Microbial Control of Insect and Mite Pests

Microbial Control of Insect and Mite Pests

From Theory to Practice

Edited by

Lawrence A. Lacey
IP Consulting International, Yakima, WA, United States

ELSEVIER

AMSTERDAM • BOSTON • HEIDELBERG • LONDON • NEW YORK • OXFORD • PARIS
SAN DIEGO • SAN FRANCISCO • SINGAPORE • SYDNEY • TOKYO

Academic Press is an imprint of Elsevier

Academic Press is an imprint of Elsevier
125 London Wall, London EC2Y 5AS, United Kingdom
525 B Street, Suite 1800, San Diego, CA 92101-4495, United States
50 Hampshire Street, 5th Floor, Cambridge, MA 02139, United States
The Boulevard, Langford Lane, Kidlington, Oxford OX5 1GB, United Kingdom

Notices
Knowledge and best practice in this field are constantly changing. As new research and experience broaden our understanding, changes in research methods, professional practices, or medical treatment may become necessary.

Practitioners and researchers must always rely on their own experience and knowledge in evaluating and using any information, methods, compounds, or experiments described herein. In using such information or methods they should be mindful of their own safety and the safety of others, including parties for whom they have a professional responsibility.

To the fullest extent of the law, neither the Publisher nor the authors, contributors, or editors, assume any liability for any injury and/or damage to persons or property as a matter of products liability, negligence or otherwise, or from any use or operation of any methods, products, instructions, or ideas contained in the material herein.

Library of Congress Cataloging-in-Publication Data
A catalog record for this book is available from the Library of Congress

British Library Cataloguing-in-Publication Data
A catalogue record for this book is available from the British Library

ISBN: 978-0-12-803527-6

For information on all Academic Press publications
visit our website at https://www.elsevier.com/

  Working together
to grow libraries in
developing countries

www.elsevier.com • www.bookaid.org

Publisher: Sara Tenney
Acquisitions Editor: Kristi Gomez
Editorial Project Manager: Pat Gonzalez
Production Project Manager: Chris Wortley
Designer: Maria Inês Cruz

Typeset by TNQ Books and Journals

Dedication

This book is dedicated to two of our finest insect pathologists, Sérgio Alves and Flávio Moscardi. Their accomplishments in microbial control and as mentors in insect pathology are detailed here by their friends and colleagues, Roberto Pereira and Pedro Neves.

Lawrence A. Lacey

Sérgio Alves (1944–2008)

If Sérgio was not the Father of Insect Pathology in Brazil, he was certainly a father figure in that he nurtured the science in its humble beginnings in Brazil, being involved in every aspect of insect pathology and microbial control, raising a new generation of insect pathologists, and making sure the science found an application in agriculture and other areas in need of insect control. Sérgio's was the first university laboratory dedicated to insect pathology and microbial control in Brazil, and he always insisted on dedicating a great portion of his efforts to the development of production systems, formulation, and application of microbial insecticides. His legacy is clearly seen in the number of products and the extensive use of microbial insecticides in Brazil and other countries in Latin America.

Sérgio's initial studies focused on the use of the fungus *Metarhizium anisopliae* against spittlebugs in pasture and sugarcane and contributed to the great expansion of microbial control of pests in Brazil and elsewhere in Latin America. His influence expanded to North America through collaboration with researchers at the University of Florida working on control of fire ants by using the fungus *Beauveria bassiana*. Sérgio's international reach included collaboration with many other international researchers and assistance in establishing microbial control laboratories and commercial microbial product factories in several countries.

His contributions include more than 180 publications, two books, 40 book chapters, the discovery of numerous insect pathogens, and scientific collaborations all over the globe. The book (S. B. Alves, (ed.) *Controle Microbiano de Insetos* FEALQ, Piracicaba, Brazil), now in its second edition, has served as the basis for training and development for scientists and practitioners of microbial control of insects. His scientific contribution is further exemplified by his latest book: *Controle Micobiano de Pragas na América Latina: avanços e desafios* (Alves and Lopes, 2008). Dr. Alves received several awards during his career including the Edilson B. Oliveira Award, the highest honor awarded by the Brazil Entomological Society. As a champion for microbial control of insects, the practical use of insect pathology in solving pest problems, during his career, Sérgio was the face, and heart, of Insect Pathology in Brazil and other countries in Latin America.

Roberto M. Pereira, PhD
University of Florida
Gainesville, FL
United States

Flávio Moscardi (1949–2012)

Flávio was born in Osvaldo Cruz, São Paulo, Brazil. He was one of the principal scientists in charge of the establishment of the biological control program against the soybean caterpillar, using the AgMNPV baculovirus, a technology that was widely implemented and used in Brazil and other Latin America countries. The virus was applied to more than 2 million ha per year in Brazil, reducing the amount of chemical pesticides used by approximately 25 million liters, and became one of the largest biological control programs worldwide and probably the largest using an entomopathogenic virus.

Moscardi graduated as an agronomic engineer from the "Escola Superior deAgricultura Luiz de Queiroz (ESALQ-USP)" and obtained his master's and PhD degrees at the University of Florida, United States. He dedicated 35 years of research with microbial control, mainly with virus in soybean for Embrapa. During his career, he authored more than 200 publications, including scientific articles published in several well-known Brazilian and international journals, book chapters, and technical notes. Moscardi was also the adviser of 23 graduate students, many who continue to work with entomopathogens and microbial control.

A member of the Brazilian Academy of Sciences, he received many awards such as the "Commendation of the National Order of Scientific Merit" granted by the Brazilian president in August 2002, "Agriculture by the Third World Academy of Sciences" in 1997, and "Distinction by the International Society of Plant Protection in 1995" and the "Frederico de Menezes Veiga" granted by Embrapa in 1991 and was honored by the Founders' lecture of the Society for Invertebrate Pathology (SIP) meeting in 2012 as presented by Daniel Sosa-Gomez. Moscardi was a regular attender at SIP meetings and had many friends and collaborators in the Society.

He was a member of the Advisory Board of Agriculture from CNPq, president of the Entomological Society of Brazil from 1998 to 2002, and president of the X CBE (Brazilian Congress of Entomology, 1984), V Siconbiol (Biological Control Symposium in 1996), VII Colloquium of Invertebrate Pathology and Microbial Control (Foz do Iguaçu-PR, 2002), and the Scientific Committee of the XXI International Congress of Entomology (2000). Besides being a great scientist, Moscardi's other great passion was fishing, where he "graduated" many fishermen by sharing his knowledge and organizing fishing expeditions.

Prof. Pedro Neves
Universidade Estadual de Londrina
Londrina PR, Brasil

Contents

4. Basic and Applied Research: Entomopathogenic Bacteria

T.R. Glare, J.-L. Jurat-Fuentes and M. O'Callaghan

5. Basic and Applied Research on Entomopathogenic Fungi

D. Chandler

6. Basic and Applied Research: Entomopathogenic Nematodes

D. Shapiro-Ilan, S. Hazir and I. Glazer

List of Contributors

J. Almeida Instituto Biológico, Laboratório de Controle Biológico, Campinas, Brazil

S.P. Arthurs University of Florida, Apopka, FL, United States

M.V. Baggio UNESP, Jaboticabal, São Paulo, Brazil

R. Bateman Imperial College, Ascot, United Kingdom

N. Becker German Mosquito Control Association (KABS), Speyer, Germany; University of Heidelberg, Heidelberg, Germany

D.J. Bruck DuPont Pioneer, Johnston, IA, United States

D. Chandler University of Warwick, Wellesbourne, Warwick, United Kingdom

E.H. Clifton Iowa State University, Ames, IA, United States

D. Conlong South African Sugarcane Research Institute, Mount Edgecombe, South Africa; University of KwaZulu-Natal, Pietermaritzburg, South Africa

S.K. Dara University of California Cooperative Extension, San Luis Obispo, CA, United States

C. Dolinski Universidade Estadual do Norte Fluminense/CCTA/LEF, Rio de Janeiro, Brazil

L.W. Duncan University of Florida, Lake Alfred, FL, United States

S. Ekesi International Centre of Insect Physiology and Ecology (icipe), Nairobi, Kenya

M. Faria EMBRAPA Genetic Resources and Biotechnology, Brasilia, Federal District, Brazil

R. Fusco Valent BioSciences Corporation (Retired), Mifflintown, PA, United States

A.J. Gassmann Iowa State University, Ames, IA, United States

T.R. Glare Lincoln University, Christchurch, New Zealand

I. Glazer ARO, Volcani Center, Israel

T. Goble Cornell University, Ithaca, NY, United States

J. Gore Mississippi State University, Stoneville, MS, United States

E.W. Gray University of Georgia, Athens, GA, United States

D. Grzywacz University of Greenwich, Chatham, Kent, United Kingdom

R. Gwynn Biorationale Limited, Duns, United Kingdom

A.E. Hajek Cornell University, Ithaca, NY, United States

R. Han Guangdong Entomological Institute, Guangzhou, Guangdong, China

S. Hazir Adnan Menderes University, Aydin, Turkey

T.A. Jackson AgResearch, Lincoln Research Center, Christchurch, New Zealand

S.T. Jaronski United States Department of Agriculture, Sidney, MT, United States

N. Jenkins Pennsylvania State University, University Park, PA, United States

J.-L. Jurat-Fuentes University of Tennessee, Knoxville, TN, United States

P.G. Koehler University of Florida, Gainesville, FL, United States

C. Kooyman Éléphant Vert, Nanyuki, Kenya

A.M. Koppenhöfer Rutgers University, New Brunswick, NJ, United States

L.A. Lacey IP Consulting International, Yakima, WA, United States

L.G. Leite Instituto Biológico, São Paulo, Brazil

J. Leland Novozymes BioAg, Salem, VA, United States

R.B. Lopes EMBRAPA Genetic Resources and Biotechnology, Brasilia, Federal District, Brazil

P. Lüthy Swiss Federal Institute of Technology, ETH Zürich, Zürich, Switzerland

N.K. Maniania International Centre of Insect Physiology and Ecology (icipe), Nairobi, Kenya

G.M. Mascarin Embrapa Rice and Beans, Santo Antônio de Goiás, Brazil

D. Moore Commonwealth Agricultural Bureaux International Europe – UK, Surrey, United Kingdom

S. Moore Citrus Research International, Port Elizabeth, South Africa; Rhodes University, Grahamstown, South Africa

S.D. Moore Citrus Research International, Port Elizabeth, South Africa; Rhodes University, Grahamstown, South Africa

M. Nakai Tokyo University of Agriculture and Technology, Tokyo, Japan

M. O'Callaghan Lincoln University, Christchurch, New Zealand; AgResearch, Christchurch, New Zealand

D.H. Oi USDA-ARS, Gainesville, FL, United States

R.M. Pereira University of Florida, Gainesville, FL, United States

A. Peters e-nema GmbH, Schwentinental, Germany

C. Prior Mount Pleasant, Bampton, Devon, United Kingdom

D. Shapiro-Ilan USDA-ARS, Southeastern Fruit and Tree Nut Laboratory, Byron, GA, United States

L.F. Solter University of Illinois, Champaign, IL, United States

D.R. Sosa-Gómez Embrapa Soja, Londrina, Paraná, Brazil

T. Su West Valley Mosquito and Vector Control District, Ontario, CA, United States

K. van Frankenhuyzen Natural Resources Canada, Sault Ste. Marie, ON, Canada

S.P. Wraight USDA-ARS, R. W. Holley Center for Agriculture and Health, Ithaca, NY, United States

S. Wu Rutgers University, New Brunswick, NJ, United States

X. Yan Guangdong Entomological Institute, Guangzhou, Guangdong, China

Preface

Since the publication of *Microbial Control of Pests and Plant Diseases* (Burges, 1981) 35 years ago, there have been very few comprehensive books published on microbial control of insect and mite pests. This book was written by an international team of 56 experts from 13 counties with a wealth of experience in a wide variety of fields in order to address the question, "which entomopathogens can be successfully employed as significant components of integrated pest management?" Although entomopathogens are seldom used as a stand-alone means of control, they have been successfully employed in all of the major types of biological control, including classical, conservation, inoculative and inundation (also referred to as augmentation control). Examples of classical and conservation biological control are presented in the first 6 chapters of the book. Most of the 19 applied chapters focus on inundation application of microbial control agents (MCAs).

The book is intended to serve as a stand-alone resource, designed to provide background on the insect pathology foundations of microbial control and how and where entomopathogens can be used for pest control. However, it is not intended to be an exhaustive review of microbial control in every crop or setting in which MCAs are used. For some of the crops in which MCAs have been applied that are not covered here, the reader is referred to the *Field Manual of Techniques in Invertebrate Pathology* (Lacey and Kaya, 2007) and *Insect Pathology* (Vega and Kaya, 2012). This book can also serve as a supplementary resource for courses in insect pathology, biological control, and integrated pest management. It provides information on basic and applied research for each of the major pathogen groups and their mass production. However most of the chapters are devoted to microbial control of a plethora of pests in major crops (corn, soybeans, sugarcane, coffee, cotton), minor crops (fruits, vegetables, tea), landscape and ornamental plants, forestry, turf, rangeland, human health, and structures.

Background on the MCAs that are employed is derived from recent original research and older studies upon which today's successes are based. The chapters focus primarily on field derived data and, where possible, application rates and efficacy data are provided to be used in implementation of control programs or as starting points for continued research. This data can also be used as the basis for determining the feasibility and cost effectiveness of implementing MCAs for specific control situations.

I sincerely appreciate the efforts and expertise of the 56 authors, 60 reviewers, and a multitude of colleagues who have furnished scientific advice and literature, graphics, and constructive comments. Thank you Cynthia Lacey for your support, grammatical tutoring, and encouragement. I am grateful to the Elsevier editorial staff, especially Pat Gonzalez and Kristi Gomez for their management and editorial skills. Inspiration for this book was provided by Steinhaus (1949, 1956), Burges and Hussey (1971), Burges (1981), Agostino Bassi, and Louis Pasteur.

Lawrence A. Lacey
April, 2016

REFERENCES

Burges, H.D. (Ed.), 1981. Microbial Control of Pests and Plant Diseases: 1970–1980. Academic Press, London, UK, 949 pp.

Burges, H.D., Hussey, N.W. (Eds.), 1971. Microbial Control of Insects and Mites. Academic Press, London, UK, 861 pp.

Lacey, L.A., Kaya, H.K. (Eds.), 2007. Field Manual of Techniques in Invertebrate Pathology: Application and Evaluation of Pathogens for Control of Insects and Other Invertebrate Pests, 2nd Edition. Springer, Dordrecht, The Netherlands, 484 pp.

Steinhaus, E., 1949. Principals of Insect Pathology. Academic Press, 757 pp.

Steinhaus, E., 1956. Living insecticides. Sci. Am. 195, 96–103.

Vega, F.E., Kaya, H.K. (Eds.), 2012. Insect Pathology, 2nd ed. Academic Press, San Diego, 490 pp.

Part I

Introduction to Insect Pathology and Microbial Control: The Tried and True and Recent Innovations

Chapter 1

Entomopathogens Used as Microbial Control Agents

L.A. Lacey

IP Consulting International, Yakima, WA, United States

1.1 INTRODUCTION TO ENTOMOPATHOGENS USED IN THE MICROBIAL CONTROL OF INSECT AND MITE PESTS

The goal of this book is to provide a current, comprehensive, and enduring reference for microbial control and the underlying principles of insect pathology that determine its success. Central to this must be a clear understanding of the nature and ecology of the microbes that cause diseases in pest species.

1.1.1 A Brief History of Entomopathogens and Their Use as Microbial Control Agents

Diseases of honeybees and silkworms have been recognized for millennia (Davidson, 2012), but the causal agents of insect diseases were not demonstrated until 1835, when Augusto Bassi associated a disease of the silkworm, *Bombyx mori*, with a fungal pathogen, later named *Beauveria bassiana* (Bassi, 1835). Bassi is known as the "Father of Insect Pathology" for demonstrating the transmission of the fungus from larva to larva and from contaminated leaves to larvae (Steinhaus, 1975; Davidson, 2012). Subsequently, a number of entomopathogens have been incriminated as the causal agents of insect disease, including Louis Pasteur's landmark demonstration of the microsporidium *Nosema bombycis* as the causal agent of pébrine disease in *B. mori* (Pasteur, 1870). According to Steinhaus (1949), in solving this problem and another entomopathogen-related silkworm disease (flacherie), Pasteur saved the French silk industry. Metchnikoff was the first to culture an entomopathogen, the fungus *Metarhizium anisopliae*, on artificial medium after isolation from the wheat cockchafer *Anisoplia austriaca* and to suggest using it as an applied microbial control agent (MCA) (Metchnikoff, 1879, 1880; Steinhaus, 1975; Lord, 2005). However, it was not until 1888 that *M. anisopliae* was evaluated as an MCA (Krassilstschik, 1888). Since then, a multitude of entomopathogens have been discovered and developed as MCAs (Steinhaus, 1963, 1975; Lord, 2005; Davidson, 2012; Lacey et al., 2015).

The first commercialization of MCAs began in France in 1938 with Sporeine, a product based on the bacterium *Bacillus thuringiensis*. The production and use of the product were discontinued, ostensibly due to World War II (Steinhaus, 1949; Davidson, 2012). The first entomopathogenic bacterium to be registered in the United States (1948) was *Paenibacillus* (*Bacillus*) *popilliae* (also known as milky spore) for control of the Japanese beetle, *Popillia japonica* (Dutky, 1963; Klein, 1981; Federici, 2005; Davidson, 2012). It was applied as an augmentative MCA to larval breeding sites of *P. japonica* larvae, especially in residential lawns and golf courses. It is no longer commercially produced due in large part to the requirement of in vivo production. Steinhaus stimulated renewed interest in the development of *B. thuringiensis* and other MCAs with the publication of "Living Insecticides" (Steinhaus, 1956). Although the first commercial *B. thuringiensis* product was registered for use in the United States in 1961, it was not until the discovery and commercial development of efficacious strains with greater virulence toward a broader range of pests that microbial control became widely used in agriculture and forestry (Kurstak, 1962; Dulmage, 1970; Glare and O'Callaghan, 2000). More detail on the development of this key MCA leading to its widespread use is presented in Chapter 4 and by Federici (2005) and by Sanchis (2011). Historical accounts of the development of other entomopathogens as MCAs are presented by Steinhaus (1949, 1975), Lord (2005), and Davidson (2012). Several significant advances in microbial control of pest insects and mites have taken place since the comprehensive reviews by Burges and Hussey (1971) and Burges (1981). Glare et al. (2012) and Lacey et al. (2015) reviewed the successes and setbacks for microbial control and made recommendations for improvements of MCAs and their future use.

Microbial Control of Insect and Mite Pests. http://dx.doi.org/10.1016/B978-0-12-803527-6.00001-9

1.2 ENTOMOPATHOGENS USED AS MCAs

1.2.1 Entomopathogenic Viruses

A wide array of viruses have been reported from a broad spectrum of insects and other arthropods (Miller, 1997; Miller and Ball, 1998). These include representatives from at least 16 families of viruses (Possee and King, 2014) including baculoviruses, cytoplasmic polyhedroviruses, entomopoxviruses, nudiviruses, and a variety of lesser known viruses that have been investigated for their potential as MCAs. The most researched for microbial control are the baculoviruses with over 600 species of insects being reported as hosts (Miller, 1997; Szewczy et al., 2006; Cory and Evans, 2007). Inundation augmentative application is the most commonly adopted strategy for their use. Of the four genera of the Baculoviridae, the *Alphabaculovirus* [nucleopolyhedroviruses (NPVs)] and *Betabaculovirus* [granuloviruses (GVs)] are the most numerous and most widely used as MCAs (Cory and Evans, 2007; Eberle et al., 2012). NPVs are predominantly found in the Lepidoptera, but also in Diptera, Hymenoptera (Symphyta), and some Crustacea while GVs infect only Lepidoptera (Miller, 1997; Eberle et al., 2012). In addition to host range, the two genera differ mainly in the manner in which virions are embedded in the infectious particles, the proteinaceous occlusion bodies (OBs). In NPVs, several viral rods are embedded together within the same large polyhedral OBs, whereas in GVs only a single virion is embedded singly in smaller OBs, sometimes called a granule. In NPVs, viral rods can be singly enveloped (SNPV) or multiply enveloped (MNPV). Morphology of GVs, MNPVs, and SNPVs is depicted in Chapter 3, Figure 3.2. MNPVs usually have broader host ranges than SNPVs. For example, the *Autographa californica* MNPV (AucaMNPV) is active against larvae of 95 species in 15 families of Lepidoptera (Vail et al., 1971, 1999; Vail and Jay, 1973; Cory and Evans, 2007). However, it is most efficacious against species in the Noctuidae (Vail et al., 1999). In contrast, another MNPV, *Lymantria dispar*, is strictly host specific (Barber et al., 1993). Selected NPVs and GVs are presented in Table 1.1, and a more extensive listing is presented in Chapter 3.

With few exceptions, baculovirus OBs must be ingested in order to invade the host insect. Virions enter the host via the midgut epithelium where they initially multiply before spreading and infecting other types of cells (fat body, tracheal matrix, hypodermis) and begin viral replication and occlusion. Detailed descriptions of the infection processes and pathogenesis are presented by Federici (1997) and in Chapter 3. The macroscopic progression of disease signs and symptoms are shown in Chapter 3, Figure 3.1. Among all entomopathogenic viruses, baculoviruses are the only ones to have been mass produced, commercialized, and applied on a widespread basis. The issues and constraints around expanding their use on a broader scale are presented in Chapter 3. Mass production of baculoviruses is covered in Chapter 7.

1.2.2 Bacteria

Although the major volume and market share of commercially produced MCAs that are used for microbial control are bacteria, relatively few species have been developed. These include gram-positive, spore-forming *B. thuringiensis* subspp. and *L. (B.) sphaericus*, the gram-negative, non–spore-forming *Serratia entomophila*, and, most recently, *Chromobacterium subtsugae*. *B. thuringiensis* subsp. are the most commonly used of all microbial pesticides worldwide, representing over 50% of all MCA sales (Glare et al., 2012; Lacey et al., 2015). The majority of *B. thuringiensis* inundative applications are for suppression of lepidopteran pests of crops followed by applications for control of forest pests and then by abatement of mosquitoes and black flies. The insecticidal activity of *B. thuringiensis* and *L. sphaericus* is mainly due to protein toxins within parasporal inclusions, which must be ingested to become larvicidal. The toxins become solubilized in the high pH of the midgut and are cleaved to the toxic moieties by gut proteolytic enzymes. These toxins recognize receptors and create pores on enterocytes, with the subsequent osmotic imbalance resulting in cell rupture, compromising integrity of the gut epithelium. Invasion of the main body cavity and subsequent septicemia caused by gut-resident bacteria leads to death of the larva soon afterward. Detailed descriptions of the biochemical nature of the toxins and their modes of action at the molecular level are presented in Chapter 4. Examples of targeted pests for inundative applications include a plethora of lepidopteran pests of field crops, tree fruit, turf, and ornamental plants and forest and dipteran pests and are shown in Table 1.1. The larvicidal activity *S. entomophila* is due to proliferation of the bacterium in the gut resulting in blockage and lack of nutrient absorption. More details on host range factors affecting larvicidal activity, safety, and production of entomopathogenic bacteria are presented in Chapters 4 and 8.

1.2.3 Fungi

Depending on the scientists reporting, between 700 and 1000 species of fungi are recognized as causal agents of disease in arthropods (Goettel et al., 2010; Vega et al., 2012). The majority of those used for microbial control of insects and mites are in the orders Hypocreales and Entomophthorales. In addition to a broad host range within the Insecta and Acarina, they are the only effective MCAs of sucking insects (Hemiptera). The only commercially produced fungi for microbial control are in the Hypocreales. These include *Beauveria bassiana*, *Metarhizium* spp., *Isaria fumosorosea*, and *Lecanicillium* spp. and are used against dozens of pests in a variety of crops.

The typical route of invasion of the host is through the cuticle. Once fungal spores have germinated on the surface of the host, the growing tip (appressorium, infection peg,

TABLE 1.1 Some Examples of Entomopathogens Used for Microbial Control of Targeted Arthropods

Pathogen Group	Pathogens	Targeted Arthropods	Crops/Habitats
Viruses: Baculoviridae			
Alphaviruses-NPV	HearSNPV, HezeSNPV	*Helicoverpa* and *Heliothis* spp.	Corn, cotton, soybean, tobacco
	SeMNPV	*Spodoptera exigua*	Vegetable, field, flower crops, and ornamentals
Betaviruses-GV	CpGV	*Cydia pomonella*	Pome fruit and walnut
	AdorGV	*Adoxophyes orana*	Apple
Bacteria			
	Bacillus thuringiensis subsp. *kurstaki* and *aizawai*	Lepidoptera	Vegetables, tree fruit, forest, stored products
	B. thuringiensis subsp. *israelensis*	Diptera: Culicidae, Simuliidae	Lotic and lentic aquatic habitats
	Lysinibacillus sphaericus	Diptera: Culicidae	Lentic aquatic habitats
	Serratia entomophila	*Costelytra zealandica*	Pasture
	Chromobacterium subtsugae	Broad host range including: *Leptinotarsa decemlineata*, Hemiptera, Acarina	Diverse crops
Fungi			
Hypocreales	*Beauveria bassiana*	Broad host range including: Lepidoptera, Coleoptera, Formicidae, Hemiptera, Acarina	Diverse crops: Cucurbits, potato, crucifers, cotton, sugarcane, and more
	Metarhizium anisopliae senso lato	Very broad host range including: Coleoptera, Diptera, Lepidoptera, Hemiptera, Isoptera, Acarina	Diverse crops/habitats and structures
	Metarhizium acridum	Grasshoppers and locusts	Diverse crops
	Isaria fumosorosea	Hemiptera	Diverse crops including: Cucurbits, cotton, crucifers, and ornamentals
	Lecanicillium spp.	Hemiptera	Diverse crops, especially greenhouse crops
Entomophthorales			
	Neozygites tanajoae	*Mononychellus tanajoa*	Cassava
	Entomophaga maimaiga	*Lymantria dispar*	Deciduous trees
Entomopathogenic Nematodes[a]			
	Steinernema carpocapsae	Lepidoptera, Coleoptera, Diptera, Hymenoptera	Field crops, orchard, ornamental and turf
	S. feltiae	Coleoptera, Lepidoptera,	Vegetables, tree fruit, ornamentals, mushrooms
	S. scapterisci	Orthoptera (*Scapteriscus* spp.)	Lawn and turf
	Heterorhabditis bacteriophora	Lepidoptera, Coleoptera	Vegetables, orchard, ornamentals, turf mushrooms
	H. megidis	Coleoptera (Scarabaeidae)	Lawn and turf

[a]*Insect orders in which naturally occurring EPN infections were found (Peters, 1996). Experimental host ranges can be narrow (H. megidis, S. scapterisci) to very broad (S. carpocapsae, S. feltiae, H. bacteriophora). See also Chapter 6.*

germ tube) penetrates the cuticle through concentrated physical energy and lytic enzymatic activity (Hajek and St. Leger, 1994). Following penetration and growth of hyphae and blastospores within the host, sporulation of the fungus takes place on the surface of the cuticle. Selected examples of the use of entomopathogenic fungi for protection of crops and urban structures are presented in Table 1.1 and in Chapter 5 and several of the applied chapters in the book. Mass production of entomopathogenic fungi is presented in Chapter 9.

1.2.4 Entomopathogenic Nematodes

A huge number of entomogenous nematodes have been reported from insects, (Welch, 1963; Poinar, 1975; Grewal et al., 2005) including those used in classic biological control (Bedding and Akhurst, 1974; Platzer, 2007; Bedding, 2009; Frank, 2009). The only nematodes currently used as MCAs for augmentative biological microbial control are entomopathogenic nematodes (EPNs) in the families Steinernematidae and Heterorhaditidae. Species of EPNs are in mutualistic and obligate associations with bacteria (*Xenorhabdis* spp. and *Photorhabdis* spp.) carried inside the gut of infective juveniles (IJs). These IJs wait for or actively seek a host insect. They gain entry into the host insect, per os, via the spiracles or in some cases by penetrating intersegmental membranes. Inside the host, the mutualistic bacteria are liberated, kill the host, and digest host tissues, and the IJs molt and begin feeding. There are normally two to three generations, although if nutrients are low, only one generation may result. Once nutrients are exhausted, IJs are produced and leave the host insect to search for or wait for new hosts. Some examples of EPNs, insect targets, and protected crops are presented in Table 1.1. Detailed information on the life cycle, host ranges, factors that enhance or retard EPN efficacy, and protected crops are given in Chapter 6. Mass production of EPNs is covered in Chapter 10.

1.3 MICROBIAL CONTROL AGENTS AND BIOLOGICAL CONTROL

Entomopathogens used as MCAs fit the three categories of biological control: classic, conservation, and augmentative, as defined by Barbosa (1998), McCrevy (2008) and Hoy (2008a,b) for natural enemies of pest arthropods. Entomopathogens have been successfully used in all three categories. Classic biological control entails introduction of naturally occurring predators, parasitoids, and entomopathogens from the native range of an invasive pest to help bring about its long-term suppression; conservation biological control stresses maintaining a habitat that is conducive to conserving natural enemies of the pest; and augmentative biological control involves enhancing the prevalence of the natural enemies to initiate increased control. In microbial control, this is accomplished with either inoculative or inundative application of MCAs. Eilenberg et al. (2001) proposed replacing augmentative biological control with two separate categories: inoculative biological control and inundative biological control.

1.3.1 Classic Biological Control

There are numerous examples of MCAs used for classic biological control of invasive insects (Hajek et al., 2005, 2007, 2009). These include entomopathogenic viruses (baculovirus, nudivirus), fungi (Entomophthorales), and parasitic and entomopathogenic nematodes (Neotylenchidae, Mermithidae, Steinernematidae). Three noteworthy examples of classical microbial control follow.

1.3.1.1 Case Study: Oryctes Nudivirus for Control of Invasive Coconut Palm Rhinoceros Beetle

One of the most impressive examples of classic biological control with an MCA is that of the Oryctes nudivirus for control of the coconut palm rhinoceros beetle, *Oryctes rhinoceros* (Hüger, 2005; Jackson et al., 2005; Jackson, 2009). The beetle is a serious exotic pest of coconut and oil palms in the southwest Pacific islands. Adults consume foliage in the crown of the palms and can reduce yield and kill trees (Bedford, 1980). Larvae develop in rotting palm logs, including the tops of standing dead palms, and in organic matter in a diversity of other sites with high organic content (sawdust, manure piles, etc.) (Bedford, 1980). Native to the coconut growing areas of the Asia–west Pacific region, the beetle was accidentally introduced into Samoa and subsequently spread to other islands in the southwest Pacific, where it became a widespread and very serious pest of coconut palms (Bedford, 1980; Jackson, 2009).

The Oryctes virus was originally collected from infected *Oryctes rhinoceros* in Malaysia by Hüger (1966) and introduced into Samoa and several southwest Pacific islands (Bedford, 1980; Hüger, 2005; Jackson, 2009). Adult beetles become chronically infected and serve as mobile reservoirs of the virus. Aggregations of mating and feeding adults in the palm crowns ensures transmission from infected to uninfected individuals and dispersal of the virus. Larval breeding sites become initially contaminated with virus via ovipositing infected females. Larvae become acutely infected after consuming virus and die within 9–25 days depending on age and temperature (25–32°C) (Hüger, 1966; Zelazny, 1972). The introduced virus has resulted in significant long-term control of the beetle and reduction of foliar damage by causing epizootics that kill larvae, curtail the life span of adults, and reduce fecundity in females (Zelazny, 1972, 1973; Bedford, 1980; Hüger, 2005; Jackson, 2009). Production and application of the virus is reviewed by Bedford (1980) and Jackson (2009). In recent years there are reports of Oryctes virus having become attenuated and therefore less effective on some of the beetle-infested islands (Jackson et al., 2005; Jackson, 2009). Jackson et al. (2005) and Jackson (2009) conclude that selection of more virulent strains of virus and improvements in application methods need to be developed to overcome this problem.

1.3.1.2 Case Study: Entomophaga maimaiga for Classic Biological Control of Gypsy Moth, L. dispar

An example of classic biological control using a fungus as an effective pathogen for long-term control of the gypsy moth, *L. dispar*, is that of the entomophthoralean *E. maimaiga* (Hajek et al., 1996; Hajek, 1997; Solter and Hajek, 2009). The moth is a native forest defoliator that outbreaks occasionally from Europe to Asia but was absent from the Western Hemisphere. From its original point of entry into the United States (Boston, Massachusetts) in 1869, it has dispersed annually farther south and west. Larvae of *L. dispar* feed on a wide range of deciduous trees and during cyclic outbreaks have been responsible for defoliation of trees in up to 2 million ha of forest (Solter and Hajek, 2009).

The fungus originated in Japan and was first released in the United States near Boston in 1910 and 1911 but was not detected in following years (Hajek et al., 1995). It was also collected in Japan in 1984 by Soper et al. (1988), and small field applications of the 1984 isolate in 1985 in New York State and in 1986 in Virginia resulted in few or no infections. It was not until 1989 that significant epizootics caused by *E. maimaiga* were reported (Andreadis and Weseloh, 1990; Hajek, 1997, 1999). The exact origin of the *E. maimaiga* causing these outbreaks is not precisely known, but it appears to have originated from areas in Japan different from the 1910–1911 collection sites (Nielsen et al., 2005). Since then, in vivo produced fungal spores (conidia and resting spores) and infected larvae have been introduced into areas on the leading edge of *L. dispar* dispersal. Subsequent infections indicate that *E. maimaiga* has become established in and spread from locations where it was not previously observed (Hajek et al., 1996; Weseloh, 2003; Solter and Hajek, 2009). More details on this continuing successful classic microbial control are presented in Chapter 21.

1.3.1.3 Case Study: Steinernema scapterisci for Control of Invasive Mole Crickets

Mole crickets, *Scapteriscus* spp., expanding northward from their centers of origin in South America, were introduced into the United States in the early 1900s (Frank, 2009). Since introduction, severe damage had been reported in turf infested by the crickets, especially in Florida, where *S. vicinus* is a serious pest of turf (Frank, 2009). Exploration for EPNs and other natural enemies of *Scapteriscus* spp. were started in South America. In 1985, an EPN from Uruguay was released in Florida. Initially, the introduction of the EPN, *S. scapterisci,* for control of the invasive mole crickets in Florida fit the classic biological control paradigm (Parkman et al., 1993). The nematode was collected in Uruguay and introduced into *S. vicinus, S. borelli,* and *S. abbreviatus* populations in Florida where it became established (Hudson et al., 1988; Parkman et al., 1993). Additionally, two imported parasitoids, also from South America, were released and became established throughout Florida. By 2000, the combined effort of the three natural enemies reduced the *Scapteriscus* populations by 95% (Frank and Walker, 2006). Since then the EPN has been applied to infestations of the cricket in several sites in Florida (Frank, 2009).

Numerous additional examples of classic microbial control are found throughout the literature; some of which are examined in this book. These include importation and establishment of the nematode *Deladenus siricidicola* (Tylenchida: Neotylenchidae) for control of the Sirex woodwasp, *Sirex noctilio* (Chapter 21); importation and establishment the of nematode *Romanomermis iyengari* (Nematoda: Mermithidae) for control of mosquitoes (Chapter 28); development of *Neozygites tanajoae* (Entomophthorales) for control of cassava greenmite in Brazil and Chapter 2 Chapter 21Benin (Chapter 2); baculoviruses in forestry (Chapter 21); and agriculture (Chapters 14 and 17). A catalog of imported exotic entomopathogens used as classic biological control agents was compiled by Hajek et al. (2005).

1.3.2 Conservation Biological Control

Epizootics caused by naturally occurring viral and fungal pathogens are often responsible for spectacular crashes of insect pest populations and are often credited with eliminating the need for further interventions (Harper, 1987; Steinkraus, 2007; Shapiro-Ilan et al., 2012). Conservation biological control relies on the protection of and improving conditions for naturally occurring MCAs to enable induction of high rates of infection in pest populations. Reliance on the natural occurrence of entomopathogens for pest management, however, can be risky due to the unpredictability of factors that govern epizootics. Because many pathogens are host-density dependent, epizootics often occur after economic injury levels have been surpassed (Odindo, 1983; Harper, 1987; Rose et al., 2000). However, agricultural practices that foster their conservation and increase their prevalence should nevertheless be considered.

A demonstration of the conservation microbial control strategy is that reported by Steinkraus (2007). By not treating cotton with broad-spectrum insecticides, populations of the cotton aphid, *Aphis gossypii*, were allowed to develop, which in turn encouraged epizootics of the fungus *Neozygites fresenii*. Conservation of naturally occurring MCAs can also be disrupted by the effect of pesticides directly on the MCAs. For example, application of fungicides and nematicides can drastically curtail the insecticidal activity of entomopathogenic fungi and EPNs, respectively.

In addition to pesticide type and timing of application, other agricultural practices [type of tillage (conventional versus conservation), irrigation, fertilizer (chemical versus organic) crop rotation, type of ground cover, etc.] can have a marked effect on the survival and pathogenicity of MCAs. Hummel et al. (2002) and Millar and Barbercheck (2002) found that tillage practices can significantly affect the survival and abundance of entomopathogenic fungi and EPNs. Hummel et al. (2002) demonstrated that detection of entomopathogens was significantly higher in conservation tilling compared with conventional tillage systems. They reported that conservation tilling (strip-till) did not affect levels of detection of *Steinernema carpocapsae*, but pesticide use significantly reduced detection of entomopathogenic fungi. Millar and Barbercheck (2002) demonstrated that tillage versus no tillage resulted in a significant negative effect of detection of *S. carpocapsae* but a positive effect on detection of *S. riobrave*. The authors surmised that the different sensitivities of the EPNs could be partly explained by differences in environmental tolerances of the two species and tendencies to disperse deeper into the soil. A number of other examples of the role of conserved EPN and fungal MCAs in microbial control are presented by Lewis et al. (1998), Ekesi et al. (2005), Meyling and Eilenberg (2007), Nielsen et al. (2007), Steinkraus (2007), Pell et al. (2010), and Campos-Herrera et al. (2010).

1.3.3 Augmentative Biological Control

Augmentation of naturally occurring MCAs can be accomplished with either inoculative or inundative applications depending on the pathogen and crop that requires protection. Inoculative microbial control is the application of smaller amounts of inoculum with the goal that the MCA will increase in the host population on its own (Eilenberg et al., 2001). This could result in outbreaks of disease earlier in the season than would otherwise occur. Examples include limited applications of *P. popilliae* and *Heterorhabditis bacteriophora* for control of *Popillia japonica* (Dutky, 1963; Klein, 1981; Klein and Georgis, 1992).

The more common strategy is the inundative application of larger amounts of MCAs to initiate widespread infections and provide immediate control of targeted pests (Steinhaus, 1949, 1963; Burges and Hussey, 1971; Burges, 1981; Lacey et al., 2001, 2015; Kaya and Lacey, 2007). In the inundative approach, the main route of infection is through the pests picking up the applied MCA rather than through secondary infections. Depending on the number of generations of the pest insect or mite and the impact of

adverse environmental factors on MCA persistence, especially exposure to ultraviolet-B radiation, one to several applications of the MCA may be required throughout the growing season.

There are scores of commercially produced microbial pesticides used for augmentative control of pest insects and mites worldwide (Glare and O'Callaghan, 2000; Federici, 2005; Alves and Lopes, 2008; Kabaluk and Gazdik, 2005; Faria and Wraight, 2007; Gwynn, 2014; Lacey et al., 2015). Some examples of the augmentative use of entomopathogenic viruses, bacteria, fungi, and nematodes are presented in Table 1.1. *B. thuringiensis* is the most widely used MCA for control of hundreds of species of insect pests (Glare and O'Callaghan, 2000; Federici, 2005; see Chapter 4). The use of MCAs against insects and mites is the predominant theme of Chapters 11–29.

1.4 ADVANTAGES AND DISADVANTAGES OF MICROBIAL CONTROL

Table 1.2 presents several advantages and disadvantages of microbial control as compiled from Steinhaus (1949), Burges and Hussey (1971), Tanada and Kaya (1993), Alves (1998), Kaya and Lacey (2007) and Kaya and Vega (2012). The most important advantages are the efficacy, specificity, and safety of MCAs. In contrast, the greatest disadvantages include the higher comparative cost, narrower spectrum of insecticidal activity, and reduced persistence relative to conventional chemical insecticides. However, benefits of environmental and food safety are increasingly having a positive effect on the sales and popularity of microbial control. Numerous studies attest to the safety of MCAs for applicators and other vertebrates, most nontarget organisms, especially honeybees and predators and parasitoids (Hokkanen and Hajek, 2003). There could be indirect effects on other natural enemies if the host insect is killed before parasitoids emerge or host and prey are significantly removed from the food chain. However, natural enemy vagility and alternative hosts and prey minimize this effect so that the overall positive effect of arthropod natural enemies on pests outweighs any potential negative effects. Publication of studies on the safety of MCAs have been reviewed by Laird et al. (1990), Glare and O'Callaghan (2000), Lacey and Siegel (2000), Goettel et al. (2001), and Hokkanen and Hajek (2003). MCAs may not always provide a stand-alone means for total pest control, yet they can be invaluable components of integrated pest management working in concert with other natural enemies, resistant plant varieties, agricultural practices, the judicious use of selective and soft pesticides (avermectin, spinosad, etc.), mating disruption, and environmental manipulation and modification (Gurr et al., 2004; Radcliffe et al., 2009; Wraight and Hajek, 2009).

TABLE 1.2 Advantages and Disadvantages of Using Microbial Control Agents (MCAs)

Advantages	Disadvantages
Specificity, minimal effects on nontarget insects	Specificity, less effective on pest complexes
Efficacious for targeted insects and mites	Short shelf-life for some unformulated MCAs
No development of resistance with most MCAs	Sensitive to inactivation by ultraviolet and other environmental conditions.
Recycling and long-term persistence of some pathogens, enabling long-term control	Multiple applications may be required due to short persistence of MCA on crops
No maximum residue limits (MRL) for MCAs on produce	May require precise timing of applications to infect early instars
Little or no secondary pest outbreaks due to preservation of natural enemies	Slower onset of death relative to chemical pesticides
Effective tool for resistance management in IPM strategies	Development of resistance of some Lepidoptera to *B. thuringiensis* and of *Cydia pomonella* to the codling moth granulovirus
Safe for applicators, other vertebrates and the food supply	In vivo production costs could be high
No preharvest spray interval	Uneconomical except for high value crops
Little or no environmental pollution	Potential for negative public perception of entomopathogens used on crops
Ease of mass production of nonobligate pathogens on inexpensive artificial media	Negative public reaction to the use of genetically modified MCAs
Good shelf-life of most formulated MCAs	
Application with conventional equipment	
Adaptable to genetic modification	

Information compiled from Steinhaus, E., 1949. Insect Pathology, Academic Press, 757 pp.; Burges, H.D., Hussey, N.W., 1971. Microbial Control of Insects and Mites. Academic Press, London, UK, 861 pp.; Tanada, Y., Kaya, H.K., 1993. Insect Pathology. Academic Press, San Diego, 666 pp.; Alves, S.B. (Ed.), 1998. Controle Microbiano de Insetos, second ed. Fundação de Estudos Agrários Luiz de Queiroz, Piracicaba, Brasil, 1163 pp.; Kaya, H.K., Lacey, L.A., 2007. Introduction to microbial control. In: Lacey, L.A., Kaya, H.K. (Eds.), Field Manual of Techniques in Invertebrate Pathology: Application and Evaluation of Pathogens for Control of Insects and Other Invertebrate Pests, second ed. Springer, Dordrecht, The Netherlands, pp. 3–7; Kaya, H.K., Vega, F.E., 2012. Scope and basic principles of insect pathology. In: Vega, F.E., Kaya, H.K. (Eds.), Insect Pathology, second ed. Academic Press, San Diego, pp. 1–12.

REFERENCES

Alves, S.B. (Ed.), 1998. Controle Microbiano de Insetos, second ed. Fundação de Estudos Agrários Luiz de Queiroz, Piracicaba, Brasil. 1163 pp.

Alves, S.B., Lopes, R.B. (Eds.), 2008. Controle Microbiano de Pragas na América Latina: avanços e desafios. Fundação de Estudos Agrários Luiz de Queiroz, Piracicaba, Brasil. , p. 414.

Andreadis, T.G., Weseloh, R.M., 1990. Discovery of *Entomophaga maimaiga* in North American gypsy moth, *Lymantria dispar* (Entomophthorales/epizootic/Lymantriidae). Proc. Natl. Acad. Sci. U. S. A. 87, 2461–2465.

Barber, K.N., Kaupp, W.J., Holmes, S., 1993. Specificity testing of the nuclear polyhedrosis virus of the gypsy moth, *Lymantria dispar* (L.) (Lepidoptera: Lymantriidae). Can. Entomol. 125, 1055–1066.

Barbosa, P. (Ed.), 1998. Conservation Biological Control. Academic Press, San Diego. 396 pp.

Bassi, A., 1835. Del mal del segno, calcinaccio o muscardino. Lodi: Tipografia Orcesi. From Davidson, 2012.

Bedding, R.A., 2009. Controlling the pine-killing woodwasp, *Sirex noctilio*, with nematodes. In: Hajek, A.E., Glare, T.R., O'Callaghan, M. (Eds.), Use of Microbes for Control and Eradication of Invasive Arthropods. Springer, Dordrecht, The Netherlands, pp. 213–235.

Bedding, R.A., Akhurst, R.J., 1974. Use of the nematode *Deladenus siricidicola* in the biological control of *Sirex noctilio* in Australia. J. Austral. Entomol. Soc. 13, 129–135.

Bedford, G.O., 1980. Biology, ecology, and control of palm rhinoceros beetles. Annu. Rev. Entomol. 25, 309–339.

Burges, H.D., 1981. Microbial Control of Pests and Plant Diseases: 1970–1980. Academic Press, London, UK. 949 pp.

Burges, H.D., Hussey, N.W., 1971. Microbial Control of Insects and Mites. Academic Press, London, UK. 861 pp.

Campos-Herrera, R., Piedra-Buena, A., Escuer, M., Montalbán, B., Gutiérrez, C., 2010. Effect of seasonality and agricultural practices on occurrence of entomopathogenic nematodes and soil characteristics in La Rioja (Northern Spain). Pedobiol. 53, 253–258.

Cory, J.S., Evans, H.F., 2007. Viruses. In: Lacey, L.A., Kaya, H.K. (Eds.), Field Manual of Techniques in Invertebrate Pathology: Application and Evaluation of Pathogens for Control of Insects and Other Invertebrate Pests, second ed. Springer, Dordrecht, The Netherlands, pp. 149–174.

Davidson, E.W., 2012. History of insect pathology. In: Vega, F.E., Kaya, H.K. (Eds.), Insect Pathology, second ed. Academic Press, San Diego, pp. 13–28.

Dulmage, H.T., 1970. Insecticidal activity of HD-1, a new isolate of *Bacillus thuringiensis* var. *alesti*. J. Invertebr. Pathol. 15, 232–239.

Dutky, S.R., 1963. The milky diseases. In: Steinhaus, E.A. (Ed.). Insect Pathology: An Advanced Treatise, vol. 2. Academic Press, New York, pp. 75–115.

Eberle, K.E., Wennmann, J.T., Kleespies, R.G., Jehle, J.A., 2012. Basic techniques in insect virology. In: Lacey, L.A. (Ed.), Manual of Techniques in Invertebrate Pathology, second ed. Academic Press, San Diego, pp. 15–74.

Eilenberg, J., Hajek, A.J., Lomer, C., 2001. Suggestions for unifying the terminology in biological control. BioControl 46, 387–400.

Ekesi, S., Shah, P.A., Clark, S.J., Pell, J.K., 2005. Conservation biological control with the fungal pathogen, *Pandora neoaphidis*: implications of aphid species, host plant and predator foraging. Agric. For. Entomol 7, 21–30.

Faria, M.R., Wraight, S.P., 2007. Mycoinsecticides and mycoacaricides: a comprehensive list with worldwide coverage and international classification of formulation types. Biol. Control 43, 237–256.

Federici, B.A., 1997. Baculovirus pathogenesis. In: Miller, L.K. (Ed.), The Baculoviruses. Plenum Press, New York, pp. 33–59.

Federici, B.A., 2005. Insecticidal bacteria: an overwhelming success for invertebrate pathology. J. Invertebr. Pathol. 89, 30–38.

Frank, J.H., 2009. *Steinernema scapterisci* as a biological control agent of *Scapteriscus* mole crickets. In: Hajek, A.E., Glare, T.R., O'Callaghan, M. (Eds.), Use of Arthropods for Control and Eradication of Invasive Arthropods. Springer BV, Netherlands, pp. 115–131.

Frank, J.H., Walker, T.J., 2006. Permanent control of pest mole crickets (Orthoptera: Gryllotalpidae: *Scapteriscus*) in Florida. Am. Entomol. 52, 138–144.

Glare, T.R., O'Callaghan, M., 2000. *Bacillus thuringiensis*: Biology, Ecology and Safety. J. Wiley and Sons, Ltd., Chichester, UK. 350 pp.

Glare, T.R., Caradus, J., Gelernter, W., Jackson, T., Keyhani, N., Kohl, J., Marrone, P., Morin, L., Stewart, A., 2012. Have biopesticides come of age? Trends Biotechnol. 30, 250–258.

Goettel, M.S., Eilenberg, J., Glare, T.R., 2010. Entomopathogenic fungi and their role in regulation of insect populations. In: Gilbert, L.I., Gill, D.S. (Eds.), Insect Control: Biological and Synthetic Agents. Academic Press, San Diego, pp. 387–431.

Goettel, M.S., Hajek, A.E., Siegel, J.P., Evans, H.C., 2001. Safety of fungal biocontrol agents. In: Butt, T., Jackson, C., Magan, N. (Eds.), Fungi as Biocontrol Agents-Progress, Problems and Potential. CABI Press, Wallingford, UK, pp. 347–375.

Grewal, P.S., Ehlers, R.-U., Shapiro-Ilan, D.I., 2005. Nematodes as Biocontrol Agents. CABI Publishing, CAB International, Wallingford, Oxfordshire, UK. 505 pp.

Gurr, G.M., Wratten, S.D., Altieri, M.A., 2004. Ecological Engineering for Pest Management: Advances in Habitat Manipulation for Arthropods. CSIRO Publishing, Collingwood, Australia. 232 pp.

Gwynn, R. (Ed.), 2014. Manual of Biocontrol Agents 5th Edition. British Crop Protection Council, Alton, UK. 520 pp.

Hajek, A.E., 1997. Fungal and viral epizootics in gypsy moth (Lepidoptera: Lymantriidae) populations in Central New York. Biol. Control 10, 58–68.

Hajek, A.E., 1999. Pathology and epizootiology of the Lepidoptera-specific mycopathogen *Entomophaga maimaiga*. Microbiol. Mol. Biol. Rev. 63, 814–835.

Hajek, A.E., St Leger, R.J., 1994. Interactions between fungal pathogens and insect hosts. Annu. Rev. Entomol. 39, 293–322.

Hajek, A.E., Humber, R.A., Elkington, J.S., 1995. The mysterious origin of *Entomophaga maimaiga* in North America. Am. Entomol. 41, 31–42.

Hajek, A.E., Elkinton, J.S., Witcosky, J.J., 1996. Introduction and spread of the fungal pathogen *Entomophaga maimaiga* along the edge of gypsy moth spread. Environ. Entomol. 25, 1235–1247.

Hajek, A.E., McManus, M.L., Delalibera Junior, I., 2005. Catalogue of Introductions of Pathogens and Nematodes for Classical Biological Control of Insects and Mites. USDA. For. Serv. FHTET-2005-05. 59 pp.

Hajek, A.E., Delalibera Jr., I., McManis, M.L., 2007. Introduction of exotic pathogens and documentation of their establishment and impact. In: Lacey, L.A., Kaya, H.K. (Eds.), Field Manual of Techniques in Invertebrate Pathology: Application and Evaluation of Pathogens for Control of Insects and Other Invertebrate Pests, second ed. Springer, Dordrecht, The Netherlands, pp. 299–325.

Hajek, A.E., Glare, T.R., O'Callaghan, M. (Eds.), 2009. Use of Microbes for Control and Eradication of Invasive Arthropods. Springer, Dordrecht, The Netherlands. 366 pp.

Harper, J.D., 1987. Applied epizootiology: microbial control of insects. In: Fuxa, J.R., Tanada, Y. (Eds.), Epizootiology of Insect Diseases. Wiley & Sons, NY, pp. 473–496.

Hokkanen, H.M.T., Hajek, A.E. (Eds.), 2003. Environmental Impacts of Microbial Insecticides: Need and Methods for Risk Assessment. Kluwer Academic Publishers, Dordrecht, The Netherlands. 269 pp.

Hoy, M.A., 2008a. Augmentative biological control. In: Capinera, J.L. (Ed.), Encyclopedia of Entomology, second ed. Springer Dordrecht, The Netherlands, pp. 327–334.

Hoy, M.A., 2008b. Classical biological control. In: Capinera, J.L. (Ed.), Encyclopedia of Entomology, second ed. Springer Dordrecht, The Netherlands, pp. 906–923.

Hudson, W.G., Frank, J.H., Castner, J.L., 1988. Biological control of *Scapteriscus* spp. mole crickets (Orthoptera: Gryllotalpidae) in Florida. Bull. Entomol. Soc. Am. 34, 192–198.

Hüger, A.M., 1966. A virus disease of the Indian rhinoceros beetle, *Oryctes rhinoceros* (Linnaeus), caused by a new type of insect virus, *Rhabdionvirus oryctes* gen. n., sp. n. J. Invertebr. Pathol. 8, 38–51.

Hüger, A.M., 2005. The *Oryctes* virus: its detection, identification, and implementation in biological control of the coconut palm rhinoceros beetle, *Oryctes rhinoceros* (Coleoptera: Scarabaeidae). J. Invertebr. Pathol. 89, 78–84.

Hummel, R.L., Walgenbach, J.F., Barbercheck, M.E., Kennedy, G.G., Hoyt, G.D., Arellano, C., 2002. Effects of production practices on soil-borne entomopathogens in Western North California vegetable systems. Environ. Entomol. 31, 84–91.

Jackson, T.A., 2009. The use of Oryctes virus for control of rhinoceros beetle in the Pacific Islands. In: Hajek, A.E., Glare, T.R., O'Callaghan, M. (Eds.), Use of Arthropods for Control and Eradication of Invasive Arthropods. Springer, The Netherlands, pp. 133–140.

Jackson, T.A., Crawford, A.M., Glare, T.R., 2005. Oryctes virus- time for a new look at a useful biocontrol agent. J. Invertebr. Pathol. 89, 91–94.

Kabaluk, T., Gazdik, K., 2005. Directory of Microbial Pesticides for Agricultural Crops in OECD Countries. http://www.organicagcentre.ca/Docs/MicrobialDirectory-English-V237-05-Revision1.pdf.

Kaya, H.K., Lacey, L.A., 2007. Introduction to microbial control. In: Lacey, L.A., Kaya, H.K. (Eds.), Field Manual of Techniques in Invertebrate Pathology: Application and Evaluation of Pathogens for Control of Insects and Other Invertebrate Pests, second ed. Springer, Dordrecht, The Netherlands, pp. 3–7.

Kaya, H.K., Vega, F.E., 2012. Scope and basic principles of insect pathology. In: Vega, F.E., Kaya, H.K. (Eds.), Insect Pathology, second ed. Academic Press, San Diego, pp. 1–12.

Klein, M.G., 1981. Advances in the use of *Bacillus popilliae* for pest control. In: Burges, H.D. (Ed.), Microbial Control of Pests and Plant Diseases 1970–1980. Academic Press, San Diego, pp. 183–192.

Klein, M.G., Georgis, R., 1992. Persistence of control of Japanese beetle (Coleoptera: Scarabaeidae) larvae with steinernematid and heterorhabditid nematodes. J. Econ. Entomol. 85, 727–730.

Krassiltstchik, I.M., 1888. La production industrielle des parasites végétaux pour la destruction des insects nuisibles. Bull. Sci. Fr. 19, 461–472.

Kurstak, E., 1962. Données sur l'epizootie bacterienne naturelle provoquée par un *Bacillus* du type *Bacillus thuringiensis* var. *alesti* sur *Ephestia kuhniella* Zeller. Entomophaga Mem. Hors. Ser. 2, 245–247.

Lacey, L.A., Siegel, J.P., 2000. Safety and ecotoxicology of entomopathogenic bacteria. In: Charles, J.-F., Delécluse, A., Nielsen-LeRoux, C. (Eds.), Entomopathogenic Bacteria: From Laboratory to Field Application. Kluwer Academic Publishers, Dordrecht, The Netherlands, pp. 253–273.

Lacey, L.A., Frutos, R., Kaya, H.K., Vail, P., 2001. Insect pathogens as biological control agents: do they have a future? Biol. Control 21, 230–248.

Lacey, L.A., Grzywacz, D., Shapiro-Ilan, D.I., Frutos, R., Goettel, M.S., Brownbridge, M., 2015. Insect pathogens as biological control agents: back to the future. J. Invertebr. Pathol. 132, 1–41.

Laird, M., Davidson, E.W., Lacey, L.A. (Eds.), 1990. Safety of Microbial Insecticides. CRC Press, Boca Raton, FL. USA. 259 pp.

Lewis, E.E., Campbell, J.F., Gaugler, R., 1998. A conservation approach to using entomopathogenic nematodes in turf and ornamentals. In: Barbosa, P. (Ed.), Conservation Biological Control. Academic Press, San Diego, pp. 235–253.

Lord, J.C., 2005. From Metchnikoff to Monsanto and beyond: the path of microbial control. J. Invertebr. Pathol. 89, 19–29.

McCrevy, K.W., 2008. Conservation biological control. In: Capinera, J.L. (Ed.), Encyclopedia of Entomology, second ed. Springer Dordrecht, The Netherlands, pp. 1021–1023.

Metchnikoff, E., 1879. O boleznach litchinok khlebnogo zhuka. In: Zapiski Imperatorskogo Obschestva Sel' Skogo Khoziaistva Iuzhnoi Rossi. Odessa, pp. 21–50From Steinhaus, 1975.

Metchnikoff, E., 1880. Zur Lehre über Insektenkrankheiten. Zool. Anz 3, 44–47 From Lord, 2005.

Meyling, N., Eilenberg, J., 2007. Ecology of the entomopathogenic fungi *Beauveria bassiana* and *Metarhizium anisopliae* in temperate agroecosystems: potential for conservation biological control. Biol. Control 43, 145–155.

Millar, L.C., Barbercheck, M.E., 2002. Effects of tillage practices on entomopathogenic nematodes in a corn agroecosystems. Biol. Control 25, 1–11.

Miller, L.K. (Ed.), 1997. The Baculoviruses. Plenum Press, New York. 447 pp.

Miller, L.K., Ball, L.K., 1998. The Insect Viruses. Plenum Press, New York. 413 pp.

Nielsen, C., Milgroom, M.G., Hajek, A.E., 2005. Genetic diversity in the gypsy moth fungal pathogen *Entomophaga maimaiga* from founder populations in North America and source populations in Asia. Mycol. Res. 109, 941–950.

Nielsen, C., Jensen, A.B., Eilenberg, J., 2007. Survival of entomophthoralean fungi infecting aphids and higher flies during unfavourable conditions and implications for conservation biological control. In: Ekesi, S., Maniania, N.K. (Eds.), Use of Entomopathogenic Fungi in Biological Pest Management. Research Signpost, Kerala, India, pp. 13–38.

Odindo, M.O., 1983. Epizootiological observations on a nuclear polyhedrosis of the African armyworm *Spodoptera exempta* (Walk.). Insect Sci. Appl. 4, 291–298.

Parkman, J.P., Hudson, W.G., Frank, J.H., Nguyen, K.B., Smart, G.C., 1993. Establishment and persistence of *Steinernema scapterisci* (Rhabditida: Steinernematidae) in field populations of *Scapteriscus* mole crickets (Orthoptera: Grillotalpidae). J. Entomol. Sci. 8, 182–190.

Pasteur, L., 1870. Études sur la maladie de ver à soir. Tome I et II. Gauthier-Villars, Paris.

Pell, J.K., Hannam, J.J., Steinkraus, D.C., 2010. Conservation biological control using fungal entomopathogens. BioControl 55, 187–198.

Peters, A., 1996. The natural host range of *Steinernema* and *Heterorhabditis* spp. and their impact on insect populations. Biocontrol Sci. Technol. 6, 389–402.

Platzer, E.G., 2007. Mermithid nematodes. In: Floore, T.G. (Ed.), Biorational Control of Mosquitoes. Amer. Mosq. Contr. Assoc. Bull. 7, 58–64.

Poinar Jr., G.O., 1975. Entomogenous Nematodes. A Manual and Host List of Insect-Nematode Associations. E. J. Brill, Leiden, The Netherlands. 317 pp.

Possee, R.D., King, L.A., 2014. Inset Viruses. Wiley and Sons. http://dx.doi.org/10.1002/9780470015902.a0020712.pub2.

Radcliffe, E.B., Hutchison, W.B., Cancelado, R.E., 2009. Integrated Pest Management: Concepts, Tactics, Strategies and Case Studies. Cambridge University Press, New York. 529 pp.

Rose, D.J.W., Dewhurst, C.F., Page, W.W., 2000. The African Armyworm Handbook, second ed. Natural Resources Institute, Greenwich, UK. 304 pp.

Sanchis, V., 2011. From microbial sprays to insect-resistant transgenic plants: history of the biospesticide *Bacillus thuringiensis*. A review. Agron. Sustain. Dev. 31, 217–231.

Shapiro-Ilan, D.I., Bruck, D., Lacey, L.A., 2012. Principles of epizootiology and microbial control. In: Vega, F.E., Kaya, H.K. (Eds.), Insect Pathology, second ed. Academic Press, San Diego, pp. 29–71.

Solter, L., Hajek, A., 2009. Control of gypsy moth, *Lymantria dispar*, in North America since 1878. In: Hajek, A.E., Glare, T.R., O'Callaghan, M. (Eds.), Use of Arthropods for Control and Eradication of Invasive Arthropods. Springer BV, Netherlands, pp. 181–212.

Soper, R.S., Shimazu, M., Humber, R.A., Ramos, M.E., Hajek, A.E., 1988. Isolation and characterization of *Entomophaga maimaiga* sp. nov., a fungal pathogen of gypsy moth, *Lymantria dispar*, from Japan. J. Invertebr. Pathol. 51, 229–241.

Steinhaus, E., 1949. Insect Pathology. Academic Press. 757 pp.

Steinhaus, E., 1956. Living insecticides. Sci. Am. 195, 96–103.

Steinhaus, E., 1963. Insect Pathology: An Advanced Treatise, vol. 2. Academic Press, New York. 689 pp.

Steinhaus, E., 1975. Disease in a Minor Chord. Ohio State University Press, Columbus. 488 pp.

Steinkraus, D.C., 2007. Documentation of naturally occurring pathogens and their impact in agroecosystems. In: Lacey, L.A., Kaya, H.K. (Eds.), Field Manual of Techniques in Invertebrate Pathology: Application and Evaluation of Pathogens for Control of Insects and Other Invertebrate Pests, second ed. Springer, Dordrecht, The Netherlands, pp. 267–281.

Szewcyk, B., Hoyos-Carvajal, L., Paluszek, M., Skrzecz, I., Lobo de Souza, M., 2006. Baculoviruses re-emerging biopesticides. Biotechnol. Adv. 24, 143–160.

Tanada, Y., Kaya, H.K., 1993. Insect Pathology. Academic Press, San Diego. 666 pp.

Vail, P.V., Jay, D.L., 1973. Pathology of the nuclear polyhedrosis virus of the alfalfa looper in alternate hosts. J. Invertebr. Pathol. 21, 198–204.

Vail, P.V., Jay, D.L., Hunter, D.K., 1971. Cross infectivity of a nuclear polyhedrosis virus isolated from the alfalfa looper, Autographa californica. Proc. IVth Int. Colloq. Insect Pathol. 297–304.

Vail, P.V., Hostetter, D.L., Hoffmann, F., 1999. Development of multinucleocapsid polyhedroviruses (MNPVs) infectious to loopers as microbial control agents. Int. Pest Manag. Rev. 4, 231–257.

Vega, F.E., Meyling, N.V., Luangsa-Ard, J.J., Blackwell, M., 2012. Fungal entomopathogens. In: Vega, F.E., Kaya, H.K. (Eds.), Insect Pathology, second ed. Academic Press, San Diego, pp. 171–220.

Welch, H.E., 1963. Nematode infections. In: Steinhaus, E. (Ed.). Insect Pathology: An Advanced Treatise, vol. 2. Academic Press, New York, pp. 363–392.

Weseloh, R.M., 2003. Short and long range dispersal of the gypsy moth (Lepidoptera: Lymantriidae) fungal pathogen, Entomophaga maimaiga (Zygomycetes: Entomophthorales). Environ. Entomol. 32, 111–122.

Wraight, S.P., Hajek, A.E., 2009. Manipulation of arthropod pathogens for IPM. In: Radcliffe, E.B., Hutchison, W.B., Cancelado, R.E. (Eds.), Integrated Pest Management: Concepts, Tactics, Strategies and Case Studies. Cambridge University Press, New York, pp. 131–150.

Zelazny, B., 1972. Studies on Rhabdionvirus rhinoceros. I. Effects on larvae of Oryctes rhinoceros and inactivation of the virus. J. Invertebr. Pathol. 20, 235–241.

Zelazny, B., 1973. Studies on Rhabdionvirus rhinoceros. II. Effects on adults of Oryctes rhinoceros. J. Invertebr. Pathol. 22, 122–126.

Chapter 2

Exploration for Entomopathogens

L.F. Solter[1], A.E. Hajek[2], L.A. Lacey[3]

[1]University of Illinois, Champaign, IL, United States; [2]Cornell University, Ithaca, NY, United States; [3]IP Consulting International, Yakima, WA, United States

2.1 INTRODUCTION

In addition to the large number of commercially available microbial control products, foreign and domestic exploration for entomopathogens of invertebrate pests, including insects, mites, nematodes, and slugs, continues to provide candidate microbial control isolates and species with host range and environmental tolerance appropriate for the target pests and system, greater virulence, and other beneficial traits. Several historic and important discoveries were made while addressing pathogen problems in insect colonies, and some of the most significant discoveries of microbial control agents were made while conducting field research unrelated to insect pathology. Purposeful exploration for biological control agents initially was focused on the more easily observable predators and parasites (DeBach, 1974), but as the importance of diseases as primary factors in invertebrate population dynamics became more clearly understood, pathogens were also targeted (Steinhaus, 1949, 1963; Anderson and May, 1981). More recently, efforts have focused on exploration for pathogen isolates that are adapted to the climatic conditions of the target host in the area of origin, as well as those with higher infectivity and virulence (Lacey and Solter, 2012; Hajek et al., 2007).

2.2 GENERAL SEARCH FOR ENTOMOPATHOGENS

Although this chapter primarily concerns purposeful exploration for microbial control agents, managed colonies of insects yielded some of the first pathogens that, although they were serious problems in the colonies they infected, eventually were recognized as valuable microbial control agents. For example, silkworm colonies were the source of the discovery of *Beauveria bassiana* by Agostino Bassi (1835) and of *Bacillus thuringiensis* (originally *B. sotto*) by Ishiwata (1901). Seventy years later, Dulmage (1970) isolated *B. thuringiensis* subsp. *alesti* (later renamed *B. thuringiensis* subsp. *kurstaki*) from mass-produced pink bollworm, *Pectinophora gossypiella*.

Field research unrelated to exploration has also yielded some important microbial control agents. Insects infected with bacteria, viruses, fungi, and nematodes are frequently found during sampling of insect pests because of their distinct signs and symptoms such as color changes, unusual deterioration, and odor (Lacey and Solter, 2012). An example is the granulovirus of codling moth *Cydia pomonella*, a serious pest of apples, which was collected by L. E. Caltagirone (Division of Biological Control, University of California, Berkeley) during a research trip in Chihuahua, Mexico (Tanada, 1964). Since discovery of the "Mexican strain" of the virus, CpGV, numerous basic and applied studies have resulted in commercial development of the pathogen and worldwide use by growers for inundative control of the moth (Lacey et al., 2008a).

2.2.1 Discoveries of Entomopathogens From Regional Native Pests

Entomopathogens have been discovered in numerous pest insects indigenous to North America. Some examples include the entomopathogenic fungi (EPF) *Neozygites fresenii* in the cotton aphid *Aphis gossypii* (Steinkraus et al., 1991), *B. bassiana* in the Colorado potato beetle *Leptinotarsa decimlineata* (Bajan and Kmitowa, 1977), *Furia gastropachae* in forest tent caterpillars *Malacosoma disstria* (Filotas et al., 2003); the entomopathogenic nematodes (EPN) *Heterorhabditis hepialius* infecting the ghost moth *Hepialis californicusa* (Stock et al., 1996), *Steinernema riobrave* infecting cotton bollworm *Helicoverpa zea* (Cabanillos et al., 1994); and the nucleopolyhedrovirus of the alfalfa looper *Autographa californica* (Vail et al., 1971; Vail and Jay, 1973). Additional examples, including those from other locations globally, are presented in Table 2.1.

2.2.2 Notable Discoveries of Entomopathogens of Exotic Pest Insects and Mites Outside of the Putative Center of Origin

Several important entomopathogens were discovered in regions where the hosts were exotic invasive pests and were subsequently developed and used as microbial control agents

Microbial Control of Insect and Mite Pests. http://dx.doi.org/10.1016/B978-0-12-803527-6.00002-0

TABLE 2.1 Examples of Entomopathogens Originally Collected in the Same Region or Country Where a Pest Is Endemic or Has Been Established for More Than 100 Years

Pathogen	Host	Country	Original and Supplementary References
Virus			
AucaMNPV	Alfalfa looper, *Autographa californica*	USA	Vail et al. (1971), Vail and Jay (1973)
AnfaMNPV, synonymous with RaouMNPV	Celery looper, *Anagrapha falcifera*, Grey looper moth, *Rachoplusia ou*	USA	Hostetter and Putler (1991), Harrison and Bonning (1999)
NeabSNPV	Balsam fir sawfly, *Neodiprion abietis*	Canada	Moreau et al. (2005), Lucarotti et al. (2007)
AdorGV	*Adoxophyes orana*	Japan, Europe	Oho et al. (1974), Blommers et al. (1987)
CpGV[a]	Codling moth, *Cydia pomonella*	Mexico	Tanada (1964), Lacey et al. (2008a)
Bacteria			
Bacillus thuringiensis	Mediterranean flour moth, *Ephestia kuhniella*	Germany	Berliner (1915)
B. t. subsp. kurstaki[a]	*E. kuhniella, Pectinophora gossypiella*	Canada, USA	Kurstak (1962), Dulmage (1970)
B. t. subsp. morrisoni var. tenebrionis[a]	Colorado potato beetle, *Leptinotarsa decemlineata*	Germany	Kreig et al. (1983)
B. t. subsp. israelensis[a, b]	*Culex quinquefasciatus*	Israel	Goldberg and Margolit (1977)
Lysinibacillus sphaericus 1593, 2362[a, b]	dead mosquito larvae, and adult *Simulium damnosum*	Indonesia, Nigeria	Singer (1973), Weiser (1984)
Paenibacillus popilliae, P. lentamorbis	Japanese beetle, *Popillia japonica*	USA	Dutky (1940, 1963)
Serratia entomophila	Pasture grub, *Entomophila zealandica*	New Zealand	Trought et al. (1982), Stucki et al. (1984), Jackson et al. (1992)
Chromobacterium subtsugae	Soil	USA	Martin et al. (2007)
Fungi			
Isaria fumosorosea[a] Apopka strain 97	Sweet potato whitefly, *Bemisia tabaci* Biotype B	Florida, USA	Osborne et al. (1990), Lacey et al. (2008b)
Beauveria bassiana[a] GHA	Spotted cucumber beetle, *Diabrotica undecimpunctata*	Oregon, USA	ARSEF (2014)
Beauveria brongniartii	*Melolontha melolontha*	Switzerland	Keller (1978), Zimmermann (1992)
Metarhizium anisopliae[a] Met 52	*C. pomonella*	Austria	Zimmermann et al. (2013)
Metarhizium acridum Green Muscle	*Ornithacris cavroisi*	Niami, Niger	Prior et al. (1992), Lomer et al. (2001, 1997)
Lecanicillium muscarium[a]	Whiteflies, thrips	UK and Sri Lanka	Cuthbertson and Walters (2005), Lacey et al. (2008b) Humber, personal communication
Lecanicillium longisporum[a]	Aphids	UK and Sri Lanka	Goettel et al. (2008) Humber, personal communication
Neozygites tanajoae	Cassava green mite, *Mononychellus tanajoa*	Brazil	Delalibera et al. (1992)

TABLE 2.1 Examples of Entomopathogens Originally Collected in the Same Region or Country Where a Pest Is Endemic or Has Been Established for More Than 100 Years—cont'd

Pathogen	Host	Country	Original and Supplementary References
Nematodes			
Steinernema carpocapsae[a]	C. pomonella	Czechoslovakia, USA	Weiser (1955), Dutky and Hough (1955)
S. glaseri	Japanese beetle, Popillia japonicum	New Jersey, USA	Glaser and Fox (1930), Gaugler et al. (1992)
S. riobrave[a]	Helicoverpa zea	Texas, USA	Cabanillas et al. (1994)
Heterorhabditis megadis[a]	P. japonicum	Ohio, USA	Poinar et al. (1987)
H. bacteriophora[a]	Heliothis punctigera	Brecon, South Australia	Poinar (1976)
H. marelatus	Galleria mellonella, soil baiting	Oregon, USA	Liu and Berry (1996)

[a]Worldwide distribution subsequent to commercialization.
[b]These species were collected in countries where the pest insect was already present, however, they have been exported worldwide for control of hundreds of pest insects: L. bacillus for control of dozens of culicid species; Bti for a far greater number of pests including: all Culicidae and Simuliidae, and certain species in the Chironomidae, Tipulidae (Tipula paludosa), and Sciaridae (Lycoriella spp.).

(MCAs). The original collections of the EPNs *Steinernema* (*Neoaplectana*) *carpocapsae* (Weiser, 1955) and *Steinernema carpocapsae* DD-136 (Dutky and Hough, 1955) from *Cydia pomonella* were made while conducting general pathogen searches in apple orchards in Czechoslovakia and the United States, respectively. Although *C. pomonella* invaded Europe and North America centuries ago, it is nevertheless an exotic species originating from Asia Minor. Because all infections of EPNs are lethal to *C. pomonella* larvae within a few days after penetration by the infective juvenile (IJ), it is unlikely that the nematode accompanied the moth from its center of origin to the United States and Czechoslovakia. Similarly, Nguyen and Duncan (2002) discovered *Steinernema diaprepesi*, an effective MCA for control of the invasive citrus root weevil *Diaprepes abbreviatus* (McCoy et al., 2007), when prospecting for EPNs in Florida.

Prospecting for natural enemies of the Japanese beetle, *Popillia japonica*, an established exotic pest in the United States, led to the discovery and development of *Paenibacillus popilliae* and *P. lentimorbis*, the causal agents of milky disease (Dutky, 1940, 1963). Likewise, the efficacious EPNs *Steinernema glaseri* and *Heterorhabditis megadis* were isolated from *P. japonica* during surveys by Glaser and Fox (1930) and Poinar et al. (1987), respectively.

The use of bait insects such as larvae of the wax moth *Galleria mellonella* for general prospecting of fungi and EPNs has resulted in the discovery of a multitude of fungi and EPNs that attack an even larger number of insect pests (Bedding and Akhurst, 1975; Zimmermann, 1986; Stock and Goodrich-Blair, 2012). Using microbiological methods, including selective media containing antibiotics,

numerous strains and new varieties of *Bacillus thuringiensis* continue to be found in a diversity of habitats, including soil, the phylloplane, grain dust, water, and diseased insects (Sanchis, 2011; Fisher and Garczynski, 2012). These and other methods used for search and initial processing of insects infected with entomopathogens and from habitats associated with the pest were reviewed by Lacey and Solter (2012), Eberle et al. (2012), Fisher and Garczynski (2012), Inglis et al. (2012), and Stock and Goodrich-Blair (2012).

2.3 FOREIGN EXPLORATION FOR ENTOMOPATHOGENS OF EXOTIC INVERTEBRATE PESTS

2.3.1 Criteria for Selecting Exploration Sites

When exotic pests are introduced into a new region without their natural enemies, massive economically devastating outbreaks can occur. Such outbreaks have often encouraged exploration for natural enemies of the pest, including entomopathogens. Several discoveries made during foreign exploration have resulted in classic biological control of exotic and other pest invertebrates (Table 2.2; and Hajek, 2009; Hajek et al., 2005, 2007). Considerations for selection of sites in which to conduct foreign and domestic exploration include identifying the center of origin of the exotic host and matching crops and climate to that of the area invaded.

2.3.1.1 Purported Center of Origin of Pest

For most invasive species, although not all, the center of origin is known or can be approximated using museum and

TABLE 2.2 Examples of Entomopathogens Collected During Foreign Exploration for Release and Establishment in Countries With Invasive Insects

Entomopathogen	Targeted Pest Insect	Country in Which Collection Was Made	Countries in Which Entomopathogen Was Released	Original and Supplemental References
Virus				
Neodiprion sertifer SNPV	Pine sawfly, *Neodiprion sertifer*	Sweden	Canada, USA, Scotland	Bird (1953), Bird and Whalen (1953)
Oryctes Nudivirus	Coconut palm rhinoceros beetle, *Oryctes rhinoceros*	Malaysia	Samoa, Fiji, Mauritius, and other islands	Hüger (1966, 2005), Bedford (1980)
Bacteria[a]				
Fungi				
Zoophthora radicans	Spotted alfalfa aphid, *Therioaphis trifolii*	Israel	Australia	Milner et al. (1982)
Entomophaga maimaiga	Gypsy moth, *Lymantria dispar*	Japan	USA	Soper et al. (1988)
Neozygites tanajoae	Cassava green mite, *Mononychellus tanajoa*	Brazil	Benin	Prior et al. (1992), Delalibera et al. (1992), Delalibera (2009)
Nosema portugal (Microsporidia)	*L. dispar*	Europe	USA	Jeffords et al. (1989), Solter and Hajek (2009)
Nematodes				
Deladenus siricidicola (Tylenchida: Neotylenchidae)	Wood wasp, *Sirex noctilio*	Europe	Australia, Brazil, North America	Bedding (1968, 2009), Bedding and Akhurst (1974)
Romanomermis iyengari (Mermithida: Mermithidae)[b]	*Anopheles* spp., *Aedes* spp., *Culex* spp.	India	Azerbaijan, Cuba, Brazil, Mexico, Benin	Gajanana et al. (1978), Santamarina Mijares (1994), Pérez-Pacheco et al. (2005), Abagli et al. (2012)
R. culicivorax	*Anopheles albimanus*, *An. punctipennis*	USA	El Salvador	Petersen et al. (1978), Petersen (1985)
Steinernema scapterisci (Rhabditida: Steinernematidae)	Mole crickets, *Scapteriscus* spp.	Uruguay	Florida, USA	Hudson et al. (1988), Nguyen and Smart (1990), Parkman et al. (1993)

[a]*See Table 2.2 for Bacillus thuringiensis subsp. israelensis and Lysinibacillus sphaericus. These species were collected in countries where the pest insect was already present; however, they were exported worldwide for control of hundreds of pest insects and therefore can also be considered as foreign, with importation into countries where diverse pests include both exotics and endemics. The range of susceptible species for L. bacillus includes dozens of culicid species; for Bti a far greater number of susceptible pests including all Culicidae and Simuliidae, and certain species in the Chironomidae, Tipulidae (Tipula paludosa), and Sciaridae (Lycoriella spp.).*
[b]*Romanomermis iyengari was introduced into neotropical countries in which most mosquito species were not exotic. Some species such as Aedes aegypti are exotic but have been established for centuries. Other exotic species such as Ae. albopictus have become more recently established.*

historic records. In addition, molecular methods can be used to search for locations with greatest genetic variability, which is often associated with the area of endemicity of an invasive. The center of origin is the first location to investigate when searching for pathogens with potential to control an invasive pest because natural enemy diversity in a host species is presumed to be highest in the host's center of origin where the host and its pathogens and parasites have a shared evolutionary history (Mitchell and Power, 2003; Li et al., 2012; Bing et al., 2014). Typically, even if the host is an outbreak species in the area of origin, its natural enemies provide a check on populations. Introduced pest species, however, may be missing key regulators of their populations (Hajek et al., 2007). An example is *Entomophaga maimaiga,* a fungal pathogen of gypsy moth, which was found in Japan where the gypsy moth was thought to have evolved before spreading west to Europe. *E. maimaiga* was eventually introduced (intentionally and possibly accidentally) to North America, where control of the invasive gypsy was insufficient to prevent frequent outbreaks.

2.3.1.2 Climate Matching

Like their hosts, pathogens may be adapted to specific climatic conditions. Annual variables used to match the climate in areas of invasion with those in the area of origin include (1) extreme minimum and maximum temperatures, (2) number of frost-free days, (3) degree days (function of time that varies with temperature), (5) precipitation annual total, (6) average temperature of the 3 warmest months of the year, (7) potential evapo-transpiration (PET) for the 3 warmest months of the year, and (8) precipitation for the 3 warmest months of the year. Tools for obtaining climate information in regions where the target insect is found include historical climate records, reports and publications from colleagues, and climate software (Climex, PC Climate, NAPPFAST). Basing selection of sites on historical temperature (mean, median, range) and precipitation should take into account seasonal variations based on cyclical climate such as the El Niño effect and more recent occurrences (within the past 5 years) such as unusual droughts or excessive precipitation that may occur more frequently due to climate warming.

2.3.1.3 Crop Selection

Polyphagous species such as the sweetpotato whitefly *Bemisia tabaci* (Hemiptera: Aleyrodidae), brown marmorated stink bug *Halyomorpha halys* (Hemiptera: Pentatomidae), and several lepidopteran and coleopteran species attack a broad range of crops. *B. tabaci* collections were made over a broad geographic and host plant range (Lacey et al., 1993, 2008b; Kirk et al., 2008), and initial exploration was conducted for natural enemies in Western Asia (India, Nepal, Pakistan), the putative center of origin for this species, on a variety of crops (cotton, cucurbits, sweet potato, eggplant and other Solanaceae) and other host plants including some *Euphorbia* spp. such as poinsettia, *Euphorbia pulcherrima.* Likewise, pines in Europe, Asia and North Africa were surveyed for most of a decade when searching for the nematode parasite of *Sirex noctilio* (Spradbery and Kirk, 1978).

2.3.1.4 Collecting From Closely Related Invertebrates

Pathogens have been collected from host species that are closely related to the intended target; for example, EPF from *Trialeurodes vaporariorum, Aleyrodes* spp. and species in the family Coccidae were isolated as potential MCAs of *Bemisia tabaci* (Lacey et al., 2008b). When searching for nematodes

to parasitize *Sirex noctilio,* hosts in three genera of siricids were evaluated (Bedding and Akhurst, 1978). Neoclassic biological control (use of natural enemies of other, sometimes related, hosts to control a pest) has not been particularly successful, especially for pathogens with relatively narrow host ranges, most likely due to subtle biological and ecological differences between the natural host and the target host that prevent optimal reproduction and transmission (e.g., Solter and Maddox, 1998; Solter et al., 2005).

2.3.1.5 Other Activities Associated With Foreign Exploration

Obtaining permission to collect and export entomopathogens from host countries, usually through the host country government agencies and university colleagues, is essential. In recent years, the Convention on Biological Diversity (CBD) determined that countries have sovereign rights over their genetic resources. Based on the 2010 Nagoya Protocol, agreements governing access to these resources and sharing of the benefits from their use must now be established between involved parties (Cock et al., 2010). While this protocol is largely in place so that profits derived from genetic resources belonging to a country are shared, this also applies to species collected for use in biological control, which often results in little profit. Although in the recent past these new rules have impacted the ability to collect natural enemies in some countries, there now is an attempt to have biological control agents excluded from this protocol.

Government (federal, state, provincial) clearance to release pathogens in the country where the exotic pest is invasive is almost always required. Specificity testing should be conducted to ensure that pathogens will not impact native nontarget species and beneficial organisms and can be initiated in the host country. Specificity testing is usually conducted in quarantine in the country where the pathogen will be released. Several additional activities associated with foreign exploration are provided by Hajek et al. (2007).

2.4 CASE STUDIES OF SUCCESSFUL INTRODUCTIONS OF NONINDIGENOUS ENTOMOPATHOGENS FOR CONTROL OF INSECT AND MITE PESTS AND PARASITIC NEMATODES FOR MOSQUITO CONTROL

2.4.1 Classic Biological Control of Cassava Green Mite With an Entomopathogenic Fungus

The cassava green mite, *Mononychellus tanajoa,* is an indigenous pest of cassava, *Manihot esculenta,* in the Neotropics.

The mite became a major pest of cassava across equatorial Africa following introduction in the early 1970s (Bellotti et al., 1999). Exploration for potential natural enemies to be introduced into Africa began in Brazil in 1988 and resulted in the discovery of the fungus, *Neozygites tanajoae* (Zygomycetes: Entomophthorales) (Delalibera, 2009). *N. tanajoae* is specific to the mite, is native to South America (Hountondji et al., 2002b; Delalibera and Hajek, 2004). Delalibera et al. (2000), and demonstrated the capability of *N. tanajoae* to produce epizootics and cause high mortality in *M. tanajoa* in Brazil. A Brazilian isolate was introduced into Benin, produced epizootics, and became established (Hountondji et al., 2002a; Delalibera, 2009; Agboton et al., 2011). Delalibera (2009) provided a detailed account of this classic microbial control success.

2.4.2 Foreign Exploration for an Effective Classic Biological Control Agent of Gypsy Moth *Lymantria dispar* for Long-Term Control

The gypsy moth is a pest across Europe and Asia but was absent from the Western Hemisphere until it was imported into the United States in 1869. From its original point of entry near Boston, Massachusetts, it has dispersed south along the Appalachian Ridge and west to Illinois and Minnesota (Tobin and Blackburn, 2007). Larvae of *L. dispar* feed on a wide range of deciduous trees, and during cyclic outbreaks have been responsible for defoliation of trees in up to 2 million ha of forest (Solter and Hajek, 2009). The fungal pathogen *Entomophaga maimaiga* was collected in Japan and first released in the United States near Boston in 1910 and 1911 but was not detected in following years (Hajek et al., 1995). It was also collected in Japan by Soper et al. (1988). Small field applications of the fungus in 1985 in New York State and in 1986 in Virginia resulted in few or no apparent infections. It was not until 1989 that significant epizootics caused by *E. maimaiga* were reported (Andreadis and Weseloh, 1990; Hajek, 1997, 1999). The exact origin of the *E. maimaiga* isolate causing these outbreaks is not precisely known (Hajek et al., 1995). Since the first outbreaks occurred, in vivo produced fungal spores (conidia and resting spores) and infected larvae have been introduced into areas on the leading edge of *L. dispar* dispersal. Subsequent infections indicate that *E. maimaiga* has become established in and spread from locations where it was not previously observed (Hajek et al., 1996; Weseloh, 2003; Solter and Hajek, 2009). In 1999, *E. maimaiga* was successfully introduced into a gypsy moth population in Bulgaria. Recent investigations suggest that the fungus is quickly spreading in Europe (Zùbrik et al., 2016). More detail on this continuing successful classic microbial control is presented by Hajek et al. (2015) and in Chapter 21.

2.4.3 Release of the *Neodiprion sertifer* Nucleopolyhedrovirus for Control of the European Pine Sawfly in North America

The European pine sawfly *Neodiprion sertifer* (Hymenoptera: Diprionidae) is endemic to the Palearctic Region, and during periodic outbreaks damage to pines can be significant. In the United States and Canada, it is an exotic pest of pine that was introduced from Sweden in 1925 (Geri, 2013). As part of the effort to control the sawfly, the *N. sertifer* nucleopolyhedrovirus (*NeseSNPV*) was introduced from Sweden into Canada (Hajek et al., 2007) and spread rapidly through the cohort and ultimately the population (Dowden and Girth, 1953; Bird, 1955). The U.S. Forest Service subsequently produced Neochek-S, a formulation of NeseSNPV, shown to provide significant control in field trials (Podgwaite et al., 1984). Establishment of NeseSNPF has resulted in long-term suppression of *N. sertifer* (Cunningham and Entwistle, 1981; Cunningham, 1998); see also Chapter 21 for additional details on *NeseSNPV*.

2.4.4 Classic Biological Control of Mosquito Larvae With Imported *Romanomermis iyengari* (Mermithidae)

Romanomermis iyengari is a tropical nematode species indigenous to India (Gajanana et al., 1978) and is successful in controlling mosquitoes breeding in natural habitats, rice fields, flooded grassland, and other habitats (Chandrahas and Rajagopalan, 1979; Paily and Jayachandran, 1987; Paily et al., 1994; Chandhiron and Paily, 2015). The ability of this nematode to cycle and persist in a variety larval habitats, including those with extensive agricultural practices and after long periods of drought (Chandhiron and Paily, 2015), has elicited interest in using *R. iyengari* as a classic biological control agent in Brazil, Cuba, Mexico, and Benin (Platzer, 2007, Chapter 28). The host range in India includes several *Anopheles* spp., *Culex* spp., and *Aedes* spp. (Gajanana et al., 1978; Paily and Balaraman, 2000; Chandhiron and Paily, 2015).

Samples of *R. iyengari* were imported into Cuba and initial field studies revealed the potential to control a number of pest species including *Anopheles albimanus*, *An. pseudopunctipennis*, *Aedes albopictus*, *Culex quinquefasciatus*, *Culex nigripalpus*, and other culicid species (Santamarina, 1994; Santamarina and Pérez-Pacheco, 1997). Methods for mass production were developed by Santamarina Mijares and Bellini (2000), Pérez-Pacheco and Flores (2005), and Alavo et al. (2015), allowing larger-scale testing and research in Cuba, Mexico, Brazil, and Benin (Santamarina et al., 1999; Pérez-Pacheco et al., 2005; Santamarina Mijares and Bellini, 2000; Abagli et al., 2012). Additionally, Alirzaev et al. (1990) reported on its efficaciousness and persistence of *R. iyengari* in mosquito habitats in Azerbaijan.

Additional notable examples of classic biological control with imported entomopathogens include introduction of the fungus *Zoophthora radicans* collected in Israel for control of the spotted alfalfa aphid, *Therioaphis trifolii* in Australia (Milner et al., 1982), release of the neotylencid nematode *Deladenus siricidicola* from Europe into Australia for control of *Sirex noctilio* (Bedding, 1968, 2009; Bedding and Akhurst, 1974), importation of the *Oryctes* Nudivirus from Malaysia into Western Samoa, Fiji, other islands in the region, and Mauritius for control of the coconut palm rhinoceros beetle *Oryctes rhinoceros* (Hüger, 1966, 2005; Bedford, 1980), importation of *Steinernema scapterisci* collected in Uruguay into Florida, USA, for control of mole crickets *Scapteriscus* spp. (Hudson et al., 1988; Frank, 2009), and others in reviews by Hajek et al. (2005, 2007, 2009), Hajek and Delalibera (2010), and Shapiro-Ilan et al. (2012).

2.5 CONCLUSIONS AND RECOMMENDATIONS

Every year, invertebrate pests of agricultural crops and human and domestic animal habitations, as well as pests of public health concern and of native species in natural areas, invade new regions without their entourage of natural enemies to keep their populations in check. Exploration, both domestic and foreign, for natural enemies of these pests can yield environmentally safe and targeted control of pests to tolerable levels. Scores of entomopathogens discovered over the past 100 years have provided efficacious microbial control. Considering the diversity of insects and mites and their host plants and habitats that have not yet been surveyed for pathogens, it is highly likely that far more species and varieties will be discovered. In addition, new molecular and analytic techniques used in metagenomics that can detect multiple pathogen groups in pooled samples and quickly determine genetic relationships should, in the future, provide increased success in the identification and isolation of potential MCAs.

REFERENCES

Abagli, A.Z., Alavo, T.B.C., Platzer, E.G., 2012. Efficacy of the insect parasitic nematode, *Romanomermis iyengari*, for malaria vector control in Benin West Africa. Malar. J. 11, 5.

Agboton, B.V., Hanna, R., Tiedemann, A.V., 2011. Molecular detection of establishment and dispersal of Brazilian isolates of *Neozygites tanajoae* in Benin (West Africa) a fungus pathogenic to cassava green mite. Exp. Appl. Acarol. 53, 235–244.

Alavo, T.B.C., Abagli, A.Z., Pérez-Pacheco, R., Platzer, E.G., 2015. Large-scale production of the malaria vector biological control agent *Romanomermis iyengari* (Nematoda: Mermithidae) in Benin, West Africa. Malar. World J. 6, 1–5.

Alirzaev, G.U., Pridantseva, E.P., Vladimirova, V.V., Alekseev, A.N., 1990. The prospects for using *Romanomermis culicivorax* and *R. iyengari* (Nematoda: Mermithidae) for mosquito control in Azerbaijan. Med. Parazitol. Mosk. 1, 11–15 (in Russian).

Anderson, R.M., May, R.M., 1981. The population dynamics of microparasites and their invertebrate hosts. Philos. Trans. Roy. Soc. B 291, 451–524.

Andreadis, T.G., Weseloh, R.M., 1990. Discovery of *Entomophaga maimaiga* in North American gypsy moth, *Lymantria dispar* (Entomophthorales/epizootic/Lymantriidae). Proc. Natl. Acad. Sci. U.S.A. 87, 2461–2465.

ARSEF, 2014. USDA Agricultural Research Service Entomopathogenic Collection. http://www.ars.usda.gov/Main/docs.htm?docid=12125.

Bajan, C., Kmitowa, K., 1977. Contribution of entomopathogenic fungi to the natural winter reduction of Colorado potato beetle adults. Pol. Ecol. Stud. 3, 107–114.

Bassi, A., 1835. Del mal del segno, calcinaccio o moscardina, malattia che affigge i bachi da seta e sul modo di liberarne le bigattaie anche le piu infextate. Parte prima. Teoria, Orcesi, Lodi. 67 pp. Rpt. *in* Opere de Agostino Bassi. Cooperativa di Pavia. 1925. Pp. 11–67. Trans. By A.C. Ainsworth and P.J. Yarrow *in* Phytopathological classics. No. 10. Monumental Printing Co., Baltimore. 49 pp.

Bedding, R.A., 1968. *Deladenus wilsoni* n. sp. and *Deladenus siricidicola* n. sp. (Neotylenchidae) entomophagous nematodes parasitic in siricid woodwasps. Nematologica 14, 515–525.

Bedding, R.A., 2009. Controlling the pine killing woodwasp, *Sirex noctilio*, with nematodes. In: Hajek, A.E., Glare, T.R., O'Callaghan, M. (Eds.), Use of Microbes for Control and Eradication of Invasive Arthropods. Springer, Dordrecht, The Netherlands, pp. 213–235.

Bedding, R.A., Akhurst, R.J., 1974. Use of the nematode *Deladenus siricidicola* in the biological control of *Sirex noctilio* in Australia. J. Aust. Entomol. 13, 129–135.

Bedding, R.A., Akhurst, R.J., 1975. A simple technique for detection of insect parasitic rhabditid nematodes in soil. Nematologica 21, 109–110.

Bedding, R.A., Akhurst, R.J., 1978. Geographical distribution and host preferences of *Deladenus* species (Nematoda: Neotylenchidae) parasitic in siricid woodwasps and associated hymenopterous parasitoids. Nematologica 24, 286–294.

Bedford, G.O., 1980. Biology, ecology, and control of palm rhinoceros beetles. Annu. Rev. Entomol. 25, 309–339.

Bellotti, A.C., Smith, L., Lapointe, S.L., 1999. Recent advances in cassava pest management. Annu. Rev. Entomol. 44, 343–370.

Berliner, E., 1915. Ueber der schlaffsucht der *Ephestia kuhniella* und *Bac. thuringiensis* n. sp. Z. Anz. Entomol. 2, 21–56.

Bing, X.-L., Xja, W.-Q., Gui, J.-D., Yan, G.-H., Wang, X.-W., Liu, S.-S., 2014. Diversity and evolution of the Wolbachia endosymbionts of *Bemisia* (Hemiptera: Aleyrodidae) whiteflies. Ecol. Evol. 4, 2714–2737. http://dx.doi.org/10.1002/ece3.1126.

Bird, F.T., 1953. The use of a virus disease in the biological control of the European pine sawfly, *Neodiprion sertifer*. Can. Entomol. 85, 437–445.

Bird, F.T., 1955. Virus diseases of sawflies. Can. Entomol. 87, 124–127.

Bird, F.T., Whalen, M.M., 1953. A virus disease of the European pine sawfly, *Neodiprion sertifer* (Geoffr.). Can. Entomol 85, 433–437.

Blommers, L., Vaal, F., Freriks, J., Helsen, H., 1987. Three years of specific control of summer fruit tortrix and codling moth on apple in the Netherlands. J. Appl. Entomol. 104, 353–371.

Cabanillas, H.E., Poinar, G.O., Raulston, J.R., 1994. *Steinernema riobravis* n. sp. (Rhabditida: Steinernematidae) from Texas. Fund. Appl. Nematol. 17, 123–131.

Chandhiron, K., Paily, K.P., 2015. Natural parasitism of *Romanomermis iyengari* (Welch) (Nematoda: Mermithidae) on various species of mosquitoes breeding in rice fields in Pondicherry, India. Biol. Control 83, 1–6.

Chandrahas, R.K., Rajagopalan, P.K., 1979. Mosquito breeding and the natural parasitism of larvae by a fungus *Coelomomyces* and mermithid nematode *Romanomermis* in paddy fields in Pondicherry. Indian J. Med. Res. 69, 63–70.

Cock, M.J.W., van Lenteren, J.D., Brodeur, J., Barratt, B.I.P., Bigler, F., Bolckmans, K., Cônsoli, F.L., Haas, F., Mason, P.G., Parra, J.R.P., 2010. Do new access and benefit sharing procedures under the convention on biological diversity threaten the future of biological control? BioControl 55, 199–218.

Cunningham, J.C., 1998. North America. In: Hunter-Fujita, F.R., Entwistle, P.F., Evans, H.F., Crook, N.E. (Eds.), Insect Viruses and Pest Management. Wiley, Chichester, UK, pp. 313–331.

Cunningham, J.C., Entwistle, P.F., 1981. Control of sawflies by baculovirus. In: Burges, H.D. (Ed.), Microbial Control of Pests and Plant Diseases 1970–1980. Academic Press, London, pp. 379–407.

Cuthbertson, A.G.S., Walters, K.F.A., 2005. Pathogenicity of the entomopathogenic fungus, *Lecanicillium muscarium*, against the sweetpotato whitefly *Bemisia tabaci* under laboratory and glasshouse conditions. Mycopathol 160, 315–319.

DeBach, P., 1974. Biological Control by Natural Enemies. Cambridge University Press, pp. 89–194.

Delalibera Jr., I., 2009. Biological control of the cassava green mite in Africa with Brazilian isolates of the fungal pathogen *Neozygites tanajoae*. Prog. Biol. Control 6, 259–269.

Delalibera, I., Hajek, A.E., 2004. Pathogenicity and specificity of *Neozygites tanajoae* and *Neozygites floridana* (Zygomycete: Entomophthorales) isolates pathogenic to the cassava green mite. Biol. Control 30, 608–616.

Delalibera Jr., I., Sosa-Gomez, D.R., de Morães, G.J., de Alencar, J.A., Garias, W., 1992. Infection of *Mononychellus tanajoa* (Acari: Tetranychidae) by the fungus *Neozygites* sp. (Entomophthorales) in Brazil. Fla. Entomol. 75, 145–147.

Delalibera Jr., I., de Morães, G.J., Lapointe, S.L., da Silva, C.A.D., Tamai, M.A., 2000. Temporal variability and progression of *Neozygites* sp. (Zygomycetes: Entomophthorales) in populations of *Mononychellus tanajoa* (Bondar) (Acari: Tetranychidae). An. Soc. Entomol. Bras. 29, 523–535.

Dowden, P.B., Girth, H.B., 1953. Use of a virus disease to control European pine sawfly. J. Econ. Entomol. 46, 525–526.

Dulmage, H.T., 1970. Insecticidal activity of HD-1, a new isolate of *Bacillus thuringiensis* var. *alesti*. J. Invertebr. Pathol. 15, 232–239.

Dutky, S.R., 1940. Two new spore-forming bacteria causing milky diseases of Japanese beetle larvae. J. Agric. Res. 61, 57–68.

Dutky, S.R., 1963. The milky diseases. In: Steinhaus, E.A. (Ed.), Insect Pathology: An Advanced Treatise, vol. 2. Academic Press, New York, pp. 75–115.

Dutky, S.R., Hough, W.S., 1955. Note on a parasitic nematode from codling moth larvae *Carpocapsa pomonella* (Lepidoptera: Olethreutidae). Proc. Entomol. Soc. Wash 57, 244.

Eberle, K.E., Wennmann, J.T., Kleespies, R.G., Jehle, J.A., 2012. Basic techniques in insect virology. In: Lacey, L.A. (Ed.), Manual of Techniques in Invertebrate Pathology. Academic Press, San Diego, pp. 15–74.

Filotas, M.J.F., Hajek, A.E., Humber, R.A., 2003. Prevalence and biology of *Furia gastropachae* (Zygomycetes: Entomophthorales) in populations of the forest tent caterpillar (Lepidoptera: Lasiocampidae). Can. Entomol. 135, 359–378.

Fisher, T.W., Garczynski, S.F., 2012. Isolation, culture, preservation, and identification of entomopathogenic bacteria of the Bacilli. In: Lacey, L.A. (Ed.), Manual of Techniques in Invertebrate Pathology. Academic Press, San Diego, pp. 75–99.

Frank, J.H., 2009. *Steinernema scapterisci* as a biological control agent of *Scapteriscus* mole crickets. In: Hajek, A.E., Glare, T.R., O'Callaghan, M. (Eds.), Use of Arthropods for Control and Eradication of Invasive Arthropods. Springer BV, Netherlands, pp. 115–131.

Gajanana, A., Kazmin, S.J., Bheema Rao, U.S., Suguna, S.G., et al., 1978. Studies on a nematode parasite (*Romanomermis* sp.: Mermithidae) of mosquito larvae isolated in Pondicherry. Indian J. Med. Res. 68, 242–247.

Gaugler, R., Campbell, J.F., Selvan, S., Lewis, E.E., 1992. Large-scale inoculative releases of the entomopathogenic nematode, *Steinernema glaseri*: assessment 50 years later. Biol. Control 2, 181–187.

Geri, C., 2013. Pine sawfly in central France. In: Berryman, A.A. (Ed.), Dynamics of Forest Insect Populations: Patterns, Causes, Implications. Springer, NY, pp. 378–407.

Glaser, R.W., Fox, H., 1930. A nematode parasite of the Japanese beetle (*Popillia japonica* Newm.). Science 71, 16–17.

Goettel, M.S., Koike, M., Kim, J.J., Aiuchi, D., Shinya, R., Brodeur, J., 2008. Potential of *Lecanicillium* spp. for management of insects, nematodes and plant diseases. J. Invertebr. Pathol. 98, 256–261.

Goldberg, L.J., Margalit, J., 1977. A bacterial spore demonstrating rapid larvicidal activity against *Anopheles sergentii, Uranotaenia unguiculata, Culex univitattus, Aedes aegypti* and *Culex univitattus*. Mosq. News 37, 355–358.

Hajek, A.E., 1997. Fungal and viral epizootics in gypsy moth (Lepidoptera: Lymantriidae) populations in Central New York. Biol. Contro l10, 58–68.

Hajek, A.E., 1999. Pathology and epizootiology of the Lepidoptera-specific mycopathogen *Entomophaga maimaiga*. Microbiol. Mol. Biol. Rev. 63, 814–835.

Hajek, A.E., 2009. Invasive arthropods and approaches to their microbial control. In: Hajek, A.E., Glare, T.R., O'Callaghan, M. (Eds.), Use of Arthropods for Control and Eradication of Invasive Arthropods. Springer BV, Netherlands, pp. 3–18.

Hajek, A.E., Delalibera Jr., L., 2010. Fungal pathogens as classical biological control agents against arthropods. BioControl 55, 147–158.

Hajek, A.E., Humber, R.A., Elkinton, J.S., 1995. The mysterious origin of *Entomophaga maimaiga* in North America. Am. Entomol. 41, 31–42.

Hajek, A.E., Elkinton, J.S., Witcosky, J.J., 1996. Introduction and spread of the fungal pathogen *Entomophaga maimaiga* along the edge of gypsy moth spread. Environ. Entomol. 25, 1235–1247.

Hajek, A.E., McManus, M.L., Delalibera Jr., I., 2005. Catalogue of Introductions of Pathogens and Nematodes for Classical Biological Control of Insects and Mites. USDA. Forest Serv. FHTET-2005-05, 59 pp.

Hajek, A.E., Delalibera Jr., I., McManus, L., 2007. Introduction of exotic pathogens and documentation of their establishment and impact. In: Lacey, L.A., Kaya, H.K. (Eds.), Field Manual of Techniques in Invertebrate Pathology: Application and Evaluation of Pathogens for Control of Insects and Other Invertebrate Pests, second ed. Springer, Dordrecht, The Netherlands, pp. 299–325.

Hajek, A.E., Glare, T.R., O'Callaghan, M. (Eds.), 2009. Use of Microbes for Control and Eradication of Invasive Arthropods. Springer, Dordrecht, The Netherlands. 366 pp.

Hajek, A.E., Tobin, P.C., Haynes, K.J., 2015. Replacement of a dominant viral pathogen by a fungal pathogen does not alter the synchronous collapse of a forest insect outbreak. Oecologia 177, 785–797.

Harrison, R.L., Bonning, B.C., 1999. The nucleopolyhedroviruses of *Rachiplusia ou* and *Anagrapha falcifera* are isolates of the same virus. J. Gen. Virol. 80, 2793–2798.

Hostetter, D.L., Puttler, B., 1991. A new broad host spectrum nuclear polyhedrosis virus isolated from a celery looper, *Anagrapha falcifera* (Kirby), (Lepidoptera: Noctuidae). Environ. Entomol. 20, 1480–1488.

Houtondji, F.C.C., Lomer, C.J., Hanna, R., Cherry, A.J., Dara, S.K., 2002a. Field evaluation of Brazilian isolates of *Neozygites floridana* (Entomophthorales: Neozygitaceae) for the microbial control of cassava green mite in Benin, West Africa. Biocontrol Sci. Tech. 12, 361–370.

Houtondji, F.C.C., Yaninek, J.S., Moraes, G.J., Oduor, G.I., 2002b. Host specificity of cassava green mite pathogen *Neozygites floridana*. BioControl 47, 61–66.

Hudson, W.G., Frank, J.H., Castner, J.L., 1988. Biological control of *Scapteriscus* spp. mole crickets (Orthoptera: Gryllotalpidae) in Florida. Bull. Entomol. Soc. Am. 34, 192–198.

Hüger, A.M., 1966. A virus disease of the Indian rhinoceros beetle, *Oryctes rhinoceros* (Linnaeus), caused by a new type of insect virus, *Rhabdionvirus oryctes gen.* n., sp. n. J. Invertebr. Pathol. 8, 38–51.

Hüger, A.M., 2005. The Oryct*es* virus: its detection, identification, and implementation in biological control of the coconut palm rhinoceros beetle, *Oryctes rhinoceros* (Coleoptera: Scarabaeidae). J. Invertebr. Pathol. 89, 78–84.

Inglis, G.D., Enkerli, J., Goettel, M.S., 2012. Laboratory techniques used for entomopathogenic fungus: Hypocreales. In: Lacey, L.A. (Ed.), Manual of Techniques in Invertebrate Pathology. Academic Press, San Diego, pp. 189–253.

Ishiwata, S., 1901. On a type of severe flacherie (sotto disease). Dainihon Sanshi Kaiho 114, 1–5 (in Japanese).

Jackson, T.A., Pearson, J.F., O'Callaghan, M., Mahanty, H.K., Wilcocks, M.J., 1992. Pathogen to product – development of *Serratia entomophila* (Enterobacteriacea) as a commercial biological agent for the New Zealand grass grub (*Costelytra zealandica*). In: Jackson, T.A., Glare, T.R. (Eds.), Use of Pathogens in Scarab Pest Management. Intercept Press, pp. 191–198. Andover.

Jeffords, M.R., Maddox, J.V., McManus, M.L., Webb, R.E., Wieber, A., 1989. Evaluation of the overwintering success of two European microsporidia inoculatively released into gypsy moth populations. J. Invertebr. Pathol. 53, 235–240.

Keller, S., 1978. Infektionsversuche mit dem Pilz *Beauveria tenella* and adulten Maikäfern (*Melolontha melolontha* L.). Mitt. Schweiz. Entomol. Gesell. 51, 13–19.

Kirk, A.A., Lacey, L.A., Goolsby, J.A., 2008. Foreign exploration for insect natural enemies of *Bemisia* for use in biological control in the USA: a successful program. In: Gould, J., Hoelmer, K., Goolsby, J. (Eds.), Classical Biological Control of *Bemisia Tabaci* in the United States: A Review of Interagency Research and Implementation. Springer, Dordrecht, The Netherlands, pp. 17–31.

Kreig, A., Huger, A.M., Langenbruch, G.A., Schnetter, W., 1983. *Bacillus thuringiensis* var. *tenebrionis*: Ein neuer, gegenüber Larven von Coleopteren wirksamer Pathotyp. Z. Ang. Entomol. 96, 500–508.

Kurstak, E., 1962. Données sur l'epizootie bacterienne naturelle provoquée par un *Bacillus* du type *Bacillus thuringiensis* var. *alesti* sur *Ephestia kuhniella* Zeller. Entomophaga Mem. Hors. Ser. 2, 245–247.

Lacey, L.A., Solter, L., 2012. Initial handling and diagnosis of diseased invertebrates. In: Lacey, L.A. (Ed.), Manual of Techniques in Invertebrate Pathology. Academic Press, San Diego, pp. 1–14.

Lacey, L.A., Kirk, A.A., Hennessey, R.D., 1993. Foreign exploration for natural enemies of *Bemisia tabaci* and implementation in integrated control programs in the United States. Proc. ANPP Int. Conf. Pests Agric. 1, 351–360.

Lacey, L.A., Thomson, D., Vincent, C., Arthurs, S.P., 2008a. Codling moth granulovirus: a comprehensive review. Biocontrol Sci. Technol. 18, 639–663.

Lacey, L.A., Wraight, S.P., Kirk, A.A., 2008b. Entomopathogenic fungi for control of *Bemisia tabaci* biotype B: foreign exploration, research and implementation. In: Gould, J., Hoelmer, K., Goolsby, J. (Eds.), Classical Biological Control of *Bemisia Tabaci* in the United States: A Review of Interagency Research and Implementation. Springer, Dordrecht, The Netherlands, pp. 33–69.

Li, J., Chen, W., Wu, J., Peng, W., An, J., Schmid-Hempel, P., Schmid-Hempel, R., 2012. Diversity of *Nosema* associated with bumblebees (*Bombus* spp.) from China. Int. J. Parasitol. 42, 49–61.

Liu, J., Berry, R.E., 1996. *Heterorhabditis marelatus* n. sp. (Rhabditida: Heterorhabditidae) from Oregon. J. Invertebr. Pathol. 67, 48–54.

Lomer, C.J., Prior, C., Kooyman, C., 1997. Development of *Metarhizium* spp. for the control of grasshoppers and locusts. In: Goettel, M.S., Johnson, D. (Eds.), Microb. Control Grasshopp. LocustsMem. Entomol. Soc. Can. 171, 265–286.

Lomer, C.J., Bateman, R.P., Johnson, D.L., Langewald, J., Thomas, M., 2001. Biological control of locusts and grasshoppers. Annu. Rev. Entomol. 46, 667–702.

Lucarotti, C.J., Moreau, G., Kettela, E.G., 2007. Abietiv', a viral biopesticide for control of the balsam fir sawfly. In: Vincent, C., Goettel, M.S., Lazarovits, G. (Eds.), Biological Control: A Global Perspective. CAB International, Oxfordshire, UK, pp. 353–361.

Martin, P.A., Gundersen-Rindal, D., Blackburn, M., Buyer, J., 2007. *Chromobacterium subtsugae* sp. nov., a betaproteobacterium toxic to Colorado potato beetle and other insect pests. Int. J. Sys. Evol. Microbiol. 57, 993–999.

McCoy, C.W., Stuart, R.J., Duncan, L.W., Shapiro-Ilan, D.I., 2007. Application and evaluation of entomopathogens for citrus pest control. In: Lacey, L.A., Kaya, H.K. (Eds.), Field Manual of Techniques in Invertebrate Pathology: Application and Evaluation of Pathogens for Control of Insects and Other Invertebrate Pests, second ed. Springer, Dordrecht, The Netherlands, pp. 567–581.

Milner, R.J., Soper, R.S., Lutton, G.G., 1982. Field release of an Israeli strain of the fungus, *Zoopthora radicans*, (Brefeld) Batko for biological control of *Therioaphis trifolii* (Monell) f. *maculata* spotted alfalfa aphid, pest of leguminous pastures. J. Aust. Entomol. Soc. 21, 113–118.

Mitchell, C.E., Power, A.G., 2003. Release of invasive plants from fungi and viral pathogens. Nature 421, 625–627.

Moreau, G., Lucarotti, C.J., Kettela, E.G., Thurston, G.S., Holmes, S., Weaver, C., Levin, D.B., Morin, B., 2005. Aerial application of nucleopolyhedrovirus induces decline in increasing and peaking populations of *Neodiprion abietis*. Biol. Control 33, 65–73.

Nguyen, K.B., Duncan, L.W., 2002. *Steinernema diaprepesi* n. sp. (Rhabditida: Steinernematidae), a parasite of the citrus root weevil *Diaprepes abbreviatus* (L) (Coleoptera: Curculionidae). J. Nematol. 34, 159–170.

Nguyen, K.B., Smart Jr., G.C., 1990. *Steinernema scapterisci* n. sp. (Rhabditida: Steinernematidae). J. Nematol. 22, 187–199.

Oho, N., Yamada, H., Nakazawa, H., 1974. A granulosis virus of the smaller tea tortrix, *Adoxophyes orana* Fischer von Roslerstamm (Lepidoptera; Tortricidae). Mushi 48, 15–20.

Osborne, L.S., Storey, G.K., McCoy, C.W., Walter, J.F., 1990. Potential for controlling the sweetpotato whitefly, *Bemisia tabaci*, with the fungus *Paecilomyces fumosoroseus*. In: Proceedings of Vth International Colloquium on Invertebrate Pathology and Microbial Control, Adelaide, Australia, August 20–24, 1990. Society for Invertebrate Pathology, pp. 386–390.

Paily, K.P., Balaraman, K., 2000. Susceptibility of ten species of mosquito larvae to the parasitic nematode *Romanomermis iyengari* and its development. Med. Vet. Entomol. 14, 426–429.

Paily, K.P., Jayachandran, S., 1987. Factors inhibiting parasitism of mosquito larvae by the mermithid nematode (*Romanomermis iyengari*) in a polluted habitat. Indian J. Med. Res. 86, 469–474.

Paily, K.P., Arunachalam, M., Reddy, C.M.R., Balaraman, K., 1994. Effect of field application of *Romanomermis iyengari* (Nematoda: Mermithidae) on the larvae of *Culex tritaeniorhynchus* and *Anopheles subpictus* breeding in grassland. Trop. Biomed. 11, 23–29.

Parkman, J.P., Hudson, W.G., Frank, J.H., Nguyen, K.B., Smart, G.C., 1993. Establishment and persistence of *Steinernema scapterisci* (Rhabditida: Steinernematidae) in field populations of *Scapteriscus* mole crickets (Orthoptera: Grillotalpidae). J. Entomol. Sci. 8, 182–190.

Pérez-Pacheco, R., Flores, G., 2005. Mass production of mermithid nematode parasites of mosquito larvae in Mexico. J. Nematol. 37, 388.

Pérez-Pacheco, R., Rodríguez-Hernández, C., Lara-Reyna, J., Montez-Belmont, R., Ruiz-Vega, J., 2005. Control of the mosquito *Anopheles pseudopunctipennis* (Diptera: Culicidae) with *Romanomermis iyengari* (Nematoda: Mermithidae) in Oaxaca, Mexico. Biol. Control 32, 137–142.

Petersen, J.J., 1985. Nematode parasites. In: Chapman, H.C. (Ed.), Biol. Control Mosq. Am. Mosq. Control Assoc. Bull., 6, 110–122.

Petersen, J.J., Chapman, H.C., Willis, O.R., Fukuda, T., 1978. Release of *Romanomermis culicivorax* for the control of *Anopheles albimanus* in El Salvador II. Application of the nematode. Am. J. Trop. Med. Hyg. 27, 1268–1273.

Platzer, E.G., 2007. Mermithid nematodes. J. Am. Mosq. Control Assoc. Bull. 7, 58–64.

Podgwaite, J.D., Rush, P., Hall, D., Walton, G.S., 1984. Efficacy of the *Neodiprion sertifer* (Hymenoptera: Diprionidae) nucleopolyhedrosis virus (baculovirus) product, Neochek-S. J. Econ. Entomol. 77, 525–528.

Poinar Jr., G.O., 1976. Description and biology of a new insect parasitic rhabditoid. *Heterorhabditis bacteriophora* n. gen., n. sp. (Heterorhabditidae, n. fam.) (Rhabditida [Oerley]). Nematologica 21, 463–470.

Poinar, G.O., Jackson, T., Klein, M., 1987. *Heterorhabditis megidis* sp. n. (Rhabditida: Heterorhabditidae) parasitic in the Japanese beetle, *Popillia japonica* (Scarabaeidae: Coleoptera), in Ohio. Proc. Helminthol. Soc. Wash. 54, 53–59.

Prior, C., Lomer, C.J., Herren, H., Paraiso, A., Kooyman, C., Smit, J.J., 1992. The IIBC/IITA.DFPV collaborative research programme on the biological control of locusts and grasshoppers. In: Lomer, C.J., Prior, C. (Eds.), Biological Control of Locusts and Grasshoppers. CAB International in Association With the International Institute of Tropical Agriculture, pp. 8–18.

Sanchis, V., 2011. From microbial sprays to insect-resistant transgenic plants: history of the biospesticide *Bacillus thuringiensis*. A review. Agron. Sus. Devel. 31, 217–231.

Santamarina, M.A., 1994. Actividad parasitaria de *Romanomermis iyengari* (Nematoda, Mermithidae) en criaderos naturales de larvas de mosquito. Misc. Zool 17, 59–65.

Santamarina, A.,M., Bellini, A.C., 2000. Producción masiva de *Romanomermis iyengari* (Nematoda: Mermithidae) y su aplicación en criaderos de anofelinos en Boa Vista (Roraima), Bras. Rev. Panam. Sal. Pub. 7, 155–161.

Santamarina, A.M., Pérez-Pacheco, R.P., 1997. Reduction of mosquito larval densities in natural sites after introduction of *Romanomermis culicivorax* (Nematoda: Mermithidae) in Cuba. J. Med. Entomol. 34, 1–4.

Santamarina, A.M., Pérez-Pacheco, R., Martinez, S.H., Cantón, L.E., Flores-Ambrosio, G.F., 1999. The *Romanomermis iyengari* parasite for *Anopheles pseudopunctipennis* suppression in natural habitats in Oaxaca State, Mexico. Rev. Panam. Sal. Pub. 5, 23–28.

Shapiro-Ilan, D.I., Bruck, D., Lacey, L.A., 2012. Principles of epizootiology and microbial control. In: Vega, F.E., Kaya, H.K. (Eds.), Insect Pathology, second ed. Academic Press, San Diego, pp. 29–71.

Singer, S., 1973. Insecticidal activity of recent bacterial isolates and their toxins against mosquito larvae. Nature 244 (5411), 110.

Solter, L., Hajek, 2009. Control of gypsy moth, *Lymantria dispar*, in North America since 1878. In: Hajek, A.E., Glare, T.R., O'Callaghan, M. (Eds.), Use of Arthropods for Control and Eradication of Invasive Arthropods. Springer BV, Netherlands, pp. 181–212.

Solter, L.F., Maddox, J.V., 1998. Physiological host specificity of microsporidia as in indicator of ecological host specificity. J. Invertebr. Pathol. 71, 207–216.

Solter, L.F., Maddox, J.V., Vossbrinck, C.R., 2005. Physiological host specificity: a model using the European corn borer *Ostrinia nubilalis* (Hübner) (Lepidoptera: Crambidae) and microsporidia of row crop and other stalk-boring insects. J. Invertebr. Pathol. 90, 127–130.

Soper, R.S., Shimazu, M., Humber, R.A., Ramos, M.E., Hajek, A.E., 1988. Isolation and characterization of *Entomophaga maimaiga* sp. nov., a fungal pathogen of gypsy moth, *Lymantria dispar*, from Japan. J. Invertebr. Pathol. 51, 229–241.

Spradbery, J.P., Kirk, A.A., 1978. Aspects of the ecology of siricid woodwasps (Hymenoptera: Siricidae) in Europe, North Africa and Turkey with special reference to the biological control of *Sirex noctilio* F. in Australia. Bull. Entomol. Res. 68, 341–359.

Steinhaus, E., 1949. Insect Pathology. Academic Press. 757 pp.

Steinhaus, E., 1963. Insect Pathology: An Advanced Treatise, vol. 2. Academic Press, New York. 689 pp.

Steinkraus, D.C., Kring, T.J., Tugwell, N.P., 1991. *Neozygites fresenii* in *Aphis gossypii* on cotton. Southwest. Entomol. 16, 118–122.

Stock, S.P., Goodrich-Blair, H., 2012. Nematode parasites, pathogens and associates of insects and invertebrates of economic importance. In: Lacey, L.A. (Ed.), Manual of Techniques in Invertebrate Pathology. Academic Press, San Diego, pp. 373–426.

Stock, P.S., Strong, D.R., Gardner, S.L., 1996. Identification of *Heterorhabditis* (Nematoda: Heterorhabditidae) from California with a new species isolated from the larvae of the ghost moth *Hepialis californicus* (Lepidoptera: Hepialidae) from the Bodega Bay natural reserve. Fun. Appl. Nematol. 19, 585–592.

Stucki, G., Jackson, T.A., Noonan, M.J., 1984. Isolation and characterization of *Serratia* strains pathogenic for larvae of the New Zealand grass grub, *Costelytra zealndica*. N. Z. J. Sci. 27, 255–260.

Tanada, Y., 1964. A granulosis virus of the codling moth, *Carpocapsae pomonella* (Linnaeus) (Olethreutidae, Lepidoptera). J. Insect Pathol. 6, 378–380.

Tobin, P.C., Blackburn, L.M. (Eds.), 2007. Slow the Spread: A National Program to Manage the Gypsy Moth, 109 pp. http://www.nrs.fs.fed.us/pubs/gtr/gtr_nrs6.pdf.

Trought, T.T.T., Jackson, T.A., French, R.A., 1982. Incidence and transmission of a disease of grass grub (*Costelytra zealandica*) in Canterbury. N. Z.eal. J. Exp. Agric 10, 79–82.

Vail, P.V., Jay, D.L., 1973. Pathology of the nuclear polyhedrosis virus of the alfalfa looper in alternate hosts. J. Invertebr. Pathol. 21, 198–204.

Vail, P.V., Jay, D.L., Hunter, D.K., 1971. Cross infectivity of a nuclear polyhedrosis virus isolated from the alfalfa looper, *Autographa californica*. Proc. IVth Int. Colloq. Insect Pathol. 297–304.

Weiser, J.A., 1955. *Neoaplectana carpocapsae*, sp. n. (Anguillulata, Steinernematinae), novy cizopasnikhousenek obaleče jablečnéého, *Carpocapsae pomonella* L. Věstnik Českoslov. Zool. Spol. 19, 44–52.

Weiser, J.A., 1984. A mosquito-virulent *Bacillus sphaericus* in adult *Simulium damnosum* from Northern Nigeria. Zent. Mikrobiol. 139, 57–60.

Weseloh, R.M., 2003. Short and long range dispersal of the gypsy moth (Lepidoptera: Lymantriidae) fungal pathogen, *Entomophaga maimaiga* (Zygomycetes: Entomophthorales). Environ. Entomol. 32, 111–122.

Zimmermann, G., 1986. The 'Galleria bait method' for detection of entomopathogenic fungi in soil. J. Appl. Entomol 102, 213–215.

Zimmermann, G., 1992. Use of the fungus, *Beauveria brongniatii*, for control of European cockchafers, *Melolontha* spp., in Europe. In: Jackson, T.A., Glare, T.R. (Eds.), Use of Pathogens in Scarab Pest Management. Intercept Press, Andover, pp. 199–208.

Zimmermann, G., Huger, A.M., Kleespies, R.G., 2013. Occurrence and prevalence of insect pathogens in populations of the codling moth, *Cydia pomonella* L.: a long-term diagnostic survey. Insects 4, 425–446.

Zùbrik, M., Hajek, A., Pilarska, D., Spilda, I., Georgiev, G., Hrasovec, B., Hirka, A., Goertz, D., Hoch, G., Barta, M., Saniga, M., Kunca, A., Nikolov, C., Vakula, J., Galko, J., Pilarski, P., Csoka, G., 2016. The potential for *Entomophaga maimaiga* to regulate gypsy moth *Lymantria dispar* (L.) (Lepidoptera: Erebidae) in Europe. J. Appl. Entomol. 140, 1–15.

Part II

Basic and Applied Research

Chapter 3

Basic and Applied Research: Baculovirus

D. Grzywacz

University of Greenwich, Chatham, Kent, United Kingdom

3.1 INTRODUCTION TO ENTOMOPATHOGENIC VIRUSES

Insect viruses are important pathogens of many arthropod species. They have been recorded from a wide range of insects (Miller and Ball, 1998), and their association with these hosts is long, perhaps for more than 200 million years. While there are more than 1100 reported viruses that infect over 20 different families of insects, the largest number described to date, over 600, are from the family Baculoviridae (Miller, 1997; Eberle et al., 2012a,b), though it has been suggested many thousands more may remain to be identified from the Lepidoptera alone (Federici, 1997).

3.2 BACULOVIRIDAE

The baculoviruses are viral pathogens that are currently classified as occurring only in three orders of insects, though older classifications include a wider host range (Herniou et al., 2012). It is on the baculoviruses that the majority of effort to develop insect viruses as practical microbial control agents (MCAs) has been focused with up to 30 baculovirus species being either developed as biopesticides or under development, and they account for majority of the commercial viral insecticides in use (Lacey et al., 2015; Table 3.1). Baculoviruses are often important factors in the population dynamics of the hosts they infect (Evans, 1986; Cory et al., 1997; Fuxa, 2004; Cory and Evans, 2007), and their biology has been extensively studied, both for their pest control potential and as model systems for studying virus pathology and biology (Adams and McLintock, 1991; Miller, 1997; Harrison and Hoover, 2012).

Baculoviruses are comparatively fast acting among biological control agents and are lethal pathogens of some of the most globally important pest species today, such as the diamondback moth (*Plutella xylostella*) and the *Heliothis/Helicoverpa* species that are characterized by the ability to develop high levels of resistance to chemical insecticides and whose control has therefore become a challenge to farmers. There is an extensive body of evidence regarding their specificity to insects that underpins the widespread consensus that they pose no significant safety issues when used for pest control (OECD, 2002; Mudgal et al., 2013).

3.2.1 Description of the Family Baculoviridae

The family Baculoviridae are viruses with double-stranded DNA circular genomes of 80–180 kbp, that are packaged in rod-shaped infective particles or nucleocapsids. These are found within crystal-like proteinaceous bodies called occlusion bodies (OBs) (Miller and Ball, 1998; Harrison and Hoover, 2012). The current taxonomy of the baculoviruses was proposed by Jehle et al. (2006) and has been confirmed by the subsequent IICTV reports (Herniou et al., 2012). This divides the family into four genera on the basis of genome sequence analysis and host classification, replacing earlier systems based more on morphological traits. The Alphabaculoviruses are all Lepidoptera-specific nucleopolyhedroviruses (NPVs), and their OBs are the classic many-sided (polyhedral) shape as seen in the type species *Autographa californica* multiple NPV (AcMNPV). The OB is composed of a 25- to 33-kDa polypeptide called polyhedrin formed into a crystalline matrix commonly ranging from 0.4 to 5 μm. Within the OB are found many individual infectious entities or virions embedded in this crystalline matrix. The Betabaculoviruses include the old grouping the granuloviruses (GVs); all are Lepidoptera-specific and the type species is *Cydia pomonella* GV (CpGV). The OBs of Betabaculoviruses are typically rod shaped and much smaller than the OBs of Alphabaculoviruses at 130–250 μm across by 300–500 μm in length and contain only a single virion in each OB that is composed of a polypeptide, very similar to polyhedrin, called granulin. The Gammabaculoviruses are Hymenoptera-specific NPVs such as the *Neodiprion lecontei* SNPV (singly enveloped NPV) with polyhedral OBs with a size in the range of 0.4–1.1 μm containing multiple singly enveloped nucleocapsids. Finally, the Deltabaculoviruses are Diptera-specific NPVs with crystalline OBs of 0.5–15 μm containing many virions. Most species of baculovirus are found within the alpha and beta baculoviruses. The largest number of the over 600 known species are from the Lepidoptera (Eberle et al., 2012a,b).

Microbial Control of Insect and Mite Pests. http://dx.doi.org/10.1016/B978-0-12-803527-6.00003-2

TABLE 3.1 Examples of Baculoviruses That Have Been or Are Being Developed as Microbial Control Agents

Baculoviruses	Host Insects	Crops	Selected References
Betabaculoviruses (granuloviruses, GV)			
CpGV	*Cydia pomonella* codling moth	Pome fruit and walnuts	Eberle and Jehle (2006), Lacey and Shapiro-Ilan (2008), Lacey et al. (2008)
PhopGV	*Phthorimaea operculella*, potato tuber moth	Potato	Sporleder and Kroschel (2008), Arthurs et al. (2008), Lacey and Kroschel (2009)
AdorGV	*Adoxophyes orana*, summer fruit tortrix	Apple, pear	Cross et al. (1999), Nakai (2009)
PlxyGV	*Plutella xylostella*, diamondback moth	Brassicas, cabbage, kale, collard, canola	Grzywacz et al. (2004)
CrleGV	Thaumatotibia (*Cryptophlebia leucotreta*) False codling moth	Citrus, cotton, avocado, macadamia nuts	Moore et al. (2004, 2015)
HomaGV	*Homona magnanima*, tea tortrix	Tea, flowers, pome fruit	Kunimi (2007), Nakai (2009)
AdhoGV	*Adoxophyes honmai*, smaller tea tortrix	Tea, ornamental trees and shrubs	Nakai (2009)
Alphabaculoviruses and gammabaculoviruses [nucleopolyhedroviruses (NPVs)]			
HearNPV	*Helicoverpa armigera*, cotton bollworm, podborer, Old World bollworm	Wide range of crops including maize, soy, cotton, vegetables, legumes	Rabindra and Jayaraj (1995), Buerger et al. (2007)
HzSNPV	*Helcoverpa zea*, corn earworm, tomato fruitworm, tobacco budworm	Wide range of crops including corn, cotton, tomato, tobacco	Ignoffo (1999)
SpexMNPV	*Spodoptera exempta*, African armyworm	Pasture grass, wheat, barley, maize, rice	Grzywacz et al. (2008)
SeMNPV	*Spodoptera exigua* beet armyworm	Wide variety of vegetables and ornamentals including asparagus, celery, beets, tomato, lettuce, as well as cotton, cereals, and oilseeds	Kolodny-Hirsch et al. (1997), Lasa et al. (2007)
SpliNPV	*Spodoptera littoralis*, African cotton leafworm, Egyptian cotton leafworm	Cotton, wide variety of vegetables, fruits, and flowers	Jones et al. (1994)
SpltNPV	*Spodoptera litura*, tobacco armyworm, tobacco cutworm, taro caterpillar	Wide range of crops including, cotton, tobacco, vegetables, ornamentals, and brassicas	Nakai and Cuc (2005)
AgMNPV	*Anticarsia gemmatalis*, velvet bean caterpillar	Soya bean, other legumes	Moscardi (2007), Panazzi (2013)
LdMNPV	*Lymantria dispar*, gypsy moth	Wide variety of deciduous trees	Podgwaite (1999)
NeleNPV	*Neodiprion lecontei* Red-headed pine sawfly	Pine species, Norway spruce	Cunningham (1995)
OpMNPV	*Orygia pseudotsugata* Douglas-fir tussock moth	Douglas fir, firs, spruce and ornamental trees	Martignoni (1999)

TABLE 3.1 Examples of Baculoviruses That Have Been or Are Being Developed as Microbial Control Agents—cont'd

Baculoviruses	Host Insects	Crops	Selected References
NeabNPV	*Neodiprion abietis* Balsam fir sawfly	Balsam fir, spruce, larch	Lucarotti et al. (2007, Moreau and Lucarotti (2007)
AcMNPV	*Autographa californica*, alfalfa looper	Wide variety of vegetable and forage crops, alfalfa, peas, brassicas	Vail et al. (1999), Yang et al. (2012)
TrniSNPV	*Trichoplusia ni, cabbage looper*	Brassicas	Vail et al. (1999)
HpNPV	*Hyblea puera, teak defoliator*	Teak	Nair et al. (1996)

FIGURE 3.1 Illustration of *Helicoverpa armigera* in various stages of baculovirus infection. (A) healthy larva, (B) NPV-infected dead larva hanging in typical wilt pose, (C) dead larva whose cuticle has ruptured releasing bodily fluid containing progeny virus occlusion bodies (OB), and (D) HearNPV OB seen under phase contrast ×400. Baculovirus OB seen as bright refractile bodies. *Photos (A–C) S. Moore, (D) D. Grzywacz.*

The most noticeable and distinctive characteristic of the baculoviruses is the OB, which is large enough to be seen and identified with a light microscope under phase contrast light at ×400. The NPV is distinctive (Fig. 3.1D) and quantifiable although the smaller GV OB may require more skill and dark field illumination to identify and count. The OB's crystalline protein matrix provides environmental protection for the virions. In the OB, these are the occlusion derived virion (ODV) forms (Fig. 3.2), and within the envelope of each virion are contained the infectious entities, the rod-like nucleocapsids

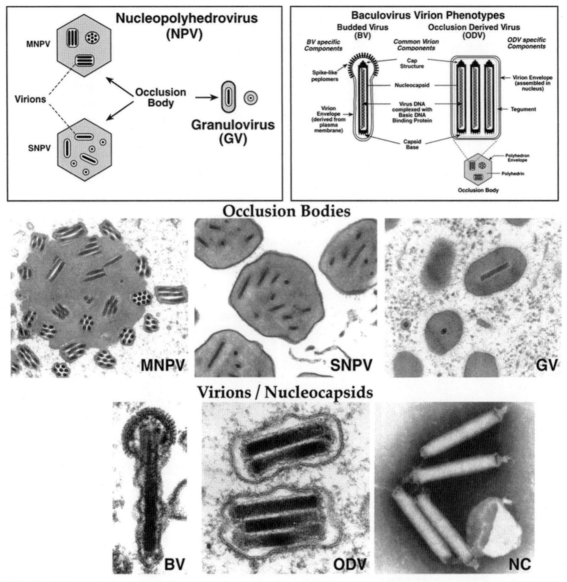

FIGURE 3.2 Baculovirus occlusion bodies, virions and nucleocapsids. (Upper left) The structures of occlusion bodies from baculoviruses in the genera Alphabaculovirus (nucleopolyhedrovirus, NPV) and Betabaculovirus (granulovirus, GV) are illustrated. Virions embedded in nucleopolyhedrovirus occlusion bodies may contain multiple nucleocapsids (MNPV) or single nucleocapsids (SNPV). (Upper right) The two baculovirus virion phenotypes are illustrated as diagrams with shared and phenotype-specific components. *(From Blissard, G.W., 1996. Baculovirus–insect cell interactions. Cytotechnology 20, 73–93.)* (Bottom) Transmission electron micrographs of occlusion bodies (MNPV, SNPV and GV), virion phenotypes BV (budded virions), ODV (occlusion-derived virions) and nucleocapsids (NC). Nucleopolyhedrovirus occlusion bodies of the MNPV (*Autographa californica* MNPV, top left) and SNPV (*Trichoplusia ni* SNPV, top middle) types are compared to granulovirus occlusion bodies (*Estigmine acrea* GV, top right). Transmission electron micrographs of virions of the BV (*Lymantria dispar* MNPV, bottom left) and ODV (*Autographa californica* MNPV, bottom centre) phenotypes are shown beside negatively stained nucleocapsids (*Autographa californica* MNPV, bottom right). (Electron micrographs courtesy of J.R. Adams [LdMNPV BV virion] and R. Granados [all others].) *Printed with permission of the publisher. Herniou, E.A., Arif, B.M., Becnel, J.J., Blissard, G.W., Bonning, B., Harrison, R., Jehle, J.A., Theilmann, D.A., Vlak, J.M., 2012. Family Baculoviridae. In: King, A.M.Q., Adams, M.J., Carstens, E.B., Lefkowitz, E.J. (Eds.), Virus Taxonomy, Classification and Nomenclature of Viruses, Ninth Report of the International Committee on Taxonomy of Viruses. Elsevier Academic Press, Amsterdam, pp. 163–173, Figure 1.*

in which the DNA of the genome is packaged with its associated nucleoproteins. These are enveloped by a trilaminar membrane and each ODV may contain either a single nucleocapsid (SNPV) or multiple nucleocapsids (MNPVs). In SNPVs and GVs, a single nucleocapsid is found within an envelope, but in MNPVs the ODV may contain from 1 to 29 nucleocapsids (Adams and McClintock, 1991; Herniou et al., 2012). Most NPV OBs fall in the range of 0.6–2.5 μm, but sizes up to 15 μm are reported. The OB is surrounded by a viral envelope or calyx composed of polysaccharides; its function is not fully clarified but it does play a role in maintaining OB integrity and in infection.

3.2.2 Pathology and Infection Process

The pathology of baculovirus infections has long been the subject of detailed scientific investigation and a number of substantive reviews (Adams and McLintock, 1991; Federici, 1997; Rohrmann, 2013), while a number of other more recent ones cover specific aspects of this topic (Passarelli, 2011; Lapointe et al., 2012). The process of infection in baculoviruses is, in most species, a two-stage process with a primary infection and amplification stage in the midgut followed by a systemic infection phase by which the infection spreads throughout the body and which concludes with the massive production of OBs. This two-stage system in the Alphabaculoviruses and some Betabaculoviruses, such as CpGV, where there is widespread systemic tissue infection, is highly efficient and can produce death in 4–10 days (Federici, 1997; Lacey et al., 2008); producing progeny OBs and rates of $1–10 \times 10^6$ OBs per mg host tissue are common in host insects (Evans, 1986; Shieh, 1989; Cherry et al., 1997). In Gammabaculoviruses and Deltabaculoviruses, where infections are restricted to midgut tissues, the production of progeny OBs is much lower.

Infections occur following ingestion of OBs by a susceptible larva, usually during feeding on OB-contaminated food such as foliage. The OBs can only initiate infection once having gained entry to the gut through ingestion, and as such baculoviruses are overwhelmingly pathogens of larval stages. Once ingested, the OB is carried to the alkaline (pH 8–10) larval midgut region where the OB dissolves as the polyhedrin solubilizes to release the ODV within minutes (Fig. 3.3). If OB are ingested by birds and mammals, in which gut pH ranges 1–7, some OBs can pass through the gut retaining infectivity; this can be an important mechanism for spreading baculovirus infections to new host populations (Lapointe et al., 2012). The released ODV then needs to pass through the peritrophic membrane (PM), found in the midgut, to access the midgut epithelial cells in order to establish infection. Baculoviruses have evolved specific mechanisms to disrupt PM structure to allow ODV passage. One is the possession of metalloproteases called enhancins, found on OBs and ODVs of GVs and some NPVs, and these contribute to successful oral infection in these viruses (Del Rincón-Castro and Ibarra, 2005; Slavicek, 2012). Once the PM is passed, virions then attach to and enter the midgut epithelial cells facilitated by a suite of virally encoded proteins called per os infectivity factors (PIF) that complex and fuse with the epithelial membrane, allowing the nucleocapsids to enter the cells. Within 30 min of the virions entering host cells, there is evidence of active transfer of the nucleocapsids to the nucleus, which they enter through the nuclear pores (Cohen et al., 2011) where the DNA commences to unwind and viral replication commences (Lapointe et al., 2012).

Viral replication in the nucleus of midgut epithelial cells results in the appearance of many progeny nucleocapsids that move toward the base of the cells and bud through the basal lamellar membrane, picking up an envelope of host plasma membrane and emerging as budded virus (BV) (Federici, 1997). The BVs are optimized for cell-to-cell transmission of the virus, and most have baculovirus attachment envelope and fusion proteins (EPFs) that enable these virions to rapidly spread the infection throughout the host (Wang et al., 2014). In alpha and beta baculoviruses, the emerging baculovirus appear 4–12 hours postinfection (hpi) and begin to infect other tissues, through the tracheal and hemolymph. Tracheal cells and tracheoblasts are among the first tissues to be infected at 12–24 hpi (Adams and McClintock, 1991). By moving through the network of trachea and by translocation within the motile tracheoblasts, the virus is able to rapidly spread through the host tissues colonizing hemolymph cells by 36 hpi, fat body 48 hpi, and most larval tissues including gonad, hypodermis, muscles, nerve ganglia, and pericardial cells (Keddie et al., 1989; Harrison and Hoover, 2012). It is the infection and destruction of these tissues that bring about eventual larval death. In some Gammabaculoviruses hosts such as Hymenoptera sawflies, the baculovirus only replicates in midgut tissues and no systemic infection follows after the primary infection. The OBs are produced in the midgut epithelial cells only while the larva is alive. These then infect other larvae through contact and consumption of the OB-laden diarrhea. In Deltabaculoviruses, infection is again confined to cells of the midgut and gastric ceca. Here, again, OBs are produced rapidly in midgut cells at 14–48 hpi but are mainly released following larval death (Jehle et al., 2006).

After primary virus replication, the midgut recovers by sloughing off and replacing infected cells, thus enabling feeding and growth to continue. This allows continued larval growth, thus increasing the pool of cells that the virus can eventually infect. This process is assisted by the expression of the baculovirus-encoded ecdysteroid UDP glucosyltransferase (*egt*) gene. Ecdysteroids are larval hormones that control molting and feeding and viral production of *egt* encoded enzyme inactivates larval ecdysteroids; this is associated with greater larval growth postinfection and the production of higher titers of OBs in hosts where this occurs (Hughes, 2013). During the early stages of infection, the first 3 days, the larvae show few symptoms but from day 4 onwards, changes in the responsivity to stimuli, a decline, and eventually a cessation of feeding become more apparent. The external appearance can also be seen to change as the cuticle becomes swollen and glossy due to the presence of OBs so that in some species a distinct white or creamy coloration can be perceived (Federici, 1997).

Baculoviruses are most infective to neonate larvae and early instars. A single OB may be sufficient to infect a neonate, though estimates for GVs seem to be higher (Rohrmann, 2013). Neonate infections can rapidly kill

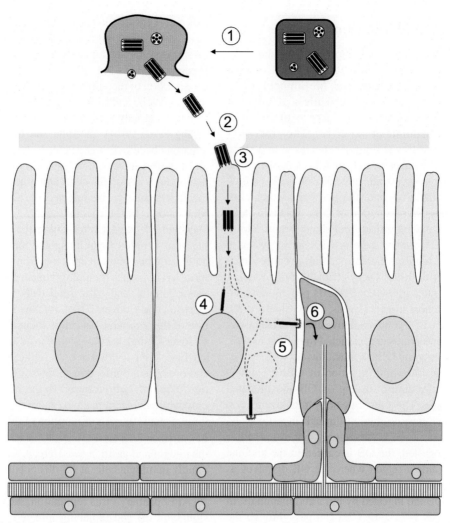

FIGURE 3.3 A model for baculovirus primary infection of the midgut and secondary infection of the tracheal system generalized from studies on lepidopteran larvae infected with multiple nucleopolyhedrovirus (MNPV). Occlusions ingested by larvae dissolve in the alkaline environment of the host midgut lumen, liberating the occlusion derived virions (ODVs) (*1*). These ODV move through the peritrophic matrix (*yellow*) lining the midgut epithelium (*2*), in some cases with the assistance of matric degrading baculovirus-encoded proteases called enhancins. ODV bind to microvilli of the midgut columnar epithelial cells, and nucleocapsids enter the microvilli after fusion between the ODV envelope and the cytoplasmic membrane (*3*). Nucleocapsids move through the cytoplasm via actin polymerization (*red lines*). Some nucleocapsids translocate to the nucleus (*4*), uncoat their DNA, and express early genes encoding the major budded virus (BV) envelope glycoprotein (GP64 or F protein). Other nucleocapsids translocate to the basolateral domain of the cell and bud through the plasma membrane (*5*), forming BV with envelope glycoproteins expressed from the subpopulation of nucleocapsids that entered the nucleus of the cell. Expression of a viral fibroblast growth factor leads to turnover of the basement membrane surrounding the tracheoblasts, which facilitates the infection of the tracheoblasts with BV from the midgut epithelial cells (*6*). Infection then spreads through the tracheal epidermal cells to other tissues beyond the midgut sheath. *Printed with permission of the publisher, Harrison, R., Hoover, K., 2012. Baculoviruses and other occluded insect viruses. In: Vega, F., Kaya, H. (Eds.), Insect Pathology. Elsevier, Amsterdam, pp. 73–131.*

within 72h (Federici, 1997). However, as larvae develop and progress through instars, they become increasingly resistant and may require dosages/concentrations of virus 3–5 orders of magnitude greater to infect fifth instars successfully (Payne et al., 1981; Jones, 2000). One mechanism of larval resistance with increasing age is that older larvae more rapidly slough infected midgut cells, which plays a large part in resistance to infection. In addition, there is evidence for an increase in the systemic immune response of older larvae (Engelhard and Volkman, 1995), and although

this has been most clearly characterized in LdMNPV infections in *Lymantria dispar* (Hoover and Grove, 2009; McNeil et al., 2010), it is probably much more widespread. Hosts' immune response to virus infection is mediated through both cellular responses such as phagocytosis and melanization/encapsulation/isolation of infected cells and through humoral responses involving the synthesis and release of antimicrobial factors into the blood and hemocele. Overall, the complexities of the insect immune system remain to be explored, and many mechanisms are as yet poorly understood. One specific mechanism for host resistance that is more precisely characterized is apoptosis, that is the programmed death of infected or defective cells. It is initiated by the host when a cell has been detected to be defective or infected, and by destroying infected cells the host can prevent viral replication and bring incipient infections under control. Thus, apoptosis represents one of the most powerful defense responses of hosts, impairing viruses' ability to replicate and proliferate within a host (Clem, 2005). The very extensive and complete infection of tissues can produce extremely efficient viral amplification with larvae of 200–300 mg producing $1–5 \times 10^9$ OBs per insect (Shapiro, 1982; Cherry et al., 1997; Hunter-Fujita et al., 1998). The primary production of OBs is in fat body, epidermis, and trachea, and this may represent up to 15% of the final dry weight of the larva. This high efficiency of those species of baculoviruses that produce systemic infections is a major positive factor in their success as pathogens. In the final stages of infection, the nuclei of infected cells become swollen with the proliferating OBs (Federici, 1997). These cells eventually lyse to release the OBs into the body; a process accelerated in many baculoviruses by the late expression encoded proteases, cathepsin and chitinases. These help to lyse larval tissue and integument to facilitate the release of OBs from the host into the environment. A notable specific behavior, seen in late infection involving virally expressed *egt* gene, is the positive geotropism that causes infected larvae to climb (Goulson, 1997). Larvae in the late stage of infection climb to the top of the plant/tree where they become anchored by the prolegs and die hanging in the distinctive V-shaped head-down posture (Fig. 3.1B) (Federici, 1997; Rebolledo et al., 2015). This climbing behavior is a combination of a positive phototactic response due to *egt* expression and hyperactivity triggered by a protein tyrosine phosphatase gene (van Houte et al., 2012, 2014). Once at the top of plants, the cadavers rupture so that the OBs are released (Fig. 3.1C) to infect larvae feeding down. This also places OBs directly on newly growing foliage at the tip of stems/branches that is often the most nutritious and attractive to early instars, thus promoting transmission. It also makes infected larvae more visible to predators such as birds who may vector longer range dispersal of the virus (Entwistle et al., 1993). Baculoviruses can kill hosts within 3–7 days in rapidly growing tropical species, but in cooler conditions with slower growing temperate species, such as the gypsy moth, *L. dispar*, death may take 10–15 days or longer (Shapiro, 1986). This is a speed of action intermediate between the rapid toxin action of *Bacillus thuringiensis* (Bt) and the generally slower infection process of fungal entomopathogens. This pathology is that of the classic acute lethal baculovirus infection seen in many NPVs and some GVs such as CpGV and *Plutella xylostella* GV (PlxyGV). Some GVs though demonstrate a more prolonged chronic infection (Tanada and Hess, 1991). In GVs such as *Trichoplusia ni*

GV, only midgut and fat body are sites of viral replication, and larvae live longer than uninfected larvae in the same instar, continue to feed, and can grow much larger than uninfected larvae, only showing symptoms shortly before death and, thus these viruses are not seen as useful control agents (Federici, 1997; Hatem et al., 2011). Even in Baculovirus species commonly encountered as acute infections, there is an alternate "unapparent" or "latent" pathology seen in otherwise apparently healthy insects, now more usually termed "covert infections." The existence of these covert infections was only proved once more sensitive polymerase chain reaction (PCR) techniques were developed (Hughes et al., 1993), and its characteristics were explored (Burden et al., 2003). It is now clear that covert infections involve the active replication of the baculoviruses in host tissues though at a very low level with no apparent pathology. In this way, a baculovirus can persist from generation to generation (vertical transmission) (Kukan, 1999), only later to become activated and emerge as a classic acute lethal infection able to spread from insect to insect (horizontal transmission). The advantage of this is that it can enable a virus genotype to persist when population densities are very low when acute pathology would be a poor survival strategy (Vilaplana et al., 2010). The ability to exist as a covert infection is also widespread and shared by many genotypes within a species, so it may represent a very common baculovirus life strategy (Virto et al., 2013). Additionally, baculovirus disease progression from covert to overt may also be influenced by the presence of other microbes commonly present in insects such as *Wolbachia* spp. that have been found to impact on baculovirus infection outcome (Graham et al., 2010).

To summarize, the pathology of baculoviruses offers us insights into why they have both strengths and weaknesses as biological control agents. The two-phase replication cycle enables baculoviruses to achieve a high degree of viral multiplication, but it also constrains the speed with which a natural baculovirus can kill. While baculoviruses are among the faster killing entomopathogens, they still take several days to kill, much slower than most, though not all pesticides, which explains the interest in using GM technology to improve this key control characteristic (Lacey et al., 2001, 2015).

3.3 HOST RANGE

A key facet of the use of baculoviruses as MCAs relates to the host range of the species. Baculoviruses are in general much more specific than some other MCAs such as entomopathogenic fungi. Many species are specific to a single species or infect only members of a genus or closely related group of genera, such as HearNPV and HzSNPV that can infect a range of *Heliothine* species (Rowley et al., 2011). Only a limited number of baculovirus species such as *Autographa californica* MNPV (AcMNPV) have the ability to infect a wide range of host species. The NPV species from the Lymantridae so far investigated tend to be very narrow in host range, while those isolated from the Noctuidae are more variable with some species showing extensive host ranges, for example among the family Noctuinae (Vail et al., 1999). AcMNPV has been reported to infect up to 95 species of Lepidoptera from a range of families. However, for many of these the required dosage of virus to produce infection can be very high, so may not be biologically relevant (Cory, 2003). A confounding issue in identifying host range is that some older studies may in fact represent cases where the attempted infection by a baculovirus resulted not in a successful infection by the challenging virus but initiated the infection of a covert infection that the host was already harbouring (Hughes et al., 1993; Harrison and Hoover, 2012).

3.4 ECOLOGY OF BACULOVIRUSES AND MICROBIAL CONTROL

3.4.1 Baculovirus Ecology

In viewing baculoviruses as crop protection agents, it must always be borne in mind that they have evolved as diverse ecological entities to exploit niches in complex ecological systems (Fuxa, 2004), in contrast to chemical pesticides, which are simple toxic agents. The abilities of baculovirus to infect, multiply, spread (both horizontally and vertically), and persist on plants or in soil are important properties that can have a significant impact on the crop protection function and give baculovirus the potential to act as more than simple chemical pesticide substitutes. There are a number of substantive reviews of the fascinating ecology of baculoviruses to which the interested reader is recommended for further detail (Evans, 1986; Harper, 1987; Cory et al., 1997; Cory and Myers, 2003; Fuxa, 2004; Myers and Cory, 2015) as only a limited discussion of how baculovirus ecology directly impacts on the theory and practice of microbial control is presented here.

The overwhelming majority of baculovirus use has been as short-term applications designed to produce cessation of crop damage through short-term population reductions (Moscardi, 1999; Cory, 2003). This insecticidal use of baculoviruses is often described as inundative and relying on the impact of infection directly from the virus applied with little or no emphasis on the impact of horizontal or secondary infections or vertical infections (Cunningham, 1995). Even where OB production through horizontal transmission is substantial, the degree of suppression may not meet the high level of pest suppression required in modern crop protection, so that control may still depend on repeated reapplication by growers to ensure new foliage or fruiting bodies are adequately protected (Harper, 1987). In general, it may be said that ecological deployment of baculoviruses through

classic control, augmentation or conservation, as with most biological agents, is much more likely to be successful in perennial crops, orchard crops, and forestry where there is a stable ecosystem facilitating natural spreading, persistence, and virus cycling, than in annual or row crops. In seasonal or annual crops, the ecosystem is inherently unstable, and this may prevent optimal horizontal transmission and disrupt the accumulation of adequate virus reservoirs on plants or in the soil. In modern annual crops, growing cycles are kept as short as possible and plants are rapidly cleared and destroyed after harvesting as part of crop hygiene or rotation cycles, which would reduce any plant based reservoir of virus. Soil and organic leaf litter remain very important reservoirs for baculovirus OBs and after outbreaks levels of infective OBs are reported to build up to $2–7 \times 10^4$ OB per g of soil (Harper, 1987). In most soils pH and temperature permit OBs to remain infective for weeks or even years. However, for these to initiate new infections, such OBs need to be translocated to larval feeding sites (Payne, 1982). In pastures precipitation has been identified as an important mechanism in initiating infections, while in forest systems windblown dust can be a major route for relocating soil borne OBs into the leaf canopy where larvae feed (Fuxa, 2004). Despite this, the precise mechanisms and quantification of the role of soil-borne baculovirus in the dynamics of natural baculovirus epizootics remain poorly understood in most systems.

3.4.2 Ecology and Microbial Control Strategies

In considering more ecological strategies than simple inundative use of baculoviruses, several different alternative strategies have been described: (1) limited application of an agent which then spreads horizontally and persists resulting in acceptable pest control without the need for further applications; this is a strategy often described as classical biological control or inoculation; (2) augmentation or the application to initiate epizootics that produces control of pest outbreaks but may not be permanent so that some reapplication is needed to maintain acceptable control; and (3) conservation, which through environmental manipulation seeks to conserve or reactivate baculovirus in the environment to facilitate epizootics.

The capacity of baculoviruses for rapid and high multiplication in hosts makes them perhaps more suitable for controlling insect pests, that are themselves overwhelmingly r-selected species, than some other biological control agents such as many predators that have longer replication times. They have the ability to maximize horizontal transmission through climbing and wandering behaviors that place progeny OBs precisely on those parts of the plant where susceptible neonate larvae aggregate and feed, thus horizontal transmission to the next generations of larvae

(Harper, 1987; Rebolledo et al., 2015). Finally, the ability of the OB itself to persist on suitable microhabitats on foliage for days or weeks and in soil for many years is crucial to their long-term ecological success. However, examples of baculoviruses that have been deployed successfully as classical biological control agents are rather limited (Hajek, 2004; Hajek and Tobin, 2010). One probably accidental example that fits is that of *Gilpinia hercyniae*, the European spruce sawfly in North America. Here a baculovirus initially appeared as an accidental introduction some years after the pest first appeared (Brown, 1982; Harper, 1987), and since then a combination of this virus and parasitoids has kept this pest under adequate control (Moreau and Lucarotti, 2007). While examples of the deliberate use of baculoviruses in classic biological control programs are limited, the cost-benefit ratios of a successful program would seem to be very high so that it remains a highly attractive potential control option in cases where suitable pest species suddenly appear in new areas. The application of the augmentation strategy for pest control is more widespread and has been mostly used in forestry and plantation crops. In North American forests, the programs for controlling the Douglas-fir tussock moth (*Orygia pseudosugata*) and the gypsy moth are examples of a baculoviruses being produced and applied to supplement the natural outbreaks of virus (Cunningham, 1995). Here the programs were backed up by a body of research both on the ecology of the pests and to develop the production and application of the virus (Doane and McManus, 1981; Shapiro, 1982). In the case of gypsy moth MNPV (LdMNPV), the virus alone did not achieve control adequate enough to prevent either the spread of gypsy moth or severe defoliation (Hajek, 2007). Subsequently a fungal pathogen *Entomophaga maimaiga*, probably introduced accidentally, emerged and became established as the major pathogen in gypsy moth populations (Hajek et al., 2015). A more recent example of a virus being produced for augmentation is the use of NPV to control the Balsam fir sawfly (*Neodiprion abietis*) in Canadian forests (Lucarotti et al., 2007; Graves et al., 2012). Other cases include the programs to control two other sawflies *N. sertifer* and *Neodiprion lecontei* with their homologous NPVs. These were introduced into the USA and Canada to initiate epizootics and to this end *N. sertifer* NPV (NeseSNPV) and *N. lecontei* NPV (NeleNPV) were both mass produced and widely evaluated. The sawfly NPVs were both produced through in-field production by collecting infected larvae, from either naturally infected outbreaks or from heavily infected plantations that had been artificially inoculated. NeleNPV was registered as a control product for augmentation and dispersal to new areas (Harper, 1987; Cunningham, 1995). However, in recent years, the commercial production and use of baculoviruses appear to have reduced as applications of Bt are deemed more flexible and cost-effective (Moreau and Lucarotti, 2007).

3.5 ENVIRONMENTAL FACTORS AFFECTING BACULOVIRUS EFFICACY AND PERSISTENCE

3.5.1 Persistence in Different Crops

In the field, the outcome of the application of a baculovirus in terms of control is the result not merely of the nature and quantity of the baculovirus applied but is influenced by a number of environmental factors, biotic and abiotic, that influence the efficacy of the baculovirus applied and its persistence. Sunlight, and specifically the ultraviolet (UV) part of the spectrum, has long been recognized as probably the most significant factor in limiting the persistence of baculovirus (Ignoffo, 1992; Burges and Jones, 1998; Tamez-Guerra et al., 2000; Lacey and Arthurs, 2005; Cory and Evans, 2007). On leaves, active OBs may persist for less than a day on crops with open architecture and subject to high UV exposure (Cherry et al., 2000), while given a more favorable environment, they may persist up to a month or even longer on foliage in temperate forests (Fuxa, 2004). The loss of infectivity is not usually accompanied by physical degradation or by a loss of OBs but due to UV light, which induces changes in the viral DNA that destroy its integrity and functioning thereby inactivating the virus (Payne, 1982). While the OB plays an important function in protecting virions from many environmental challenges and facilitating its persistence, it does not seem to provide any substantive protection from UV in sunlight. It is this limitation that has driven the many efforts to develop baculovirus formulations, adjuvants, and tank mixes intended to provide improved UV protection for baculovirus sprays (Burgess and Jones, 1998; Behle and Birthisel, 2014).

The location of OBs is also an important factor in persistence. When applied to upper surfaces of leaves, OBs lose activity much faster than those on lower leaf surfaces, primarily due to the action of UV (Jones et al., 1993; Peng et al., 1999). Targeting virus applications to maximize deposition on under-leaf surfaces has been a goal for many control situations as this ensures maximal persistence. Also, under-leaf surfaces are commonly the preferred feeding sites for the neonates and early instars that are the most sensitive to infection and whose rapid infection can reduce subsequent crop damage most. A similar targeting issue may occur where crop damage occurs when larvae feed on key parts such as flowers or fruiting bodies where any damage leads to crop loss. Cryptic feeding by larvae in protected sites that are hard to spray can also prevent successful application of OBs. While OBs are temperature sensitive and can be inactivated rapidly once temperatures exceed 50°C, this factor does not seem to be an important one in restricting baculovirus efficacy in most field situations (McLeod et al., 1977; Ignoffo, 1992).

There has been a considerable body of work to improve application of insect viruses (Hunter-Fujita et al., 1998;

Mierzejewski et al., 2008). A number of sprayers or attachments for sprayers were developed to better target OBs to under-leaf surfaces, but in practice there has been little adoption by growers of such specialized equipment and it is probably unrealistic to expect they would buy specific microbial sprayers until these productions build a more significant market share (Gan-Mor and Mathews, 2003; Chapple et al., 2008; Lacey et al., 2015). Wind and rain can lead to translocation of infective OBs, removing them from target areas where susceptible young instars of pests feed to lower portions of the plant or onto the soil where transmission is less likely. Once OB sprays have dried onto foliage, even in waxy species they seem to be relatively rain-fast (Payne, 1982), but any precipitation occurring before this drying is completed can lead to control failures as OBs are removed from target foliage.

The crop architecture can also have an impact on efficacy and persistence of baculovirus as shading by foliage can protect OBs from full sunlight, enabling activity to persist longer (Biever and Hostetter, 1985). The generally excellent results reported for using NPV on crops such as tomato may well be a result of longer persistence in part due to the higher degree of shading experienced by OB on leaf surfaces of this crop (Arrizubieta et al., 2015). It may be that the appearance of new foliage is the deciding factor in determine the need for reapplication of baculovirus, commonly at least every 7–10 days regardless of the inherent persistence of the baculovirus on the crop. Finally, the persistence of baculovirus on different crops can vary profoundly in response to the chemical nature of the plants. Host plants can severely degrade the persistence of baculovirus or reduce infectivity where they secrete or contain plant defense chemicals that can inactivate baculovirus (Duffey et al., 1995; Cory and Hoover, 2006), though on a few host plants it has been reported that baculovirus can become more potent when the host is feeding on plant tissue (Keating et al., 1988). Specific phytochemical mechanisms that interfere with baculovirus infectivity on crops have been identified in cotton, where two different mechanisms have been reported (Elleman and Entwistle, 1985; Hoover et al., 1998a,b), and more recently another different chemical inactivation mechanism in chickpea has been described (Stevenson et al., 2010). In the case of cotton, it was originally identified that OBs on the leaf surfaces were rapidly inactivated (Young and Yearian, 1974; McLeod et al., 1977). This was identified as due to Mg^{2+} and Ca^{2+} cations in the leaf glandular exudates (Elleman and Entwistle, 1985). A second inactivation mechanism in cotton leaves was identified later (Hoover et al., 1998a); here, the effect due to the action of free radicals during digestion of foliage and OBs that damage ODVs directly and/or stimulated the infected cells in the gut to be sloughed off reducing infection (Hoover et al., 2000; Cory and Hoover, 2006). Quinone or phenolic binding to OB may also reduce OB infectivity

as it may impair release of ODVs from the OBs (Felton and Duffey, 1990). However, there is also evidence that with some host plants feeding on foliage with high levels of tannins and acids can increase virus infectivity (Keating et al., 1988), while high foliage palatability, through increasing consumption of OB can increase larval mortality (Bixby-Brosa and Potter, 2010). Host behavior is an important factor in baculovirus ecology and transmission in *N. sertifer*, while the NPV replicates only in the gut and so only limited multiplication occurs, it is gregarious, and commonly a single infection leads to the infection of all larvae in a colony passed on through feces and regurgitation (Cunningham, 1995). In contrast, where larvae feed only for short periods after hatching on the surface, before borrowing into fruiting bodies or stems (Payne, 1982), this poses a real challenge for successful baculovirus control as spraying has to be very precisely timed to ensure adequate infection rates. Nonetheless, this is not an insuperable barrier to commercial success of a baculovirus as one of the most widely used of all baculovirus CpGV targets just such a pest (Lacey et al., 2008).

3.6 BACULOVIRUS USE ON CROPS AND CASE STUDIES OF BACULOVIRUS PRODUCTS

A key factor in cases of successful baculovirus use on crops is related to the application rate needed to control target pests and how this equates to cost. Arguably the most successful case of baculovirus adoption has been the Brazilian program to control the velvet bean caterpillar (*Anticarsia gemmatalis*) with the *A. gemmatalis* NPV (AgMNPV). Here the low application rate needed (1.5×10^{11} OBs/ha), played a major role in making this economically viable. In most NPVs used to control pests, rates need to be much higher; at least 10^{12} OBs/ha for HearNPV (Cherry et al., 2000; Arrizubieta et al., 2015), HzSNPV (Ignoffo, 1999), SpexNPV (Grzywacz et al., 2008), and SeMNPV (Lasa et al., 2007). Granulovirus products contain only a single virion per OB and require higher application rates; effective rates are reported as $2–3 \times 10^{13}$ OBs/ha for field applications of CrleGV (Moore et al., 2015) and PlxyGV (Grzywacz et al., 2004), though on some orchard crops the use of 10^{13} OBs/ha seems to be the accepted rate for CpGV (Lacey and Shapiro-Ilan, 2008; Lacey et al., 2008) and AdorGV (Blommers et al., 1987; Nakai, 2009). In forests where the objective is to initiate local epizootics, much lower application rates can be used to give acceptable control. For the sawfly *Neodiprion abietis* application rates of NeabSNPV as low as 10^9 OBs/ha are reported, while with NeseSNPV and LdMNPV, application rates of 5×10^9 OBs/ha seem to be adequate (Cunningham, 1995; Graves et al., 2012). Despite these successes, it must be remembered that in most globally important field/row crops such as cotton, maize, rice, etc., despite sometimes extensive research efforts to develop and evaluate baculovirus (Lacey et al., 2001, 2015), adoption remains minimal. A few of the key factors in baculovirus successes can be summarized as (1) availability of a highly pathogenic isolate; (2) mass production at a cost that the market can bear; (3) a crop situation in which baculovirus can be readily applied to pest feeding sites; (4) crop and pest environment that facilitates good baculovirus horizontal transmission; and (5) a use where due to incompatibility with an established IPM system or MRL regulations or environmental drivers, chemical pesticides are not acceptable.

3.6.1 Baculovirus Products and Crops

Baculoviruses, although they currently form a minor part of the crop protection sector, approximately 10% of the biological pesticides market (Glare et al., 2012), are used on a very wide variety of crops. This is in part because some of the target pests for commercially available baculoviruses are pests of many crops and are of global importance and distribution as pests, eg, *Helicoverpa/Heliothis* spp. and *Spodoptera* spp. (Gowda, 2005). Table 3.1 provides a list of baculoviruses in current commercial use or under active commercial development (Gywnn, 2014; Lacey et al., 2015); some of the ones still under precommercial development are not listed here but can be found elsewhere (Haase et al., 2015; Lacey et al., 2015). Baculovirus products have in the main been developed to meet crop pest needs not readily met by chemical insecticides, either for reasons of pest resistance to chemicals, environmental or residue concerns, or IPM incompatibility. These species are also pests in a wide range of crops, giving baculovirus potential market niches in both field and protected crops grown in polytunnels and glasshouses (Arrizubieta et al., 2015; Lacey et al., 2015). In recent years, the niche markets for baculovirus, as for all biopesticides, have also expanded with the increasing withdrawal of registrations for chemical pesticides due to MRL issues. However, while the potential market opportunities for baculovirus have grown due to their high compatibility with IPM systems, lack of MRL issues, and safety, they are still constrained by their generally higher cost, narrower host spectrum of activity, and shorter persistence than for chemical pesticides.

The cropping systems in which baculoviruses have been most widely adopted fall into several general categories including high-value horticultural crops such as vegetables and fruit, high-value orchard crops, vegetables, organic crops, and environmentally sensitive agriculture.

- High-value horticulture has become a key market for baculovirus products. These intensive cropping horticulture systems can absorb higher costs of using baculovirus products and handle more sensitive timing of applications and more frequent reapplication of baculovirus products. An important subsector of potential importance is in the glasshouse crops where IPM systems reliant on predators and parasitoids have become well established. If new pests arrive in these systems, any control for these novel pests needs to be predator and parasitoid compatible, thus ruling out chemical insecticides but opening a niche for specific biopesticides like baculoviruses.

- The rise of the organic sector in horticulture has also had an important role in expanding use of baculoviruses, such as organic apple production, which is a key market for CpGV products in Europe and the United States.

- Orchard crops such as apples are often ecologically stable cropping systems with a significant degree of natural pest control and thus present a supportive environment for specific biopesticides such as baculoviruses.

- Finally, in forestry, many major forest pests are Lepidoptera or Hymenoptera for which baculoviruses are known (Cunningham, 1995; Moreau and Lucarotti, 2007).

3.6.2 Case Studies of Baculovirus Use

Looking at a number of specific cases where baculovirus products have been deployed widely can usefully illustrate the crop factors that can favor the successful use of baculovirus biopesticides and some of the issues that can constrain their use.

3.6.2.1 Codling Moth Control With CpGV

One of the earliest and most widely used of baculovirus products are those based on the codling moth granulovirus CpGV used for control of codling moth on apples. It is now widely used in Europe and North and South America. Codling moth is a serious pest of pome fruit, and in response to serious problems of secondary pest outbreaks after use of broad-spectrum chemical pesticides from the 1960s onwards, there was a search for a more specific sustainable control agent. An isolate of CpGV was early on identified from Mexican codling moth and was found to be a very virulent baculovirus (Tanada, 1964; Payne, 1982) and subject to a major program of research in USA and Europe. Extensive field testing in North America and Europe established protocols for successful application timing and application rates, as well as the scouting and decision-making criteria for treatment (Lacey and Shapiro-Ilan, 2008). Several products based on the Mexican GV isolate (CpGV-M) were registered by 1995 (Hüber, 1998) and use in Europe had reached 100,000 ha with some use in North America mainly on organic orchards (Lacey and Shapiro-Ilan, 2008).

The situation for CpGV use became more complicated when resistance to the CpGV-M was reported from areas where it has been used for 20 years or more (Eberle and

Jehle, 2006). It was rapidly determined that this resistance can be overcome by using baculovirus product containing different CpGV isolates rather than the original CpGV-M isolate (Eberle et al., 2008), and new products incorporating the new isolates have now been brought to market (Schmitt et al., 2013; Gwynn, 2014). This illustrates that while resistance potential of pests to baculovirus may be lower than that with simple chemical insecticides, it is not negligible and that it remains advisable practice to use a baculovirus only as one component in an IPM strategy (Lacey and Shapiro-Ilan, 2008).

In some ways the success of these CpGV baculovirus products is surprising as codling moth larvae feed inside the apple for much of their larval cycle only wandering and feeding on the surface for a short period after hatching, so that natural infection rates are low. This short window for infection, often only hours, combined with the short persistence of the GV with a half-life of 24–72 h (Glen and Payne, 1984; Arthurs and Lacey, 2004) and the presence of multiple generations of codling moth over a prolonged growing season would appear a serious challenge to effective control with the CpGV (Lacey and Shapiro-Ilan, 2008). Nonetheless, if application can be timed to coincide with peak fruit entry by the first instar then the baculovirus can rapidly be acquired by a high proportion of the larvae before any significant damage occurs (Ballard et al., 2000; Lacey and Shapiro-Ilan, 2008). Although some early season fruit scaring may be an issue, regular reapplication of CpGV every 7–14 days can bring about population reduction sufficient to limit this to usually levels (Arthurs et al., 2005). However, CpGV is not viable as a stand-alone control, either where codling moth pest pressure is very high or where other pests are prominent, so that it needs to be combined with other "soft" IPM components. Thus, it frequently needs to be combined with the implementation of cultural controls, crop sanitation, pheromone mating disruption, parasitoid and predator conservation with some use of "soft" chemistry to achieve effective IPM (Lacey and Shapiro-Ilan, 2008). This case study of CpGV illustrates that even a less than ideal baculovirus can be successful, but it needs always to be considered as part of an overall IPM system and that includes a focus on maintaining good insect resistance management practice.

3.6.2.2 Anticarsia gemmatalis *MNPV*

The development and use of AgMNPV in Brazil for the control of *A. gemmatalis*, velvet bean caterpillar, on soya bean can be seen as a model program for the public/private development and deployment of a successful baculovirus (Moscardi, 1999). To avoid reliance on chemical pesticides and possibility of environmental damage, the Brazilian research institute EMBRAPA commissioned the development of a soya IPM program in the 1970s. As part of this program, a team led by Dr. Flavio Moscardi identified the potential of AgMNPV. Subsequently after extensive successful field testing of the virus the program moved into developing an AgMNPV product in the 1980s. One very favorable aspect of using AgMNPV was the very low application rate of 1.5×10^{11} OBs/ha, well below the rate required for many other NPVs (Moscardi, 1999). This combined with the adoption of a field-based production system (see Chapter 7) was crucial to reducing the production cost of AgMNPV to US$1.20–$1.50 ha^{-1}, making it more economic than chemical alternatives (Moscardi, 1999; Moscardi and Sosa-Gomez, 2007). In addition, the early adoption of nonliquefying isolates of AgMNPV was very important in improving the viability of hand collecting infected larvae as part of the field production system (Moscardi and Sosa-Gomez, 2007). An interesting finding was that while host strains resistant to AgMNPV could readily be generated in laboratory experiments (Abot et al., 1996), no resistance problems appeared in the field to threaten control, despite the extensive use of this virus on up to 2 million ha (Moscardi and Sosa-Gomez, 2007).

The adoption of AgMNPV by farmers once the product was commercialized was rapid, and by 2002–2003 it was used on 2 million ha of soya. However, after 2003 use of the AgMNPV declined precipitously as soya farmers switched to a no-tillage soya cropping systems that involved the routine prophylactic use of a combined herbicide and broad spectrum insecticide reducing the market for the virus product. In consequence, AgMNPV is now used on less than 200,000 ha (Moscardi et al., 2011; Panazzi, 2013), though there has been some expansion of use up to 100,000 ha in Paraguay and a start to its use in Mexico (Haase et al., 2015).

3.6.2.3 PhopGV and Potato Tuber Moth

Another pioneering baculovirus pest control system used mainly in South America is the use of a GV for the control of potato tuber moth (PTM), *Phthorimaea operculella*. Work in the late 1960s had identified a GV from *P. operculella* (PhopGV), and research subsequently assessed its potential use for PTM control (Raman et al., 1987). Field trials of PhopGV in potatoes showed variable results and the high cost meant it was not adopted for field use (Lacey and Kroschel, 2009). PhopGV however was developed into a product for use in rustic stores to protect seed potatoes and tubers for consumption. Pilot production of PhopGV as a talc formulation (Matapol) was established at a number of sites in Bolivia, Ecuador, Colombia, and Peru and still continues in some locations (Moscardi, 1999; Haase et al., 2015). The product is specifically packaged and aimed at smallholder subsistence farmers. The use of PhopGV in stores meant that rapid knock down was not needed, and under the dark and dry conditions the poor UV stability of the GV was not an issue (Sporleder and Kroschel, 2008). A very interesting aspect of using the talc formulation was that it greatly increased the efficacy of control compared with virus only suspensions so that much of the initial kill was due to the insecticidal effect of the talc carrier (Arthurs et al., 2008; Mascarin and Delalibera, 2012). Talc and various other diatomaceous earths are well-known physical insecticides widely used in stored products. The Matapol-type products may, with some truth, be described more as diatomaceous earths with added baculovirus than baculovirus products per se. Research efforts to expand the use of PhopGV have more recently involved testing in the United States (Arthurs et al., 2008) and in refrigerated potato stores (Lacey et al., 2010).

3.6.2.4 North America Forest Pests

Foresters in North America from the 1950s became aware that using broad spectrum chemical pesticides to control pests was rarely cost-effective, as the disruption of natural pest control in forests through use of these pesticides frequently lead to serious secondary pest issues. Forestry and baculovirus research has a long historical association as the "wilt" and "tree top diseases" recognized in forest pests in the 19th century in both North America and Europe were caused by baculoviruses. In North America, it is government agencies that manage forests, facilitating a strategic approach to pest control including that of invasive pests that are potentially the most devastating.

It was against this background that the forest services in the United States and Canada investigated baculoviruses as potential control agents from the 1960s for control of defoliating Lepidoptera (Cunningham, 1995; also see Chapter 22). The development of baculovirus for forestry use drove much valuable research on baculovirus ecology, application technology, and production in North America on gypsy moth (Doane and McManus, 1981; Moscardi, 1999), sawflies (Lucarotti et al., 2007), and a variety of other forest-defoliating pests (Cunningham, 1995; van Frankenhuyzen et al., 2007; also see Chapter 22). The North American forestry research investigated and evaluated a wide variety of different baculoviruses and their hosts over a prolonged period from the 1950s, leading to the registration of five products based on four baculoviruses: LdMNPV, NeleNPV, OpMNPV, and NeabNPV (Moreau and Lucarotti, 2007). Despite the technical success much of this work, the current use of baculoviruses in forestry in North America has remained marginal as they have been supplanted by Bt (Moreau and Lucarotti, 2007; van Frankenhuyzen et al., 2007). Though registration of some of these has now lapsed, development continues with NeabNPV (Graves et al., 2012) and commercial sale of LdMNPV and NeabNPV continues (Gwynn, 2014).

3.6.2.5 Baculovirus Programs That Did Not Progress to Adoption

While examination of "success" stories for baculovirus can throw interesting insights into factors contributing to successful baculovirus-based IPM, one should also learn from the case studies where baculoviruses have been developed but not adopted.

A major focus for baculovirus product development has in the past been on controlling *Heliothis/Helicoverpa* spp. on cotton, as these pests have long been the most prominent pests of this global crop. HzSNPV was the first baculovirus insecticide to be developed and brought to registration in the 1970s (Ignoffo, 1999). However, shortly after its appearance the first modern photostable pyrethroids were released and these with their combination of low-cost and broad-spectrum activity led production of the HzSNPV product (Virion H, later Elcar) to cease in 1982. Subsequently, the widespread appearance of pyrethroid resistance in cotton pests in the 1980 and 1990s revived interest in HzSNPV leading to the registration of Gemstar in the United States. Widespread development of Heliothine resistance to pyrethroids and other pesticides in the late 1980s (Gowda, 2005) also drove the development and registration of other local HzSNPV and HearSNPV products in India, Australia, and China where the NPVs were seen as useful integrated resistance management tools (Buerger et al., 2007; Sun and Peng, 2007; Sun, 2015). With the arrival of insect resistant genetically modified (GM) cotton in the late 1990s the market for baculovirus in cotton largely disappeared. The HzSNPV and HearNPV products have since found new niche markets in crops such as vegetables, oilseeds, legumes, and sorghum, especially where *Helicoverpa armigera* is a problem and whose geographical range is expanding (Franzmann et al., 2008).

3.6.2.6 Genetically Modified Baculoviruses

Another example of failure to move to product has been the very substantial research effort to develop GM baculoviruses to overcome some of the constraints associated with "wild type" isolates. This included developing GM baculovirus incorporating inserted genes or deletions that speed up cessation of feeding and death. The genes evaluated include insect specific toxins from the scorpions *Androctonus australis* and *Leiurus quinquestriatus*, the spider *Tengeneria agrestis*, the itch mite *Pyemotes tritici* and juvenile hormone esterases. Many GM baculoviruses were successfully developed and tested in laboratory and some in field trials (Bonning et al., 2003; Inceoglu et al., 2006; Szewcyk et al., 2006).

Despite the technical successes of the GM baculoviruses, commercial development of GM baculovirus products has not progressed. An important constraint has undoubtedly been the rising costs of registering and deploying GM technology, currently up to $130 million per new GM trait (McDougall, 2011). This is particularly crucial for baculovirus which currently has narrow host ranges and therefore more limited markets than their chemical equivalents. Other difficulties lie in developing large-scale in vitro production, necessary for most of the GM baculoviruses (see Chapter 7). However, as the GM approach can potentially resolve some of the serious limitations around baculoviruses, its use is still seen by its protagonists as a technology whose time will definitely come (Kamita et al., 2010).

In summary, baculovirus products have taken off in a limited number of niches where a number of separate factors, biological, agronomic, and economic, need to come together positively to produce a control technology that can achieve acceptable pest damage limitation within the users' acceptability and cost constraints. The biological environment—pest ecology, plant species, and cropping system—needs to be favorable for effective virus application, persistence and spread. However, the cases such as that of CpGV, have shown that even if the virus–pest dynamics do not seem ideal, a successful product can be developed if it is deployed as a component of a strong IPM system. To stand any chance of adoption by farmers any baculovirus needs first to travel the long and uncertain road to reach the status of a commercially available product, which is in itself a complex and often poorly understood process.

3.7 OVERVIEW OF BACULOVIRUS PRODUCT DEVELOPMENT AND COMMERCIALIZATION

"It is a truth universally acknowledged that a successful baculovirus in possession of a good pathogenicity must be in want of a commercial partner" (Austen, 1813). In truth, with apologies for subverting the immortal prose of Ms. Austen, the courtship of baculovirus and commercial producers is a process fraught with as many uncertainties and generally as much ignorance as any more conventional courtship. There is a significant problem that although in recent years new commercial baculovirus products have been appearing on the markets, essentially these are developed from baculovirus that have all been known for the last 30 years and products based on novel baculoviruses are not appearing on the market (R. Gwynn, personal communication). Why is the supply of new baculovirus to the commercial sector so poor when the opportunities for biological control sector have never been brighter (Glare et al., 2012; Lacey et al., 2015)? One possible hypothesis for this dearth of new product development in baculovirus might be the shift in research in the last two decades from applied microbial control to more emphasis on fundamental genetic and GM-related research. The concept behind this shift was that the development of GM baculovirus alone could significantly improve key

baculovirus characteristics. However, with the failure to get GM baculovirus to market, these potentialities have not been realized (Inceoglu et al., 2006; Reid et al., 2014). Now, with new market opportunities generating a pull toward new baculovirus products, this dearth may be reversed, but it will take time to make up for the lost decades. In order for research to generate new products, it is not merely that there has to be a greater focus on applied biopesticides research or a need for more funding of this topic. Arguably there also needs to be a better understanding among researchers and science funders of the paradigm that is "research for product" and to recognize that this is crucially different from "research for knowledge." The issue of why more biopesticides research has not led to products emerging has been identified and discussed by Lisansky (1997) and Gaugler (1997). A central theme in these studies on biopesticide commercialization is that without public sector scientists having a clear comprehensive understanding of what biopesticide product development involves, they are not likely to conduct the type of research that will facilitate commercial uptake and exploitation of the organisms they are investigating. This can be exacerbated because for most public sector scientists the primary priority is to publish scientific papers; thus the research is driven primarily by the perceived priorities of high-impact journal reviewers rather than the goal of product development. The need for a comprehensible framework for biopesticide development was addressed by the landmark study of Ravensberg (2011), which was a unique in-depth analysis of the roadmap to the development of biocontrol products. This topic is too complex and detailed to give it justice in the space available here, but this study should be essential reading for any biopesticide scientist whose work may have commercial applications.

There is something of a consensus in the commercial biopesticides sector that many academic research programs simply fail to produce the type of research knowledge that facilitates commercial uptake (M. Brown, personal communication) and that this needs to be considered when formulating research, not as a poststudy afterthought. The priority interests of the commercial sector are (1) conduct research on commercially significant pests; (2) include evidence of comparative efficacy against alternatives; and (3) include data on replicability and reliability. While it is often very tempting to conduct research on model systems if that pest/pathogen is of no real commercial importance, this approach significantly reduces the potential commercial interest. If the choice is taken to focus research on a significant pest species, even then it can require some effort to identify a suitable species as pest status can change greatly over time, something not always picked up by academic researchers. The key issue is to identify what pests the industry identifies as a problem currently or in the future. Commercial organizations in looking at efficacy are always interested on how it compares to existing methods of control rather than against a no treatment control. Thus, trials that include no chemical insecticide or other industry standard control will often, for good reason, fail to excite much commercial interest.

Finally, in doing such comparative trials, it is always highly desirable to collect the types of data that are of true commercial significance. To this end, data on pest numbers need to be supported by data on economically significant crop damage if the trial is to be evaluated comprehensively. This can be a real dilemma where the specific biocontrol (pheromone or pathogen) may have very effectively reduced the target pest, but in the absence of a broad spectrum action, this allows another nontarget pest to florish and cause serious crop/produce damage. In this situation, full disclosure is advised as the best policy in the long term, as commercial developers are well aware of specificity issues and can cope with this aspect if they have confidence in the transparency and honesty of the research.

In any trials, thought should be given to collecting the type and quality of data that could fit into a registration dossier. Conducting trials to good laboratory practice and/or good experimental practice, as standard hugely increases confidence in the result and enhances its value to the industry. Examples are those published by the European and Mediterranean Plant Protection Organisation (2012). In addition, when conducting a trial, thought should be given from the start in how the new technology would be integrated on farms with existing cropping or pest control practices as farmers cannot adopt a pest control technology that requires resources they do not have or that the economic value of the crop cannot support. Nor can they change the basics of cropping, cultivation, and rotation just to facilitate the better efficacy of a specific biocontrol agent. Thus, trial design should reflect what is practicable and possible rather than what is ideal! A key issue for biopesticides has been the issue of reliability (Lisansky, 1997), so any set of results needs to be based on robust and meaningful replication. To judge the viability of a control, there is an absolute need to know how it works in different years and on different locations; without such data, commercial developers are right to exercise caution in interpreting results. There is no easy resolution to this issue, given that much publically funded research is in the form of 3-year doctorates, and for these projects it can be beyond their means to conduct multiyear and multilocation replications. If a program of research on a novel biocontrol approach never moves forward to build such a body of data, then a lack of commercial interest should not be surprising.

Finally, there is a clear potential conflict in the desire of researchers/funders to rush to press and publish as soon as possible and a commercial imperative not to disclose data, results, and ideas that might have significant intellectual property value before this can be accurately assessed. This needs to be very prudently managed and publication

carefully planned if a potentially valuable commercial opportunity is not to be lost. Here early and close liaison with a commercial partner established in crop protection and the researchers own institutional intellectual property advisor is invaluable in guiding the process.

3.8 OVERVIEW OF FUTURE DEVELOPMENT AND NEEDS

In looking to the future of baculovirus in pest control, the major question is what research and product development issues need to be resolved in order to expand their attractiveness to users and allow them to play a full role in future crop protection. In one way, this is definitely an exercise in "back to the future" as this issue and the needs of baculovirus research have remained remarkably similar over the last two decades (Lisansky, 1997; Lacey et al., 2001, 2015). Overall, scientists in the field have been readily able to recognize and agree on the research needs; it is the solutions that have been a bit slower to come!

1. Reduction in cost of baculovirus products. There is no doubt that a major factor in the success of AgMNPV in Brazil was due to its low cost of <$5 ha^{-1}. Reid et al. (2014) argued that to expand baculovirus use significantly into major field crops a farm product price of $20 ha^{-1} needs to be achieved. There are several potential routes to achieve this. It could come through lowering production cost or developing more active strains or formulations (see Chapter 7).

2. Scaling up baculovirus production. While in vivo production has delivered the volume needed for the current niche markets (Grzywacz et al., 2014), there are questions about its ability to deliver the quantities needed for major field crops (Reid et al., 2014). The argument is that this could only be met through the development of commercially viable large scale (<20,000 L) in vitro systems of production (see Chapter 7).

3. Faster speed of kill and earlier cessation of crop damage. The slower killing speed of baculovirus remains a significant constraint (Szewcyk et al., 2006). This may be as much an issue of adverse farmer perceptions as a biological reality, but finding natural isolates with faster action could be a desirable way forward. The other route to address this issue is through GM of baculovirus, as discussed earlier.

4. Improving reliability of baculovirus products. Many users do not use baculovirus products because they are perceived of as unreliable and erratic (Lisansky, 1997). Surveys of farmers have revealed that again perceived lack of reliability is as important as higher cost in deterring users (Marrone, 2007).

5. Expanding baculovirus host range. Baculovirus products would be much more commercially attractive if products had a wider host range. Research here on the role of enhancins and other infectivity factors (Slavicek, 2012) may be the key to understanding the mechanisms that underlie host specificity.

6. Increasing baculovirus persistence. A substantial advance would be to extend the persistence of baculovirus, especially its UV persistence. One approach would be through creating or identifying UV-resistant isolates, but probably more practical would be the development of UV-resistant formulation technology that worked in the field, which could then be applied to any baculovirus/biopesticide (Behle and Birthisel, 2014). Significant progress might come if a way were found to not merely add the UV blocker to the product but bind it intimately to the viral OB without impairing infectivity (see Chapter 7).

7. Overcoming phyto-inactivation of baculoviruses on key crops such as cotton and chickpea (Cory and Hoover, 2006; Stevenson et al., 2010) would be a useful advance if baculoviruses are to win markets in these major crops.

8. The development of baculovirus storable formulations with a shelf-life comparable to chemical pesticides (>2 years) at ambient temperatures would be another useful advance, especially in tropical cropping systems.

9. The development of better biologically based IPM for systems that used the particular merits of all MCAs, virus, fungi, bacteria, and nematodes, while minimizing their biological and physical limitations would help to promote baculovirus use (Lacey et al., 2015). These would aim to integrate MCAs more effectively and move away from insecticidal application to proactive ecological use.

10. Finally, the streamlining and rationalization of registration systems to fast-track and facilitate the registration of MCAs like baculovirus of accepted safety (OECD, 2002; Mudgal et al., 2013). The current system in place in the European Union (EU) and many other jurisdictions is widely seen as a barrier to new products (Chandler et al., 2011; Ehlers, 2011). Streamlining has occurred to some extent in North America (Kabaluk et al., 2010) so that registering an MCA product costs $500,000 but in the EU, despite significant attempts to rationalize it (Chandler et al., 2011), registration still costs around €1.3 million (US$1.5 million) and can take several years, acting as a serious barrier to product innovation (Ravensberg, 2011).

In summary, baculoviruses are useful tools in the biological control agent armory with a place in the current biopesticides market. Currently they are very much niche products operating in the high value or environmentally sensitive market sectors; they could play a much larger role in crop protection in future, but only if key constraints to their efficacy, marketing, and registration are addressed.

REFERENCES

Abot, A.R., Moscardi, F., Fuxa, J.R., Sosa-Gómez, D.R., Richter, A.R., 1996. Development of resistance by *Anticarsia gemmatalis* from Brazil and United States to a nuclear polyhedrosis virus under laboratory selection pressure. Biol. Control 7, 126–130.

Adams, J.R., Mclintock, J.T., 1991. Baculoviridae. Nuclear polyhedrosis viruses. Part 1 Nuclear polyhedrosis viruses of insects. In: Adams, J.R., Bonami, J.R. (Eds.), Atlas of Invertebrate Viruses. CRC Press, Boca Raton, pp. 87–204.

Arrizubieta, M., Simón, O., Torres-Vila, L.M., Figueiredo, E., Mendiola, F.J., Mexia, A., Williams, T., 2015. Insecticidal efficacy and persistence of a co-occluded binary mixture of *Helicoverpa armigera* nucleopolyhedrovirus (HearNPV) variants in protected and field-grown tomato crops in the Iberian Peninsula. Pest. Manag. Sci. 72, 660–670. http://dx.doi.org/10.1002/ps.4035.

Arthurs, S.P., Lacey, L.A., 2004. Field evaluation of commercial formulations of the codling moth granulovirus: persistence of activity and success of seasonal applications against natural infestations of codling moth in Pacific Northwest apple orchards. Biol. Control 31, 388–397.

Arthurs, S.P., Lacey, L.A., Fritts Jr., R., 2005. Optimizing the use of the codling moth granulovirus: effects of application rate and spraying frequency on control of codling moth larvae in Pacific Northwest apple orchards. J. Econ. Entomol. 98, 1459–1468.

Arthurs, S.P., Lacey, L.A., Pruneda, J.N., Rondon, S., 2008. Semi-field evaluation of a granulovirus and *Bacillus thuringiensis* ssp. *kurstaki* for season-long control of the potato tuber moth, *Phthorimaea operculella*. Entomol. Exp. Appl. 129, 276–285.

Austen, J., 1813. Pride and Prejudice. Thomas Egerton, London.

Ballard, J., Ellis, D.J., Payne, C.C., 2000. Uptake of granulovirus from the surface of apples and leaves by first instar larvae of the codling moth *Cydia pomonella* L. (Lepidoptera: Olethreutidae). Biocontrol Sci. Technol. 10, 617–625.

Behle, R., Birthisel, T., 2014. Formulation of entomopathogens as bioinsecticides. In: Morales-Ramos, J.A., Guadalupe Rojas, M., Shapiro-Ilan, D.L. (Eds.), Mass Production of Beneficial Organisms. Elsevier, Amsterdam, pp. 483–517.

Biever, K.D., Hostetter, D.L., 1985. Field persistence of *Trichoplusia ni* (Lepidoptera: Noctuidae) single-embedded nuclear polyhedrosis virus on cabbage foliage. Environ. Entomol. 14, 579–581.

Bixby-Brosi, A.J., Potter, D.A., 2010. Evaluating a naturally occurring baculovirus for extended biological control of the black cutworm (Lepidoptera: Noctuidae) in golf course habitats. J. Econ. Entomol. 103, 1555–1563.

Blommers, L., Vaal, F., Freriks, J., Helsen, H., 1987. Three years of specific control of summer fruit tortrix and codling moth on apple in the Netherlands. J. Appl. Entomol. 104, 353–371.

Bonning, B.C., Boughton, J.A., Jin, H., Harrison, R.L., 2003. Genetic enhancement of baculovirus insecticides. In: Upadhyay, K. (Ed.), Advances in Microbial Control of Insect Pests. Kluwer Academic, New York, pp. 109–126.

Brown, D.A., 1982. Two naturally occurring nuclear polyhedrosis virus variants of *Neodiprion sertifer* Geoffr. (Hymenoptera: Diprionidae). Appl. Environ. Microbiol. 43, 65–69.

Buerger, P., Hauxwell, C., Murray, D., 2007. Nucleopolydrovirus introduction in Australia. Virol. Sin. 22, 173–179.

Burden, J.P., Nixon, C.P., Hodgkinson, A.E., Posse, R.P., Sait, S.M., King, L.A., Hails, R.S., 2003. Covert infections as a mechanism for long-term persistence of baculoviruses. Ecol. Lett. 6, 524–531.

Burges, H.D., Jones, K.A., 1998. Formulation of bacteria, viruses and protozoa to control insects. In: Burges, H.D. (Ed.), Formulation of Microbial Biopesticides. Kluwer Academic Publishers, Dordrecht, pp. 33–127.

Chapple, A.C., Downer, R.A., Bateman, R.P., 2008. Theory and practice of microbial insecticide application. In: Lacey, L.A., Kaya, H.K. (Eds.), Field Manual of Techniques in Invertebrate Pathology: Application and Evaluation of Pathogens for Control of Insects and Other Invertebrate Pests. Springer, Dordrecht, pp. 9–36.

Chandler, D., Bailey, A.S., Tatchell, G.M., Davidson, G., Greaves, J., Grant, W.P., 2011. The development, regulation and use of biopesticides for integrated pest management. Philos. Trans. R. Soc. B Biol. Sci. 366, 1987–1998.

Cherry, A.J., Parnell, M., Grzywacz, D., Brown, M., Jones, K.A., 1997. The optimization of in vivo nuclear polyhedrosis virus production in *Spodoptera exempta* (Walker) and *Spodoptera exigua* (Hubner). J. Invertebr. Pathol. 70, 50–58.

Cherry, A.J., Rabindra, R.J., Parnell, M.A., Geetha, N., Kennedy, J.S., Grzywacz, D., 2000. Field evaluation of *Helicoverpa armigera* nucleopolyhedrovirus formulations for control of the chickpea pod borer, *H. armigera* (Hubn.), on chickpea (*Cicer arietinum* var. Shoba) in southern India. Crop Prot. 19, 51–60.

Clem, R.J., 2005. The role of apoptosis in defence against baculovirus infection in insects. Curr. Top. Microbiol. Immunol. 289, 113–129.

Cohen, S., Au, S., Panté, N., 2011. How viruses access the nucleus. BBA-Mol. Cell. Res. 1813, 1634–1645.

Cory, J.S., 2003. Ecological impacts of virus insecticides: host range and non-target organisms. In: Hokkanen, H.M.T., Hajek, A. (Eds.), Environmental Impacts of Microbial Insecticides Progress in Biological Control 1, pp. 73–91.

Cory, J.S., Myers, J.H., 2003. The ecology and evolution of insect baculoviruses. Annu. Rev. Ecol. Evol. Syst. 34, 239–272.

Cory, J.S., Hoover, K., 2006. Plant mediated effects in insect-pathogen interactions. Trends Ecol. Evol. 21, 278–286.

Cory, J.S., Evans, H.F., 2007. Viruses. In: Lacey, L.A., Kaya, H.K. (Eds.), Field Manual of Techniques in Invertebrate Pathology: Application and Evaluation of Pathogens for Control of Insects and Other Invertebrate Pests, second ed. Springer, Dordrecht, Netherlands, pp. 149–174.

Cory, J.S., Hails, R.S., Sait, S.M., 1997. Baculovirus ecology. In: Miller, L. (Ed.), The Baculoviruses. Plenum Press, New York, pp. 301–330.

Cross, J.V., Solomon, M.G., Chandler, D., Jarrett, P., Richardson, P.N., Winstanley, D., Bathon, H., Hüber, J., Keller, B., Langenbruch, G.A., Zimmermann, G., 1999. Biocontrol of pests of apples and pears in Northern and Central Europe: 1. Microbial agents and nematodes. Biocontrol Sci. Technol. 9, 125–149.

Cunningham, J.C., 1995. Baculoviruses as microbial pesticides. In: Reuveni, R. (Ed.), Novel Approaches to Integrated Pest Management. Lewis, Boca Raton, pp. 261–292.

Del Rincón-Castro, M.C., Ibarra, J.E., 2005. Effect of a nuclepolyhedrovirus of *Autographa californica* expressing the enhancin gene of *Trichoplusia ni* granulovirus on *T. ni* larvae. Biocontrol Sci. Technol. 15, 701–710.

Doane, C.C., McManus, M.L., 1981. The Gypsy Moth: Research Towards Integrated Pest Management. Forest Service, U.S. Department of Agriculture, Technical Bulletin 1584. USDA, Washington DC, p. 757.

Duffey, S.S., Hoover, K., Bonning, B., Hammock, B.D., 1995. The impact of host plant on the efficacy of baculoviruses. In: Roe, M., Kuhr, R. (Eds.), Reviews in Pesticide Toxicology. CTI Toxicology Communications, pp. 137–275.

Eberle, K.E., Jehle, J.A., 2006. Field resistance of codling moth against *Cydia pomonella* granulovirus (CpGV) is autosomal and incompletely dominant inherited. J. Invertebr. Pathol. 93, 201–206.

Eberle, K.E., Asser-Kaiser, S., Sayed, S.M., Nguyen, H.T., Jehle, J.A., 2008. Overcoming the resistance of codling moth against conventional *Cydia pomonella* granulovirus (CpGV-M) by a new isolate CpGV-I12. J. Invertebr. Pathol. 93, 293–298.

Eberle, K.E., Wennmann, J.T., Kleespies, R.G., Jehle, J.A., 2012a. Basic techniques in insect virology. In: Lacey, L.A. (Ed.), Manual of Techniques in Invertebrate Pathology, second ed. Academic Press, San Diego, pp. 15–74.

Eberle, K.E., Jehle, J.A., Hüber, J., 2012b. Microbial control of crop pests using insect viruses. In: Abrol, D.P., Shankar, U. (Eds.), Integrated Pest Management: Principles and Practice. CABI Publishing, Wallingford, pp. 281–298.

Ehlers, R. (Ed.), 2011. Regulation of Biological Control Agents. Springer, Dordrecht. 416 pp.

Elleman, C.J., Entwistle, P.F., 1985. Inactivation of a nuclear polyhedrosis-virus on cotton by the substances produced by the cotton leaf surface glands. Ann. Appl. Biol. 106, 83–92.

Engelhard, E.K., Volkman, L.E., 1995. Developmental resistance in 4th-instar *Trichoplusia ni* orally inoculated with *Autographa californica* M nuclear polyhedrosis virus. Virology 209, 384–389.

Entwistle, P.F., Forkner, A.C., Green, B.M., Cory, J.S., 1993. Avian dispersal of nuclear polyhedrosis viruses after induced epizootics in the pine beauty moth, *Panolis flammea* (Lepidoptera: Noctuidae). Biol. Control 3, 61–69.

European and Mediterranean Plant Protection Organisation, 2012. Standard PP 1/181 conduct and reporting of efficacy evaluation trials, including good experimental practice. Bull. OEPP/EPPO Bull. 42, 382–393. ISSN:0250-8052. http://dx.doi.org/10.1111/epp. 2611.

Evans, H.F., 1986. Ecology and epizootiology of baculoviruses. In: Granados, R.R., Federici, B.A. (Eds.), The Biology of Baculoviruses Volume II: Practical Application for Insect Control. CRC Press, Boca Raton, pp. 89–132.

Federici, B.A., 1997. Baculovirus pathogenesis. In: Miller, L.K. (Ed.), The Baculoviruses. Plenum, New York, pp. 33–60.

Felton, G.W., Duffey, S.S., 1990. Inactivation of baculovirus by quinones formed in insect-damaged plant-tissues. J. Chem. Ecol. 16, 1221–1236.

Franzmann, B.A., Hardy, A.T., Murray, D.A.H., Henzell, R.G., 2008. Host plant resistance and biopesticides: ingredients for successful integrated pest management (IPM) in Australian sorghum production. Aust. J. Exp. Agric. 48, 1594–1600.

Fuxa, J.R., 2004. Ecology of nucleopolyhedroviruses. Agr. Ecosyst. Environ. 103, 27–43.

Gan-Mor, S., Matthews, G.A., 2003. Recent developments in sprayers for application of biopesticides – an overview. Bio. Syst. Eng. 84, 119–125.

Gaugler, R., 1997. Alternative paradigms for commercialising biopesticides. Phytoparasitica 25, 179–182.

Glare, T.R., Caradus, J., Gelernter, W., Jackson, T., Keyhani, N., Kohl, J., Marrone, P., Morin, L., Stewart, A., 2012. Have biopesticides come of age? Trends Biotechnol. 30, 250–258.

Glen, D.M., Payne, C.C., 1984. Production and field evaluation of codling moth granulosis virus for control of *Cydia pomonella* in the United Kingdom. Ann. Appl. Biol. 104, 87–98.

Goulson, D., 1997. Wipfelkrankheit: modification of host behaviour during baculoviral infection. Oecologia 109, 219–228.

Gowda, C.L.L., 2005. *Helicoverpa* the global problem. In: Sharma, C.H. (Ed.), *Heliothis/Helicoverpa* Management: Emerging Trends and Strategies for Future Research. Oxford and IBH Publishing, New Delhi, pp. 1–6.

Graves, R., Lucarotti, C.J., Quiring, D., 2012. Spread of a *Gammabaculovirus* within larval populations of its natural balsam fir sawfly (*Neodiprion abietis*) host following its aerial application. Insects 3, 912–929.

Graham, R.I., Grzywacz, D., Mushobozi, W., Wilson, K., 2010. *Wolbachia* in a major African crop pest increases susceptibility to viral disease rather than protects. Ecol. Lett. 15, 993–1000.

Grzywacz, D., Parnell, D., Kibata, G., Odour, G., Ogutu, O.O., Miano, D., Winstanley, D., 2004. The development of endemic baculoviruses of *Plutella xylostella* (diamondback moth, DBM) for control of DBM in East Africa. In: Endersby, N., Ridland, P.M. (Eds.), The Management of Diamondback Moth and Other Crucifer Pests. Proceedings of the 4th International Workshop, 26–29 Nov 2001. Melbourne, Victoria, Australia, pp. 197–206.

Grzywacz, D., Mushobozi, W.L., Parnell, M., Jolliffe, F., Wilson, K., 2008. The evaluation of *Spodoptera exempta* nucleopolyhedrovirus (SpexNPV) for the field control of African armyworm (*Spodoptera exempta*) in Tanzania. Crop Prot. 27, 17–24.

Grzywacz, D., Moore, D., Rabindra, R.J., 2014. Mass production of entomopathogens in less industrialized countries. In: Juan, A., Morales-Ramos, M., Guadalupe Rojas, M., Shapiro-Ilan, D.I. (Eds.), Mass Production of Beneficial Organisms. Elsevier, Amsterdam, pp. 519–553.

Gwynn, R., 2014. Manual of Biocontrol Agents 5th Edition. British Crop Protection Council, Alton. 520 pp.

Haase, S., Sciocco-Cap, A., Romanowski, V., 2015. Baculovirus insecticides in Latin America: historical overview, current status and future perspectives. Viruses 7, 2230–2267.

Hajek, A.E., 2004. Natural Enemies. An Introduction to Biological Control. Cambridge University Press, Cambridge. 366 pp.

Hajek, A.E., 2007. Introduction of a fungus into North America for control of gypsy moth. In: Vincent, C., Goettel, M.S., Lazarovits, G. (Eds.), Biological Control: A Global Perspective. CAB International, Wallingford, pp. 53–62.

Hajek, A.E., Tobin, P.C., 2010. Micro-managing arthropod invasions: eradication and control of invasive arthropods with microbes. Biol. Invasions 12, 2895–2912.

Hajek, A.E., Tobin, P.C., Haynes, K.J., 2015. Replacement of a dominant viral pathogen by a fungal pathogen does not alter the collapse of a regional forest insect outbreak. Oecologia 177, 785–797.

Hatem, A. El-S., Aldebris, H.K., Vargas-Osuna, E.V., 2011. Effects of the *Spodoptera littoralis* granulovirus on the development and reproduction of cotton leafworm *S. littoralis*. Biol. Control 59, 192–199.

Harrison, R., Hoover, K., 2012. Baculoviruses and other occluded insect viruses. In: Vega, F., Kaya, H. (Eds.), Insect Pathology. Elsevier, Amsterdam, pp. 73–131.

Harper, J.D., 1987. Applied epizootiology: microbial control of insects. In: Fuxa, J.R., Tanada, Y. (Eds.), Epizootiology of Insect Diseases. Wiley, New York, pp. 473–496.

Herniou, E.A., Arif, B.M., Becnel, J.J., Blissard, G.W., Bonning, B., Harrison, R., Jehle, J.A., Theilmann, D.A., Vlak, J.M., 2012. Family Baculoviridae. In: King, A.M.Q., Adams, M.J., Carstens, E.B., Lefkowitz, E.J. (Eds.), Virus Taxonomy, Classification and Nomenclature of Viruses, Ninth Report of the International Committee on Taxonomy of Viruses. Elsevier Academic Press, Amsterdam, pp. 163–173.

Hoover, K., Stout, M.J., Alaniz, S.A., Hammock, B.D., Duffey, S.S., 1998a. Influence of induced plant defenses in cotton and tomato on the efficacy of baculoviruses on noctuid larvae. J. Chem. Ecol. 24, 253–271.

Hoover, K., Yee, J.L., Schultz, C.M., Rocke, D.M., Hammock, B.D., Duffey, S.S., 1998b. Effects of plant identity and chemical constituents on the efficacy of a baculovirus against *Heliothis virescens*. J. Chem. Ecol. 24, 221–252.

Hoover, K., Washburn, J.O., Volkman, L.E., 2000. Midgut-based resistance of *Heliothis virescens* to baculovirus infection mediated by phytochemicals in cotton. J. Insect Physiol. 6, 999–1007.

Hoover, K., Grove, M.J., 2009. Specificity of developmental resistance in gypsy moth (*Lymantria dispar*) to two DNA-insect viruses. Virol. Sin. 24, 493–500.

Hüber, J., 1998. Western Europe. In: Hunter-Fujita, F.R., Entwistle, P.F., Evans, H.F., Crook, N.E. (Eds.), Insect Viruses and Pest Management. Wiley and Sons, New York, pp. 201–215.

Hughes, A., 2013. Origin of ecdysosteroid UDP-glycosyltransferases of baculoviruses through horizontal gene transfer from Lepidoptera. Coevolution 1, 1–7.

Hughes, D.S., Posse, R.D., King, L.A., 1993. Activation and detection of a latent baculovirus resembling *Mamestra brassicae* nuclear polyhedrosis virus in *M. brassicae* insects. Virology 194, 604–615.

Hunter-Fujita, F.R., Entwistle, P.F., Evans, H.F., Crook, N.E., 1998. Insect Viruses and Pest Management. John Wiley & Sons Ltd., Chichester. 620 pp.

Inceoglu, A.B., Kamita, S., Hammock, B.D., 2006. Genetically modified baculoviruses: historical overview and future outlook. In: Bonning, B.C. (Ed.), Advances in Virus Research. Insect Viruses: Biotechnological Applications, vol. 68. Academic Press, San Diego, pp. 109–126.

Ignoffo, C.M., 1992. Environmental factors affecting persistence of entomopathogens. Fla. Entomol. 75, 516–525.

Ignoffo, C.M., 1999. The first viral pesticide: past present and future. J. Ind. Microbiol. Biotechnol. 22, 407–417.

Jehle, J.A., Blissard, G.W., Bonning, B.C., Cory, J.S., Herniou, E.A., Rohrmann, G.R., Theilmann, D.A., Theim, S.M., Vlak, J.M., 2006. On the classification and nomenclature of baculoviruses: a proposal for revision. Arch. Virol. 151, 1257–1266.

Jones, K.A., 2000. Bioassays of entomopathogenic viruses. In: Navon, A., Ascher, K.R.S. (Eds.), Bioassays of Entomopathogenic Microbes and Nematodes. CAB Publishing, Wallingford, pp. 95–140.

Jones, K.A., Moawad, G., McKinley, D.J., Grzywacz, D., 1993. The effect of natural sunlight on *Spodoptera littoralis* nuclear polyhedrosis virus. Biocontrol Sci. Technol. 3, 189–197.

Jones, K.A., Irving, N.S., Moawad, G., Grzywacz, D., Hamid, A., Farghaly, A., 1994. Field trials with NPV to control *Spodoptera littoralis* on cotton in Egypt. Crop Prot. 13, 337–340.

Kabaluk, T., Svircev, A., Goettel, M., Woo, S.G. (Eds.), 2010. Use and Regulation of Microbial Pesticides in Representative Jurisdiction's Worldwide. IOBC Global. 99 pp.

Kamita, S.G., Kang, K.D., Inceoglu, A.B., Hammock, B.D., 2010. A10 addendum: genetically modified baculoviruses for pest insect control. In: Gilbert, L.I., Gill, S. (Eds.), Insect Control Biological and Synthetic Agents. Academic Press, Amsterdam, pp. 383–387.

Keating, S.T., Yendol, W.G., Schultz, J.C., 1988. Relationship between susceptibility of gypsy-moth larvae (Lepidoptera, Lymantriidae) to a baculovirus and host plant foliage constituents. Environ. Entomol. 17, 952–958.

Keddie, B.A., Aponte, G.W., Volkman, L.E., 1989. The pathway of infection of *Autographa californica* M nuclear polyhedrosis virus in an insect host. Science 243, 1728–1730.

Kolodny-Hirsch, D.M., Sitchawat, T., Jansiri, T., Chenrchaivachirakul, A., Ketunuti, U., 1997. Field evaluation of a commercial formulation of the *Spodoptera exigua* (Lepidoptera: Noctuidae) nuclear polyhedrosis virus for control of beet armyworm on vegetable crops in Thailand. Biocontrol Sci.Technol. 7, 475–488.

Kukan, B., 1999. Vertical transmission of nucleopolyhedrovirus in insects. J. Invert. Pathol. 74, 103–111.

Kunimi, Y., 2007. Current status and prospects on microbial control in Japan. J. Invertebr. Pathol. 95, 181–186.

Lacey, L.A., Arthurs, S.P., 2005. New method for testing solar sensitivity of commercial formulations of the granulovirus of codling moth (*Cydia pomonella*, Tortricidae: Lepidoptera). J. Invertebr. Pathol. 90, 85–90.

Lacey, L.A., Kroschel, J., 2009. Microbial control of the potato tuber moth (Lepidoptera: Geleciidae). Fruit Veg. Cereal Sci. Biotechnol. 3, 46–54.

Lacey, L.A., Shapiro-Ilan, D.I., 2008. Microbial control of insect pests in temperate orchard systems: potential for incorporation into IPM. Annu. Rev. Entomol. 53, 121–144.

Lacey, L.A., Frutos, R., Kaya, H.K., Vail, P., 2001. Insect pathogens as biological control agents: do they have a future? Biol. Control 21, 230–248.

Lacey, L.A., Thomson, D., Vincent, C., Arthurs, S.P., 2008. Codling moth granulovirus: a comprehensive review. Biocontrol Sci. Technol. 18, 639–663.

Lacey, L.A., Headrick, H.L., Horton, D.R., Schriber, A., 2010. Effect of granulovirus on the mortality and dispersal of potato tuber worm (Lepidoptera: Gelechiidae) in refrigerated storage warehouse conditions. Biocontrol Sci. Technol. 20, 437–447.

Lacey, L.A., Grzywacz, D., Shapiro-Ilan, D.I., Frutos, R., Brownbridge, M., Goettel, M.S., 2015. Insect pathogens as biological control agents: back to the future. J. Invertebr. Pathol. 132, 1–41.

Lapointe, R., Thumbi, D.K., Lucarotti, C.J., 2012. Recent advances in our knowledge of baculovirus molecular biology and its relevance for the registration of baculovirus-based products for insect pest population control. In: Soloneski, S., Larramendy, M.L. (Eds.), Integrated Pest Management and Pest Control. In Tech Open Access Publisher, Rijeka, Croatia, pp. 481–522.

Lasa, R.C., Pagola, I., Ibanez, J.E., Belda, J.E., Caballero, P., Williams, T., 2007. Efficacy of *Spodoptera exigua* multiple nucleopolyhedrovirus (SeMNPV) as a biological insecticide for beet armyworm in greenhouses in Southern Spain. Biocontrol Sci. Technol. 17, 221–232.

Lisansky, S., 1997. Microbial biopesticides. In: Evans, H.F. (Ed.), Microbial Insecticides; Novelty or Necessity. BCPC Symposium Proceedings No. 68. British Crop Protection Council, Farnham, UK, pp. 3–11.

Lucarotti, C.J., Moreau, G., Kettela, E.G., 2007. Abietiv™, a viral biopesticide for control of the balsam fir sawfly. In: Vincent, C., Goettel, M.S., Lazarovits, G. (Eds.), Biological Control: A Global Perspective. CAB International, Wallingford, pp. 353–361.

Mascarin, G.M., Delalibera Jr., I., 2012. Insecticidal activity of the granulosis virus in combination with neem products and talc powder against the potato tuber worm *Phthorimaea operculella* (Zeller) (Lepidoptera: Gelechiidae). Neotrop. Entomol. 41, 223–231.

Marrone, P.G., 2007. Barriers to adoption of biological control agents and biological pesticides. In: Radcliffe, E.B., Hutchison, W.D., Cancelado, R.E. (Eds.), Integrated Pest Management Concepts, Tactics, Strategies and Case Studies. Cambridge University Press, Cambridge, pp. 163–178.

Martignoni, M.E., 1999. History of TM BioControl: the first registered virus based product for insect control of a forest insect. Am. Entomol. 45, 30–37.

McDougall, P., 2011. The Cost and Time Involved in the Discovery, Development and Authorization of a New Plant Biotechnology Derived Trait. Crop Life International. http://www.croplife.org/view_document.aspx?docId=3338.

McLeod, P.J., Yearian, W.C., Young, S.Y., 1977. Inactivation of *Baculovirus heliothis* by ultraviolet irradiation, dew, and temperature. J. Invertebr. Pathol. 30, 237–241.

McNeil, J., Cox-Foster, D., Gardner, M., Slavicek, J., Thiem, S., Hoover, K., 2010. Pathogenesis of *Lymantria dispar* multiple nucleopolyhedrovirus in *L. dispar* and mechanisms of developmental resistance. J. Gen. Virol. 91, 1590–1600.

Mierzejewski, K., Reardon, R.C., Dubois, N., 2008. Conventional application equipment: aerial application. In: Lacey, L.A., Kaya, H.K. (Eds.), Field Manual of Techniques in Invertebrate Pathology; Application and Evaluation of Pathogens for Control of Insects and Other Invertebrate Pests, second ed. Springer, Dordrecht, Netherlands, pp. 99–126.

Miller, L.K. (Ed.), 1997. The Baculoviruses. Plenum Press, New York. 477 pp.

Miller, L., Ball, L.A. (Eds.), 1998. The Insect Viruses. Plenum Press, New York. 411 pp.

Moore, S.D., Kirkman, W., Richards, G.I., Stephen, P., 2015. The *Cryptophlebia leucotreta* granulovirus – 10 years of commercial field use. Viruses 7, 1284–1312.

Moore, S.D., Kirkman, W., Stephen, P., 2004. Crytogran: a virus for biological control of false codling moth. S. Afr. Fruit. J. 7, 56–60.

Moreau, G., Lucarotti, C.J., 2007. A brief review of the past use of baculoviruses for the management of eruptive forest defoliators and recent developments on a sawfly virus in Canada. For. Chron. 83, 105–112.

Moscardi, F., 1999. Assessment of the application of baculoviruses for the control of Lepidoptera. Annu. Rev. Entomol. 44, 257–289.

Moscardi, F., 2007. Development and use of the nucleopolyhedrovirus of the velvetbean caterpillar in soybeans. In: Vincent, C., Goettel, M.S., Lazarovits, G. (Eds.), Biological Control: A Global Perspective. CAB International, Wallingford, UK, pp. 344–353.

Moscardi, F., Sosa-Gomez, D., 2007. Microbial control of insect pests of soybean. In: Lacey, L.A., Kaya, H.K. (Eds.), Field Manual of Techniques in Invertebrate Pathology: Application and Evaluation of Pathogens for Control of Insects and Other Invertebrate Pests, second ed. Springer, Dordrecht, The Netherlands, pp. 411–426.

Moscardi, F., de Souza, M.L., de Castro, M.E.B., Moscardi, M.L., Szewczyk, B., 2011. Baculovirus pesticides: present state and future perspectives. In: Ahmad, I., Ahmad, F., Pichtel, J. (Eds.), Microbes and Microbial Technology. Springer, Dordrecht, pp. 415–445.

Mudgal, S., De Toni, A., Tostivint, C., Hokkanen, H., Chandler, D., 2013. Scientific Support, Literature Review and Data Collection and Analysis for Risk Assessment on Microbial Organisms Used as Active Substance in Plant Protection Products – Lot 1 Environmental Risk Characterization. EFSA Supporting Publications. European Food Standards Agency, EN-518. 149 pp. Available online: www.efsa.europa.eu/publications.

Myers, J.H., Cory, J.S., 2015. Ecology and evolution of pathogens in natural populations of Lepidoptera. Evol. Appl. 9, 213–247.

Nakai, M., 2009. Biological control of tortricidae in tea fields in Japan using insect viruses and parasitoids. Virol. Sin. 24, 323–332.

Nakai, M., Cuc, N.T.T., 2005. Field application of an insect virus in the Mekong Delta: effects of a Vietnamese nucleopolyhedrovirus on *Spodoptera litura* (Lepidoptera: Noctuidae) and its parasitic natural enemies. Biocontrol Sci. Technol. 15, 443–453.

Nair, K.S.S., Babjan, B., Sajeev, T.V., Sudheendrakumar, V.V., Mohamed-Ali, M.I., Varma, R.V., Mohandas, K., 1996. Field efficacy of nuclear polyhedrosis virus for protection of teak against the defoliator *Hyblea puera* Cramer (Lepidorptera: Hyblaeidae). J. Biol. Control 10, 79–85.

O.E.C.D, 2002. Consensus Document on Information Used in Assessment of Environmental Applications Involving Baculoviruses. Series on harmonisation of regulatory oversight in biotechnology, No. 20. ENV/JM/MONO(2002)1 OECD.

Panazzi, A.R., 2013. History and contemporary perspectives of the integrated pest management of soybean in Brazil. Neotrop. Entomol. 42, 119–127.

Passarelli, A.L., 2011. Barriers to success; how baculoviruses establish efficient systemic infections. Virology 411, 383–392.

Payne, C.C., 1982. Insect viruses as control agents. Parasitology 84, 35–77.

Payne, C.C., Thatchell, G.M., Williams, C.F., 1981. The comparative susceptibilities of *Pieris brassicae* and *P. rapae* to a granulosis virus from *P. brassicae*. J. Invertebr. Pathol. 38, 273–280.

Peng, F., Fuxa, J.R., Richter, A.R., Johnson, S.J., 1999. Effects of heat-sensitive agents, soil type, moisture, and leaf surface on persistence of *Anticarsia gemmatalis* (Lepidoptera: Noctuidae) nucleopolyhedrovirus. Environ. Entomol. 28, 330–338.

Podgwaite, J.D., 1999. Gypchek: biological insecticide for the gypsy moth. J. For. 97, 16–19.

Rabindra, R.J., Jayaraj, S., 1995. Management of *Helicoverpa armigera* with nuclear polyhedrosis virus on cotton using different spray equipment and adjuvants. J. Biol. Control 9, 34–36.

Raman, K.V., Booth, R.H., Palacios, M., 1987. Control of potato tuber moth *Phthorimaea operculella* (Zeller) in rustic potato stores. Trop. Sci. 27, 175–194.

Ravensberg, W.J., 2011. A Roadmap to the Successful Development and Commercialization of Microbial Pest Control Products for Control of Arthropods. Progress in Biological Control, vol. 10. Springer, Dordrecht, The Netherlands. 383 pp.

Rebolledo, D., Lasa, R., Guevara, R., Murillo, R., Williams, T., 2015. Baculovirus-induced climbing behavior favors intraspecific necrophagy and efficient disease transmission in *Spodoptera exigua*. PLoS One 10, e0136742. http://dx.doi.org/10.1371/journal.pone.0136742.

Reid, S., Chan, L., Van Oers, M., 2014. Production of entomopathogenic viruses. In: Juan, A., Morales-Ramos, M., Guadalupe Rojas, M., Shapiro-Ilan, D.I. (Eds.), Mass Production of Beneficial Organisms. Elsevier, Amsterdam, pp. 437–482.

Rohrmann, G.F., 2013. Baculovirus Molecular Biology, third ed. National Library of Medicine (US), National Centre for Biotechnology Information, Bethesda (MD). http://www.ncbi.nlm.nih.gov/books/NBK114593/.

Rowley, D.L., Popham, H.J.R., Harrison, R.L., 2011. Genetic variation and virulence of nucleopolyhedroviruses isolated worldwide from the heliothine pests *Helicoverpa armigera*, *Helicoverpa zea*, and *Heliothis virescens*. J. Invertebr. Pathol. 107, 112–126.

Schmitt, A., Bisutti, I.L., Ladurner, E., Benuzzi, M., Sauphanor, B., Kienzle, J., Zing, D., Undorf-Spahn, D., Fritsch, E., Hüber, J., Jehle, J.A., 2013. The occurrence and distribution of resistance of codling moth to *Cydia pomonella* granulovirus in Europe. J. Appl. Entomol. 137, 641–649.

Shapiro, M., 1982. In vivo mass production of insect viruses. In: Kustaki, E. (Ed.), Microbial and Viral Pesticides. Dekker, New York, pp. 463–492.

Shapiro, M., 1986. In vivo production of Baculoviruses. In: Granados, R.R., Federici, B.A. (Eds.), The Biology of Baculoviruses Volume II Practical Application for Insect Control. CRC Press, Boca Raton, pp. 32–61.

Shieh, T.R., 1989. Industrial production of viral pesticides. Adv. Virus Res. 36, 315–343.

Slavicek, J.M., 2012. Baculovirus enhancins and their role in viral pathogenicity. In: Adoga, M.P. (Ed.), Molecular Virology. Intech, Rijecka, pp. 147–155.

Sporleder, M., Kroschel, J., 2008. The potato tuber moth granulovirus (PoGV): use, limitations and possibilities for field applications. In: Kroschel, J., Lacey, L.A. (Eds.), Integrated Pest Management for the Potato Tuber Moth, *Phthorimaea Operculella* (Zeller) – A Potato Pest of Global Importance. Advances in Crop Research, vol. 10. Margraf Publishers, Weikersheim, Germany, pp. 49–71.

Stevenson, P.C., D'Cunha, R.F., Grzywacz, D., 2010. Inactivation of baculovirus by the isoflavenoids on chickpea (*Cicer arietinum*) leaf surfaces reduces the efficacy of nucleopolyhedrovirus against *Helicoverpa armigera*. J. Chem. Ecol. 36, 227–235.

Sun, X., Peng, H., 2007. Recent advances in control of insect pests by using viruses in China. Virol. Sin. 22, 158–162.

Sun, X., 2015. History and current status of development and use of viral insecticides in China. Viruses 7, 306–319.

Szewcyk, B., Hoyos-Carvajal, L., Paluszek, M., Skrzecz, I., Lobo de Souza, M., 2006. Baculoviruses re-emerging biopesticides. Biotechnol. Adv. 24, 143–160.

Tamez-Guerra, P., McGuire, M.R., Behle, R.W., Hamm, J.J., Sumner, H.R., Shasha, B.S., 2000. Sunlight persistence and rain fastness of spray-dried formulations of baculovirus isolated from *Anagrapha falcifera* (Lepidoptera: Noctuidae). J. Econ. Entomol. 93, 210–218.

Tanada, Y., 1964. A granulosis virus of the codling moth, *Carpocapsae pomonella* (Linnaeus) (Olethreutidae, Lepidoptera). J. Insect Pathol. 6, 378–380.

Tanada, Y., Hess, R.T., 1991. Baculoviridae. Granulosis viruses. In: Adams, J.R., Bonami, J.R. (Eds.), Atlas of Invertebrate Viruses. CRC Press, Boca Raton, pp. 227–258.

Vail, P.V., Hostetter, D.L., Hoffmann, F., 1999. Development of multinucleocapsid polyhedroviruses (MNPVs) infectious to loopers as microbial control agents. Integr. Pest Manage. Rev. 4, 231–257.

van Frankenhuyzen, K., Reardon, R.C., Dubois, N.R., 2007. Forest defoliators. In: Lacey, L.A., Kaya, H.K. (Eds.), Field Manual of Techniques in Invertebrate Pathology: Application and Evaluation of Pathogens for Control of Insects and Other Invertebrate Pests. Springer, Dordrecht, The Netherlands, pp. 481–504.

van Houte, S., Ros, V.I., Mastenbroek, T.G., Vendrig, N.J., Hoover, K., Spitzen, J., van Oers, M.M., 2012. Protein tyrosine phosphatase-induced hyperactivity is a conserved strategy of a subset of baculoviruses to manipulate lepidopteran host behavior. PLoS One 7, e46933. http://dx.doi.org/10.1371/journal.pone.0046933.

van Houte, S., van Oers, M.M., Han, Y., Vlak, J.M., Ros, V.I., 2014. Baculovirus infection triggers a positive phototactic response in caterpillars to induce 'tree-top' disease. Biol. Lett. 10 (12), 20140680.

Vilaplana, L., Wilson, K., Redman, E.M., Cory, J.S., 2010. Pathogen persistence in migratory insects: high levels of vertically-transmitted virus infection in field populations of the African armyworm. Evol. Ecol. 24, 147–160.

Virto, C., Navarroc, D., Mar Tellez, M., Herrerod, S., Williams, T., Murillo, R., Caballero, P., 2013. Natural populations of *Spodoptera exigua* are infected by multiple viruses that are transmitted to their offspring. J. Invertebr. Pathol. 122, 22–27.

Wang, M., Wang, J., Yin, F., Tan, Y., Deng, F., Chen, X., Jehle, J.A., Vlak, J.M., Hu, Z., Wang, H., 2014. Unravelling the entry mechanism of baculoviruses and its evolutionary implications. J. Virol. 88, 2301–2311.

Yang, M.M., Meng, L.L., Zang, Y.A., Wang, Y.Z., Qu, L.J., Wang, Q.H., Ding, J.Y., 2012. Baculoviruses and insect pest control in China. Afr. J. Microbiol. Res. 214–218.

Young, S.Y., Yearian, W.C., 1974. Persistence of *Heliothis* NPV on foliage of cotton, soybean, and tomato. Environ. Entomol. 3, 253–255.

Chapter 4

Basic and Applied Research: Entomopathogenic Bacteria

T.R. Glare[1], J.-L. Jurat-Fuentes[2], M. O'Callaghan[1,3]

[1]*Lincoln University, Christchurch, New Zealand;* [2]*University of Tennessee, Knoxville, TN, United States;* [3]*AgResearch, Christchurch, New Zealand*

4.1 INTRODUCTION

Microbial biopesticides based on entomopathogenic bacteria dominate the biopesticide market worldwide due to their cost-effective mass production, specificity, persistence in the environment, and safety. Importantly, the rapid cessation of feeding observed in infected insects reduces plant damage and contributes to maintaining pest populations below economic thresholds. However, bacterial insecticides may, in some cases, not be competitive against synthetic pesticides due to a more narrow activity spectrum, rapid environmental degradation, and lower efficacy.

Bacteria are unicellular prokaryotic microorganisms lacking a nuclear membrane and other defined intracellular membrane–enclosed organelles. For the purpose of this chapter, we have used classifications based on the "List of Prokaryotic Names with Standing in Nomenclature" database (Euzéby, 1997) as a reference. While virtually all insect life stages interact with bacteria in diverse symbiotic relationships, the focus of this chapter is entomopathogenic bacteria with relevance to insect control. Commercialized bacterial entomopathogens have obligate or facultative relationships with their hosts or produce toxins that can be exploited for insect control. Obligate bacterial entomopathogens like *Paenibacillus* spp. complete their life cycles within the insect host, while facultative pathogens such as *Serratia* spp. can also grow in the environment outside the host. Bacteria such as *Bacillus thuringiensis* (Bt) and *Chromobacterium subtsugae* produce toxins that can be used without the live bacteria, resulting in stable, noninfectious biopesticides.

Bacterial entomopathogens first enter the host hemocoel and avoid defensive insect responses to proliferate and produce virulence factors to cause disease, which ultimately kills the host. In most cases, these virulence factors are encoded by genes located in operons controlling fast expression of functionally related genes. Pathogenicity islands acquired through horizontal gene transfer and containing virulence genes have been identified in commercially relevant entomopathogenic bacteria, such as *Photorhabdus luminescens* (Waterfield et al., 2004) and *Xenorhabdus nematophila* (Brown et al., 2004). In other cases, virulence factors are localized on plasmids that may be transferred through conjugation as in the case of crystal toxin genes from Bt (Held et al., 1982). Upon host death, bacteria use the carcass as a nutrient resource until formation of dormant life stages, such as spores in the case of *Bacillus* spp., or they infect a new host.

4.2 CONCISE HISTORY AND KEY SPECIES DESCRIPTION

The most commercially successful bacterial entomopathogens are gram-positive bacteria in the order Bacillales, more specifically the genera *Bacillus*, *Paenibacillus*, and *Lysinibacillus*. Members of these genera form endospores under adverse conditions that in some cases are accompanied by proteinaceous parasporal bodies. Examples of microbial insecticides based on non–spore-forming bacteria are few due to their instability in storage. Improvements in formulation resulting in expanded shelf-life under ambient conditions have allowed a strain of *Serratia entomophila* to be commercialized. Next, we briefly describe each relevant bacterium and summarize their commercial development as pesticides.

4.2.1 Bacillaceae

4.2.1.1 Bacillus thuringiensis

Despite being first isolated from diseased silkworms in Japan in 1901 by S. Ishiwata, Bt was formally described and named by Ernst Berliner after isolation from diseased Mediterranean flour moth (*Ephestia küehniella*) larvae (Berliner, 1915). Subsequent isolation and screening efforts have allowed the identification of Bt strains with activity against diverse insect species in six taxonomic orders (van Frankenhuyzen, 2009).

Microbial Control of Insect and Mite Pests. http://dx.doi.org/10.1016/B978-0-12-803527-6.00004-4

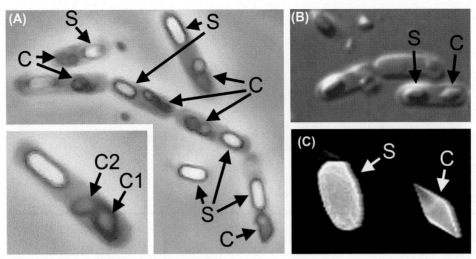

FIGURE 4.1 (A) Phase contrast micrograph of sporulated culture of *B. thuringiensis* subsp. *kurstaki* strain HD1, one of the most widely used Bt isolates in products for control of lepidopteran pests. Spores (S) and protein crystals (C) can be observed inside sporangia. The HD1 strain produces two crystals, the first is composed of Cry1A toxins and is rhombohedral in shape (insert in A, C1), while the second is smaller, round, and composed of Cry2Aa toxin (C2). (B) Differential interference contrast (DIC) micrograph of a culture of *B. thuringiensis* subsp. *kurstaki* strain HD73. The spore and rhombohedral crystal produced by this strain and composed of Cry1Ac toxin are visible inside sporangia. (C) Electron microscope micrograph of a spore and crystal of *B. thuringiensis* subsp. *kurstaki* strain HD73.

While Bt produces a number of virulence factors, specificity is mostly determined by the production of toxins, most of which are synthesized and packaged during sporulation into a parasporal crystal (Fig. 4.1). The presence of this crystal is a phenotypic trait used to differentiate Bt from other spore-forming *Bacilli*. Two main types of toxins, Cry (from *Cry*stal) and Cyt (from *Cyt*olytic), may be present in these parasporal bodies, while Vip (from *Vegetative Insecticidal Protein*) toxins are produced and secreted by vegetative Bt cells. The Cry, Cyt, and Vip toxins are named and classified according to protein sequence similarity (Crickmore et al., 1998); >300 Cry, 11 Cyt, and 32 Vip toxin holotypes have been described (Crickmore et al., 2015). All these Bt toxins target midgut epithelial cells in the host to compromise epithelial integrity and facilitate bacterial invasion of the hemocoel. Once Bt cells invade the hemocoel, they grow and multiply until nutritional resources are depleted, then enter sporulation phase (Raymond et al., 2010).

While diverse morphological, biochemical, and antigenic methods have been proposed, the most widely accepted classification system has been based on grouping together into subspecies those Bt strains that share serotyping of the H flagellar antigen in vegetative cells (de Barjac and Bonnefoi, 1968). In recent years, whole genome sequencing and genomic techniques have been used to describe and classify new Bt isolates, and serotyping has been rarely used in part due to lack of a reliable source of antisera. While 85 Bt serotypes or subspecies have been distinguished by serotyping (Jurat-Fuentes and Jackson, 2012), strains from within only five of these subspecies are currently used as active ingredient in pesticides registered with the U.S. Environmental Protection Agency. These Bt strains are natural isolates active against insects in the orders Lepidoptera (subsp. *kurstaki*, *aizawai*, and *galleriae*), Coleoptera (subsp. *morrisoni* biovar *tenebrionis*), and Diptera (subsp. *israelensis*). In addition, pesticides based on recombinant Bt (Raven, Lepinox, Crymax), or recombinant strains of *Pseudomonas fluorescens* (MVPII, Mattch) and *Clavibacter xyli* subsp. *cynodontis* (InCide) producing Bt toxins, have also been commercialized to improve efficacy and expand host range (Glare and O'Callaghan, 2000).

4.2.1.2 *Paenibacillus* spp.

Entomopathogenic *Paenibacillus* species were previously classified as *Bacillus* and include *P. larvae*, *Paenibacillus popilliae*, and *P. lentimorbus*. Although the *P. larvae* group includes causative agents of devastating bee diseases (Genersch, 2010), they will not be discussed further given their lack of potential as microbial pesticides. Bacteria in the *Paenibacillus* group are characterized as being catalase negative, having a relatively large ellipsoidal spore that deforms the sporangium in a central or subterminal location (Fig. 4.2A), and being unable to grow in nutrient broth or agar. This lack of growth on nutritive media has hindered the cost-effective development of *P. popilliae* and *P. lentimorbus* as microbial pesticides and distinguishes this genus from *Bacillus*.

The first documented successful insect control program based on a bacterial entomopathogen was the use of *Paenibacillus popilliae* to control larvae of the Japanese beetle (*Popillia japonica*) in turf (Klein, 1988). This bacterium and *P. lentimorbus* were recognized as the causative agents of milky disease in scarab beetles (Dutky, 1940) and are

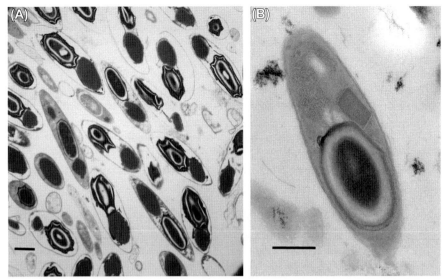

FIGURE 4.2   (A) Transmission electron micrograph of *Paenibacillus popilliae* sporangium containing spore and associated protein body (bar = 1 μm). (B) Transmission electron micrograph of *Brevibacillus laterosporus* sporangium containing spore and associated protein body (bar = 0.5 μm). *Photo courtesy of Marsha Ormskirk and Duane Harland.*

obligate pathogens only found as resistant spores outside hosts. While initially considered members of the same species, *P. popilliae* and *P. lentimorbus* have been confirmed as distinct species at the molecular level (Rippere et al., 1998). Morphological differences between the species include the spindle-shaped sporangium of *P. popilliae* versus the lemon-shaped sporangium of *P. lentimorbus* (Fig. 4.2) and, more importantly, the presence of a parasporal body in *P. popilliae* that is absent from sporulated *P. lentimorbus* (Dutky, 1940).

Milky disease has been detected affecting scarab larvae in most continents and many islands (Klein and Jackson, 1992). Once spores are ingested by the scarab larvae, they germinate in the alkaline midgut fluids and the vegetative cells penetrate the midgut epithelium to reach the luminal side of the basal membrane, where they undergo a primary multiplication cycle (Splittstoesser et al., 1973). Toxins present in the parasporal body have been proposed to facilitate disruption of the gut epithelial barrier and invasion of the hemocoel (Zhang et al., 1997), where multiplication continues. However, lack of clear differences in pathogenesis between *P. popilliae* and paraspore-free *P. lentimorbus* supports a limited role for parasporal toxins in infection. Importantly, specificity of infection seems to depend on the ability to disrupt the gut epithelial layer to invade the hemocoel as injection of spores overcomes strain specificity (Franken et al., 1996). The infected larvae remain actively feeding, with successive waves of sporulation and germination in the hemolymph until the last stages of infection when refractive spores dominate and confer on the hemolymph, a characteristic milky-white appearance. In contrast, infection of overwintering scarab larvae by *P. lentimorbus* induces clotting of hemolymph and a brown

discoloration of the diseased larvae. The infected larva fails to molt, becomes moribund, and dies, possibly due to depletion of lipid reserves (Sharpe and Detroy, 1979). The spores are released from the larval carcass and remain viable and infective in soil.

4.2.1.3 Lysinibacillus sphaericus

The defining phenotype of *Lysinibacillus sphaericus*, bacteria previously named as *Bacillus sphaericus* (Ahmed et al., 2007), is the production of a spherical spore located in a terminal position within a swollen sporangium. Phylogenetic analyses based on 16S ribosomal DNA support the existence of seven subgroups within *L. sphaericus sensu lato*, including *L. sphaericus* (*sensu stricto*), *L. fusiformis*, and four possibly new species (Nakamura, 2000). Within these groups, the characteristic mosquitocidal activity of *L. sphaericus sensu lato* strains is confined to a group close to the *L. fusiformis* cluster but not to the *L. sphaericus sensu stricto* group. The mosquitocidal activity of these strains depends on the production of Bin (from *Bin*ary) and/or Mtx (from *M*osquitocidal *T*oxin) toxins that target midgut cells in the host larvae (reviewed in Silva-Filha et al., 2014). At sporulation, some strains also produce a binary toxin (Cry48 and Cry49) related to the crystal toxins of Bt (Jones et al., 2007). While the Bin and Cry48/49 toxins are produced and stored as parasporal bodies in the sporangium, Mtx toxins are produced during the vegetative stage (Park et al., 2010). Additional nonmosquitocidal insecticidal toxins have also been reported from *L. sphaericus* strains (Nishiwaki et al., 2007), although mosquitoes are the main host and target for commercial *L. sphaericus* pesticides. Isolation of new strains with improved efficacy against *Culex* *quinquefasciatus* larvae (Park et al., 2007; Hire et al., 2010) suggests potential for the development of improved products.

The *L. sphaericus* bacterium is commonly isolated from soil and aquatic habitats (Guerineau et al., 1991), and an enriching medium may be used for isolation of pathogenic *L. sphaericus* strains (Yousten et al., 1985). Mosquitocidal activity is highly variable depending on the specificity of the Bin and MTX toxins produced and the mosquito species (Silva-Filha et al., 2014), with *Culex* mosquitoes being most sensitive followed by *Anopheles, Mansonia*, and some *Aedes* spp. (Lacey, 2007; Berry, 2012). Apart from mosquito larvae, *L. sphaericus* products are also applied to control black flies and nonbiting midges.

Shortly after ingestion of the *L. sphaericus* spores, the infected host stops feeding and the bacteria appear to be confined within the peritrophic matrix where they are digested by the host, possibly contributing to release of toxins (Davidson, 1979). The Bin and Cry toxins present in parasporal inclusions bind to receptors on midgut cells and induce hypertrophy, increased presence of lysosomes, apocrine secretion, and complete destruction of the epithelium (Singh and Gill, 1988; Oliveira et al., 2009). After midgut disruption, neural and skeletal tissue cells are also damaged, probably inducing the onset of paralysis observed in infected larvae. Once at the hemocoel, spores germinate and vegetative cells grow and the pathogenic cycle ends with sporulation and formation of toxin crystals, thereby amplifying spore numbers in the environment (Charles and Nicolas, 1986) and increasing persistence of mosquitocidal activity. It is possible that synergism between vegetative (Mtx) toxins and parasporal Bin toxins may have a role in the natural progression of the infection (Berry, 2012).

4.2.1.4 *Brevibacillus laterosporus*

Within the *Brevibacillus* spp., only *Brevibacillus laterosporus* (reclassified from *Bacillus* by Shida et al., 1996) is reported as an insect pathogen. It is commonly found in water and soil and was first described by Laubach (1916) as an aerobic, gram-positive, endospore-forming bacterium that can also be a facultative anaerobe. The production of crystalline inclusions within the sporangia is common for this species (Orlova et al., 1998; de Oliveira et al., 2004; Huang et al., 2005; Smirnova et al., 2011) (Fig. 4.2B).

B. laterosporus has been recorded as a pathogen to species in three insect orders. Susceptible species of Coleoptera include the corn rootworm (*Diabrotica* spp.) (Schnepf et al., 2002; Boets et al., 2011) and the Mexican cotton boll weevil (*Anthonomus grandis*) (de Oliveira et al., 2004). Among the Diptera, activity has been reported against larvae of black flies (*Simulium vittatum*), mosquitoes (*Cx. quinquefasciatus* and *Aedes aegypti*) (Favret and Yousten, 1985; Rivers et al., 1991), and houseflies (*Musca domestica*) (Ruiu et al., 2006). Susceptibility to *B. laterosporus* in Lepidoptera has been reported in larvae of the velvetbean caterpillar (*Anticarsia gemmatalis*) (de Oliveira et al., 2004) and the diamondback moth (*Plutella xylostella*) (Glare et al., 2012). The extent of cross-order toxicity is not well determined, although activity of isolates to both Diptera and Coleoptera (Rivers et al., 1991) and Diptera and Lepidoptera (Glare et al., 2014) has been described. Activity has been demonstrated for other invertebrate groups, including freshwater snails (*Biomphalaria glabrata*) (de Oliveira et al., 2004), and phytoparasitic (*Heterodera glycines*) and zooparasitic (*Trichostrongylus colubriformis*) nematodes (Bone and Singer, 1991).

The involvement of the sometimes canoe-shaped crystalline inclusion bodies (similar to those in Bt) of *B. laterosporus* in invertebrate toxicity is still debated. The main insecticidal activity of crystal-containing strains against larvae of *Aedes aegypti* and *An. stephensi* mosquitoes was reported to be within the parasporal crystals (Orlova et al., 1998; Zubasheva et al., 2010). However, other studies have reported similar toxicity in isolates without parasporal crystals (Favret and Yousten, 1985; Rivers et al., 1991). Larvae of *Anticarsia gemmatalis* were shown to be twice as susceptible to sporangia compared to vegetative cells (de Oliveira et al., 2004). Several patents on *B. laterosporus* toxins have identified "insecticidal secreted proteins", possibly resembling Vip proteins of Bt (Schnepf et al., 2002; Boets et al., 2011). A number of potential invertebrate toxin-encoding genes have been revealed by genome sequencing, including *cry* genes (Djukic et al., 2011; Sharma et al., 2012), although their role in toxicity has not been demonstrated.

4.2.2 Enterobacteriaceae

4.2.2.1 *Serratia entomophila* and *S. proteamaculans*

Lack of production of an environmentally stable life stage has limited commercial production and application of non–spore-forming bacteria. An exception to this rule is the bacterium *Serratia entomophila* (strain 154) (Enterobacteriaceae) which was commercialized to control an economically damaging soil-dwelling scarab pest of pasture, the New Zealand grass grub (*Costelytra zealandica*) (Jackson et al., 1992) (Fig. 4.3). *S. entomophila* was described as a new species (Grimont et al., 1988) following its isolation from larvae with an unusual appearance that were found during unexplained population collapses in the field (Trought et al., 1982).

The bacterium specifically causes amber disease in *C. zealandica* larvae following a unique chronic pathology. Upon ingestion of the bacteria by larvae feeding on organic matter and plant roots, feeding stops within a couple of days and clearing of the larval midgut results in the characteristic coloration of amber disease. Diseased larvae can remain

FIGURE 4.3 The development of *Serratia entomophila* for control of grass grub in New Zealand. (A) Pasture damage due to *Costelytra zealandica*; (B) *S. entomophila* colonies on a medium; (C) fermentation of *S. entomophila*; (D) dry formulations containing *S. entomophila* loaded into a seed drill; and (E) application by a seed drill to pasture.

active in soil without feeding for several weeks by self-consumption of fat bodies, leading to structural tissues becoming weakened and allowing bacteria to enter the hemocoel and causing lethal septicemia (Jackson et al., 1993).

The ability of two *Serratia* species to cause amber disease was explained by the identification of a 155-kb plasmid designated pADAP containing two clusters of genes associated with amber disease (Grkovic et al., 1995). The *sep* (*Serratia entomophila pathogenicity*) gene cluster contains three genes (*sepA*, *sepB*, and *sepC*) also detected in plasmids from *S. proteamaculans*, *S. liquefaciens*, and *Yersinia frederiksenii* (Dodd et al., 2006; Hurst et al., 2011c), suggesting that they are part of a discrete horizontally mobile region. The Sep proteins are similar to Tc (from *t*oxin *c*omplex) proteins from *Photorhabdus luminescens* (Hurst et al., 2007a), and *sep* gene expression in the larval gut causes gut clearance and development of amber coloration in diseased larvae (Hurst et al., 2007a). The second gene cluster (*afp*) is composed of 18 ORFs that code for a novel *a*nti-*f*eeding *p*rophage (Hurst et al., 2007b) that is a bullet-shaped toxin-delivery apparatus similar to the R-pyocins of *Pseudomonas aeruginosa* (Heymann et al., 2013). The role of some of the ORFs in assembly of this novel phage-tail like particle has recently been elucidated (Rybakova et al., 2013). Both the Sep and Afp virulence factors are essential for manifestation of amber disease; Afp causes rapid cessation of feeding upon ingestion, while subsequent induction and release of Sep toxins causes expulsion of gut contents allowing bacterial colonization of cuticular membranes of the foregut and hindgut from where constitutive expression of the toxins is maintained.

While *Serratia entomophila* was originally described in New Zealand, an isolate (strain Mor4.1) pathogenic to several species in the genera *Phyllophaga* and *Anomala* has been reported from Mexico (Eugenia Nunez-Valdez et al., 2008). In the Mor4.1 strain, lipopolysaccharide has been shown to be one of the critical virulence factors active in the hemocoel of *Phyllophaga blanchardi* larvae (Rodriguez-Segura et al., 2012). Despite the broader range of activity compared to New Zealand isolates, the Mor4.1 strain has not been commercialized to date.

4.2.2.2 Other *Serratia* spp.

Other species of *Serratia* are ubiquitous in the environment, and several of the 14 species in this genus have been found associated with diseased and dead insects (Grimont and Grimont, 1978). In particular, *S. marcescens* has been reported as a potential or facultative pathogen and following oral ingestion can cause disease in the blow fly *Lucilia sericata* (O'Callaghan et al., 1996) and May beetles (*Melolontha melolontha*) (Jackson and Zimmermann, 1996). A serralysin metalloprotease secreted by *S. marcescens* has been demonstrated to increase the pathogenicity to insects by suppressing cellular immunity (Ishii et al., 2014). Reports of human infections caused by *S. marcescens* has limited research on this bacterium as a microbial control agent, but recent identification of a new strain with activity against diamondback moth (Jeong et al., 2010), a new species (*S. nematodiphila*) symbiotically associated with an entomopathogenic *Heterorhabditinoides* sp. nematode (Zhang et al., 2009), and identification of a putative new species of *Serratia* associated with the nematode *Caenorhabditis briggsae* (Petersen and Tisa, 2012) suggest further useful microbial control agents will be identified from this genus.

4.2.2.3 Yersinia spp.

The genus *Yersinia* (Enterobacteriaceae) has undergone extensive diversification during the course of its evolution and includes well-known pathogenic species such as *Y. pestis*, the causative agent of bubonic plague, and species that have no known pathogenic effects (Sulakvelidze, 2000). Insecticidal activity of *Yersinia* spp. has been recognized only relatively recently; it was first demonstrated for *Y. enterocolitica* and expression of toxin complex (Tc) proteins was necessary for insecticidal activity (Bresolin et al., 2006). Since then, *Tc* genes have been detected in genomes of several other *Yersinia* species, including *Y. mollaretii*, *Y. pestis*, and *Y. pseudotuberculosis*. Variable oral insecticidal activity of strains from different *Yersinia* species was reported against *Galleria mellonella* and *Manduca sexta* (Fuchs et al., 2008).

The novel species *Yersinia entomophaga*, which, like *Serratia entomophila*, was isolated from a diseased larva of *C. zealandica* in New Zealand (Hurst et al., 2011b), is highly pathogenic to a wide range of important insect pests in the orders Coleoptera, Lepidoptera, and Orthoptera (Hurst et al., 2011a). Its potential utility as a microbial control agent has been tested against several pest species endemic to New Zealand, including the porina caterpillar (*Wiseana* spp.) (Brownbridge et al., 2008) and the geometrid caterpillar *Scopula rubraria* (Jones et al., 2015), but bioassays have also shown activity against *C. zealandica*, the diamondback moth (*Plutella xylostella*), the small white butterfly (*Pieris rapae*), locust (*Locusta migratoria*), and the cotton bollworm (*Helicoverpa armigera*) (Hurst et al., 2011a). Following ingestion of *Y. entomophaga* by *C. zealandica* larvae, there is rapid cessation of feeding, followed by regurgitation of gut contents and degradation of the gut epithelial membranes to allow invasion of the hemocoel, septicemia, and death usually within 2–5 days of ingestion (Marshall et al., 2012).

Yersinia entomophaga produces an orally active proteinaceous toxin complex (Yen-Tc) composed of ABC toxins that were first identified and characterized in the bacterium *Photorhabdus luminescens* (Bowen et al., 1998). Release of Tc in *Y. entomophaga* seems to be temperature dependent, with large amounts released when the bacterium is cultured at 25°C (Hurst et al., 2011b). This Tc is composed of seven subunit proteins: two TcA-like proteins (YenA1, YenA2), a TcB-like protein (YenB), two TcC-like proteins (YenC1, YenC2), and two chitinases (Chi1, Chi2), which combine to form the insect active Tc (Hurst et al., 2011b).

4.2.2.4 Other Enterobacteriaceae

A number of other gram-negative bacteria have been recorded as pathogenic to insects (Vodovar et al., 2005). Three species of *Proteus* (*P. mirabilis*, *P. vulgaris*, and *P. rettifgeri*) have been listed as potential pathogens in insects (Tanada and Kaya, 1993). However, these appear opportunistic and *P. mirabilis* causes 90% of all *Proteus* infections in humans, which suggests *Proteus* spp. are more generalist pathogens, limiting their potential for development as microbial insecticides.

4.2.3 Pseudomonadaceae

Pseudomonas entomophila is an entomopathogenic bacterium that can kill insects from at least three orders (Dieppois et al., 2015). The mode of action in *Drosophila*, where most of the studies have taken place, includes destruction of the gut epithelium. Genome sequencing has shown secretion system and toxins that are likely responsible for the toxicity (Vodovar et al., 2006). The ability of some plant-associated *Pseudomonas* to infect and kill insects has been reported (Stavrinides et al., 2009; Kupferschmied et al., 2013). A number of *Pseudomonas* are known to have insecticidal toxins similar to nematode-vectored insect pathogens, including *P. chlororaphis* (Ruffner et al., 2015) and *P. taiwanensis*, a broad-spectrum entomopathogen (Chen et al., 2014). *Pseudomonas putida* was described as a pathogen of the Colorado potato beetle, *Leptinotarsa decemlineata* (Muratoglu et al., 2011).

4.2.4 Clostridiaceae

Clostridium bifermentans is an anaerobic, spore-forming, gram-positive bacillus, infrequently recorded as infectious to humans (Edagiz et al., 2015). The *C. bifermentans* subsp. *malaysia* isolate is known to produce a range of toxins with activity against mosquitoes and blackflies but showed no short-term toxicity to vertebrates (Edagiz et al., 2015). Inoculation of the bacterium using diverse routes demonstrated the safety of bacterial cells to laboratory mammals and goldfish (Thiéry et al., 1992). The strain produces Cry proteins, eventually classified as Cry16A and Cry17A (Barloy et al., 1996, 1998), and other potential toxins similar to hemolysins, which individually did not cause mortality in mosquitoes (Juarez-Perez and Delecluse, 2001). However, using the four toxins that are encoded on a single operon (Cry16A, Cry17A, Cbm17.1, and Cbm17.2), toxicity was found for *Aedes* spp., while *Anopheles* mosquitoes were more susceptible to whole cultures of *C. bifermentans* subsp. *malaysia* (Qureshi et al., 2014).

4.2.5 Neisseriaceae

The violet-pigmented bacterium *Chromobacterium subtsugae* is a motile, gram-negative bacterium orally toxic to Colorado potato beetle larvae, the Southern green stink bug (*Nezara viridula*), corn rootworm *Diabrotica* spp., diamondback moth (*Plutella xylostella*), and the small hive beetle *Aethina tumida* (Martin et al., 2007a,b; Stamm et al., 2013).

The insecticidal activity results from the expression of stable bioactive factors. Extracts of the bacterium have been commercialized by Marrone Bio Innovations, and the company advertises toxicity to thrips, two-spotted spider mites whiteflies, caterpillars, psyllids, and beetles. After ingestion of bacterial bioactives, death can take between 2 and 7 days. An undescribed *Chromobacterium* sp. was isolated from the midgut of the mosquito *Aedes aegypti* and shown to reduce the survival of larvae and adults, and mosquitoes were less likely to carry the malaria causing *Plasmodium* and dengue fever virus (Ramirez et al., 2014).

4.2.6 Other Proteobacteria

Bacteria from the genus *Burkholderia* are often symbiotic and sometimes have negative effects on their hosts, such as affecting the number of eggs and oviposition time of Hemiptera (Kil et al., 2014). The new species *Burkholderia rinojensis* killed *S. exigua* and two-spotted spider mites (*Tetranychus urticae*) both through contact and ingestion (Cordova-Kreylos et al., 2013). Members of the *Burkholderia cepacia* complex are known to infect a range of hosts, including insects (Uehlinger et al., 2009).

Rickettsiaceae, such as *Rickettsia* and *Wolbachia*, have very diverse lifestyles including obligate, symbiotic, and pathogenic associations with insects (Lukasik et al., 2013; Nikoh et al., 2014). *Wolbachia pipientis* (Rickettsiales: Rickettsiaceae) is a maternally inherited endosymbiont of many arthropods that affect host reproductive biology (Moretti and Calvitti, 2013). True insect pathogens appear to be limited to the genus *Rickettsiella* (Leclerque et al., 2012). The genus is classified in the Coxiellaceae in the order Legionellales of the Gammaproteobacteria (Garrity et al., 2005). A number of insect pathogenic species are currently recognized, including *Rickettsiella popilliae*, *R. grylli*, *R. chironomi*, and *R. stethorae*, with numerous more specific pathotype names also used (Leclerque et al., 2012). There is little known about the mode of action of insect pathogenic *Rickettsiella*. The infective propagules are ingested, and infection is usually in the fat bodies or hemolymph cells. Mortality can take months to occur after ingestion, which limits their potential as microbial insecticides.

4.2.7 Actinobacteria

Streptomyces (Streptomycetaceae) contains isolates that show some pathogenicity to insects, whether their metabolites are used or viable cells are applied directly. Fifteen isolates of *Streptomyces* were consistently pathogenic to *Helicoverpa armigera*, *S. litura*, and *Chilo partellus* (Vijayabharathi et al., 2014). *Streptomyces albus*, an endophyte of grass, showed high toxicity to the cotton aphid *Aphis gossypii* (Shi et al., 2013). A number of insecticidal compounds are produced

by *Streptomcyes*, including flavensomycin, antimycin A, piericidins, macrotetralides, and prasinons (Ruiu et al., 2013). Avermectin and other macrocyclic lactone derivatives such as milbemycins and doramectin have been shown to display potent anthelmintic and insecticidal properties and can be produced by *S. avermitilis* (Wang et al., 2011) and *S. hygroscopicus* (Takiguchi et al., 1980). Avermectin has been successfully used in commercial insecticides, and the discoverers, Satoshi Ōmura and William Campbell, were awarded the 2015 Nobel Prize in Medicine.

Another important metabolite group from Actinobacteria are the spinosyns produced by *Saccharopolyspora spinosa*. The species, first described by Mertz and Yao (1990), was shown to produce unique insecticidal metabolites during aerobic fermentation. Spinosyns are macrolides—molecules that contain a 12-member macrocyclic lactone in a unique tetracyclic ring. This family of metabolites was identified and developed by Dow AgroSciences as selective, environmentally friendly group of insecticides known as spinosads (Sparks et al., 1998).

4.3 MODE OF ACTION

4.3.1 Toxicity and Toxins

Among the bacteria just described, several produce insecticidal toxins as the main mechanism for virulence. In some cases, these bacteria may not be truly infective and do not always colonize the host insect cadaver. *Bacillus thuringiensis*, *Chromobacterium subtsugae*, and *B. laterosporus* are examples of bacteria where toxins are needed for insect death. Major families of insecticidal proteins produced by these bacteria are the Cry, Cyt, Vip, and Bin toxins (Djukic et al., 2011; Adang et al., 2014). Among these insecticidal toxins, the Cry toxins are the most thoroughly studied and characterized.

4.3.1.1 Cry Toxins

The Cry toxins are not unique to the parasporal body of Bt but have also been identified from other bacteria (Barloy et al., 1996) as secreted proteins (Varani et al., 2013) or encoded by cryptic genes (Crickmore et al., 1994). Amino acid sequence identity is currently used as the basis for nomenclature and classification of Cry toxins, with over 70 Cry groups currently described at the primary rank (less than 40% identity) represented by a specific numeral (eg, Cry1, Cry2, Cry3, and so on) (Crickmore et al., 2015). Further divisions within these Cry groups based on less than 70% identity are represented by different letters (ie, Cry1A, Cry1B, Cry2A, etc.). Despite their range in size from 40 to approximately 148 kDa, the majority of Cry proteins follow a consistent (Li et al., 1991; Grochulski et al., 1995) contain at least a subset of five conserved sequence blocks (Höfte and Whiteley, 1989), which are critical for structural stability and toxin functionality (Adang et al., 2014). In this structure, domain I is a bundle of seven or eight α-helices with the hydrophobic α-helix five located in the center. Structural similarities with pore-forming domains in other bacterial toxins (Li et al., 1991) and currently available data suggest that domain I is involved in membrane insertion and pore formation. In contrast, domain II is made up of three antiparallel β sheets with a Greek key motif arranged in a triangular β prism, while domain III is a β-sandwich composed of two antiparallel β-sheets compressed into a "jelly roll" topology. Both β-sheet domains display high structural similarity to carbohydrate-binding domains described in lectins (Burton et al., 1999), which initially suggested their involvement in determining specificity. Two other structural classes found within Cry toxins are the predominately β-sheet Cry toxins with similarity to *Clostridium perfringens* epsilon toxin, and the Bin toxins resembling the mosquitocidal binary toxin from *Lysinibacillus sphaericus*.

Current evidence supports a mode of action for all three-domain Cry toxins (Fig. 4.4) involving formation of toxin pores in target cells. Upon ingestion by a susceptible host, the parasporal Bt crystal is solubilized by the gut physicochemical conditions and processed by Bt proteases and/or proteases in the host gut fluids to an active toxin core. Both solubilization and activation can influence specificity (Du et al., 1994; Walters et al., 2008), and their alteration may result in resistance (Oppert et al., 1997). During activation of the ~120-kDa protoxin, about 500 amino acids from the C terminus and 43 from the N-terminus are sequentially digested to a 65- to 55-kDa toxin core. Shorter Cry protoxins lack the C-terminus and are mostly processed at the N-terminus. Processing of the N-terminus is critical to formation of toxin oligomers and insecticidal activity (Bravo et al., 2002). The activated Cry toxin may associate with carbohydrate moieties on the peritrophic matrix (Hayakawa et al., 2004) with circulating lipids (Ma et al., 2012) or recognize specific binding sites on the brush border membrane of midgut cells. This binding step is necessary, yet not sufficient, for susceptibility (Wolfersberger, 1990). A number of proteins and glycoconjugates have been identified as containing binding sites for Cry toxins (Adang et al., 2014), but data directly supporting receptor functionality have only been provided for cadherin-like proteins (Tsuda et al., 2003),

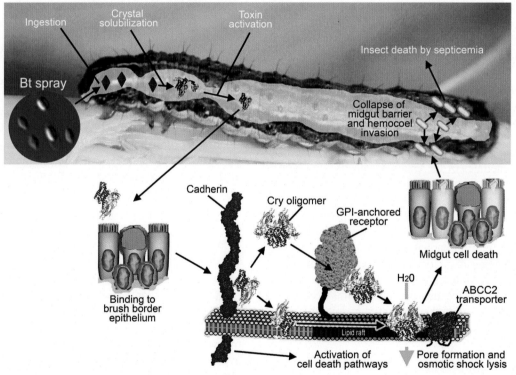

FIGURE 4.4 Diagram of the current models of Cry toxin action. After ingestion of toxin crystals and spores present in Bt pesticides, the Cry proteins are solubilized and processed by proteases in the host digestive fluids to a toxin core. After traversing the peritrophic matrix, the Cry toxin core binds with high affinity to cadherin receptors on the brush border membrane of the midgut epithelial cells. This binding event may activate intracellular cell death pathways and/or be conducive to further toxin processing resulting in formation of a Cry toxin oligomer. These Cry oligomers display high binding affinity to aminopeptidase-N and alkaline phosphatase, which preferentially locate to lipid rafts and are attached to the cell membrane by a glycosyl-phosphatidylinositol (GPI) anchor. Binding to these GPI-anchored receptors and potential interaction with ABCC2 transporter proteins is proposed to facilitate insertion of the oligomer and pore formation in the membrane, leading to osmotic cell lysis. There is also evidence for membrane insertion of Cry toxin monomers and oligomerization occurring in the membrane. Enterocyte death results in disruption of the midgut epithelium barrier, allowing for gut bacteria to invade the hemocoel, where they proliferate to cause septicemia and death of the insect. The same model applies to Cry toxins produced by transgenic Bt plants, although these proteins are usually produced by the plants as truncated (activated) forms and thus would surpass the solubilization and activation steps.

aminopeptidase-N (Sivakumar et al., 2007), alkaline phosphatase (Fernandez et al., 2006), and proteins in the ATP binding cassette (ABC) transporter family (Tanaka et al., 2013).

Mechanistic details of the post-binding events leading to Cry toxicity remain controversial (Vachon et al., 2012). The most accepted model includes sequential interaction with cadherin followed by aminopeptidase or alkaline phosphatase (Pardo-Lopez et al., 2013). This "sequential binding" model (Fig. 4.4) was developed for Cry1A toxin–receptor interactions and supports initial reversible binding of Cry1A activated toxins to abundant aminopeptidase and alkaline phosphatase proteins, favoring higher affinity toxin interactions with less abundant cadherin proteins (Pacheco et al., 2009). Binding of toxin monomers to the membrane-proximal region of cadherin promotes proteolytic removal of alpha helix 1 and formation of a toxin oligomer that displays high affinity for binding to aminopeptidase and alkaline phosphatase proteins (Bravo et al., 2004). Binding to these proteins facilitates accumulation of toxin oligomers on specialized membrane regions called lipid rafts (Zhuang et al., 2002), favoring insertion of the oligomer and formation of a toxin pore that leads to cell death by osmotic shock. There is also evidence (as reviewed by Vachon et al., 2012) that Cry toxins form oligomers by association of monomers already inserted in the membrane. An alternative model for Cry cytotoxicity that is currently limited to cultured insect cells (Fig. 4.4) proposes that binding to cadherin activates intracellular oncotic cell death pathways responsible for enterocyte death and that pore formation is not involved in cytotoxicity (Zhang et al., 2008).

Most of the available evidence suggests that formation of pores by Cry toxins follows an "umbrella" model in which binding to receptors results in a conformational change in the toxin so that a hydrophobic hairpin composed of alpha helices 4 and 5 of domain I inserts in the membrane (Schwartz et al., 1997). In this model the alpha helix 4 lines the Cry toxin pore and controls the passage of ions through the pore (Kumar and Aronson, 1999), yet pore properties are also dependent on midgut epithelium components (Schwartz et al., 1997; Peyronnet et al., 2001) and ionic composition (Fortier et al., 2005). The position of domains II and III during insertion is controversial. There is evidence for the whole toxin molecule being confined to the membrane (Aronson, 2000; Nair and Dean, 2008) and for the toxin remaining associated with receptors on the membrane surface (Pardo-López et al., 2006; Fortier et al., 2007). Massive enterocyte death results in a compromised midgut epithelial barrier, allowing Bt and other bacteria to invade the hemocoel, causing septicemia and insect death (Johnston and Crickmore, 2009).

In contrast to the three-domain Cry toxins, other Cry toxins (Cry51Aa, Cry15, Cry23, Cry33, Cry45, and Cry46) present a single beta-stranded domain structure with similarities to aerolysin-type beta-pore forming toxins. Some of these proteins have been shown to display unique toxicity against hemipteran and coleopteran pests (Baum et al., 2012; Xu et al., 2015). While structural similarities suggest commonalities to aerolysin-type toxins, the mode of action of these Cry toxins has not been experimentally characterized.

4.3.1.2 Vip Proteins

The low sequence and structural similarity of Vip proteins or binary Cry toxins support differences in the intoxication process when compared with the three-domain Cry toxins. In addition, the fact that Vip proteins are produced by vegetative cells, which mostly occur in the host during the later stages of infection, may suggest that while active by ingestion, the main target of Vip proteins may be located in the hemocoel. Data available for Vip toxins suggest that Vip1 and Vip2 function as A/B binary toxins (Barth et al., 2004) while Vip3 toxins probably depend on pore formation for cytotoxicity (Lee et al., 2003). Homology to other binary toxins suggests that the Vip1 protein is the binding and translocation (or "B") component, while Vip2, which displays similarity to actin-ADP-ribosylating toxins, represents the toxic (or "A") component in the binary toxin (Jucovic et al., 2008). Both binding to insect midgut proteins (Sattar and Maiti, 2011) and formation of pores in lipid membranes (Leuber et al., 2006) have been reported for Vip1. The Vip1 pore would allow Vip2 to penetrate the cell and employ ribosylating activity to inhibit actin polymerization (Han et al., 1999). In contrast, the mode of action of Vip3 toxins seems to mirror the Cry intoxication process and includes processing by midgut proteases, binding to receptors, and formation of pores (Lee et al., 2003; Liu et al., 2011). Interestingly, and in contrast to Cry proteins, Vip3 toxins do not seem to have a protease-resistant toxin core. A ribosomal protein has been reported as functional Vip3A binding protein and receptor in Sf21 insect cell cultures (Singh et al., 2010), yet the mechanism leading to enterocyte death remains largely unknown.

4.3.1.3 Bin Toxins

The Bin toxins produced by *Lysinibacillus sphaericus* are produced as a single parasporal crystal containing equimolar amounts of two proteins of 42 kDa (BinA or P42) and 51 kDa (BinB or P51) (Broadwell et al., 1990). The *binA* and *binB* genes are highly conserved among *L. sphaericus* strains and are contained in a ~35-kb operon that is located both on the chromosome and on the pBsph plasmid (Hu et al., 2008). The two toxins have low overall sequence similarity but share several regions of identity that are essential for toxicity (Clark and Baumann, 1991). Variability in small regions allows classification of the Bin toxins produced by *L. sphaericus* strains used in field applications into four groups (Bin1 to Bin4) (Humphreys and Berry, 1998).

Both BinA and BinB are processed to 40-kDa (BinA) and 43-kDa (BinB) proteins by midgut fluids (Broadwell et al., 1990). The crystal structure of the BinB protein has been recently resolved and includes a lectin-like N-terminal domain and a C-terminal domain with similarity to aerolysin-type toxins (Srisucharitpanit et al., 2014). The lectin-like N-terminal region of the BinB protein is critical for binding to the larval gut (Singkhamanan et al., 2010), while the C-terminus is responsible for interacting with BinA (Oei et al., 1992). Binding of BinA/BinB to receptors results in formation of pores with a β-barrel structure (Boonserm et al., 2006), which allow internalization of BinA to exert its toxic properties. The receptor for BinB in the mosquito *Culex pipiens* is a glycosyl-phosphatidylinositol (GPI) anchored 60-kDa α-glucosidase (Cpm1) on the midgut brush border membrane (Silva-Filha et al., 1999). While loss of Bin binding has been reported to associate with resistance to *L. sphaericus* in *Culex* spp. (Nielsen-Leroux et al., 1995), experiments with cultured mosquito cells support no direct link between toxicity and Bin toxin binding or insertion and that the precise mechanism of BinA cytotoxicity may depend on postinternalization processes (Schroeder et al., 1989).

Information on the mode of action of binary Cry toxins is mostly focused on the Cry34/35 complex (Ellis et al., 2002). Both proteins appear as monomers in solution and their structure has been resolved (Kelker et al., 2014). The β sandwich fold of Cry34 resembles proteins of the Aegerolysin family capable of membrane pore formation, while the mostly β sheet structure of Cry35 resembles members of the Toxin-10 family forming β sheet pores. These structures are in agreement with previous reports (Masson et al., 2004) and the Cry34/35 binary toxin is expected to exert its cytotoxicity through pore formation. The Cry35 protein contains structural lectin folds that may dictate binding interactions, which are greatly promoted by Cry34 in brush border membrane protein vesicles of *D. virgifera* (Li et al., 2013). A role for Cry34 in binding has not been ruled out (Kelker et al., 2014), and no Cry34/35 or Cry35 receptors have been identified to date.

4.3.1.4 Mtx Toxins

Several strains of *L. sphaericus* produce mosquitocidal toxins (Mtx) as soluble proteins during the vegetative stage (reviewed in Carpusca et al., 2006). The mature Mtx proteins contain an N-terminal 27-kDa fragment with ADP-ribosyltransferase activity (Thanabalu et al., 1993) and a C-terminal 70-kDa fragment with sequence similarity to the lectin-like binding component of ricin (Hazes and Read, 1995). It is hypothesized that after binding through interactions between the 70-kDa fragment and as yet unidentified receptors, the toxin is internalized into endosomes, where exposure to low pH may favor translocation of the 27-kDa fragment into the cytosol to ADP-ribosylate proteins (Schirmer et al., 2002).

Combination of Mtx and Cry toxins from Bt subsp. *israelensis* showed moderate synergy against *Culex quinquefasciatus* (Wirth et al., 2014).

4.3.1.5 Toxin Complex

A growing number of bacterial entomopathogens have been shown to produce proteins in the insecticidal-like Tc family. Typically, Tc toxins are composed of three protein subunits, TcA, TcB, and TcC, which combine to form the insect-active complex (ffrench-Constant et al., 2007). The details of the mechanism of action in Tc toxins have recently been elucidated. The TcC subunit contains the ADP-ribosyl transferase activity responsible for cytotoxicity, which is located inside a large hollow structure formed by the TcB and TcC subunits (Busby et al., 2013). This TcB/TcC heterodimer is attached to a pentamer formed by TcA subunits which is involved in binding to as yet unknown receptors on the target cell surface. After binding, the complex is endocytosed and subsequently the TcA pentamer forms a vuvuzela-shaped, pH-triggered channel (Gatsogiannis et al., 2013), through which the cytotoxic domain of TcC is translocated in the cytoplasm through a syringe-like mechanism of protein translocation (Meusch et al., 2014).

4.3.2 Infections

A desirable attribute of most bacterial insecticides is the quick cessation of feeding and rapid death, which reduces loss of foliage and usually allows the insect population to be maintained below economic thresholds. Bacterial growth during the disease process follows the three main phases (lag, exponential, and stationary) observed in culture. During the lag phase, vegetative cells adapt to the new host and initiate rapid proliferation by binary fission leading to an exponential or log phase of growth. For most entomopathogenic bacteria, this log growth phase occurs in the host hemocoel, resulting in septicemia and death of the insect host. Accumulation of secondary metabolites and nutrient depletion result in a stationary growth phase that may lead to sporulation. These bacterial spores do not exhibit metabolic activity and are highly resistant against adverse environments.

The primary route of entry for commercialized entomopathogenic bacteria is ingestion during feeding. Once in the oral cavity, the bacteria have to overcome the host's defensive mechanisms which can include production of reactive oxygen and other antibacterial compounds (Musser et al., 2005) and structures in the digestive system acting as filters of microbes in food (Sturtevant and Revell, 1953; Glancy et al., 1981). The physicochemical conditions in the host gut inhibit entomopathogenic bacterial growth yet they usually activate bacterial pathogenesis, such as production of Sep toxins by *Serratia entomophila* (Hurst et al., 2007a) or solubilization and activation of Cry toxins. The gut flora may also limit successful infection by producing antimicrobial compounds (Yoshiyama and Kimura, 2009). The peritrophic matrix may contribute to reduced susceptibility (Rees et al., 2009), yet several bacteria secrete chitinases during infection to overcome this protective layer of the digestive system (Wiwat et al., 2000; Thamthiankul et al., 2001). Bacterial toxins and enzymes target enterocytes to disrupt the epithelial barrier and allow bacteria invasion of the hemocoel.

It is interesting that the bacteria that cause chronic infections, such as *Serratia entomophila* and *Paenibacillus popilliae*, have also been found to have genes from well-described insect toxin families. The disease processes of these slow infectious bacteria appears very different from the toxin-induced mortality of Bt or *Chromobacterium subtsugae*, but genome and plasmid sequencing has shown the presence of Cry (Zhang et al., 1998), Tc (Hurst et al., 2000), and other toxin-encoding regions.

4.4 NATURAL ECOLOGY

Entomopathogenic bacteria occupy many niches, functioning as rhizosphere and phyllosphere colonizers, both obligate and facultative pathogens, and as endophytes of plants. The main ecological niche of some species is primarily as pathogen, surviving between hosts as environmentally resistant spores.

Particularly puzzling is the niche of *B. thuringiensis* as it is a relatively poor colonizer of insects despite its ability to produce potent insecticidal toxins. The lack of an obvious invertebrate host in soil where Bt has been predominantly isolated has given rise to the suggestion that Bt is not primarily an insect pathogen. It has been proposed that Bt evolved in commensal relationships with plants to provide defense against insect attack (Smith and Couche, 1991). This "bodyguard theory" states that microbe–plant symbiosis has evolved so that microbes that are insect antagonistic are recruited and/or maintained by the plant for protection and has been supported by recent ecological and molecular studies on Bt demonstrating rhizosphere competence (Vidal-Quist et al., 2013), endophytic ability (Tao et al., 2014), and even plant-mediated volatile attraction. More recently, it has been proposed that nematodes are the missing vector species for Bt virulence (Roan et al., 2015). The rarity of Bt epizootics may be the result of rapid invasion and displacement from the hemocoel of toxin-producing strains by non–toxin-producing Bt during pathogenesis (Raymond et al., 2012). In a review considering evidence from many studies, it was concluded that Bt is a pathogen as (1) it has a large array of insecticidal toxins, enzymes, and antibiotics, which reflect a pathogenic lifestyle; (2) the toxins act through the gut epithelium; (3) Bt can multiply in the hemocoel; and (4) horizontal infections do occur (Raymond et al., 2010).

The association of Bt with plants may also be a factor in their success as pathogens.

Bacterial entomopathogens carry a range of insecticidal encoding genes rather than producing a single effective toxin. A clear example is the association of Vip and Cry toxins to the vegetative and spore life stages of Bt, respectively, which suggests diverse relevance of the toxins during different stages of the infection. This strategy may be related to the constant evolutionary pressure on the host, which is constantly exposed to the microbes, and competing bacteria (Castagnola and Stock, 2014). Interestingly, diverse bacterial pathogens (eg, *Photorhabdus*, *Serratia*, and *Yersinia*) often share similar insecticidal virulence cassettes, possibly through horizontal gene transfer.

Some species have been shown to be capable of endophytic colonization of plants and may also enhance the growth of plants in the absence of insects, suggesting an additional role as biofertilizers. The importance of endophytic lifestyles in the entomopathogenic ability of these species is unknown but is an area of active investigation. Species recorded as endophytes include *B. laterosporus* (Ormskirk et al., 2015), *Pseudomonas putida*, and *Bacillus* spp. (eg, Jasim et al., 2015; Kumar et al., 2015; Li et al., 2015). *B. thuringiensis* has been found to naturally colonize cotton plants, and infected cotton and cabbage plants had activity against *Spodoptera frugiperda* and *Plutella xylostella*, respectively (Monnerat et al., 2009). The multiple effects of some of these putative insecticidal bacteria was demonstrated in the recent publication that a strain of *B. laterosporus* also induces systematic resistance in tobacco, although insect toxicity was not mentioned for the specific strain (Wang et al., 2015).

New discoveries about the social lives of bacteria are increasing our understanding of entomopathogenic bacteria. For example, there is evidence that social dynamics are driving loss of pathogenic function in some human pathogens (Andersen et al., 2015). There are similar interactions in the ecology of entomopathogenic bacteria (Raymond et al., 2012). For example, the role of quorum sensing in pathogen virulence is being explored with Bt, with evidence of cooperation between bacteria found (Zhou et al., 2014).

4.5 HOST SUSCEPTIBILITY RANGE AND EFFICACY

The range of insects affected by the entomopathogenic bacteria varies from the apparently single species of scarab infected by the amber disease-causing *Serratia* spp. to the multiple orders susceptible to toxins produced by *Chromobacterium subtsugae*, *B. thuringiensis*, and *Y. entomophaga*. The discovery of broad host range species in recent years points to the future identification of undiscovered pathogens and reflects increasing effort to find novel strains for use as biopesticides. In some cases, the broad host range reflects the production of families of similar toxins, such as the Tc family (Hurst et al., 2011b; Chen et al., 2014).

B. thuringiensis is reported to be toxic to over 3000 invertebrates (Glare and O'Callaghan, 2000), but no strain or single toxin has such a wide target range. Some Cry toxins may be active against a few insects in several orders (eg, Lepidoptera and Diptera), but in general the toxins appear more specific (van Frankenhuyzen, 2009). This specificity is mostly dictated by domains II and III of the three-domain Cry toxins recognizing host receptors (Bravo et al., 2013). Comparative to Bt, *B. laterosporus* has been reported from only a few hosts, but they cover Coleoptera, Diptera, Lepidoptera, and other invertebrates. As with most pathogens described herein, individual strains within a species do not affect all insects known to be susceptible to a species.

The infective bacteria (as opposed to toxin producing), those that demonstrate cell multiplication within the live insect and have a more chronic effect, provide contrasting modes of action and host specificity. *Paenibacillus popilliae* and *P. lentimorbus* are pathogens of various scarab larvae, including the Japanese beetle. Interestingly, the amber disease-causing *Serratia entomophila* and some *S. proteamaculans* are also restricted to a single scarab species, *Costelytra zealandica*, perhaps reflecting the need for specialist pathogens to overcome well-developed defense mechanisms in soil-dwelling insects, due to their constant exposure to bacteria.

The development of resistance to toxin-based entomopathogenic bacteria is a constant issue, especially with the increasing use of such toxins in transgenic plants. The evolution of field resistance to several Bt crops follows from experimental demonstration of the ease that resistance to Cry toxins can be developed in the laboratory (Siegwart et al., 2015). At least nine insect species have been found to develop resistance to Bt toxins, mainly in the laboratory (Siegwart et al., 2015) and there are also reports of Vip3Aa20-resistant alleles in populations of *Spodoptera* and *Helicoverpa* (Bernardi et al., 2015). In the field, only two lepidopteran species have been reported to develop resistance to Bt pesticides, *Plutella xylostella* and *T. ni*. Considering that alterations in Bt toxin receptors is the most common mechanism resulting in high levels of resistance, the low number of insects with field-evolved resistance could be explained by the combination of toxins with different modes of action (ie, recognizing different midgut receptors) present in Bt pesticides. However, in these field-evolved resistance cases alterations in homologous *ABCC2* transporter genes serving as Cry toxin receptors have been genetically linked to resistance (Baxter et al., 2011). At least in *P. xylostella*, altered expression of *ABCC2* genes is *trans*-regulated by a gene in the mitogen-activated protein kinase signaling pathway (Guo et al., 2015). In turn, functional field-evolved resistance to Bt crops producing a single Cry toxin has been reported for *Busseola fusca*

(corn stalk borer) in South Africa, *Spodoptera frugiperda* in Puerto Rico, *Pectinophora gossypiella* (pink bollworm) in India, and *D. virgifera* in the continental United States (Tabashnik et al., 2013). However, when tested these insects remained susceptible to Bt pesticides (Jakka et al., 2014). One method to overcome resistance could be the use of mixtures of pathogens, since in theory it would be more difficult for insects to develop resistance to two agents. However, Raymond et al. (2013) found that *P. xylostella* evolved resistance at similar rate when Bt mixtures were used instead of single Bt strains, suggesting that mixtures can have unexpected effects on resistance selection.

Field-evolved resistance has also been reported for *Lysinibacillus sphaericus* toxins in mosquitoes from a number of countries (Charles and Nielsen-LeRoux, 2000). As with Bt toxins, most cases of resistance to *L. sphaericus* involve alterations in binding of the Bin toxin to midgut receptors (Silva-Filha et al., 2014).

4.6 PRACTICAL APPLICATIONS AND COMMERCIALIZATION

The first reported attempts to use bacteria as inundative microbial control agents were in Mexico and South America using the locust bacterium *Coccobacillus acridiorum* (D'Herelle, 1912). Practical application of Bt for pest control was first realized when a strain isolated from diseased *E. küehniella* was shown to be effective in controlling *Ostrinia nubilalis* larvae (Husz, 1928). The first pesticide product based on Bt (Sporeine) was developed and commercialized in France in 1938 to control flour moths, but the development of potent synthetic pesticides and World War II stalled research on Bt as biopesticide (Sanchis, 2011). Combinations of polyhedrosis virus and Bt were shown to address limitations of individual microbes to control the alfalfa caterpillar (*Colias philodice*) in California (Steinhaus, 1951). The continuous use of Bt sprays on a limited commercial scale to control diverse defoliating pests and reports of lack of detrimental environmental effects provided evidence of the potential use of Bt as an environmentally safe biopesticide (Heimpel and Angus, 1960). Increased interest led to subsequent isolation of multiple spore-forming entomopathogenic bacteria.

Paenibacillus popilliae, *P. lentimorbus*, and Bt were registered in the United States in 1958. The Bt subsp. *kurstaki* used in most Bt pesticides against lepidopteran larvae was first registered in the United States in 1961 (Sundh and Goettel, 2013). A large range of Bt products were developed subsequently (Glare and O'Callaghan, 2000). *Lysinibacillus sphaericus* serotype H5a5b was first registered in the United States in 1991. The most used *L. sphaericus* larvicides in the United States and Southeast Asia are based on highly active strains 2362 and C3-41, respectively (Park et al., 2010). These strains display very similar flagellar serotypes and identical Bin toxins (Yuan et al., 1999). Main pesticidal attributes of

L. sphaericus compared to Bt subsp. *israelensis* are higher persistence and reduced sedimentation in mosquito breeding sites (Silapanuntakul et al., 1983; Yousten et al., 1992). Higher persistence may be associated with recycling of the bacterium in the environment as new spores are produced in cadavers of infected larvae (Charles and Nicolas, 1986).

Cost-effective production is an essential part of development of the bacterium as a biopesticide (Visnovsky et al., 2008). Commercialization is also sometimes hindered by lack of efficacious application methods, especially in the case of non–spore-forming bacteria, such as *S. entomophila*. The first product based on *S. entomophila* was a liquid formulation, marketed as Invade, which was applied into grooves in the soil using a modified seed drill with high water requirements. Further research led to development of a granular formulation (Bioshield) with greatly improved shelf-life under ambient conditions and delivery using conventional machinery (Johnson et al., 2001; Townsend et al., 2004) and seed coating (eg, Young et al., 2009, 2010). Seeds are an ideal delivery mechanism for the introduction of entomopathogenic microbes into the plant rhizosphere as they allow targeted contact with and infection of insect pests can introduce high populations of viable cells.

Of the increasing global market for pesticides, biopesticides are the fastest growing segment, estimated to be expanding at twice the rate of synthetic chemicals (Ruiu, 2015). The market and product range is dynamic, and lists of biopesticides are out of date almost as soon as written, as companies launch new products and some are withdrawn from the market. There are recent reviews of biopesticides (Kabaluk et al., 2010) and manuals produced by several private companies yearly, such as BCC research. Kabaluk et al. (2010) listed 149 registrations of bacteria-based insecticides in over 15 countries, but the active agents were only from three species.

The major pesticide companies around the world have been very active in developing and acquiring biopesticide products, including microbial-based biopesticides. The company Agraquest and the fungal production company Prophyta were acquired by Bayer CropScience; Becker Underwood was acquired by BASF; and the producer of nematocides based on bacteria *Pasteuria*, Bioscience Inc., was acquired by Syngenta. Whether this new activity in biopesticides will lead to more stable markets and products or will drive more diversity in the product range is uncertain. One of the new type of products is Grandevo (Marrone BioInnovations), based on *Chromobacterium subtsugae*, where the bioactive metabolites are used rather than an infective agent.

4.7 ENVIRONMENTAL AND MAMMALIAN SAFETY

While microbial control agents are generally regarded as safe alternatives to chemical insecticides, registration of products based on entomopathogenic bacteria requires thorough assessment of their environmental and mammalian safety. This may include testing against fish and amphibians, birds, and small mammals such as mice, rats, or rabbits using standardized protocols. Numerous studies on vertebrate safety (reviewed in Glare and O'Callaghan, 2000) indicate very little safety risk from direct exposure to Bt and, despite many years of broadscale application of Bt-based products, very few human health problems have been reported. Monitoring of public health of communities exposed to Bt spraying has also raised few proven concerns. Because Bt is part of the *B. cereus* complex and some strains of Bt can produce *B. cereus*–type diarrheal enterotoxins (Damgaard, 1995), quantification of enterotoxins produced by Bt strains for commercial use is assessed, even though growth conditions for Bt production are not conducive for expression of these toxins.

Genome sequencing is providing new understanding of the genetic basis of pathogenicity to invertebrates and how different species within the same genus can either cause disease in humans or invertebrates but not both. Increasing recognition of the fluidity of the bacterial genome and the potential for acquisition of virulence factors coded for on phages, pathogenicity islands, and plasmids points to the need to thoroughly investigate the genetic basis and regulation of virulence factors in entomopathogens with commercial potential.

Ecological risk assessment requires consideration of several different aspects of the biology and ecology of the entomopathogenic bacterium and the environment into which it will be introduced, such that the risks of its potential impact on nontarget species can be fairly assessed. The key areas that must be considered in ecological/environmental impact assessment of bacterial biopesticides are both direct and indirect effects on nontarget species (including sublethal effects); establishment, persistence, and dispersal in the environment; potential for ecosystem disruption (including both aquatic and terrestrial as appropriate); and horizontal gene transfer (Glare and O'Callaghan, 2003). Development of more specific genetic markers is needed to improve the rigor of environmental monitoring and impact assessments; this is becoming possible through advancement and cost reductions in whole genome sequencing.

Also essential is a detailed understanding of the mode of action and host range of the bacterium as this will guide selection of nontarget species that may be at risk following field application. The high level of specificity of *S. entomophila* ensures it has no nontarget effects and a good environmental safety profile; more extensive testing of nontarget species relevant to the potential receiving environment of entomopathogenic bacteria with broad host ranges will be needed. Candidate biopesticidal microorganisms are typically tested in the first instance against beneficial species (earthworms, pollinators, predators, and parasites), genera and species most closely related to the target pest, and species likely to be present in the receiving environment. There may also be key iconic (often endemic) species of particular conservation or cultural value that will be tested early in the assessment process. Nontarget testing in laboratory assays allows rapid assessment of direct effects. However, in many cases test species may show sensitivity in laboratory assays, which are often maximum challenge (high dose) studies, but are not affected at a population level following field application where they are likely to be exposed to lower doses. There are many reasons for the observed differences between susceptibility of species tested in the laboratory and field, and the limitations of laboratory assays must be considered when drawing conclusions about the environmental safety of the bacterium (Glare and O'Callaghan, 2003).

Field studies are valuable in proving the safety of entomopathogens, but evaluation of non-target effects in the field presents many challenges, not least of which is the decision about what organisms or ecosystem functions should be measured. The most extensive studies on ecosystem-level effects of bacterial application have been carried out with Bt subsp. *kurstaki* and Bt subsp. *israelensis* based products. Field application would obviously be expected to impact nontarget lepidopteran species and declines in lepidopteran populations are seen in the year of application, but populations of nontarget species generally recovered in subsequent years (Glare and O'Callaghan, 2003). Similarly, little impact has been reported as a result of Bt subsp. *israelensis* applications in aquatic systems (Lacey and Merritt, 2003).

Some of the newly identified entomopathogens discussed earlier have some way to being confirmed as having low environmental impact and some with broad host range will require extensive testing of nontarget effects appropriate to the target environment/location. Predictive models of interactions between entomopathogens and invertebrates would be useful in streamlining environmental impact assessment of candidate biopesticidal bacteria.

4.8 REGISTRATION

The process of registration of bacteria-based biopesticides differs between countries and regions (Kabaluk et al., 2010, see Chapter 30). Few countries have separate guidelines for biological agents as opposed to synthetic pesticides, which can result in inappropriate evaluation methods (Sundh and Goettel, 2013). Registration may be required of both the active substance and the eventual product. There are many guidelines and publications covering requirements in different jurisdictions, which can rapidly become out of date. In the United States, the Environmental Protection Agency (EPA) recognizes biologically based control products with separate registration pathways (Leahy et al., 2014). The EPA has three classes of biopesticides, microbial pesticides; plant-incorporated protectants, and biochemical pesticides, which

cover the bacterial entomopathogens. The European Union has been attempting to promote the use of nonchemical pest control approaches through legislation specifically favoring integrated pest management and revised regulations for plant protection products from 2011 (Villaverde et al., 2014).

4.9 FUTURE

As our understanding of the multiple bacterial entomopathogens increases, so does the potential for their utilization in new and interesting ways. The increase in biopesticides sales around the world is not based on *B. thuringiensis* but on a range of other species. Toxin-producing bacteria are particularly attractive as active agents in new products as often they have broader spectra of activity and formulation and application benefits. New bacterial insect pathogens are also being discovered, such as *Chromobacterium subtsugae* and *Yersinia entomophaga*. The future for biopesticides is certainly looking bright, with the increasing withdrawal of chemical insecticides, low numbers of new chemicals being developed, and public pressure for safe pesticides. However, for the potential to be realized, agents including bacteria will need to meet the requirements of modern agriculture. These were summarized in Glare et al. (2012) as increased field persistence (ie, 21 days on foliage), better understanding of bioactive molecules from the microbes, more strategic selection of target pests and markets, more investment in research and development, and suitable registration and legislative environments.

REFERENCES

Adang, M., Crickmore, N., Jurat-Fuentes, J.L., 2014. Diversity of *Bacillus thuringiensis* crystal toxins and mechanism of action. In: Dhadialla, T.S., Gill, S. (Eds.), Advances in Insect Physiology. Insect Midgut and Insecticidal Proteins, vol. 47. Academic Press, San Diego, pp. 39–87.

Ahmed, I., Yokota, A., Yamazoe, A., Fujiwara, T., 2007. Proposal of *Lysinibacillus boronitolerans* gen. nov. sp. nov., and transfer of *Bacillus fusiformis* to *Lysinibacillus fusiformis* comb. nov. and *Bacillus sphaericus* to *Lysinibacillus sphaericus* comb. nov. Intern. J. Syst. Evol. Microbiol. 57, 1117–1125.

Andersen, S.B., Marvig, R.L., Molin, S., Krogh Johansen, H., Griffin, A.S., 2015. Long-term social dynamics drive loss of function in pathogenic bacteria. Proc. Natl. Acad. Sci. 112, 10756–10761.

Aronson, A., 2000. Incorporation of protease K into larval insect membrane vesicles does not result in disruption of integrity or function of the pore-forming *Bacillus thuringiensis* delta-endotoxin. Appl. Environ. Microbiol. 66, 4568–4570.

Barloy, F., Delecluse, A., Nicolas, L., Lecadet, M.M., 1996. Cloning and expression of the first anaerobic toxin gene from *Clostridium bifermentans* subsp. *malaysia*, encoding a new mosquitocidal protein with homologies to *Bacillus thuringiensis* delta-endotoxins. J. Bacteriol. 178, 3099–3105.

Barloy, F., Lecadet, M.M., Delecluse, A., 1998. Cloning and sequencing of three new putative toxin genes from *Clostridium bifermentans* CH18. Gene 211, 293–299.

Barth, H., Aktories, K., Popoff, M.R., Stiles, B.G., 2004. Binary bacterial toxins: biochemistry, biology, and applications of common *Clostridium* and *Bacillus* proteins. Microbiol. Mol. Biol. Rev. 68, 373–402.

Baum, J.A., Sukuru, U.R., Penn, S.R., Meyer, S.E., Subbarao, S., Shi, X., Flasinski, S., Heck, G.R., Brown, R.S., Clark, T.L., 2012. Cotton plants expressing a hemipteran-active *Bacillus thuringiensis* crystal protein impact the development and survival of *Lygus hesperus* (Hemiptera: Miridae) nymphs. J. Econ. Entomol. 105, 616–624.

Baxter, S.W., Badenes-Perez, F.R., Morrison, A., Vogel, H., Crickmore, N., Kain, W., Wang, P., Heckel, D.G., Jiggins, C.D., 2011. Parallel evolution of *Bacillus thuringiensis* toxin resistance in Lepidoptera. Genetics 189, 675–679.

Berliner, E., 1915. Uber die schlaffsucht der mehlmottenraupe (*Ephestia kuhniella* Zell.) und ihren erreger *B. thuringiensis* n. sp. Z. Angew. Entom. 2, 29–56.

Bernardi, O., Bernardi, D., Ribeiro, R.S., Okuma, D.M., Salmeron, E., Fatoretto, J., Medeiros, F.C.L., Burd, T., Omoto, C., 2015. Frequency of resistance to Vip3Aa20 toxin from *Bacillus thuringiensis* in *Spodoptera frugiperda* (Lepidoptera: Noctuidae) populations in Brazil. Crop Prot. 76, 7–14.

Berry, C., 2012. The bacterium, *Lysinibacillus sphaericus*, as an insect pathogen. J. Invertebr. Pathol. 109, 1–10.

Boets, A., Arnaut, G., Van Rie, J., Damme, N., 2011. Toxins. Patent No. US 7919609B2.

Bone, L.W., Singer, S., 1991. Control of Parasitic Nematode Ova/Larvae with a *Bacillus laterosporus*. US Patent No. 5045314.

Boonserm, P., Moonsom, S., Boonchoy, C., Promdonkoy, B., Parthasarathy, K., Torres, J., 2006. Association of the components of the binary toxin from *Bacillus sphaericus* in solution and with model lipid bilayers. Biochem. Biophys. Res. Comm. 342, 1273–1278.

Bowen, D., Rocheleau, T.A., Blackburn, M., Andreev, O., Golubeva, E., Bhartia, R., ffrench-Constant, R.H., 1998. Insecticidal toxins from the bacterium *Photorhabdus luminescens*. Science 280, 2129–2132.

Bravo, A., Sanchez, J., Kouskoura, T., Crickmore, N., 2002. N-terminal activation is an essential early step in the mechanism of action of the *Bacillus thuringiensis* Cry1Ac insecticidal toxin. J. Biol. Chem. 277, 23985–23987.

Bravo, A., Gómez, I., Conde, J., Muñoz-Garay, C., Sánchez, J., Miranda, R., Zhuang, M., Gill, S.S., Soberón, M., 2004. Oligomerization triggers binding of a *Bacillus thuringiensis* Cry1Ab pore-forming toxin to aminopeptidase N receptor leading to insertion into membrane microdomains. Biochim. Biophys. Acta 1667, 38–46.

Bravo, A., Gomez, I., Porta, H., Garcia-Gomez, B.I., Rodriguez-Almazan, C., Pardo, L., Soberon, M., 2013. Evolution of *Bacillus thuringiensis* Cry toxins insecticidal activity. Microb. Biotechnol. 6, 17–26.

Bresolin, G., Morgan, J.A.W., Ilgen, D., Scherer, S., Fuchs, T.M., 2006. Low temperature-induced insecticidal activity of *Yersinia enterocolitica*. Mol. Microbiol. 59, 503–512.

Broadwell, A.H., Baumann, L., Baumann, P., 1990. The 42-kilodalton and 51-kilodalton mosquitocidal proteins of *Bacillus sphaericus* 2362-construction of recombinants with enhanced expression and *in vivo* studies of processing and toxicity. J. Bacteriol. 172, 2217–2223.

Brown, S.E., Cao, A.T., Hines, E.R., Akhurst, R.J., East, P.D., 2004. A novel secreted protein toxin from the insect pathogenic bacterium *Xenorhabdus nematophila*. J. Biol. Chem. 279, 14595–14601.

Brownbridge, M., Ferguson, C., Saville, D.J., Swaminathan, J., Hurst, M.R.H., Jackson, T.A., 2008. Potential for biological control of porina (*Wiseana* spp.) with a novel insecticidal bacterium, *Yersinia* n. sp. (MH96) EN65 strain. N.Z. Plant Prot. 61, 229–235.

Burton, S.L., Ellar, D.J., Li, J., Derbyshire, D.J., 1999. N-acetylgalactosamine on the putative insect receptor aminopeptidase N is recognised by a site on the domain III lectin-like fold of a *Bacillus thuringiensis* insecticidal toxin. J. Mol. Biol. 287, 1011–1022.

Busby, J.N., Panjikar, S., Landsberg, M.J., Hurst, M.R., Lott, J.S., 2013. The BC component of ABC toxins is an RHS-repeat-containing protein encapsulation device. Nature 501, 547–550.

Carpusca, I., Jank, T., Aktories, K., 2006. *Bacillus sphaericus* mosquitocidal toxin (MTX) and pierisin: the enigmatic offspring from the family of ADP-ribosyltransferases. Mol. Microbiol. 62, 621–630.

Castagnola, A., Stock, S., 2014. Common virulence factors and tissue targets of entomopathogenic bacteria for biological control of lepidopteran pests. Insects 5, 139.

Charles, J.F., Nicolas, L., 1986. Recycling of *Bacillus sphaericus* 2362 in mosquito larvae: a laboratory study. Ann. Inst. Pasteur. Microbiol. 137, 101–111.

Charles, J.F., Nielsen-LeRoux, C., 2000. Mosquitocidal bacterial toxins: diversity, mode of action and resistance phenomena. Mem. Inst. Oswaldo Cruz 95, 201–206.

Chen, W.J., Hsieh, F.C., Hsu, F.C., Tasy, Y.F., Liu, J.R., Shih, M.C., 2014. Characterization of an insecticidal toxin and pathogenicity of *Pseudomonas taiwanensis* against insects. PLoS Pathog. 10, 14.

Clark, M.A., Baumann, P., 1991. Modification of the *Bacillus sphaericus* 51-kilodalton and 42-kilodalton mosquitocidal proteins – effects of internal deletions, duplications, and formation of hybrid proteins. Appl. Environ. Microbiol. 57, 267–271.

Cordova-Kreylos, A.L., Fernandez, L.E., Koivunen, M., Yang, A., Flor-Weiler, L., Marrone, P.G., 2013. Isolation and characterization of *Burkholderia rinojensis* sp. nov., a non-*Burkholderia cepacia* complex soil bacterium with insecticidal and miticidal activities. Appl. Environ. Microbiol. 79, 7669–7678.

Crickmore, N., Wheeler, V.C., Ellar, D.J., 1994. Use of an operon fusion to induce expression and crystallisation of a *Bacillus thuringiensis* delta-endotoxin encoded by a cryptic gene. Mol. Gen. Genet. 242, 365–368.

Crickmore, N., Zeigler, D.R., Feitelson, J., Schnepf, E., Van Rie, J., Lereclus, D., Baum, J., Dean, D.H., 1998. Revision of the nomenclature for the *Bacillus thuringiensis* pesticidal crystal proteins. Microbiol. Mol. Biol. Rev. 62, 807–813.

Crickmore, N., Baum, J., Bravo, A., Lereclus, D., Narva, K., Sampson, K., Schnepf, E., Sun, M., Zeigler, D.R., 2015. *Bacillus thuringiensis* Toxin Nomenclature. www.btnomenclature.info.

D'Herelle, F., 1912. Sur la propagation, dans la Republique Argentine, de l'epizootie des sauterelles du Mexique. Comptes Rendus Acad. Sci. Ser. D 154, 623–625.

Damgaard, P.H., 1995. Diarrheal enterotoxin production by strains of *Bacillus thuringiensis* isolated from commercial *Bacillus thuringiensis*-based insecticides. FEMS Immun. Med. Microbiol. 12, 245–249.

Davidson, E.W., 1979. Ultrastructure of midgut events in the pathogenesis of *Bacillus sphaericus* strain SSII-1 infections of *Culex pipiens quinquefasciatus* larvae. Can. J. Microbiol. 25, 178–184.

de Barjac, H., Bonnefoi, A., 1968. A classification of strains of *Bacillus thuringiensis* Berliner with a key to their differentiation. J. Invertebr. Pathol. 11, 335–347.

de Oliveira, E.J., Rabinovitch, L., Monnerat, R.G., Passos, L.K.J., Zahner, V., 2004. Molecular characterization of *Brevibacillus laterosporus* and its potential use in biological control. Appl. Environ. Microbiol. 70, 6657–6664.

Dieppois, G., Opota, O., Lalucat, J., Lemaitre, B., 2015. Pseudomonas entomophila: a versatile bacterium with entomopathogenic properties. In: Ramos, J.-L., et al. (Ed.), *Pseudomonas*: New Aspects of *Pseudomonas* Biology. Springer, pp. 25–34.

Djukic, M., Poehlein, A., Thurmer, A., Daniel, R., 2011. Genome sequence of *Brevibacillus laterosporus* LMG 15441, a pathogen of invertebrates. J. Bacteriol. 193, 5535–5536.

Dodd, S.J., Hurst, M.R.H., Glare, T.R., O'Callaghan, M., Ronson, C.W., 2006. Occurrence of sep insecticidal toxin complex genes in *Serratia* spp. and *Yersinia frederiksenii*. Appl. Environ. Microbiol. 72, 6584–6592.

Du, C., Martin, P.A.W., Nickerson, K.W., 1994. Comparison of disulfide contents and solubility at alkaline pH of insecticidal and noninsecticidal *Bacillus thuringiensis* protein crystals. Appl. Environ. Microbiol. 60, 3847–3853.

Dutky, S.R., 1940. Two new spore-forming bacteria causing milky diseases of Japanese beetle larvae. J. Agric. Res. 61, 57–68.

Edagiz, S., Lagace-Wiens, P., Embil, J., Karlowsky, J., Walkty, A., 2015. Empyema caused by *Clostridium bifermentans*: a case report. Can. J. Infect. Dis. Med. Microbiol. 26, 105–107.

Ellis, R.T., Stockhoff, B.A., Stamp, L., Schnepf, H.E., Schwab, G.E., Knuth, M., Russell, J., Cardineau, G.A., Narva, K.E., 2002. Novel *Bacillus thuringiensis* binary insecticidal crystal proteins active on western corn rootworm, *Diabrotica virgifera virgifera* LeConte. Appl. Environ. Microbiol. 68, 1137–1145.

Eugenia Nunez-Valdez, M., Calderon, M.A., Aranda, E., Hernandez, L., Ramirez-Gama, R.M., Lina, L., Rodriguez-Segura, Z., Gutierrez, M.D.C., Villalobos, F.J., 2008. Identification of a putative Mexican strain of *Serratia entomophila* pathogenic against root-damaging larvae of Scarabaeidae (Coleoptera). Appl. Environ. Microbiol. 74, 802–810.

Euzéby, J.P., 1997. List of bacterial names with standing in nomenclature. Intern. J. Syst. Bacteriol. 47, 590–592. http://www.bacterio.net.

Favret, M.E., Yousten, A.A., 1985. Insecticidal activity of *Bacillus laterosporus*. J. Invertebr. Pathol. 45, 195–203.

Fernandez, L.E., Aimanova, K.G., Gill, S.S., Bravo, A., Soberon, M., 2006. A GPI-anchored alkaline phosphatase is a functional midgut receptor of Cry11Aa toxin in *Aedes aegypti* larvae. Biochem. J. 394, 77–84.

ffrench-Constant, R.H., Dowling, A., Waterfield, N.R., 2007. Insecticidal toxins from *Photorhabdus* bacteria and their potential use in agriculture. Toxicon 49, 436–451.

Fortier, M., Vachon, V., Kirouac, M., Schwartz, J.L., Laprade, R., 2005. Differential effects of ionic strength, divalent cations and pH on the pore-forming activity of *Bacillus thuringiensis* insecticidal toxins. J. Membr. Biol. 208, 77–87.

Fortier, M., Vachon, V., Marceau, L., Schwartz, J.L., Laprade, R., 2007. Kinetics of pore formation by the *Bacillus thuringiensis* toxin Cry1Ac. Biochim. Biophys. Acta. Biomembr. 1768, 1291–1298.

Franken, E., Krieger, L., Schnetter, W., 1996. *Bacillus popilliae*: a difficult pathogen. Bull. OILB/SROP 19, 40–45.

Fuchs, T.M., Bresolin, G., Marcinowski, L., Schachtner, J., Scherer, S., 2008. Insecticidal genes of *Yersinia* spp.: taxonomical distribution, contribution to toxicity towards *Manduca sexta* and *Galleria mellonella*, and evolution. BMC Microbiol. 8.

Garrity, G., Brenner, D.J., Krieg, N.R., Staley, J.R.E., 2005. Bergey's Manual® of Systematic Bacteriology. The Proteobacteria, Part B: The Gammaproteobacteria, vol. 2. Springer, US.

Gatsogiannis, C., Lang, A.E., Meusch, D., Pfaumann, V., Hofnagel, O., Benz, R., Aktories, K., Raunser, S., 2013. A syringe-like injection mechanism in *Photorhabdus luminescens* toxins. Nature 495, 520–523.

Genersch, E., 2010. American foulbrood in honeybees and its causative agent, *Paenibacillus larvae*. J. Invertebr. Pathol. 103 (Suppl.), S10–S19.

Glancy, B.M., VanderMeer, R.K., Glover, A., Lofgren, C.S., Vinson, S.B., 1981. Filtration of microparticles from liquids ingested by the red imported fire ant *Solenopsis invicta* Buren. Insect. Soc. 28, 395–401.

Glare, T., Caradus, J., Gelernter, W., Jackson, T., Keyhani, N., Kohl, J., Marrone, P., Morin, L., Stewart, A., 2012. Have biopesticides come of age? Trends Biotechnol. 30, 250–258.

Glare, T.R., Hampton, J.G., Cox, M.P., Bienkowski, D.A., 2014. Biocontrol Compositions. Patent application PCT/IB2013/055157/WO2014045131 A1.

Glare, T.R., O'Callaghan, M., 2000. *Bacillus thuringiensis*: Biology, Ecology and Safety. Wiley. 350 pp.

Glare, T.R., O'Callaghan, M., 2003. Environmental impacts of bacterial biopesticides. In: Hokkanen, H.M.T., Hajek, A.E. (Eds.), Environmental Impacts of Microbial Insecticides: Need and Methods for Risk Assessment. Kluwer, pp. 119–149.

Grimont, P.A., Grimont, F., 1978. The genus *Serratia*. Ann. Rev. Microbiol. 32, 221–248.

Grimont, P.A.D., Jackson, T.A., Ageron, E., Noonan, M.J., 1988. *Serratia entomophila* sp. nov. associated with amber disease in the New Zealand grass grub *Costelytra zealandica*. Intern. J. Syst. Evol. Microbiol. 38, 1–6.

Grkovic, S., Glare, T.R., Jackson, T.A., Corbett, G.E., 1995. Genes essential for amber disease in grass grubs are located on the large plasmid found in *Serratia entomophila* and *Serratia proteamaculans*. Appl. Environ. Microbiol. 61, 2218–2223.

Grochulski, P., Masson, L., Borisova, S., Pusztai-Carey, M., Schwartz, J.L., Brousseau, R., Cygler, M., 1995. *Bacillus thuringiensis* CryIA(a) insecticidal toxin: crystal structure and channel formation. J. Mol. Biol. 254, 447–464.

Guerineau, M., Alexander, B., Priest, F.G., 1991. Isolation and identification of *Bacillus sphaericus* strains pathogenic for mosquito larvae. J. Invertebr. Pathol. 57, 325–333.

Guo, Z., Kang, S., Chen, D., Wu, Q., Wang, S., Xie, W., Zhu, X., Baxter, S.W., Zhou, X., Jurat-Fuentes, J.L., Zhang, Y., 2015. MAPK signaling pathway alters expression of midgut ALP and ABCC genes and causes resistance to *Bacillus thuringiensis* Cry1Ac toxin in diamondback moth. PLoS Genet. 11, e1005124.

Han, S., Craig, J.A., Putnam, C.D., Carozzi, N.B., Tainer, J.A., 1999. Evolution and mechanism from structures of an ADP-ribosylating toxin and NAD complex. Nat. Struct. Biol. 6, 932–936.

Hayakawa, T., Shitomi, Y., Miyamoto, K., Hori, H., 2004. GalNAc pretreatment inhibits trapping of *Bacillus thuringiensis* Cry1Ac on the peritrophic membrane of *Bombyx mori*. FEBS Lett. 576, 331–335.

Hazes, B., Read, R.J., 1995. A mosquitocidal toxin with a ricin-like cell-binding domain. Nat. Struct. Biol. 2, 358–359.

Heimpel, A.M., Angus, T.A., 1960. Bacterial insecticides. Bacteriol. Rev. 24, 266–288.

Held, G.A., Bulla Jr., L.A., Ferrari, E., Hoch, J., Aronson, A.I., Minnich, S.A., 1982. Cloning and localization of the lepidopteran protoxin gene of *Bacillus thuringiensis* subsp. *kurstaki*. Proc. Natl. Acad. Sci. U.S.A. 79, 6065–6069.

Heymann, J.B., Bartho, J.D., Rybakova, D., Venugopal, H.P., Winkler, D.C., Sen, A., Hurst, M.R.H., Mitra, A.K., 2013. Three-dimensional structure of the toxin-delivery particle antifeeding prophage of *Serratia entomophila*. J. Biol. Chem. 288, 25276–25284.

Hire, R.S., Hadapad, A.B., Vijayalakshmi, N., Dongre, T.K., 2010. Characterization of highly toxic indigenous strains of mosquitocidal organism *Bacillus sphaericus*. FEMS Microbiol. Lett. 305, 155–161.

Höfte, H., Whiteley, H.R., 1989. Insecticidal crystal proteins of *Bacillus thuringiensis*. Microbiol. Rev. 53, 242–255.

Hu, X., Fan, W., Han, B., Liu, H., Zheng, D., Li, Q., Dong, W., Yan, J., Gao, M., Berry, C., Yuan, Z., 2008. Complete genome sequence of the mosquitocidal bacterium *Bacillus sphaericus* C3-41 and comparison with those of closely related *Bacillus* species. J. Bacteriol. 190, 2892–2902.

Huang, X.W., Tian, B.Y., Niu, Q.H., Yang, J.K., Zhang, L.M., Zhang, K.Q., 2005. An extracellular protease from *Brevibacillus laterosporus* G4 without parasporal crystals can serve as a pathogenic factor in infection of nematodes. Res. Microbiol. 156, 719–727.

Humphreys, M.J., Berry, C., 1998. Variants of the *Bacillus sphaericus* binary toxins: implications for differential toxicity of strains. J.Invertebr. Pathol. 71, 184–185.

Hurst, M.R.H., Glare, T.R., Jackson, T.A., Ronson, C.W., 2000. Plasmid-located pathogenicity determinants of *Serratia entomophila*, the causal agent of amber disease of grass grub, show similarity to the insecticidal toxins of *Photorhabdus luminescens*. J. Bacteriol. 182, 5127–5138.

Hurst, M.R., Jones, S.M., Tan, B., Jackson, T.A., 2007a. Induced expression of the *Serratia entomophila* Scp proteins shows activity towards the larvae of the New Zealand grass grub *Costelytra zealandica*. FEMS Microbiol. Lett. 275, 160–167.

Hurst, M.R.H., Beard, S.S., Jackson, T.A., Jones, S.M., 2007b. Isolation and characterization of the *Serratia entomophila* antifeeding prophage. FEMS Microbiol. Lett. 270, 42–48.

Hurst, M.R.H., Becher, S.A., Young, S.D., Nelson, T.L., Glare, T.R., 2011a. *Yersinia entomophaga* sp. nov., isolated from the New Zealand grass grub *Costelytra zealandica*. Intern. J. Syst. Evol. Microbiol. 61, 844–849.

Hurst, M.R.H., Jones, S.A., Binglin, T., Harper, L.A., Jackson, T.A., Glare, T.R., 2011b. The main virulence determinant of *Yersinia entomophaga* MH96 is a broad-host-range toxin complex active against insects. J. Bacteriol. 193, 1966–1980.

Hurst, M.R.N., Becher, S.A., O'Callaghan, M., 2011c. Nucleotide sequence of the *Serratia entomophila* plasmid pADAP and the *Serratia proteamaculans* pU143 plasmid virulence associated region. Plasmid 65, 32–41.

Husz, B., 1928. *Bacillus thuringiensis* Berl., a bacterium pathogenic to corn borer larvae. A preliminary report. Int. Corn. Borer Invest. Sci. Rep. 1927–1928, 191–193.

Ishii, K., Adachi, T., Hamamoto, H., Sekimizu, K., 2014. *Serratia marcescens* suppresses host cellular immunity via the production of an adhesion-inhibitory factor against immunosurveillance cells. J. Biol. Chem. 289, 5876–5888.

Jackson, T.A., Zimmermann, G., 1996. Is there a role for *Serratia* spp. in the biocontrol of *Melolontha* spp. Bull. OILB/SROP 19, 47–53.

Jackson, T.A., Pearson, J.F., O'Callaghan, M., Mahanty, H.K., Willcocks, M.J., 1992. Pathogen to product – development of *Serratia entomophila* (Enterobacteriaceae) as a commercial biological agent for the New Zealand grass grub *Costelytra zealandica*. In: Jackson, T.A., Glare, T.R. (Eds.), Use of Pathogens in Scarab Pest Management. Intercept, Andover, pp. 191–198.

Jackson, T.A., Huger, A.M., Glare, T.R., 1993. Pathology of amber disease in the New Zealand grass grub *Costelytra zealandica* (Coleoptera, Scarabaeidae). J. Invertebr. Pathol. 61, 123–130.

Jakka, S.R., Knight, V.R., Jurat-Fuentes, J.L., 2014. *Spodoptera frugiperda* (J. E. Smith) with field-evolved resistance to Bt maize are susceptible to Bt pesticides. J. Invertebr. Pathol. 122, 52–54.

Jasim, B., Geethu, P.R., Mathew, J., Radhakrishnan, E.K., 2015. Effect of endophytic *Bacillus* sp. from selected medicinal plants on growth promotion and diosgenin production in *Trigonella foenum-graecum*. Plant Cell Tissue Organ Cult. 122, 565–572.

Jeong, H.U., Mun, H.Y., Oh, H.K., Kim, S.B., Yang, K.Y., Kim, I., Lee, H.B., 2010. Evaluation of insecticidal activity of a bacterial strain, *Serratia* sp. EML-SE1 against diamondback moth. J. Microbiol. 48, 541–545.

Johnson, V.W., Pearson, J.F., Jackson, T.A., 2001. Formulation of *Serratia entomophila* for biological control of grass grub. N.Z. Plant Prot. 54, 125–127.

Johnston, P.R., Crickmore, N., 2009. Gut bacteria are not required for the insecticidal activity of *Bacillus thuringiensis* toward the tobacco hornworm, *Manduca sexta*. Appl. Environ. Microbiol. 75, 5094–5099.

Jones, G.W., Nielsen-LeRoux, C., Yang, Y., Yuan, Z., Dumas, V.F., Monnerat, R.G., Berry, C., 2007. A new Cry toxin with a unique two-component dependency from *Bacillus sphaericus*. FASEB J. 21, 4112–4120.

Jones, S.A., Ferguson, C.M., Philip, B.A., Koten, C.V., Hurst, M.R.H., 2015. Assessing the potential of *Yersinia entomophaga* to control plantain moth in a laboratory assay. N.Z. Plant Prot. 68, 146–150.

Juarez Perez, V., Delecluse, A., 2001. The Cry toxins and the putative hemolysins of *Clostridium bifermentans* ser. *malaysia* are not involved in mosquitocidal activity. J. Invertebr. Pathol. 78, 57–58.

Jucovic, M., Walters, F.S., Warren, G.W., Palekar, N.V., Chen, J.S., 2008. From enzyme to zymogen: engineering Vip2, an ADP-ribosyltransferase from *Bacillus cereus*, for conditional toxicity. Protein Eng. Des. Sel. 21, 631–638.

Jurat-Fuentes, J.L., Jackson, T.A., 2012. Bacterial entomopathogens. In: Vega, F.E., Kaya, H.K. (Eds.), Insect Pathology. Academic Press, San Diego, pp. 265–349.

Kabaluk, J.T., Svircev, A.M., Goettel, M.S., Woo, S.G.E., 2010. The Use and Regulation of Microbial Pesticides in Representative Jurisdictions Worldwide. IOBC Global.

Kelker, M.S., Berry, C., Evans, S.L., Pai, R., McCaskill, D.G., Wang, N.X., Russell, J.C., Baker, M.D., Yang, C., Pflugrath, J.W., Wade, M., Wess, T.J., Narva, K.E., 2014. Structural and biophysical characterization of *Bacillus thuringiensis* insecticidal proteins Cry34Ab1 and Cry35Ab1. PLoS One 9, e112555.

Kil, Y.J., Seo, M.J., Kang, D.K., Oh, S.N., Cho, H.S., Youn, Y.N., Yasunaga-Aoki, C., Yu, Y.M., 2014. Effects of enterobacteria (*Burkholderia* sp.) on development of *Riptortus pedestris*. J. Fac. Agric. Kyushu Univ. 59, 77–84.

Klein, M.G., 1988. Pest management of soil-inhabiting insects with microorganisms. Agric. Ecosyst. Environ. 24, 337–349.

Klein, M.G., Jackson, T.A., 1992. Bacterial diseases of scarabs. In: Jackson, T.A., Glare, T.R. (Eds.), Use of Pathogens in Scarab Pest Management. Intercept, Andover, UK, pp. 43–61.

Kumar, A.S., Aronson, A.I., 1999. Analysis of mutations in the pore-forming region essential for insecticidal activity of a *Bacillus thuringiensis* delta-endotoxin. J. Bacteriol. 181, 6103–6107.

Kumar, V., Kumar, A., Pandey, K.D., Roy, B.K., 2015. Isolation and characterization of bacterial endophytes from the roots of *Cassia tora* L. Ann. Microbiol. 65, 1391–1399.

Kupferschmied, P., Maurhofer, M., Keel, C., 2013. Promise for plant pest control: root-associated pseudomonads with insecticidal activities. Front. Plant Sci. 4, 17.

Lacey, L.A., 2007. *Bacillus thuringiensis* serovariety *israelensis* and *Bacillus sphaericus* for mosquito control. In: Floore, T.G. (Ed.), Biorational Control of Mosquitoes. Am. Mosq. Control Assoc. Bull., vol. 7, pp. 133–163.

Lacey, L.A., Merritt, R.W., 2003. The safety of bacterial microbial agents used for black fly and mosquito control in aquatic environments. In: Hokkanen, H.M.T., Hajek, A.E. (Eds.), Environmental Impacts of Microbial Insecticides: Need and Methods for Risk Assessment. Kluwer, pp. 151–168.

Laubach, C.A., 1916. Studies on aerobic spore-bearing non-pathogenic bacteria: part II spore-bearing bacteria in dust. J. Bacteriol. 1, 493.

Leahy, J., Mendelsohn, M., Kough, J., Jones, R., Berckes, N., 2014. Biopesticide oversight and registration at the US environmental protection agency. In: Gross, A.D., et al. (Ed.), Biopesticides: State of the Art and Future Opportunities. Amer. Chem. Soc., Washington, pp. 3–18.

Leclerque, A., Kleespies, R.G., Schuster, C., Richards, N.K., Marshall, S.D.G., Jackson, T.A., 2012. Multilocus sequence analysis (MLSA) of *Rickettsiella costelytrae* and *Rickettsiella pyronotae*, intracellular bacterial entomopathogens from New Zealand. J. Appl. Microbiol. 113, 1228–1237.

Lee, M.K., Walters, F.S., Hart, H., Palekar, N., Chen, J.S., 2003. The mode of action of the *Bacillus thuringiensis* vegetative insecticidal protein Vip3A differs from that of Cry1Ab delta-endotoxin. Appl. Environ. Microbiol. 69, 4648–4657.

Leuber, M., Orlik, F., Schiffler, B., Sickmann, A., Benz, R., 2006. Vegetative insecticidal protein (Vip1Ac) of *Bacillus thuringiensis* HD201: evidence for oligomer and channel formation. Biochemistry 45, 283–288.

Li, H., Olson, M., Lin, G., Hey, T., Tan, S.Y., Narva, K.E., 2013. *Bacillus thuringiensis* Cry34Ab1/Cry35Ab1 interactions with western corn rootworm midgut membrane binding sites. PLoS One 8, e53079.

Li, H.Y., Soares, M.A., Torres, M.S., Bergen, M., White, J.F., 2015. Endophytic bacterium, *Bacillus amyloliquefaciens*, enhances ornamental host resistance to diseases and insect pests. J. Plant Interact. 10, 224–229.

Li, J.D., Carroll, J., Ellar, D.J., 1991. Crystal structure of insecticidal delta-endotoxin from *Bacillus thuringiensis* at 2.5 A resolution. Nature 353, 815–821.

Liu, J.G., Yang, A.Z., Shen, X.H., Hua, B.G., Shi, G.L., 2011. Specific binding of activated Vip3Aa10 to *Helicoverpa armigera* brush border membrane vesicles results in pore formation. J. Invertebr. Pathol. 108, 92–97.

Lukasik, P., van Asch, M., Guo, H.F., Ferrari, J., Godfray, H.C.J., 2013. Unrelated facultative endosymbionts protect aphids against a fungal pathogen. Ecol. Lett. 16, 214–218.

Ma, G., Rahman, M.M., Grant, W., Schmidt, O., Asgari, S., 2012. Insect tolerance to the crystal toxins Cry1Ac and Cry2Ab is mediated by the binding of monomeric toxin to lipophorin glycolipids causing oligomerization and sequestration reactions. Dev. Comp. Immunol. 37, 184–192.

Marshall, S.D.G., Hares, M.C., Jones, S.A., Harper, L.A., Vernon, J.R., Harland, D.P., Jackson, T.A., Hurst, M.R.H., 2012. Histopathological effects of the Yen-Tc toxin complex from *Yersinia entomophaga* MH96 (Enterobacteriaceae) on the *Costelytra zealandica* (Coleoptera: Scarabaeidae) larval midgut. Appl. Environ. Microbiol. 78, 4835–4847.

Martin, P.A.W., Gundersen-Rindal, D., Blackburn, M., Buyer, J., 2007a. *Chromobacterium subtsugae* sp. nov., a betaproteobacterium toxic to Colorado potato beetle and other insect pests. Intern. J. Syst. Evol. Microbiol. 57, 993–999.

Martin, P.A.W., Hirose, E., Aldrich, J.R., 2007b. Toxicity of *Chromobacterium subtsugae* to southern green stink bug (Heteroptera: Pentatomidae) and corn rootworm (Coleoptera: Chrysomelidae). J. Econ. Entomol. 100, 680–684.

Masson, L., Schwab, G., Mazza, A., Brousseau, R., Potvin, L., Schwartz, J.L., 2004. A novel *Bacillus thuringiensis* (PS149B1) containing a Cry34Ab1/Cry35Ab1 binary toxin specific for the western corn rootworm *Diabrotica virgifera virgifera* LeConte forms ion channels in lipid membranes. Biochemistry 43, 12349–12357.

Mertz, F.P., Yao, R.C., 1990. *Saccharopolyspora spinosa* sp. nov. isolated from soil collected in a sugar mill rum still. Intern. J. Syst. Bacteriol. 40, 34–39.

Meusch, D., Gatsogiannis, C., Efremov, R.G., Lang, A.E., Hofnagel, O., Vetter, I.R., Aktories, K., Raunser, S., 2014. Mechanism of Tc toxin action revealed in molecular detail. Nature 508, 61–65.

Monnerat, R.G., Soares, C.M., Capdeville, G., Jones, G., Martins, E.S., Praca, L., Cordeiro, B.A., Braz, S.V., dos Santos, R.C., Berry, C., 2009. Translocation and insecticidal activity of *Bacillus thuringiensis* living inside of plants. Microb. Biotech. 2, 512–520.

Moretti, R., Calvitti, M., 2013. Male mating performance and cytoplasmic incompatibility in a wPip *Wolbachia* trans-infected line of *Aedes albopictus* (*Stegomyia albopicta*). Med. Vet. Entomol. 27, 377–386.

Muratoglu, H., Demirbag, Z., Sezen, K., 2011. An entomopathogenic bacterium, *Pseudomonas putida*, from *Leptinotarsa decemlineata*. Turk. J. Biol. 35, 275–282.

Musser, R.O., Kwon, H.S., Williams, S.A., White, C.J., Romano, M.A., Holt, S.M., Bradbury, S., Brown, J.K., Felton, G.W., 2005. Evidence that caterpillar labial saliva suppresses infectivity of potential bacterial pathogens. Arch. Insect Biochem. Physiol. 58, 138–144.

Nair, M.S., Dean, D.H., 2008. All domains of Cry1A toxins insert into insect brush border membranes. J. Biol. Chem. 283, 26324–26331.

Nakamura, L.K., 2000. Phylogeny of *Bacillus sphaericus*-like organisms. Intern. J. Syst. Evol. Microbiol. 50, 1715–1722.

Nielsen-Leroux, C., Charles, J.F., Thiery, I., Georghiou, G.P., 1995. Resistance in a laboratory population of *Culex quinquefasciatus* (Diptera: Culicidae) to *Bacillus sphaericus* binary toxin is due to a change in the receptor on midgut brush-border membranes. Eur. J. Biochem. 228, 206–210.

Nikoh, N., Hosokawa, T., Moriyama, M., Oshima, K., Hattori, M., Fukatsu, T., 2014. Evolutionary origin of insect-*Wolbachia* nutritional mutualism. Proc. Natl. Acad. Sci. U.S.A. 111, 10257–10262.

Nishiwaki, H., Nakashima, K., Ishida, C., Kawamura, T., Matsuda, K., 2007. Cloning, functional characterization, and mode of action of a novel insecticidal pore-forming toxin, sphaericolysin, produced by *Bacillus sphaericus*. Appl. Environ. Microbiol. 73, 3404–3411.

O'Callaghan, M., Garnham, M.L., Nelson, T.L., Baird, D., Jackson, T.A., 1996. The pathogenicity of *Serratia* strains to *Lucilia sericata* (Diptera: Calliphoridae). J. Invertebr. Pathol. 68, 22–27.

Oei, C., Hindley, J., Berry, C., 1992. Binding of purified *Bacillus sphaericus* binary toxin and its deletion derivatives to *Culex quinquefasciatus* gut: elucidation of functional binding domains. J. Gen. Microbiol. 138, 1515–1526.

Oppert, B., Kramer, K.J., Beeman, R.W., Johnson, D., McGaughey, W.H., 1997. Proteinase-mediated insect resistance to *Bacillus thuringiensis* toxins. J. Biol. Chem. 272, 23473–23476.

Oliveira, C.D., Tadei, W.P., Abdalla, F.C., 2009. Occurrence of apocrine secretion in the larval gut epithelial cells of *Aedes aegypti* L., *Anopheles albitarsis* Lynch-Arribalzaga and *Culex quinquefasciatus* Say (Diptera: Culicidae): a defense strategy against infection by *Bacillus sphaericus* Neide? Neotrop. Entomol. 38, 624–631.

Orlova, M.V., Smirnova, T.A., Ganushkina, L.A., Yacubovich, V.Y., Azizbekyan, R.R., 1998. Insecticidal activity of *Bacillus laterosporus*. Appl. Environ. Microbiol. 64, 2723–2725.

Ormskirk, M., Narciso, J., Glare, T., 2015. *Brevibacillus laterosporus* as potential endophyte in Brassica production system. In: 9th International Symposium on Fungal Endophytes of Grasses, Melbourne, Australia, pp. 58–60.

Pacheco, S., Gómez, I., Arenas, I., Saab-Rincon, G., Rodríguez-Almazán, C., Gill, S.S., Bravo, A., Soberón, M., 2009. Domain II loop 3 of *Bacillus thuringiensis* Cry1Ab toxin is involved in a "ping pong" binding mechanism with *Manduca sexta* aminopeptidase-N and cadherin receptors. J. Biol. Chem. 284, 32750–32757.

Pardo-López, L., Gómez, I., Rausell, C., Sanchez, J., Soberón, M., Bravo, A., 2006. Structural changes of the Cry1Ac oligomeric pre-pore from *Bacillus thuringiensis* induced by N-acetylgalactosamine facilitates toxin membrane insertion. Biochemistry 45, 10329–10336.

Pardo-Lopez, L., Soberon, M., Bravo, A., 2013. *Bacillus thuringiensis* insecticidal three-domain Cry toxins: mode of action, insect resistance and consequences for crop protection. FEMS Microbiol. Rev. 37, 3–22.

Park, H.W., Bideshi, D.K., Federici, B.A., 2010. Properties and applied use of the mosquitocidal bacterium, *Bacillus sphaericus*. J. Asia-Pacific Entomol. 13, 159–168.

Park, H.W., Mangum, C.M., Zhong, H.E., Hayes, S.R., 2007. Isolation of *Bacillus sphaericus* with improved efficacy against *Culex quinquefasciatus*. J. Am. Mosq. Control Assoc. 23, 478–480.

Petersen, L.M., Tisa, L.S., 2012. Influence of temperature on the physiology and virulence of the insect pathogen *Serratia* sp. strain SCBI. Appl. Environ. Microbiol. 78, 8840–8844.

Peyronnet, O., Vachon, V., Schwartz, J.L., Laprade, R., 2001. Ion channels induced in planar lipid bilayers by the *Bacillus thuringiensis* toxin Cry1Aa in the presence of gypsy moth (*Lymantria dispar*) brush border membrane. J. Membr. Biol. 184, 45–54.

Qureshi, N., Chawla, S., Likitvivatanavong, S., Lee, H.L., Gill, S.S., 2014. The Cry toxin operon of *Clostridium bifermentans* subsp. *malaysia* is highly toxic to *Aedes* larval mosquitoes. Appl. Environ. Microbiol. 80, 5689–5697.

Ramirez, J.L., Short, S.M., Bahia, A.C., Saraiva, R.G., Dong, Y.M., Kang, S., Tripathi, A., Mlambo, G., Dimopoulos, G., 2014. *Chromobacterium* Csp-P reduces malaria and dengue infection in vector mosquitoes and has entomopathogenic and *in vitro* anti-pathogen activities. PLoS Pathog 10.

Raymond, B., Johnston, P.R., Nielsen-LeRoux, C., Lereclus, D., Crickmore, N., 2010. *Bacillus thuringiensis*: an impotent pathogen? Trends Microbiol. 18, 189–194.

Raymond, B., West, S.A., Griffin, A.S., Bonsall, M.B., 2012. The dynamics of cooperative bacterial virulence in the field. Science 337, 85–88.

Raymond, B., Wright, D.J., Crickmore, N., Bonsall, M.B., 2013. The impact of strain diversity and mixed infections on the evolution of resistance to *Bacillus thuringiensis*. Proc. Roy. Soc. B Biol. Sci. 280, 9.

Rees, J.S., Jarrett, P., Ellar, D.J., 2009. Peritrophic membrane contribution to Bt Cry delta-endotoxin susceptibility in Lepidoptera and the effect of calcofluor. J. Invertebr. Pathol. 100, 139–146.

Rippere, K.E., Tran, M.T., Yousten, A.A., Hilu, K.H., Klein, M.G., 1998. *Bacillus popilliae* and *Bacillus lentimorbus*, bacteria causing milky disease in Japanese beetles and related scarab larvae. Int. J. Syst. Bacteriol. 48, 395–402.

Rivers, D.B., Vann, C.N., Zimmack, H.L., Dean, D.H., 1991. Mosquitocidal activity of *Bacillus laterosporus*. J. Invertebr. Pathol. 58, 444–447.

Roan, L.F., Crickmore, N., Peng, D.H., Sun, M., 2015. Are nematodes a missing link in the confounded ecology of the entomopathogen *Bacillus thuringiensis*? Trends Microbiol. 23, 341–346.

Rodriguez-Segura, Z., Chen, J., Villalobos, F.J., Gill, S., Eugenia Nunez-Valdez, M., 2012. The lipopolysaccharide biosynthesis core of the Mexican pathogenic strain *Serratia entomophila* is associated with toxicity to larvae of *Phyllophaga blanchardi*. J. Invertebr. Pathol. 110, 24–32.

Ruffner, B., Pechy-Tarr, M., Hofte, M., Bloemberg, G., Grunder, J., Keel, C., Maurhofer, M., 2015. Evolutionary patchwork of an insecticidal toxin shared between plant-associated pseudomonads and the insect pathogens *Photorhabdus* and *Xenorhabdus*. BMC Genomics 16, 14.

Ruiu, L., 2015. Insect pathogenic bacteria in integrated pest management. Insects 6, 352.

Ruiu, L., Delrio, G., Ellar, D.J., Floris, I., Paglietti, B., Rubino, S., Satta, A., 2006. Lethal and sublethal effects of *Brevibacillus laterosporus* on the housefly (*Musca domestica*). Entomol. Expert Appl. 118, 137–144.

Ruiu, L., Satta, A., Floris, I., 2013. Emerging entomopathogenic bacteria for insect pest management. Bull. Insectol. 66, 181–186.

Rybakova, D., Radjainia, M., Turner, A., Sen, A., Mitra, A.K., Hurst, M.R.H., 2013. Role of antifeeding prophage (Afp) protein Afp16 in terminating the length of the Afp tailocin and stabilizing its sheath. Mol. Microbiol. 89, 702–714.

Sanchis, V., 2011. From microbial sprays to insect-resistant transgenic plants: history of the biopesticide *Bacillus thuringiensis*. A review. Agron. Sustainable Dev. 31, 217–231.

Sattar, S., Maiti, M.K., 2011. Molecular characterization of a novel vegetative insecticidal protein from *Bacillus thuringiensis* effective against sap-sucking insect pest. J. Microbiol. Biotechnol. 21, 937–946.

Schirmer, J., Just, I., Aktories, K., 2002. The ADP-ribosylating mosquitocidal toxin from *Bacillus sphaericus*: proteolytic activation, enzyme activity, and cytotoxic effects. J. Biol. Chem. 277, 11941–11948.

Schnepf, H., Narva, K., Stockhoff, B., Lee, S., Walz, M., Sturgis, B., 2002. Pesticidal Toxins and Genes from *Bacillus Laterosporus* Strains. USA Patent No. US 2002/0120114A1.

Schroeder, J.M., Chamberlain, C., Davidson, E.W., 1989. Resistance to the *Bacillus sphaericus* toxin in cultured mosquito cells. In Vitro Cell. Dev. Biol. 25, 887–891.

Schwartz, J.L., Lu, Y.J., Soehnlein, P., Brousseau, R., Masson, L., Laprade, R., Adang, M.J., 1997. Ion channels formed in planar lipid bilayers by *Bacillus thuringiensis* toxins in the presence of *Manduca sexta* midgut receptors. FEBS Lett. 412, 270–276.

Sharma, V., Singh, P.K., Midha, S., Ranjan, M., Korpole, S., Patil, P.B., 2012. Genome sequence of *Brevibacillus laterosporus* strain GI-9. J. Bacteriol. 194, 1279.

Sharpe, E.S., Detroy, R.W., 1979. Fat body depletion, a debilitating result of milky disease in Japanese beetle larvae. J. Invertebr. Pathol. 34, 92–94.

Shi, Y.W., Zhang, X.B., Lou, K., 2013. Isolation, characterization, and insecticidal activity of an endophyte of drunken horse grass, *Achnatherum inebrians*. J. Insect Sci. 13, 12.

Shida, O., Takagi, H., Kadowaki, K., Komagata, K., 1996. Proposal for two new genera, *Brevibacillus* gen. nov. and *Aneurinibacillus* gen. nov. Intern. J. Syst. Bacteriol. 46, 939–946.

Siegwart, M., Graillot, B., Lopez, C.B., Besse, S., Bardin, M., Nicot, P.C., Lopez-Ferber, M., 2015. Resistance to bio-insecticides or how to enhance their sustainability: a review. Front. Plant Sci. 6.

Silapanuntakul, S., Pantuwatana, S., Bhumiratana, A., Charoensiri, K., 1983. The comparative persistence and toxicity of *Bacillus sphaericus* strain 1593 and *Bacillus thuringiensis* serotype H-14 against mosquito larvae in different kinds of environments. J. Invertebr. Pathol. 42, 387–392.

Silva-Filha, M.H., Nielsen-Leroux, C., Charles, J.F., 1999. Identification of the receptor for *Bacillus sphaericus* crystal toxin in the brush border membrane of the mosquito *Culex pipiens* (Diptera: Culicidae). Insect Biochem. Mol. Biol. 29, 711–721.

Silva-Filha, M.H.N.L., Berry, B., Regis, L., 2014. *Lysinibacillus sphaericus*: toxins and mode of action, applications for mosquito control and resistance management. Adv. Insect Physiol. 47, 89–176.

Singh, G., Sachdev, B., Sharma, N., Seth, R., Bhatnagar, R.K., 2010. Interaction of *Bacillus thuringiensis* vegetative insecticidal protein with ribosomal S2 protein triggers larvicidal activity in *Spodoptera frugiperda*. Appl. Environ. Microbiol. 76, 7202–7209.

Singh, G.J.P., Gill, S.S., 1988. An electron microscope study of the toxic action of *Bacillus sphaericus* in *Culex quinquefasciatus* larvae. J. Invertebr. Pathol. 52, 237–247.

Singkhamanan, K., Promdonkoy, B., Chaisri, U., Boonserm, P., 2010. Identification of amino acids required for receptor binding and toxicity of the *Bacillus sphaericus* binary toxin. FEMS Microbiol. Lett. 303, 84–91.

Sivakumar, S., Rajagopal, R., Venkatesh, G.R., Srivastava, A., Bhatnagar, R.K., 2007. Knockdown of aminopeptidase-N from *Helicoverpa armigera* larvae and in transfected Sf21 cells by RNA interference reveals its functional interaction with *Bacillus thuringiensis* insecticidal protein Cry1Ac. J. Biol. Chem. 282, 7312–7319.

Smirnova, T.A., Shevlyagina, N.V., Nikolaenko, M.A., Sorokin, V.V., Zubasheva, M.V., Azizbekyan, R.R., 2011. Characterization of *Brevibacillus laterosporus* spores and crystals. Biotekhnologiya 29–37.

Smith, R.A., Couche, G.A., 1991. The phylloplane as a source of *Bacillus thuringiensis* variants. Appl. Environ. Microbiol. 57, 311–315.

Sparks, T.C., Thompson, G.D., Kirst, H.A., Hertlein, M.B., Larson, L.L., Worden, T.V., Thibault, S.T., 1998. Biological activity of the spinosyns, new fermentation derived insect control agents, on tobacco budworm (Lepidoptera: Noctuidae) larvae. J. Econ. Entomol. 91, 1277–1283.

Splittstoesser, C.M., Tashiro, H., Lin, S.L., Steinkraus, K.H., Fiori, B.J., 1973. Histopathology of the European chafer *Amphimallon majalis* infected with *Bacillus popilliae*. J. Invertebr. Pathol. 22, 161–167.

Srisucharitpanit, K., Yao, M., Promdonkoy, B., Chimnaronk, S., Tanaka, I., Boonserm, P., 2014. Crystal structure of BinB: a receptor binding component of the binary toxin from *Lysinibacillus sphaericus*. Proteins 82, 2703–2712.

Stamm, M.D., Prochaska, T.J., Matz, N.A., Baxendale, R.W., 2013. Efficacy of *Chromobacterium subtsugae* for control of southern masked chafers. Arthrop. Manage.Test 38.

Stavrinides, J., McCloskey, J.K., Ochman, H., 2009. Pea aphid as both host and vector for the phytopathogenic bacterium *Pseudomonas syringae*. Appl. Environ. Microbiol. 75, 2230–2235.

Steinhaus, E.A., 1951. Possible use of *Bacillus thuringiensis* Berliner as an aid in the biological control of the alfalfa caterpillar. Hilgardia 20, 359–381.

Sturtevant, A.P., Revell, I.L., 1953. Reduction of *Bacillus larvae* spores in liquid food of honey bees by action of the honey stopper, and its relation to the development of American foulbrood. J. Econ. Entomol. 46, 855–860.

Sulakvelidze, A., 2000. Yersiniae other than *Y. enterocolitica*, *Y. pseudotuberculosis*, and *Y. pestis*: the ignored species. Microbes Infect. 2, 497–513.

Sundh, I., Goettel, M.S., 2013. Regulating biocontrol agents: a historical perspective and a critical examination comparing microbial and macrobial agents. Biocontrol 58, 575–593.

Tabashnik, B.E., Brevault, T., Carriere, Y., 2013. Insect resistance to Bt crops: lessons from the first billion acres. Nat. Biotechnol. 31, 510–521.

Takiguchi, Y., Mishima, H., Okuda, M., Terao, M., 1980. Milbemycins, a new family of macrolide antibiotics – fermentation, isolation and physicochemical properties. J. Antibiot. 33, 1120–1127.

Tanada, Y., Kaya, H.K., 1993. Insect Pathology. Springer. 666 pp.

Tanaka, S., Miyamoto, K., Noda, H., Jurat-Fuentes, J.L., Yoshizawa, Y., Endo, H., Sato, R., 2013. The ATP-binding cassette transporter subfamily C member 2 in *Bombyx mori* larvae is a functional receptor for Cry toxins from *Bacillus thuringiensis*. FEBS J. 280, 1782–1794.

Tao, A.L., Pang, F.H., Huang, S.L., Yu, G.M., Li, B., Wang, T., 2014. Characterisation of endophytic *Bacillus thuringiensis* strains isolated from wheat plants as biocontrol agents against wheat flag smut. Biocontrol Sci. Technol. 24, 901–924.

Thamthiankul, S., Suan-Ngay, S., Tantimavanich, S., Panbangred, W., 2001. Chitinase from *Bacillus thuringiensis* subsp. *pakistani*. Appl. Microbiol. Biotechnol. 56, 395–401.

Thanabalu, T., Berry, C., Hindley, J., 1993. Cytotoxicity and ADP-ribosylating activity of the mosquitocidal toxin from *Bacillus sphaericus* SSII-1: possible roles of the 27- and 70-kilodalton peptides. J. Bacteriol. 175, 2314–2320.

Thiéry, I., Hamon, S., Cosmao Dumanoir, V., de Barjac, H., 1992. Vertebrate safety of *Clostridium bifermentans* serovar *malaysia*, a new larvicidal agent for vector control. J. Econ. Entomol. 85, 1618–1623.

Townsend, R.J., Ferguson, C.M., Proffitt, J., Slay, M.W.A., Swaminathan, J., Day, S., Gerard, E., O'Callaghan, M., Johnson, V.W., Jackson, T.A., 2004. Establishment of *Serratia entomophila* after application of a new formulation for grass grub control. N.Z. Plant Prot. 57, 310–313.

Trought, T.E.T., Jackson, T.A., French, R.A., 1982. Incidence and transmission of a disease of grass grub (*Costelytra zealandica*) in Canterbury. N.Z. J. Exp. Agric. 10, 79–82.

Tsuda, Y., Nakatani, F., Hashimoto, K., Ikawa, S., Matsuura, C., Fukada, T., Sugimoto, K., Himeno, M., 2003. Cytotoxic activity of *Bacillus thuringiensis* Cry proteins on mammalian cells transfected with cadherin-like Cry receptor gene of *Bombyx mori* (silkworm). Biochem. J. 369, 697–703.

Uehlinger, S., Schwager, S., Bernier, S.P., Riedel, K., Nguyen, D.T., Sokol, P.A., Eberl, L., 2009. Identification of specific and universal virulence factors in *Burkholderia cenocepacia* strains by using multiple infection hosts. Infect. Immun. 77, 4102–4110.

Vachon, V., Laprade, R., Schwartz, J.L., 2012. Current models of the mode of action of *Bacillus thuringiensis* insecticidal crystal proteins: a critical review. J. Invertebr. Pathol. 111, 1–12.

van Frankenhuyzen, K., 2009. Insecticidal activity of *Bacillus thuringiensis* crystal proteins. J. Invertebr. Pathol. 101, 1–16.

Varani, A.M., Lemos, M.V., Fernandes, C.C., Lemos, E.G., Alves, E.C., Desiderio, J.A., 2013. Draft genome sequence of *Bacillus thuringiensis* var. *thuringiensis* strain T01–328, a Brazilian isolate that produces a soluble pesticide protein, Cry1Ia. Genome Announce. 1 (5).

Vidal-Quist, J.C., Rogers, H.J., Mahenthiralingam, E., Berry, C., 2013. *Bacillus thuringiensis* colonises plant roots in a phylogeny-dependent manner. FEMS Microbiol. Ecol. 86, 474–489.

Vijayabharathi, R., Kumari, B.R., Sathya, A., Srinivas, V., Abhishek, R., Sharma, H.C., Gopalakrishnan, S., 2014. Biological activity of entomopathogenic actinomycetes against lepidopteran insects (Noctuidae: Lepidoptera). Can. J. Plant Sci. 94, 759–769.

Villaverde, J.J., Sevilla-Moran, B., Sandin-Espana, P., Lopez-Goti, C., Alonso-Prados, J.L., 2014. Biopesticides in the framework of the European pesticide regulation (EC) No. 1107/2009. Pest Manage. Sci. 70, 2–5.

Visnovsky, G.A., Smalley, D.J., O'Callaghan, M., Jackson, T.A., 2008. Influence of culture medium composition, dissolved oxygen concentration and harvesting time on the production of *Serratia entomophila*, a microbial control agent of the New Zealand grass grub. Biocontrol Sci. Technol. 18, 87–100.

Vodovar, N., Vallenet, D., Cruveiller, S., Rouy, Z., Barbe, V., Acosta, C., Cattolico, L., Jubin, C., Lajus, A., Segurens, B., Vacherie, B., Wincker, P., Weissenbach, J., Lemaitre, B., Medigue, C., Boccard, F., 2006. Complete genome sequence of the entomopathogenic and metabolically versatile soil bacterium *Pseudomonas entomophila*. Nat. Biotechnol. 24, 673–679.

Vodovar, N., Vinals, M., Liehl, P., Basset, A., Degrouard, J., Spellman, P., Boccard, F., Lemaitre, B., 2005. Drosophila host defense after oral infection by an entomopathogenic *Pseudomonas* species. Proc. Natl. Acad. Sci. U.S.A. 102, 11414–11419.

Walters, F.S., Stacy, C.M., Lee, M.K., Palekar, N., Chen, J.S., 2008. An engineered chymotrypsin/cathepsin G site in domain I renders *Bacillus thuringiensis* Cry3A active against Western corn rootworm larvae. Appl. Environ. Microbiol. 74, 367–374.

Wang, H.Q., Yang, X.F., Guo, L.H., Zeng, H.M., Qiu, D.W., 2015. PeBL1, a novel protein elicitor from *Brevibacillus laterosporus* Strain A60, activates defense responses and systemic resistance in *Nicotiana benthamiana*. Appl. Environ. Microbiol. 81, 2706–2716.

Wang, X.J., Zhang, J., Wang, J.D., Huang, S.X., Chen, Y.H., Liu, C.X., Xiang, W.S., 2011. Four new doramectin congeners with acaricidal and insecticidal activity from *Streptomyces avermitilis* NEAU1069. Chem. Biodivers. 8, 2117–2125.

Waterfield, N.R., Daborn, P.J., Ffrench-Constant, R.H., 2004. Insect pathogenicity islands in the insect pathogenic bacterium *Photorhabdus*. Physiol. Entomol. 29, 240–250.

Wirth, M.C., Berry, C., Walton, W.E., Federici, B.A., 2014. Mtx toxins from *Lysinibacillus sphaericus* enhance mosquitocidal cry-toxin activity and suppress cry-resistance in *Culex quinquefasciatus*. J. Invertebr. Pathol. 115, 62–67.

Wiwat, C., Thaithanun, S., Pantuwatana, S., Bhumiratana, A., 2000. Toxicity of chitinase-producing *Bacillus thuringiensis* ssp. *kurstaki* HD-1 (G) toward *Plutella xylostella*. J. Invertebr. Pathol. 76, 270–277.

Wolfersberger, M.G., 1990. The toxicity of two *Bacillus thuringiensis* delta-endotoxins to gypsy moth larvae is inversely related to the affinity of binding sites on midgut brush border membranes for the toxins. Experientia 46, 475–477.

Xu, C., Chinte, U., Chen, L., Yao, Q., Meng, Y., Zhou, D., Bi, L.J., Rose, J., Adang, M.J., Wang, B.C., Yu, Z., Sun, M., 2015. Crystal structure of Cry51Aa1: a potential novel insecticidal aerolysin-type β-pore-forming toxin from *Bacillus thuringiensis*. Biochem. Biophys. Res. Commun. 462, 184–189.

Yoshiyama, M., Kimura, K., 2009. Bacteria in the gut of Japanese honeybee, *Apis cerana japonica*, and their antagonistic effect against *Paenibacillus larvae*, the causal agent of American foulbrood. J. Invertbr. Pathol. 102, 91–96.

Young, S.D., Townsend, R.J., O'Callaghan, M., 2009. Bacterial entomopathogens improve cereal establishment in the presence of grass grub larvae. N.Z. Plant Prot. 62, 1–6.

Young, S.D., Townsend, R.J., Swaminathan, J., O'Callaghan, M., 2010. *Serratia entomophila*-coated seed to improve ryegrass establishment in the presence of grass grubs. N.Z. Plant Prot. 63, 229–234.

Yousten, A.A., Fretz, S.B., Jelley, S.A., 1985. Selective medium for mosquito-pathogenic strains of *Bacillus sphaericus*. Appl. Environ. Microbiol. 49, 1532–1533.

Yousten, A.A., Genthner, F.J., Benfield, E.F., 1992. Fate of *Bacillus sphaericus* and *Bacillus thuringiensis* serovar *israelensis* in the aquatic environment. J. Am. Mosq. Control Assoc. 8, 143–148.

Yuan, Z., Neilsen-LeRoux, C., Pasteur, N., Delecluse, A., Charles, J.F., Frutos, R., 1999. Cloning and expression of the binary toxin genes of *Bacillus sphaericus* C3-41 in a crystal minus *B. thuringiensis* subsp. *israelensis*. Acta Microbiol. Sin. 39, 29–35.

Zhang, C.-X., Yang, S.-Y., Xu, M.-X., Sun, J., Liu, H., Liu, J.-R., Liu, H., Kan, F., Sun, J., Lai, R., Zhang, K.-Y., 2009. *Serratia nematodiphila* sp. nov., associated symbiotically with the entomopathogenic nematode *Heterorhabditidoides chongmingensis* (Rhabditida: Rhabditidae). Intern. J. Syst. Evol. Microbiol. 59, 2646.

Zhang, J., Hodgman, T.C., Krieger, L., Schnetter, W., Schairer, H.U., 1997. Cloning and analysis of the first cry gene from *Bacillus popilliae*. J. Bacteriol. 179, 4336–4341.

Zhang, J.B., Schairer, H.U., Schnetter, W., Lereclus, D., Agaisse, H., 1998. *Bacillus popilliae* cry18Aa operon is transcribed by sigma(E) and sigma(K) forms of RNA polymerase from a single initiation site. Nucleic Acids Res. 26, 1288–1293.

Zhang, X., Griko, N.B., Corona, S.K., Bulla, L.A., 2008. Enhanced exocytosis of the receptor BT-R-1 induced by the Cry1Ab toxin of *Bacillus thuringiensis* directly correlates to the execution of cell death. Comp. Biochem. Physiol. B Biochem. Mol. Biol. 149, 581–588.

Zhou, L.Q., Slamti, L., Nielsen-LeRoux, C., Lereclus, D., Raymond, B., 2014. The social biology of quorum sensing in a naturalistic host pathogen system. Curr. Biol. 24, 2417–2422.

Zhuang, M., Oltean, D.I., Gómez, I., Pullikuth, A.K., Soberón, M., Bravo, A., Gill, S.S., 2002. *Heliothis virescens* and *Manduca sexta* lipid rafts are involved in Cry1A toxin binding to the midgut epithelium and subsequent pore formation. J. Biol. Chem. 277, 13863–13872.

Zubasheva, M.V., Ganushkina, L.A., Smirnova, T.A., Azizbekyan, R.R., 2010. Larvicidal activity of crystal-forming strains of *Brevibacillus laterosporus*. Appl. Biochem. Microbiol. 46, 755–762.

Chapter 5

Basic and Applied Research on Entomopathogenic Fungi

D. Chandler

University of Warwick, Wellesbourne, Warwick, United Kingdom

5.1 INTRODUCTION

In this chapter, the biology of entomopathogenic fungi (EPF) is described and a brief introduction is given to their use in pest management. The Fungi are a kingdom of heterotrophic eukaryotes characterized by the possession of cell walls containing chitin and glucan. They are typically filamentous and multicellular, with simple morphologies constructed from tube-like hyphae that give rise to modular forms of indeterminate growth, although some species have a unicellular vegetative form. Nutrition is obtained via the secretion of degradative enzymes into the external environment and reabsorption of breakdown products into fungal cells. The EPF include any member of the fungal kingdom that causes infection in an insect or other arthropod leading to an observable disease (the term "entomo" is used *sensu lato* (*s.l.*) to encompass insects and other terrestrial arthropods, because some fungal species pathogenic to insects can also infect mites, ticks, and spiders). We are concerned specifically with parasitic EPF, that is, fungi that have evolved symbiotic, nonmutualistic relationships with arthropods that are detrimental to the fitness of the arthropod host and are beneficial to the fitness of the fungus. Nonparasitic fungi that occasionally cause opportunistic infections in arthropods are not included. Most of the chapter is given to the biology of the two main fungal groups that are used for biological pest control: the Hypocreales and the Entomophthoromycota.

5.2 ENTOMOPATHOGENIC FUNGI: EVOLUTIONARY RELATIONSHIPS

Entomopathogenicity has evolved independently in at least seven fungal phyla/subphyla, and there are estimated to be over 700 EPF species (Roberts and Humber, 1981). This represents less than 1% of the total number of described fungal species (McLaughlin et al., 2009). However, while entomopathogenicity itself is a rare life-history trait in the fungal kingdom, it is arguably a successful one as EPF are abundant in terrestrial environments and the majority of insect and arachnid species are, in principle, susceptible to fungal infections. As a consequence, EPF are important drivers of evolution in arthropods via the host–pathogen "arms race" in which the development of virulence-related traits in fungal pathogens induces the development of defensive countermeasures in arthropods over evolutionary time scales (Pedrini et al., 2015).

For the basal fungi, EPF are found in three main groups: the Entomophthoromycota, the Blastocladiomycota, and the Microsporidia (Humber, 2008). The largest number of EPF species in the basal fungi occurs in the phylum Entomophthoromycota. Most species within this group are obligate entomopathogens, occurring in five of the six known families (Humber, 2008). These fungi are often highly specialized parasites that show distinct ecomorphological adaptations to the life cycles of their hosts, including the production of conidia (spores produced from asexual reproduction) discharged actively from host cadavers and timed to take advantage of favorable environmental conditions, manipulation of host behavior, and the production of resting spores to survive adverse conditions. The phylum Blastocladiomycota contains the dipteran-pathogenic genera *Coelomomyces* and *Coelomycidium*. *Coelomomyces* contains over 70 species that are pathogens of a range of dipteran families, including Culicidae, Psychodidae and Chironomidae (Scholte et al., 2004). These fungi have a complex life cycle involving intermediate copepod hosts and cause natural epizootics in larval populations of aquatic dipterans. Although they have been investigated as biological control agents of mosquito larvae, infection rates were variable, which was attributed partly to the challenges of mass production (Scholte et al., 2004; Bukhari et al., 2013). The Microsporidia are ubiquitous, obligate intracellular pathogens of vertebrates and invertebrates that have lost many metabolic functions and show significant genome size reduction as an adaptation to their parasitic life cycles (Corradi, 2015). Previously considered to be protozoans, but now known to be fungi, some are pathogens of beneficial arthropods, for example *Nosema ceranae*, an invasive pathogen of honeybees (Higes et al., 2008). *Paranosema locustae* has potential as a control agent of locusts and grasshoppers (Guo et al., 2012) although mass production is difficult and has prevented its widespread use (Jenkins and Goettel, 1997).

Within the higher fungi (Dikarya), the majority of EPF are located in the order Hypocreales of the class Sordariomycetes (phylum Ascomycota, subphylum Pezizomycotina). This order contains morphologically distinct sexual and asexual EPF forms that until recently were believed to be evolutionarily separate taxa. Entomoparasites in the Ascomycota are also located within the Laboulbeniomycetes (obligate biotrophic parasites that colonise insect cuticle without normally causing death or other significant disease effects (Weir and Blackwell, 2001)) and within the genus *Ascosphaera* (Pezizomycotina, Eurotiales), which are pathogens of bees. Finally, there are some members of the Septobasidiales (Basidiomycota, Pucciniomycotina) that are parasites of scale insects (Henk and Vilgalys, 2007).

5.3 BIOLOGY OF THE TWO MAIN GROUPS OF ENTOMOPATHOGENIC FUNGI USED AS BIOLOGICAL CONTROL AGENTS OF ARTHROPOD PESTS: THE HYPOCREALES AND THE ENTOMOPHTHOROMYCOTA

The entomopathogenic Hypocreales and the Entomophthoromycota infect arthropods by direct penetration of the host integument, using spores that attach to the epicuticle, germinate on it, and then are able to grow through the procuticle (which presents a significant physical and chemical barrier to infection) using a combination of mechanical pressure from hyphal tips and the secretion of proteolytic enzymes. This method of topical infection is common among filamentous fungal pathogens of plants and animals, and they are probably predisposed to infecting hosts with a rigid external surface (this stands in contrast to bacterial and viral entomopathogens, which infect predominantly *per os*). EPF in both the Hypocreales and the Entomophthoromycota generally exhibit high virulence (mortality rates of up to 100% are not uncommon) associated with high intrahost growth rates and transmission to new susceptible hosts. In addition to their ability to overcome the immune defenses of their arthropod hosts, some species are also able to manipulate host behavior, which is thought to be an adaptation to increasing the probability of transmission to new susceptible hosts. Traditionally, these two groups of fungi have been used in different ways for pest management. The anamorphic forms of hypocrealean EPF are often suitable for industrial-scale production, and they are used as inundative microbial biopesticides, in which large amounts of fungal biomass—usually in the

form of conidia—are applied to the target pest population. The conidia are formulated as sprays, dusts, or granules and applied to soil, foliage, or (in the case of insect vector control such as mosquitoes) surfaces in domestic settings such as bed-nets. The Entomophthoromycota are often difficult to mass produce and so are not used to any significant extent as biopesticides, but they are important because of their capacity to develop epizootics in insect and mite populations. They have been studied as components of conservation biological control, with the aim of conserving or enhancing natural fungal populations in the landscape (Steinkraus, 2007).

The largest number of EPF species is found within the Hypocreales. This order is ecologically diverse and contains saproptrophs, entomopathogens, pathogens of nematodes and other microinvertebrates, plant pathogens, endophytes, yeast-like symbionts, and mycoparasites. A multigene phylogenetic analysis and ancestral character state reconstruction of the Hypocreales has enabled us to understand the evolutionary origins of the different nutritional modes that exist within the order (Vega et al., 2009). The Hypocreales originated from ancestors that functioned as plant parasites and its species have evolved subsequently through interkingdom host jumping. A minimum of five to eight unidirectional interkingdom host jumps have occurred within the order: three to five to fungi, one or two to animals, and one to plants (Spatafora et al., 2007). The most diverse genus of entomopathogens in the Hypocreales is *Cordyceps s.l.*, which is estimated to have more than 400 species. These species were previously assigned to the family Clavicipitaceae. The previous classification was based on morphological criteria, such as the possession of stalked stromata in entomopathogenic taxa, which have subsequently been shown by DNA-based phylogenetic analysis to be the result of convergent evolution. In the latest taxonomic revision, based on a multilocus phylogeny (Sung et al., 2007), *Cordyceps* spp. have been shown to be distributed among three monophyletic families: the Clavicipitaceae *sensu stricto* (*s.s.*), and two new families—the Cordycipitaceae and the Ophiocordycipitaceae. The Clavicipitaceae and Ophiocordycipitaceae are sister taxa that have evolved from a common lineage of entomopathogens. The Cordycipitaceae are all entomopathogenic, while the Clavicipitaceae *s.s.* and the Ophiocordycipitaceae contain taxa that can obtain their nutrition from animals, plants, or fungi. These three families contain fungal species that exist in either or both teleomorphic and anamorphic states. These states are phenotypically very different from each other. Teleomorphs are sexually reproducing forms characterized by the production of ascospores in cylindrical asci borne on conspicuous stromata, while anamorphs are morphologically simple filamentous fungi that reproduce asexually by the production of conidia on hyphae. Currently, the teleomorphs and anamorphs have different scientific names (ie, an anamorph and a teleomorph that belong to the same genetic/evolutionary lineage are nonetheless assigned to different genera), reflecting the fact that until recently they were considered to be evolutionarily separate entities. However, as the anamorph–teleomorph connections are resolved, the scientific names will be unified as appropriate (eg, Kepler et al., 2014). The main entomopathogenic genera in the three hypocrealean families are as follows (Humber, 2008):

> Clavicipitaceae. Entomopathogenic teleomorphs: *Hypocrella, Metacordyceps, Regiocrella, Torrubiella*. Entomopathogenic anamorphs: *Aschersonia, Metarhizium, Nomuraea*, some *Paecilomyces*-like fungi excluded from *Isaria s.s.*
> Cordycipitaceae. Entomopathogenic teleomorphs: *Cordyceps, Torrubiella*. Entomopathogenic anamorphs: *Beauveria, Microhilum, Engyodentium, Isaria, Mariannaea*-like species, *Lecanicillium, Simplicillium*.
> Ophiocordycipitaceae. Entomopathogenic teleomorphs: *Ophiocordyceps, Elaphocordyceps*. Entomopathogenic anamorphs: *Haptocillium, Harposporium, Hirsutella, Hymenostilbe*, some *Paecilomyces*-like species, *Paraisaria, Sorosporella, Syngliocladium, Tolypocladium*.

The phylogenies of the most agronomically important genera in the Hypocreales are now being resolved. *Beauveria* has been shown to be monophyletic, and a multilocus phylogeny has resolved the genus into 10 species (Rehner et al., 2011). *Beauveria bassiana s.l.*, one of the most widely used EPF species for biological control (Faria and Wraight, 2007), is really a species complex composed of an undetermined number of cryptic lineages including *Beauveria bassiana s.s.* Different species within the complex occur together in farmed and wild habitats to form multispecies assemblages although some lineages show a preference for particular microhabitats within the landscape (see later) (Rehner and Buckley, 2005; Meyling et al., 2009; Rehner et al., 2011). Similarly, the *Metarhizium anisopliae* spp. complex has been resolved into nine separate species based on multilocus phylogeny and including *M. anisopliae s.s.* (Bischoff et al., 2009).

The Entomophthoromycota consists of over 250 species, most of which are obligate entomopathogens, although the phylum also contains a number of species that are saprotrophs or facultative entomopathogens, pathogens of soil-dwelling nematodes and tardigrades, pathogens of algae or ferns, facultative pathogens of vertebrates, or commensals resident in the guts of reptiles and amphibians (Gryganskyi et al., 2013a). The latest multigene phylogenies have resolved the phylum into three classes and six families, with the entomopathogenic genera within them being as follows (Humber, 2012; Gryganskyi et al., 2012, 2013a):

Class Basidiobolomycetes: order Basidiobolales; family Basidiobolaceae; *Basidiobolus*
Class Neozygitomycetes: order Neozygitales; family Neozygitaceae; *Apterivorax, Neozygites, Thaxterosporium*
Class Entomophthoromycetes: order Entomophthorales; (1) family Ancylistaceae; *Conidiobolus*; (2) family Entomophthoraceae; (a) subfamily Erynioideae: *Erynia, Eryniopsis* (in part), *Furia, Orthomyces, Pandora, Strongwellsea, Zoophthora*; (b) subfamily Entomophthoroideae: *Batkoa, Entomophaga, Entomophthora, Eryniopsis* (in part), *Massospora*

5.4 ENTOMOPATHOGENIC FUNGI: ECOLOGY AND HOST–PATHOGEN INTERACTIONS

The natural occurrence of EPF depends on the availability and density of hosts, the presence of alternative sources of nutrition, habitat type, and environmental conditions. The EPF in the Entomophthoromycota are observed as infections in single individuals or as spectacular epizootics in host arthropod populations, generally in epigeal habitats. The Hypocreales contain some entomopathogenic genera that are mainly associated with above-ground insects; these include *Lecanicillium*, which is often associated with hemipteran insects (eg, scales, whiteflies, and aphids) although some strains have a broader host range (Zare and Gams, 2001) and *Aschersonia* (teleomorph *Hypocrella*), which is a specialist pathogen of whiteflies and scale insects (Liu et al., 2006). The teleomorphic hypocrealean species tend to be found in habitats free of human disturbance, with the main centers for their occurrence being tropical and subtropical areas, particularly in East and Southeast Asia (Sung et al., 2007). The anamorphic hypocrealean genera *Beauveria, Isaria*, and *Metarhizium* are ubiquitous components of the soil microbiota in temperate regions including agroecosystems and forests and woodlands (Chandler et al., 1997; Keller et al., 2003). Soil is a favorable environment because it is occupied by large numbers of potential arthropod hosts, while free carbon from root exudates is also present as a potential source of energy for growth and survival (St Leger, 2008). Soil itself confers protection to fungal mycelium and spores against damage from UV radiation and extremes of heat and cold, and water availability is generally not limiting to fungal growth and infection (Vega et al., 2009). Population levels of these fungi are affected by agricultural practices including tillage and pesticide use (Mietkiewski et al., 1997; Hummel et al., 2002), and the distribution of particular species and subspecies clades is influenced strongly by habitat type (Bidochka et al., 2001, 2002; Meyling et al., 2009). The relationship between habitat disturbance and occurrence of sexual versus asexual reproductive modes in the Hypocreales may be associated with the way in which spore release takes place in these EPF. Teleomorphs produce actively discharged, sexually produced ascospores from stromata on the host cadaver. Stromata may persist on the host cadaver for many months and hence are vulnerable to disturbance (Boomsma et al., 2014). In contrast, the anamorphs produce very large numbers of asexually generated haploid conidia that are dispersed passively from the insect cadaver for a relatively short period, often falling into the soil. Similar to the hypocrealean teleomorphs, the Entomophthoromycota actively discharge conidia from infected cadavers, but the period of production is short and is often complete within 24–36h of host death (Mullens and Rodriguez, 1985; Glare and Milner, 1991).

Some EPF show clear specialization for parasitism of particular arthropod groups, while others are generalists. While some EPF in the Entomophthoromycota are generalist or opportunist pathogens, most are specialists of particular groups of arthropod hosts. For example, *Entomophthora muscae s.l.* is a species complex that specializes on adult dipterans, with individual subspecies/species within the complex varying in their host specificities (Gryganskyi et al., 2013b); *Pandora neoaphidis* is a specialist parasite of aphids and is frequently found causing epizootics on aphid pests of crops (Scorsetti et al., 2010); and *Neozygites floridana* is a specialist of tetranychid mites (Chandler et al., 2000). Most of the teleomorphic hypocrealean EPF function as ecologically obligate parasites of arthropods and have preferences for particular arthropod orders although some show a greater level of specialization and have host ranges restricted to individual arthropod families, genera, or even at species level (Kepler et al., 2012). The host ranges of the anamorphic hypocrealean EPF vary according to the particular fungal genus/species. Some genera are specialists (eg, *Lecanicillium*, *Aschersonia*, see earlier), while other genera (eg, *Metarhizium*) represent a continuum, and contain species that are specialists, generalists, or which have intermediate host ranges. A recent genomic analysis of seven species within the genus *Metarhizium* found that generalist species (*M. anisopliae*, *Metarhizium brunneum*, *Metarhizium robertsii*) have evolved from specialists via transitional species with intermediate host ranges (Hu et al., 2014). The evolution of the generalist species was associated with reproductive isolation (ie, loss of ability to recombine sexually), horizontal gene transfer from bacteria, and expansion of protein families associated with fungal virulence (Hu et al., 2014). These generalist species also exhibit phenotypic plasticity; they have the ability to utilize plants for growth and survival, for example through endophytic colonization of plant tissue and growth in the rhizosphere. In a study from Canada, *M. robertsii* was preferentially associated with forested habitats and had an ability for cold active growth (8°C), while *M. brunneum* was associated more with agricultural land, was able to grow at 37°C and was more resistant to UV radiation (Bidochka et al., 2001).

M. robertsii also appeared to operate more as an endophyte, while *M. brunneum* functioned more as a rhizosphere colonizer (Bruck, 2005; Behie et al., 2015). The genus *Beauveria* also exhibits a distinct relationship between phylogeny and habitat type: in a study of different genetic groups of *Beauveria* in soils collected from contrasting hedgerow and tilled field habitats, seven distinct phylogenetic groups were obtained from hedgerows, while only one group (*B. bassiana s.s.*) was associated with tilled soils (Meyling et al., 2009).

5.5 ENTOMOPATHOGENIC FUNGI LIFE CYCLES

EPF in the Hypocreales and the Entomophthoromycota face an identical set of challenges to successfully complete their life cycles. They must first come into contact with a susceptible host, adhere to the host cuticle, and then penetrate it. They must then overcome and/or avoid host immune defenses to obtain nutrition and proliferate. Finally, they must be transmitted to new hosts. The infection process itself is a co-evolved interaction between the fungal pathogen and its host in which both partners respond to natural selection to maximize their respective fitness. There is natural selection on the fungus for traits that result in better exploitation of the host, while selection will operate on the host to favor traits that prevent or avoid infection (Roy et al., 2006). Historically, much of the research on fungal infections has been about identifying and describing the different stages that constitute the fungal infection cycle rather than addressing specific hypotheses about host pathogen co-evolution (Boomsma et al., 2014). The most-detailed studies have been done with the anamorphic hypocrealean genera that are used for biological pest control, especially *Beauveria* and *Metarhizium*. Our knowledge of the mechanisms underpinning the interaction between the pathogen and its host still has many unanswered questions, particularly with respect to how the host immune system responds to infection and how the fungus has evolved to overcome host immune defenses. Fortunately, there are good opportunities to translate detailed knowledge about the molecular biology of the immune system in the model organism *Drosophila melanogaster* into studies involving EPF infections of *Drosophila* and other arthropod hosts (Gottar et al., 2006; Dong et al., 2012; Ouyang et al., 2015; Paparazzo et al., 2015).

5.5.1 Infection of Arthropods: Attachment to Hosts and Penetration of the Cuticle

Infection starts with fungal spores (conidia or ascospores) contacting, and then adhering to, the cuticle of a susceptible host. The anamorphic hypocrealean fungi generally require a high threshold number of conidia necessary to cause infection that, measured in terms of conidia concentration in a liquid suspension applied to the host, ranges from 10^2 to $10^9 \, \text{mL}^{-1}$ (Hesketh et al., 2010). In contrast, the EPF in the Entomophthoromycota tend to cause infection at lower conidia dosage, which could be as low as the equivalent of one or two conidia per host in some cases (Oduor et al., 1997). The attachment of spores on the epicuticle of the host, and the subsequent germination and prepenetration growth, are critical phases of the infection process (Ortiz-Urquiza and Keyhani, 2013). Most of the hypocrealean EPF produce hydrophobic conidia, presumably an adaptation to enable rapid attachment to the waxy epicuticle. In *B. bassiana s.l.* and *M. anisopliae s.l.*, binding of the conidia is promoted by hydrophobin proteins that form rodlet layers on the conidium surface (Holder et al., 2007). The initial attachment is followed by the secretion of an adhesive mucus (Boucias and Pendland, 1991). The Entomophthoromycota produce conidia that are ready formed with a mucus layer, deposited on the outside of the conidium or contained underneath a fissile outer wall, and which allows firm adhesion to the host cuticle (Eilenberg et al., 1986).

Environmental factors such as relative humidity, temperature, and solar radiation have a profound effect on germination and hyphal growth. Spore germination requires the presence of water and oxygen and is characterized by rapid swelling as a result of hydration. Imbibition of water into the conidia by osmosis increases its internal hydrostatic pressure and is followed by the formation of a thin-walled germ tube. Germination requires a minimum equilibrium relative humidity (ERH) of 93% at the site of infection (Andersen et al., 2006). The conidia of the hypocrealean EPF are small (<8 μm) and have limited nutrient reserves, and although the imbibition of water occurs without an exogenous nutrient being present, germ tube formation itself will only occur if a suitable carbon source is available (Dillon and Charnley, 1990). In addition to limitations imposed on them by environmental conditions, fungal spores also have to overcome a range of antifungal defense mechanisms that represent the first line of defense for the arthropod against infection. The chemical and physical composition of the cuticle changes as the fungal hypha grows through it, and a series of biochemical and morphological adaptations have evolved in fungal parasites that enable them to respond to this variable environment (Hajek and St Leger, 1994). Hydrated conidia release proteases during the pregermination period, presumably for nutrient acquisition and to aid penetration (Qazi and Khachatourians, 2007). Developing fungal hyphae utilize epicuticular waxes and lipids for growth, and studies with *B. bassiana s.l.*, for example, have identified at least 16 fungal enzymes involved in the oxidative degradation and assimilation of epicuticular lipids (Pedrini et al., 2010). However, some cuticular lipids have antifungal properties, and their synthesis on the cuticle has probably evolved as a defense mechanism to inhibit attachment and/or germination (Sosa-Gomez et al., 1997) while some insect

species are known also to produce glandular secretions with different chemical constituents that have antifungal activity (Gross et al., 1998). These actively synthesized defenses, combined with the fact that the surface of the cuticle is relatively poor in readily accessible nutrients, can significantly extend the median germination time to approximately 30 h (Butt et al., 1995; Andersen et al., 2006). In contrast, the conidia of EPF in the Entomophthoromycota are relatively large (20–30 μm in diamater) (Keller, 2007) and germinate quickly; the mean time for infection by *Pandora neoaphidis* is only 4.5 h at 20°C, for example (Glare and Milner, 1991).

5.5.2 Penetration of the Cuticle and Interaction With the Host Immune Response

The procuticle is a major physical barrier to infection and has to be breached before the fungus can multiply within the host (Hajek and St Leger, 1994). The transition from lateral hyphal extension over the epicuticle surface to the formation of penetration structures is a thigmotropic process: it requires a certain amount of physical contact between the fungus and the cuticle to induce the formation of fungal penetration structures (St Leger et al., 1991). Some fungal genera, such as *Metarhizium*, form appressoria on the cuticle surface, which concentrates the production of degradative enzymes over a small area as well as providing a higher physical pressure for penetration by an infection peg that forms beneath the appressorium (Hajek and St Leger, 1994). Penetration of the cuticle is achieved using a combination of mechanical pressure from growing hyphal tips and enzymatic degradation. Insect procuticle consists of layers of chitin embedded within a protein matrix, and in *M. anisopliae s.l.*, for example, the serine protease Pr1 is an important virulence determinant, responsible for solubilization of proteins in the procuticle to facilitate penetration and provide nutrients to support fungal growth (St Leger et al., 1987). Penetration can be direct, or hyphal growth may include lateral extension via penetrant plates that can fracture the cuticle (Hajek and St Leger, 1994). Developing fungal hyphae grow within the interstices of the cuticle until they enter the hemocoel, which then triggers two interlinked and rapid immune reactions; encapsulation of invading fungal cells and melanization of fungal tissue and surrounding areas. Insect hemocytes are central to these reactions and are composed of three functional groups: (1) plasmatocytes, which form the majority of hemocyte cells circulating in larvae and which act as phagocytes of bacteria and other small microbial cells; (2) crystal cells, which are involved in the melanization reaction and produce prophenoloxidases (proPOs); and (3) lamellocytes, which are involved in the encapsulation of large infecting objects that are too big to be engulfed by phagocytes (Lemaitre and Hoffmann, 2007). Detection of an invading object results in a large increase in the synthesis of lamellocytes from precursors, which then form a surrounding capsule within which melanization takes place (Williams, 2007). The reaction occurs following activation of a serine protease cascade that terminates in the cleavage of the zymogen proPO into active phenoloxidase (PO), which in turn catalyzes the oxidation of phenolic compounds into melanin, a polymer of orthoquinones. Melanization also takes place in the cuticle (where it is the most obvious symptom of host defense against a fungal infection) (Dubovskiy et al., 2013) and as a free reaction in the hemolymph where it happens within seconds of an invading pathogen being detected (Chambers et al., 2012).

The EPF in the Hypocreales and Entomophthoromycota have both evolved strategies that enable them to evade or overcome the antifungal cellular defense system of their hosts, including undertaking a morphological change upon entering the hemocoel. For the Hypocreales, this involves switching from hyphal growth to the formation of small thin-walled, yeast-like hyphal bodies (aka blastospores). These structures facilitate rapid dispersal in the hemolymph and avoid detection by pathogen recognition molecules (see later) although they are still susceptible to phagocytosis by plasmatocytes (Jiang et al., 2010). Virulent fungal strains are also able to overcome the hemocyte response by causing a reduction in the numbers of granulocytes in the days immediately after infection (Hou and Chang, 1985; Hung et al., 1993). Many of the EPF in the Entomophthoromycota grow as protoplasts (not unlike blastospores) within the hemocoel of their hosts, which also helps them to evade detection by hemocytes (Butt et al., 1996).

Invasion of the hemocoel also initiates the systemic/humoral immune response, which results in the synthesis of a range of antifungal host immune effectors. The systemic immune response takes longer to be initiated than the cellular response, with effector synthesis peaking at 6–24h postinfection in *D. melanogaster* (Ganesan et al., 2011). It probably represents the last line of immune defense, and it has been proposed that it evolved as a defense mechanism against persistent infections (Haine et al., 2008). The antifungal component of the systemic immune response is mediated by the Toll pathway. This is a complex cascade reaction that starts with sensing and detection of invading fungal cells and ends in the activation and translocation of transcription factors from the cytoplasm to the nucleus, resulting in the synthesis and secretion of antimicrobial peptides (AMPs) from the fat body into the hemolymph. The Toll pathway is also responsible for mediating the systemic defense against gram-positive bacteria, as well as playing a key role in signal transduction for programmed insect development. The two other main immune signaling pathways are the immune deficiency (Imd) pathway and Janus kinase/signal transduction and activator of transcription (JAK/STAT) pathway. These two pathways operate in a similar way to Toll, inducing the synthesis of

antibacterial AMPs and stress/injury response proteins, respectively. There is cross-talk between the three pathways, and the whole system enables arthropods to detect and discriminate between different types of pathogen. At the same time, there are multiple levels of regulation to prevent their aberrant activation, which can result in insect death (Georgel et al., 2001). The insect Toll pathway commences with a serine proteolytic cascade involving soluble proenzymes circulating in the hemolymph that undergo sequential zymogen activation and that leads to the processing and maturation of an insect cytokine, Spatzle, by the Spatzle processing enzyme (SPE). In infections of *D. melanogaster* by the hypocrealean entomopathogens *Beauveria* and *Metarhizium*, the cascade is a two-step process initiated by the fungal serine protease Pr1, which cleaves an insect protease, Persephone, which in turn activates SPE (Valanne et al., 2011). Once Spatzle has been activated, it binds to Toll receptors, which are signal transducing proteins that span the membranes of insect cells, and this initiates a second, intracellular signaling cascade reaction that results in the translocation of the nuclear factor–κB (NF-κB) transcription factor Dif from the cytoplasm into the nucleus. Dif induces the expression of a range of target genes including those encoding AMPs. These effectors are small (<10kDa) and membrane active and remain in the hemolymph for several days after pathogen challenge (Lemaitre and Hoffmann, 2007). AMPs are found in all metazoans but they are not evolutionarily conserved. Different taxonomic groups of insects have their own families of AMPs, and within particular insect taxa, AMP genes show high rates of gene family expansion and contraction (Juneja and Lazzaro, 2009). In *Drosophila*, about 20 AMPs have been identified in seven different classes (Lemaitre and Hoffmann, 2007), with antifungal AMPs occurring within three classes of peptides: metchnikowin, drosomycin, and cecropin (one member of which has both antifungal and antibacterial activity) (Ekengren and Hultmark, 1999; Lematire and Hoffmann, 2007). Metchnikowin and drosomycin are specific to *Drosophila*, whereas cecropins are synthesized by many insect groups. Infection of *D. melanogaster* by *B. bassiana s.l.* results in induced expression of *drosomycin* and *metchnikowin* genes via the Toll pathway (Lemaitre et al., 1997). *D. melanogaster* mutants defective in the Toll pathway have a higher susceptibility to infection by *B. bassiana s.l.*, but resistance cannot be restored by ectopic expression of *drosomycin* (Tzou et al., 2002). In *D. melanogaster*, the AMPs that are synthesized through the Toll pathway are active within the hemocoel, and they do not appear to be induced in epithelial cells, for example those underlying the cuticle (Lemaitre and Hoffmann, 2007).

Hypocrealean EPF avoid and disable the host systemic immune system although the detailed mechanisms by which this occurs are still unclear. Avoidance of detection probably occurs via the morphological switch by the invading fungus from hyphal growth to hyphal bodies within the hemocoel, presumably to reduce detection of β-glucans (Jiang et al., 2010). Production of protoplasts by the Entomophthoromycota may fulfill the same function (see earlier). EPF may also be resistant to insect antifungal defense compounds. For example, a strain of *B. bassiana s.l.* that showed high levels of virulence against *D. melanogaster* was not susceptible to cecropins in an in vitro growth assay, whereas a strain of *M. anisopliae s.l.* that had low virulence against *D. melanogaster* was susceptible to cecropin-induced growth inhibition (Ekengren and Hultmark, 1999). Evidence of general pathogen interference in immune signaling pathways comes from observations that the core proteins in both the Toll and Imd signaling pathways exist as strict orthologs in different groups of insects but show high levels of amino acid divergence, with particular evidence of adaptive evolution in *Drosophila* spp. (Juneja and Lazzaro, 2009). It has been proposed that this divergence has evolved in response to selection by pathogens interfering with immune signaling pathways as part of the infection process (Begun and Whitley, 2000).

The hypocrealean EPF synthesize an array of secondary metabolites that are thought to aid infection and host colonization or to suppress the growth of other pathogenic microorganisms including opportunistic, secondary infections (Zimmermann, 2007a,b); nevertheless, the exact contribution of individual compounds to pathogenesis can be hard to ascertain experimentally (Rohlfs and Churchill, 2011). The destruxins, a family of cyclic hexadepsipeptides produced by *Metarhizium* sp., cause insect cell paralysis and impair the normal function of muscle and organs (Samuels et al., 1988). The cyclic depsipeptide beauvericin (synthesized by *B. bassiana s.l.* and *Isaria* spp.) may operate as an ionophore and also has insecticidal, cytotoxic, and antibiotic properties (Zimmermann, 2007a). At least seven other secondary metabolites have been chemically described from *Beauveria* spp.: bassianin and tenellin are nonpeptide metabolites with toxic effects on erythrocyte membranes; bassianolide is a cyclic depsipetide with antibiotic and ionophore properties; bassiacridin is an insecticidal protein active against locusts; beauveriolides and beauverolides are depsipeptides investigated as lead compounds in human medicine although their natural function in insect infection is not known; and oosporein is a dihydroxybenzoquinone with activity against gram-positive bacteria (Zimmermann, 2007a). Some of the secondary metabolites of hypocrealean EPF may be involved in active suppression of the host immune response. For example, injection of destruxin A from *M. anisopliae s.l.* into *D. melanogaster* caused downregulation of AMP gene expression including *diptericin, cecropin, attacin,* and *metchnikowin* (Pal et al., 2007), while destruxins A and D inhibited the attachment ability of plasmatocytes from larvae of the lepidopteran *Galleria mellonella*, a finding that was also observed with plasmatocytes isolated from

M. anisopliae–infected larvae (Vilcinskas et al., 1997). Most of these secondary metabolites have been studied using in vitro assays or by injecting them into insects (often at high concentrations) and observing their effects. However, to understand their precise role in pathogenesis, loss of function studies in different EPF species and subspecies are required. To date, only a small number of these studies have been done, and while some have shown that knocking out the ability of a fungal strain to synthesize a particular metabolite reduces its virulence, other studies have found no effect. For example, *B. bassiana s.l.* strains that had undergone targeted disruption of genes responsible for production of beauvericin and bassianolide were less virulent to *Spodoptera frugiperda, G. mellonella,* and *Helicoverpa zea* than unmodified controls (Xu et al., 2008, 2009). In contrast, targeted gene knockout of destruxin production in *M. robertsii* had no effect on virulence of the fungus to larvae of *Spodoptera exigua, G. mellonella,* or *Tenebrio molitor* (Donzelli et al., 2012) Furthermore, gene disruption of production of the peptide serinocyclin in *M. robertsii* had no effect on virulence against *S. exigua* or *Leptinotarsa decemlineata* (Moon et al., 2008). It might be that secondary metabolites only make a significant contribution to pathogenesis in certain host species that possess the specific targets for those metabolites; alternatively, it may be that secondary metabolites supplement primary determinants of virulence when there is redundancy in virulence factors; and/or that secondary metabolites play subtle roles in infection that are hard to detect using standard laboratory bioassays of fungal virulence (Donzelli et al., 2012). It has also been proposed that—from an evolutionary perspective—there is a trade-off between the action of toxic fungal metabolites and the amount of fungal growth in the host, since highly active metabolites could result in host death before the fungus has produced sufficient growth for spore production and transmission (Boomsma et al., 2014).

Little is known about the molecular mechanisms underlying the infection processes of the Entomophthoromycota. The current lack of whole genome sequences for these fungi prevents routine transcriptomics studies of the infection process. One exception concerns *Pandora formicae*, an entomophthoralean pathogen of the European red wood ant, *Formica polyctena* (Malagocka et al., 2015). A transcriptional analysis of *P. formicae*–infected ants collected from the wild, and in different stages of fungus-associated behavioral modification (see later), identified expression of a wide range of fungal genes encoding pathogenicity-associated proteins (Malagocka et al., 2015). These included subtilisin-like serine proteases, trypsin-like serine proteases, a homolog of a zinc-dependent metalloprotease from *Zoophthora radicans*, as well as numerous fungal lipases associated with nutrient acquisition and signaling pathways regulating fungal morphogenesis (Malagocka et al., 2015).

5.5.3 Host Death and Transmission to New Arthropod Hosts

EPF-induced host death occurs as a result of physical damage and loss of normal function following fungal colonization of tissues and organs, the action of fungal metabolites, loss of water, and starvation. The time to death of the host, which varies from 3–20 days after inoculation, depends on a range of factors including: arthropod species; the physiological state of the host (starvation, food type including consumption of plant secondary compounds, overcrowding, and other forms of stress can affect the host susceptibility to infection); the spore dose received; the species and strain of fungus; and environmental conditions (Hesketh et al., 2010). There are often premortality effects such as reduction in feeding or flight activity, which can start to occur within 1–4 days of infection (Moore et al., 1992; Seyoum et al., 1994; Roy et al., 2006). In successful infections, the entirety of the soft tissue of the insect is replaced with fungal biomass. Following the death of the host, the fungus grows out through the cuticle to produce spores (conidia or ascospores) for transmission to new hosts. In hard-bodied insects, this external growth usually occurs through the parts of the body where the integument is thinnest, such as intersegmental membranes, whereas for soft-bodied insects, the fungus will grow through the cuticle at most positions on the body. External growth and the production of spores require favorable environmental temperature and humidity (Arthurs and Thomas, 2001). Under unfavorable conditions, the host cadaver can remain in the form of a "mummy" for an extended period until conditions favorable for spore production return.

Transmission of the fungus to new hosts determines its fitness, and there will be natural selection on traits that increase the success of the transmission process, such as spore production, the ability of resting spores to survive for prolonged periods between host availability, and spore dispersal mechanisms (Roy et al., 2006). There are clear differences in the mechanisms of spore production between the teleomorphic and anamorphic states of the entomopathogenic Hypocreales and the EPF in the Entomophthoromycota. The anamorphic hypocrealeans produce conidia that are released passively over a short period, usually just a few days, and large reservoirs of conidia can accumulate in soil. Under natural circumstances, susceptible hosts acquire conidia by direct physical contact with an infected, conidiating cadaver or indirectly by coming into contact with conidia in soil or on foliage (Hesketh et al., 2010). Because of their small size, the conidia produced by the anamorphs can be distributed by water splash or wind, facilitating medium- to long-range dispersal in the environment as an aid to indirect transmission, while dispersal can also occur when nonhost organisms such as springtails (*Collembola*) transport conidia attached to their cuticle or that have been eaten in soil (Dromph, 2001). Passive acquisition of spores from the environment probably constitutes a "sit and wait" strategy for EPF in which the chances of successful transmission will be positively correlated with the number of spores released in the environment. Indeed, the anamorphic fungi tend to produce conidia in massive numbers (10^6 or more) on cadavers although the number is dependent upon the size of the host: *Hirsutella thompsonii*, a pathogen of eriophyoid mites, only produces few conidia on its hosts as there is insufficient host biomass to support the production of more (Humber, 2008). Conidia can also persist for long periods of time if conditions are favorable, which allows large reservoirs of conidia to build up in some environments (Meyling and Eilenberg, 2007; Meyling et al., 2011). In contrast, the ascospores produced by teleomorphic states are actively discharged and ascospore production occurs over prolonged periods of time from fruiting bodies. This "trickle release" strategy might reduce the requirement for ascospores to be able to persist for long periods in the environment in the absence of a host. As stated previously, EPF in the Entomophthoromycota are transmitted to new hosts through the production of ballistic, primary conidia that are actively discharged from the host body for about 24 h. These conidia are formed from conidiophores that grow out through the host cuticle very shortly after death, and the process normally requires high humidity (Newman and Carner, 1975; Glare et al., 1986). Some species discharge conidia out of the leaf boundary layer where they can be dispersed by air currents (Eilenberg and Pell, 2007). For many species, if a primary conidium does not land on an arthropod host, it will germinate to produce a secondary ballistic conidium, and this process may repeat itself a number of times until a host is contacted or the conidium runs out of energy reserves (McDonald and Nolan, 1995). The transmission cycles of many of the entomopathogenic Entomophthoromycota are synchronized with diurnal fluctuations in environmental conditions, so that discharge of conidia occurs when temperature and humidity are favorable for infection, often at dusk or during the night (Brown and Hasibuan, 1995).

The use of hypocrealean EPF as inundative biological control agents bypasses most of the "natural" methods of transmission, and conidia are acquired by host insects in the form of spray droplets, from granules or powder formulations applied to soil, or by secondary pick up of conidia deposited in spray droplets on plant surfaces. The amount of movement of the insect can influence the amount of inoculum acquired, and some biocontrol strategies have used treatments such as alarm pheromones or sublethal doses of insecticide to increase pest movement and thereby increase secondary pickup of conidia (Roditakis et al., 2000). Transmission can also occur during biocontrol by physical contact between fungus-treated hosts and uncontaminated hosts. Transmission of a lethal dose in this way can occur with only a single contact between a new host and a cadaver (Kreutz et al., 2004). Transmission during biocontrol can also take place between male and female hosts during mating, the efficiency of which can depend on whether it is the male or female that has been treated with fungal inoculum (Kaaya and Okech, 1990; Quesada-Moraga et al., 2008; Baverstock et al., 2010).

Some arthropod hosts undergo a change in their behavior shortly before death, moving into positions that are often assumed to be more conducive to transmission. For example, grasshoppers infected with *Entomophaga grylli* often climb to the tops of plants as a final act, a behavior that is known as "summit disease" (Roy et al., 2006). However, it is not always clear whether the behavioral change results from manipulation of the host by the fungus or vice versa, and so the exact effects on the host's exclusive fitness may not be clear [for example, it could be a response of the host to remove itself from related individuals to reduce the probability of pathogen transmission or to move into a warmer location to raise body temperature as a defense mechanism against infection (Roy et al., 2006)]. There is some evidence from laboratory studies that the anamorphic hypocrealean fungi manipulate host behavior, but it is not known how widespread this is in nature. In laboratory bioassays, larvae of vine weevil (*Otiorhynchus sulcatus*) were attracted to plant roots treated with conidia of *M. brunneum* (Kepler and Bruck, 2006), which might be an adaptation by the fungus to increase the chances of infection occurring. The best-studied system—where there is clear, mechanistic evidence of pathogen manipulation of host behavior—concerns *Ophiocordyceps unilateralis* infections of ants, where distinct mechanisms have evolved for manipulating the behavior of infected individuals as an aid to transmission, and in which host behavior functions as an extended phenotype of the fungus (Hughes et al., 2011; Boomsma et al., 2014). Infection of tropical carpenter ants *Camponotus leonardi* by *O. unilateralis* causes diseased individuals to descend from nests in the forest canopy to understory vegetation where conditions are best for growth of stromata and ascospore production on cadavers. Behaviorally affected ants exhibit repeated convulsions that cause them to fall from vegetation and stop them from climbing back into the canopy. Fungal colonization of the head, associated with muscular atrophy in the mandibles, causes ants to bite into abaxial leaf veins where they become fixed and die, usually in high-density aggregations termed "graveyards" (Pontoppidan et al., 2009). Ants foraging on the ground are therefore vulnerable to acquiring ascospores showering down from infected cadavers above them (Hughes et al., 2011). Potential pre-death behavioral manipulation has also evolved in the entomophthoralean fungus *P. formicae*, which is associated with summit disease in infected European red wood ants, *Formica polyctena* (Malagocka et al., 2015). This is the only known example of infection of a eusocial insect by a member of the Entomophthoromycota (Malagocka et al., 2015).

Detection and avoidance behaviors of insects towards EPF have been documented in five insect orders (Baverstock et al., 2010). Avoidance behavior is not a common trait, however, and studies suggest that most insect species are not able to detect the presence of an entomopathogenic fungus, or can detect it but do not show avoidance behavior (Baverstock et al., 2010). The exception to this appears to be with social insects where—as we have just seen—selection for behaviors that prevent or minimize the probability of infection would appear to be under particularly strong selection pressure. Social insects, such as termites, exhibit strong avoidance behaviors as well as allogrooming to remove conidia from the cuticle of infected individuals (Chouvenc et al., 2008).

5.5.4 Growth and Survival Outside of the Arthropod Host

Most of the teleomorphic hypocrealean EPF are believed to function ecologically as obligate parasites of arthropods. However, some hypocrealean EPF can also operate in more than one econutritional mode and can use different resources depending on their availability (Vega et al., 2009; Ownley et al., 2010). This includes endophytism of both above- and below-ground tissues, growth within the rhizosphere, and mycoparasitism. Some species of *Hypocrella* (teleomorphic *Clavicipitaceae*) start their development as parasites of scale insects and then switch to parasitism of the insect's host plant, enabling them to gain sufficient nutrition to produce very large stromata, which would be impossible were they to rely solely on the insect tissue as an energy source for fruiting body production (Humber, 2008). Strains of *B. bassiana s.l.* and *Lecanicillium* spp. are known to be antagonistic to plant pathogenic fungi although the modes of action differ (Ownley et al., 2010). *B. bassiana s.l.* is reported to antagonize soil-borne plant pathogenic fungi, including *Pythium*, *Fusarium*, and *Rhizoctonia*, as well as some foliar pathogens such as grapevine downy mildew (*Plasmopara viticola*): modes of action include mycoparasitism, competition, antibiosis, and induced systemic resistance depending on the species of plant pathogen (Lee et al., 1999; Ownley et al., 2008 a,b; Ownley et al., 2010; Jaber, 2015). *Lecanicillium* is antagonistic toward powdery mildew and rust fungi and functions primarily through mycoparasitism (Whipps, 1993; Askary et al., 1998; Kim et al., 2007, 2008). Following pioneering research by Bing and Lewis (1991, 1992) that demonstrated the ability of *B. bassiana s.l.* to grown as an endophyte in maize, endophytism has now also been reported in *Lecanicillium* and *Metarhizium* and has been documented in at least 15 species of crop plants (Russo et al., 2015; Vidal and Jaber, 2015). The natural occurrence of endophytic entomopathogens varies according to geographic region: in an assessment of the occurrence of endophytes from wild grasses, forbs, and sedges in Canada, both *B. bassiana s.l.* and *Metarhizium* sp. were isolated in sample sites in Ontario, but only *B. bassiana s.l.* was isolated from samples collected in Newfoundland (Behie et al., 2015). Moreover, whereas *B. bassiana s.l.* was found to occur in all plant tissues, 99% of the *Metarhizium* spp. isolations were from plant roots, and the majority of these (95%) were identified by RFLP as *M. robertsii*, with the remainder being *M. brunneum* (4.3%) and *M. guizhouense* (0.7%) (Behie et al., 2015). *M. brunneum* shows enhanced growth in the presence of developing roots of certain plant species, and hence it may have evolved to be rhizosphere competent rather than endophytic. For example, it exhibited enhanced growth in the rhizosphere of container grown *Picea abies*, accumulating significantly more biomass than in the bulk soil (Bruck, 2005), although growth was not observed in the rhizosphere of *Taxus baccata* (Bruck, 2010). Phenotypic plasticity in *Metarhizium*, and in particular the ability to grow in plants or in the root zone, is a trait associated with the generalist species *M. robertsii* and *M. brunneum*, consistent with their evolution from species that are specific for particular insect groups (Hu et al., 2014). There might be multiple benefits to plants from these associations although more evidence is needed to elucidate the mechanisms of interaction and the effects on plant and fungal fitness. *M. robertsi* has been shown to stimulate plant root development (Sasan and Bidochka, 2012) as well as translocate nitrogen from fungus-killed insects to plants (Behie et al., 2012), while endophytic *B. bassiana s.l.* can enhance plant growth (Lopez and Sword, 2015). Endophytic growth of both *Beauveria* and *Metarhizium* is reported to have negative effects on the performance of a range of insect herbivores, including lepidopteran and coleopteran larvae that tunnel into plant stems and roots, and aphids feeding on leaves and stems (Powell et al., 2009; Lopez et al., 2014; Lopez and Sword, 2015; Mantzoukas et al., 2015; Vidal and Jaber, 2015). However, establishment of fungal infection in the insect host appears to be rare (Vidal and Jaber, 2015). The outcomes of the relationship for fungal fitness are not yet clear. A priori, the fungus may benefit by being able to grow on an alternative source of nutrition in the absence of suitable arthropod hosts. However, it is not yet clear how important these other associations are to fungal survival compared to entomopathogenicity and whether they present a strategy for long-term survival and reproduction or are used as a "stop gap" measure if suitable host arthropods are not present in the environment.

5.6 USE OF ENTOMOPATHOGENIC FUNGI FOR ARTHROPOD PEST MANAGEMENT

EPF have been investigated widely for the biological control of crop and forestry pests, as well as arthropod vectors of human and animal disease (Chandler et al., 2000; Lacey et al., 2001, 2015; Shah and Pell, 2003; Blanford et al.,

2011; Fernandes et al., 2012; Paula et al., 2013; Nana et al., 2015). They are used in augmentation (inundation and inoculation), classic, and conservation biological control. They pose minimal risk to beneficial organisms including bees, earthworms, and parasitoid wasps, while they also present minimal human safety concerns (Goettel et al., 2001; Zimmermann, 2007a,b; Garrido-Jurado et al., 2011) and hence are likely to be classified as "low-risk" substances by government regulators (Chandler et al., 2011). EPF biopesticides can usually be applied with existing spray equipment (Feng et al., 1994; Price et al., 1997; Pu et al., 2005), and they have potential to reproduce on or in close vicinity to the target pest, which can give an element of self-perpetuating control (Lomer et al., 2001). As alternatives to conventional chemical pesticides, they can help reduce the selection pressure for the evolution of pesticide resistance in pest populations, and they are be able to stop the expression of resistance once it has evolved (Farenhorst et al., 2009, 2010).

The vast majority of work on biological control with EPF concerns the use of anamorphic hypocrealean fungi as inundative biopesticides. Over 170 different EPF biopesticides have been developed since the 1960s, of which three-quarters are currently commercially available, undergoing registration or have active authorizations but are not currently sold (Faria and Wraight, 2007). Most are based on the genera *Beauveria* and *Metarhizium*, but products are also available based on *Isaria*, *Lecanicillium*, and others. About half have been developed in South and Central America, 20% in North America, 12% in Europe and Asia, and 3% in Africa (Faria and Wraight, 2007). The target pests include species within the Hemiptera, Coleoptera, Lepidoptera, Diptera, Orthoptera, and Acari. It is worth pointing out that EPF are the only entomopathogens currently available for the control of hemipteran pests. The majority of products are intended for use in horticultural crop production systems (greenhouse edible and ornamental crops, fruit crops, and field vegetables); however, there are also examples of successful use on broad acre crops. The largest single area of use is probably for *M. anisopliae s.l.* against spittlebug pests on sugarcane (*Mahanarva* spp.) and pasture (*Brachiaria* spp.) in Brazil (see Chapter 20). The fungus is mass-produced on rice grains and is sold to farmers as a conidia powder or as fungus-colonized substrate, being used on about 750,000 ha of sugarcane and 250,000 ha of grassland annually (Li et al., 2010). This biopesticide program was developed using government funding for research and development, which, alongside the use of rice grains as a low-cost method for mass production, has kept the price of the fungal product competitive with that of chemical insecticides (Li et al., 2010). In Africa and Australia, *Metarhizium acridum* has been developed as a biopesticide for large-area control of locusts and grasshoppers, including the desert locust *Schistocerca gregaria*, as an alternative to blanket sprays of synthetic chemical pesticides (Milner and Prior, 1994; Lomer et al., 2001; see also Chapter 23). Conidia are mass produced on rice grains, harvested, dried, and formulated in mineral oil for low-volume spray application; this allows the fungus to work under conditions of low ambient humidity, and it can give up to 90% control within 14–20 days (Bateman et al., 1993; Thomas et al., 1996; Lomer et al., 2001). Unlike broad-spectrum conventional chemical pesticides, the fungus is considered to pose minimal risk to nontarget organisms. In addition to its high level of virulence, it has good persistence on the soil and vegetation, and it has some capacity to spread within the host population via the production of new conidia on infected cadavers (Thomas et al., 1995; Lomer et al., 2001). The most successful example of EPF for inoculation control is the use of *Beauveria brongniartii* applied against European cockchafer beetles, *Melolontha melolontha*, which is an important pest of pastureland, orchards, and forest trees in central Europe. Application of the fungus can give useful levels of control for up to 9 years (Keller et al., 1997).

As described by Jaronski (2010), successful use of EPF biopesticides essentially comes down to "winning the numbers game" to ensure that individual target pests acquire sufficient conidia to kill them under prevailing environmental conditions. To determine the optimal timing, frequency, and rate of biopesticide application, the IPM practitioner needs to know: (1) the effective conidia concentration, that is, the number of conidia per unit surface area (for foliar applications) or per unit volume of soil/growing medium (for biopesticides used below ground) required to cause infection and (2) the effect of environmental conditions on infectivity and persistence of activity, that is, the effect of temperature, UV radiation, humidity, rainfall, leaf chemistry and topography, and agronomic practices such as application of pesticides that can be harmful to EPF (Jaronski, 2010). A number of studies have been done to systematically investigate the optimum frequency and rate of EPF applications on crops (Poprawski et al., 1997; Wraight et al., 2000; Wraight and Ramos, 2002; Castrillo et al., 2010; Gatarayiha et al., 2011), but, in general, this important aspect of biopesticide development has not received sufficient attention (Jaronski, 2010).

Temperature directly determines the rate of EPF infection and has indirect effects through its influence on the activity of the target pest and host plant (Thomas and Blanford, 2003). It is important to know not only the optimum temperature for fungal activity but also the thermal tolerance (ie, the upper and lower temperatures at which the EPF will work). Determining the effect of temperature on (1) arthropod growth or development and (2) fungal activity (ie, growth, germination) is relatively straightforward. However, given the complex, nonlinear interaction of temperature, host, and fungus, these may not be good predictors of the effect of temperature on the rate of disease progression

(Thomas and Blanford, 2003). Almost all of the published studies on EPF thermal biology are done under constant temperature conditions (eg, Bugeme et al., 2008; Guzman-Franco et al., 2008; Cabanillas and Jones, 2009; Kikankie et al., 2010; Evans et al., 2015), with very few studies investigating how fungal activity changes under fluctuating temperatures (Bouamama et al., 2010). The response of fungal activity to temperature is a bell-shaped curve skewed to the lower temperatures, sometimes with a rapid decline in activity as temperatures increase past the optimum, which means that even small changes in temperature can have a profound effect on fungal activity rate (Davidson et al., 2003; Smits et al., 2003). This nonlinear response to temperature can be modeled (Davidson et al., 2003; Smits et al., 2003) but this is not yet being used to predict changes in EPF performance under fluctuating temperatures in field environments.

All EPF require freely accessible water at the site of infection to germinate and grow (Hallsworth and Magan, 1995; Andersen et al., 2006), and hence crop canopy humidity can have a strong effect on fungal activity. Infection can occur under low ambient humidity conditions providing the humidity of the microclimate at the spore–arthropod interface is favorable, which could occur, for example, with small arthropods that sit within the leaf boundary layer (Ferro and Southwick, 1984). The use of oil formulations of *M. acridum* against desert locust *S. gregaria* enabled infection at very low ambient humidity conditions (35% r.h.). There are two potential reasons for this: oil enables not only more efficient attachment of conidia to the host cuticle, resulting in a far steeper dose–response curve compared to a water-based application (Bateman et al., 1993), but also the formulation may promote conidia deposition at important penetration sites for EPF such as intersegmental membranes (Prior et al., 1988).

EPF conidia are particularly sensitive to UV radiation and are generally inactivated after just a few hours of direct exposure to sunlight in tropical or temperate regions (Braga et al., 2001; Fernandes et al., 2007). Use of oil-based formulations and incorporation of sunscreens can significantly reduce UV damage to conidia (Inglis et al., 1995; Thompson et al., 2006; Cohen and Joseph, 2009). There are also opportunities to select fungal strains with some UV tolerance, and in general strains from regions closer to the equator are more tolerant of UV exposure than are strains from temperate latitudes (Fernandes et al., 2009). The molecular basis for UV tolerance is starting to be elucidated in *Beauveria* and *Metarhizium*, which may provide a better mechanism for selecting UV-tolerant strains or to improve tolerance through genetic engineering (Fang and St Leger, 2012; Xie et al., 2012; Ying et al., 2014; Li et al., 2015; Lovett and St Leger, 2015; Zhang et al., 2015). Rainfall can also cause rapid and significant loss of inoculum. For example, 15 min of moderate rainfall in a simulator resulted in about 90% removal of *B. bassiana s.l.* conidia from foliage (Inglis et al., 2000). When mustard beetles, *Phaedon cochleariae*, were placed onto oilseed rape foliage that had been sprayed with *M. anisopliae s.l.* and subjected to 1 h of simulated rainfall, mortality was reduced by 42–57% compared with sprayed foliage that had not been subjected to rainfall; the formulation used for the fungus also influenced the results (Inyang et al., 2000).

The combined effect of these environmental factors means that, without the mitigation provided by appropriate formulations and application techniques, EPF biopesticides do not persist for long on foliage in the field. *B. bassiana s.l.* is reported to persist for only 6 days on field-grown strawberries (Sabbahi et al., 2008), and *M. anisopliae s.l.* for 3 days on maize (Pilz et al., 2011) and 7 days on *Amaranthus* (Guerrero-Guerra et al., 2013). Greenhouses confer protection for EPF against adverse environments, meaning that they persist for substantially longer than under field conditions. For example, *B. bassiana s.l.* persisted for more than 3 weeks on a range of crop types grown in a greenhouse under natural light conditions including maize, beans, cucumber, eggplant, and tomato (Gatarayiha et al., 2010).

Loss of conidia viability on foliage by adverse environmental conditions could have a significant effect on the persistence of activity of EPF biopesticides against target pests that are mobile and acquire conidia from the environment rather than by being a direct target for sprays (Ugine et al., 2005; Behle, 2006). Unfavorable environmental conditions may delay pest mortality until favorable conditions return, for example by the pest moving to a location that has a microclimate conducive to fungal activity. In a detailed series of field experiments (Wraight and Ramos, 2015), Colorado potato beetle (CPB) larvae, *Leptinotarsa decemlineata*, were found to be readily infected by foliar applications of *B. bassiana s.s.* but fungal development and host death were delayed until larvae had entered the soil to pupate where they were protected from UV radiation, high temperatures, and the effects of low humidity at the foliar surface. Despite the delay in the time to death, a single application of the fungus still provided an average 60% control of first generation adult beetles, while potato yields were 18% higher than controls in a multispray program, suggesting that the fungus has considerable potential against CPB in an IPM program (Wraight and Ramos, 2015).

Changes to the way biopesticides are applied to plant surfaces can give improved control by ensuring better coverage, by targeting the biopesticide more efficiently to where the pest is located and reducing the amount of product lost through run-off (Glare et al., 2012). In principle, changes to spray application method could provide significant improvements to the performance of EPF biopesticides by optimizing the size of droplets or concentrating spray applications to small, targeted zones. For example, switching to a sprayer with up-pointing hydraulic drop nozzles provided up to a 30-fold increase in the deposition of *B.*

bassiana s.l. conidia to the undersides of leaves compared with that achieved using standard nozzles (Wraight and Ramos, 2002). Delivery method is also important for soil application; spray application can result in greater conidia concentrations at the base of plants compared to the application of a granular formulation, for example, although the use of cereal grain–based granules can result in localized increases in fungal density as the EPF grows on the granule (Boetel et al., 2012). Despite the clear importance of application technology to biopesticide efficacy, relatively few studies have been done in this area with EPF, particularly for identifying practical measures for optimizing application that can be adopted readily by farmers and growers (including nozzle size, operating pressure, tank system, etc.) (Gan-Mor and Matthews, 2003; Pu et al., 2005; Saito, 2005; Ugine et al., 2007).

Beauveria and *Metarhizium* persist for substantially longer in soil than on foliar surfaces. Following application to soil, there may even be a short lived increase in numbers caused by reproduction of the fungus on host arthropods, followed by decline to background over time. A meta-analysis by Scheepmaker and Butt (2010) showed that the rate of decline in soil to background levels varies according to the species of fungus, ranging from 0.5 to 1.5 years for *B. bassiana s.l.*, about 4 years for *B. brongniartii*, and longer than 10 years for *M. anisopliae s.l.* (Scheepmaker and Butt, 2010). The rate of decline in soil is also thought to depend on the "quality" of the EPF product (the way it was mass produced, formulated, and stored prior to application), physical attributes of the environment (temperature, soil type, water availability), the presence of micro- and macrofauna that feed on fungi, and agricultural practices including tillage (which causes a marked reduction in levels) and use of chemical pesticides (Scheepmaker and Butt, 2010).

Some pesticides—in particular, fungicides—are antagonistic or lethal to EPF, and hence it is important to determine their compatibility if they are to be used together in an IPM program (Fabrice et al., 2013; Pelizza et al., 2015). Manufacturers of microbial biopesticides provide technical information to farmers and growers that list which chemical pesticides are compatible/incompatible with the microbial agent, but they generally do not provide details of the methods used to test compatibility. Compatibility studies are normally done using in vitro assays of microbial growth or survival during exposure to a test pesticide (eg, Jaros-Su et al., 1999; Bruck, 2009; Schumacher and Poehling, 2012; Ribeiro et al., 2014). These tests are often designed to mimic the types of exposure that will occur in the field. The type of test influences the results, so it is usually a good idea to conduct more than one type of test. In some cases there can be false-positive results in which inhibition observed in vitro does not translate to field-scale effects. This can occur as a result of compartmentalization of the chemical pesticide within plant tissue, lower pesticide concentrations encountered under field conditions, or drying of pesticide residues on foliar surfaces (Inglis et al., 2001; Cuthbertson et al., 2005). Even if a chemical pesticide is seemingly incompatible with an EPF, it may be possible to prevent negative effects on the fungus by careful choice of formulation. For example laboratory studies suggest that formulating conidia in oils can protect them from fungicide damage (Lopes et al., 2011).

As mentioned previously, most biopesticides are developed as stand-alone treatments, and the ways in which they interact with other pest management tools in an IPM context are often overlooked (Lacey and Shapiro-Ilan, 2008). A number of EPF biopesticides work well as single treatment solutions, such as *M. acridum* targeted against locusts and grasshoppers (Lomer et al., 2001). However, in most circumstances EPF biopesticides are best suited to use in an integrated system where they can contribute to incremental improvements in sustainable pest control when combined with other pest management tools (Lacey and Shapiro-Ilan, 2008). Unfortunately, there has been a lack of published research in this area although this is likely to change in the future as IPM approaches become used more widely in the world's main agricultural regions. In Europe and North America, production of greenhouse edible crops such as tomato, pepper, and cucurbits relies heavily on use of "macro" biocontrol agents (predators and parasitoids) for pest management. Pests such as western flower thrips (*Frankliniella occidentalis*) and two-spotted spider mite (*Tetranychus urticae*) are routinely managed using trickle release applications of predatory mites. However, there are often periods in the season when control breaks down. For example, on tomato crops, predator movement and prey location is impaired by the action of the dense layer of glandular trichomes on the leaves, which can result in unsatisfactory levels of control. In the past, synthetic chemical pesticides have been used as supplementary treatments during periods of poor predator activity, but their use has become limited because of the evolution of resistance in the target pest populations. EPF biopesticides seem particularly well suited to fill this gap. For example, high levels (>75%) of control of western flower thrips populations were obtained on greenhouse cucumber crops using consecutive sprays of commercial preparations of *B. bassiana s.l.*, while populations of predatory mites, *Amblyseius cucumeris* were unaffected (Jacobson et al., 2001). Similarly, sprays of *B. bassiana s.l.* gave up to 97% control of two-spotted spider mite eggs, nymphs, and adults on a tomato crop within 7 days of application, while having minimal effect on the predator *Phytoseiulus persimilis* (Chandler et al., 2005). Elsewhere, commercial products based on *M. anisopliae s.l.*, *B. bassiana s.l.*, and *Isaria fumosorosea* incorporated individually in 8-week rotation programs with commonly used chemical insecticides were found to give the same level of control of western flower thrips on greenhouse chrysanthemum crops as a rotation program just containing the chemical insecticides (Kivett et al., 2015). Incorporating the EPF into the rotation was proposed as a way of reducing selection pressure for resistance development in western flower thrips populations, and also enabled cost savings of up to 34% compared to insecticide-only programs, depending on the type of biopesticide used (Kivett et al., 2015). Incorporating EPF with other selective insecticides may have merit for controlling more than one pest affecting a crop, provided that the application of individual agents in the IPM system is timed to match periods of peak activity for each pest. An IPM system comprised of—in order—one application each of petroleum spray oil, *B. bassiana s.l.*, azadirachtin, and *Bacillus thuringiensis*, each applied at an interval of 15 days, significantly reduced populations of red spider mite, *Tetranychus marianae*, and tomato fruitworm, *Helicoverpa armigera*, on tomato crops in Guam, USA (Reddy and Miller, 2014). This IPM program of sprays gave better control than the standard grower practice of 13 consecutive sprays of carbaryl or malathion or six consecutive sprays of *B. bassiana s.l.* or *B. thuringiensis* (Reddy and Miller, 2014).

EPF are not yet used widely as classical biocontrol agents or in conservation control; however, there are examples of successful exploitation of both strategies. Four genera in the Entomophthoromycota, five genera in the Hypocreales, and five genera in the Microsporidia have been used in classical control programs, with 67 documented releases against pests in the Thysanoptera, Hemiptera, Orthoptera, Coleoptera, Diptera, Lepidoptera, and Acari (Hajek and Delalibera, 2010). The establishment success rate was estimated at around 60%, and there are no records of any of these releases having a negative effect on nontarget species in the release environment, although few of the programs appear to have contributed to suppression of the target pest population (Hajek et al., 2007), exceptions being the introduction of *Zoophthora radicans* into Australia for control of spotted alfalfa aphid, *Therioaphis trifolii* (Milner et al., 1982) and regulation of gypsy moth, *Lymantria dispar*, populations by *Entomophthora maimaiga* in the United States (Hajek et al., 1996). Ongoing classic biological control programs include investigation of *Neozygites tanajoae*, which naturally regulates populations of cassava green mite *Mononychellus tanajoae* in Brazil, and is being studied as a classic control agent for use in Africa where the pest is invasive and causes significant damage to cassava production (Delalibera et al., 1992; Hountondji et al., 2007).

Although EPF are common components of the natural enemy complex in many agro-ecosystems, their role in the regulation of pest populations in agricultural crops has often been overlooked. Altering crop management practices to enhance natural EPF activity could be a valuable contribution to IPM programs through conservation biological control (CBC) although a lack of fundamental knowledge about EPF ecology may be impairing the development and implementation of CBC (Meyling and Eilenberg, 2007). Research has shown that reducing or avoiding fungicide sprays that were antagonistic to *Neozygites floridana* improved the natural activity of this entomophthoralean fungus against two-spotted spider mite, *Tetranychus urticae*, on broad acre crops (Smitley et al., 1986). *Pandora neoaphidis,* which regularly causes epizootics in aphid populations on crops, has been investigated for its CBC potential in the United Kingdom by manipulating field margins, which act as a refuge for the fungus by providing alternative aphid hosts on wild plants (Ekesi et al., 2005). In the southeast United States, fungal epizootics in populations of cotton aphids, *Aphis gossypii*, are caused by *Neozygites fressenii* (Steinkraus et al., 1991, 1995). These disease outbreaks occur at the same time of year over wide areas and reduce aphid populations by 80% within 4 days. Epizootics are likely to occur if the fungus reaches a threshold prevalence level of 15% within an aphid population, while fungus-infected aphids can first be detected at low levels up to 10 days in advance of the epizootics. This has formed the basis of a prediction service set up for cotton farmers in 10 U.S. states. Farmers send in samples of aphids to the service for diagnosis and are informed about the likelihood of an epizootic occurring, based on the 15% threshold level. If an epizootic is likely, they have the opportunity to hold back on spraying insecticides, which is estimated to save millions of dollars in reduced spray costs every year (Hollingsworth et al., 1995; Pell et al., 2010).

5.7 FUTURE OPPORTUNITIES AND CHALLENGES

EPF and other entomopathogens have been investigated as microbial control agents for many years, but the number of commercial products available to farmers and growers has always been relatively small compared to the availability of conventional chemical pesticides (Marrone, 2007). However, the loss of large numbers of conventional pesticides, through the evolution of resistance in target pests or market withdrawal following government safety reviews, is creating a new market for biopesticides. If the current level of new biopesticide product authorizations is maintained, then numbers of marketed biopesticide products will surpass those of conventional pesticides in some regions of the world in the next few years. Since the development costs for biopesticides are markedly less than those for conventional chemical pesticides, it should be possible to develop biopesticide products tailored for specific markets in a profitable way. As far as EPF are concerned, the challenge is to develop new products that meet farmers' needs, which must include the development of IPM strategies that take advantage of the attractive properties of EPF. This will require a shift of emphasis on the way that EPF biopesticides have been developed, particularly by making better use of new knowledge about their life histories, ecology and evolution. Traditionally, EPF biopesticides have been used as chemical pesticide analogues, with the emphasis on achieving rapid kill of pest populations but with little regard to the ecological properties of the fungus. However recent discoveries about the ability of some EPF species to function as endophytes and rhizosphere colonizers presents new opportunities for using them in the field, for example applying them as seed dressings or soil inoculants to colonize crop plants. We should also be able to use the detailed EPF phylogenies that are now coming through, in combination with new knowledge on fungal phenotypes, to select EPF species and strains with better intrinsic effectiveness and which may be finely tuned to particular crop protection situations, such as better tolerance of UV radiation for prolonged persistence on plant surface for outdoor crops, or selection of specialists for individual pest genera or species that enable nontarget effects to be kept to the minimum. As more arthropod and fungal genomes become available from high throughput sequencing programs, we will have a better understanding of the mechanisms by which host arthropods defend themselves against infection, and the measures that have evolved in EPF to avoid or disable the host immune system. This should also lead to new opportunities for improving EPF biopesticide effectiveness, for example through gene editing to improve strain performance, or the identification of potentiating compounds that impede or delay the arthropod immune response to fungal infection and which could be applied as EPF co-formulants. Finally, better knowledge of the ecology of the Entomophthoromycota and specialist hypocrealean fungi, which cause natural outbreaks in pest populations, should lead to new opportunities for conservation biocontrol, to make agricultural ecosystems more resistant to the development of pest populations.

REFERENCES

Andersen, M., Magan, N., Mead, A., Chandler, D., 2006. Development of a population-based threshold model of conidial germination for analysing the effects of physiological manipulation on the stress tolerance and infectivity of insect pathogenic fungi. Environ. Microbiol. 8, 1625–1634.

Arthurs, S.P., Thomas, M.B., 2001. Effects of temperature and relative humidity on sporulation of *Metarhizium anisopliae* var. *acridum* in mycosed cadavers of *Schistocerca gregaria*. J. Invertebr. Pathol. 78, 59–65.

Askary, H., Carriere, Y., Belanget, R.R., Brodeur, J., 1998. Pathogenicity of the fungus *Verticillium lecanii* to aphids and powdery mildew. Biocontrol Sci. Technol. 8, 23–32.

Bateman, R.P., Carey, M., Moore, D., Prior, C., 1993. The enhanced infectivity of *Metarhizium flavoviride* in oil formulations to desert locusts at low humidities. Ann. App. Biol. 122, 145–152.

Baverstock, J., Roy, H.E., Pell, J.K., 2010. Entomopathogenic fungi and insect behaviour: from unsuspecting hosts to targeted vectors. BioControl 55, 89–102.

Begun, D.J., Whitley, P., 2000. Adaptive evolution of relish, a Drosophila NF-kappaB/IkappaB protein. Genetics 154, 1231–1238.

Behie, S.W., Zelisko, P.M., Bidochka, M.J., 2012. Endophytic insect-parasitic fungi translocate nitrogen directly from insects to plants. Science 336, 1576–1577.

Behie, S.W., Jones, S.J., Bidochka, M.J., 2015. Plant tissue localization of the endophytic insect pathogenic fungi *Metarhizium* and *Beauveria*. Fungal Ecol. 13, 112–119.

Behle, R.W., 2006. Importance of direct spray and spray residue contact for infection of *Trichoplusia ni* larvae by field applications of *Beauveria bassiana*. J. Econ. Entomol. 99, 1120–1128.

Bidochka, M.J., Kamp, A.M., Lavender, T.M., Dekoning, J., De Croos, J.N.A., 2001. Habitat association in two genetic groups of the insect-pathogenic fungus *Metarhizium anisopliae*: uncovering cryptic species? Appl. Environ. Microb. 67, 1335–1342.

Bidochka, M.J., Menzies, F.V., Kamp, A.M., 2002. Genetic groups of the insect-pathogenic fungus *Beauveria bassiana* are associated with habitat and thermal growth preferences. Arch. Microbiol. 178, 531–537.

Bing, L., Lewis, L.C., 1991. Suppression of *Ostrinia nubilalis* (Hubner) (Lepidoptera: Pyralidae) by endophytic *Beauveria bassiana* (Balsamo) Vuill. Environ. Entomol. 20, 1207–1211.

Bing, L., Lewis, L.C., 1992. Endophytic *Beauveria bassiana* (Balsamo) Vuill in corn: the influence on the plant growth stage and *Ostrinia nubilalis* (Hubner). Biocontrol Sci. Technol. 1, 29–47.

Bischoff, J.F., Rehner, S.A., Humber, R.A., 2009. A multilocus phylogeny of the *Metarhizium ansiopliae* lineage. Mycologia 101, 512–530.

Blanford, S., Shi, W.P., Christian, R., Marden, J.H., Koekemoer, L.L., Brooke, B.D., Coetzee, M., Read, A.F., Thomas, M.B., 2011. Lethal and pre-lethal effects of a fungal biopesticide contribute to substantial and rapid control of malaria vectors. PLoS One 6, e23591. http://dx.doi.org/10.1371/journal.pone.0023591.

Boetel, M.A., Majumdar, A., Jaronski, S.T., Horsley, R.D., 2012. Cover crop and conidia delivery system impacts on soil persistence of *Metarhizium anisopliae* (Hypocreales: Clavicipitaceae) in sugarbeet. Biocontrol Sci. Technol. 22, 1284–1304.

Boomsma, J.J., Jensen, A.B., Meyling, N.V., Eilenberg, J., 2014. Evolutionary interaction networks of insect pathogenic fungi. Annu. Rev. Entomol. 59, 467–485.

Bouamama, N., Vidal, C., Fargues, J., 2010. Effects of fluctuating moisture and temperature regimes on the persistence of quiescent conidia of *Isaria fumosorosea*. J. Invertebr. Pathol. 105, 139–144.

Boucias, D.G., Pendland, J.C., 1991. Attachment of mycopathogens to cuticle: the initial event of mycosis in arthropod hosts. In: Cole, G.T., Hoch, H.C. (Eds.), The Fungal Spore and Disease Initiation in Plants and Animals. Plenum, New York, USA, pp. 101–128.

Braga, G.U.L., Flint, S.D., Miller, C.D., Anderson, A.J., Roberts, D.W., 2001. Both solar UVA and UVB radiation impair conidial culturability and delay germination in the entomopathogenic fungus *Metarhizium anisopliae*. Photochem. Photobiol. 74, 734–739.

Brown, G.C., Hasibuan, R., 1995. Conidial discharge and transmission efficiency of *Neozygites floridana*, an entomopathogenic fungus infecting two-spotted spider mites under laboratory conditions. J. Invertebr. Pathol. 65, 10–16.

Bruck, D.J., 2005. Ecology of *Metarhizium anisopliae* in soilless potting media and the rhizosphere: implications for pest management. Biol. Control 32, 155–163.

Bruck, D.J., 2009. Impact of fungicides on *Metarhizium anisopliae* in the rhizosphere, bulk soil and *in vitro*. Biocontrol 54, 597–606.

Bruck, D.J., 2010. Fungal entomopathogens in the rhizosphere. BioControl 55, 103–112.

Bugeme, D.M., Maniania, N.K., Knapp, M., Boga, H.I., 2008. Effect of temperature on virulence of *Beauveria bassiana* and *Metarhizium anisopliae* isolates to *Tetranychus evansi*. Exp. Appl. Acarol. 46, 275–285.

Bukhari, T., Takken, W., Koenraadt, C.J.M., 2013. Biological tools for control of larval stages of malaria vectors – a review. Biocontrol Sci. Technol. 23, 987–1023.

Butt, T.M., Ibrahim, L., Clark, S.J., Beckett, A., 1995. The germination behavior of *Metarhizium anisopliae* on the surface of aphid and flea beetle cuticles. Mycol. Res. 99, 945–950.

Butt, T.M., Hajek, A.E., Humber, R.A., 1996. Gypsy moth immune defenses in response to hyphal bodies and natural protoplasts of entomophthoralean fungi. J. Invertebr. Pathol. 68, 278–285.

Cabanillas, H.E., Jones, W.A., 2009. Effects of temperature and culture media on vegetative growth of an entomopathogenic fungus *Isaria* sp. (Hypocreales: Clavicipitaceae) naturally affecting the whitefly, *Bemisia tabaci* in Texas. Mycopathologia 167, 263–271.

Castrillo, L.A., Griggs, M.H., Liu, H.P., Bauer, L.S., Vandenberg, J.D., 2010. Assessing deposition and persistence of *Beauveria bassiana* GHA (Ascomycota: Hypocreales) applied for control of the emerald ash borer, *Agrilus planipennis* (Coleoptera: Buprestidae), in a commercial tree nursery. Biol. Control 54, 61–67.

Chambers, M.C., Lightfield, K.L., Schneider, D.S., 2012. How the fly balances its ability to combat different pathogens. PLoS Pathog. 8, e1002970. http://dx.doi.org/10.1371/journal.ppat.1002970.

Chandler, D., Hay, D.B., Reid, A.P., 1997. Sampling and occurrence of entomopathogenic fungi and nematodes in UK soils. Appl. Soil Ecol. 5, 133–141.

Chandler, D., Davidson, G., Pell, J.K., Ball, B.V., Shaw, K., Sunderland, K.D., 2000. Fungal biocontrol of Acari. Biocontrol Sci. Technol. 10, 357–384.

Chandler, D., Davidson, G., Jacobson, R.J., 2005. Laboratory and glasshouse evaluation of entomopathogenic fungi against the two-spotted spider mite, *Tetranychus urticae* (Acari: Tetranychidae) on tomato, *Lycopersicon esculentum*. Biocontrol Sci. Technol. 15, 37–54.

Chandler, D., Bailey, A.S., Tatchell, G.M., Davidson, G., Greaves, J., Grant, W.P., 2011. The development, regulation and use of biopesticides for integrated pest management. Philos. Trans. R. Soc. B 366, 1987–1998.

Chouvenc, T., Su, N.Y., Elliott, M.L., 2008. Interaction between the subterranean termite *Reticulitermes flavipes* (Isoptera: Rhinotermitidae) and the entomopathogenic fungius *Metarhizium anisopliae* in foraging arenas. J. Econ. Entomol. 101, 885–893.

Cohen, E., Joseph, T., 2009. Photostabilization of *Beauveria bassiana* conidia using anionic dyes. Appl. Clay Sci. 42, 569–574.

Corradi, N., 2015. Microsporidia: Eukaryotic intracellular parasites shaped by gene loss and horizontal gene transfers. Ann. Rev. Microbiol. 69, 167–183.

Cuthbertson, A.G.S., Walters, K.F.A., Deppe, C., 2005. Compatibility of the entomopathogenic fungus *Lecanicillium muscarium* and insecticides for eradication of sweetpotato whitefly, *Bemisia tabaci*. Mycopathologia 160, 35–41.

Davidson, G., Phelps, K., Sunderland, K.D., Pell, J.K., Ball, B.V., Shaw, K.E., Chandler, D., 2003. Study of temperature-growth interactions of entomopathogenic fungi with potential for control of *Varroa destructor* using a nonlinear model of poikilotherm development. J. Appl. Microbiol. 94, 816–825.

Delalibera Jr., I., Sosa Gomez, D.R., De Moraes, G.J., Alencar, J.A., Farias Araujo, W., 1992. Infection of *Mononychellus tanajoa* (Acari: Tetranychidae) by the fungus *Neozygites* sp. (Entomophthorales) in northeastern Brazil. Fla. Entomol. 75, 145–147.

Dillon, R.J., Charnley, A.K., 1990. Initiation of germination in conidia of the entomopathogenic fungus, *Metarhizium anisopliae*. Mycol. Res. 94, 299–304.

Dong, Y.M., Morton, J.C., Ramirez, J.L., Souza-Neto, J.A., Dimopoulos, G., 2012. The entomopathogenic fungus *Beauveria bassiana* activate toll and JAK-STAT pathway-controlled effector genes and anti-dengue activity in *Aedes aegypti*. Insect Biochem. Mol. Biol. 42, 126–132.

Donzelli, B.G.G., Krasnoff, S.B., Sun-Moon, Y., Churchill, A.C.L., Gibson, D.M., 2012. Genetic basis of destruxin production in the entomopathogen *Metarhizium robertsii*. Curr. Genet. 58, 105–116.

Dromph, K.M., 2001. Dispersal of entomopathogenic fungi by collembolans. Soil Biol. Biochem. 33, 2047–2051.

Dubovskiy, I.M., Whitten, M.A., Kryukov, V.Y., Yaroslavtseva, O.N., Grizanova, E.V., Greig, C., Mukherjee, K., Vilcinskas, A., Mitkovets, P.V., Glupov, V.V., Butt, T.M., 2013. More than a colour change: insect melanism, disease resistance and fecundity. Proc. R. Soc. Lond. B Biol. Sci. 280, 20130584. http://dx.doi.org/10.1098/rspb.2013.0584.

Eilenberg, J., Pell, J.K., 2007. Ecology. In: Keller, S. (Ed.), Arthropod Pathogenic Entomophthorales: Biology, Ecology, Identification. European Cooperation in the Field Scientific and Technical Research (COST) Food and Agriculture, EUR 22829, Brussels, pp. 7–26.

Eilenberg, J., Bresciani, J., Latge, J.P., 1986. Ultrastructural studies of primary spore formation and discharge in the genus *Entomophthora*. J. Invertebr. Pathol. 48, 318–324.

Ekengren, S., Hultmark, D., 1999. *Drosophila* cecropin as an antifungal agent. Insect Biochem. Mol. Biol. 29, 965–972.

Ekesi, S., Shah, P.A., Clark, S.J., Pell, J.K., 2005. Conservation biological control with the fungal pathogen *Pandora neoaphidis*: implications of aphid species, host plant and predator foraging. Agr. For. Entomol. 7, 21–30.

Evans, B.G., Jordan, K.S., Brownbridge, M., Hallett, R.H., 2015. Effect of temperature and host life stage on efficacy of soil entomopathogens against the swede midge (Diptera: Cecidomyiidae). J. Econ. Entomol. 108, 473–483.

Fabrice, C.E.S., Tonussi, R.L., Orlandelli, R.C., Lourenco, D.A.L., Pamphile, J.A., 2013. Compatibility of entomopathogenic fungus *Metarhizium anisopliae* (Metschnikoff) Sorokin with fungicide thiophanate-methyl assessed by germination speed parameter. J. Food Agri. Environ. 11, 368–372.

Fang, W.G., St Leger, R.J., 2012. Enhanced UV resistance and improved killing of malaria mosquitoes by photolyase transgenic entomopathogenic fungi. PLoS One 7, e43069. http://dx.doi.org/10.1371/journal.pone.0043069.

Farenhorst, M., Mouatcho, J.C., Kikankie, C.K., Brooke, B.D., Hunt, R.H., Thomas, M.B., Koekemoer, L.L., Knols, B.G.J., Coetzee, M., 2009. Fungal infection counters insecticide resistance in African malaria mosquitoes. Proc. Natl. Acad. Sci. U.S.A. 106, 17443–17447.

Farenhorst, M., Knols, B.G.J., Thomas, M.B., Howard, A.F.V., Takken, W., Rowland, M., N'Guessan, R., 2010. Synergy in efficacy of fungal entomopathogens and permethrin against West African insecticide-resistant *Anopheles gambiae* mosquitoes. PLoS One 5, 1–10. http://dx.doi.org/10.1371/journal.pone.0012081 e12081.

Faria, M.R., Wraight, S.P., 2007. Mycoinsecticides and mycoacaricides: a comprehensive list with worldwide coverage and international classification of formulation types. Biol. Control 43, 237–256.

Feng, M.G., Poprawski, T.J., Khachatourians, G.G., 1994. Production, formulation and application of the entomopathogenic fungus *Beauveria bassiana* for insect control: current status. Biocontrol Sci. Technol. 4, 3–34.

Fernandes, E.K.K., Rangel, D.E.N., Moraes, A.M.L., Bittencourt, V.R.E.P., Roberts, D.W., 2007. Variability in tolerance to UV-B radiation among *Beauveria* spp. isolates. J. Invertebr. Pathol. 96, 237–243.

Fernandes, E.K.K., Moraes, A.M.L., Pacheco, R.S., Rangel, D.E.N., Miller, M.P., Bittencourt, V.R.E.P., Roberts, D.W., 2009. Genetic diversity among Brazilian isolates of *Beauveria bassiana*: comparisons with non-Brazilian isolates and other *Beauveria* species. J. App. Microbiol. 107, 760–774.

Fernandes, E.K.K., Bittencourt, V.R.E.P., Roberts, D.W., 2012. Perspectives on the potential of entomopathogenic fungi in biological control of ticks. Exp. Parasitol. 130, 300–305.

Ferro, D.N., Southwick, E.E., 1984. Microclimates of small arthropods: estimating humidity within the leaf boundary layer. Environ. Entomol. 13, 926–929.

Ganesan, S., Aggarwal, K., Paquette, N., Silverman, N., 2011. NF-kappa B/Rel proteins and the humoral immune responses of *Drosophila melanogaster*. Curr. Top. Microbiol. 39, 25–60.

Gan-Mor, S., Matthews, G.A., 2003. Recent developments in sprayers for application of biopesticides – an overview. Biosyst. Eng. 84, 119–125.

Garrido-Jurado, I., Ruano, F., Campos, M., Quesada-Moraga, E., 2011. Effects of soil treatments with entomopathogenic fungi on soil dwelling non-target arthropods at a commercial olive orchard. Biol. Control 59, 239–244.

Gatarayiha, M.C., Laing, M.D., Miller, R.M., 2010. Effects of crop type on persistence and control efficacy of *Beauveria bassiana* against the two spotted spider mite. Biocontrol 55, 767–776.

Gatarayiha, M.C., Laing, M.D., Miller, R.M., 2011. Field evaluation of *Beauveria bassiana* efficacy for the control of *Tetranychus urticae* Koch (Acari: Tetranychidae). J. Appl. Entomol. 135, 582–592.

Georgel, P., Naitza, S., Kappler, C., Ferrandon, D., Zachary, D., Swimmer, C., Kopczynski, C., Duyk, G., Reichart, J.M., Hoffmann, J.A., 2001. Drosophila immune deficiency (IMD) is a death domain protein that activates antibacterial defense and can promote apoptosis. Dev. Cell 1, 503–514.

Glare, T.R., Milner, R.J., 1991. Ecology of entomopathogenic fungi. In: Arora, D.K., Ajello, L., Mukerji, K.G. (Eds.), Handbook of Applied Mycology. Humans, Animals and Insects, vol. 2. Dekker, New York, USA, pp. 547–612.

Glare, T.R., Milner, R.J., Chilvers, G.A., 1986. The effect of environmental factors on the production, discharge, and germination of primary conidia of *Zoophthora phalloides* Batko. J. Invertebr. Pathol. 48, 275–283.

Glare, T., Caradus, J., Gelernter, W., Jackson, Y., Keyhani, N., Kohl, J., Marrone, P., Morin, L., Stewart, A., 2012. Have biopesticides come of age? Trends Biotechnol. 30, 250–258.

Goettel, M.S., Hajek, A.E., Siegel, J.P., Evans, H.C., 2001. Safety of fungal biocontrol agents. In: Butt, T., Jackson, C., Magan, N. (Eds.), Fungi as Biocontrol Agents – Progress Problems and Potential. CABI Press, Wallingford, UK, pp. 347–375.

Gottar, M., Gobert, V., Matskevich, A.A., Reichhart, J.M., Wang, C.S., Butt, T.M., Belvin, M., Hoffmann, J.A., Ferrandon, D., 2006. Dual detection of fungal infections in *Drosophila* via recognition of glucans and sensing of virulence factors. Cell 127, 1425–1437.

Gross, J., Muller, C., Vilcinskas, A., Hilker, M., 1998. Antimicrobial activity of exocrine glandular secretions, haemolymph and larval regurgitate of the mustard leaf beetle *Phaedon cochleariae*. J. Invertebr. Pathol. 72, 296–303.

Gryganskyi, A.P., Humber, R.A., Miadlikovska, J., Smith, M.E., Wu, S., Voigt, K., Walther, G., Anishchenko, I.M., Vilgalys, R., 2012. Molecular phylogeny of the *Entomophthoromycota*. Mol. Phylogen. Evol. 65, 682–694.

Gryganskyi, A.P., Humber, R.A., Smith, M.E., Hodge, K., Huang, B., Voigt, K., Vilgalys, R., 2013a. Phylogenetic lineages in *Entomophthoromycota*. Persoonia 30, 94–105.

Gryganskyi, A.P., Humber, R.A., Stajich, J.E., Mullens, B., Anishchenko, I.M., Vilgalys, R., 2013b. Sequential utilization of hosts from different fly families by genetically distinct, sympatric populations within the *Entomophthora muscae* species complex. PLoS One 8, e71168. http://dx.doi.org/10.1371/journal.pone.0071168.

Guerrero-Guerra, C., Reyes-Montes, M.R., Toriello, C., Hernández-Velázquez, V., Santiago-López, I., Mora-Palomino, L., Calderón-Segura, M.E., Fernández, S.D., Calderón-Ezquerro, C., 2013. Study of the persistence and viability of *Metarhizium acridum* in Mexico's agricultural area. Aerobiologia 29, 249–261.

Guo, Y.Y., An, Z., Shi, W.P., 2012. Control of grasshoppers by combined application of *Paranosema locustae* and an insect growth regulator (IGR) (Cascade) in Rangelands in China. J. Econ. Entomol. 105, 1915–1920.

Guzman-Franco, A.W., Clark, S.J., Alderson, P.G., Pell, J.K., 2008. Effect of temperature on the *in vitro* radial growth of *Zoophthora radicans* and *Pandora blunckii*, two co-occurring fungal pathogens of the diamondback moth *Plutella xylostella*. Biocontrol 53, 501–516.

Haine, E.R., Moret, Y., Siva-Jothy, M.T., Rolff, J., 2008. Antimicrobial defense and persistent infection in insects. Science 322, 1257–1259.

Hajek, A.E., Delalibera Jr., L., 2010. Fungal pathogens as classical biological control agents against arthropods. Biocontrol 55, 147–158.

Hajek, A.E., St Leger, R.J., 1994. Interactions between fungal pathogens and insect hosts. Annu. Rev. Entomol. 39, 293–322.

Hajek, A.E., Elkinton, J.S., Witcosky, J.J., 1996. Introduction and spread of the fungal pathogen *Entomophaga maimaiga* (Zygomycetes: Entomophthorales) along the leading edge of gypsy moth (Lepidoptera: Lymantriidae) spread. Environ. Entomol. 25, 1235–1247.

Hajek, A.E., Delalibera Jr., I., McManis, L., 2007. Introduction of exotic pathogens and documentation of their establishment and impact. In: Lacey, L.A., Kaya, H.K. (Eds.), Field Manual of Techniques in Invertebrate Pathology: Application and Evaluation of Pathogens for Control of Insects and Other Invertebrate Pests, second ed. Springer, Dordrecht, The Netherlands, pp. 299–325.

Hallsworth, J.E., Magan, N., 1995. Manipulation of intracellular glycerol and erythritol enhances germination of conidia at low water availability. Microbiology 141, 1109–1115.

Henk, D.A., Vilgalys, R., 2007. Molecular phylogeny suggests a single origin of insect symbiosis in the Pucciniomycetes with support for some relationships within the genus *Septobasidium*. Am. J. Bot. 94, 1515–1526.

Hesketh, H., Roy, H.E., Eilenberg, J., Pell, J.K., Hails, R.S., 2010. Challenges in modelling complexity of fungal entomopathogens in seminatural populations of insects. BioControl 55, 55–73.

Higes, M., Martin-Hernandez, R., Botias, C., Bailon, E.G., Gonzalez-Porto, A.V., Barrios, L., del Nozal, M.J., Bernal, J.L., Jimenez, J.J., Palencia, P.G., Meana, A., 2008. How natural infection by *Nosema ceranae* causes honeybee colony collapse. Environ. Microbiol. 10, 2659–2669.

Holder, D.J., Kirkland, B.H., Lewis, M.H., Keyhani, N.O., 2007. Surface characteristics of the entomopathogenic fungus *Beauveria* (*Cordyceps*) *bassiana*. Microbiology 153, 3448–3457.

Hollingsworth, R.G., Steinkraus, D.C., McNew, R.W., 1995. Sampling to predict fungal epizootics in cotton aphids (Homoptera: Aphididae). Environ. Entomol. 24, 1414–1421.

Hou, R.F., Chang, J., 1985. Cellular defense response to *Beauveria bassiana* in the silkworm *Bombyx mori*. Appl. Entomol. Zool. 20, 118–125.

Hountondji, F.C.C., Hanna, R., Cherry, A.J., Sabelis, M.W., Agboton, B., Korie, S., 2007. Scaling-up tests on virulence of the cassava green mite fungal pathogen *Neozygites tanajoae* (Entomophthorales: Neozygitaceae) under controlled conditions: first observations at the population level. Exp. Appl. Acarol. 41, 153–168.

Hu, X., Xiao, G., Zheng, P., Shang, Y., Su, Y., Zhang, X., Liu, X., Zhan, S., St Leger, R.J., Wang, C., 2014. Trajectory and genomic determinants of fungal-pathogen speciation and host adaptation. Proc. Natl. Acad. Sci. U.S.A. 111, 16796–16801.

Hughes, D.P., Andersen, S.B., Hywel-Jones, N.L., Himaman, W., Billen, J., Boomsma, J.J., 2011. Behavioural mechanisms and morphological symptoms of zombie ants dying from fungal infection. BMC Ecol. 11, 13.

Humber, R.A., 2008. Evolution of entomopathogenicity in fungi. J. Invertebr. Pathol. 98, 262–266.

Humber, R.A., 2012. Entomophthoromycota: a new phylum and reclassification for entomophthoroid fungi. Mycotaxon 120, 477–492.

Hummel, R.L., Walgenbach, J.F., Barbercheck, M.E., Kennedy, G.G., Hoyt, G.D., Arellano, C., 2002. Effects of production practices on soil-borne entomopathogens in western North Carolina vegetable systems. Environ. Entomol. 31, 84–91.

Hung, S.-Y., Boucias, D.G., Vey, A., 1993. Effect of *Beauveria bassiana* and *Candida albicans* on the cellular defense response of *Spodoptera exigua*. J. Invertebr. Pathol. 61, 179–187.

Inglis, G.D., Goettel, M.S., Johnson, D.L., 1995. Influence of ultraviolet light protectants on persistence of the entomopathogenic fungus, *Beauveria bassiana*. Biol. Control 5, 581–590.

Inglis, G.D., Ivie, T.J., Duke, G.M., Goettel, M.S., 2000. Influence of rain and conidial formulation on persistence of *Beauveria bassiana* on potato leaves and Colorado potato beetle larvae. Biol. Control 18, 55–64.

Inglis, G.D., Goettel, M.S., Butt, T.M., Strasser, H., 2001. Use of hyphomycetous fungi for managing insect pests. In: Butt, T.M., Jackson, C., Magan, N. (Eds.), Fungi as Biocontrol Agents: Progress Problems and Potential. CABI Publishing, Wallingford, UK, pp. 23–70.

Inyang, E.N., McCartney, H.A., Oyejola, B., Ibrahim, L., Pye, B.J., Archer, S.A., Butt, T.M., 2000. Effect of formulation, application and rain on the persistence of the entomogenous fungus *Metarhizium anisopliae* on oilseed rape. Mycol. Res. 104, 653–661.

Jaber, L.R., 2015. Grapevine leaf tissue colonization by the fungal entomopathogen *Beauveria bassiana s.l.* and its effect against downy mildew. BioControl 60, 103–112.

Jacobson, R.J., Chandler, D., Fenlon, J., Russell, K.M., 2001. Compatibility of *Beauveria bassiana* (Balsamo) Vuillemin with *Amblyseius cucumeris* Oudemans (Acarina: Phytoseiidae) to control *Frankliniella occidentalis* Pergande (Thysanoptera: Thripidae) on cucumber plants. Biocontrol Sci. Technol. 11, 381–400.

Jaronski, S.T., 2010. Ecological factors in the inundative use of fungal entomopathogens. BioControl 55, 159–185.

Jaros-Su, J., Groden, E., Zhang, J., 1999. Effects of selected fungicides and the timing of fungicide application on *Beauveria bassiana*-induced mortality of the Colorado potato beetle (Coleoptera: Chrysomelidae). Biol. Control 15, 259–269.

Jenkins, N.E., Goettel, M.S., 1997. Methods for mass-production of microbial control agents of grasshoppers and locusts. Mem. Entomol. Soc. Can. 171, 37–48.

Jiang, H., Vilcinskas, A., Kanost, M.R., 2010. Immunity in lepidopteran insects. In: Soderhall, K. (Ed.), Invertebrate Immunity. Advances in Experimental Medicine and Biology, vol. 708. Springer US, pp. 181–204.

Juneja, P., Lazzaro, B.P., 2009. Population genetics of insect immune responses. In: Rolff, J., Reynolds, S.E. (Eds.), Insect Function and Immunity: Evolution, Ecology and Mechanisms. Oxford University Press, Oxford, UK, pp. 206–224.

Kaaya, G.P., Okech, M.A., 1990. Horizontal transmission of mycotic infection in adult tsetse, *Glossina morsitans-morsitans*. Entomophaga 35, 589–600.

Keller, S., 2007. Fungal structures and biology. In: Keller, S. (Ed.), Arthropod Pathogenic Entomophthorales: Biology, Ecology, Identification. European Cooperation in the Field Scientific and Technical Research (COST) Food and Agriculture, EUR 22829, Brussels, pp. 27–54.

Keller, S., Schweizer, C., Keller, E., Brenner, H., 1997. Control of white grubs (*Melolontha melolontha* L.) by treating adults with the fungus *Beauveria brongniartii*. Biocontrol Sci. Technol. 7, 105–116.

Keller, S., Kessler, P., Schweizer, C., 2003. Distribution of insect pathogenic soil fungi in Switzerland with special reference to *Beauveria brongniartii* and *Metarhizium anisopliae*. BioControl 48, 307–319.

Kepler, R.M., Bruck, D.J., 2006. Examination of the interaction between the black vine weevil (Coleoptera: Curculionidae) and an entomopathogenic fungus reveals a new tritrophic interaction. Environ. Entomol. 35, 1021–1029.

Kepler, R.M., Sung, G.H., Ban, S., Nakagiri, A., Chen, M.J., Huang, B., Li, Z., Spatafora, J.W., 2012. New teleomorph combinations in the entomopathogenic genus *Metacordyceps*. Mycologia 104, 182–197.

Kepler, R.M., Humber, R.A., Bischoff, J.F., Rehner, S.A., 2014. Clarification of generic and species boundaries for *Metarhizium* and related fungi through multigene phylogenetics. Mycologia 106, 811–829.

Kikankie, C.K., Brooke, B.D., Knols, B.G.J., Koekemoer, L.L., Farenhorst, M., Hunt, R.H., Thomas, M.B., Coetzee, M., 2010. The infectivity of the entomopathogenic fungus *Beauveria bassiana* to insecticide-resistant and susceptible *Anopheles arabiensis* mosquitoes at two different temperatures. Malar. J. 9. http://dx.doi.org/10.1186/1475-2875-9-71.

Kim, J.J., Goettel, M.S., Gillespie, D.R., 2007. Potential of *Lecanicillium* species for dual microbial control of aphids and the cucumber powdery mildew fungus, *Sphaerotheca fuliginea*. Biol. Control 40, 327–332.

Kim, J.J., Goettel, M.S., Gillespie, D.R., 2008. Evaluation of *Lecanicillium longisporum*, Vertalec® for simultaneous suppression of cotton aphid, *Aphis gossypii*, and cucumber powdery mildew, *Sphaerotheca fuliginea*, on potted cucumbers. Biol. Control 45, 404–409.

Kivett, J.M., Cloyd, R.A., Bello, N.M., 2015. Insecticide rotation programs with entomopathogenic organisms for suppression of western flower thrips (Thysanoptera: Thripidae) adult populations under greenhouse conditions. J. Econ. Entomol. 108, 1936–1946.

Kreutz, J., Zimmermann, G., Vaupel, O., 2004. Horizontal transmission of the entomopathogenic fungus *Beauveria bassiana* among the spruce bark beetle, *Ips typographus* (Col., Scolytidae) in the laboratory and under field conditions. Biocontrol Sci. Technol. 14, 837–848.

Lacey, L.A., Shapiro-Ilan, D.I., 2008. Microbial control of insect pests in temperate orchard systems: potential for incorporation into IPM. Annu. Rev. Entomol. 53, 121–144.

Lacey, L.A., Frutos, R., Kaya, H.K., Vail, P., 2001. Insect pathogens as biological control agents: do they have a future? Biol. Control 21, 230–248.

Lacey, L.A., Grzywacz, D., Shapiro-Ilan, D.I., Frutos, R., Brownbridge, M., Goettel, M.S., 2015. Insect pathogens as biological control agents: back to the future. J. Invertebr. Pathol. 132, 1–41.

Lee, S.M., Yeo, W.H., Jee, H.J., Shin, S.C., Moon, Y.S., 1999. Effect of entomopathogenic fungi on growth of cucumber and *Rhizoctonia solani*. J. For. Sci. 62, 118–125.

Lemaitre, B., Hoffmann, J., 2007. The host defense of *Drosophila melanogaster*. Annu. Rev. Immunol. 25, 697–743.

Lemaitre, B., Reichhart, J., Hoffmann, J., 1997. Drosophila host defense: differential induction of antimicrobial peptide genes after infection by various classes of microorganisms. Proc. Natl. Acad. Sci. U.S.A. 94, 14614–14619.

Li, Z.Z., Alves, S.B., Roberts, D.W., Fan, M., Delalibera Jr., I., Tang, J., Lopes, R.B., Faria, M., Rangel, D.E.N., 2010. Biological control of insects in Brazil and China: history, current programs and reasons for their successes using entomopathogenic fungi. Biocontrol Sci. Technol. 20, 117–136.

Li, F., Shi, H.Q., Ying, S.H., Feng, M.G., 2015. Distinct contributions of one Fe- and two Cu/Zn-cofactored superoxide dismutases to antioxidation, UV tolerance and virulence of *Beauveria bassiana*. Fungal Genet. Biol. 81, 160–171.

Liu, M., Chaverri, P., Hodge, K.T., 2006. A taxonomic revision of the insect biocontrol fungus *Aschersonia aleyrodis*, its allies with white stromata and their *Hypocrella* sexual states. Mycol. Res. 110, 537–554.

Lomer, C.J., Bateman, R.P., Johnson, D.L., Langewald, J., Thomas, M., 2001. Biological control of locusts and grasshoppers. Annu. Rev. Entomol. 46, 667–702.

Lopes, R.B., Pauli, G., Mascarin, G.M., Faria, M., 2011. Protection of entomopathogenic conidia against chemical fungicides afforded by an oil-based formulation. Biocontrol Sci. Technol. 21, 125–137.

Lopez, D.C., Sword, G.A., 2015. The endophytic fungal entomopathogens *Beauveria bassiana* and *Purpureocillium lilacinum* enhance the growth of cultivated cotton (*Gossypium hirsutum*) and negatively affect survival of the cotton bollworm (*Helicoverpa zea*). Biol. Control 89, 53–60.

Lopez, D.C., Zhu-Salzman, K., Ek-Ramos, M.J., Sword, G.A., 2014. The entomopathogenic fungal endophytes *Purpureocillium lilacinum* (formerly *Paecilomyces lilacinus*) and *Beauveria bassiana* negatively affect cotton aphid reproduction under both greenhouse and field conditions. PLoS One 9, e103891. http://dx.doi.org/10.1371/journal.pone.0103891.

Lovett, B., St Leger, R.J., 2015. Stress is the rule rather than the exception for *Metarhizium*. Curr. Genet. 61, 253–261.

Malagocka, J., Grell, M.N., Lange, L., Eilenberg, J., Jensen, A.B., 2015. Transcriptome of an entomophthoralean fungus (*Pandora formicae*) shows molecular machinery adjusted for successful host exploitation and transmission. J. Invertebr. Pathol. 128, 47–56.

Mantzoukas, S., Chondrogiannis, C., Grammatikopoulos, G., 2015. Effects of three endophytic entomopathogens on sweet sorghum and on the larvae of the stalk borer *Sesamia nonagrioides*. Entom. Exp. Appl. 154, 78–87.

Marrone, P.G., 2007. Barriers to adoption of biological control agents and biological pesticides. CAB Rev. Perspect. Agric. Vet. Sci. Nutr. Nat. Resour. 2 (51) CABI Publishing, Wallingford, UK, 12pp.

McDonald, D.M., Nolan, R.A., 1995. Effects of relative humidity and temperature on *Entomophaga aulicae* conidium discharge from infected eastern hemlock looper larvae and subsequent conidium development. J. Invertebr. Pathol. 65, 83–90.

McLaughlin, D.J., Hibbett, D.S., Lutzoni, F., Spatafora, J.W., Vilgalys, R., 2009. The search for the fungal tree of life. Trends Microbiol. 17, 488–497.

Meyling, N.V., Eilenberg, J., 2007. Ecology of the entomopathogenic fungi *Beauveria bassiana* and *Metarhizium anisopliae* in temperate agroecosystems: potential for conservation biological control. Biol. Control 43, 145–155.

Meyling, N.V., Lubeck, M., Buckley, E.P., Eilenberg, J., Rehner, S.A., 2009. Community composition, host range and genetic structure of the fungal entomopathogen *Beauveria* in adjoining agricultural and seminatural habitats. Mol. Ecol. 6, 1282–1293.

Meyling, N.V., Thorup-Kristensen, K., Eilenberg, J., 2011. Below and aboveground abundance and distribution of fungal entomopathogens in experimental conventional and organic cropping. Biol. Control 59, 180–186.

Mietkiewski, R.T., Pell, J.K., Clark, S.J., 1997. Influence of pesticide use on the natural occurrence of entomopathogenic fungi in arable soils in the UK: field and laboratory comparisons. Biocontrol Sci. Technol. 7, 565–575.

Milner, R.J., Prior, C., 1994. Susceptibility of the Australian plague locust, *Chortoicetes terminifera*, and the wingless grasshopper, *Phaulacridium vittatum*, to the fungi *Metarhizium* spp. Biol. Control 4, 132–137.

Milner, R.J., Soper, R.S., Lutton, G.G., 1982. Field release of an Israeli strain of the fungus *Zoophthora radicans* (Brefeld) Batko for biological-control of *Therioaphis trifolii* (Monell) f. *maculata*. J. Aust. Entomol. Soc. 21, 113–118.

Moon, Y.S., Donzelli, B.G., Krasnoff, S., Mclane, H., Griggs, M.H., Cooke, P., Vandenberg, J.D., Gibson, D.M., Churchill, A.C.L., 2008. Agrobacterium-mediated disruption of a nonribosomal peptide synthetase gene in the invertebrate pathogen *Metarhizium anisopliae* reveals a peptide spore factor. Appl. Environ. Microb. 74, 4366–4380.

Moore, D., Reed, M., Lepatourel, G., Abraham, Y.J., Prior, C., 1992. Reduction of feeding by the desert locust, *Schistocerca gregaria*, after infection with *Metarhizium flavoviride*. J. Invertebr. Pathol. 60, 304–307.

Mullens, B.A., Rodriguez, J.L., 1985. Dynamics of *Entomophthora muscae* (Entomophthorales, Entomophthoraceae) conidial discharge from *Musca domestica* (Diptera, Muscidae) cadavers. Environ. Entomol. 14, 317–322.

Nana, P., Nchu, F., Ekesi, S., Boga, H.I., Kamtchouing, P., Maniania, N.K., 2015. Efficacy of spot spray application of *Metarhizium anisopliae* formulated in emulsifiable extract of *Calpurnia aurea* in attracting and infecting adult *Rhipicephalus appendiculatus* ticks in semi-field experiments. J. Pest Sci. 88, 613–619.

Newman, G.G., Carner, G.R., 1975. Environmental factors affecting conidial sporulation and germination of *Entomophthora gammae*. Environ. Entomol. 4, 615–618.

Oduor, G.I., Yaninek, J.S., De Moraes, G.J., Van Der Geest, L.P.S., 1997. The effect of pathogen dosage on the pathogenicity of *Neozygites floridana* (Zygomycetes: Entomophthorales) to *Mononychellus tanajoa* (Acari: Tetranychidae). J. Invertebr. Pathol. 70, 127–130.

Ortiz-Urquiza, A., Keyhani, N.O., 2013. Action on the surface: entomopathogenic fungi versus the insect cuticle. Insects 4, 357–374.

Ouyang, L., Xu, X.X., Freed, S.F., Gao, Y.F., Yu, J., Wang, S., Ju, W.Y., Zhang, Y.Q., Jin, F.L., 2015. Cecropins from *Plutella xylostella* and their interaction with *Metarhizium anisopliae*. PLoS One 10, e0142451. http://dx.doi.org/10.1371/journal.pone.0142451.

Ownley, B.H., Dee, M.M., Gwinn, K.D., 2008a. Effect of conidia seed treatment rate of entomopathogenic *Beauveria bassiana* 11-98 on endophytic colonization of tomato seedlings and control of *Rhizoctonia* disease. Phytopathology 98, S118.

Ownley, B.H., Griffin, M.R., Klingeman, W.E., Gwinn, K.D., Moulton, J.K., Pereira, R.M., 2008b. *Beauveria bassiana*: endophytic colonization and plant disease control. J. Invertebr. Pathol. 98, 267–270.

Ownley, B.H., Gwinn, K.D., Vega, F.E., 2010. Endophytic fungal entomopathogens with activity against plant pathogens: ecology and evolution. BioControl 55, 113–128.

Pal, S., St Leger, R.J., Wu, L.P., 2007. Fungal peptide destruxin A plays a specific role in suppressing the innate immune response on *Drosophila melanogaster*. J. Biol. Chem. 282, 8969–8977.

Paparazzo, F., Tellier, A., Stephan, W., Hutter, S., 2015. Survival rate and transcriptional response upon infection with the generalist parasite *Beauveria bassiana* in a world wide sample of *Drosophila melanogaster*. PLoS One 10, e0132129.

Paula, A.R., Carolino, A.T., Silva, C.P., Pereira, C.R., Samuels, R.I., 2013. Testing fungus impregnated cloths for the control of adult *Aedes aegypti* under natural conditions. Parasit. Vectors 6. http://dx.doi.org/10.1186/1756-3305-6-256.

Pedrini, N., Zhang, S., Juarez, M.P., Keyhani, N.O., 2010. Molecular characterization and expression analysis of a suite of cytochrome P450 enzymes implicated in insect hydrocarbon degradation in the entomopathogenic fungus *Beauveria bassiana*. Microbiology 156, 2549–2557.

Pedrini, N., Ortiz-Urquiza, A., Huarte-Bonnet, C., Fan, Y., Juárez, M., Keyhani, N.O., 2015. Tenebrionid secretions and a fungal benzoquinone oxidoreductase form competing components of an arms race between a host and pathogen. Proc. Natl. Acad. Sci. U.S.A. 112, 3651–3660.

Pelizza, S.A., Scorsetti, A.C., Fogel, M.N., Pacheco-Marino, S.G., Stenglein, S.A., Cabello, M.N., Lange, C.E., 2015. Compatibility between entomopathogenic fungi and biorational insecticides in toxicity against *Ronderosia bergi* under laboratory conditions. Biocontrol 60, 81–91.

Pell, J.K., Hannam, J.J., Steinkraus, D.C., 2010. Conservation biological control using fungal entomopathogens. Biocontrol 55, 187–198.

Pilz, C., Enkerli, J., Wegensteiner, R., Keller, S., 2011. Establishment and persistence of the entomopathogenic fungus *Metarhizium anisopliae* in maize fields. J. Appl. Entomol. 135, 393–403.

Pontoppidan, M.B., Himaman, W., Hywel-Jones, N.L., Boomsma, J.J., Hughes, D.P., 2009. Graveyards on the move: the spatio-temporal distribution of dead ophiocordyceps-infected ants. PLoS One 4 (3), e4835. http://dx.doi.org/10.1371/journal.pone.0004835.

Poprawski, T.J., Carruthers, R.I., Speese, J., Vacek, D.C., Wendel, L.E., 1997. Early-season applications of the fungus *Beauveria bassiana* and introduction of the hemipteran predator *Perillus bioculatus* for control of Colorado potato beetle. Biol. Control 10, 48–57.

Powell, W.A., Klingeman, W.E., Ownley, B.H., Gwinn, K.D., 2009. Evidence of endophytic *Beauveria bassiana* in seed-treated tomato plants acting as a systemic entomopathogen to larval *Helicoverpa zea* (Lepidoptera: Noctuidae). J. Entomol. Sci. 44, 391–396.

Price, R.E., Bateman, R.P., Brown, H.D., Butler, E.T., Müller, E.J., 1997. Aerial spray trials against brown locust (*Locustana pardalina*, Walker) nymphs in South Africa using oil-based formulations of *Metarhizium flavoviride*. Crop Prot. 16, 345–351.

Prior, C., Jollands, P., le Patourel, G., 1988. Infectivity of oil and water formulations on *Beauveria bassiana* (Deuteromycotina: Hyphomycetes) to the cocoa weevil pest *Pantrhytes plutus* (Coleoptera: Curculionidae). J. Inverteb. Pathol. 52, 66–72.

Pu, X.Y., Feng, M.G., Shi, C.H., 2005. Impact of three application methods on the field efficacy of a *Beauveria bassiana*-based mycoinsecticide against the false-eye leafhopper, *Empoasca vitis* (Homoptera: Cicadellidae) in the tea canopy. Crop Prot. 24, 167–175.

Qazi, S.S., Khachatourians, G.G., 2007. Hydrated conidia of *Metarhizium ansiopliae* release a family of metalloproteases. J. Invertebr. Pathol. 95, 48–59.

Quesada-Moraga, E., Martin-Carballo, I., Garrido-Jurado, I., Santiago-Alvarez, C., 2008. Horizontal transmission of *Metarhizium anisopliae* among laboratory populations of *Ceratitis capitata* (Widermann) (Diptera: Tephriditae). Biol. Control 47, 115–124.

Reddy, G.V.P., Miller, R.H., 2014. Biorational versus conventional insecticides – comparative field study for managing red spider mite and fruit borer on tomato. Crop Prot. 64, 88–92.

Rehner, S.A., Buckley, E.A., 2005. *Beauveria* phylogeny inferred from nuclear ITS and EF1-alpha sequences: evidence for cryptic diversification and links to *Cordyceps* teleomorphs. Mycologia 97, 84–98.

Rehner, S.A., Minnis, A.M., Sung, G.H., Luangsa-ard, J.J., Devotto, L., Humber, R.A., 2011. Phylogeny and systematics of the anamorphic, entomopathogenic genus *Beauveria*. Mycologia 103, 1055–1073.

Ribeiro, L.D., Mota, L.H.C., D'alessandro, C.P., Vendramim, J.D., Delalibera, I., 2014. *In vitro* compatibility of an acetogenin-based bioinsecticide with three species of entomopathogenic fungi. Fla. Entomol. 97, 1395–1403.

Roberts, D.W., Humber, R.A., 1981. Entomogenous fungi. In: Cole, G.T., Kendrick, B. (Eds.), Biology of Conidial Fungi. Academic Press, New York, USA, pp. 201–236.

Roditakis, E., Cousin, I.D., Balrow, K., Franks, N.R., Charnley, A.K., 2000. Improving secondary pick up of insect fungal pathogen conidia by manipulating host behaviour. Ann. Appl. Biol. 137, 329–335.

Rohlfs, M., Churchill, A.C.L., 2011. Fungal secondary metabolites as modulators of interactions with insects and other arthropods. Fungal Genet. Biol. 48, 23–34.

Roy, H.E., Steinkraus, D.C., Eilenberg, J., Hajek, A.E., Pell, J.K., 2006. Bizarre interactions and endgames: entomopathogenic fungi and their arthropod hosts. Annu. Rev. Entomol. 51, 331–357.

Russo, M.L., Pelizza, S.A., Cabello, M.N., Stenglein, S.A., Scorsetti, A.C., 2015. Endophytic colonisation of tobacco, corn, wheat and soybeans by the fungal entomopathogen *Beauveria bassiana* (Ascomycota, Hypocreales). Biocontrol Sci. Technol. 25, 475–480.

Sabbahi, R., Merzouki, A., Guertin, C., 2008. Efficacy of *Beauveria bassiana* (Bals.) Vuill. against the tarnished plant bug, *Lygus lineolaris* L., in strawberries. J. Appl. Entomol. 132, 124–134.

Saito, T., 2005. Preliminary experiments to control the silverleaf whitefly with electrostatic spraying of a mycoinsecticide. App. Entomol. Zool. 40, 289–292.

Samuels, R.I., Charnley, A.K., Reynolds, S.E., 1988. The role of destruxins in the pathogenicity of 3 strains of *Metarhizium anisopliae* for the tobacco hornworm *Manduca sexta*. Mycopathologia 104, 51–58.

Sasan, R.K., Bidochka, M.J., 2012. The insect pathogenic fungus *Metarhizium robertsii* (Clavicipitaceae) is also an endophyte that stimulates plant root development. Am. J. Bot. 99, 101–107.

Scheepmaker, J.W.A., Butt, T.M., 2010. Natural and released inoculum levels of entomopathogenic fungal biocontrol agents in soil in relation to risk assessment and in accordance with EU regulations. Biocontrol Sci. Technol. 20, 503–552.

Scholte, E.J., Knols, B.G.J., Samson, R.A., Takken, W., 2004. Entomopathogenic fungi for mosquito control: a review. J. Insect Sci. 4, 1–24.

Schumacher, V., Poehling, H.M., 2012. *In vitro* effect of pesticides on the germination, vegetative growth, and conidial production of two strains of *Metarhizium anisopliae*. Fungal Biol. 116, 121–132.

Scorsetti, A.C., Maciá, A., Steinkraus, D.C., López Lastra, C.C., 2010. Prevalence of *Pandora neoaphidis* (Zygomycetes: Entomophthorales) infecting *Nasonovia ribisnigri* (Hemiptera: Aphididae) on lettuce crops in Argentina. Biol. Control 52, 46–50.

Seyoum, E., Moore, D., Charnley, A.K., 1994. Reduction in flight activity and food consumption by the desert locust, *Schistocerca gregaria* Forskal (Orth, Cyrtacanthacrinae), after infection with *Metarhizium flavoviride*. J. Appl. Entomol. 118, 310–315.

Shah, P.A., Pell, J.K., 2003. Entomopathogenic fungi as biological control agents. Appl. Microbiol. Biotechnol. 61, 413–423.

Smitley, D.R., Kennedy, G.G., Brooks, W.M., 1986. Role of the entomogenous fungus, *Neozygites floridana*, in population declines of the twospotted mite, *Tetranychus urticae*, on field corn. Entomol. Exp. Appl. 41, 255–264.

Smits, N., Briere, J.F., Fargues, J., 2003. Comparison of non-linear temperature-dependent development rate models applied to in vitro growth of entomopathogenic fungi. Mycol. Res. 107, 1476–1484.

Sosa-Gomez, D.R., Boucias, D.G., Nation, J.L., 1997. Attachment of *Metarhizium anisopliae* to the southern green stink bug *Nezara viridula* cuticle and fungistatic effect of cuticular lipids and aldehydes. J. Invertebr. Pathol. 69, 211–217.

Spatafora, J.W., Sung, G.-H., Sung, J.-M., Hywel-Jones, N.L., White Jr., J.F., 2007. Phylogenetic evidence for an animal pathogen origin of ergot and the grass endophytes. Mol. Ecol. 16, 1701–1711.

Steinkraus, D.C., 2007. Documentation of naturally occurring pathogens and their impact in agroecosystems. In: Lacey, L.A., Kaya, H.K. (Eds.), Field Manual of Techniques in Invertebrate Pathology: Application and Evaluation of Pathogens for Control of Insects and Other Invertebrate Pests, second ed. Springer, Dordrecht, The Netherlands, pp. 267–281.

Steinkraus, D.C., Kring, T.J., Tugwell, N., 1991. *Neozygites fresenii* (Entomophthorales, Neozygitaceae) in *Aphis gossypii* (Homoptera, Aphididae) on cotton. Southwest Entomol. 16, 118–122.

Steinkraus, D.C., Hollingsworth, R.G., Slaymaker, P.H., 1995. Prevalence of *Neozygites fresenii* (Entomophthorales, Neozygitiaceae) on cotton aphids (Homoptera, Aphididae) in Arkansas cotton. Environ. Entomol. 24, 465–474.

St Leger, R.J., 2008. Studies on adaptation of *Metarhizium anisopliae* to life in the soil. J. Invertebr. Pathol. 98, 271–276.

St Leger, R.J., Charnley, A.K., Cooper, R.M., 1987. Characterization of cuticle-degrading proteases produced by the entomopathogen *Metarhizium anisopliae*. Arch. Biochem. Biophys. 253, 221–232.

St Leger, R.J., Goettel, M., Roberts, D.W., Staples, R.C., 1991. Pre-penetration events during infection of host cuticle by *Metarhizium anisopliae*. J. Invertebr. Pathol. 58, 168–179.

Sung, G.-H., Hywel-Jones, N.L., Sung, J.M., Luangsa-ard, J.J., Shrestha, B., Spatafora, J.W., 2007. Phylogenetic classification of *Cordyceps* and the clavicipitaceous fungi. Stud. Mycol. 57, 5–59.

Thomas, M.B., Blanford, S., 2003. Thermal biology in insect parasite interactions. Trends Ecol. Evol. 18, 344–350.

Thomas, M.B., Wood, S.N., Lomer, C.J., 1995. Biological-control of locusts and grasshoppers using a fungal pathogen – the importance of secondary cycling. Proc. R. Soc. Lond. B 259, 265–270.

Thomas, M.B., Langewald, J., Wood, S.N., 1996. Evaluating the effects of a biopesticide on populations of the variegated grasshopper, *Zonocerus variegatus*. J. Appl. Ecol. 33, 1509–1516.

Thompson, S.R., Brandenburg, R.L., Arends, J.J., 2006. Impact of moisture and UV degradation on *Beauveria bassiana* (Balsamo) Vuillemin conidial viability in turfgrass. Biol. Control 39, 401–407.

Tzou, P., Reichhart, J.M., Lemaitre, B., 2002. Constitutive expression of a single antimicrobial peptide can restore wild-type resistance to infection in immune-deficient *Drosophila* mutants. Proc. Natl. Acad. Sci. U.S.A. 99, 2152–2157.

Ugine, T.A., Wraight, S.P., Sanderson, J.P., 2005. Acquisition of lethal doses of *Beauveria bassiana* conidia by western flower thrips, *Frankliniella occidentalis*, exposed to foliar spray residues of formulated and unformulated conidia. J. Invertbr. Pathol. 90, 10–23.

Ugine, T.A., Wraight, S.P., Sanderson, J.P., 2007. Effects of manipulating spray-application parameters on efficacy of the entomopathogenic fungus *Beauveria bassiana* against western flower thrips, *Frankliniella occidentalis*, infesting greenhouse impatiens crops. Biocontrol Sci. Technol. 17, 193–219.

Valanne, S., Wang, J.H., Ramet, M., 2011. The *Drosophila* toll signaling pathway. J. Immunol. 186, 649–656.

Vega, F.E., Goettel, M.S., Blackwell, M., Chandler, D., Jackson, M.A., Keller, S., Koike, M., Maniania, N.K., Monzón, A., Ownley, B.H., Pell, J.K., Rangel, D.E.N., Roy, H.E., 2009. Fungal entomopathogens: new insights on their ecology. Fungal Ecol. 2, 149–159.

Vidal, S., Jaber, L.R., 2015. Entomopathogenic fungi as endophytes: plant-endophyte-herbivore interactions and prospects for use in biological control. Curr. Sci. 109, 46–54.

Vilcinskas, A., Matha, V., Gotz, P., 1997. Effects of the entomopathogenic fungus *Metarhizium anisopliae* and its secondary metabolites on morphology and cytoskeleton of plasmatolytes isolated from *Galleria mellonella*. J. Insect Physiol. 43, 1149–1159.

Weir, A., Blackwell, M., 2001. Molecular data support the *Laboulbeniales* as a separate class of *Ascomycota, Laboulbeniomycetes*. Mycol. Res. 105, 1182–1190.

Whipps, J.M., 1993. A review of white rust (*Puccinia horiana* Henn.) disease on chrysanthemum and the potential for its biological control with *Verticillium lecanii* (Zimm.) Viegas. Ann. Appl. Biol. 122, 173–187.

Williams, M.J., 2007. Drosophila hemopoiesis and cellular immunity. J. Immunol. 178, 4711–4716.

Wraight, S.P., Ramos, M.E., 2002. Application parameters affecting field efficacy of *Beauveria bassiana* foliar treatments against Colorado potato beetle *Leptinotarsa decemlneata*. Biol. Control 23, 164–178.

Wraight, S.P., Ramos, M.E., 2015. Delayed efficacy of *Beauveria bassiana* foliar spray applications against Colorado potato beetle: impacts of number and timing of applications on larval and next-generation adult populations. Biol. Control 83, 51–67.

Wraight, S.P., Carruthers, R.I., Jaronski, S.T., Bradley, C.A., Garza, C.J., Galaini-Wraight, S., 2000. Evaluation of the entomopathogenic fungi *Beauveria bassiana* and *Paecilomyces fumosoroseus* for microbial control of the silverleaf whitefly, *Bemisia argentifolii*. Biol. Control 17, 203–217.

Xie, X.Q., Li, F., Ying, S.H., Feng, M.G., 2012. Additive contributions of two manganese-cored superoxide dismutases (mnsods) to antioxidation, UV tolerance and virulence of *Beauveria bassiana*. PLoS One 7, e30298. http://dx.doi.org/10.1371/journal.pone.

Xu, Y., Orozco, R., Wijeratne, E.M.K., Gunatilaka, A.A.L., Stock, S.P., Molnar, I., 2008. Biosynthesis of the cyclooligomer depsipeptide beauvericin, a virulence factor of the entomopathogenic fungus *Beauveria bassiana*. Chem. Biol. 15, 898–907.

Xu, Y., Orozco, R., Wijeratne, E.M.K., Espinosa-Artiles, P., Gunatilaka, A.A.L., Stock, S.P., Molnar, I., 2009. Biosynthesis of the cyclooligomer depsipeptide bassianolide, an insecticidal virulence factor of *Beauveria bassiana*. Fungal Genet. Biol. 46, 353–364.

Ying, S.H., Ji, X.P., Wang, X.X., Feng, M.G., Keyhani, N.O., 2014. The transcriptional co-activator multiprotein bridging factor 1 from the fungal insect pathogen, *Beauveria bassiana*, mediates regulation of hyphal morphogenesis, stress tolerance and virulence. Environ. Microbiol. 16, 1879–1897.

Zare, R., Gams, W., 2001. A revision of *Verticillium* section *Prostrata*. IV. The genera *Lecanicillium* and *Simplicillium* gen. nov. Nova Hedwig. 73, 1–50.

Zhang, L.B., Tang, L., Ying, S.H., Feng, M.G., 2015. Subcellular localization of six thioredoxins and their antioxidant activity and contributions to biological control potential in *Beauveria bassiana*. Fungal Genet. Biol. 76, 1–9.

Zimmermann, G., 2007a. Review on safety of the entomopathogenic fungi *Beauveria bassiana* and *Beauveria brongniartii*. Biocontrol Sci. Technol. 17, 553–596.

Zimmermann, G., 2007b. Review on safety of the entomopathogenic fungus *Metarhizium anisopliae*. Biocontrol Sci. Technol. 17, 879–920.

Chapter 6

Basic and Applied Research: Entomopathogenic Nematodes

D. Shapiro-Ilan[1], S. Hazir[2], I. Glazer[3]

[1]USDA-ARS, Southeastern Fruit and Tree Nut Laboratory, Byron, GA, United States; [2]Adnan Menderes University, Aydin, Turkey; [3]ARO, Volcani Center, Israel

6.1 INTRODUCTION AND BASIC BIOLOGY

The objective of this chapter is to review the basic biology of entomopathogenic nematodes (EPNs) in the context of microbial control. Nematodes (Phylum Nematoda) are non-segmented round worms also known as eelworms or threadworms. Life histories of nematodes vary from free-living to parasitic. EPNs have been defined as parasitic nematodes that are mutualistically associated with bacterial symbionts and all life stages of the nematode, except for the dauer stage, are found exclusively inside the insect host (Grewal et al., 2005). Historically, this definition of EPNs has referred to the families Steinernematidae and Heterorhabditidae. Recently, the definition was expanded to include other nematodes such as certain species of the genus *Oscheius* (Dillman et al., 2012a). Nonetheless, this chapter will focus solely on the genera *Heterorhabditis* and *Steinernema* because they are the only EPNs that have been developed for microbial control. More than 100 species of steinernematids and heterorhabditids have been described to date (at least 90 steinernematids and 20 heterorhabditids) (Tables 6.1 and 6.2).

6.1.1 EPN Association With Symbiotic Bacteria

Steinernematid nematodes are associated with *Xenorhabdus* spp. and heterorhabditids with *Photorhabdus* spp. (Poinar, 1990; Lewis and Clarke, 2012). In the dauer juvenile, commonly referred to as the infective juvenile (IJ), the symbiotic bacteria are harbored in a specialized bacterial receptacle in the anterior part of the intestine for steinernematids and in the gut mucosa for heterorhabditids (Bird and Akhurst, 1983; Ciche and Ensign, 2003; Martens and Goodrich-Blair, 2005; Ciche et al., 2008). The association is highly specific. Each nematode species is associated with only one bacterial species although a bacterial species may be associated with more than one nematode species (Fischer-Le Saux et al., 1999; Boemare, 2002; Stock and Goodrich-Blair, 2008).

The relationship between the nematode and bacterium is truly mutualistic because the nematode is dependent upon the bacterium for (1) quickly killing its insect host, (2) creating a suitable environment for nematode development by producing antibiotics that suppress competing microorganisms, (3) transforming the host tissues into a food source, and (4) serving as a food resource. The bacteria require the nematodes for (1) dissemination from one insect to another, (2) protection from the external environment, (3) penetration into the host's hemocoel, and (4) inhibition of the host's antibacterial proteins (Akhurst and Boemare, 1990; Forst and Clarke, 2002; Hazir et al., 2003; Stock and Goodrich-Blair, 2008).

A generalized EPN life cycle is depicted in Fig. 6.1. Infection occurs when the IJs enter into their insect hosts through natural apertures (oral cavity, anus, and spiracles) or, in some cases, through the cuticle. After penetrating the hemocoel, the IJs release their symbiotic bacteria. Once the bacteria are released into the hemocoel, they contribute to the killing of the insect host and multiply to a high density in the cadaver. Depending on the size of the insect host, two or more nematode generations can occur in the cadaver although small hosts may only have one generation. As nutrients become limiting, second stage nematodes develop into the IJ stage where they reinitiate the symbiosis by sequestering bacterial cells. The IJs then exit host cadaver to search for new hosts to infect.

The number of IJs produced per insect host varies with host and nematode species and can be influenced by environmental factors (Shapiro-Ilan et al., 2014a). To some extent, variation in IJ yield per host among nematode species is inversely proportional to IJ size, yet some species simply have innately high reproductive capacities, such as *H. indica* and *S. riobrave*. For example, in *Galleria mellonella*, yields of *S. riobrave* (average body length of IJ = 622 mm) may exceed 300,000 IJs per larva, whereas for *S. glaseri* (average body length of IJ = 1133 mm) yields do

TABLE 6.1 List of *Steinernema* Species

S. abbasi, Elawad, Ahmad, and Reid	*S. kushidai* Mamiya
S. aciari Qiu, Hu, Zhou, Mei, Nguyen, and Pang	*S. lamjungense* Khatri-Chhetri, Waeyenberge, Spiridonov, Manandhar, and Moens
S. affine (Bovien)	*S. loci* Phan, Nguyen, and Moens
S. arenarium (Artyukhovsky)	*S. leizhouense* Nguyen, Qiu, Zhou, and Pang
S. anatoliense Hazir, Stock, and Keskin	*S. longicaudum* Shen and Wang
S. ashiuense Phan, Takemoto, and Futai	*S. masoodi* Ali, Shaheen, Pervez, and Hussain
S. asiaticum Anis, Shahina, Reid, and Rowe	*S. minutum* Maneesakorn, Grewal, and Chandrapatya
S. apuliae Triggiani, Mracek, and Reid	*S. monticolum* Stock, Choo, and Kaya
S. australe Edgington, Buddie, Tymo, Hunt, Nguyen, France, Merino, and Moore	*S. neocurtillae* Nguyen and Smart
S. backanense Phan, Spiridonov, Subbotin, and Moens	*S. nepalense* Khatri-Chhetri, Waeyenberge, Spiridonov, Manandhar, and Moens
S. balochiense, Fayyaz, Khanum, Ali, Solangi, Gulsher, and Javed	*S. nyetense* Kanga, Trinh, Waeyenberge, Spiridonov, Hauser, and Moens
S. bifurcatum Fayyaz, Yan, Qiu, Han, Gulsher, Khanum, and Javed	*S. oregonense* Liu and Berry
S. boemarei, Lee, Sicard, Skeie, and Stock,	*S. pakistanense* Shahina, Anis, Reid, Rowe, and Maqbool
S. brazilense Nguyen, Ginarte, Leite, Santos, and Haracava	*S. phyllophagae* Nyguyen and Buss
S. bicornutum Tallosi, Peters, and Ehlers	*S. poinari*, Mrácek, Puza, and Nermut
S. cameroonense Kanga, Trinh, Waeyenberge, Spiridonov, Hauser, and Moens	*S. puertoricense* Roman and Figueroa
S. carpocapsae (Weiser)	*S. pui* Qiu, Zhao, Wu, Lv, and Pang
S. caudatum Xu, Wang, and Li	*S. puntauvense* Uribe-Lorio, Mora, and Stock
S. ceratophorum Jian, Reid, and Hunt	*S. rarum* (de Doucet)
S. changbaiense Ma, Chen, De Clercq, Han, and Moens	*S. riobrave* Cabanillas, Poinar, and Raulston
S. cholashanense Nguyen, Puza, and Mracek	*S. ritteri* de Doucet and Doucet
S. citrae Stokwe, Malan, Nguyen, Knoetze, and Tiedt	*S. robustispiculum* Phan, Subbotin, Waeyenberge, and Moens
S. costaricense Uribe-Lorio, Mora, and Stock	*S. sacchari*, Nthenga, Knoetze, Berry, Tiedt, and Malan
S. colombiense Lopez-Nunez, Plichta, Gongora-Botero, and Stock	*S. sangi* Phan, Nguyen, and Moens
S. cubanum Mracek, Hernandez, and Boemare	*S. sasonense* Phan, Spiridonov, Subbotin, and Moens
S. cumgarense Phan, Spiridonov, Subbotin, and Moens	*S. scapterisci* Nguyen and Smart
S. diaprepesi Nguyen and Duncan	*S. scarabaei* Stock and Koppenhöfer
S. eapokense Phan, Spiridonov, Subbotin, and Moens	*S. schliemanni* Spiridonov, Wayenberge, and Moens
S. ethiopiense Tamiru, Waeyenberge, Hailu, Ehlers, Půža, and Mráček	*S. seemae* Ali, Shaheen, Pervez, and Hussain
S. everestense Khatri-Chhetri, Waeyenberge, Spiridonov, Manandhar, and Moens	*S. serratum* Liu
S. feltiae (Filipjev)	*S. siamkayai* Stock, Somsook, and Kaya
S. glaseri (Steiner)	*S. sichuanense* Mracek, Nguyen, Tailliez, Boemare, and Chen
S. guangdongense Qiu, Fan, Zhou, Pang, and Nguyen	*S. silvaticum* Sturhan, Spiridonov, and Mracek
S. hebeiense Chen, Li, Yan, Spiridonov, and Moens	*Steinemema surkhetense* Khatri-Chhetri, Waeyenberge, Spiridonov, Manandhar, and Moens
S. hermaphroditum Stock, Griffin, and Chaerani	*S. tami* Pham, Nguyen, Reid, and Spiridonov
S. ichnusae Tarasco, Mrácek, Nguyen, and Triggiani	*S. thanhi* Phan, Nguyen, and Moens
S. innovationi, Çimen, Lee, Hatting, Hazir, and Stock	*S. thermophilum* Ganguly and Singh
S. intermedium (Poinar)	*S. tielingense* Ma, Chen, Li, Han, Khatri-Chhetri, and De Clercq
S. jeffreyense, Malan, Knoetze, and Tiedt	*S. tophus*, Çimen, Lee, Hatting, Hazir, and Stock
S. jollieti Spiridonov, Krasomil-Osterfeld, and Moens	*S. unicornum* Egington, Buddie Tymo, France, Merino, and Hunt
S. karii Waturu, Hunt and Reid	*S. vulcanicum* Clausi, Longo, Rappazzo, Tarasco, and Vinciguerra
S. huense Phan, Mráček, Půža, Nermů, and Jarošová	*S. websteri* Cutler and Stock
S. khoisanae Nguyen, Malan, and Gozel	*S. weiseri* Mracek, Sturhan, and Reid
S. kraussei (Steiner)	*S. xinbinense* Ma, Chen, De Clercq, Wayenberge, Han, and Moens
	S. xueshanense Mracek, Qi-Zhi, and Nguyen
	S. yirgalemense Nguyen, Tesfamariam, Gozel, Gaugler, and Adams

not exceed 50,000 IJs in the same host (Grewal et al., 1994; Stock and Hunt, 2005).

6.1.2 Survival Physiology and Immune Response

Prior to locating an insect host, IJs must survive based on their energy reserves, which are also critical to their ability to infect (Lewis et al., 1995; Patel et al., 1997; Patel and Wright, 1997; Wright and Perry, 2002). Quantitative and qualitative analyses of biochemical reserves of EPNs indicate that lipid content and fatty acid composition are critical and are dependent on media or host components (Selvan et al., 1993; Abu Hatab et al., 1998).

Once the IJ reaches the target insect, it faces the host's first line of defense, that is physical structural barriers, such

TABLE 6.2 List of *Heterorhabditis* Species

H. amazonensis Andaló, Nguyen, and Moino	*H. indica* Poinar, Karunakar, and David
H. atacamensis Egington, Buddie Moore, France, Merino, and Hunt	*H. marelata* Liu and Berry
	H. megidis Poinar, Jackson, and Klein
H. bacteriophora Poinar	*H. mexicana* Nguyen, Shapiro-Ilan, Stuart, McCoy, James, and Adams
H. baujardi Phan, Subbotin, Nguyen, and Moens	
H. beicherriana Li, Liu, Nermut, Puza, and Mracek	*H. noenieputensis*, Malan, Knoetze, and Tiedt
H. brevicaudis Liu	*H. poinari* Kakulia and Mikaia
H. downesi Stock, Burnell, and Griffin	*H. safricana* Malan, Nguyen, De Waal, and Tiedt
H. floridensis Nguyen, Gozel, Koppenhöfer, and Adams	*H. sonorensis* Stock, Rivera-Orduno, and Flores-Lara
H. georgiana Nguyen, Shapiro-Ilan, and Mbata	*H. taysearae* Shamseldean, El-Sooud, Abd-Elgawad, and Saleh
H. gerrardi Plitchta, Joyce, Clarke, Waterfield, and Stock	*H. zealandica* Poinar

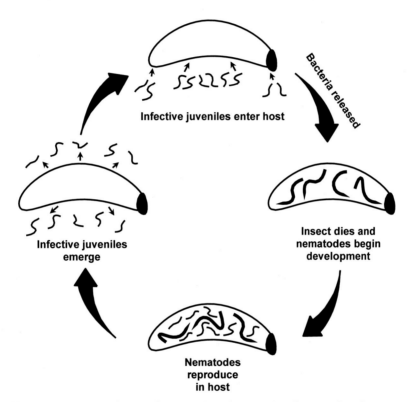

FIGURE 6.1 A generalized life cycle of entomopathogenic nematodes. *Illustration by Bill Joyner, USDA.*

as the cuticle and the peritrophic membrane of the midgut and the respective epithelia. Once these barriers are breached, the insect defends itself using both cellular and humoral immune responses (Castillo et al., 2011). Cellular immune reactions are mediated by different types of immune cells, which may contribute to phagocytosis, nodule formation and encapsulation (Li et al., 2007). The humoral effectors recruited by host insects include the inducible antimicrobial peptides (AMPs), cell adhesion molecules, clotting components, and the prophenoloxidase system (Li et al., 2007; Lemaitre and Hoffmann, 2007; Wang et al., 2010). To overcome the host immune system and cause death, the invading IJs or their symbiotic bacteria release various virulence factors including proteases, phenoloxidase inhibitors, and toxins (Burman, 1982; Laumond et al., 1989; Forst et al.,

1997; Simões et al., 2000; Balasubramanian et al., 2009, 2010; Waterfield et al., 2009; Jing et al., 2010).

6.1.3 Foraging and Infection Dynamics

Substantial attention has been given to the ability of the IJs to locate and invade insect hosts. Indeed, host-seeking ability is an important factor in biocontrol efficacy. In terms of host-seeking strategies, EPNs have been classified as a continuum between ambushers and cruiser type foragers (Lewis et al., 1992; Campbell and Lewis, 2002; Lewis, 2002). The ambushers, such as *Steinernema carpocapsae* and *Steinernema scapterisci*, are characterized by relative low motility and a tendency to stay near the soil surface where they may stand on their tails waiting for a host to pass. The cruiser nematodes, such as, *S. glaseri* and *Heterorhabditis bacteriophora*, are highly mobile and actively seek hosts throughout the soil profile. However, most nematode species are positioned between these two extremes and are considered intermediate in their foraging strategy, such as *S. riobrave* and *S. feltiae* (Campbell and Gaugler, 1997). Moreover, there are many instances in which cruiser or ambusher foraging strategists have effectively controlled various insects regardless of host mobility or location in the soil profile (Shapiro-Ilan et al., 2014a). Furthermore, accumulating data and reports suggest that the environment can have an effect on nematodes' host-seeking ability (Kruitbos ct al., 2010).

A key component of host finding is olfaction, such as sensing of volatile cues (Hallem et al., 2011; Dillman et al., 2012b). EPNs respond differentially to the odor blends emitted by live hosts. Further, IJs use CO_2 as a general attractant, but the relative importance of CO_2 versus host-specific odorants varies for different nematode–host combinations and for different host-seeking behaviors (Hallem et al., 2011; Dillman et al., 2012b).

Other factors affecting IJ navigation and host seeking include physical cues such as vibration (Torr et al., 2004), electromagnetic signals (Ilan et al., 2013), plant volatiles, and group behavior. Rasmann et al. (2005) reported an insect-induced belowground plant signal, (*E*)-*b*-caryophyllene, which acts as a "call for help" by attracting EPNs to maize roots damaged by corn rootworm, *Diabrotica virgifera virgifera*. Other studies have also indicated plant volatiles can direct EPN dispersal and may increase biocontrol activity in various environments (Ali et al., 2012; Hiltpold and Turlings, 2012; Willett et al., 2015). Another driver of EPN foraging and infection behavior stems from a tendency of IJs to move and infect in aggregate, that is, displaying group behavior (Fushing et al., 2008; Shapiro-Ilan et al., 2014b).

6.2 FACTORS AFFECTING SURVIVAL AND EFFICACY

EPNs have been tested against a large number of insect pest species in the laboratory, greenhouse, and field with varying results (Koppenhöfer, 2000; Grewal et al., 2005). Under field conditions, applications of EPNs have resulted in poor to excellent control of the target pest depending on a variety of biotic and abiotic factors that affect EPN survival and efficacy (Shapiro et al., 2006).

6.2.1 Biotic Factors

Many biotic factors can influence the efficacy of nematodes as biocontrol agents, but matching the biology and ecology of both the nematode and the target pest is perhaps the most crucial step toward successful application. Proper matching of the nematode to the host includes virulence, host finding, and environmental tolerance. If a nematode species does not possess a high level of virulence toward the target pest, there are minimal chances of success; however, in some cases, persistence in the environment might overcome a shortcoming in virulence (Shapiro-Ilan et al., 2002, 2006). Host-seeking ability may also be an important factor in biocontrol efficacy. As indicated earlier, EPN host-finding strategies are classified as a continuum between ambushers and cruiser type, and these foraging behaviors may influence the choice of nematode to be used.

6.2.1.1 Rate of Application

Although only one nematode is needed to infect and kill an insect host, in reality, a minimum number needs to be applied for effective biocontrol. Thus, as a general rule EPNs should usually be applied to soil at rates of 2.5×10^9 IJs/ha (=25 IJs/cm^2) or higher (Georgis et al., 1995; Shapiro-Ilan et al., 2002). In cases where the pest is particularly susceptible or in controlled conditions such as in the greenhouse, lower application rates may provide efficacy (Levine and Oloumi-Sadeghi, 1992; Shapiro-Ilan et al., 2002). Conversely, some insect pests are less susceptible to nematodes or occur deeper in the soil profile which may require higher application rates to achieve sufficient efficacy (McCoy et al., 2000; Shapiro-Ilan et al., 2002; Demir et al., 2015).

6.2.1.2 Interactions With Other Organisms

EPNs in the soil environment interact with different organisms such as virus, bacteria, fungi, protozoans, turbellarians, other nematodes, collembolans, mites, tardigrades, oligochaete, parasitoids, and vertebrate and invertebrate scavengers (Kaya, 2002; Ulug et al., 2014). Some of these organisms can have detrimental effects on EPN applications. For example, nematophagous fungi (Jaffee and Strong, 2005; El-Borai et al., 2009, 2011) and mites are known as effective predators of EPNs (Epsky and Capinera, 1988; Karagoz et al., 2007; Ekmen et al., 2010a,b; Cakmak et al., 2013).

Various interactions, such as antagonism, additivity, or synergy, can occur between EPNs and other entomopathogens or nonmicrobial biocontrol agents. Examples of antagonists include *Beauveria bassiana* (Barbercheck and Kaya, 1990, 1991; Brinkman and Gardner, 2000) and *Isaria fumosorosea* (Shapiro-Ilan et al., 2004). However, EPNs have also been reported to act synergistically with certain entomopathogens, such as with *Paenibacillus popilliae* (Thurston et al., 1994), *Bacillus thuringiensis* (Koppenhöfer and Kaya, 1997), and the entomopathogenic fungus, *Metarhizium anisopliae* (Ansari et al., 2004). The nature of the interaction depends on nematode species, host species, or application rates and times. Neutral or competitive interactions between parasitic Hymenoptera and EPNs have been reported (Lacey et al., 2003; Everard et al., 2009). Interestingly, Mbata and

Shapiro-Ilan (2010) reported that IJs preferentially infected parasitized host larvae compared with nonparasitized host larvae, yet the parasitoids did not different between EPN-infected and noninfected hosts.

During the period that EPN-infected hosts are in the soil, the cadavers are at risk of being consumed by arthropod scavengers. However, the symbiotic bacteria associated with EPNs produce compound(s) that deter various scavengers such as wasps, cockroaches, crickets, predator insects, collembolans, and ants (Zhou et al., 2002; Gulcu et al., 2012); the chemical compounds produced responsible for this deterrent activity are called "scavenger deterrent factors (SDFs)" (Gulcu et al., 2012). However, the scavenging mite, *S. polyphyllae*, and some ant species will feed on EPN-killed insects, demonstrating that not all scavengers are deterred by SDFs (Ekmen et al., 2010a,b; Ulug et al., 2014).

6.2.1.3 Recycling Potential

Generally, following soil applications, EPN populations remain high enough to provide effective control for 2–8 weeks (Duncan and McCoy, 1996; Kaya, 1990; Shapiro-Ilan et al., 2002). Some recycling commonly occurs after nematode application, but it is usually not adequate for achieving multiseason control. However, in some cases, effective pest control may be obtained for more than one season (Klein and Georgis, 1992; Parkman et al., 1994; Shields et al., 1999). Factors such as soil type, ground cover, insect species and host density, and the nematode species affect the persistence, infectivity, and motility of IJs and their ability to recycle (Kaya, 1990; Klein and Georgis, 1992; Shapiro-Ilan et al., 2002).

6.2.2 Abiotic Factors Affecting Survival and Efficacy

6.2.2.1 Soil Moisture

Nematodes are highly sensitive to desiccation and require a water film to disperse (Kaya, 1990; Shapiro-Ilan et al., 2006, 2012); adequate moisture levels are critical for nematode survival and efficacy (Koppenhöfer et al., 1995). Insect mortality is very low or absent at extremely low and high moisture (nearly saturated) soils (Grant and Villani, 2003). In some cases, IJs may survive and become less active at low moisture conditions, and subsequently virulence can be restored by rehydrating the soil (Glazer, 2002; Grant and Villani, 2003). Koppenhöfer and Fuzy (2007) reported that nematode infectivity was highest at moderate soil moistures (−10 to −100 kPa), and tended to be lower in wet (−1 kPa) and moderately dry (−1000 kPa) soil. Soil moisture ranges for infectivity and persistence vary among EPN species.

6.2.2.2 Temperature

Temperature greatly affects EPN activity (Kaya, 1990). Extended exposure to temperature extremes (<0°C or >40°C) is lethal to most EPN species (Brown and Gaugler, 1996). Optimum temperatures for survival, infection, and reproduction vary among EPN species and strains (Grewal et al., 1994; Hazir et al., 2001). Some nematodes, such as *S. siamkayai*, *S. glaseri*, *S. riobrave*, *H. indica*, and *H. floridensis*, are relatively heat tolerant and efficacy is maintained at temperatures of >30°C (Shapiro-Ilan et al., 2012), whereas others are relatively cold-adapted and efficacy is maintained at <15°C, such as *H. megidis*, *S. feltiae*, and *H. marelata* (Berry et al., 1997; Grewal et al., 1994; Kung et al., 1991; Hazir et al., 2001).

6.2.2.3 Soil Characteristics

Soil texture is an important factor that affects IJ efficacy (Barbercheck, 1992; Kaya, 1990). Generally, heavy soils, such as clay soils, are the least conducive to EPN efficacy and sandy soils are most conducive; however, significant exceptions have been observed (Georgis and Gaugler, 1991; Shapiro et al., 2000). In addition to soil texture, organic matter level and dissolved solids are also important for nematode survival and efficacy (Kaspi et al., 2010). Soil pH in most agroecosystems, having a range of 4–8, does not have a strong effect on IJ survival, but a pH of 10 or higher could be detrimental (Kung et al., 1990). The effect of soil parameters including soil texture, pH, and organic matter may vary with nematode species (Koppenhöfer and Fuzy, 2006).

6.2.2.4 Ultraviolet Radiation

Exposure to UV radiation can adversely affect EPN survival, virulence, or reproduction (Gaugler et al., 1992; Fujiie and Yokoyama, 1998). Shapiro-Ilan et al. (2015a) indicated significant variation in UV tolerance among EPN strains and species, with *Steinernema carpocapsae* exhibiting higher levels of UV tolerance than other strains or species. Moreover, a lack of impact on viability did not necessarily translate into a lack of nematode virulence. Therefore, it is recommended that IJ applications to soil be made early in the morning or evening to avoid direct exposure to UV light (Shapiro-Ilan et al., 2012). In addition to the timing of application, the adverse effects of UV radiation can further be minimized by using sufficient amounts of water to wash the IJs into the soil, which also can minimize the effects of desiccation and exposure to high temperatures.

6.3 APPLICATION AND FORMULATION TECHNOLOGY

EPNs can be applied with almost all types of agricultural application equipment including pressurized sprayers, mist blowers, and electrostatic sprayers or as aerial sprays (Wright et al., 2005; Shapiro-Ilan et al., 2006, 2012). The appropriate application equipment depends on the cropping system, and in each case there are a variety of handling considerations including volume, agitation, nozzle type, pressure and recycling time, system environmental conditions, and spray distribution pattern (Fife et al., 2003, 2005; Wright et al., 2005; Shapiro-Ilan et al., 2006, 2012). It is important to provide adequate agitation during application. For small plot applications, hand-held equipment, such as water cans or back-pack sprayers, may be suitable. When nematodes are applied to larger plots, an appropriate spraying apparatus such as a boom sprayer should be considered. Other viable methods include microjet irrigation systems, trunk-sprayers, subsurface injection, or baits (Wright et al., 2005; Shapiro-Ilan et al., 2015b). Various formulations for EPNs may be used for application in aqueous suspension including activated charcoal, alginate and polyacrylamide gels, clay, diatomaceous earth, paste, peat, polyurethane sponge, vermiculite, and water-dispersible granules (WDGs) (Grewal, 2002; Shapiro-Ilan et al., 2012).

6.4 COMMERCIALIZATION AND TARGETS

EPNs possess numerous positive attributes as microbial control agents (Shapiro and Grewal, 2008; Shapiro-Ilan et al., 2014a). EPNs used in biological control are safe to humans and are generally safe to other nontarget organisms and the environment (Akhurst and Smith, 2002; Ehlers, 2005). The high level of safety associated with EPNs has resulted in exclusion from pesticide registration requirements in many countries such as the United States and nations in the European Union (Ehlers, 2005). With few exceptions, such as *Steinernema scarabaei* (Koppenhöfer and Fuzy, 2003) and *Steinernema scapterisci* (Nguyen and Smart, 1991), EPNs have a wide host range. Indeed, some EPN species can infect dozens of insect species across four or more orders (Poinar, 1979; Klein, 1990; Table 6.3), and a number of EPN species are used commercially against 12 or more insect species (see Table 6.3). Entomopathogenic nematodes are amenable to mass production in vivo in insects or in vitro by solid or liquid fermentation (Shapiro-Ilan and Gaugler, 2002; Shapiro-Ilan et al., 2014a, Chapter 10 this volume).

Based on these attributes, EPNs have been cultured commercially for more than 30 years as microbial control agents (Georgis, 2002; Poinar and Grewal, 2012). Currently, EPNs are produced by more than a dozen companies worldwide, and, to date, at least 13 species have reached commercial development: *Heterorhabditis bacteriophora, H. indica, H. marelata, H. megidis, H. zealandica, Steinernema carpocapsae, S. feltiae, S. glaseri, S. kushidai, S. kraussei, S. longicaudum, S. riobrave*, and *Steinernema scapterisci* (Kaya et al., 2006; Lacey et al., 2015).

EPNs can suppress a wide variety of economically important pests, many of which are targeted commercially (Table 6.3, also see Shapiro-Ilan and Gaugler, 2002; Grewal et al., 2005; Georgis et al., 2006; Lacey and Georgis, 2012; Shapiro-Ilan et al., 2014a). The use of EPNs against the insect pests listed in Table 6.3 is supported by refereed journal articles indicating field efficacy of at least 75% suppression of the target pest. Efficacy for a number of other commercially targeted pests, such as the cat flea (*Ctenocephalides felis*), chinch bugs (Hemiptera: Blissidae), and wireworms (Coleoptera: Elateridae), may be supported by proprietary data. Due to cost considerations and sensitivity to environmental extremes, most of the successful commercial applications have been restricted to soil applications in relatively high-value crops (Grewal et al., 2005; Campos-Herrera, 2015). However, in some cases EPNs have successfully controlled target pests above ground or in cryptic habitats (Arthurs et al., 2004; Grewal et al., 2005).

6.5 METHODS TO IMPROVE EFFICACY

One of the primary goals of research activity related to EPNs is their improvement as biological control agents. Biocontrol efficacy can be achieved by improving the organism itself, the application method or formulation, or through environmental manipulation.

6.5.1 Improve the Organism

Biocontrol efficacy can be improved through strain discovery, selection, hybridization, or transgenic manipulation (Gaugler, 1987). Initially, the potential EPN candidate species and strains should be tested for biocontrol efficacy to the particular target pest of interest. If existing strains are not sufficiently efficacious, a straightforward method to improve biocontrol is to simply find a superior strain or species of EPN. This can be accomplished through surveys and subsequent screening of EPNs that have been isolated relative to existing strains (Shapiro-Ilan et al., 2014a).

If existing or recently isolated strains do not exhibit sufficient biocontrol potential, then genetic means may be employed to enhance efficacy (Gaugler, 1987; Fodor et al., 1990; Burnell and Dowds, 1996; Segal and Glazer, 1998, 2000). Directed selection for specific traits important to biocontrol is one approach. Using genetic selection techniques, EPNs have been improved for traits such as host finding (Gaugler and Campbell, 1989, 1991; Salame et al., 2010), dispersal (Bal et al., 2014), efficacy (Tomalak, 1994a,b), nematicide resistance (Glazer et al., 1997), environmental tolerance to heat and or desiccation (Ehlers et al., 2005; Salame et al., 2010), and sensitivity to volatile cues (Hiltpold et al., 2010). However, it must be noted that directed selection for one trait may inadvertently select for an inferior level

TABLE 6.3 Pest Targeted Commercially With Entomopathogenic Nematodes (EPNs)

Pest Common Name[a]	Pest Scientific Name	Key Crop(s) Targeted	Primary EPNs Used
Artichoke plume moth	*Platyptilia carduidactyla* (Riley)	Artichoke	Sc
Armyworms	Lepidoptera: Noctuidae	Vegetables	Sc, Sf, Sr
Banana moth	*Opogona sachari* Bojer	Ornamentals	Hb, Sc
Banana root borer	*Cosmopolites sordidus* (Gemar)	Banana	Sc, Sf, Sg
Billbug	*Sphenophorus* spp. (Coleoptera: Curculionidae)	Turf	Hb,Sc
Black cutworm	*Agrotis ipsilon* (Hufnagel)	Turf, vegetables	Sc
Black vine weevil	*Otiorhynchus sulcatus* (F.)	Berries, ornamentals	Hb, Hd, Hm, Hmeg, Sc, Sg
Borers (eg, peach tree borer)	*Synanthedon* spp. and other sesiids	Fruit trees and ornamentals	Hb, Sc, Sf
Citrus root weevil	*Pachnaeus* spp. (Coleoptera: Curculionidae	Citrus, ornamentals	Sr, Hb
Codling moth	*Cydia pomonella* (L.)	Pome fruit	Sc, Sf
Corn earworm	*Helicoverpa zea* Boddie	Vegetables	Sc, Sf, Sr
Corn rootworm	*Diabrotica* spp.	Vegetables	Hb, Sc
Cranberry girdler	Chrysoteuchia topiaria (*Zeller*)	Cranberries	Sc
Crane fly	Diptera: Tipulidae	Turf	Sc
Diaprepes root weevil	*Diaprepes abbreviatus* (L.)	Citrus, ornamentals	Hb, Sr
Fungus gnats	Diptera: Sciaridae	Mushrooms, greenhouse	Sf, Hb
Iris borer	*Macronoctua onusta* Grote	Iris	Hb, Sc
Large pine weevil	*Hylobius albietis* (L.)	Forest plantings	Hd, Sc
Leafminers	*Liriomyza* spp. (Diptera: Agromyzidae)	Vegetables, ornamentals	Sc, Sf
Mole crickets	*Scapteriscus* spp.	Turf	Sc, Sr, Scap
Navel orange worm	*Amyelois transitella* (Walker)	Nut and fruit trees	Sc
Plum curculio	*Conotrachelus nenuphar* (Herbst)	Fruit trees	Sr, Sf
Scarab grubs	Coleoptera: Scarabaeidae	Turf, ornamentals	Hb, Sc, Sg, Ss, Hz
Shore flies	*Scatella* spp.	Ornamentals	Sc, Sf
Strawberry root weevil	*Otiorhynchus ovatus* (L.)	Berries	Hm, Sc
Sweet potato weevil	*Cylas formicarius* (Summers)	Sweet potato	Hb, Sc, Sf
Western flower thrips	*Frankliniella occidentalis* (Pergande)	Greenhouse, flowers	Sc, Sf

[a]*At least one scientific paper reported ≥75% suppression of these pests in the field; not meant to be an exhaustive list. Hb=Heterorhabditis bacteriophora, Hd=H. downesi, Hm=H. marelata, Hmeg=H. megidis, Hz=H. zealandica, Sc=Steinernema carpocapsae, Sf=S. feltiae, Sg=S. glaseri, Sk=S. kushidai, Sr=S. riobrave, Sscap=S. scapterisci, Ss=S. scarabaei.*

of another trait (Gaugler et al., 1990). Another approach to improving the organism is through hybridization, that is, controlled crosses and subsequent screening of progeny for superior biocontrol populations. Hybridization has been demonstrated for both heterorhabditids (Shapiro et al., 1997) and steinernematids (Shapiro-Ilan et al., 2005). Mukuka et al. (2010) combined both selection and hybridization techniques

to enhance environmental tolerance in *Heterorhabditis bacteriophora*.

Genetic manipulation has been demonstrated as a potential mechanism for EPN improvement. Hashmi et al. (1995, 1998) used genetic engineering as a means of improving heat tolerance of *H. bacteriophora*, and subsequently a field release of the nematode was made (Gaugler et al., 1997).

Additionally, Vellai et al. (1999) reported successful transformation of the yeast desiccation–related gene encoding trehalose-6-phosphate synthase into *S. feltiae*. In a new direction for studying EPN genomics and manipulation, Ciche and Sternberg (2007) developed the use of RNAi in heterorhabditid nematodes. Moshayov et al. (2013) used this approach to silence genes presumably related to the recovery process from the infective to the parasitic stage. Genetic approaches hold great promise for improvement of EPNs, and both classic and advanced genetic techniques can significantly enhance EPN performance. Advances in genetic improvement as well as other avenues to improved efficacy will be leveraged by genome sequencing of EPNs and their symbionts (Duchaud et al., 2003; Bai et al., 2013; Dillman et al., 2015).

Once a suitable EPN strain is selected for biocontrol purposes, it is critical that the genetic stability of that population is secured. Attenuation of beneficial traits, which may result from repeated subculturing, can jeopardize biocontrol efforts. Trait changes can be genetically based, such as inbreeding, drift, inadvertent selection, or arise from nongenetic factors, such as disease or nutrition (Chaston et al., 2011; Shapiro-Ilan et al., 2014a). Studies have indicated deterioration under laboratory conditions for diverse traits of EPNs such as virulence, environmental tolerance, reproductive capacity, and host finding (Shapiro et al., 1996; Wang and Grewal, 2002; Bai et al., 2005; Bilgrami et al., 2006). Trait deterioration in *Heterorhabditis bacteriophora* was genetically based with inbreeding depression being the prominent issue (Bai et al., 2005; Adhikari et al., 2009; Chaston et al., 2011). Avoidance of trait deterioration can be pursued through cryopreservation, but several shortcomings exist with that approach such as genetic bottle necking or potential for mechanical failure (Bai et al., 2005; Shapiro-Ilan et al., 2014a). Alternatively, creation of homozygous inbred lines on solid media was demonstrated as a method for deterring EPN trait deterioration (Bai et al., 2005). Following this research, Anbesse et al. (2013) reported that multiple heterorhabditid inbred lines can be automatically created during liquid culture because heterorhabditids cannot mate in liquid broth; thus, all progeny are produced by selfing. There are advantages and disadvantages in producing inbred lines singly in solid culture dishes versus *en masse* in liquid culture. Only heterorhabditids can automatically produce inbred lines, whereas both heterorhabditids and steinernematids are amenable to inbred line generation on solid culture plates (Shapiro-Ilan et al., 2014a). Nonetheless, the creation of inbred lines appears to be a highly valuable approach to maintaining strain stability.

6.5.2 Improve Formulation or Application

Enhanced efficacy in EPN applications can be facilitated through improved formulation. Substantial progress has been made in recent years in developing EPN formulations, particularly for above-ground applications. For example, mixing EPNs with a surfactant and polymer has led to improved aboveground efficacy (Schroer and Ehlers, 2005). Improved efficacy may also be achieved by the addition of surfactants to increase leaf coverage (Williams and Walters, 2000). Furthermore, *Steinernema carpocapsae* applications for control of the lesser peach tree borer, *Synanthedon pictipes*, were greatly improved by a follow-up application of a sprayable gel; Barricade is a sprayable gel commonly used for protecting structures from fire (Shapiro-Ilan et al., 2010a). Barricade can also provide protection to EPNs when applied in a single spray at lower concentrations (unpublished data) and was used successfully in soil EPN applications to retain soil moisture in lieu of irrigation (Shapiro-Ilan et al., 2015b). Other improved formulations for aboveground application of EPNs have included those based on chitosan (Llàcer et al., 2009), a wood flour foam (Lacey et al., 2010), or other adjuvants (van Niekerk and Malan, 2015).

EPN efficacy can also be enhanced through improved application equipment or approaches. Despite well-established procedures, equipment can be improved further, such as optimizing spray systems like nozzles, pumps, and spray distribution for superior pathogen survival and dispersion (Shapiro-Ilan et al., 2006, 2012; Brusselman et al., 2012). Bait formulations might increase EPN persistence and reduce the quantity of microbial agents required per unit area (Grewal, 2002); use of baits have thus far been limited in practical application, but could conceivably be developed further for applications. Another novel application approach that has gained attention is delivery of EPNs in their infected host cadavers (Jansson et al., 1993; Shapiro and Glazer, 1996; Dolinski et al., 2015). Advantages to the cadaver application approach relative to standard application in aqueous suspension include increased nematode dispersal (Shapiro and Glazer, 1996), infectivity (Shapiro and Lewis, 1999), survival (Perez et al., 2003), and efficacy (Shapiro-Ilan et al., 2003), whereas other studies did not detect a benefit in the cadaver approach (Bruck et al., 2005; Ramalingam et al., 2015). EPNs applied in host cadavers were also effective and persistent when added to bags of potting media for subsequent distribution to target pest sites (Deol et al., 2011). Application of cadavers may be facilitated through formulations that have been developed to protect cadavers from rupture and improve ease of handling (Shapiro-Ilan et al., 2001, 2010b; Del Valle et al., 2009), and development of mechanized equipment for field distribution (Zhu et al., 2011). In a related approach, living preinfected hosts can carry the next generation of emerging IJs to hard-to-reach target sites, and thereby facilitate enhanced control in cryptic habitats; this method was recently demonstrated when using EPNs against the goat moth *Cossus cossus*, which bores deep into the trunks of chestnut trees (Gumus et al., 2015).

Application of EPNs can be improved by leveraging interactions with other control agents. Certain combinations of EPNs with other agents are synergistic and such combinations can be used to enhance pest control. Synergy can be obtained through combinations with chemical agents (Koppenhöfer and Grewal, 2005; Koppenhöfer and Fuzy, 2008; Morales-Rodriguez and Peck, 2009; Mbata and Shapiro-Ilan, 2013) or biotic agents (Koppenhöfer and Grewal, 2005; Ansari et al., 2010). Yet it must be emphasized that variation in response and biocontrol outcome exists among EPN and host species as well as timing and application rates.

In addition to combinations with other control agents, EPNs may be combined with phoretic agents, such as earthworms (Shapiro-Ilan and Brown, 2013). Possibly, earthworms and EPNs could be sold a unit for small-scale applications, such as gardens, nurseries, potted plants, and greenhouses. From a single package or combined application, the grower could obtain the added value of improved soil conditions due to the earthworms as well as superior biocontrol based on the earthworm–nematode relationship. Moreover, on a larger scale, it may be feasible to increase field populations of earthworms or other phoretic agents through cultural practices, such as addition of organic matter (Berry and Karlen, 1993). This tactic could lead to improved biocontrol by native EPNs or improve EPN distribution following inundative or inoculative applications.

6.5.3 Improve the Environment: Environmental Manipulation to Enhance Biocontrol

Physical or biotic manipulation of the environment at the target site can increase biocontrol efficacy through a variety of mechanisms. For example, approaches that decrease exposure to harmful biotic or abiotic factors or enhance EPN reproduction, virulence, persistence, and exposure to the host will expand biocontrol utility. For example, the persistence of efficacy EPNs can be enhanced through making the soil environment more conducive to EPN survival, such as changing soil pH or other parameters, or via the addition of soil amendments such as mulch or crop residues (Shapiro et al., 1999; Lacey et al., 2006; de Waal et al., 2011). Alternatively, the soil environment can be altered on a microlevel: citrus trees were planted in islands of soil that was more conducive to EPN control than endemic soils; 68% more adult *Diaprepes abbreviatus* weevils were captured from native soil relative to the imported sandy soil which favors EPN persistence and efficacy (Duncan et al., 2013).

6.6 CONCLUSION: FUTURE DEVELOPMENTS AND NEEDS

EPNs can be excellent biocontrol agents for insect pests and for this reason and others outlined in this chapter, a viable commercial industry has developed. Nonetheless, a number of factors have prevented EPNs from being employed on a

wider scale. It is critical to match the right nematode species to the target pest. Furthermore, biotic agents, including pathogens and predators of EPN, other soil organisms, as well as abiotic factors such as ultraviolet radiation, soil moisture and relative humidity, and temperature can affect the efficacy of applied EPNs. Recent, improvements in EPN formulation, application equipment or approaches, and strain improvement have been made to enhance application efficacy. Additional research toward lowering product costs, increasing product availability, and improving efficacy and carryover effect will stimulate wider use of EPNs in biocontrol. Major areas of investigation that are likely to yield substantial results include genetic improvement and stabilization of EPN strains, expansion of formulation technology particularly focused on aboveground applications, and conservation biocontrol. Also, discovery of new EPN species and strains as well as development of new target pests will lead to expanded use. New EPN species are being described at a high rate although the biocontrol potential of most of these nematodes has yet to be determined (Lacey et al., 2015). Another avenue to wider utilization is to employ the symbiotic bacteria of EPN or their metabolites or byproducts as control materials for arthropod pests (Mohan et al., 2003; Bussaman et al., 2006; ffrench-Constant et al., 2007; Da Silva et al., 2013) or plant pathogens (Isaacson and Webster, 2002; Böszörmènyi et al., 2009; Shapiro-Ilan et al., 2014c). With advances such as these, EPNs will continue to help reduce reliance on chemical inputs in agriculture and enhance sustainability.

REFERENCES

Abu Hatab, M., Gaugler, R., Ehlers, R.U., 1998. Influence of culture method on *Steinernema glaseri* lipids. J. Parasitol. 84, 215–221.

Adhikari, B.N., Chin-Yo, L., Xiaodong, B., Ciche, T.A., Grewal, P.S., Dillman, A.R., Chaston, J.M., Shapiro-Ilan, D.I., Bilgrami, A.L., Gaugler, R., Sternberg, P.W., Adams, B.J., 2009. Transcriptional profiling of trait deterioration in the insect pathogenic nematode *Heterorhabditis bacteriophora*. BMC Genomics 10, 609.

Akhurst, R.J., Boemare, N.E., 1990. Biology and taxonomy of *Xenorhabdus*. In: Gaugler, R., Kaya, H.K. (Eds.), Entomopathogenic Nematodes in Biological Control. CRC Press, Boca Raton, FL, pp. 75–90.

Akhurst, R., Smith, K., 2002. Regulation and safety. In: Gaugler, R. (Ed.), Entomopathogenic Nematology. CABI, Wallingford, UK, pp. 311–332.

Ali, J.G., Alborn, H.T., Campos-Herrera, R., Kaplan, F., Duncan, L.W., et al., 2012. Subterranean, herbivore-induced plant volatile increases biological control activity of multiple beneficial nematode species in distinct habitats. PLoS One 7 (6), e38146. http://dx.doi.org/10.1371/journal.pone.0038146.

Anbesse, S., Sumaya, H.N., Dörfler, V.A., Strauch, O., Ehlers, R.U., 2013. Stabilization of heat tolerance traits in *Heterorhabditis bacteriophora* through selective breeding and creation of inbred lines in liquid culture. BioControl 58, 85–93.

Ansari, M.A., Tirry, L., Moens, M., 2004. Interaction between *Metarhizium anisopliae* CLO 53 and entomopathogenic nematodes for the control of *Hoplia philanthus*. Biol. Control 31, 172–180.

Ansari, M.A., Shah, F.A., Butt, T.M., 2010. The entomopathogenic nematode *Steinernema kraussei* and *Metarhizium anisopliae* work synergistically in controlling overwintering larvae of the black vine weevil, *Otiorhynchus sulcatus*, in strawberry growbags. Biocontrol Sci. Technol. 20, 99–105.

Arthurs, S., Heinz, K.M., Prasifka, J.R., 2004. An analysis of using entomopathogenic nematodes against above-ground pests. Bull. Entomol. Res. 94, 297–306.

Bai, C., Shapiro-Ilan, D.I., Gaugler, R., Hopper, K.R., 2005. Stabilization of beneficial traits in *Heterorhabditis bacteriophora* through creation of inbred lines. Biol. Control 32, 220–227.

Bai, X., Adams, B.J., Ciche, T.,A., Clifton, S., Gaugler, R., Kwi–suk, K., Spieth, J., Sternberg, W.P., Wilson, K.R., Grewal, S.P., 2013. A lover and a fighter: the genome sequence of an entomopathogenic nematode *Heterorhabditis bacteriophora*. PLoS One 8, e69618. http://dx.doi.org/10.1371/journal.pone.0069618.

Bal, K.H., Michel, P.A., Grewal, P.S., 2014. Genetic selection of the ambush foraging entomopathogenic nematode, *Steinernema carpocapsae* for enhanced dispersal and its associated trade–offs. Evol. Ecol. 28, 923–939.

Balasubramanian, N., Hao, Y.D., Toubarro, D., Nascimento, D., Simões, N., 2009. Purification, biochemical and molecular analysis of a chymotrypsin protease with prophenoloxidase suppression activity from the entomopathogenic nematode *Steinernema carpocapsae*. Int. J. Parasitol. 39, 975–984.

Balasubramanian, N., Toubarro, D., Simões, N., 2010. Biochemical study and in vitro insect immune suppression by a trypsin-like secreted protease from the nematode *Steinernema carpocapsae*. Parasite Immunol. 32, 165–175.

Barbercheck, M.E., 1992. Effect of soil physical factors on biological control agents of soil insect pests. Fla. Entomol. 75, 539–548.

Barbercheck, M.E., Kaya, H.K., 1990. Interactions between *Beauveria bassiana* and the entomogenous nematodes *Steinernema feltiae* and *Heterorhabditis heliothidis*. J. Invertebr. Pathol. 55, 225–234.

Barbercheck, M.E., Kaya, H.K., 1991. Effect of host condition and soil texture on host finding by the entomogenous nematodes *Heterorhabditis bacteriophora* (Rhabditida: Heterorhabditidae) and *Steinernema carpocapsae* (Rhabditida: Steinernematidae). Environ. Entomol. 20, 582–589.

Berry, E.C., Karlen, D.L., 1993. Comparison of alternate farming systems II. Earthworm population and density and species diversity. Am. J.Altern. Agric. 8, 21–26.

Berry, R.E., Liu, J., Groth, E., 1997. Efficacy and persistence of *Heterorhabditis marelatus* (Rhabditida: Heterorhabditidae) against root weevils (Coleoptera: Curculionidae) in strawberry. Environ. Entomol. 26, 465–470.

Bilgrami, A.L., Gaugler, R., Shapiro-Ilan, D.I., Adams, B.J., 2006. Source of trait deterioration in entomopathogenic nematodes *Heterorhabditis bacteriophora* and *Steinernema carpocapsae* during *in vivo* culture. Nematology 8, 397–409.

Bird, A.F., Akhurst, R.J., 1983. The nature of the intestinal vesicle in nematodes of the family Steinernematidae. Int. J. Parasitol. 13, 599–606.

Boemare, N., 2002. Biology, taxonomy, and systematics of *Photorhabdus* and *Xenorhabdus*. In: Gaugler, R. (Ed.), Entomopathogenic Nematology. CABI, Wallingford, UK, pp. 35–56.

Böszörmènyi, E., Ersek, T., Fodor, A., Fodor, A.M., Foldes, L.S., Hevesi, M., Hogan, J.S., Katona, Z., Klein, M.G., Kormany, A., Pekar, S., Szentirmai, A., Sztaricskai, F., Taylor, R.A.J., 2009. Isolation and activity of *Xenorhabdus* antimicrobial compounds against the plant pathogens *Erwinia amylovora* and *Phytophthora nicotianae*. J. Appl. Microbiol. 107, 746–759.

Brinkman, M.A., Gardner, W.A., 2000. Possible antagonistic activity of two entomopathogens infecting workers of the red imported fire ant (Hymenoptera: Formicidae). J. Entomol. Sci. 35, 205–207.

Brown, I.M., Gaugler, R., 1996. Cold tolerance of steinernematid and heterorhabditid nematodes. J. Thermal Biol. 21, 115–121.

Bruck, D.J., Shapiro-Ilan, D.I., Lewis, E.E., 2005. Evaluation of application technologies of entomopathogenic nematodes for control of the black vine weevil, *Otiorhynchus sulcatus*. J. Econ. Entomol. 98, 1884–1889.

Brusselman, E., Beck, B., Pollet, S., Temmerman, F., Spanoghe, P., Moens, M., Nuyttens, D., 2012. Effect of the spray application technique on the deposition of entomopathogenic nematodes in vegetables. Pest Manage. Sci. 68, 444–453.

Burman, M., 1982. *Neoaplectana carpocapsae*: toxin production by axenic insect parasitic nematodes. Nematologica 28, 62–70.

Burnell, A.M., Dowds, B.C.A., 1996. The genetic improvement of entomopathogenic nematodes and their symbiotic bacteria: phenotypic targets, genetic limitations and an assessment of possible hazards. Biocontrol Sci. Technol. 6, 435–447.

Bussaman, P., Sermswan, R.W., Grewal, P.S., 2006. Toxicity of the entomopathogenic bacteria *Photorhabdus* and *Xenorhabdus* to the mushroom mite (*Luciaphorus* sp.; Acari: Pygmephoridae). Biocontrol Sci. Technol. 16, 245–256.

Cakmak, I., Hazir, S., Ulug, D., Karagoz, M., 2013. Olfactory response of *Sancassania polyphyllae* (Acari: Acaridae) to its phoretic host larva killed by the entomopathogenic nematode, *Steinernema glaseri* (Rhabditida: Steinernematidae). Biol. Control 65, 212–217.

Campbell, J.F., Gaugler, R., 1997. Inter-specific variation in entomopathogenic nematode foraging strategy: dichotomy or variation along a continuum? Fundam. Appl. Nematol. 20, 393–398.

Campbell, J.F., Lewis, E.E., 2002. Entomopathogenic nematode host search strategies. In: Lewis, E.E., Campbell, J.F., Sukhdeo, M.V.K. (Eds.), The Behavioural Ecology of Parasites. CABI Publishing, Wallingford, UK, pp. 13–38.

Campos-Herrera, R. (Ed.), 2015. Nematode Pathogenesis of Insects and Other Pests. Spinger, Cham, Switzerland. 531 pp.

Castillo, J.C., Reynolds, S.E., Eleftherianos, I., 2011. Insect immune responses to nematode parasites. Trends Parasitol. 27, 537–547.

Chaston, J.M., Dillman, A.R., Shapiro-Ilan, D.I., Bilgrami, A.L., Gaugler, R., Hopper, K.R., Adams, B.J., 2011. Outcrossing and crossbreeding recovers deteriorated traits in laboratory cultured *Steinernema carpocapsae* nematodes. Int. J. Parasitol. 41, 801–809.

Ciche, T.A., Ensign, J.C., 2003. For the insect pathogen *Photorhabdus luminescens*, which end of a nematode is out? Appl. Environ. Microbiol. 69, 1890–1897.

Ciche, T.A., Sternberg, P.W., 2007. Postembryonic RNAi in *Heterorhabditis bacteriophora*: a nematode insect parasite and host for insect pathogenic symbionts. BMC Dev. Biol. 7, 101–111.

Ciche, T.A., Kim, K., Kaufmann-Daszczuk, B., Nguyen, K.C.Q., Hall, D.H., 2008. Cell invasion and matricide during *Photorhabdus luminescens* transmission by *Heterorhabditis bacteriophora* nematodes. Appl. Environ. Microbiol. 74, 2275–2287.

da Silva, O.S., Prado, G.R., Da Silva, J.L.R., Silva, C.E., Da Costa, M., Heermann, R., 2013. Oral toxicity of *Photorhabdus luminescens* and *Xenorhabdus nematophila* (Enterobacteriaceae) against *Aedes aegypti* (Diptera: Culicidae). Parasitol. Res. 112, 2891–2896.

De Waal, J.Y., Malan, A.P., Addison, M.F., 2011. Evaluating mulches together with *Heterorhabditis zealandica* (Rhabditida: Heterorhabditidae) for the control of diapausing codling moth larvae, *Cydia pomonella* (L.) (Lepidoptera: Tortricidae). Biocontrol Sci. Technol. 20, 255 271.

Del Valle, E.E., Dolinksi, C., Barreto, E.L.S., Souza, R.M., 2009. Effect of cadaver coatings on emergence and infectivity of the entomopathogenic nematode *Heterorhabditis baujardi* LPP7 (Rhabditida: Heterorhabditidae) and the removal of cadavers by ants. Biol. Control 50, 21–24.

Demir, S., Karagoz, M., Hazir, S., Kaya, H.K., 2015. Evaluation of entomopathogenic nematodes and their combined application against *Curculio elephas* and *Polyphylla fullo* larvae. J. Pest Sci. 88, 163–170.

Deol, Y.S., Jagdale, G.B., Cañas, L., Grewal, P.S., 2011. Delivery of entomopathogenic nematodes directly through commercial growing media via the inclusion of infected host cadavers: a novel approach. Biol. Control 58, 60–67.

Dillman, A.R., Chaston, J.M., Adams, B.J., Ciche, T.A., Goodrich-Blair, H., Stock, S.P., Sternberg, P.W., 2012a. An entomopathogenic nematode by any other name. PLoS Pathog. 8, e1002527.

Dillman, A.R., Guillermin, M.L., Lee, J.H., Kim, B., Sternberg, P.W., Hallem, E.A., 2012b. Olfaction shapes host–parasite interactions in parasitic nematodes. PNAS. http://dx.doi.org/10.1073/pnas.1211436109.

Dillman, A.R., Macchietto, M., Porter, C.F., Rogers, A., Williams, B., Antoshechkin, I., Lee, M.M., Goodwin, Z., Lu, X., Lewis, E.E., Goodrich-Blair, H., Stock, S.P., Adams, B.J., Sternberg, P.W., Mortazavi, A., 2015. Comparative genomics of *Steinernema* reveals deeply conserved gene regulatory networks. Genome Biol. 16, 200. http://dx.doi.org/10.1186/s13059-015-0746-6.

Dolinski, C., Shapiro-Ilan, D.I., Lewis, E.E., 2015. Insect cadaver applications: pros and cons. In: Campos-Herrera, R. (Ed.), Nematode Pathogenesis of Insects and Other Pests – Ecology and Applied Technologies for Sustainable Plant and Crop Protection. Springer Publishing, Cham, Switzerland, pp. 207–230.

Duchaud, E., Rusniok, C., Frangeul, L., Buchrieser, C., Givaudan, A., Taourit, S., Bocs, S., Boursaux-Eude, C., Chandler, M., Charles, J.F., Dassa, E., Derose, R., Derzelle, S., Freyssinet, G., Gaudriault, S., Medigue, C., Lanois, A., Powell, K., Siguier, P., Vincent, R., Wingate, V., Zouine, M., Glaser, P., Boemare, N., Danchin, A., Kunst, F., 2003. The genome sequence of the entomopathogenic bacterium *Photorhabdus luminescens*. Nat. Biotechnol. 21, 1307–1313.

Duncan, L.W., McCoy, C.W., 1996. Vertical distribution in soil, persistence, and efficacy against citrus root weevil (Coleoptera: Curculionidae) of two species of entomogenous nematodes (Rhabditida: Steinernematidae: Heterorhabditidae). Environ. Entomol. 25, 174–178.

Duncan, L.W., Stuart, R.J., El-Borai, F.E., Campos-Herrera, R., Pathak, E., Giurcanu, M., Graham, J.H., 2013. Modifying orchard planting sites conserves entomopathogenic nematodes, reduces weevil herbivory and increases citrus tree growth, survival and fruit yield. Biol. Control 64, 26–36.

Epsky, N., Capinera, J.L., 1988. Efficacy of the entomogenous nematode *Steinernema feltiae* against a subterranean termite, *Reticulitermes tibialis* (Isoptera: Rhinotermitidae). J. Econ. Entomol. 81, 1313–1317.

Ehlers, R.-U., 2005. Forum on safety and regulation. In: Grewal, P.S., Ehlers, R.-U., Shapiro-Ilan, D.I. (Eds.), Nematodes as Biocontrol Agents. CABI, Wallingford, UK, pp. 107–114.

Ehlers, R.U., Oestergaard, J., Hollmer, S., Wingen, M., Strauch, O., 2005. Genetic selection for heat tolerance and low temperature activity of the entomopathogenic nematode–bacterium complex *Heterorhabditis bacteriophora–Photorhabdus luminescens*. BioControl 50, 699–716.

Ekmen, Z.I., Hazir, S., Cakmak, I., Ozer, N., Karagoz, M., Kaya, H.K., 2010a. Potential negative effects on biological control by *Sancassania polyphyllae* (Acari: Acaridae) on an entomopathogenic nematode species. Biol. Control 54, 166–171.

Ekmen, Z.I., Cakmak, I., Karagoz, M., Hazir, S., 2010b. Food preference of *Sancassania polyphyllae* (Acari: Acaridae): living entomopathogenic nematodes or insect tissues. Biocontrol Sci. Technol. 20, 553–566.

El-Borai, F.E., Bright, D.B., Graham, J.H., Stuart, R.J., Cubero, J., Duncan, L.W., 2009. Differential susceptibility of entomopathogenic nematodes to nematophagous fungi from Florida citrus orchards. Nematology 11, 233–243.

El-Borai, F.E., Campos-Herrera, R., Stuart, R.J., Duncan, L.W., 2011. Substrate modulation, group effects and the behavioral responses of entomopathogenic nematodes to nematophagous fungi. J. Invertebr. Pathol. 106, 347–356.

Everard, A., Griffin, C.T., Dillon, A.B., 2009. Competition and intraguild predation between the braconid parasitoid *Bracon hylobii* and the entomopathogenic nematode *Heterorhabditis downesi*, natural enemies of the large pine weevil, *Hylobius abietis*. Bull. Entomol. Res. 99, 151–161.

ffrench-Constant, R.H., Dowling, A., Waterfield, N.R., 2007. Insecticidal toxins from *Photorhabdus* bacteria and their potential use in agriculture. Toxicon 49, 436–451.

Fife, J.P., Derksen, R.C., Ozkan, H.E., Grewal, P.S., 2003. Effects of pressure differentials on the viability and infectivity of entomopathogenic nematodes. Biol. Control 27, 65–72.

Fife, J.P., Ozkan, H.E., Derksen, R.C., Grewal, P.S., Krause, C.R., 2005. Viability of a biological pest control agent through hydraulic nozzles. Trans. ASAE 48, 45–54.

Fischer-Le Saux, M., Arteaga-Hernández, E., Mrácek, Z., Boemare, N., 1999. The bacterial symbiont *Xenorhabdus poinarii* (Enterbacteriaceae) is harbored by two phylogenetic related host nematodes: the entomopathogenic species *Steinernema cubanum* and *Steinernema glaseri* (Nematoda: Steinernematidae). FEMS Microbiol. Ecol. 29, 149–157.

Fodor, A., Vecseri, G., Farkas, T., 1990. *Caenorhabditis elegans* as a model for the study of entomopathogenic nematodes. In: Gaugler, R., Kaya, H.K. (Eds.), Entomopathogenic Nematodes in Biological Control. CABI, Wallingford, UK, pp. 249–269.

Forst, S., Clarke, D., 2002. Bacteria-nematode symbioses. In: Gaugler, R. (Ed.), Entomopathogenic Nematology. CABI Publishing, Wallingford, UK, pp. 57–77.

Forst, S., Dowds, B., Boemare, N., Stackebrandt, E., 1997. *Xenorhabdus* and *Photorhabdus* spp.: bugs that kill bugs. Annu. Rev. Microbiol. 51, 47–72.

Fujiie, A., Yokoyama, T., 1998. Effects of ultraviolet light on the entomopathogenic nematode, *Steinernema kushidai* and its symbiotic bacterium, *Xenorhabdus japonicus*. Appl. Entomol. Zool. 33, 263–269.

Fushing, H.L., Zhu, L., Shapiro-Ilan, D.I., Campbell, J.F., Lewis, E.E., 2008. State-space based mass event-history model I: many decision-making agents with one target. Ann. Appl. Stat. 2, 1503–1522.

Gaugler, R., 1987. Entomogenous nematodes and their prospects for genetic improvement. In: Maramorosch, K. (Ed.), Biotechnology in Invertebrate Pathology and Cell Culture. Academic Press, San Diego, CA, US, pp. 457–484.

Gaugler, R., Campbell, J.F., 1989. Selection for host–finding in *Steinernema feltiae*. J. Invertebr. Pathol. 54, 363–372.

Gaugler, R., Campbell, J.F., 1991. Selection for enhanced host–finding of scarab larvae (Coleoptera: Scarabaeidae) in an entomopathogenic nematode. Environ. Entomol. 20, 700–706.

Gaugler, R., Campbell, J.F., McGuire, T.R., 1990. Fitness of a genetically improved entomopathogenic nematode. J. Invertebr. Pathol. 56, 106–116.

Gaugler, R., Bednarek, A., Campbell, J.F., 1992. Ultraviolet inactivation of heterorhabditid and steinernematid nematodes. J. Invertebr. Pathol. 59, 155–160.

Gaugler, R., Wilson, M., Shearer, P., 1997. Field release and environmental fate of a transgenic entomopathogenic nematode. Biol. Control 9, 75–80.

Georgis, R., 2002. The Biosys experiment: an insider's perspective. In: Gaugler, R. (Ed.), Entomopathogenic Nematology. CABI, Wallingford, UK, pp. 357–372.

Georgis, R., Gaugler, R., 1991. Predictability in biological control using entomopathogenic nematodes. J. Econ. Entomol. 84, 713–720.

Georgis, R., Dunlop, D.B., Grewal, P.S., 1995. Formulation of entomopathogenic nematodes. In: Hall, F.R., Barry, J.W. (Eds.), Biorational Pest Control Agents: Formulation and Delivery. American Chemical Society, Washington, DC, pp. 197–205.

Georgis, R., Koppenhöfer, A.M., Lacey, L.A., Bélair, G., Duncan, L.W., Grewal, P.S., Samish, M., Tan, L., Torr, P., van Tol, R.W.H.M., 2006. Successes and failures in the use of parasitic nematodes for pest control. Biol. Control 38, 103–123.

Glazer, I., 2002. Survival biology. In: Gaugler, R. (Ed.), Entomopathogenic Nematology. CABI, Wallingford, UK, pp. 169–187.

Glazer, I., Salame, L., Segal, D., 1997. Genetic enhancement of nematicidal resistance of entomopathogenic nematodes. Biocontrol Sci. Technol. 7, 499–512.

Grant, J.A., Villani, M.G., 2003. Soil moisture effects on entomopathogenic nematodes. Environ. Entomol. 32, 80–87.

Grewal, P.S., 2002. Formulation and application technology. In: Gaugler, R. (Ed.), Entomopathogenic Nematology. CABI, Wallingford, pp. 265–288.

Grewal, P.S., Selvan, S., Gaugler, R., 1994. Thermal adaptation of entomopathogenic nematodes-niche breadth for infection, establishment and reproduction. J. Therm. Biol. 19, 245–253.

Grewal, P.S., Ehlers, R.-U., Shapiro-Ilan, D.I. (Eds.), 2005. Nematodes as Biological Control Agents. CABI, Wallingford. 505 pp.

Gulcu, B., Hazir, S., Kaya, H.K., 2012. Scavenger deterrent factor (SDF) from symbiotic bacteria of entomopathogenic nematodes. J. Invertebr. Pathol. 110, 326–333.

Gumus, A., Karagoz, M., Shapiro-Ilan, D.I., Hazir, S., 2015. A novel approach to biocontrol: release of live insect hosts pre-infected with entomopathogenic nematodes. J. Invertebr. Pathol. 130, 56–60.

Hallem, E.A., Dillman, A.R., Hong, A.V., Zhang, Y., Yano, J.M., DeMarco, S.F., Sternberg, P.W., 2011. A sensory code for host seeking in parasitic nematodes. Curr. Biol. 21, 377–383.

Hashmi, S., Hashmi, G., Gaugler, R., 1995. Genetic transformation of an entomopathogenic nematode by microinjection. J. Invertebr. Pathol. 66, 293–296.

Hashmi, S., Hashmi, G., Glazer, I., Gaugler, R., 1998. Thermal response of *Heterorhabditis bacteriophora* transformed with the *Caenorhabditis elegans hsp70* encoding gene. J. Exp. Zool. 281, 164–170.

Hazir, S., Stock, S.P., Kaya, H.K., Koppenhöfer, A.M., Keskin, N., 2001. Developmental temperature effects on five geographic isolates of the entomopathogenic nematode *Steinernema feltiae* (Steinernematidae). J. Invertebr. Pathol. 77, 243–250.

Hazir, S., Kaya, H.K., Stock, S.P., Keskin, N., 2003. Entomopathogenic nematodes (Steinernematidae and Heterorhabditidae) for biological control of soil pests. Turk. J. Biol. 27, 181–202.

Hiltpold, I., Turlings, T.C.J., 2012. Manipulation of chemically mediated interactions in agricultural soils to enhance the control of crop pests and to improve crop yield. J. Chem. Ecol. 38, 641–650.

Hiltpold, I., Baroni, M., Toepfer, S., Kuhlmann, U., Turlings, T.C.J., 2010. Selection of entomopathogenic nematodes for enhanced responsiveness to a volatile root signal helps to control a major root pest. J. Exp. Biol. 213, 2417–2423.

Ilan, T., Kim-Shapiro, D.B., Bock, C., Shapiro-Ilan, D.I., 2013. The impact of magnetic fields, electric fields and current on the directional movement of *Steinernema carpocapsae*. Int. J. Parasitol. 43, 781–784.

Isaacson, P.J., Webster, J.M., 2002. Antimicrobial activity of *Xenorhabdus* sp. RIO (Enterobacteriaceae) symbiont of the entomopathogenic nematode, *Steinernema riobrave* (Rhabditida: Steinernematidae). J. Invertebr. Pathol. 79, 146–153.

Jaffee, B.A., Strong, D.R., 2005. Strong bottom-up and weak top-down effects in soil: nematode-parasitized insects and nematode-trapping fungi. Soil Biol. Biochem. 37, 1011–1021.

Jansson, R.K., Lecrone, S.H., Gaugler, R., 1993. Field efficacy and persistence of entomopathogenic nematodes (Rhabditida: Steinernematidae, Heterorhabditidae) for control of sweetpotato weevil (Coleoptera: Apionidae) in southern Florida. J. Econ. Entomol. 86, 1055–1063.

Jing, Y., Toubarro, D., Hao, Y., Simões, N., 2010. Cloning, characterisation and heterologous expression of an astacin metalloprotease, Sc-AST, from the entomoparasitic nematode *Steinernema carpocapsae*. Mol. Biochem. Parasitol. 174, 101–108.

Karagoz, M., Gulcu, B., Cakmak, I., Kaya, H.K., Hazir, S., 2007. Predation of entomopathogenic nematodes by *Sancassania* sp. (Acari: Acaridae). Exp. Appl. Acarol. 43, 85–95.

Kaspi, R., Ross, A., Hodson, A.K., Stevens, G.N., Kaya, H.K., Lewis, E.E., 2010. Foraging efficacy of the entomopathogenic nematode *Steinernema riobrave* in different soil types from California citrus groves. Appl. Soil Ecol. 45, 243–253.

Kaya, H.K., 1990. Soil ecology. In: Gaugler, R., Kaya, H.K. (Eds.), Entomopathogenic Nematodes in Biological Control. CRC Press, Boca Raton, FL, pp. 93–116.

Kaya, H.K., 2002. Natural enemies and other antagonists. In: Gaugler, R. (Ed.), Entomopathogenic Nematology. CABI, Wallingford, pp. 189–204.

Kaya, H.K., Aguillera, M.M., Alumai, A., Choo, H.Y., de la Torre, M., Foder, A., Ganguly, S., Hazir, S., Lakatos, T., Pye, A., Wilson, M., Yamanaka, S., Yang, H., Ehlers, R.-U., 2006. Status of entomopathogenic nematodes and their symbiotic bacteria from selected countries or regions of the world. Biol. Control 38, 134–155.

Klein, M.G., 1990. Efficacy against soil-inhabiting insect pests. In: Gaugler, R., Kaya, H.K. (Eds.), Entomopathogenic Nematodes in Biological Control. CRC Press, Boca Raton, FL, pp. 195–214.

Klein, M.G., Georgis, R., 1992. Persistence of control of Japanese beetle (Coleoptera: Scarabaeidae) larvae with steinernematid and heterorhabditid nematodes. J. Econ. Entomol. 85, 727–730.

Koppenhöfer, A.M., 2000. Nematodes. In: Lacey, L.A., Kaya, H.K. (Eds.), Field Manual of Techniques in Invertebrate Pathology. Kluwer, Dordrecht, The Netherlands, pp. 283–301.

Koppenhöfer, A.M., Fuzy, E.M., 2003. Ecological characterization of *Steinernema scarabaei*, a scarab-adapted entomopathogenic nematode from New Jersey. J. Invertebr. Pathol. 83, 139–148.

Koppenhöfer, A.M., Fuzy, E.M., 2006. Effect of soil type on infectivity and persistence of the entomopathogenic nematodes *Steinernema scarabaei*, *Steinernema glaseri*, *Heterorhabditis zealandica*, and *Heterorhabditis bacteriophora*. J. Invertebr. Pathol. 92, 11–22.

Koppenhöfer, A.M., Fuzy, E.M., 2007. Soil moisture effects on infectivity and persistence of the entomopathogenic nematodes *Steinernema scarabaei*, *S. glaseri*, *Heterorhabditis zealandica*, and *H. bacteriophora*. Appl. Soil Ecol. 35, 128–139.

Koppenhöfer, A.M., Fuzy, E.M., 2008. Effect of the anthranilic diamide insecticide, chlorantraniliprole, on *Heterorhabditis bacteriophora* (Rhabditida: Heterorhabditidae) efficacy against white grubs (Coleoptera: Scarabaeidae). Biol. Control 45, 93–102.

Koppenhöfer, A.M., Grewal, P.S., 2005. Compatibility and interactions with agrochemicals and other biocontrol agents. In: Grewal, P.S., Ehlers, R.-U., Shapiro-Ilan, D.I. (Eds.), Nematodes as Biological Control Agents. CABI Publishing, Wallingford, pp. 363–381.

Koppenhöfer, A.M., Kaya, H.K., 1997. Additive and synergistic interactions between entomopathogenic nematodes and *Bacillus thuringiensis* for scarab grub control. Biol. Control 8, 131–137.

Koppenhöfer, A.M., Kaya, H.K., Taormino, S.P., 1995. Infectivity of entomopathogenic nematodes (Rhabditida: Steinernematidae) at different soil depths and moistures. J. Invertebr. Pathol. 65, 193–199.

Kruitbos, L.M., Heritage, S., Hapca, S., Wilson, M.J., 2010. The influence of habitat quality on the foraging strategies of the entomopathogenic nematodes *Steinernema carpocapsae* and *Heterorhabditis megidis*. Parasitology 137, 303–309.

Kung, S., Gaugler, R., Kaya, H.K., 1990. Influence of soil pH and oxygen on persistence of *Steinernema* spp. J. Nematol 22, 440–445.

Kung, S., Gaugler, R., Kaya, H.K., 1991. Effects of soil temperature, moisture, and relative humidity on entomopathogenic nematode persistence. J. Invertebr. Pathol. 57, 242–249.

Lacey, L.A., Georgis, R., 2012. Entomopathogenic nematodes for control of insect pests above and below ground with comments on commercial production. J. Nematol 44, 218–225.

Lacey, L.A., Unruh, T.R., Headrick, H.L., 2003. Interactions of two parasitoids (Hymenoptera: Ichneumonidae) of codling moth (Lepidoptera: Tortricidae) with the entomopathogenic nematode *Steinernema carpocapsae* (Rhabditida: Steinernematidae). J. Invertebr. Pathol. 83, 230–239.

Lacey, L.A., Arthurs, S.P., Granatstein, D., Headrick, H., Fritts Jr., R., 2006. Use of entomopathogenic nematodes (Steinernematidae) in conjunction with mulches for control of codling moth (Lepidoptera: Tortricidae). J. Entomol. Sci. 41, 107–119.

Lacey, L.A., Shapiro-Ilan, D.I., Glenn, G.M., 2010. The effect of post-application anti-desiccant agents and formulated host-cadavers on entomopathogenic nematode efficacy for control of diapausing codling moth larvae (Lepidoptera: Tortricidae). Biocontrol Sci. Technol. 20, 909–921.

Lacey, L.A., Grzywacz, D., Shapiro-Ilan, D.I., Frutos, R., Brownbridge, M., Goettel, M.S., 2015. Insect pathogens as biological control agents: back to the future. J. Invertebr. Pathol. 132, 1–41.

Laumond, C., Simões, N., Boemare, N., 1989. Toxins of entomoparasitic nematodes. Pathogenicity of *Steinernema carpocapsae* – prospectives of genetic engineering. Conte Rendu Acad. Agric. France 75, 135–138.

Lemaitre, B., Hoffmann, J., 2007. The host defense of *Drosophila melanogaster*. Annu. Rev. Immunol. 25, 697–743.

Levine, E., Oloumi-Sadeghi, H., 1992. Field evaluation of *Steinernema carpocapsae* (Rhabditida: Steinernematidae) against the black cutworm (Lepidoptera: Noctuidae) larvae in field corn. J. Entomol. Sci. 27, 427–435.

Lewis, E.E., 2002. Behavioral ecology. In: Gaugler, R. (Ed.), Entomopathogenic Nematology. CABI Publishing, Wallingford, UK, pp. 205–224.

Lewis, E.E., Clarke, D.J., 2012. Nematode parasites and entomopathogens. In: Vega, F.E., Kaya, H.K. (Eds.), Insect Pathology, second ed. Elsevier, Amsterdam, pp. 395–424.

Lewis, E.E., Gaugler, R., Harrison, R., 1992. Entomopathogenic nematode host finding: response to host contact cues by cruise and ambush foragers. Parasitology 105, 309–319.

Lewis, E.E., Grewal, P.S., Gaugler, R., 1995. Hierarchical order of host cues in parasite foraging strategies. Parasitology 119, 207–213.

Li, X.Y., Cowles, R.S., Cowles, E.A., Gaugler, R., Cox-Foster, D.L., 2007. Relationship between the successful infection of entomopathogenic nematodes and the host immune response. Int. J. Parasitol. 37, 365–374.

Llàcer, E., Martinez de Altube, M.M., Jacas, J.A., 2009. Evaluation of the efficacy of *Steinernema carpocapsae* in a chitosan formulation against the red palm weevil, *Rhynchophorus ferrugineus*, in *Phoenix canariensis*. BioControl 54, 559–565.

Martens, E.C., Goodrich-Blair, H., 2005. The *Steinernema carpocapsae* intestinal vesicle contains a subcellular structure with which *Xenorhabdus nematophila* associates during colonization initiation. Cell Microbiol. 7, 1723–1735.

Mbata, G.N., Shapiro-Ilan, D.I., 2010. Compatibility of *Heterorhabditis indica* (Rhabditida: Heterorhabditidae) and *Habrobracon hebetor* (Hymenoptera: Braconidae) for biological control of *Plodia interpunctella* (Lepidoptera: Pyralidae). Biol. Control 54, 75–82.

Mbata, G.N., Shapiro-Ilan, D.I., 2013. The potential for controlling *Pangaeus bilineatus* (Say) (Heteroptera: Cydnidae) using a combination of entomopathogens and an insecticide. J. Econ. Entomol. 106, 2072–2076.

McCoy, C.W., Shapiro, D.I., Duncan, L.W., Nguyen, K., 2000. Entomopathogenic nematodes and other natural enemies as mortality factors for larvae of *Diaprepes abbreviatus* (Coleoptera: Curculionidae). Biol. Control 19, 182–190.

Mohan, S., Raman, R., Gaur, H.S., 2003. Foliar application of *Photorhabdus luminescens*, symbiotic bacteria from entomopathogenic nematode *Heterorhabditis indica*, to kill cabbage butterfly *Pieris brassicae*. Curr. Sci. 84, 1397.

Morales-Rodriguez, A., Peck, D.C., 2009. Synergies between biological and neonicotinoid insecticides for the curative control of the white grubs *Amphimallon majale* and *Popillia japonica*. Biol. Control 51, 169–180.

Moshayov, A., Koltai, H., Glazer, I., 2013. Molecular characterization of the recovery process in the entomopathogenic nematode *Heterorhabditis bacteriophora*. Int. J. Parasitol. 43, 843–852.

Mukuka, J., Strauch, O., Ehlers, R.-U., 2010. Improvement of heat and desiccation tolerance in *Heterorhabditis bacteriophora* through cross–breeding of tolerant strains and successive genetic selection. BioControl 55, 511–521.

Nguyen, K.B., Smart Jr., G.C., 1991. Pathogenicity of *Steinernema scapterisci* to selected invertebrates. J. Nematol. 23, 7–11.

Parkman, J.P., Frank, J.H., Nguyen, K.B., Smart Jr., G.C., 1994. Inoculative release of *Steinernema scapterisci* (Rhabditida: Steinernematidae) to suppress pest mole crickets (Orthoptera: Gryllotapidae) on golf courses. Environ. Entomol. 23, 1331–1337.

Patel, M.N., Wright, D.J., 1997. Glycogen: its importance in the infectivity of aged juveniles of *Steinernema carpocapsae*. Parasitology 114, 591–596.

Patel, M.N., Stolinski, M., Wright, D.J., 1997. Neutral lipids and the assessment of infectivity in entomopathogenic nematodes: observations on four *Steinernema* species. Parasitology 114, 489–496.

Perez, E.E., Lewis, E.E., Shapiro-Ilan, D.I., 2003. Impact of host cadaver on survival and infectivity of entomopathogenic nematodes (Rhabditida: Steinernematidae and Heterorhabditidae) under desiccating conditions. J. Invertebr. Pathol. 82, 111–118.

Poinar Jr., G.O., 1979. Nematodes for Biological Control of Insects. CRC Press, Boca Raton, FL. 277 pp.

Poinar Jr., G.O., 1990. Biology and taxonomy of Steinernematidae and Heterorhabditidae. In: Gaugler, R., Kaya, H.K. (Eds.), Entomopathogenic Nematodes in Biological Control. CRC Press, Boca Raton, FL, pp. 23–62.

Poinar Jr., G.O., Grewal, P.S., 2012. History of entomopathogenic nematology. J. Nematol. 44, 153–161.

Ramalingam, K.R., Hazir, C., Gumus, A., Asan, C., Karagoz, M., Hazir, S., 2015. Efficacy of the entomopathogenic nematode, *Heterorhabditis bacteriophora*, using different application methods in the presence or absence of a natural enemy. Turk. J. Agric. For. 39, 277–285.

Rasmann, S., Köllner, T.G., Degenhardt, J., Hiltpold, I., Toepfer, S., Kuhlmann, U., Gershenzon, J., Turlings, T.C.J., 2005. Recruitment of entomopathogenic nematodes by insect-damaged maize roots. Nature 434, 732–737.

Salame, L., Glazer, I., Chubinishvilli, M.T., Chkhubianishvili, T., 2010. Genetic improvement of the desiccation tolerance and host–seeking ability of the entomopathogenic nematode *Steinernema feltiae*. Phytoparasitica 38, 359–368.

Schroer, S., Ehlers, R.-U., 2005. Foliar application of the entomopathogenic nematode *Steinernema carpocapsae* for biological control of diamondback moth larvae (*Plutella xylostella*). Biol. Control 33, 81–86.

Segal, D., Glazer, I., 1998. Genetic approaches for enhancing beneficial traits in entomopathogenic nematodes. Jpn. J. Nematol. 28, 101–107.

Segal, D., Glazer, I., 2000. Genetics for improving biological control agents: the case of entomopathogenic nematodes. Crop Prot. 19, 685–689.

Selvan, S., Gaugler, R., Lewis, E.E., 1993. Biochemical energy reserves of entomopathogenic nematodes. J. Parasitol. 79, 167–172.

Shapiro-Ilan, D.I., Brown, I., 2013. Earthworms as phoretic hosts for *Steinernema carpocapsae* and *Beauveria bassiana*: implications for enhanced biological control. Biol. Control 66, 41–48.

Shapiro-Ilan, D.I., Gaugler, R., 2002. Production technology for entomopathogenic nematodes and their bacterial symbionts. J. Ind. Microbiol. Biotechnol. 28, 137–146.

Shapiro-Ilan, D.I., Glazer, I., 1996. Comparison of entomopathogenic nematode dispersal from infected hosts versus aqueous suspension. Environ. Entomol. 25, 1455–1461.

Shapiro-Ilan, D.I., Grewal, P.S., 2008. Entomopathogenic nematodes and insect management. In: Capinera, J.L. (Ed.), Encyclopedia of Entomology, second ed. Springer, Dordrecht, The Netherlands, pp. 1336–1340.

Shapiro-Ilan, D.I., Lewis, E.E., 1999. Comparison of entomopathogenic nematode infectivity from infected hosts versus aqueous suspension. Environ. Entomol. 28, 907–911.

Shapiro-Ilan, D.I., Glazer, I., Segal, D., 1996. Trait stability and fitness of the heat tolerant entomopathogenic nematode *Heterorhabditis bacteriophora* IS5 strain. Biol. Control 6, 238–244.

Shapiro-Ilan, D.I., Glazer, I., Segal, D., 1997. Genetic improvement of heat tolerance in *Heterorhabditis bacteriophora* through hybridization. Biol. Control 8, 153–159.

Shapiro, D.I., Obrycki, J.J., Lewis, L.C., Abbas, M., 1999. The effects of fertilizers on black cutworm, *Agrotis ipsilon*, (Lepidoptera: Noctuidae) suppression by *Steinernema carpocapsae*. J. Nematol. 31, 690–693.

Shapiro, D.I., McCoy, C.W., Fares, A., Obreza, T., Dou, H., 2000. Effects of soil type on virulence and persistence of entomopathogenic nematodes in relation to control of *Diaprepes abbreviatus*. Environ. Entomol. 29, 1083–1087.

Shapiro-Ilan, D.I., Lewis, E.E., Behle, R.W., McGuire, M.R., 2001. Formulation of entomopathogenic nematode-infected-cadavers. J. Invertebr. Pathol. 78, 17–23.

Shapiro-Ilan, D.I., Gouge, D.H., Koppenhöfer, A.M., 2002. Factors affecting commercial success: case studies in cotton, turf and citrus. In: Gaugler, R. (Ed.), Entomopathogenic Nematology. CABI Publishing, Wallingford, UK, pp. 333–356.

Shapiro-Ilan, D.I., Lewis, E.E., Tedders, W.L., Son, Y., 2003. Superior efficacy observed in entomopathogenic nematodes applied in infected host cadavers compared with application in aqueous suspension. J. Invertebr. Pathol. 83, 270–272.

Shapiro-Ilan, D.I., Jackson, M., Reilly, C.C., Hotchkiss, M.W., 2004. Effects of combining an entomopathogenic fungi or bacterium with entomopathogenic nematodes on mortality of *Curculio caryae* (Coleoptera: Curculionidae). Biol. Control 30, 119–126.

Shapiro-Ilan, D.I., Stuart, J.R., McCoy, W.C., 2005. Targeted improvement of *Steinernema carpocapsae* for control of the pecan weevil, *Curculio caryae* (Horn) (Coleoptera: Curculionidae) through hybridization and bacterial transfer. Biol. Control 34, 215–221.

Shapiro-Ilan, D.I., Gouge, D.H., Piggott, S.J., Patterson Fife, J., 2006. Application technology and environmental considerations for use of entomopathogenic nematodes in biological control. Biol. Control 38, 124–133.

Shapiro-Ilan, D.I., Cottrell, T.E., Mizell III, R.F., Horton, D.L., Behle, B., Dunlap, C., 2010a. Efficacy of *Steinernema carpocapsae* for control of the lesser peachtree borer, *Synanthedon pictipes*: improved aboveground suppression with a novel gel application. Biol. Control 54, 23–28.

Shapiro-Ilan, D.I., Morales-Ramos, J.A., Rojas, M.G., Tedders, W.L., 2010b. Effects of a novel entomopathogenic nematode–infected host formulation on cadaver integrity, nematode yield, and suppression of *Diaprepes abbreviatus* and *Aethina tumida* under controlled conditions. J. Invertebr. Pathol. 103, 103–108.

Shapiro-Ilan, D.I., Han, R., Dolinksi, C., 2012. Entomopathogenic nematode production and application technology. J. Nematol 44, 206–217.

Shapiro-Ilan, D.I., Han, R., Qiu, X., 2014a. Production of entomopathogenic nematodes. In: Morales-Ramos, J., Rojas, G., Shapiro-Ilan, D.I. (Eds.), Mass Production of Beneficial Organisms: Invertebrates and Entomopathogens. Academic Press, San Diego, pp. 321–356.

Shapiro-Ilan, D.I., Lewis, E.E., Schliekelman, P., 2014b. Aggregative group behavior in insect parasitic nematode dispersal. Int. J. Parasitol. 44, 49–54.

Shapiro-Ilan, D.I., Bock, C.H., Hotchkiss, M.W., 2014c. Suppression of pecan and peach pathogens on different substrates using *Xenorhabdus bovienii* and *Photorhabdus luminescens*. Biol. Control 77, 1–6.

Shapiro-Ilan, D.I., Hazir, S., Leite, L., 2015a. Viability and virulence of entomopathogenic nematodes exposed to ultraviolet radiation. J. Nematol. 47, 184–189.

Shapiro-Ilan, D.I., Cottrell, T.E., Mizell III, R.F., Horton, D.L., Abdo, Z., 2015b. Field suppression of the peachtree borer, *Synanthedon exitiosa*, using *Steinernema carpocapsae*: effects of irrigation, a sprayable gel and application method. Biol. Control 82, 7–12.

Shields, E.J., Testa, A., Miller, J.M., Flanders, K.L., 1999. Field efficacy and persistence of the entomopathogenic nematodes *Heterorhabditis bacteriophora* 'Oswego' and *H. bacteriophora* 'NC' on Alfalfa snout beetle larvae (Coleoptera: Curculionidae). Environ. Entomol. 28, 128–136.

Simões, N., Caldas, C., Rosa, J.S., Bonifassi, E., Laumond, C., 2000. Pathogenicity caused by high virulent and low virulent strains of *Steinernema carpocapsae* to *Galleria mellonella*. J. Invertebr. Pathol. 75, 47–54.

Stock, S.P., Goodrich-Blair, H., 2008. Entomopathogenic nematodes and their bacterial symbionts: the inside out of a mutualistic association. Symbiosis 46, 65–75.

Stock, S.P., Hunt, D.J., 2005. Morphology and systematics of nematodes used in biocontrol. In: Grewal, P.S., Ehlers, R.U., Shapiro-Ilan, D.I. (Eds.), Nematodes as Biocontrol Agents. CABI, Wallingford, pp. 3–43.

Thurston, G.S., Kaya, H.K., Gaugler, R., 1994. Characterizing the enhanced susceptibility of milky disease-infected scarabaeid grubs to entomopathogenic nematodes. Biol. Control 4, 67–73.

Tomalak, M., 1994a. Selective breeding of *Steinernema feltiae* (Filipjev) (Nematoda: Steinernematidae) for improved efficacy in control of a mushroom fly, *Lycoriella solani* Winnertz (Diptera: Sciaridae). Biocontrol Sci. Technol. 4, 187–198.

Tomalak, M., 1994b. Genetic improvement of *Steinernema feltiae* for integrated control of the Western Flower Thrips, *Frankliniella occidentalis*. IOBC/WPRS Bull. 17, 17–20.

Torr, P., Heritage, S., Wilson, M.J., 2004. Vibrations as a novel signal for host location by parasitic nematodes. Int. J. Parasitol. 34, 997–999.

Ulug, D., Hazir, S., Kaya, H.K., Lewis, E.E., 2014. Potential natural enemies of entomopathogenic nematodes (Steinernematidae and Heterorhabditidae). J. Ecol. Entomol. 39, 462–469.

van Niekerk, S., Malan, A.P., 2015. Adjuvants to improve aerial control of the citrus mealybug *Planococcus citri* (Hemiptera: Pseudococcidae) using entomopathogenic nematodes. J. Helminthol. 89, 189–195.

Vellai, T., Molnár, A., Lakatos, L., Bánfalvi, Z., Fodor, A., Sáringer, G., 1999. Transgenic nematodes carrying a cloned stress resistant gene from yeast. In: Glazer, I., Richardson, P., Boemare, N., Coudert, F. (Eds.), Survival of Entomopathogenic Nematodes. European Commission Publications, Luxemburg, pp. 105–119.

Wang, X., Grewal, P.S., 2002. Rapid genetic deterioration of environmental tolerance and reproductive potential of an entomopathogenic nematode during laboratory maintenance. Biol. Control 23, 71–78.

Wang, Z., Wilhelmsson, C., Hyrsl, P., Loof, T.G., Dobes, P., Klupp, M., et al., 2010. Pathogen entrapment by transglutaminase a conserved early innate immune mechanism. PLoS Pathog. 6, 1000763 PMID:20169185.

Waterfield, N.R., Ciche, T., Clarke, D., 2009. *Photorhabdus* and a host of hosts. Annu. Rev. Microbiol. 63, 557–574.

Willett, D.S., Alborn, H.T., Duncan, L.W., Stelinski, L.L., 2015. Social networks of educated nematodes. Scientific Rep. 5, 14388.

Williams, E.C., Walters, K.F.A., 2000. Foliar application of the entomopathogenic nematode *Steinernema feltiae* against leafminers on vegetables. Biocontrol Sci. Technol. 10, 61–70.

Wright, D.J., Perry, R.N., 2002. Physiology and biochemistry. In: Gaugler, R. (Ed.), Entomopathogenic Nematology. CABI, Wallingford, UK, pp. 145–168.

Wright, D.J., Peters, A., Schroer, S., Fife, J.P., 2005. Application technology. In: Grewal, P.S., Ehlers, R.-U., Shapiro-Ilan, D.I. (Eds.), Nematodes as Biocontrol Agents. CABI, New York, NY, pp. 91–106.

Zhou, X.S., Kaya, H.K., Heungens, K., Goodrich–Blair, H., 2002. Response of ants to a deterrent factor(s) produced by the symbiotic bacteria of entomopathogenic nematodes. Appl. Environ. Microbiol. 68, 6202–6209.

Zhu, H., Grewal, P.S., Reding, M.E., 2011. Development of a desiccated cadaver delivery system to apply entomopathogenic nematodes for control of soil pests. App. Eng. Agric. 27, 317–324.

Part III

Mass Production, Formulation and Quality Control

Chapter 7

Production, Formulation, and Bioassay of Baculoviruses for Pest Control

D. Grzywacz[1], S. Moore[2,3]

[1]University of Greenwich, Chatham, Kent, United Kingdom; [2]Citrus Research International, Port Elizabeth, South Africa; [3]Rhodes University, Grahamstown, South Africa

7.1 INTRODUCTION

The need to produce a reliable baculovirus product in adequate quantities, of suitable quality, and at an affordable cost is a fundamental requirement, which must be met if the product is to become a viable pest control agent. Baculoviruses, as all viruses, can only be produced through propagation in living host cells of a susceptible insect species (Shapiro, 1986; Eberle et al., 2012a). These cells may be growing in intact insects (in vivo) or in tissue cell cultures (in vitro). Each baculovirus isolate/host system has individual requirements, and while some baculoviruses only replicate in a single insect species, others have a wider host range and may successfully replicate in several host species and cell lines. Successful baculovirus production involves not only the successful replication of the virus but also crucially the mass production of viable infectious particles occluded within protective protein matrices, the occlusion body (OB).

In vivo production is a sequential process involving the rearing of sufficient healthy host insects and then infecting them to multiply the virus so as to produce an adequate yield of infective mature OBs (Shapiro, 1986; Cherry et al., 1997; van Beek and Davis, 2009). Baculovirus production in insect larvae is a simple process that can readily be undertaken without the need for complex or expensive production systems. Consequently, baculovirus production can be undertaken at a laboratory scale or by small enterprises as a low-cost operation without the need for major capital expenditure (Grzywacz et al., 2014).

As an alternative to relying on mass reared insects, in vitro production produces baculoviruses in cell culture systems (Granados et al., 2007; Claus et al., 2012; Reid et al., 2014). This is seen as a potentially more controllable system for the automated production of large amounts of high-quality OBs without high labor costs. To date no in vitro production has been commercialized, and a number of issues need to be resolved and research advances made before this approach is likely to be commercially feasible.

Baculoviruses can be very effectively used for pest control as simple aqueous suspensions (Moore et al., 2015) of infected insects, but the cost effectiveness, application, storage, and handling can all be improved by formulation so that most commercial products in use today involve some element of formulation. A significant advantage of baculoviruses as biopesticides is that the OB is stable and robust, making its storage, processing, and formulation relatively straightforward. To be acceptable as a pest control product, it is absolutely essential that batches of product reliably achieve certain standards of efficacy. This can only be determined and assured through a rigorous application of quality control in which bioassay is the key element (Jenkins and Grzywacz, 2000).

At present a number of different baculoviruses (>15) are produced commercially, and the number of commercial baculovirus products worldwide exceeds 80 (Kabaluk et al., 2010; Gwynn, 2014; Sun, 2015; Haase et al., 2015). The current generation of products use baculoviruses from the genera Alphabaculovirus and Gammabaculovirus, the nucleopolyhedroviruses (NPVs), and from the genus Betabaculovirus, the granuloviruses (GVs). Commercial production uses in vivo production systems of varying degrees of technological complexity (Eberle et al., 2012a; Grzywacz et al., 2014; Lacey et al., 2015). However, if the need for baculovirus products expands beyond the current niche markets into providing pest control for large-scale field crops, our current in vivo production systems will need to be radically scaled up, or alternatively, and perhaps more challengingly, commercially viable in vitro production methods will need to be developed.

7.2 BACULOVIRUS PRODUCTION IN CELL CULTURE

In vivo mass production is the only viable means of mass producing baculoviruses for microbial control products (Szewcyk et al., 2006; Reid et al., 2014), but it has long been

Microbial Control of Insect and Mite Pests. http://dx.doi.org/10.1016/B978-0-12-803527-6.00007-X

recognized that this has limitations and that the development of practicable, economically viable cell culture-based mass production systems could have a number of important potential advantages (Black et al., 1997; Moscardi, 1999). However, despite extensive efforts, it has not been possible to develop suitable reactors needed for large-scale production of insect cells nor to develop medium at a sufficiently low cost (Claus et al., 2012; Reid et al., 2014). Issues of baculovirus quality also remain a challenge in baculovirus cell culture, as does low cell yield (Nguyen et al., 2011).

Adopting tissue culture-based production theoretically could make for a cheaper and more flexible production system, as a single bioreactor supported by a suite of different frozen cell lines can be used in principle to produce many different baculoviruses as needed, without the cost of maintaining multiple insect cultures and scaling them up to meet new production targets as the market requirement changes (Reid et al., 2014).

Another argument for developing cell culture of baculoviruses is that it is the only option for producing genetically modified (GM) baculoviruses (Black et al., 1997; Inceoglu et al., 2006). GM baculoviruses, capable of killing insects more quickly, invariably interfere with the two-stage baculovirus, and in vivo production of GM strains can be <10% of that in wild type baculoviruses. Thus it has long been a major thrust of research into baculoviruses to develop viable large-scale bioreactor production, and this has produced a number of detailed reviews of the issues to which the reader is directed for a more detailed discussion (Black et al., 1997; Granados et al., 2007; Claus et al., 2012; Reid et al., 2014). The essential elements for tissue culture mass production are insect cell lines that can support viral replication, a cell medium that can support cell culture, and suitable bioreactors for large-scale cell culture. Other key supporting components are the knowledge base and expertise on how to manage and optimize cell and virus replication as well as suitable stable virus lines that can replicate reliably in vitro.

Since the development of the first insect cell lines in the 1960s many cell lines from Lepidoptera that are capable of supporting viral replication have been developed (Lynn, 2007). Key properties in cell lines needed for mass production include a short replication time (<24 h) and the ability to produce >200 OB per cell (Reid et al., 2014). While a few such cells lines are now available for some well-researched baculoviruses, such as *Autographa californica* MNPV, *Spodoptera frugiperda* MNPV, and *Helicoverpa armigera* SNPV, we still do not have such cell lines for other baculoviruses of high commercial interest, such as *Plutella xylostella* GV.

Media capable of supporting the demanding nutritional requirements of insect cell growth were originally modeled on the composition of insect hemolymph. In order to achieve this, it was necessary to include a chemically defined mixture of salts, amino acids, vitamins, and carbohydrates with complex natural supplements, such as fetal bovine serum (FBS) (Reid et al., 2014). This approach had a number of drawbacks in that batches of FBS were chemically undefined and varied between batches; they included proteins that could impact subsequent processing, and the protein content led to foaming in reactors (Claus et al., 2012). However, the main issue was the high cost of FBS, up to 90% of the total media cost, which made mass production of baculoviruses for insecticides economically unviable. Some of these issues were resolved with the development of serum-free media in which purified amino acids, up to 21, replaced FBS and other natural supplements. However, the cost of these media is still too high for them to be considered as viable media for commercial biopesticide production. It has been estimated that viable mass media need to be $1.00–$2.50 US per liter, which is about 1/30th of the current cost.

The key factors in the development of bioreactors for the mass multiplication of insect cells are the high oxygen requirement of insect cells (these require dissolved oxygen at 40–70% maximum) and the large size and fragility of insect cells. Insect cells are amenable for small-scale culture in standard static suspended culture systems, but these are too expensive to consider for mass biopesticide production. To meet industry needs for biopesticides, it has been stated that reactors of 10–20,000 L are the realistic minimal size (Black et al., 1997). Possible configurations for insect cell bioreactors include those using stirred tank reactors and airlift reactors. The key issue is balancing the need for high oxygenation to support cell growth while avoiding the physical stresses that destroy the fragile insect cells. This issue becomes increasingly difficult as the size of the reactors increases. The use of turbines or impellors can create shear forces that damage cells while the bubbling and the bubble bursting that occur with airlift systems can also be lethal to cells (Claus et al., 2012). The stirred tank designs have been reported as viable at up to 600 L, but there is clearly a need for a quantum leap in scale if industry requirements are to be met (Reid et al., 2014).

Small-scale batch production of high levels of functional OBs has been reliably achieved (Lua and Reid, 2003). However, the reliable production of OBs in larger scale continuous culture is more problematic, and a reduction in yields of OBs and virulence are serious problems. In cell culture, there is no need for a functional occlusion derived virions (ODV) phenotype to ensure continued replication, so the appearance and proliferation of mutants that lack proteins crucial for infection or that do not produce OB-containing virions can readily occur (Eberle et al., 2012a). Thus the proliferation of phenotypes lacking the ability to orally infect hosts can become a serious problem, including the possibility of reduced virulence (Chakraborty et al., 1999) and a lower speed of kill (Bonning and Hammock,

1996). Clearly, to render cell culture OB production systems productive and reliable, systems and protocols need to be developed to avoid the proliferation of nonfunctional OBs. In addition, the ability to ensure that viral inoculum can be screened to detect and subsequently remove defective mutants would also be a useful development, though difficult to achieve. Tissue culture production of baculoviruses for insect pest control has been under development for 30 years, and while some progress has been made, the production costs would need to be at or below $12.00 US per 10^{12} OBs in order to be commercially viable as a direct competitor to in vivo production of a wild type virus (Reid et al., 2014).

7.3 BACULOVIRUS PRODUCTION IN VIVO

Detailed descriptions of commercial production systems for baculoviruses are scarce (Gryzwacz et al., 2014), probably because of the perceived proprietary nature of these systems by the producers. However, valuable information on production and production systems has been well summarized by Shapiro et al. (1981), Shapiro (1986), Shieh (1989), Hunter-Fujita et al. (1998), Grzywacz et al. (2004), van Beek and Davies (2009), and Gryzwacz et al. (2014).

Over 50 baculovirus products have been used against different insect pests worldwide, and all have been produced in vivo (Szewczyk et al., 2006; Moscardi et al., 2011; Reid et al., 2014). By far the majority of commercial baculovirus production is in laboratory-type production facilities with cultured insects on an artificial diet, rather than in the field with wild collected larvae and a natural diet (van Beek and Davis, 2009; Moscardi, 2007). Although field production in some specific cases may be cheaper (Moscardi, 2007; Grzywacz et al., 2014), it is the greater product predictability, and hence product consistency, that make production facilities more attractive. However, in order to ensure that production facilities are indeed a feasible and attractive option, production processes must be as practical and hence as manageable as possible, production volumes must be acceptable, so that the outcome justifies the effort, and the processes must not be too costly, so that baculovirus products can be competitive in the market place.

7.3.1 The Production Facility

For the optimization of a logical production flow, ease of production, and risk minimization, baculovirus production facilities should be custom designed and purpose specific. The facility should consist of two major sections: one for host rearing and one for virus production. These sections must be distinctly separated, at different sites if necessary, and there must be no movement from the virus production side to the insect rearing side, so as to eliminate any risk of the produced virus getting back to contaminate the mother

(host) culture. Although an existing building can be modified and used in order to save initial establishment costs, if the design of the structure is not ideal, even greater difficulties than usual could be experienced in establishing and maintaining a healthy and contaminant-free host culture.

Each production facility is unique, and there is no standardized facility design that can be applied to all. However, there are features that should be incorporated into all production facilities to ensure successful production (Fisher, 2009a). For example, sanitation and hygiene are of extreme importance, and the facility must be designed with these in mind (Grzywacz et al., 2004). The host production section should, therefore, have distinct "dirty" and "clean" zones. Within the "dirty" zone would be, for example, the entry and equipment preparation room(s), possibly an office, a quarantine room, the diet preparation room (in some facilities this would fall within the clean zone), and the washing room (Grzywacz et al., 2004). The layout of a small rearing insectary is shown in Fig. 7.1. The implication is not that it is acceptable for these rooms to be dirty but rather that they by nature cannot be subject to the same sanitary extremes as the other rooms. Within the "clean" zone could be a sterilization room, an egg room, a larval rearing room (Fig. 7.2a), and an adult moth room. The level of cleanliness and sanitation in these rooms must be unmitigated in order to minimize the possibility of contamination with and transmission of insect pathogens (Hunter-Fujita et al., 1998). Within the virus production section of the facility would be a virus inoculation room; an incubation room; possibly a separate harvesting room; a product preparation room(s) (where the virus is homogenized, filtered, sampled for quality control, formulated, and packaged); and a wash room.

Production facilities can be built out of bricks and mortar or can be modular units assembled on a concrete foundation underneath a supported roof structure. The modular units could be erected from laminated insulated-core (eg, polystyrene) panels, such as are used in the construction of cold-rooms. This solution is attractive, as it can be highly flexible, and the scale of production can more easily be adjusted to meet changing market needs or incorporate innovations in production systems. For examples of production facility design, see Shapiro et al. (1981), Shapiro (1986), Shieh (1989), Hunter-Fujita et al. (1998), and Grzywacz et al. (2004).

Although low-tech production is possible, and is often the best way to begin, there are certain pieces of equipment that are essential. These include autoclaves for the sterilization of equipment and possibly for diet preparation, a laminar flow or biohazard chamber (for surface sterilization of eggs and inoculation onto diet [Fig. 7.2b]), a digital balance for the accurate measurement of dietary ingredients, air conditioners, humidifiers, and temperature/humidity monitors with data recorders. A more comprehensive list of the essential pieces of equipment required for insect rearing

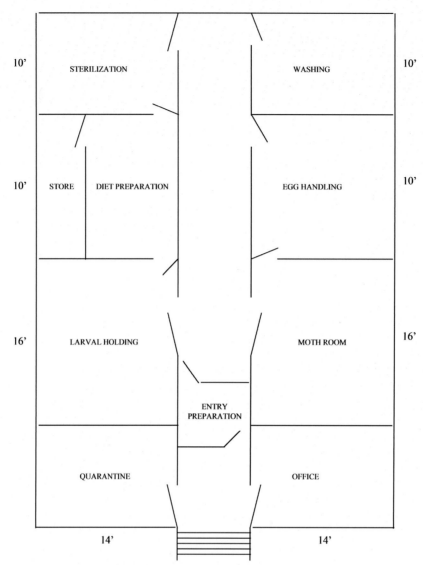

FIGURE 7.1 Plan of insectary for the pilot-scale production of insects for baculovirus production showing an arrangement of rooms and facilities.

is given in Grzywacz et al. (2004) and Fisher (2009a). For virus processing, essential pieces of equipment include a homogenizer, a filter of some sort (for semipurifying OBs from insect body parts and larval diet particles [Fig. 7.2c]), a compound light microscope equipped with phase contrast and dark-field capabilities, a 400× objective, and a bacterial counting chamber. In addition there will be the equipment for producing the specific types of formulations needed, such as refrigerated agitated dairy tanks for liquid formulation (Fig. 7.2d) or spray-driers for powder formulations.

Nothing is more important than the selection and training of personnel to run the facility and perform the rearing and production processes. Staff must be properly trained and absolutely committed to what they are doing so that no problems go unnoticed and quality standards are strictly adhered to. Separate personnel must be appointed for host

rearing and for virus production, and there must be no intermingling of the two during the day (Gryzwacz et al., 2004). Fisher (2009b) provides a very detailed outline of the attributes, qualifications, and training of insectary managers and personnel.

Detailed protocols must be drawn up for every activity in the rearing and production process. The purpose of these protocols is to provide standard operating procedures, quality control standards, operator health and safety procedures, and minimization of contamination risk. Consequently, protocols must be clearly documented and displayed in the relevant rooms within the production facility. Personnel must consult these daily and record operations, observations, production volumes, and quality control measurements relative to the documented standards. Any protocol must be considered a dynamic document, subject to change and

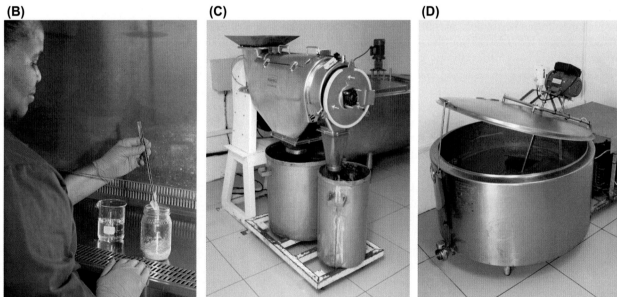

FIGURE 7.2 Equipment and laboratory facilities in a commercial baculovirus production facility. (A) Rearing jars in insectary. (B) Placing eggs. (C) Russell Finex filter for separating virus suspension from debris (insect and diet particles). (D) Refrigerated dairy tank for the storage of processed and formulated baculovirus before bottling. *Photographer: Craig Chambers.*

hence improvements as experience and the data collection grow. Fisher (2009b) can be consulted for details of such procedures on host rearing.

7.3.2 Host Production

A supply of healthy, and as far as possible disease-free insects, is the first requirement of any program to produce insect viruses (Gryzwacz et al., 2004). Hence, insects being introduced to the facility first must be quarantined to check for disease before being released into the main facility (Hunter-Fujita et al., 1998). Diseased individuals must be

removed and the disease-causing agent (pathogen) identified. If evidence of a high-risk agent (eg, a microsporidian) is obtained, or if the number of diseased individuals is considered too great, the remainder of the quarantined insects are destroyed, and a new batch is brought in for observation. Even if no overt signs of disease are observed, covert baculovirus infections could be present. In fact, covert baculovirus infections are very common among lepidopteran populations of most species, and ridding a laboratory culture of a covert virus is extremely difficult or even impossible if the virus is transmitted transovarially. Eastwell et al. (1999) determined that up to 70% of individuals in a laboratory culture of codling moth (*Cydia pomonella*), being commercially reared for a sterile insect release program, were infected (albeit usually covertly) with CpGV. Although they were unable to rid the insect culture of the CpGV, good management prevented debilitating epizootics from disrupting production (Cossentine et al., 2005).

Populations of the relevant insect species from certain geographic regions may be more suitable for virus production than those from other regions. For example, Opoku-Debrah et al. (2014) determined that the false codling moth (*Thaumatotibia leucotreta*) from the Western Cape of South Africa exhibited significantly lower pupal mass, female fecundity, egg hatch, pupal survival, adult eclosion, and the longest duration in larval and pupal development compared to cultures of other populations investigated. This rendered insects from this population far less suitable for commercial production of CrleGV (Moore et al., 2015). In addition, genetic changes can become established within an insect culture as the result of environment and rearing methods (Caprio, 2009), but these changes should be detected through monitoring key biological traits in each generation. If such changes are detected and found to be detrimental, the situation can be remedied by introducing new disease-free insects into, or even replacing entirely, the laboratory culture.

The search for a suitable host for CpGV production included an attempt to use *T. leucotreta*, which is far easier and more efficiently reared than the closely related *C. pomonella* (Moore et al., 2015; Chambers, 2015). The attempt by Hoechst in Germany was unsuccessful, as CrleGV soon outcompeted CpGV in the culture (Cross et al., 1999). These results illustrate not only the virus latency problem in apparently uninfected insect cultures (CrleGV in this case) but also reveal significant differences between the two insect species with regard to their ability to support infection by the two different GVs. Nonetheless, this strategy is again being attempted by Chambers (2015) with a renewed hope of success due to improved techniques (based on qPCR) for rapid differentiation between the two viruses. Successful replication of virus has been achieved in heterologous hosts in a few cases (Kunimi et al., 1996; Kelly and Entwistle, 1998). However, these were noncommercial programs, and heterologous host production was not adopted commercially.

Regardless of the suitability of a particular population of insects for laboratory rearing and baculovirus production, it normally takes a number of generations of culturing in order to successfully establish a healthy, sustainable insect culture in the laboratory (Gryzwacz et al., 2004). Once this has happened, the optimization of rearing parameters must be established through a rigorous execution of laboratory experiments, measuring biological fitness and performance (Leppla, 2009). This pertains to factors, such as temperature and humidity regimes for larvae and adults, light–dark

cycles, oviposition substrates, egg handling, dietary ingredients and preparation. The provision of a hygienic, cost-effective insect diet is important for successful rearing of the host. The preferred option for insect rearing is an artificial diet made from standardized ingredients and preferably heat sterilized. Details of diets for various species are available (Singh, 1977; Singh and Moore, 1985; Anderson and Leppla, 1993). However, each new program can establish the details of the most efficient and cost-effective diets for its particular insect and system.

7.3.3 Virus Production

For many species of Lepidoptera, for which virus production is being pursued, a number of different isolates of the same species of virus are available. Virulence of isolates can be compared in laboratory bioassays. However, genetic heterogeneity between field populations of the host may lead to differences in virulence to the different host populations (Opoku-Debrah et al., 2014), meaning that virulence is a host–pathogen relationship, rather than simply a measurement of the pathogen's ability. In addition, it is considered unlikely that detectable differences in virulence in the laboratory bioassays will indeed translate to detectable differences in efficacy in the field, unless dramatic, as was recorded with different CrleGV isolates against *T. leucotreta* in the field (Moore et al., 2015), which had been shown to significantly differ in virulence in laboratory bioassays (Opoku-Debrah et al., 2013). Therefore, the final selection of production isolates must always be based on the results of representative field trial data in combination with an assessment of each isolate's productivity and suitability for mass production.

A stock suspension of the chosen isolate must be prepared as an inoculum for the production of all future batches of virus. The virus must be purified according to standard protocols (Hunter-Fujita et al., 1998; Eberle et al., 2012b) and must be genetically characterized in order to be able to audit future batches of inoculum and to genetically reconcile production batches with the inoculum (see Section 7.4 for more details). Selected strains should be lodged in separate culture collection facilities, both as a product reference and as a backup in case of accidents at the main facility (Grzywacz et al., 2004). The most appropriate inoculation dosage, the most appropriate larval instar to be inoculated, and the most appropriate incubation period and conditions must be established. A number of studies have determined that the optimal age or weight of larvae is crucial in production efficacy for individual species (Cherry et al., 1997; Grzywacz et al., 1998; Senthil Kumar et al., 2005). If larvae are inoculated when either too small or too large, the production of OB is reduced. Incubation temperature can also influence both the quantity and quality of the virus produced (Subramanian et al., 2006). For production of a virus, a suspension of the virus can either be incorporated into the artificial diet or simply be applied by spraying on the diet surface (Shapiro et al., 1981). However, application to the diet surface appears to be the generally preferred method for mass production, as it was found that less inoculum was required to attain infection from surface application, and virus production per larva was dramatically higher (Shapiro et al., 1981).

Effective multiplication of viruses in vivo is usually dependent upon the provision and maintenance of appropriate rearing conditions for the host. Temperature and humidity need to be kept within defined limits to avoid stressing infected insects such that they die before virus replication is completed (Grzywacz et al., 2014). The incubation period and hence time of harvesting are also critical, not only in ensuring the maximum possible virus load but also in ensuring that microbial contamination remains below acceptable levels (Cherry et al., 1997; Grzywacz et al., 1997; Senthil Kumar et al., 2005). Biological activity of the virus harvested can also be affected by harvesting dead rather than moribund larvae; greater activity has been reported for OBs harvested from dead larvae (Ignoffo and Shapiro, 1978; Bell, 1991).

The productivity of in vivo baculoviruses production varies with different host/virus systems. The productivity of NPV in permissive hosts seems to range between 3×10^6 and 6×10^6 OBs per mg larval weight (Cherry et al., 1997), thus NPVs in commercial production are reported as yielding between 1×10^9 and 5×10^9 OBs per larva (Evans, 1986; Ignoffo, 1999; Grzywacz et al., 2014). Available figures for the productivity in specific species include 2×10^9 OBs per larva for the NPV of *Spodoptera exigua*, *Spodoptera exempta*, and *Spodoptera littoralis* produced in homologous hosts (Smits and Vlak, 1988; Cherry et al., 1997; Grzywacz et al., 1998); for HearSNPV and HzSNPV yields of $(2-5) \times 10^9$ OBs per larva were reported (Shieh, 1989; Ignoffo, 1999; Arrizubieta et al., 2014). The productivity of GVs produced commercially is generally higher than for NPVs but is also more variable, being reported at $3.5 \times 10^6 - 2.3 \times 10^9$ per mg larva for various different host/virus systems (Moore, 2002; Grzywacz et al., 2004; Cuartas et al., 2014). In commercially produced species, reported productivity has ranged from 4.5×10^9 to 1.1×10^{11} OBs per larva (Evans, 1986; Grzywacz et al., 2004; Moore, 2002) with reported specific productivities including 4×10^{10} OBs per larva for *P. xylostella* GV in *P. xylostella* (Grzywacz et al., 2004), 9×10^9 and 8×10^{10} OBs per larva for *C. pomonella* GV in *C. pomonella* and *T. leucotreta* (Glen and Payne, 1984; Chambers, 2015), up to 2.3×10^{10} OBs per larva for *Phthorimaea operculella* GV (PhopGV; Arthurs et al., 2008; Cuartas et al., 2014), and 1.1×10^{11} OBs per larva for CrleGV in *T. leucotreta* (Moore, 2002).

Another means of measuring virus productivity is the productivity ratio (PR), calculated as the number of OBs yielded per larva divided by the number of infecting OBs (Hunter-Fujita et al., 1998), which has been reported to range between 8.4×10^2 for *Panolis flammea* NPV in *Mamestra brassica* (Kelly and Entwistle, 1988) and 1.2×10^6 for *S. exigua* NPV in *S. exigua* (Smits and Vlak, 1988). Hunter-Fujita et al. (1998) tabulated PRs for several NPVs. However, PRs reported for GVs, such as for CpGV and CrleGV, also fall within the range reported for NPVs (Moore, 2002; Chambers, 2015).

Depending on the nature of the host-virus interactions, either only dead larvae or the entire diet including the larvae are harvested. The decision depends on whether larvae can be harvested intact or if they rupture before harvesting due to the expression of viral chitinase and/or cathepsin genes (Hawtin et al., 1997). It also depends on whether the larvae are surface feeders or cryptic (within the diet) feeders. Even though symptomatically infected cryptic feeding larvae tend to move out of the diet, lying on top of the diet or hanging from the sides or lids of virus incubation containers, there is often an unacceptably high percentage of infected larvae that remain within the diet (Moore, 2002) and would be lost if the entire diet was not harvested. Different densities of larvae is a factor that could affect the OB yields per larva. Yields per larva often decrease with increasing larval density, as was found for LdMNPV produced in gypsy moth, *Lymantria dispar* (Shapiro et al., 1981), SeMNPV in *S. exigua* (Cherry et al., 1997), and CpGV in *C. pomonella* (Chambers, 2015). However, one also needs to be cognizant of the OB yield per container and the greater labor and expense required in rearing fewer larvae per container. For example, as *S. exigua* larval density increased from 25 to 200 larvae per container, the OB yield per larva decreased, but the OB yield per container still increased (Cherry et al., 1997). To rear only 25 larvae per container, Cherry et al. (1997) would have had to use 5.4 times as many containers as they would if they had reared 200 larvae per container. This could signify a substantial increase in cost and labor.

Once harvested, larvae or the larval-diet mixture can be stored frozen until processing takes place, or it can be processed immediately. This would involve homogenization of the harvest and filtration of the homogenate. Protocols for this can be obtained from Hunter-Fujita et al. (1998), Moore (2002), or Grzywacz et al. (2004), among others. Once samples of the filtered homogenate have been subjected to the necessary quality control tests, formulation and packaging of the final product can take place.

7.4 QUALITY CONTROL

A key constraint to the wider adoption of baculoviruses has been the wide perception that these products, like most other biopesticides, are erratic in field performance (Lisansky, 1997). In authors' experience, issues of inconsistent quality and failure to achieve standard levels of product quality have been among the most important problems in scaling up and maintaining baculovirus production in both public and private sector programs (Moscardi, 2007). All too often, while expanding the volume of production has been a high priority, too few resources are focused on implementing a fit for purpose, real-time quality control system, which is the key to achieving and maintaining reliable large-scale production (Jenkins and Grzywacz, 2000).

A properly defined and implemented quality control system is central to ensuring standardized product performance, product safety, and reliable production, all of which are essential to user confidence and the economic well-being of any production organization, private or public. Such a system is central to early detection and avoidance of the problems that can derail production. The key elements that require monitoring to assure product quality include baculovirus identity, activity, and product purity. However, to support and sustain production, the health of the host insect lines and the quality of the baculovirus inoculum also require active monitoring.

Quality control in baculovirus production starts with the raw materials, which for in vivo production must focus on the health of the live host insects. It is useful to maintain detailed records of egg laying, hatching rates, pupal size, and larval/pupal mortality rates as well as rates of pupal deformation (Moore et al., 1985). It can also be very useful to carry out regular microscopic examinations on any "sickly" insects for the presence of entomopathogens; especially problematic are chronic pathogens, such as GVs, Microsporidia, and Cypoviruses, whose initial presence causes few obvious problems but whose uncontrolled spread through a culture can seriously contaminate and ultimately lead to the collapse of baculovirus production.

7.4.1 Inoculum Quality

A reliable standardized starting inoculum of the desired baculovirus that has been purified and whose identity has been validated by DNA profiling is essential to sustainable production. The use of a sucrose gradient, purified inoculum should wherever practicable be adopted as standard, and production batches should routinely be sampled and checked for identity. Standard protocols can be found in Hunter-Fujita et al. (1998) and Grzywacz et al. (2004).

7.4.2 Quantification of Occlusion Body

Quantifying or counting the baculoviruses is fortunately relatively straightforward and requires little beyond an appropriate microscope and a bacterial cell counter, as these occluded viruses are unusual in being visible with a light microscope. The OBs of NPVs are between 2 and 15 μm and clearly identifiable as bright refractile bodies with a phase contrast illumination at ×400. The OBs of GV are

more challenging, as they are smaller ovoids $0.2 \times 0.5 \, \mu m$ and are demanding to count under phase contrast, so that some workers find dark field incident illumination more accurate. The basic protocols for microscopic counting of baculovirus suspensions are available in Evans and Shapiro (1997), Hunter-Fujita et al. (1998), Jones (2000), Grzywacz et al. (2004), and Eberle et al. (2012b). Other counting techniques for OB, as dry films on leaves, are discussed by Jones (2000). While OB counting is generally straightforward it does require the right equipment, care, and proper training. The authors have seen a number of cases where researchers and quality control staff have failed to conduct accurate counts, usually overestimating OBs, because they have been working with nonphase contrast or optically poor microscopes or because they have been inadequately trained. Staff involved in counting baculovirus should practice counting initially on pure samples before graduating to counting semipure and then finally the crude impure samples they mainly need to work with. If at all possible, staff should receive training at a laboratory or with researchers already proficient in OB counting to ensure that a satisfactory standard is reached.

7.4.3 Quantification of Activity

In quantifying baculovirus activity, bioassay remains the key tool for determining the success of production and quality standardization (Jones, 2000; Eberle et al., 2012b). While the microscope can be used to quantify the number of OBs, that does not equate to infectivity, and it is always necessary to assess the infectivity separately, which requires a bioassay. The reason OB counts may not correspond to activity are various and may include early harvesting before OBs have matured and contain sufficient virions or are fully infectious. It may also be an issue that batches of OBs may have been mishandled and lost activity.

The most important point about bioassays is the need to conduct them to internationally accepted protocols and to a standard and consistent methodology. The use of LD_{50} as the key bioassay parameter is essential to measure activity, compare samples against the background of natural variation, and get useful estimates of activity for comparison between samples and against historical data. There are a number of sources for specific detailed protocols for baculovirus bioassays (Evans and Shapiro, 1997; Hunter-Fujita et al., 1998; Jones, 2000; Grzywacz et al., 2004; Eberle et al., 2012b). For routine monitoring of production, neonate assays are often the most convenient, as they utilize large numbers of standard larvae produced at a relatively low cost. Either droplet or surfacing dosing assays can be used, depending upon which suits the host species better. Bioassays on artificial diet are preferred, as this reduces potential variability seen with natural food sources but may be impractical for some species. The replication, recording, and statistical analysis of bioassay data need to be to internationally accepted standards (Marcus and Eaves, 2000; Robertson et al., 2007). The exact choice of logit, probit, or other analysis models varies between laboratories but probably makes little difference compared to other factors, so may be chosen as convenient. However, particular care needs to be paid to proper sample sizes and adequate replication to ensure that results are accurate enough to detect any real differences and so that data sets are acceptable to regulators if needed. While every effort must be made to ensure replicability, some degree of natural variation in insect bioassays is unavoidable and needs to be taken into account when evaluating or comparing results (Robertson et al., 1995).

7.4.4 Contamination

It is a standard part of any biopesticide's registration that products sold or released for field use have been checked for microbiological safety (Podgewaite et al., 1983; United States Environmental Protection Agency, 1996). Some degree of microbial contamination with in vivo production systems seems unavoidable, as microbes routinely live as commensal organisms on larvae or in the gut, and their removal during processing would be prohibitively expensive. Some microbes may indeed be essential for healthy growth, and a variety of species may multiply either in host insects during life or on the cadavers or feces. Baculovirus-infected insects harvested dead are commonly found to contain microbes at levels of up to 2×10^8 colony forming units (CFU) per larva, and these consist mainly of symbiotic *Enterococcus* species with lower levels of yeasts and some *Enterobacteriacae* (Podgwaite et al., 1983; Grzywacz et al., 1997; Lasa et al., 2008). Lower levels ($<1 \times 10^5$ CFU) can be achieved when larvae are harvested before death, but if harvesting is delayed after death by 24 h these can reach up to 10^9 CFU per larva, as saprophytic *Bacillus* species multiply on the cadavers (Grzywacz et al., 1997; Jenkins and Grzywacz, 2000). A target of 1×10^8 CFU/mL for liquid formulations or 5×10^8 CFU/g for dry powders has been proposed as both safe and attainable (Jenkins and Grzywacz, 2000). So far there have been no reports of human pathogens in baculovirus production, but checking for the presence of specific pathogens, such as *Shigella* spp., *Salmonella* spp., *Vibrio* spp., and *Escherichia coli* and assurance of their absence are mandatory in most biopesticide registration regimes (Jenkins and Grzywacz, 2003; Anon., 2011).

7.5 FORMULATION

Jones and Burges (1998) list the basic functions of formulation, which can be summarized as shelf life, usability, and field performance (efficacy and persistence). However, other important factors influencing formulation decisions are cost; the nature of the market (pertaining to factors like the availability of cool-chain transport and storage, need for transportation, and subjective preference); and the nature of the target crop(s) and target organism. Primary or shelf formulation takes place in the production facility and is the eventual product as it is sold, whereas tank formulation is made by the user immediately before product application and in line with the producer's registered recommendations.

The most common preparation of a baculovirus biopesticide is as a liquid formulation, invariably as a suspension concentrate. Dry formulations can be applied dry, as dusts, or can be sprayed with water, as wettable powder or granular formulations. The application of dusts is uncommon and is generally used only for postharvest application of stored products, such as with the PhopGV for control of potato tuber moth (*P. operculella*) on stored potatoes (Lagnaoui et al., 1997). Dry baculovirus formulations are generally prepared through air-drying, freeze-drying (or lyophilization) spray drying, coprecipitation with lactose, or microencapsulation.

As with details on commercial production systems, there is a dearth of available information on the specifics of formulations used in commercial products, as such information is viewed as proprietary by manufacturers and, therefore, does not reach the public domain. However, there is a body of published information in the public domain, generated by public research scientists in the field.

7.5.1 Shelf Life

The longest period of time in the life of a product elapses during storage, ie, between manufacture and use in the field. Virus must remain viable during this time, with minimal loss of virulence, retention of formulation properties, and with a minimum suggested shelf life of 18 months (Jones and Burges, 1998). However, it may be noted that chemical insecticides have a shelf life of at least 2–4 years and that matching this would be highly desirable (Jones and Burges, 1998). The principal causes of product deterioration are high temperatures, length of time of exposure to causative factors, presence of free water, adverse pH, enzymes, surfactants, and combinations of these factors (Burges and Jones, 1998).

Liquid formulations (suspension concentrates in the case of viruses) tend to be preferred to solid formulations by the end user, simply due to their greater ease of use. However, in order for their shelf life to compete with or even supersede that of dry formulation, pH must be adjusted to the ideal level, and a bacterial suppressant is required (Hunter-Fujita et al., 1998). Although the ideal pH for suppressing the replication of many microorganisms is <4.0 to 4.5, such a low pH could be detrimental to the virus, so a near neutral pH is preferable (Burges and Jones, 1998). Compounds, such as potassium sorbate, sorbic acid, or glycerol (Behle and Birthisel, 2014) can be used to suppress contaminant microorganisms in liquid formulations. Both liquid and dry formulations need to be refrigerated (at around 4°C) for optimal shelf life, with no indication that dry formulations are superior in this department (Behle et al., 2003). The storage of virus at below freezing can ensure little to no loss of virulence for many years (Behle et al., 2003).

7.5.2 Ease of Use

Whether prepared as a liquid or dry formulation, ease of use in the field is of immeasurable importance, sometimes superseding price and efficacy as the most important factor in product choice by the farmer. Therefore, the product should be usable in a manner that is familiar to the farmer, and no special equipment or preparation procedures should be required. The product should also be devoid of large particles so that spray nozzles will not be blocked. Good production should minimize any strong unpleasant product odors, such as can occur in poorly produced products (Ranga Rao and Meher, 2004), and dry formulations should not produce dust clouds when decanted and should not clump or cake when added to the tank (Jones and Burges, 1998). Liquid formulations generally need to be prepared with thickeners or suspenders to maintain even distribution of the virus in the carrier (Jones and Burges, 1998), and product packaging should be user-friendly. These are just some examples of what is important in the usability or perceived usability of a product.

7.5.3 Field Performance: Efficacy

Although primary formulation of a product is important for the reasons given thus far, it has been proposed that it is unlikely that shelf formulation alone can significantly improve field performance (Jones et al., 1997). Numerous studies have succeeded in demonstrating improved performance due to the addition of formulation ingredients in laboratory bioassays; however, many of these are subsequently found to not make a significant difference to the efficacy recorded in the field (Vail et al., 1999). Thus in practice shelf formulation more often serves as a marketing tool for the manufacturer rather than making a real and observable difference to the end user. Examples are given below.

The most important factor in ensuring that a baculovirus product is efficacious (apart from the manner in which it is applied) is that the registered concentration to be applied is adequate to achieve an acceptable level of pest control. The virus concentration on the plant surface on which the target pest is feeding must be sufficient to ensure a high probability of a lethal dose being ingested. Moore et al. (2015) demonstrated that the minimum number of CrleGV OBs that should be applied per ha of mature citrus trees for control of *T. leucotreta* is approximately 2×10^{13} OBs.

It was noted that one of the commercial CrleGV products was consistently more efficacious than the other, and it was concluded that the most likely reason for this difference in efficacy was that the more efficacious product, when used at the recommended rate, resulted in 7.6 times more OBs per ha than the less efficacious product (based on the registered concentrations of the two products).

Although it has been proposed that shelf formulation may not be capable of markedly improving the efficacy of a baculovirus product, this is certainly not so for tank formulation. Certain compounds, when added to the tank mix in sufficient amounts, have been shown to consistently improve the efficacy of the product, usually by acting as a phagostimulant. There exists a large body of work on phagostimulants and baits outlined by Hunter-Fujita et al. (1998), and a fairly comprehensive list of those used with baculoviruses is given by Burges and Jones (1998). Possibly one of the most commonly used and successful phagostimulants has been molasses. Ballard et al. (2000) demonstrated that the addition of molasses at 10% or 15% to tank mixes of CpGV either significantly reduced *C. pomonella* deep damage to apples or enabled a significant reduction in dosage rate applied, without loss of efficacy. Moore et al. (2015) demonstrated a similar phenomenon with the addition of molasses to CrleGV, substantially and sometimes significantly enhancing efficacy against *T. leucotreta* on citrus even at concentrations as low as 0.5% and even 0.25%.

Another potential means to enhance the efficacy of baculoviruses is through the addition of a synergist. In this regard, optical brighteners have attracted more research focus than any other synergists, as adding these to baculoviruses enhances activity in bioassays (Shapiro and Robertson, 1992; Caballero et al., 2009). They appear to act by blocking of the sloughing of infected primary target cells in the midgut (Washburn et al., 1998), and subsequently, Dougherty et al. (2006) reported that apoptosis (cell death) of infected midgut cells was inhibited, preventing the loss of infected cells and thereby allowing the virus to initiate infection more efficiently. However, this does not always translate into a significant improvement in efficacy in the field to balance increased cost (McGuire et al., 2001), and as yet these are not used in commercial baculovirus products.

There are a few examples of other effective synergists, such as salicin (a phenolic glycoside present in aspen foliage) with Gypchek (LdMNPV; Cook et al., 2003) and boric acid combined with a SfMNPV phagostimulant formulation (Cisneros et al., 2002). Another far more recent example is the combination of CpGV and mutualistic yeasts isolated from *C. pomonella* larvae that significantly increased the mortality of neonate *C. pomonella* larvae in both laboratory and field trials, particularly when brown cane sugar was also added (Knight et al., 2015).

A range of other additives can be used in shelf formulations or added to the tank. These can be antievaporants; humectants; thickeners; or surfactants (spreaders, wetters, emulsifiers, stickers). However, it is not clear whether the addition of any of these makes a significant difference to efficacy in the field.

7.5.4 Field Performance: Persistence

Apart from the need to achieve as high a level of efficacy as possible, it is equally important to ensure that the efficacy of the baculovirus persists for as long as possible on the plant surface, particularly considering the environmental constraints of baculoviruses. Although they appear to be relatively rainfast on certain plant surfaces (Jaques, 1972; Kirkman, 2007), they are known to degrade rapidly in direct sunlight. Sunlight is the most destructive of the environmental factors for baculovirus (Shapiro, 1995) and consequently has the most influence on whether the efficacy of a baculovirus product will persist in the field or not. In the field, the half-life of baculoviruses varies greatly, from as little as 10 h to 10 days, though in general, the half-life in full sunlight without protective screens is around 24 h (Burges and Jones, 1998). However, the need for UV protection of virus is determined by the plant architecture and where on the plant the target pest feeds. Moore (2002) determined that degradation of virus on the northern (sunny) side of citrus trees in South Africa was dramatically more rapid than on the southern (shady) side of trees and that the 17-week persistence of CrleGV at around 80% efficacy, recorded in a Navel orange orchard, was in part attributable to the UV protection of baculovirus in the shaded navel end of fruit, precisely where the majority of neonate larvae attempted to penetrate the fruit (Moore et al., 2015).

A large number of sunscreens have been tested in numerous studies conducted since the early 1970s. Space restrictions prevent us from going into detail on these, and for more detail, Burges and Jones (1998) and Hunter-Fujita et al. (1998) can be consulted. Burges and Jones (1998), after reviewing all of the available literature, concluded that the most effective UV protectants in approximate descending order were melanin, insect remains, optical brighteners, Orzan LS (a lignosulphate), Coax, molasses, clay and flour carriers, and lignin. However, they concluded that in general, UV protection was not considered sufficiently dramatic to translate into meaningful differences in the field. In concluding that insect remains (ie, unpurified baculovirus suspensions) were one of the most effective UV protectants, Jones (1988) found that none of 12 recognized sunscreens gave better protection than the debris in unpurified *S. littoralis* NPV. Shapiro (1992) demonstrated the protective power of a large number of optical brighteners with LdMNPV in laboratory bioassays. Thus, similarly to formulations that improved product efficacy in laboratory

bioassays, the effect of these UV protectants sometimes translates into an improvement in persistence in the field (Webb et al., 1999); however, this is not always the case (Arthurs et al., 2006), so as yet these UV protectants have not been utilized commercially.

Tamez-Guerra et al. (2000) and McGuire et al. (2001) found that formulations containing lignin retained activity significantly longer than unformulated virus. Arthurs et al. (2006) found that lignin-based CpGV formulations provided solar protection at relatively high virus application rates, but in season-long orchard tests, lignin formulations did not significantly improve the control of *C. pomonella* on apples. Kirkman (2007) found that although by incorporating lignin sulphate with CrleGV, efficacy in the field was significantly more persistent, by applying sprays in the evening, rather than the day, this difference disappeared. In addition, the lignin sulphate carrier used was expensive and had to be applied at rates that were too high to make it an attractive option.

This appears to be the problem with a number of the potential UV protectants, making their addition to the shelf formulation impractical or even impossible and their addition to the tank formulation unattractive. The volumes of these UV protectants required to make a meaningful difference inhibit or even prohibit their usability and affordability for commercial field use. Burges and Jones (1998) stated that the use of UV protectants at 1–10% in low- or high-volume tank mixes is wasteful and that microencapsulation would be a more sensible option. To date, encapsulation has most commonly been conducted for UV protection. However, in one of the only studies to conduct trials beyond the laboratory, Gómez et al. (2013) showed no improvement in performance relative to nonencapsulated virus in greenhouse trials. Consequently, as with so many of the innovations producing positive results in the laboratory, none of the currently available commercial baculovirus formulations are known to be encapsulated.

A simple method for improving persistence in the field is to apply higher rates of virus. Moore et al. (2015) argued that the registered application rate of CrleGV could be reduced from 5×10^{13} to 2×10^{13} OBs per ha without any immediate loss of efficacy. However, the lower rate did result in less effective control as the effective virus concentration more rapidly fell below the critical level after application due to the effect of UV inactivation.

7.5.5 Cost

As baculovirus-based biopesticides are often not as immediately efficacious as the chemical insecticides against which they are competing (particularly in speed of kill), it is critical that they be as competitive as possible in as many other aspects as possible. The whole question of the cost and efficacy of baculoviruses and their relative role in modern integrated pest management (IPM) is a complex one (see Chapter 3), and many factors play a role (Lacey et al., 2015), but cost is always a key issue for users. It is, therefore, of great importance that the formulation of baculoviruses is not prohibitively expensive. The costs relative to the benefits provided by formulation ingredients must be properly evaluated through robust and representative field studies (Behle and Birthisel, 2014). Formulation costs are not a major component of commercial product costs, as those proven formulation additives are low cost while the more expensive potential additives have yet to have the degree of demonstrable efficacy that would justify their inclusion.

7.5.6 Worker Safety

Baculoviruses are obligate parasites that can replicate only in arthropods or arthropod cell lines and so are inherently safe to humans (Mudgal et al., 2013). There are certain lepidopteran larvae, however, eg, gypsy moth, which have urticaceous setae that can be irritable to people and even result in some respiratory problems (Hunter-Fujita et al., 1998). Where necessary these hairs can be removed to acceptable levels through centrifugation in the production process, though this increases production costs (Kelly et al., 1989). All materials considered for inclusion in baculovirus formulations, for whatever purpose, should also be tested for human, veterinary, and environmental safety before incorporation.

7.6 FIELD PRODUCTION OF BACULOVIRUSES

Almost all baculovirus products produced today and historically are produced in vivo from insects that originate in laboratory cultures that are specifically reared for baculovirus production (Shapiro, 1986; Hunter-Fuijita et al., 1998; Grzywacz et al., 2014). This approach alone is felt by most producers to ensure that the larvae for production are available in the continuity, quantity, and quality necessary to maximize productivity. However, there have been and continue to be situations where the insects used in production are either harvested from the field for inoculation in a production facility, or the whole process of production is carried out in the field. In the latter, either naturally virus-infected insects are harvested, or more often natural pest outbreaks are artificially inoculated through spraying with baculovirus, then later harvested when virus multiplication has occurred, to provide the bulk baculovirus for the production of a product.

Many host species are not suitable for this technique, as they do not occur in the density or predictability necessary to make it feasible, but it has been adopted in the case of a few species for the commercial production of virus (Moscardi, 1999; Grzywacz et al., 2014). It is particularly attractive where host insects occur as high-density outbreaks of larvae of a similar age, as with some univoltine insects, where hatching is synchronized, perhaps closely coordinated with the seasonal appearance of new foliage on host plants. It may seem most feasible in those species, such as some lymantrid forest defoliators that form communal tents or nests and may overwinter in these and thus can be readily harvested "en mass," eg, *Neodiprion sertifer* (Shieh and Bohmfalk, 1980). Field harvesting may be the only feasible option for hosts where no viable or affordable artificial diet has been developed, as for some sawflies (Moreau and Lucarotti, 2007). It may also be the only economically viable option for univoltine species or ones whose long larval rearing period makes continuous, year-round laboratory-based production impractical or too expensive.

Field production may be used as an entry point for baculovirus product development to enable the pilot-scale production of the baculovirus to provide sufficient material for large-scale evaluation trials without the commitment of large capital sums for a custom built rearing and infection facility but prior to eventually moving to laboratory culture production. Field production of baculovirus has also evolved in some developing countries where the need to meet the low cost required by poor farmers has stimulated the low-cost production of a number of baculoviruses (Ranga Rao and Meher, 2004; Grzywacz et al., 2014).

7.7 OVERVIEW OF FUTURE DEVELOPMENTS AND NEEDS

Although IPM and biological control are age-old phenomena and have been implemented successfully for decades in a number of agricultural sectors, their implementation has essentially been by choice. What has now changed over the last decade, and is continuing to change, is the pressure being placed on the use of chemical insecticides (Chandler et al., 2011; Eberle et al., 2012a). Pesticide review programs, increased stringency in the regulation of maximum residue levels, arbitrary retailer residue restrictions, and growing public sentiment are contributing to this pressure (Lacey et al., 2015). This is creating opportunities for biological pesticides, including baculovirus-based products, seen in the estimated more than 20-fold growth in the global biopesticide market from 2000 to 2010 (Ravensberg, 2011; Glare et al., 2012).

To facilitate the anticipated continued growth in demand for baculovirus-based pesticides, production systems need to improve. Whether these remain in vivo or if in vitro production does eventually become a commercial reality, production systems need to be improved to supply greater volumes at more competitive prices. A target product price below $20 per ha has been one that has been suggested as a figure that could unlock larger scale use (Reid et al.,

2014). However, the price that the end user would be prepared to pay would depend on the value of the crop and the status of the pest. For example, the user would be prepared to pay more to control a phytosanitary pest on an export crop destined for a residue-sensitive market, than for a cosmetic pest on a crop destined for processing. This increase in production will require not only increased automation of production, but maybe novel innovations, such as in vivo production in heterologous hosts, that may be cheaper and easier to rear and may facilitate greater productivity. Such a breakthrough will require either a reliable and rapid technique to determine whether the virus being produced is indeed the one being inoculated or a method to rid a laboratory population of covert virus infection, thus eliminating the risk of contamination of the product with another virus.

Much work has been conducted on the formulation of baculovirus products, often demonstrating significant improvement in the laboratory. However, this has rarely translated into meaningfully improved field efficacy, as explained throughout this chapter. It is generally accepted that the greatest environmental shortcoming of baculoviruses is their degradation in sunlight. Consequently, an innovative breakthrough in UV protection is required, as the current generation of available UV protectants seem not to translate into an acceptable level of increased UV protection when used in the field. Again, it is imperative that this be an affordable and practically usable option. This might not only be achieved through improved formulation but possibly also genetic selection of more UV-resistant strains or even genetic recombination.

As a result of the pressure on chemical pesticides and the growth in biopesticides (measured by the proportion they make up of the global pesticide market), multinational agricultural chemical companies have actively purchased biopesticide companies over the past few years (Moore, 2015). If this shift is indeed sincere, we could see dramatically increased investment into the improvement of baculovirus production and formulation technologies and into new and innovative related technologies. Calls for in vitro production systems, as the solution to mass-scale commercial baculovirus production, have been made for years now (Black et al., 1997; Hunter-Fujita et al., 1998; Moscardi, 1999; Szewczyk et al., 2006; Reid et al., 2014). The entry (or reentry) of these multinational companies into the biopesticide arena may be the injection required to make the in vitro production of baculoviruses a reality.

In vitro production systems will also open the door for genetic modification. Although the often perceived benefit of genetic modification of improved speed of kill is questionable (Eberle et al., 2012a), there may be other benefits, such as the improved UV tolerance already mentioned, overcoming host barriers to baculovirus infection, and broadening of host range (Eberle et al., 2012a). If public sentiment and regulation resist this avenue, these types of

improvements can be pursued through the selection of the most appropriate species and isolates of baculoviruses and the exploitation of synergistic possibilities with other pesticides, biological or chemical.

Finally, registration laws will need to change to facilitate easier, more affordable, and more rapid registration of biologicals (Chandler et al., 2011; Glare et al., 2012). This will be achieved through the relevant regulators recognizing the unique qualities and particularly the safety of biopesticides, especially baculoviruses, and modifying registration requirements accordingly, so that they are clear, proportionate, and predictable. The harmonization of registration systems would also help with the economics of extending product registrations to new countries. These registration laws have already started to change in certain countries (Kabaluk et al., 2010; Glare et al., 2012), thus beginning to remove one of the biggest hurdles to the commercialization and uptake of baculovirus-based biopesticides.

REFERENCES

Anderson, T.E., Leppla, N.C. (Eds.), 1993. Advances in Insect Rearing for Research and Pest Management. Westview Press Inc., Boulder, USA, 519 pp.

Anon., 2011. OECD Issue Paper on Microbial Contamination Limits for Microbial Pest Control Products, Series on Pesticides, No. 65. Online at: http://www.oecd.org/officialdocuments/publicdisplaydocumentpdf/?cote=env/jm/mono%282011%2943&doclanguage=en.

Arthurs, S.P., Lacey, L.A., Behle, R.W., 2006. Evaluation of spray-dried lignin-based formulations and adjuvants as solar protectants for the granulovirus of the codling moth, *Cydia pomonella* (L). J. Invertebr. Pathol. 93, 88–95.

Arthurs, S.P., Lacey, L.A., Pruneda, J.N., Rondon, S., 2008. Semi-field evaluation of a granulovirus and *Bacillus thuringiensis* ssp. *kurstaki* for season-long control of the potato tuber moth, *Phthorimaea operculella*. Entomol. Exp. Appl. 129, 276–285.

Arrizubieta, M., Williams, T., Caballero, P., Simón, O., 2014. Selection of a nucleopolyhedrovirus isolate from *Helicoverpa armigera* as the basis for a biological insecticide. Pest Manag. Sci. 70, 967–976.

Ballard, J., Ellis, D.J., Payne, C.C., 2000. The role of formulation additives in increasing the potency of *Cydia pomonella* granulovirus for codling moth larvae, in laboratory and field experiments. Biocontrol Sci.Technol. 10, 627–640.

Behle, R., Birthisel, T., 2014. Formulation of entomopathogens as bioinsecticides. In: Morales-Ramos, J.A., Guadalupe Rojas, M., Shapiro-Ilan, D.I. (Eds.), Mass Production of Beneficial Organisms. Elsevier, Amsterdam, pp. 483–517.

Behle, R.W., Tamez-Guerra, P., McGuire, M.R., 2003. Field activity and storage stability of *Anagrapha falcifera* nucleopolyhedrovirus (AfMNPV) in spray-dried lignin-based formulations. J. Econ. Entomol. 96, 1066–1075.

Bell, M.R., 1991. *In vivo* production of a nuclear polyhedrosis virus utilizing tobacco budworm and a multicellular larval rearing container. J. Entomol. Sci. 26, 69–75.

Black, B.C., Brennan, L.A., Dierks, P.M., Gard, I.E., 1997. Commercialization of baculovirus insecticides. In: Miller, L.K. (Ed.), The Baculoviruses. Plenum Press, New York, pp. 341–387.

Bonning, B.C., Hammock, B.D., 1996. Development of recombinant baculoviruses for insect control. Annu. Rev. Entomol. 41, 191–210.

Burges, H.D., Jones, K.A., 1998. Formulation of bacteria, viruses and protozoa to control insects. In: Burges, H.D. (Ed.), Formulation of Microbial Biopesticides. Kluwer Academic Publishers, Dordrecht, The Netherlands, pp. 33–127.

Caballero, P., Murillo, R., Munoz, D., Williams, T., 2009. The nucleopolyhedrovirus of *Spodoptera exigua* (Lepidoptera: Noctuidae) as a biopesticide: analysis of recent advances in Spain. Rev. Colomb. Entomol. 35, 105–115.

Caprio, M.A., 2009. Genetic considerations and strategies for rearing high quality insects. In: Schneider, J.C. (Ed.), Principles and Procedures for Rearing High Quality Insects. Mississippi State University, Mississippi, USA, pp. 87–95.

Chakraborty, S., Monsour, C., Teakle, R., Reid, S., 1999. Yield, biological activity, and field performance of a wild-type *Helicoverpa* nucleopolyhedrovirus produced in *H. zea* cell cultures. J. Invertebr. Pathol. 73, 199–205.

Chambers, C., 2015. Production of *Cydia pomonella* Granulovirus (CpGV) in a Heterologous Host, *Thaumatotibia leucotreta* (Meyrick) (False Codling Moth) (Ph.D. thesis). Rhodes University, Grahamstown, South Africa, 222 pp.

Chandler, D., Bailey, A.S., Tatchell, G.M., Davidson, G., Greaves, J., Grant, W.P., 2011. The development, regulation and use of biopesticides for integrated pest management. Philos. Trans. R. Soc. Lond. Ser. B 366, 1987–1998.

Cherry, A.J., Parnell, M.A., Grzywacz, D., Jones, K.A., 1997. The optimization of *in vivo* nuclear polyhedrosis virus production in *Spodoptera exempta* (Walker) and *Spodoptera exigua* (Hübner). J. Invertebr. Pathol. 70, 50–58.

Cisneros, J., Pérez, J.A., Penagos, D.I., Ruiz, J., Goulson, D., Caballero, P., Cave, R.D., Williams, T., 2002. Formulation of a nucleopolyhedrovirus with boric acid for control of *Spodoptera frugiperda* (Lepidoptera: Noctuidae) in maize. Biol. Control 23, 87–95.

Claus, J.D., Gioria, V.V., Micheloud, G.A., Visnovsky, G., 2012. Production of insecticidal baculoviruses in insect cell cultures: potential and limitations. In: Soloneski, S., Larramendy, M. (Eds.), Insecticides—Basic and Other Applications. InTech, Rijeka, Croatia, pp. 127–152.

Cook, S.P., Webb, R.E., Podgwaite, J.D., Reardon, R.C., 2003. Increased mortality of gypsy moth *Lymantria dispar* (L.) (Lepidoptera: Lymantriidae) exposed to gypsy moth nuclear polyhedrosis virus in combination with the phenolic gycoside salicin. J. Econ. Entomol. 96, 1662–1667.

Cossentine, J.E., Jensen, L.B.M., Eastwell, K.C., 2005. Incidence and transmission of a granulovirus in a large codling moth [*Cydia pomonella* L. (Lepidoptera: Tortricidae)] rearing facility. J. Invertebr. Pathol. 90, 187–192.

Cross, J.V., Solomon, M.G., Chandler, D., Jarrett, P., Richardson, P.N., Winstanley, D., Bathon, H., Hüber, J., Keller, B., Langenbruch, G.A., Zimmerman, G., 1999. Biocontrol of pests of apples and pears in northern and central Europe: 1. Microbial agents and nematodes. Biocontrol Sci. Technol. 9, 125–149.

Cuartas, P., Barrera, G., Barreto, E., Villamizar, L., 2014. Characterisation of a Colombian granulovirus (Baculoviridae: *Betabaculovirus*) isolated from *Spodoptera frugiperda* (Lepidoptera: Noctuidae) larvae. Biocontrol Sci. Technol. 24, 1265–1285.

Dougherty, E.M., Narang, N., Loeb, M., Lynn, D.E., Shapiro, M., 2006. Fluorescent brightener inhibits apoptosis in baculovirus-infected gypsy moth larval midgut cells in vitro. Biocontrol Sci. Technol. 16, 157–168.

Eastwell, K.C., Cossentine, J.E., Bernardy, M.G., 1999. Characterisation of *Cydia pomonella* granulovirus from codling moths in a laboratory colony and in orchards of British Columbia. Ann. Appl. Biol. 134, 285–291.

Eberle, K.E., Jehle, J.A., Hüber, J., 2012a. Microbial control of crop pests using insect viruses. In: Abrol, D.P., Shankar, U. (Eds.), Integrated Pest Management: Principles and Practice. CAB International, Wallingford, pp. 281–298.

Eberle, K.E., Wennmann, J.T., Kleespies, R.G., Jehle, J.A., 2012b. Basic techniques in insect virology. In: Lacey, L.A. (Ed.), Manual of Techniques in Invertebrate Pathology. Academic Press, San Diego, pp. 16–75.

Evans, H.F., 1986. Ecology and epizootiology of baculoviruses. In: Granados, R., Federici, B.A. (Eds.), Biology of Baculoviruses. Practical Application for Insect Pest Control, Vol. 2. CRC Press, Boca Raton, pp. 89–132.

Evans, H.F., Shapiro, M., 1997. Viruses. In: Lacey, L.A. (Ed.), Manual of Techniques in Insect Pathology. Academic Press, San Diego, pp. 17–53.

Fisher, W.R., 2009a. Insectary design and construction. In: Schneider, J.C. (Ed.), Principles and Procedures for Rearing High Quality Insects. Mississippi State University, Mississippi, USA, pp. 9–42.

Fisher, W.R., 2009b. The insectary manager. In: Schneider, J.C. (Ed.), Principles and Procedures for Rearing High Quality Insects. Mississippi State University, Mississippi, USA, pp. 43–70.

Glare, T.R., Caradus, J., Gelernter, W., Jackson, T., Keyhani, N., Kohl, J., Marrone, P., Morin, L., Stewart, A., 2012. Have biopesticides come of age? Trends Biotechnol. 30, 250–258.

Glen, D.M., Payne, C.C., 1984. Production and field evaluation of codling moth granulosis virus for control of *Cydia pomonella* in the United Kingdom. Ann. Appl. Biol. 104, 87–98.

Gómez, J., Guevara, J., Cuartas, P., Espinel, C., Villamizar, L., 2013. Microencapsulated *Spodoptera frugiperda* nucleopolyhedrovirus: insecticidal activity and effect on arthropod populations in maize. Biocontrol Sci. Technol. 23, 829–846.

Granados, R.R., Li, G., Blissard, G.W., 2007. Insect cell culture and biotechnology. Virol. Sin. 22, 83–93.

Grzywacz, D., McKinley, D., Jones, K.A., Moawad, G., 1997. Microbial contamination in *Spodoptera littoralis* nuclear polyhedrosis virus produced in insects in Egypt. J. Invertebr. Pathol. 69, 151–156.

Grzywacz, D., Jones, K.A., Moawad, G., Cherry, A., 1998. The in vivo production of *Spodoptera littoralis* nuclear polyhedrosis virus. J. Virol. Methods 71, 115–122.

Grzywacz, D., Rabindra, R.J., Brown, M., Jones, K.A., Parnell, M., 2004. *Helicoverpa armigera* Nucleopolyhedrovirus Production Manual. Natural Resources Institute. 107 pp. http://www.fao.org/docs/eims/upload/agrotech/2011/HaNPVmanual-pt1.pdf.

Grzywacz, D., Moore, D., Rabindra, R.J., 2014. Mass production of entomopathogens in less industrialized countries. In: Morales-Ramos, J.A., Rojas, M.G., Shapiro-Ilan, D.I. (Eds.), Mass Production of Beneficial Organisms. Elsevier, Amsterdam, pp. 519–553.

Gwynn, R., 2014. Manual of Biocontrol Agents, fifth ed. British Crop Protection Council, Alton, 520 pp.

Haase, S., Sciocco-Cap, A., Romanowski, V., 2015. Baculovirus insecticides in Latin America: historical overview, current status and future perspectives. Viruses 7, 2230–2267.

Hawtin, R.E., Zarkowska, T., Arnold, K., Thomas, C.J., Gooday, G.W., King, L.A., Kuzio, J.A., Possee, R.D., 1997. Liquefaction of *Autographa californica* nucleopolyhedrovirus-infected insects is dependent on the integrity of virus-encoded chitinase and cathepsin genes. Virology 238, 243–253.

Hunter-Fuijita, F.R., Entwistle, P.F., Evans, H.F., Crook, N.E., 1998. Insect Viruses and Pest Management. John Wiley & Sons Ltd, Chichester, 620 pp.

Ignoffo, C.M., 1999. The first viral pesticide: past present and future. J. Ind. Microbiol. Biotechnol. 22, 407–417.

Ignoffo, C.M., Shapiro, M., 1978. Characteristics of baculovirus preparations processed from living and dead larvae. J. Econ. Entomol. 71, 186–188.

Inceoglu, A.B., Kamita, S.G., Hammock, B.D., 2006. Genetically modified baculoviruses: a historical overview and future outlook. Adv. Virus Res. 68, 323–360.

Jaques, R.P., 1972. The inactivation of foliar deposits of viruses of *Trichoplusia ni* (Lepidoptera: Noctuidae) and *Pieris rapae* (Lepidoptera: Pieridae) and tests on protectant additives. Can. Entomol. 104, 1985–1994.

Jenkins, N.E., Grzywacz, D., 2000. Quality control of fungal and viral biocontrol agents: assurance of product performance. Biocontrol Sci. Technol. 10, 753–777.

Jenkins, N.E., Grzywacz, D., 2003. Towards the standardization of quality control of fungal and viral biocontrol agents. In: van Lenteren, J.C. (Ed.), Quality Control and Production of Biological Control Agents: Theory and Testing Procedures. CAB International, Wallingford, pp. 247–263.

Jones, K.A., 1988. Studies on the Persistence of *Spodoptera littoralis* Nuclear Polyhedrosis Virus on Cotton in Egypt (Ph.D. thesis). University of Reading, UK, 340 pp.

Jones, K.A., 2000. Bioassays of entomopathogenic viruses. In: Navon, A., Ascher, K.R.S. (Eds.), Bioassays of Entomopathogenic Microbe and Nematodes. CABI Publishing, Wallingford, Oxon, UK, pp. 95–140.

Jones, K.A., Burges, H.D., 1998. Technology of formulation and application. In: Burges, H.D. (Ed.), Formulation of Microbial Biopesticides. Kluwer Academic Publishers, Dordrecht, The Netherlands, pp. 2–30.

Jones, K.A., Cherry, A.J., Grzywacz, D., Burges, H.D., 1997. Formulation: is it an excuse for poor application? In: BCPC Symposium Proceedings, No. 68. British Crop Protection Council, Farnham, UK, pp. 173–180.

Kabaluk, J.T., Svircev, A.M., Goettel, M.S., Woo, S.G. (Eds.), 2010. The Use and Regulation of Microbial Pesticides in Representative Jurisdictions Worldwide. IOBC Global, 99 pp.

Kelly, P.M., Entwistle, P.F., 1988. *In vivo* mass production in the cabbage moth (*Mamestra brassicae*) of a heterologous (*Panolis*) and a homologous (*Mamestra*) nuclear polyhedrosis virus. J. Virol. Methods 19, 249–256.

Kelly, P.M., Speight, M.R., Entwistle, P.F., 1989. Mass production and purification of *Euproctis chrysorrhoea* (L.) nuclear polyhedrosis virus. J. Virol. Methods 25, 93–99.

Kirkman, W., 2007. Understanding and Improving the Residual Efficacy of the *Cryptophlebia leucotreta* Granulovirus (CRYPTOGRAN) (M. Sc. thesis). Rhodes University, Grahamstown, South Africa, 115 pp.

Knight, A.L., Basoalto, E., Witzgall, P., 2015. Improving the performance of the granulosis virus of codling moth (Lepidoptera: Tortricidae) by adding the yeast *Saccharomyces cerevisiae* with sugar. Environ. Entomol. 44, 252–259.

Kunimi, Y., Fuxa, J.R., Hammock, B.D., 1996. Comparison of wild type and genetically engineered nuclear polyhedrosis viruses of *Autographa californica* for mortality, virus replication and polyhedra production in *Trichoplusia nilarvae*. Entomol. Exp. Appl. 81, 251–257.

Lagnaoui, A., Ben Salah, H., El-Bedewy, R., 1997. Integrated management to control potato tuber moth in North Africa and the Middle East. CIP Circ. 22, 10–15.

Lacey, L.A., Grzywacz, D., Shapiro-Ilan, D.I., Frutos, R., Brownbridge, M., Goettel, M.S., 2015. Insect pathogens as biological control agents: back to the future. J. Invertebr. Pathol. 132, 1–41.

Lasa, R., Williams, T., Caballero, P., 2008. Insecticidal properties and microbial contaminants in a *Spodoptera exigua* multiple nucleopolyhedrovirus (Baculoviroidae) formulation stored at different temperatures. J. Econ. Entomol. 101, 42–49.

Leppla, N.C., 2009. The basics of quality control for insect rearing. In: Schneider, J.C. (Ed.), Principles and Procedures for Rearing High Quality Insects. Mississippi State University, Mississippi, USA, pp. 289–306.

Lisansky, S., 1997. Microbial biopesticides. In: Evans, H.F. (Ed.), Microbial Insecticides; Novelty or Necessity. BCPC Symposium Proceedings No. 68. British Crop Protection Council, Farnham, UK, pp. 3–11.

Lua, L.H., Reid, S., 2003. Growth, viral production and metabolism of a *Helicoverpa zea* cell line in serum-free culture. Cytotechnology 42, 109–120.

Lynn, D.E., 2007. Available lepidopteran insect cell lines. In: Murhammer, D.W. (Ed.), Baculovirus and Insect Cell Expression Protocols. Methods in Molecular Biology, vol. 338. Humana Press Inc., Towata, pp. 117–137.

McGuire, M.R., Tamez-Guerra, P., Behle, R.W., Streett, D.A., 2001. Comparative field stability of selected entomopathogenic virus formulations. J. Econ. Entomol. 94, 1037–1044.

Marcus, R., Eaves, D.M., 2000. Statistical and computation analysis of bioassay data. In: Navon, A., Ascher, K.R.S. (Eds.), Bioassays of Entomopathogenic Microbes and Nematodes. CAB International Wallingford, pp. 249–294.

Moore, R.F., Odell, T.M., Calkins, C.O., 1985. Quality assessment in laboratory reared insects. In: Singh, P., Moore, R.F. (Eds.), Handbook of Insect Rearing. Amsterdam, vol. 1. Elsevier, pp. 107–135.

Moore, S.D., 2002. The Development and Evaluation of *Cryptophlebia leucotreta* Granulovirus (CrleGV) as a Biological Control Agent for the Management of False Codling Moth, *Cryptophlebia leucotreta*, on Citrus (Ph.D. thesis). Rhodes University, Grahamstown, South Africa, 311 pp.

Moore, S.D., 2015. The future of microbial pesticides: fantasy or reality? In: Proceedings of the Joint 19th ESSA and 37th ZSSA Congress, Grahamstown, South Africa, 12–17 July 2015, p. 72.

Moore, S.D., Kirkman, W., Richards, G.I., Stephen, P., 2015. The *Cryptophlebia leucotreta* granulovirus – 10 years of commercial field use. Viruses 7, 1284–1312.

Moreau, G., Lucarotti, C.J., 2007. A brief review of the past use of baculoviruses for the management of eruptive forest defoliators and recent developments on a sawfly virus in forest. Chronicle 83, 105–112.

Moscardi, F., 1999. Assessment of the application of baculoviruses for the control of Lepidoptera. Annu. Rev. Entomol. 44, 257–289.

Moscardi, F., 2007. Development and use of the nucleopolyhedrovirus of the velvetbean caterpillar in soybeans. In: Vincent, C., Goettel, M.S., Lazarovits, G. (Eds.), Biological Control: A Global Perspective. CAB International, Wallingford, UK, pp. 344–353.

Moscardi, F., de Souza, M.L., de Castro, M.E.B., Moscardi, M.L., Szewczyk, B., 2011. Baculovirus pesticides: present state and future perspectives. In: Ahmad, I., Ahmad, F., Pichtel, J. (Eds.), Microbes and Microbial Technology. Springer, Dordrecht, The Netherlands, pp. 415–445.

Moscardi, F., Leite, L.G., Zamataro, C.E., 1997. Production of nuclear polyhedrosis virus of *Anticarsia gemmatalis* Hübner (Lepidoptera: Noctuidae): effect of virus dosage, host density and age. An. Soc. Entomológica do Bras. 26, 121–132.

Mudgal, S., De Toni, A., Tostivint, C., Hokkanen, H., Chandler, D., 2013. Scientific Support, Literature Review and Data Collection and Analysis for Risk Assessment on Microbial Organisms Used as Active Substance in Plant Protection Products –Lot 1 Environmental Risk Characterization. EFSA Supporting Publications. European Food Standards Agency, EN-518, 149 pp. Available online: www.efsa. europa.eu/publications.

Nguyen, Q., Qi, Y.M., Wu, Y., Chan, L.C.L., Nielsen, L.K., Reid, S., 2011. In vitro production of *Helicoverpa baculovirus* biopesticides—automated selection of insect cell clones for manufacturing and systems biology studies. J. Virol. Methods 175, 197–205.

Opoku-Debrah, J.K., Moore, S.D., Hill, M.P., Knox, C., 2013. Characterisation of novel CrleGV isolates for false codling moth control-lessons learnt from codling moth resistance to CpGV. In: Proceedings of the Insect Pathogens and Entomoparasitic Nematodes IOBC-WPRS Bulletin, Zagreb, Croatia, 16–20 June 2013, vol. 90, pp. 155–159.

Opoku-Debrah, J.K., Hill, M.P., Knox, C., Moore, S.D., 2014. Comparison of the biology of geographically distinct populations of the citrus pest, *Thaumatotibia leucotreta* (Meyrick) (Lepidoptera: Tortricidae), in South Africa. Afr. Entomol. 22, 530–537.

Podgewaite, J.D., Bruen, R.B., Shapiro, M., 1983. Micro-organisms associated with production lots of the nuclear polyhedrosis virus of the gypsy moth *Lymantria dispar*. Entomophaga 28, 9–16.

Ranga Rao, G.V., Meher, K.S., 2004. Optimization of in vivo production of *Helicoverpa armigera* NPV and regulation of malador associated with the process. Indian J. Plant Protect. 32, 15–18.

Ravensberg, W.J., 2011. A Roadmap to the Successful Development and Commercialization of Microbial Pest Control Products for Control of Arthropods. Springer Science and Business Media, Netherlands, p. 386.

Reid, S., Chan, L., van Oers, M., 2014. Production of entomopathogenic viruses. In: Morales-Ramos, J.A., Guadalupe Rojas, M., Shapiro-Ilan, D.I. (Eds.), Mass Production of Beneficial Organisms. Elsevier, Amsterdam, pp. 437–482.

Robertson, J.L., Russell, R.M., Preisler, H.K., Savin, N.E., 2007. Bioassay with Arthropods, second ed. CRC Press Inc., Boca Raton, p. 199.

Robertson, J.L., Preisler, H.K., Ng, S.S., Hickle, L.A., Gelernter, W.D., 1995. Natural variation: a complicating factor in bioassays with chemical and microbial pesticides. J. Econ. Entomol. 88, 1–10.

Senthil Kumar, C.M., Sathiah, N., Rabindra, R.J., 2005. Optimizing the time of harvest of nucleopolyhedrovirus infected *Spodoptera litura* (Fabricius) larvae under in vivo production systems. Curr. Sci. 88, 1682–1684.

Shapiro, M., 1986. *In vivo* production of baculoviruses. In: Granados, R.R., Federici, B.A. (Eds.), The Biology of Baculoviruses Volume II: Practical Application for Insect Control. CRC Press, Inc., Boca Raton, pp. 32–61.

Shapiro, M., 1992. Use of optical brighteners as radiation protectants for the gypsy moth (Lepidoptera: Lymantriidae) nuclear polyhedrosis virus. J. Econ. Entomol. 85, 1682–1686.

Shapiro, M., 1995. Radiation protection and activity enhancement of viruses. In: Biorational Pest Control Agents: Formulation and Delivery. Am. Chem. Soc. Symp, vol. 595, pp. 153–164.

Shapiro, M., Robertson, J.L., 1992. Enhancement of gypsy moth (Lepidoptera: Lymantriidae) baculovirus activity by optical brighteners. J. Econ. Entomol. 85, 1120–1124.

Shapiro, M., Bell, R.A., Owens, C.D., 1981. *In vivo* mass production of gypsy moth nucleopolyhedrovirus. In: Doane, C.C., MacManus, M.L. (Eds.), The Gypsy Moth: Research Towards Pest Management. Forest Service Technical Bulletin 1584. United States Department of Agriculture, Washington, pp. 633–655.

Shieh, T.R., 1989. Industrial production of viral pesticides. Adv. Virus Res. 36, 315–343.

Shieh, T.R., Bohmfalk, G.T., 1980. Production and efficacy of baculoviruses. Biotechnol. Bioeng. 22, 1357–1375.

Singh, P., 1977. Artificial Diets for Insects, Mites, and Spiders. Plenum Publishing Corporation, New York, USA, 594 pp.

Singh, P., Moore, R.F. (Eds.), 1985. Handbook of Insect Rearing. Elsevier, Amsterdam.

Smits, P.H., Vlak, J.M., 1988. Quantitative and qualitative aspects in the production of a nuclear polyhedrosis virus in *Spodoptera exigua* larvae. Ann. Appl. Biol. 112, 249–257.

Subramanian, S., Santharam, G., Sathiah, N., Kennedy, J.S., Rabindra, R.J., 2006. Influence of incubation temperature on productivity and quality of *Spodoptera litura* nucleopolyhedrovirus. Biol. Control 37, 367–374.

Sun, X., 2015. History and current status of development and use of viral insecticides in China. Viruses 7, 306–319.

Szewcyk, B., Hoyos-Carvajal, L., Paluszek, M., Skrzecz, I., Lobo de Souza, M., 2006. Baculoviruses re-emerging biopesticides. Biotechnol. Adv. 24, 143–160.

Tamez-Guerra, P., McGuire, M.R., Behle, R.W., Hamm, J.J., Sumner, H.R., Shasha, B.S., 2000. Sunlight persistence and rainfastness of spray-dried formulations of baculovirus isolated from *Anagrapha falcifera* (Lepidoptera: Noctuidae). J. Econ. Entomol. 93, 210–218.

United States Environmental Protection Agency, 1996. Microbial Pesticide Test Guidelines: OPPTS 885.1300 Discussion of Formation of Unintentional Ingredients [EPA 712–C–96–294]. Environmental Protection Agency, United States. http://www.epa.gov/ocspp/pubs/frs/publications/Test_Guidelines/series885.htm.

Vail, P.V., Hoffmann, D.F., Tebbets, J.S., 1999. Influence of fluorescent brighteners on the field activity of the celery looper nucleopolyhedrovirus. Southwest. Entomol. 24, 87–98.

van Beek, N., Davis, D.C., 2009. Baculovirus production in insect larvae. In: Murhammer, D.W. (Ed.), Baculovirus and Insect Cell Expression Protocols. Methods in Molecular Biology, vol. 338. Humana Press, Towata, NJ, pp. 367–378.

Washburn, J.O., Kirkpatrick, B.A., Haas-Stapleton, E., Volkman, L.E., 1998. Evidence that the stilbene-derived optical brightener M2R enhances *Autographa californica* M nucleopolyhedrovirus infection of *Trichoplusia ni* and *Heliothis virescens* by preventing sloughing of infected midgut epithelial cells. Biol. Control 11, 58–69.

Webb, R.E., Thorpe, K.W., Podgwaite, J.D., Reardon, R.C., White, G.B., Talley, S.E., 1999. Field evaluation of an improved formulation of Gypchek (a nuclear polyhedrosis virus product) against the gypsy moth (Lepidoptera: Lymantriidae). J. Entomol. Sci. (USA) 34, 72–83.

Chapter 8

Entomopathogenic Bacteria: Mass Production, Formulation, and Quality Control

T.A. Jackson
AgResearch, Lincoln Research Center, Christchurch, New Zealand

8.1 INTRODUCTION

Bacterial biopesticides comprise the major proportion of the microbial pesticide market and have become widely available as commercial products registered around the world (Kabaluk and Gazdik, 2005). This is not surprising, as bacteria are closely associated with insects in forms as diverse as natural symbionts, infective disease-causing agents, and bearers of insecticidal toxins. The very diversity of bacteria with natural and manipulated recombinants, transferrable plasmids, and a near infinite diversity in nature provides an excellent resource for those seeking to develop biopesticides. From early beginnings in the 1930s and 1940s a range of strains and specific bacterial isolates have been developed as biopesticides and form the most important group of organisms commercialized as microbial control agents (MCAs; Lacey et al., 2015).

The spore-forming bacilli of the family Bacillaceae are by far the most widely used of the bacteria for the production of biopesticides, as this bacterial family contains a wide range of toxin-producing isolates, mostly belonging to *Bacillus thuringiensis* (Bt), and some with activities against lepidopteran, dipteran, and coleopteran pests that have been developed as commercial biopesticides (Fisher and Garczynski, 2012; Jurat-Fuentes and Jackson, 2012, Chapter 4). While many strains or products based on the Bt subspecies *kurstaki* have been used for lepidopteran control, Bt subspecies *israelensis* (Bti) and *Lysinibacillus* (*Bacillus*) *sphaericus* have also been used for dipteran control in public health vector control, and Bt biovar. *tenebrionis* has been used for the control of some coleopteran pests.

Although many nonspore-forming bacteria have shown insecticidal activity, few have performed adequately and consistently in the field to become microbial products. One exception is the bacterium *Serratia entomophila* (Enterobacteriaceae), a specific pathogen of the New Zealand grass grub *Costelytra zealandica* (Jackson et al., 2001), which has been developed as a live biological control product for the pest (Jackson, 2007). Another nonspore-forming bacterium *Chromobacterium subtsugae* (Betaproteobacteria) is pathogenic to a range of insect pests (Martin et al., 2007a) and has been developed for sale as a product containing live bacteria and fermentation residues. Many other nonspore-formers such as *Yersinia entomophaga* (Hurst et al., 2011) and *Pseudomonas entomophaga* (Vodovar et al., 2006) have shown excellent insecticidal activity but have yet to be developed as biopesticide products.

Undoubtedly, the most important bacterium commercialized to date in biopesticides is Bt, with a global market at the turn of the century estimated at more than US$110 million per year in 2003 and increased to more than US$210 million per year in 2010 (Whalon and Wingerd, 2003; Lacey et al., 2015) with continued steady growth. Federici (2007) considered that a major reason for the success of Bt as commercial biopesticides has been the ease of production. This, together with scalability of the fermentation production process, has allowed production of the volumes required for insect pest control in agriculture and public health.

Despite their success, bacterial biopesticides represent only a small proportion of the total pesticide market, mainly limited to organic agriculture or for specialized roles within integrated pest management programs (IPMs), as they have difficulty competing with low-cost chemical pesticides. Thus the challenge for bacterial insecticides and MCAs in general is to increase efficacy by increasing toxicity through better isolates or improved strains (Chapter 4) or improved application techniques and strategies (several chapters in this volume) or to reduce costs by increasing production per unit volume and formulation technologies to reduce postproduction losses of virulence or viability. These issues were recognized by Lacey et al. (2015) who considered that more research was needed on production, formulation, and application to make MCAs more acceptable and user-friendly.

Microbial Control of Insect and Mite Pests. http://dx.doi.org/10.1016/B978-0-12-803527-6.00008-1

A number of thorough reviews and texts of bacterial production and formulation have been published, including a production handbook (Lisansky et al., 1993), a review (Couch and Jurat-Fuentes, 2014), and a text on microbial formulation (Burges and Jones, 1998), which are excellent resources for further information on these topics. This chapter reviews the processes of production and formulation of bacterial biopesticides considering efforts that have been made to increase production efficiency, reduce production costs, and minimize losses during storage. Questions of quality control will also be considered and opportunities for new developments identified.

8.2 MASS PRODUCTION OF BACTERIA AS THE ACTIVE INGREDIENTS OF BIOPESTICIDES

The first step in production of a microbial insecticide will be the production of sufficient cells to meet the needs of the end user for an economical reduction in pest numbers. A number of methods have been used for the production of sufficient bacteria for use on a wide scale and will be described below.

8.2.1 In Vivo Production

Paenibacillus (*Bacillus*) *popilliae* was the first bacterium to be registered in the United States and was used on a large scale for control of the Japanese beetle (*Popillia japonica*; Klein, 1992). The bacterium, cause of milky disease in larvae of *P. japonica*, was produced in vivo. The method was described in detail by Fleming (1968) and is outlined in Koppenhöfer et al. (2012). Healthy larvae were collected from the field and inoculated

with 33 µl of sterile spore suspension with a dose of approximately 1×10^7 cells/larva and then placed in soil held at 30°C for up to 20 days until most larvae showed symptoms of infection. Infected larvae were held in cool water before macerating with a meat chopper and mixing with calcium carbonate and talc to produce a powder containing 1×10^8 spores/g. Fleming (1968) reported that, from 1939 to 1953, more than 80 tons of standardized spore powder were produced in the United States Japanese Beetle Laboratory for control of the beetle, which was then spreading throughout the eastern United States. A colonization approach was adopted where spore powder was applied at 160,000 sites throughout the infested region to initiate infections, until natural infections became common and the program was terminated. From Fleming's data it can be estimated that about 5 million Japanese beetle larvae were collected, infected, and processed during the program. Seeking to overcome the laborious production process led to many attempts at artificial production of *P. popilliae*, but none have been successful for commercial production (Stahley and Klein, 1992). In vivo production of milky spore powder is still used by a company producing "Milky Spore" for the home lawn and organics markets (Fig. 8.1; http://1274ym4fbva545shrh1 giu44.wpengine.netdna-cdn.com/wp-content/uploads/milky-spore-label.pdf).

8.2.2 Solid Substrate Fermentation

Solid substrates, particularly grains, provide a good medium for the production of microorganisms in the food and beverages industries and can also be used for the production of biopesticides (Mitchell et al., 2002). Solid state fermentation (SFF) involves the production of microorganisms

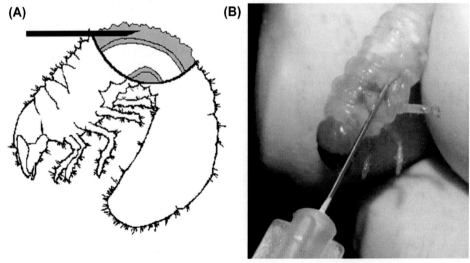

FIGURE 8.1 Inoculation of scarab larvae for the production of *Paenibacillus popilliae* by intracoelomic injection with a 30-gauge needle inserted dorsally between the second and third abdominal segments (A, *from Jackson, T.A., Saville, D.J., 2000. Bioassays of replicating bacteria against soil-dwelling pests. In: Navon, A., Ascher, K.R.S. (Eds.), Bioassays of Entomopathogenic Microbes and Nematodes. CAB International, pp. 73–94. B, modified from a figure previously reproduced by Elsevier in Koppenhöfer, A.M., Jackson, T.A., Klein, M.G., 2012. Bacteria for use against soil-inhabiting insects. In: Lacey, L.A. (Ed.), Manual of Techniques in Invertebrate Pathology, second ed. Academic Press, Amsterdam, pp. 129–149).*

on moist solid substrates in the absence of free water and has been applied on a large scale to the production of fermented foods, enzymes, and mycoinsecticides (Chapter 9). The principles of SFF have been described by Mitchell et al. (2002). Briefly, an SFF vessel (bioreactor) contains a substrate bed of discrete, solid particles with interparticular spaces allowing gas transfer. For microbial production, the substrate usually consists of moistened grains or other porous materials amended with nutrients. Bacteria can be inoculated and mixed into the substrate where they grow as biofilms on the moist particulate surfaces until nutrients have been fully utilized or other factors limit growth. SFF vessels can be aerated or agitated to vary the production process. The production of Bti by SFF has been described by Dulmage et al. (1990). Bacteria were inoculated into a matrix of wheat bran, perlite, and nutrients with a moisture content of 60% by weight. An aerated bin system was used with incubation for about 36h before harvesting. To avoid excessive heating and air channeling, the matrix was stirred at least once during the fermentation. The fermentation produced an average of 8.2×10^9 spores/g and 820 international units (IU)/mg dry wt (assay insect *Estigmene acrea*). A wide range of low-cost, coarse materials, such as brans, husks, and perlite, which are suitable for mixing and will allow air flow in static beds, have been tested in the solid substrate production of Bt (Beegle et al., 1991). A range of fermenter designs to assist air flow and mixing, from static beds to rotary drums and continuous screw fermenters, are described by Chisti (1999). SFF production can be scaled up, and Aranda et al. (2000) report that 900 tons of Bt were produced in a pilot plant at Hubei Academy of Agricultural Sciences, China, in 1990. In Brazil, SFF has been used for the production of Bt where studies on SSF were initiated to avoid the need for large aerated fermenters (Capalbo, 1995). Production was focused on the availability of low-cost agricultural by-products as substrates, and yields of $>10^{10}$ spores/g could be obtained with the fermenter product showing high activity against insect pests. In further studies, Bt subspecies *tolworthi* was produced in 50g batches on nutrient-amended rice sealed in polypropylene bags producing more than 10^9 bacterial CFU/g for the control of fall armyworm (*Spodoptera frugiperda*) in maize (Capalbo et al., 2001). The method has been adopted and is used in a small biofactory in northeastern Brazil (Valicente, pers. comm). Foda et al. (2015) tested a range of substrates for the production of *Lysinibacillus sphaericus* and found that coarse wheat bran with yeast extract produced the highest spore yields and maximum mosquitocidal activity.

8.2.3 Liquid Fermentation

Liquid fermentation of bacteria is the most common method of mass production of bacteria for biopesticides, as it can be scaled from small flasks to industrial fermenters, allowing large volumes to be produced. The basics of fermentation technology have been described in texts such as McNeil and Harvey (2008) and Demain and Davies (1999). Liquid fermentation involves the production of biomass (cells and by-products) from soluble nutrients. Bacteria require carbon, nitrogen, and trace elements for growth, and these are dissolved in water and sterilized before inoculation with the desired bacterial strain. After inoculation, cell growth can be defined in four phases: (1) Lag phase, where the cells adapt to the physicochemical environment with very little replication. (2) Log phase, where the cells grow and divide in an exponential fashion. (3) Stationary phase, where growth slows down or stops as the substrate is metabolized or toxic metabolites accumulate. (4) Death phase, where the energy reserves of the cells are exhausted, and the culture dies over time. The shape of the growth curve will be affected by media and fermentation conditions, and the fermentation strategy will usually be to maximize the production of cells or metabolites as required in the final product.

A sterile technique is essential and becomes even more critical as the production volume is scaled up. General methods for the small-scale production of bacteria are provided by Hilton (1999), which are appropriate for both spore-forming and nonspore-forming bacteria to be used as bioinsecticides. In general, certified starter cultures should be produced from an initial characterized "standard" strain, which has been selected for its pest control properties and production characteristics. The "standard" should be maintained in small aliquots under stable conditions (either ultra deep-freeze or lyophilization as a seed bank). Aliquots of the standard will be taken from the bank and used to produce a starter culture, which will be grown to reach the top of the log phase before inoculation of the fermenter seed tank (or scaled up through a series of seed tanks) where it will grow to the same point at the late log phase before introduction into the main fermenter. The number of transfers (subcultures) from the seed stock to the fermenter should be kept to a minimum to reduce the risk of plasmid loss, other genetic rearrangements, or contamination during culturing. On no account should starter cultures be kept on the bench, under refrigeration at 4°C, or regularly transferred to fresh agar plates. Isolate/aliquot and batch numbers should be allocated and continue through each stage of the process following the procedures for Good Laboratory Practice.

Success of a fermentation will depend on the fermenter design, fermentation medium, operation, and strategy. Industrial fermenters (bioreactors) are large agitated vessels that can be sterilized in situ prior to inoculation and must provide sufficient mixing and aeration as required for growth of the cells. Principles of bioreactor design are provided by Charles and Wilson (2002), and factors to consider are heat transfer, mixing, and aeration. During growth, bacterial cells are highly oxygen-demanding, and insufficient aeration will limit growth rate. Air lift fermenters, containing an internal air sparged draft tube, use differential hydrostatic forces to provide better mixing and oxygen transfer than traditional stirred tank fermenters (Merchuk et al., 2009). Commercial bioreactors for production of Bt can vary from a few thousands liters to >10^6 L, and Couch (2000) considered that fermenters of less than 3×10^4 L were "pilot plants." Problems of mixing and aeration increase with scale and with increasing viscosity of the bacterial culture through the growth phase (Charles and Wilson, 2002).

In a bioreactor, bacteria can be produced by batch, fed batch, or continuous flow culture (McNeil and Harvey, 2008). A batch fermentation will proceed through stages from the stored pure stock culture to the cell concentrate product. Aliquots of pure, characterized cells are taken from storage and transferred aseptically into a small volume, usually a shake flask, for growth on a rotary shaker until the culture is well into the log growth phase. The dividing cells are then introduced into a small seed fermenter, which will be 1–5% of the volume of the final commercial fermenter. While the culture in the seed fermenter is still in the log growth phase it will be transferred to inoculate the main fermenter where growth will continue until the culture reaches the species/product-dependent characteristics required for harvesting. In a fed batch fermentation (Macauley-Patrick and Finn, 2008), extra nutrients are added to extend the log growth stage, and the process can lead to very high cell yields, but the process is more complicated than batch fermentation and may interfere with processes such as sporulation on nutrient depletion required for the bacilliform, biopesticidal bacteria.

In the fermenter, the bacteria produced for biopesticides are "oxygen hungry" (Lisansky et al., 1993), and in the oxygen-demanding log growth stage dissolved oxygen levels should be maintained above 20%. This is achieved by increased air flow or agitation, both of which produce foam, which can be disastrous in the confined space of the fermenter (Dulmage et al., 1990). Under pressure, excess foam will flow into the exhaust filters, and the fermentation residue will be lost into the drains. Foaming can be controlled by the addition of chemical antifoams or by mechanical control. Most large-scale fermenters will introduce antifoams when the foam level reaches a "high level" detection probe.

A significant threat to the large-scale production of bacteria is infection by bacteriophages, which can destroy the culture through lysis of the cells (Beegle et al., 1991). Wu et al. (2014), using proteomics, have shown how bacteriophage infection of Bt depressed host energy metabolism and hijacked the host translational machinery to damage the cells and suppress the growth of the culture. Osman et al. (2013) reported on the selection of phage-resistant strains to overcome this problem. *S. entomophila* is also host to a wide range of bacteriophages that can cause culture collapse (Wilson et al., 1993).

8.2.3.1 *Liquid Fermentation of* Bacillus thuringiensis

Production of Bt dwarfs that of all other microbial biopesticides, as large volumes of Bt products have been marketed for more than 50 years. Specific methods for the production of insecticidal Bt have been described by Dulmage et al. (1990), Beegle et al. (1991), Lisansky et al. (1993), Couch (2000), and Couch and Jurat-Fuentes (2014). Bt is typically grown under batch fermentation at 30°C with an air feed of not less than one VVM (volume of air per volume of liquid per minute) and an initial concentration of glucose of 18 g/L (Couch and Jurat-Fuentes, 2014; Couch, 2000; Lisansky et al., 1993). The stock culture is inoculated aseptically into shake flasks where the cell culture is grown through the lag phase into the log phase when cells are motile, elongate, and actively dividing. This active culture is used to inoculate the seed fermenter where growth rapidly enters the log phase allowing for harvesting and transfer to the main fermenter after 12 h. After a further 12 h from seeding the main fermenter, the bacteria will be in the log growth phase with a high demand for oxygen, which Beegle et al. (1991) considered best met by increasing the agitation rate, increasing the back pressure in the fermentation vessel, or lowering the fermentation temperature to increase the amount of dissolved oxygen in the medium. Glucose may be fed into the fermentation through the log growth phase (Couch and Jurat-Fuentes, 2014). Bt cultures reach their maximum cell yield by about 18 h but require considerable time to complete the fermentation while the cells sporulate, produce insecticidal crystals (parasporal inclusions), and lyse. Dulmage et al. (1990) report that for Bti the log phase is completed by 16–18 h, sporulation by 20–24 h, and cell lysis by 35–40 h, after which the culture can be harvested. The total fermentation time for lepidopteran-active Bt isolates is longer, and the reported times to complete the fermentation are variable; 40–44 h (Beegle et al., 1991), 57 h (Lisansky et al., 1993); from seed tank to completion, 62–92 h (Couch and Jurat-Fuentes, 2014), but the length of time emphasizes the need for full sporulation to produce the maximum quantity of accompanying toxin-bearing crystals. Once the Bt fermentation is complete, it is cooled to 4°C, and the pH is adjusted to 4.5 with the fermentation yielding 6–8% solids with 1–3% spores and crystals (Couch, 2000).

8.2.3.2 *Liquid Fermentation of* Lysinibacillus sphaericus

The mosquitocidal *L. sphaericus* can be produced in similar batch fermentations as described previously for Bt but shows some differences in optimal production temperature and media components (Dulmage et al., 1990). Lacey (1984) summarized the knowledge on the growth requirements of *L. sphaericus* and commented that the bacterium will grow between 25°C and 40°C, but sporulation and toxin production will be inhibited at higher temperatures. Media is differentiated from that for the production of Bt, as *L. sphaericus* cannot use carbohydrates as carbon sources (Beegle et al., 1991), and carbon is best provided from partially hydrolyzed proteins. Lacey (1984, 1990) and El-Bendary (2006) reported on various substrates for *L. sphaericus* production, ranging from animal proteins to plant meal and waste products. Prabakaran et al. (2007) produced *L. sphaericus* to a high cell density (3.7×10^9/mL) in a 100-L fermenter using a medium incorporating soybean flower and peanut cake as carbon sources. Using a fed batch strategy and a corn-steep-based medium, Sasaki et al. (1998) achieved a cell dry weight of 13.0 g/L and a spore yield of 1.64×10^9/mL after 56 h of culture. Continuous flow fermentation has also been used for the production of large quantities of *L. sphaericus* (Lacey, 1984). As with Bt, *L. sphaericus* fermentation is terminated with full sporulation and crystal toxin formation.

8.2.3.3 Liquid Fermentation of Serratia entomophila

Growth media and production of the nonspore-forming bacterium *S. entomophila* by liquid fermentation is described by Visnovsky et al. (2008). The bacterium grew to higher cell densities on carbohydrates rather than proteins and grew well on a range of sugars with the highest yields from media containing sucrose as the primary carbon source. The addition of yeast extract increased the cell yield, which could be further improved using a fed batch strategy. A typical batch fermentation for *S. entomophila* is carried out at 30°C, aerated to maintain dissolved oxygen levels >20% and completed in 26–28 h when the fermentation is well into the stationary growth phase. A completed fermentation will produce a cell yield of $>5 \times 10^{10}$ CFU/mL corresponding to >30 g/L dry cell weight. Unlike Bt, there are no morphological changes such as sporulation to indicate completion of fermentation. Visnovsky et al. (2008) found that cells harvested early in the fermentation, through the log and early stationary phases, had a poor survival in storage, while those harvested after 20 h, well into the stationary growth phase, showed high levels of survival in storage under refrigeration. A high throughput of air and constant agitation in the *S. entomophila* fermentation can lead to the production of large quantities of foam, which led Surrey et al. (1996) to develop a pneumatic antifoam delivery system to overcome the problem. The threat of bacteriophages (Wilson et al., 1993) was minimized through the selection of phage-resistant strains for production (O'Callaghan et al., 1992a,b; Jackson et al., 1992) after episodes of bacteriophage-induced culture collapse.

8.2.3.4 Fermentation of Chromobacterium subtsugae

Another nonspore-forming bacterium produced by fermentation is *C. subtsugae*. The bacterium shows optimum growth at 25°C, pH 6.5–8.0 with 0–1.5% NaCl added to the medium (Martin et al., 2007b). The highest insect mortality followed treatment with a combination of live cells and the insecticidal toxins produced in the stationary growth phase.

8.2.4 Objectives of Fermentation for Spore-Forming and Nonspore-Forming Bacteria

The objective of fermentation will be to produce the active ingredient of the biopesticide in the most cost-effective manner. This is achieved differently when comparing between the fermentation production of spore-forming and nonspore-forming bacteria. For the spore-former Bt, maximizing the toxin yield is of greatest importance, while for nonspore-forming *S. entomophila* and other live bacteria, the objective will be to produce the maximum density of robust, live cells within the fermenter. Perani and Bishop (2000) showed that sporulation and toxin production by Bt on minimal media varied between strains, and Monnerat et al. (2004, 2007) were able to select high toxin-producing/virulent strains from a wild type culture collection. Strain selection also had a significant effect on the production and postharvest viability of *S. entomophila* (Pearson et al., 1993, 1996).

The economics of Bt production will depend on the cell density per unit volume and the cost of raw materials. Cell and spore production by Bt and *L. sphaericus* after batch fermentation on refined media are generally in the range of 1 to 5×10^9 cells/mL (El-Bendary, 2006; Beegle et al., 1991), but Poopathi (2010) considered the costs of refined media to be an excessive expense. Lachhab et al. (2001) showed that similar yields to those on refined media (potency of 1.3×10^4 IU/µL) could be obtained using low-cost wastewater sludge as a nutrient source. Chicken feather and rice husk wastes have also been found to provide effective low-cost substrates for *L. sphaericus* and Bti production (Poopathi and Abidha, 2009; Poopathi, 2010). In further experiments, Poopathi and Archana (2012) found that coconut cake powder (CCP), an industrial by-product, was also an effective medium for the production of Bti. CCP (5%) was added to tap water, boiled, and filtered before being inoculated with bacteria and incubated in an agitated vessel at 30°C for 72 h before harvesting. The CCP medium produced 6–8% biomass that was marginally enhanced by the addition of 1% $MnCl_2$.

The nonspore-forming, smaller cell of *S. entomophila* can be produced at high densities in the bioreactor where commercial production at 4×10^{10} cells/mL equilibrates to 1 L of fermenter product applied per ha (Jackson et al., 1992). The high yields obtained are representative of a high-density cell culture (HDCC) with a dry weight yield of 3%, but with advanced HDCC fermentation techniques there is considerable scope to further increase the yield (Visnovsky et al., 2008).

8.3 HARVESTING

Once fermentation is complete, cell mass must be harvested to preserve insecticidal activity through formulation, storage, and application. Solid substrate products are washed from the substrate or simply dried on the media and used as granules. Brar et al. (2006) considered that while much attention had been given to the production and media costs for bioreactor-produced Bt, the true constraints to efficient production are embedded in the harvesting and formulation costs. The Bt broth consists of spores, crystals, and other materials, which must be harvested in an efficient manner to be used in the formulation step. As these cell products comprise only 1–3% of the total broth they must be concentrated to produce an active ingredient that can be formulated into a final product. While lactose–acetone precipitation was once used to assist cell concentration, this process resulted in a loss of bioactivity and has been replaced by centrifugation and filtration to separate the solid active ingredients from the soluble liquid fraction (Brar et al., 2006). Centrifugation at >8000×g can be used for recovery and with removal of the liquid phase can concentrate the total solids from 4–6% to 15–30% of the volume (Couch and Jurat-Fuentes, 2014). It is increasingly realized that, as well as the endotoxins, Bt secretes large amounts of other "virulence factors" into the culture medium, including Vips, Cry proteins, proteases, and chitinase enzymes (Jurat-Fuentes and Jackson, 2012). Jallouli et al. (2014) showed that these components could be conserved with ultrafiltration, resulting in a higher toxicity than the centrifuged product. In contrast with Bt fermentation, *S. entomophila* is produced as a high-density cell culture with a use rate of <1 L fermentation broth/ha. Therefore no cell concentration step is needed for product formulation.

8.4 FORMULATION OF BACTERIA FOR STORAGE AND APPLICATION

Brar et al. (2006) considered that formulation is the crucial link between production and application, and the objectives of pesticide formulation can be defined as follows: "To provide the user with a convenient, safe product which will not deteriorate over time and to obtain the maximum activity inherent in the active ingredient" (Knowles, 2008). The same principles apply to bacterial formulations but recognize the sensitivity of microbes to environmental conditions both in storage and beyond. The main objectives of formulation for

biopesticides have been defined by Jones and Burges (1998) and Ravensberg (2011) and broadly cover the following: (1) stabilization of the cells that have been mass produced (and concentrated) to be packed, stored, and shipped to the end user with minimal losses; (2) production of a user-friendly product that can be applied effectively and economically to the target pest; (3) protection of the bacterial cell to enhance survival, persistence, and activity at the target site; and (4) minimizing the exposure of workers during packaging and application and of the consumer of food crops.

Due to the multiple objectives of formulation and the wide range of materials and formulation processes, microbial formulation is often considered more art than science with information often held by companies as trade secrets and confidential know-how (Behle and Birthisel, 2014). However, the general principles can be described even though the level of detailed information will vary among commercial products.

At the end of the cell production process, fermentation, and concentration, the bacterial culture is in a stable phase, either as spores or stationary phase cells, but it is held in an aqueous phase surrounded by enzymes and nutrients, which can have a deleterious effect on the cells. Subsequent actions will depend on the characteristics of the cells. Spore-former concentrates are generally spray dried into a "technical grade active ingredient (TGAI)," a dry powder with a moisture content of less than 8%, which can be standardized and characterized for potency and stored before further formulation (Couch, 2000). Nonspore-formers are more sensitive to drying and long-term storage but can be stored for periods as aqueous cell concentrates under refrigeration.

Formulation types have been classified and codified by CropLife International (http://www.cipac.org/cumindex/formcode_right.htm) and summarized by Knowles (2008) and Gašić and Tanović (2013). Formulations for biopesticides can be divided into dry and liquid products (Jones and Burges, 1998). Dry products used for bacterial biopesticides include wettable powders (WPs), water dispersible granules (WGs), and granules for dry application (Fig. 8.2). WPs consist of the active ingredient, which will be the dried TGAI, a filler, a free flow agent, a dispersant, and a wetter (Lisansky et al., 1993). The fillers and free flow agents can be clays, talcs, or milk powder. WPs are designed to be mixed with water as a carrier for application. Fillers and dispersants should be hydrophilic to aid mixing and prevent balls of dry material from forming and clogging the application equipment. With all microbial applications it is important to avoid chlorinated water sources. WGs can also be used to formulate the microbe prior to mixing and their dispersal in water. WG formulations are better for users, as there is less dust on handling, and they are easier to mix into the liquid. Granules can be formed by coating the active ingredient onto an absorbent carrier such as clays

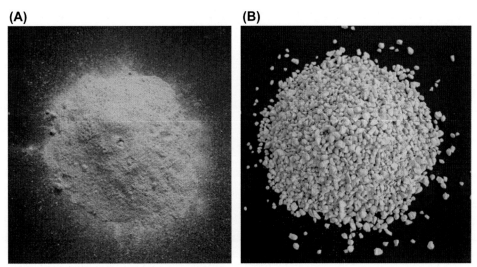

FIGURE 8.2 (A) *Bacillus thuringiensis* formulated as a wettable powder (Dipel). (B) *Serratia entomophila* formulated as a zeolite-coated granule (Bioshield).

and silicas or vegetable materials such as ground corn cobs or coconut shells (Knowles, 2008). For applications to soil, bacteria can be coated onto seed as a granule core (Wright et al., 2005). Stickers may be used to aid coverage and resins, and polymers can be used to delay release after application. Granules can also be produced by extrusion (Johnson et al., 2001) where the microbe is mixed with a clay or other carrier to form a paste and extruded under pressure to form discrete pellets. Dry formulations are formed either by mixing dry materials or applying the microbe in aqueous suspension to a dry substrate and subsequent drying to form the final flowable product. An extreme form of granule is the briquette, a floating granule produced for mosquito control.

Liquid formulations are also described in Jones and Burges (1998), and many are produced as suspension concentrates (SC) where the microbe is suspended in a carrier liquid (water or oil) together with a suspending agent to provide a viscosity that limits settling of the suspended microbial particles. Emulsion with a surfactant can be used to produce an emulsifiable concentrate (EC). Most biopesticide EC formulations have the microbial active ingredient suspended in an oil-based solvent which, when mixed with water and an appropriate emulsifying agent, forms a milky emulsion. Microbial cells are held in oil droplets, the dispersed phase, which are suspended in water, the continuous phase. Invert emulsions, where water droplets are suspended in oil, are more viscous and less likely to separate into the different phases.

Microbes can also be formulated using encapsulation to coat the organism with a protective material such as a polymer. Microcapsules can be formed using phase separation, interfacial reactions, and physical methods such as multiorifice centrifugation and electrostatic encapsulation (Jones and Burges, 1998). A form of encapsulation has been achieved by genetic engineering to produce Bt toxins in an alternative bacterium, which can be subsequently dried to protect the cell. Once produced, microcapsules must be formulated into a dry or liquid formulation, as described above.

The benefits and disadvantages of the alternatives of liquid and dry formulations are discussed by Burges and Jones (1998), Hynes and Boyetchko (2006), Brar et al. (2006), Couch and Jurat-Fuentes (2014), and others but the final choice for the end product will depend on a compromise between the differing formulation objectives: cell survival and/or preservation of insecticidal activity; ease of production and application; postapplication protection of the product; and minimal worker/consumer exposure.

8.4.1 Cell Stabilization

While it has long been considered that biopesticides require a shelf life of >18 months at ambient temperatures (Couch and Ignoffo, 1981), strategies can be adopted to adapt the supply chain for less robust products. Lactic products with probiotic bacteria, for example, have a short shelf life at ambient temperatures and are usually distributed with refrigeration. While the instability of nonspore-forming rhizobial legume inoculants poses problems, they are still widely used, as they are an essential part of the production of pulses such as soybean (Temperano et al., 2002). Ravensberg (2011) considered that, while industry would like a biopesticide to have a shelf life of more than 2 years at ambient temperature, for the greenhouse (protected cultivation) industries, products with less than 6 months of shelf life under refrigerated storage are acceptable. Behle and Birthisel (2014) considered that >90% survival of microbial agents was acceptable, but this too will vary with the organism and final product. Thus the requirements of formulation for bacterial biopesticides will depend on the end use of the product, and the final formulation will be a compromise between the intrinsic characteristics of the organism and the needs of the market place.

Spores and crystals are very stable in dry conditions, which has encouraged the production of dry formulations for the spore-forming bacteria. For the in vivo production of *P. popilliae* the spore-filled macerate of dead insects was mixed with talc and clays in a dry formulation, which could be stored and remained viable for several years (Fleming, 1968; Klein, 1992). Similarly, Behle and Birthisel (2014) reported that dried Bt products retained activity for several years, even when stored at room temperature. However, moisture content must be considered. Couch (2000) considered that Bt products should have a moisture content of 6–8%, as further drying will lead to a loss of potency. Lisansky et al. (1993) warned against greater moisture content, as it can encourage contaminant growth. Liquid suspensions, in contrast, are subject to deterioration during long storage (Brar et al., 2006). As a precautionary measure, the manufacturers of the aqueous Foray 4B (Valent Biosciences) recommend that the product should be stored at temperatures of 0–25°C and used within 12 months of the date of manufacture (http://forestry.valent-biosciences.com/docs/forestry-resources-library/foray-48b-label---canada). Valent Biosciences further showed that dry flowable granules could be stored at 25°C with minimal losses in potency, while water-based Bt products lost nearly 50% of their potency in the first 12 months of storage (http://cropprotection.valentbiosciences.com/valent-biosciences-corporation-biorational-crop-protection/insecticides/products/dipel/dry-flowable-granule).

Lacking resistant spores, nonspore-forming bacteria are much more sensitive to postproduction environmental conditions than spore-formers, and the difficulty of maintaining viability from production to use has limited the development of this group of organisms. Cool storage is an option. Due to rapid deterioration at ambient temperatures, unformulated bacterial broth cultures of *S. entomophila* had to be maintained under refrigeration to preserve cell viability (Pearson et al., 1993), but even then, during early development of the bacterium, unformulated bacterial broth cultures were used, which could only be maintained in a viable state for 3 weeks when held under refrigeration (Jackson et al., 1992). With improvement in fermentation strategy and conditions (see above), a fermentate was produced, which could be stored under refrigeration (4°C) for more than 6 months without significant losses (Pearson et al., 1993), providing enough stability for the product, Invade, to be manufactured and distributed through cool store networks. This 6-month period was sufficient to treat a seasonal pest like the New Zealand grass grub *C. zealandica* (Jackson et al., 1992), but the need for cool temperature storage limited

distribution of the product. To provide a more stable product the bacterium was incorporated into a biopolymer gel and mixed into a clay paste before extrusion as pellets. The pellets could be stored at ambient temperatures for 6 months with minimal losses of viable bacteria (Johnson et al., 2001). In a further development, the bacteria/biopolymer mix was coated onto a zeolite core to form a granule, which maintained bacterial survival for up to a year in ambient conditions (Swaminathan and Jackson, 2007). Interestingly, seed coating formulations retained bacterial viability at 4°C but had significant losses at 20°C (Wright et al., 2005). The bacterium *C. subtsugae* is another nonspore-forming bacterium, which has been commercialized as the bioinsecticide Grandevo, containing live bacteria, spent fermentation media, and formulating ingredients in a WP format. The material is reported as stable if stored in cool conditions.

8.4.2 Production of an Efficacious, User-Friendly, Economical Product

For a biopesticide to be successful it needs to be efficacious, easy to apply, and provide an economic solution to the target pest. The efficacy of biopesticides is reported in scientific papers and product literature and so will not be covered in this chapter. However, the formulation will play a role in efficacy by delivering an infective/toxic microbe to the target in a viable and competent form. Viability was addressed in the previous section, but the formulation will also have a major role in delivery to the target. Bacterial biopesticides to be sprayed on a crop must be fully compatible with the application machinery, and so dispersants and wetting agents are often added to prevent machinery blockages and obtain full coverage of the crop.

The formulation selection will depend on the target to be treated. When the New Zealand city of Auckland was invaded by the exotic white spotted tussock moth *Orgyia thyellina*, Bt subsp. *kurstaki*, as the product Foray 48B, was chosen as the main control agent as an aqueous suspension that could be applied as a low-volume concentrate by air for broad coverage of the city (Hosking et al., 2003). Suspension concentrates of Bt are also widely used in forestry, agriculture, and horticulture, as they are compatible with the foresters' and farmers' existing spray machinery and can be easily incorporated into the regular pest control programs. Concealed, soil, and water pests will need a more targeted approach. The corn borers (*Diatraea* spp.) are important pests of maize and can be controlled by Bt if the biopesticide can reach them as they feed on the stem tissue. The solution was to apply corn grit granules coated with Bt suspended in oil into the whorls of the developing plants. Couch (2000) reported that the corn grit granule was the most common form of Bt granule in use until it was superseded by transgenic Bt corn. Bt remains widely used for agricultural, forestry, and plantation pests but mainly as dry, dispersible granules due to their better storage and distribution characteristics.

The control of public health pests poses different problems. Mosquito larvae, for example, develop in water and can emerge as problems from wide areas such as lakes and paddy fields or in small containers of water and drains in urban environments. Bti can be used as a larvicide to prevent adult emergence, but clearly there is a place for formulations to address the different breeding conditions. For large-scale application for mosquito control on flood plains in the Rhine valley, where $600\,km^2$ are infested, Becker (1997) reported that 1000 tons of Bti granules (Bti powder in oil coating over quartz sand) was applied by helicopter at a dose of 10–20 kg/ha between 1981 and 1996. In a successful campaign for mosquito control on a smaller scale ($4.5\,km^2$) in Kenya, Fillinger and Lindsay (2006) found that WG formulations of Bti subsp. *israelensis* and *L. sphaericus* were most convenient for backpack sprayer applications but that corn grit formulations could also be used for hand applications in areas that were hard to access. Formulations can also be designed for the best impact on specific target species (Lacey, 1990; Lacey et al., 1984, 1988). The mosquito larval targets for biopesticide control have distinctive feeding habits, with anopheline mosquito larvae feeding at the surface of ponds, while *Aedes* spp., culicidae, and black flies feed at the lower depths. Delivery to the target insects has been resolved by formulation. Surface feeders are treated with sprays or floating granules while those in the depths are treated with dense granules that sink to the bottom. Valent BioSciences produce a range of formulations of Bti and *L. sphaericus* from suspension concentrates to treat open waterways to granules designed to be applied by air to penetrate foliage and sink into the water (see Chapter 27; http://publichealth.valentbiosciences.com). VectoBac 12AS is an aqueous suspension concentrate formulation of Bti (strain AM65-52) for the control of mosquitoes, black flies, and closely related fly larvae that is designed to be sprayed onto open waterways. VectoMax WSP is a water-soluble pouch containing Bti (strain AM65-52) and *L. sphaericus* 2362 (strain ABTS 1743) in a formulation designed for the slow-release treatment of small volume breeding sites for control of *Culex* spp. vectors of West Nile virus (WNV) in urban and suburban areas. Bactimos PT is formulated as a dense pellet that will sink to the bottom of ponds for the control of chironomid midges. The evolution of the Vectobac granule demonstrates the relationship between the ballistics of the granule, improved by increased weight, versus the need to obtain effective coverage of the target zone. The first granule, Vectobac G, was large, comprised of 65 granules/g, and provided coverage of $32.5\,granules/m^2$ at a use rate of 5.6 kg/ha. The Vectobac GS product with 485 granules/g produces a coverage of nearly $250\,granules/m^2$ at the same application rate while

Vectobac GR with 830 granules/g produces a coverage of more than $400\,granules/m^2$. Where the objective is contact with a randomly dispersed insect, the highest probability of contact will be with the greatest number of granules, but the trade-off is a loss of precision in the delivery of the smaller granules.

Application to soil involves similar compromises. Application of *S. entomophila* as a liquid SC involved the dilution of 1 L of bacterial concentrate in 100 L of water/ha to obtain adequate distribution when drilled into the soil with a modified seed drill (Jackson et al., 1992). Better product stability led to the adoption of a granular formulation that could be applied through standard seed drills, but in order to achieve an adequate distribution of granules through the soil the use rate was 30 kg/ha (Townsend et al., 2004).

Formulation can give rise to a wide range of products to address end user requirements and provide options for delivery. The traditional Bt insecticide Dipel (Btk ABTS-351) is sold by Valent BioSciences in dry flowable, emulsifiable suspension, and granular formulations (www.valent.com). The dry flowable formulation contains 54% fermentation solids, spores, and toxins. The alternative formulations, emulsifiable suspension and granules, contain 23.7% and 2.3%, respectively. Sometimes formulations can go beyond the traditional liquid and dry formats. A novel formulation of Bti was produced by Becker (2003), who incorporated Bti and *L. sphaericus* into ice pellets using a special ice-making machine. The ice floated on water and melted, releasing the toxins on the surface, in contrast to quartz granule formulations, which sank quickly in the water. The ice formulation also reduced wear on spray nozzles, and a greater spray width was achieved than with quartz granules.

The Brazilian research institute, Embrapa, has focused on the development of endemic spore-forming bacteria as commercial products. A large number of bacterial isolates were collected, characterized, and added to the Embrapa Collection of Entomopathogenic Bacteria (http://plataformarg.cenargen.embrapa.br/rede-microbiana/colecoes-de-culturas/colecao-de-bacterias-de-invertebrados/colecao-de-bacterias-de-invertebrados). Bacterial isolates from the collection were selected and tested against key agricultural and public health pests. Strains were then selected for commercialization on the basis of toxicity to the target pest and efficiency of production. These include an endemic strain of the Bt subspecies *kurstaki*, developed as the product Ponto.Final for the control of lepidopteran larvae in agriculture and horticulture, and strains of Bti developed as the products Bt-horus and Fim da picada for the control of biting black flies (*Simulium* spp.) and *Aedes aegypti* mosquitos causing dengue fever. Sphaerus, incorporating *L. sphaericus* for the control of malaria-transmitting *Anopheles* spp. mosquitos, has also been developed (bthek.com.br).

Bt commercial products in Brazil are produced by aerobic liquid fermentation on a medium containing yeast extract, proteins, and salts for 30 h to obtain a cell paste of 2.5 GDW/L (fermentation conditions: temperature 30°C, dissolved oxygen 20%, pH 7.0; Monnerat and Soares, pers. comm.). Bt and *L. sphaericus* products in Brazil are sold in liquid form as suspension concentrates containing UV protectants, emulsifiers, preservatives, and inert carriers with 1.2% a.i. (active ingredient) Bt and 2.5% a.i. *L. sphaericus* (bthek.com.br).

8.4.3 Protection of the Bacterial Cell at the Target Site

Formulation can also provide protection for the microbe/toxins after they have been applied to the target site. Bacteria sprayed onto leaf surfaces will be rapidly degraded by the effects of UV light and desiccation. The use of sunscreens was extensively reviewed by Burges and Jones (1998), who listed positive protectant effects from UV light for bacteria protected with vitamins, cosmetics, dyes, proteins, starches, and melanin. The best protection was provided by melanins, which are dark brown to black pigments that provide protection from the damaging effects of UV light to organisms ranging from humans to microbes. Melanin synthesis in free-living microbes is thought to provide a survival advantage in the environment (Nosanchuk and Casadevall, 2006), and *Streptomyces lividans*-synthesized melanin provided excellent photoprotection from UV light when mixed with spore/crystals of Bti (Liu et al., 1993). Sansinenea and Ortiz (2015) review the use of mutagenesis to obtain Bt mutants producing melanin, and Sansinenea et al. (2015) report on a wild type Bt isolate with UV resistance naturally producing melanin. UV protection for *Bacillus* sp. using lignins and optical brighteners has also been described by Schisler et al. (2004). Ratnakar et al. (2013) reduced degradation of *C. subtsugae* by sunlight by formulating with sodium benzoate.

Incorporation into granules can also protect the bacterium at the target site. Granules and pouches for mosquito control (see above) provide extended control by protecting the bacteria/toxins within the formulation and delivering them into the surrounding water through a slow degradation of the formulation substrate. Granular formulations applied to the soil can also protect the microbes contained within them. O'Callaghan and Gerard (2005) found that a granular application of *S. entomophila* in granules resulted in better long-term persistence of the bacterium in soil than from an application in liquid concentrates.

8.4.4 Minimizing Worker/Consumer Exposure

Bacterial biopesticides have been extensively tested for mammalian safety and have been registered as posing no health risk by regulatory authorities in many parts of the world. This lack of toxicity and hazard is one of the great benefits of the registered bacterial biopesticides, meaning that there are no withholding periods between application and harvesting and that these products can be used to control pests late in the crop cycle. However, as with all manufacturing processes and agricultural production, there is a need to avoid unnecessary exposure for workers and consumers. Exposure to dust is probably the main issue in handling dry materials for formulation. Work sites should be properly ventilated, and production workers should wear protective clothing and masks for work in dusty conditions. Aerosols should also be avoided. Registered products will be accompanied by directions for use and a Material Safety Data Sheet containing information on potential hazards. Allergies, skin sensitization, and eye irritation are recognized hazards often resulting from incorrect use that leads to unnecessary human exposure.

8.5 QUALITY CONTROL

Quality control begins at the start of the production process using guidelines for Good Manufacturing Process and Standard Operating Procedures (SOP) for each step of the production process, from fermentation to formulation and packing to ensure a consistent end product. The final product will then be checked to ensure that it meets the label requirements for the end user. Where the bioinsecticide active ingredient is a live bacterium, as in the case of *P. popilliae* or *S. entomophila*, the biopesticide product can be standardized on the basis of the number of live cells and formulations, adjusted accordingly. In the production of Milky Spore, a concentrate spore suspension was prepared, quantified by microscopy, and diluted with an appropriate quantity of talc to produce a standard product containing 1×10^8 spores/g. For *S. entomophila* products to control the grass grub, the level of infection is consistently related to the dose of live cells (Jackson et al., 2001). Quality control of *S. entomophila*-based products is assured by tests for purity of the strain and virulence of the starter culture followed by viable cell quantification, purity, and virulence testing of the final product (Pearson and Jackson, 1995).

For Bt products, verifying quality control is more complex due to a lack of a reliable relationship between spore count and potency (Beegle et al., 1991), which results from the large number of factors involved in toxin production and the difficulty of maintaining the toxic components through the downstream processing after fermentation. The use of Bt before 1970 was plagued by low potency and variability in efficacy, which led to calls for standardization in order to provide a definable level of quality assurance for consumers (Beegle and Yamamoto, 1992). A standardization system, based on the "potency" of the Bt product as determined by bioassay, was proposed and adopted by the industry. To date, no reliable chemical or biochemical method has been developed to determine the potency of Bt preparations, and so the measurement of potency continues to rely on bioassay

(Couch and Jurat-Fuentes, 2014). However, bioassays are complicated (Navon and Ascher, 2000), and even with the best SOPs, they will vary over time and between laboratories. To resolve these discrepancies, a set of spore/toxin standards was identified, which could be used for comparative assays so that the potency of the test material could be defined in relation to the standard (Beegle et al., 1991). The process of determining potency is described in detail by Dulmage et al. (1990). Briefly, the test material and standard are tested in replicated assays over a range of doses to determine the LC_{50} (the lethal concentration of toxin that will kill 50% of the test population) of each material. The potency of the test material as expressed in International Units (IU)/mg of test material is calculated from the LC_{50} of the standard divided by the LC_{50} of the test material to create a relative proportion of activity with respect to the original standard. This ratio is multiplied by the assigned potency value for the standard preparation, as indicated by the formula

$$\text{Potency of test material, IU/mg} = \frac{LC_{50} \text{ Standard}}{LC_{50} \text{ Test material}} \times \text{Potency of Standard}$$

The defined potency will be dependent on the insect under test and the assay conditions. A new standard will be allocated an arbitrary potency value. The first Bt standard (E−61 from the Institute Pasteur, Paris) was assigned a potency of 1000 IU/mg but was later replaced by HD-1-S-1971 (potency 18,000 IU/mg) and HD-1-S-80 (potency 16,000 IU/mg; Navon, 2000). Standards also exist for Bti and *L. sphaericus*. Potency is a useful measure for comparison between products. Bt subsp. *kurstaki* strain ABTS-351, the active ingredient of the Dipel products, has a potency of 32,000 Cabbage Looper Units (CLU) per mg in the dry flowable formulation; 17,600 CLU/mg in the emulsifiable suspension; and 1600 CLU/mg in the granule (http://www.valent.com).

The use of standards, however, has not been without problems. Without proper storage, the potency of the standard will decline over time (Beegle et al., 1991), and methods and conditions vary between laboratories, making comparisons difficult (Navon, 2000). It is not surprising, therefore, that some companies do not report potency, and the standard label for Bt products in the United States includes the following caution statement: "The percent active ingredient does not indicate product performance and potency measurements are not federally standardized."

8.6 CONCLUSIONS AND FUTURE NEEDS

As shown in Chapter 4 and Jurat-Fuentes and Jackson (2012), bacteria possess a wide range of identified insecticidal activities, and there are certainly new activities to be discovered. However, to be useful for the control of insect pest species these organisms need to be converted into products that are economical and convenient for the end user. Success has been achieved with the production and commercialization of Bt products through large-scale fermentation and drying, but this success seems to have limited investigations of other production methods, even though other systems have been successful in specific situations. When the Japanese beetle was an invasive, acute problem in the eastern United States, and there was no in vitro method to produce *P. popilliae*, in vivo production was carried out successfully at an industrial scale to colonize the infested area with the biocontrol bacterium (Fleming, 1968). The in vivo production of *P. popilliae* was highly labor intensive and hence expensive and could not be justified for anything other than an inoculative process with a microorganism that could not be cultured by other means.

For resource-poor farmers in northeastern Brazil, solid substrate fermentation of Bt on rice has provided a feasible and economical method for controlling their key pests of maize but relies on a good labor supply and inexpensive substrates for production (Capalbo, 1995). The success of these products may be related to the relatively small scale of the operation. Solid substrate fermentation is subject to problems of scale up: heat transfer limits the size of bag or tray production, and the process is labor intensive and hence costly (Ravensburg, 2011). Even with fungal biopesticides, solid substrate production has been most successful when limited to community or local industry (eg, glasshouse, sugar cane estate) production and has not been able to penetrate broad-scale agriculture. As Grzywacz et al. (2014) have pointed out, insect pest control is a problem over large areas, and the amount of grain that would be needed to meet the pest control demand through solid substrate production would be staggering. Given these constraints it is likely that solid state production of bacteria on grains will remain a niche activity confined to local consumption in small communities or used for the production of small batches of specific organisms.

As the demand for organic or low-chemical residue food products has increased, Bt products produced by large-scale liquid fermentation have been able to meet the demand and provide the necessary insect control. Bt products account for about 70% of the total biopesticide sales with the overall market estimated to have a value of more than US$200 million (Ravensberg, 2011). Economies of scale are achieved by companies fermenting and processing large volumes of bacteria to bring standardized products to the market place. Fermentation can also be contracted to specialized companies to allow for smaller scale production of niche products, as has been the case with the production of *S. entomophila*.

Producing a bacterial biopesticide will involve accumulated costs from media, labor, and capital. High capital costs cannot be justified for a small end market although it may be possible to lease equipment or contract out fermentation.

Where labor costs are high, the labor intensive in vivo and solid state production systems are unlikely to be cost efficient. The value of investing in media amendments to increase yield or less expensive ingredients to reduce media costs will need to be considered carefully and will depend on their cost in proportion to that of the whole fermentation. When comparing biopesticide production with other fermentation industries, raw material costs for fermentation media may be as little as 5% for high-value products, such as pharmaceuticals, but can be as high as 50% of production costs for ethanol fermentation (Dahod, 1999). For Bt products, Lisansky et al. (1993) estimated raw material costs at about 25% of the total production, allowing for some benefit to be obtained from a reduction in culture media costs. Low-cost raw material media for the production of Bt, from chicken waste to sewage sludge, have been described above, but the costs saved must be measured against reduced activity per unit volume of the fermenter. For large-scale commercial Bt producers, production methods are optimized to maximize insecticidal proteins per unit volume of the fermentation media (Couch and Jurat-Fuentes, 2014) and minimize the cost of labor with an increased scale of production (Lisansky et al., 1993). Visnovsky et al. (2008) also demonstrated that the economics of production of *S. entomophila* could be markedly improved by increasing cell density within the fermenter.

Fitting production capacity to the market demand will also be an issue. Excessive production resulting in unused perishable product will become a sunk cost and threaten the economics of the production system. For this reason, commercial companies have insisted on a long shelf life for microbial products. While product stability is highly desirable and suitable for the distribution of biopesticides through conventional marketing systems, alternative systems can be developed for less robust products. Localized production, where the biopesticide producer is close to the user, or refrigerated supply chains are two such options.

In conclusion, bacterial biopesticides have been successful in a number of different formats, as they have found niches in the agricultural and public health pest control markets. As selective agents, they will usually need to be incorporated into integrated management systems for control of the target pests. As safe agents they will have the flexibility to be used on food crops and in the home environment. Bacterial biopesticides complement efforts to reduce chemical pesticide use and contamination of the environment and have a positive future within programs for pest and disease control.

ACKNOWLEDGMENTS

My thanks to the Microbial Products Team, AgResearch (past and present), and collaborators for useful discussions and contributions to the development of bacterial biopesticides. Special thanks to Rose Monnerat for providing information on biopesticide production in Brazil and Robert Behle and Terry Couch for useful comments on an earlier draft of the manuscript.

REFERENCES

Aranda, E., Lorence, A., Refugio Trejo, M., 2000. Rural production of *Bacillus thuringiensis* by solid state fermentation. In: Charles, J.F., Delecluse, A., Nielsen-Leroux, C. (Eds.), Entomopathogenic Bacteria; from Laboratory to Field Application. Kluwer, The Netherlands, pp. 317–332.

Becker, N., 1997. Microbial control of mosquitoes: management of the upper rhine mosquito population as a model programme. Parasitol. Today 13, 485–487.

Becker, N., 2003. Ice granules containing endotoxins of microbial agents for the control of mosquito larvae – a new application technique. J. Am. Mosq. Control Assoc. 19, 63–66.

Beegle, C.C., Yamamoto, T., 1992. History of *Bacillus thuringiensis* Berliner research and development. Can. Entomol. 124, 587–616.

Beegle, C.C., Rose, R.I., Ziniu, Y., 1991. Mass production of *Bacillus thuringiensis* and *B. sphaericus* for microbial control of insect pests. In: Maramorosch, K. (Ed.), Biotechnology for Biological Control of Pests and Vectors. CRC Press, Boca Raton, pp. 195–216.

Behle, R., Birthisel, T., 2014. Formulation of entomopathogens as bioinsecticides. In: Morales-Ramos, J.A., Guadalupe Rojas, M., Shapiro-Ilan, D. (Eds.), Mass Production of Beneficial Organisms. Academic Press, San Diego, pp. 483–517.

Brar, S.K., Verma, M., Tyagi, R.D., Valero, J.R., 2006. Recent advances in downstream processing and formulations of *Bacillus thuringiensis* based biopesticides. Process Biochem. 41, 323–342.

Burges, H.D., Jones, K.A., 1998. Formulation of bacteria, viruses and protozoa to control insects. In: Burges, H.D. (Ed.), Formulation of Microbial Biopesticides: Beneficial Microorganisms, Nematodes and Seed Treatments. Kluwer Academic Publishers, Dortrecht, pp. 33–127.

Capalbo, D.M.F., 1995. *Bacillus thuringiensis*: fermentation process and risk assessment. A short review. Mem. Inst. Oswaldo Cruz 90, 135–138.

Capalbo, D.M.F., Valicente, F.H., Moraes, I.O., Pelizer, L.H., 2001. Solid state fermentation of *Bacillus thuringiensis tolworthi* to control fall armyworm in maize. EJB Electron. J. Biotech. http://dx.doi.org/10.2225/vol4-issue2-fulltext-5.

Charles, M., Wilson, J., 2002. Fermenter design. In: Flickinger, M.C., Drew, S.W. (Eds.). Flickinger, M.C., Drew, S.W. (Eds.), Encyclopedia of Bioprocess Technology: Fermentation, Biocatalysis, and Bioseparation, vol. 5. Wiley, New York, pp. 1157–1189.

Chisti, Y., 1999. Solid substrate fermentations, enzyme production, food enrichment. In: Flickinger, M.C., Drew, S.W. (Eds.). Flickinger, M.C., Drew, S.W. (Eds.), Encyclopedia of Bioprocess Technology: Fermentation, Biocatalysis, and Bioseparation, vol. 5. Wiley, New York, pp. 2446–2462.

Couch, T.L., 2000. Industrial fermentation and formulation of entomopathogenic bacteria. In: Charles, J.F., Delecluse, A., Nielsen-Leroux, C. (Eds.), Entomopathogenic Bacteria; from Laboratory to Field Application. Kluwer, The Netherlands, pp. 297–316.

Couch, T.L., Ignoffo, C.M., 1981. Formulation of insect pathogens. In: Burges, H.D. (Ed.), Microbial Control of Pests and Plant Diseases 1970–80. Academic Press, London, pp. 621–634.

Couch, T.L., Jurat-Fuentes, J.L., 2014. Commercial production of entomopathogenic bacteria. In: Morales-Ramos, J.A., Guadalupe Rojas, M., Shapiro-Ilan, D. (Eds.), Mass Production of Beneficial Organisms. Academic Press, San Diego, pp. 415–435.

Dahod, S.K., 1999. Raw material selection and medium development for industrial fermentation processes. In: Demain, A.L., Davies, J.E. (Eds.), Manual of Industrial Microbiology and Biotechnology. American Society of Microbiology. ASM Press, Washington, DC, pp. 213–220.

Demain, A.L., Davies, J.E. (Eds.), 1999. Manual of Industrial Microbiology and Biotechnology. American Society of Microbiology. ASM Press, Washington, DC.

Dulmage, T., Yousten, A.A., Singer, S., Lacey, L.A., 1990. Guidelines for Production of *Bacillus thuringiensis* H-14 and *Bacillus sphaericus*. UNDP/World Bank/WHO. 58 pp.

El-Bendary, M., 2006. *Bacillus thuringiensis* and *Bacillus sphaericus* biopesticides production. J. Basic Microbiol. 46, 158–170.

Federici, B.A., 2007. Bacteria as biological control agents for insects: economics, engineering and environmental safety. In: Vurro, M., Gressel, J. (Eds.), Novel Biotechnologies for Biocontrol Agents Enhancement and Management. Springer, Dortrecht, pp. 25–51.

Fleming, W.E., 1968. Biological Control of the Japanese Beetle. Agricultural Research Service, Tech. Bull. No. 1383. USDA, Washington. 78 pp.

Fillinger, U., Lindsay, S.W., 2006. Suppression of exposure to malaria vectors by an order of magnitude using microbial larvicides in rural Kenya. Trop. Med. Int. Health 11, 1629–1642.

Fisher, T.W., Garczynski, S.F., 2012. Isolation, culture, preservation, and identification of entomopathogenic bacteria of the Bacilli. In: Lacey, L.A. (Ed.), Manual of Techniques in Invertebrate Pathology, second ed., Academic Press, San Diego, pp. 75–99.

Foda, M.S., El-Beih, F.M., Moharam, M.E., El-Gamal, N.N.A., 2015. High efficiency production of mosquitocidal toxin by a novel *Bacillus sphaericus* isolate from Egyptian soils on local agroindustrial byproducts. J. Am. Sci. 6, 761–769.

Gašić, S., Tanović, B., 2013. Biopesticide formulations, possibility of application and future trends. Pestic. Phytomed. Belgr. 28, 97–102.

Grzywacz, D., Moore, D., Rabindra, R.J., 2014. Mass production of entomopathogens in less industrialised countries. In: Morales-Ramos, J.A., Guadalupe Rojas, M., Shapiro-Ilan, D. (Eds.), Mass Production of Beneficial Organisms. Academic Press, San Diego, pp. 518–561.

Hilton, M.D., 1999. Small scale liquid fermentations. In: Demain, J.E., Davies, A.L. (Eds.), Manual of Industrial Microbiology and Biotechnology. American Society for Microbiology, Washington, pp. 49–60.

Hosking, G., Clearwater, J., Handiside, J., Kay, M., Ray, J., Simmons, N., 2003. Tussock moth eradication – a success story from New Zealand. Int. J. Pest Manag. 49, 17–24.

Hurst, M.R., Becher, S.A., Young, S.D., Nelson, T.L., Glare, T.R., 2011. *Yersinia entomophaga* sp. nov., isolated from the New Zealand grass grub *Costelytra zealandica*. Int. J. Syst. Evol. Microbiol. 61, 844–849.

Hynes, R.K., Boyetchko, S.M., 2006. Research initiatives in the art and science of biopesticide formulations. Soil Biol. Biochem. 38, 845–849.

Jackson, T.A., 2007. A novel bacterium for control of grass grub. In: Vincent, C., Goettel, M.S., Lazarovits, G. (Eds.), Biological Control: A Global Perspective. CABI Publishing, Wallingford, UK, pp. 160–168.

Jackson, T.A., Boucias, D.G., Thaler, J.O., 2001. Pathobiology of amber disease, caused by *Serratia* spp., in the New Zealand grass grub, *Costelytra zealandica*. J. Invertebr. Pathol. 78, 232–243.

Jackson, T.A., Pearson, J.F., O'Callaghan, M., Mahanty, H.K., Willocks, M., 1992. Pathogen to product – development of *Serratia entomophila* (Enterobacteriacae) as a commercial biological control agent for the New Zealand grass grub (*Costelytra zealandica*). In: Jackson, T.A., Glare, T.R. (Eds.), Use of Pathogens in Scarab Pest Management. Intercept, Andover, UK, pp. 191–198.

Jallouli, W., Sellami, S., Sellami, M., Tounsi, S., 2014. Impact of liquid formulation based on ultrafiltration-recovered bioactive components on toxicity of *Bacillus thuringiensis* subsp. *kurstaki* strain BLB1 against *Ephestia kuehniella*. Process Biochem. 49, 2010–2015.

Johnson, V.W., Pearson, J.F., Jackson, T.A., 2001. Formulation of *Serratia entomophila* for biological control of grass grub. N. Z. Plant Prot. 54, 125–127.

Jones, K., Burges, D., 1998. Technology of formulation and application. In: Burges, H.D. (Ed.), Formulation of Microbial Biopesticides: Beneficial Microorganisms, Nematodes and Seed Treatments. Kluwer Academic Publishers, Dortrecht, pp. 7–30.

Jurat-Fuentes, J.L., Jackson, T.A., 2012. Bacterial pathogens. In: Vega, F.E., Kaya, H.K. (Eds.), Insect Pathology 2nd Edition. Academic Press, Elsevier, London, pp. 265–349.

Kabaluk, T., Gazdik, K., 2005. Directory of Microbial Pesticides for Agricultural Crops in OECD Countries. Agriculture and Agri-Food Canada. 99 pp.

Klein, M.G., 1992. Use of *Bacillus popilliae* in Japanese beetle control. In: Jackson, T.A., Glare, T.R. (Eds.), Use of Pathogens in Scarab Pest Management. Intercept, Andover, UK, pp. 179–189.

Knowles, A., 2008. Recent developments of safer formulations of agrochemicals. Environmentalist 28, 35–44.

Koppenhöfer, A.M., Jackson, T.A., Klein, M.G., 2012. Bacteria for use against soil-inhabiting insects. In: Lacey, L.A. (Ed.), Manual of Techniques in Invertebrate Pathology, 2nd Edition. Academic Press, Amsterdam, pp. 129–149.

Lacey, L.A., 1984. Production and formulation of *Bacillus sphaericus*. Mosq. News 44, 153–159.

Lacey, L.A., 1990. Persistence and formulation of *Bacillus sphaericus*. In: de Barjac, H., Sutherland, D. (Eds.), Bacterial Control of Mosquitoes and Blackflies: Biochemistry, Genetics and Applications of *Bacillus thuringiensis israelensis* and *Bacillus sphaericus*. Rutgers Univ Press, New Brunswick, NJ, pp. 284–294.

Lacey, L.A., Urbina, M.J., Heitzman, C., 1984. Sustained release formulations of *Bacillus thuringiensis* (H-14) for control of container-breeding *Culex quinquefasciatus*. Mosq. News 44, 26–32.

Lacey, L.A., Ross, D.H., Lacey, C.M., Inman, A., Dulmage, H.T., 1988. Experimental formulations of *Bacillus sphaericus* for the control of anopheles and culicine larvae. J. Indust. Microb. 2, 39–47.

Lacey, L.A., Grzywacz, D., Shapiro-Ilan, D., Frutos, R., Brownbridge, M., Goettel, M.S., 2015. Insect pathogens as biocontrol agents: back to the future. J. Invertebr. Pathol. 132, 1–41.

Lachhab, K., Tyagi, R.D., Valéro, J.R., 2001. Production of *Bacillus thuringiensis* biopesticides using wastewater sludge as a raw material: effect of inoculum and sludge solids concentration. Process Biochem. 37, 197–208.

Lisansky, S.G., Quinlan, R., Tassoni, G., 1993. The *Bacillus thuringiensis* Production Handbook: Laboratory Methods, Manufacturing, Quality Control, Registration. CPL Press, Newberry, UK. pp. 124.

Liu, Y.T., Sui, M.J., Ji, D.D., Wu, I.H., Chou, C.C., Chen, C.C., 1993. Protection from ultraviolet irradiation by melanin of mosquito activity of *Bacillus thuringiensis* var *israelensis*. J. Invertebr. Path 62, 131–136.

Macauley-Patrick, S., Finn, B., 2008. Modes of fermenter operation. In: McNeil, B., Harvey, L.M. (Eds.), Practical Fermentation Technology. John Wiley & Sons Ltd. 388 pp.

Martin, P.A.W., Gundersen-Rindal, D., Blackburn, M.B., Buyer, J., 2007a. *Chromobacterium subtsugae* sp. nov. a betaproteobacterium toxic to Colorado potato beetle and other insect pests. Int. J. Syst. Evol. Microbiol. 57, 993–999.

Martin, P.A.W., Shropshire, A.D.S., Gundersen-Rindal, D., Blackburn, M.B., 2007b. *Chromobacterium substugae* sp. nov. For Control of Insect Pests. U.S. Patent 7244607 B2.

McNeil, B., Harvey, L.M., 2008. Practical Fermentation Technology. John Wiley & Sons Ltd.

Merchuk, J.C., Garcia Camacho, F., Flickinger, M.C., 2009. Bioreactors: Airlift Reactors. Encyclopedia of Industrial Biotechnology. John Wiley & Sons Inc., pp. 851–912.

Mitchell, D.A., Stuart, D.M., Tanner, R.D., 2002. Solid State Fermentation, Microbial Growth Kinetics. Encyclopedia of Bioprocess Technology. John Wiley & Sons Inc., pp. 2407–2429.

Monnerat, R., Da Silva, S.F., Dias, D.S., Martins, É.S., Praça, L.B., Jones, G.W., Soares, C.M., de Souza Dias, J.M.C., Berry, C., 2004. Screening of Brazilian *Bacillus sphaericus* strains for high toxicity against *Culex quinquefasciatus* and *Aedes aegypti*. J. Appl. Ent. 128, 469–473.

Monnerat, R.G., Batista, A.C., de Medeiros, P.T., Martins, É.S., Melatti, V.M., Praça, L.B., Dumas, V.F., Morinaga, C., Demo, C., Gomes, A.C.M., Falcão, R., Siqueira, C.B., Silva-Werneck, J.O., Berry, C., 2007. Screening of Brazilian *Bacillus thuringiensis* isolates active against *Spodoptera frugiperda*, *Plutella xylostella* and *Anticarsia gemmatalis*. Biol. Control 41, 291–295.

Navon, A., 2000. Bioassays of *Bacillus thuringiensis*. 1A. Bioassay of *Bacillus thuringiensis* products used against agricultural pests. In: Navon, A., Ascher, K.R.S. (Eds.), Bioassays of Entomopathogenic Microbes and Nematodes. CAB Publishing, Wallingford, pp. 1–24.

Navon, A., Ascher, K.R.S. (Eds.), 2000. Bioassays of Entomopathogenic Microbes and Nematodes. CAB Publishing, Wallingford. 324 pp.

Nosanchuk, J.D., Casadevall, A., 2006. Impact of melanin on microbial virulence and clinical resistance to antimicrobial compounds. Antimicrob. Agents Chemother. 50, 3519–3528.

O'Callaghan, M., Jackson, T.A., Mahanty, H.K., 1992a. Selection, development and testing of phage resistant strains of *Serratia entomophila* for grass grub control. Biocontrol Sci. Technol. 2, 297–305.

O'Callaghan, M., Gerard, E.M., 2005. Establishment of *Serratia entomophila* in soil from a granular formulation. N. Z. Plant Prot. 58, 122–125.

O'Callaghan, M., Pearson, J.F., Jackson, T.A., Mahanty, H.K., 1992b. Fermentation failure overcome with the selection of phage resistant bacterial strains. In: Proc. 10th Australian Biotechnol. Conf., pp. 251–252.

Osman, Y.A., Mohamedin, A.H., El-Kafrawy, R.E., 2013. Molecular characterisation of *Bacillus thuringiensis* mutant resistant to bacteriophage. Egypt. J. Biol. Pest Control 23, 57–63.

Pearson, J.F., Jackson, T.A., 1995. Quality control management of the grass grub microbial control product, Invade®. Proc. Agron. Soc. N.Z. 25, 51–53.

Pearson, J.F., Young, S.D., Jackson, T.A., 1993. Longevity of high density *Serratia entomophila* cultures under refrigerated and ambient conditions. In: Proc. 46th N.Z. Plant Prot. Soc. Conf., pp. 237–238.

Pearson, J.F., Young, S.D., Surrey, M.R., Jackson, T.A., 1996. Choice of bacterial strain affects production parameters and characteristics of a commercial biocontrol product. In: Jackson, T.A., Glare, T.R. (Eds.), Proc. 3rd Int. Workshop on Microbial Control of Soil Dwelling Pests, pp. 165–170.

Perani, M., Bishop, A.H., 2000. Effects of media composition of delta-endotoxin production and morphology of *Bacillus thuringiensis* in wild types and spontaneously mutated strains. Microbios 101, 47–66.

Poopathi, S., 2010. Novel fermentation media for production of mosquito pathogenic bacilli in mosquito control. In: Mendez-Vilas, A. (Ed.), Current Research Topics in Applied Microbiology and Microbial Biotechnology, pp. 349–358.

Poopathi, S., Abidha, S., 2009. A medium for the production of biopesticides (*Bacillus sphaericus* and *Bacillus thuringiensis* subsp. *israelensis*) in mosquito control. J. Econ. Entomol. 102, 1423–1430.

Poopathi, S., Archana, B., 2012. A novel cost-effective medium for the production of *Bacillus thuringiensis* subsp. *israelensis* for mosquito control. Trop. Biomed. 29, 81–91.

Prabakaran, G., Balaraman, K., Hoti, S.L., Manonmani, A.M., 2007. A cost-effective medium for the large-scale production of *Bacillus sphaericus* H5a5b (VCRC B42) for mosquito control. Biol. Control 41, 379–383.

Ratnakar, A., Namnath, J., Marrone, P., 2013. *Chromobacterium* Formulations, Compositions, Metabolites and Their Uses. Patent WO213062977 A1.

Ravensberg, W.J., 2011. A roadmap to the successful development and commercialization of microbial pest control products for control of arthropods. Prog. Biol. Control 10 Springer, Dordrecht, 383 pp.

Sansinenea, E., Ortiz, A., 2015. Melanin: a photoprotection for *Bacillus thuringiensis* based biopesticides. Biotech. Lett. 37, 483–490.

Sansinenea, E., Salazar, F., Ramirez, M., Ortiz, A., 2015. An ultra-violet tolerant wild-type strain of melanin-producing *Bacillus thuringiensis*. Jund. J. Microbiol. 8 (7), e20910. http://dx.doi.org/10.5812/jjm.20910v2.

Sasaki, K., Jiavinyaboonya, S., Rogers, P.L., 1998. Enhancement of sporulation and crystal toxin production by cornsteep liquor feeding during intermittent fed-batch culture of *Bacillus sphaericus* 2362. Biotech. Lett. 20, 165–168.

Schisler, D.A., Slininger, P.J., Behle, R.W., Jackson, M.A., 2004. Formulation of *Bacillus* spp. for biological control of plant diseases. Phytopath 94, 1267–1271.

Stahley, D.P., Klein, M.G., 1992. Problems with in vitro production of spores of *Bacillus popilliae* for use in biological control of the Japanese beetle. J. Invertebr. Pathol. 60, 283–291.

Surrey, M.R., Hoefakker, P.M.C., Reader, S.L., Kennedy, M.J., Davies, R.J., 1996. The culture and processing of biological control agents by Industrial Research Limited. In: Jackson, T.A., Glare, T.R. (Eds.), Proc. 3rd Int. Workshop on Microbial Control of Soil Dwelling Pests. AgResearch, Lincoln, pp. 159–165.

Swaminathan, J., Jackson, T.A., 2007. Agent Stabilisation and Delivery Process and Product. Patent No. NZ 560574, PCT/NZ2008/000299, WO2009061221A2.

Temperano, F.J., Albareda, M., Camacho, M., Daza, A., Santamaria, C., Rodriguez-Navarro, D.N., 2002. Survival of several *Rhizobium/Bradyrhizobium* strains on different inoculant formulations and inoculated seeds. Int. Microbiol. 5, 81–86.

Townsend, R.J., Ferguson, C.M., Proffitt, J.R., Slay, M.W.A., Swaminathan, J., Day, S., Gerard, E., O'Callaghan, M., Johnson, V.W., Jackson, T.A., 2004. Establishment of *Serratia entomophila* after application of a new formulation for grass grub control. N.Z. Plant Prot. 57, 310–313.

Visnovsky, G.A., Smalley, D.J., O'Callaghan, M., Jackson, T.A., 2008. Influence of culture medium composition, dissolved oxygen concentration and harvesting time on the production of *Serratia entomophila*, a microbial control agent of the New Zealand grass grub. Biocontrol Sci. Tech. 18, 87–100.

Vodovar, N., Vallenet, D., Cruveiller, S., Rouy, Z., Barbe, V., Acosta, C., Cattolico, L., Jubin, C., Lajus, A., Segurens, B., Vacherie, B., Wincker, P., Weissenbach, J., Lemaitre, B., Medigue, C., Boccard, F., 2006. Complete genome sequence of the entomopathogenic and metabolically versatile soil bacterium *Pseudomonas entomophila*. Nat. Biotechnol. 24, 673–679.

Whalon, M.E., Wingerd, B.A., 2003. Bt: mode of action and use. Arch. Insect Biochem. Physiol. 54, 200–211.

Wilson, C.J., Jackson, T.A., Mahanty, H.K., 1993. Preliminary characterisation of bacteriophages of *Serratia entomophila*. J. Appl. Bact. 74, 484–489.

Wright, D.A., Swaminathan, J., Blaser, M., Jackson, T.A., 2005. Carrot seed coating with bacteria for seedling protection from grass grub damage. N.Z. Plant Prot. 58, 229–233.

Wu, D., Yuan, Y., Liu, P., Wu, Y., Gao, M., 2014. Cellular responses in *Bacillus thuringiensis* CS33 during bacteriophage BtCS33 infection. J. Proteomics 101, 192–204.

Chapter 9

Mass Production of Fungal Entomopathogens

S.T. Jaronski[1], G.M. Mascarin[2]

[1]United States Department of Agriculture, Sidney, MT, United States; [2]Embrapa Rice and Beans, Santo Antônio de Goiás, Brazil

9.1 INTRODUCTION

Mycoinsecticides have seen a greatly increased popularity during the past two decades. Faria and Wraight (2007) recorded 110 commercial products based on entomopathogenic fungi (EPF), of which 40% were based on *Beauveria bassiana* and 39% on *Metarhizium anisopliae* sensu lato. The remainder of products incorporated *Beauveria brongniartii*, *Isaria fumosorosea*, *Isaria farinosus*, *Lecanicillium longisporum*, and *Lecanicillium muscarium*. Since that publication, several products have disappeared from the international marketplace, but a considerable number have been introduced, especially on the Indian subcontinent, in China and Latin America (Kabaluk et al., 2010; Lacey et al., 2015). For example, in Ecuador alone, there are more than a dozen new products with *B. bassiana*, *M. anisopliae*, *I. fumosorosea*, or *Lecanicillium lecanii* (Jaronski, unpublished observations). In addition, several multinational agricultural chemical companies, such as Bayer, BASF, Monsanto, DuPont, and Arysta, have moved into the microbial biopesticide market and are developing EPF for their portfolios (Ravensberg, 2015).

In beginning the development of a mycoinsecticide, one has a number of choices about the form of mass production. Specific target markets, such as foliar insect pests vs. soil pests, size of those markets, technical capabilities, amount of capital investment, and human labor rates, are all components to be considered. Ravensberg (2011) discussed these at length. A large number of infectious propagules, whether aerial conidia or blastospores, are often necessary when these fungi are typically used as an inundative treatment, especially with foliar application. Quantities of spores (in the generic sense) on the order of 2×10^{12}–5×10^{13}/ha are required for typical, acute, insecticidal mode of action in outdoor field crops. Some of the reasons for such large quantities were discussed by Jaronski (2010). Multiplied by a commercially attractive area, such as 10,000–100,000 ha, such application rates mean that efficient and cost-effective mass production is paramount for commercial success.

Jaronski (2013) provided a detailed review of the "state of the art" for mass production of each of the different EPF. This chapter will describe general production and formulation strategies that focus on developing Hypocrealean fungal propagules designed for insect control with an emphasis on the newest advances. While Entomophthoralean and Oomycete fungi can be important microbial control agents, their production will not be discussed here. Instead, the reader is referred to Jaronski (2013).

9.2 SELECTING FUNGI AND FUNGAL PROPAGULE DEVELOPMENT

The development process of a mycoinsecticide can be complex and can be conceived as a jigsaw puzzle. The selection process must evaluate not just virulence as a measure of potential efficacy but also the potential of the fungal isolate to form a stable propagule that can be economically mass-produced: one that has good long-term stability ("shelf-life"), is amenable to available application technologies, has acceptable environmental and toxicology profiles, and is capable of consistent efficacy under environmental and ecological conditions typical for the target insect(s) (Jackson et al., 2010). A complication can arise in the form of multiple targets for the commercial success of one product. The considerable cost of registration in North America or the European Union (EU) for each strain of the same fungus species can necessitate use of a single strain against multiple targets in multiple crops. A single insect target may not be commercially feasible. Often, a compromise among characteristics is necessary, a compromise between fundamental efficacy and the other criteria, especially mass production.

In nature, the typical infectious propagule of the entomopathogenic Ascomycetes is the aerial conidium. Other propagules of possible exploitation are blastospores, submerged (microcycle) conidia, sporogenically competent mycelia, and microsclerotia (Fig. 9.1). Choice of the appropriate propagule includes consideration of the projected use, good virulence,

Microbial Control of Insect and Mite Pests. http://dx.doi.org/10.1016/B978-0-12-803527-6.00009-3
2017 Published by Elsevier Inc.

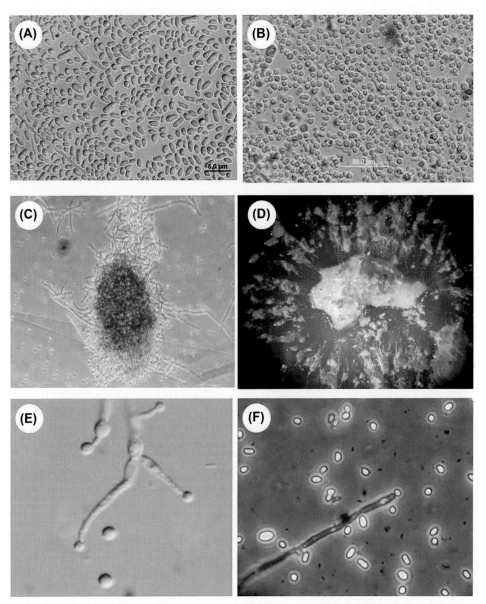

FIGURE 9.1 Entomopathogenic fungal propagules produced by liquid culture fermentation. Blastospores of *Beauveria bassiana* with an oblong shape produced in low-glucose medium (A) and spherical blastospores produced in high-glucose medium (B). Microsclerotia of *Metarhizium robertsii* (C) and granule of microsclerotia sporulated upon rehydration on water agar medium (D). Submerged conidia of *B. bassiana* (E) and *Metarhizium acridum* (F).

desiccation tolerance, thermal tolerance, speed of germination and infection, environmental stability and reproduction, and UV tolerance, as well as the inherent ability of the chosen fungus to produce that propagule (Jackson et al., 1997, 2010; Vega et al., 1999; Fernandes et al., 2015). The selection of a fungal entomopathogen that economically produces a stable propagule, which provides consistent insect control under varied field conditions, is the ultimate goal of the selection process. If the mycoinsecticide is to be applied as a spray (ie, a "contact" biopesticide), then the production method must yield high numbers of discrete, infective propagules. For this reason, most commercial enterprises use solid substrate fermentation to produce aerial conidia.

The ability of fungal spores to infect insects through the cuticle, by contact, rather than by ingestion is their greatest advantage, yet this ability also represents one of the major constraints to the commercial use of these biocontrol agents. Frequent molting by some insects, such as aphids, necessitates faster-than-normal spore germination. Blastospores are vegetative, hydrophilic fungal propagules, which are the preferred mode of growth for many EPF within the hemocoel of infected insects (Pendland and Boucias, 1997; Humber, 2008). The yeast-like growth allows the fungus better access to the nutrients within the insect and is also considered a virulence factor because blastospores possess some ability to deceive the insect's immune system (Wang and St Leger, 2006; Humber, 2008; Boomsma et al., 2014). Numerous entomopathogens of the genera *Isaria, Beauveria, Lecanicillium, Hirsutella, Nomuraea,* and *Metarhizium* can be induced to grow in a "yeast-like" fashion in submerged liquid culture. The use of vegetative, yeast-like cells (blastospores) that germinate and infect insects more rapidly than conidia has been proposed as suitable "contact" mycoinsecticides for spray applications in such instances (Jackson et al., 1997; Kleespies and Zimmermann, 1998; Mascarin et al., 2015a). Blastospores have a rapid germination rate (eg, >90% germination in 6 h for *I. fumosorosea* blastospores versus 16–24 h for conidia), which can make them useful as a contact mycoinsecticide under certain circumstances (Vega et al., 1999). Due to this accelerated germination process, faster germination may also reduce the exposure period to deleterious environmental stresses (UV, low humidity) after application, compared with aerial conidia. At the same time, blastospores are more intolerant of environmental stresses than aerial conidia and, so far, have reduced viability after processing, a much shorter shelf-life, especially under ambient conditions, and persistence after application. This disparity between aerial conidia and blastospores is of considerable importance in situations where secondary acquisition of spores by an insect, such as during movement across treated surfaces, rather than direct spray is of equal or greater importance in efficacy. One example is the use of *Metarhizium acridum* against locusts (Van der Valk, 2007). Formulation components may ameliorate this defect to some extent by partially protecting the spores and extending their persistence. Some species are less amenable to exploitation as blastospores than others. *M. anisopliae* s.l. and *Hirsutella thompsonii* blastospores have limited stability and are of lower virulence than conidia (Adamek, 1963; Winkelhoff and McCoy, 1984). Another propagule, the microcycle conidium, is produced in a liquid medium directly by a germinating aerial conidium without the intervention of hyphal growth (Anderson and Smith, 1971). These microcycle conidia have ultrastructural and morphological differences from aerial conidia. They lack a layer in the spore wall and have some different physical properties (Hegedus et al., 1990). The speed of microcycle conidium germination is intermediate between blastospores and aerial conidia.

Granular mycoinsecticide formulations for use in soil require the production of a persistent fungal propagule that is capable of delivering an infective inoculum to the insect host when required. Historically, such granules have consisted of aerial conidia on a solid carrier, even "spent solid substrate" (the remaining culture substrate after loose conidia have been removed). It is possible to formulate a nutritive carrier for use in soil, one that will support conidial germination on rehydration, vegetative outgrowth, and sporulation or simply sustain the viability of conidia long enough for the target insect to contact the granule and facilitate spore transfer to the insect. The recently discovered *Metarhizium* microsclerotia (Jackson and Jaronski, 2009) offer an alternative to conidia on a carrier granule. These sclerotial forms are desiccation tolerant and germinate sporogenically in soil to produce conidia in situ that were shown to infect and kill susceptible soil-dwelling insects (Jaronski and Jackson, 2008).

9.3 IMPORTANT GENERAL PRODUCTION PARAMETERS

9.3.1 Culture Maintenance

As a critical starting point, a genetically uniform culture for subsequent mass production must be identified. Single spore- or single-colony isolation of a fungus from the insect host is essential because simultaneous presence of several genotypes within a naturally infected host is possible (Jaronski, unpublished data). A typical strategy is to isolate the fungus from an insect to make a genetically uniform, primary "mother culture" from which all subsequent production is made. Reisolation of fungus can be from a target insect species, but a surrogate, such as larval *Tenebrio* or *Galleria*, is suitable. On an annual or semiannual basis, a part of the primary stock is removed from storage and used to create sufficient number of subcultures on agar medium to support all subsequent mass production for the next interval. Caution must be taken to ensure that the preserved fungus has not been subcultured several times (≥10 times consecutively) on artificial media, since successive passages through artificial media *may* result in variation in the normal pattern of morphogenesis (Butt et al., 2006) and/or virulence attenuation (Shah et al., 2007). Regular periodic subculture of a fungus is not recommended because genetic changes can occur, attenuating virulence or sporulation ability during repeated subculture (Wang et al., 2002, 2005; Ansari and Butt, 2011). Ideally, the primary culture should be no more than four or five in vitro passages from an insect host. The fungus must remain viable during storage, while the preservation method must inhibit genetic variation. Low-temperature storage (liquid nitrogen, −80°C), or under desiccation (freeze-drying, storage of dry spores with silica desiccant) is the norm. Stock cultures can be preserved as sporulated fungus on small agar pieces placed in 10% glycerol and stored at −80°C. A commercial version of this practice (Microbank, Pro-Lab Diagnostics) is used at some laboratories. For more detailed descriptions of methodologies for the preservation of fungal cultures, readers can consult the review by Humber (2012). Generally, it is advised to have more than one method of preservation of a fungus, and the culture storage should be replicated in another institution or culture collection to serve as a backup. Morphological and molecular characterization of the fungus and other details related to date of collection, site of collection, and insect host or substratum and notation of a code for each isolate are crucial information for maintaining fungal stock cultures. Photographs of the original culture as radial colonies on agar medium are useful to detect subsequent changes in the strain.

9.3.2 Process Sterility

To avoid process contamination, the fermentation medium, air, and equipment must be sterilized. It is necessary to start by eliminating all the native microorganisms present in the raw materials and equipment. If this is not done satisfactorily, the contaminants could quickly outnumber the production strain, and no production will occur or the end product will be unacceptably contaminated. Sterilization can be achieved by heat, special filters, or, in some cases, gamma-radiation.

9.3.3 Nutrients

Nutrients are key elements to support fungal growth as they provide the building blocks, energy source and co-factors for biochemical reactions. Carbon, nitrogen, oxygen, hydrogen, minerals, and vitamins are required in different concentrations depending on fungal species and strain. Selection of optimal media ingredients and concentrations can be somewhat time-consuming and laborious, especially for liquid fermentation, but it is extremely important to define the appropriate nutritional conditions for maximum propagule yield in the shortest fermentation time. Nutrients can also affect fungal morphogenesis, propagule formation, specific growth rate, and propagule quality and fitness for use in biological control (Jackson, 1997). Therefore, nutritional studies are of great relevance to develop cost-effective production media for EPF.

Dissolved oxygen is often the limiting factor in aerobic fermentation of these filamentous fungal entomopathogens. A sufficient oxygen supply is a fundamental requirement for successful cultivation of these fungi. Industrial filamentous fungal fermentations are typically operated in fed-batch mode, in which the supply rate of a growth-limiting substrate is kept constant by regular addition to the fermentation. For this reason, oxygen control represents an important operational challenge due to the varying biomass concentration. Mushroom bags and oxygen-rich air supply are commonly used to provide sufficient oxygenation for promotion of optimal growth and sporulation on solid-substrate fermentations or in stationary liquid fermentation, where the aim is devoted to aerial conidia production. Similarly, oxygen-enriched cultures along with high agitation speeds are also means of improving media oxygenation and quicker and greater biomass growth of fungi by liquid culture fermentation. However, high concentrations of dissolved oxygen in liquid cultures or oxygen in the atmosphere ($>21\% O_2$)

can be harmful to fungal growth, as cultures undergo an oxidative stress that can limit balanced growth and reduce biomass dry weight and cell viability. An oxidative growth environment is also dangerous to the normal physiological and biochemical functions of fungal fermentation. Because not all genotypes or strains are equally responsive to the same oxygen availability in the growth environment, studies on oxygen rate consumption are required to optimize oxygen levels required by a specific strain (Garza-López et al., 2012; Tlecuitl-Beristain et al., 2010).

Most often, determination of optimal media and fermentation conditions has been accomplished in a laborious stepwise fashion. More recent response surface methodology (RSM) has been used to efficiently determine the best parameters. For example, Bhanu Prakash et al. (2008) used a "2^3 full factorial central composite design" and RSM to simultaneously optimize the production variables of pH, moisture content, and yeast extract supplement concentration on sorghum, rice, and barley for *M. anisopliae* solid substrate production.

9.3.4 Strain Selection

Selection of the most appropriate strain, as well as fungus species, is also critical in developing fermentation processes. The inherent spore production of any given strain has genetic determinants that respond to specific fermentation conditions. In general, a "good" medium provides satisfactory but not necessarily optimal propagule yield for several fungi. For example, rice is a solid substrate on which most of the hypocrealean EPF (eg, *Beauveria*, *Metarhizium*, *Isaria*, and *Lecanicillium*) will grow and form aerial conidia. However, some strains within the same fungal species are less or even unresponsive to rice and, as a result, will attain poor yields. On a species level, this is typically the case for *Aschersonia*, *Hirsutella*, and *Nomuraea*, which produce very low yields of spores when cultivated on moistened rice grains (Jaronski, 2013). There can also be significant differences among strains of a species (Jaronski, 2013). Therefore, the selection of a virulent and genetic stable strain factors into the development of the fermentation medium.

9.4 SOLID-STATE AND BIPHASIC FERMENTATION

Solid substrate fermentation production of aerial conidia is the primary production method in current use, both by small-to-medium-sized enterprises and large multinational companies. The primary end product is the dry aerial conidium. The process has several names: solid-state fermentation and solid-substrate fermentation. "Biphasic fermentation" is a subsidiary term related to use of a liquid fermentation inoculum rather than a conidial suspension to initiate the solid substrate phase, but the overall process is the same. Fundamentals of solid substrate production of the Hypocreales up to 1984 are given in Bartlett and Jaronski (1988). Jaronski (2013) later reviewed the updated "state of the art," while protocol details for small-scale solid-substrate production are presented in Jaronski and Jackson (2012). The basic concept is fungal colonization of a sterile, solid, nutritive substrate, typically a cereal grain, with subsequent vegetative growth and conidiation. Typical conidial yields can be on the order of 4×10^{12}–10^{13}/kg for both *B. bassiana* and *M. anisopliae* s.l. (Bradley et al., 1992, 2002; Lopez-Perez et al., 2015). Having chosen solid substrate as the production mode, one is then faced with additional choices: substrate, fermentation containers, aeration system, and a number of other factors, such as initial substrate moisture level, source of inoculum, degree and manner of aeration, and addition of supplementary nutrition. Chen (2013) discusses many aspects of solid-substrate fermentation theory and practice in general. One inherent factor is the basic genetic propensity for excellent sporulation by the fungus strain under consideration. Fermentation parameters can be adjusted for modest gains with a given strain, but major improvements with a poorly sporulating (mycelial) strain are not usually possible (Jaronski, unpublished data). Prediction of conidiation ability on a cereal substrate rarely parallels that on an agar medium (Jaronski, unpublished data). Small-scale (100–500-g substrate) evaluations of candidate strains are usually a necessary component of strain selection.

Cereal grains, such as rice, barley, rye, wheat, sorghum, or corn are suitable substrates. Surface volume ratio, particle size as it affects head space between particles, and substrate surface preparation (abrading, flaking) are other aspects that need to be considered for optimal spore production. A porous inert substrate (eg, Floor-Dry, Solid-A-Sorb, or Celatom MP diatomaceous earth granules, EP Minerals LLC, Reno NV, USA) saturated with an appropriate liquid medium can be a substitute for cereal grains. Advantages of this approach include tailoring the medium for optimal spore production by a fungus strain and the ability to recycle the inert substrate after removal of conidia and reprocessing.

Fermentation containers can range from sterilizable plastic bags to large aerated chambers used in mushroom production. Most small-scale production involves autoclavable plastic bags that either contain integral venting (eg, Unicorn bags, Unicorn Imp. & Mfg. Corp., Plano TX, USA; SacO2 bags, Microsac, Nevele, Belgium) or have their open ends plugged with cotton or foam. Hypocrealean vegetative growth and conidiation demand excellent gas exchange, thus bag venting; insufficient aeration leads to greatly reduced conidiation in favor of mycelial growth. For a larger, mechanized scale, the commercial mushroom industry provides a variety of chambers and environmental control systems that can be adapted to hypocrealean spore production. One American biopesticide manufacturer boasts of a solid-substrate batch size of 10,000 kg and a routine yield of 10^{13} conidia/kg substrate (Jaronski, unpublished observation). The patent literature contains several intermediate production systems, such as Prophyta Biologischer Pflanzenschutz GMBH (1999), Suryanarayan and Mazumdar (2003), and Kulkarni and Gorwalla (2010) and can be a source of ideas for mechanized production methods.

Regardless of the substrate and container, sterilization by some means is necessary. Steam sterilization at appropriate temperature, pressure, and time is typical. Tyndallization is practiced for very large batches of substrate. Gamma-ray radiation sterilization is possible but not widely practiced. Typical production cycle is 7–14 days, after which the culture must be dried, usually within 1–4 days, and spores removed via mechanical means. Drying conidia to an end point of $<0.3\,a_w$ ($<7\%$ gravimetric moisture) is critical for good shelf-life. Jaronski and Jackson (2012) and Jaronski (2013) should be consulted for more details.

9.5 LIQUID FERMENTATION

Efficient submerged liquid fermentation for production of entomopathogenic Hypocreales has been a long-sought goal. The appeal of submerged liquid fermentation is based on its amenability to massive scaleup, using existing commercial liquid fermentation equipment, closer control of environmental variables, and shorter process times (ie, hours rather than days as for solid-substrate fermentation). At the same time, commercial-scale production requires considerable investment in fermentation equipment, while solid-substrate fermentation can be accomplished with simple, though labor-intensive, methods. Liquid fermentation can be classified into two categories: submerged liquid fermentation, where the fungus is submerged in a constantly agitated and aerated liquid medium to form blastospores, microcycle conidia, or microsclerotia, and stationary liquid fermentation, in which sporulation takes place on a still, liquid surface, producing mycelium and aerial conidia. The end-products of submerged fermentation can be blastospores, "microcycle" conidia, stabilized mycelial products, or microsclerotia (*Metarhizium* and *Nomuraea* only) but not aerial conidia. Until recently, most attempts to produce satisfactory titers of pure blastospores or pure microcycle conidia have been disappointing from an industrial perspective. The majority of the anamorphic hypocrealean fungi will grow simultaneously as blastopores and hyphal filaments. Mixed cell types dilute the desired spore, requiring larger fermentation volumes for a given quantity of the desired spore. Harvesting and purification of such mixtures can be difficult. The physical properties of the harvested and stabilized product can be subsequently incompatible with formulation. While *I. fumosorosea* and *Lecanicilium* spp. have been routinely produced in submerged fermentation, because neither genus readily lends itself to solid-substrate fermentation, *Metarhizium* and *Beauveria* have, until recently, eluded successful large-scale production because of this situation. Recent successes, notably with *Beauveria* (Mascarin et al., 2015a,b) and *Metarhizium* (Kleespies and Zimmermann, 1992, 1998), may have increased the commercial attractiveness of submerged fermentation production of blastospores. Because of the recent developments, which are not covered in any current reviews, we will focus more closely on submerged fermentation in the following subsections.

9.5.1 Blastospores

Cost-effective methods for producing high, relatively pure concentrations of desiccation tolerant blastospores of *Beauveria* and *Isaria* with short fermentation times have been successfully achieved (Jackson et al., 1997; Mascarin et al., 2015a). Blastospore-based mycoinsecticides of *L. lecanii* (Mycotal and Vertalec) and *I. fumosorosea* (PFR97 and No-Fly) are currently produced commercially (Faria and Wraight, 2007).

In general, nutrient-rich media containing high concentrations of carbon and nitrogen sources are more likely to produce larger quantities of these vegetative propagules (blastospores, hyphal bodies, mycelium, and microsclerotia). In medium optimization schemes, emphasis is placed on the maximization of fungal biomass as a requirement for commercial development of a mycoinsecticide.

Blastospores of *B. bassiana* and *I. fumosorosea* have been obtained in liquid cultures amended with high concentrations of glucose, under high aeration rates, and an appropriate source of nitrogen to yield high concentrations of blastospores. For example, a low-cost liquid medium (<U.S. 20¢/L) composed of high carbon concentration (≥80-g glucose/L) supplemented with 25 g/L of cottonseed flour, as a substitute of casamino acids, supports rapid production of high concentrations of desiccation-tolerant blastospores that remain viable for at least 1 year under refrigerated conditions (Jackson, 2012; Mascarin et al., 2015a). Desiccation tolerance of blastospores, necessary for producing a shelf-stable, sprayable material, can be improved if an appropriate source (eg, casamino acids) and concentration of nitrogen are provided for both *B. bassiana* and *I. fumosorosea* (Cliquet and Jackson, 1999; Jackson et al., 2003; Mascarin et al., 2015a). Such blastospores have been shown to survive desiccation (air-drying), to remain stable under refrigerated storage (4°C), but not ambient temperatures, and to infect whiteflies better than aerial conidia. The influence of different sources of nitrogen on blastospore virulence and speed of germination remains elusive.

Propagule formation, quality, and fitness for biological control can be affected by the carbon-to-nitrogen ratio

(Jackson, 1997; Jackson et al., 1997). Cliquet and Jackson (2005) found that high amounts of nitrogen and carbon (80 g/L glucose and 13.2 g/L casamino acids) in the liquid medium supported higher blastospore yields with improved desiccation tolerance (75% survival) after freeze-drying. This desiccation tolerance was associated with the higher protein content found in blastospores rather than lipids and carbohydrates.

Romero-Rangel et al. (2012) identified a liquid medium composed of casamino acids to be the best for producing *Hirsutella* sp. with a yield up to 3.8×10^7 blastospores/mL after 14 days incubation at 26°C and 250 rpm. Little information with regards to liquid culture production of blastospores of *Lecanicillium* spp. has been documented; most is proprietary. A liquid medium containing glucose and yeast extract was reported by Latgé et al. (1986) that yielded >5×10^9 blastospores/mL. For production of *Metarhizium* spp. blastospores, Kleespies and Zimmermann (1992) reported blastospore yields of up to 10^8/mL. In another study, Kleespies and Zimmermann (1998) noticed that addition of 5% lecithin increased blastospore production to 1.9×10^8/mL. In contrast, Vega et al. (2003) observed that, irrespective of medium composition, *M. anisopliae* failed to produce high quantities of blastospores; the highest yield in their studies was 3.9×10^7/mL. Similarly, Ypsilos and Magan (2004, 2005) reported that, depending on the media, *M. anisopliae* produced blastospores in the range 1.5×10^6–5×10^7/mL. A cost-effective liquid production method for blastospores of *I. fumosorosea* was accomplished in 1997 (Jackson et al., 1997; Jackson, 1999).

Early published methods for producing blastospores of *Beauveria* using liquid culture fermentation processes required long fermentation times (6–8 days) to maximize blastospore yields. Samsinikova (1966) reported a yield of about 6.5×10^8 hyphal bodies/mL in *B. bassiana* cultures produced in 25 g/L glucose and 25 g/L starch supplemented with 20 g/L corn steep liquor. The resulting blastospores did not survive desiccation or storage (Samsinakova, 1966). Rombach (1989) reported maximum yields of *B. bassiana* hyphal bodies (eg, blastospores) of 7.5×10^8/mL after 3 days of incubation at 150 rpm in a liquid medium consisting of 25 g/L sucrose and 25 g/L yeast extract. No desiccation tolerance or storage stability studies were carried out. Humphreys et al. (1989) grew *B. bassiana* cultures in liquid media containing 56 mmol/L glucose and 0.6 g/L ammonium sulfate or 10 g/L glucose and 20 g/L yeast extract incubated at 200 rpm and 25°C to produce <10^8 *B. bassiana* blastospores/mL. Lane et al. (1991) grew *B. bassiana* in a liquid medium enriched with different concentrations of glucose and ammonium sulfate that resulted in poor yield of blastospores (≤5×10^6/mL) after 80 h of incubation at 25°C and 200 rpm. Vidal et al. (1998) reported maximum yield of approximately 5.5×10^8 *B. bassiana* blastospores/mL using a low aeration rate (90 rpm) after 3 days of fermentation period in a liquid medium developed by Jackson (1999). Desiccation tolerance or storage stability studies were not evaluated. Vega et al. (2003) observed up to 1.0×10^9 B. bassiana blastospores/mL but only after 7 days of incubation at 28°C and 300 rpm, using a liquid medium enriched with 25 g/L casamino acids and 80 g/L glucose. Pham et al. (2009) developed a liquid medium containing 30 g/L corn meal, 20 g/L corn steep powder, and 2 g/L rice bran that yielded a maximum spore concentration of 8.54×10^8/mL after 8 days, at 25°C and 200 rpm. Chong-Rodriguez et al. (2011) reported 6.4×10^9 submerged spores/mL within a 6-day fermentation period in a liquid medium containing sucrose (50 g/L), glucose (5 g/L), and corn steep liquor (20 g/L) (26°C and 300 rpm), but the spores had poor survival after 1-month storage at 4°C. (Spores produced in this study were predominantly submerged conidia, instead of blastospores.) More recently, Lohse et al. (2014) documented a fermentation process for B. bassiana using inexpensive liquid media yielding 1.23×10^9 blastospores/mL by 72 h at 25°C and 200–600 rpm. In contrast to these reports, higher concentrations of Beauveria, Isaria, and Lecanicillium blastospores [$(1–5) \times 10^9$ blastospores/mL] were achieved in a shorter incubation time (2–3 days) using a liquid culture medium containing high glucose concentrations (>100 g/L) and appropriate amounts of nitrogen sources (25–30 g/L) and very high aeration rates (Jackson et al., 2003; Mascarin et al., 2015a,b). In addition, the blastospores produced by this optimized liquid medium survived air-drying (<20% viability loss) and remained stable over a year at 4°C and over 6 months at 28°C in vacuum packaging (Mascarin et al., unpublished results). It is possible that the earlier studies on blastospore liquid culture production could have been limited by an insufficient oxygen supply in the broth, unsuitable carbon and nitrogen sources, inappropriate concentrations of these macronutrients (C:N ratio), low inoculum density, lack of precultures to shorten or avoid the lag-phase of growth, pH, or water availability in the liquid medium.

This fermentation process requires the coupling of high aeration rates, high concentrations of glucose or other compatible solutes to reduce the water activity of the medium, as well as the use of appropriate sources of nitrogen for the rapid production of large numbers of blastospores (Mascarin et al., 2015a,b). Higher aeration rates afford better oxygen transfer to the liquid cultures of EPF, and this is translated into higher concentrations of blastospores in shorter fermentations times (2–3 days) with good desiccation tolerance and shelf-life (Issaly et al., 2005; Jackson, 2012; Mascarin et al., 2015a,b). Oxygen is one of the most important nutrients in aerobic fermentation of EPF. Since oxygen comprises in average 20% of total biomass in microorganisms, it is the second most important element in biomass composition (Zabriskie et al., 2008). The excess or limited oxygen supply in fungal liquid cultures can generate oxidative stress and hypoxia, respectively, causing unbalanced growth and subsequent impact on propagule yield and quality. It has been proposed that inhibition of alternative oxidative enzymes in fungal mitochondria, such as NADH:ubiquinone oxidoreductases (NADH dehydrogenases), has led to significantly enhanced specific growth rate, substrate uptake, carbon dioxide evolution, higher protein content, and more efficient use of substrates by Aspergillus niger (Voulgaris et al., 2012). By blocking the electron flow via these enzymes, flux through the main respiratory pathway rises, leading to enhanced ATP generation. Understanding the mechanisms associated with alternative respiratory pathways in aerobic liquid fermentations of EPF may shed light into new means of improving biomass accumulation along with substrate utilization and the role of antioxidative enzymes in coping with oxidative stress growth conditions.

Several studies have demonstrated that there are some substances, produced under very low titers by some other dimorphic fungi in liquid fermentation that apparently act as a morphogenetic autoregulatory substances (a.k.a. quorum-sensing molecule) regulating the switch of cell morphologies between mycelium and yeast-like forms (Sato et al., 2004). This phenomenon is controlled by the inoculum density and temperature used in the liquid culture of some filamentous fungi (Ramirez-Martinez, 1971; Hornby et al., 2004). There are no reports of these factors regulating dimorphism in EPF so far.

9.5.2 Submerged (Microcycle) Conidia

Submerged conidia of B. bassiana have also been shown to arise directly from blastospores in a process known as microcycle conidiation (Thomas et al., 1987). Formation of submerged conidia in liquid cultures is associated with the carbon source and its concentration (Feng et al., 1994; Jackson, 1997). Nutrient-limited media, with low carbon concentration (8 g/L) and high C:N ratio with supports rapid differentiation of hyphae to form conidia, if the fungus is capable of conidiation in liquid culture. Both B. bassiana and M. acridum, but not M. anisopliae s.l., are capable of producing submerged or microcycle conidia under certain liquid fermentation conditions (Thomas et al., 1987; Jenkins and Prior, 1993; Kassa et al., 2004). These submerged conidia are not hydrophobic, unlike aerial conidia, and thus present different challenges in formulation and use. The microcycle conidia of B. bassiana are produced after 96 h of fermentation only in the presence of inorganic nitrogen, as nitrate, and with very high levels of carbohydrate. Furthermore, B. bassiana grown in liquid medium amended with glutamic acid as the nitrogen source (3% total nitrogen in the medium) induced submerged conidia form with fewer blastospores in 2–3-day-old cultures (Mascarin et al., unpublished data). Submerged conidia are morphologically different from aerial conidia, lacking one layer to their cell walls (Hegedus et al., 1990). Germination speed for submerged conidia is intermediate between aerial conidia and blastospores. Submerged conidia of *M. acridum* were produced on structures very similar to aerial phialides and were morphologically indistinguishable from aerial conidia, although they possessed different physical properties (Leland et al., 2005). Nitrogen, in the form of Brewer's yeast, in the presence of excess sucrose was found to be essential for the production of submerged conidia by *M. acridum* cultures.

Formation of submerged conidia of *B. bassiana* was induced by glucose instead of citrate, lactose, sorbitol, starch, fructose, maltose or glycerol, with an increased conidia: blastospore ratio (Thomas et al., 1987). Hegedus et al. (1990) verified that growth liquid media containing *N*-acetyl-D-glucosamine (GlcNAc) proved to be more suitable for production of a larger quantity of submerged conidia of *B. bassiana* rather than blastospores. When *B. bassiana* was grown in medium supplemented with chitin, the fungus produced 86.3% of the spores as submerged conidia exceeding 10^6/mL after 48 h. In addition, phosphate limitation resulted in an increased percentage of submerged conidia (Hegedus et al., 1990). It was noted that *B. bassiana* was capable of producing up to 10^9 conidia/mL when grown in submerged liquid cultures (Kondryatiev et al., 1971). More recently, Lohse et al. (2014) optimized a cost-effective liquid medium to produce total spores (submerged conidia plus blastospores) of an endophytic entomopathogenic *B. bassiana* isolate ATP-02 using shake flask cultures. Most of these spores were submerged conidia with the highest concentration reaching 1.33×10^9 total spores/mL after 10 days of fermentation in liquid medium containing TKI broth (Thomas et al., 1987) with 5% sugar beet molasses, which consists of 50% sucrose as a carbon source, with conditions set to 25°C and 200 rpm. The scale-up to a 2-L stirred tank reactor was carried out at 25°C, 200–600 rpm and one volume per volume per minute (vvm) airflow at pH 5.5 and, as a result, this fermentation process yielded a concentration of 1.29×10^9 total spores/mL with more than 95% of these propagules consisting of blastospores after only 72 h. Later, most of these propagules were submerged conidia. Thus, the maximum concentration obtained was 0.84×10^9 viable spores/mL at 168 h after inoculation, which corresponded to 10^8 viable submerged conidia/mL. With regard to the culture medium, the cost of 10^{12} total spores was estimated as 0.24€. Torre and Cárdenas-Cota (1996) documented an interesting process of achieving high numbers of submerged conidia through microcycle conidiation of *I. fumosorosea* grown in liquid medium. The liquid medium presented a C:N ratio of 25:1 and cultures were grown for at 37°C in the first 24 h prior to incubation for 3 days at 30°C with 12-h light regimen. This fermentation process was able to yield 10^{11} conidia/L. Production of submerged conidia of *M. acridum* have been optimized by adding polyethylene glycol (PEG 200) up to 150 g/L to the modified Jenkins's medium, containing 40 g/L waste Brewer's yeast and 50 g/L lecithin, with yields reaching 1.2×10^9 in 6 days of fermentation at 24°C and 150 rpm (Leland et al., 2005). Despite efforts in production and formulation of submerged conidia (Kassa et al., 2004; Leland et al., 2005; Lohse et al., 2014), submerged conidia have not been commercially developed. The biggest problem is contamination with blastospores and hyphal fragments (Jaronski, unpublished information).

9.5.3 Microsclerotia

Microsclerotia are small, compact hyphal aggregates that are often melanized and desiccation tolerant and produce secondary compounds that can be antibacterial, antifungal, or deter insect feeding (Cooke, 1983). This microsclerotium is considered a survival structure produced by a number of fungi to overcome adverse nutritional or environmental conditions, remaining dormant until conditions are suitable for growth. Microsclerotia germinate hyphally and sporogenically to produce infective conidia. A number of *Metarhizium* species, (but not *M. acridum*), are capable of forming microsclerotia using shake flask cultures as well as in deep-tank fermentation (Jaronski and Jackson, 2008; Jackson and Jaronski, 2009, unpublished data; Mascarin et al., 2013). *Beauveria bassiana* does not seem to have microsclerotial potential, based on an analysis of 12 strains (Jaronski, unpublished data). For control of soil-dwelling insects, incorporation of microsclerotial formulations into the soil provides a feasible and innovative strategy for delivering stable fungal propagules that produce infective conidial inoculum in situ. Cottonseed flour and soybean meal as nitrogen sources rendered high yields of microsclerotia in deep-tank fermentation, and the resultant microsclerotia presented similar stability under room storage conditions and virulence to mealworm, *Alphitobius diaperinus*, compared with microsclerotia obtained using pure casamino acids (Behle and Jackson, 2014).

Song et al. (2013) reported that *Nomuraea rileyi* is also capable of forming microsclerotia in liquid media, This ability is not surprising because the green-spored species of *Nomuraea* have now been transferred into the ascomycete genus *Metacordyceps* along with *Metarhizium* by Kepler et al. (2014) and thus are closely related to the latter fungus. A liquid medium composition of glucose (32 g/L) and ammonium citrate (2 g/L), fortified with 0.15 g/L of ferrous sulfate, provided the highest yields of microsclerotia (2.19×10^5/mL) after 6 days of fermentation at 200 rpm and 28°C (Song et al., 2015). The microsclerotia obtained from this optimized culture medium exhibited virulence and greater thermal tolerance than solid-substrate produced conidia, and maintained 86.4% viability after 1-year room temperature storage. This fungal structure could have great importance for the control of the fall armyworm, *Spodoptera frugiperda* by *N. rileyi*. The typically high humidity in this insect's microenvironment could be conducive for germination and sporulation of microsclerotial granules, resulting in considerable production of an infective inoculum in situ.

9.5.4 Aerial Conidia by Liquid Surface Fermentation

Aerial conidia can also be produced on stationary liquid medium, the process being termed surface liquid culture or stationary cultivation. The process allows the maximization of the surface–volume ratio to promote profuse and rapid growth. For production of *B. bassiana* and *Lecanicillium* spp., the system consists of 30-cm polyethylene tubes sealed on both sides and partially filled with a fungus suspension cultured in submerged nutritive broth (8 g/L peptone and 10 g/L sorbitol). The culture broth forms a 1-cm-thick layer in these tubes and sterile airflow is flushed inside this container to provide sufficient aeration for fungal growth and sporulation. Therefore, mycelial growth takes place in the liquid medium, while conidial production occurs on the surface, which resembles what happens when the fungus infects and grows out from a host insect. After 12–16 days of incubation, it was possible to harvest up to 10^9 conidia/L or /cm^2 of media surface for *B. bassiana* (Kybal and Vlcek, 1976; Samsinakova et al., 1981). Commercial production of several EPF, including a *B. bassiana*, Boverol, in the Czech Republic uses this method (Fytovita spol, 2015).

9.5.5 Affecting Propagule Attributes by Environmental Manipulation

Fungal spores have a range of environmental factors (temperature, humidity) that affects their efficacy. Screening programs aim to identify strains with suitable attributes to meet a particular situation. Formulations can be devised to compensate for the limitations of a strain to a variable level of success. Other researchers have altered fungus properties through genetic modifications. Several recent studies have identified an alternative approach in manipulation of fermentation conditions to improve environmental stress tolerance (reviewed by Jaronski, 2013). Physical, chemical, and nutritional conditions can be altered during fungal growth to manipulate endogenous reserves for production of propagules with improved stress tolerance to abiotic factors and virulence to insect pests (Jackson, 1997; Magan, 2001; Rangel et al., 2015).

Generally, it has been shown that growth of fungi under stress conditions is associated with production of conidia with increased stress tolerance but also accompanied with very low conidial yields that do not favor commercial production (Rangel et al., 2015). Low yields pose a major constraint to the commercial physiological manipulation. Reducing water availability in solid substrate media has

been explored as a strategy to increase endogenous reserves, such as sugar alcohols (ie, polyols), in aerial conidia grown at least on agar media; whether the same phenomenon exists with solid substrate fermentation remains to be determined. The fungus grown under water stress conditions, for example, is capable of accumulating more polyols, which, in turn, have been associated with higher virulence and speed of germination even under low relative humidity (Magan, 2001; Andersen et al., 2006; Rangel et al., 2008). Despite the limitation in the scaled-up production of physiologically manipulated conidia with solid-substrate fermentation, a parallel approach has been proposed with liquid fermentation, one in which the concentration of glucose or other ionic or nonionic osmolytes are increased to reduce water availability. Although osmotic pressure in the liquid medium becomes high (>1.0 MPa) as water availability is reduced, there seems to be an unexpected enhancement in "yeast-like" growth for some fungal entomopathogens, which is translated into higher yields of blastospores in relation to filamentous growth. This phenomenon has been recently observed for *B. bassiana* (Mascarin et al., 2015b) and *I. fumosorosea* (Mascarin, unpublished data). In addition, high glucose concentrations in the liquid media promoted better oxygen transfer rate or oxygenation in liquid cultures of *B. bassiana*, resulting in a less viscous broth and more blastospores. This situation suggests that oxygenation and osmotic pressure of the culture broth play an important role in blastospore yield.

The reduction in water activity of the liquid media has been associated with an increase in spore production by various EPF (Humphreys et al., 1989; Inch and Trinci, 1987; Kleespies and Zimmermann, 1998; Ypsilos and Magan, 2005). Highly aerated cultures (350 rpm with shake flask cultures) of *B. bassiana* grown in the presence of water stress imposed by high glucose concentration (140 g/L) rendered considerably smaller, spherical blastospores with improved virulence (lower LC_{50} and LT_{50}) to whitefly nymphs in comparison with the larger, oblong blastospores produced under low osmotic pressure (40 g glucose/L) (Mascarin et al., 2015b). Although not yet proven, it is reasonable to link these results with the intracellular accumulation of polyols in blastospores, similar to what has been observed for aerial conidia grown under water stress conditions, as a response to the increase of glucose concentration in the medium in order to restore the osmotic balance.

Studies showing the influence of carbon concentration and osmotic pressure on production of intracellular polyols by fungal cells in *Aspergillus* revealed that glycerol is strictly linked to an increase in the osmotic pressure of the medium, while erythritol is coupled to both high osmotic pressure and high level of carbons (Diano et al., 2009). Another theory arises from the fact that glucose is the most preferable carbon source assimilated by fungi (Ronne, 1995). The excess glucose in liquid medium is associated with carbon catabolite repression, but how this relates to yeast–mycelium morphogenesis and virulence in EPF remains elusive. More importantly, adaptation to hyperosmotic stress is a required ability for survival in all cellular organisms, especially free-living filamentous fungi. These adaptive responses are mostly governed by the high osmolarity glycerol pathway, which is conserved in diverse fungal species, yet very little is known regarding this aspect for fungal entomopathogens (Duran et al., 2010). More studies are required to shed light onto the role of carbon sources and osmotic pressure in propagule yield and fitness of blastospores. There are some data suggesting that light conditions during growth may improve heat tolerance and UV radiation (Rangel et al., 2011, 2015), but these observations could be treated with caution because they were made using agar media in Petri dishes, not regular solid-substrate, and have not been repeated by others.

9.6 QUALITY CONTROL

Consistent purity is critical for commercially acceptable formulations. Aside from regulatory prohibitions of contaminant microorganisms, the titer of the produced "technical grade active ingredient" (tgai) must be relatively consistent in quantity and viability. The number of conidia or blastospores per gram of tgai and their viability determine the physical amount that must be added to a formulation. For example, if *Beauveria* tgai has 1.2×10^{11} conidia/g with a viability of 95% (typical of a high-purity *Beauveria* spore powder), an oil formulation of 2.2×10^{13} conidia/L (eg, BotaniGard ES) requires 193 g/L of conidia; but if the tgai has only 80% viability, then 229 g tgai/L will be required, resulting in a thicker suspension. Oil-based formulations of *Metarhizium* are even more susceptible to such a problem because the conidia are larger and the tgai has a lower titer, typically $(4–5) \times 10^{10}$ conidia/g. Thus, more solids per liter are required for a given final concentration. An emulsifiable formulation of 5×10^{12} conidia/L, such as Met52 EC, would optimally require 100 g spores (at 5×10^{10} conidia/g and 100% viability), but as much as 166 g/L if titer was 4×10^{10}/g and viability 75%. The rheological properties of a formulation can be significantly changed in such situations, as well as there being a major departure from the official formula submitted to regulators. European and North American regulatory systems require a specific statement of formula with precise narrow limits for each ingredient, such as ±5% for a component representing 1–20% of the formulation (USEPA, 2010). Some of the variability can be compensated by blending production batches of different titers to a predetermined level but such a solution requires accumulation of appropriate batches.

9.6.1 Viability Testing and Spore Counts

With conidia or blastospores, standard quality control measures involve hemocytometer counts of diluted spore suspensions and determination of viability by germination on an agar medium (Inglis et al., 2012). While estimation of viable propagules per unit can be done via suspension, dilution, and plating on agar media to obtain colony forming units, this method can give underestimates of as much as 20% because of incomplete dispersion of individual propagules (Oliveira et al., 2015; Jaronski, unpublished data). Assessment of microsclerotial granules can be done by incubating a sample on water agar for several days, then washing the resulting spores off the agar and counting in a hemocytometer (Jackson and Jaronski, 2009). A simpler method is to merely determine the proportion of granules exhibiting full conidiation after incubation on water agar or moistened filter paper.

Recently, Faria et al. (2015) proposed short incubation periods during germination tests for assessing viability as an indicator of product quality. These authors observed that debilitated conidia exhibiting slow-germination (requiring >16 h to germinate) seem to be less virulent than vigorous conidia exhibiting fast germination (requiring ≤16 h to germinate). James and Jaronski (2000), however, observed that the live conidia in a preparation of 50% viability were just as infectious for whiteflies as fresh conidia. The incorporation of propagule vigor into quality control protocols may be a valuable parameter for quality assurance of conidial preparations.

9.6.2 Moisture Content

Another aspect that can affect virulence and must be taken into account in mass production relates to moisture content of conidia after drying. For extended shelf-life, conidia should be dried to less than 5% moisture or a water activity $(a_w) < 0.3$. As a rule of thumb, the drier, colder and lower oxygen availability, the longer survivorship of fungal propagules can be achieved. It has been shown that very dry conidia $(a_w < 0.1)$ can suffer from imbibitional damage due to rapid rehydration in cold water (Faria et al., 2009). These very dry conidia should be rehydrated slowly under water vapor or by using water with temperature >25°C, even though imbibitional damage might vary among fungal species and strains. Interestingly, the viability of such very dry conidia is preserved if they are suspended in oil or oil-based formulation.

9.6.3 Bioassay as a Quality Assurance Tool

The ultimate determination of experimental batch or product efficacy will be through bioassay. Due to the vast number of insect and mite species that are susceptible to infection by hypocrealean fungi, the methodology varies considerably among and within orders. Several bioassay protocols for EPF are presented by Butt and Goettel (2000), Hajek et al. (2012), and Inglis et al. (2012). A caveat about bioassay LC/LD_{50} values is their frequently limited sensitivity (large 95% confidence intervals) because the log dose–probit/logit response slope in fungal bioassays is very low, <2, often 0.5–1, especially with immersion protocols but even with careful spray methods (eg, Wraight et al., 2010). Because the standard deviation of the LC/LD_{50}, thus the 95% confidence interval, is directly proportional to the slope of the log dose–response regression divided by $n-1$ insects used, a large number of insects and 4–10 doses are needed for good precision to detect small changes in efficacy (Robertson et al., 2007). Median survival time (MST) is often used in bioassays to gauge virulence as an alternative to multiple-dose bioassays and can be computed with Kaplan–Meier survivorship analysis (Kleinbaum, 2012). Probit-logit analysis is inappropriate unless mortality is determined over time with independent groups (Robertson et al., 1995). The MST also has a limitation to sensitivity, however, unless observations are made at half-day or even quarter-day intervals (Bateman et al., 1996). Daily observations have a precision of 1 day, whereas differences among lots of fungi are often less. In addition, independent replication of bioassays, regardless of method, is essential because of the natural variation among cohorts of the same insect over time, as well as variation in preparation and delivery of doses (Robertson et al., 1995). Thus, except for a cursory indication of the relative efficacy of a production lot of fungus as a measure of quality, considerable effort will be necessary for quality assurance bioassays. A very valuable reference that should be consulted before fully engaging in such bioassays is Robertson et al. (2007).

9.7 FORMULATION OF MYCOINSECTICIDES

Practical and efficacious mycopesticide formulations have a number of requirements. Fungal propagules must be must be kept dormant, but alive, during storage before use (storage should be ideally at ambient temperatures, not refrigeration). The propagules need to be easily suspended in some sort of carrier, typically aqueous, for most spray applications. The propagules need to be kept alive on the treated substrate (plant canopy, soil surface, building surface) as long as possible, especially in the face of UV-A and UV-B radiation, and rainfall. Fungal "behavior" during germination and infection must be not be adversely affected by adjuvants.

Typical technical-grade (unformulated) fungal propagules (conidia, dried blastospores, dried submerged conidia) are not very "user friendly." Aerial conidia, especially, are very hydrophobic and thus extremely difficult to suspend in aqueous carriers. Spray dried blastospores are more hydrophilic, but they often require wetting agents for proper dispersion in an aqueous carrier. Fungal propagules must be incorporated with various wetting agents, suspension agents, dispersants, antifoamers, emulsifiers, and

spreaders. At a minimum, all three forms of spores require a nonionic wetting agent for proper suspension. The hydrophobic aerial conidia are, however, readily suspended in oil formulations, which renders them suitable to low- and ultralow-volume applications. Further, a number of studies indicate that conidia applied in oil have superior infectivity to pure aqueous application (Prior et al., 1988; Bateman et al., 1993; Inglis et al., 1996). A wide variety of formulation types can be adapted for various mycoinsecticide uses. Ultra-low volume suspensions (SU) and oil flowable concentrates (OF), oil dispersions (emulsifiable suspensions) (OD), and flowable suspensions (SC), are vegetable- or paraffinic-oil–based. Additional formulation types include wettable powders (WP), water-dispersible granules (WDG), foams, contact powders (CP), granules (G), and baits. Commercial mycoinsecticide formulations in many parts of the world are generally emulsifiable suspensions or wettable powders, although in some countries unformulated spores are sold as such, along with a small separate quantity of wetting agent (Jaronski, unpublished observations).

Although there are opportunities for provision of fresh mycoinsecticides to the user, without storage, in most commercial situations stability in storage is critical, especially at non-refrigerated temperatures. This situation also requires additives. For example, optimal storage of dry conidia can be achieved by incorporation of desiccants and oxygen scavengers in water- and air-impermeable packaging (Jin et al., 1993). Historically, formulation development has been empirical, such as Mascarin et al. (2014), with most research proprietary, but more rational approaches have been identified recently (eg, Jin et al., 1993; Luke et al., 2015). Burges (1998) provides much detail about formulation of mycoinsecticides. The recent development of carnauba and candelilla wax dusts as carriers for aerial conidia (Meikle et al., 2008; Meikle and Nansen, 2012; Exosect, 2015), enhancing delivery to and contact with target insects, opens new use possibilities for these fungi, especially with stored product insects. Carnauba wax dusts may also have potential against foliar pests, but this aspect has not been evaluated.

There is very little information concerning practical formulation of blastospores, in comparison with aerial conidia. Blastospores are hydrophilic and vegetative in nature and are considered more sensitive to desiccation and less amenable to storage conditions than aerial conidia. Kim et al. (2013) described an oil-based formulation of blastospores containing isotridecyl alcohol ethoxylated-3EO (TDE-3) and sodium alginate (SA) as adjuvants. This formulation showed enhanced efficacy against whiteflies (95.7% mortality), versus the SA-free oil formulation (72.8%) after 10 days. With the recent development of practical submerged fermentation protocols for the various Hypocreales (Jackson et al., 1997; Issaly et al., 2005; Mascarin et al., 2015a), there is considerable need for blastospore formulation research.

Use of EPF against soil insects is best accomplished with granular formulations. The advantages of granules over soil drenches are numerous and were summarized by Jaronski (2007, 2010). The simplest granular formulations can be made by coating a suitably sized inert or nutritive granular material with conidia using a water miscible binder. An advantage of using a nutritive granule, such as coarse maize meal, is that the conidia will germinate on the granule when it becomes rehydrated, the fungus will briefly grow out using the granules as a nutritive source, then sporulate, creating a focus of a large number of conidia. An insect need only briefly contact a sporulated granule to acquire a lethal dose of spores. A variant formulation consists of the spent solid substrate after most of the conidia have been removed. The product Met52 G (Novozymes Biologicals) is such a formulation, consisting of spent broken rice solid-substrate, but the size and shape distribution of the particles limit their use in field agriculture; in North America and the EU, planting granule application systems require granules of 0.5–1.5 mm in diameter. In the EU, a *Metarhizium* granular formulation was developed in which the fungus is allowed only to colonize barley grains, not sporulate, at which point the grains are dried. The dried, colonized grains are applied to the soil in pastures using an air drill planter. Moisture is drawn from the soil into the grain, reactivating the fungus, which then grows out and sporulates, creating a focus of fresh, infective conidia.

The recent discovery that *Metarhizium* spp. can produce storage-stable microsclerotia (Jackson and Jaronski, 2009) may greatly facilitate the use of this genus in soil (Jaronski and Jackson, 2008). In addition, microsclerotial granules formulated with hydromulch as a sticker agent have been used for the control of the invasive Asian longhorned beetle, *Anoplorus glabripennis*, by spraying this formulation onto the trunks of host trees (Globe et al., 2015). This approach has marked the first attempt of applying microsclerotia for control of an arboreal insect pest. Additional formulations were described by Jackson et al. (2010) and Li et al. (2010).

9.8 SUMMARY

During the past two decades, EPF have undergone a great increase in popularity. New companies are being continually established for commercial development. Public domain research efforts for implementation have undergone an explosion of effort. Mass production and formulation are critical for this process. Optimization of fermentation processes, recovery of the biomass from the fermentation broth, and formulation of the final product are all research areas important in the commercial developmental of a mycoinsecticide. One key area for research is a better understanding of how environmental factors during fermentation can improve the quality and attributes of conidia and blastospores (Section 9.5.5). With further breakthroughs

in this area, the entomopathogenic Hypocreales could see additional advances in their commercial implementation.

Mention of trade names or commercial products in this publication is solely for the purpose of providing specific information and does not imply recommendation or endorsement by the U.S. Department of Agriculture (USDA). The USDA is an Equal Opportunity provider and employer.

REFERENCES

Adamek, L., 1963. Submersed cultivation of the fungus *Metarhizium anisopliae* (Metsch.). Folia Microbiol. 10, 255–257.

Anderson, J.G., Smith, J.E., 1971. The production of conidiophores and conidia by newly germinated conidia of *Aspergillus niger*. J. Gen. Microbiol. 69, 185–197.

Andersen, M., Magan, N., Mead, A., Chandler, D., 2006. Development of a population-based threshold model of conidial germination for analysing the effects of physiological manipulation on the stress tolerance and infectivity of insect pathogenic fungi. Environ. Microbiol. 8, 1625–1634.

Ansari, M.A., Butt, T.M., 2011. Effects of successive subculturing on stability, virulence, conidial yield, germination and shelf-life of entomopathogenic fungi. J. Appl. Microbiol. 110, 1460–1469.

Bartlett, M.C., Jaronski, S.T., 1988. Mass production of entomogenous fungi for biological control of insects. In: Burge, M.N. (Ed.), Fungi in Biological Control Systems. Manchester University Press, Manchester, UK, pp. 61–85.

Bateman, R., Carey, M., Moore, D., Prior, C., 1993. The enhanced infectivity of *Metarhizium flavoviride* in oil formulations to desert locusts at low humidities. Ann. Appl. Biol. 122, 145–152.

Bateman, R., Carey, M., Batt, D., Prior, C., Abraham, Y., Moore, D., Jenkins, N., Fenlon, J., 1996. Screening for virulent isolates of entomopathogenic fungi against the desert locust, *Schistocerca gregaria* (Forskål). Biocontrol Sci. Technol. 6, 549–560.

Behle, R.W., Jackson, M.A., 2014. Effect of fermentation media on the production, efficacy and storage stability of *Metarhizium brunneum* microsclerotia formulated as a prototype granule. J. Econ. Entomol. 107, 582–590.

Bhanu Prakash, G.V.S., Padmaja, V., Siva Kiran, R.R., 2008. Statistical optimization of process variables for the large-scale production of *Metarhizium anisopliae* conidiospores in solid-state fermentation. Bioresour. Technol. 99, 1530–1537.

Boomsma, J.J., Jensen, A.B., Meyling, N.V., Eilenberg, J., 2014. Evolutionary interaction networks of insect pathogenic fungi. Annu. Rev. Entomol. 59, 467–485.

Bradley, C.A., Black, W.E., Kearns, R., Wood, P., 1992. Role of production technology in mycoinsecticide development. In: Leatham, G.E. (Ed.), Frontiers in Industrial Microbiology. Chapman & Hall, New York, pp. 160–173.

Bradley, C.A., Wood, P.P., Black, W.E., Kearns, R.D., Britton, J., 2002. Solid Culture Substrate Including Barley. U.S. Patent Application 2002/0006650 A1.

Burges, H.D., 1998. Formulation of mycoinsecticides. In: Burges, H.D. (Ed.), Formulation of Microbial Pesticides: Beneficial Microorganisms, Nematodes and Seed Treatments. Kluwer Academic Publishers, Dordrecht, The Netherlands, pp. 132–185.

Butt, T.M., Goettel, M.S., 2000. Bioassays of entomogenous fungi. In: Navon, A., Ascher, K.R.S. (Eds.), Bioassays of Entomopathogenic Microbes and Nematodes. CABI Publishing, Wallingford, Oxon, UK, pp. 141–195.

Butt, T.M., Wang, C., Shah, F.A., Hall, R., 2006. Degeneration of ento-mogenous fungi. In: Eilenberg, J., Hokkanen (Eds.), An Ecological and Societal Approach to Biological Control. Springer, Dordrecht, pp. 213–226.

Chen, H., 2013. Modern Solid State Fermentation Theory and Practice. Springer Verlag Dordrecht, The Netherlands. 324 pp.

Chong-Rodriguez, M.J., Maldonado-Blanco, M.G., Hernandez-Escareno, J.J., Galan-Wong, L.J., Sandoval-Coronado, C.F., 2011. Study of *Beauveria bassiana* growth, blastospore yield, desiccation-tolerance, viability and toxic activity using different liquid media. Afr. J. Bio-technol. 10, 5736–5742.

Cliquet, S., Jackson, M.A., 1999. Influence of culture conditions on pro-duction and freeze-drying tolerance of *Paecilomyces fumosoroseus* blastospores. J. Ind. Microbiol. Biotechnol. 23, 97–102.

Cliquet, S., Jackson, M.A., 2005. Impact of carbon and nitrogen nutrition on the quality, yield and composition of blastospores of the bioinsec-ticidal fungus *Paecilomyces fumosoroseus*. J. Ind. Microbiol. Biotech-nol. 32, 204–210.

Cooke, R., 1983. Morphogenesis of sclerotia. In: Smith, J.E. (Ed.), Fungal Differentiation: A Contemporary Synthesis. Marcel Dekker Inc., New York, pp. 397–418.

Diano, A., Peeters, J., Dynesen, J., Nielsen, J., 2009. Physiology of *Aspergillus niger* in oxygen-limited continuous cultures: influence of aeration, carbon source concentration and dilution rate. Biotechnol. Bioeng. 103, 956–965.

Duran, R., Cary, J.W., Calvo, A.M., 2010. Role of the osmotic stress regu-latory pathway in morphogenesis and secondary metabolism in fila-mentous fungi. Toxins 2, 367–381.

Exosect, 2015. Entostat Delivers Microbial Control Agents. https://www.exos-ect.com/technology/entostat/delivers-microbial-control-agents.aspx.

Faria, M.R., Wraight, S.P., 2007. Mycoinsecticides and mycoacaricides: a comprehensive list with worldwide coverage and international clas-sification of formulation types. Biol. Control 43, 237–256.

Faria, M., Hajek, A.E., Wraight, S.P., 2009. Imbibitional damage in conidia of the entomopathogenic fungi *Beauveria bassiana*, *Metarhizium acridum*, and *Metarhizium anisopliae*. Biol. Control 51, 346–354.

Faria, M., Lopes, R.B., Souza, D.A., Wraight, S.P., 2015. Conidial vigor vs. viability as predictors of virulence of entomopathogenic fungi. J. Invertebr. Pathol. 125, 68–72.

Feng, M.G., Poprawski, T.J., Khachatourians, G.G., 1994. Production, for-mulation and application of the entomopathogenic fungus *Beauveria bassiana* for insect control: current status. Biocontrol Sci. Technol. 4, 3–34.

Fernandes, É.K.K., Rangel, D.E.N., Braga, G.U.L., Roberts, D.W., 2015. Tolerance of entomopathogenic fungi to ultraviolet radiation: a review on screening of strains and their formulation. Curr. Genet. 61, 427–440.

Fytovita spol, 2015. About us, http://www.fytovita.cz/o-nas.html.

Garza-López, P.M., Konigsberg, M., Gómez-Quiroz, L.H., Loera, O., 2012. Physiological and antioxidant response by *Beauveria bassiana* Bals (Vuill.) to different oxygen concentrations. World J. Ind. Micro-biol. Biotechnol. 28, 353–359.

Globe, T.A., Hajek, A.E., Jackson, M.A., Gardescu, S., 2015. Microsclero-tia of *Metarhizium brunneum* F52 applied in hydromulch for control of Asian longhorned beetles (Coleoptera: Cerambycidae). J. Econ. Entomol. 108, 433–443.

Hajek, A.E., Papierok, B., Eilenberg, J., 2012. Methods for study of the Entomophthorales. In: Lacey, L.A. (Ed.), Manual of Techniques in Invertebrate Pathology, second ed. Academic Press, San Diego, pp. 285–316.

Hegedus, D.D., Bidochka, M.J., Khachatourians, G.G., 1990. *Beauveria bassiana* submerged conidia production in a defined medium contain-ing chitin, two hexosamines or glucose. Appl. Microbiol. Biotechnol. 33, 641–647.

Hornby, J.C., Jacobitz-Kizzier, S.M., McNeel, D.J., Jensesn, E.C., Treves, D.S., Nickerson, K.W., 2004. Inoculum size effect in dimorphic fungi: extracellular control of yeast-mycelium dimorphism in *Ceratocystis ulmi*. Appl. Environ. Microbiol. 70, 1356–1359.

Humber, R.A., 2008. Evolution of entomopathogenicity in fungi. J. Inver-tebr. Pathol. 98, 262–266.

Humber, R.A., 2012. Preservation of entomopathogenic fungal cultures. In: Lacey, L.A. (Ed.), Manual of Techniques in Insect Pathology, sec-ond ed. Academic Press, San Diego, pp. 317–327.

Humphreys, A.M., Matewele, P., Trinci, A.P.J., 1989. Effects of water activ-ity on morphology, growth and blastospore production of *Metarhizium anisopliae*, *Beauveria bassiana* and *Paecilomyces farinosus* in batch and fed-batch culture. Mycol. Res. 92, 257–264.

Inch, J.M., Trinci, A.P.J., 1987. Effects of water activity on growth and sporulation of *Paecilomyces farinosus* in liquid and solid media. J. Gen. Microbiol. 113, 247–252.

Inglis, G.D., Johnson, D.L., Goettel, M.S., 1996. Effect of bait substrate and formulation on infection of grasshopper nymphs by *Beauveria bassiana*. Biocontrol Sci. Technol. 6, 35–50.

Inglis, D.G., Enkerli, J., Goettel, M.S., 2012. Laboratory techniques used for entomopathogenic fungi: Hypocreales. In: Lacey, L.A. (Ed.), Manual of Techniques in Invertebrate Pathology, second ed. Academic Press, San Diego, pp. 189–253.

Issaly, N., Chauveau, H., Aglevor, F., Fargues, J., Durand, A., 2005. Influence of nutrient, pH and dissolved oxygen on the production of *Metarhizium flavoviride* Mf189 blastospores in submerged batch cul-ture. Process Biochem. 40, 1425–1431.

Jackson, M.A., 1997. Optimizing nutritional conditions for the liquid cul-ture production of effective fungal biological control agents. J. Ind. Microbiol. Biotechnol. 19, 180–187.

Jackson, M.A., 1999. Method for Producing Desiccation Tolerant *Paecilo-myces fumosoroseus* Spores. U.S. Patent Number 5968808.

Jackson, M.A., 2012. Dissolved oxygen levels affect dimorphic growth by the entomopathogenic fungus *Isaria fumosorosea*. Biocontrol Sci. Technol. 22, 67–79.

Jackson, M.A., Jaronski, S.T., 2009. Production of microsclerotia of the fungal entomopathogen *Metarhizium anisopliae* and their use as a bio-control agent for soil-inhabiting insects. Mycol. Res. 113, 842–850.

Jackson, M.A., McGuire, M.R., Lacey, L.A., Wraight, S.P., 1997. Liquid culture production of desiccation tolerant blastospores of the bioinsec-ticidal fungus *Paecilomyces fumosoroseus*. Mycol. Res. 101, 35–41.

Jackson, M.A., Cliquet, S., Iten, L.B., 2003. Media and fermentation pro-cess for the rapid production of high concentrations of stable blas-tospores of the bioinsecticidal fungus *Paecilomyces fumosoroseus*. Biocontrol Sci. Technol. 13, 23–33.

Jackson, M.A., Dunlap, C.A., Jaronski, S.T., 2010. Ecological consider-ations in producing and formulating fungal entomopathogens for use in insect biocontrol. BioControl 55, 129–145.

James, R.R., Jaronski, S.T., 2000. Effect of low viability on infectivity of *Beauveria bassiana* conidia toward the silverleaf whitefly. J. Invertebr. Pathol. 76, 227–228.

Jaronski, S.T., 2007. Soil ecology of the entomopathogenic ascomycetes: a critical examination of what (we think) we know. In: Maniana, K., Ekesi, S. (Eds.), Use of Entomopathogenic Fungi in Biological Pest Management. Research Signposts, Trivandrum, India, pp. 91–144.

Jaronski, S.T., 2010. Role of fungal ecology in the inundative use of entomopathogenic fungi. BioControl 55, 159–185.

Jaronski, S.T., 2013. Mass production of entomopathogenic fungi: state of the art. In: Morales-Ramos, J.A., Rojas, M.G., Shapiro-Ilan, D.J. (Eds.), Mass Production of Beneficial Organisms. Elsevier Inc., Amsterdam, pp. 357–415.

Jaronski, S.T., Jackson, M.A., 2008. Efficacy of *Metarhizium anisopliae* microsclerotial granules. Biocontrol Sci. Technol. 18, 849–863.

Jaronski, S.T., Jackson, M.A., 2012. Mass production of entomopathogenic Hypocreales. In: Lacey, L.A. (Ed.), Manual of Techniques in Invertebrate Pathology, second ed. Academic Press, San Diego, pp. 255–284.

Jenkins, N.A., Prior, C., 1993. Growth and formation of try conidia by *Metarhizium flavoviridae* in a simple liquid medium. Mycol. Res. 97, 1489–1494.

Jin, X., Grigas, K.E., Johnson, C.A., Perry, P., Miller, D.W., 1993. Method for Storing Fungal Conidia. U.S. Patent 5989898.

Kabaluk, J.T., Svircev, A.M., Goettel, M.S., Woo, S.G. (Eds.), 2010. The Use and Regulation of Microbial Pesticides in Representative Jurisdictions Worldwide. IOBC Global. 99 pp. Available online through: www.IOBC-Global.org.

Kassa, A., Vidal, S.D., Zimmermann, G., 2004. Production and processing of *Metarhizium anisopliae* var. *acridum* submerged conidia for locust and grasshopper control. Mycol. Res. 108, 93–100.

Kepler, R.M., Humber, R.M., Bischoff, J.F., Rehner, S.A., 2014. Clarification of generic and species boundaries for *Metarhizium* and related fungi through multigene phylogenetics. Mycologia 106, 811–829.

Kim, J.S., Je, Y.H., Skinner, M., Parker, B.L., 2013. An oil-based formulation of *Isaria fumosorosea* blastospores for management of greenhouse whitefly *Trialeurodes vaporariorum* (Homoptera: Aleyrodidae). Pest Manag. Sci. 69, 576–581.

Kleespies, R.G., Zimmermann, G., 1992. Production of blastospores by three strains of *Metarhizium anisopliae* (Metch.) Sorokin in submerged culture. Biocontrol Sci. Technol. 2, 127–135.

Kleespies, R.G., Zimmermann, G., 1998. Effect of additives on the production, viability and virulence of blastospores of *Metarhizium anisopliae*. Biocontrol Sci. Technol. 8, 207–214.

Kleinbaum, D.G., 2012. Survival Analysis: A Self-learning Text, third ed. Springer Science and Business Media LLC, New York. 700 pp.

Kondryatiev, N.N., Alioshina, O.A., Il'icheva, S.N., Perikhanova, A.G., Sinitsina, L.P., Oupenskaia, A.A., Chagov, E.M., 1971. Method for Obtaining the Entomopathogenic Material from the Fungus *Beauveria bassiana*. USSR Patent No. 313531 (in Russian).

Kulkarni, S.S., Gorwalla, E.P., 2010. Stacked Basket Bioreactor for Solid State Fermentation. PCT World Patent Application 2010/032260 A4.

Kybal, J., Vlcek, V., 1976. A simple device for stationary cultivation of microorganisms. Biotechnol. Bioeng. 18, 1713–1718.

Lacey, L.A., Grzywacz, D., Shapiro-Ilan, D.I., Frutos, R., Brownbridge, M., Goettel, M.S., 2015. Insect pathogens as biological control agents: back to the future. J. Invertebr. Pathol. 132, 1–41.

Lane, B.S., Trinci, A.P.J., Gillespie, A.T., 1991. Endogenous reserves and survival of blastospores of *Beauveria bassiana* harvested from carbon- and nitrogen-limited batch cultures. Mycol. Res. 95, 821–828.

Latgé, J.P., Hall, R.A., Cabrera-Cabrera, R.I., Kerwin, J.C., 1986. Liquid fermentation of entomopathogenic fungi. In: Samson, R.A., Vlak, J.M., Peters, D. (Eds.), Fundamental and Applied Aspects of Invertebrate Pathology, Wageningen: Foundation of the 4th International Colloquium of Invertebrate Pathology, pp. 603–604.

Leland, J.E., Mullins, D.E., Vaughan, L.J., Warren, H.L., 2005. Effects of media composition on submerged culture spores of the entomopathogenic fungus, *Metarhizium anisopliae* var. *acridum*, part 1: comparison of cell wall characteristics and drying stability among three spore types. Biocontrol Sci. Technol. 15, 379–392.

Li, Z., Alves, S.B., Roberts, D.W., Fan, M., Delalibera, I., Tang, J., Lopes, R.B., Faria, M., Rangel, D., 2010. Biological control of insects in Brazil and China: history, current programs and reasons for their successes using entomopathogenic fungi. Biocontrol Sci. Technol. 20, 117–136.

Lohse, R., Jakobs-Schönwandt, D., Patel, A.V., 2014. Screening of liquid media and fermentation of an endophytic *Beauveria bassiana* strain in a bioreactor. AMB Express 4, 47.

Lopez-Perez, M., Rodriguez-Gomez, D., Loera, O., 2015. Production of conidia of *Beauveria bassiana* in solid-state culture: current status and future perspectives. Crit. Rev. Biotechnol. 35, 334–341.

Luke, B., Faull, J., Bateman, R., 2015. Using particle size analysis to determine the hydrophobicity and suspension of fungal conidia with particular relevance to formulation of biopesticide. Biocontrol Sci. Technol. 25, 383–398.

Magan, N., 2001. Physiological approaches to improving the ecological fitness of fungal biocontrol agents. In: Butt, T.M., Jackson, C.W., Magan, N. (Eds.), Fungi as Biocontrol Agents. CAB Publishing, Oxon, pp. 239–251.

Mascarin, G.M., Kobori, N.N., Vital, R.C.J., Jackson, M.A., Quintela, E.D., 2013. Production of microsclerotia by Brazilian strains of *Metarhizium* spp. using submerged liquid culture fermentation. World J. Microbiol. Biotechnol. 30, 1583–1590.

Mascarin, G.M., Kobori, N.N., Quintela, E.D., Arthurs, S.P., Delalibera Jr., I., 2014. Toxicity of non-ionic surfactants and interactions with fungal entomopathogens toward *Bemisia tabaci* biotype B. BioControl 59, 111–123.

Mascarin, G.M., Jackson, M.A., Kobori, N.N., Behle, R.W., Delalibera Jr., I., 2015a. Liquid culture fermentation for rapid production of desiccation tolerant blastospores of *Beauveria bassiana* and *Isaria fumosorosea* strains. J. Invertebr. Pathol. 127, 11–20.

Mascarin, G.M., Jackson, M.A., Kobori, N.N., Behle, R.W., Dunlap, C.A., Delalibera Jr., I., 2015b. Glucose concentration alters dissolved oxygen levels in liquid cultures of *Beauveria bassiana* and affects formation and bioefficacy of blastospores. Appl. Microbiol. Biotechnol. 99, 6653–6665.

Meikle, W.G., Nansen, C., 2012. Biocontrol of Varroa Mites with *Beauveria bassiana*. U.S. Patent 8226938 B1.

Meikle, W.G., Mercadier, G., Holst, N., Girod, V., 2008. Impact of two treatments of a formulation of *Beauveria bassiana* (Deuteromycota: Hyphomycetes) conidia on Varroa mites (Acari: Varroidae) and on honeybee (Hymenoptera: Apidae) colony health. Exp. Appl. Acarol. 46, 105–117.

Oliveira, D.G., Pauli, G., Mascarin, G.M., Delalibera Jr., I., 2015. A protocol for determination of conidial viability of the fungal entomopathogens *Beauveria bassiana* and *Metarhizium anisopliae* from commercial products. J. Microbiol. Methods 119, 44–52.

Pendland, J.C., Boucias, D.G., 1997. In vitro growth of the entomopathogenic hyphomycete *Nomuraea rileyi*. Mycologia 89, 66–71.

Pham, T.A., Kim, J.J., Kim, K., 2009. Production of blastospore of entomopathogenic *Beauveria bassiana* in a submerged batch culture. Mycobiology 37, 218–224.

Prior, C., Jollands, P., Le Patourel, G., 1988. Infectivity of oil and water formulations of *Beauveria bassiana* (Deuteromycotina: Hyphomycetes) to the cocoa weevil pest *Pantorhytes plutus* (Coleoptera: Curculionidae). J. Invertebr. Pathol. 52, 66–72.

Prophyta Biologischer Pflanzenschutz GMBH, 1999. Solid-State Fermenter and Method for Solid-State Fermentation. World PCT Patent 99/57239.

Ramirez-Martinez, J.R., 1971. *Paracoccidioides brasiliensis*: conversion of yeast-like forms into mycelia in submerged culture. J. Bacteriol. 105, 523–526.

Rangel, D.E.N., Alston, D.G., Roberts, D.W., 2008. Effects of physical and nutritional conditions during mycelial growth on conidial germination speed, adhesion to host cuticle, and virulence of *Metarhizium anisopliae*, an entomopathogenic fungi. Mycol. Res. 112, 1355–1361.

Rangel, D.E.N., Fernandes, E.K.K., Braga, G.U.L., Roberts, D.W., 2011. Visible light during mycelial growth and conidiation of *Metarhizium robertsii* produces conidia with increased stress tolerance. FEMS Microbiol. Lett. 315, 81–86.

Rangel, D.E.N., Braga, G.U.L., Fernandes, E.K.K., Keyser, C.A., Hallsworth, J.E., Roberts, D.W., 2015. Stress tolerance and virulence of insect-pathogenic fungi are determined by environmental conditions during conidial formation. Curr. Genet. 61, 383–404.

Ravensberg, W.J., 2011. Mass production and product development of microbial pest control products for control of arthropods. In: Ravensberg, W.J. (Ed.), A Roadmap to the Successful Development and Commercialization, Progress in Biological Control, vol. 10. Springer Science+Business Media B.V., pp. 59–127.

Ravensberg, W., 2015. Crop protection in 2030: towards a natural, efficient, safe and sustainable approach. In: International Symposium on Biopesticides, Swansea University, September 7–9, 2015. www.ibma-global.org/upload/documents/201509wrpresentationinswansea.pdf.

Robertson, J.L., Preisler, H.K., Ng, S.S., Hickle, L.A., Gelernter, W.D., 1995. Natural variation: a complicating factor in bioassays with chemical and microbial pesticides. J. Econ. Entomol. 88, 1–10.

Robertson, J.L., Russell, R.M., Preisler, H.K., Savin, N.E., 2007. Bioassays with Arthropods Second Edition. CRC Press, Boca Raton. 199 pp.

Rombach, M.C., 1989. Production of *Beauveria bassiana* (Deutoromycotina, Hyphomycetes) sympoduloconidia in submerged culture. Entomophaga 34, 45–52.

Romero-Rangel, O., Maldonado-Blanco, M.G., Aguilar-López, C.C., Elías-Santos, M., Rodríguez-Guerra, R., López-Arroyo, J.I., 2012. Production of mycelium and blastospores of *Hirsutella* sp. in submerged culture. Afr. J. Biotechnol. 11, 15336–15340.

Ronne, H., 1995. Glucose repression in fungi. Trends Genet. 11, 12–17.

Samsinakova, A., 1966. Growth and sporulation of submerged cultures of the fungus *Beauveria bassiana* in various media. J. Invertebr. Pathol. 8, 395–400.

Samsinakova, A., Kalalovam, S., Vlcek, V., Kybal, J., 1981. Mass production of *Beauveria bassiana* for regulation of *Leptinotarsa decemlineata* populations. J. Invertebr. Pathol. 38, 169–174.

Sato, T., Watanabe, T., Mikami, T., Matsumoto, T., 2004. Farnesol, a morphogenetic autoregulatory substance in the dimorphic fungus *Candida albicans*, inhibits hyphae growth through suppression of a mitogen-activated protein kinase cascade. Biol. Pharm. Bull. 27, 751–752.

Shah, F.A., Allen, N., Wright, C.J., Butt, T.M., 2007. Repeated in vitro subculturing alters spore surface properties and virulence of *Metarhizium anisopliae*. FEMS Microbiol. 276, 60–66.

Song, Z.Y., Yin, Y.P., Jiang, S.S., Liu, J.J., Chen, H., Wang, Z.K., 2013. Comparative transcriptome analysis of microsclerotia development in *Nomuraea rileyi*. BMC Genome 14, 411.

Song, Z.Y., Yin, Y.P., Jiang, S.S., Liu, J., Wang, Z., 2015. Optimization of culture medium for microsclerotia production by *Nomuraea rileyi* and analysis of their viability for use as a mycoinsecticide. BioControl 59, 597–605.

Suryanarayan, S., Mazumdar, K., 2003. Solid State Fermentation. U.S. Patent 6664095.

Thomas, K.C., Khachatourians, G.G., Ingledew, W.M., 1987. Production and properties of *Beauveria bassiana* conidia cultivated in submerged culture. Can. J. Microbiol. 33, 12–20.

Tlecuitl-Beristain, S., Viniegra-Gonzalez, G., Diaz-Godinez, G., Loera, O., 2010. Medium selection and effect of higher oxygen concentration pulses on *Metarhizium anisopliae* var. *lepidiotum* conidial production and quality. Mycopathologia 169, 387–394.

Torre, M., Cárdenas-Cota, H.M., 1996. Production of *Paecilomyces fumosoroseus* conidia in submerged culture. Entomophaga 41, 443–453.

US Environmental Protection Administration, 2010. How to Prepare a Confidential Statement of Formula (CSF) for Biochemical and Microbial Pesticides. http://www2.epa.gov/pesticide-registration/biopesticides-confidential-statement-formula.

Van der Valk, H., 2007. Review of the Efficacy of *Metarhizium anisopliae* var. *acridum* against the Desert Locust Plant Production and Protection Division, Locusts and Other Migratory Pest Group N. AGP/DL/TS/34. 81 pp. http://www.fao.org/ag/locusts/common/ecg/1295/en/TS34e.pdf.

Vega, F.E., Jackson, M.A., McGuire, M.R., 1999. Germination of conidia and blastospores of *Paecilomyces fumosoroseus* on the cuticle of the silverleaf whitefly, *Bemisia argentifolii*. Mycopathologia 147, 33–35.

Vega, F.E., Jackson, M.A., Mercadier, M., Poprawski, T.J., 2003. The impact of nutrition on spore yields for various fungal entomopathogens in liquid culture. World J. Microbiol. Biotechnol. 19, 363–368.

Vidal, C., Fargues, J., Lacey, L.A., Jackson, M.A., 1998. Effect of various liquid culture media on morphology, growth, propagule production, and pathogenic activity to *Bemisia argentifolii* of the entomopathogenic hyphomycete, *Paecilomyces fumosoroseus*. Mycopathologia 143, 33–46.

Voulgaris, I., O'Donnell, A., Harvey, L., McNeil, B., 2012. Inactivating alternative NADH dehydrogenases: enhancing fungal bioprocesses by improving growth and biomass yield? Sci. Rep. 2 (322). http://dx.doi.org/10.1038/srep00322.

Wang, C., St Leger, R., 2006. A collagenous protective coat enables *Metarhizium anisopliae* to evade insect immune responses. Proc. Natl. Acad. Sci. 103, 6647–6652.

Wang, C., Typas, M.A., Butt, T.A., 2002. Detection and characterisation of pr1 virulent gene deficiencies in the insect pathogenic fungus *Metarhizium anisopliae*. FEMS Microbiol. Lett. 213, 251–255.

Wang, C., Butt, T.A., St Leger, R., 2005. Colony sectorization of *Metarhizium anisopliae* is a sign of ageing. Microbiology 151, 3223–3236.

Winkelhoff, A.J.V., McCoy, C.W., 1984. Conidiation of *Hirsutella thompsonii* var. *synnematosa* in submerged culture. J. Invertebr. Pathol. 41, 59–68.

Wraight, S.P., Ramos, M.E., Avery, P.B., Jaronski, S.T., Vandenberg, J.D., 2010. Comparative virulence of *Beauveria bassiana* isolates against lepidopteran pests of vegetable crops. J. Invertebr. Pathol. 103, 186–199.

Ypsilos, I.K., Magan, N., 2004. Impact of water-stress and washing treatments on production, synthesis and retention of endogenous sugar alcohols and germinability of *Metarhizium anisopliae* blastospores. Mycol. Res. 108, 1337–1345.

Ypsilos, I.K., Magan, N., 2005. Characterisation of optimum cultural environmental conditions for the production of high numbers of *Metarhizium anisopliae* blastospores with enhanced ecological fitness. Biocontrol Sci. Technol. 15, 683–699.

Zabriskie, D.W., Armiger, W.B., Phillips, D.H., Albano, P.A., 2008. Traders' Guide to Fermentation Media Formulation. Traders Protein, Memphis, TN, USA. 60 pp.

Chapter 10

Production of Entomopathogenic Nematodes

A. Peters[1], R. Han[2], X. Yan[2], L.G. Leite[3]

[1]e-nema GmbH, Schwentinental, Germany; [2]Guangdong Entomological Institute, Guangzhou, Guangdong, China; [3]Instituto Biológico, São Paulo, Brazil

10.1 INTRODUCTION

Mass production of microbial control agents (MCAs) of reliable quality is the backbone of the biopesticide industry. Today, the market for biopesticides is dominated by antagonistic arthropods, nematodes, and microorganisms (bacteria and fungi). While arthropods are propagated in large-scale insect and mite farms, microorganisms are grown in submerged or solid-state cultures in bioreactors. Entomopathogenic nematodes (EPNs) of the genera *Steinernema* and *Heterorhabditis* take an intermediate position. They can be mass-produced on host insects, but, since they essentially feed on bacteria rather than insects, they may also be grown without insects on artificial solid or liquid media after preculturing the symbiotic bacteria.

The suitability for industrial mass production is a prerequisite for the successful commercialization of EPNs. Additionally, they have a moderately wide host range, allowing for applications against a variety of different insect groups but still minimizing adverse effects to nontarget insects (Bathon, 1996). The infective juvenile (IJ), which is the only free living stage, may be stored for one to several months. They are sufficiently small to pass through standard spraying equipment, and they actively move in suitable moist environments like the soil or galleries in wood or leaves (Wright et al., 2005). Soon after Krausse discovered the first *Steinernema* species (Krausse, 1917), the nematode *Steinernema glaseri*, produced on an artificial medium based on dog food, was used in controlling the scarabaeid *Popillia japonica* in the United States (Glaser, 1940). At that time, the symbiotic relation to bacteria was unknown and, subsequently, mass production on artificial media collapsed after a few cycles probably due to contaminating bacteria, which did not support growth and propagation of the nematode. The renaissance of using *Steinernema* and *Heterorhabditis* started in the mid-1980s fostered by an increasing public concern about the use of chemical crop protection products and the widespread adoption of biological pollination in greenhouses.

10.2 BIOLOGY OF ENTOMOPATHOGENIC NEMATODES

The species of EPNs used in inundative biocontrol agents belong to the genera *Steinernema* (Steinernematidae) and *Heterorhabditis* (Rhabditidae). Descending from nonspecific bacteria feeding nematodes, a symbiotic relation to entomopathogenic bacteria evolved in these two genera. When looking closely, the differences between the two genera are overwhelming evidence for a convergent evolution: In the genus *Heterorhabditis*, a protruding tooth can be found on the head of the infective juvenile (IJ), which is used to rupture the integument of the host insect. IJs of *Heterorhabditis* spp. usually are ensheathed in the cuticle of the preinfective J2 stage and only shed during penetration inside the host insect. The clean surface of the freshly shed nematode is less likely recognized by the non–self-response system in the insect's hemocoel (Peters et al., 1997). *Heterorhabditis* spp., if deprived from their symbionts, are nonpathogenic to insects (Han and Ehlers, 2000), whereas symbiont-free *Steinernema* spp. will kill their insect hosts, although it takes them about 10 times longer to do so (Ehlers et al., 1997). Inside the hemocoel, IJs of *Heterorhabditis* release the symbionts by regurgitation, while bacteria are defecated by *Steinernema* spp. (Ciche et al., 2006). With the exception of *Steinernema hermaphroditum* (Griffin et al., 2001), *Steinernema* spp. are gonochoristic showing just the amphimictic mode of reproduction. In *Heterorhabditis* spp., the IJs develop into automictic hermaphrodites. The offspring of these will develop to amphimictic males and females or to IJs and subsequently automictic hermaphrodites, depending on the availability of food (Strauch et al., 1994), which has important consequences for mass production strategies. *Steinernema* spp. (clade IV: *Strongyloidea*) are associated with *Xenorhabdus* spp. and *Heterorhabditis* spp. (clade V: *Rhabditidae*) with *Photorhabdus* spp. (Blaxter et al., 1998; Dorris et al., 2002). Interestingly, the symbionts,

Xenorhabdus and *Photorhabdus*, form a monophyletic clade in the *Enterobacteriaceae* (Rainey et al., 1995).

Both nematode genera have similar life cycles starting with an IJ, which vectors entomopathogenic bacteria inside the hemocoel of a suitable host. After penetration of the host, the nematode and the symbiotic bacteria act together in overcoming the host insect's immune response (Dowds and Peters, 2002). Subsequently, the bacteria multiply in the insect's hemocoel, the insect is killed, and the nematode propagates feeding on the bacteria. When the food in the insect cadaver is depleted, an enduring third-stage juvenile (Dauer stage), the IJ, is formed and leaves the cadaver if conditions outside are sufficiently moist and adequately warm (Brown and Gaugler, 1997).

The IJs carry their symbiotic bacteria in the intestine while searching for an insect host. They only have one chance to release them and have hence evolved mechanisms to ensure that it does not happen before they have successfully entered the insect's hemocoel. A heat- and protease-resistant, low-molecular-weight component from the insect's hemocoel (Ciche et al., 2006) triggers the development of IJs and the release of the symbiotic bacteria. A spontaneous release of the bacteria does not occur, not even in the artificial production medium typically based on soybean protein, yeast extract, and vegetable oil.

Since 1995, another member of the *Rhabditidae* has advanced as a potent biological control agent. *Phasmarhabditis hermaphrodita* is used to control slugs and snails (Wilson and Grewal, 2005). Unlike species of *Steinernema* and *Heterorhabditis*, this species is not associated with one specific bacterium. It can feed and reproduce on a large variety of different bacterial species, and it could be shown that pathogenicity to slugs largely depends on which bacteria they were fed on (Wilson et al., 1994). *Phasmarhabditis hermaphrodita* can be viewed as a model for the ancestors of *Steinernema* and *Heterorhabditis*, which apparently have independently picked up insect pathogenic bacteria and developed mechanisms to ensure the exclusive transmission of their symbionts.

10.3 PRODUCTION METHODS

10.3.1 Production in Insects

For early field testing, in vivo production techniques were used to produce EPNs. Nutrilite Corporation in Lakeview, CA, used larvae of the wax moth, *Galleria mellonella*, to produce Biotrol NCS-DD-136 in 1970 for experimental use. In 1981, "The Nematode Farm" in Berkeley produced several EPNs (*Steinernema carpocapsae*, *Steinernema glaseri*, and *Heterorhabditis bacteriophora*) in *G. mellonella* for commercial use against garden pests. Also in 1981, B&R Supply in Exeter, CA, raised *S. carpocapsae* on crickets and packaged a product called Neocide for use against

the carpenter worm (Poinar and Grewal, 2012). In Italy, Bioerre had based their EPN production successfully on *G. mellonella* until the early 1990s (Deseö et al., 1990).

In vivo production of EPNs remains a cottage industry, generating a relatively low-volume output but requiring only moderate capital outlay. The default strategy is to rear nematodes in insect hosts that are highly reliable, exceptionally susceptible, and relatively inexpensive. Although *Tenebrio molitor* are substantially cheaper (Blinova and Ivanova, 1987), with rare exceptions (eg, crickets by B&R Supply in the mid-1980s), the last instar of the greater wax moth, *Galleria mellonella*, is the host of choice. *G. mellonella* are reared by more companies than any other insect. No insect has been found to be more susceptible to a wider range of nematode species. Yields per larva range from 1 to 3.5×10^5 IJs (Dutky et al., 1964; Milstead and Poinar, 1978). Only a few EPNs exhibit relatively poor reproduction in *G. mellonella*. *Steinernema kushidai* and *S. scarabaei* appear to be especially adapted to hosts in the family Scarabaeidae (order Coleoptera), and *S. scapterisci* appears to be especially adapted to the order Orthoptera (Mamiya, 1989; Nguyen and Smart, 1990; Kaya and Stock, 1997; Grewal et al., 1999; Koppenhofer and Fuzy, 2003; Stock and Goodrich-Blair, 2012).

10.3.1.1 Laboratory Production

White (1927) laid the foundation for in vivo culture in a classic article on the collection of parasitic nematodes from dung. The concept was adapted by Dutky et al. (1964) to harvest IJs, taking advantage of their natural migration away from the host cadaver on emergence. The White trap consists of a dish or tray on which the cadavers are set; the dish is surrounded by water, which is contained by a larger arena (Fig. 10.1). Every EPN laboratory uses the White trap albeit with some modifications (Lindegren et al., 1993), especially for production for experiments and for maintenance of nematode collections.

10.3.1.2 Large-Scale Production

To mass produce nematodes, scale-up is necessary, which is the process of taking production from the bench to larger dimensions and volume. Temperature is the easiest parameter to be completely controllable along the process, while humidity and aeration are more difficult to be keep constant with increasing scale. Spraying insect hosts with IJs can accelerate inoculation, or they can be dipped into a nematode suspension. However, the most common method is inoculating the substrate around the insect. Host density and inoculation rate should be optimized for each particular nematode and host species (Shapiro et al., 2002). A dosage that is too low results in low host mortality. In contrast, a dosage that is too high has been suggested to potentially result in flawed infections due to competition with secondary invaders (Woodring

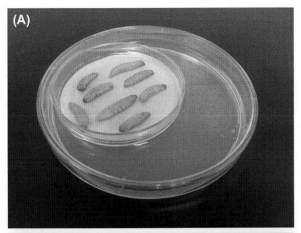

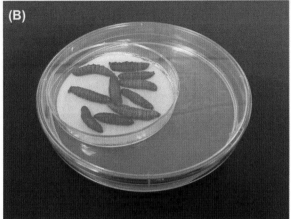

FIGURE 10.1 White trap containing *Galleria mellonella* larvae infected with *Steinernema carpocapsae* (A) and *Heterorhabditis megidis* (B).

and Kaya, 1988). Thus, intermediate dosages should be used to maximize yield (Boff et al., 2000). For example, rates of approximately 25–200 IJs per insect are usually sufficient (depending on nematode species and method of inoculation) for infecting *G. mellonella*, whereas higher rates are generally needed to infect *Tenebrio molitor* (eg, 100–600 IJs per insect). Crowding of hosts can lead to oxygen deprivation or buildup of ammonia, which suppresses nematode yield (Shapiro et al., 2002).

Nematode yield also varies proportionally to host size (Blinova and Ivanova, 1987; Flanders et al., 1996) and is inversely proportional to nematode size (Grewal et al., 1994; Hominick et al., 1997; Shapiro and Gaugler, 2002). This has resulted in small nematodes being chosen, but the choice of host species and nematode for in vivo production should ultimately rest on nematode yield per cost of insect and the suitability of the nematode for the pest target (Blinova and Ivanova, 1987; Shapiro et al., 2002). Nematode yield varies depending on several environmental factors including temperature, aeration, and moisture (Burman and Pye, 1980; Woodring and Kaya, 1988; Friedman, 1990; Grewal et al., 1994; Shapiro et al., 2002; Dolinski et al., 2007).

Temperature is critical during the rearing process as it affects both yield and the time to emergence (Grewal et al., 1994). Optimum rearing temperature in *G. mellonella* varied from 18 to 28°C dependent on the nematode species: 18.5°C for certain strains of *Steinernema feltiae* (a cold-tolerant species), 28°C for *S. riobrave* (a heat-tolerant species), and 25°C for *S. carpocapsae* (intermediate). In addition to appropriate temperatures, adequate aeration and humidity are important environmental factors that must be maintained throughout the production cycle (Woodring and Kaya, 1988). Aeration should be as high as possible to avoid accumulation of ammonia or other harmful gases. Humidity should be controlled to be nonsaturated and not too dry. Saturated levels may induce adults of the nematode to leave the cadaver, favoring the growth of contaminants, such as fungi and bacteria. If the insect is contaminated, the cadavers may attract phorid flies, whose larvae are scavengers and may destroy all the infected cadavers. To maintain the humidity, the entire system is usually kept closed but not sealed to allow aeration, or sealed with particulate air filters to avoid infestation by flies. A harvest apparatus described by Carne and Reed (1964) consists of a disk supporting cadavers resting in the mouth of a large funnel. The water level in the funnel is maintained at cadaver height by a constant-leveling device. Emerging IJs migrate through the disk and settle to the funnel bottom, where they are collected by opening the stopcock. The central innovation of the system is the perforated disk in the center of the funnels, through which nematodes pass to the water reservoir, reducing the need for significant lateral migration. There are no reports that the design was used in in vivo mass production. Some countries of Latin America have commercially produced nematodes on *G. mellonella*, using a technique that was developed in Cuba. *G. mellonella* larvae are reared in about 100-L metal buckets, fed with diet based on corn flour. Fully grown larvae are dipped in the nematode suspension and held inside Petri dishes to be incubated at room temperature. After the insect dies, cadavers are transferred to large trays containing modified White traps (J. E. Almeida, personal communication).

10.3.1.2.1 Mechanization

The primary expenses for in vivo production include the costs of insect hosts and labor. Therefore, the economics of in vivo production can be improved substantially by producing the insect hosts "in-house" and mechanizing the process to reduce labor cost (Shapiro-Ilan et al., 2014a). Various steps in the in vivo production can be mechanized (eg, inoculation, harvest, concentration). Several approaches to mechanization of nematode inoculation and harvest have been developed or proposed. Gaugler and Brown (2001) have proposed an in vivo production system, named "LOTEK," that does not rely on migration into a stagnant water reservoir. The system uses shallow perforated holding trays for inoculation, incubation, and harvest. Each tray holding several 100 hosts is dipped into a nematode suspension and then stacked in a high-humidity environment (Gaugler et al., 2002). Before IJs start leaving the cadavers, the tray is moved to a harvester, which conceptually resembles the mist chambers used to extract nematodes from soil (Southey, 1986). Each tray is suspended under two rigid plastic pipes equipped with atomizer nozzles. A timer connected to a water supply provides periodic misting cycles (3 min at 6-h intervals) for the tray; air movement is also important to moderate humidity and potential anoxia. Exposure to free water induces IJs to emerge. The nematodes are rinsed with the runoff water into an angled drip tray. Each drip tray serves 10 or more holding trays arrayed overhead and directs nematodes by gravity flow into a collecting pipe and then to a central storage tank, which is aerated. Harvest is completed in 48h with a 97% cadaver extraction efficiency. Cadavers are then discarded and the holding tray replaced for in situ cleaning with detergents and disinfectants delivered via the misting nozzles.

Because the system is largely automated, it is projected to reduce labor costs. Regardless of the means of harvest, collected IJs must be concentrated and separated from culture residue, bacteria, and undesired life stages before formulation. Insufficient separation inevitably results in inferior product shelf-life due to microbial activity. Nematode concentration and separation have been costly for companies practicing in vivo methods. The small producers tend to let suspensions settle and then decant the supernatant as waste. Other than its inherent inefficiency, this crude method's primary deficiency is that nematodes are held in a stressful low-oxygen environment. Use of a linear flow of liquid across a filter membrane (eg, in LOTEK) moves particles from the surface and keeps the membrane clean, and thus solves the separation issue as has been suggested by Gaugler and Brown (2001). They also advocate a simple and inexpensive continuous-deflection separation method for capturing nematodes from storm water (Schwarz and Wells, 1999). This approach tested on a lab bench scale removes 98% of wastewater in two passes with less than 5% loss of nematodes. In an encouraging study, the first of its kind, Young et al. (2002) described the physical properties of nematode culture components, with the long-term goal of providing a theoretical underpinning for the design of new separation procedures. The point is that new, scale-appropriate, separation technology is urgently needed for producers who are currently dependent on gravity settling.

10.3.1.2.2 Streamlining

Shapiro-Ilan et al. (2014b) introduced a single-tray design where infected host cadavers are placed on a substrate intended as the formulation carrier (eg, a gel) within the final package. Once IJs emerge, the cadavers are removed and the final nematode product is ready for shipment or storage. In this approach, the process is streamlined substantially because nematodes emerge directly into their formulation and package; thus, additional concentration and packaging steps are removed from the process.

Another approach to streamline in vivo production may be through production and application of EPNs in infected hosts. Using this method, infected-host cadavers are applied to the target site and pest suppression is subsequently obtained by the emerging IJs. This approach is likely to reduce production costs substantially because labor-intensive steps of harvest and concentration are avoided (Shapiro-Ilan et al., 2001). Application of nematodes in infected-host cadavers has been shown to provide significant pest control (Jansson et al., 1993; Jansson and Lecrone, 1994). Additionally, studies indicate that nematodes emerging into soil directly from the host cadaver can be more infective and disperse more readily than nematodes applied in aqueous formulations/suspensions (Shapiro and Glazer, 1996; Shapiro and Lewis, 1999). Application of nematode-infected hosts has not been commercialized possibly due to problems in storage, transport, and application of the fragile cadavers. To overcome these hindrance, Shapiro-Ilan et al. (2001) found that infected hosts can be formulated with a coating (eg, starch and clay), thereby increasing desiccation tolerance and storability and preventing cadaver rupture or adhesion during handling. Alternatively, hard-bodied insects such as *Tenebrio molitor* can be used, because the smooth rigid cuticles are naturally resistant to rupture or sticking together. In greenhouse trials, superior efficacy (vs. *Diaprepes abbreviatus* and *Otiorhynchus sulcatus*) was observed when nematodes were applied in infected *T. molitor* compared with aqueous application (Shapiro-Ilan et al., 2003).

10.3.1.2.3 Grower-Based Production

More recently, research focused on developing producer-friendly nematode rearing and application techniques and teaching producers and agribusiness professionals new techniques for rearing and applying nematodes to control *Otiorhynchus ligustici* in alfalfa (Shields, 2012). In this program, the farmer buys the insect *G. mellonella* for sale in almost every pet shop and obtains the nematode inoculum from the university. The farmer inoculates 11 pots containing sawdust plus around 250 waxmoth larvae with IJs ordered from the university, covers the pots with perforated lids, and incubates them at room temperature for 17 days. During this period, the larvae die and nematodes leave the cadaver as IJs, remaining in the sawdust. The farmer washes each cup over a screen, allowing the nematodes to pass through the screen into a container while sawdust and wax worms remain in the pot. The IJ suspension is then taken to the field and applied to the soil for insect control. A mix of *S. carpocapsae* and *S. feltiae* is recommended to control *O. ligustici*.

10.3.2 Production on Artificial Media

The IJs usually only start to develop when sensing specific compounds of the insect hemocoel. Where does the start signal in an artificial medium come from? Adding the trigger substance from the insect's hemocoel would be a tedious and expensive way to start the development of nematodes in artificial media. The bacteria also produce a signal component triggering development, which has been partly characterized (Aumann and Ehlers, 2001). The component is unstable and hence the timing for adding the IJs to the bacteria preculture is of crucial importance for a high and synchronous development of the IJs in artificial media (Johnigk et al., 2004).

The symbiotic bacteria not only trigger the onset of IJ development, they also provide the essential food. *Steinernema* and *Heterorhabditis* spp. feed on their symbiotic bacteria and grow poorly when fed on other bacteria genera, even if closely related. It is therefore crucial to establish and maintain a monoxenic culture when growing nematodes in artificial media. When cultured in vivo, the non–self-response system of a freshly infected insect creates a sterile environment for the invaded nematode, and once it has released its symbiotic bacteria, which are not recognized as non-self (Dowds and Peters, 2002), a monoxenic culture is established in the insect hemocoel. Later, when the intestine of the cadaver collapses, other bacteria will manage to grow and might even be found between the first- and second-stage cuticle (Gouge and Snyder, 2006).

10.3.2.1 Establishing Monoxenic Cultures

To obtain monoxenic culture for in vitro culture, the symbiotic bacteria are isolated from freshly infected cadavers. Two methods are described to obtain axenic nematodes for introduction to the pure bacteria culture. One is to surface sterilize the IJs with bleach (NaOCl 1%) or hyamine (0.4%) before adding them to the symbiotic bacterial culture (Akhurst, 1980; Wouts, 1981). This method may not always work, because surface sterilization fails to kill all bacteria beneath the IJs' cuticle. Another method is to sterilize the eggs (Lunau et al., 1993). Eggs are collected by rupturing gravid females (*Steinernema* spp.) or hermaphrodites (*Heterorhabditis* spp.) in sterile Ringer's solution (0.9% NaCl, 0.042% KCl, 0.048% CaCl$_2$ and 0.02% NaHCO$_3$). The eggs are separated by sieving and surface-sterilized in a disinfectant (2.5 mL of 4 mol/L NaOH, 0.5 mL of 12% NaOCl, and 21.5 mL of distilled water) for 5 min before rinsing twice in sterile Ringer's solution. The sterilized eggs are then transferred to sterile lysogeny broth (LB) and incubated until the first-stage juveniles (J1) hatch. Two days later, if no bacteria are detected in the LB, nematodes are combined with a 1-day-old culture of their symbiotic bacteria on a fortified lipid agar (1.6% nutrient broth, 1.2% agar, and 1% corn oil). Nematodes propagate on these solid-state monoxenic

cultures, and the resulting IJs can be used to start solid-state or liquid cultures, where the symbiotic bacteria have been precultured until the stationary phase.

10.3.2.2 Solid State Culture

The very first nematodes reported to be used in biocontrol were produced on solid-state culture (Glaser, 1940). At that time, the symbiotic association with bacteria was not known and the nematode was simply placed on dog food. Modern solid-state culture systems are based on monoxenic cultures where nematodes are added to a pure culture of their symbiotic bacteria in a nutritive medium. The in vitro solid-state production of EPN involves four steps: preparation of culture media, culture and inoculation of symbiotic bacteria, inoculation with nematodes, and harvest.

Solid-state media used for EPN production started with agar media (House et al., 1965; Wouts, 1981; Dunphy and Webster, 1989), and now the three-dimensional media with nutrients in polyether-polyurethane sponge (Fig. 10.2) (Bedding, 1981, 1984) are used for mass production. Bedding cultures consist of crumbled polyurethane foam sponge thinly coated with poultry offal homogenate. Because the poultry entrails cannot be standardized and may lead to unreliable results, a medium based on yeast extract, corn oil, corn starch, and dried egg solids was developed (Wouts, 1981). An improved medium containing 15% soy flour, 5% wheat flour and corn oil, 1% yeast extract and egg yolk flour, and 10% crumbled polyether polyurethane foam (Han et al., 1992, 1993) is now widely used for solid culture of EPN in developing countries. Containers used for EPN solid culture can be flasks or culture vessels comprising a tray with side walls and overlapping lids that allow gas exchange through a layer of foam (Bedding et al., 1991) or bags with gas-permeable strips for ventilation (Gaugler and Han, 2002).

Composition of the culture media affects the yield of the IJs. Different EPN isolates require specific compositions of the production medium. Medium ingredients that may affect nematode yield include lipid, water, yeast extract, egg yolk, lard, soy flour, salts, and proteins (Dunphy and Webster, 1989; Han et al., 1992; Ehlers and Shapiro-Ilan, 2005; Salma and Shahina, 2012; Shapiro-Ilan et al., 2014a).

The symbiotic bacteria of different isolates of EPN can easily be isolated from nematode-infected insect larvae or surface-sterilized IJs. The symbiotic bacteria are well known to have two phase variants. The primary variant (phase I) is conducive to the growth of IJs and tends to be retained by them (Han and Ehlers, 2001). Phase I variant can be selected on nutrient bromothymol blue agar (NBTA) (0.5% peptone, 0.3% beef extract, 1.5% agar, 0.025% bromothymol blue, and 0.004% 2,3,5-triphenyltetrazolium chloride) or MacConkey media (1.7% peptone 17, 0.3% proteose peptone, 1% lactose, 0.15% bile salts, 0.5% sodium

FIGURE 10.2 Production in sponge media. (A) Sponge media inoculated with symbiotic bacteria (left: *Steinernema glaseri*; right: *Heterorhabditis bacteriophora*). (B) *H. indica* infective juveniles grown in sponge media. (C) *S. feltiae* infective juveniles grown in sponge media.

chloride, 0.003% neutral red, and 1.35% agar) (Akhurst, 1980). Symbiotic bacteria can be cultured in lysogeny broth (LB) (0.5% NaCl 5 g, 1% tryptone, 0.5% yeast extract) or yeast salts (YS) broth (0.05% $NH_4H_2PO_4$, 0.05% K_2HPO_4, 0.02% $MgSO_4\cdot7H_2O$, 0.5% NaCl, and 0.5% yeast extract) (Hirao and Ehlers, 2009). Symbiotic bacteria stock can be prepared by mixing fresh bacteria liquid culture with glycerol to a final concentration of 15% (v/v) and storing the vials at −80°C. A freshly grown bacterial culture should be inoculated into the solid media 1–4 days before nematode inoculation (Fig. 10.2A) as the symbiotic bacteria need time to convert the culture medium into a suitable medium for nematode development and reproduction (Forst et al., 2002). Bacterial inoculum size does not appear to be important in yield determination (Han et al., 1992, 1993).

Monoxenic cultures obtained as described earlier can be scaled-up by inoculation of the agar with nematodes to the sponge media. Inoculation of monoxenic nematodes to new culture media can be accomplished either by transferring the sponge media with IJs directly or by washing the IJs out of the media before inoculation under aseptic conditions, depending on different culture containers used. Nematode inoculum concentration (IJs per unit of media) may impact yields of nematodes for some, but not all, nematode species. For example, *S. carpocapsae* Agriotos produced optimum yields at inoculum amount of 2000 IJs/g (Han et al., 1993), whereas *S. carpocapsae* CB2B and *Heterorhabditis bacteriophora* H06 were not affected by inoculum amount (Han et al., 1992). However, increasing inoculum size may accelerate nematode propagation and decrease culture time.

After 2–5 weeks of culture, nematodes can be harvested (Bedding, 1981, 1984) (Fig. 10.2B and C). IJs can be extracted from solid media with centrifugal sifters or by washing nematodes out of the sponge in washing machines and then separating the IJs via sedimentation, centrifugation, or sieving (Ehlers and Shapiro-Ilan, 2005). Before formulation and storage, the separated IJs should be washed two or more times to remove impurities, which would fuel the growth of undesired fungal or bacterial contaminants in the final product.

Yields of IJs depend on nematode species and strains, which have different innate reproductive capacities and sizes. For example, *H. indica* LN2 (9.3×10^5 IJs/g medium) yielded much more than *H. bacteriophora* H06 (5.1×10^5 IJs/g medium) under the same culture conditions (Shapiro-Ilan et al., 2014a).

Culture temperature and culture time are also important parameters in in vitro solid-state production. Optimal temperature for maximum production in solid-state culture varies among different species and strains. For example, 27°C supported optimal production for *S. carpocapsae* CB2B and 25°C for *H. bacteriophora* H06 (Han et al., 1992) while *H. bacteriophora* (ie, *heliothidis*) grew the best at 30°C on lipid agar (Dunphy and Webster, 1989). When culturing different species of EPNs from Pakistan, it was found that most species obtained maximum yields at 32±2°C, whereas *Steinernema feltiae* produced the highest yields at 20±2°C (Salma and Shahina, 2012). Usually, longer culture time can provide higher yields, but nematode mortality may also increase with time (Han et al., 1992, 1993).

10.3.2.3 Advantages and Disadvantages

Solid-state production of EPN is labor intensive, so it is superior to liquid culture production in developing countries where labor cost is low (Bedding, 1990; Ehlers et al., 2000). Solid-state culture has some advantages. The effect of symbiotic bacteria phase variation on the yields is less detrimental than in liquid cultures (Han and Ehlers, 2001). Little investment in biotechnology equipment is necessary and the water and energy consumption is lower than in liquid culture systems. The risk for process failure is partitioned over several smaller production units. But when it comes to large-scale production, solid-state production is vulnerable to contamination during upstream and downstream processing and the uneven distribution of the nematodes in the medium prevents systematic sampling (Ehlers and Shapiro-Ilan, 2005). Solid-state cultures cannot be incubated in layers higher than approximately 20 cm since the weight of the medium would compress the sponge structure and subsequently cause anoxia in the bottom part of the medium, which would impede scaling up of this production system.

10.3.3 Liquid Culture

The main advantage of liquid culture is the economies of scale. The production units may be increased from a few mL to several m^3 with hardly any increase in labor cost. Figuratively, nematodes are grown in an artificial insect cadaver of several tons in weight. The bioreactor is a stainless-steel tank with probes for various physical parameters and devices for controlling these parameters (Fig. 10.3). Typically, the temperature, pH, and dissolved oxygen (DO) is measured and controlled. The aeration rate (L/min of sterile air supplied to the liquid) and the stirrer speed are used to control the DO. The pH is not controlled but monitored to identify switches in the metabolic activity of the symbiotic bacteria (Fig. 10.4). As with the production in solid-state culture, a monoxenic culture of nematodes and their symbiotic bacteria is absolutely required for the successful liquid culture of EPNs.

A liquid culture process starts with the inoculation of a pure culture of the symbiotic bacteria in its exponential growth phase into a suitable sterilized liquid medium at a concentration of 0.01–0.1% v/v. The bacteria grow and exhibit typical metabolic switches, which are reflected by sudden changes in the oxygen uptake rate (OUR) and the pH. The timing of adding nematodes affects the recovery, that is, the proportion of IJs leaving their arrested stage, and the final nematode yield (Johnigk et al., 2004). By relating the timing of nematode inoculation to the last pH minimum during the bacteria preculture, processes of *Heterorhabditis bacteriophora* can be started reproducibly (Fig. 10.4). The nematodes recover from the arrested Dauer stage and start feeding on the symbiotic bacteria. They develop to adults in

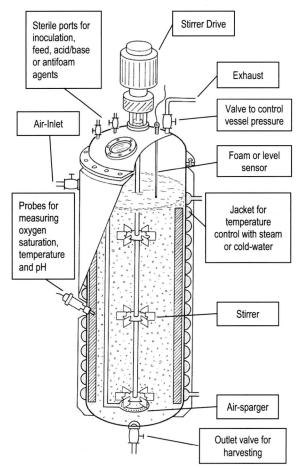

FIGURE 10.3 Sketch of a bioreactor. The bioreactor is filled with sterile medium or sterilized in situ. The inlet air is passed through a sterile filter.

a few days and produce offspring. When the hermaphrodites produce eggs, their metabolic activity is highest, which is reflected by another increase in the OUR. Compared to the pure bacteria culture, the maximum stirrer speed is lowered after nematodes are added to avoid shear-stress, which may kill the fragile hermaphrodites. The increased OUR can therefore not be adequately compensated by increasing the stirrer speed and the DO% drops below the set point, which does not, however, affect the nematode vitality.

The main considerations in scaling up the production are to ensure a good quality of bacteria to trigger nematode development, to provide a suitable food source, and to synchronize nematode development in order to obtain a pure IJ suspension at the end of the process. The symbiotic bacteria of both genera (*Xenorhabdus* and *Photorhabdus*) may switch between different morphologically and metabolically distinct phases (Akhurst, 1980; Boemare et al., 1997), some of which were shown to produce inferior yields in liquid culture processes of *Heterorhabditis* (Han and Ehlers, 2001) and *Steinernema* (Hirao and Ehlers, 2009). Secondary phase bacteria do result in a lower proportion of IJs

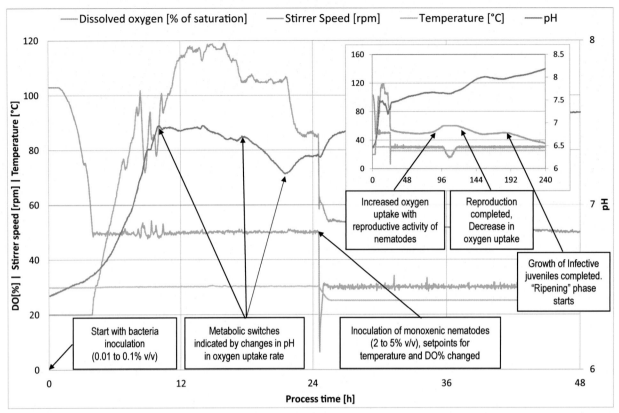

FIGURE 10.4 Changes of typical process parameters of a process in 8000 L for producing *Heterorhabditis bacteriophora*. The oxygen uptake rate (OUR) during the bacteria preculture is much higher than after nematode inoculation and a higher stirrer speed is needed. Set values for temperature and oxygen concentration are decreased when nematodes are added to the culture. The process is finished when all nematodes have developed to infective juveniles at approximately 240 h.

developing than primary form bacteria (Hirao and Ehlers, 2009). The phase variation is by definition not correlated to any changes in bacterial DNA (Owuama, 2001; Park et al., 2007). Phase variation was shown to be induced by osmotic stress, heat treatment or by culturing bacteria at low oxygen levels (Krasomil-Osterfeld, 1994, 1995). To avoid phase variation, bacteria should be cultured in the same media and oxygen stress should be avoided in any scaling-up step. Fortunately, in most cases, bacteria clones can be selected, which are less prone to phase variation. The bacteria should then be grown to the late stationary phase before nematodes are added (Johnigk et al., 2004). High bacteria concentrations increased recovery and finally yield in *S. carpocapsae* and *S. feltiae* (Hirao and Ehlers, 2009).

The recovery from the arrested IJ state is a deciding step for the population dynamics in liquid culture. If the recovery is low, the number of adults in the first generation will be low and few bacteria will be consumed. The offspring of the first generation encounter a relatively high food density and therefore do not develop to IJs but to adults. By this advanced time of the process, the bacteria are beyond the stationary phase and thus of inferior quality and the adults of the second generation will produce hardly any offspring. The final yield in these two-generation

processes is often lower than in one-generation processes, where the bacteria were depleted by a high concentration of synchronously occurring adults in the parental generation (Hirao and Ehlers, 2010). Moreover, the process time is prolonged in two-generation processes, and the higher proportion of stages other than IJs impedes the cleaning procedure when the nematodes are harvested. Residues of dead noninfective stages in the final product will decrease shelf-life.

In *Heterorhabditis*, the consequences of a poor recovery are even more problematic, which is due to the extraordinary set of developmental stages occurring (see earlier). The automictic hermaphrodites are the only stages that can produce offspring in liquid culture. Males of *Heterorhabditis* copulate by attaching only the ventral side of their tail to the females via a suction cup, the *bursa copulatrix*. This behavior requires a gentle controlled approaching of males and females. In an aqueous suspension, nematodes cannot move in a directed way and mating will never occur (Strauch et al., 1994). Therefore, the amphimictic males and females are dead-ends in liquid culture. Nematode offspring only arises from automictic hermaphrodites, and the production process must be steered to push the population to a synchronous development of IJs to

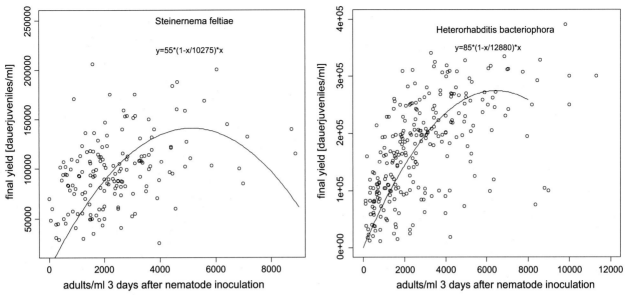

FIGURE 10.5 Relation between adult nematodes at day 3 after nematode inoculation and final yield of infective juveniles in nematode liquid cultures at meter-cubed scale. Data from production processes from 2000 to 2015 at e-nema GmbH for *Steinernema feltiae* and *Heterorhabditis bacteriophora*. The two-parameter model assumes a maximum apparent fecundity of r (IJs per adult) and a critical upper density of adults (N_{crit}). The apparent fecundity is assumed to decrease linearly with increasing adult densities: $IJ = r \left(1 - \dfrac{N}{N_{crit}} \right) N$.

hermaphrodites, which feed voraciously on the bacteria, leaving a scarce food supply for the synchronously hatching juveniles (Johnigk et al., 2004). Only a small proportion of eggs will be laid, but the rest will develop inside the uterus of the hermaphrodites and feed on eggs and tissue of the mother, a phenomenon called endotokia matricida (Johnigk and Ehlers, 1999). In a successful process, the starving first-stage juveniles will exclusively develop into IJs and can be harvested about 10 days after the juveniles have been added to the bacteria.

The attentive reader may now wonder how the obligate amphimictic *Steinernema* species are produced in liquid culture. The answer is that they can copulate even in moving liquids due to a different mating behavior. The males coil around females, which they manage to do rather quickly when randomly encountering a female, even in an aqueous suspension. The coupling is sufficiently tight to withstand water velocities and shear forces in liquid culture (Strauch et al., 1994).

There are numerous parameters that can be changed to optimize the production process. Besides the obvious parameters like DO, pH, and temperature, there are less obvious parameters like the timing of nematode inoculation and the inoculum concentration. Johnigk et al. (2004) have shown that the timing of nematode inoculation has a significant impact on recovery (ie, the proportion of inoculated nematodes recovering from the arrested IJ stage). Recovery has been shown to be dependent on the CO_2 concentration in the medium (Jessen et al., 2000), which can be increased by decreasing the air flow and increasing the vessel pressure.

The density of the bacteria should be as high as possible for maximizing nematode yield. In a batch culture, there are natural limits to the amount of fresh bacteria that can be provided. When females of *S. riobrave* were constantly supplied with fresh bacteria in droplets attached to a cover-slide [hanging drops (Stock and Goodrich-Blair, 2012)], approximately 1000 offspring were obtained per female (Addis et al., 2014), but the net reproduction rate in shaken flasks never surpassed 300 per female (Addis et al., 2015). The recovery of *S. feltiae* and *S. carpocapsae* decreased at lower bacterial densities (Hirao and Ehlers, 2009). If the concentration of adult nematodes feeding on the bacteria is too low, a larger proportion of the F1 generation will develop into adults and start a second generation. If the proportion is too high, there will not be enough bacteria left to support the formation of progeny. An analysis of the relation between the final yield and the concentration of adults 72 h after nematode inoculation for production processes on an industrial scale at e-nema illustrate this phenomenon (Fig. 10.5). The fitting function is based on the assumption that an optimal apparent fecundity r can be reached with least crowding of adults, and this fecundity is reduced with more competition for food at higher adult densities. At an upper-critical density (N_{crit}), no offspring would be produced at all, since the adults would not be able to complete development. For obvious reasons, the data for high adult concentrations at the start of the process are scarce, since IJs are the source of income for the companies and should not be wasted for inoculation. While the model fits the data well on the left side of the graph, the fit on the right part is poor, which could be attributed to insufficient data for this region.

In experimental studies in shaken flasks, adult concentrations up to 6000/mL were tested (Hirao and Ehlers, 2010), and while the model fit the data well on the left side, the IJ yield did not decline to 0 even with very high adult densities. Apparently, the adults are able to compensate for the food deficiency under highly crowded conditions.

10.4 FORMULATION

IJs may either be delivered inside the infected cadavers (see earlier), or they are suspended in water and separated from the growing medium via sedimentation, centrifugation, or sieving techniques. The separation characteristics of IJs of *P. hermaphrodita*, *S. feltiae*, and *H. megidis* were described by Young et al. (2002). The size and the density of the nematodes are mainly explored in separation techniques at a commercial scale. Techniques based on the migration activity of nematodes, while often used to harvest pure IJ suspensions in laboratory scale (Dutky et al., 1964), cannot easily be scaled up to commercial scale. Ultimately, the IJs are stored in water and amended with salts (Strauch et al., 2000). They need to be stirred and aerated in these aqueous suspensions and therefore cannot be shipped this way. Formulations for EPNs must allow sufficient gas exchange and humidity (water activity >0.96). To produce such formulations, excess water is first removed via various sieving techniques. The resulting nematode paste is then mixed with suitable binders (polyurethane sponge, clay, diatomaceous earth, superabsorbant gels) to turn the slurry into a moist powder. This moist powder allows for sufficient gas exchange and at the same time buffers the water content. The high water activity required enables the growth of bacterial and fungal contaminants in the formulation. The addition of antimicrobial compounds is problematic since the products are often used in organic farming and the symbiotic bacteria might be impaired by generic antimicrobial compounds. The nematodes should therefore be cleaned carefully and packed under aseptic conditions (Grewal and Peters, 2005). Detailed recipes for formulating EPNs can be found in Peters (2016).

10.5 QUALITY ASSURANCE AND ASSESSMENT

The quality of a final nematode product is assured by adhering to a standard production protocol. Production should be started from uniform stock cultures. While bacteria clones can easily be stored at −80°C or lyophilized in quantities allowing production over decades, maintaining a stable stock of nematodes is not as easy. Only small amounts may be stored in liquid nitrogen and genetic deterioration by founder effects can hardly be avoided when starting cultures from cryopreserved nematodes (Wang and Grewal, 2002).

On a commercial scale, genetic variability in production can be assured by regularly sourcing monoxenic cultures from nematode populations kept under natural selection conditions where they would propagate on a target pest insect.

Genetic deterioration can be minimized by creating homozygous inbred lines (Bai et al., 2005). No decline in beneficial traits was observed in inbred lines serially cultured in vivo while these traits declined by 30% when the foundation population was cultured in vivo. Since in liquid culture, *Heterorhabditis* sp. can only reproduce by self-fertilization, inbreeding can be accelerated compared to in vivo or solid-state culture. Furthermore, it has been demonstrated that genetic deterioration in inbred lines selected for desiccation tolerance is retarded in liquid culture compared to in vivo culture (Anbesse et al., 2013). Since mating cannot occur in liquid culture, a population cultured exclusively in liquid culture will automatically consist of multiple homozygous inbred lines after a few generations, which is likely to retard genetic deterioration.

The final assessment of a product based on EPNs is typically done with a holistic approach (Grewal and Peters, 2005). The numerous traits influencing the performance of nematodes, like mobility, lipid reserves, and retention of symbiotic bacteria, are not assessed independently, but the performance is checked in a bioassay where a dose close to the LD_{50} is applied to suitable host insects in an arena similar to the target environment (sand or peat). The variability of assay results proved to be lower when larvae of the lesser mealworm (*Tenebrio molitor*) were used instead of *G. mellonella* (Peters, 2000). A detailed description of standardized bioassays for nematodes can be found in Glazer and Lewis (2000).

10.6 CONCLUSIONS

The extent of the use of EPNs in biocontrol would not be thinkable without efficient mass-rearing techniques. The development and the increase in scale of the liquid culture resulted in a reduction of production costs by more than 75% in the past 30 years, and this process has not finished yet. As a result, the use of EPNs in low-value crops is now at hand. The efficacy and the price of using *H. bacteriophora* in maize in Europe against the Western corn rootworm (*Diabrotica virgifera virgifera*) can compete with that of chemical pesticides (Toepfer et al., 2009, 2010).

Another important prerequisite for the success of EPNs as biocontrol agents is the low level of regulation. Nematodes are macro-organisms and, like the equally widely used beneficial arthropods, do not need to be registered as plant protection agents in most countries. It is probably due to this liberal legislation that the nematode production business is versatile and various culture techniques coexist. This coexistence is important to sustain the innovative potential of EPNs. The worldwide concentration of the liquid culture business in three principal companies has resulted in a reduction of species produced. There are more than 100 species (Nguyen, 2015) known to science but less than five produced in significant numbers. While some nematode species are not produced due to a limited market size, others seem to be not adapted for in vitro culturing, such as *S. scarabaei*, a species with outstanding activity against grubs (Koppenhofer et al., 2008). The alternative production methods, probably on easily reared scarabaeid larvae (*Pachnoda* sp.) rather than waxmoth larvae, can overcome this limitation and, subsequently, contribute to the success story of EPNs.

Despite limitations in cost, efficiency, and scale, in vivo production is a cottage industry that has managed to sustain itself throughout the development of numerous larger companies producing nematodes in vitro (Gaugler et al., 2000; Gaugler and Han, 2002). In vivo production bases its success especially on small-market niches to control susceptible insect targets in smaller areas or specific sites of the field, such as fungus gnats in nursery or green house conditions, or small hive beetle in beehives.

In vivo and solid-state production is likely to continue as local businesses for niche markets where competition by in vitro producers is limited, and in developing countries where labor is inexpensive. Another area for which in vivo production may be highly suitable is local production by growers or grower cooperatives (Gaugler, 1997). A significant benefit of a more local production is the avoidance of long delivery chains at controlled temperature, which is a major cost driver for worldwide delivery of centrally produced EPNs. We anticipate that innovations to improve efficiency will enable in vivo production to play an expanded role in pest management programs.

REFERENCES

Addis, T., Teshome, A., Strauch, O., Ehlers, R.-U., 2014. Life history trait analysis of the entomopathogenic nematode *Steinernema riobrave*. Nematology 16, 929–936.

Addis, T., Teshome, A., Strauch, O., Ehlers, R.-U., 2015. Life history trait analysis of the entomopathogenic nematode *Steinernema feltiae* provides the basis for prediction of dauer juvenile yields in monoxenic liquid culture. Appl. Microbiol. Biotechnol. http://dx.doi.org/10.1007/s00253-015-7220-y.

Akhurst, R.J., 1980. Morphological and functional dimorphism in *Xenorhabdus* spp., bacteria symbiotically associated with the insect pathogenic nematodes, *Neoaplectana* and *Heterorhabditis*. J. Gen. Microbiol. 121, 303–309.

Anbesse, S., Sumaya, N.H., Dörfler, A.V., Strauch, O., Ehlers, R.-U., 2013. Selective breeding for desiccation tolerance in liquid culture provides genetically stable inbred lines of the entomopathogenic nematode *Heterorhabditis bacteriophora*. Appl. Microbiol. Biotechnol. 97, 731–739.

Aumann, J., Ehlers, R.-U., 2001. Physico-chemical properties and mode of action of a signal from the symbiotic bacterium *Photorhabdus luminescens* inducing dauer juvenile recovery in the entomopathogenic nematode *Heterorhabditis bacteriophora*. Nematology 3, 849–853.

Bai, C., Shapiro-Ilan, D.-I., Gaugler, R., Hopper, K.-R., 2005. Stabilization of beneficial traits in *Heterorhabditis bacteriophora* through creation of inbred lines. Biol. Control 32, 220–227.

Bathon, H., 1996. Impact of entomopathogenic nematodes on non-target hosts. Biocontrol Sci. Technol. 6, 421–434.

Bedding, R.A., 1981. Low cost in vitro mass production of *Neoaplectana* and *Heterorhabditis* species (Nematoda) for field control of insect pests. Nematologica 27, 109–114.

Bedding, R.A., 1984. Large scale production, storage and transport of the insect parasitic nematodes *Neoaplectana* spp. and *Heterorhabditis* spp. Ann. Appl. Biol. 104, 117–120.

Bedding, R.A., 1990. Logistics and strategies for introducing entomopathogenic nematode technology into developing countries. In: Poinar, G.O. (Ed.), Entomopathogenic Nematodes in Biological Control. CRC Press, Boca Raton, FL, pp. 233–246.

Bedding, R.A., Stanfield, M.A., Crompton, G., 1991. Apparatus and Method for Rearing Nematodes, Fungi, Tissue Cultures and the Like, and for Harvesting Nematodes. World Patent. WO 91/15569.

Blaxter, M.L., De Ley, P., Garey, J.R., Liu, L.X., Scheldemann, P., Vierstraete, A., Vanfleteren, J.R., Mackey, L.Y., Dorris, M., Frisse, L.M., Vida, J.T., Thomas, W.K., 1998. A molecular evolutionary framework for the phylum Nematoda. Nature 392, 71–75.

Blinova, S.L., Ivanova, E.S., 1987. Culturing the Nematode-Bacterial Complex of *Neoaplectana carpocapsae* in Insects. Helminths of Insects. Amerind Publishing Company, New Delhi, pp. 13–21.

Boemare, N., Givaudan, A., Brehelin, M., Laumond, C., 1997. Symbiosis and pathogenicity of nematode-bacterium complexes. Symbiosis 22, 21–45.

Boff, M.I.C., Wiegers, G.L., Gerritsen, L.J.M., Smits, P.H., 2000. Development of the entomopathogenic nematode *Heterorhabditis megidis* strain NLH-E 87.3 in *Galleria mellonella*. Nematology 2, 303–308.

Brown, I.M., Gaugler, R., 1997. Temperature and humidity influence emergence and survival of entomopathogenic nematodes. Nematologica 43, 363–375.

Burman, M., Pye, A.E., 1980. *Neoaplectana carpocapsae*: Respiration of infective juveniles. Nematologica 26, 214–219.

Carne, P.B., Reed, E.M., 1964. A simple apparatus for harvesting infective stage nematodes emerging from their insect hosts. Parasitol 54, 551–553.

Ciche, T.-A., Darby, C., Ehlers, R.-U., Forst, S., Goodrich-Blair, H., 2006. Dangerous liaisons: the symbiosis of entomopathogenic nematodes and bacteria. Biol. Control 38, 22–46.

Deseö, K.V., Ruggeri, L., Lazzari, G., 1990. Mass-production and quality control of entomopathogenic nematodes in *Galleria mellonella* L. larvae. Proc. Int. Colloq. Invertebr. Pathol. Microb. Control 5, 250.

Dolinski, C., Del Valle, E.E., Burla, R.S., Machado, I.R., 2007. Biological traits of two native Brazilian entomopathogenic nematodes (Heterorhabditidae: Rhabditidae). Nematol. Bras. 31, 180–185.

Dorris, M., Viney, M.E., Blaxter, M.L., 2002. Molecular phylogenetic analysis of the genus *Strongyloides* and related nematodes. Int. J. Parasitol. 32, 1507–1517.

Dowds, B.C.A., Peters, A., 2002. Virulence mechanisms. In: Gaugler, R. (Ed.), Entomopathogenic Nematology. CABI Publishing, Wallingford, pp. 79–98.

Dunphy, G.B., Webster, J.M., 1989. The monoxenic culture of *Neoaplectana carpocapsae* DD-136 and *Heterorhabditis heliothidis*. Rev. Nématol. 12, 113–123.

Dutky, S.R., Thompson, J.V., Cantwell, G.E., 1964. A technique for the mass propagation of the DD-136 nematode. J. Insect Pathol. 6, 417–422.

Ehlers, R.-U., Shapiro-Ilan, D.I., 2005. Mass production. In: Grewal, P.S., Ehlers, R.-U., Shapiro-Ilan, D.I. (Eds.), Nematodes as Biocontrol Agents. CABI Publishing, Wallingford, UK, pp. 65–78.

Ehlers, R.-U., Wulff, A., Peters, A., 1997. Pathogenicity of axenic *Steinernema feltiae*, *Xenorhabdus bovienii*, and the bacto-helminthic complex to larvae of *Tipula oleracea* (Diptera) and *Galleria mellonella* (Lepidoptera). J. Invertebr. Pathol. 69, 212–217.

Ehlers, R.-U., Niemann, I., Hollmer, S., Strauch, O., Jende, D., Shanmugasundaram, M., Mehta, U.K., Easwaramoorthy, S.K., Burnell, A., 2000. Mass production potential of the bacto-helminthic biocontrol complex *Heterorhabditis indica-Photorhabdus luminescens*. Biocontrol Sci. Technol. 10, 607–616.

Flanders, K.L., Miller, J.M., Shields, E.J., 1996. *In vivo* production of *Heterorhabditis bacteriophora* 'Oswego' (Rhabditida: Heterorhabditidae), a potential biological control agent for soil-inhabiting insects in temperate regions. J. Econ. Entomol. 89, 373–380.

Forst, S., Clarke, D., Gaugler, R., 2002. Bacteria-nematode symbiosis. In: Gaugler, R. (Ed.), Entomopathogenic Nematology. CABI Publishing, New York, pp. 57–77.

Friedman, M.J., 1990. Commercial production and development. In: Poinar, G.O. (Ed.), Entomopathogenic Nematodes in Biological Control. CRC Press, Boca Raton, FL, pp. 153–172.

Gaugler, R., 1997. Alternative paradigms for commercializing biopesticides. Phytoparasitica 25, 179–182.

Gaugler, R., Brown, I., 2001. Apparatus and Method for Mass Production of Insecticidal Nematodes. US Patent No. 09/533, 180.

Gaugler, R., Han, R., 2002. Production technology. In: Gaugler, R. (Ed.), Entomopathogenic Nematology. CABI Publishing, Oxon, UK, pp. 289–310.

Gaugler, R., Grewal, P., Kaya, H.K., Smith Fiola, D., 2000. Quality assessment of commercially produced entomopathogenic nematodes. Biol. Control 17, 100–109.

Gaugler, R., Brown, I., Shapiro, I.D., Atwa, A., 2002. Automated technology for in vivo mass production of entomopathogenic nematodes. Biol. Control 24, 199–206.

Glaser, R.W., 1940. Continued culture of a nematode parasitic in the Japanese beetle. J. Exp. Zool. 84, 1–12.

Glazer, I., Lewis, E.E., 2000. Bioassays for entomopathogenic nematodes. In: Navon, A., Ascher, K.R.S. (Eds.), Bioassays of Entomopathogenic Microbes and Nematodes. CABI Publishing, Wallingford, Oxon, UK, pp. 229–247.

Gouge, D.H., Snyder, J.L., 2006. Temporal association of entomopathogenic nematodes (Rhabditida : Steinernematidae and Heterorhabditidae) and bacteria. J. Invertebr. Pathol. 91, 147–157.

Grewal, P.S., Peters, A., 2005. Formulation and quality. In: Grewal, P.S., Ehlers, R.-U., Shapiro-Ilan, D.I. (Eds.), Nematodes as Biocontrol Agents. CABI Publishing, Wallingford, pp. 79–90.

Grewal, P.S., Selvan, S., Gaugler, R., 1994. Thermal adaptation of entomopathogenic nematodes: niche breadth for infection, establishment, and reproduction. J. Therm. Biol. 19, 245–253.

Grewal, P.S., Converse, V., Georgis, R., 1999. Influence of production and bioassay methods on infectivity of two ambush foragers (Nematoda: Steinernematidae). J. Invertebr. Pathol. 73, 40–44.

Griffin, C.T., O'Callaghan, K.M., Dix, I., 2001. A self-fertile species of *Steinernema* from Indonesia: further evidence of convergent evolution amongst entomopathogenic nematodes? Parasitology 122, 181–186.

Han, R., Ehlers, R.-U., 2001. Effect of *Photorhabdus luminescens* phase variants on the *in vivo* and *in vitro* development and reproduction of the entomopathogenic nematodes *Heterorhabditis bacteriophora* and *Steinernema carpocapsae*. FEMS Microbiol. Ecol. 35, 239–247.

Han, R., Cao, L., Liu, X., 1992. Relationship between medium composition, inoculum size, temperature and culture time in the yields of *Steinernema* and *Heterorhabditis* nematodes. Fundam. Appl. Nematol. 15, 223–229.

Han, R., Cao, L., Liu, X., 1993. Effects of inoculum size, temperature and time on *in-vitro* production of *Steinernema carpocapsae* Agriotos. Nematologica 39, 366–375.

Han, R.C., Ehlers, R.-U., 2000. Pathogenicity, development, and reproduction of *Heterorhabditis bacteriophora* and *Steinernema carpocapsae* under axenic *in vivo* conditions. J. Invertebr. Pathol. 75, 55–58.

Hirao, A., Ehlers, R.-U., 2009. Influence of cell density and phase variants of bacterial symbionts (*Xenorhabdus* spp.) on dauer juvenile recovery and development of biocontrol nematodes *Steinernema carpocapsae* and *S. feltiae* (Nematoda: Rhabditida). Appl. Microbiol. Biotechnol. 84, 77–85.

Hirao, A., Ehlers, R.-U., 2010. Influence of inoculum density on population dynamics and dauer juvenile yields in liquid culture of biocontrol nematodes *Steinernema carpocapsae* and *S. feltiae* (Nematoda: Rhabditida). Appl. Microbiol. Biotechnol. 85, 507–515.

Hominick, W.M., Briscoe, B.R., Del Pino, F.G., Heng, J., Hunt, D.J., Kozodoy, E., Mracek, Z., Nguyen, K.B., Reid, A.P., Spiridonov, S., Stock, P., Sturhan, D., Waturu, C., Yoshida, M., 1997. Biosystematics of entomopathogenic nematodes: current status, protocols and definitions. J. Helminthol. 71, 271–298.

House, H.L., Welch, H.E., Cleugh, T.R., 1965. Food medium of prepared dog biscuit for the mass-production of the nematode DD-136 (Nematoda: Steinernematidae). Nature 206, 847.

Jansson, R.K., Lecrone, S.H., 1994. Application methods for entomopathogenic nematodes (Rhabditida: Heterorhabditidae) aqueous suspensions versus infected cadavers. Fla. Entomol. 77, 281–284.

Jansson, R.K., Lecrone, S.H., Gaugler, R., 1993. Field efficacy and persistence of entomopathogenic nematodes (Rhabditida: Steinernematidae, Heterorhabditidae) for control of sweetpotato weevil (Coleoptera: Apionidae) in southern Florida. J. Econ. Entomol. 86, 1055–1063.

Jessen, P., Strauch, O., Wyss, U., Luttmann, R., Ehlers, R.U., 2000. Carbon dioxide triggers recovery from dauer juvenile stage in entomopathogenic nematodes (*Heterorhabditis* spp.). Nematology 2, 319–324.

Johnigk, S.A., Ehlers, R.-U., 1999. Juvenile development and life cycle of *Heterorhabditis bacteriophora* and *H. indica* (Nematoda: Heterorhabditidae). Nematology 1, 251–260.

Johnigk, S.A., Ecke, F., Poehling, M., Ehlers, R.-U., 2004. Liquid culture mass production of biocontrol nematodes, *Heterorhabditis bacteriophora* (Nematoda: Rhabditida): improved timing of dauer juvenile inoculation. Appl. Microbiol. Biotechnol. 64, 651–658.

Kaya, H.K., Stock, S.P., 1997. Techniques in insect nematology. In: Lacey, L.A. (Ed.), Manual of Techniques in Insect Pathology. Academic Press, San Diego, pp. 281–324.

Koppenhofer, A.M., Fuzy, E.M., 2003. Biological and chemical control of the Asiatic garden beetle, *Maladera castanea* (Coleoptera: Scarabaeidae). J. Econ. Entomol. 96, 1076–1082.

Koppenhofer, A.-M., Rodriguez-Saona, C.-R., Polavarapu, S., Holdcraft, R.-J., 2008. Entomopathogenic nematodes for control of *Phyllophaga georgiana* (Coleoptera: Scarabaeidae) in cranberries. Biocontrol Sci. Technol. 18, 21–31.

Krasomil-Osterfeld, K., 1994. Phase variation in *Photorhabdus*, *Xenorhabdus* and other bacteria. A review. In: Genetics of Entomopathogenic Nematode – Bacterium Complexes. European Commission Directorate-General XII, Science, Research and Development Environment Research Programme, Luxembourg, Belgium, pp. 70–79.

Krasomil-Osterfeld, K.C., 1995. Influence of osmolarity on phase shift in *Photorhabdus luminescens*. Appl. Environ. Microbiol. 61, 3748–3749.

Krausse, A., 1917. Forstentomologische Exkursionen ins Eggegebirge zum Studium der Massenvermehrung der *Cephaleia abietis* L. Arch. Naturgesch. 6, 46–49.

Lindegren, J.E., Valero, K.A., Mackey, B.E., 1993. Simple *in vivo* production and storage methods for *Steinernema carpocapsae* infective juveniles. J. Nematol. 25, 193–197.

Lunau, S., Stoessel, S., Schmidt-Peisker, A.J., Ehlers, R.-U., 1993. Establishment of monoxenic inocula for scaling-up in vitro cultures of the entomopathogenic nematodes *Steinernema* spp. and *Heterorhabditis* spp. Nematologica 39, 385–399.

Mamiya, Y., 1989. Comparison of the infectivity of *Steinernema kushidai* (Nematoda: Steinernematidae) and other steinernematid and heterorhabditid nematodes from three different insects. Appl. Entomol. Zool. 24, 302–308.

Milstead, J.E., Poinar Jr., G.O., 1978. A new entomophagous nematode for pest management systems. Calif. Agric. 32, 12.

Nguyen, K.B., 2015. Species of *Steinernema*, Species of *Heterorhabditis*. http://entnemdept.ufl.edu/nguyen/morph/steinsp1.htm.

Nguyen, K.B., Smart, G.C.J.R., 1990. *Steinernema scapterisci*, new species (Rhabditida: Steinernematidae). J. Nematol. 22, 187–199.

Owuama, C.I., 2001. Entomopathogenic symbiotic bacteria, *Xenorhabdus* and *Photorhabdus* of nematodes. World J. Microbiol. Biotechnol. 17, 505–515.

Park, Y., Herbert, E.-E., Cowles, C.-E., Cowles, K.-N., Menard, M.-L., Orchard, S.-S., Goodrich-Blair, H., 2007. Clonal variation in *Xenorhabdus nematophila* virulence and suppression of *Manduca sexta* immunity. Cell. Microbiol. 9, 645–656.

Peters, A., 2000. Insect based assay for entomopathogenic nematode infectiousness: definitions, guidelines, problems. IOBC/WPRS Bull. 23, 109–114.

Peters, A., 2016. Formulation of nematodes. In: Glare, T.A., Moran-Diez, M.E. (Eds.), Microbial-Based Pesticides – Methods and Protocols. Springer, Netherlands (in press).

Peters, A., Gouge, D.H., Ehlers, R.U., Hague, N.G.M., 1997. Avoidance of encapsulation by *Heterorhabditis* spp. infecting larvae of *Tipula oleracea*. J. Invertebr. Pathol. 70, 161–164.

Poinar, G.O., Grewal, P.S., 2012. History of entomopathogenic nematology. J. Nematol. 44, 153–161.

Rainey, F.A., Ehlers, R.-U., Stackebrandt, E., 1995. Inability of the polyphasic approach to systematics to determine the relatedness of the genera *Xenorhabdus* and *Photorhabdus*. Int. J. Syst. Bacteriol. 45, 379–381.

Salma, J., Shahina, F., 2012. Mass production of eight Pakistani strains of entomopathogenic nematodes (Steinernematidae and Heterorhabditidae). Pak. J. Nematol. 30, 1–20.

Schwarz, T.S., Wells, S.A., 1999. Continuous deflection separation of stormwater particulates. Adv. Filtr. Sep. Technol. 12, 219–226.

Shapiro-Ilan, D.I., Gaugler, R., 2002. Production technology for entomopathogenic nematodes and their bacterial symbionts. J. Ind. Microbiol. Biotechnol. 28, 137–146.

Shapiro-Ilan, D.I., Glazer, I., 1996. Comparison of entomopathogenic nematode dispersal from infected hosts versus aqueous suspension. Environ. Entomol. 25, 1455–1461.

Shapiro-Ilan, D.I., Lewis, E.E., 1999. Comparison of entomopathogenic nematode infectivity from infected hosts versus aqueous suspension. Environ. Entomol. 28, 907–911.

Shapiro-Ilan, D.I., Lewis, E.E., Behle, R.W., McGuire, M.R., 2001. Formulation of entomopathogenic nematode-infected cadavers. J. Invertebr. Pathol. 78, 17–23.

Shapiro-Ilan, D.I., Gaugler, R., Tedders, W.L., Brown, I., Lewis, E.E., 2002. Optimization of inoculation for *in vivo* production of entomopathogenic nematodes. J. Nematol. 34, 343–350.

Shapiro-Ilan, D.I., Lewis, E.E., Son, Y., Tedders, W.L., 2003. Superior efficacy observed in entomopathogenic nematodes applied in infected-host cadavers compared with application in aqueous suspension. J. Invertebr. Pathol. 83, 270–272.

Shapiro-Ilan, D.I., Han, R., Qiu, X., 2014a. Production of entomopathogenic nematodes. In: Morales-Ramos, J.A., Rojas, M.G., Shapiro-Ilan, D.I. (Eds.), Mass Production of Beneficial Organisms. Academic Press, San Diego, pp. 321–355.

Shapiro-Ilan, D.I., Tedders, W.L., Morales-Ramos, J.A., Rojas, M.G., 2014b. System and Method for Producing Beneficial Parasites. US Patent. 1 13/217,956, DN 172.07.

Shields, E., 2012. Management of Alfalfa Snout Beetle. http://www.nnyagdev.org/wp-content/uploads/2012/01/Shields_ASBPamphlet_FINAL.pdf.

Southey, J.F., 1986. Laboratory Methods for Work with Plant and Soil Nematodes. Ministry of Agriculture, Fisheries and Food. HMSO, London.

Stock, S.P., Goodrich-Blair, H., 2012. Nematode parasites, pathogens and associates of insects and invertebrates of economic importance. In: Lacey, L.A. (Ed.), Manual of Techniques in Insect Pathology, second ed. Academic Press, San Diego, pp. 373–426.

Strauch, O., Stoessel, S., Ehlers, R.-U., 1994. Culture conditions define automictic or amphimictic reproduction in entomopathogenic rhabditid nematodes of the genus *Heterorhabditis*. Fundam. Appl. Nematol. 17, 575–582.

Strauch, O., Niemann, I., Neumann, A., Schmidt, A.J., Peters, A., Ehlers, R.U., 2000. Storage and formulation of the entomopathogenic nematodes *Heterorhabditis indica* and *H. bacteriophora*. BioControl 45, 483–500.

Toepfer, S., Burger, S., Ehlers, R.-U., Peters, A., Kuhlmann, U., 2009. Controlling western corn rootworm larvae with entomopathogenic nematodes: effect of application techniques on plant-scale efficacy. J. Appl. Entomol. 134, 467–480.

Toepfer, K., Knuth, P., Peters, A., Burger, R., 2010. Insektenpathogene Nematoden gegen Wurzelbohrer. Mais 2, 68–70.

Wang, X., Grewal, P.S., 2002. Rapid genetic deterioration of environmental tolerance and reproductive potential of an entomopathogenic nematode during laboratory maintenance. Biol. Control 23, 71–78.

White, G.F., 1927. A method for obtaining infective nematode larvae from cultures. Science 66, 302–303.

Wilson, M.J., Glen, D.M., Pearce, J.D., Rodgers, P.B., 1994. Effects of different bacterial isolates on growth and pathogenicity of the slug-parasitic nematode *Phasmarhabditis hermaphrodita*. In: Proceedings Brighton Crop Protection Conference, Pests and Diseases, vol. 3, pp. 1055–1060.

Wilson, M.J., Grewal, P.S., 2005. Biology, production and formulation of slug-parasitic nematodes. In: Grewal, P.S., Ehlers, R.-U., Shapiro-Ilan, D.I. (Eds.), Nematodes as Biocontrol Agents. CABI Publishing, Wallingford, pp. 421–430.

Woodring, J.L., Kaya, H.K., 1988. Steinernematid and Heterorhabditid Nematodes: A Handbook of Techniques. Arkansas Agricultural Experiment Station, Fayetteville, Arkansas.

Wouts, W.M., 1981. Mass production of the entomogenous nematode *Heterorhabditis heliothidis* (Nematoda: Heterorhabditidae) on artificial media. J. Nematol. 13, 467–469.

Wright, D.J., Peters, A., Schroer, S., Fife, J.P., 2005. Application technology. In: Grewal, P.S., Shapiro-Ilan, D., Ehlers, R.-U. (Eds.), Nematodes as Biocontrol Agents. CAB International, pp. 91–106.

Young, J.M., Dunnill, P., Pearce, J.D., 2002. Separation characteristics of liquid nematode cultures and the design of recovery operations. Biotechnol. Prog. 18, 29–35.

Part IV

Applied Research and Implementation of Microbial Control Agents for Pest Control

Chapter 11

Current and Potential Applications of Biopesticides to Manage Insect Pests of Maize

A.J. Gassmann, E.H. Clifton

Iowa State University, Ames, IA, United States

11.1 INTRODUCTION

Maize is widely grown in many countries and attacked by a diverse array of insect pests (Steffey et al., 1999). Some of the most injurious pests of maize are Lepidoptera, which occupy several feeding niches on maize plants. These include foliage feeders (eg, fall armyworm *Spodoptera frugiperda)*, stalk borers (eg, European corn borer *Ostrinia nubilalis* and maize stalk borer *Busseola fusca*), and ear feeders (eg, corn earworm *Helicoverpa zea*). Multiple species of Coleoptera in the genus *Diabrotica* feed on the roots of maize as larvae (and on maize silk and pollen as adults). Additionally, there are secondary pests that feed on newly planted seeds (eg, seedcorn maggot *Delia platura* and wireworms in the genus *Agriotes* [Coleoptera: Elateridae]) and seedlings (eg, black cutworm, *Agrotis ipsilon*).

Traditionally, the management of maize pests has focused on cultural control and conventional insecticides. However, a wide variety of microbial control agents (MCAs) have the potential to enhance pest management in maize. Because some key pests complete their injurious stage below the ground, the study of interactions with entomopathogens may be complicated. Two such pests are the western corn rootworm *Diabrotica virgifera virgifera* and the northern corn rootworm *Diabrotica barberi*. Soil-borne pests may often interact with multiple species of entomopathogens creating complex multitrophic interactions. Either inundative or conservation biological control with entomopathogens may aid in suppressing soil-borne pests.

Beginning the 1990s, maize that was genetically engineered to produce insecticidal toxins from the bacterium *Bacillus thuringiensis* (Bt) was brought to the market. Bt maize has provided an effective tool to manage several lepidopteran and coleopteran pests although recent cases of pest resistance illustrate the limitations of this technology. It is clear that the development of more integrated approaches for the management of maize pests will be needed to increase the sustainability of maize production. In this chapter we describe the potential application of MCAs for the management of maize pests, the current uses of Bt maize and challenges posed by pest resistance, and how biopesticides may be combined to enhance pest suppression and preserve yield.

11.2 FUNGI

11.2.1 Hypocreales

Fungal entomopathogens, in particular the hypocreallean Ascomycetes *Metarhizium* and *Beauveria*, can occur naturally in maize fields (Bruck and Lewis, 2001; Pilz et al., 2008; Rudeen et al., 2013) and in some cases may contribute to the suppression of insect pests (St Leger, 2008). Additionally, studies have shown that entomopathogenic fungi (EPF) can form endophytic symbioses in the rhizosphere (Hu and St Leger, 2002). Fungi that could persist in proximity to maize roots, or even within the root tissues, may function to protect plants from feeding injury. Kabaluk and Ericsson (2007) found higher concentrations of *Metarhizium brunneum* F52 within the rhizosphere over the course of the growing season following the inoculation of seeds (3.8×10^8 conidia/seed; 5×10^{13} conidia/ha), suggesting the potential application of EPF as a seed treatment (ie, coating) for the management of maize pests.

Many isolates of *Metarhizium* and *Beauveria* have been screened as potential MCAs for maize pests. Laboratory assays with *M. brunneum* achieved up to 47% infection of western corn rootworm when larvae were directly exposed to suspensions of the fungus (1×10^7 conidia/mL; Pilz et al., 2007). Ansari et al. (2009) exposed wireworms (*Agriotes lineatus*) to suspensions of fungi (1×10^8 conidia/mL) and found that some strains of *M. brunneum* were pathogenic, but *Beauveria bassiana* was not. In another laboratory study, Hoffmann et al. (2014) found that soil inoculated with *B. bassiana* or *M. brunneum* (1×10^4 to 1×10^7 conidia/g soil) increased the mortality of western corn rootworm larvae. It should be noted that the type of soil used in these bioassays is important. Bioassays evaluating *B. bassiana* against southern corn rootworm *Diabrotica undecimpunctata howardi* showed that larval mortality was significantly higher in sterile soil compared to nonsterile soil, suggesting that other soil microbiota could be fungistatic and inhibit infections (Jaronski, 2007). Furthermore, Jaronski (2007) showed that isolates of *B. bassiana* varied in efficacy against southern corn rootworm and that the efficacy of strains often differed between soil types, indicating both an effect of soil type and strain.

Because maize is grown in many different regions and soil types around the world, it is unlikely that one isolate will perform consistently across locations. Extensive reviews of the ecological factors that may impede the efficacy of fungal entomopathogens in the field are provided by Jaronski (2007, 2010). An important conclusion of this work is that the effective use of EPF in pest management requires careful consideration of EPF biology and the biotic and abiotic factors that govern efficacy and that EPF should not be considered simply as another chemical insecticide.

Krueger and Roberts (1997) applied strains of *M. anisopliae* and *B. bassiana* to field plots of maize infested with southern corn rootworm, using either a low rate of inoculum (2.5–4.6×10^{10} CFU/m row) or a high rate (2.5–4.6×10^{11} CFU/m row). In both years of their experiment, the high rate of *M. anisopliae* achieved a significantly lower percentage of goosenecked plants (treated plots = 3.3% and 21.3% and control plots = 40.2% and 45.9% in 1988 and 1989, respectively) and significantly lowered root injury (1–6 rating scale; treated plots = 1.60 and 2.41 and control plots = 3.05 and 3.44 in 1988 and 1989, respectively). By contrast, both the low and high rates of *B. bassiana* failed to reduce the percentage of goosenecked plants, and only the high rate of *B. bassiana* achieved significantly lower root injury (treated plots = 1.95 and 2.94 and control plots = 3.05 and 3.44 in 1988 and 1989, respectively). Pilz et al. (2009) achieved a 31% reduction of western corn rootworm emergence and a 23% reduction of root injury in maize fields after applying *M. brunneum* to the soil at a rate of 4–7×10^{13} spores/ha using colonized barley kernels. However, the reduction in emergence of rootworm adults and root injury was less than that achieved with the chemical insecticides tefluthrin and clothianidin (Pilz et al., 2009).

Rather than inoculating soil, some studies have investigated coating the epidermis of maize seeds with EPF. Kabaluk and Ericsson (2007) treated maize seeds with *M. brunneum* F52 (3.8×10^8 conidia/seed; 5×10^{13} conidia/ha) prior to planting, which resulted in a significantly greater stand density of 77.9% in the treated plots compared to

66.7% in the controls and yielded 9.6 Mg/ha compared to 7.6 Mg/ha in the controls. The fungus sporulating from larval wireworm (*Agriotes obscurus*) cadavers in the treated plots was identified as *M. brunneum* F52, suggesting that EPF persisted at virulent levels. It is important to note that wireworm larvae feed directly on newly germinated maize seeds, a feeding niche that would bring them into direct contact with the treated seeds. Nonetheless, these results point to the potential utility of treating seeds with EPF as a management option for early season pests of maize. Additionally, evidence suggests that EPF may translocate nitrogen and other resources from insect cadavers to plants (Behie et al., 2012).

Past research also has addressed the potential of EPF to suppress aboveground pests. This work has focused primarily on Lepidoptera, which comprise most of the economically important aboveground pests of maize. In some cases, the suppression of aboveground maize pests may occur from natural populations of EPF. For example, *B. bassiana* can sometimes induce an epizootic in European corn borer. However, the proportion of larvae killed in these events is often low and takes place later in the season after the crop has already sustained injury (Lewis and Cossentine, 1986). Additionally, *B. bassiana* was found to be a mortality factor of southwestern corn borer, *Diatraea grandiosella*, in maize fields in Texas (Knutson and Gilstrap, 1990).

More commonly, research on aboveground pests of maize has focused on inundative applications of EPF. *B. bassiana* has been tested in several countries against European corn borer and in some cases reduced larval populations by more than 90% using rates of 1×10^{13} conidia/ha (Bartlett and Lefebvre, 1934; Ferron, 1981; Hussey and Tinsley, 1981; Riba, 1984). A field study by Bing and Lewis (1991) found that tunneling by European corn borer larvae was reduced to 4.0 and 5.6 cm/plant (compared to 6.8 cm/plant in the untreated control) when maize plants received *B. bassiana* as either granules applied to the whorl at a rate of 3.0×10^{10} to 2.5×10^{12} conidia/ha or a suspension (1 mL at 1.0×10^6 to 4.55×10^7 conidia/mL) injected into the base of the plant. It should be noted that the timing of EPF applications to maize is a critical factor. In general, foliar applications of EPF targeting European corn borer should be applied shortly after peak hatch when larvae are feeding on the foliage (Mason et al., 1996). If applied later, larvae may have already tunneled into the interior of the maize plant where they can be sheltered from encountering the fungal conidia.

In addition to infecting pests through direct contact following foliar application, *B. bassiana* may form an endophytic relationship with maize plants, possibly infecting European corn borer larvae via ingested plant tissues (Bing and Lewis, 1991; Wagner and Lewis, 2000). *B. bassiana* is likely well suited to persist in the humid maize stalk tunnels formed by boring insect larvae where infection could be enhanced, because propagules would be sheltered from ultraviolet light and desiccation.

Other aboveground maize pests targeted with EPF include fall armyworm and adult corn rootworm. A lab study treated fall armyworm larvae with isolates of either *M. brunneum* or *B. bassiana* (containing 1×10^8 conidia/mL) in the form of either topical applications or feeding with inoculated plant tissues (foliage was dipped in the same concentration of conidia), resulting in >70% mortality for both methods of application with all the isolates of *M. brunneum* and *B. bassiana* (control mortality = 30%; Sánchez-Peña et al., 2007). Bruck and Lewis (2002) applied foliar applications of *B. bassiana* granules (ca. 8.4×10^7 conidia/plant) to maize at various phenological stages to manage adult western corn rootworm and in 1 year of the study, found that pollen-shed applications infected 51% of the beetles compared to a 6.0% infection in control plots. However, infection was less pronounced for whorl stage applications across all years.

It is likely that strains of EPF differ in their efficacy against maize pests and the extent to which they are adapted to the maize fields. Wraight et al. (2010) conducted a comprehensive screen of *B. bassiana* isolates selected from the USDA-ARS Collection of Entomopathogenic Fungi and found that the widely available GHA strain had higher LC_{50} values than most other strains when tested against larval European corn borer, black cutworm, and fall armyworm, but the GHA strain had a lower LC_{50} for larval corn earworm. The same study found that one of the strains evaluated (strain 1200) was more virulent against several maize pests compared to GHA. Similarly, Rudeen et al. (2013) detected ca. 28% mortality of western corn rootworm larvae with two field-collected strains of *Metarhizium* spp. applied at 6.1×10^6 conidia/g soil compared to ca. 5% mortality with the widely used *M. brunneum* F52 strain at the same rate. These results highlight the potential to improve management of maize pests with EPF through proper strain selection.

11.2.2 Microsporidia

Among the microsporidia, *Nosema pyrausta* is the most studied for insect pests of maize and one of the few pathogens in this group that has shown efficacy in the field. In the 1920s, the microsporidium, *Perezia pyraustae*, was found in European corn borer in France (Paillot, 1927), and *N. pyrausta* was identified from European corn borer in Russia (Kotlán, 1928). Steinhaus (1952) suggested that the two pathogens may actually be the same species due to a high similarity in morphology and pathogenicity. Most of the literature after 1975 refers only to the genus *Nosema*, because *Perezia* was no longer recognized taxonomically (Lewis et al., 2009). In the 1950s, researchers collected similar microsporidia from European corn borer populations in North America

(Hall, 1952). Tokarev et al. (2015) employed rRNA sequencing to determine that the Russian isolate of *N. pyrausta* was approximately 99.7% similar to isolates from China and the United States, yet was a distinct haplotype.

N. pyrausta appears to be an important natural pathogen of European corn borer populations in the United States (Lewis et al., 2009). While it is difficult to estimate the level of mortality inflicted on European corn borer by *N. pyrausta*, it is prevalent in natural populations, and it can be spread both horizontally and vertically. A survey of European corn borer larvae by Zimmack et al. (1954) detected *N. pyrausta* in more than 80% of the counties in several Midwestern states. Field surveys in Illinois recovered *N. pyrausta* from 13% of living pupae and from 98% of dead pupae (Kramer, 1959). While an infection of European corn borer by *N. pyrausta* is not always fatal, sublethal effects such as reduced longevity and fecundity have been reported (Zimmack et al., 1954; Windels et al., 1976).

Some field research has been conducted testing the potential efficacy of foliar applications of *N. pyrausta* as an MCA of European corn borer. Lewis and Lynch (1978) achieved 63.8% and 97.2% infection against first- and second-generation European corn borer, respectively, with a single foliar application of *N. pyrausta* spore suspensions to maize plants at a rate of ca. 2.3×10^8 spores/plant 4 days after plants were infested with egg masses. The same study by Lewis and Lynch (1978) achieved reductions in larval abundance of as much as 48.4% and 17.2% in the first and second generation, respectively. The study also applied suspensions of *N. pyrausta* to maize plots to suppress a natural infestation of second-generation European corn borer larvae, and the authors were able to infect 75.0% and 58.3% of larvae in treatments of 2.43×10^8 spores/mL at a rate of 27.5 mL/plant and 15.3 mL/plant, respectively.

11.3 ENTOMOPATHOGENIC NEMATODES

Because of their soil-based habitat, entomopathogenic nematodes (EPNs) have been studied primarily for the management of belowground pests of maize and pests that inhabit the surface of the soil or contact the soil when searching for a site to pupate or diapause (Poinar et al., 1983; Jackson, 1985; Arthurs et al., 2004). EPNs may be introduced to a soil environment for the management of soil-borne pests (Journey and Ostlie, 2000; Toepfer et al., 2008), but they also have been found to occur naturally within maize cropping systems (Pilz et al., 2009). Additionally, research has found evidence for chemically mediated tritrophic interactions among maize, EPNs, and larval rootworms (Rasmann et al., 2005), and this may be applied to improve pest management in the field (Hiltpold et al., 2012).

One of the primary maize pests studied for management with EPNs is larval corn rootworm. Thurston and Yule (1990) applied steinernematid nematodes in furrow at rates of 1.0×10^4 or 1.0×10^5 infective juveniles (IJs)/m at the time of planting and recovered 2.3 to 2.5 northern corn rootworm larvae/plant compared to 4.3 larvae/plant in the untreated control, but applications of EPNs did not improve plant height or yield. In a 5-year field study, Jackson (1996) applied strains of *Heterorhabditis bacteriophora* and *Steinernema carpocapsae* to the soil furrow at a rate of either 2.0×10^5 IJs/30.5 cm of row or 6.0×10^5 IJs/plant, and for 2 of the 5 years found a reduction in root injury (1–9 scale; treatment = 3.8 to 5.5; control = 6.2 to 6.8) and emergence of adult western corn rootworm (treatment = 0.4 to 16.1 adults/emergence cage; control = 20.5 to 21.1 adults/cage). Similarly, Toepfer et al. (2008) applied EPNs at a rate of 2.8×10^9 IJs/ha and reported significant reductions in the emergence of adult western corn rootworm (32–81%) and reduced root injury (54–91%) in some field locations.

For the effective management of maize pests with EPNs, it appears that the timing of the application is important. Journey and Ostlie (2000) applied varying rates of *S. carpocapsae* (1.0×10^4 to 1.0×10^7 IJs/30.5 cm of row) to field plots with different developmental stages of western corn rootworm and found that targeting the third instar larvae provided the greatest reductions in root injury (1–6 scale; treatment = 2.5; control = 4.0). The type of soil in maize fields also may influence the efficacy. Barbercheck (1992) showed that sandy soil provided good conditions for EPNs to seek and infect hosts. However, Toepfer et al. (2010) found a greater suppression of western corn rootworm using *H. bacteriophora* at a rate of 1.3×10^9 IJs/ha in soils with a high clay and silt content (ca. 58% reduction in adult emergence) compared to sandy soils (ca. 33% reduction). Additionally, *Heterorhabditis megidis* and *Steinernema feltiae* showed a similar pattern in the study (Toepfer et al., 2010).

Evidence suggests a chemical communication between maize and EPNs in response to root feeding by insect pests. Rasmann et al. (2005) found that maize roots release the volatile (E)-β-caryophyllene (EβC) when fed on by western corn rootworm larvae. Interestingly, EβC is not released by all maize lines, with many North American lines not emitting the compound, while it does occur more commonly in European lines of maize (Rasmann et al., 2005).

Hiltpold and Turlings (2008) built upon this work by studying the diffusion of EβC and other maize volatiles in sand and soil at varying moisture levels (0–10%) and found a two-fold greater amount of horizontal diffusion of EβC through soil with 1% versus 10% moisture. In addition, Hiltpold et al. (2010) tested three different EPNs (*H. bacteriophora*, *Heterorhabditis megidis*, and *S. feltiae*) at a rate of 2.1×10^5 IJs/m of row in maize plots containing maize cultivars with or without genes for EβC emission. The authors found that all EPN strains reduced root injury (ca. 35–80%) from western corn rootworm and reduced adult emergence (ca. 20–80%). *H. megidis* and *S. feltiae* achieved significantly greater reductions in root injury for

the EβC-emitting cultivar than for the nonemitting cultivar (EβC emitting = ca. 25–45% reduction versus control = ca. 35% reduction), and *H. megidis* reduced adult emergence significantly more with the EβC-emitting cultivar (ca. 50% reduction) compared to the nonemitting cultivar (ca. 20% reduction).

Additionally, it may be possible to use the interaction between maize allelochemicals and western corn rootworm to improve the efficacy of western corn rootworm management with EPNs. Hiltpold et al. (2012) tested capsules containing *H. bacteriophora* that were coated with synthetic blends of maize kairomones targeting larval corn rootworm. The authors found that the nematodes could readily survive inside the capsules and that western corn rootworm larvae were as attracted to the capsules as they were to maize roots.

Other maize pests including corn earworm and black cutworm also may be targeted by EPNs. Even though the larval stage of corn earworm feeds above ground, this insect enters the soil to pupate, and black cutworm larvae reside on the soil surface and feed on maize seedlings. Cabanillas and Raulston (1995) tested *Steinernema riobrave* against corn earworm and found that in-furrow irrigation (2.0×10^5 IJs/m²) infected 95% of corn earworm compared to 56% and 84% infectivity when EPN were applied before and after irrigation, respectively. Levine and Oloumi-Sadeghi (1992) used *S. carpocapsae* (1.25–2.5×10^9 IJs/ha) against black cutworm in field plots and reduced the percentage of cut plants (treatment = 11%; control = 48.9%).

Some research has addressed the application of EPNs for the management of pests on maize foliage. In a field experiment, Ben-Yakir et al. (1998) sprayed an *S. carpocapsae* suspension (50,000 IJs/plant) on maize ears and reduced the number of European corn borer larvae by 42–82%. Bong (1986) studied single and double applications of *S. carpocapsae* at varying concentrations (4×10^3 and 4×10^4 IJs/ear) to protect maize ears from corn earworm and reported a significant reduction in corn earworm larvae per ear for all concentrations, with the highest concentration, applied twice, reducing the number of larvae from 1.84/ear in the untreated control to 0.59 larvae/ear. Similarly, Richter and Fuxa (1990) applied *S. feltiae* to maize foliage at rates of 2.5×10^8 or 2.5×10^{10} IJs/ha to manage fall armyworm. Infection percentages varied by year and location, but on average, up to 33–43% of fall armyworm larvae were infected. The study also applied either 4×10^4 or 8×10^4 IJs/ear to manage both corn earworm and fall armyworm and found that the high dose could reduce the number of insects per ear 4 days after application (treatment = ca. 0.7 insects/5 ears; control = 1.5 insects/5 ears).

The selection of the correct species and strain of EPN is important. For example, a greater mortality of corn earworm was achieved in field plots with applications of EPNs at a rate of 2.0×10^5 IJs/m² using *S. riobrave* compared to

S. carpocapsae (Cabanillas and Raulston, 1996). Additionally, the formulation of all MCAs, and EPNs in particular, can have pronounced effects on efficacy. The functions of formulation include stabilization for greater shelf life, ease of handling and application, environmental persistence, and improved efficacy (Jones and Burges, 1998). EPNs have been formulated for storage on moist substrates (sponge, vermiculite), immobilized in gel matrixes, and as desiccated infective juveniles (Georgis and Kaya, 1998). The latter provides the longest preapplication survival due to a reduction of energy utilization.

11.4 BACULOVIRUSES

Research on entomopathogenic baculoviruses has focused on foliage and ear feeding pests of maize. This research may have been stimulated in part by early reports of nucleopolyhedrovirus (NPV) infecting fall armyworm in maize (Chapman and Glaser, 1915; Allen, 1921). Fuxa (1982) determined that the NPV infecting fall armyworm was prevalent in some regions of North America, infecting as much as 68% of populations found in pastures and sometimes infecting a similar proportion in maize fields after mid-July. Because of the susceptibility of fall armyworm to NPVs and the yield losses inflicted by fall armyworm, NPVs have been investigated as a potential MCA. Hamm and Hare (1982) tested *S. frugiperda* (SpfrMNPV) at rates of $5–15\times10^{11}$ occlusion bodies (OBs)/ha applied one to three times with overhead irrigation. Three applications at 5×10^{11} OBs/ha resulted in 63.2% mortality, which was greater than the two applications at 7.5×10^{11} (55.7%) and one application at 15×10^{11} (42.1%). Hamm and Shapiro (1992) were able to increase the virulence of SpfrMNPV against fall armyworm by nine-fold when they added an optical brightener to the virus formulation, thereby slowing the degradation of the virus from ultraviolet radiation.

Ignoffo et al. (1965) used *H. zea* SNPV (HezeSNPV) foliar sprays containing ca. 1.5×10^{12} OBs/ha with a wetting agent and were able to reduce the populations of corn earworm larvae on maize by 68% and 86% using five and eight applications, respectively. Hamm and Young (1971) investigated applications of both SpfrMNPV and HezeSNPV at a rate of ca. 1.5×10^{13} OBs/ha to manage *S. frugiperda* and *H. zea*. Virus occlusion bodies applied singly at the early tassel stage and applied five times at the silk stage were shown to significantly reduce damage to maize (8.4 mean damage index compared to 10.0 in the untreated control).

An MNPV was isolated and characterized from black cutworm (Boughton et al., 1999). The isolate, AgipMNPV, showed a high virulence in lab experiments with black cutworm ($LC_{50}=260$ OBs/µL). Used in the greenhouse, AgipMNPV applied at 7×10^{8} OBs/m^2 significantly reduced black cutworm injury to maize seedlings (1–5 scale; treatment = 3.0; control = 4.8; Boughton et al., 2001).

11.5 BACTERIA

The application of bacteria for the management of maize pests has focused on Bt and lepidopteran pests. Research on Bt formulations for maize have focused on commercial products (Beegle and Yamamoto, 1992), including *B. thuringiensis* subsp. *kurstaki* (Btk) in Valent's Dipel ES (emulsifiable suspension) and Mycogen's MVP (Mycogen Vegetable Product). Foliar applications of Btk at 0.079 kg AI/ha reduced late instar European corn borer larvae to 38.5 larvae per plant compared to 81.8 larvae in the control (Linduska, 1990; Bartels and Hutchison, 1995). The addition of oils to these Bt products may improve the coverage of maize foliage and increase pest mortality. Hazzard et al. (2003) compared mineral oil, corn oil, MVP, and a mixture of corn oil and MVP for the treatment of maize silks in the field and found that the combination of corn oil and MVP (108 g delta endotoxin/kg oil suspension) resulted in 9% ears infested with *S. frugiperda* and *H. zea* larvae compared to 86% infestations in the control and as a result increased the marketability of the maize ears to 95% in the treated plots compared to 48% marketable ears in the control plots.

Aside from Bt for the management of Lepidoptera, little has been done to investigate entomopathogenic bacteria for the management of maize pests. One exception is the work by Martin et al. (2007), who tested the bacterium *Chromobacterium subtsugae* on both southern and western corn rootworm adults with a baited formulation and achieved 80% mortality 6 days after feeding.

11.6 *BACILLUS THURINGIENSIS* MAIZE

Genetically engineered maize producing insecticidal Bt proteins entered commercial cultivation in 1996 and was first planted in the United States where it covered less than 200,000 ha (James, 1996). By 2012, Bt maize was planted in at least 15 countries and covered over 40 million ha (James, 2012). The dramatic increase in the planting of Bt maize illustrates the importance and utility of this technology for managing insect pests. Initially, Bt maize produced single Bt toxins; however, over time, multiple Bt toxins were combined within a single hybrid (EPA, 2011). Bt toxins may either by stacked, where each toxin has a unique spectrum of target pests, or pyramided, which occurs when multiple toxins target the same pest.

Most of the Bt toxins that target insect pests of maize are crystalline toxins (ie, Cry toxins), and these target lepidopteran and coleopteran pests. Examples include Cry1Ab, Cry1F and Cry2Ab2 for lepidopteran pests, and Cry3Bb1 and Cry34/35Ab1 for coleopteran pests (Vaughn et al., 2005; Storer et al., 2006; Reisig et al., 2015). In some cases the structure of the Cry protein is modified to increase activity against a target pest, for example modified Cry3A (ie, mCry3A) targeting western corn rootworm (Walters et al., 2008).

Additionally, domains for different toxins may be combined to enhance biological activity. An example of this is eCry3.1Ab for the management of western corn rootworm, which replaces part of domain 3 from Cry3A with that of Cry1Ab (Walters et al., 2010; Frank et al., 2015). In addition to Cry toxins, vegetative Bt toxin (ie, Vip toxins) also can be engineered into transgenic maize, as is the case with Vip3Aa20 for lepidopteran pests (Bowen et al., 2014).

In general, Bt maize has been highly effective against target pests, including pests that have historically presented management challenges. For example, European corn borer larvae will feed within the stalk and ear of maize plants where they are sheltered from conventional insecticides (Mason et al., 1996). However, the systemic production of lepidopteran active Bt toxins, including Cry1F and Cry1Ab, provides an effective means to kill this pest, even when it feeds within the maize plant (Siegfried and Hellmich, 2012). Before the commercialization of Bt maize, European corn borer was among the most significant pests of maize and imposed yield losses and management costs in excess of $1 billion US/year (Mason et al., 1996). However, the widespread planting of Bt maize has been associated with the regional suppression of European corn borer in several Midwest states, and the planting of Bt maize targeting European corn borer is estimated to have provided farmers in the United States with nearly $7 billion US in additional profits between 1996 and 2009 (Hutchison et al., 2010).

Moreover, the planting of Bt maize has enabled farmers to rely less on conventional insecticides. For example, in fields with a continuous cultivation of maize, the management of western and northern corn rootworm has traditionally relied on soil-applied insecticides at planting (Levine and Oloumi-Sadeghi, 1991). The commercialization of Bt maize targeting rootworm larvae has replaced some of the need for soil-applied insecticides in the management of rootworm larvae (Gray et al., 2009). The cost of managing corn rootworm larvae with Bt maize versus soil-applied insecticides can be similar (Dunbar and Gassmann, 2013). However, in general, managing pests through the planting of Bt maize has fewer nontarget effects compared to pest management with conventional insecticides (Marvier et al., 2007).

Because of the high levels of pest mortality imposed by Bt maize, there is intense selective pressure on pest populations to evolve resistance, and the development of Bt resistance has the potential to diminish the benefits achieved through the commercialization of Bt maize. To date there are several documented cases of pests evolving resistance to Bt maize in the field, and in these instances, resistance was associated with a reduced efficacy of Bt maize against a target pest, ie, practical resistance as defined by Tabashnik et al. (2014). These cases of field-evolved resistance include western corn rootworm resistance to mCry3A and Cry3Bb1 maize in the United States (Gassmann et al., 2011,

2012, 2014), fall armyworm resistance to Cry1F maize in Puerto Rico and Brazil (Storer et al., 2010; Farias et al., 2014a, 2014b), and maize stem borer *B. fusca* resistance to Cry1Ab maize in South Africa (van Rensburg, 2007). Strategies to mitigate the effects of Bt resistance have included discontinuing the use of some types of Bt maize in certain geographies (Storer et al., 2010), increasing the use of conventional insecticides (Farias et al., 2014a), and pyramiding multiple Bt toxins to compensate for the loss of efficacy of a Bt trait (Gassmann, 2012).

The management of resistance to Bt maize has focused on the refuge strategy (Gould, 1998). Refuges of non-Bt maize, or other suitable larval host plants, enable the survival of Bt-susceptible insects. Mating between Bt-susceptible insects from refuges and resistant individuals from Bt crops produces heterozygous progeny, and to the extent that these heterozygous progeny have lower fitness on Bt maize than a homozygous-resistant genotype, the evolution of Bt resistance may be delayed (Carrière et al., 2010). When resistance is a functionally recessive trait (ie, heterozygous-resistant individuals are unable to survive on Bt maize), delays in resistance are greater than when resistance is non-recessive (Tabashnik et al., 2008). It is possible to achieve recessive inheritance of resistance when Bt maize produces a high dose of toxin. A high dose is defined as 25 times the amount of Bt toxin required to kill a susceptible individual or a dose that is capable of killing 99.99% of susceptible individuals (EPA, 1998). Some of the first Bt toxins commercialized for the management of maize pests met this assumption of high dose. These included Cry1Ab and Cry1F toxins for the management of European corn borer (Huang et al., 2011). However, in other cases, such as Bt maize targeting western corn rootworm, the assumption of high dose was not satisfied (Gassmann, 2012). It is notable that after over a decade and a half of commercial use, and extensive cultivation within the United States, available data indicate a lack of field-evolved resistance by European corn borer to Bt maize (Siegfried and Hellmich, 2012). By contrast, field-evolved resistance to Bt maize by western corn rootworm was found after just 7 years of cultivation (Gassmann et al., 2011).

Pyramiding of multiple Bt toxins also may be used in conjunction with the refuge strategy to delay Bt resistance (Roush, 1998). Unlike the high-dose strategy, which delays resistance by making resistance a functionally recessive trait, pyramiding delays resistance by using multiple toxins with independent modes of action that target the same pest. In this manner, individuals with resistance to one toxin in a pyramid are killed by the other toxin and vice versa. Thus, for a high-dose trait, the survival of heterozygotes on a Bt crop (ie, dominance) is the critical factor affecting the rate of resistance evolution, but for pyramided crops, it is the mortality of susceptible individuals that is the most important (Roush, 1998). Because pyramiding does not require a high dose to delay resistance, it is useful for Bt traits that do not achieve a high dose. Examples of pyramided Bt traits that are less than high dose include Bt maize targeting corn rootworm and corn earworm (Lutrell and Jackson, 2012; Keweshan et al., 2015).

For the refuge strategy to delay resistance, either alone, with high-dose events, or with pyramided events, the refuge population must consist of primarily Bt-susceptible individuals. Because Bt crops select for Bt resistance, resistance alleles, which increase in frequency in Bt fields, may move into refuge populations through pest dispersal (Comins, 1977). The accumulation of resistance alleles within refuge populations will diminish the effectiveness of the refuge to delay resistance and lead to a rapid increase in the frequency of resistance alleles within a population. Fitness costs of Bt resistance arise in the absence of Bt toxins when individuals with resistance alleles have lower fitness than Bt-susceptible individuals (Gassmann et al., 2009). Fitness costs remove resistance alleles from refuge populations and can delay the onset of resistance (Gassmann et al., 2008). For insect pests of maize, evidence indicates that fitness costs of Bt resistance may be present in some cases but not others. For example, resistance to both Cry1F and Cry1Ab by European corn borer often appears to have associated fitness costs (Crespo et al., 2010; Pereira et al., 2011). Along with the high dose of toxin production by Bt maize, these fitness costs may also contribute to the observed lack of field-evolved resistance in this species. By contrast, resistance to CryBb1 maize by western corn rootworm appears to have few accompanying fitness costs (Petzold-Maxwell et al., 2012b; Hoffmann et al., 2014, 2015; Ingber and Gassmann, 2015), which may have contributed to some of the current management challenges associated with field-evolved resistance to Cry3Bb1 maize and mCry3A maize in this species (Gassmann et al., 2014). In general there is a positive relationship between the magnitude of resistance and the presence of fitness costs (Gassmann et al., 2009). For European corn borer targeted by Cry1Ab and Cry1F, the magnitude of resistance is far greater than for western corn rootworm targeted by Cry3Bb1, and this may contribute to the more pronounced fitness costs of Bt resistance in European corn borer compared to western corn rootworm (Gassmann et al., 2014; Ingber and Gassmann, 2015).

In general, the case of western corn rootworm and Bt maize offers insights into the potential shortcomings of Bt maize for the management of insect pests. There are several features about the interaction between western corn rootworm and Bt maize that likely contributed to the rapid evolution of resistance in this species. Field data indicate that resistance to Cry3Bb1 maize by western corn rootworm arose following as few as three generations of selection (which is equivalent to 3 years in the case of this univoltine insect) and that additional years of selection were associated with increased levels of resistance (Gassmann et al., 2011, 2014).

Factors contributing to resistance likely included the lack of a high dose for Bt maize targeting western corn rootworm, nonrecessive inheritance of resistance, and few accompanying fitness costs (Gassmann, 2012; Petzold-Maxwell et al., 2012b; Ingber and Gassmann, 2015). This suggests that for Bt crops that achieve less than a high dose, resistance may develop rapidly and, due to a lack of fitness costs, may persist once selected. In general, when Bt maize does not achieve a high dose against its target pest, there appears to be a substantially elevated risk of resistance. Pyramiding of multiple Bt toxins is one strategy to delay resistance. Additionally, it will be important to use these Bt crops in a broader context of integrated pest management (Romeis et al., 2008). Among these approaches, it may be beneficial to consider combining Bt maize with other MCAs (Petzold-Maxwell et al., 2013).

11.7 COMBINING TACTICS

In some instances it may be possible to enhance the management of maize pests by combining multiple MCAs. This approach may increase pest suppression and better preserve yield. Additionally, it may impose less selection for resistance; however, such multispecies dynamics are complex, and although resistance evolution may be delayed in some cases, in other cases, the rate of resistance evolution actually may be accelerated, depending on how the presence of MCAs affects the differential fitness between pest genotypes. For example, Gould et al. (1991) applied a computer model to test how natural enemies could affect the rate at which insect pests developed resistance to host-plant resistance traits, which were achieved either through biotechnology (ie, Bt crops) or traditional breeding. The authors found that in some cases pests could develop resistance to Bt plants more rapidly and in other cases more slowly and that in general, these differences arose from whether or not natural enemies imposed greater mortality on the Bt-resistant genotype or the Bt-susceptible genotype. To date, most research has focused on combining Bt maize with other MCAs for the management of corn rootworm (Meissle et al., 2009; Petzold-Maxwell et al., 2012a). However, data also exist on the potential application of multiple MCAs for the management of aboveground lepidopteran pests.

Research on the management of aboveground lepidopteran pests of maize with multiple MCAs has centered on European corn borer. The first instars of European corn borer larvae were significantly more susceptible to Btk when larvae were infected by *N. pyrausta* ($LC_{50} = 0.006$ mg Btk/kg diet) compared to larvae not infected with *N. pyrausta* ($LC_{50} = 0.027$ mg Btk/kg diet), and results suggested that these MCAs acted additively (Pierce et al., 2001). Compared to a treatment of *B. bassiana* alone, combining *B. bassiana* and *N. pyrausta* (ingested from a diet containing 1.0×10^6 spores/mL) against European corn borer decreased larval LC_{50}s by 42–21%, in an additive manner, although some evidence for synergism was found (Rahman et al., 2010). By contrast, Lewis and Bing (1991) tested combinations of Bt (4.8×10^8 and 9.6×10^8 infective units (IUs)/ha) and *B. bassiana* (5.9×10^9 and 1.2×10^{10} conidia/ha) against European corn borer and found that an additional suppression of stalk tunneling was not achieved when using both pathogens in tandem.

The case of Bt maize and western corn rootworm provides an example of how transgenic crops may be used in conjunction with MCAs. Because Bt maize targeting western corn rootworm does not achieve a high dose of toxin, some injury occurs to Bt maize roots even when a rootworm population has not developed Bt resistance (Gassmann, 2012). As a result, using additional MCAs against corn rootworm can further reduce root injury to Bt maize. While research to date has focused on Bt maize and corn rootworm, such work can serve as a model for other cases where Bt maize achieves less than a high dose against a target pests (eg, corn earworm or fall armyworm). In these cases, it may also be possible to augment a transgenic pest management approach by incorporating other MCAs such as viruses, fungi, or nematodes. Both laboratory and field experiments have found complementary and additive effects when Bt maize is combined with other MCAs.

In a laboratory study, Meissle et al. (2009) combined Cry3Bb1 maize and *M. brunneum* in bioassays and found additive mortality from both agents against western corn rootworm larvae. Similarly, laboratory experiments by Petzold-Maxwell et al. (2012a) combined Cry34/35Ab1 maize with a blend of MCAs (*M. brunneum*, *B. bassiana*, *S. carpocapsae*, *Steinernema glaseri*, and *H. bacteriophora*) and found that mortality of western corn rootworm larvae was increased in an additive manner compared to either a blend of MCAs or Bt maize alone.

The benefit of combining Bt maize with entomopathogens for the management of corn rootworm also has been demonstrated in the field. Petzold-Maxwell et al. (2013) studied Cry34/35Ab1 maize in combination with a blend of MCAs. The EPF *M. brunneum* was applied to the soil at planting as microsclerotial granules (ca. 5.0×10^{13} conidia/ha), and the EPNs *S. carpocapsae* and *H. bacteriophora* were applied to the soil (ca. 1.6×10^9 IJs/ha) when plants were at an early vegetative stage. In a year of high rootworm abundance, injury to Cry34/35Ab1 maize was significantly less when entomopathogens were present, and in a year of low rootworm abundance, injury to non-Bt maize was diminished by the presence of entomopathogens. Overall, the yield was 11% greater for plots that received MCAs.

11.8 FUTURE PROSPECTS

Perhaps the biggest challenge of pest management in maize is the large area over which the crop is grown and the low profit margin. The one notable exception may be the production of maize as a vegetable rather than a grain. However, in general, for pest management strategies to be palatable to farmers, they must be simple to apply over a large area and carry a low accompanying cost of application. Bt maize meets both of these requirements and is the greatest success to date in the application of biopesticides for the management of maize pests. In many instances, the planting of Bt maize has provided effective management of pests, in particular when pests are targeted by multiple high-dose events. However, in other cases, the development of pest resistance has substantially diminished the benefits conferred by Bt maize. This has arisen primarily when pests were targeted by less than high-dose events that were present singly in Bt maize. Future pest management in maize may well include the development of new Bt transgenic traits. Of particular value would be the development of high-dose Bt traits against rootworm and lepidopteran pests that pose challenges from resistance evolution, and these would include western corn rootworm, fall armyworm, and maize stalk borer. Ideally, new high-dose events would be brought to the market as pyramids before being deployed singly in the field. In conjunction with non-Bt refuges, such an approach should reduce the rate of resistance evolution and provide a more sustainable approach to managing some key pests of maize.

For belowground pests of maize and pests that attack seedlings, it may be possible to use EPF that colonize the rhizosphere or are endophytic. Such EPF could be applied to maize seeds as a seed treatment prior to planting. To date, seed treatments with *B. bassiana* and *Metarhizium* spp. have been tested with crops including tomato, wheat, and cotton (Prabhukarthikeyan et al., 2013; Lopez et al., 2014; Keyser et al., 2015). In a study with tomato, Qayyum et al. (2015) found that *B. bassiana* applied to tomato roots could disperse within the plant, and this was associated with an increased infection of corn earworm *Helicoverpa armigera* larvae feeding on leaves. EPF also can induce the expression of plant defense genes and inhibit the development of antagonistic plant pathogens such as *Fusarium* spp. (Prabhukarthikeyan et al., 2013). Additionally, plants may benefit from the presence of EPFs through the enhanced growth of roots (Sasan and Bidochka, 2012). Future technology that improves EPF in the form of seed treatments may provide a way to protect maize plants from insect pests and plant pathogens, and increase yield.

Aside from seed treatments, large-scale applications of MCAs in maize fields are likely not practical. However, there is the potential to bolster the abundance of entomopathogens in maize fields through conservation biological control. Past research provides ample evidence that maize fields are suitable habitats for EPNs and EPF (Lewis et al., 2006; Pilz et al., 2008; Rudeen et al., 2013). Additionally, these pathogens can impose mortality of maize pests. Past work also has demonstrated that agricultural practices, such as organic versus conventional farming, affect the abundance of natural enemies in agricultural systems, including entomopathogens (Bengtsson et al., 2005; Crowder et al., 2010; Garratt et al., 2011). For example, organic systems can have a high abundance of EPF compared to conventional systems (Klingen et al., 2002; Clifton et al., 2015). Additionally, specific farming practices, such as the use of organic fertilizers and an absence of tillage, can increase the abundance of EPF in the soil (Clifton et al., 2015). The adoption of agricultural practices that bolster the abundance of entomopathogens may help to maintain pests below the economic injury level, reducing the need for other pest management tactics, including conventional insecticides, and resulting in a more diversified and sustainable production system.

REFERENCES

Allen, H.W., 1921. Notes on a bombylid parasite and a polyhedral disease of the southern grass worm, *Laphygma frugiperda*. J. Econ. Entomol. 14, 510–511.

Ansari, M.A., Evans, M., Butt, T.M., 2009. Identification of pathogenic strains of entomopathogenic nematodes and fungi for wireworm control. Crop Prot. 28, 269–272.

Arthurs, S., Heinz, K.M., Prasifka, J.R., 2004. An analysis of using entomopathogenic nematodes against above-ground pests. B. Entomol. Res. 94, 297–306.

Barbercheck, M.E., 1992. Effect of soil physical factors on biological control agents of soil insect pests. Fla. Entomol. 75, 539–548.

Bartels, D.W., Hutchison, W.D., 1995. On-farm efficacy of aerially applied *Bacillus thuringiensis* for European corn borer (Lepidoptera: Pyralidae) and corn earworm (Lepidoptera: Noctuidae) control in sweet corn. J. Econ. Entomol. 88, 380–386.

Bartlett, K.A., Lefebvre, C.L., 1934. Field experiments with *Beauveria bassiana* (Bals.) Vuill., a fungus attacking the European corn borer. J. Econ. Entomol. 27, 1147–1157.

Beegle, C.C., Yamamoto, T., 1992. History of *Bacillus thuringiensis* Berliner research and development. Can. Entomol. 124, 587–616.

Behie, S.W., Zelisko, P.M., Bidochka, M.J., 2012. Endophytic insect-parasitic fungi translocate nitrogen directly from insects to plants. Science 336, 1576–1577.

Ben-Yakir, D., Efron, D., Chen, M., Glazer, I., 1998. Evaluation of entomopathogenic nematodes for biocontrol of the European corn borer, *Ostrinia nubilalis*, on sweet corn in Israel. Phytoparasitica 26, 101–108.

Bengtsson, J., Ahnström, J., Weibull, A.-C., 2005. The effects of organic agriculture on biodiversity and abundance: a meta-analysis. J. Appl. Ecol. 42, 261–269.

Bing, L.A., Lewis, L.C., 1991. Suppression of *Ostrinia nubilalis* (Hübner) (Lepidoptera: Pyralidae) by endophytic *Beauveria bassiana* (Balsamo) Vuillemin. Environ. Entomol. 20, 1207–1211.

Bong, C.F.J., 1986. Field control of *Heliothis zea* (Boddie) (Lepidoptera: Noctuidae) using a parasitic nematode. Int. J. Trop. Insect Sci. 7, 23–25.

Boughton, A.J., Harrison, R.L., Lewis, L.C., Bonning, B.C., 1999. Characterization of a nucleopolyhedrovirus from the black cutworm, *Agrotis ipsilon* (Lepidoptera: Noctuidae). J. Invertebr. Pathol. 74, 289–294.

Boughton, A.J., Lewis, L.C., Bonning, B.C., 2001. Potential of *Agrotis ipsilon* nucleopolyhedrovirus for suppression of the black cutworm (Lepidoptera: Noctuidae) and effect of an optical brightener on virus efficacy. J. Econ. Entomol. 94, 1045–1052.

Bowen, K.L., Flanders, K.L., Hagan, A.K., Ortiz, B., 2014. Insect damage, aflatoxin content, and yield of Bt corn in Alabama. J. Econ. Entomol. 107, 1818–1827.

Bruck, D.J., Lewis, L.C., 2001. Adult *Diabrotica* spp. (Coleoptera: Chrysomelidae) infection at emergence with indigenous *Beauveria bassiana* (Deuteromycotina: Hyphomycetes). J. Invertebr. Pathol. 77, 288–289.

Bruck, D.J., Lewis, L.C., 2002. Whorl and pollen-shed stage application of *Beauveria bassiana* for suppression of adult western corn rootworm. Entomol. Exp. Appl. 103, 161–169.

Cabanillas, H.E., Raulston, J.R., 1995. Impact of *Steinernema riobravis* (Rhabditida: Steinernematidae) on the control of *Helicoverpa zea* (Lepidoptera: Noctuidae) in corn. J. Econ. Entomol. 88, 58–64.

Cabanillas, H.E., Raulston, J.R., 1996. Evaluation of *Steinernema riobravis*, *S. carpocapsae*, and irrigation timing for the control of corn earworm, *Helicoverpa zea*. J. Nematol. 28, 75–82.

Carrière, Y., Crowder, D.W., Tabashnik, B.E., 2010. Evolutionary ecology of insect adaptation to Bt crops. Evol. Appl. 3, 561–573.

Chapman, J.W., Glaser, R.W., 1915. A preliminary list of insects which have wilt, with a comparative study of their polyhedra. J. Econ. Entomol. 8, 140–149.

Clifton, E.H., Jaronski, S.T., Hodgson, E.W., Gassmann, A.J., 2015. Abundance of soil-borne entomopathogenic fungi in organic and conventional fields in the Midwestern USA with an emphasis on the effect of herbicides and fungicides on fungal persistence. PLoS One 10 (7), e0133613. http://dx.doi.org/10.1371/journal.pone.0133613.

Comins, H.N., 1977. The development of insecticide resistance in the presence of migration. J. Theor. Biol. 64, 177–197.

Crespo, A.L.B., Spencer, T.A., Tan, S.Y., Siegfried, B.D., 2010. Fitness costs of Cry1Ab resistance in a field-derived strain of *Ostrinia nubilalis*. J. Econ. Entomol. 103, 1386–1393.

Crowder, D.W., Northfield, T.D., Strand, M.R., Snyder, W.E., 2010. Organic agriculture promotes evenness and natural pest control. Nature 466, 109–112.

Dunbar, M.W., Gassmann, A.J., 2013. Abundance and distribution of western and northern corn rootworm (*Diabrotica* spp.) and prevalence of rotation resistance in eastern Iowa. J. Econ. Entomol. 106, 168–180.

EPA, 1998. Final Report of the FIFRA Scientific Advisory Panel Subpanel on *Bacillus Thuringiensis* (Bt) Plant-Pesticides and Resistance Management. Accessed at http://archive.epa.gov/scipoly/sap/meetings/web/pdf/finalfeb.pdf.

EPA, 2011. Bt11 x 59122–7 x MIR 604 x 1507 (Cry1Ab x Cry34/35 x mCry3A x Cry1F). Accessed at: http://www.epa.gov/ingredients-used-pesticide-products/current-previously-registered-section-3-plant-incorporated.

Farias, J.R., Horikoshi, R.J., Santos, A.C., Omoto, C., 2014a. Geographical and temporal variablity in susceptiblity to Cry1F toxin from *Bacillus thuringiensis* in *Spodoptera frugiperda* (Lepidoptera: Noctuidae) populations in Brazil. J. Econ. Entomol. 107, 2182–2189.

Farias, J.R., Andow, D.A., Horikoshi, R.J., Sorgatto, R.J., Fresia, P., dos Santos, A.C., Omoto, C., 2014b. Field-evolved resistance to Cry1F maize by *Spodoptera frugipera* (Lepidoptera: Noctuidae) in Brazil. Crop Prot. 64, 150–158.

Ferron, P., 1981. Pest control by the fungi *Beauveria* and *Metarhizium*. In: Burges, H.D. (Ed.), Microbial Control of Pests and Plant Diseases, 1970–1980. Kluwer Academic Publishers, New York & London, pp. 465–482.

Frank, D.L., Kurtz, R., Tinsley, N.A., Gassmann, A.J., Meinke, L.J., Moellenbeck, D., Gray, M.E., Bledsoe, L.W., Krupke, C.H., Estes, R.E., Weber, P., Hibbard, B.E., 2015. Effect of seed mixtures and soil-insecticide on the emergence of western and northern corn rootworm (Coleoptera: Chrysomelidae) from transgenic Bt maize expressing the mCry3A plus eCry3.1Ab proteins. J. Econ. Entomol. 108, 1260–1270.

Fuxa, J.R., 1982. Prevalence of viral infections in populations of fall armyworm, *Spodoptera frugiperda*, in southeastern Louisiana. Environ. Entomol. 11, 239–242.

Garratt, M.P.D., Wright, D.J., Leather, S.R., 2011. The effects of farming system and fertilisers on pests and natural enemies: a synthesis of current research. Agric. Ecosyst. Environ. 141, 261–270.

Gassmann, A.J., 2012. Field-evolved resistance to Bt maize by western corn rootworm: predictions from the laboratory and effects in the field. J. Invertebr. Pathol. 110, 287–293.

Gassmann, A.J., Stock, S.P., Sisterson, M.S., Carrière, Y., Tabashnik, B.E., 2008. Synergism between entomopathogenic nematodes and *Bacillus thuringiensis* crops: integrating biological control and resistance management. J. Appl. Ecol. 45, 957–966.

Gassmann, A.J., Carrière, Y., Tabashnik, B.E., 2009. Fitness costs of insect resistance to *Bacillus thuringiensis*. Annu. Rev. Entomol. 54, 147–163.

Gassmann, A.J., Petzold-Maxwell, J.L., Keweshan, R.S., Dunbar, M.W., 2011. Field-evolved resistance to Bt maize by western corn rootworm. PLoS One 6 (7), e22629. http://dx.doi.org/10.1371/journal.pone.0022629.

Gassmann, A.J., Petzold-Maxwell, J.L., Keweshan, R.S., Dunbar, M.W., 2012. Western corn rootworm and Bt maize: challenges of pest resistance in the field. GM Crops Food 3, 235–244.

Gassmann, A.J., Petzold-Maxwell, J.L., Clifton, E.H., Dunbar, M.W., Hoffmann, A.M., Ingber, D.A., Keweshan, R.S., 2014. Field-evolved resistance by western corn rootworm to multiple *Bacillus thuringiensis* toxins in transgenic maize. Proc. Natl. Acad. Sci. USA 111, 5141–5146.

Georgis, R., Kaya, H.K., 1998. Formulation of entomopathogenic nematodes. In: Burges, H.D. (Ed.), Formulation of Microbial Biopesticides. Springer, Berlin, pp. 289–308.

Gould, F., 1998. Sustainability of transgenic insecticidal cultivars: integrating pest genetics and ecology. Annu. Rev. Entomol. 43, 701–726.

Gould, F., Kennedy, G.G., Johnson, M.T., 1991. Effects of natural enemies on the rate of herbivore adaptation to resistant host plants. Entomol. Exp. Appl. 58, 1–14.

Gray, M.E., Sappington, T.W., Miller, N.J., Moeser, J., Bohn, M.O., 2009. Adaptation and invasiveness of western corn rootworm: intensifying research on a worsening pest. Annu. Rev. Entomol. 54, 303–321.

Hall, I.M., 1952. Observations on *Perezia pyraustae* Paillot, a microsporidian parasite of the European corn borer. J. Parasitol. 38, 48–52.

Hamm, J.J., Hare, W.W., 1982. Application of entomopathogens in irrigation water for control of fall armyworms and corn earworms (Lepidoptera: Noctuidae) on corn. J. Econ. Entomol. 75, 1074–1079.

Hamm, J.J., Shapiro, M., 1992. Infectivity of fall armyworm (Lepidoptera: Noctuidae) nuclear polyhedrosis virus enhanced by a fluorescent brightener. J. Econ. Entomol. 85, 2149–2152.

Hamm, J.J., Young, J.R., 1971. Value of virus presilk treatment for corn earworm and fall armyworm control in sweet corn. J. Econ. Entomol. 64, 144–146.

Hazzard, R.V., Schultz, B.B., Groden, E., Ngollo, E.D., Seidlecki, E., 2003. Evaluation of oils and microbial pathogens for control of lepidopteran pests of sweet corn in New England. J. Econ. Entomol. 96, 1653–1661.

Hiltpold, I., Turlings, T.C.J., 2008. Belowground chemical signaling in maize: when simplicity rhymes with efficiency. J. Chem. Ecol. 34, 628–635.

Hiltpold, I., Toepfer, S., Kuhlmann, U., Turlings, T.C.J., 2010. How maize root volatiles affect the efficacy of entomopathogenic nematodes in controlling the western corn rootworm? Chemoecology 20, 155–162.

Hiltpold, I., Hibbard, B.E., French, B.W., Turlings, T.C., 2012. Capsules containing entomopathogenic nematodes as a Trojan horse approach to control the western corn rootworm. Plant Soil 358, 11–25.

Hoffmann, A.M., French, B.W., Jaronski, S.T., Aaron, J., 2014. Effects of entomopathogens on mortality of western corn rootworm (Coleoptera: Chrysomelidae) and fitness costs of resistance to Cry3Bb1 maize. J. Econ. Entomol. 107, 352–360.

Hoffmann, A.M., French, B.W., Hellmich, R.L., Lauter, N., Gassmann, A.J., 2015. Fitness costs of resistance to Cry3Bb1 maize by western corn rootworm. J. Appl. Entomol. 139, 403–415.

Hu, G., St Leger, R.J., 2002. Field studies using a recombinant mycoinsecticide (*Metarhizium anisopliae*) reveal that it is rhizosphere competent. Appl. Environ. Microbiol. 68, 6383–6387.

Huang, F., Andow, D.A., Buschman, L.L., 2011. Success of the high-dose/refuge resistance management strategy after 15 years of Bt crop use in North America. Entomol. Exp. Appl. 140, 1–16.

Hussey, N.W.R., Tinsley, T.W., 1981. Impressions of insect pathology in the People's Republic of China. In: Burges, H.D. (Ed.), Microbial Control of Pests and Plant Diseases, 1970–1980. Academic Press, New York & London, pp. 785–795.

Hutchison, W., Burkness, E., Mitchell, P., Moon, R., Leslie, T., Fleischer, S., Abrahamson, M., Hamilton, K., Steffey, K., Gray, M., Hellmich, R., Kaster, L., Hunt, T., Wright, R., Pecinovsky, R., Rabaey, T., Flood, B., Raun, E., 2010. Areawide suppression of European corn borer with Bt maize reaps savings to non-Bt maize growers. Science 330, 222–225.

Ignoffo, C.M., Chapman, A.J., Martin, D.F., 1965. The nuclear-polyhedrosis virus of *Heliothis zea* (Boddie) and *Heliothis virescens* (Fabricius). J. Invertebr. Pathol. 7, 227–235.

Ingber, D.A., Gassmann, A.J., 2015. Inheritance and fitness costs of resistance to Cry3Bb1 corn by western corn rootworm (Coleoptera: Chrysomelidae). J. Econ. Entomol. 108, 2421–2432.

Jackson, J.J., 1985. Parasitism of the Western Corn Rootworm with the Nematode, *Steinernema Feltiae* (Ph.D. dissertation). University of Minnesota, St. Paul.

Jackson, J.J., 1996. Field performance of entomopathogenic nematodes for suppression of western corn rootworm (Coleoptera: Chrysomelidae). J. Econ. Entomol. 89, 366–372.

James, C., 1996. Global Review of the Field Testing and Commercialization of Transgenic Plants, 1986 to 1995: The First Decade of Crop Biotechnology. ISAAA Briefs No. 1. ISAAA, Ithaca.

James, C., 2012. Global Status of Commercialized Biotech/GM Crops: 2012. ISAAA Brief No. 44. ISAAA, Ithaca, New York.

Jaronski, S.T., 2007. Soil ecology of the entomopathogenic Ascomycetes: a critical examination of what we (think) we know. In: Ekesi, S., Maniania, N.K. (Eds.), Use of Entomopathogenic Fungi in Biological Pest Management. Research SignPosts, Trivandrum, India, pp. 91–144.

Jaronski, S.T., 2010. Ecological factors in the inundative use of fungal entomopathogens. BioControl 55, 159–185.

Jones, K.A., Burges, H.D., 1998. Technology of formulation and application. In: Burges, H.D. (Ed.), Formulation of Microbial Biopesticides. Springer, Netherlands, pp. 7–30.

Journey, A.M., Ostlie, K.R., 2000. Biological control of the western corn rootworm (Coleoptera: Chrysomelidae) using the entomopathogenic nematode, *Steinernema carpocapsae*. Environ. Entomol. 29, 822–831.

Kabaluk, J.T., Ericsson, J.D., 2007. *Metarhizium anisopliae* seed treatment increases yield of field corn when applied for wireworm control. Agron. J. 99, 1377–1381.

Keweshan, R.S., Head, G.P., Gassmann, A.J., 2015. Effects of pyramided Bt corn and blended refuges on western corn rootworm and northern corn rootworm (Coleoptera: Chrysomelidae). J. Econ. Entomol. 108, 720–729.

Keyser, C.A., Jensen, B., Meyling, N.V., 2015. Dual effects of *Metarhizium* spp. and *Clonostachys rosea* against an insect and a seed-borne pathogen in wheat. Pest Manag. Sci. 72 (3), 517–526.

Klingen, I., Eilenberg, J., Meadow, R., 2002. Effects of farming system, field margins and bait insect on the occurrence of insect pathogenic fungi in soils. Agric. Ecosyst. Environ. 91, 191–198.

Knutson, A.E., Gilstrap, F.E., 1990. Seasonal occurrence of *Beauveria bassiana* in the southwestern corn borer (Lepidoptera: Pyralidae) in the Texas High Plains. J. Kans. Entomol. Soc. 63, 243–251.

Kotlán, A., 1928. A double parasitic infection of a larva of *Pyrausta nubilalis* Hb. Int. Corn Borer Invest. Sci. Rep. 1927–1928, 174–178.

Kramer, J.P., 1959. Observations on the seasonal incidence of microsporidiosis in European corn borer populations in Illinois. Entomophaga 4, 37–42.

Krueger, S.R., Roberts, D.W., 1997. Soil treatment with entomopathogenic fungi for corn rootworm (*Diabrotica* spp.) larval control. Biol. Control 9, 67–74.

Levine, E., Oloumi-Sadeghi, H., 1991. Management of diabroticite rootworms in corn. Annu. Rev. Entomol. 36, 229–255.

Levine, E., Oloumi-Sadeghi, H., 1992. Field evaluation of *Steinernema carpocapsae* (Rhabditida: Steinernematidae) against black cutworm (Lepidoptera: Noctuidae) larvae in field corn. J. Entomol. Sci. 27.

Lewis, L.C., Bing, L.A., 1991. *Bacillus thuringiensis* Berliner and *Beauveria bassiana* (Balsamo) Vuillimen for European corn borer control: program for immediate and season-long suppression. Can. Entomol. 123, 387–393.

Lewis, L.C., Cossentine, J.E., 1986. Season long intraplant epizootics of entomopathogens, *Beauveria bassiana* and *Nosema pyrausta*, in a corn agroecosystem. Entomophaga 31, 363–369.

Lewis, L.C., Lynch, R.E., 1978. Foliar application of *Nosema pyrausta* for suppression of populations of European corn borer. Entomophaga 23, 83–88.

Lewis, L.C., Sumerford, D.V., Bing, L.A., Gunnarson, R.D., 2006. Dynamics of *Nosema pyrausta* in natural populations of the European corn borer, *Ostrinia nubilalis*: a six-year study. Biocontrol 51, 627–642.

Lewis, L.C., Bruck, D.J., Prasifka, J.R., Raun, E.S., 2009. *Nosema pyrausta*: its biology, history, and potential role in a landscape of transgenic insecticidal crops. Biol. Control 48, 223–231.

Linduska, J.L., 1990. Silk sprays to control corn earworms, dusky sap beetle and European corn borers in sweet corn, 1989. Insectic. Acaric. Tests 15, 108.

Lopez, D.C., Zhu-Salzman, K., Ek-Ramos, M.J., Sword, G.A., 2014. The entomopathogenic fungal endophytes *Purpureocillium lilacinum* (formerly *Paecilomyces lilacinus*) and *Beauveria bassiana* negatively affect cotton aphid reproduction under both greenhouse and field conditions. PLoS One 9 (8), e103891. http://dx.doi.org/10.1371/journal.pone.0103891.

Lutrell, R.G., Jackson, R.J., 2012. *Helicoverpa zea* and Bt cotton in the United States. GM Crops Food 3, 213–227.

Martin, P.A.W., Hirose, E., Aldrich, J.R., 2007. Toxicity of *Chromobacterium subtsugae* to southern green stink bug (Heteroptera: Pentatomidae) and corn rootworm (Coleoptera: Chrysomelidae). J. Econ. Entomol. 100, 680–684.

Marvier, M., McCreedy, C., Regetz, J., Kareiva, P., 2007. A meta-analysis of effects of Bt cotton and maize on nontarget invertebrates. Science 316, 1475–1477.

Mason, C.E., Rice, M.E., Calvin, D.D., Van Duyn, J.W., Showers, W.B., Hutchison, W.D., Witkowski, J.F., Higgins, R.A., Onstad, D.W., Dively, G.P., 1996. Ecology and Management of European Corn Borer. Iowa State University, Ames.

Meissle, M., Pilz, C., Romeis, J.R., 2009. Susceptibility of *Diabrotica virgifera virgifera* (Coleoptera: Chrysomelidae) to the entomopathogenic fungus *Metarhizium anisopliae* when feeding on *Bacillus thuringiensis* Cry3Bb1-expressing maize. Appl. Environ. Microbiol. 75, 3937–3943.

Paillot, A., 1927. Sur deux protozoaires nouveaux parasites des chenilles de *Pyrausta nubilalis* Hb. CR Acad. Sci. 185, 673–675.

Pereira, E.J.G., Storer, N.P., Siegfried, B.D., 2011. Fitness costs of Cry1F resistance in laboratory-selected European corn borer (Lepidoptera: Crambidae). J. Appl. Entomol. 135, 17–24.

Petzold-Maxwell, J.L., Jaronski, S.T., Gassmann, A.J., 2012a. Tritrophic interactions among Bt maize, an insect pest and entomopathogens: effects on development and survival of western corn rootworm. Ann. Appl. Biol. 160, 43–55.

Petzold-Maxwell, J.L., Cibils-Stewart, X., French, B.W., Gassmann, A.J., 2012b. Adaptation by western corn rootworm (Coleoptera: Chrysomelidae) to Bt maize: inheritance, fitness costs and feeding preference. J. Econ. Entomol. 105, 1407–1418.

Petzold-Maxwell, J.L., Jaronski, S.T., Clifton, E.H., Dunbar, M.W., Jackson, M.A., Gassmann, A.J., 2013. Interactions among Bt maize, entomopathogens, and rootworm species (Coleoptera: Chrysomelidae) in the field: effects on survival, yield, and root injury. J. Econ. Entomol. 106, 622–632.

Pierce, C.M., Solter, L.F., Weinzierl, R.A., 2001. Interactions between *Nosema pyrausta* (Microsporidia: Nosematidae) and *Bacillus thuringiensis* subsp. *kurstaki* in the European corn borer (Lepidoptera: Pyralidae). J. Econ. Entomol. 94, 1361–1368.

Pilz, C., Wegensteiner, R., Keller, S., 2007. Selection of entomopathogenic fungi for the control of the western corn rootworm *Diabrotica virgifera virgifera*. J. Appl. Entomol. 131, 426–431.

Pilz, C., Wegensteiner, R., Keller, S., 2008. Natural occurrence of insect pathogenic fungi and insect parasitic nematodes in *Diabrotica virgifera virgifera* populations. BioControl 53, 353–359.

Pilz, C., Keller, S., Kuhlmann, U., Toepfer, S., 2009. Comparative efficacy assessment of fungi, nematodes and insecticides to control western corn rootworm larvae in maize. BioControl 54, 671–684.

Poinar, G., Evans, J.S., Schuster, E., 1983. Field test of the entomogenous nematode, *Neoaplectana carpocapsae*, for control of corn rootworm larvae (*Diabrotica* spp., Coleoptera). Prot. Ecol. 5, 337–342.

Prabhukarthikeyan, R., Saravanakumar, D., Raguchander, T., 2013. Combination of endophytic *Bacillus* and *Beauveria* for the management of *Fusarium* wilt and fruit borer in tomato. Pest Manag. Sci. 70, 1742–1750.

Qayyum, M.A., Wakil, W., Arif, M.J., Dunlap, C.A., 2015. Infection of *Helicoverpa armigera* by endophytic *Beauveria bassiana* colonizing tomato plants. Biol. Control 90, 200–207.

Rahman, K.M.A., Barta, M., Cagáň, L., 2010. Effects of combining *Beauveria bassiana* and *Nosema pyrausta* on the mortality of *Ostrinia nubilalis*. Cent. Eur. J. Biol. 5, 472–480.

Rasmann, S., Köllner, T.G., Degenhardt, J., Hiltpold, I., Toepfer, S., Kuhlmann, U., Gershenzon, J., Turlings, T.C.J., 2005. Recruitment of entomopathogenic nematodes by insect-damaged maize roots. Nature 434, 732–737.

Reisig, D.D., Akin, D.S., All, J.N., Bessin, R.T., Brewer, M.J., Buntin, D.G., Catchot, A.L., Cook, D., Flanders, K.L., Huang, F.N., Johnson, D.W., Leonard, B.R., McLeod, P.J., Porter, R.P., Reay-Jones, F.P.F., Tindall, K.V., Stewart, S.D., Troxclair, N.N., Youngman, R.R., Rice, M.E., 2015. Lepidoptera (Crambidae, Noctuidae, and Pyralidae) injury to corn containing single and pyramided Bt traits, and blended or block refuge, in the Southern United States. J. Econ. Entomol. 108, 157–165.

Riba, G., 1984. Field plot tests using an artificial mutant of the entomopathogenic fungus, *Beauveria bassiana* (Hyphomycetes) against the European corn borer, *Ostrinia nubilalis* (Lepidoptera: Pyralidae). Entomophaga 29, 41–48.

Richter, A.R., Fuxa, J.R., 1990. Effect of *Steinernema feltiae* on *Spodoptera frugiperda* and *Heliothis zea* (Lepidoptera: Noctuidae) in corn. J. Econ. Entomol. 83, 1286–1291.

Romeis, J., Shelton, A.M., Kennedy, G.G. (Eds.), 2008. Integration of Insect-Resistant Genetically Modified Crops within IPM Programs. Springer, New York.

Roush, R.T., 1998. Two-toxin strategies for management of insecticidal transgenic crops: can pyramiding succeed where pesticide mixtures have not? Phil. Trans. R. Soc. B 353, 1777–1786.

Rudeen, M.L., Jaronski, S.T., Petzold-Maxwell, J.L., Gassmann, A.J., 2013. Entomopathogenic fungi in cornfields and their potential to manage larval western corn rootworm *Diabrotica virgifera virgifera*. J. Invertebr. Pathol. 114, 329–332.

Sánchez-Peña, S.R., Casas-De-Hoyo, E., Hernandez-Zul, R., Wall, K.M., 2007. A comparison of the activity of soil fungal isolates against three insect pests. J. Agric. Urban Entomol. 24, 43–48.

Sasan, R.K., Bidochka, M.J., 2012. The insect-pathogenic fungus *Metarhizium robertsii* (Clavicipitaceae) is also an endophyte that stimulates plant root development. Am. J. Bot. 99, 101–107.

Siegfried, B.D., Hellmich, R.L., 2012. Understanding successful resistance management: the European corn borer and Bt corn in the United States. GM Crops Food 3, 184–193.

St Leger, R.J., 2008. Studies on adaptations of *Metarhizium anisopliae* to life in the soil. J. Invertebr. Pathol. 98, 271–276.

Steffey, K.L., Rice, M.E., All, J., Andow, D.A., Gray, M.E., Van Duyn, J.W. (Eds.), 1999. Handbook of Corn Insects. The Entomological Society of America, Lanham.

Steinhaus, E.A., 1952. Microbial infections in European corn borer larvae held in the laboratory. J. Econ. Entomol. 45, 48–51.

Storer, N.P., Babcock, J.N., Edwards, J.M., 2006. Field measures of western corn rootworm (Coleoptera: Chrysomelidae) mortality caused by Cry34/Cry35Ab1 proteins expressed in maize event 59122 and implications for trait durability. J. Econ. Entomol. 99, 1381–1387.

Storer, N.P., Babcock, J.M., Schlenz, M., Meade, T., Thompson, G.D., Bing, J.W., Huchaba, R.M., 2010. Discovery and characterization of field resistance to Bt maize: *Spodoptera frugiperda* (Lepidoptera: Noctuidae) in Puerto Rico. J. Econ. Entomol. 103, 1031–1038.

Tabashnik, B.E., Gassmann, A.J., Crowder, D.W., Carrière, Y., 2008. Insect resistance to Bt crops: evidence versus theory. Nat. Biotechnol. 26, 199–202.

Tabashnik, B.E., Mota-Sanchez, D., Whalon, M.E., Hollingworth, R.M., Carrière, Y., 2014. Defining terms for proactive management of resistance to Bt crops and pesticides. J. Econ. Entomol. 107, 496–507.

Thurston, G.S., Yule, W.N., 1990. Control of larval northern corn rootworm (*Diabrotica barberi*) with two steinernematid nematode species. J. Nematol. 22, 127–131.

Toepfer, S., Peters, A., Ehlers, R.U., Kuhlmann, U., 2008. Comparative assessment of the efficacy of entomopathogenic nematode species at reducing western corn rootworm larvae and root damage in maize. J. Appl. Entomol. 132, 337–348.

Toepfer, S., Kurtz, B., Kuhlmann, U., 2010. Influence of soil on the efficacy of entomopathogenic nematodes in reducing *Diabrotica virgifera virgifera* in maize. J. Pest Sci. 83, 257–264.

Tokarev, Y.S., Malysh, J.M., Kononchuk, A.G., Seliverstova, E.V., Frolov, A.N., Issi, I.V., 2015. Redefinition of *Nosema pyrausta* (*Perezia pyraustae* Paillot 1927) basing upon ultrastructural and molecular phylogenetic studies. Parasitol. Res. 114, 759–761.

van Rensburg, J.B.J., 2007. First report of field resistance by stem borer, *Busseola fusca* (Fuller) to Bt-transgenic maize. S. Afr. J. Plant Soil 24, 147–151.

Vaughn, T., Cavato, T., Brar, G., Coombe, T., DeGooyer, T., Ford, S., Groth, M., Howe, A., Johnson, S., Kolacz, K., Pilcher, C., Purcell, J., Romano, C., English, L., Pershing, J., 2005. A method of controlling corn rootworm feeding using *Bacillus thuringiensis* protein expressed in transgenic maize. Crop Sci. 45, 931–938.

Wagner, B.L., Lewis, L.C., 2000. Colonization of corn, *Zea mays*, by the entomopathogenic fungus *Beauveria bassiana*. Appl. Environ. Microbiol. 66, 3468–3473.

Walters, F.S., Stacy, C.M., Lee, M.K., Palekar, N., Chen, J.S., 2008. An engineered chymotrypsin/cathepsin G site domian I renders *Bacillus thuringiensis* Cry3A active against western corn rootworm larvae. Appl. Environ. Microbiol. 74, 367–374.

Walters, F.S., deFontes, C.M., Hart, H., Warren, G.W., Chen, J.S., 2010. Lepidopteran-active variable-region sequence imparts coleopteran activity in eCry3.1Ab, an engineered *Bacillus thuringiensis* hybrid insecticidal protein. Appl. Environ. Microbiol. 76, 3082–3088.

Windels, M.B., Chiang, H.C., Furgala, B., 1976. Effects of *Nosema pyrausta* on pupa and adult stages of the European corn borer *Ostrinia nubilalis*. J. Invertebr. Pathol. 27, 239–242.

Wraight, S.P., Ramos, M.E., Avery, P.B., Jaronski, S.T., Vandenberg, J.D., 2010. Comparative virulence of *Beauveria bassiana* isolates against lepidopteran pests of vegetable crops. J. Invertebr. Pathol. 103, 186–199.

Zimmack, H.L., Arbuthnot, K.D., Brindley, T.A., 1954. Distribution of the European corn borer parasite *Perezia pyraustae*, and its effect on the host. J. Econ. Entomol. 47, 641–645.

Chapter 12

Microbial Control of Insect and Mite Pests of Cotton

J. Leland[1], J. Gore[2]

[1]*Novozymes BioAg, Salem, VA, United States;* [2]*Mississippi State University, Stoneville, MS, United States*

12.1 INTRODUCTION

Cultivated cotton, *Gossypium hirsutum* and *G. barbadense*, is a unique crop from an insect management standpoint. They were derived from a perennial shrub but have been domesticated and are now planted as an annual crop (Mauney, 1986). Although cotton is grown as an annual crop, it maintains numerous traits of a perennial plant. Most notably from an insect pest management standpoint, the indeterminate growth habit of cotton is very important. Cotton begins reproductive development about 40 days after planting when the first flower bud appears (Mauney, 1986). Unlike other annual crops such as field corn, *Zea mays*, and most soybean, *Glycine max*, varieties, cotton continues vegetative growth throughout the reproductive stages (Landivar and Benedict, 1996). As a result, the reproductive stages of cotton occur over a 9- to 10-week period. This makes cotton attractive to insects and susceptible to yield losses over a longer period of time compared to many other crops.

12.1.1 Thrips

Thrips species are usually the first insects to infest cotton. Planting starts when soil temperatures reach or exceed 15.6°C, and seedlings emerge in 3–7 days (Gipson, 1986). Seedling growth is slow, making cotton more susceptible to injury from early season insect pests than many other crops. During the seedling stage, numerous thrips species infest cotton and can cause significant yield losses if left uncontrolled (Leigh et al., 1996). As a result, preventative control is often recommended in all regions where cotton is grown (Cook et al., 2011). The most effective controls have included at-planting insecticides applied as systemic in-furrow granules and sprays, or seed treatments. The most common thrips species that infest cotton during the seedling stage include tobacco thrips, *Frankliniella fusca* (Hinds), and Western flower thrips, *F. tritici* (Fitch). More comprehensive lists of thrips species that infest cotton can be found in Stewart et al. (2013) and Cook et al. (2011).

12.1.2 Aphids

Initial infestations of cotton aphid, *Aphis gossypii* Glover, occur between the seedling and early reproductive stages of cotton (Slosser et al., 1989). However, populations do not usually reach treatable levels until later in the growing season, following insecticide applications for other pests (Johnson et al., 1996). In general, cotton aphid populations are maintained below treatable levels by the actions of natural enemies unless sprays targeting other pests have been made that disrupt the natural enemy complex (Weathersbee and Hardee, 1994). In general, cotton aphid is an indirect pest that feeds on the underside of leaves and can produce large amounts of honeydew. There is considerable debate about the pest status of cotton aphid and its impact on cotton yields. However, the accumulation of honeydew in arid regions such as the southwestern United States can result in sticky cotton. This interferes with harvest and reduces lint quality. In the southeastern United States, frequent rainfall in the fall limits the accumulation of honeydew except in dry years. Insecticide sprays targeting cotton aphid occur in most regions, especially in arid regions to preserve lint quality and improve milling quality.

12.1.3 Hemipteran Pests

Several species of true bugs have emerged as important pests of cotton (Leigh et al., 1996). The complex varies significantly by region. The tarnished plant bug, *Lygus lineolaris*, and the Western tarnished plant bug, *Lygus hesperus*, are the most important pests in the mid-southern and western United States, respectively (Fig. 12.1). In parts of Texas, the complex of true bugs infesting cotton includes multiple *Lygus* spp., cotton flea hopper, *Pseudatomoscelis seriatus*, and the green mirid, *Creontiades dilutes*. In the southeastern United States, the stink bug species, *Euschistus servus*, *Acrosternum hilare*, and *Nezra viridula*, are the most important insect pests of cotton on an annual basis. Multiple species of true bugs infest cotton worldwide. The true bugs are direct pests that feed on flower buds and bolls.

Microbial Control of Insect and Mite Pests. http://dx.doi.org/10.1016/B978-0-12-803527-6.00012-3

FIGURE 12.1 Cotton plots in which the Western tarnished plant bug, *Lygus hesperus*, was controlled (left) and not controlled (right).

12.1.4 Spider Mites

Several species of spider mites infest cotton throughout the world. The most common species on a global scale is the two-spotted spider mite, *Tetranychus urticae*. This species can cause significant yield losses in cotton (Gore et al., 2013) and has a long history of rapidly developing resistance to synthetic pesticides used for their control (Beers et al., 1998). Spider mites feed on the underside of leaves. In general, injury occurs from reduced photosynthesis at moderate population densities and defoliation at higher population densities (Furr and Pfrimmer, 1968).

12.1.5 Whiteflies

Several species of whitefly can occur in cotton, but the sweet potato whitefly, *Bemisia tabaci*, is the most important (Leigh et al., 1996). Multiple biotypes of this species exist with varying levels of insecticide resistance. Whiteflies are primarily a pest in arid regions or where vegetable production occurs in close proximity to cotton. Similar to spider mites and cotton aphid, whiteflies feed on the underside of leaves. Injury can occur from defoliation or from the accumulation of honeydew that reduces lint quality.

12.1.6 Lepidoptera

Numerous insect pests in the order Lepidoptera infest cotton worldwide. Prior to the introduction of Bt cotton (cotton that has been genetically modified to produce insecticidal proteins from *Bacillus thuringiensis*), caterpillars were the most important insect pests of cotton in many regions. In most regions, multiple species of Lepidoptera occur in a sequence or as a complex throughout the season. They include both direct pests that feed on the developing fruit and indirect pests that injure cotton through defoliation.

12.1.6.1 Tobacco Budworm

Historically, the tobacco budworm, *Heliothis virescens*, was the most important lepidopteran pest of cotton in many areas of the Americas prior to the introduction of Bt cotton (Leigh et al., 1996). However, this species is highly susceptible to the Bt proteins expressed in current Bt cotton varieties and no sprays have been needed for tobacco budworm in Bt cotton since its introduction into commercial varieties (Luttrell et al., 1999) (Fig. 12.2). In areas where tobacco budworm was the primary direct pest, adoption of Bt cotton has been greater than 95% and this species is rarely considered with current integrated pest management (IPM) plans.

12.1.6.2 Bollworm

In those areas where tobacco budworm was the primary lepidopteran pest, the bollworm (ie, corn earworm), *Helicoverpa zea*, has become the primary economically important species (Gore et al., 2008). It is much less susceptible to the proteins expressed in Bt cotton than tobacco budworm (Luttrell et al., 1999), and supplemental foliar insecticide sprays are often needed in Bt cotton to prevent yield losses (Gore et al., 2008).

12.1.6.3 Pink Bollworm

The pink bollworm, *Pectinophora gossypiella*, is an important insect pest of cotton in many areas of the world (Leigh et al., 1996). Similar to tobacco budworm, this species is highly susceptible to the proteins in current Bt cotton varieties. Because of the success of sterile release programs in the western United States and the introduction of cotton varieties expressing Bt toxins, it is no longer considered a major pest in the United States.

FIGURE 12.2 Seed cotton harvested from plots of non–Bt cotton (right) and Bt cotton (left) where tobacco budworm, *Heliothis virescens*, and bollworm, *Helicoverpa zea*, were the primary lepidopteran species present.

12.1.6.4 Old World Bollworm

The Old World bollworm, *Helicoverpa armigera*, is the most important caterpillar pest of cotton in many areas of the world, including Asia, Australia, Africa, and Europe. Recently, this species was identified in South America and appears to be migrating north. The Old World bollworm is relatively tolerant to the proteins in Bt cotton and has a long history of rapidly developing resistance to foliar insecticides (Armes et al., 1996).

12.1.6.5 Armyworms

Numerous *Spodoptera* species occur as pests of cotton worldwide (Sparks, 1979; Hendricks et al., 1995). Some of the more common species include *S. frugiperda*, *S. exigua*, *S. eridania*, *S. latifascia*, and *S. ornithogalli*. *Spodoptera* spp. can be devastating pests of cotton in many areas, and foliar sprays are often needed to prevent economic losses. Most of these species are indirect pests that feed on the foliage except *S. frujiperda*, which feeds directly on the fruit of cotton.

12.1.6.6 Loopers

The soybean looper, *Chrysodeixis includes*, and the cabbage looper, *Trichoplusia ni*, both noctuids, are the most common species of looper in cotton. Of these species, *C. includens* is the most important because it is relatively tolerant to some of the proteins in Bt cotton and has developed resistance to numerous foliar insecticides (Leonard et al., 1990). Similar to many of the armyworm species, loopers are indirect pests that feed on the foliage of cotton.

12.2 MICROBIAL BIOPESTICIDES FOR USE IN THE COTTON CROP FOR RESISTANCE MANAGEMENT

12.2.1 Sprayable Bt and Baculovirus Product Use—Impact of Transgenic Bt Cotton

Use of sprayable Bt and baculovirus products on cotton were increasing during the 1990s in response to resistance of lepidopteran pests to insecticides. However, with the introduction of transgenic Bt cotton and its mass adoption on the majority of cotton-growing hectares globally, this role for microbial biopesticides has declined.

In the United States, there were significant efforts toward developing the nucleopolyhedrovirus, HezeSNPV, to control *Helicoverpa zea* and *Heliothis virescens* in wild host plants around cotton fields of the Mississippi Delta as part of an areawide management program (Bell and Hardee, 1994a, 1994b, 1995; Hardee and Bell, 1995; Street et al., 1997; Hardee et al., 1999). With the introduction of Bt cotton, these programs were deprioritized with a shift in focus toward the areawide management of *Lygus* species, as will be described in Section 12.3.1.This same scenario has also played out in other regions of the world. A second example of this is described in the expansion and then decline of Baculovirus product use in Australia as described in Buerger et al. (2007), where Elcar (HezeSNPV) was first introduced in the 1970s but at the time was competing with pyrethroids. As resistance began to develop to pyrethoids in the 1990s, Gemstar (HezeSNPV) was then introduced, which was followed by ViVUS Gold (HearNPV) in 2004 in order to use a native Baculovirus strain. Strict import regulations in Australia make it beneficial to work with a native strain that is locally produced. With the widespread use of Bt cotton in Australia, Baculovirus products have primarily found a fit in other crops. Baculovirus has become an established component of IPM programs where *Helicoverpa armigera* is a major pest (eg, soybeans, cotton, corn, sorghum, sweet corn, lettuce, and tomato) and in systems where it is more of a minor pest because its selectivity does not disrupt natural enemies. However, transgenic crops such as cotton have reduced the use of all foliar insecticides, including microbials (Buerger et al., 2007). Despite *H. armigera's* long history of developing resistance to chemical insecticides, resistance to Baculovirus has not been documented to date regardless of its widespread use in sorghum and cotton in Australia (Buerger et al., 2007).

China has also put significant effort into Baculovirus development and currently produces approximately 1600 metric tons of viral insecticide formulations annually (Sun, 2015). This represents 57 products from 11 viruses for about 0.2% of total insecticide input in China (Sun, 2015). NPVs are applied on more than 10,000 ha annually in China over a range of crops (Yang et al., 2012 cited by Beas-Catena et al., 2014). HearNPV was first commercialized in China in 1993. Although GM cotton has also affected the need for insecticide sprays, in 2012, 968 metric tons of HearNPV produced by 10 companies were used against cotton bollworm (Sun, 2015). *Helicoverpa armigera* remains resistant to several classes of insecticides in China despite their decreased use in Bt cotton (Yang et al., 2013). HearNPV products are also available in South Africa primarily targeting *H. armigera* on other crops but there is potential for its expanded use in cotton in South Africa (Knox et al., 2015).

12.2.2 Efforts to Improve Baculovirus Efficacy

In order for baculoviruses to be a significant option for control of Heliothines in cotton, some challenges will have to be overcome. These include (1) high cost of in vivo production; (2) slow speed of kill; (3) poor persistence due to UV inactivation; and (4) getting adequate coverage to target secluded feeders, a feature they share with Bt sprays (Steinkraus et al., 2007).

Significant efforts have gone into genetic modifications, formulations, and insecticide combinations of baculoviruses to improve their speed of kill and persistence. These efforts have been covered in other reviews and will not be fully described here. In brief, recombinant viruses have improved speed of kill by addition of insect hormones or enzymes, or insect specific toxins, expression of Bt cry 1–5 crystal protein, and binding peptides for nanomaterials to improve persistence (Baes-Catena et al., 2014; Sun, 2015). Despite extensive laboratory work, there has been

limited work on demonstrating improved field efficacy of GM baculoviruses. One study demonstrated significantly higher (22%) lint yield based on mature boll counts from plots treated with HearSNPV-AaIT than from wild-type virus (Sun et al., 2009).

12.2.3 Efforts to Improve Efficacy of Bt and Entomopathogenic Fungi

For advances made with MCAs that could make them more competitive in inundative control strategies for cotton and other insect pests, see Chapters 4 and 5. In addition to the work described for baculoviruses, it is worth mentioning here some of the recent efforts to improve the efficacy of entomopathogenic fungi (EPFs) and Bt that could one day make these classes of MCAs more competitive alternatives to chemical insecticides in cotton. Much of this research has focused on the development of genetically modified organisms. A Bt cotton strain (WG-001) with significantly increased crystal toxin production and enhanced insecticidal activity was approved in China and commercial production and application are now under way (Huang et al., 2007). Efforts to improve virulence of EPFs were reviewed by Wang and Feng (2014). They included overexpression of chitinase or protease enzymes; introduction of exogenous toxins such as from spider, scorpions, or Bt; and improved tolerance to UV, oxidation, and heat.

It remains to be seen if another technology is on the horizon that will be as impactful as Bt cotton was in the 1990s toward changing the landscape of cotton IPM. Due to the increased significance of sap sucking insect pests in cotton, there is a lot of focus on these targets of the next transformative step. However, efforts to develop transgenic crops with activity against sap sucking insects have met with many challenges (Chougule and Bonning, 2012). Low levels of toxicity have been observed for Bt δ-endotoxins against aphids, in part because many studies used toxin crystals or spore suspensions that would be difficult to solubilize in the acidic aphid midgut rather than presolubilized toxin (Payne and Cannon, 1993; Walters and English, 1995; Cristofoletti et al., 2003 cited by Chougule and Bonning, 2012). However, even solubilized toxins have showed low levels of activity (Porcar et al., 2009 cited by Chougule and Bonning, 2012). One vegetative insecticidal protein (Vip) purified from Bt has also shown activity against *Aphis gossypii*, and receptor binding results suggest the toxin may have aphid specificity (Sattar and Maiti, 2011 cited by Chougule and Bonning, 2012). Hemiptera may not be susceptible to Bt toxins due to their feeding habit. Evolutionarily, Bt is a soil bacterium that contaminates leaves and thus is ingested by soil feeding insects. Hemiptera, by feeding on internal fluids within the plant, simply have not been exposed to Bt toxins. The type and abundance of proteolytic enzymes and pH may be inappropriate for proteolytic activation. Given the challenge in expanding the spectrum of Bt cotton beyond control of Lepidoptera, it is likely that there will be a continued need for managing sap sucking insects pests for the foreseeable future.

12.3 BIOPESTICIDES FOR USE IN NON–BT REFUGE COTTON AND ALTERNATE HOST PLANTS FOR RESISTANCE MANAGEMENT

In part due to the higher cost of microbial biopesticides, strategies have been proposed to use them in a focused manner on either non–Bt refuge cotton or alternate host plants. One example of this is the proposed use of EPFs against tarnished plant bugs (*Lygus* spp.) in wild host plants in intensively managed areas of the Mississippi Delta or in the San Joaquin Valley of California, USA (see Section 12.3.1). As described in Section 12.2.1, a similar proposal using baculoviruses against lepidopteran pests in wild host plants around cotton fields in the Mississippi Delta predated the work on tarnished plant bug but was deemphasized with the introduction of Bt cotton. A second example was the proposed use of entomopathogenic nematodes (EPNs) in non–Bt cotton refuges adjacent to Bt cotton fields to reduce the frequency of resistance genes establishing in these refuges by imposing a selective fitness cost against pink bollworm harboring resistance genes. These targeted approaches described below rely on the unique mode of action of the MCAs and a strategic understanding of the migration of pest insects among various host plants to maximize the potential utility of biopesticides in resistance management.

12.3.1 Fungi in Wild Host Plants for *Lygus* Species

Steinkraus et al. (2007) provided an excellent review of the earlier work on the use of *Beauveria bassiana* against *L. lineolaris* in the Mississippi Delta and *L. hesperus* in the San Joaquin Valley of California. In brief, native isolates of *B. bassiana* were collected from the two regions and then selected based on activity toward the two *Lygus* spp., environmental stability (eg, UV tolerance), and production potential in solid state fermentation (McGuire, 2002; Leland and Snodgrass, 2004; Leland, 2005; Leland et al., 2005; McGuire et al., 2005). One native *B. bassiana* isolate from each region and a commercial isolate (*B. bassiana* GHA) were then evaluated in field trials in each region using isolated plots (Fig. 12.3), caged insects, and insects returned to the laboratory to evaluate incidence of infection based on sporulation from cadavers after incubation at room temperature (Leland and McGuire, 2006; McGuire et al., 2006). Other work has focused on using *B. bassiana* in combination with trap crops (Lund et al., 2006). Since this initial work, the majority of follow-up research has focused on understanding the impact of *B. bassiana* on *Lygus* spp. survival and fecundity at various temperatures in order to guide best practices around seasons for application to wild host plants. Spurgeon (2010) evaluated mortality of *L. hesperus* at three temperatures (12.8°C, 18.3°C, and 23.9°C) following exposure to five *B. bassiana* isolates (one commercial and four native isolates at 6.8×10^3 conidia/cm^2). Mortality rates were slower at the lower temperatures over the 21-day bioassay. However, the proportion of cadavers exhibiting mycosis was approximately 80% for all temperatures. There was not enough difference among the isolates to use mortality rates at lower temperature as criteria for

FIGURE 12.3 Plots (0.025 ha) of mustard isolated by four rows of corn to limit migration among plots. These were used to study the impact of *Beauveria bassiana* applications on the Western tarnished plant bug, *Lygus hesperus*, and beneficial insects on alternate host plants as part of a cotton areawide management program.

isolate selection. Ugine (2011) also addressed the issue of low temperatures on efficacy of *B. bassiana* toward *L. lineolaris*. Bioassays on adult females over a range of 200–400 conidia/mm^2 were conducted at 18°C, 21°C, 25°C, 30°C, and 32°C to determine the number of eggs laid until death. There was a significant reduction in total eggs laid at all temperatures by *B. bassiana*, although it did not affect the daily egg production rate. Estimates of the impact of temperature and *B. bassiana* exposure on calculated population increase and population doubling times indicated that *B. bassiana* should be used to target *L. lineolaris* in diapause or during cool temperatures to minimize reproduction prior to mortality. Certainly ecological factors beyond just temperature will impact the efficacy of EPFs in cotton. A review of the ecological factors impacting the inundative use of EPFs is provided by Jaronski (2010). In the phylloplane, sunlight, humidity, and temperature impact both immediate efficacy and persistence. Some specific references to cotton in this review included suggestions for the use of a horizontal bar preceding the spray boom to bend the cotton and improve application to the underside of leaves. A second suggestion noted that evapotranspiration can decrease temperatures at the cotton leaf surface under hot desert conditions, thus explaining the efficacy of *B. bassiana* against whiteflies in Arizona cotton (Jaronski et al., 1997).

Additional follow-up work has focused on the susceptibility of different life stages of *L. lineolaris* to a *B. bassiana* isolate native to Mississippi (NI8). Portilla et al. (2014) evaluated mortality of all second through fifth instars of *L. lineolaris* and adults when exposed to *B. bassiana* (NI8). Among these life stages, the second instar was the least susceptible. Bioassays were conducted using a newly developed nonautoclaved diet, which could be a useful tool for additional work with *Lygus* spp. rather than relying solely on plant material for rearing and assays. Jin et al. (2008) developed formulations using optimal hydrophilic–lipophilic balance to select surfactants that improved suspension of *B. bassiana* (NI8 and GHA) conidia in water and subsequent germination. The new formulations did not result in a measurable increase in mortality in laboratory bioassays.

Similar programs are beginning to address the growing pest status of the Miridae in cotton growing regions of China. Field trials over 10 years in northern China have demonstrated an increase in the pest status of mirids in cotton and other crops (Lu et al., 2010). The increases show correlation with lower insecticide use in Bt cotton. On an areawide basis, there has been a reduction in cotton bollworm (*Helicoverpa armigera*) due to the impact of Bt cotton on cotton bollworm populations. The case is different for mirids. They build up in early season host plants before moving into managed crops. Cotton is a preferred host crop in mid to late June because it is one of the few that is flowering at that time. Prior to the introduction of Bt cotton, mirids entering the cotton fields would be killed by insecticide application, thus reducing populations on a regional basis. As a result, where cotton was previously a sink for mirid populations due to insecticide use, it is now acting as a source. The opposite is true for aphids that were a primary pest in cotton in the mid 1970s because of intensive insecticide use against *H. armigera* but are now at low densities because natural enemies have become more abundant with the reduction of insecticide use (Lu et al., 2010). Efforts to mitigate this have included early spring weeding to reduce habitats, interspersing cotton with alfalfa and mung bean to attract and provide natural enemy habitat, releases of natural enemies such as *Peristenus spretus* (Hymenoptera: Braconidae) (Luo et al., 2011 cited by Luo et al., 2014), and application of *B. bassiana* during the warm and rainy season (Tong et al., 2010 cited by Luo et al., 2014). Tong et al. (2010) screened seven strains of *B. bassiana* against the three dominant mirid pests in China: *Apolygus locorum*, *Adelphocoris suturalis*, and *Adelphocoris lineolatus*. Relative activity of the strains against the three mirid species was strain dependent rather than a single isolate showing a higher level of activity to all species. For the most active strain (C-1) against *A. locorum*, activity was inversely correlated with instar stage based on LT$_{50}$ values.

12.3.2 EPNs in Refuge Plots for Pink Bollworm Control

Dual-gene Bt (*Cry1Ac* and *Cry2Ab*) cottons were commercialized in 2004, and they were used to successfully eradicate *Pectinophora gossypiella* from the United States. It remains a pest of cotton in India where it has developed resistance to Bt *Cry1Ac* cotton (Bagla, 2010; Dhurua and Gujar, 2011, cited by Gassmann et al., 2012). In the southwestern United States, growers were required to provide refuges of non–Bt cotton near Bt crops in order to allow nonselected populations of pink bollworm to exist that could mate with potentially resistant individuals developed in Bt crops to delay selection of Bt-resistant individuals. Prior to eradication in the southwestern United States, a strategy was proposed to introduce EPNs in refuge crops to impose an additional fitness cost on pink bollworms containing alleles that imparted resistance to Bt toxins. Pink bollworm pupates in the top 1.3 cm of soil. *Steinernema riobrave* is an intermediate strategist (ambusher and seeker), thus having an ability to move into infested plant material or soil to locate the host. Whereas *S. carpocapsae* is more prone to ambush its host and is more appropriate for targeting mobile larvae near the soil surface (Shapiro-Ilan et al., 2006). Timing for EPN application could be against prepupae during the hot midseason or against diapausing larvae presowing along with irrigation. Applications presowing against diapausing larvae would be more convenient, but targeting prepupae midseason makes more sense on an areawide basis due to pink bollworm's high mobility

(Steinkraus et al., 2007). *Steinernema riobrave* is more heat tolerant than *S. carpocapsae*, which should be considered when applying EPNs during the hot midseason of cotton in the southwestern United States (Shapiro-Ilan et al., 2006). Several studies have evaluated the potential for EPNs to impose a fitness cost on pink bollworms carrying resistant alleles to *Cry1Ac* Bt cotton (*Cry1Ac*). Gassmann et al. (2006) demonstrated that *S. riobrave* imposed a fitness cost on resistance to *Cry1Ac* in pink bollworm. Gassmann et al. (2008, 2009) followed up by demonstrating that *S. riobrave* imposed a recessive fitness cost on resistance to *Cry1Ac* in pink bollworm. Further work demonstrated that *Heterorhabditis bacteriophora* imposed a fitness cost affecting heterozygous resistance to *Cry1Ac* in pink bollworm reared on cotton bolls (Gassmann et al., 2009) but not homozygous resistant individuals nor on pink bollworms reared on diet with or without gossypol (Gassmann et al., 2008). Hannon et al. (2010) demonstrated that *S. riobrave* imposed a dominant fitness cost on resistance to *Cry1Ac* in pink bollworm; *S. carpocapsae*, *Steinernema* sp. (ML18 strain), and *H. sonorensis* imposed no fitness cost. Gassmann et al. (2012) mimicked high dose/refuge scenarios in the lab on diet and greenhouse. In greenhouse experiments, *S. riobrave* did not delay resistance. In the laboratory experiments, *S. riobrave* delayed resistance after two generations but not after four. *Steinernema riobrave* imposed a 20% fitness cost on homozygous resistant pink bollworm but not fitness cost on heterozygous resistant pink bollworm. These studies indicate that fitness costs imposed by *S. riobrave* would be most effective at delaying resistance when resistance allele frequency is low and post populations remain sufficiently high. The nature of the fitness cost appears to differ between genotypes of pink bollworm being tested, adding complexity to this strategy. When considering using EPNs for these strategies, it is worth reviewing the impacts of application and environmental conditions on efficacy. A review of application and environmental considerations for biological control with EPNs is provided by Shapiro-Ilan et al. (2006). The review describes considerations such as operating pressures for application equipment, typical minimum rates for efficacy, environmental factors such as UV, soil moisture, and temperature, and compatibility with other agricultural products.

12.4 POTENTIAL OF EPFs AS ROOT COLONIZERS AND ENDOPHYTES – REDUCED RATES REQUIRED AND ALTERNATE MODES OF ACTION

The role of EPFs as rhizosphere colonizers, endophytes, plant pathogen antagonists, and potentially plant growth–promoting agents has drawn increased attention in the past decade, building on work in the 1980s and 1990s such as

that with *B. bassiana* and corn done by Les Lewis (Vega, 2008). For broad acreage crops such as cotton where inputs are more price sensitive than higher-value crops such as vegetables, the strategy for incorporating MCAs into the cropping system could help overcome cost limitations associated with the inundative application of MCAs. As long as the pathogens bring growers a return on investment for their inputs by way of yield protection or reduced inputs, these uses could be justified, even if a high level of efficacy against a specific insect target is not achieved. Several studies have demonstrated colonization of cotton by applied EPFs that led to positive results with regard to mitigating damage by insect pests and plant pathogens. Three EPFs (*B. bassiana*, *Lecanicillium lecanii*, and *Aspergillus parasiticus*) were shown to colonize the leaves of six crop plants including cotton when inoculated as conidia (Gurulingappa et al., 2010). Of these fungi, only *A. parasiticus* was able to colonize cotton leaves and roots through the soil. However, its endophytic presence reduced cotton growth, whereas colonization of cotton leaves by either *B. bassiana* or *L. lecanii* reduced aphid (*Aphis gossypii*) reproduction. Follow-up studies indicated that filtrates from liquid culture of *L. lecanii* reduced aphid reproduction, and methanolic fraction of extracts from the liquid culture filtrate of *B. bassiana* caused significant aphid mortality (Gurulingappa et al., 2011). Lopez et al. (2014) also demonstrated that endophytic colonization of cotton by *B. bassiana* or *Purpureocillium lilacinum* was correlated with reduced aphid reproduction both in the greenhouse and on the field. In the greenhouse studies, impacts on aphid reproduction were significantly negatively correlated with presence of endophytic colonization of the cotton plants following seed treatment. In the field studies, the presence of endophytic colonization of plants from seed treatments was not confirmed, but nonetheless seed treatments with *B. bassiana* significantly reduced aphid reproduction.

In addition to effects on insects, effects on plant pathogens have been demonstrated. Ownley et al. (2008) demonstrated that seed application of *B. bassiana* (strain 11–98) resulted in endophytic colonization of cotton and protection against the plant pathogen *Rhizoctonia solani*. *B. bassiana* was recovered from surface sterilized roots, stems, and leaves and recovery was positively correlated with seed treatment rate. Scanning electron microscope images of the colonization were provided (Ownley et al., 2008). Evidence of protection from *Rhizoctonia* was provided by positive correlation of seedling growth with *B. bassiana* seed treatment concentrations in the presence of *Rhizoctonia* with no effect observed in the absence of *Rhizoctonia*. *B. bassiana* also protected against *Xanthamonas axonopodis* pv. *malvacearum* (bacterial blight) as evidenced by reduced disease severity. These authors proposed that a mechanism for protection from the fungal plant pathogens was competition for space and induced systemic resistance for protection from *X. axonopodis* pv. *malvacearum* (Ownley et al., 2008). This hypothesis was further supported by Griffin et al. (2006) who evaluated *B. bassiana* in an induced systemic resistance (ISR) screening bioassay in cotton against *X. axonopodis* pv. *malvacearum*. The assay demonstrated a similar level of protection from *X. axonopodis* pv. *malvacearum* as that produced by the chemical ISR inducer 2,6-dichloro-isonicotinic acid (INA). Some of the work with *Lecanicillium* spp. has taken this a step further using protoplast fusion to create strains with improved biological control characteristics (Goettel et al., 2008). This was initially done using commercial bioinsecticide strains (from Vertalec and Mycotal) with a strain (B-2) that had high epiphytic ability (Koike et al., 2004 cited by Goettel et al., 2008). Additional scientists selected hybrid strains optimized for activity against aphids and whiteflies with foliar survival under low humidity conditions (Aiuchi et al., 2007 cited by Goettel et al., 2008) and combined activity against cotton aphids and powdery mildew (Kim et al., 2007, 2008 cited by Goettel et al., 2008). Protoplast fusion was also used to develop more efficient *Lecanicillium* spp. strains for control of soybean cyst nematode (Shinya et al., 2008a, 2008b; 2008c; Koike et al., 2004 cited by Goettel et al., 2008). It is envisioned that hybrids of *Lecanicillium* spp. could be created with sufficient efficacy and host range improvements to provide some level of control for plant diseases, pest insects, and plant parasitic nematodes (Goettel et al., 2008). Although much of this work has been done on plants other than cotton, cotton aphid is a significant cotton pest and the principles have potential application toward extending the utility of fungal based biopesticides in cotton systems.

Cory and Erricsson (2010) turned this EPF–plant interaction view around considering the impact of plant-mediated effects on EPFs. Plant volatiles and plant surface chemistry have effects on the EPFs. Plant secondary chemicals can inhibit fungal growth. These plant-mediated effects may directly influence infection of the insect or feeding on the plant by the insect host may indirectly affect the lifecycle of the fungus. Some specific cotton examples given in this review were decreased susceptibility of *B. argentifolii* to *Isaria fumosorosea* and *B. bassiana* when reared on cotton rather than melon (a preferred host plant), which was suggested to be related to a secondary chemical produced by cotton (Poprawski and Jones, 2001; Poprawski et al., 2000).

The negative effects of cotton on activity is not unique to EPFs. For example, this has also been observed for baculoviruses. Mortality rates were shown to be 2.5 time higher for *Heliothis virescens* fed diet or lettuce relative to those fed cotton treated with AcMNPV (Hoover et al., 2000). The resistance to infection correlates with foliar peroxidase levels, which may cause midgut cell sloughing leading to resistance (Hoover et al., 1998, 2000).

12.5 CONSERVATION BIOLOGICAL CONTROL: INFLUENCING NATURAL EPIZOOTICS

The role of *Neozygites fresenii* in managing cotton aphid populations in the southeastern United States was previously reviewed (Steinkraus et al., 2007). In brief, this naturally occurring EPF tends to cause epizootics in cotton aphid populations, reducing them below economic injury levels. A scouting service has been a useful tool for growers to determine if insecticide applications are needed to manage the populations or if *N. fresenii* epizootics are likely to occur in the near future. Insect predators, in particular ladybird beetles (Coccinellidae), also play a significant role in managing cotton aphid populations. A detailed 3-year study on the interaction of cotton aphid natural enemies and cotton aphid populations was reported by Abney et al. (2008). Fungal epizootics of *N. fresenii* occurred in mid-July in all 3 years and reduced cotton aphid populations below treatment thresholds. Coccinellid species (especially *Scymnus* spp.) were the most important predators, but other generalists such as spiders, fire ants, hemipterans, and neuropterans were also present. Results from arthropod exclusion cages suggested that the predators or parasitoids had little impact on cotton aphid populations prior to fungal epizootics but may have played a more significant role in suppressing aphid populations after epizootics in the late season. Simelane et al. (2008) investigated the interaction of cotton aphids, *N. fresenii* infection and predation rate, and development of *Coccinella septempunctata*. Predation rates of cotton aphids by fourth-instar and adult *C. septempunctata* were not impacted by early stage infection of the cotton aphids by *N. fresenii*. Although the number of prey consumed by *C. semptempunctata* was not impacted by the cotton aphids being infected with *N. fresenii*, other developmental and reproductive factors were impacted negatively, reducing overall fitness. These impacts included longer stadia, higher mortality rates, smaller adult body size, and reduction in numbers of eggs laid. New action thresholds for cotton aphid insecticide applications take into account the development of the cotton aphid population and the presence of natural enemies such as *N. fresenii* and ladybird beetles (Conway et al., 2006; Greene, 2006 cited by Simelane et al., 2008).

Other entomopathogens may be playing a role in managing cotton insect pests even if the effects are not as dramatic as observed in cotton aphids. Recently a new fungus, *Pandora heteropterae*, was isolated from *L. lineolaris* in Arkansas in 2010 and previously had only been reported once from an unidentified host species in Poland (Hannam and Steinkraus, 2010). Prevalence in the wild *L. lineolaris* population was very low (0.32%). In laboratory bioassays, it was found to infect seven hemipterans from the families Miridae, Coreidae, Lygaeidae, and Pentatomidae. Further work is needed to determine if this fungus may cause epizootics under different environmental conditions, and it may easily go unrecognized due to its rapid kill and decomposition of the host or if the epizootics occur in populations outside the cotton growing season.

12.6 OTHER COTTON PESTS

12.6.1 Boll Weevils

Research on alternative control options for boll weevils has declined in the United States since the advent of boll weevil eradication programs. However, boll weevil remains a pest of cotton in Brazil, Paraguay, Argentina, and Bolivia. Nussenbaum and Lecuona (2012) screened 28 *M. anisopliae* isolates and 66 *B. bassiana* isolates against boll weevil. Two *M. anisopliae* isolates were the most virulent. Sublethal effects on feeding and weight gain were also noted. Overwintering adults in leaf litter and feeding and ovipositing on cotton can be targeted (Nussenbaum and Lecuona, 2012).

12.6.2 Stink Bugs

There has been little success in developing MCAs against stink bugs. Most work has focused on EPFs but the stink bug's fungistatic natural defenses may limit their potential, particularly in certain species like *Nezara virudula* and generally more significant in older instars and adults (Sosa-Gómez et al., 1996; Sosa-Gómez and Moscardi, 1998; Lopes et al., 2015; Silva et al., 2015; Raafat et al., 2015). Despite the potential for natural defenses mitigating pathogenicity in older nymphs and adults of some species, some isolates have demonstrated activity against specific stink bug species. Examples of these include *Tibraca limbateventris* (Quintela et al., 2013; Silva et al., 2015), *Plautia crossota stali* and *Glaucias subpunctatus* (Ihara et al., 2008), *Dichelops melacanthus* (Lopes et al., 2015), and *Halymorpha halys* (Parker et al., 2015). Although these examples of pentatomid susceptibility are not directly related to cotton trials, they underscore the value in considering susceptibility to the specific stink bug complex in the target region and varying susceptibility of life stages when considering EPFs as a control option in a given cropping system.

12.6.3 Thrips, Aphids, Whiteflies, and Mites

There has been little adoption of EPFs in cotton for control of these pests. For thrips, it has been difficult to compete with systemic in-furrow insecticidal granules and sprays or seed treatments in terms of efficacy and cost. Therefore, it has not been a focus of development. As described earlier, the major role of microbial control of aphids is in monitoring and incorporating *Neozygites fresenii* epizootics into management practices. Spider mites and whiteflies also

have EPF options available but these have not been adopted on a wide scale. Increasing regulatory pressure reducing chemical pesticides available to growers and resistance development to available chemical insecticides will continue to put pressure on developing new control options. These pressures may in the future increase the need for MCA products targeting these pests in cotton. As mentioned earlier, cotton production in China has had a greater focus on EPFs for sucking insect pests. For example, Shi et al. (2008) evaluated two *Beauveria bassiana* and two *M. anisopliae* isolates against spider mites in cotton in China. Applications were made as emulsified oil formulation at 1.5×10^{13} and 1.05×10^{13} conidia/ha for *B. bassiana* and *M. anisopliae*, respectively. During the trials, they were able to obtain desirable control over a period of 30–35 days with an efficacy of greater than 80%. High humidity in the canopy and moderate daily mean temperatures were likely advantageous toward control. One available resource for future development of EPF for whitefly control in cotton is a large EPF collection made by U.S. Department of Agriculture (USDA) at the European Biological Control Laboratory (EBCL) in Montpellier, France, and in Weslaco, TX (Lacey et al., 2008). This is primarily held at the USDA-ARS Entomopathogenic Fungal (ARSEF) collection in Ithaca, NY. Collections were made over 1990 to 1996 from 18 countries representing the Mediterranean, Middle East, Western Asia, Southeast Asia, Latin America, North America, and the South Pacific. Activity against *B. tabaci* was demonstrated for isolates of *Isaria* (*Paecilomyces*) spp., *Lecanicillium lecanii*, and *B. bassiana*, and *Aschersonia* spp. *Isaria fumosorosea* was the species most often found to be attacking *Bemisia tabaci* in the field and epizootics were observed during certain humid seasons in Pakistan, India, and Nepal (Lacey et al., 1993, 2008, 1996). Field results for *B. bassiana* and *I. fumosorosea* isolates from this program have been variable, with some achieving moderate to high levels of control (50–87%) and others showing little success (Jaronski and Lord, 1996; Wraight et al., 1996; Akey and Henneberry, 1998; Liu et al., 1999; Lacey et al., 2008). Possible explanations for this variability include the impact of allelochemicals from cotton as discussed at the end of Section 12.4 and the impact of irrigation on moisture at the leaf boundary layer as reviewed by Jaronski (2010).

12.7 CONCLUSIONS

Despite significant efforts toward developing MCAs, there has been little adoption in cotton. Three areas where MCAs could play a more significant role in the future cotton pest management are natural epizootics, resistance management, and plant colonization. The value of natural epizootics has been recognized for the cotton aphid fungus, but insect pathogens are likely playing other roles. Developing a sufficient understanding to adjust thresholds and conserve or augment these epizootics may play a valuable role in pest management. MCAs provide alternative modes of action for IRM. Strategically using these tools to target populations concentrated on alternate host plants may help manage resistance in the population and limit the number of acres requiring treatment. There is a growing body of literature describing the dual role of insect pathogens as plant colonizers either in the rhizosphere or as endophytes. This literature is further pointing toward benefits of these plant/microbe associations beyond insect control. If these benefits can be realized at relatively low application rates, it may provide an avenue for wider-scale adoption of MCAs in broad acre crops. The future MCAs in general looks very promising (Glare et al., 2012; Lacey et al., 2015), and the associated increased investments and advancements will likely open up new options to all growers, including cotton growers. Cotton will continue to be an intensively managed crop under significant pest pressure. There will always be a need for new pest management tools as old tools are lost due to resistance or regulatory pressures, and MCAs can play a valuable role in this future.

REFERENCES

Abney, M.R., Ruberson, J.R., Herzog, G.A., Krink, T.J., Steinkraus, D.C., Roberts, P.M., 2008. Rise and fall of cotton aphid (Hemiptera: Aphididae) populations in southeastern cotton production systems. J. Econ. Entomol. 101, 23–35.

Aiuchi, D., Baba, Y., Inami, K., Shinya, R., Tani, M., Kuramochi, K., Horie, S., Koike, M., 2007. Screening of *Verticillium lecanii* (*Lecanicillium* spp.) hybrid strains based on evaluation of pathogenicity against cotton aphid and greenhouse whitefly, and viability on the leaf surface. Jpn. J. Appl. Entomol. Zool. 51, 205–212.

Akey, D.H., Henneberry, T.J., 1998. Control of silverleaf whitefly with the entomopathogenic fungi *Paecilomyces fumosoroseus* and *B. bassiana* in upland cotton in Arizona. In: Proc. Beltwide Cotton Conf, pp. 1073–1077.

Armes, N.J., Jadhav, D.R., DeSouza, K.R., 1996. A survey of insecticide resistance in *Helicoverpa armigera* in the Indian sub-continent. Bull. Entomol. Res. 86, 499–514.

Bagla, P., 2010. Hardy cotton-munching pests are latest blow to GM crops. Science 327, 1439.

Beas-Catena, A., Sánchez-Mirón, A., García-Camacho, F., Contreras-Gómez, A., Molina-Grima, E., 2014. Baculovirus biopesticide: an overview. J. Anim. Plant Sci. 24, 362–373.

Beers, E.H., Riedl, H., Dunley, J.E., 1998. Resistance to abamectin and reversion to susceptibility to fenbutin oxide in spider mite (Acari: Tetranychidae) populations in the Pacific Northwest. J. Econ. Entomol. 91, 352–360.

Bell, M.R., Hardee, D.D., 1994a. Early season application of a baculovirus for area-wide management of Heliothis/Helicoverpa (Lepidoptera: Noctuidae): 1992 field trial. J. Entomol. Sci. 29, 192–200.

Bell, M.R., Hardee, D.D., 1994b. Tobacco budworm: possible use of various entomopathogens in large area pest management. In: Proc. Beltwide Cotton Conf, pp. 1168–1170.

Bell, M.R., Hardee, D.D., 1995. Tobacco budworm and cotton bollworm: methodology for virus production and application in large-area management trials. In: Proc. Beltwide Cotton Conf, pp. 857–858.

Buerger, P., Hauxwell, C., Murray, D., 2007. Nucleopolyhedrovirus introduction in Australia. Virol. Sin. 22, 173–179.

Chougule, N.P., Bonning, B.C., 2012. Toxins for transgenic resistance to hemipteran pests. Toxins 4, 405–429.

Conway, H.E., Steinkraus, D.C., Ruberson, J.R., Kring, T.J., 2006. Experimental treatment threshold for the cotton aphid (Homoptera: Aphididae) using natural enemies in Arkansas cotton. J. Entomol. Sci. 41, 361–373.

Cook, D., Herbert, A., Akin, D.S., Reed, J., 2011. Biology, crop injury, and management of thrips (Thysanoptera: Thripidae) infesting cotton seedlings in the United States. J. Integr. Pest Manag. 2, 1–9. http://dx.doi.org/10.1603/IPM10024.

Cory, J.S., Ericsson, J.D., 2010. Fungal entomopathogens in a tritrophic context. BioControl 55, 75–88.

Cristofoletti, P.T., Ribeiro, A.F., Deraison, C., Rahbe, Y., Terra, W.R., 2003. Midgut adaptation and digestive enzyme distribution in a phloem feeding insect, the pea aphid *Acyrthosiphon pisum*. J. Insect Physiol. 49, 11–24.

Dhurua, S., Gujar, G.T., 2011. Field-evolved resistance to Bt toxin Cry1Ac in the pink bollworm, *Pectinophora gossypiella* (Saunders) (Lepidoptera: Gelechiidae) from India. Pest Manag. Sci. 67, 898–903.

Furr, R.E., Pfrimmer, T.R., 1968. Effects of early-, mid-, and late-season infestations of two-spotted spider mites on the yield of cotton. J. Econ. Entomol 61, 1446–1447.

Gassmann, A.J., Stock, S.P., Carriére, Y., Tabashnik, B.E., 2006. Effect of entomopathogenic nematodes on the fitness cost of resistance to Bt toxin Cry1Ac in pink bollworm (Lepidoptera: Gelechiidae). J. Econ. Entomol. 99, 920–926.

Gassmann, A.J., Stock, S.P., Sisterson, M.S., Carriére, Y., Tabashnik, B.E., 2008. Synergism between entomopathogenic nematodes and *Bacillus thuringiensis* crops: integrating biological control and resistance management. J. Appl. Ecol. 45, 957–966.

Gassmann, A.J., Fabrick, J.A., Sisterson, M.S., Hannon, E.R., Stock, S.P., Carriére, Y., Tabashnik, B.E., 2009. Effects of pink bollworm resistance to *Bacillus thuringiensis* on phenoloxidase activity and susceptibility to entomopathogenic nematodes. J. Econ. Entomol. 102, 1224–1232.

Gassmann, A.J., Hannon, E.R., Sisterson, M.S., Stock, S.P., Carriére, Y., Tabashnik, B., 2012. Effects of entomopathogenic nematodes on evolution of pink bollworm resistance to *Bacillus thuringiensis* toxin Cry1Ac. J. Econ. Entomol. 105, 994–1005.

Gipson, J.R., 1986. Temperature effects on growth, development, and fiber properties. In: Mauney, J.R., Stewart, J.M. (Eds.), Cotton Physiology. The Cotton Foundation, Memphis, TN, pp. 47–56.

Glare, T., Caradus, J., Gelernter, W., Jackson, T., Keyhani, N., Kohl, J., Marrone, P., Morin, L., Stewart, A., 2012. Have biopesticides come of age? Trends Biotechnol. 30, 250–258.

Goettel, M.S., Koike, M., Kim, J.J., Aiuchi, D., Shinya, R., Brodeur, J., 2008. Potential of *Lecanicillium* spp. for management of insects, nematodes and plant diseases. J. Invertebr. Pathol. 98, 256–261.

Gore, J., Adamczyk Jr., J.J., Catchot, A., Jackson, R., 2008. Yield response of dual-toxin Bt cotton to *Helicoverpa zea* infestations. J. Econ. Entomol. 101, 1594–1599.

Gore, J., Cook, D.R., Catchot, A.L., Musser, F.R., Stewart, S.D., Leonard, B.R., Lorenz, G., Studebaker, G., Akin, D.S., Tindall, K.V., Jackson, R.E., 2013. Impact of twospotted spider mite (Acari: Tetranychidae) infestation timing on cotton yields. J. Cotton Sci. 17, 34–39.

Greene, J.K., 2006. Insecticide Recommendations for Arkansas. University of Arkansas, Div. Agriculture. Coop. Ext. Serv. Misc. Pub. 144, 253.

Griffin, M.R., Ownley, B.H., Klingeman, W.E., Pereira, R.M., 2006. Evidence of induced systemic resistance with *Beauveria bassiana* against *Xanthomonas* in cotton. Phytopathology 96, S42.

Gurulingappa, P., Sword, G.A., Murdoch, G., McGee, P.A., 2010. Colonization of crop plants by fungal entomopathogens and their effects on two insect pests when *in planta*. Biol. Control 55, 34–41.

Gurulingappa, P., McGee, P.A., Sword, G., 2011. Endophytic *Lecanicillium lecanii* and *Beauveria bassiana* reduce the survival and fecundity of *Aphis gossypii* following contact with conidia and secondary metabolites. Crop Prot. 30, 349–353.

Hannam, J.J., Steinkraus, D.C., 2010. The natural occurrence of *Pandora heteropterae* (Zygomycetes: Entomophthorales) infecting *Lygus lineolaris* (Hemiptera: Miridae). J. Invertebr. Pathol. 103, 96–102.

Hannon, E.R., Sisterson, M.S., Stock, S.P., Carrière, Y., Tabashnik, B.E., Gassmann, A.J., 2010. Effects of four nematode species on fitness costs of pink bollworm resistance to *Bacillus thuringiensis* toxin Cry1Ac. J. Econ. Entomol. 103, 1821–1831.

Hardee, D.D., Bell, M.R., 1995. Area-wide management of Heliothis/Helicoverpa in the delta of Mississippi. In: Constable, G.A., Forrester, N.W. (Eds.), Challenging the Future: Proceed. World Cotton Res. Conf.-1, Brisbane Australia, February 14–17, 1994. CSIRO, Melbourne, Australia, pp. 434–436.

Hardee, D.D., Bell, M.R., Street, D.A., 1999. A review of area-wide management of *Helicoverpa* and *Heliothis* (Lepidoptera: Noctuidae) with pathogens (1987–1997). Southwest. Entomol. 24, 62–75.

Hendricks, D.E., Hubbard, D.W., Hardee, D.D., 1995. Occurrence of beet armyworm moths in the lower Mississippi river delta as indicated by numbers caught in traps in 1994. Southwest. Entomol. 20, 157–164.

Hoover, K., Stout, M.J., Alaniz, S.A., Hammock, B.D., Duffey, S.S., 1998. Influence of induced plant defenses in cotton and tomato on the efficacy of baculoviruses on noctuid larvae. J. Chem. Ecol. 24, 253–271.

Hoover, K., Washburn, J.O., Volkman, L.E., 2000. Midgut-based resistance of *Heliothis virescens* to baculovirus infection mediated by phytochemicals in cotton. J. Insect Physiol. 6, 999–1007.

Huang, D.F., Zhang, J., Song, F.-P., Lang, Z.-H., 2007. Microbial control and biotechnology research on *Bacillus thuringiensis* in China. J. Invertebr. Pathol. 95, 175–180.

Ihara, F., Toyama, M., Mishiro, K., Yaginuma, K., 2008. Laboratory studies on the infection of stink bugs with *Metarhizium anisopliae* strain FRM515. Appl. Entomol. Zool. 43, 503–509.

Jaronski, S.T., 2010. Ecological factors in the inundative use of fungal entomopathogens. BioControl 55, 159–185.

Jaronski, S.T., Lord, J.C., 1996. Evaluation of *Beauveria bassiana* (Mycotrol WP) for control of whitefly in spring cantaloupes, 1995. Arthropod Manag. Tests 21, 103.

Jaronski, S.T., Rosinska, J., Brown, C., Osterlind, R., Staten, R., Craft, R., Antilla, L., 1997. Impact of *Beauveria bassiana* Mycotech strain GHA, buprofezin, and pyriproxyfen on whitefly predators in Arizona cotton. In: Silverleaf Whitefly 1997 Supplement to the 5-Year National Research and Action Plan: Progress Review, Technology Transfer and New Research and Action Plan (1997–2001), p. 49.

Jin, X., Streett, D.A., Dunlap, C.A., Lyn, M.E., 2008. Application of hydrophilic–lipophilic balance (HLB) number to optimize a compatible non-ionic surfactant for dried aerial conidia of *Beauveria bassiana*. Biol. Control 46, 226–233.

Johnson, D.R., Caron, R.E., Head, R.B., Jones, F.G., Tynes, J.S., 1996. Insect and mite pest management in the mid-SouthBeauveria bassiana. Biol. Control 46In: King, E.J., Phillips, J.R., Coleman, R.J. (Eds.), Cotton Insects and Mites: Characterization and Management. The Cotton Foundation, Memphis, TN, pp. 673–693.

Kim, J.J., Goettel, M.S., Gillespie, D.R., Jones, F.G., Tynes, J.S., 2007. Potential of *Lecanicillium* species for dual microbial control of aphids and the cucumber powdery mildew fungus, *Sphaerotheca fuliginea*. Biol. Control 40,King, E.J., Phillips, J.R., Coleman, R.J. (Eds.), Cotton Insects and Mites: Characterization and Management. The Cotton Foundation, Memphis, TN 327–332.

Kim, J.J., Goettel, M.S., Gillespie, D.R., 2008. Evaluation of *Lecanicillium longisporum*, Vertalec for simultaneous suppression of cotton aphid, *Aphis gossypii*, and cucumber powdery mildew, *Sphaerotheca fuliginea*, on potted cucumbers. Biol. Control 45, 404–409.

Knox, C., Moore, S.D., Luke, G.A., Hill, M.P., 2015. Baculovirus-based strategies for the management of insect pests: a focus on development and application in South AfricaLecanicillium longisporumAphis gossypiiSphaerotheca fuliginea. Biocontrol Sci. Technol. 25, 1–20.

Koike, M., Higashio, T., Komori, A., Akiyama, K., Kishimoto, N., Masuda, E., Sasaki, M., Yoshida, S., Tani, M., Kuramochi, K., Sugimoto, M., Nagao, H., 2004. *Verticillium lecanii* (*Lecanicillium* spp.) as epiphyte and their application to biological control of pest and disease in a glasshouse and a field. IOBC/WPRS Bull. 27 (8), 41–44.

Lacey, L.A., Kirk, A.A., Hennessey, R.D., Akiyama, K., Kishimoto, N., Masuda, E., Sasaki, M., Yoshida, S., Tani, M., Kuramochi, K., Sugimoto, M., Nagao, H., December 7–9, 1993. Foreign exploration for natural enemies of *Bemisia tabaci* and implementation in integrated control programs in the United StatesLecanicillium. IOBC/WPRS Bull. 27 (8)In: Proceed. ANPP Int. Conf. on Pests in Agriculture, vol. 1. Association Nacionale de Protection des Plantes, Paris France, Montpellier, France, pp. 351–360.

Lacey, L.A., Fransen, J.J., Carruthers, R., 1996. Global distribution of naturally occurring fungi of *Bemisia*, their biologies and use as biological control agents. In: Gerling, D., Mayer, R. (Eds.), Bemisia 1995: Taxonomy, Biology, Damage, and Management. Proceed. ANPP Int. Conf. on Pests in Agriculture, vol. 1. Intercept, Andover, UK, pp. 401–433.

Lacey, L.A., Wraight, S.P., Kirk, A.A., 2008. Entomopathogenic fungi for control of *Bemisia tabaci* biotype B Foreign exploration, research and implementation. In: Gould, J., Hoelmer, K., Goolsby, J. (Eds.), Classical Biological Control of *Bemisia Tabaci* in the United States – a Review of Interagency Research and Implementation. Progress in Biological Control, vol. 4. Springer, Dordrecht, The Netherlands, pp. 33–69.

Lacey, L.A., Grzywacz, D., Shapiro-Ilan, D.I., Frutos, R., Brownbridge, M., Goettel, M.S., 2015. Insect pathogens as biological control agents: back to the future. J. Invertebr. Pathol 132, 1–41.

Landivar, J.A., Benedict, J.H., Kirk, A.A., 1996. Monitoring System for the Management of Cotton Growth and FruitingBemisia tabaci. Texas Agric. Exp. Station Bull. B02, College Station, TXIn: Gould, J., Hoelmer, K., Goolsby, J. (Eds.), Classical Biological Control of *Bemisia Tabaci* in the United States – a Review of Interagency Research and Implementation. Progress in Biological Control, vol. 4. Springer, Dordrecht, The Netherlands, pp. 33–69. 25 pp.

Leigh, T.F., Roach, S.H., Watson, T.G., Frutos, R., Brownbridge, M., Goettel, M.S., 1996. Biology and ecology of important insect and mite pests of cotton. J. Invertebr. Pathol. 132In: King, E.G., Phillips, J.R., Coleman, R.J. (Eds.), Cotton Insects and Mites: Characterization and Management. The Cotton Foundation, Memphis, TN, pp. 17–85.

Leland, J.E., 2005. Characteristics of Beauveria bassiana isolated from Lygus lineolaris populations of Mississippi. J. Agric. Urban Entomol 22, 57–71.

Leland, J.E., McGuire, M.R., 2006. Effects of different *Beauveria bassiana* isolates on field populations of *Lygus lineolaris* in pigweed (*Amaranthus* spp.). Biol. Control 39, 272–281Texas Agric. Exp. Station Bull. B02, College Station, TX. 25 pp.Leigh, T.F., Roach, S.H., Watson, T.G., 1996. Biology and ecology of important insect and mite pests of cotton. In: King, E.G., Phillips, J.R., Coleman, R.J. (Eds.), Cotton Insects and Mites: Characterization and Management. The Cotton Foundation, Memphis, TN, pp. 17–85.

Leland, J.E., Snodgrass, G.L., 2004. Prevalence of naturally occurring *Beauveria bassiana* in *Lygus lineolaris* (Heteroptera: Miridae) populations from wild host plants of MississippiAmaranthus. J. Agric. Urban Entomol. 23, 157–163.

Leland, J.E., McGuire, M.R., Grace, J.A., Jaronski, S.T., Ulloa, M., Park, Y.H., Plattner, R.D., 2005. Strain selection of a fungal entomopathogen, *Beauveria bassiana*, for control of plant bugs (*Lygus* spp.) (Heteroptera: Miridae). Biol. Control 35, 104–114.

Leonard, B.R., Boethel, D.J., Sparks Jr.J.A., A.N., Layton, M.B., Mink, J.S., Pavlov, A.M., Burris, E., Graves, J.B., 1990. Variations in response of soybean looper (Lepidoptera: Noctuidae) to selected insecticides in LouisianaBeauveria bassianaLygus. J. Econ. Entomol. 83, 27–34.

Leland, J.E., 2005. Characteristics of *Beauveria bassiana* isolated from *Lygus lineolaris* populations of Mississippi. J. Agric. Urban Entomol. 22, 57–71.

Liu, T.-X., Stansly, P.A., Sparks Jr., A.N., Knowles, T.C., Chu, C.C., Pavlov, A.M., Burris, E., Graves, J.B., 1999. Application of Mycotrol and Naturalis-L (*Beauveria bassiana*) for management of *Bemisia argentifolii* (Homoptera: Aleyrodidae) on vegetables, cotton and ornamentals in the southern United States. Subtropical Plant Sci. J. Rio Gd. Val. Hort. Soc. 51, 41–48.

Lopes, R.B., Laumann, R.A., Blassioli-Morães, M.C., Borges, M., Faria, M., 2015. The fungistatic and fungicidal effects of volatiles from metathoracic glands of soybean-attacking stink bugs (Heteroptera: Pentatomidae) on the entomopathogen *Beauveria bassiana*. J. Invertebr. Pathol. 132, 77–85.

Lopez, D.C., Zhu-Salzman, K., Ek-Ramos, Jr., M.J., Sword, G.A., Chu, C.C., 2014. The entomopathogenic fungal endophytes *Purpureocillium lilacinum* (Formerly *Paecilomyces lilacinus*) and *Beauveria bassiana* negatively affect cotton aphid reproduction under both Greenhouse and field conditions. PLoS One 9 (8), 1–8.

Lu, Y.H., Wu, K.M., Jiang, Y.Y., Xia, B., Li, P., Feng, H.Q., Wyckhuys, K.A.G., Guo, Y.Y., 2010. Mirid bug outbreaks in multiple crops correlated with wide-scale adoption of Bt cotton in ChinaBeauveria bassiana. Science 328, 1151–1154.

Lund, J., Teague, T.G., Steinkraus, D.C., Leland, J.E., 2006. Control of the tarnished plant bug (*Lygus lineolaris*) in midsouth cotton using the entomopathogenic fungus (*Beauveria bassiana*) and the insect growth regulator Diamond®Beauveria bassiana. AAES Res. Ser. 552 (8), 169–175.

Luo, S.P., Haye, T., Lu, Y.H., Li, H.M., Zhang, F., Kuhlmann, U., Wyckhuys, K.A.G., Guo, Y.Y., 2011. An Artificial Rearing Method of *Peristenus Spretus*. Science 328, 1151–1154Patent, application number: 201110092491.3 (China).

Luo, S., Naranjo, S.E., Wu, K., Leland, J.E., 2014. Biological control of cotton pests in ChinaLygus lineolarisBeauveria bassiana. Biol. Control 68, 6–14.

Luttrell, R.G., Wan, L., Knighten, K., Li, H.M., Zhang, F., Kuhlmann, U., 1999. Variation in susceptibility of Noctuid (Lepidoptera) larvae attacking cotton and soybean to purified endotoxin proteins and commercial formulations of *Bacillus thuringiensis*. J. Econ. Entomol. 92, 21–32Patent, application number: 201110092491.3 (China).

Mauney, J.R., Naranjo, S.E., Wu, K., 1986. Vegetative growth and development of fruiting sites. Biol. Control 68In: Mauney, J.R., Stewart, J. McD. (Eds.), Cotton Physiology. The Cotton Foundation, Memphis, TN, pp. 11–28.

McGuire, M.R., 2002. Prevalence and distribution of naturally occurring Beauveria bassiana in San Joaquin Valley populations of Lygus hesperus (Heteroptera: Miridae). J. Agric. Urban Entomol 19, 237–246.

McGuire, M.R., Ulloa, M., Park, Y.H., Hudson, N., 2005. Biological and molecular characteristics of *Beauveria bassiana* isolates from California *Lygus hesperus* (Hemiptera: Miridae) populations. Biol. Control 33, 307–314.

McGuire, M.R., Leland, J.E., Dara, S., Park, Y.H., Ulloa, M., 2006. Effect of different isolates of *Beauveria bassiana* on field populations of *Lygus hesperus*. Biol. Control 38,Mauney, J.R., Stewart, J.McD. (Eds.), Cotton Physiology. The Cotton Foundation, Memphis, TN 390–396.

Nussenbaum, A.L., Lecuona, R.E., Park, Y.H., Hudson, N., 2012. Selection of *Beauveria bassiana* sensu lato and *Metarhizium anisopliae* sensu lato isolates as microbial control agents against the boll weevil (*Anthonomus grandis*) in Argentina. J. Invertebr. Pathol. 11, 1–7.

Ownley, B.H., Griffin, M.R., Klingeman, W.E., Gwinn, K.D., Moulton, J.K., Pereira, R.M., 2008. *Beauveria bassiana*: endophytic colonization and plant disease controlLygus hesperus. J. Invertebr. Pathol. 98, 267–270.

Parker, B.L., Skinner, M., Gouli, S., Gouli, V., Kim, J.S., 2015. Virulence of BotaniGard® to second instar brown marmorated stink bug, *Halyomorpha halys* (Stål) (Heteroptera: Pentatomidae)Lygus hesperus. Insects 6, 319–324.

Payne, J.R., Cannon, R.J.C., November 16, 1993. Use of *Bacillus thuringiensis* Isolates for Controlling Pests in the Family AphididaeMetarhizium anisopliaeAnthonomus grandis. US Patent 5262159J. Invertebr. Pathol. 11, 1–7.

Poprawski, T.J., Jones, W.J., Klingeman, W.E., Gwinn, K.D., Moulton, J.K., Pereira, R.M., 2001. Host plant effects on activity of the mitosporic fungi *Beauveria bassiana* and *Paecilomyces fumosoroseus* against two populations of *Bemesia* whiteflies (Homoptera: Aleyrodidae). Mycopathologia 151, 11–20.

Poprawski, T.J., Greenberg, S.M., Ciomperlik, M.G., Gouli, V., Kim, J.S., 2000. Effect of host plant on *Beauveria bassiana*- and *Paecilomyces fumosoroseus*-induced mortality of *Trialeurodes vaporariorum* (Homoptera: Aleyrodidae). Environ. Entomol. 29, 1048–1053.

Porcar, M., Grenier, A.M., Federici, B., Rahbe, Y., 2009. Effects of *Bacillus thuringiensis* δ-endotoxins on the pea aphid (*Acyrthosiphon pisum*). US Patent 5262159Appl. Environ. Microbiol. 75, 4897–4900.

Portilla, M., Snodgrass, G., Luttrell, R., 2014. A novel bioassay to evaluate the potential of *Beauveria bassiana* strain NI8 and the insect growth regulator novaluron against *Lygus lineolaris* on a non-autoclaved solid artificial dietBemesia. J. Insect Sci. 14, 1–13.

Quintela, E.D., Mascarin, G.M., Silva, R.A., Barrigossi, J.A.F., Martins, J.F.S., 2013. Enhanced susceptibility of *Tibraca limbativentris* (Heteroptera: Pentatomidae) to *Metarhizium anisopliae* with sublethal doses of chemical insecticidesTrialeurodes vaporariorum. Biol. Control 66, 56–64.

Raafat, I., Meshrif, W.S., El Husseiny, E.M., El-Hariry, M., Seif, A.I., 2015. *Nezara viridula* (Hemiptera: Pentatomidae) cuticle as a barrier for *Beauveria bassiana* and *Paecilomyces* sp. infection. Afr. Entomol. 23, 75–87.

Sattar, S., Maiti, M.K., Luttrell, R., 2011. Molecular characterization of a novel vegetative insecticidal protein from *Bacillus thuringiensis* effective against sap-sucking insect pestLygus lineolaris. J. Microbiol. Biotechnol. 21, 937–946.

Shapiro-Ilan, D., Gouge, D.H., Piggott, S.J., Fife, J.P., Martins, J.F.S., 2006. Application technology and environmental considerations for use of entomopathogenic nematodes in biological controlTibraca limbativentrisMetarhizium anisopliae. Biol. Control 38, 124–133.

Shi, W.-B., Zhang, L.-L., Feng, M.-G., El-Hariry, M., Seif, A.I., 2008. Field trials of four formulations of *Beauveria bassiana* and *Metarhizium anisopliae* for control of cotton spider mites (Acari: Tetranychidae) in the Tarim Basin of ChinaPaecilomyces. Biol. Control 45, 48–55.

Shinya, R., Aiuchi, D., Kushida, A., Tani, M., Kuramochi, K., Kushida, A., Koike, M., 2008. Effects of fungal culture filtrates of *Verticillium lecanii* (*Lecanicillium* spp.) hybrid strains on *Heterodera glycines* egg and juveniles. J. Invertebr. Pathol. 97, 291–297.

Shinya, R., Aiuchi, D., Kushida, A., Tani, M., Kuramochi, K., Koike, M., 2008. Pathogenicity and its mode of action in different sedentary stages of *Heterodera glycines* (Tylenchida: Heteroderidae) by *Verticillium lecanii* hybrid strains. Appl. Entomol. Zool. 43, 227–233.

Shinya, R., Watanabe, A., Aiuchi, D., Tani, M., Kuramochi, K., Kushida, A., Koike, M., 2008. Potential of *Verticillium lecanii* (*Lecanicillium* spp.) hybrid strains as biological control agents for soybean cyst nematode: is protoplast fusion an effective tool for development of plant-parasitic nematode control agents? Jpn. J. Nematol. 38 (10), 9–18.

Silva, R.A., Quintela, E.D., Mascarin, G.M., Pedrini, N., Lião, L.M., Ferri, P.H., Koike, M., 2015a. Unveiling chemical defense in the rice stalk stink bug against the entomopathogenic fungus *Metarhizium anisopliae*LecanicilliumHeterodera glycines. J. Invertebr. Pathol. 127, 93–100.

Simelane, D.O., Steinkraus, D.C., Kring, T.J., Tani, M., Kuramochi, K., Koike, M., 2008b. Predation rate and development of *Coccinella septempunctata* L. influenced by *Neozygites fresenii*-infected cotton aphid prey. Biol. Control 44, 128–135.

Slosser, J.E., Pinchak, W.E., Rummel, D.R., Tani, M., Kuramochi, K., Kushida, A., Koike, M., 1989c. A review of known and potential factors affecting the population dynamics of the cotton aphidVerticillium lecaniiLecanicillium. Southwest. Entomol. 14 (10), 302–312.

Sosa-Gómez, D.R., Moscardi, F., Mascarin, G.M., Pedrini, N., Lião, L.M., Ferri, P.H., 1998. Laboratory and field studies on the infection of stink bugs, *Nezara viridula*, *Peizodorus guildinii*, and *Euschistus heros* (Hemiptera: Pentatomidae) with *Metarhizium anisopliae* and *B. bassiana* in Brazil. J. Invertebr. Pathol. 71, 115–120.

Sosa-Gómez, D.R., Boucias, D.G., Nation, J.L., 1996. Attachment of *Metarhizium anisopliae* to southern green stink bug *Nezara viridula* cuticle and fungistatic effect of cuticular lipids and aldehydes. J. Invertebr. Pathol. 69, 31–39.

Sparks, A.N., Pinchak, W.E., Rummel, D.R., 1979. A review of the biology of fall armyworm. Fla. Entomol. 62, 82–87.

Spurgeon, D.W., Moscardi, F., 2010. Efficacy of *Beauveria bassiana* against *Lygus hesperus* (Hemiptera: Miridae) at low temperaturesEuschistus herosMetarhizium anisopliaeB. bassiana. J. Entomol. Sci. 45, 211–219.

Steinkraus, D.C., Young, S.Y., Gouge, D.H., Leland, J.E., 2007. Microbial insecticide application and evaluation: cotton. In: Lacey, L.A., Kaya, H.K. (Eds.), Manual of Techniques in Invertebrate Pathology, second ed. Springer, Dordrecht, pp. 427–455.

Stewart, S.D., Akin, D.S., Reed, J., Bacheler, J., Catchot, A., Cook, D., Gore, J., Greene, J., Herbert, A., Jackson, R.E., Kerns, D.L., Leonard, B.R., Lorenz, G.M., Micinski, S., Reisig, D., Roberts, P., Studebaker, G., Tindall, K., Toews, M., 2013. Survey of thrips species infesting cotton across the southern U.S. cotton belt. J. Cotton Sci. 17, 263–269.

Street, D.A., Bell, M.R., Hardee, D.D., 1997. Update on the area-wide budworm/bollworm management program with virus: is it a cost effective insurance program?Beauveria bassianaLygus hesperus J. Entomol. Sci. 45In: Proc. Beltwide Cotton Conf, pp. 1148–1150.

Sun, X., 2015. History and current status of development and use of viral insecticides in China. Viruses 7, 306–319.

Sun, X.L., Wu, D., Sun, X.C., Jin, L., Ma, Y., Bonning, B.C., Peng, H.Y., Hu, Z.H., 2009. Impact of *Helicoverpa armigera* nucleopolyhedroviruses expressing a cathepsin L-like protease on target and nontarget insect species on cotton. Biol. Control 49,Lacey, L.A., Kaya, H.K. (Eds.), Manual of Techniques in Invertebrate Pathology, second ed. Springer, Dordrecht 77–83.

Tong, Y.J., Wu, K.M., Lu, Y.H., Gao, X.W., Catchot, A., Cook, D., Gore, J., Greene, J., Herbert, A., Jackson, R.E., Kerns, D.L., Leonard, B.R., Lorenz, G.M., Micinski, S., Reisig, D., Roberts, P., Studebaker, G., Tindall, K., Toews, M., 2010. Pathogenicity of *Beauveria* spp. strains to three species of mirids, *Apolygus lucorum*, *Adelphocoris suturalis* and *Adelphocoris lineolatus*. Acta Phytophylacica Sin. 37, 172–176.

Ugine, T.A., Bell, M.R., Hardee, D.D., 2011. The effect of temperature and exposure to *Beauveria bassiana* on tarnished plant bug *Lygus lineolaris* (Heteroptera: Miridae) population dynamics, and the broader implications of treating insects with entomopathogenic fungi over a range of temperatures. Biol. Control 59,Proc. Beltwide Cotton Conf 373–383.

Vega, F.E., Wu, D., Sun, X.C., Jin, L., Ma, Y., Bonning, B.C., Peng, H.Y., Hu, Z.H., 2008. Insect pathology and fungal endophytesHelicoverpa armigera. J. Invertebr. Pathol. 98, 277–279.

Walters, F.S., English, L.H., 1995. Toxicity of *Bacillus thuringiensis* δ-endotoxins toward the potato aphid in an artificial diet bioassay. Entomol. Exp. Appl. 77, 211–216.

Wang, C., Feng, M.-G., Lu, Y.H., Gao, X.W., 2014. Advances in fundamental and applied studies in China of fungal biocontrol agents for use against arthropod pestsBeauveriaApolygus lucorumAdelphocoris suturalisAdelphocoris lineolatus. Biol. Control 68, 129–135.

Weathersbee III, A.A., Hardee, D.D., 1994. Abundance of cotton aphids (Homoptera: Aphididae) and associated biological control agents on six cotton cultivarsBeauveria bassianaLygus lineolaris. J. Econ. Entomol. 87, 258–265.

Wraight, S.P., Carruthers, R.I., Bradley, C.A., June 9–14, 1996. Development of entomopathogenic fungi for microbial control of whiteflies of the *Bemisia tabaci* complex. J. Invertebr. Pathol. 98In: Proc. V Siconbiol Simpósio de Controle Biológico, Foz do Iguaçu, Brazil. Embresa Brasileira de Pesquisa Agropecuaria – Centro Nacional de Pesquisa de Soja, Londrina, PR, Brazil, pp. 28–34. EMBRAPA/CNPSo, Foz do Iguaçú, Brazil.

Yang, M.M., Meng, L.L., Zhang, Y.A., Wang, Y.Z., Qu, L.J., Wang, Q.H., Ding, J.Y., 2012. Baculoviruses and insect pests control in ChinaBacillus thuringiensis. Afr. J. Microbiol. Res. 6, 214–218.

Yang, Y., Li, Y., Wu, Y., 2013. Current status of insecticide resistance in *Helicoverpa armigera* after 15 years of Bt cotton planting in China. J. Econ. Entomol. 106, 375–381.

Chapter 13

Microbial Control of Soybean Pest Insects and Mites

D.R. Sosa-Gómez

Embrapa Soja, Londrina, Paraná, Brazil

13.1 INTRODUCTION

The top-producing soybean countries worldwide are the United States, Brazil, Argentina, China, India, Paraguay, Canada, and Uruguay. Countries in North and South America plant 89.55 million ha of this commodity annually, which is ~76% of the total soybean cultivation area in the world (USDA, 2015, http://apps.fas.usda.gov/psdonline/circulars/production.pdf). Soybean crops are attacked by more than 180 invertebrate species around the world (Biswas, 2013; Sosa-Gómez et al., 2014; http://www.ent.iastate.edu/soybeaninsects/). The United States harbors some soybean pests that do not occur in South American countries, such as *Aphis glycines* and *Hypena scabra*, as well as other minor pests, such as *Dectes texanus* and *Popillia japonica*. Other species are the same across the Nearctic and Neotropical regions, such as the soybean looper (*Chrysodeixis includens*), the velvetbean caterpillar (*Anticarsia gemmatalis*), the whitefly (*Bemisa tabaci*), two spotted mite (*Tetranychus urticae*), and stink bugs (*Piezodorus guildinii* and *Nezara viridula*).

Among the *Spodoptera* species, *Spodoptera cosmioides*, *Spodoptera eridania*, and *Spodoptera albula* are usually associated with soybean Brazilian fields. In the United States, *Spodoptera exigua*, *Spodoptera ornithogalli*, and the fall armyworm, *Spodoptera frugiperda*, have been reported as soybean pests. In contrast, the main lepidopteran pests that are widely distributed across the soybean cultivation area in Brazil are, in decreasing order of importance, *C. includens*, *A. gemmatalis*, and the old world bollworm, *Helicoverpa armigera*. Despite their widespread distribution, *H. armigera* is more important in the northeastern and northern central regions, whereas it occurs at low levels and has little economic impact in the southern region. In the majority of instances soybean pests from South America are common across all countries in this region, with a few exceptions of relative geographical importance. For example, the Neotropical Brown Stink Bug (*Euschistus heros*), which is a key pest and is economically significant in Brazil and Paraguay, is not a major pest in Argentina or Uruguay.

In India, the most important lepidopteran soybean pests are *Spodoptera litura*, *Chrysodeixis acuta*, *Thysanoplusia orichalcea* (cited as *Diachrysia orichalcea*; Noctuidae: Plusiinae), and *H. armigera*. Other lepidopteran pests of regional importance are *Spilosoma obliqua*, *Omiodes indicata* (cited as *Hedylepta indicate*), and *S. exigua* (Hubner) (Biswas, 2013). Among them, most of the lepidopteran species are susceptible to entomopathogenic diseases and are controlled, or potentially can be controlled, by microbial agents. A new scenario in Brazilian soybean fields will be large areas with *Bacillus thuringiensis* (Bt) soybean expressing Cry1Ac toxin. During the soybean season 2014–15, the area with Bt soybean reached 6.3 million ha with a potential for rapid and exponential expansion of this area. Therefore, interactions between key entomopathogenic diseases and this new scenario should be determined.

Stink bug species that occur on soybean can be afflicted by fungal infections (Sosa-Gómez et al., 1997). However, stink bugs usually require a high inoculum load to become infected with entomopathogenic fungi because of their low susceptibility to mycoses. Chronic diseases caused by viruses (picorna-like virus, rhabdovirus, reovirus, and unclassified viruses) are known as *N. viridula* and *E. heros*, which are both important soybean stink bug pests (Williamson and von Wechmar, 1992; Liu et al., 2015), but their potential as control agents remains unexplored. Also, a high prevalence of trypanosomatid flagellates has been found in populations of stink bugs of soybean (Fuxa et al., 2000; Sosa-Gómez et al., 2005). In addition, the impact of viruses and protozoa on stink bug biology, behavior, and damage-causing capacity is unknown.

13.2 ENTOMOPATHOGENIC FUNGI

13.2.1 Natural Occurrence of Entomopathogenic Fungi in Soybean

In soybean systems, entomopathogenic and mite-pathogenic fungi play important roles in regulating the arthropod populations of important pests (Sosa-Gómez et al., 2010). The most prevalent entomopathogenic diseases of the lepidopteran species include mycoses caused by *Metarhizium (Nomuraea) rileyi*, *Zoophthora radicans* s.l., *Pandora gammae*, and *Isaria tenuipes*. These fungi can cause impressive epizootics in *C. includens*, *C. acuta*, *A. gemmatalis*, *Rachiplusia nu*, *S. frugiperda*, *S. litura*, *H. scabra*, *Crocidosema aporema*, and *Spilosoma* spp. populations. Similarly, *Beauveria bassiana* s.l. can cause epizootics in coleopteran species and in *Diabrotica speciosa* and *Aracanthus mourei* populations, but epizootics in these species are rare. Epizootics of *Pandora neoaphidis* have been reported in *A. glycines* in the United States (Nielsen and Hajek, 2005; Koch et al., 2010). In Brazilian soybean fields, the fungus *Batkoa apiculata* can be found in adults of the coleopteran *Lagria villosa* when cold nights during the summer provide dew formation; infection with this fungus results in dead insects with open wings. Additionally, the mite-pathogenic fungus, *Neozygites floridana*, in a less easily perceived manner, regulates populations of *T. urticae*, *Mononychellus planki*, and other tetranychid mites (Roggia et al., 2009).

Our knowledge of the differentiation, phylogeny, virulence, nutritional requirements, environmental stressors, and microecology of individual species of arthropod pathogenic fungi has increased in the past 20 years (Castrillo and Humber, 2009; Rehner, 2009; Roy et al., 2010; Kepler et al., 2014). The need for knowledge of ethological studies for a number of pathogenic species, production costs, formulation, and microenvironmental requirements have limited their commercial application. An underestimated factor is the susceptibility of the target arthropod to the pathogen due to the lack of bioassay data relative to the precise dosage to which the target arthropod is exposed. In field test experiments, there is often a lack of knowledge regarding how many infective units are necessary to kill a proportion of a healthy population. This information is necessary to infer the potential to control the insect pests under field conditions. Possibly the most important and useful control approach is the conservation of naturally occurring entomopathogenic fungi. For example, fungicide applications on soybean crops through the suppression of the natural inoculum increases the probability of arthropod population outbreaks (Stansly and Orellana, 1990; Sosa-Gómez et al., 2003; Koch et al., 2010).

Some pathogens, such as *I. tenuipes*, appear during unusually wet seasons, and their synnemata can be found forming spore masses on Plusiinae pupae on the leaves where these species form cocoons or arise from the soil, as in the case of *A. gemmatalis*, which pupates at 2–4 cm underground. Despite *I. tenuipes* being prevalent in Plusiinae caterpillars (such as *C. includens* and *R. nu*) during wet seasons, conservative measures are of reduced importance in the short term, because this pathogen causes chronic infections. Infection usually takes place during the larval stage, but with mortality occurring at the pupal stage, this pathogen does not suppress larval-induced plant damage.

13.2.2 Impact of Agrochemicals on Entomopathogenic Fungi

Soybean rust is caused by the fungus *Phakopsora pachyrhizi* and is considered a serious soybean disease in Asia, Africa, and the Americas. Very active fungicide compounds directed against soybean rust, especially early in the growing season, can cause the suppression of entomopathogenic inoculum, reducing the intensity of epizootic diseases and delaying their occurrence (Stansly and Orellana, 1990; Sosa-Gómez et al., 2003; Koch et al., 2010). Unfortunately, the most effective fungicides applied against soybean rust, such as strobilurin, pyrazole, carboxamide, and triazol chemical groups, as well as their commercial formulations (mixtures such as trifloxystrobin + prothioconazole, pyraclostrobin + epoxiconazole + fluxapyroxad, azoxystrobin + benzovindiflupyr, and bixafen + prothioconazole + trifloxystrobin), are very toxic to *Metarhizium rileyi* and other entomopathogenic fungi (Sosa-Gómez, unpublished). Under cultivation conditions in Brazil, soybean rust usually requires, depending on the severity of the disease, two to four applications when the recommended fungicide threshold is adopted. However, the fungicide treatments can increase to four to eight applications in regions having two soybean seasons in one year with control measures being applied during the second soybean season (Godoy, personal communication).

Fungicides also may impact mite-pathogenic fungi, such as *N. floridana*, which can operate in epizootic fashion against phytophagous tetranychid mite populations, such as *M. planki* and *T. urticae*; both species having wide distributions in southern Brazilian, Paraguayan, and Bolivian soybean fields (Sosa-Gómez, unpublished). Also, strobilurin and triazole fungicides can affect the prevalence of *P. neoaphidis* on soybean aphid populations (Koch et al., 2010). Fungicide application timing is a determining factor that can influence epizootic intensity. Early fungicide applications that coincide with the appearance of the first disease cases when the initial inoculum is being produced can lead to inoculum depletion, delaying epizootic initiation or lowering its cumulative prevalence (Sosa-Gómez et al., 2003; Koch et al., 2010).

In addition to fungicides, other agrochemicals used over very large areas, such as glyphosate formulations, can interfere with the survival of entomopathogenic fungi. For example, important soybean entomopathogens, including *N. floridana* and *M. rileyi*, can exhibit reduced spore germination and/or mycelial growth in response to various herbicide formulations (Morjan et al., 2002). Significantly, over the years the prevalence of *M. rileyi* epizootics in *A.*

gemmatalis and *C. includens* populations in Brazilian soybean fields has diminished. In samples taken during three concurrent soybean seasons from fungicide- and glyphosate-treated areas (most soybean growers apply these treatments), we detected a <3% mortality rate for *M. rileyi* on *A. gemmatalis* (*n* = 980) and *C. includens* (*n* = 1259) populations. These same areas had shown infection rates ranging from 20% to 90% from early January through early February (Sujii et al., 2002), when fungicide applications were uncommon.

13.2.3 Microbial Control of Soybean Pests With Fungal Entomopathogens

Without doubt, the most conspicuous and spectacular entomopathogenic fungi epizootics in soybean systems are caused by the dimorphic fungus *M. rileyi*. This species was moved to the *Metarhizium* genus by Kepler et al. (2014). Early basic and applied studies were performed on *M. rileyi* by Boucias et al. (2000a, 2000b), Suwannakut et al. (2005), Ignoffo (1981, 1992), and Ignoffo and Boucias (1992). However, the main bottleneck in producing *M. rileyi*-based commercial products has been the mass-scale production costs. Field applications of the entomopathogenic fungus *M. rileyi* in soybean agroecosystems have been made against *H. scabra*, *Heliothis zea*, *C. includens* (Ignoffo et al., 1976, 1978), *A. gemmatalis*, *C. includens* (Stansly and Orellana, 1990), and *S. litura* (Kulkarni and Lingappa, 2002; Patil and Abhilash, 2014). In addition, the potential to control *E. heros*, *P. guildinii*, and *N. viridula* with applications of *Metarhizium anisopliae* and *B. bassiana* have been evaluated, but they require unusually high dosages to control these pest populations (Sosa-Gómez and Moscardi, 1998).

Artificial epizootics were successfully created using applications of *M. rileyi* conidia in the range of $2–5.4 \times 10^{13}$ conidia/ha. The most effective rates reduced the larval populations of *H. zea* and *H. scabra*, on average, by 77% and 82.5%, respectively (Ignoffo et al., 1976, 1978). A smaller reduction (29–54%) of a *S. litura* larval population was obtained by Kulkarni and Lingappa (2002). Differences in the efficacy of *M. rileyi* could be due to different susceptibilities among species (Puttler et al., 1976).

Interestingly, Vega-Aquino et al. (2010) observed that the efficiency of oil-formulated conidia of *I. tenuipes* and *M. rileyi* was enhanced, depending on the type of oil and the targeted noctuid species. Oil formulations also improved the dispersion of *B. bassiana* in field applications when compared with powder applications; however, a similar improvement was not observed for *M. anisopliae* conidia (Guinossi et al., 2012). Despite the substantial improvement of oil formulations, there have been no reports on the efficacy of these formulations against soybean pests.

The commercial applicability of the most studied fungi, *B. bassiana* and *M. anisopliae*, on soybean systems has been downgraded because of the natural tolerance of soybean pests against these pathogens. Stink bugs require high conidial dosages to acquire mycoses, unless they are stressed or at the end of their lifespan (Sosa-Gómez and Moscardi, 1998). Studies on using entomopathogenic fungi to control *B. tabaci* have been well reviewed by Lacey et al. (2008) and Faria and Wraight (2001), and the constraining factors cited by the authors are slow rate of kill, a dependence on favorable environmental conditions, and the need for less costly products.

13.3 ENTOMOPATHOGENIC BACULOVIRUSES

Several Baculoviruses have reached commercial status; however, few are applied extensively in soybean crops. Among the United States, India, and the countries of South America, Brazil treats the greatest area of soybean cultivation with microbial agents. Commercial products based on Baculoviruses, such as the nucleopolyhedroviruses (NPVs) of *A. gemmatalis* (AgMNPV) and *H. zea* (HzSNPV), have been registered and extensively applied (Table 13.1).

13.3.1 Viruses Associated With *Anticarsia gemmatalis*

Reported viruses in *A. gemmatalis* belong to the Baculoviridae family, Alphabaculovirus genus, Group I (AgMNPV), and to the Iridoviridae family (Kinard et al., 1995). Most studies have been conducted in the United States and Brazil because *A. gemmatalis* is only a pest in the Americas. The genome of AgMNPV has been sequenced (Oliveira et al., 2006), which made phylogenetic and evolutionary studies possible and enabled us to understand why this virus can be produced and harvested from relatively intact insects. Increasing population densities of *C. includens* after the 2003–04 growing season required the use of broader spectrum insecticides than AgMNPV. Mixed infestations of *C. includens* and *A. gemmatalis* have been controlled with carbamates, organophosphates, and, more recently, diamides. Thus, there was a gradual reduction in AgMNPV applications, from 394,000 ha treated in 2007–09 (Moscardi et al., 2011) to <100,000 ha during the 2013–14 season (Santos, personal communication). Another factor responsible for this reduction has been the reluctance of companies to produce the virus because of the widespread adoption of soybean expressing the Cry1Ac toxin, which has a high activity against both defoliators, leading to a restricted market. However, Baculovirus-based insecticides could be a promising alternative to control these pests in refuge areas as part of a resistance management program to avoid the selection of Cry1Ac protein-resistant phenotypes.

The AgMNPV formulation was improved by using heteroflocculation with titanium dioxide, resulting in protection against UV light. When larvae were fed a diet containing

TABLE 13.1 Entomopathogenic Viruses Used, or Registered to Be Used, on Brazilian Soybean Crops; Registered Baculovirus Products With Possible Applications on Soybean Crops in Argentina, China, India, and Australia Can Be Found in Kabaluk et al. (2010) and Chapter 3

Virus Name	Commercial Name	Company	Brazilian Distributer
HzSNPV	Gemstar	Certis	Ihara and Biocontrole
	Gemstar LC	Certis	Ihara and Biocontrole
	Gemstar Max	Certis	Mitsui
HearNPV	Diplomata	Andermatt Biocontrol	Koppert from Brazil
HzSNPV	HzNPV-CCAB	Ag BiTech Pty Ltd.	CCAB[a]
HearMNPV	Helicovex	Andermatt Biocontrol Ag	FMC Brazil
AgMNPV	Baculovirus Soja WP	BSbio Produtos (Bosquiroli & Santos Ltd.)	BSbio Produtos
AgMNPV	Baculo-Soja	Novozymes BioAg	Novozymes BioAg
AgMNPV	Baculovirus AEE	AEE[b]/CNPSoja	AEE/CNPSoja
AgMNPV	Grap Baculovirus	Agrocete	Agrocete
AgMNPV	Protégé	Adama Brasil	Adama Brasil

AgMNPV, Anticarsia gemmatalis multiple nucleopolyhedrovirus; HearMNPV, Helicoverpa armigera multiple nucleopolyhedrovir; HzSNPV, Heliothis zea single nuclear polyhedrosis virus.
[a]*Consortium of Agriculture and Stockbreeding Brazilian Cooperatives.*
[b]*Association of Employees from Embrapa.*

irradiated Baculovirus, the average larval mortality rate was 5.0% in contrast to ~60%, as observed for the diet that contained nonirradiated occlusion bodies (OBs). For larvae fed a diet containing the titanium-protected virus, the mortality rate reached the same level as the OB-fed larvae (Medugno et al., unpublished). The virus could also be used, along with the *C. includens* single-nucleocapsid nucleopolyhedrovirus (PsinSNPV), to create a commercial product with a dual function: the control of the velvetbean caterpillar and the soybean looper. These viruses act independently, without the benefit of synergism or additive effects, either on velvetbean caterpillar or on soybean looper (Sosa-Gómez, unpublished).

In Brazil, AgMNPV can be applied when soybean fields are colonized by populations of velvetbean caterpillar without the simultaneous occurrence of soybean loopers. However, since the 2002–03 soybean growing season, *C. includens* has been the predominant pest species, reducing the need for AgMNPV. Although cross-infectivity studies suggest that AgMNPV could infect *C. includens*, *H. zea*, and *Heliothis virescens* (Carner et al., 1979), simultaneous field applications of AgMNPV and PsinSNPV did not cause a higher mortality of *C. includens* when compared with the application of PsinSNPV alone, possibly because of the high dose of AgMNPV required to cause mortality in soybean looper (Sosa-Gómez, unpublished).

AgMNPV is commercially available as a wettable powder formulation with at least 7×10^9 OB/g, which results in

the application of ~22 g commercial formulation/ha. The shelf life of this formulation is 6 months at room temperature, 1 year at 4°C, and 3 years at −18°C (Sosa-Gómez and Moscardi, unpublished). The protocol to formulate this product has been described by Moscardi and Sosa-Gómez (2007). The virus should be applied at a recommended rate of 1.5×10^{11} OB/ha when 20 small velvetbean caterpillar larvae (<1.5 cm in length) per meter of row are found, and the defoliation is ≤30% at the vegetative stage and <15% during the reproductive stage (Moscardi and Sosa-Gómez, 1992). If small (<1.5 cm) and large larvae (≥1.5 cm) are found, the threshold should not be higher than 15 small larvae and 5 large larvae per meter of row.

13.3.2 Viruses Associated With Heliothinae

Most economically important species of Heliothinae are susceptible NPVs (Ignoffo and Couch, 1981), granuloviruses (GVs; Falcon et al., 1967; Gitay and Polson, 1971), cytoplasmic polyhedrosis viruses (Li et al., 2006), iridoviruses (Carey et al., 1978), and various RNA viruses (Brooks et al., 2002). Among these groups, the NPV and the RNA viruses have been explored in field trials, with NPVs reaching the commercial phase. The NPV virus from *H. virescens* was the first virus to be registered and commercialized as a microbial insecticide in the early 1970s (Ignoffo and Couch, 1981). Interest in this virus was renewed after the introduction of *H. armigera* in Brazil. Field trials on

H. armigera with a tetravirus had promising results when the virus was applied against early instars (Christian et al., 2005). Various NPVs and GVs can infect members of the Heliothinae subfamily; however, their virulence can differ among hosts (Ignoffo et al., 1983; Ardisson-Araújo et al., 2015). However, in certain cases baculoviruses can infect multiple hosts; for example, the HzSNPV can provide effective field control against both *H. zea* and *H. virescens* (Ignoffo et al., 1965).

Interactions between the cytoplasmic polyhedrosis virus from *H. armigera* and Cry toxin may indicate the potential to control its host with this virus when applied to transgenic crops with the *Cry1Ac* gene. Exposure to the two agents simultaneously caused higher mortality rates and debilitating effects on *H. armigera* than either agent alone (Marzban et al., 2009). On the contrary, a Densovirus appears to have a protective effect against Baculovirus infections with a low viral load and against toxemia induced by low doses of Cry1Ac toxin (Xu et al., 2012, 2014).

Registered products based on the viral control of *H. armigera* with HearSNPV are available in China, India, Europe, and Australia (Kabaluk et al., 2010). After the first official report of *H. armigera* in Brazil to the Ministry of Brazilian Agriculture in March of 2013 (Folha de São Paulo, 2013; Specht et al., 2013), growers mobilized to request the emergency registration of Baculoviruses against *H. armigera*. This pest represents a serious risk to corn, soybean, cotton, and vegetable crops in South America. At present, the most important Baculoviruses used on Brazilian soybean are HzNPV and AgMNPV (see Section 13.3.1); their main targets are the old world bollworm and the velvetbean caterpillar, respectively.

Outbreaks of *H. armigera*, in the intensive agricultural systems located west of Bahia and in the central region of Brazil, have been the impetus for applications of Heliothinae-active baculoviruses in various crops, including soybean, cotton, and corn. In the past two seasons (2013–15), their viral applications were expanded and intensified, resulting in the commercial formulations produced by Certis, Andermatt, and Ag BiTech to be completely sold out during the 2013–14 season (Oliveira, personal communication). During the 2013–14 season, the commodities (soybean, cotton, and corn) areas treated with commercial Baculovirus against *H. armigera* reached 1.3 million ha. As of 2015, control measures against *H. armigera* seem to be restricted to the northeast and central regions of Brazil. To date, the *H. armigera* populations in other regions during the soybean season have remained low.

HzSNPV targets the first three instar stages (less than 1.5 cm in length) of *H. armigera* and should be applied when the mean larval density is four larvae (<1.5 cm in length) per meter of row during the vegetative stage and two larvae (<1.5 cm in length) per meter of row during the reproductive stage. Additionally, defoliation should be <30% and <15% during the vegetative and reproductive stages, respectively. Sampling from 1 m of a row should be performed using the ground cloth method (Bueno and Sosa-Gómez, 2014) and complemented with visual observations, because small larvae can be hidden and firmly attached inside unopened leaflets. Twilight or nocturnal applications are recommended to avoid damage of viral particles by UV light. When populations are high and apical buds of small plants are under attack, insecticides with rapid modes of action need to replace the slower acting viral controls.

In soybean the registered rates of application for these virus formulations vary between 200 and 375 mL/ha for Gemstar and Gemstar LC, 40 and 75 mL/ha for Gemstar Max, and 50 and 200 mL/ha for Helicovex. Rates of Gemstar Max are lower because the commercial product contains 10×10^9 OB/mL, which is fivefold the amount of OB present in the Gemstar formulation. Usually, HzSNPV or HearSNPV are sold as liquid concentrated suspensions with a shelf life at room temperature of up to 2 months. Storage at −18°C preserves their integrity for at least 2 years.

13.3.3 Viruses Associated With Plusiinae, With an Emphasis on *Chrysodeixis includens*

Many viruses have been reported in *C. includens* and other soybean-associated Plusiinae, such as picornaviruses (Chao et al., 1986), icosahedral viruses (Chao et al., 1985a), densonucleosis viruses (Chao et al., 1985b), and GVs (Carner and Barnett, 1975). Of the viruses that target Plusiinae, the most studied virus is PsinSNPV (McLeod et al., 1982; Livingston et al., 1980), with more recent studies focusing on its genetic differentiation in geographic isolates (Alexandre et al., 2010), virulence (Alexandre et al., 2010), gene variability (Craveiro et al., 2013), and its genome sequence (Ribeiro, personal communication).

Despite the economic importance of the soybean looper as a key pest in Brazil, the virus that targets this species, PsinSNPV, has not been used commercially. This is mainly because of the lower virulence of this virus compared with other viruses and because of the difficulties in implementing mass rearing of *C. includens* through more than 10–12 generations because of inbreeding depression (Sosa-Gómez, unpublished). Geographical isolates of PsinSNPV show an LC$_{50}$ of 2500–9274 OB/mL of artificial diet (Alexandre et al., 2010). Therefore, the most virulent isolates have a biological activity that is six- to eightfold lower than AgMNPV and 2.6-fold lower than HzSNPV against their respective hosts (Fig. 13.1). A comparative analysis of restriction patterns from PsinSNPV geographical isolates revealed wide variation in DNA fragments, particularly when the restriction analysis was performed with *Eco*RI endonuclease (Alexandre et al., 2010). Additionally, a high variability in *pif-2* has been found in PsinSNPV isolates

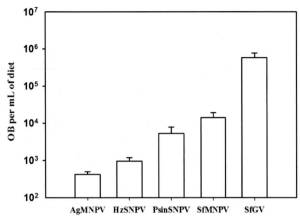

FIGURE 13.1 Susceptibility of Lepidoptera to their own Baculoviruses. LC_{50} values are from laboratory bioassays performed at Embrapa Soybean using a bioassay method that relies on an OB suspension in an artificial diet (Sosa-Gómez, unpublished). Vertical axes in log scale, *error bars* are Fiducial Limits 95%. The granulovirus and nucleopolyhedroviruses were individually assayed on their own Noctuidae hosts, except for *HzSNPV*, which was assayed on the Brazilian strain of *Helicoverpa armigera*. *AgMNPV*, *Anticarsia gemmatalis* multiple nucleopolyhedrovirus; *HzSNPV*, *Helicoverpa zea* single nucleopolyhedrovirus; *OB*, occlusion body; *PsinSNPV*, *Pseudoplusia includens* single nucleopolyhedrovirus; *SfGV*, *Spodoptera frugiperda* granulovirus; *SfMNPV*, *Spodoptera frugiperda* multiple nucleopolyhedrovirus.

when compared with other viruses. This gene is expressed in the late infection phase and is an important factor for oral infectivity. It is part of a protein complex of the occlusion-derived virus membrane involved in per os infections of the host. Thus the authors postulated that this variability could be linked to the different virulence levels observed among the isolates (Craveiro et al., 2013).

Field assays in the United States and Brazil showed that this virus still needs improvement to optimize its performance. Soybean field experiments in Louisiana, with 247 larval equivalents/ha, showed that virus applications resulted in epizootics and a reduction of 60% of the larval population 1 week after treatment (McLeod et al., 1982). Moreover, field tests carried out with 3×10^{12} OB/ha by Livingston et al. (1980) resulted in a reduced larval population of soybean loopers in the NPV-treated plots, but treatment with the insecticide methomyl resulted in the only significant yield increase over untreated control plots in that study. In addition, Brazilian soybean field assays of PsinSNPV, with application rates ranging from 0.5×10^{12} to 3×10^{12} OB/ha, produced a slight reduction in the number of larvae >1.5 cm. However, rates of 6×10^{12} OB/ha reduced populations of soybean looper significantly after 7 days of application. In the PsinSNPV-treated plots defoliation ranged from 10% to 18%, whereas in the control plots it ranged from 14% to 31%, exceeding approximately twofold the threshold damages for pest control, which is 15% of defoliation at the reproductive stage (Sosa-Gómez, unpublished).

Another soybean looper, *R. nu*, is restricted to South America, being most common in the southern Brazilian region and Argentina. This species is an important pest of sunflower as well as soybean. Two morphologically different viruses, approximately tetrahedral (SNPV) and polyhedral (MNPV) in shape, were reported on *R. nu* in Argentina (Young and Yearian, 1983; Rodriguez et al., 2012). The RanuMNPV was characterized using restriction fragment length polymorphisms and *p74* and *polyhedrin* sequences (Rodriguez et al., 2012).

13.3.4 Viruses Associated With *Spodoptera frugiperda*

Spodoptera frugiperda is usually a secondary pest on soybean, acting as a cutworm and defoliator pod borer. When *S. frugiperda* is found on soybean, the preceding crop is usually a typical host, such as oat or *Bt* corn, or a susceptible host weed, such as *Amaranthus* sp., *Agrostis* sp., *Cyperus* spp., *Cenchrus* sp., *Digitaria* spp., *Echinocloa* sp., *Ipomoea* spp., or *Sorghum halepense*.

The most important pest in Brazilian corn fields is the fall armyworm, which damages even *Bt* corn (Farias et al., 2014). The fall armyworm has undergone selection for insecticide resistance, and their increasing populations in soybean suggest that they could become a problem. Several viruses have been associated with *S. frugiperda*, including viruses from the family Ascoviridae, which lead to both chronic and fatal diseases (Federici and Govindarajan, 1990; Xue and Cheng, 2011), and a Rhabdovirus, which is associated with Sf9 cell lines (Ma et al., 2014). The most prevalent viruses associated with this species are *S. frugiperda* nucleopolyhedrovirus (SfMNPV) and a GV. The SfMNPV has been used more frequently to target *S. frugiperda* because of its greater virulence compared with the GV (Fig. 13.1). SfGV infection leads to a chronic disease that takes more than 2 weeks to kill when inoculated into neonate larvae. In Brazil, SfMNPV commercial products have been developed to treat maize crops.

A serious limitation to producing SfMNPV is the cannibalistic behavior of *S. frugiperda* during mass rearing and virus-induced lysis (melting) of infected larvae (Vieira et al., 2012). Naturally occurring isolates that have an impaired capacity to liquefy larval integument have been found. Sequencing the chitinase A (v-chiA) gene revealed that it had a frameshift mutation that reduced the size of the putative enzyme. An LC_{50} comparison with highly virulent isolates did not show significant differences, but the defective virus took a longer period (16.9 h difference of mean survival time) to kill its host (Vieira et al., 2012). Commercial products based on SfMNPV are currently in the Brazilian federal registration process by Vitae Rural Biotech, Mina Gerais, Brazil. The microbial control of *S.*

frugiperda on corn requires doses ranging from 4×10^{12} OB/ha to 8.8×10^{12} OB/ha (Valicente and Cruz, 1991). However, detailed studies on this insect/virus system need be conducted to define the insect thresholds, proper timing, and application rates on soybean.

13.3.5 Viruses in Lepidopteran Species of Minor Importance to Soybean

The most common Brazilian species of *Spodoptera* that attack soybean, in decreasing order of economic importance, are *S. eridania*, *S. cosmioides*, *S. albula*, and *S. frugiperda* (see Section 13.3.4). Among these species, we found natural occurrences of NPVs in *S. eridania* and *S. cosmioides* as well as in *R. nu*. Entomopathogenic viruses applicable for the control of *Spodoptera* species in soybean have been registered in India for *S. litura* (Ramanujam et al., 2014) and in Mexico for *S. albula* (Grzywacz et al., 2014), but the main targets of these products are pests of nonsoybean crops. We also diagnosed an NPV in *Helicoverpa gelotopoeon* from the Argentinian laboratory populations established at the Estación Experimental Obispo Colombres, San Miguel de Tucumán, Argentina. This virus has the potential to be used against *H. gelotopoeon*, which is one of the most prevalent soybean pests in Argentina (Murúa et al., 2014).

13.4 ENTOMOPATHOGENIC BACTERIA

Microbial pesticides formulated with Bt subspecies (*Bt kurstaki* and *Bt aizawai*) are used by soybean growers to control *A. gemmatalis*, *H. virescens*, *H. zea*, *H. armigera*, *T. orichalcea*, *C. includens*, *S. exigua*, and *Spodoptera littoralis*. However, some species, like *C. includens*, when controlled by Bt HD-1 strain, require higher field rates compared with *A. gemmatalis* (Morales et al., 1995; Sosa-Gómez, unpublished).

Microbial products based on Bt gained special relevance after outbreaks of *H. armigera* in Brazil during the 2013–14 soybean season. At this time, there was an exponential increase of Bt use, and the Brazilian market absorbed most of the Bt production from important companies such as Valent Bioscience and Certis. The Brazilian soybean area treated with Bt products in 2012–13 was approximately 800,000 ha. In 2013–14 it was around 3,500,000 ha, and in 2014–15 it was close to 2,200,000 ha (Oliveira, personal communication). This increase was due to difficulties in controlling *H. armigera* with chemical insecticides and the susceptibility of this pest to HD-1 strain products. However, in the 2014–15 season, their use decreased due to a low occurrence of this pest.

Even though the use of Bt-based products is desirable, their use in countries with large acreage of Bt soybean should be carefully evaluated under the optics of a resistant management system. The elements to analyze are the contribution of the toxins present in the commercial product to bioactivity (mortality), susceptibility of the pest to the toxin or complex of toxins expressed by the host plant, level of toxin expression in the Bt culture, and the probability of cross-resistance. This is a key issue for countries where the Bt crop covers a large area. For example, in Brazil the area of soybean expressing Cry1Ac and glyphosate tolerance has reached 6.3 million ha (20% of the soybean area; Céleres, 2015).

13.5 GENETIC MODIFICATIONS TO ENHANCE SOYBEAN PEST CONTROL

Genetic modifications of Baculoviruses could significantly affect integrated pest management programs. Examples include AgMNPV, which was modified to inactivate the ecdysteroid glucosyltransferase (*egt*) gene, which encodes the enzyme ecdysteroid UDP glucosyltransferase, inactivation of which interferes with larval development and molting and reduces survival time. The recombinant virus, vAgEGTΔ-*lacZ*, had an LC_{50} that was 3.9-fold lower than that of wild type AgMNPV when bioassayed in third-instar *A. gemmatalis* larvae, and it killed in a significantly faster fashion (~1–2.8 days earlier) than did the wild type strain (Pinedo et al., 2003).

Naturally occurring viruses, as observed in SfMNPV populations with genomic deletions or disruptions of the *egt* gene, can also possess fast-killing genotypes (Harrison et al., 2008; Simon et al., 2012). In addition, insecticidal activity can be enhanced by using recombinant viruses containing the cathepsin-L gene (*ScathL*) from *Sarcophaga peregrina* and the *keratinase* gene from *Aspergillus fumigatus*. The recombinant viruses reduced the mean time to death of third-instar *S. frugiperda* larvae from 5.2 days with the wild type virus to 2.6 days with the vSynScathL virus and 3.6 days with the vSynKerat virus (Gramkow et al., 2010).

Other examples include the insertions of chitinase and cathepsin genes from the *Choristoneura fumiferana* defective nucleopolyhedrovirus into the genome of AgMNPV. The modified virus showed a higher insecticidal activity against velvetbean caterpillar and caused mortality in a shorter time when compared with wild type AgMNPV. Interestingly, an unexpected effect was that the transformed virus also showed a higher production of OBs than did the wild type strain (Lima et al., 2013). Genetically modified viruses can be effective control agents of agricultural pests (Kamita et al., 2010). However, these genetically modified viruses require approval by environmental protection agencies for commercial distribution and practical use. For this reason, most virus experiments have been limited to the laboratory and experimental plots.

13.6 CONCLUSIONS

Alphabaculoviruses (HzSNPV and AgMNPV) from the Baculoviridae family are the most used bioinsecticides, in terms of treated area, in soybean agroecosystems. Selective insecticides that cause a minimal disruption and no pest resurgence in agricultural systems are desirable products. Because of their specificity, Baculoviruses have a potential use in refuge areas as a resistance management strategy in Bt soybean. Based on the Brazilian experience, growers are prone to use them when faced with insecticide resistance problems or unusually high populations, as observed after *H. armigera* outbreaks in 2012–13 in the state of Bahia and when applications of conventional insecticides were ineffective control measures. Other entomopathogenic agents have not been applied on a large scale or as commercial products in soybean. The main factors required to increase viral-based insecticide use are technologies to reduce production costs, improved and less expensive commercial formulations, and increased shelf life at room temperature. This last point, product stability, is especially challenging in tropical countries. A key to increasing broader use will be the establishment of custom economic threshold levels, appropriate timing for field applications present, and/or improved microbial-based products. Finally, the, promotion of the beneficial environmental effects of using selective microbial controls to the general public is necessary.

REFERENCES

Alexandre, T.M., Maria, Z., Ribeiro, A., Craveiro, S.R., Cunha, F., Cristina, I., Fonseca, B., Moscardi, F., Castro, M.E.B., 2010. Evaluation of seven viral isolates as potential biocontrol agents against *Pseudoplusia includens* (Lepidoptera: Noctuidae) caterpillars. J. Invertebr. Pathol. 105, 98–104.

Ardisson-Araújo, D.M.P., Sosa-Gómez, D.R., Melo, F.L., Báo, S.N., Ribeiro, B.M., 2015. Characterization of *Helicoverpa zea* single nucleopolyhedrovirus in Brazil during the first old world bollworm (Noctuidae: *Helicoverpa armigera*) Nationwide outbreak. Virus Rev. Res. 20, 1–4.

Boucias, D.G., Stokes, C., Suazo, A., Funderburk, J., 2000a. AFLP analysis of the entomopathogen *Nomuraea rileyi*. Mycologia 92, 638–648.

Boucias, D.G., Tigano, M.S., Sosa-Gómez, D.R., Glare, T.R., Inglis, P.W., 2000b. Genotypic properties of the entomopathogenic fungus *Nomuraea rileyi*. Biol. Control 19, 124–138.

Biswas, G.C., 2013. Insect pests of soybean (*Glycine max* L.) their nature of damage and succession with the crop stages. J. Asiat. Soc. Bangladesh Sci. 39 (1), 1–8.

Brooks, E.M., Gordon, K.H., Dorrian, S.J., Hines, E.R., Hanzlik, T.N., 2002. Infection of its lepidopteran host by the *Helicoverpa armigera* stunt virus (Tetraviridae). J. Invertebr. Pathol. 80, 97–111.

Bueno, A.F., Sosa-Gómez, D.R., 2014. The old world bollworm in the Neopropical region: the experience of Brazilian growers with *Helicoverpa armigera*. Outlook Pest Manag. 261–264.

Carey, G.P., Lescott, T., Robertson, J.S., Spencer, L.K., Kelly, D.C., 1978. Three African isolates of small iridescent viruses: type 21 from *Heliothis armigera* (Lepidoptera: Noctuidae), type 23 from *Heteronychus arator* (Coleoptera: Scarabaeidae), and type 28 from *Lethocerus columbiae* (Hemiptera, Heteroptera: Belostomatidae). Virology 85, 307–309.

Carner, G.R., Barnett, O.W., 1975. A granulosis virus of green cloverworm. J. Invertebr. Pathol. 25, 269–271.

Carner, G.R., Hudson, J.S., Barnett, O.W., 1979. The infectivity of a nuclear polyhedrosis virus of the velevetbean caterpillar for eight noctuid hosts. J. Invertebr. Pathol. 33, 211–216.

Castrillo, L.A., Humber, R.A., 2009. Molecular methods for identification and diagnosis of fungi. In: Stock, S.P., Vandenberg, J., Glazer, I., Boemare, N. (Eds.), Insect Pathogens: Molecular Approaches and Techniques. CABI, pp. 50–70.

Céleres, June 2015. Informativo Biotecnologia. pp. 1–7. http://www.celeres.com.br/docs/biotecnologia/IB1501_150611.pdf.

Chao, Y.C., Young, S.Y., Kim, K.S., 1985a. Cytopathology of the soybean looper, *Pseudoplusia includens*, infected with the *Pseudoplusia includens* icosahedral virus. J. Invertebr. Pathol. 45, 16–23.

Chao, Y.C., Young III, S.Y., Kim, K.S., Scott, H.A., 1985b. A newly isolated densonucleosis virus from *Pseudoplusia includens* (Lepidoptera: Noctuidae). J. Invertebr. Pathol. 46, 70–82.

Chao, Y.C., Young III, S.Y., Kim, K.S., 1986. Characterization of a picornavirus isolated from *Pseudoplusia includens* (Lepidoptera: Noctuidae). J. Invertebr. Pathol. 47, 247–257.

Christian, P.D., Murray, D., Powell, R., Hopkinson, J., Gibb, N.N., Hanzlik, T.N., 2005. Effective control of a field population of *Helicoverpa armigera* by using the small RNA virus *Helicoverpa armigera* Stunt Virus (Tetraviridae: Omegatetravirus). J. Econ. Entomol. 98, 1839–1847.

Craveiro, S.R., Melo, F.L., Ribeiro, Z.M.A., Ribeiro, B.M., Báo, S.N., Inglis, P.W., Castro, M.E., 2013. *Pseudoplusia includens single nucleopolyhedrovirus*: genetic diversity, phylogeny and hypervariability of the pif-2 gene. J. Invertebr. Pathol. 114, 258–267.

Falcon, L.A., Kane, W.R., Etzel, L.K., Leutenegger, R., 1967. Isolation of a granulosis virus from the noctuid *Heliothis zea*. J. Invertebr. Pathol. 9, 134–136.

Faria, M., Wraight, S.P., 2001. Biological control of *Bemisia tabaci* with fungi. Crop Prot. 20, 767–778.

Farias, J.R., Andow, D.A., Horikoshi, R.J., Sorgatto, R.J., Fresia, P., dos Santos, A.C., Omoto, C., 2014. Field-evolved resistance to Cry1F maize by *Spodoptera frugiperda* (Lepidoptera: Noctuidae) in Brazil. Crop Prot. 64, 150–158.

Federici, B.A., Govindarajan, R., 1990. Comparative histopathology of three ascovirus isolates in larval noctuids. J. Invertebr. Pathol. 56, 300–311.

Folha de São Paulo, October 23, 2013. Embrapa identifica espécie da nova lagarta. Folha de São Paulo, São Paulo. Caderno Mercado, p. 9.

Fuxa, J.E., Fuxa, J.R., Richter, A.R., Weidner, E.H., 2000. Prevalence of a trypanosomatid in the southern green stink bug, *Nezara viridula*. J. Eukaryot. Microbiol. 47, 388–394.

Gitay, H., Polson, A., 1971. Isolation of a granulosis virus from *Heliothis armigera* and its persistence in avian feces. J. Invertebr. Pathol. 17, 288–290.

Gramkow, A.W., Perecmanis, S., Sousa, R.L.B., Noronha, E.F., Felix, C.R., Nagata, T., Ribeiro, B.M., 2010. Insecticidal activity of two proteases against *Spodoptera frugiperda* larvae infected with recombinant viruses. Virol. J. 7, 143.

Grzywacz, D., Moore, D., Rabindra, R.J., 2014. Mass production of entomopathogens in less industrialized countries. In: Morales-Ramos, J.A., Rojas, M.G., Shapiro-Ilan, D.I. (Eds.), Mass Production of Beneficial Organisms: Invertebrates and Entomopathogens. Academic Press, Elsevier, Oxford, UK, pp. 519–564.

Guinossi, H.M., Moscardi, F., Oliveira, M.C., Sosa-Gómez, D.R., 2012. Spatial dispersal of *Metarhizium anisopliae* and *Beauveria bassiana* in soybean fields. Trop. Plant Pathol. 37, 44–49.

Harrison, R.L., Puttler, B., Popham, H.J.R., 2008. Genomic sequence analysis of a fast killing isolate of *Spodoptera frugiperda* multiple nucleopolyhedrovirus. J. Gen. Virol. 89, 775–790.

Ignoffo, C.M., 1981. The fungus *Nomuraea rileyi* as a microbial insecticide. In: Burges, H.D. (Ed.), Microbial Control of Pests and Plant Diseases 1970–1908. Academic Press, New York, pp. 513–538.

Ignoffo, C.M., 1992. Environmental factors affecting persistence of entomopathogens. Fla. Entomol 75, 516–525.

Ignoffo, C.M., Boucias, D.G., 1992. Relative activity of geographical isolates of *Nomuraea* bioassayed against the cabbage looper and velvetbean caterpillar. J. Invertebr. Pathol. 59, 215–217.

Ignoffo, C.M., Couch, T.L., 1981. The nucleopolyhedrosis virus of *Heliothis* species as a microbial insecticide. In: Burges, H.D. (Ed.), Microbial Control of Pests and Plant Diseases 1970–1908. Academic Press, New York, pp. 328–362.

Ignoffo, C.M., Chapman, A.J., Martin, D.F., 1965. The nuclear-polyhedrosis virus of *Heliothis zea* (Boddie) and *Heliothis virescens* (Fabricius). III. Effectiveness of the virus, against field population of *Heliothis* on cotton, corn, and grain sorghum. J. Invertebr. Pathol. 7, 227–235.

Ignoffo, C.M., Marston, N.L., Hostetter, D.L., Puttler, B., Bell, J.V., 1976. Natural induced epizootics of *Nomuraea rileyi* in soybean caterpillars. J. Invertebr. Pathol. 27, 191–198.

Ignoffo, C.M., Hostetter, D.L., Biever, K.D., Garcia, C., Thomas, G.D., Dickerson, W.A., Pinnell, R.E., 1978. Evaluation of an entomopathogenic bacterium, fungus, and virus for control *Heliothis zea* on soybean. J. Econ. Entomol. 71, 165–168.

Ignoffo, C.M., McIntosh, A.H., Garcia, C., 1983. Susceptibility of larvae of *Heliothis zea*, *H. virescens*, and *H. armigera* (Lep.: Noctuidae) to 3 baculoviruses. Entomophaga 28, 1–8.

Kabaluk, J.T., Svircev, A.M., Goettel, M.S., Woo, S.G., 2010. The Use and Regulation of Microbial Pesticides in Representative Jurisdictions Worldwide. IOBC Global. 99 pp. Available online: www.IOBC-Global.org.

Kamita, S.G., Kang, K.-D., Hammock, B.D., 2010. Genetically modified baculoviruses for insect pest control. In: Gilbert, L.I., Gill, S.S. (Eds.), Insect Control. Biological and Synthetic Agents, first ed. Academic Press, London, England, pp. 331–385.

Kepler, R.M., Humber, R.A., Bischoff, J.F., Rehner, S.A., 2014. Clarification of generic and species boundaries for *Metarhizium* and related fungi through multigene phylogenetics. Mycologia 106, 811–829.

Kinard, G.R., Barnett, O.W., Carner, G.R., 1995. Characterization of an iridescent virus isolated from the velvetbean caterpillar, *Anticarsia gemmatalis*. J. Invertebr. Pathol. 66, 258–263.

Koch, K.A., Potter, B., Ragsdale, D.W., 2010. Non-target impacts of soybean rust fungicides on the fungal entomopathogens of soybean aphids. J. Invertebr. Pathol. 103, 156–164.

Kulkarni, N.S., Lingappa, S., 2002. Bioefficacy of entomopatogenic fungus, *Nomuraea rileyi* (Farlow) Samson on *Spodoptera litura* and *Cydia ptychora* in soybean and on *S. litura* in potato. Karnataka J. Agric. Sci. 15, 47–55.

Lacey, L.A., Wraight, S.P., Kirk, A.A., 2008. Entomopathogenic fungi for control of *Bemisia* spp.: foreign exploration, research and implementation. In: Gould, J.K., Hoelmer, K., Goolsby, J. (Eds.), Classical Biological Control of *Bemisia tabaci* in the USA: A Review of Interagency Research and Implementation. Springer, Dordrecht, The Netherlands, pp. 33–70.

Li, Y., Tan, L., Li, Y., Chen, W., Zhang, J., Hu, Y., 2006. Identification and genome characterization of *Heliothis armigera* cypovirus types 5 and 14 and *Heliothis assulta* cypovirus type 14. J. Gen. Virol. 87, 387–394.

Lima, A.A., Aragão, C.W.S., de Castro, M.E.B., de Castro Oliveira, J.V., Sosa Gómez, D.R., Ribeiro, B.M., 2013. A recombinant *Anticarsia gemmatalis* MNPV harboring chiA and v-cath genes from *Choristoneura fumiferana* defective NPV induce host liquefaction and increased insecticidal activity. PLoS One 8, e74592.

Liu, S., Chen, Y., Bonning, B.C., 2015. RNA virus discovery in insects. Curr. Opin. Insect Sci. 8. http://dx.doi.org/10.1016/j.cois.2014.12.005.

Livingston, J.M., McLeod, P.J., Yearian, W.C., Young III, S.Y., 1980. Laboratory and field evaluation of a nuclear polyhedrosis virus of the soybean looper, *Pseudoplusia includens*. J. Ga. Entomol. Soc. 15, 194–199.

Ma, H., Galvin, T.A., Glasner, D.R., Shaheduzzaman, S., Khan, A.S., 2014. Identification of a novel Rhabdovirus in *Spodoptera frugiperda* cell lines. J. Virol. 88, 6576–6585.

Marzban, R., He, Q., Liu, X., Zhang, Q., 2009. Effects of *Bacillus thuringiensis* toxin Cry1Ac and cytoplasmic polyhedrosis virus of *Helicoverpa armigera* (Hübner) (HaCPV) on cotton bollworm (Lepidoptera: Noctuidae). J. Invertebr. Pathol. 101, 71–76.

McLeod, P.J., Young, S.Y., Yearian, W.C., 1982. Application of a baculovirus of *Pseudoplusia includens* to soybean: efficacy and seasonal persistence. Environ. Entomol. 11, 412–416.

Morales, L., Moscardi, F., Kastelic, J.G., Sosa-Gómez, D.R., Paro, F.R., Soldorio, I.L., 1995. Suscetibilidade de *Anticarsia gemmatalis* Hubner e *Chrysodeixis includens* (Walker) (Lepidoptera: Noctuidae), a *Bacillus thuringiensis* (Berliner). Ann. Soc. Entomol. Bras. 24, 593–598.

Morjan, W.E., Pedigo, L.P., Lewis, L., 2002. Fungicidal effects of glyphosate formulations on four species of entomopathogenic fungi. Environ. Entomol. 31, 1206–1212.

Moscardi, F., Sosa-Gómez, D.R., 1992. Use of viruses against soybean caterpillars in Brazil. In: Copping, L.G., Green, M.B., Rees, R.T. (Eds.), Pest Management in Soybean. Published for SCI by Elsevier Applied Science, pp. 98–109.

Moscardi, F., Sosa-Gómez, D.R., 2007. Microbial control of insect pests of soybean. In: Lacey, L.A., Kaya, H.K. (Eds.), Field Manual of Techniques in Invertebrate Pathology. Application and Evaluation of Pathogens for Control of Insects and Other Invertebrate Pests, second ed. Springer, Dordrecht, The Netherlands, pp. 411–426.

Moscardi, F., Souza, M.L., Castro, M.E.B., Moscardi, M., Szewczyk, B., 2011. Baculovirus pesticides: present state and future perspectives. In: Ahmad, I., Ahmad, F., Pichtel, J. (Eds.), Microbes and Microbial Technology Agricultural and Environmental Applications. Springer, Dordrecht, The Netherlands, pp. 415–445.

Murúa, M.G., Scalora, F.S., Navarro, F., Cazado, L.E., Casmuz, A., Villagrán, M.E., Lobos, E., Gastaminza, G., 2014. First record of *Helicoverpa armigera* (Lepidoptera: Noctuidae) in Argentina. Fla. Entomol. 97, 854–856.

Nielsen, C., Hajek, A.E., 2005. Control of invasive soybean aphid, *Aphis glycines* (Hemiptera: Aphididae), populations by existing natural enemies in New York State, with emphasis on entomopathogenic fungi. Environmental Entomology 34, 1036–1047.

Oliveira, J.V., Wolff, J.L., Garcia-Maruniak, A., Ribeiro, B.M., de Castro, M.E., de Souza, M.L., Moscardi, F., Maruniak, J.E., Zanotto, P.M., 2006. Genome of the most widely used viral biopesticide: *Anticarsia gemmatalis* multiple nucleopolyhedrovirus. J. Gen. Virol. 87, 3233–3250.

Patil, R.H., Abhilash, C., 2014. *Nomuraea rileyi* (Farlow) Samson: a biopesticide IPM component for the management of leaf eating caterpillars in soybean ecosystem. In: International Conference on Biological, Civil and Environmental Engineering (BCEE-2014) March 17–18, 2014 Dubai (UAE. http://dx.doi.org/10.15242/IICBE.C0314069.

Pinedo, F.J.R., Moscardi, F., Luque, T., Olszewski, J.A., Ribeiro, B.M., 2003. Inactivation of the ecdysteroid UDP-glusosyltransferase (egt) gene of *Anticarsia gemmatalis* nucleopolyhedrovirus (AgMNPV) improves its virulence towards its insect host. J. Invertebr. Pathol. 27, 336–344.

Puttler, B., Ignoffo, C.M., Hostetter, D.L., 1976. Relative susceptibility of nine caterpillar species to the fungus *Nomuraea rileyi*. J. Invertebr. Pathol. 27, 269–270.

Ramanujam, B., Rangeshwaran, R., Sivakmar, G., Mohan, M., Yandigeri, M., 2014. Management of insect pests by microorganisms. Proc. Indian Natl. Acad. Sci. 80, 455–471.

Rehner, S.A., 2009. Molecular systematics of entomopathogenic fungi. In: Stock, S.P., Vandenberg, J., Glazer, I., Boemare, N. (Eds.), Insect Pathogens: Molecular Approaches and Techniques. CABI, Oxfordshire, UK, pp. 145–165.

Rodriguez, V.A., Belaich, M.N., Quintana, G., Sciocco Cap, A., Ghiringhelli, P.D., 2012. Isolation and characterization of a nucleopolyhedrovirus from *Rachiplusia nu* (Guenée) (Lepidoptera: Noctuidae). Int. J. Virol. Mol. Biol. 1, 28–34.

Roggia, S., Guedes, J.V.C., Kuss-Roggia, R.C.R., Vasconcelos, G.J.N., Navia, D., Delalibera Jr., I., 2009. Ácaros predadores e o fungo *Neozygites floridana* associados a tetraniquídeos em soja no Rio Grande do Sul. Pesq. Agropec. Bras. 44, 107–110.

Roy, H.E., Brodie, E.L., Chandler, D., Goettel, M.S., 2010. Deep space and hidden depths: understanding the evolution and ecology of fungal entomopathogens. Biocontrol 55, 1–6.

Simon, S., Williams, T., López-Ferber, M., Caballero, P., 2012. Deletion of egt is responsible for the fast-killing phenotype of natural deletion genotypes in a *Spodoptera frugiperda* multiple nucleopolyhedrovirus population. J. Invertebr. Pathol. 111, 260–263.

Sosa-Gómez, D.R., Moscardi, F., 1998. Laboratory and field studies on the infection of stink bugs, *Nezara viridula, Piezodorus guildinii*, and *Euschistus heros* (Hemiptera: Pentatomidae) with *Metarhizium anisopliae* and *Beauveria bassiana* in Brazil. J. Invertebr. Pathol. 71, 115–120.

Sosa-Gómez, D.R., Boucias, D.G., Nation, J.L., 1997. Attachment of *Metarhizium anisopliae* to the southern green stink bug *Nezara viridula* cuticle and fungistatic effect of cuticular lipids and aldehydes. J. Invertebr. Pathol. 69, 31–39.

Sosa-Gómez, D.R., Delpin, K.E., Moscardi, F., Nozaki, M.H., 2003. The impact of fungicides on *Nomuraea rileyi* (Farlow) Samson epizootics and on populations of *Anticarsia gemmatalis* Hubner (Lepidoptera: Noctuidae), on soybean. Neotrop. Entomol. 32, 287–291.

Sosa-Gómez, D.R., Borges, E., Viera, I.H.T.L., Costa, F., Oliveira, C.N., 2005. Trypanosomatid prevalence in *Nezara viridula* (L.), *Euschistus heros* (Fabricius) and *Piezodorus guildinii* (Westwood) (Heteroptera: Pentatomidae) populations in Northern Paraná, Brazil. Neotrop. Entomol. 34, 341–347.

Sosa-Gómez, D.R., López Lastra, C.C., Humber, R.A., 2010. An overview of arthropod-associated fungi from Argentina and Brazil. Mycopathologia 170, 61–76.

Sosa-Gómez, D.R., Côrrea-Ferreira, B.S., Hoffmann-Campo, C.B., Corso, I., Oliveira, L., Moscardi, F., Panizzi, A.R., Bueno, A.F., Hirose, E., Roggia, S., 2014. Manual de identificação de insetos e outros invertebrados da cultura da soja, third ed. Embrapa Soja Documentos, Londrina. 100 pp.

Specht, A., Sosa-Gómez, D.R., Paula-Moraes, S.V., Yano, S.A.C., 2013. Identificação morfológica e molecular de *Helicoverpa armigera* (Lepidoptera: Noctuidae) e ampliação de seu registro de ocorrência no Brasil. Pesq. Agropec. Bras. 48, 689–692.

Stansly, P.A., Orellana, G.J., 1990. Field manipulations of *Nomuraea rileyi* (Moniliales: Moniliacea): effects on soybean defoliators in Costal Ecuador. J. Econ. Entomol. 83, 2193–2195.

Sujii, E.R., Tigano, M.S., Sosa-Gómez, D., 2002. Simulação do impacto do fungo *Nomuraea rileyi* em populações da lagarta da soja, *Anticarsia gemmatalis*. Pesq. Agropec. Bras. 37, 1551–1558.

Suwannakut, S., Boucias, D.G., Wiwat, C., 2005. Genotypic analysis of *Nomuraea rileyi* collected from various noctuid hosts. J. Invertebr. Pathol. 90, 169–176.

USDA, 2015. World Agricultural Production. Foreign Agricultural Service. Office of Global Analysis. Circular Series. WAP, pp. 1–15.

Valicente, F.H., Cruz, I., 1991. Controle biológico da lagarta-do-cartucho, *Spodoptera frugiperda*, com o baculovírus. Embrapa Milho e Sorgo. Circ. Técnica 15.

Vega-Aquino, P., Sanchez-Peña, S., Blanco, C.A., 2010. Activity of oil-formulated conidia of the fungal entomopathogens *Nomuraea rileyi* and *Isaria tenuipes* against lepidopterous larvae. J. Invertebr. Pathol. 103, 145–149.

Vieira, C.M., Tuelher, E.S., Valicente, F.H., Caldas-Wolff, J.L., 2012. Characterization of a *Spodoptera frugiperda* multiple nucleopolyhedrovirus isolate that does not liquefy the integument of infected larvae. J. Invertebr. Pathol. 111, 189–192.

Williamson, C., von Wechmar, M.B., 1992. Two novel viruses associated with severe disease symptoms of the green stinkbug *Nezara viridula*. J. Gen. Virol. 73, 2467–2471.

Xu, P., Cheng, P., Liu, Z., Li, Y., Murphy, R.W., Wu, K., 2012. Complete genome sequence of a monosense densovirus infecting the bollworm, *Helicoverpa armigera*. J. Virol. 86, 10909.

Xu, P., Liu, Y., Graham, R.I., Wilson, K., Wu, K., 2014. Densovirus is a mutualistic symbiont of a global crop pest (*Helicoverpa armigera*) and protects against a baculovirus and Bt biopesticide. PLoS Pathog. 10, e1004490.

Xue, J.L., Cheng, X.W., 2011. Comparative analysis of a highly variable region within the genomes of *Spodoptera frugiperda* ascovirus 1d (SfAV-1d) and SfAV-1a. J. Gen. Virol. 92, 2797–2802.

Young, S.Y., Yearian, W.C., 1983. Pathology of a nuclear polyhedrosis virus of *Rachiplusia nu* in *Rachiplusia ou* (Lepidoptera: Noctuidae). J. Invertebr. Pathol. 42, 410–412.

Chapter 14

Microbial Control of Arthropod Pests in Small Fruits and Vegetables in Temperate Climate

S.K. Dara
University of California Cooperative Extension, San Luis Obispo, CA, United States

14.1 INTRODUCTION

Vegetable and small fruit crops have a wide range of arthropod pests that are susceptible to a variety of entomopathogenic bacteria, fungi, nematodes, and viruses (Collier et al., 2001; Bruck et al., 2005; Booth et al., 2007; Vandenberg et al., 2007; Robinson, 2009; Beas-Catena et al., 2014). In many cropping systems around the world, pest management is mainly dependent on chemical pesticides because they are generally less expensive, easy to store and apply, and perceived to be more efficacious compared to microbial pesticides. However, various factors, including difficulty in controlling some pests with chemical pesticides and associated resistance problems, increasing concern for environmental safety and preference for organically grown fruits and vegetables, and new formulation technologies that make biopesticides more effective and competitive, have contributed to a recent boom in the biopesticide industry (USDA-NASS, 2010; Glare et al., 2012; Sinha, 2012; Dunham, 2014; Lacey et al., 2015). Emerging partnerships between major chemical companies and biopesticide companies give impetus to the biopesticide industry and emphasize the importance of microbial control agents (MCAs) in pest management.

Several coleopteran, dipteran, hemipteran, lepidopteran, acarina, and other pests attack small fruits and vegetables causing significant yield losses. The major entomopathogens that are used by growers or evaluated by researchers for pest management in these crops include the bacterium *Bacillus thuringiensis* (Bt) subsp. *kurstaki*; the fungi: *Beauveria bassiana, Isaria fumosorosea* (formerly *Paecilomyces fumosoroseus*), and *Metarhizium brunneum* (formerly *M. anisopliae*); the entomopathogenic nematodes (EPNs): *Steinernema* spp. and *Heterorhabditis* spp.; and the baculoviruses, granuloviruses (GV) and nucleopolyhedroviruses (NPV) (Cranshaw, 1998; Faria and Wraight, 2001; Priyatno and Ibrahim, 2004; Booth et al., 2007; Vandenberg et al., 2007; Koppenhöfer et al., 2008; Dara, 2013, 2014a; Zahn et al., 2013; Beas-Catena et al., 2014; Correa-Cuadros et al., 2014). Several of these entomopathogens are commercially available in multiple formulations. Among these entomopathogens, Bt is the most successful and commonly used biopesticide in many parts of the world. Growing conditions of some of the small fruits and vegetables in temperate climate are environmentally conducive for entomopathogens and integrated pest management (IPM) could benefit from the uses of MCAs. Because of the limited information on the efficacy of microbial pesticides in several cropping systems or lack of effective strategies for integrating them with other control practices, the full potential of MCAs is not yet explored in many small fruit and vegetable growing regions.

About 30% of the insecticides used worldwide are targeted against pests of vegetables and fruits (Shelton et al., 2008). If MCAs can be substituted for one or more applications of broad-spectrum chemical insecticides, this practice can lead to a significant reduction in the use of synthetic chemicals and the associated risks. Among different entomopathogens, fungi are well suited for arthropods with piercing and sucking mouthparts and bacteria and viruses are ideal for those with chewing mouthparts. EPNs are generally used for soil-dwelling pests. While some MCAs alone provide good control as stand-alone materials, combining them with compatible botanical and chemical pesticides can be even more effective. Because of their relative specificity and little or no effect on nontarget organisms, especially natural enemies, MCAs are ideal components of vegetable and small fruit IPM.

Microbial Control of Insect and Mite Pests. http://dx.doi.org/10.1016/B978-0-12-803527-6.00014-7

14.2 MICROBIAL CONTROL IN VEGETABLES IN TEMPERATE CLIMATE

14.2.1 *Bacillus thuringiensis*

Bt subspp. *kurstaki* (Btk) and *aizawai* (Bta) are commonly used against lepidopteran pests on a wide range of vegetable crops. Bt products dominate the biopesticide industry because they are generally effective, selective, and affordable (Lacey et al., 2015). Unlike some other entomopathogens where the organism has to cause an infection and develop in the target host, the toxins or insecticidal crystal proteins in Bt products have a specific mode of action where the toxin breaches the insect gut and results in loss of osmotic balance, rupturing of cells, and death from septicemia. Bt products are somewhat similar to chemical pesticides in their efficacy and affordability and have the advantage of being safely applied until the time of harvest.

Btk and Bta are used to manage several species of lepidopteran pests on multiple vegetable crops worldwide. Recommended label rates of Bt products for these pests are 0.5–1.7 kg or 1.2–4.7 L/ha depending on the formulation, and they can be applied one or more times during the cropping season as demanded by pest infestations and economic injury levels.

Compatibility or synergistic interaction with other biocontrol agents also makes Bt products suitable for use in IPM programs. For example, the predatory stink bug, *Podisus nigrispinus*, consumed more larvae of the diamondback moth, *Plutella xylostella*, in the presence of Bt sprays, thus improving overall control (Magalhães et al., 2015). Although resistance development to Bt in some insects such as *P. xylostella* was a problem in the 1980s and 1990s in some regions where it was used excessively (Tabashnik et al., 1990), Bt products are still a popular choice in California for controlling *P. xylostella*, the imported cabbageworm, *Pieris rapae*, and the cabbage looper, *Trichoplusia ni*, on cole crops and the beet armyworm, *Spodoptera exigua*, on lettuce (Natwick et al., 2010a,b).

14.2.1.1 Some Notable Examples of Insect Control With Bt in Vegetables

Tomato leaf miner, *Tuta absoluta*, is a serious pest of tomatoes in South America and is rapidly spreading to other parts of the world including temperate agro-ecosystems (NAPPO, 2012). It can also cause significant damage to potato, eggplant, and other solanaceous hosts. Laboratory, greenhouse, and field studies conducted in Spain showed that commercial formulations of Bt caused a significant reduction in *T. absoluta* numbers and prevented yield loss (González-Cabrera et al., 2011). Each plant was sprayed with about 20 mL of 2 g of commercial formulation (Dipel DF, Turex, and Costar) in a liter of water, and maximum recommended application rates of these formulations ranged between 50 and 200 g/100 L.

Several authors have reported fair to good control of potato tubermoth (PTM also known as potato tuberworm), *Phthorimaea operculella*, with Btk on potato foliage (original citations provided by Kroschel and Lacey, 2008; Sporleder and Lacey, 2013). Arthurs et al. (2008a) reported fairly good control of very high PTM populations with Btk, applied 1.12 kg/ha, but several applications were needed throughout the growing season. They reported that alternating Btk and PTM granulovirus (PhopGV) or applying PhopGV alone at 10^{13} occlusion bodies (OBs)/ha was significantly more effective than using Btk alone.

Applications of Bt to potato tubers in rustic stores for control of PTM has been reported by a multitude of authors (cited in Kroschel and Lacey, 2008; Sporleder and Lacey, 2013). For example, in Tunisia, an integrated control approach comprising Bt applied at the beginning of the storage period in combination with cultural control (early harvest) eliminated the reliance on parathion sprays (von Arx et al., 1987). Formulation of Bt with various carriers has been reported to have improved Bt activity in PTM infested tubers in rustic stores. Raman et al. (1987) reported that Btk (DiPel) was effective in reducing feeding damage in storage in Peru when applied as a dust formulation. Btk mixed with fine sand dust containing quartz (40 g Btk, 960 g sand per ton of tubers) provided effective control in tuber storage in the Republic of Yemen (Kroschel and Koch, 1996). In addition to killing larvae attempting to enter tubers, the formulation also controlled 96% of larvae that were already inside tubers. Arthurs et al. (2008b) demonstrated that tubers treated with 37.5 mg Btk WP mixed in talcum or diatomaceous earth/kg of tubers before infestation resulted in 99% PTM larval mortality.

A study in the United Kingdom reported that thermal fogging of Bt in glasshouse tomatoes provided good control of the moth, bright-line brown-eye, *Lacanobia oleracea*, at 1, 2, and 4 kg/ha (Burges and Jarrett, 1980). Nearly 100% control was achieved at 4 kg/ha rate and prevented foliar and fruit damage. In a later study, Jarrett and Burgess (1982) also reported that high volume applications of Bt at 0.04% and 0.12% w/v reduced foliar and fruit damage from *L. oleracea* by 77–99% in glasshouse tomato. They suggested timely application of Bt were as a good to chemical pesticides. Bt is also used to manage several species of lepidopteran pests on multiple vegetable crops in Florida (Maynard et al., 1999; Mosslet and Dunn, 2008) and California (Natwick et al., 2010c) in the United States and South Korea (Jee, 2007).

Colorado potato beetle (CPB), *Leptinotarsa decemlinata*, is the most important defoliator of potato worldwide (Wraight et al., 2009). Since the isolation of Bt var. *tenebrionis*, a variety of Bt with elevated larvicidal activity for CPB by Krieg et al. (1983), numerous studies reporting good efficacy for commercial formulations of Btt have been conducted (Zehnder et al., 1992; Wraight and Ramos,

2002, 2005; Wraight et al., 2009). In other field trials in Washington State, USA, a commercial formulation of Btk, genetically engineered to overproduce a coleopteran-active toxin, *Cry 3Bb*, applied at 1.17 and 7.0 L/ha against CPB resulted in 33 and 40 tons/ha yield, respectively compared to 24 tons in the untreated control and 46 tons in aldicarb treatment (3.37 kg a.i./ha) (Lacey et al., 1999a). For maximum larvicidal activity, Btt must be applied when larvae are as young as possible (Zehnder et al., 1992). In addition to their inherent larvicidal activity, Btt and the hypocrealian fungus, *B. bassiana*, exhibit synergistic activity (Wraight and Ramos, 2005). Combination of Btt and *B. bassiana* products improved the control efficacy by 6–35% compared to the control provided by individual products against CPB. One disadvantage of Btt is that it is sensitive to UV degradation and may be washed off by rain or irrigation requiring multiple applications. Wraight et al. (2007) discussed that lower virulence of Btt to older larvae, short residual activity of foliar applications, lack of recycling ability in the pest habitat, and extended periods of egg laying limited the efficacy of Btt against CPB.

Despite the risks of resistance with pests like *Plutella xylostella* (Shelton et al., 1993) or *T. ni* (Janmat and Myers, 2003), Bt *Cry* genes in transgenic crops such as eggplant, potato, corn, and soybean also contribute to pest management, reduction in chemical pesticide use, and related costs in many countries and ensure food safety (Reed et al., 2001; Carpenter et al., 2002; Shelton et al., 2002; Ghislain et al., 2003; Benbrook, 2004; Krishna and Qaim, 2008). In addition to field evaluation and development of Bt potatoes engineered to express the *Cry 3B* toxin for control of CPB, Bt potato has also been engineered with a modified *Bt-cry1Ia1* gene (often referred to as *cry5*) to transform it for resistance to both the CPB and lepidopterans (Douches et al., 2002, 2011)

14.2.2 Bacterial Byproducts

Although not an entomopathogen, spinosad, derived from the bacterium *Saccharopolyspora spinosa*, is a commercially available active ingredient that is used against a variety of vegetable pests. Spinosad is an important management recommendation for leafminers (*Liriomyza* spp.) in cole crops and lettuce (Natwick et al., 2010a,b), several lepidopteran pests in spinach (Koike et al., 2009) and thrips, and lepidopteran pests in tomato (Natwick et al., 2013) in California. However, spinosad and other spinosyns can adversely affect beneficial arthropods (Biondi et al., 2012), and their overuse can lead to pesticide resistance (Bielza et al., 2007; Lebedev et al., 2013).

A broad spectrum of vegetable pest insects including larval CPB, adults of corn rootworms, *Diabrotica* spp., and the southern green stinkbug, *Nezara viridula*, are killed by fermentation byproducts and spent fermentation medium of

another bacterium, *Chromobacter subtsugae* (Martin et al., 2007a,b). Live bacteria were not needed for toxicity to *N. viridula* adults (Martin et al., 2007b). It is also reported by Marrone Bio Innovations (Ratnakar et al., 2013) to have multiple effects such as reducing fecundity and oviposition, deterring feeding and acting as a stomach poison on aphids, psyllids, whiteflies, *Lygus*, mealybugs, thrips, and phytophagous mites.

14.2.3 Fungi

Entomopathogenic fungi (EPF) infect the host through conidial or blastospore germination and formation of appressoria enabling them to penetrate the exoskeleton, making them especially useful for control of sucking insects (eg, aphids, stink bugs, scales, whiteflies). Several formulations of entomopathogenic fungi in the order Hypocreales are commercially available for use in organic and conventional agriculture. *Lecanicillium lecanii* s.l. applied at 10^7 conidia/mL was effective in controlling the brown stink bug, *Riptortus linearis*, a major pest of soybean in Indonesia (Prayogo, 2014). *Lecanicillium lecanii* s.l. alone and in combination with *Annona squamosa* or *Jatropha curcas* seed powder caused a significant reduction in egg hatch and seed and pod damage to soybean in field studies. *Beauveria bassiana* is used to manage flea beetles on organically produced vegetables in the Pacific Northwest of the United States (Parker et al., 2012).

EPF provide immediate control options to growers and home gardeners who are looking for nonchemical solutions. In different field studies in California, a *B. bassiana*–based formulation at 2.2 kg/ha provided moderate control similar to some of the chemical pesticides against the Western flower thrips, *Frankliniella occidentalis*, on lettuce and the cabbage aphid, *Brevicoryne brassicae* and *Myzus persicae*, on broccoli (Dara, 2013).

Some hemipteran pests are difficult to manage in general, but EPF could play an important role in controlling both native and invasive species. For example, the Bagrada bug, *Bagrada hilaris*, which is a new invasive pest on cole crops and other hosts in California and other states can be controlled by the combination of EPF and azadirachtin (Dara, personal observation; Martin et al., 2013). In another study in California, the combination of *B. bassiana* (3.5 L/ha) and azadirachtin (0.5 L/ha) caused a 62% reduction in the rice root aphid, *Rhopalosiphum rufiabdominale*, and the honeysuckle aphid, *Hyadaphis foeniculi*, numbers compared to pretreatment counts (Dara, 2015c). James (2003) and Islam et al. (2010) also reported improved control of whiteflies when EPF were combined with azadirachtin. Application of *I. fumosorosea* (2×10^8 conidia/mL) and azadirachtin (Neemix 4.5 at 60 µg/mL) caused 90% mortality in *B. argentifolii* nymphs (James, 2003). Similarly, the combination of *B. bassiana* (10^7 conidia/mL) and azadirachtin (Azadirachtin EC at 5 mL/L of water) also caused higher mortality in eggs (29.5%) and nymphs (97.2%) of *Bemisia tabaci*, which was more than the mortality cause by the fungus or azadirachtin at higher rates (Islam et al., 2010).

The whitefly, *B. tabaci* biotype B (also referred to as *B. argentifolii*), is an extremely important polyphagous pest throughout the world (Gerling and Mayer, 1996). Several species and strains of fungi and parasitoids have been collected and evaluated for its control (Lacey et al., 1993, 2008; Kirk and Lacey, 1996). The most promising control agent was *I. fumosorosea* (Osborne et al., 1990; Vidal et al., 1998b; Lacey et al., 2008). Biotic and abiotic factors affecting its insecticidal activity for *B. tabaci* were reported by Fargues et al. (1996), Vidal et al. (1997a,b, 1998a,b), Wraight et al. (1998), Lacey and Mercadier (1998), and Lacey et al. (1999b, 2008).

Wraight et al. (2000) reported that multiple applications of 5×10^{13} conidia/ha of *I. fumosorosea* or *B. bassiana* in 180 L water at 4- to 5-day intervals resulted in 90% control of third- and fourth-instar nymphs of *B. argentifolii* on cucumbers and cantaloupe. A lower application rate (1.25×10^{13} conidia/ha), applied at the same intervals, reduced numbers of large nymphs by 85% in cantaloupe. Targeting applications to the underside of leaves where the nymphs remain sessile for most of their development is a key factor for effective use of EPFs against this species.

EPF also provide good control options for multiple pests of potato (Lacey and Kroschel, 2009; Wraight et al., 2009; Lacey et al., 2011). Both *M. brunneum* and *I. fumosorosea* reduced eggs and nymphs of the potato psyllid, *Bactericera cockerelli*, plant damage, and the symptoms of zebra chip disease and increased the tuber yields in three field studies conducted in Texas (Lacey et al., 2011). Application of *M. brunneum* at 0.5, 1.1 and 2.2 L/ha caused 45%, 59%, and 67% reduction in eggs and nymphs of *B. cockerelli*, respectively in one study. In another study, *I. fumosorosea* was applied with and without neem oil, resulting in 76% and 78% reduction in *B. cockerelli* numbers, respectively. In the third study, *M. brunneum* at 1.1 and 2.2 L/ha and *I. fumosorosea* at 1.1 and 2.2 kg/ha caused 62%, 62%, 66%, and 65% reduction in *B. cockerelli*, respectively.

Poprawski et al. (1997) reported that four early-season application of unformulated *B. bassiana* applied at 5×10^{13} conidia/ha caused a significant reduction (77%) in mature larvae of CPB compared to the insecticide treatments (50%). While these reductions did not translate into yield improvements in this study, the potential of *B. bassiana* in reducing larval stages that cause most of the defoliation is promising. In a multi-year Canadian field study, a granular formulation of a noncommercial isolate of *M. anisopliae*, applied as broadcast preplant incorporation to obtain 2.5×10^{14} conidia/ha reduced the wireworm (*Agriotes obscurus*) damage (Kabaluk et al., 2005).

EPF could also control plant parasitic nematodes and promote plant growth. *Paecilomyces lilacinus* and cow manure application reduced eggs of the root knot nematode, *Meloidogyne incognita*, and improved the plant biomass of tomato and lettuce (Machado et al., 2010). Improved growth, survival, nutrient absorption, and biomass production of cabbage plants grown with limited water and lighting was seen with application of *B. bassiana*, *I. fumosorosea*, and *M. brunneum* (Dara et al., 2014) shedding light on potential new uses for EPF in crop production. It is possible that these fungi, through mycorrhiza-like interaction, helped the plants to absorb water and nutrients and promoted their growth. This is especially important in Brassica and other crops that do not form typical mycorrhizal relationships. With such a role, EPF could also be useful for improved plant production in addition to their typical use in plant protection.

14.2.4 Entomopathogenic Nematodes

EPNs are effectively used against soil-inhabiting stages of various coleopteran, dipteran, and lepidopteran pests. Banded cucumber beetle, *Diabrotica balteata*, is a pest of vegetable and other crops throughout the southern United States and elsewhere (Miret and Rodriguez, 1984; Del C. Bandas et al., 2004; Capinera, 2007). In potted plant studies, *Heterorhabditis* sp. caused nearly 100% reduction in larval numbers when used at 5000 or 20,000 infective juveniles (IJs)/pot (Creighton and Fassuliotis, 1985). In another study, *Steinernema feltiae* controlled the cutworm, *Agrotis segetum*, as effectively as endosulfan in lettuce under field conditions at 2.5×10^5 nematodes/m² in sandy soil and at 10^6 IJs/m² in loamy soil (Lössbroek and Theunissen, 1985). Similarly, steinernematid nematodes caused 40–77% mortality in early instars of the cabbage maggot, *Delia radicum* (Nielsen, 2003). Heterorhabditid and two steinernematid nematodes provided 66–100% control of CPB, in a study conducted in small potato plots (Wright et al., 1987). Dauer-stage larvae of *Heterorhabditis heliothidis*, *S. feltiae*, and *S. bibionis* at 5×10^5/m² reduced black cutworm, *Agrotis ipsilon*, damage by more than 50% in a study conducted using corn seedlings (Capinera et al., 1988). Lewis et al. (2007) discussed controlling several species of lepidopteran pests on corn that cause damage to roots, stalks, ears, and leaves by using *Heterorhabditis bacteriophora*, *Steinernema carpocapsae*, *S. feltiae*, and *S. riobrave*. Different EPNs are recommended for the control of moths, borers, weevils, armyworm, cutworm, earworm, and rootworm on vegetables and small fruits in the Pacific Northwest in the United States (Miles et al., 2012; see also Chapter 6). EPNs are also considered as potential tools for controlling various potato pests (Lacey et al., 2009).

EPNs can also play an important role in managing plant parasitic nematodes. For example, both heterorhabditid and steinernematid species suppressed plant pathogenic root-knot nematodes, *Meloidogyne mayaguensis* and *M. incognita*, and root galls formed in tomato (Pérez and Lewis, 2002; Molina et al., 2007; Maru and Siddiqui, 2012).

14.2.5 Viruses

Like Bt, baculoviruses need to be ingested by the insect and are well suited for larvae of lepidopteran pests. Baculoviruses such as GVs and NPVs are available as biopesticides and are seen as safe alternatives to chemical pesticides in multiple and niche cropping systems (Szewczyk et al., 2009; del Rincón-Castro and Ibarra, 2011; see also Chapter 3). PTM is an important pest of potatoes and, in some cases tomato, around the world and is associated with potato tubermoth GV (PhopGV) (Lacey and Kroschel, 2009). Several field studies reviewed by Lacey and Kroschel (2009) showed PhopGV as a fair to good MCA of PTM. Significant reduction in PTM populations was achieved with potato PhopGV alone applied at 10^{12} OBs/ha and in combination with Bt in studies conducted under semifield conditions in Washington (Arthurs et al., 2008a; see Section 2.1.2 in this chapter). The efficacy of PhopGV in protecting rustic stores of tubers is reviewed by Sporleder and Lacey (2013). In simulated commercial storage, treatment of infested and noninfested tubers with suspensions of PhopGV (10^9 OBs/mL) before placing them in refrigerated storage for 5 weeks at 10°C significantly reduced the dispersal of larvae from infested to noninfested tubers and reduced damage within infested tubers. Mortality of PTM larvae in treated tubers was 87% compared to 37% in controls (Lacey et al., 2010).

Multiyear field studies in California showed that a multinucleocapsid NPV (MNPV) was as effective as methomyl and permethrin in controlling *S. exigua* on head lettuce (Gelernter et al., 1986). In an Indian study, Subramanian et al. (2010) showed that *Plutella xylostella* granulovirus (PlxyGV) was as effective as quinalphos application for managing *P. xylostella* in cauliflower. In a review on the use of baculoviruses in Brazil, Moscardi (1989) reported that AgNPV use against the soybean caterpillar, *Anticarsia gemmatalis*, increased from 2000 ha in 1982–1983 to 500,000 ha in 1987–1988. Field application rates of 10–320 larval equivalents (LE)/ha caused 70–100% larval mortality in various studies.

In field studies conducted in Thailand, *S. exigua* NPV provided significant protection from *S. exigua* damage in Chinese kale (at 5–60 × 10^{11} OBs/ha) when applied at 4-day intervals during the growing season and in garden pea (at 3.1–12.5 × 10^{11} OBs/ha) after the second application (Kolodny-Hirsch et al., 1997). HzNPV is used for the control of *Heliothis* and *Helicoverpa* spp. on different crops in various countries at a rate of 100–200 LE/ha (Federici, 1999). Different ways of improving the efficacy of

viruses were studied by combining them with other MCAs or certain enhancement agents. For example, the combination of *S. carpocapsae* and *S. exigua* MNPV caused a significantly higher mortality of the beet armyworm, *S. exigua*, than either pathogen alone in field grown soybean (Gothama et al., 1996). Boughton et al. (2001) reported that using an optical brightener improved the efficacy of *A. ipsilon* MNPV in laboratory studies but not in greenhouse or field studies. However, they found significant reduction in feeding damage by third-instar *A. ipsilon* to corn seedlings.

The need for in vivo production increases costs and generally limits the use of entomopathogenic viruses to high-value crops or in environmentally sensitive ecosystems. However, these examples demonstrate that entomopathogenic viruses can be as effective as chemical pesticides in controlling several insect pests and play an important role in IPM.

14.3 MICROBIAL CONTROL IN SMALL FRUITS

14.3.1 Bacteria

Bt products are used to manage a variety of lepidopteran pests in blueberries, caneberries, grapes, and strawberries in Canada, Europe, and the United States (Raine, 1965; Yarborough and Collins, 1997; Gajek and Jörg, 2003; Booth et al., 2007; Ramanaidu et al., 2011; Haviland, 2014; Zalom et al., 2014; Bolda and Bettiga, 2015; Varela et al., 2015). Multiple species of leafrollers in caneberries, grapes, and strawberries, the Western grapeleaf skeletonizer, *Harrisina brillians*, in grapes, the light brown apple moth, *Epiphyas postvittana*, in blueberries and grapes, the blueberry spanworm, *Itame agrillacearia*, in wild blueberry, and various lepidopteran pests in strawberries are some of the pests where Bt is an important control option around the world.

Bt products were also found effective against other small fruit pests such as the European grapevine moth, *Lobesia botrana*, in Greece (Ifoulis and Savopoulou-Soultani, 2004) and Turkey (Ataç et al., 1990), the grape leaffolder, *D. funeralis*, in California (AliNiazee and Jensen, 1973), and the oblique-banded leaf roller, *Choristoneura rosaceana*, in Canada (Li and Fritzpatrik, 1996). *L. botrana* is a very important pest of grapes in Europe and North and South America, and Bt products are considered as important and effective management options (Scalco et al., 1997; Shahini et al., 2010; Varela et al., 2010). Bt was effective in reducing fruit damage in the laboratory and numbers of *L. botrana* in multiple field studies in Greece (Roditakis, 1986). In another field study in Greece, a dust and a liquid formulation of Bt reduced *L. botrana* numbers by 68% and 78%, respectively compared to untreated control (Ifoulis and Savopoulou-Soultani, 2004). Efficacy varied among cultivars depending on the color of grapes and the compactness of bunches. Multiple Bt formulations were as effective as carbaryl in providing 95% or more control of *D. funeralis* under heavy infestations in a California vineyard (AliNiazee and Jensen, 1973). Bt formulations, Dipel WP and Foray 48B, were effective against the larvae of *C. rosaceana* in field-grown raspberries in Canada at high application rates and low spray volume (Li and Fritzpatrick, 1996).

In a California field study, biopesticides based on fermentation byproducts of *C. substugae* strain PRAA4-1 and heat-killed *Burkholderia* spp. strain A396 showed efficacy similar to that of chemical miticides against the two-spotted spider mite, *Tetranychus urticae*, in strawberries (Dara, 2015a). While chemical miticides reduced the number of eggs and mobile stages of *T. urticae* by 31–48%, *C. substugae* caused a 25% reduction at a rate of 2.2 kg/ha and *Burkholderia* spp. caused a 48.3% reduction at 18.8 L/ha. Mite infestations were fairly high during this study. While treatments provided only moderate control, *Burkholderia* spp. was as effective as some of the chemical miticides.

In California, spinosad is recommended against the citrus thrips, *Scirtothrips citri* and *Epiphyas postvittana* in blueberry (Haviland, 2014), and multiple species of thrips and lepidopteran pests in grape (Varela et al., 2015) and strawberry (Zalom et al., 2014).

14.3.2 Fungi

Natural infections of EPF are important mortality factors of small fruit pests. For example, several species of hypocrealean fungi and the entomophthoralean fungus *Entomophaga aulicae* were found infecting pests of low-bush blueberry in eastern Canada, and *B. bassiana* contributed to about 10% of the infections (Strongman et al., 1997). *B. bassiana* applied at 4.74×10^{13} conidia/ha was very effective against the blueberry spanworm, *Prochoerodes lineola*, and the blueberry flea beetle, *Altica Sylvia*, in field studies in Maine (Collins and Drummond, 1998a,b).

Fruit flies are important pests of small fruits such as grapes, bush and cane fruit, and various vegetables in different parts of the world (Booth et al., 2007). Some African isolates of *B. bassiana* and *M. anisopliae* caused high mortality (72–100%) to three species of fruit fly (*Ceratitis* spp.) in laboratory studies that used an autoinoculation device (Dimbi et al., 2003). Using entomopathogenic fungi with autoinoculation devices could improve microbial control of fruit flies in the field.

Spotted wing drosophila, *Drosophila suzukii* is an invasive pest in the United States, threatening blueberry, cherry, caneberry, grape, and strawberry industries (Hoenisch, 2010; Walsh et al., 2011). Laboratory and field studies show that EPF and EPNs could be potential MCAs for this pest. Among many pathogens evaluated, *I. fumosorosea*, *M. anisopliae*, and *S. carpocapsae* were found effective against adult and larval stages of *D. suzukii* (Bolda, personal communication; Naranjo-Lázaro et al., 2014; Woltz et al., 2015). A study conducted in Spain showed that *M. brunneum* in bait traps caused 100% mortality of adult *D. suzukii* that were attracted to the traps (Fernández-Bravo and Quesada-Moraga, 2014).

Grape phylloxera, *Daktulosphaira vitifoliae*, is a global pest of grapes, and *B. bassiana* and *M. anisopliae* are potential control agents in the United States (Garnett et al., 2001) and Europe (Kirchmair et al., 2004). These studies with local isolates indicate the microbial control potential of the fungi, however, commercializing those virulent isolates and conducting additional studies with registered formulations will benefit practical agriculture.

The use of multiple entomopathogens can sometimes increase control efficacy compared to individual pathogens even when they are applied at lower rates. Laboratory and greenhouse studies in the United Kingdom showed a synergy between *M. anisopliae* (1.1×10^9 conidia/L peat compost) and EPNs, *H. bacteriophora* (12 IJs/cup), *S. feltiae* (37 IJs/cup), and *S. kraussei* (25 IJs/cup), resulting in 100% mortality of the black vine weevil *Otiorhynchus sulcatus* larvae (Ansari et al., 2008). Similarly, combining or rotating entomopathogens with botanical or chemical pesticides can also be an important IPM strategy in reducing chemical pesticide use without compromising pest control efficacy. Recent studies in California strawberries showed that *B. bassiana* or *M. brunneum* combined with azadirachtin or low rates of chemical pesticides can be as effective as some of the chemical pesticides against *T. urticae* and the Western tarnished plant bug, *Lygus hesperus* (Dara, 2014a, 2015a,b).

Lygus hesperus is a major pest of strawberries in California, and its management with chemical pesticides is a challenge for growers (Zalom et al., 2014). In a large-scale field study against *L. hesperus*, none of the treatments reduced pest populations, but only limited the buildup (Dara, 2015b). In a 3-week treatment program, the application of a new chemical, sulfoxaflor (0.32 L/ha) limited the population buildup to 14%, while the application of diatomaceous earth (79 kg/ha) dust followed by *B. bassiana* (the low label rate of 2.4 L/ha) plus acetamiprid (0.22 kg/ha), and *M. brunneum* (1.2 L/ha) plus azadirachtin (0.59 L/ha) limited the population buildup to 17% posttreatment. Targeting adults with EPF and nymphal stages with azadirachtin provided control similar to that achieved by their chemical counterparts (adulticides and an insect growth regulator). A more recent study showed that using *B. bassiana* and *M. brunneum* with azadirachtin or pyrethrum in a rotation program with chemical or mechanical control options is a good IPM strategy for managing *L. hesperus* in strawberries (Dara, 2016a). In a different study, a combination of low label rates of *B. bassiana* (2.4 L/ha) and bifenazate (0.84 kg/ha) provided 35% control of *T. urticae*, whereas bifenazate alone at the high label rate (1.1 kg/ha) provided 31% control and the control from other chemical, botanical, and microbial treatments varied between 25% and 49% (Dara, 2015a). When control with individual products is difficult, combining multiple products can be effective by taking advantage of different modes of action to control the pests. For example, while EPF can infect all life stages of *L. hesperus*, molting can reduce the chances of infection in nymphal stages. When the botanical insect growth regulator azadirachtin, which is also an insecticide, antifeedant, and repellent, is used with EPF, different life stages of *L. hesperus* can be targeted by the combination. EPF are also considered important in managing different species of weevils in strawberries in Poland (Labanowska and Olszak, 2003). In the Polish study, *B. bassiana* reduced *O. sulcatus* and the strawberry root weevil, *O. ovatus*, numbers by 59% under field conditions.

Recent studies suggest that EPF could also impact herbivores through endophytic colonization of the host plants (Vega, 2008). *B. bassiana* colonized various strawberry plant parts through soil application (Dara et al., 2013) and negatively impacted *Myzus persicae* that were feeding on treated plants in greenhouse studies (Dara, 2016b). Similarly, compared to untreated strawberry plants, plants that received soil application of *M. brunneum* withstood *T. urticae* infestations in a greenhouse study (Dara and Dara, 2015).

EPF could promote plant growth and health in addition to their role in pest management (Liao et al., 2014). A 2013 field study in California showed that inoculating the roots of strawberry transplants with *B. bassiana* promoted the growth and health of the plants during the next few months (Dara, 2014b). Another study conducted in a commercial strawberry field in 2014 also showed that *B. bassiana* could positively influence plant growth and yield (Dara, 2014c). Preliminary results from a 2015 field study in commercial strawberries indicate that soil treatment with EPF could offer protection from plant diseases such as powdery mildew and botrytis fruit rot (Dara, 2015d). *B. bassiana*, *I. fumosorosea*, and *M. brunneum* are among various beneficial fungal and bacterial microbes evaluated in this study by treating the transplants at the time of planting followed by periodical application. In Germany, *M. anisopliae* inhibited the growth of plant pathogens, *Verticillium* spp. and *Phytophthora* spp, in strawberries (Bisutti et al., 2013). Multipurpose use of EPF to promote plant growth while protecting from diseases and arthropod pests can enhance their appeal and increase their use in temperate fruits and vegetable crops.

14.3.3 Entomopathogenic Nematodes

Small fruits are infested by a variety of weevils, worms, and borers, which are excellent candidates for control with EPNs. Both heterorhabditid and steinernematid EPNs are used to manage *O. sulcatus* and other root weevils in strawberry and other small fruits including currants and gooseberries in New England in the United States (Grubinger, 2002; Cowles et al., 2005; Abbey and Concklin, 2011). A high level of weevil (*Otiorhynchus* spp.) control was achieved with *Heterorhabditis* spp. and *S. feltiae* in cranberries (Shanks and Agudelo-Silva, 1990; Simser and Roberts, 1994; Berry and Liu, 1999; Booth et al., 2002) and strawberries (Backhaus, 1994). For example, *H. bacteriophora* at 3.9×10^{10} IJs/ha caused a 70% reduction in *O. sulcatus* populations in cranberry bogs 10 months after application (Shanks and Agudelo-Silva, 1990). Similarly, by using a high application pressure for spraying over cranberry bogs, *H. bacteriophora* and *S. carpocapsae* caused 89% and 98% mortality in *O. ovatus*, respectively (Simser and Roberts, 1994). Application rates, application techniques, irrigation practices, and favorable environmental conditions were important for the success of EPNs in these studies. For example, increased spray pressure or warmer seasonal temperatures appeared to improve EPN efficacy. Different species of white grubs (Coleoptera: Scarabaeidae) infest blueberries and cranberries, but not all of them are equally susceptible to EPNs. Using the right nematode at a susceptible life stage of the grubs is necessary to achieve effective control (Cowles et al., 2005).

Raspberry crown borer, *Pennisetia marginata*, is a pest of blackberries, boysenberries, loganberries, raspberries, salmonberries, and thimbleberries (Raine, 1962). *Steinernema feltiae* at 6×10^4 IJs/plant in 150 mL of water caused a 33–67% reduction in *P. marginata* populations within 3–5 days after application (Capinera et al., 1986). In field studies conducted in Florida, *H. bacteriophora* and *H. megidis* at 2.44×10^7 IJs/ha caused a 54–80% reduction in the pupae of the grape root borer, *Vitacea polistiformis* compared to a 47–66% reduction from the chemical pesticide chlorpyrifos (Hix, 2008, 2009). *Steinernema carpocapsae* is recommended for controlling the branch and twig borer, *Melalgus confertus*, in California grapes (Varela et al., 2015). As seen in other cases, using multiple products can improve the control efficacy. In a strawberry field study in Egypt, *H. bacteriophora* and *S. carpocapsae* applied at 2000 IJs/mL (2 L/4200 m² area) caused 80% and 70% mortality in *A. ipsilon* larvae, respectively (Fetoh et al., 2009). Adding spinosad (0.125 mg/L) to EPNs improved the mortality to 92% by *H. bacteriophora* and to 75% by *S. carpocapsae*.

A recent survey conducted in British Columbia, Canada, indicated that several conventional blueberry and cranberry growers use EPNs for managing various pests, which demonstrates their importance in conventional agriculture (Ferris and Dessureault, 2013).

14.3.4 Baculoviruses

Compared to other entomopathogens, baculoviruses are less commonly used as control options for small fruit pests because of the host specificity of most of the viruses and high cost. However, there are a few successful examples of pest management through biopesticide applications of GVs and NPVs in addition to natural outbreaks of baculoviruses. The corn earworm, *Helicoverpa zea*, is a pest of strawberries in different parts of the world (Zalom et al., 2014), and it is one of the few examples where a commercially produced baculovirus (HzSNPV) is used (Reid, 2015; CDPR, 2015a,b). *Harrisina brillians* populations are kept under control by the natural occurrence of *H. brillians* GV (HabrGV) and a tachinid parasitoid in California (Stark et al., 1999). Using ELISA, it was determined that HabrGV caused up to 97% of mortality in field collected pupae and up to 47% of the adults emerging from overwintering pupae also had HabrGV infections. Stern and Federici (1990) reported the classic biological control potential of HabrGV by horizontal and vertical transmission in *H. brillians*. Infected males transmit the virus to females during mating and infected females transmit to their offspring affecting larval survival and the fecundity of adult females of the offspring.

14.4 CONCLUSIONS

Entomopathogens play a significant role in management of small fruit and vegetable pests in temperate climate. Bt and baculoviruses for lepidopteran pests, EPF for hemipteran pests, and EPNs for coleopteran, dipteran, and lepidopteran pests are good candidates in multiple cropping systems. Small fruits and vegetables have a variety of pests that are good targets for one or more entomopathogens. Rotating and combining MCAs with chemical or botanical pesticides can provide effective pest control while reducing the reliance of chemical pesticides and the associated risk of pesticide resistance. Continuing field studies and developing IPM strategies that include microbial control as an important component will contribute to sustainable management practices for small fruit and vegetable industries.

REFERENCES

Abbey, T., Concklin, M., 2011. Beneficial Nematodes. University of Connecticut IPM Publication. http://ipm.uconn.edu/documents/raw2/Beneficial%20Nematodes/Beneficial%20Nematodes.php?aid=184.

AliNiazee, M.T., Jensen, F.L., 1973. Microbial control of the grape leaffolder with different formulations of *Bacillus thuringiensis*. J. Econ. Entomol. 66, 157–158.

Ansari, M.A., Shah, F.A., Butt, T.M., 2008. Combined use of entomopathogenic nematodes and *Metarhizium anisopliae* as a new approach for black vine weevil, *Otiorhynchus sulcatus*, control. Entomol. Exp. Appl. 129, 340–347.

Arthurs, S.P., Lacey, L.A., Pruneda, J.N., Rondon, S.I., 2008a. Semi-field evaluation of a granulovirus and *Bacillus thuringiensis* ssp. *kurstaki* for season-long control of the potato tuber moth, *Phthorimaea operculella*. Entomol. Exp. Appl. 129, 276–285.

Arthurs, S.P., Lacey, L.A., de La Rosa, F., 2008b. Evaluation of a granulovirus (PoGV) and *Bacillus thuringiensis* subsp. *kurstaki* for control of the potato tuberworm (Lepidoptera: Gelechiidae) in stored tubers. J. Econ. Entomol. 101, 1540–1546.

Ataç, O., Bulut, H., Çevik, T., 1990. Investigations on the effectiveness of *Bacillus thuringiensis* alone and in combination with low doses of carbaryl against European grape berry moth (*Lobesia botrana* Den. et Schiff). In: Proc. Sec. Turk. Nat. Cong. Biol. Con., pp. 127–135.

Backhaus, G.F., 1994. Biological control of *Otiorhynchus sulcatus* F. by use of entomopathogenic nematodes of the genus *Heterorhabditis*. Acta Hort. 364, 131–142.

Beas-Catena, A., Sánchez-Mirón, A., Garcia-Camacho, F., Contreras-Gómez, A., Molina-Grima, E., 2014. Baculovirus biopesticides: an overview. J. Anim. Plant Sci. 24, 362–373.

Benbrook, C.M., 2004. Genetically Engineered Crops and Pesticide Use in the United States: The First Nine Years. Technical Paper 7, p. 53. https://www.organic-center.org/reportfiles/Full_first_nine.pdf.

Berry, R.L., Liu, J., 1999. Biological control of cranberry insects with beneficial nematodes. In: Deziel, G., Hogan, M. (Eds.), Cranberry Research Compilation. Progress and Final Reports on Cranberry Research Conducted in 1998. Cranberry Institute, Wareham, Massachusetts, pp. 67–70.

Bielza, P., Quinto, V., Contreras, J., Torné, M., Martin, A., Espinosa, P.J., 2007. Resistance to spinosad in the western flower thrips, *Frankliniella occidentalis* (pergande), in greenhouses of south-eastern Spain. Pest Manag. Sci. 63, 682–687.

Biondi, A., Mommaerts, V., Smagghe, G., Viñuela, E., Zappalà, L., Desneux, N., 2012. The non-target impact of spinosyns on beneficial arthropods. Pest Manag. Sci. 68, 1523–1536.

Bisutti, I.L., Steen, C., Stephan, D., 2013. Does *Metarhizium anisopliae* influence strawberries in presence of pest and disease? In: 46th Annual Meetings of the Society for Invertebrate Pathology, 11–15 August, Pittsburgh, PA, USA, p. 88.

Bolda, M.P., Bettiga, L.J., 2015. UC IPM Pest Management Guidelines: Caneberries. University of California Statewide Integrated Pest Management Program. UC ANR Publication 3437, Oakland.

Booth, S.R., Tanigoshi, L.K., Shanks Jr., C.H., 2002. Evaluation of entomopathogenic nematodes to manage root weevil larvae in Washington State cranberry, strawberry, and red raspberry. Environ. Entomol. 31, 895–902.

Booth, S.R., Drummond, F.A., Groden, E., 2007. Small fruits. In: Lacey, L.A., Kaya, H.K. (Eds.), Field Manual of Techniques in Invertebrate Pathology. Springer, The Netherlands, pp. 583–598.

Boughton, A.J., Lewis, L.C., Bonning, B.C., 2001. Potential of *Agrotis ipsilon* nucleopolyhedrovirus for suppression of the black cutworm (Lepidoptera: Noctuidae) and effect of an optical brightener on virus efficacy. J. Econ. Entomol. 94, 1045–1052.

Bruck, D.J., Shapiro-Ilan, D.I., Lewis, E.E., 2005. Evaluation of application technologies of entomopathogenic nematodes for control of the black vine weevil. J. Econ. Entomol. 98, 1884–1889.

Burges, H.D., Jarrett, P., 1980. Application and distribution of *Bacillus thuringiensis* for control of tomato moth in glasshouses. In: Proc. 1979 Brit. Crop Protec. Conf. Pests and Dis., Brighton, England, pp. 433–439.

California Department of Pesticide Regulation (CDPR), 2015a. Summary of Pesticide Use Report Data 2013 Indexed by Commodity. Sacramento http://www.cdpr.ca.gov/docs/pur/pur13rep/comrpt13.pdf#page=131.

California Department of Pesticide Regulation (CDPR), May, 2015b. Summary of Pesticide Use Report Data 2013, Indexed by Commodity. 754 pp. http://www.cdpr.ca.gov/docs/pur/pur13rep/comrpt13.pdf.

Capinera, J.L., 2007. Banded Cucumber Beetle, *Diabrotica Balteata* LeConte (Insecta: Coleoptera: Chrysomelidae). University of Florida. Featured Creatures. Ext. Pub. No. EENY-93 http://entnemdept.ufl.edu/creatures/veg/bean/banded_cucumber_beetle.htm.

Capinera, J.L., Cranshaw, W.S., Hughes, H.G., 1986. Suppression of raspberry crown borer *Pennisetia marginata* (Harris) (Lepidoptera: Sesiidae) with soil applications of *Steinernema feltiae* (Rhabditida: Steinernematidae). J. Invertebr. Pathol. 48, 257–258.

Capinera, J.L., Pelissier, D., Menuout, G.S., Epsky, N.D., 1988. Control of black cutworm, *Agrotis ipsilon* (Lepidoptera: Noctuidae) with entomogenous nematodes (Nematoda: Steinernematidae, Heterorhabditidae). J. Invertebr. Pathol. 52, 427–435.

Carpenter, J., Felsot, A., Goode, T., Hammig, M., Onstad, D., Sankula, D., 2002. Comparative Environmental Impacts of Biotechnology-Derived and Traditional Soybean, Corn, and Cotton Crops. Council for Agricultural Science and Technology, Ames, Iowa. 199 pp. http://www.cast-science.org/download.cfm?PublicationID=2895&File=103033ce57ea286231a17b4738617785411dTR.

Collier, R.H., Finch, S., Davies, G., 2001. Pest insect control in organically-produced crops of field vegetables. Meded. Rijksuniv. Gent Fak. Landbouwkd. Toegep. Biol. Wet. 66, 259–267.

Collins, J.A., Drummond, F.A., 1998a. Blueberry spanworm control, 1997. Art. Manage. Tests 23, 47.

Collins, J.A., Drummond, F.A., 1998b. Blueberry flea beetle control, 1997. Art. Manage. Tests 23, 48.

Correa-Cuadros, J.P., Rodriquez-Bocanegra, M.X., Sáenz-Aponte, A., 2014. Susceptibility of *Plutella xylostella* (Lepidoptera: Plutellidae; Linnaeus 1758) to *Beauveria bassiana* Bb9205, *Metarhizium anisopliae* Ma9236 and *Heterorhabditis bacteriophora* HNI0100. Univ. Sci. 19, 277–285.

Cowles, R.S., Polavarapu, S., WIlliams, R.N., Thies, A., Ehlers, R.-U., 2005. Soft fruit applications. In: Grewal, P.S., Ehlers, R.-U., Shapiro-Ilan, D.I. (Eds.), Nematodes as Biocontrol Agents. CABI, Oxfordshire, UK, pp. 231–254.

Cranshaw, W., 1998. Pests of the West: Prevention and Control for Today's Garden and Small Farm. Fulcrum Publishing, Canada. 249 pp.

Creighton, C.S., Fassuliotis, F., 1985. *Heterorhabditis* sp. (Nematoda: Heterorhabditidae): a nematode isolated from the banded cucumber beetle *Diabrotica balteata*. J. Nematol. 17, 150–153.

Dara, S.K., April 2013. Field trials for managing aphids on broccoli and western flower thrips on lettuce. CAPCA Adviser 16, 29–32.

Dara, S.K., February 2014a. New strawberry IPM studies with chemical, botanical, and microbial solutions. CAPCA Adviser 17, 34–37.

Dara, S.K., 2014b. A beneficial fungus promoting strawberry plant growth and health. Veg. West 18, 18–19.

Dara, S.K., 2014c. *Beauveria bassiana* and California strawberries: endophytic, mycorrhizal, and entomopathogenic interactions. In: 47th Annual Meetings of the Society for Invertebrate Pathology, 3–7 August, Mainz, Germany, p. 148.

Dara, S.K., August 2015a. Twospotted spider mite and its management in strawberries. CAPCA Adviser 18, 56–58.

Dara, S.K., February 2015b. Integrating chemical and non-chemical solutions for managing lygus bug in California strawberries. CAPCA Adviser 18, 40–44.

Dara, S.K., 2015c. Reporting the Occurrence of Rice Root Aphid and Honeysuckle Aphid and Their Management in Organic Celery. University of California Cooperative Extension eNewsletter. Strawberries and Vegetables http://ucanr.edu/blogs/blogcore/postdetail.cfm?postnum=18740.

Dara, S.K., 2015d. Non-entomopathogenic role of entomopathogenic fungi in strawberry production. In: 48th Annual Meetings of the Society for Invertebrate Pathology, 9–13 August, Vancouver, Canada, p. 103.

Dara, S.K., 2016a. IPM solutions for insect pests in California strawberries: efficacy of botanical, chemical, mechanical, and microbial options. CAPCA Adviser 19, 40–46.

Dara, S.K., 2016b. Endophytic *Beauveria bassiana* negatively impacts green peach aphids on strawberries. University of California Cooperative Extension eNewsletter. Strawberries and Vegetables http://ucanr.edu/blogs/blogcore/postdetail.cfm?postnum=21711.

Dara, S.K., Dara, S.R., 2015. Soil Application of the Entomopathogenic Fungus *Metarhizium brunneum* Protects Strawberry Plants from Spider Mite Damage. University of California Cooperative Extension eNewsletter. Strawberries and Vegetables http://ucanr.edu/blogs/blogcore/postdetail.cfm?postnum=16821.

Dara, S.K., Dara, S.R., Dara, S.S., 2013. Endophytic colonization and pest management potential of *Beauveria bassiana* in strawberries. J. Berry Res. 3, 203–211.

Dara, S.K., Dara, S.S.R., Dara, S.S., 2014. Entomopathogenic fungi as plant growth enhancers. In: 47th Annual Meeting of the Society for Invertebrate Pathology, 3–7 August, Mainz, Germany, p. 103.

Del C. Bandas, L., Dario, C.P., Guillermo, C.S., 2004. Efecto de la asociación patilla (*Citrullus lanatus*) con maíz (*Zea mays*) sobre la población y el daño causado por tres insectos plaga y el rendimiento de estos cultivos en la Ciénaga Grande de Lorica, Córdoba. Rev. Colomb. Entomol. 30, 161–169.

Dimbi, S., Maniania, N.K., Lux, S.A., Ekesi, S., Mueke, J.K., 2003. Pathogenicity of *Metarhizium anisopliae* (Metsch.) Sorokin and *Beauveria bassiana* (Balsamo) Vuillemin, to three adult fruit fly species: *Ceratitis capitata* (Weidemann), *C. rosa* var. *fasciventris* Karsch and *C. cosyra* (Walker) (Diptera: Tephritidae). Mycopathol. 156, 375–382.

Douches, D.S., Li, W., Zarka, K., Coombs, J., Pett, W., Grafius, E., El-Nasr, T., 2002. Development of Bt-cry5 insect resistant potato lines "Spunta-G2" and "Spunta-G3". HortScience 37, 1103–1107.

Douches, D.S., Coombs, J., Lacey, L.A., Felcher, K., Pett, W., 2011. Choice and no-choice evaluations of transgenic potatoes for resistance to potato tuber moth (*Phthorimaea operculella* Zeller) in the laboratory and field. Am. J. Potato Res. 88, 91–95.

Dunham, B., 2014. Microbial Pesticides: A Key Role in the Multinational Portfolio. New Ag International, pp. 32–36. Sep/Oct issue.

Fargues, J., Goettel, M.S., Smits, N., Ouedraogo, A., Vidal, C., Lacey, L.A., Lomer, C.J., Rougier, M., 1996. Variability in susceptibility to simulated sunlight of conidia among isolates of entomopathogenic Hyphomycetes. Mycopathol. 135, 171–181.

Faria, M., Wraight, S.P., 2001. Biological control of *Bemisia tabaci* with fungi. Crop Prot. 20, 767–778.

Federici, B.A., 1999. A perspective on pathogens as biological control agents for insect pests. In: Bellows, T.S., Fisher, T.W. (Eds.), Handbook of Biological Control. Academic Press, San Diego, pp. 517–548.

Fernández-Bravo, M., Quesada-Moraga, E., 2014. An experimental autoinoculation device to control an invasive Asiatic pest, *Drosophila suzukii*. In: 47th Annual Meetings of the Society for Invertebrate Pathology, 3–7 August, Mainz, Germany, p. 114.

Ferris, K., Dessureault, M., 2013. Organic Blueberry and Cranberry Production in British Columbia: Gaps Analysis. Project report to the Organic Sector Development Program, Fraserland Organics, British Columbia Blueberry Council, 29 pp. http://certifiedorganic.bc.ca/programs/osdp/I-176_Berry_GAP_Analysis_Final_Report.pdf.

Fetoh, B.E.A., Khaled, A.S., El-Nagar, T.F.K., 2009. Combined effect of entomopathogenic nematodes and biopesticides to control the greasy cut worm, *Agrotis ipsilon* (Hufn.) in the strawberry fields. Egypt. Acad. J. Biol. Sci. 2, 227–236.

Gajek, D., Jörg, E., 2003. Status of integrated production of soft fruit in Europe. In: Gordon, S.C., Cross, J.V. (Eds.), Integrated Plant Protection in Orchards – Soft FruitsIOBC/WPRS Bull., 26, pp. 1–6.

Garnett, J., Walker, M.A., Kocsis, L., Omer, A.D., 2001. Biology and management of grape phylloxera. Annu. Rev. Entomol. 46, 387–412.

Gelernter, W.D., Toscano, N.C., Kido, K., Federici, B.A., 1986. Comparison of a nuclear polyhedrosis virus and chemical insecticides for control of the beet armyworm (Lepidoptera: Noctuidae) on head lettuce. J. Econ. Entomol. 79, 714–717.

Gerling, D., Mayer, R. (Eds.), 1996. *Bemisia* 1995: Taxonomy, Biology, Damage, and Management. Intercept, Andover. 702 pp.

Ghislain, M., Lagnaoui, A., Walker, T., 2003. Fulfilling the promise of Bt potato in developing countries. J. New Seeds 5, 93–113.

Glare, T., Caradus, J., Gelernter, W., Jackson, T., Keyhani, N., Köhl, J., Marrone, P., Morin, L., Stewart, A., 2012. Have biopesticides come of age? Trends in Biotechnol. 30, 250–258.

González-Cabrera, J., Mollá, O., Montón, H., Urbaneja, A., 2011. Efficacy of *Bacillus thuringiensis* (Berliner) in controlling the tomato borer, *Tuta absoluta* (Meyrick) (Lepidoptera: Gelechiidae). BioControl 56, 71–80.

Gothama, A.A., Lawrence, G.W., Sikorowski, P.P., 1996. Activity and persistence of *Steinernema carpocapsae* and *Spodoptera exigua* nuclear polyhedrosis virus against *S. exigua* larvae on soybean. J. Nematol. 28, 68–74.

Grubinger, V., 2002. Beneficial Nematodes for Black Vine Weevil Control in Strawberry. University of Vermont Ext. Pub. http://www.uvm.edu/vtvegandberry/factsheets/blackvineweevil.html.

Haviland, D.R., 2014. UC IPM Pest Management Guidelines: Blueberry. University of California Statewide Integrated Pest Management Program. UC ANR Publication 3542, Oakland.

Hix, R.L., 2008. Grape root borer control with nematodes, 2007. Art. Manage. Tests 33. http://dx.doi.org/10.1093/amt/33.1.C8.

Hix, R.L., 2009. Grape root borer control with nematodes, 2008. Art. Manage. Tests 34. http://dx.doi.org/10.4182/amt.2009.C17.

Hoenisch, R., October 2010. Spotted Wing Drosophila Found in California, Oregon, Washington, and British Columbia. Foundation Plant Services Grape Program Newsletter. http://iv.ucdavis.edu/files/71831.pdf.

Ifoulis, A.A., Savopoulou-Soultani, M., 2004. Biological control of *Lobesia botrana* (Lepidoptera: Tortricidae) larvae by using different formulations of *Bacillus thuringiensis* in 11 vine cultivars under field conditions. J. Econ. Entomol. 97, 340–343.

Islam, M.T., Castle, S.J., Ren, S., 2010. Compatibility of the insect pathogenic fungus *Beauveria bassiana* with neem against sweetpotato whitefly, *Bemisia tabaci*, on eggplant. Entomol. Exp. Appl. 134, 28–34.

James, R.R., 2003. Combining azadirachtin and *Paecilomyces fumosoroseus* (Deuteromycotina: Hyphomycetes) to control *Bemisia argentifolii* (Homoptera: Aleyrodidae). J. Econ. Entomol. 6, 25–30.

Janmaat, A.F., Myers, J., 2003. Rapid evolution and the cost of resistance to *Bacillus thuringiensis* in greenhouse populations of cabbage looper, *Trichoplusia ni*. Proc. Biol. Sci. 270, 2263–2270.

Jarrett, P., Burges, H.D., 1982. Control of tomato moth *Lacanobia oleracea* by *Bacillus thuringiensis* on glasshouse tomatoes and the influence of larval behaviour. Entomol. Exp. Appl. 31, 239–244.

Jee, H.J., 2007. Current Status of Bio-fertilizers and Bio-pesticides Development, Farmer's Acceptance and Their Utilization in Korea. Food and Fertilizer Technology Center (FFTC), p. 601. http://en.fftc.org.tw/htmlarea_file/library/20110712072318/eb601.pdf.

Kabaluk, T., Goettel, M., Erlandson, M., Ericsson, J., Duke, G., Vernon, B., 2005. *Metarhizium anisopliae* as a biological control for wireworms and a report of some other naturally-occurring parasites. IOBC/WPRS Bull. 28, 109–115.

Kirchmair, M., Huber, L., Porten, M., Rainer, J., Strasser, H., 2004. *Metarhizium anisopliae*, a potential agent for the control of grape phylloxera. BioControl 49, 295–303.

Koike, S.T., LeStrange, M., Chaney, W.E., 2009. UC IPM Pest Management Guidelines: Spinach. UC ANR Publication 3467.

Kirk, A.A., Lacey, L.A., 1996. A systematic approach to foreign exploration for natural enemies of *Bemisia*. In: Gerling, D., Mayer, R. (Eds.), *Bemisia* 1995: Taxonomy, Biology, Damage, and Management. Intercept, Andover, pp. 531–533.

Kolodny-Hirsch, D.M., Sitchawat, T., Jansiri, T., Chenrchaivachirakul, A., Ketunuti, U., 1997. Field evaluation of a commercial formulation of the *Spodoptera exigua* (Lepidoptera: Noctuidae) nuclear polyhedrosis virus for control of beet armyworm on vegetable crops in Thailand. Biocontrol Sci. Tech. 7, 475–488.

Koppenhöfer, A.M., Rodriguez-Saona, C.R., Polavarapu, S., Holdcraft, R.J., 2008. Entomopathogenic nematodes for control of *Phyllophaga georgiana* (Coleoptera: Scarabaeidae) in cranberries. Biocontrol Sci. Tech. 18, 21–31.

Krieg, A., Huger, A.M., Langenbruch, G.A., Schnetter, W., 1983. *Bacillus thuringiensis* var. *tenebrionis*: Ein neuer, gegenüber larven von coleopteren wirksamer pathotyp. Z. Angew. Entomol. 96, 500–508.

Krishna, V.V., Qaim, M., 2008. Potential impacts of Bt eggplant on economic surplus and farmers' health in India. Agri. Econ. 38, 167–180.

Kroschel, J., Koch, W., 1996. Studies on the use of chemicals, botanicals and *Bacillus thuringiensis* in the management of the potato tuber moth in potato stores. Crop Prot. 15, 197–203.

Kroschel, J., Lacey, L.A. (Eds.), 2008. Integrated Pest Management for the Potato Tuber Moth: A Potato Pest of Global Proportion, Tropical Agriculture 20, Advances in Crop Research 10. Margraf Publishers, Weikersheim, Germany. 147 pp.

Labanowska, B.H., Olszak, R.W., 2003. Soil pests and their chemical and biological control on strawberry plantations in Poland. In: Gordon, S.C., Cross, J.V. (Eds.), Integrated Plant Protection in Orchards – Soft FruitsIOBC/WPRS Bull., 26, pp. 93–99.

Lacey, L.A., Kroschel, J., 2009. Microbial control of the potato tuber moth (Lepidoptera: Gelechiidae). Fruit Veg. Cereal Sci. Biotechnol. 3, 46–54.

Lacey, L.A., Mercadier, G., 1998. The effect of selected allelochemicals on germination of conidia and blastospores and mycelial growth of the entomopathogenic fungus, *Paecilomyces fumosoroseus* (Deuteromycotina: Hyphomycetes). Mycopathologia 142, 17–25.

Lacey, L.A., Kirk, A.A., Hennessey, R.D., 1993. Foreign exploration for natural enemies of *Bemisia tabaci* and implementation in integrated control programs in the United States. In: Proc. Internat. Conf. on Pests in AgricultureAssoc. Nat. Protect. Plantes, vol. 1, pp. 351–360.

Lacey, L.A., Horton, D.R., Chauvin, R.L., Stocker, J.M., 1999a. Comparative efficacy of *Beauveria bassiana*, *Bacillus thuringiensis*, and aldicarb for control of Colorado potato beetle in an irrigated desert agroecosystem and their effects on biodiversity. Entomol. Exp. Appl. 93, 189–200.

Lacey, L.A., Kirk, A.A., Millar, L., Mercadier, G., Vidal, C., 1999b. Ovicidal and larvicidal activity of conidia and blastospores of *Paecilomyces fumosoroseus* (Deuteromycotina: Hyphomycetes) against *Bemisia argentifolii* (Homoptera: Aleyrodidae) with a description of a bioassay system allowing prolonged survival of control insects. Biocontrol Sci. Technol. 9, 9–18.

Lacey, L.A., Wraight, S.P., Kirk, A.A., 2008. Entomopathogenic fungi for control of *Bemisia tabaci* biotype B: foreign exploration, research and implementation. In: Gould, J., Hoelmer, K., Goolsby, J. (Eds.), Classical Biological Control of *Bemisia tabaci* in the United States – A Review of Interagency Research and Implementation. Springer, Dordrecht, The Netherlands, pp. 33–69.

Lacey, L.A., Kroschel, J., Wraight, S.P., Goettel, M.S., 2009. An introduction to microbial control of insect pests of potato. Fruit, Veg. Cereal Sci. Biotechnol. 3, 20–24.

Lacey, L.A., Headrick, H.L., Horton, D.R., Schriber, A., 2010. Effect of granulovirus on the mortality and dispersal of potato tuber worm (Lepidoptera: Gelechiidae) in refrigerated storage warehouse conditions. Biocontrol Sci. Technol. 20, 437–447.

Lacey, L.A., Liu, T.-X., Buchman, J.E., Munyaneza, J.E., Goolsby, J.A., Horton, D.R., 2011. Entomopathogenic fungi (Hypocreales) for control of potato psyllid, *Bactericera cockerelli* (Šulc) (Hemiptera: Triozidae) in an area endemic for zebra chip disease of potato. Biol. Control 56, 271–278.

Lacey, L.A., Grzywacz, D., Shapiro-Ilan, D.I., Frutos, R., Goettel, M.S., Brownbridge, M., 2015. Insect pathogens as biological control agents: back to the future. J. Invertebr. Pathol. 132, 1–41.

Lebedev, G., Abo-Moch, F., Gafni, G., Ben-Yakir, D., Ghanim, M., 2013. High-level of resistance to spinosad, emamectin benzoate and carbosulfuran in populations of *Thrips tabaci* collected in Israel. Pest Manag. Sci. 69, 274–277.

Lewis, L.C., Bruck, D.J., Jackson, J.J., 2007. Microbial control of insect pests of corn. In: Lacey, L.A., Kaya, H.K. (Eds.), Field Manual of Techniques in Invertebrate Pathology. Springer, The Netherlands, pp. 375–392.

Li, S.Y., Fitzpatrick, S.M., 1996. The effects of application rate and spray volume on efficacy of two formulations of *Bacillus thuringiensis* Berliner var. *kurstaki* against *Choristoneura rosaceana* (Harris) (Lepidoptera: Tortricidae) on raspberries. Can. Entomol. 128, 605–612.

Liao, X., O'Brien, T.R., Fang, W., St Leger, R.J., 2014. The plant beneficial effects of *Metarhizium* species correlate with their association with roots. Appl. Microbiol. Biotechnol. 98, 7089–7096.

Lössbroek, T.G., Theunissen, J., 1985. The entomogenous nematode *Neoaplectana bibionis* as a biological control agent of *Agrotis segetum* in lettuce. Entomol. Exp. Appl. 39, 261–264.

Machado, J.C., Vieira, B.S., Lopes, E.A., Canedo, E.J., 2010. *Paecilomyces lilacinus* and cow manure for the control of *Meloidogyne incognita* in tomato and lettuce. Nematol. Bras. 34, 231–235.

Magalhães, G.O., Vacari, A.M., Laurentis, V.L., De Bortoli, S.A., Planczyk, R.A., 2015. Interactions of *Bacillus thuringiensis* bioinsecticides and the predatory stink bug *Podisus nigrispinus* to control *Plutella xylostella*. J. Appl. Entomol. 139, 123–133.

Martin, P.A., Gundersen-Rindal, D., Blackburn, M., Buyer, J., 2007a. *Chromobacterium subtsugae* sp. nov., a betaproteobacterium toxic to Colorado potato beetle and other insect pests. Int. J. Syst. Evol. Microbiol. 57, 993–999.

Martin, P.A., Hirose, E., Aldrich, J.R., 2007b. Toxicity of *Chromobacterium subtsugae* to southern green stink bug (Heteroptera: Pentatomidae) and corn rootworm (Coleoptera: Chrysomelidae). J. Econ. Entomol. 100, 680–684.

Martin, T.A., Palumbo, J.C., Dara, S.K., Natwick, E.T., October 2013. Learning more about Bagrada bug management in cole crops. CAPCA Adviser 16, 36–45.

Maru, A.K., Siddiqui, A.U., 2012. Management of Root-Knot Nematode Using Entomopathogenic Nematodes. LAP LAMBERT Academic Publishing. 104 pp.

Maynard, D.N., Hochmuth, G.J., Vavrina, C.S., Stall, W.M., Kucharek, T.A., Stansly, P.A., Taylor, T.G., Smith, S.A., Smajstrla, A.G., 1999. Lettuce, Endive, Escarole Production in Florida. University of Florida. Coop. Ext. Serv. HS728, 10 pp. http://university.uog.edu/cals/people/pubs/lfygrns/cv12600.pdf.

Miles, C., Blethen, C., Gaugler, R., Shapiro-Ilan, D., Murray, T., 2012. Using Entomopathogenic Nematodes for Crop Insect Pest Control. Washington State University. Pacific Northwest Ext. Pub. No. PNW544 http://cru.cahe.wsu.edu/CEPublications/PNW544/PNW544.pdf.

Miret, R., Rodriguez, M., 1984. Incidencia de plagas y enfermedades en 8 géneros de leguminosas. Pastos Forrajes 7, 177–188.

Molina, J.P., Dolinski, C., Souza, R.M., Lewis, E.E., 2007. Effect of entomopathogenic nematodes (Rhabditida: Steinernematidae and Heterorhabditidae) on *Meloidogyne mayaguensis* Rammah and Hirschmann (Tylenchida: Meloidoginidae) infection in tomato plants. J. Nematol. 39, 338–342.

Moscardi, F., 1989. Use of viruses for pest control in Brazil: the case of the nuclear polyhedrosis virus of the soybean caterpillar, *Anticarsia gemmatalis*. Mem. Inst. Oswaldo Cruz 84, 51–56.

Mosslet, M.A., Dunn, E., 2008. Florida Crop/Pest Management Profile: Lettuce. UF IFAS Extension. CIR1460 http://edis.ifas.ufl.edu/pdffiles/PI/PI07000.pdf.

Naranjo-Lázaro, J.M., Mellin-Rosas, M.A., González-Padilla, V.D., Sánchez-González, J.A., Moreno-Carrillo, G., Arredondo-Bernal, H.C., 2014. Susceptibility of *Drosophila suzukii* Matsumura (Diptera: Drosophilidae) to entomopathogenic fungi. Southwest. Entomol. 39, 201–203.

Natwick, E.T., Bentley, W.J., Chaney, W.E., Toscano, N.C., 2010a. UC IPM Pest Management Guidelines: Cole Crops. University of California Statewide Integrated Pest Management Program. UC ANR Publication 3442, Oakland.

Natwick, E.T., Chaney, W.E., Toscano, N.C., 2010b. UC IPM Pest Management Guidelines: Lettuce. University of California Statewide Integrated Pest Management Program. UC ANR Publication 3450, Oakland.

Natwick, E.T., Summers, C.G., Haviland, D.R., Godfrey, L.D., 2010c. UC IPM Pest Management Guidelines: Sugarbeet. University of California Statewide Integrated Pest Management Program. UC ANR Publication 3469, Oakland.

Natwick, E.T., Stoddard, C.S., Zalom, F.G., Trumble, J.T., Miyao, G., Stapleton, J.J., 2013. UC IPM Pest Management Guidelines: Tomato. University of California Statewide Integrated Pest Management Program. UC ANR Publication 3470, Oakland.

Nielson, O., 2003. Susceptibility of *Delia radicum* to steinernematid nematodes. BioControl 48, 431–446.

North American Plant Protection Organization (NAPPO), 2012. Surveillance Protocol for the Tomato Leaf Miner, *Tuta Absoluta*, for NAPPO Member Countries. 18 pp. http://www.aphis.usda.gov/import_export/plants/plant_exports/downloads/Tuta_absoluta_surveillanceprotocol_08-06-2012-e.pdf.

Osborne, L.S., Storey, G.K., McCoy, C.W., Walter, J.F., 1990. Potential for controlling the sweetpotato whitefly, *Bemisia tabaci*, with the fungus, *Paecilomyces fumosoroseus*. In: Proc. Vth Int. Colloq. Invertebr. Pathol. Microb. Control, Adelaide, Australia, pp. 386–390.

Parker, J., Miles, C., Murray, T., Snyder, W., 2012. Organic Management of Flea Beetles. Washington State University. Pacific Northwest Ext. Pub. No. PNW640 http://cru.cahe.wsu.edu/CEPublications/PNW640/PNW640.pdf.

Pérez, E.E., Lewis, E.E., 2002. Use of entomopathogenic nematodes to suppress *Meloidogyne incognita* on greenhouse tomatoes. J. Nematol. 34, 171–174.

Poprawski, T.J., Carruthers, R.I., Speese III, J., Vacek, D.C., Wendel, L.E., 1997. Early-season applications of the fungus *Beauveria bassiana* and introduction of the hemipteran predator *Perillus bioculatus* for control of Colorado potato beetle. Biol. Control 10, 48–57.

Priyatno, T.P., Ibrahim, Y.B., 2004. Pathogenicity of *Paecilomyces fumosoroseum* (Wise) Brown & Smith, *Beauveria bassiana* (Bals.) Vuill. and *Metarhizium anisopliae* (Metsch.) Sorokin on the striped flea beetle *Phyllotreta striolata* F. (Coleoptera: Chrysomelidae). Pertanika J. Trop. Agric. Sci. 27, 171–177.

Prayogo, S.Y., 2014. Integration of botanical pesticide and entomopathogenic fungi to control the brown stink bug *Riptortus linearis* F. (Hemiptera: Alydidae) in soybean. J. HPT Trop. 14, 41–50.

Raine, J., 1962. Life history and behavior of the raspberry crown borer *Bembecia marginata* (Harr.) (Lepidotera: Aegeriidae). Can. Entomol. 94, 1216–1222.

Raine, J., 1965. Control of *Dasystoma salicellum*, a new pest of blueberries in British Columbia. Can. J. Plant Sci. 45, 243–245.

Raman, K.V., Booth, R.H., Palacios, M., 1987. Control of potato tuber moth *Phthorimaea operculella* (Zeller) in rustic potato stores. Trop. Sci. 27, 175–194.

Ramanaidu, K., Hardman, J.M., Percival, D.C., Cutler, G.C., 2011. Laboratory and field susceptibility of blueberry spanworm (Lepidoptera: Geometridae) to conventional and reduced-risk insecticides. Crop Prot. 30, 1643–1648.

Ratnakar, A., Namnath, J., Marrone, P., 2013. *Chromobacterium* Formulations, Compositions, Metabolites and Their Uses. Patent WO213062977 A1.

Reed, G.L., Jensen, A.S., Riebe, J., Head, G., Duan, J.J., 2001. Transgenic Bt potato and conventional insecticides for Colorado potato beetle management: comparative efficacy and non-target impacts. Entomol. Exp. Appl. 100, 89–100.

Reid, A., 2015. Using Pesticides in Strawberry Production – Your Responsibilities as a Grower. Department of Agriculture and Food, Government of Western Australia. https://www.agric.wa.gov.au/strawberries/using-pesticides-strawberry-production-%E2%80%93-your-responsibilities-grower.

del Rincón-Castro, M.C., Ibarra, J.E., 2011. Entomopathogenic viruses. In: Rosas-Garcïa, N.M. (Ed.), Biological Control of Insect Pests. CABI, Oxfordshire, UK, pp. 29–64.

Robinson, J., 2009. Insect management. In: Masabni, J.G., Dainello, F.J. (Eds.), Texas Vegetable Growers Handbook. , fourth ed. Texas A&M University, College Station, TX. http://aggie-horticulture.tamu.edu/vegetable/guides/texas-vegetable-growers-handbook/chapter-vi-insect-management/.

Roditakis, N.E., 1986. Effectiveness of *Bacillus thuringiensis* Berliner var. kurstaki on the grape berry moth *Lobesia botrana* Den. and Shiff. (Lepidoptera, Tortricidae) under field and laboratory conditions in Crete. Entomol. Hell. 4, 31–35.

Scalco, A., Charmillot, P.J., Pasquier, D., Antonin, P., 1997. Comparaison de produits a base de *Bacillus thuringiensis* dans la lutte contre les vers de la grappe: du laboratoire au vignoble. Rev. Duisse Vitic. Arboric. Hortic. 29, 345–350.

Shahini, S., Kullaj, E., Cakalli, A., Lazarevska, S., Pfeiffer, D.G., Gumeni, F., 2010. Population dynamics and biological control of European grapevine moth (*Lobesia botrana*: Lepidoptera: Tortricidae) in Albania using different strains of *Bacillus thuringiensis*. Int. J. Pest Manag. 56, 281–286.

Shanks Jr., C.H., Agudelo-Silva, F., 1990. Field pathogenicity and persistence of heterorhabditid and steinernematid nematodes (Nematoda) infecting black vine weevil larvae (Coleoptera: Curculionidae) in cranberry bogs. J. Econ. Entomol. 83, 107–110.

Shelton, A.M., Robertson, J.L., Tang, J.D., Perez, C., Eigenbrode, S.D., Preisler, H.K., Wilsey, W.T., Cooley, R.J., 1993. Resistance of diamondback moth (Lepidoptera: Plutellidae) to *Bacillus thuringiensis* subspecies in the field. J. Econ. Entomol. 86, 697–705.

Shelton, A.M., Zhao, J.-Z., Roush, R.T., 2002. Economic, ecological, food safety, and social consequences of the deployment of BT transgenic plants. Annu. Rev. Entomol. 47, 845–881.

Shelton, A.M., Fuchs, M., Shotkoski, F.A., 2008. Transgenic vegetables and fruits for control of insects and insect-vectored pathogens. In: Romeis, J., Shelton, A.M., Kennedy, G.G. (Eds.), Integration of Insect-Resistant Genetically Modified Crops within IPM Programs. Springer, The Netherlands, pp. 249–271.

Simser, D., Roberts, S., 1994. Suppression of strawberry root weevil, *Otiorhynchus ovatus*, in cranberries by entomopathogenic nematodes (Nematoda: Steinernematidae and Heterorhabditidae). Nematologica 40, 456–462.

Sinha, B., 2012. Global biopesticide research trends: a bibliometric assessment. Ind. J. Agric. Sci. 82, 95–101.

Sporleder, M., Lacey, L.A., 2013. Biopesticides. In: Giordanengo, P., Vincent, C., Alyokhin, A. (Eds.), Insect Pests of Potato: Global Perspectives on Biology and Management. Academic Press, Amsterdam, pp. 463–497.

Stark, D.M., Purcell, A.H., Mills, N.J., 1999. Natural occurrence of *Ametadoria misella* (Diptera: Tachinidae) and the granulovirus of *Harrisina brillians* (Lepidoptera: Zygaenidae) in California. Environ. Entomol. 28, 868–875.

Stern, V.M., Federici, B.A., 1990. Granulosis virus: biological control for western grapeleaf skeletonizer. Cal. Agri. 44, 21–22.

Strongman, D., MacKenzie, K., Dixon, P., 1997. Entomopathogenic fungi in lowbush blueberry fields. In: Proc. Sixth Intl. Symp. *Vaccinium* Cult. Acta Hort., vol. 446, pp. 465–476.

Subramanian, S., Rabindra, R.J., Sathiah, N., 2010. Economic threshold for the management of *Plutella xylostella* with granulovirus in cauliflower ecosystem. Phytoparasitica 38, 5–17.

Szewczyk, B., Rabalski, L., Krol, E., Sihler, W., de Souza, M.L., 2009. *Baculovirus* biopesticides – a safe alternative to chemical protection of plants. J. Biopestic. 2, 209–216.

Tabashnik, B.E., Cushing, N.L., Finson, N., Johnson, M.W., 1990. Field development of resistance to *Bacillus thuringiensis* in diamondback moth (Lepidoptera: Plutellidae). J. Econ. Entomol. 83, 1671–1676.

United States Department of Agriculture-National Agricultural Statistics Service (USDA-NASS), 2010. Organic Production Survey (2008). 2007 Census of Agriculture. http://www.agcensus.usda.gov/Publications/2007/Online_Highlights/Organics/ORGANICS.pdf.

Vandenberg, J.D., Wraight, S.P., Shelton, A.M., 2007. Application and evaluation of entomopathogens in crucifers and cucurbits. In: Lacey, L.A., Kaya, H.K. (Eds.), Field Manual of Techniques in Invertebrate Pathology, second ed. Springer, The Netherlands, pp. 361–374.

Varela, L.G., Smith, R.J., Cooper, M.L., Hoenisch, R.W., March/April 2010. European Grapevine Moth, *Lobesia Botrana*, in Napa Valley Vineyards. Practical Winery & Vineyard, pp. 1–5.

Varela, L.G., Haviland, D.R., Bentley, W.J., Zalom, F.G., Bettiga, L.J., Smith, R.J., Daane, K.M., 2015. UC IPM Pest Management Guidelines: Grape. University of California Statewide Integrated Pest Management Program. UC ANR Publication 3448, Oakland.

Vega, F.E., 2008. Insect pathology and fungal endophytes. J. Invertebr. Pathol. 98, 277–279.

Vidal, C., Lacey, L.A., Fargues, J., 1997a. Pathogenicity of *Paecilomyces fumosoroseus* (Deuteromycotina: Hyphomycetes) against *Bemisia argent ifolii* (Homoptera: Aleyrodidae) with a description of a bioassay method. J. Econ. Entomol. 90, 765–772.

Vidal, C., Fargues, J., Lacey, L.A., 1997b. Intraspecific variability of *Paecilomyces fumosoroseus*: effect of temperature on vegetative growth. J. Invertebr. Pathol. 70, 18–26.

Vidal, C., Fargues, J., Lacey, L.A., Jackson, M.A., 1998a. Effect of various liquid culture media on morphology, growth, propagule production, and pathogenic activity to *Bemisia argentifolii* of the entomopathogenic *Paecilomyces fumosoroseus*. Mycopathologia 143, 33–46.

Vidal, C., Osborne, L.S., Lacey, L.A., Fargues, J., 1998b. Effect of host plant on the potential of *Paecilomyces fumosoroseus* (Deuteromycotina: Hyphomycetes) for controlling the silverleaf whitefly, *Bemisia argentifolii* (Homoptera: Aleyrodidae) in greenhouses. Biol. Control 12, 191–199.

von Arx, R., Goueder, J., Cheikh, M., Temime, A.B., 1987. Integrated control of potato tubermoth *Phthorimaea operculella* (Zeller) in Tunisia. Insect Sci. Appl. 8, 989–994.

Walsh, D.B., Bolda, M.P., Goodhue, R.E., Dreves, A.J., Lee, J., Bruck, D.J., Walton, V.M., O'Neal, S.D., Frank, G.Z., 2011. *Drosophila suzukii* (Diptera: Drosophilidae): invasive pest of ripening soft fruit expanding its geographic range and damage potential. Int. Pest Manag. 106, 289–295.

Woltz, J.M., Donahue, K.M., Bruck, D.J., Lee, J.C., 2015. Efficacy of commercially available predators, nematodes and fungal entomopathogens for augmentative control of *Drosophila suzukii*. J. Appl. Entomol. 139, 759–770. http://dx.doi.org/10.1111/jen.12200.

Wraight, S.P., Ramos, M.E., 2002. Application parameters affecting field efficacy of *Beauveria bassiana* foliar treatments against Colorado potato beetle *Leptinotarsa decemlineata*. Biol. Control 23, 164–178.

Wraight, S.P., Ramos, M.E., 2005. Synergistic interaction between *Beauveria bassiana*- and *Bacillus thuringiensis tenebrionis*-based biopesticides applied against field populations of Colorado potato beetle larvae. J. Invertebr. Pathol. 90, 139–150.

Wraight, S.P., Carruthers, R.I., Bradley, C.A., Jaronski, S.T., Lacey, L.A., Wood, P., Galaini-Wraight, S., 1998. Pathogenicity of the entomopathogenic fungi *Paecilomyces* spp. and *Beauveria bassiana* against the silverleaf whitefly, *Bemisia argentifolii*. J. Invertebr. Pathol. 71, 217–226.

Wraight, S.P., Carruthers, R.I., Jaronski, S.T., Bradley, C.A., Garza, C.J., Galaini-Wraight, S., 2000. Evaluation of the entomopathogenic fungi *Beauveria bassiana* and *Paecilomyces fumosoroseus* for microbial control of the silverleaf whitefly, *Bemisia argentifolii*. Biol. Control 17, 203–217.

Wraight, S.P., Sporleder, M., Poprawski, T.J., Lacey, L.A., 2007. Application and evaluation of entomopathogens in potato. In: Lacey, L.A., Kaya, H.K. (Eds.), Field Manual of Techniques in Invertebrate Pathology. Springer, The Netherlands, pp. 329–359.

Wraight, S.P., Lacey, L.A., Kabaluk, J.T., Goettel, M.S., 2009. Potential for microbial biological control of coleopteran and hemipteran pests of potato. Fruit Veg. Cereal Sci. Biotechnol. 3, 25–38.

Wright, R.J., Agudelo-Silva, F., Georgis, R., 1987. Soil applications of steinernematid and heterorhabditid nematodes for control of Colorado potato beetles, *Leptinotarsa decemlineata* (Say). J. Nematol. 19, 201–206.

Yarborough, D.E., Collins, J.A., 1997. Insect Control Guide for Wild Blueberries. Wild Blueberry Fact Sheet 209, Bulletin 2001. University of Maine Cooperative Extension.

Zahn, D.K., Haviland, D.R., Stanghellini, M.E., Morse, J.G., 2013. Evaluation of *Beauveria bassiana* for management of citrus thrips (Thysanoptera: Thripidae) in California blueberries. J. Econ. Entomol. 106, 1986–1995.

Zalom, F.G., Bolda, M.P., Dara, S.K., Joseph, S., 2014. UC IPM Pest Management Guidelines: Strawberry. University of California Statewide Integrated Pest Management Program. UC ANR Publication 3468, Oakland.

Zehnder, G.W., Ghidiu, G.M., Speese, J., 1992. Use of the occurrence of peak Colorado potato beetle (Coleoptera: Chrysomelidae) egg hatch for timing of *Bacillus thuringiensis* spray applications in potatoes. J. Econ. Entomol. 85, 281–288.

Chapter 15

Microbial Control of Insect Pests of Tea and Coffee

M. Nakai[1], L.A. Lacey[2]

[1]Tokyo University of Agriculture and Technology, Tokyo, Japan; [2]IP Consulting International, Yakima, WA, United States

15.1 INTRODUCTION

The two most common infusion beverages in the world, tea (*Camellia sinensis*, Fig. 15.1) and coffee (*Coffea arabica*, Fig. 15.2), are grown in the tropics and subtropics worldwide. Approximately of 5.3 million tonnes of tea were grown annually in 50 countries in 2013 (FAO STAT). Tea tree presumably originated in China, but its cultivation has spread to the parts of the world with an appropriate growing climate. The top five producers by export tonnage are China (1,924,457), India (1,208,780), Kenya (432,400), Sri Lanka (340,230), and Vietnam (214,300) (FAO STAT, 2013), which are mainly located in Asia and Africa (Fig. 15.3A). Besides Asia and Africa, Argentina and Brazil are production centers in South America. Considerably less is produced in Europe, but some is grown in the Azorean archipelago (Portugal).

Coffee is cultivated in the tropics by some 20 million farmers in more than 80 countries in Africa, Asia, and Latin America and generated over US$173 billion in 2012 (ICO, 2014 cited by Vega et al., 2015). The global production of coffee in 2014 was 8,808,062 tonnes (Fig. 15.3B). The top five producers by export tonnage are Brazil (2,720,520), Vietnam (1,650,000), Colombia (696,000), Indonesia (411,000), and Ethiopia (6000) (ICO, 2015).

A plethora of pests (plant pathogens and parasites, insects, and mites) attack tea and coffee (Waller et al., 2007; Chen and Chen, 1989). An integrated approach for their control has been adopted by many growers as a consequence of insecticide resistance, food safety issues, and environmental concerns (Hazarika et al., 2009; Aristizábal et al., 2012; Benavides et al., 2012). In this chapter, we will cover those significant pests of tea and coffee in which entomopathogens have been reported or where microbial control has been demonstrated.

15.2 INSECT PESTS OF TEA AND THEIR MICROBIAL CONTROL

Tea is a perennial evergreen plant and has a unique structure with dense leaves on the picking surface, but the interior has very little foliage. This structure allows various pests to inhabit tea plants. There are 1031 species of arthropods associated with tea, 3% of which are considered pests (Chen et al., 1989; Ye et al., 2014). Tea pests and their control have been well documented by Lehmann-Danzinger (2000), Hazarika et al. (2009), and Roy and Muraleedharan (2014). The major pests are mirids (Hemiptera: Miridae), tea tortricid (Lepidoptera: Tortricidae), shot hole borers (Coleoptera: Curculionidae: Scolytinae), and mites, which include *Tetranychus kanzawai*, *Brevipalpus phoenicis*, *Acaphylla theae*, *Calcarus carinatus*, and *Oligonychus coffeae*. *Bacillus thuringiensis* is the most widely used microbial control agent (MCA) against tea pests, mainly lepidopteran species.

15.2.1 Defoliators of Tea

15.2.1.1 Tea Tortricids

Tea tortricids (Lepidoptera: Tortricidae) include *Homona magnanima*, *Homona coffearia*, *Adoxophyes honmai*, *Adoxophyes dubia*, and *Archips insulanus*, which are common pests of tea throughout Asia. Larvae roll the leaves as their retreats and damage them by feeding, resulting in reduction of the harvestable product (the leaf). Leaf quality is also reduced because the rolled leaves contain feces and silk. Rolled leaves visually attract farmers' attention, resulting in the likelihood of the application of chemical pesticides, which might cause an increase of resistance to pesticides. For example, *A. honmai* has acquired resistance to various chemical pesticides, including carbamates,

FIGURE 15.1 Tea plantation in Taiwan. *Photograph by Madoka Nakai.*

FIGURE 15.2 Coffee plantation in Vietnam. *Photograph by Shugo Hama.*

organophosphates, pyrethroids, benzoylureas, diacylhydrazines, and diamides. Diamide is known as a novel chemical ingredient, but the resistance of *A. honmai* to it was reported within four years of field application (Uchiyama and Ozawa, 2014). Further, preference of consumers for less chemical residue helps in the promotion and development of biological control of tea pests.

15.2.1.2 Microbial Control of Tortricids With Viruses

A field study in Tsukuba, Japan, was conducted to monitor the prevalence of parasitic natural enemies of *A. honmai* larvae. An entomopoxvirus (AHEV) and an endoparasitoid, *Ascogaster reticulatus* (Hymenoptera: Braconidae), are the most important natural enemies of *A. honmai* larvae (Nakai et al., 1997). A maximum infection rate of AHEV was 60% among larval *A. honmai* populations. AHEV has a broad host range, which includes *Homona magnanima* and *A. insuralis* (Takatsuka et al., 2010). An EPV is also highly prevalent in *Homona coffearia* in Sri Lanka (Nakai et al., unpublished data). The mechanism responsible for high prevalence of EPVs in tortricids in tea systems is unknown but could be attributed to the slow killing trait of the virus, which could enable transmission among generations. One of the cues for this speculation is based on results of the following field experiments. A nucleopolyhedrovirus (NPV) isolated from *A. honmai* (AdhoNPV) also shows a slow killing trait with death occurring as long as 15 days after inoculation of neonate larvae. Another isolate of NPV from *A. orana* (AdorNPV) (an apple pest in the United Kingdom) is genetically closely related to AdhoNPV in Japan but has a faster rate of kill (survival time of neonate inoculation is 5.8 days). The transmission rate to the next generation in the field was higher in AdhoNPV than in AdorNPV, whereas the slow-killing NPV exhibited higher damage of tea leaves compared with fast-killing AdorNPV (Takahashi et al.,

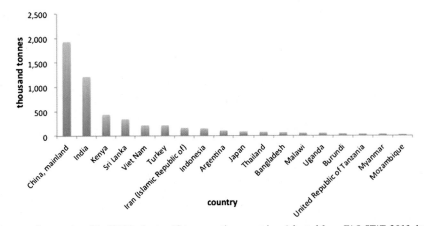

FIGURE 15.3A Amount of tea produced in 2013 in the top 18 tea exporting countries. *Adapted from FAO STAT, 2013. http://faostat3.fao.org/.*

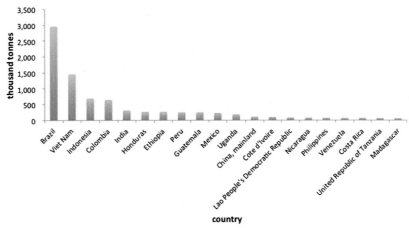

FIGURE 15.3B Amount of coffee produced in 2013 in the top 20 coffee exporting countries. *FAO STAT, 2013. http://faostat3.fao.org/.*

2015). Both *A. honmai* and *H. magnanima* have four or five generations per year in Japan without overlapping generations. The slow-killing trait of viruses has adapted to the occurrence of the tortricid larvae using the perennial crop system as tea fields.

Granuloviruses (GVs) were developed to control *A. honmai* and *Homona magnanima* larvae in Kagoshima, Japan, since the 1980s as a government project. Details were described in Nishi and Nonaka (1996) and Nakai (2009). The GVs isolated from *A. orana* and *H. magnanima* were propagated in vivo using egg dipping in a suspension of AdhoGV occlusion bodies (OBs) (Nishi and Nonaka, 1996). Using this method, the neonate larvae consume the OBs when they egress from the egg. The GVs are also slow killing, taking longer than three weeks, with infected larvae dying in the final instar. The dead larvae were homogenized and filtered with a cloth to prepare a semipurified OB suspension. An equivalent of 2000 dead larvae for *A. honmai* and 1000 larvae for *H. magnanima* were sufficient to control the pests in 1 ha tea fields. The two GVs are produced separately with each host, because GV isolated from *A. orana* (AdorGV) and from *H. magnanima* (HomaGV) are not infectious to each other's hosts. Most farmers evaluated the GVs as efficient control agents for tea leafrollers, and approximately 80% of farmers used the GVs for control of tea tortricids.

15.2.1.3 Microbial Control of Tea Pests With Other Viruses

Geometrids are important defoliators. Among them, the common tea looper caterpillar, *Buzura supressaria* (Lepidoptera: Geometridae), is one of the most important pests in China and India. In China, NPVs from *Buzura suppressaris*, *Ectopis oblique* (Lepidoptera: Geometridae), *Ectopis grisescens* (Lepidoptera: Geometridae), *Euproctis pseudonspersa* (Lepidoptera: Lymantriidae), and *Iragoides fasciata* (Lepidoptera: Limacodidae) and GVs from *Andraca bipuncta* (Lepidoptera: Endromidae) were evaluated

as potential bioinsecticides in tea plantations (Sun, 2015). Among them, *Euproctis pseudonspersa NPV* and *Ectopis grisescens NPV* were authorized as commercial insecticides by the Ministry of Agriculture of China (Sun, 2015). The LC_{50} (lethal to 50% of treated individuals) of a strain of BusuNPV was assessed as 1.66×10^4 OBs/mL. As the result of field application in China, 97% and 94% of mortality was assessed for first and second generations, respectively, 10 days after application of BusuNPV (Peng et al., 1998).

15.2.1.4 Bacillus thuringiensis

Bacillus thuringiensis (Bt) is known to be one of the most effective agents to control lepidopterans and dipteran tea pests (Hazarika et al., 2005, 2009). Efficacy of various commercialized Bt agents against *H. magnanima*, *A. honmai*, and *A. theae* (Diptera: Agromyzidae) was reported by Kariya (1977). Leaf-dipping laboratory bioassays and field tests determined that Bt agents are effective against those species, although the mortality varied among different agents (Kariya, 1977). Application of Bt for control of other lepidopteran pests including *B. suppressaria*, *Dasychira baibarana* (Lepidptera: Lymantriidae), and *E. pseudoconspersa* resulted in about 90% mortality in small-scale field studies. However, Bt treatment of *E. obliqua* only produced about 50% mortality in some cases (reviewed by Ye et al., 2014).

15.2.2 Stem and Phloem Feeders

15.2.2.1 Entomopathogenic Fungi Against Whiteflies

Disease caused by entomopathogenic fungi (EPF) naturally occurs in tea pests. *Zoophthora (Erynia) radicans* (Zygomycota: Entomophthorales) was reported as responsible for 16% mortality of *H. magnanima* pupae in tea fields near Tokyo (Mao and Kunimi, 1991). Epizootics of *Aegerita webberi* (Basidiomycota: Polyporales) were observed in China in an *Aleurocanthus spiniferus* (Homoptera: Aleyrodidae) population resulting in 90% mortality (Ye et al., 2014). *Isaria (Paecilomyces) cinnamomeus* (Ascomycota: Eurotiales) was isolated from the camellia whitefly, *Aleurocanthus camelliae* (Hemiptera: Aleyrodidae), and first recorded in Japan in 2004 (Saito et al., 2012). The isolate of *I. cinnamomeus* caused greater than 50% and 90% infection in whitefly nymphs when applied at 10^6 and 10^7 conidia/mL, respectively. However, the control efficacy of commercialized fungal insecticides Preferb (*Isaria fumosorosea*) and Mycotal (*Lecanicillium muscarium*) produced less than 10% mortality at the recommended dosages of 5×10^6 and 9×10^6 conidia/mL, respectively, following field applications. This suggests that *I. cinnamomeus* can be a more effective MCA than the current commercially available EPF products in the Japanese market.

15.2.2.2 Entomopathogenic Fungi Against Leafhoppers (Homoptera: Cicadellidae)

Good efficacy of *B. bassiana* was reported for control of nymph and adult stages of the tea leafhopper, *Empoasca flavescent*, and false-eye leafhopper, *Empoasca vitis* (reviewed by Ye et al., 2014). An oil-based emulsion of *B. bassiana* conidia and its feasibility to be incorporated into imidacloprid control were examined against *E. vitis* population in tea fields in southern in China (Feng et al., 2004). An oil-based emulsifiable formulation (2×10^7 conidia/mL) of *B. bassiana* containing 1% imidacloprid in 10% WP (wettable powder) produced 69% mortality in autumn weather conditions, though *B. bassiana* alone was not effective (Feng et al., 2004). Feng et al. (2004) recommended improvement of spraying technique of fungal agents because the tea leafhoppers actively move below the tea canopy rather than above. Conventional hand sprayers do not sufficiently enable the conidia to attach to insects. Moreover, the authors suggested that a temperate weather pattern in Yunnan, China, with rain from May through September is suitable for using fungal application (Feng et al., 2004).

15.2.3 Entomopathogenic Fungi for Control of Tea Weevil

Between 1988 and 1994, *B. bassiana* resulted 1–21% mortality of larval and pupal *Myllocerinus aurolineatus* in tea fields in China (Ye et al., 2014). Those fungi showing higher field prevalence were considered to be suitable MCAs for tea pests. For example, *B. bassiana* isolated from the tea weevil *M. aurolineatus* were applied in 1,200 ha in Fujian Province, China (Sun et al., 1993). Laboratory experiments showed that mortality of adult weevils was 100% seven days following immersion in a suspension of 10^8 conidia/mL (Wang et al., 2013). Use of a powder formulation of *B. bassiana* was recommended for soil application. Control of

the weevils reached an average of 80%, and the following year was also reduced by an average of 70% by using EPF products (Sun et al., 1993).

15.2.4 Chafer

The yellow elongate chafer *Heptophylla picea* (Coleoptera: Scarabaeidae) feeds on the roots of tea plants during the larval stage. Plant damage inflicted by third instar larvae suppresses the growth of first budding of the leaves the following spring. Because the chemical pesticides that are used do not reach into deep soil where the susceptible young larvae are located, large amounts of chemical pesticide application was required to control old, less-susceptible adults on the surface of the soil. According to a study to determine a suitable procedure to control *H. picea* using *Beauveria amorpha*, the fungi should be applied into the soil in October to November when the soil temperature is above 15°C (Yaginuma et al., 2004). Because *B. amorpha*-infected larvae move to the soil surface (0–10 cm) from a depth of 20–30 cm and the average soil temperature was above 15°C in October, infections of *H. picea* were observed. *Beauveria amorpha* showed little infectivity below 15°C (Hiromori et al., 2004). Mortality of third-instar *H. picea* larvae were 100% at 25°C at 17 and 28 days after treatment at 5.0×10^7 and 1.0×10^7 conidia/mL, respectively (Hiromori et al., 2004), though *B. amorpha* is not effective against adults. Yaginuma et al. (2006) isolated *Beauveria brongniartii* as a potential MCA of *H. picea* adults. The strain of *B. brongniartii* produced 100% mortality of *H. picea* adults at a concentration of 10^7 conidia/mL with a short LT_{50} of 8 days for females and 7 days for males (Yaginuma et al., 2006). Fungal control of *H. picea* can be efficient if it is used in combination of two fungal strains, *B. amorpha* and *B. brongniartii*.

15.2.5 Termite Control With *Bacillus thuringiensis*

Bacillus thuringiensis subsp. *israelensis* (Bti) was examined for its pathogenicity against tea termites, *Microtermes obesi* (Blattodea: Termitidae) and *Microcerotermes beesoni* (Blattodea: Termitidae). Bti applied at 10^4 cells/mL resulted in more than 90% mortality against the worker caste of both termite species, and the LT_{50} for *M. obesi* and *M. beesoni* was 2.8 and 2.5 days, respectively (Singha et al., 2010).

15.3 INSECT AND MITE PESTS OF COFFEE AND THEIR MICROBIAL CONTROL

There are approximately 3000 species of insects and mites associated with coffee, more than 850 of which feed on leaves, stems, fruit/beans, and roots of the plant (Pelley, 1973; Waller et al., 2007; Barrera, 2008; Bustillo, 2015).

However, considerably fewer are regarded as significant pests (Barrera, 2008). Bustillo (2015) lists the 40 most economically important mites and insects, including three species of tetranychid mites, five families of Hemiptera, three lepidopteran families, and four families of Coleoptera. The number and severity of pests vary by country, but all coffee-growing countries share seven significant pest species: the coffee green scale (*Coccus viritis*), black citrus aphid (*Toxoptera aurantii*), wooly whitefly (*Aleurothrixus floccosus*), citrus and pineapple mealy bugs (*Planococcus citri* and *Dysmicoccus brevipes*), coffee berry borer (CBB; *Hypothenemus hampei*), and coffee bean weevil (*Araecerus fasciculatus*). Left unchecked or not well controlled, insect and mite pests can inflict insurmountable losses to farmers. Globally these pests cause billions of U.S. dollars in losses due to feeding on leaves, stems, and most of all fruit (Gaitan et al., 2015).

15.3.1 Fruit-Feeding Insect Pests of Coffee

15.3.1.1 Coffee Berry Borer, H. hampei (Coleoptera: Curculionidae: Scolytinae)

The CBB is by far the most economically important pest of coffee worldwide (Pelley, 1973; Damon, 2000; Barrera, 2008; Vega et al., 2009, 2015; Gaitán et al., 2015). It originated in Africa but is now distributed wherever coffee is grown. The life cycle of this pest is presented in detail by Vega et al. (2009, 2015), and an extensive bibliography on CBB biology, ecology, control, and economic importance was published by Pérez et al. (2015). Larvae cause damage by feeding within the fruit, resulting in downgrading of the quality of beans or rendering the coffee bean unusable. Oliveira et al. (2013) determined that the economic injury level (EIL) is reached when infestations of the berries are equal to or higher than 7.9–23.7% for high- and average-yield conventional crops, respectively. However, when qualitative and quantitative losses caused by CBB are considered together, the EIL was 4.3% of bored berries (Oliveira et al., 2013). Yearly losses caused by CBB have been estimated for Brazil alone at US$215–358 million (Oliveira et al., 2013; Infante et al., 2014; Vega et al., 2015).

15.3.1.1.1 Microbial Control of Coffee Berry Borer With Entomopathogenic Fungi

Beauveria bassiana is the most-studied fungus for control of coffee pests with the majority of research on the control of CBB. It is reported as a naturally occurring pathogen of adult CBB around the world (Balakrishnan et al., 1994; Kumar et al., 1994; Alves, 1998; Alves et al., 2008; Vega et al., 1999; Daman, 2000; Bustillo et al., 2002; Costa et al., 2002; Monzón et al., 2008; Sampedro-Rosas et al., 2008; Benavides et al., 2012; Bustillo, 2015, and several

additional reports by country summarized by Vega et al., 2015). Epizootics of this fungus in CBB populations have been reported as being responsible for up to 70% infection in India, Africa, and Latin America (Vázquez Moreno et al., 2010; Méndez et al., 2014; Vega et al., 2015). However, occurrences of epizootics are not accurately predictable and should not be relied upon for maintaining CBB and other pest insect populations below the EIL. The inundative application of *B. bassiana* for CBB control, on the other hand, can be timed for effective reduction of the pest population based on time of berry invasion and optimal environmental conditions. Field assessments of its utility as an MCA have been conducted across Latin America, Africa, and India with variable efficacy. Typically, rates of $(1–2) \times 10^{12}$ conidia/ha in aqueous suspensions are applied to foliage. Field trials of inundatively applied *B. bassiana* suspensions producing mortality from 14% to 90% are reported by Vélez-Arango (1997), Bustillo et al. (1999), de la Rosa et al. (2000), Haraprasad et al. (2001), Samuels et al. (2002), Jaramillo et al. (2006), Alves et al. (2008), Irulandi et al. (2008), Vera et al. (2010), Aristizábal et al. (2012), Benavides et al. (2012), Suáez and Mejía (2012), Díaz-Vicente et al. (2014), and others summarized by Vega et al. (2009, 2015).

A variety of factors could be responsible for the observed variation in control including viability of conidia, isolate, temperature, altitude, humidity, formulation, application equipment, timing and rate of application (both in terms of environmental conditions and location of CBB—inside or outside of fruit), UV radiation, compatibility with agrochemicals and other natural enemies, and host factors. As with most other EPFs, *B. bassiana* is most effective when humidity is high (>90%), temperature is moderate (25°C) and when and where UV radiation is low. Shaded plantations can facilitate this range of conditions. Unfortunately, some pests and plant pathogens of coffee also benefit from shaded environments. Inglis et al. (2001) and Vega et al. (2012, 2015) provide additional detail to the biotic and abiotic factors that limit or enhance the activity of several EPFs.

Metarhizium anisopliae s.l. has also been evaluated for insecticidal activity against CBB in laboratory and field conditions (Lecuona et al., 1986; Bernal et al., 1999; de la Rosa et al., 2000; Samuels et al., 2002; Alves et al., 2008). This species demonstrated fair adulticidal activity, but in side-by-side tests with *B. bassiana*, it was less efficacious the *B. bassiana*.

Although several other fungi have been reported from CBB (Barrera, 2008; Vega et al., 2015), with the exception of those EPFs living endophytically in coffee plants, few have been further studied for microbial control.

15.3.1.1.2 Endophytic Fungi

Fungi and bacteria that live within the tissues of asymptomatic plants are defined as endophytes (Vega, 2008). Exploitation of the endophytic EPFs as possible control agents of CBB and other pests has been proposed by Posada et al. (2007) and Vega (2008). Benefits to the host plant are deterrence of arthropod pests and disease-causing agents (Vega, 2008). Vega et al. (2008) reported several genera of fungi (*Acremonium*, *Beauveria*, *Cladosporium*, *Clonostachys*, and *Isaria*) isolated from coffee plants collected in Colombia, Hawaii, and Puerto Rico. Two isolates, *B. bassiana* and *Cladosporium rosea*, were insecticidal for CBB. Species that are also known as free living entomopathogens have been reported, including *B. bassiana* (Vega et al., 2008). *Beauveria bassiana* was established in coffee seedlings after applying suspensions of the fungus as foliar sprays, stem injections, and soil drenches (Posada and Vega, 2006; Posada et al., 2007). Injection resulted in the highest endophytic recovery. However, recovery of the fungus declined over time (Vega, 2008).

15.3.1.1.3 *Bacillus thuringiensis*

Bt subspecies, most notably Bt subsp. *kurstaki* and Bt subsp. *aizawai*, are used for control of a wide variety of Lepidoptera on dozens of crops (Sanchis, 2011), but very little is published on their application for control of fruit-feeding coffee pests. Méndes-López et al. (2003) screened 170 Mexican Bt strains against CBB. Most of these produced little or no insecticidal activity against first-instar CBB. However, a few of these strains belonging to the Bt subspecies *israelensis* (Bti) showed good larvicidal activity against CBB. The authors conducted bioassays with purified parasporal inclusions of native and Pasteur Institute strains of Bti and estimated an LC_{50} Bti 219.5 ng/cm^2 of diet. No additional evaluations of Bti have been reported following their initial results. Prior to their findings, this Bt sub-species has been known predominantly as having larvicidal activity for dipteran species in the suborder Nematocera (Lacey and Siegel, 2000).

A diversity of Bt isolates were found by Arrieta et al. (2004) from soil leaf litter, fresh foliage, coffee beans, and insects, living and dead, that were dissected from coffee berries in six Costa Rican coffee plantations infested with CBB. Of the 202 isolates collected, 43 different genetic profiles were found: 78 strains contained the *vip3* gene, 82 the *cry2* gene, 45 the *cry1* gene, and 29 strains harbored *cry3–cry7* genes. No subsequent bioassays against CBB were reported. de la Rosa et al. (2005) bioassayed 61 Mexican Bt isolates from the same collection used by Méndez-López et al. (2003) and found toxicity levels ranging from 8% to 83%.

15.3.1.1.4 Entomopathogenic Nematodes as Potential Control Agents of Coffee Berry Borer

Two families of Rhabditid nematodes, Steinernematidae and Heterorhabditidae, comprise the vast majority of entomopathogenic nematodes (EPN) species. All species of EPN require a living host insect in which to reproduce. They rely on two foraging strategies, seeking and ambushing, or a combination of both for finding host insects. Details on the biology and ecology of EPNs can be found in Chapter 6. Applied research on EPNs for CBB control has been somewhat limited. EPNs are principally natural enemies of soil insects (Georgis et al., 2006), and as such their efficacy on leaves and fruit on the tree is considerably reduced. Hence, application to fallen CBB-infested berries on the surface of the soil would provide the best possibility for control. Laboratory studies conducted with CBB-infested berries treated with *Steinernema* sp., *Steinernema carpocapsae*, and *Heterorhabditis* have produced variable mortality in adult CBB. Those observed by Manton et al. (2012) are an example of the low end of the efficacy spectrum. Eight days following spray application of ca. 700 infective juveniles (IJs)/cm^2 of *S. carpocapsae* (an ambusher) on CBB-infested coffee berries on soil in the laboratory, they observed 23.7% and 26.6% mortality in adults and larvae respectively. In field trials, small plots (133 cm^2) with infested berries on the ground were drenched with 500 mL of an *S. carpocapsae* suspension (4600 IJs/mL) resulting in 1.73×10^4 IJs/cm^2. After 7 days, mortality was 4.7% in adults and 17.1% in larvae. Lara et al. (2004) present results on the higher end of the field efficacy spectrum. Their treatments included three concentrations of *Steinernema* sp. (species name not provided) at 1.25, 2.5, and 5.0×10^5 IJs 300 mL^{-1} of water and the same three concentrations of *Heterorhabditis* sp. (species not provided) applied to the base of coffee trees onto CBB-infested berries. Following treatment, the plots were covered with a muslin tent ($1.25 \times 1.25 \times 2.4$ m). After 1 month, they demonstrated the highest control of CBB in fallen fruit (29.7% and 28.9%) by *Steinernema* sp. at 2.5×10^5 plot^{-1} and *Heterorhabditis* sp. at 1.25×10^5 plot^{-1}, respectively.

Comparing these results with those of studies on EPNs for control of soil-inhabiting insects, the differences are astronomical. For example, Japanese beetle, *Popillia japonica*, larvae in turf sprayed with 2.5×10^9 IJs/ha (25 IJs/cm^2) of *Heterorhabditis zealandica* (a seeker) responded with 73–98% mortality (see Chapter 24). Considering the degree of control due to low efficacy and a major part of the population is not in the area being treated (ie, on the tree), relatively high cost, short shelf-life, proper EPN product storage (8°C), labor expense, and ready access to water, it is unlikely that EPNs will be adopted as a means of CBB control.

Parasitic nematodes can also affect CBB populations in nonlethal ways (Vega et al., 2005). Castillo et al. (2002) discovered a species of *Sphaerulariopsis* (Tylenchida: Sphaerularioidea) attacking immature and adult stages of *Hypothenemus hampei* in Mexico. Subsequently, Poinar et al. (2004) renamed and reclassified this nematode as *Metaparasitylenchus hypothenemi* (Nematoda: Allantonematidae). Parasitism by this species does not result in high mortality of CBB but significantly reduces the fecundity of females (Poinar et al., 2004; Vega et al., 2005). Other consequences of parasitism include reduced vagility and longevity.

15.3.1.2 Other Fruit-Feeding Pests

15.3.1.2.1 Coffee Berry Moth

Prophantis smaragdina (*Lepidoptera*: *Pyralidae*). In addition to damage caused by berry moth larvae feeding on fruit and seeds, the wounds to the berry facilitate infection by fungal disease–causing agents (Barerra, 2008; Chartier et al., 2013; Bustillo, 2015). This species has recently become a major coffee pest in East Africa and is considered as the principal cause of crop losses in Réunion Island (Chartier et al., 2013). Chartier et al. (2013) observe that Bt can be an effective MCA of *P. smaragdina* using two applications one month apart (application rate was not provided) during the dry season.

15.3.1.2.2 Antestia Coffee Bugs, *Antestiopsis lineaticolis*, *Antestia orbitalis*, and *Antestia* spp. (Hemiptera: Pentatomidae)

These species can be serious pests in East Africa. Prior to harvest, the bugs feed on berries and flowers. Postharvest, they feed on most parts of the tree. Feeding on new shoots results in no subsequent fruit development. In addition to damage to berries due to feeding, *Antestiopsis* spp. are known to transmit plant pathogenic fungi (eg, *Nematospora coryli* and other fungi), which causes rotting of beans (Bustillo, 2015). Very little research on microbial control has been reported on *Antestia* coffee bugs. Nahayo and Bayisenge (2012) conducted a laboratory bioassay in which *A. lineaticolis* nymphs were sprayed with a spore suspension made with 50 g of *B. bassiana* spore–covered rice in 200 mL of water and incubated for 5 days and observed 95% mortality. Considering the rather elevated concentration required for control, the prospects for using EPFs for *Antestia* coffee bugs are not favorable in the near future. However, detailed quantitative evaluations of hemipteran-active EPF isolates grown and prepared under optimal conditions are warranted.

15.3.2 Stem and Phloem Feeders

15.3.2.1 Spider Mites (Acari: Tetranychidae)

Infestations of spider mites in the genus *Oligonychus* (*O. coffeae*, *Oligonychus yothersi*, *O. ilicis*) are reported as significant pests of coffee (Bustillo, 2015). *Oligonychus coffeae* is an important pest of coffee plantations in the Americas, Asia, and Africa. *Oligonychus yothersi* and *O. ilicis* are known only from the Americas.

Numerous species of mites, including several tetranychids, are infected by EPFs, predominantly species in the Entomophthorales (eg, *Neozygites floridana*) and Hypocreales: *Beauveria bassiana*, *Isaria* spp. (formerly *Paecilomyces*), *Metarhizium anisopliae* s.l., *Hirsutella thompsonii*, *Cladosporium cladosporioides*, *Cephalosporium diversiphialidum*, and *Lecanicillium lecanii* s.l. (formerly *Verticillium lecanii*) (Chandler et al., 2000; van der Geest et al., 2000; Wekesa et al., 2015) and are at times responsible for spectacular epizootics (Wekesa et al., 2015) including those caused by *N. floridana* in *Oligonychus pratensis*, a pest of field corn (Dick and Buschman, 1995).

There has been very little research conducted on microbial control of *Oligonychus* spp. using EPFs in coffee plantations. de Oliveira et al. (2002) conducted bioassays of 80 isolates of *B. bassiana*, *Isaria fumosorosea*, and *M. anisopliae* s.l. against *O. yothersi* under laboratory conditions and demonstrated the mites' susceptibility to suspensions of 10^7 conidia/mL. After 5 days, mortality was determined. Nine isolates of *B. bassiana* produced the highest mortality ranging from 70% to 89%. The authors recommend additional studies to further evaluate efficacy and determine strategies for utilization of the most virulent isolates in the field.

Mamun et al. (2014) assessed the efficacy of *B. bassiana*, *M. anisopliae* s.l., *Isaria fumosorosea*, and *L. lecanii* s.l. against *O. coffeae* on tea plants. All of the WP formulations of these fungi except *L. lecanii* contained 2×10^8 conidia/g and were applied at 5 kg/ha. *Lecanicillium lecanii* s.l. containing 10^8 conidia/g spores was applied at 4.0 kg/ha. All four of the fungi significantly reduced mite populations by 61%, 67%, 80%, and 83%, respectively, one week after application. Encouragingly, they reported no mortality of phytosiid mites and other predators, such as *Stethorus gilviforns* (Coleoptera: Coccinellidae) and *Oxyopes* sp. (Arachnida: Oxyopidae). Considering the effectiveness of EPFs for control of mites on other crops (Chandler et al., 2000; Wekesa et al., 2015), evaluation in coffee plantations is warranted for classical biological control (*Neozygites* spp.) and inundative applications (Hypocreales, especially *L. lecanii*).

15.3.2.2 Coffee Green Scale, Coccus viridis (Hemiptera: Coccidae)

This species is the most serious soft-scale attacking coffee worldwide, and in some countries (eg, Papua New Guinea) it is considered to be the most economically important coffee pest. In the humid tropics, *L. lecanii* is the main mortality factor of this species (Murphy, 1997). The hemispherical scale or brown scale, *Saissetia coffeae* (Hemiptera: Coccidae) can also be an important pest in some locations. Reddy (2014) recommended application of *L. lecanii* at 10^7 conidia/mL just at the onset of the rainy season as an inundatively applied MCA for control of both coccids in coffee.

15.3.2.3 Citrus Black Aphid, Toxoptera aurantii (Hemiptera: Aphididae)

Toxoptera aurantii is a polyphagous species with 120 host plants including coffee. No detailed reports were found in the literature on microbial control of *T. aurantii* in coffee. Bustillo (2015) noted that *L. lecanii* was used for management of *T. aurantii* in coffee but dosage and frequency of application were not given. Cortez Madrigal et al. (2003) studied 15 isolates of *L. lecanii* from Mexico and Cuba with regard to conditions and media for optimal mycelial growth, sporulation, production of conidia, and virulence. The top five isolates produced 83–100% mortality and 80–97% sporulation on *T. aurantii* under laboratory conditions 4 days after spraying them with suspensions containing 2×10^7 conidia/mL. With one exception, there appeared to be no significant differences in mortality due to original host species. Only one of the top five was isolated from *T. aurantii*. The majority of research on EPFs for aphid control has been conducted on other species (Hall, 1981; Nielsen and Wraight, 2009). Their susceptibility to *L. lecanii*, other Hypocreales, and species in the Entomophthorales is an indication that several candidates are available for testing and ultimately for use in control programs. For example, Poprawski et al. (1999) conducted field evaluations of *B. bassiana* in citrus applied as an aqueous suspension of 5×10^{13} conidia/ha (Mycotrol) for control of *Toxoptera citricida*, a closely related species of *T. aurantii* and observed 98% mortality.

15.3.2.4 Wooly Whitefly (Aleurothrixus floccosus)

This whitefly is a pest of several cultivars including citrus and coffee in the Americas. Bustillo (2015) mentions that *Aschersonia aleyrodis* and *L. lecanii* have been used in the management of this species. *Aschersonia* spp. and *L. lecanii* are ubiquitous and infect several whitefly species (Fransen, 1990; Meekes et al., 2002; Lacey et al., 2008). Because of their documented efficacy on other whiteflies, their potential for control of *Aleurothrixus floccosus* should be investigated. Due to the hydrophobic nature of their waxy secretions, experimentation with wetting agents should be done concomitantly with efficacy studies.

15.3.3 Leaf-Feeding Pests Including Defoliators and Miners

15.3.3.1 Defoliators

Barrera (2008) listed 21 species of lepidopteran defoliators (Geometridae, Noctuidae, Pyralidae) of coffee in the American tropics. *Spodoptera* spp. and *Agrotis* spp. are defoliators and cutworms, respectively, found on a variety of crops, and are pests of coffee in some areas (Barrera, 2008). Bustillo (2015) considers Lepidoptera in the family Geometridae as some of the most important defoliators. Some plantations have reported near-complete defoliation

of plants. Although the plants may recover, that season's crop may be lost. *Ascotis selenaria*, *Oxydia* spp., *Paragonia procidaria*, and *Glena bisulca* are some of the most injurious species where there is excessive use of insecticides (Bustillo, 2015).

15.3.3.1.1 Baculovirus

Nucleopolyhedroviruses (NPVs) and granuloviruses (GVs) have been developed for control of several lepidopteran pests (see Chapter 3, also Vail et al., 1999), some of which can be pests of coffee (eg, *Spodoptera* spp., *Agrotis* spp., *Trichoplusia ni*, and other noctuids) (Barrera, 2008). Baculoviruses have not yet been evaluated on coffee for control of pest Lepidoptera. However, they have been developed and commercialized for control of the above species and several others, predominantly in the Noctuidae (Vail et al., 1999). Their evaluation is warranted where noctuid pests cause significant damage to coffee plants or seedlings.

15.3.3.1.2 *Bacillus thuringiensis*

The Bt subspecies are used for control of a wide variety of Lepidoptera, but very little is published on their application for control of defoliating coffee pests. Polanczyk et al. (2012) mentioned that the imperial moth (*Eacles imperialis magnifica*) a pest that is capable of completely defoliating coffee plants in Brazil, is controlled with Bt, but no specifics on rate and frequency of application were provided. Surprisingly, the treatment of coffee leaves with a suspension of Bt induced local and systemic resistance to the rust *Hemileia vastatrix* within 14–18 days of application (Guzzo and Martins, 1996).

15.3.3.2 Coffee Leaf Miner, Leucoptera coffeella (Lepidoptera: Lyonetiidae)

Leucoptera coffeella attacks coffee foliage throughout Latin America. Tunneling by feeding larvae can result in reduction of berry production proportionate to the amount of tunneling. Extensive infestation can result in up to a 61% loss of leaves and roots and 50% reduction in photosynthetic activity of the remaining leaves (Bustillo, 2015).

15.3.3.2.1 Transgenic Plants for Leaf Miner Control

Leroy et al. (2000) produced and evaluated genetically modified coffee plants expressing the Bt *cry1Ac* gene for resistance to *L. coffeella*. Although the results of laboratory and field tests indicated good control of this pest (Leroy et al., 2000; Perthuis et al., 2005), further testing has not been reported.

15.4 CONCLUSIONS AND FUTURE DEVELOPMENT

Tea is grown intensively with high input of chemical fertilizer and pesticides in particular countries including Japan and Taiwan; however, it is cultivated extensively in most major tea producing countries in large-scale plantations. The differences in farming systems affect the occurrences of both insect pests and plant diseases (Altieri and Nicholls, 2003). In the case of intensive tea cultivation, the annual rate of nitrogen input is 500–600 kg/ha in Shuzuoka Prefecture, Japan, and conventional chemical pesticides are applied more than 10 times a year (Nakai, 2009). This high input of chemical fertilizer and chemical pesticide triggers pest outbreaks. For example, development of Kanzawa spider mite, *Tetranychus kanzawai* (Acari; Tetranychidae), was stimulated by chemical fertilizer (Kitaoka et al., 2013). Development of resurgence and resistance were also recorded in tea cultivation due to high input of chemical pesticides (Hazarika et al., 2009; Ye et al., 2014). Because the chemical pesticides kill the natural enemies including insect predators and parasitoids, pest density can be increased due to exclusion of those natural mortality factors. On the other hand, extensive farming, often seen in India, Sri Lanka, Kenya, and Turkey, uses less chemicals and fertilizers and the pest problem is less than that in intensive farming. Tea is cultivated many countries, whereas the evolving problem is not only due to climate and distribution but also due to type of farming system. Microbial control is, however, rather compatible with chemical control and can be incorporated into integrated pest management (IPM) programs in any farming system.

The majority of applied research for microbial control of coffee pests has understandably been on CBB, the most serious pest of coffee. CBB has a broad diversity of natural enemies (predators, parasitoids, entomopathogens) (Bustillo et al., 2002; Bustillo Pardey, 2005; Vega et al., 2009, 2015; Benavides et al., 2012). Despite their suppressive effects on the beetle, augmentation of natural enemies is often required to maintain CBB populations below EIL. Inundative application of *B. bassiana* for its control is increasingly being used by more farmers in lieu of broad-spectrum chemical pesticides, particularly in Colombia. The benefits of using MCAs, especially for the conservation of other natural enemies for the abatement of secondary pests, are not fully appreciated. Despite the amount of time required before insects die, they may feed less and have reduced fecundity (Inglis et al., 2001). Subsequent sporulation under the appropriate environmental conditions becomes another source of inoculum for pest insects and mites feeding in or passing through the vicinity of this additional source of spores. Many species of coffee pest mites and insects are susceptible to EPFs that are used for CBB suppression. Generalist MCAs, such as several Hypocreales, especially *B. bassiana*, may concomitantly infect aphids, scales, coffee bugs, thrips, and additional coleopteran and lepidopteran pests. The benefits of sharing this information with growers by researchers and agricultural extension

personnel is evidenced by successful IPM programs in Colombia (Bustillo Pardey, 2005; Aristizábal et al., 2012; Benavides et al., 2012).

Although the cost:benefit ratio is the major factor in determining the acceptance of MCAs for CBB control, logistical considerations may be an even bigger issue. According to Fernando Vega (personal communication), anything that needs to be sprayed is a serious problem due to the topography of the majority of coffee plantations (hills, mountains), which involves carrying backpack sprayers weighing 20 kg up and down hills, the lack of ready access to water, and the fact that a living organism is being used. To further complicate the picture, the insect lives inside the coffee berry, making it a more difficult target for EPF that need to come into contact with the pest.

Improvement of delivery of entomopathogens to difficult-to-treat pest populations will be required before microbial control of mite and insect pests of coffee and tea will be accepted by growers. This will be especially the case for treatment of CBB with EPF. Successful auto-dissemination of EPF (*B. bassiana*, *M. anisopliae*, *Zoophthora radicans*) and other entomopathogens using devices baited with attractant substances (kairomones and pheromones) has been reported for control of several insect pests by Vega et al. (2007). Development of this strategy for CBB and other coffee and tea pests is warranted. Candidate devices and attractants for CBB have been reported by Ortiz et al. (2004), and Dufour and Frérot (2008). Further improvements in control of CBB and other coffee and tea pests with MCAs such as exploration for more virulent entomopathogens and improved formulation and delivery systems are warranted.

ACKNOWLEDGMENTS

We are grateful for the literature and suggestions provided by Prof. Zhihong (Rose) Hu, Wuhan Institute of Virology, and Dr. Fernando Vega, USDA, Agricultural Research Service. We also thank Dr. Vega for review of the manuscript and his insights into the role of MCAs in control of coffee pests.

REFERENCES

Altieri, M.A., Nicholls, C.I., 2003. Soil fertility management and insect pests: harmonizing soil and plant health in agroecosystems. Soil Tillage Res. 72, 203–211.

Alves, S.B., 1998. Fungos entomopatogênicos. In: Alves, S.B. (Ed.), Control Microbiano de Insetos. FEALQ. Piracicaba, Brazil, pp. 289–381.

Alves, S.B., Almeida, J.E., Salvo, S.D., 1998. Associação de produtos fitossanitarios com *Beauveria bassiana* no controle abroca y ferrugem do cafeeiro. Man. Int. Plag. 48, 12–24.

Alves, S.B., Lopes, R.B., Vieira, S.A., Tamai, M.A., 2008. Fungos entomopatogênicas usado no controle de pragas na América Latina. In: Alves, S.B., Lopes, R.B. (Eds.), Controle Microbiano de Pragas na América Latina. FEALQ. Piracicaba, Brazil, pp. 69–110.

Aristizábal, L.F., Lara, O., Arthurs, S.P., 2012. Implementing an integrated pest management program for coffee berry borer in a specialty coffee plantation in Colombia. J. Integ. Pest Manag. 3, 2012.

Arrieta, G., Hernández, A., Espinoza, A.M., 2004. Diversity of *Bacillus thuringiensis* strains isolated from coffee plantations infested with the coffee berry borer *Hypothenemus hampei*. Rev. Biol. Trop. 52, 757–764.

Balakrishnan, M.M., Sreedharan, K., Krishnamoorthy Bhat, P., 1994. Occurrence of the entomopathogenic fungus *Beauveria bassiana* on certain coffee pests in India. J. Coffee Res. 24, 33–35.

Barrera, J.F., 2008. Coffee pests and their management. In: Capinera, J.L. (Ed.), Encyclopedia of Entomology, second ed. Springer, Dordrecht, The Netherlands, pp. 961–998.

Benavides, P., Góngora, C., Bustillo, A., 2012. IPM program to control coffee berry borer *Hypothenemus hampei*, with emphasis on highly pathogenic mixed strains of *Beauveria bassiana*, to overcome insecticide resistance in Colombia. In: Perveen, F. (Ed.), Insecticides – Advances in Integrated Pest Management. InTech, pp. 511–540. Also available from: http://www.intechopen.com/books/insecticides-advances-in-integrated-pest-management/ipm-program-tocontrol-coffee-berry-borer-hypothenemus-hampei-with-emphasis-on-highly-pathogenic-mix.

Bernal, M.G., Bustillo, A.E., Chaves, B., Benavides, P., 1999. Efecto de *Beauveria bassiana* y *Metarhizium anisopliae* sobre poblaciones de *Hypothenemus hampei* (Coleoptera: Scolytidae) que emergen de frutos. Rev. Colom. Entomol. 25, 11–16.

Bustillo, A.E., M. G. Bernal, M.G., Benavides, P., Chaves, B., 1999. Dynamics of *Beauveria bassiana* and *Metarhizium anisopliae* infecting *Hypothenemus hampei* (Coleoptera: Scolytidae) populations emerging from fallen coffee berries. Fla. Entomol. 82, 491–498.

Bustillo, A.E., Cárdenas, R., Posada, F.J., 2002. Natural enemies and competitors of *Hypothenemus hampei* (Ferrari) (Coleoptera: Scolytidae) in Colombia. Neotrop. Entomol. 31, 635–639.

Bustillo, A.E., 2015. Part II pests. In: Gaitán, A.L., Cristancho, M.A., Castro Caisedo, B.L., Rividas, C.A., Cadena Gómez, G. (Eds.), Compendium of Coffee Diseases and Pests. American Phytopathological Society, St. Paul. 75 pp.

Bustillo Pardey, A.E., 2005. El papel del control biológico en el manejo integrado de la broca del café, *Hypothenemus hampei* (Ferrari) (Coleoptera: Curculionidae: Scolytinae). Rev. Acad. Colom. Cien. Exact. Fís. Nat. 29, 55–68.

Castillo, A., Infante, F., Barrera, J.F., Carta, L., Vega, F.E., 2002. First field report of a nematode (Tylenchida: Sphaerularioidea) attacking the coffee berry borer, *Hypothenemus hampei* (Ferrari) (Coleoptera: Scolytidae) in the Americas. J. Invertebr. Pathol. 79, 199–202.

Chandler, D., Davidson, G., Pell, J.K., Ball, B.V., Haw, K., Sunderland, K.D., 2000. Fungal biocontrol of Acari. Biocontrol Sci. Technol. 10, 357–384.

Chartier, M., Quilici, S., Frérot, B., Noirot, M., Glénac, S., Descroix, F., 2013. La pyrale du café à la Réunion: comment mieux gérer cet important ravageur des cultures du fameux café Bourbon pointu? Phytoma 660, 43–46.

Chen, Z., Chen, X., 1989. An analysis of the world tea fauna. J. Tea Sci. 9, 13–22.

Cortez Madrigal, H., Ortiz García, C.F., Alatorre Rosas, R., Bravo Mojica, H., Mora Aguilera, G., Aceves Navarro, L.A., 2003. Caracterización cultural de cepas de *Lecanicillium* (=*Verticillium) lecanii* (Zimm.) Zare y Gams y su patogenicidad sobre *Toxoptera aurantii* Boyer. Rev. Mex. Fitopatol. 21, 161–167.

Costa, J.N.M., da Silva, R.B., de Araujo Ribeiro, P., Garcia, A., 2002. Ocorrência de *Beauveria bassiana* (Bals.) Vuill. em broca-do-café (*Hypothenemus hampei*, Ferrari) no estado de Rondônia, Brasil. Acta Amaz. 32, 517–519.

Damon, A., 2000. A review of the biology and control of the coffee berry borer, *Hypothenemus hampei* (Coleoptera: Scolytidae). Bull. Entomol. Res. 90, 453–465.

de la Rosa, W., Alatorre, R., Barrera, J.F., Toriello, C., 2000. Effect of *Beauveria bassiana* and *Metarhizium anisopliae* (Deuteromycetes) upon the coffee berry borer (Coleoptera: Scolytidae) under field conditions. J. Econ. Entomol. 93, 1409–1414.

de la Rosa, W., Figueroa, M., Ibarra, J.E., 2005. Selection of *Bacillus thuringiensis* strains native to Mexico active against the coffee berry borer, *Hypothenemus hampei* (Ferrari) (Coleoptera: Curculionidae: Scolytinae). Vedalia 12, 3–9.

Díaz-Vicente, J.M., Pérez-Quintanilla, J.N., Magallanes-Cedeño, R., Pinson-Rincón, E.P., de Coss-Flores, M.E., Cabrera-Alvarado, M.E., 2014. Control biológico de la broca del café *Hypothenemus hampei* Ferrari (Coleoptera: Curculionidae) con diferentes dosis del hongo *Beauveria bassiana* (Bals.) Vuill. (Moniliales: Moniliaceae) en Unión Juárez, Chiapas, México. Vedalia 15, 15–21.

Dick, G.L., Buschman, L.L., 1995. Seasonal occurrence of a fungal pathogen, *Neozygites adjarica* (Entomophthorales: Neozygitaceae), infecting banks grass mites, *Oligonychus pratensis*, and two-spotted spider mites, *Tetranychus urticae* (Acari: Tetranychidae), in field corn. J. Kans. Entomol. Soc. 68, 425–436.

Dufour, B.P., Frérot, B., 2008. Optimization of coffee berry borer, *Hypothenemus hampei* Ferrari (Col., Scolytidae), mass trapping with an attractant mixture. J. Appl. Entomol. 132, 591–600.

FAO STAT, 2013. http://faostat3.fao.org/.

Feng, M.-G., Pu, X.-Y., Ying, S.-H., Wang, Y.-G., 2004. Field trials of an oil-based emulsifiable formulation of *Beauveria bassiana* conidia and low application rates of imidacloprid for control of false-eye leafhopper *Empoasca vitis* on tea in southern China. Crop Prot. 23, 489–496.

Fransen, J.J., 1990. Natural enemies of whiteflies: fungi. In: Gerling, D. (Ed.), Whiteflies: Their Bionomics, Pest Status and Management. Intercept, Andover, UK, pp. 187–210.

Gaitán, A.L., Cristancho, M.A., Castro Caisedo, B.L., Rividas, C.A., Cadena Gómez, G., 2015. Compendium of Coffee Diseases and Pests. Am. Phytopathol. Soc., St. Paul. 75 pp.

Georgis, R., Koppenhöfer, A.M., Lacey, L.A., Bélair, G., Duncan, L.W., Grewal, P.S., Samish, M., Tan, L., Torr, P., van Tol, R.W.H.M., 2006. Successes and failures in the use of parasitic nematodes for pest control. Biol. Control 38, 103–123.

Guzzo, S.D., Martins, E.M.F., 1996. Local and systemic induction of β-1,3-glucanase and chitinase in coffee leaves protected against *Hemileia vastatrix* by *Bacillus thuringiensis*. J. Phytopathol. 144, 449–454.

Hall, R.A., 1981. The fungus *Verticillium lecanii* as a microbial insecticide against aphids and scales. In: Burges, H.D. (Ed.), Microbial Control of Pests and Plants Disease. Academic Press, New York, pp. 483–498.

Haraprasad, N., Niranjana, S.R., Prakash, H.S., Shetty, H.S., Wahab, S., 2001. *Beauveria bassiana* – a potential mycopesticide for the efficient control of coffee berry borer, *Hypothenemus hampei* (Ferrari) in India. Biocontrol Sci. Technol. 11, 251–260.

Hazarika, L.X., Bhatacharyya, B., Kalita Das, P., 2005. Bt as a biocide and its role in management of tea pests. Int. Tea Sci 4, 7–16.

Hazarika, L.K., Bhuyan, M., Hazarika, B.N., 2009. Insect pests of tea and their management. Ann. Rev. Entomol 54, 267–284.

Hiromori, H., Yaginuma, D., Kajino, K., Hatsukade, M., 2004. The effects of temperature on the insect activity of *Beauveria amorpha* to *Heptophylla picea*. Appl. Entomol. Zool 39, 389–392.

ICO, 2014. International Coffee Organization—World Coffee Trade (1963–2013): A Review of the Markets, Challenges and Opportunities Facing the Sector. International Coffee Organization, London, ICC, pp. 111–115. http://ico.org.

ICO, 2015. Total coffee production by all exporting countries in 2014. http://ico.org/new_historical.asp.

Infante, F., Pérez, J., Vega, F.E., 2014. The coffee berry borer: the centenary of a biological invasion in Brazil. Braz. J. Biol. 74 (Suppl.), S125–S126.

Inglis, G.D., Goettel, M.S., Butt, T.M., Strasser, H., 2001. Use of hyphomycetous fungi for managing insect pests. In: Butt, T.M., Jackson, C.W., Magan, N. (Eds.), Fungi as Biocontrol Agents: Progress, Problems and Potential. CABI Publishing, Wallingford, Oxon, UK, pp. 23–70.

Irulandi, S., Rajendran, R., Samuel, S.D., Chinniah, C., Kumar, P.K.V., Sreedharan, K.S., 2008. Efficacy of *Beauveria bassiana* (Balsamo) Vuill. and an exotic parasitoid, *Cephalonomia stephanoderis* (Betrem) (Hymenoptera: Bethylidae) for the management of coffee berry borer, (Ferrari) (Coleoptera: Scolytidae). J. Biol. Control 22, 91–97.

Jaramillo, J., Borgemeister, C., Baker, P., 2006. Coffee berry borer, *Hypothenemus hampei* (Coleoptera: Curculionidae): searching for sustainable control strategies. Bull. Entomol. Res. 96, 1–12.

Kariya, A., 1977. Control of tea pests with *Bacillus thuringiensis*. JARQ 11, 173–178.

Kitaoka, D., Ohkawa-Ohtsu, N., Murase, K., Omata, R., Nakajima, K., Hayashi, A., Suzuki, S., NAKAI, M., Yokayama, T., Kimura, S.D., 2013. Effect of chemical fertilizer and poultry manure on nitrogen content of tea leaves (*Camellia sinensis* (L.) O. Kuntze var. *sinensis*) and on development of Kanzawa spider mite *Tetranychus kanzawai* Kishida (Acari: Tetranychidae) ingesting the tea leaves. Jpn. J. Org. Agric. Sci. 5, 37–44.

Kumar, P.K.V., Prakasan, C.B., Vijayalakshmi, C.K., 1994. Record of entomopathogenic fungi on *Hypothenemus hampei* (Ferrari) from South India. J. Coffee Res. 24, 119–120.

Lacey, L., Siegel, J.P., 2000. Safety and ecotoxicology of entomopathogenic bacteria. In: Charles, J.-F., Delécluse, A., Nielsen-LeRoux, C. (Eds.), Entomopathogenic Bacteria: From Laboratory to Field Application. Kluwer Academic Publishers, Dordrecht, pp. 253–273.

Lacey, L.A., Wraight, S., Kirk, A.A., 2008. Entomopathogenic fungi for control of *Bemisia tabaci* biotype B: foreign exploration, research and implementation. In: Gould, J., Hoelmer, K., Goolsby, J. (Eds.), Classical Biological Control of *Bemisia tabaci* in the United States – A Review of Interagency Research and Implementation. Springer, Dordrecht, pp. 33–69.

Lara, G.J.C., López, N.J.C., Bustillo, P.A.E., 2004. Effecto de entomonemátodos sobre poblaciones de la broca del café, *Hypothenemus hampei* (Coleoptera: Scolytidae), en frutos en el suelo. Rev. Colomb. Entomol. 30, 179–185.

Lecuona, R.E., Fernandes, P.M., Alves, S.B., Bleicher, E., 1986. Patogenicidade de *Metarhizium anisopliae* (Metsch.) Sorok., à broca-do-café *Hypothenemus hampei* (Ferrari, 1867) (Coleoptera: Scolytidae). An. Soc. Entomol. 15, 21–27.

Lehmann-Danzinger, H., 2000. Diseases and pests of tea: overview and possibilities of integrated pest and disease management. J. Agric. Trop. Subtrop. 101, 13–38.

Leroy, T., Henry, A.M., Royer, M., Altosaar, I., Frutos, R., Duris, D., 2000. Genetically modified coffee plants expressing the *Bacillus thuringiensis cry1Ac* gene for resistance to leaf miner. Plant Cell Rep. 19, 382–389.

Mamun, M.S.A., Ahmed, M., Hoque, M.M., Sikder, M.B.H., Mi, A., 2014. *In vitro* and *in vivo* screening of some entomopathogens against red spider mite, *Oligonychus coffeae* Nietner (Acarina: Tetranychidae) in tea. Tea J. Bangladesh 43, 34–44.

Mao, H., Kunimi, Y., 1991. Pupal mortality of the oriental tea tortrix, *Homona magnanima* DIAKONOFF (Lepidoptera: Tortricidae), caused by parasitoids and pathogens. Appl. Entomol. Zool. 35, 241–245.

Manton, J.L., Hollingsworth, R.G., Cabos, R.Y.M., 2012. Potential of *Steinernema carpocapsae* (Rhabditida: Steinernematidae) against *Hypothenemus hampei* (Coleoptera: Curculionidae) in Hawaii. Fla. Entomol. 95, 1194–1197.

Meekes, E.T.M., Fransen, J.J., van Lenteren, J.C., 2002. Pathogenicity of *Aschersonia* spp. against whiteflies *Bemisia argentifolii* and *Trialeurodes vaporariorum*. J. Invertebr. Pathol. 81, 1–11.

Méndez, A., García, I., Jiménez, L.C., Pozo, E., 2014. Natural epizooty of *Beauveria bassiana* (Bals.-Criv.) Vuill. In *Hypothenemus hampei* Ferrari in Mayabeque, Cuba. Rev. Prot. Veg. 29, 150.

Méndes-López, I., Basurto-Ríos, R., Ibarra, J.E., 2003. *Bacillus thuringiensis* serovar *israelensis* is highly toxic to the coffee berry borer, *Hypothenemus hampei* Ferr. (Coleoptera: Scolytidae). FEMS Microb. Lett. 226, 73–77.

Monzón, A.J., Guharay, F., Klingen, I., 2008. Natural occurrence of *Beauveria bassiana* in *Hypothenemus hampei* (Coleoptera: Curculionidae) populations in unsprayed coffee fields. J. Invertebr. Pathol. 97, 134–141.

Murphy, S.T., 1997. Coffee. In: Ben-Dov, Y., Hogson, C.J. (Eds.), Soft Scale Insects: Their Biology, Natural Enemies and Control. Elsevier Science B.V., Amsterdam, The Netherlands, pp. 367–380.

Nahayo, A., Bayisenge, J., 2012. Biological control of coffee antestia bugs (*Antestiopsis lineaticolis*) by using *Beauveria bassiana*. N.Y. Sci. J. 5, 106–113.

Nakai, M., 2009. Biological control of Tortricidae in tea fields in Japan using insect viruses and parasitoids. Virol. Sin. 24, 323–332. http://dx.doi.org/10.1007/s12250-009-3057-9.

Nakai, M., Takeda, M., Kunimi, Y., 1997. Seasonal changes in prevalence of viral disease and parasitism by parasitic insects in a larval population of the smaller tea tortrix, Adoxophyes sp. (Lepidoptera: Tortricidae) in a tea field. Appl. Entomol. Zool. 32, 609–615.

Nielsen, C., Wraight, S.P., 2009. Exotic aphid control with pathogens. In: Hajek, A.E., Glare, T.R., O'Callaghan, M. (Eds.), Use of Microbes for Control and Eradication of Invasive Arthropods. Springer, Dordrecht, pp. 93–113.

Nishi, Y., Nonaka, T., 1996. Biological control of the tea tortrix, using granulosis virus in the tea field. Agrochem. Jpn. 69, 7–10.

de Oliveira, R.C., Alves, L.F.A., Neves, P.M.O.J., 2002. Suscetibilidade de *Oligonychus yothersi* (Acari: Tetranychidae) ao fungo *Beauveria bassiana*. Sci. Agric. 59, 187–189.

Oliveira, C.M., Auad, A.M., Mendes, S.M., Frizzas, M.R., 2013. Economic impact of exotic insect pests in Brazilian agriculture. J. Appl. Entomol. 137, 1–15.

Ortiz, A., Ortiz, A., Vega, F.E., Posada, F., 2004. Volatile composition of coffee berries at different stages of ripeness and their possible attraction to the coffee berry borer *Hypothenemus hampei* (Coleoptera: Curculionidae). J. Agric. Food Chem. 52, 5914–5918.

Pelley, R.H., 1973. Coffee insects. Annu. Rev. Entomol. 18, 121–142.

Peng, H., Feng, J., et al., 1998. Efficacy analysis of viral insecticide of *Buzura supperssaria* single nuclear polyhedrosis virus (BsSNPV). Chin. J. Appl. Environ. Biol. 4, 258–262. Abstract was shown in Chin. J. Appl. Environ. Biol. http://www.cibj.com/en/oa/DArticle.aspx?type=view&id=693http://www.cabdirect.org/abstracts/19951107369.html?freeview=true.

Pérez, J., Infante, F., Vega, F.E., 2015. A coffee berry borer (Coleoptera: Curculionidae: Scolytinae) bibliography. Insect Sci. 15, 1–41. http://dx.doi.org/10.1093/jisesa/iev053.

Perthuis, B., Pradon, J.L., Montagnon, C., Dufour, M., Leroy, T., 2005. Stable resistance against the leaf miner *Leucoptera coffeella* expressed by genetically transformed *Coffea canephora* in a pluriannual field experiment in French Guiana. Euphytica 144, 321–329.

Poinar, G., Vega, F.E., Castillo, A., Chaves, I.E., Infante, F., 2004. *Metaparasitylenchus hypothenemi* n. sp. (Nematoda: Allantonematidae), a parasite of the coffee berry borer *Hypothenemus hampei* (Curculionidae). J. Parasitol. 90, 1106–1110.

Polanczyk, R.A., De Bortoli, S.A., De Bortoli, C.P., 2012. *Bacillus thuringiensis* – based biopesticides against agricultural pests in Latin America. In: Soloneski, S. (Ed.), Integrated Pest Management and Pest Control – Current and Future Tactics. InTech. Available from: http://www.intechopen.com/books/integrated-pest-management-and-pest-control-current-and-future-tactics/bacillus-thuringiensis-based-biopesticides-against-agricultural-pests-in-latin-america.

Poprawski, T.J., Parker, P.E., Tsai, J.H., 1999. Laboratory and field evaluation of hyphomycete insect pathogenic fungi for control of brown citrus aphid (Homoptera: Aphididae). Environ. Entomol. 28, 315–321.

Posada, F., Vega, F.E., 2006. Inoculation and colonization of coffee seedlings (*Coffea arabica* L.) with the fungal entomopathogen *Beauveria bassiana* (Ascomycota: Hypocreales). Mycoscience 47, 284–289.

Posada, F.J., Aime, M.C., Peterson, S.W., Rehner, S.A., Vega, F.E., 2007. Inoculation of coffee plants with the fungal entomopathogen *Beauveria bassiana* (Ascomycota: Hypocreales). Mycol. Res. 111, 748–757.

Reddy, P.P., 2014. Biointensive Integrated Pest Management in Horticultural Ecosystems. Springer, Dordrecht. 280 pp.

Roy, S., Muraleedharan, N., 2014. Microbial management of arthropod pests of tea: current state and prospects. Appl. Mirobiol. Biotechnol. 98, 5375–5386.

Saito, T., Takatsuka, J., Shimazu, M., 2012. Characterization of *Paecilomyces cinnamomeus* from the camellia whitefly, *Aleurocanthus camelliae* (Hemiptera: Aleyrodidae), infesting tea in Japan. J. Invertebr. Pathol. 110, 14–23.

Sampedro-Rosas, L., Villanueva-Arce, J., Rosas-Acevedo, J.L., 2008. Aislamiento y validación en campo de *Beauveria bassiana* (Balsamo) contra *Hypothenemus hampei* (Ferrari) en la región cafetalera del municipio de Atoyac de Álvarez, Gro. México. Rev. Latinoamer. Rec. Nat. 4, 199–202.

Samuels, R.I., Pereira, R.C., Gava, C.A.T., 2002. Infection of the coffee berry borer *Hypothenemus hampei* (Coleoptera: Scolytidae) by Brazilian isolates of the entomopathogenic fungi *Beauveria bassiana* and *Metarhizium anisopliae* (Deuteromycotina: Hyphomycetes). Biocontrol Sci. Technol. 12, 631–635.

Sanchis, V., 2011. From microbial sprays to insect-resistant transgenic plants: history of the biospesticide *Bacillus thuringiensis*. A review. Agron. Sustain. Dev. 31, 217–231.

Singha, D., Singha, B., Dutta, B.K., 2010. In vitro pathogenicity of *Bacillus thuringiensis* against tea termites. J. Biol. Control 24, 279–281.

Suáez, H.D., Mejía, J., 2012. Manejo de *Hypothenemus hampei* (Coleoptera: Scolytidae) con *Beauveria bassiana* en cafetales de Pueblo Bello (Cesar). Rev. Colomb. Microbiol. Trop. 2, 41–50.

Sun, X., 2015. History and current status of development and use of viral insecticides in China. Viruses 7, 306–319. http://dx.doi.org/10.3390/v7010306.

Sun, J.D., Wu, G.Y., Lin, A., Zeng, M.S., Wang, Q.S., Xu, D.Y., 1993. Investigation and demonstration of the integrated control the tea weevil by a mixture of pesticides and microbes. Tea Sci. Technol. Bull. 3, 32–34. Translated by CABI http://www.cabdirect.org/abstracts/19951107369.html?freeview=true.

Takahashi, M., Nakai, M., Saito, Y., Sato, Y., Ishijima, C., Kunimi, Y., 2015. Field efficacy and transmission of fast- and slow-killing nucleopolyhedroviruses that are infectious to *Adoxophyes honmai* (Lepidoptera: Tortricidae). Viruses 5, 1271–1283.

Takatsuka, J., Okuno, S., Ishii, T., Nakai, M., Kunimi, Y., 2010. Fitness-related traits of entomopoxviruses isolated from *Adoxophyes honmai* (Lepidoptera: Tortricidae) at three locations in Japan. J. Invertebr. Pathol. 105, 121–131.

Uchiyama, T., Ozawa, A., 2014. Rapid development of resistance to diamide insecticides in the smaller tea tortrix, *Adoxophyes honmai* (Lepidoptera: Tortricidae), in the tea fields of Shizuoka Prefecture, Japan. Appl. Entomol. Zool. 49, 529–534.

Vail, P.V., Hostetter, D.L., Hoffmann, D., 1999. Development of multinucleocapsid polyhedroviruses (MNPVs) infectious to loopers as microbial control agents. Integ. Pest Manag. Rev. 4, 231–257.

van der Geest, L.P.S., Elliot, S.L., Breeuwer, J.A.J., Beerling, E.A.M., 2000. Diseases of mites. Exp. Appl. Acarol. 24, 497–560.

Vázquez Moreno, L.L., Elosegui Claro, O., Leyva Cardona, L., Polanco Aballe, A., Becerra Barrios, M., Monzón, S., Rodríguez Fernández, A., Tamayo Rosales, E., Toledo Duque de Estrada, C., Navarro Lantes, A., García Hernández, M., 2010. Ocurrencia de epizootias causadas por *Beauveria bassiana* (Bals.) Vuill. en poblaciones de la broca del café (*Hypothenemus hampei* Ferrari) en las zonas cafetaleras de Cuba. Fitosanidad 14, 111–116.

Vega, F.E., 2008. Insect pathology and fungal endophytes. J. Invertebr. Pathol. 98, 277–279.

Vega, F.E., Mercadier, G., Damon, A., Kirk, A., 1999. Natural enemies of the coffee berry borer, *Hypothenemus hampei* (Ferrari) (Coleoptera: Scolytidae) in Togo and Cote D'ivoire, and other insects associated with coffee beans. Afr. Entomol. 7, 243–248.

Vega, F.E., Pava-Ripoll, M., Posada, F., Buyer, J.S., 2005. Endophytic bacteria in *Coffea arabica* L. J. Basic Microbiol. 45, 371–380.

Vega, F.E., Dowd, P.F., Lacey, L.A., Pell, J.K., Jackson, D.M., Klein, M.G., 2007. Dissemination of beneficial microbial agents by insects. In: Lacey, L.A., Kaya, H.K. (Eds.), Field Manual of Techniques in Invertebrate Pathology: Application and Evaluation of Pathogens for Control of Insects and Other Invertebrate Pests, second ed. Springer, Dordrecht, pp. 127–146.

Vega, F.E., Posada, F., Aime, M.C., Pava-Ripoll, M., Infante, F., Rehner, S.A., 2008. Entomopathogenic fungal endophytes. Biol. Control 46, 72–82.

Vega, F.E., Infante, F., Castillo, A., Jaramillo, J., 2009. The coffee berry borer, *Hypothenemus hampei* (Ferrari) (Coleoptera: Curculionidae): a short review, with recent findings and future research directions. Terr. Arthropod Rev. 2, 129–147.

Vega, F.E., Meyling, N.V., Luangsa-ard, J.J., Blackwell, M., 2012. Fungal entomopathogens. In: Vega, F.E., Kaya, H.K. (Eds.), Insect Pathology, second ed. Academic Press, San Diego, pp. 171–220.

Vega, F.E., Infante, F., Johnson, A.J., 2015. The genus *Hypothenemus*, with emphasis on *H. hampei*, the coffee berry borer. In: Beetles, B. (Ed.), Biology and Ecology of Native and Invasive Species. Academic Press, San Diego, pp. 427–494.

Vélez-Arango, P.E., 1997. Evaluación de formulaciones en aceite y en agua del hongo *Beauveria bassiana* (Balsamo) Vuillemin en campo. Rev. Colomb. Entomol. 23, 59–64.

Vera, J.T., Montoya, E.C., Benavides, P., Góngora, C.E., 2010. Evaluation of *Beauveria bassiana* (Ascomycota: Hypocreales) as a control of the coffee berry borer *Hypothenemus hampei* (Coleoptera: Curculionidae Scolytinae) emerging from fallen, infested coffee berries on the ground. Biocontrol Sci. Technol. 21, 1–14.

Wang, D.-F., Liu, F.-J., Li, H.-L., et al., 2013. Time-dose-mortality model analysis of *Beauveria bassiana* XJBb3005 against *Myllocerinus aurolineatus*. Fujian J. Agricul. Sci. 28, 807–811.

Waller, J.M., Bigger, M., Hillocks, R.J., 2007. Coffee Pests, Diseases and Their Management. CABI Publishing, Wallingford, Oxfordshire, UK. 434 pp.

Wekesa, V.W., Hountondji, F.C., Dara, S.K., 2015. Mite pathogens and their use in biological control. In: Carrillo, D., Moraes, G.J., Peña, J.E. (Eds.), Prospects for Biological Control of Plant Feeding Mites and Other Harmful Organisms. Springer International Publishing, Dordrecht, pp. 309–328.

Yaginuma, D., Hiromori, H., Hatsukade, M., 2004. Relationship between virulence and adhesion of entomopathogenic fungus *Beauveria amorpha* (Strain: HpBa-1) to the yellowish elongate chafer, *Heptophylla picea* (Motschulsky) (Coleoptera: Scarabaeidae). Jpn. J. Appl. Entomol. Zool. 48, 101–108 (in Japanese with English summary).

Yaginuma, D., Hiromori, H., Hatsukade, M., 2006. Virulence of the entomopathogenic fungus *Beauveria brongniartii* to several life stages of the yellowish elongate chafer *Heptophylla picea* Motschulsky (Coleoptera: Scarabaeidae). Appl. Entomol. Zool. 41, 287–293.

Ye, G.-Y., Xiao, Q., Chen, M., Chen, X.-X., Yuan, Z.-J., Stanley, D.W., Hu, C., 2014. Tea: biological control of insect and mite pests in China. Biol. Control 68, 73–79.

Chapter 16

Microbial Control of Mite and Insect Pests of Greenhouse Crops

S.P. Wraight[1], R.B. Lopes[2], M. Faria[2]

[1]USDA-ARS, R. W. Holley Center for Agriculture and Health, Ithaca, NY, United States; [2]EMBRAPA Genetic Resources and Biotechnology, Brasilia, Federal District, Brazil

16.1 GENERAL INTRODUCTION

The production of vegetable and ornamental crops in greenhouses and other protected environments has experienced strong growth in recent decades (Chang et al., 2013). The reasons are numerous, but they relate primarily to the potential for year-round production, regardless of ambient environmental conditions. Under the highly favorable growing conditions possible in these protected environments, productivity and profitability can be substantially greater compared with open-field systems, particularly where high-quality products can be offered to meet specific market demands. Despite the still considerable investments in infrastructure, the continued emergence of more-efficient technologies, equipment, and management systems has contributed to the growth of this farming system. Greenhouse production worldwide rose from 0.7 million ha two decades ago to 3.7 million in 2010. Growth has been greatest in China, Spain, South Korea, Japan, and Turkey, which together account for 96% of global protected-crop cultivation (Chang et al., 2013).

Greenhouse production (greenhouse hereafter referring to all forms of protected culture) presents many challenges, and among these, pest management has occupied a prominent position. In the same way that plants benefit from the optimal conditions offered by greenhouses, so do many plant pathogens and arthropod pests. Constant availability of host plants, often in monoculture, in combination with favorable environmental conditions can lead to pest population explosions. Most insects and mites attacking greenhouse crops have high reproductive capacities, producing numerous, overlapping generations in a single crop cycle. Management of these pests is further complicated by the fact that most are highly polyphagous and many, such as leaf miners, thrips, and mealybugs, reside in cryptic habitats such as soil, flowers, abaxial leaf surfaces, leaf nodes, and leaf galleries.

16.2 MAJOR GROUPS OF GREENHOUSE ARTHROPOD PESTS

Greenhouse crops are attacked by a great diversity of insects and other arthropod pests. Presented here is a brief introduction to the major groups of these pests and the types of crop damage they cause.

16.2.1 Sucking Arthropods

The most important arthropod pests of greenhouse crops worldwide feed on and damage their host plants by piercing cells or tissues with their needle-like mouthparts and sucking out liquid food materials. This group includes many species of thrips, whiteflies, aphids, mealybugs, scales, and mites; most are highly polyphagous. The tomato/potato psyllid *Bactericera cockerelli* and a few species of leafhoppers and heteropteran bugs are pests of field crops that invade greenhouses in some regions, causing greatest damage to fruit and vegetable crops (Malais and Ravensberg, 2003; Abdullah et al., 2008).

16.2.1.1 Thrips (Thysanoptera)

Many species of thrips (nearly all of the family Thripidae) infest greenhouse crops. The most commonly encountered pests include the Western flower thrips (WFT) *Frankliniella occidentalis*, onion thrips *Thrips tabaci*, melon thrips *Thrips palmi*, greenhouse thrips *Heliothrips haemorrhoidalis*, poinsettia or impatiens thrips *Echinothrips americanus*, and chilli thrips *Scirtothrips dorsalis*. Thrips are highly cryptic in habit, feeding on leaf undersides or in new leaf clusters enclosing meristem tissues of vegetative host plants or in developing flowers of mature plants where they are protected from direct inoculation via spray applications. These insects pierce plant epidermal cells and extract the liquid contents. Feeding produces chlorotic scarring of leaf

Microbial Control of Insect and Mite Pests. http://dx.doi.org/10.1016/B978-0-12-803527-6.00016-0

tissues, and injury to developing leaves and fruit results in deformities. Thrips are vectors of plant viruses (Tospoviridae and others), a problem that has impeded the adoption of Integrated Pest Management (IPM) practices.

16.2.1.2 Whiteflies (Hemiptera: Aleyrodidae)

Whiteflies are among the most important pests of greenhouse vegetable and ornamental crops. Economic damage is attributable primarily to two species, the greenhouse whitefly *Trialeurodes vaporariorum* and sweet potato whitefly *Bemisia tabaci*, although *B. tabaci* is widely viewed as a complex of morphologically indistinguishable sibling species, generally referred to as biotypes. Whiteflies are phloem-sap feeders, excreting copious quantities of excess water and sugars (honeydew) that support growth of molds; the consequent cosmetic damage to ornamentals, fruits, and vegetables is intolerable. *Bemisia* whiteflies inject toxic substances during feeding that cause chlorosis, fruit deformations, and uneven fruit ripening. However, greatest damage associated with whitefly infestations results from transmission of plant viruses (Jones, 2003).

16.2.1.3 Aphids (Hemiptera: Aphididae)

Greenhouse crops are damaged by numerous aphid pests, but a few highly polyphagous species, including green peach aphid *Myzus persicae*, melon aphid *Aphis gossypii*, potato aphid *Macrosiphum euphorbiae*, and foxglove aphid *Aulacorthum solani*, are of primary importance (Chow and Heinz, 2004). Reproduction in these pests is largely asexual, and development is rapid, leading to explosive population increases. Heavy infestations cause stunting of host plants; toxic substances injected during feeding also cause abnormal growth. As phloem feeders, aphids produce large amounts of honeydew. However, as with thrips and whiteflies, greatest damage from these pests may be caused by transmission of plant viruses.

16.2.1.4 Mealybugs and Scale Insects (Hemiptera)

Though generally considered minor pests of greenhouse crops, established infestations of these pests can be difficult to control. Important mealybug pests in greenhouses include the citrus mealybug *Planococcus citri*, obscure mealybug *Pseudococcus affinis*, and long-tailed mealybug *Pseudococcus longispinus*, all of the family Pseudococcidae (Tingle and Copland, 1988). Soft scales (Coccidae), most notably brown soft scale *Coccus hesperidum*, hemispherical scale *Saissetia coffeae*, and black scale *Saissetia oleae*, are common scale-insect pests of greenhouse crops (Gillespie and Costello, 2004). Mealybugs and soft scales are phloem-sap feeders, and the principal damage to greenhouse crops results from excretion of honeydew. A few

species of armored scales (Diaspididae), including oleander scale *Aspidiotus nerii*, can damage greenhouse plants.

16.2.1.5 Mites

Protected crops are damaged by numerous species of mites, primarily of the order Trombidiformes (suborder Prostigmata). These mites are fluid feeders, and the agricultural pests of the group pierce plant cells and suck out the contents. Affected leaves and stems turn chlorotic. During feeding, some species inject substances that interfere with plant growth, resulting in deformities similar to those caused by virus infections. The most important mite pests of greenhouse crops include the two-spotted spider mite *Tetranychus urticae*, carmine spider mite *T. cinnabarinus*, tomato red spider mite *Tetranychus evansi*, broad mite *Polyphagotarsonemus latus*, cyclamen mite *Phytonemus pallidus*, and tomato russet mite *Aculops lycopersici*.

16.2.2 Chewing Insects

The larvae of many insects of the orders Diptera, Hymenoptera, Coleoptera, and Lepidoptera can cause severe damage to greenhouse ornamental and horticultural crops. Many species are secondary pests, causing only minor damage or becoming serious problems only sporadically. However, these groups include many primary pests of regional specialty crops.

16.2.2.1 Diptera

The larval stages of some flies tunnel and feed within leaf tissues. A number of polyphagous species of the family Agromyzidae (particularly members of the genera *Liriomyza* and *Chromatomyia*) are important pests of greenhouse ornamental and vegetable crops worldwide (Capinera, 2001; Chow and Heinz, 2004). Some midges of the family Cecidomyiidae also infest greenhouse crops as gall formers and leaf miners. The cryptic larvae of leaf-mining pests are not readily targeted with control agents. Fungus gnats (Sciaridae) and shore flies (Ephydridae) are ubiquitous pests of greenhouse systems. Adults of these flies are important mechanical vectors of various plant pathogens and cause cosmetic damage to ornamentals. *Lycoriella* fungus gnats are pests of mushroom production; larvae feed on the compost substrate, rendering it unsuitable for production and also cause direct feeding damage to mushroom mycelia and fruiting bodies. Larvae of fungus gnats, including *Bradysia*, *Sciara*, and *Corynoptera* spp., feed on the roots and tunnel into the stems of many crop plants. Damage to large plants is generally minor, but damage to seedlings and cuttings can be severe. In protected environments, many overlapping generations occur each year, and control is difficult due to the cryptic (tunneling) behavior of these pests. Root maggots of the genus *Delia* (Anthomyiidae) are important pests of ornamental and vegetable crops in some regions. Larvae of most species bore into the roots and basal stems of host plants, but a few tunnel into leaves and shoots.

16.2.2.2 Coleoptera

Although many species of beetles damage greenhouse crops, few are considered primary pests. One exception is the black vine weevil *Otiorhynchus sulcatus*, a pest of many ornamental and horticultural crops throughout the temperate regions of the world (Moorhouse et al., 1992). The larvae feed primarily on plant roots but also girdle the crowns of their host plants. Unlike most key greenhouse pests, this insect has only one generation per year; however, because greenhouse infestations generally arise from multiple invasions, all stages of the pest are commonly present. Other curculionids, including the strawberry root weevil *O. ovatus* and the pepper weevil *Anthonomus eugenii*, can be equally damaging to certain specialty crops.

16.2.2.3 Lepidoptera

The most important lepidoperan pests of greenhouse crops are members of the family Noctuidae. Most noctuid pests are polyphagous and attack both open and protected crops. In the latter, these insects rarely attain key pest status, although larvae of species such as the tomato looper *Chrysodeixis chalcites*, tomato moth *Lacanobia oleracea*, beet armyworm *Spodoptera exigua*, cabbage looper *Trichoplusia ni*, cabbage moth *Mamestra brassicae*, and silver-Y moth *Autographa gamma* frequently cause considerable damage to various greenhouse vegetable and ornamental crops, including tomato, pepper, brassicas, lettuce, chrysanthemums, carnations, roses, and gerbera. Tortricid leaf rollers and gelechiid and pyralid borers are also common pests of many of the aforementioned crops and of soft fruits such as strawberry and raspberry; larvae of many species tunnel into plant stems and fruits or roll leaves with webbing, becoming much more damaging and difficult to control.

16.3 MICROBIAL BIOCONTROL AGENTS AND THEIR USE IN GREENHOUSE PEST MANAGEMENT

16.3.1 Biological Control of Arthropod Pests in Protected Crops

A number of recent reviews and other publications have addressed the biological control of greenhouse pests, including use of predators, parasitoids, entomogenous nematodes, and entomopathogenic microorganisms (eg, Van Driesche and Heinz, 2004; Heinz et al., 2004; Shipp et al., 2007; van Lenteren, 2009). In virtually all cases, these works have addressed deployment of these agents as components of IPM systems. With respect to our coverage of microbes used as biological control agents, we restrict our definition to living microbes, in accordance with the definition of biological control recommended by Eilenberg et al. (2001). Thus, our use of the terms microbial biological control or microbial biocontrol excludes preparations containing entomotoxic microbial byproducts alone (purified toxins) or microbes whose insecticidal activity derives solely from toxic metabolites (no difference in activity between live and dead preparations). We furthermore restrict our definition of biopesticides to formulations of living microbes. The bacterium *Bacillus thuringiensis* (Bt) produces potent toxins that, in many cases, are solely responsible for insecticidal activity. However, we treat this pathogen as a microbial biological control agent in light of the fact that nearly all Bt-based biopesticides contain living spores, and these spores and the vegetative cells they generate are known to contribute significantly to pathogenesis in some insects (Jurat-Fuentes and Jackson, 2012). Entomopathogenic nematodes (EPNs), though macrobial agents, act in concert with microbial symbionts (bacteria) that are directly responsible for death of the insect host, and we include a brief discussion of these agents.

The pest control efficacy of microbial control agents is highly dependent on specific environmental conditions, and environment can be manipulated to a greater degree in protected crops than in open fields. For example, some greenhouse glazings can reduce the intensity of UV-B radiation by about 90% (Lasa et al., 2007), promoting microbe survival and persistence. Unexpected or severe rain wash is not a factor in protected crops, and temperature and humidity conditions can often be regulated (though not without caveats, as discussed below). Because high-value vegetable and ornamental crops may require frequent, even daily harvests, there are obvious economic advantages to the use of biological control agents with minimal reentry and preharvest intervals. The unique modes of action of microbial control agents also make them potentially useful in pesticide rotation schemes for resistance management. Many greenhouse crops (particularly ornamental plants) are of higher value than field crops. This allows more frequent applications and applications at greater rates and using highly efficient (eg, hand-targeted) application equipment. There is potential also for efficient application via sophisticated greenhouse irrigation systems. Greenhouse environments support use of a broad range of beneficial macroorganisms (including predators, parasitoids, EPNs, and pollinators), and use of many microbial biocontrol agents is compatible with these organisms. Examples of insect pathogens applied against important pest species from these groups in greenhouses are cited in Table 16.1.

TABLE 16.1 Major Greenhouse Pests and Examples of Potential Microbial Biocontrol Agents

	Pest Arthropods	Main Greenhouse Crops	Microbial Control Agents	Selected Studies[a]
Thrips				
Western flower thrips	*Frankliniella occidentalis*	Ornamental and vegetable crops	*Beauveria bassiana, Metarhizium anisopliae, Steinernema feltiae*	Maniania et al. (2001), Buitenhuis and Shipp (2005), Ugine et al. (2007), and Wraight et al. (2016)
Onion thrips	*Thrips tabaci*	Ornamental and vegetable crops	*B. bassiana, Lecanicillium sp., M. anisopliae*	Pourian et al. (2010), Wu et al. (2013), and Annamalai et al. (2015)
Melon thrips	*Thrips palmi*	Ornamental and vegetable crops (mainly Cucurbitaceae and Solanaceae)	*B. bassiana*	Castineiras et al. (1996)
Chilli thrips	*Scirtothrips dorsalis*	Ornamental and vegetable crops (mainly pepper)	*B. bassiana, Isaria fumosorosea, Metarhizium brunneum*	Arthurs et al. (2013)
Whiteflies				
Greenhouse whitefly	*Trialeurodes vaporariorum*	Ornamental and vegetable crops	*Isaria fumosorosea, Lecanicillium muscarium, B. bassiana*	Shipp et al. (2003), Feng et al. (2004), and Fargues et al. (2005)
Sweetpotato whitefly	*Bemisia tabaci*	Ornamental and vegetable crops	*B. bassiana, I. fumosorosea, Lecanicillium muscarium, S. feltiae*	Cuthbertson and Walters (2005), Saito (2005), Lacey et al. (2008), Qiu et al. (2008), and Liu and Stansly (2009)
Aphids				
Cotton or melon aphid	*Aphis gossypii*	Ornamental and vegetable crops (mainly Cucurbitaceae)	*Lecanicillium longisporum*	Kim et al. (2010)
Green peach aphid	*Myzus persicae*	Ornamental and vegetable crops (mainly Brassicaceae and Solanaceae)	*B. bassiana*	Kapongo et al. (2008)
Cabbage aphid	*Brevicoryne brassicae*	Vegetable crops (Brassicaceae)	*Heterorhabditis bacteriophora, Steinernema carpocapsae*	Kaya et al. (1995)
Potato aphid	*Macrosiphum euphorbiae*	Ornamental and vegetable crops (mainly Solanaceae)	*B. bassiana*	Shah et al. (2000)
Foxglove aphid	*Aulacorthum solani*	Ornamental and vegetable crops (mainly tomato and pepper)	*B. bassiana, L. longisporum, Lecanicillium attenuatum, M. anisopliae*	Kim et al. (2007) and Jandricic et al. (2014)
Mealybugs				
Cryptic or citriculus mealybug	*Pseudococcus cryptus*	Tomato	*M. anisopliae*	Panyasiri et al. (2007)

Dipterans

Leafminers	*Liriomyza trifolii*	Ornamental and vegetable crops	*I. fumosorosea, S. carpocapsae*	Broadbent and Olthof (1995) and Wekesa et al. (2011)
	L. huidobrensis	Ornamental and vegetable crops	*S. feltiae, Heterorhabditis megidis*	Weintraub and Horowitz (1995)
Fungus gnats	*Lycoriella* spp.	Mushrooms	*S. feltiae, Bacillus thuringiensis* subsp. *israelensis*	Fenton et al. (2002) and Shamshad (2010)
	Bradysia spp.	Mushrooms, ornamentals	*S. feltiae, H. bacteriophora, Heterorhabditis indica*	Harris et al. (1995) and Shamshad (2010)
Root-feeding flies	*Delia* spp.	Vegetable crops (Brassicaceae)	*Metarhizium* spp., *S. feltiae, Steinernema afine*	Schroder et al. (1996), Willmott et al. (2002), and Chandler and Davidson (2005)

Lepidopterans

Cabbage looper	*Trichoplusia ni*	Tomato, sweet (bell) pepper, cucumber	*B. thuringiensis* subsp. *kurstaki, Trichoplusia ni* SNPV, *Autographa californica* MNPV	Erlandson et al. (2007) and Janmaat et al. (2007)
Beet armyworm	*Spodoptera exigua*	Ornamentals, sweet pepper	*Spodoptera exigua* MNPV	Bianchi et al. (2000) and Caballero et al. (2009)
Tomato moth	*Lacanobia olearacea*	Tomato	*B. thuringiensis* subsp. *kurstaki*	Jarrett and Burges (1982)
South American tomato pinworm	*Tuta absoluta*	Tomato	*B. thuringiensis, H. bacteriophora, S. carpocapsae, S. feltiae*	González-Cabrera et al. (2011) and Urbaneja et al. (2012)

Coleopterans

Black vine weevil	*Otiorhynchus sulcatus*	Ornamentals, strawberry	*Metarhizium* spp. *H. bacteriophora, S. feltiae, Steinernema kraussei*	Ansari et al. (2008b, 2010)

Mites

Two-spotted spider mite	*Tetranychus urticae*	Ornamental and vegetable crops	*B. bassiana*	Chandler et al. (2005)
Tomato red spider mite	*Tetranychus evansi*	Vegetable crops (Solanaceae)	*Neozygites floridana*	Navajas et al. (2013)
Broad mite	*Polyphagotarsonemus latus*	Ornamental and vegetable crops	*B. bassiana, Hirsutella thompsonii, Isaria* sp.	Peña et al. (1996)

[a]Except for *A. solani* and *P. latus*, all cited studies with microbial control agents were conducted under greenhouse conditions.

16.3.2 Entomopathogenic Bacteria

Microbial biocontrol of greenhouse pests using entomopathogenic bacteria is based on a single species, Bt. New products based on *Chromobacterium subtsugae* and heat-killed *Burkholderia* spp. have been registered for control of greenhouse pests, including aphids, whiteflies, thrips, and mites; however, the insecticidal activity of these bacteria is attributed to toxins and apparently not dependent on or enhanced by live cells (Martin et al., 2007). Here, we refer to these agents as biorational chemical insecticides (insecticides based on natural product chemistries).

16.3.2.1 Lepidoptera

In greenhouse production systems, spray applications of Bt have successfully controlled a number of Lepidoptera species. Commercial Bt-based products currently available for management of greenhouse pests, and their targets are presented in Table 16.2. Most products with activity against Lepidoptera are based on strains of *Bacillus thuringiensis* subsp. *kurstaki* (Btk) or *B. thuringiensis* subsp. *aizawai*

(Bta). Shelf-life of most dry or oil-based formulations of these pathogens is at least 2 years when stored below 25–30°C; shelf-life is shorter for water-based formulations (Devisetty et al., 1998).

Bt spray applications against Lepidoptera spp. infesting greenhouse crops have been most successful against early-instar larvae of noctuid pests, including *Lacanobia oleracea*, *Autographa gamma*, *Mamestra brassicae*, and *T. ni* (Burges and Jarrett, 1978; Jarrett and Burges, 1982; Gillespie and Costello, 2004). Product labels of dry formulations of Btk and Bta currently marketed in the United States call for applications at rates of 0.25–2 lb per ac or 100 gal (0.28–2.24 kg/ha or 378.5 L) for control of lepidopteran pests. Heavy reliance on Btk for control of *T. ni* in Canadian greenhouses has resulted in resistance, a problem attributed, in part, to the enhanced persistence of this entomopathogen in the greenhouse environment (Janmaat et al., 2007). Additionally, Bt is less effective against pests of the families Tortricidae, Gelechiidae, and Crambidae (eg, carnation leaf roller *Caecimorpha pronubana*, tomato pinworm *Keiferia lycopersicella*, and European pepper moth

TABLE 16.2 Representative Bt- and Baculovirus-Based Larvicides Currently Available for Management of Insect Pests in Greenhouse Crops

Microorganism		Target	Commercial Product	Manufacturer	References
Bacteria[a]	*Bacillis thuringiensis* subsp. *kurstaki*	*Plutella xylostella*, *Spodoptera* spp., *Helicoverpa* spp., *Heliothis* spp.	CoStar	Certis, USA	Website: certisusa.com
		P. xylostella, *T. ni*	Dipel	Valent BioSciences, USA	Website: microbials. valentbiosciences.com
		P. xylostella, *Spodoptera* spp., *T. ni*	BMP 123	Becker Microbial Products, USA	Websiste: beckermicrobial-productsinc.com
	B. thuringiensis subsp. *aizawai*	*P. xylostella*, *Spodoptera* spp., *T. ni*	Agree	Certis, USA	Website: certisusa.com
		P. xylostella, *Spodoptera* spp.	XenTari	Valent BioSciences, USA	Website: microbials. valentbiosciences.com
	B. thuringiensis subsp. *israelensis*	Fungus gnats (Diptera: Sciaridae)	Gnatrol	Valent BioSciences, USA	Website: microbials. valentbiosciences.com
Baculoviruses	*Autographa californica nucelopolyhedrovirus* (AcNPV)	*Trichoplusia ni*	Loopex	Sylvar Technologies, Canada	Websites: biocontrol.ch; sylvar.ca
	Spodoptera exigua nucleopolyhedrovirus (SeNPV)	*Spodoptera exigua*	Spod-X	Certis, USA	Website: certisusa.com
			Vir-ex	Biocolor S.L., Spain	Website: biocolor-tec.es
			Spexit	Andermatt Biocontrol, Switzerland	Website: biocontrol.ch

[a]Most Bt-based products are available in more than one formulation type and are marketed in a considerable number of countries. Listed products are usually recommended for control of outdoor pests (not listed) or control of listed pests in open field crops and sometimes are marketed in different countries under different trade names. Information on Bt-based products was updated from Roh et al. (2007).

Duponchelia fovealis) because the larvae of these insects are cryptic and do not feed on exposed (treated) plant surfaces. Although larvae of most species of Gelechiidae feed internally on parts of susceptible hosts, González-Cabrera et al. (2011) demonstrated effective control of tomato borer *Tuta absoluta* in a commercial greenhouse; Btk (formulated as CoStar WG) applied weekly at dosages ranging from 45 to 181 million international units (IU)/L in a volume of 2000 L/ha using a high-pressure hand-gun protected tomato fruit from damage.

16.3.2.2 Diptera

B. thuringiensis subsp. *israelensis* (Bti) is applied for control of larval fungus gnats of the genera *Bradysia* and *Lycoriella* (Diptera: Sciaridae). In one test performed under greenhouse conditions, Osborne et al. (1985) applied Bti against *B. coprophila* infesting potted *Hoya carnosa*. Pots were drenched with three applications of Bti at 8- to 10-day intervals at rates of 1.4 or 4.1 million IU/L; control at the respective rates was 64% and 80%. As with Lepidoptera, Bt is not highly effective against late-instar fungus gnat larvae, and infestations generally arise as completely mixed-age populations, necessitating frequent applications (Shamshad, 2010). Also, the larvae cause considerable damage by boring into the roots and stems of plant cuttings and seedlings and into the spawn mycelia, stems, and caps of mushrooms, making them difficult targets for control. Results have consequently been inconsistent.

Bt varieties exhibiting activity against agromyzid pests have been reported (Payne et al., 1994); however, to our knowledge no commercial biopesticides are registered for agromyzid leaf-miner control.

16.3.3 Entomopathogenic Viruses

These microbial biocontrol agents can be highly effective against susceptible lepidopteran pests. For example, greater than 90% control of early-instar *Spodoptera exigua* larvae has been consistently achieved in greenhouse crops treated with SeMNPV applied at rates of $1–3 \times 10^{11}$ OB/ha (Bianchi et al., 2000; Lasa et al., 2007).

Baculoviruses are completely safe to nontarget organisms as a result of their narrow host ranges, making them compatible with other biological control agents used for greenhouse pest management. Like Bt, they are extremely sensitive to solar radiation, and so their persistence and efficacy can be enhanced in certain protected crop environments. Commercial baculovirus-based products currently available on the world market for management of greenhouse pests and their targets are presented in Table 16.2. Few of the primary pests of greenhouse crops are potential targets for insect pathogenic viruses, and these pathogens have played only a minor role in greenhouse pest management.

16.3.4 Entomopathogenic Fungi

The most important arthropod pests of greenhouse crops (including the thrips, whiteflies, aphids, mealybugs, and mites) are susceptible to fungi that infect their hosts via direct penetration of the body wall. This mode of action is advantageous in that it broadens the host range of these pathogens, but the infection process is highly dependent on environmental conditions of moisture, temperature, and insolation.

16.3.4.1 Commercial Fungus-Based Products

In a survey conducted nearly 10 years ago, 52 of 171 mycoinsecticides and mycoacaricides (about 30%) identified from the literature and websites had been developed for control of key greenhouse pests such as whiteflies, aphids, thrips, and tetranychid mites (de Faria and Wraight, 2007). Many of these products are no longer available or were designed to control these pests in field crops; however, a recent survey shows that at least 28 mycoinsecticides and/or mycoacaricides are currently marketed worldwide for use in protected environments (Table 16.3). These products are based on *Lecanicillium* spp. (*Lecanicillium muscarium* and possibly other species), *Isaria* spp. (*Isaria javanica* and possibly other species), *Beauveria bassiana* sensu lato, and *Metarhizium anisopliae* sensu lato. *Lecanicillium longisporum*, formerly identified as *Verticillium lecanii* and formulated as Vertalec, is no longer marketed (product discontinued due to limited market size). Unlike most of the conidia-based mycoinsecticides that have been used for control of pests in large-hectarage annual crops and rangelands, mycoinsecticides developed for greenhouse applications have been based on either conidia or yeast-like hyphal bodies called blastospores, which are produced via liquid fermentation. Despite numerous advantages, development of blastospore-based biopesticides has lagged because blastospores are more difficult to harvest, stabilize, and formulate than conidia and because even moderate shelf-life of desiccation-stabilized blastospores (about 9–14 months) is dependent on refrigeration (Mascarin et al., 2015). This contrasts sharply with conidia-based formulations, which if hermetically packaged with desiccants and oxygen scavengers can exhibit even longer shelf-lives at temperatures of 25°C or greater (Faria et al., 2012).

16.3.4.2 Abiotic and Biotic Constraints

The essential need for a near–moisture-saturated microenvironment and moderate temperatures account for much of the inconsistency in efficacy that has hindered the development of fungus-based biopesticides. With respect to environmental moisture, the required near-saturation conditions in the spore microenvironment required for germination can obviously arise from the ambient environment (Vidal et al., 1998)

TABLE 16.3 Representative Mycoinsecticides Currently Adopted for Control of Greenhouse Pest Insects and Mites

Country(ies) Where Registered/Marketed	Trade Name	Key Greenhouse Pests Targeted (Claimed)[a]	Manufacturer	Source(s)
***Beauveria bassiana* Sensu Stricto**				
USA, Mexico, Colombia, Denmark, Sweden, Italy, Spain, Japan	BotaniGard 22 WP	Whiteflies, thrips, aphids, psyllids, mealybugs, weevils	LAM International, USA	Product label; Website: laverlamintl.com
USA, Mexico, Denmark, Sweden, Italy, Spain, Japan	BotaniGard ES	Whiteflies, thrips, aphids, psyllids, mealybugs	LAM International, USA	Product label; Website: laverlamintl.com
USA, Mexico, Colombia, Denmark, Sweden, Italy	Mycotrol O	Whiteflies, thrips, aphids, psyllids, mealybugs	LAM International, USA	Product label; Website: laverlamintl.com
USA, Mexico, Greece, Italy, Spain, Switzerland	Naturalis-L	Whiteflies, thrips	Troy Biosciences (USA)	Product label
***B. bassiana* Sensu Lato**				
Brazil	Ballvéria	Whiteflies	Ballagro Agro Tecnologia (Brazil)	Website: agricultura.gov.br
Brazil	Bovebio	Whiteflies, mites	Biofungi – Industria e Comercio de Defensivos Biológicos e Inoculantes (Brazil)	Website: agricultura.gov.br
Brazil	Boveril WP	Mites	Koppert Brasil	Website: koppert.com.br
Colombia	BioExpert	Whiteflies, thrips	Live Systems Technology (Colombia)	Website: ica.gov.co
Colombia	Cadete SC	Whiteflies, thrips	Mycros Internacional (Colombia)	Website: ica.gov.co
Colombia	Adral	Aphids	Bio-Crop (Colombia)	Website: ica.gov.co
Colombia	Bassicore SC	Whiteflies	Core Biotechnology (Colombia)	Website: ica.gov.co
South Africa	Broadband	Whiteflies, thrips, mites, diamondback moth	BASF South Africa	Product label; Website: agro.basf.co.za
Lecanicillium muscarium				
Denmark, Finland, Italy, UK, Netherlands, Italy, Turkey, Switzerland, Japan, France	Mycotal	Whiteflies, thrips	Koppert Biological Systems (Netherlands)	Website: koppert.com
***Lecanicillium* spp.**				
Honduras, El Salvador, Nicaragua, Jamaica	Verzam	Whiteflies, aphids, thrips, mites	Escuela Agrícola Panamericana (Honduras)	R. Trabanino (personal communication)
Colombia	Vercani WP	Whiteflies	Natural Control (Colombia)	Website: ica.gov.co
Uruguay	Lecafol	Whiteflies	Lage y Cía. S.A., Uruguay	Website: lageycia.com

Metarhizium brunneum

Country/Region	Product	Pests	Company	Source
USA	Met52 EC	Whiteflies, thrips, weevils	Novozymes Biologicals (USA)	Product label; Websites: met52ec.com; bioag.novozymes.com
USA, Europe (>10 countries)	Met52 Granular	Weevils, thrips	Novozymes Biologicals (USA)	Product label; Websites: met52ec.com; bioag.novozymes.com
Netherlands	Bio1020	Weevils	Novozymes Biologicals France	Website: ctb.agro.nl

Metarhizium anisopliae Sensu Lato

Country/Region	Product	Pests	Company	Source
Kenya	Achieve	Mites	Real IPM (Kenya)	Website: realipm.com; Product label
Kenya, South Africa, Ethiopia, Ghana	Real Metarhizium 69 OD	Mealybugs, thrips, whiteflies, weevils	Real IPM (Kenya)	Website: realipm.com; Product label

Isaria javanica

Country/Region	Product	Pests	Company	Source
USA, Mexico	PFR-97 20% WDG	Whiteflies, aphids, thrips, mealybugs, psyllids, weevils leafminers, mites	Certis, Inc. (USA)	Product label; Website: certisusa.com
Europe (many countries)	PreFeRal WG	Whiteflies	Biobest (Belgium)	Website: biobestgroup.com

Isaria spp.

Country/Region	Product	Pests	Company	Source
USA	Nofly	Whiteflies, aphids, thrips, psyllids, mealybugs, weevils, fungus gnats	Novozymes Biologicals (USA)	Product label
Mexico	Pae-Sin	Whiteflies	Agrobiológicos del Noroeste – Agrobionsa (Mexico)	Website: agrobionsa.com
Colombia	Successor SC	Whiteflies, aphids, thrips, mites	Live Systems Technology (Colombia)	Website: ica.gov.co L. Cortés (personal communication)
Colombia	Acarex	Mites	Bio-Crop (Colombia)	Website: ica.gov.co
Colombia	Sporada SC	Mites	Mycros Internacional (Colombia)	Website: ica.gov.co

[a]Many of the listed products are also used to control nongreenhouse pests (not included in the list).
Updated and adapted from de Faria, M.R., Wraight, S.P., 2007. Mycoinsecticides and mycoacaricides: a comprehensive list with worldwide coverage and international classification of formulation types. Biol. Control 43, 237–256.

but also from niches on the host's body with thin cuticle through which water vapor can readily pass and establish a humid boundary layer; from similar niches/boundary layers surrounding leaves or other parts of the host plant (habitable by small or minute insects); or from the insect–host plant association or interaction. Examples of the latter include the humid microclimate established between a plant phylloplane and a sessile insect and around the head and mouthparts of an insect chewing on succulent plant materials (a mode of feeding that contrasts sharply with the piercing/sucking mode, which generates little moisture for exploitation by fungal pathogens).

Many of the primary pests of greenhouse crops have the potential to develop rapidly (passing quickly through immature stages), and populations of these insects grow as multiple overlapping generations (resulting in mixed-age populations). This means that at any given time, a significant number of individuals in the population is nearing a molt, reducing their chances of becoming infected. On reaching adulthood, this protection is lost, but adults gain a measure of protection by being less intimately associated with humid boundary layers of the host plant and in some cases by development of a heavily sclerotized cuticle. Adults of these pests also have short prereproductive periods (generally less than 2 days), and the reproductive system is one of the last to be affected by fungal infection (Hall, 1976). Thus, infected insects continue to oviposit or larviposit at nearly normal rates until the day of death, which even under optimal temperature conditions for fungal activity does not generally occur until after 3–4 days (Ugine et al., 2006; Jandricic et al., 2014). Many of the key greenhouse pests are vectors of highly destructive plant viruses and other pathogens, which can translate into very low economic thresholds.

Efforts to control difficult pest such as aphids and thrips have included intensive spray programs of frequent, high-dose applications of fungal pathogens, usually exceeding 10^{13} propagules/ha. Many mycoinsecticide labels indicate that application every 5 days or even more frequently may be necessary to achieve control of some pests. Even for high-value specialty crops, however, such control programs are costly and generally not sustainable, greatly increasing risks of resistance development and creating strong economic incentive for pest managers to seek alternatives. Initially encouraging studies in chrysanthemum suggested that greenhouse environments could be readily manipulated to support epizootic activity of fungal pathogens (Hall and Burges, 1979; Helyer et al., 1992). Helyer et al. (1992) concluded that epizootics of *Lecanicillium* sp. (identified as *Verticillium lecanii*) producing effective control of aphids could be induced by maintaining high humidity for 4 consecutive nights following fungal applications or by a cycle of 2 nights high humidity/2 nights ambient.

However, this approach has been countered by concerns of plant pathologists, who advocate keeping greenhouse environments as dry as possible to prevent disease outbreaks (Elad and Shtienberg, 1995). Dry conditions are generally maintained by active ventilation and drip-irrigation systems. Studies to identify the levels of moisture required for the activity of fungal pathogens against greenhouse pests and manipulation of greenhouse environments to support activity of insect pathogens without leaving crops at unacceptable risk from plant diseases has remained an active area of research (Fargues et al., 2005; Shipp et al., 2003; Vidal et al., 2003). Shipp et al. (2003) concluded from laboratory assays of *Beauveria bassiana* conidia that "optimal climatic conditions should be maintained for 2 d post-spray for controlling *Frankliniella occidentalis*." This conclusion was confirmed by Mukawa et al. (2011) and Wraight et al. (2016).

16.3.4.3 Entomopathogenic Fungi Versus Whiteflies

Despite the challenges associated with use of fungal pathogens discussed earlier, the advantages of microbial biocontrol have led to adoption of fungus-based products as components of greenhouse IPM and insecticide resistance management programs in many countries. A notable example is the use of fungi for control of *Bemisia* and *Trialeurodes* whiteflies. When effectively targeted, the scale-like nymphs of these insects are highly vulnerable to fungal infection, even under fluctuating or low ambient moisture conditions (Vidal et al., 1998; Wraight et al., 2000; Fargues et al., 2003, 2005; Wraight et al., 2016), and once infected, the nymphs die before reaching reproductive maturity (under temperature conditions favorable to fungal development). In a series of small-scale trials for control of young *Trialeurodes vaporariorum* larvae in greenhouse tomatoes under moderate humidity and variable ventilation conditions, two applications of the *Lecanicillium muscarium*-based Mycotal product made 4–5 days apart using an air-assisted backpack sprayer (each application at a rate of 2×10^{13} viable propagules in a spray volume of 2000 L/ha) resulted in mortality rates as high as 89–96% (Fargues et al., 2005). It was concluded that microclimatic conditions prevailing in the targeted insect microhabitat, the abaxial leaf-surface boundary layer, were "disconnected" from ambient, an observation corroborated by Shipp et al. (2011). In similar tests of *Beauveria bassiana* (formulated as Naturalis L), good efficacy (>75% control) was observed in both dry and humid greenhouses treated at a rate of about 8×10^{11} colony-forming units/ha (Fargues et al., 2003). Additional tests with whitefly-infested greenhouse tomatoes revealed that applications of Mycotal, Naturalis, and the *Isaria javanica* (previously identified as *Isaria fumosorosea*)-based product PreFeRal were compatible with beneficial insects (Hamdi et al., 2011).

16.3.4.4 Entomopathogenic Fungi Versus Thrips

Shipp et al. (2003) reported 90% or greater control of WFT on cucumber treated with weekly sprays of *Beauveria bassiana* (BotaniGard ES) at a rate of 10^8 conidia/mL in a greenhouse maintained at constant moderate temperature and high humidity (23–25°C and 86–92% RH). Shipp et al. (2003) also reported greater than 80% infection of thrips exposed to direct sprays on cucumber-leaf undersides and then caged on the treated leaves in commercial greenhouses with moderate temperatures (means ranging from 17 to 21°C) and mean RH as low as 73%. In greenhouse tests with mean temperature of 20°C and RH of 78%, Jacobson et al. (2001) also recorded good control of WFT (pest population reduction of 87%) on cucumber exposed to weekly high-volume (complete coverage) sprays of *B. bassiana* strain GHA (BotaniGard WP) applied at a rate of 5.5×10^7 conidia/mL. The same program of sprays of *B. bassiana* strain ATCC 74040 (Naturalis-L) applied at a rate of 10^5 conidia/mL reduced thrip numbers by 73%. These results are encouraging, but are based on a single crop plant of the family Cucurbitaceae. Cucumber produces large leaves that, when mature, tend to have more or less concave lower (abaxial) surfaces with abundant stomata and stout trichomes. These traits produce a protected, cool, humid microenvironment favored by thrips and highly favorable to activity of entomopathogenic fungi (Bolckmans et al., 1995; Wraight et al., 2000). In contrast to the aforementioned results in cucumber, high-volume spray applications of the BotaniGard WP product at a rate of 5×10^7 conidia/mL produced less than 50% control of WFT on flowering impatiens, even in greenhouses averaging greater than 80% RH (Ugine et al., 2007). On these plants, most thrips (more than 70%) were found in the single-form flowers where adults feeding on pollen were exposed to ambient conditions.

In recent greenhouse-cage tests, chilli thrips (*Scirtothrips dorsalis*) on sweet pepper were highly susceptible to fungal entomopathogens (Arthurs et al., 2013). Three applications of *B. bassiana* (BotaniGard WP) or *Metarhizium brunneum* (Met52 EC) made at 7- to 14-day intervals at respective rates of 10^8 and 10^7 conidia/mL (sprayed to runoff and also applied to the soil surface at the base of each plant) produced 81–94% reductions in thrips populations. Humidity inside the cages was elevated to higher than 95% for 12h following each application and averaged 79%.

Laboratory studies indicate that applications of spore suspensions targeting the prepupal and pupal stages of WFT in "soil" may be more effective than foliar applications

(Ansari et al., 2007, 2008a; Helyer et al., 1995). In one set of experiments, pots infested with *Frankliniella occidentalis* larvae and transferred to a greenhouse were treated with *M. anisopliae* conidia as a drench (20 mL of spore suspension applied to the surface of a 125-mL volume of growing medium) or premixed into the medium (dry conidia mixed into dry medium), both treatments equaling 10^{10} conidia/L of medium. Mortality from both treatments reached 84–93% within 12 days (Ansari et al., 2008a). In this particular test, ambient temperature reached 37°C. Skinner et al. (2012) have demonstrated an alternative approach of inoculating the soil surface with mycelial granules that undergo intensive sporulation postapplication, but implementation of this technology would have to address issues of increased risks of inhalation exposure to greenhouse workers. Thrip pupae are easily targeted with such applications as they remain at or near the soil surface. Most recently, formulations of *Metarhizium* and *Beauveria* spp. have been developed that are based on an alternative type of propagule from liquid fermentation called microsclerotia (Jackson and Jaronski, 2009), which can be incorporated into soils. Assays have shown the potential for control of the WFT (Wang et al., 2013), but additional trials with this and other soil-dwelling insects are still needed.

16.3.4.5 Entomopathogenic Fungi Versus Aphids

Aphids exhibit the fastest developmental rates and greatest fecundity of any of the primary greenhouse pests. These traits make them difficult targets for microbial biocontrol using the slow-acting, generalist fungal pathogens currently formulated as biopesticides. Indeed, the fungi that naturally regulate aphid populations in field environments, the Entomophthorales, are specialists with remarkable speeds of action and epizootic potential. Unfortunately, the Entomophthorales have proved far less amenable to mass production and formulation and have not been successfully commercialized (Milner, 1997; Wraight et al., 2003). A discussion of the microbial control challenge posed by aphids is presented by Jandricic et al. (2014). Despite these challenges many currently registered mycoinsecticides are labeled for use against aphids and are capable of controlling aphids under manipulated high-humidity conditions (Kim et al., 2010; and see Section 16.3.4.2) and of suppressing aphid populations under a broad range of conditions (Olson and Oetting, 1999; Shah et al., 2000; Kapongo et al., 2008). Shipp et al. (2003) recorded 61–72% infection of *Aphis gossypii* nymphs and adults caged on cucumber leaves and treated with a single, very high dose (10^8 conidia/mL) of *B. bassiana* (BotaniGard ES) under conditions of 70–77% relative humidity; infection was 88–96% when humidity was increased to 86–92%. Olson and Oetting (1999) observed 69%, 84%, and 94% reductions in numbers of *Myzus persicae* on chrysanthemum sprayed to runoff with BotaniGard ES at rates of 2.6, 9.2, and 15.8×10^7 conidia/mL, respectively (greenhouse environmental conditions not reported). Kapongo et al. (2008) reported about 30% infection of nymphal *M. persicae* populations exposed to bumble bee-vectored *B. bassiana*.

16.3.4.6 Entomopathogenic Fungi Versus Mealybugs

Few evaluations of entomopathogenic fungi for control of the primary species of mealybugs infesting greenhouse crops (*Planococcus* and *Pseudococcus* spp.) have been reported. *M. anisopliae* has been identified as a promising pathogen of *Pseudococcus cryptus* and other mealybug species in laboratory bioassays and greenhouse cage tests. Panyasiri et al. (2007) reported an LC_{50} of *M. anisolpliae* strain Ma65 of 2.35×10^6 conidia/mL applied against screen-caged *P. cryptus* on tomato in a tropical greenhouse.

16.3.4.7 Entomopathogenic Fungi Versus Root-Feeding Insects

Insects that feed on roots have been targeted by incorporation of fungal spores or mycelial granules into the soil at planting. The most prominent examples have used suspensions of *M. anisopliae* s.l. conidia as soil drenches or preparations comprising conidia of *Metarhizium brunneum* on a grain matrix (spent culture substrate) for control of *Otiorhynchus* larvae (Moorhouse et al., 1993; Bruck and Donahue, 2007; see also Chapter 24). In potted ornamentals, single, prophylactic soil drenches of *M. anisopliae* s.l. applied at a rate of 10^7 conidia/mL prior to infestation with *Otiorhynchus sulcatus* eggs produced 78% control within 5 weeks and 90% within 17 weeks; incorporation of the conidial suspension into the potting medium at the rate of 5 mL/L of medium produced similar results (Moorhouse et al., 1993). Fungus gnat larvae are another potential target for soil applications of fungal pathogens in greenhouses; however, larvae of *Bradysia impatiens* were found refractory to infection by *Beauveria bassiana* GHA and *M. brunneum* F-52 (E. Sensenbach and S. Wraight, unpublished data).

16.3.4.8 Entomopathogenic Fungi Versus Mites

Numerous laboratory studies have identified various fungi, including *Beauveria*, *Metarhizium*, and *Isaria* spp., as virulent against phytophagous mites, including the egg stage. Most studies have focused on *Tetranychus* spider mites, particularly *Tetranychus urticae*, *Tetranychus evansi*, and *T. cinnabarinus*. Only a few tests have been conducted under greenhouse conditions (Chandler et al., 2005; Erler et al., 2013; Gatarayiha et al., 2010), and results have been

highly variable. Chandler et al. (2005) reported 93% control of total *T. urticae* (including eggs) on greenhouse tomato treated with two weekly applications of *Beauveria bassiana* (Naturalis-L) at a low rate of 10^5 conidia/mL sprayed to runoff. Using a similar spray program with *B. bassiana* applied against *T. urticae* on four different vegetable crops at a higher rate of 4.2×10^6 conidia/mL (formulated as Eco-Bb WP suspended with 0.01% Break-thru trisiloxane surfactant), Gatarayiha et al. (2010) observed 60–86% mortality of adults, but only low mortality of mite eggs and nymphs. Greenhouse environmental conditions were not reported for either of the earlier-described studies. Shipp et al. (2003) reported no infection of adult *T. urticae* by *B. bassiana* (BotaniGard) under very high-humidity conditions in laboratory bioassays.

16.3.5 Entomopathogenic Nematodes

EPNs have been subjects of extensive research and development as biocontrol agents of many greenhouse pests (Broadbent and Olthof, 1995; Jagdale et al., 2004; Qiu et al., 2008; Trdan et al., 2007). Most research has been on development for control of soil-inhabiting pests, including fungus gnat larvae, thrips prepupae and pupae, cutworms, and vine weevil larvae (Ebssa et al., 2001; Tomalak et al., 2005; Shamshad, 2010). But studies have also examined potential of novel formulations of these agents for control of many above-ground pests such as leaf miner larvae, whitefly nymphs, thrips larvae and adults, and Lepidoptera larvae (Arthurs et al., 2004; Shapiro-Ilan et al., 2006; Cuthbertson et al., 2007). More extensive coverage of this research is reviewed elsewhere in this book.

16.4 FUTURE OF MICROBIAL BIOLOGICAL CONTROL OF GREENHOUSE ARTHROPOD PESTS

A number of entomopathogenic microbes, most notably Bt and baculoviruses, have been proved effective as stand-alone control agents against some greenhouse crop pests, and the unique modes of action of these and other entomopathogenic microbes have made them candidates for pesticide rotation schemes in widely adopted resistance management programs. However, intensive use of Bt has led to resistance in some pest populations, and there is a recognized potential for similar problems with other pathogens. In addition, slow modes of action have limited the capacity of many microbial control agents to compete directly with chemical insecticides (including biorational chemical insecticides) or to play effective roles (to carry their weight) in rotational spray programs. This is particularly the case in ornamental- and specialty-crop production systems where cosmetic damage can significantly affect market value and in crop systems threatened by potentially devastating plant viruses vectored by insect pests (see Chapter 24). Consequently, there is broad consensus that the greatest role for most arthropod pathogens in protected-crop production lies in IPM programs based on the use of multiple biocontrol agents against pests at incipient stages of infestation. In these programs, synthetic or biorational chemical insecticides can be held in reserve to control pest population outbreaks (an approach that reduces the risk of pests developing resistance to highly effective chemicals that in many cases are practically irreplaceable). For pathogens to be effective in this approach, biological control researchers and greenhouse pest managers will need to devise biopesticide-use strategies that maximize potential for complementary or synergistic interactions and minimize negative interactions with other components of the IPM system.

Problems of slow speed of kill can sometimes be mitigated by increasing dose and targeting early life stages. At the same time, microbial biocontrol agents are highly amenable to timely applications at very high rates. This trait enables use of these agents in ways not economically feasible with many other biocontrol agents. Thus, biopesticides might substitute for chemical insecticides to control pest population outbreaks (in the approach described earlier), even if efficacy is sufficient only to slow pest population growth to a rate that ultimately enables efficacy of a lagging predator or parasite population. Potential for use of biopesticides in this role is obviously greatest in organic production where chemical insecticides are not an option or in IPM programs where available chemical insecticides are incompatible with established biocontrol agents.

It is our belief that the future of microbial biological control will depend on our capacity to use microbial control agents wisely and in ways (and with expectations) commensurate with their natural modes of action. The challenges to adoption and sustainable use of these agents in pest management, especially in the form of competition from new generations of soft but effective chemical insecticides, can seem overwhelming to biocontrol researchers and developers, but as related in the many reviews presented herein, progress during the past 25 years has been remarkable, particularly in the areas of mass production and formulation. Public demand for insecticide residue–free food has never been stronger, and growth in the greenhouse markets for microbial biocontrol agents is expected to continue.

ACKNOWLEDGMENTS

We are grateful to the following people for their valuable information on commercial products available for management of greenhouse crops: Chie Goto (NARO Agricultural Research Center, Japan), Ard van der Maarel (Koppert Biological Systems, Netherlands), Henry Wainwright (Real IPM Company, Kenya), Luis Fernando Cruz Cortés (Live Systems Technology S.A., Colombia), Catherine Dickburt (Biobest N.V., Belgium), Nikolai van Beek (Kenya Biologics, Kenya), and Rogelio Trabanino (Escuela Agrícola Panamericana, Honduras). Mention of trade names or commercial products in this publication is solely for the purpose of providing specific information. The U.S. Department of Agriculture neither guarantees nor warrants the standard of the product, and use of a name by USDA implies no recommendation of the product to the exclusion of others that may also be suitable.

REFERENCES

Abdullah, N.M.M., 2008. Life history of the potato psyllid *Bactericera cockerelli* (Homoptera: Psyllidae) in controlled environment agriculture in Arizona. Afr. J. Agric. Res. 3, 60–67.

Annamalai, M., Kaushik, H.D., Selvaraj, K., 2015. Bioefficacy of *Beauveria bassiana* (Balsamo) Vuillemin and *Lecanicillium lecanii* Zimmerman against *Thrips tabaci* Lindeman. Proc. Natl. Acad. Sci. India Sect. B (Biol. Sci.) 1–7 (online).

Ansari, M.A., Shah, F.A., Whittaker, M., Prasad, M., Butt, T.M., 2007. Control of western flower thrips (*Frankliniella occidentalis*) pupae with *Metarhizium anisopliae* in peat and peat alternative growing media. Biol. Control 40, 293–297.

Ansari, M.A., Brownbridge, M., Shah, F.A., Butt, T.M., 2008a. Efficacy of entomopathogenic fungi against soil-dwelling life stages of western flower thrips, *Frankliniella occidentalis*, in plant-growing media. Entomol. Exp. Appl. 127, 80–87.

Ansari, M.A., Shah, F.A., Butt, T.M., 2008b. Combined use of entomopathogenic nematodes and *Metarhizium anisopliae* as a new approach for black vine weevil, *Otiorhynchus sulcatus*, control. Entomol. Exp. Appl. 129, 340–347.

Ansari, M.A., Shah, F.A., Butt, T.M., 2010. The entomopathogenic nematode *Steinernema kraussei* and *Metarhizium anisopliae* work synergistically in controlling overwintering larvae of the black vine weevil, *Otiorhynchus sulcatus*, in strawberry growbags. Biocontrol Sci. Technol. 20, 99–105.

Arthurs, S., Heinz, K.M., Prasifka, J.R., 2004. An analysis of using entomopathogenic nematodes against above-ground pests. Bull. Entomol. Res. 94, 297–306.

Arthurs, S.P., Aristizábal, L.F., Avery, P.B., 2013. Evaluation of entomopathogenic fungi against chilli thrips, *Scirtothrips dorsalis*. J. Insect Sci. 13 Article 31.

Bianchi, F.J.J.A., Joosten, N.N., Vlak, J.M., van der Werf, W., 2000. Greenhouse evaluation of dose- and time-mortality relationships of two nucleopolyhedroviruses for the control of beet armyworm, *Spodoptera exigua*, on Chrysanthemum. Biol. Control 19, 252–258.

Bolckmans, K., Sterk, G., Eyal, J., Sels, B., Stepman, W., 1995. PreFeRal, (*Paecilomyces fumosoroseus* (Wize) Brown and Smith, strain Apopka 97), a new microbial insecticide for the biological control of whiteflies in greenhouses. Med. Fac. Landbouww. Univ. Gent. 60, 719–724.

Broadbent, A.B., Olthof, T.H.A., 1995. Foliar application of *Steinernema carpocapsae* (Rhabditida, Steinernematidae) to control *Liriomyza trifolii* (Diptera, Agromyzidae) larvae in chrysanthemums. Environ. Entomol. 24, 431–435.

Bruck, D.J., Donahue, K.M., 2007. Persistence of *Metarhizium anisopliae* incorporated into soilless potting media for control of black vine weevil *Otiorhynchus sulcatus* in container-grown ornamentals. J. Invertebr. Pathol. 95, 146–150.

Buitenhuis, R., Shipp, J.L., 2005. Efficacy of entomopathogenic nematode *Steinernema feltiae* (Rhabditida: Steinernematidae) as influenced by *Frankliniella occidentalis* (Thysanoptera: Thripidae) developmental stage and host plant stage. J. Econ. Entomol. 98, 1480–1485.

Burges, H.D., Jarrett, P., 1978. Caterpillar control with *Bacillus thuringiensis*. Grower 90, 589–595.

Caballero, P., Murilo, R., Munoz, D., Williams, T., 2009. The nucleopolyhedrovirus of *Spodoptera exigua* (Lepidoptera: Noctuidae) as a biopesticide: analysis of recent advances in Spain. Rev. Colomb. Entomol. 35, 105–115.

Capinera, J.L., 2001. Handbook of Vegetable Pests. Academic Press, New York.

Castineiras, A., Peña, J.E., Osborne, L., 1996. Potential of *Beauveria bassiana* and *Paecilomyces fumosoroseus* (Deuteromycotina: hyphomycetes) as biological control agents of *thrips palmi* (Thysanoptera: Thripidae). Fla. Entomol. 79, 458–461.

Chandler, D., Davidson, G., 2005. Evaluation of entomopathogenic fungus *Metarhizium anisopliae* against soil-dwelling stages of cabbage maggot (Diptera: Anthomyiidae) in glasshouse and field experiments and effect of fungicides on fungal activity. J. Econ. Entomol. 98, 1856–1862.

Chandler, D., Davidson, G., Jacobson, R.J., 2005. Laboratory and glasshouse evaluation of entoopathogenic fungi against the two-spotted spider mite, *Tetranychus urticae* (Acari: Tetranychidae), on tomato, *Lycopersicon esculentum*. Biocontrol Sci. Technol. 15, 37–54.

Chang, J., Wu, X., Wang, Y., Meyerson, L.A., Gu, B., Min, Y., Xue, H., Peng, C., Ge, Y., 2013. Does growing vegetables in plastic greenhouses enhance regional ecosystem services beyond the food supply? Front. Ecol. Environ. 11, 43–49.

Chow, A., Heinz, K.M., 2004. Biological control of leafminers in vegetable crops. In: Heinz, K.M., Van Driesche, R.G., Parrella, M.P. (Eds.), Biocontrol in Protected Culture. Ball Publishing, Batavia, IL, pp. 221–251.

Cuthbertson, A.G.S., Walters, K.F.A., 2005. Pathogenicity of the entomopathogenic fungus, *Lecanicillium muscarium*, against the sweetpotato whitefly *Bemisia tabaci* under laboratory and glasshouse conditions. Mycopathologia 160, 315–319.

Cuthbertson, A.G.S., Walters, K.F.A., Northing, P., Luo, W., 2007. Efficacy of the entomopathogenic nematode, *Steinernema feltiae*, against sweetpotato whitefly *Bemisia tabaci* (Homoptera: Aleyrodidae) under laboratory and glasshouse conditions. Bull. Entomol. Res. 97, 9–14.

Devisetty, B.N., Wang, Y., Sudershan, P., Kirkpatrick, B.L., Cibulsky, R.J., Birkhold, D., 1998. Formulation and delivery systems for enhanced and extended activity of biopesticides. In: Nalewaja, J.D., Goss, G.R., Tann, R.S. (Eds.), Pesticide Formulations and Application Systems. vol. 18. ASTM, West Conshohocken, PA, pp. 242–272.

Ebssa, L., Borgmeister, C., Berndt, O., Poehling, H.-M., 2001. Efficacy of entomopathogenic nematodes against soil-dwelling life stages of western flower thrips, *Frankliniella occidentalis* (Thysanoptera: Thripidae). J. Invertebr. Pathol. 78, 119–127.

Eilenberg, J., Hajek, A., Lomer, C., 2001. Suggestions for unifying the terminology in biological control. BioControl 46, 387–400.

Elad, Y., Shtienberg, D., 1995. *Botrytis cinerea* in greenhouse vegetables: chemical, cultural, physiological, and biological controls and their integration. Integr. Pest Manage. Rev. 1, 15–29.

Erlandson, M., Newhouse, S., Moore, K., Janmaat, A., Myers, J., Theilmann, D., 2007. Characterization of baculovirus isolates from *Trichoplusia ni* populations from vegetable greenhouses. Biol. Control 41, 256–263.

Erler, F., Ates, A.O., Bahar, Y., 2013. Evaluation of two entomopathogenci fungi, *Beauveria bassiana* and *Metarhizium anisopliae*, for the control of carmine spider mite, *Tetranychus cinnabarinus* (Boisduval) under greenhouse conditions. Egypt. J. Biol. Pest Control 23, 233–240.

Fargues, J., Vidal, C., Smits, N., Rougier, M., Boulard, T., Mermier, M., Nicot, P., Reich, P., Jeannequin, B., Ridray, G., Lagier, J., 2003. Climatic factors on entomopathogenic hyphomycetes infection of *Trialeurodes vaporiorum* (Homoptera: Aleyrodidae) in Mediterranean glasshouse tomato. Biol. Control 28, 320–331.

Fargues, J., Smits, N., Rougier, M., Boulard, T., Ridray, G., Lagier, J., Jeannequin, B., Fatnassi, H., Mermier, M., 2005. Effects of microclimate heterogeneity and ventilation systems on entomopathogenic hyphomycete infection of *Trialeurodes vaporiorum* (Homoptera: Aleyrodidae) in Mediterranean glasshouse tomato. Biol. Control 32, 461–472.

de Faria, M.R., Wraight, S.P., 2007. Mycoinsecticides and mycoacaricides: a comprehensive list with worldwide coverage and international classification of formulation types. Biol. Control 43, 237–256.

Faria, M., Hotchkiss, J.H., Wraight, S.P., 2012. Application of modified atmosphere packaging (gas flushing and active packaging) for extending the shelf life of *Beauveria bassiana* conidia at high temperatures. Biol. Control 61, 78–88.

Feng, M.G., Chen, B., Ying, S.H., 2004. Trials of *Beauveria bassiana*, *Paecilomyces fumosoroseus* and imidacloprid for management of *Trialeurodes vaporariorum* (Homoptera: Aleyrodidae) on greenhouse grown lettuce. Biocontrol Sci. Technol. 14, 531–544.

Fenton, A., Gwynn, R.L., Gupta, A., Norman, R., Fairbaim, J.P., Hudson, P.J., 2002. Optimal application strategies for entomopathogenic nematodes: integrating theoretical and empirical approaches. J. Appl. Ecol. 39, 481–492.

Gatarayiha, M.C., Laing, M.D., Ray, M., 2010. Effects of adjuvant and conidial concentration on the efficacy of *Beauveria bassiana* for the control of the two-spotted spider mite, *Tetranychus urticae*. Exp. Appl. Acarol. 50, 217–229.

Gillespie, D.R., Costello, R.A., 2004. Biological control of minor pests. In: Heinz, K.M., Van Driesche, R.G., Parrella, M.P. (Eds.), Biocontrol in Protected Culture. Ball Publishing, Batavia, IL, pp. 313–329.

González-Cabrera, J., Mollá, J., Monton, H., Urbaneja, A., 2011. Efficacy of *Bacillus thuringiensis* (Berliner) in controlling the tomato borer, *Tuta absoluta* (Meyrick) (Lepidoptera: Gelechiidae). BioControl 56, 71–80.

Hall, R.A., 1976. *Verticillium lecanii* on the aphid *Macrosiphoniella sanborni*. J. Invertebr. Pathol. 28, 389–391.

Hall, R.A., Burges, H.D., 1979. Control of aphids in glasshouses with the fungus *Verticillium lecanii*. Ann. Appl. Biol. 93, 235–246.

Hamdi, F., Fargues, J., Gilles, R., Jeannequin, B., Bonato, O., 2011. Compatibility among entomopathogenic hypocreales and two beneficial insects used to control *Trialeurodes vaporariorum* (Hemiptera: Aleyrodidae) in Mediterranean greenhouses. J. Invertebr. Pathol. 108, 22–29.

Harris, M.A., Oetting, R.D., Gardner, W.A., 1995. Use of entomopathogenic nematodes and a new monitoring technique for control of fungus gnats, *Bradysia coprophila* (Diptera, Sciaridae), in floriculture. Biol. Control 5, 412–418.

Heinz, K.M., Van Driesche, R.G., Parrella, M.P. (Eds.), 2004. Biocontrol in Protected Culture. Ball Publishing, Batavia, IL.

Helyer, N., Gill, G., Bywater, A., Chambers, R., 1992. Elevated humidities for control of chrysanthemum pests with *Verticillium lecanii*. Pestic. Sci. 36, 373–378.

Helyer, N.L., Brobyn, P.J., Richardson, P.N., Edmondson, R.N., 1995. Control of western flower thrips (*Frankliniella occidentalis* Pergande) pupae in compost. Ann. Appl. Biol. 127, 405–412.

Jackson, M.A., Jaronski, S.T., 2009. Production of microsclerotia of the fungal entomopathogen *Metarhizium anisopliae* and their potential for use as a biocontrol agent for soil-inhabiting insects. Mycol. Res. 113, 842–850.

Jacobson, R.J., Chandler, D., Fenlon, J., Russell, K.M., 2001. Compatibility of *Beauveria bassiana* (Balsamo) Vuillemin with *Amblyseius cucumeris* Oudemans (Acarina: Phytoseiidae) to control *Frankliniella occidentalis* Pergande (Thysanoptera: Thripidae) on cucumber plants. Biocontrol Sci. Technol. 11, 391–400.

Jagdale, G.B., Casey, M.L., Grewal, P.S., Lindquist, R.K., 2004. Application rate and timing, potting medium, and host plant effects on the efficacy of *Steinernema feltiae* against the fungus gnat, *Bradysia coprophila*, in floriculture. Biol. Control 29, 296–305.

Jandricic, S.E., Filotas, M.J., Sanderson, J.P., Wraight, S.P., 2014. Pathogenicity of conidia-based preparations of entomopathogenic fungi against the greenhouse pest aphids green peach aphid *Myzus persicae* (Sulzer), cotton aphid *Aphis gossypii* Glover, and foxglove aphid *Aulacorthum solani* (Kaltenbach) (Hemiptera: Aphididae). J. Invertebr. Pathol. 118, 34–46.

Janmaat, A.F., Ware, J., Myers, J., 2007. Effects of crop type on *Bacillus thuringiensis* toxicity and residual activity against *Trichoplusia ni* in greenhouses. J. Appl. Entomol. 131, 333–337.

Jarrett, P., Burges, H.D., 1982. Control of tomato moth *Lacanobia oleracea* by *Bacillus thuringiensis* on glasshouse tomatoes and the influence of larval behaviour. Entomol. Exp. Appl. 31, 239–244.

Jones, D.R., 2003. Plant viruses transmitted by whiteflies. Eur. J. Plant Pathol. 109, 195–219.

Jurat-Fuentes, J.L., Jackson, T.A., 2012. Bacterial entomopathogens. In: Vega, F.E., Kaya, H.K. (Eds.), Insect Pathology, second ed. Academic Press, San Diego, pp. 265–349.

Kapongo, J.P., Shipp, L., Kevan, P., Broadbent, B., 2008. Optimal concentration of *Beauveria bassiana* vectored by bumble bees in relation to pest and bee mortality in greenhouse tomato and sweet pepper. BioControl 53, 797–812.

Kaya, H.K., Burlando, T.M., Choo, H.Y., 1995. Integration of entomopathogenic nematodes with *Bacillus thuringiensis* or pesticidal soap for control of insect pests. Biol. Control 5, 432–441.

Kim, J.J., Goettel, M.S., Gillespie, D.R., 2007. Potential of *Lecanicillium* species for dual microbial control of aphids and the cucumber powdery mildew fungus, *Sphaerotheca fuliginea*. Biol. Control 40, 327–332.

Kim, J.J., Goettel, M.S., Gillespie, D.R., 2010. Evaluation of *Lecanicillium longisporum*, Vertalec® against the cotton aphid, *Aphis gossypii*, and cucumber powdery mildew, *Sphaerotheca fuliginea* in a greenhouse environment. Crop Prot. 29, 540–544.

Lacey, L.A., Wraight, S.P., Kirk, A.A., 2008. Entomopathogenic fungi for control of *Bemisia tabaci* biotype B: foreign exploration, research and implementation. In: Gould, J., Hoelmer, K., Goolsby, J. (Eds.), Classical Biological Control of *Bemisia tabaci* in the United States – A Review of Interagency Research and Implementation. Progress in Biological Control, vol. 4. Springer, Dordrecht, The Netherlands, pp. 33–69.

Lasa, A., Ruiz-Portero, C., Alcázar, M.D., Belda, J.E., Caballeroa, P., Williams, T., 2007. Efficacy of optical brightener formulations of *Spodoptera exigua* multiple nucleopolyhedrovirus (SeMNPV) as a biological insecticide in greenhouses in southern Spain. Biol. Control 40, 89–96.

Liu, T.-X., Stansly, P.A., 2009. Effects of relative humidity on efficacy of BotaniGuard (TM) (*Beauveria bassiana*) on nymphs of sweetpotato whitefly, *Bemisia tabaci* (Hemiptera: Aleyrodidae), on hibiscus in greenhouses. Southwest. Entomol. 34, 189–191.

Malais, M.H., Ravensberg, W.J., 2003. Knowing and Recognizing: The Biology of Glasshouse Pests and Their Natural Enemies. Koppert B.V., Berkel en Rodenrijs, The Netherlands.

Maniania, N.K., Ekesi, S., Löhr, B., Mwangi, F., 2001. Prospects for biological control of the western flower thrips, *Frankliniella occidentalis*, with the entomopathogenic fungus, *Metarhizium anisopliae*, on chrysanthemum. Mycopathologia 155, 229–235.

Martin, P.A.W., Hirose, E., Aldrich, J.R., 2007. Toxicity of *Cromobacterium subtsugae* to southern green stink bug (Heteroptera: Pentatomidae) and corn rootworm (Coleoptera: Chrysomelidae). J. Econ. Entomol. 100, 680–684.

Mascarin, G.M., Jackson, M.A., Kobori, N.N., Behle, R.W., Delalibera Jr., I., 2015. Liquid culture fermentation for rapid production of desiccation tolerant blastospores of *Beauveria bassiana* and *Isaria fumosorosea* strains. J. Invertebr. Pathol. 127, 11–20.

Milner, R.J., 1997. Prospects for biopesticides for aphid control. Entomophaga 42, 227–239.

Moorhouse, E.R., Charnley, A.K., Gillespie, A.T., 1992. A review of the biology and control of the vine weevil, *Otiorhynchus sulcatus* (Coleoptera: Curculionidae). Ann. Appl. Biol. 121, 431–454.

Moorhouse, E.R., Gillespie, A.T., Charnley, A.K., 1993. Application of *Metarhizium anisopliae* (Metsch.) Sor. conidia to control *Otiorhynchus sulcatus* (F.) (Coleoptera: Curculionidae) larvae on glasshouse pot plants. Ann. Appl. Biol. 122, 623–636.

Mukawa, S., Tooyama, H., Ikegami, T., 2011. Influence of humidity on the infection of western flower thrips, *Frankliniella occidentalis* (Thysanoptera: Thripidae), by *Beauveria bassiana*. Jpn. Soc. Appl. Entomol. Zool. 46, 255–264.

Navajas, M., de Moraes, G.J., Auger, P., Migeon, A., 2013. Review of the invasion of *Tetranychus evansi*: biology, colonization pathways, potential expansion and prospects for biological control. Entomol. Exp. Appl. 59, 43–65.

Olson, D.L., Oetting, R.D., 1999. Compatibility of insect growth regulators and *Beauveria bassiana* (Balsamo) Vuillemin in controlling green peach aphid (Homoptera: Aphididae) on greenhouse chrysanthemums. J. Entomol. Sci. 34, 286–294.

Osborne, L.S., Boucias, D.G., Lindquist, R.K., 1985. Activity of *Bacillus thuringiensis* var. *israelensis* on *Bradysia coprophila* (Diptera, Sciaridae). J. Invertebr. Pathol. 78, 922–925.

Panyasiri, C., Attathom, T., Poehling, H.-M., 2007. Pathogenicity of entomopathogenic fungi-potential candidates to control insect pests on tomato under protected cultivation in Thailand. J. Plant Dis. Prot. 114, 278–287.

Payne, J.M., Uyeda, K.A., Stalder, C.J., Michaels, T.E., 1994. *Bacillus thuringiensis* Isolates Active against Dipteran Pests. U.S. Patent 5298245A.

Peña, J.E., Osborne, L.S., Duncan, R.E., 1996. Potential of fungi as biocontrol agents of *polyphagotarsonemus latus* (Acari: Tarsonemidae). Entomophaga 41, 27–36.

Pourian, H.R., Talaei-Hassanloui, R., Kosari, A.A., Ashouri, A., 2010. Effects of *Metarhizium anisopliae* on searching, feeding and predation by *Orius albidipennis* (Hem., Anthocoridae) on *Thrips tabaci* (Thy., Thripidae) larvae. Biocontrol Sci. Technol. 21, 15–21.

Qiu, B.L., Mandour, N.S., Xu, C.X., Ren, S.X., 2008. Evaluation of the entomopathogenic nematode *Steinernema feltiae* as a biological control agent of the whitefly, *Bemisia tabaci*. Int. J. Pest Manage. 54, 247–253.

Roh, J.Y., Choy, J.Y., Li, M.S., Jin, B.R., Je, Y.H., 2007. *Bacillus thuringiensis* as a specific, safe, and effective tool for insect pest control. J. Microbiol. Biotechnol. 17, 547–559.

Saito, T., 2005. Preliminary experiments to control the silverleaf whitefly with electrostatic spraying of a mycoinsecticide. Appl. Entomol. Zool. 40, 289–292.

Schroder, P.C., Ferguson, C.S., Shelton, A.M., Wilsey, W.T., Hoffmann, M.P., Petzoldt, C., 1996. Greenhouse and field evaluations of entomopathogenic nematodes (Nematode: Heterorhabditiae and Steinernematidae) for control of cabbage maggot (Diptera: Anthomyiidae) on cabbage. J. Econ. Entomol. 89, 1109–1115.

Shah, P.A., Aebi, M., Tuor, U., 2000. Infection of *Macrosipum euphorbiae* with mycelial preparations of *Erynia neoaphidis* in a greenhouse trial. Mycol. Res. 104, 645–652.

Shamshad, A., 2010. The development of integrated pest management for the control of mushroom sciarid flies, *Lycoriella ingenua* (Dufour) and *Bradysia ocellaris* (Comstock), in cultivated mushrooms. Pest Manage. Sci. 66, 1063–1074.

Shapiro-Ilan, D.I., Gouge, D.H., Piggott, S.J., Fife, J.P., 2006. Application technology and environmental considerations for use of entomopathogenic nematodes in biological control. Biol. Control 38, 124–133.

Shipp, J.L., Zhang, Y., Hunt, D.W.A., Ferguson, G., 2003. Influence of humidity and greenhouse microclimate on the efficacy of *Beauveria bassiana* (Balsamo) for control of greenhouse arthropod pests. Environ. Entomol. 32, 1154–1163.

Shipp, L., Elliott, D., Gillespie, D., Brodeur, J., 2007. From chemical to biological control in Canadian greenhouse crops. In: Vincent, C., Goettel, M.S., Lazarovits, G. (Eds.), Biological Control: A Global Perspective. CABI, Wallingford, UK, pp. 118–127.

Shipp, L., Johansen, N., Vanninen, I., Jacobson, R., 2011. Greenhouse climate: an important consideration when developing pest management programs for greenhouse crops. Acta Hortic. 893, 133–143.

Skinner, M., Gouli, S., Frank, C.E., Parker, B.L., Kim, J.S., 2012. Management of *Frankliniella occidentalis* (Thysanoptera: Thripidae) with granular formulations of entomopathogenic fungi. Biol. Control 63, 246–252.

Tingle, C.C.D., Copland, M.J.W., 1988. Effects of temperature and host-plant on regulation of glasshouse mealybug (Hemiptera, Pseudococcidae) populations by introduced parasitoids (Hymenoptera, Encyrtidae). Bull. Entomol. Res. 78, 135–142.

Tomalak, M., Piggott, S., Jagdale, G.B., 2005. Glasshouse applications. In: Grewal, P.S., Ehlers, R.U., Shapiro-Ilan, D.I. (Eds.), Nematodes as Biocontrol Agents. CABI Pub., Wallingford, UK, pp. 147–166.

Trdan, S., Znidarcic, D., Vidrih, M., 2007. Control of *Frankliniella occidentalis* on glasshouse-grown cucumbers: an efficacious comparison of foliar application of *Steinernema feltiae* and spraying with abamectin. Russ. J. Nematol. 15, 25–34.

Ugine, T.A., Wraight, S.P., Sanderson, J.P., 2006. Influences of impatiens pollen and exposure to *Beauveria bassiana* on bionomics of western flower thrips *Frankliniella occidentalis*. Biol. Control 37, 186–195.

Ugine, T.A., Wraight, S.P., Sanderson, J.P., 2007. Effects of manipulating spray application parameters on efficacy of the entomopathogenic fungus *Beauveria bassiana* against western flower thrips, *Frankliniella occidentalis*, infesting greenhouse impatiens crops. Biocontrol Sci. Technol. 17, 193–219.

Urbaneja, A., González-Cabrera, J., Arnó, J., Gabarra, R., 2012. Prospects for the biological control of *Tuta absoluta* in tomatoes of the Mediterranean basin. Pest Manage. Sci. 68, 1215–1222.

Van Driesche, R.G., Heinz, K.M., 2004. Biological control as a component of IPM systems. In: Heinz, K.M., Van Driesche, R.G., Parrella, M.P. (Eds.), Biocontrol in Protected Culture. Ball Publishing, Batavia, IL, pp. 25–36.

van Lenteren, J.C., 2009. IPM in greenhouse vegetables and ornamentals. In: Radcliffe, E.B., Hutchison, W.D., Cancelado, R.E. (Eds.), Integrated Pest Management: Concepts, Tactics, Strategies and Case Studies. Cambridge University Press, Cambridge, UK, pp. 354–365.

Vidal, C., Osborne, L., Lacey, L.A., Fargues, J., 1998. Effect of host plant on the potential of *Paecilomyces fumosoroseus* (Deuteromycotina: hyphomycetes) for controlling the silverleaf whitefly, *Bemisia argentifolii* (Homoptera: Aleyrodidae) in greenhouses. Biol. Control 12, 191–199.

Vidal, C., Fargues, J., Rougier, M., Smits, N., 2003. Effect of air humidity on the infection potential of hyphomycetous fungi as mycoinsecticides for *Trialeurodes vaporariorum*. Biocontrol Sci. Technol. 13, 183–198.

Wang, H.H., Lei, Z.R., Reitz, S., Li, Y.P., Xu, X.N., 2013. Production of microsclerotia of the fungal entomopathogen *Lecanicillium lecanii* (Hypocreales: Cordycipitaceae) as a biological control agent against soil-dwelling stages of *Frankliniella occidentalis* (Thysanoptera: Thripidae). Biocontrol Sci. Technol. 23, 234–238.

Weintraub, P.G., Horowitz, A.R., 1995. The newest leafminer pest in Israel, *Liriomyza huidobrensis*. Phytoparasitica 23, 177–184.

Wekesa, V.W., Avery, P.B., McKenzie, C.L., Powell, C.A., Osborne, L.S., 2011. Control of *Liriomyza trifolii* (Diptera: Agromyzidae) in cut flowers using *Isaria fumosorosea* (Hypocreales: Cordycipitaceae) alone and in combination with insecticides. J. Entomol. Sci. 46, 80–84.

Willmott, D.M., Hart, A.J., Long, S.J., Richardson, P.N., Chandler, D., 2002. Susceptibility of cabbage root fly *Delia radium*, in potted cauliflower (*Brassica oleraceae* var. *botrytis*) to isolates of entomopathogenic nematodes (*Steinernema* and *Heterorhabditis* spp.) indigenous to the UK. Nematology 4, 965–970.

Wraight, S.P., Carruthers, R.I., Jaronski, S.T., Bradley, C.A., Garza, C.J., Galaini-Wraight, S., 2000. Evaluation of the entomopathogenic fungi *Beauveria bassiana* and *Paecilomyces fumosoroseus* for microbial control of the silverleaf whitefly, *Bemisia argentifolii*. Biol. Control 17, 203–217.

Wraight, S.P., Galaini-Wraight, S., Carruthers, R.I., Roberts, D.W., 2003. *Zoophthora radicans* (Zygomycetes: Entomophthorales) conidia production from naturally-infected *Empoasca kraemeri* and dry-formulated mycelium under laboratory and field conditions. Biol. Control 28, 60–77.

Wraight, S.P., Ugine, T.A., Ramos, M.E., Sanderson, J.P., 2016. Efficacy of spray applications of entomopathogenic fungi against western flower thrips infesting greenhouse impatiens under variable moisture conditions. Biol. Control 97, 31–47.

Wu, S.Y., Gao, Y.L., Xu, X.N., Zhang, Y.P., Wang, J., Lei, Z.R., Smagghe, G., 2013. Laboratory and greenhouse evaluation of a new entomopathogenic strain of *Beauveria bassiana* for control of the onion thrips *Thrips tabaci*. Biocontrol Sci. Technol. 23, 794–802.

Chapter 17

Microbial Control of Arthropod Pests of Orchards in Temperate Climates

D. Shapiro-Ilan[1], S.P. Arthurs[2], L.A. Lacey[3]

[1]*USDA-ARS, Southeastern Fruit and Tree Nut Laboratory, Byron, GA, United States;* [2]*University of Florida, Apopka, FL, United States;* [3]*IP Consulting International, Yakima, WA, United States*

17.1 INTRODUCTION

Temperate orchards consist of perennial tree plantings including pome fruits, stone fruits, and tree nuts. A number of economically important pests occur in these orchards. Although chemical insecticides are the mainstay for controlling many temperate orchard pests, the orchards possess attributes that make them amenable to the use of entomopathogens. These include hosts available through most of the year, favorable soil conditions with minimal disturbance, significant canopy shade and their perennial nature that allows compatible beneficial insect populations to build up. Furthermore, temperate orchard commodities are of relatively high value, which facilitates economic feasibility of microbial control applications. Consequently, microbial control agents (MCAs) have been evaluated against multiple orchard pests, and some associated MCAs have become commercially successful. With increased environmental and regulatory issues limiting the use of broad spectrum insecticides, opportunities for implementation and integration of alternative methods (such as MCAs) are being created. Here, we review progress of microbial control used on orchard pests including successes and failures, research needs, and potential for the future. Many insect and mite pests of temperate orchards have been targeted with entomopathogens (Table 17.1). Due to space limitations, we focus our discussion on those that have been researched in depth and have supporting data outside of laboratory settings.

17.2 POME FRUIT

17.2.1 Codling Moth

Cydia pomonella (Lepidoptera: Tortricidae) remains the most serious pest of apple production worldwide (Pajač

et al., 2011). It is also a significant pest of pears, quince, and walnuts. Mated females oviposit on fruit or nearby leaves. Neonate larvae penetrate and develop inside fruits, while fifth instars exit fruits to seek cryptic habitats in which to pupate. One to four generations of codling moth (CM) are produced depending on latitude, altitude, and climate. The diapause-bound larvae overwinter in their hibernacula, pupating and emerging the following spring (Higbee et al., 2001).

The use of organophosphate insecticides, in particular azinphos methyl, was the predominant means of control of CM for almost six decades. Due to pesticide resistance, environmental concerns, and regulatory issues, several conventional broad spectrum insecticides have been eliminated and thus alternatives must be developed. Mating disruption (MD) is one such tactic (Vickers and Rothschild, 1991; Barnes et al., 1992; Calkins, 1999). This tactic works best when CM populations are low (Vickers and Rothschild, 1991). Application of MCAs for CM control can complement MD while preserving natural enemies for CM and other orchard pests.

17.2.1.1 CM Granulovirus

The *Cydia pomonella* granulovirus (CpGV) has proved to be a highly successful microbial control product demonstrating high virulence to CM (Lacey et al., 2005a) and lack of effects on nontargets (Arthurs et al., 2007). Lacey et al. (2008b) reviewed the commercial development of CpGV following its discovery, and subsequent role in integrated pest management (IPM) programs for CM worldwide. CpGV was widely tested prior to and following its commercial development, starting in Europe in the 1970s. Several CpGV products were launched in North America in 2003 and proved popular among organic growers (Arthurs and Lacey, 2004). Within 5 years, CpGV was used on more

TABLE 17.1 Insect Pests of Temperate Climate Tree Fruit and Nuts on Which Microbial Control Agents Have Been Used or Are Under Development for Orchard Crop Protection

Targeted Pest(s)	Crop	Microbial Control Agents	Efficacy/Rate	Reference(s)
Hymenoptera: Tenthredinidae				
Apple sawfly, *Hoplocampa testudinea*	Apple	*Steinernema carpocapsae*, *S. feltiae*, *Heterorhabditis bacteriophora*	Up to 72% control at 40–80 IJs/cm²	Vincent and Belair (1992)
Lepidoptera: Tortricidae				
Codling moth, *Cydia pomonella*	Apple, pear, and walnut	Granulovirus (CpGV)	≈10¹³ OB/ha at 7–10 day intervals provide >90% control	Jaques et al. (1994), Arthurs and Lacey (2004), Arthurs et al. (2005), Lacey et al. (2008b), and Vail et al. (1991)
	Apple and pear	*S. carpocapsae*, *S. feltiae*, *H. zealandica*	≥70% control of diapausing larvae at ≈10⁶ IJs/tree	Kaya et al. (1984), Unruh and Lacey (2001), Cornale et al. (2006), Lacey et al. (2006a), Kienzle et al. (2008), De Waal et al. (2011), and Nevaneethan et al. (2010)
	Fruit bins	*S. carpocapsae*, *S. feltiae*, *H. zealandica*	Up to 97% control of diapausing larvae at 10–25 IJs/mL	Lacey and Chauvin (1999), Cossentine et al. (2002), Lacey et al. (2005b), and De Waal et al. (2010)
Filbertworm, *Cydia latiferreana*	Hazel nuts	*H. marelata*, *S. carpocapsae*, *S. kraussei*	80–95% mortality in field at 40–200 IJs/cm²	Bruck and Walton (2007) and Chambers et al. (2010)
Oriental fruit moth, *Grapholita molesta*	Stone fruit, apple	CpGV	Low susceptibility, LC₉₅ > 5 × 10⁴ OB/10 µL	Lacey et al. (2005a)
		S. carpocapsae, *S. feltiae*, *H. marelata*	Up to 82% mortality at 25 IJs/mL	Riga et al. (2006)
Light brown apple moth, *Epiphyas postvittana*	Apple, pear, grape	Btk	4–48% control at 0.2–1.0 kg/ha	Baily et al. (1996) and Suckling and Brockerhoff (2010)
		EppoNPV	Naturally occurring epizootics	MacCollom and Reed (1971), Briese et al. (1980), and Longworth and Singh (1980)
Summer fruit tortrix, *Adoxophyes orana*	Apple	AdorGV	Four applications eliminated fruit damage over 2 years	Kocourek et al. (2007)
		AdorNPV	Maintained below economic thresholds	Blommers et al. (1987)
Oblique banded leaf roller, *Choristoneura rosaceana*	Apple	Btk	Label rates (2–16 g/L) caused mortality up to 21 days; applications during bloom recommended	Côté et al. (2001) and Cossentine et al. (2003)
		Beauveria bassiana	23 isolates (10⁷ conidia/mL) caused 66–89% larval and pupal mortality	Todorova et al. (2002)

Pest	Crop	Agent	Result	Reference
Fruittree leafroller, *Archips* spp.	Apple and pear	GV and NPV	Up to 24% natural infection	Goyer et al. (2001)
		Btk	Field applications reduced infestation and early fruit damage	Sorenson and Falcon (1980) and Pasqualini (1996)
Pandemis spp.		GV	Greenhouse tests at 3×10^6 OBs/mL prevented pupation	Pfannenstiel et al. (2004) and Unruh et al. (2012)
		Btk	Dipel not effective	De Reede et al. (1985)
False codling moth, *Thaumatotibia leucotreta*	Stone fruit, citrus	*Cryptophlebia leucotreta* GV (CrleGV)	Between 30 and 92% reduction at 2×10^{13} OB/ha applied with molasses	Moore et al. (2015); see also Chapter 19
Lepidoptera: Pyralidae				
Navel orangeworm *Amyelois transitella*	Almonds, pistachios, walnuts	*S. carpocapsae*, *S. feltiae*	Up to 75% mortality at 10^5 IJs/m^2	Siegel et al. (2004, 2006)
Lepidoptera: Sesiidae				
Peachtree borer, *Synanthedon exitiosa*	Peach & other stone fruits	*H. bacteriophora*, *S. carpocapsae*	3×10^5–1.5×10^6 IJs/tree gave 78–100% control (equal to chlorpyrifos)	Cossentine et al. (1990) and Shapiro-Ilan et al. (2009a, 2015)
Lesser peachtree borer, *S. pictipes*	Peach & other stone fruits	*S. carpocapsae* + Barricade gel	10^6 IJs/wound gave 70–100% control (equal to chlorpyrifos)	Shapiro-Ilan et al. (2010, 2016)
Apple clearwing, *Synanthedon myopaeformis*	Rosaceae family	*S. carpocapsae*, *S. feltiae*	10^6 IJs/300mL gave 63–75% infection	Parvizi (2003)
	various	*B. bassiana* and *Metarhizium brunneum*	LC50's of 2.9×10^5 and 3.4×10^5 spores/mL	Cossentine et al. (2010)
Lepidoptera: Yponomeutidae				
Apple fruit moth, *Argyresthia conjugella*	Apple	*Isaria fumosorosea*	70% reduced spring emergence by 10^7 conidia/cm^2 soil	Vänninen and Hokkanen (1997)
Coleoptera				
Plum curculio, *Conotrachelus nenuphar*	Pome and stone fruits	*S. riobrave*, *S. feltiae*	*S. riobrave* gave 80–100% control at 40–400 IJs/cm^2; *S. feltiae* efficacy lower (<60%)	Shapiro-Ilan et al. (2004, 2008a, 2013a), Alston et al. (2005), and Pereault et al. (2009)
Pecan weevil, *Curculio caryae*	Pecan	*S. carpocapsae*, *B. bassiana*, *Chromobacterium subtsugae*	>80% control (EPNs 50–100 IJs/cm^2; fungi 2.5×10^9 spores/tree *C. subtsugae* 3.4 kg/ha)	Shapiro-Ilan et al. (2008b, 2013b) and Shapiro-Ilan and Gardner (2012)
Hazelnut weevil, *Curculio nucum*	Hazelnut	*B. bassiana*, *B. brongniartii*, *Metarhizium* spp.	Up to 99% reduction by 2×10^7 spores/mL versus 64% in controls	Paparatti and Speranza (2005)
		H. bacteriophora, *S. feltiae*	23–88% control by 5×10^5–10^6 IJs/m^2	Sarraquigne et al. (2009) and Batalla-Carrera et al. (2013)
Flat-headed rootborer, *Capnodis tenebrionis*	Stone fruit, seed fruit	*S. feltiae*	88–97% control by 10^6 IJs/tree	Morton and García-Del-Pino (2008)

Continued

TABLE 17.1 Insect Pests of Temperate Climate Tree Fruit and Nuts on Which Microbial Control Agents Have Been Used or Are Under Development for Orchard Crop Protection—cont'd

Targeted Pest(s)	Crop	Microbial Control Agents	Efficacy/Rate	Reference(s)
Diptera				
Fruit flies, *Rhagoletis* spp.	Cherries	*B. bassiana, I. fumosorosea, M. brunneum*	>90% infection of adult flies at in the laboratory; infested fruits reduced by 65% in field tests	Yee and Lacey (2005) and Daniel and Wyss (2009, 2010)
		Steinernema and *Heterorhabditis* spp.	62–100% larval mortality by 50–100 IJs/cm²	Yee and Lacey (2003) and Kepenekci et al. (2015)
Peach fruit flies, *Bactrocera zonata*	Peach or other stone fruits	*B. bassiana*	Spores might be vectored in sterile release programs (SIT)	Sooker et al. (2014)
Hemiptera				
Pear psylla, *Cacopsylla pyricola*	Pear	*B. bassiana, I. fumosorosea, M. brunneum*	Variable mortality (18–88%) in field applications	Puterka (1999) and Erler et al. (2014)
Black pecan aphid, *Melanocallis caryaefoliae*	Pecan	*I. fumosorosea* (combined with a eucalyptus extract)	Up to 82% mortality in field applications	Shapiro-Ilan et al. (2013b)

than 4000 ha annually in North America, while the Carpovirusine formulation is used on 100,000 ha in Europe (Lacey et al., 2008b).

In general, CpGV concentrates are diluted in the range of 10^{13} OB/ha and are applied with the beginning of CM egg hatch in spring to target neonate larvae before they enter fruits (Lacey et al., 2008b). Infected larvae typically die within a few days leaving a small (0.25 cm) brown mark on the fruit skin, which prevents significant downgrading of the fruit especially for processed markets (Fig. 17.1). The time frame of egg hatch requires that growers time virus applications against each larval generation, using regional phenology models. Research in the Pacific Northwest showed that several applications at 7- to 14-day intervals starting at about 3% egg hatch controlled greater than 90% CM populations (Arthurs et al., 2005). CpGV products have stable storage characteristics (Lacey et al., 2008a), which are limiting factors with some microbial insecticides.

Due to the limited persistence of baculoviruses following aboveground field applications, several researchers have tested sunscreens and other formulations additives; reviewed by Lacey et al. (2008b). Modest improvements in CpGV residual activity have been achieved with various additives, including molasses, sucrose and skimmed milk powder (Lacey and Arthurs, 2005), and zinc oxide and titanium dioxide (Wu et al., 2015). Encapsulation of CpGV with lignins (Arthurs et al., 2006) and production of protective microparticles (Pemsel et al., 2010) have also been tested. Most recently, Knight and Witzgall (2013) demonstrated that mixing CpGV with mutualistic yeasts isolated from larvae of codling moth significantly increased mortality of neonate codling moth larvae, compared with CpGV alone. A field trial showed that fruit injury and larval survival were significantly reduced when apple trees were sprayed with CpGV, yeast, and sugar. Efforts to develop

genetically improved CpGV isolates (Winstanley et al., 1998), and modified formulations have not progressed significantly.

Starting in 2005, high levels of resistance to CpGV (greater than 1000-fold) were detected in Europe among isolated CM populations that were treated with the virus over an extended period (Asser-Kaiser et al., 2007). CpGV-resistant populations now occur in several European countries (Schmitt et al., 2013). Several new isolates (such as Madex Plus and Madex I12) have been approved for use (Eberle et al., 2008) and appear to be successful (Zingg et al., 2011; Kutinkova et al., 2012). The authors recommend that CpGV should be combined with mating disruption and other effective insecticides such as oils to reduce resistance pressure.

17.2.1.2 Entomopathogenic Nematodes for CM Control

Overwintering fifth-instar CM are difficult targets for control due to location of their hibernacula in tree bark, prop piles, wooden fruit bins, and other cryptic habitats. Entomopathogenic nematodes (EPNs) can naturally seek and find stages of orchard pests that are not easily targeted by conventional chemical insecticides. A number of biotic and abiotic factors affect EPN efficacy in orchards (Lacey et al., 2006a, 2007). Important considerations include maintenance of moisture (Unruh and Lacey, 2001), temperatures that permit host seeking (ie, 15–25°C, depending on EPN species), and application rate ($\geq 10^6$ infective juveniles [IJs]/tree; $\geq 10^9$ IJs/ha) (Cornale et al., 2006; Lacey et al., 2006a; Kienzle et al., 2008). Larvicidal activity has also been improved with adjuvants and mulches that decrease evaporation and postapplication irrigation with spray equipment configured for optimal deposition (Lacey et al., 2006a,b, 2010;

FIGURE 17.1 Golden Delicious apple showing "shallow stings" resulting from CpGV-infected codling moth. The neonates start feeding but generally die before fully penetrating fruits.

FIGURE 17.2 Airblast sprayer configured to apply entomopathogenic nematodes for codling moth control in apple orchards. The high-volume, large-droplet spray is targeted at mulch and lower tree limbs to target overwintering stages in the fall or spring before larvae pupate.

Kienzle et al., 2008; Nevaneethan et al., 2010; De Waal et al., 2011) (Fig. 17.2). Lacey et al. (2006a) reported that *S. feltiae* was more effective than *Steinernema carpocapsae* and produced up to 97% mortality of CM (applied at 10^6 IJs/tree). Similar results for control of diapausing CM larvae were reported from South Africa, Italy, and Germany using *H. zealandica* and *Steinernema* spp. (Cornale et al., 2006; Kienzle et al., 2008; De Waal et al., 2010; Nevaneethan et al., 2010; De Luca et al., 2015).

EPNs have also been used to treat CM larval cocoons in wooden fruit bins that become infested with diapause-bound larvae when placed in orchards prior to harvest. Cocooned larvae in crevices of bins were readily infected by EPNs when bins were briefly submerged in water containing IJs (Lacey and Chauvin, 1999). Subsequent tests were performed in tanks used for flotation of fruit (Lacey et al., 2005b). A concentration of 10–25 IJs/mL of *S. carpocapsae* or *S. feltiae* provided greater than 50% mortality (Lacey and Chauvin, 1999; Cossentine et al., 2002; Lacey et al., 2005b). Addition of an adjuvant (Silwet L 77) to increase penetration of wood surfaces and a humectant gel (Stockosorb) to prolong moisture, further increased efficacy. Subsequently, De Waal et al. (2010) demonstrated maximal CM control in infested bins treated with *H. zealandica*. IJs was attained when bins were pre-wet for at least 1 min and maintained at greater than 95% RH pos-treatment for at least 3 days to ensure nematode survival.

17.2.2 Leafrollers (Lepidoptera: Tortricidae)

Tortricid leafrollers are among the most economically important defoliators in temperate orchards (van der Geest and Evenhuis, 1991; Blommers, 1994). Key pests in western North America include *Choristoneura rosaceana*, *Adoxophyes orana*, *Pandemis heparana*, *P. pyrusana*, *Archips* spp., and a recent invader from Australia, the light brown apple moth, *Epiphyas postvittana*, can be destructive in pome fruit and grape (Suckling and Brockerhoff, 2010).

17.2.2.1 Baculoviruses

A granulovirus reported from the summer fruit tortrix, *Adoxophyes orana* (AdorGV) (Flückiger, 1982; Sekita et al., 1984; Kocourek et al., 2007), was commercially developed and used in Europe (Cross et al., 1999). The AdorGV product Capex 2 effectively eliminated fruit damage in commercial apple orchards in the Czech Republic following 2 years of targeting both overwintering and spring larval generations (Kocourek et al., 2007). Blommers et al. (1987) also reported good efficacy for a nucleopolyhedrovirus of *A. orana* (AdorNPV) for control of *A. orana* over 3 years. Application of AdorNPV alone and in combination with mating disruption maintained *A. orana* below

economic thresholds, with a concomitant decline in additional leafrollers in the same orchards. Due to the specificity of the virus, the reduction in secondary leafroller species was ascribed to predators and parasitoids. The paucity of *A. orana* ostensibly necessitated exploitation of other hosts. Endemic baculoviruses of other leafrollers have been documented producing periodic epizootics in different regions of the world (MacCollom and Reed, 1971; Briese et al., 1980; Longworth and Singh, 1980; Pfannenstiel et al., 2004; Suckling and Brockerhoff, 2010), but have not been developed as commercial products.

17.2.2.2 Bacillus thuringiensis

Leafrollers have been routinely controlled for decades with products based on Bt subsp. *kurstaki* (Btk) and, to a lesser extent, Bt subsp. *aizawa* (Bta) by both conventional and organic growers alike (Blommers, 1994; Baily et al., 1996; Pasqualini, 1996; Cross et al., 1999; Cossentine et al., 2003; Lacey et al., 2007; Lacey and Shapiro-Ilan, 2008). Bt has several advantages for controlling caterpillar pests, including safety for predators and parasitoids and lack of negative environmental impacts (Lacey and Siegel, 2000; Sanahuja et al., 2011; Chapter 1). Several authors (Shapiro-Ilan et al., 2007; Lacey and Unruh, 2005) demonstrate that Bt contributes to orchard IPM by allowing natural enemies to survive while suppressing both primary and secondary pests. The short preharvest and restricted entry intervals of Bt products provides logistical benefits for growers.

Due to a variety of factors, variable efficacy of Bt has been reported by several researchers and practitioners. The toxic protein moieties (δ-endotoxins) in the parasporal crystalline inclusions produced by the bacterium at the time of sporulation must be ingested in order to be insecticidal. Therefore, good coverage while pests are actively feeding is needed for maximum control efficacy of Bt. Adverse environmental conditions, including intense sunlight and heavy rains, will quickly reduce Bt persistence. Timing of applications to coincide with younger larvae and warmer weather ($\geq 15°C$) improves activity of Bt. For example, when larvae of leafroller species emerge from diapause, temperature is often suboptimal for larvicidal activity. In lieu of using chemical pesticides, organic growers increase the dosage of Bt to obtain acceptable mortality when lower temperatures prevail. The effectiveness of Bt toxins declines as larvae grow, thus requiring a progressively larger dose. Sublethal dosage of Bt can result in a shorter lifespan, reduced feeding, and reduction of oviposition (Li et al., 1995; Harris et al., 1997; Knight et al., 1998). Protocols for the application and evaluation of Bt against orchard pest insects are presented by Lacey et al. (2007). Descriptions of the Bt larvicidal toxins and their mode of action are presented in Chapter 4.

17.2.3 Pear Psylla

The hemipteran pear psylla species complex includes key pests of pear throughout most pear-growing regions of the world (Beers and Brunner, 2015). The pear psylla, *Cacopsylla pyri* (Hemiptera: Psyllidae), offers an opportunity for slower-acting fungal entomopathogens because it is a foliage feeder that does not directly damage fruit, allowing time for spores to germinate, infect, and kill psyllids before damage occurs. Several indigenous and exotic fungal isolates were tested against *C. pyri* in a range of tests by Puterka (1999). In field trials, psyllid mortalities obtained by single applications of *Beauveria bassiana* 2860 or *Isaria fumosorosea* Pf 2658 at spore concentrations of 5.4×10^{13} spores/ha plus 0.1% emulsifiable oil or acrylic polymer ranged from 18% to 37%. A commercial formulation (Naturalis L.) performed similarly. More recently, Erler et al. (2014) tested an emulsifiable concentrate formulation of *Metarhizium brunneum* F52 containing 5.5×10^9 conidia/mL, against *C. pyri* in Turkey. The product reduced psyllid eggs and young nymphs by up to 79% and 88%, respectively, 7 days after treatment, comparable with that of a standard pesticide, Novaluron, over 2 years of testing. However, the fungus was less active against older nymphs (third to fifth instars) with ≈50% mortality, suggesting that timing treatments is important.

17.2.4 Aphids

Aphids are common pests in temperate orchards, especially where natural enemies are displaced (van Emden and Harrington, 2007). Under favorable environmental conditions, aphid outbreaks are controlled naturally by obligate entomophthoralean fungi (Zygomycota), such as *Zoophthora* and *Pandora* spp. (Nielsen et al., 2007). Aphids, however, may be considered generally poor targets for microbial control with the mass-produced fungi from the Hypocreales (Ascomycota) due to lack of pathogenicity, rapid molting, and unsuitable environmental conditions for infection (Steinkraus, 2006; also see Chapter 9). Nevertheless, species of *Lecanicillium* have been reported as pathogens of several aphids (Diaz et al., 2009). Currently, only *L. longisporum* and *L. muscarium* have been formulated and commercialized in Europe as Vertalec and Mycotal to control aphids and whiteflies in protected crops. Inundative applications of *Lecanicillium* spp. have not been tested against aphids in pome fruits. Entomopathogenic fungi (EPF) might be enhanced in IPM programs where other natural enemies occur. Bird et al. (2004) reported that workers of the black ant, *Lasius niger* vectored *L. longisporum* conidia to the rosy apple aphid, *Dysaphis plantaginea*, while tending their colonies in an organic apple orchard. However, only relatively minor infections in aphids (≈3%) were observed during the trial, possibly limited by ants grooming or removing infected aphids. More promise for microbial control of aphids in apple orchards may come from new products, such as *Chromobacterium subtsugae* and heat-killed *Burkholderia* spp. that have been registered for control of aphids, whiteflies, thrips, and mites (Martin et al., 2007).

17.3 STONE FRUIT

17.3.1 Plum Curculio

The plum curculio, *Conotrachelus nenuphar* (Coleoptera: Curculionidae), is a key pest of pome and stone fruit in eastern North America (Racette et al., 1992; Horton and Johnson, 2005); we will, however, address *C. nenuphar* control in this section under stone fruit. Adult weevils enter orchards from overwintering sites in the spring to feed, and oviposit in fruit. Attacked fruit aborts or is damaged rendering it nonsaleable. Larvae develop in fallen fruit, exit as fourth instars, and burrow into the soil (1–8 cm) to pupate (Racette et al., 1992). After emergence, adults feed on fruit and migrate to overwintering sites in the orchard or surrounding area (Racette et al., 1992; Olthof and Hagley, 1993). In the southern United States, a second generation may occur prior to overwintering (Horton and Johnson, 2005).

Laboratory research indicates that *C. nenuphar* is susceptible to EPF (*Beauveria* spp. and *Metarhizium* spp.) (Alston et al., 2005; Tedders et al., 1982) and to various EPN strains and species (Olthof and Hagley, 1993; Shapiro-Ilan et al., 2002, 2011; Alston et al., 2005). Field trials have primarily tested these MCAs against the ground-dwelling stages of *C. nenuphar*. Overall, *Steinernema riobrave* has proved to be the most effective MCA for control of *C. nenuphar*. In field trials conducted in peach and apple orchards across the eastern United States, soil applications of *S. riobrave* consistently averaged greater than 92% control (Shapiro-Ilan et al., 2004, 2008a, 2013a). Additionally, Pereault et al. (2009) observed 80–89% reduction of *C. nenuphar* in Michigan apple and cherry orchards. In contrast, *S. feltiae* applications failed (Shapiro-Ilan et al., 2004), produced low (22–39%) (Alston et al., 2005) or intermediate (58%) (Shapiro-Ilan et al., 2013a) levels of control. Applications of *Heterorhabditis bacteriophora*, *B. bassiana*, or *Metarhizium brunneum* have been ineffective (Jenkins et al., 2006; Pereault et al., 2009).

Soil applications of MCAs targeting *C. nenuphar* do not control adults that fly into the orchard's canopy (soil applications prevent subsequent damage resulting from below-ground stages of the pest) (Shapiro-Ilan et al., 2005). Therefore, the use of MCAs in conjunction with other tactics (applied to the canopy) is likely to be more effective. The use of EPNs (eg, *Steinernema riobrave*) as a component of an integrated program that targets multiple stages of *C. nenuphar* may be feasible. One integrated approach entails using volatile lures to attract adult *C. nenuphar* to selected sentinel trees on the orchard perimeter; the canopies of sentinel trees are then sprayed with adult-killing insecticides while the other trees remain pesticide free. Restricting treatments to sentinel trees can reduce insecticide treatment by more than 90% (Leskey et al., 2008). Given that particularly high populations of *C. nenuphar* can be expected in sentinel trees, some damage in these trees is expected. Once the infested fruit in the sentinel trees drops, EPNs are applied to suppress the ground-dwelling stages (Shapiro-Ilan et al., 2013a; unpublished data).

17.3.2 Borers

17.3.2.1 Peachtree Borer

Several borer pests of stone fruits have shown susceptibility to entomopathogens (Barnett et al., 1993; Kain and Agnello, 1999; Grewal et al., 2005). Clearwing moth larvae (Lepidoptera: Sesiidae) in the genus *Synanthedon* are generally very susceptible to EPNs (Grewal et al., 2005). The peachtree borer, *Synanthedon exitiosa*, is a major pest of *Prunus* spp. including peach (Horton and Johnson, 2005). Moths emerge and mate during late summer and early fall. Mated females oviposit on or near the bark of host plants. Hatched larvae bore into the trunk of stone fruit trees near the soil surface and tunnel toward roots. Larvae feed at the crown and on major roots.

Current management for *S. exitiosa* generally relies on post-harvest applications of chemical insecticides, such as chlorpyrifos, in the late summer to prevent or limit damage (Horton et al., 2014). However, multiple field trials have consistently demonstrated that applications of the EPN *S. carpocapsae* are as effective as the chemical insecticide standards (Shapiro-Ilan et al., 2009a, 2015). One approach to using EPNs for *S. exitiosa* control is to apply them in the same fashion as chemical standards, that is, as a preventative treatment during egg-laying period. For example, using this approach, a 78–100% reduction in *S. exitiosa* infestation was achieved by applying 1.5 million *S. carpocapsae* to soil surrounding the tree base during egg-laying in the late summer or early fall (Shapiro-Ilan et al., 2009a). Shapiro-Ilan et al. (2015) further demonstrated that *S. carpocapsae* can be effectively applied for *S. exitiosa* control using various equipment including a boom sprayer, trunk sprayer, handgun, or watering can. Since supplemental irrigation is needed to maximize the effectiveness of EPN applications in peach orchards, Shapiro-Ilan et al. (2015) reported that a sprayable gel (Barricade) can be used to protect nematodes from desiccation in lieu of irrigation (eg, for growers that lack irrigation).

Another strategy to using EPNs for *S. exitiosa* management is to apply the treatments curatively to existing infestations in the spring. This approach was first successfully

demonstrated in Canada using *Heterorhabditis bacteriophora*, where an 80% reduction in moth emergence was observed (Cossentine et al., 1990). Subsequently, Cottrell and Shapiro-Ilan (2006) observed an 88% reduction in moth emergence following a late-spring application of *S. carpocapsae* to soil surrounding the tree base at a rate of 3×10^5 IJs per tree. In more recent trials (Shapiro-Ilan et al., unpublished data), *S. carpocapsae* applied at 1 million IJs per tree using various equipment (boom sprayer, trunk sprayer, watering can, handgun) caused high levels of curative control (up to 87%), which exceeded control from chlorpyrifos treatments. In summary, for both preventative and curative control approaches, the high levels of efficacy observed with *S. carpocapsae*, coupled with relatively low rates of application (less than 300 million IJs/ha), suggest that the tactics could successfully be incorporated into an IPM program for *S. exitiosa*.

17.3.2.2 Lesser Peachtree Borer

The lesser peachtree borer, *Synanthedon pictipes*, is an important pest of *Prunus* spp. in the eastern United States (Horton and Johnson, 2005). In the southeastern United States, adult emergence typically begins in March and peaks in April and May, with a second emergence between July and September. Adult moths lay eggs on the trunk and scaffold limbs usually in cracks in the tree's bark or near injured areas (Horton and Johnson, 2005). Larvae bore into the inner bark and cambium where they feed and develop thereby causing severe damage. Larvae overwinter in the tunnels.

Similar to *S. exitiosa*, current control recommendations for *Synanthedon pictipes* rely on intensive use of chemical insecticides (eg, chlorpyrifos). However, field applications of EPNs may be a viable alternative. Laboratory studies were conducted to determine innate virulence to *S. pictipes* larvae among 10 EPN strains and species (Shapiro-Ilan and Cottrell, 2006; Cottrell et al., 2011). Overall, *S. carpocapsae* was most effective, especially against older larvae in these laboratory studies. However, when EPNs were applied in aqueous suspension to *S. pictipes*–infested wounds on peach trees, the treatments failed, presumably due to EPN exposure to UV radiation and/or desiccation (Shapiro-Ilan et al., 2010, unpublished data). Therefore, an improved formulation was needed to facilitate EPN survival and efficacy in these aboveground habitats.

Shapiro-Ilan et al. (2010) compared several strategies to enhance EPN survival on peach tree limbs, including addition of an acrylic resin to the tank, or postapplication coverings of a diaper, or latex paint. The most effective treatment was Barricade, a fire gel commonly used to protect houses and other structures from wild fire. The Barricade plus *S. carpocapsae* treatment (1 million IJs per infested wound) resulted in 0% *S. pictipes* survival in 1 year and 30% survival

FIGURE 17.3 Barricade fire gel applied as a protective formulation in a peach orchard. Entomopathogenic nematodes (*Steinernema carpocapsae*) were applied to wounds infested by *Synanthedon pictipes*. The gel was applied separately after the nematodes to provide protection from UV and desiccation.

in a second year (Shapiro-Ilan et al., 2010). Initially, *S. carpocapsae* was applied to *S. pictipes* infested wounds first with Barricade applied after as a covering (Fig. 17.3). The Barricade was applied separately because a full concentration is too viscous to be applied by normal agricultural sprayers. However, more recently Shapiro-Ilan et al. (2016) discovered that diluted Barricade can be applied in a tank mix with the nematodes. A 2% v/v formulation of Barricade plus *S. carpocapsae* was equally effective as the full rate (approximately 4%) of Barricade plus *S. carpocapsae* and suppressed *S. pictipes* at the same level as chlorpyrifos (Shapiro-Ilan et al., 2016). The use of sprayable gels (such as Barricade) or mulches may have wide applicability to improve the efficacy of above-ground EPN applications in various cropping systems.

17.3.3 Fruit Flies

EPFs and EPNs have been evaluated against fruit flies (Diptera: Tephritidae) that infest cultivated cherries in North America and elsewhere. In laboratory assays, Daniel and Wyss (2009) reported that isolates of *B. bassiana* and *Isaria fumosorosea* caused 90–100% mortality of European cherry fruit fly, *R. cerasi* adult flies at 10^7 conidia/mL but were far less effective (less than 25% mortality) against mature larvae and pupae. Similar differential susceptibility among life stages was observed by Yee and Lacey (2005), who evaluated the *Metarhizium brunneum* against the Western cherry fruit fly (*R. indifferens*). Adult flies were readily infected in vials (100% infection at 4.6×10^8 spores/10 flies), with slightly lower rates of infection (15–68%) observed for teneral adults emerging from treated soil, yet third instars were not infected. In field trials, Daniel and Wyss (2010) reported that *B. bassiana* (product Naturalis-L) applied against *R.*

cerasi at 7-day intervals during the flight period reduced the number of infested fruits at harvest by an average of 65% over 2 years.

Compared with fungi, nematodes appear to be more effective against larval stages of fruit flies. Kepenekci et al. (2015) reported that *S. feltiae* caused 95% mortality of *R. cerasi* larvae, followed by *H. marelata* (82%) and *Heterorhabditis bacteriophora* (76%), at 1000 IJs/larva. Yee and Lacey (2003) reported that *R. indifferens* larvae were the most susceptible stage to *Steinernema* spp. in laboratory bioassays. Nematodes also infected 11–53% of adults that emerged. The use of EPFs and EPNs has also been recommended against other major tephritid pests that may infest peaches or other stone fruits (Soliman et al., 2014).

Overall, studies on microbial control of fruit flies suggest that EPNs have potential for targeting last-instar larval populations, thus decreasing the adult population the following spring; emerging adults may be further targeted with EPF. In Europe, commercial *B. bassiana* formulations are used to control adult *R. cerasi* in organic cherry production (Daniel and Grunder, 2012). The availability of highly effective baits based on spinosad (Gazit et al., 2013) may reduce the commercial demand for microbial control of fruit flies generally. However, MCAs might be deployed in wider IPM applications for fruit flies. For example, Sookar et al. (2014) suggested that sterile male peach fruit flies, *Bactrocera zonata* could be used as vectors of *B. bassiana* to suppress pest populations as part of a sterile insect release program. Additional information on tropical fruit flies is presented in Chapter 18.

17.4 NUT CROPS

17.4.1 Pecan Weevil

The pecan weevil, *Curculio caryae* (Coleoptera: Curculionidae), is a key pest of pecans throughout the southeastern United States and several other states (Shapiro-Ilan et al., 2007). *C. caryae* have a 2- or 3-year life cycle. Adults emerge from soil in late July to early August, feed on developing nuts, and oviposit into the nuts after dough stage (Shapiro-Ilan et al., 2007). Larvae develop within the nut and fourth instars drop to the soil, where they burrow to a depth of 8–25 cm, form a soil cell, and overwinter. The following year, approximately 90% of the larvae pupate and spend the next 9 months in the soil as adults (Shapiro-Ilan et al., 2007). The remaining 10% spend 2 years in the soil as larvae emerging as adults in the third year (Shapiro-Ilan et al., 2007). Control recommendations for *C. caryae* currently consist of multiple above-ground applications of chemical insecticides (primarily carbaryl and pyrethroids) to suppress adults.

EPNs, EPFs, and entomopathogenic bacteria (*C. subtsugae*) applied during periods of *C. caryae's* life cycle have shown high levels of efficacy (Lacey and Shapiro-Ilan, 2008; Shapiro-Ilan et al., 2013b). Microbial control of *C. caryae* using nematodes or fungi is targeted against larvae after they drop from the nut, or against adults while they are in the soil or on emergence. *C. subtsugae* may also be used as an effective canopy spray for control of *C. caryae* adults (Shapiro-Ilan et al., 2013b).

In laboratory screenings, *C. caryae* larvae had low to moderate susceptibility to nine EPN species, yet adults were highly susceptible, particularly to *S. carpocapsae* (Shapiro-Ilan, 2001a,b; Shapiro-Ilan et al., 2003a). Initial field trials targeting adults, *S. carpocapsae* (applied to soil under the canopy) resulted in 60% mortality, but this was considered insufficient for adequately protecting fruits (Shapiro-Ilan et al., 2006). Therefore, subsequent studies evaluated multiple seasonal applications of EPNs, targeted larvae (first year) and later the adults (second year). Field experiments using this approach showed that <1% of weevils survived in *S. carpocapsae* treated plots (Shapiro-Ilan and Gardner, 2012). Larger scale trials are underway.

Another promising group of MCAs for *C. caryae* suppression are the EPFs, particularly *B. bassiana* and *Metarhizium* spp. Initial field studies with these MCAs, however, failed to cause economic control of *C. caryae* (Gottwald and Tedders, 1983). Thus, improved methods of application were explored (Shapiro-Ilan et al., 2003b, 2008b, 2009b). Overall, the most promising methods were applications of *B. bassiana* to the trunk, or to the ground in conjunction with a cover crop (Sudan grass or clover intended to provide UV protection and conducive humidity conditions) (75% or greater adult *C. caryae* mortality at 2.5×10^{12} conidia/tree) (Shapiro-Ilan et al., 2008b, 2012a; Hudson et al., 2010). Application of regenerating trunk bands containing *Metarhizium brunneum* was also effective (Shapiro-Ilan et al., 2009b). The trunk application approach is attractive because it has a relatively small application area (from the base to 1.5 m height) and targets the majority (greater than 90%) of emerging *C. caryae* that crawl or fly to the trunk (Cottrell and Wood, 2008). Nonetheless, given that some of the weevils do fly directly to the canopy, some additional ground or foliar applications to control these weevils may be necessary as well. Conservation approaches, such as using clover cover crops to enhance persistence of endemic *B. bassiana* may also contribute to population reduction (Shapiro-Ilan et al., 2012a). One factor that may hinder efficacy against *C. caryae* is that their pupal cell possesses antibiotic properties that inhibit the fungi (Shapiro-Ilan and Mizell, 2015).

New biological insecticides based on *C. subtsugae* are proving to be highly effective at controlling *C. caryae*. Initial field trials using the product Grandevo resulted in 74.5% corrected mortality within 7 d (Shapiro-Ilan et al., 2013b). Subsequently, in large-plot field trials, four applications of Grandevo at 3 lb/acre (3.4 kg/ha) resulted in greater than 87% control, which was equal to four applications of

standard chemical insecticides (carbaryl alternated with pyrethroids) (unpublished data). Given that EPFs and EPNs are not effective as foliar sprays for *C. caryae*, *C. subtsugae* is currently the only microbial option for control of *C. caryae* control in the canopy.

Additional research is needed to optimize microbial control approaches for *C. caryae*. Conceivably, EPNs can be applied in the spring and fungi in the fall to control belowground stages, whereas *C. subtsugae* can be applied against adult weevils attacking the tree. This integrated approach has proven effective in organic orchards (eg, greater than 80% corrected control, unpublished data). Research to determine long-term effects of these microbial agents on *C. caryae* populations, as well as optimization of rates, timing, and integration with other insecticides is needed.

17.4.2 Navel Orangeworm

The navel orangeworm (NOW), *Amyelois transitella* (Lepidoptera: Pyralidae), is a key pest of almonds and pistachios in California. Larvae overwinter in nuts left after harvest, both on the ground and in trees (referred to as mummy nuts by orchardists). These larvae are a significant source of NOW in the following growing season. NOW larvae in mummy nuts are currently controlled using cultural methods such as disking or flail mowing, which is expensive and causes air pollution. Siegel et al. (2004, 2006) evaluated *S. carpocapsae* for control of NOW in six field trials in 1 m² plots conducted from November 2003 through December 2004 in pistachio orchards in the Central Valley of California. Applications of 10^5 IJs/m² (10^9/ha) in a volume of 187 mL water/m² followed by irrigation resulted in mortality as high as 74.6%.

17.5 CONCLUSIONS AND RECOMMENDATIONS

Issues of cost and efficacy, when compared with chemical insecticides, remain barriers to the successful adoption of MCAs in temperate orchards. There are exceptions; for example, we have highlighted some examples where MCAs may be equally effective as the chemical standard such as *S. carpocapsae* against *S. exitiosa* or granulovirus against codling moth. To generate wider adoption of MCAs, options are to improve the effectiveness of biocontrol organisms and/or develop novel approaches to incorporate the MCAs into integrated management programs. Microbial control efficacy can be greatly enhanced through strain discovery or improvement techniques (eg, selection, hybridization, or transgenics), or via improved formulation and application methods (Shapiro-Ilan et al., 2012b). Microbial control can be integrated with other tactics through targeting specific pest stages and through conservation biocontrol approaches such as habitat manipulation. Another important avenue to wider MCA adoption that we advocate is to factor in the added benefits of MCAs (eg, environmental and human safety) when weighing recommendation of chemical insecticides versus MCAs.

Substantial research and exploration needs remain for developing MCAs in temperate orchards. While the range of pests conceivably targeted with MCAs is large, research is still needed to investigate their potential under operational conditions. Laboratory-based studies should aim to provide useful context for more focused and thoughtful scaled up evaluation in the field. For example, there is a need to better understand interactions between microbial controls, other natural enemies, and chemical pesticides (especially fungicides); once interactions are discovered in the laboratory, they will need to be confirmed in the field. We also note that there are a variety of MCAs or biological-based insecticide products (some relatively new) that have not been tested against many important pests of temperate orchards.

In the past 10 years, interest in the commercial development of entomopathogens has led to new products as well as enhancements in effectiveness of existing ones. We expect the trend to continue. Evidence of future expansion is indicated by the increased interest in MCAs by large and medium-size corporations. Adoption of MCAs will also be fueled by the continued removal of broad-spectrum chemical insecticides due to environmental and regulatory concerns. On the other hand, microbial control products will have to compete with other new-generation low-risk insecticides that will make their successful adoption a challenge. Nonetheless, through advances in research, we expect biocontrol applications and integrated programs will be increasingly tailored to specific orchard systems and engineered for maximum efficiency.

REFERENCES

Alston, D.G., Rangel, D.E.N., Lacey, L.A., Golez, H.G., Kim, J.J., Roberts, D.W., 2005. Evaluation of novel fungus and nematode isolates for control of *Conotrachelus nenuphar* (Coleoptera: Curculionidae) larvae. Biol. Control 35, 163–171.

Arthurs, S.P., Lacey, L.A., 2004. Field evaluation of commercial formulations of the codling moth granulovirus (CpGV): persistence of activity and success of seasonal applications against natural infestations in the Pacific Northwest. Biol. Control 31, 388–397.

Arthurs, S., Lacey, L.A., Fritts Jr., R., 2005. Optimizing the use of the codling moth granulovirus: effects of application rate and spraying frequency on control of codling moth larvae in Pacific Northwest apple orchards. J. Econ. Entomol. 98, 1459–1468.

Arthurs, S.P., Lacey, L.A., Behle, R.W., 2006. Evaluation of spray-dried lignin-based formulations and adjuvants as ultraviolet light protectants for the granulovirus of the codling moth, *Cydia pomonella* (L.). J. Invertebr. Pathol. 93, 88–95.

Arthurs, S.P., Lacey, L.A., Miliczky, E.R., 2007. Evaluation of the codling moth granulovirus and spinosad for codling moth control and impact on non-target species in pear orchards. Biol. Control 49, 99–109.

Asser-Kaiser, S., Fritsch, E., Undorf-Spahn, K., Kienzle, J., Eberle, K.E., Gund, N.A., Reineke, A., Zebitz, C.P.W., Heckel, D.G., Huber, J., Jehle, J.A., 2007. Rapid emergence of baculovirus resistance in codling moth due to dominant, sex-linked inheritance. Science 317 (5846), 1916–1918.

Baily, P., Baker, G., Caon, G., 1996. Field efficacy and persistence of *Bacillus thuringiensis* var. *kurstaki* against *Epiphyas postvittana* (Walker) (Lepidoptera: Tortricidae) in relation to larval behaviour on grapevine leaves. Aust. J. Entomol. 35, 297–302.

Barnes, M.M., Millar, J.G., Kirsch, P.A., Hawks, D.C., 1992. Codling moth (Lepidoptera: Tortricidae) control by dissemination of synthetic female sex pheromone. J. Econ. Entomol. 85, 1274–1277.

Barnett, W.W., Edstrom, J.P., Coviello, R.L., Zalom, F.G., 1993. Insect pathogen "Bt" controls peach twig borer on fruits and almonds. Calif. Agric. 47, 4–6.

Batalla-Carrera, L., Morton, A., Garcia-del-Pino, F., 2013. Field efficacy against the hazelnut weevil, *Curculio nucum* and short-term persistence of entomopathogenic nematodes. Span. J. Agric. Res. 11, 1112–1119.

Beers, E.H., Brunner, J.F., 2015. Orchard Pest Management Online. Washington State University Tree Fruit Research and Extension Center. http://jenny.tfrec.wsu.edu/opm/.

Bird, A.E., Hesketh, H., Cross, J.V., Copland, M., 2004. The common black ant, *Lasius niger* (Hymenoptera: Formicidae), as a vector of the entomopathogen *Lecanicillium longisporum* to rosy apple aphid, *Dysaphis plantaginea* (Homoptera: Aphididae). Biocontrol Sci. Technol. 14, 757–767.

Blommers, L.H.M., 1994. Integrated pest management in European apple orchards. Annu. Rev. Entomol. 39, 213–241.

Blommers, L., Vaal, F., Freriks, J., Helsen, H., 1987. Three years of specific control of summer fruit tortrix and codling moth on apple in the Netherlands. J. Appl. Entomol. 104, 353–371.

Briese, D.T., Mende, H.A., Grace, T.D.C., Geier, P.W., 1980. Resistance to a nuclear polyhedrosis virus in the light-brown apple moth *Epiphyas postvittana* (Lepidoptera: Tortricidae). J. Invertebr. Pathol. 36, 211–215.

Bruck, D.J., Walton, V.M., 2007. Susceptibility of the filbertworm (*Cydia latiferreana*, Lepidoptera: Tortricidae) and filbert weevil (*Curculio occidentalis*, Coleoptera: Curculionidae) to entomopathogenic nematodes. J. Invertebr. Pathol. 96, 93–96.

Calkins, C.O., 1999. Review of the codling moth areawide suppression program in the Western United States. J. Agric. Entomol. 15, 327–333.

Chambers, U., Bruck, D.J., Olsen, J., Walton, V.M., 2010. Control of overwintering filbertworm (Lepidoptera: Tortricidae) larvae with *Steinernema carpocapsae*. J. Econ. Entomol. 103, 416–422.

Cornale, R., Reggiani, A., Ladurner, E., Fiorentini, F., 2006. Efficacia di bioinsetticidi a base di nematodi entomopatogeni (*Steinernema carpocapsae* e *S. feltiae*) nei confronti di larve di *Cydia pomonella*. ATTI Gior. Fitopatolog. 1, 37–42.

Cossentine, J.E., Banham, F.L., Jensen, L.B., 1990. Efficacy of the nematode, *Heterorhabditis heliothidis* (Rhabditida: Heterorhabditidae) against the peachtree borer, *Synanthedon exitiosa* (Lepidoptera: Sesiidae) in peach trees. J. Entomol. Soc. B. C. 87, 82–84.

Cossentine, J.E., Jensen, L.B., Moyls, L., 2002. Fruit bins washed with *Steinernema carpocapsae* (Rhabditida: Steinernematidae) to control *Cydia pomonella* (Lepidoptera: Tortricidae). Biocontrol Sci. Technol. 12, 251–258.

Cossentine, J.E., Jensen, L.B., Deglow, E.K., 2003. Strategy for orchard use of *Bacillus thuringiensis* while minimizing impact on *Choristoneura rosaceana* parasitoids. Entomol. Exp. Appl. 109, 205–210.

Cossentine, J.E., Judd, G.J.R., Bissett, J.D., Lacey, L.A., 2010. Susceptibility of apple clearwing moth larvae, *Synanthedon myopaeformis* (Lepidoptera: Sesiidae) to *Beauveria bassiana* and *Metarhizium brunneum*. Biocontrol Sci. Technol. 20, 703–707.

Côté, J.C., Vincent, C., Son, K.H., Bok, S.H., 2001. Persistence of insecticidal activity of novel bio-encapsulated formulations of *Bacillus thuringiensis* var. *kurstaki* against *Choristoneura rosaceana* [Lepidoptera: Tortricidae]. Phytoprotection 82, 73–82.

Cottrell, T.E., Shapiro-Ilan, D.I., 2006. Susceptibility of the peachtree borer, *Synanthedon exitiosa*, to *Steinernema carpocapsae* and *Steinernema riobrave* in laboratory and field trials. J. Invertebr. Pathol. 92, 85–88.

Cottrell, T.E., Wood, B.W., 2008. Movement of adult pecan weevils *Curculio caryae* (Coleoptera: Curculionidae) within pecan orchards. Agric. Forest Entomol. 10, 363–373.

Cottrell, T.E., Shapiro-Ilan, D.I., Horton, D.L., Mizell III, R.F., 2011. Laboratory virulence and orchard efficacy of entomopathogenic nematodes toward the lesser peachtree borer (Lepidoptera: Sesiidae). Environ. Entomol. 104, 47–53.

Cross, J.V., Solomon, M.G., Chandler, D., Jarrett, P., Richardson, P.N., Winstanley, D., Bathon, H., Huber, J., Keller, B., Langenbruch, G.A., Zimmermann, G., 1999. Biocontrol of pests of apples and pears in Northern and Central Europe: 1. Microbial agents and nematodes. Biocontrol Sci. Technol. 9, 125–149.

Daniel, C., Grunder, J., 2012. Integrated management of European cherry fruit fly *Rhagoletis cerasi* (L.): situation in Switzerland and Europe. Insects 3, 956–988.

Daniel, C., Wyss, E., 2009. Susceptibility of different life stages of the European cherry fruit fly, *Rhagoletis cerasi*, to entomopathogenic fungi. J. Appl. Entomol. 133, 473–483.

Daniel, C., Wyss, E., 2010. Field applications of *Beauveria bassiana* to control the European cherry fruit fly *Rhagoletis cerasi*. J. Appl. Entomol. 134, 675–681.

De Luca, F., Clausi, M., Troccoli, A., Curto, G., Giancarlo Rappazzo, G., Tarasco, E., 2015. Entomopathogenic nematodes in Italy: occurrence and use in microbial control strategies. In: Campos-Herrera, R. (Ed.), Nematode Pathogenesis of Insects and Other Pests: Ecology and Applied Technologies for Sustainable Plant and Crop Protection. Springer, Dordrecht, pp. 431–449.

De Reede, R.H., Gruys, P., Vaal, F., 1985. Leafrollers in apple IPM under regimes based on *Bacillus thuringiensis*, on diflubenzuron, or on epofenonane. Entomol. Exp. Appl. 37, 263–274.

De Waal, J.Y., Malan, A.P., Levings, J., Addison, M.F., 2010. Key elements in the successful control of diapausing codling moth, *Cydia pomonella* (Lepidoptera: Tortricidae) in wooden fruit bins with a South African isolate of *Heterorhabditis zealandica* (Rhabditida: Heterorhabditidae). Biocontrol Sci. Technol. 20, 489–502.

De Waal, J.Y., Malan, A.P., Addison, M.F., 2011. Efficacy of entomopathogenic nematodes (Rhabditida: Heterorhabditidae and Steinernematidae) against codling moth, *Cydia pomonella* (Lepidoptera: Tortricidae) in temperate regions. Biocontrol Sci. Technol. 20, 489–502.

Diaz, B.M., Oggerin, M., Lastra, C.C.L., Rubio, V., Fereres, A., 2009. Characterization and virulence of *Lecanicillium lecanii* against different aphid species. BioControl 54, 825–835.

Eberle, K.E., Asser-Kaiser, S., Sayed, S.M., Nguyen, H.T., Jehle, J.A., 2008. Overcoming the resistance of codling moth against conventional *Cydia pomonella* granulovirus (CpGV-M) by a new isolate CpGV-I12. J. Invertebr. Pathol. 98, 293–298.

Erler, F., Pradier, T., Aciloglu, B., 2014. Field evaluation of an entomopathogenic fungus, *Metarhizium brunneum* strain F52, against pear psylla, *Cacopsylla pyri*. Pest Manag. Sci. 70, 496–501.

Flückiger, C.R., 1982. Untersuchungen über drei Baculovirus-Isolate des Schalenwicklers, *Adoxophyes orana* F. v. R. (Lep., Tortricidae), dessen Phänologies und erste Feldversuche, als Grundlagen zur mikrobiologischen Bekämpfung dieses Obstschädlings. Mitt. Schweiz. Entomol. Gesell. 55, 241–288.

Gazit, Y., Gavriel, S., Akiva, R., Timar, D., 2013. Toxicity of baited spinosad formulations to *Ceratitis capitata*: from the laboratory to the application. Entomol. Exp. Appl. 147, 120–125.

Gottwald, T.R., Tedders, W.L., 1983. Suppression of pecan weevil (Coleoptera: Curculionidae) populations with entomopathogenic fungi. Environ. Entomol. 12, 471–474.

Grewal, P.S., Ehlers, R.-U., Shapiro-Ilan, D.I. (Eds.), 2005. Nematodes as Biological Control Agents. CABI, Wallingford. 505 pp.

Goyer, R.A., Wei, H., Fuxa, J.R., 2001. Prevalence of viral diseases of the fruittree leafroller, *Archips argyrospila* (Walker), in Louisiana. J. Entomol. Sci. 36, 17–22.

Harris, M.O., Mafile, F., Dhana, S., 1997. Behavioral responses of light brown apple moth neonate larvae on diets containing *Bacillus thuringiensis* formulations or endotoxins. Entomol. Exp. Appl. 84, 207–219.

Higbee, B., Calkins, C., Temple, C., 2001. Overwintering of codling moth (Lepidoptera: Tortricidae) larvae in apple harvest bins and subsequent moth emergence. J. Econ. Entomol. 94, 1511–1517.

Horton, D., Johnson, D. (Eds.), 2005. Southeastern peach Grower's Handbook. Univ. GA Coop. Ext. Serv., G.E.S. Handbook No. 1.

Horton, D., Brannen, P., Bellinger, B., Lockwood, D., Ritchie, D., 2014. 2014 Southeastern Peach, Nectarine, and Plum Pest Management and Culture Guide. Bulletin 1171. University of Georgia, Athens, GA.

Hudson, W.G., Shapiro-Ilan, D.I., Gardner, W.A., Cottrell, T.E., Behle, B., 2010. Biological Control of Pecan Weevils in the Southeast: A Sustainable Organic Approach. www.sare.org/publications/factsheetorg/publications/factsheet.

Jaques, R.P., Hardman, J., Laing, J., Smith, R., 1994. Orchard trials in Canada on control of *Cydia pomonella* (Lep: Tortricidae) by granulosis virus. Entomophaga 39, 281–292.

Jenkins, D.A., Mizell III, R.F., Shapiro-Ilan, D., Cottrell, T., Horton, D., 2006. Invertebrate predators and parasitoids of plum curculio, *Conotrachelus nenuphar* (Herbst) (Coleoptera: Curculionidae) in Georgia and Florida. Fla. Entomol. 89, 435–440.

Kain, D.F., Agnello, A.M., 1999. Pest status of American plum borer (Lepidoptera: Pyralidae) and fruit tree borer control with synthetic insecticides and entomopathogenic nematodes in New York State. J. Econ. Entomol. 92, 193–200.

Kaya, H.K., Joos, J.L., Falcon, L.A., Berlowitz, A., 1984. Suppression of the codling moth (Lepidoptera: Olethreutidae) with the entomogenous nematode, *Steinernema feltiae*. J. Econ. Entomol. 77, 1240–1244.

Kepenekci, I., Hazir, S., Özdem, A., 2015. Evaluation of native entomopathogenic nematodes for the control of the European cherry fruit fly *Rhagoletis cerasi* L. (Diptera: Tephritidae) larvae in soil. Turk. J. Agric. For 39, 74–79.

Kienzle, J., Zimmer, J., Volk, F., Zebitz, C.P.W., 2008. Experiences with entomopathogenic nematodes for the control of overwintering codling moth larvae in Germany (Rhabditida: Steinernematidae). In: Boos, M. (Ed.), Ecofruit – 13th International Conference on Cultivation Technique and Phytopathological Problems in Organic Fruit, pp. 277–283 (Weinsberg/Germany).

Knight, A.L., Witzgall, P., 2013. Combining mutualistic yeast and pathogenic virus — a novel method for codling moth control. J. Chem. Ecol. 39, 1019–1026.

Knight, A.L., Lacey, L.A., Stockhoff, B.A., Warner, R.L., 1998. Activity of Cry 1 endotoxins of *Bacillus thuringiensis* for four tree fruit leafroller pest species (Lepidoptera: Tortricidae). J. Agric. Entomol. 15, 92–103.

Kocourek, F., Pultar, O., Stará, J., 2007. Comparison of the efficacy of AdorGV and chemical insecticides against the Summer fruit tortrix, *Adoxophyes orana*, in commercial apple orchards in the Czech Republic. IOBC/WPRS Bull. 30 (1), 171–176.

Kutinkova, H., Samietz, J., Dzhuvinov, V., Zingg, D., Kessler, P., 2012. Successful application of the baculovirus product Madex® for control of *Cydia pomonella* (L.) in Bulgaria. J. Plant Prot. Res. 52, 205–213.

Lacey, L.A., Arthurs, S.P., 2005. New method for testing solar sensitivity of commercial formulations of the granulovirus of codling moth (*Cydia pomonella*, Tortricidae: Lepidoptera). J. Invertebr. Pathol. 90, 85–90.

Lacey, L.A., Chauvin, R.L., 1999. Entomopathogenic nematodes for control of codling moth in fruit bins. J. Econ. Entomol. 92, 104–109.

Lacey, L.A., Shapiro-Ilan, D.I., 2008. Microbial control of insect pests in temperate orchard systems: potential for incorporation into IPM. Annu. Rev. Entomol. 53, 121–144.

Lacey, L.A., Siegel, J.P., 2000. Safety and ecotoxicology of entomopathogenic bacteria. In: Charles, J.-F., Delécluse, A., Nielsen-La Roux, C. (Eds.), Entomopathogenic Bacteria: From Laboratory to Field. Springer, Dordrecht, The Netherlands, pp. 253–273.

Lacey, L.A., Unruh, T.R., 2005. Biological control of codling moth (*Cydia pomonella*, Tortricidae: Lepidoptera) and its role in Integrated Pest Management, with emphasis on entomopathogens. Vidalia 12, 33–60.

Lacey, L.A., Arthurs, S.P., Headrick, H., 2005a. Comparative activity of the codling moth granulovirus against *Grapholita molesta* and *Cydia pomonella* (Lepidoptera: Tortricidae). J. Entomol. Soc. B. C. 102, 79–80.

Lacey, L.A., Neven, L.G., Headrick, H.L., Fritts Jr., R., 2005b. Factors affecting entomopathogenic nematodes (Steinernematidae) for the control of overwintering codling moth (Lepidoptera: Tortricidae) in fruit bins. J. Econ. Entomol. 98, 1863–1869.

Lacey, L.A., Arthurs, S.P., Unruh, T.R., Headrick, H., Fritts Jr., R., 2006a. Entomopathogenic nematodes for control of codling moth (Lepidoptera: Tortricidae) in apple and pear orchards: effect of nematode species and seasonal temperatures, adjuvants, application equipment and post-application irrigation. Biol. Control 37, 214–223.

Lacey, L.A., Granatstein, D., Arthurs, S.P., Headrick, H., Fritts Jr., R., 2006b. Use of entomopathogenic nematodes (Steinernematidae) in conjunction with mulches for control of overwintering codling moth (Lepidoptera: Tortricidae). J. Entomol. Sci. 41, 107–119.

Lacey, L.A., Arthurs, S.P., Knight, A.L., Huber, J., 2007. Microbial control of lepidopteran pests of apple orchards. In: Lacey, L.A., Kaya, H.K. (Eds.), Field Manual of Techniques in Invertebrate Pathology. Springer, Dordrecht, pp. 527–546.

Lacey, L.A., Headrick, H., Arthurs, S.P., 2008a. The effect of temperature on the long-term storage of codling moth granulovirus formulations. J. Econ. Entomol. 101, 288–294.

Lacey, L.A., Thomson, D., Vincent, C., Arthurs, S.P., 2008b. Codling moth granulovirus: a comprehensive review. Biocontrol Sci. Technol. 18, 639–663.

Lacey, L.A., Shapiro-Ilan, D.I., Glenn, G.M., 2010. The effect of post-application anti-desiccant agents and formulated host-cadavers on entomopathogenic nematode efficacy for control of diapausing codling moth larvae (Lepidoptera: Tortricidae). Biocontrol Sci. Technol. 20, 909–921.

Leskey, T.C., Pinero, J.C., Prokopy, R.J., 2008. Oder-baited trap trees: a novel management tool for plum curculio (Coleoptera: Curculionidae). J. Econ. Entomol. 101, 1302–1309.

Li, S., Fitzpatrick, S., Isman, M., 1995. Susceptibility of different instars of the obliquebanded leafroller (Lepidoptera: Tortricidae) to *Bacillus thuringiensis* var. *kurstaki*. Biol. Control 88, 610–614.

Longworth, J.F., Singh, P., 1980. A nuclear polyhedrosis virus of the light brown apple moth, *Epiphyas postvittana* (Lepidoptera: Tortricidae). J. Invertebr. Pathol. 35, 84–87.

MacCollom, G.B., Reed, E.M., 1971. A nuclear polyhedrosis virus of the light brown apple moth, *Epiphyas postvittana*. J. Invertebr. Pathol. 18, 337–343.

Martin, P.A.W., Gundersen–Rindal, D., Blackburn, M., Buyer, J., 2007. *Chromobacterium subtsugae* sp. nov., a betaproteobacterium toxic to Colorado potato beetle and other insect pests. Int. J. Syst. Evol. Microbiol. 57, 993–999.

Moore, S.D., Kirkman, W., Richards, G.I., Stephen, P.R., 2015. The *Cryptophlebia leucotreta* granulovirus – 10 years of commercial field use. Viruses (Basel) 7, 1284–1312.

Morton, A., García-Del-Pino, F., 2008. Field efficacy of the entomopathogenic nematode *Steinernema feltiae* against the Mediterranean flat-headed rootborer *Capnodis tenebrionis*. J. Appl. Entomol. 132, 632–637.

Nevaneethan, T., Strauch, O., Besse, S., Bonhomme, A., Ehlers, R.U., 2010. Influence of humidity and a surfactant-polymer-formulation on the control potential of the entomopathogenic nematode *Steinernema feltiae* against diapausing codling moth larvae (*Cydia pomonella* L.) (Lepidoptera: Tortricidae). BioControl 55, 777–788.

Nielsen, C., Jensen, A.B., Eilenberg, J., 2007. Survival of entomophthoralean fungi infecting aphids and higher flies during unfavourable conditions and implications for conservation biological control. In: Ekesi, S., Maniania, N.K. (Eds.), Use of Entomopathogenic Fungi in Biological Pest Management. Research Signpost, Kerala, India, pp. 13–38.

Olthof, T.H., Hagley, E.C., 1993. Laboratory studies of the efficacy of steinernematid nematodes against the plum curculio (Coleoptera: Curculionidae). J. Econ. Entomol. 86, 1078–1082.

Pajač, I., Pejić, I., Barić, B., 2011. Codling moth, *Cydia pomonella* (Lepidoptera: Tortricidae)–major pest in apple production: an overview of its biology, resistance, genetic structure and control strategies. Agric. Conspec. Sci. 76, 87–92.

Paparatti, B., Speranza, S., 2005. Biological control of hazelnut weevil (*Curculio nucum* L., Coleoptera, Curculionidae) using the entomopathogenic fungus *Beauveria bassiana* (Balsamo) Vuill. (Deuteromycotina, Hyphomycetes). Acta Hortic. 686, 407–412.

Parvizi, R., 2003. An evaluation of the efficacy of the entomopathogenic nematode *Heterorhabditis bacteriophora* and *Steinernema* sp. in controlling immature stages of the apple clearwing, *Synanthedon myopaaeformis*. Iran. J. Agric. Sci. 34, 303–311.

Pasqualini, E., 1996. Efficacy of *Bacillus thuringiensis* var. *kurstaki* against leafrollers in Emilia-Romagna. Acta Hortic. 422, 342–343.

Pemsel, M., Schwab, S., Scheurer, A., Freitag, D., Schatz, R., Schlücker, E., 2010. Advanced PGSS process for the encapsulation of the biopesticide *Cydia pomonella* granulovirus. J. Supercrit. Fluids 53, 174–178.

Pereault, R.J., Whalon, M.E., Alston, D.G., 2009. Field efficacy of ento-mopathogenic fungi and nematodes targeting caged last-instar plum curculio (Coleoptera: Curculionidae) in Michigan cherry and apple orchards. Environ. Entomol. 38, 1126–1134.

Pfannenstiel, R.S., Szymanski, M., Lacey, L.A., Brunner, J.F., Spence, K., 2004. Discovery of a granulovirus of *Pandemis pyrusana* (Lepidoptera: Tortricidae), a leafroller pest of apples in Washington. J. Invertebr. Pathol. 86, 124–127.

Puterka, G.J., 1999. Fungal pathogens for arthropod pest control in orchard systems: mycoinsecticidal approach for pear psylla control. BioControl 44, 183–209.

Racette, G., Chouinard, G., Vincent, C., Hill, S.B., 1992. Ecology and management of plum curculio, *Conotrachelus nenuphar* [Coleoptera: Curculionidae], in apple orchards. Phytoprotection 73, 85–100.

Riga, K., Lacey, L.A., Guerra, N., Headrick, H.L., 2006. Control of the oriental fruit moth, *Grapholita molesta*, using entomopathogenic nematodes in laboratory and bin assays. J. Nematol. 38, 168–171.

Sanahuja, G., Banakar, R., Twyman, R.M., Capell, T., Christou, P., 2011. *Bacillus thuringiensis*: a century of research, development and commercial applications. Plant Biotechnol. J. 9, 283–300.

Sarraquigne, J.P., Couturié, E., Fernandez, M.M., 2009. Integrated control of hazelnut weevil (*Curculio nucum*): an evaluation of entomopathogenic nematodes and parasitic fungi. Acta Hortic. 845, 555–560.

Schmitt, A., Bisutti, I.L., Ladurner, E., Benuzzi, M., Sauphanor, B., Kienzle, J., Jehle, J.A., 2013. The occurrence and distribution of resistance of codling moth to *Cydia pomonella* granulovirus in Europe. J. Appl. Entomol. 137, 641–649.

Sekita, N., Kawashima, K., Aizu, H., Shirisaki, S., Yamada, M., 1984. A short term control of *Adoxophyes orana fasciata* Walsingham (Lepidoptera: Tortricidae) by a granulosis virus in apple orchards. Appl. Entomol. Zool. 19, 498–508.

Shapiro-Ilan, D.I., 2001a. Virulence of entomopathogenic nematodes to pecan weevil larvae *Curculio caryae* (Coleoptera: Curculionidae) in the laboratory. J. Econ. Entomol. 94, 7–13.

Shapiro-Ilan, D.I., 2001b. Virulence of entomopathogenic nematodes to pecan weevil adults (Coleoptera: Curculionidae). J. Entomol. Sci. 36, 325–328.

Shapiro-Ilan, D.I., Cottrell, T.E., 2006. Susceptibility of the lesser peachtree borer (Lepidoptera: Sesiidae) to entomopathogenic nematodes under laboratory conditions. Environ. Entomol. 35, 358–365.

Shapiro-Ilan, D.I., Gardner, W.A., 2012. Improved control of *Curculio caryae* (Coleoptera: Curculionidae) through multi-stage pre-emergence applications of *Steinernema carpocapsae*. J. Entomol. Sci. 47, 27–34.

Shapiro-Ilan, D.I., Mizell, R.F., 2015. An insect pupal cell with antimicrobial properties that suppress an entomopathogenic fungus. J. Invertebr. Pathol. 124, 114–116.

Shapiro-Ilan, D.I., Mizell, R.F., Campbell, J.F., 2002. Susceptibility of the plum curculio, *Conotrachelus nenuphar*, to entomopathogenic nematodes. J. Nematol. 34, 246–249.

Shapiro-Ilan, D.I., Stuart, R., McCoy, C.W., 2003a. Comparison of beneficial traits among strains of the entomopathogenic nematode, *Steinernema carpocapsae*, for control of *Curculio caryae* (Coleoptera: Curculionidae). Biol. Control 28, 129–136.

Shapiro-Ilan, D.I., Gardner, W.A., Fuxa, J.R., Wood, B.W., Nguyen, K.B., Adams, B.J., Humber, R.A., Hall, M.J., 2003b. Survey of entomopathogenic nematodes and fungi endemic to pecan orchards of the Southeastern United States and their virulence to the pecan weevil (Coleoptera: Curculionidae). Environ. Entomol. 32, 187–195.

Shapiro-Ilan, D.I., Mizell, R.F., Cottrell, T.E., Horton, D.L., 2004. Measuring field efficacy of *Steinernema feltiae* and *Steinernema riobrave* for suppression of plum curculio, *Conotrachelus nenuphar*, larvae. Biol. Control 30, 496–503.

Shapiro-Ilan, D.I., Duncan, L.W., Lacey, L.A., Han, R., 2005. Orchard crops. In: Grewal, P., Ehlers, R.-U., Shapiro-Ilan, D.I. (Eds.), Nematodes as Biological Control Agents. CABI Publishing, Wallingford, pp. 215–230.

Shapiro-Ilan, D.I., Cottrell, T.E., Brown, I., Gardner, W.A., Hubbard, R.K., Wood, B.W., 2006. Effect of soil moisture and a surfactant on entomopathogenic nematode suppression of the pecan weevil, *Curculio caryae*. J. Nematol. 38, 474–482.

Shapiro-Ilan, D.I., Lacey, L.A., Siegel, J.P., 2007. Microbial control of insect pests of stone fruit and nut crops. In: Lacey, L.A., Kaya, H.K. (Eds.), Field Manual of Techniques in Invertebrate Pathology. Springer, Dordrecht, pp. 547–565.

Shapiro-Ilan, D.I., Mizell III, R.F., Cottrell, T.E., Horton, D.L., 2008a. Control of plum curculio, *Conotrachelus nenuphar* with entomopathogenic nematodes: effects of application timing, alternate host plant, and nematode strain. Biol. Control 44, 207–215.

Shapiro-Ilan, D.I., Gardner, W.A., Cottrell, T.E., Behle, R.W., Wood, B.W., 2008b. A comparison of application methods for suppressing the pecan weevil (Coleoptera: Curculionidae) with *Beauveria bassiana* under field conditions. Environ. Entomol. 37, 162–171.

Shapiro-Ilan, D.I., Cottrell, T.E., Mizell III, R.F., Horton, D.L., Davis, J., 2009a. A novel approach to biological control with entomopathogenic nematodes: Prophylactic control of the peachtree borer, *Synanthedon exitiosa*. Biol. Control 48, 259–263.

Shapiro-Ilan, D.I., Gardner, W.A., Cottrell, T.E., Leland, L., Behle, R.W., 2009b. Mortality and mycosis of adult *Curculio caryae* (Coleoptera: Curculionidae) following application of *Metarhizium anisopliae*: laboratory and field trials. J. Entomol. Sci. 44, 24–36.

Shapiro-Ilan, D.I., Cottrell, T.E., Mizell III, R.F., Horton, D.L., Behle, B., Dunlap, C., 2010. Efficacy of *Steinernema carpocapsae* for control of the lesser peachtree borer, *Synanthedon pictipes*: improved aboveground suppression with a novel gel application. Biol. Control 54, 23–28.

Shapiro-Ilan, D.I., Leskey, T.C., Wright, S.E., 2011. Virulence of entomopathogenic nematodes to plum curculio, *Conotrachelus nenuphar*: effects of strain, temperature, and soil type. J. Nematol. 43, 187–195.

Shapiro-Ilan, D.I., Gardner, W.A., Wells, L., Wood, B.W., 2012a. Cumulative impact of a clover cover crop on the persistence and efficacy of *Beauveria bassiana* in suppressing the pecan weevil (Coleoptera: Curculionidae). Environ. Entomol. 41, 298–307.

Shapiro-Ilan, D.I., Bruck, D.J., Lacey, L.A., 2012b. Principles of epizootiology and microbial control. In: Vega, F.E., Kaya, H.K. (Eds.), Insect Pathology, second ed. Elsevier, Amsterdam, pp. 29–72.

Shapiro-Ilan, D.I., Wright, S.E., Tuttle, A.F., Cooley, D.R., Leskey, T.C., 2013a. Using entomopathogenic nematodes for biological control of plum curculio, *Conotrachelus nenuphar*: effects of irrigation and species in apple orchards. Biol. Control 67, 123–129.

Shapiro-Ilan, D.I., Cottrell, T.E., Jackson, M.A., Wood, B.W., 2013b. Control of key pecan insect pests using biorational pesticides. J. Econ. Entomol. 106, 257–266.

Shapiro-Ilan, D.I., Cottrell, T.E., Mizell III, R.F., Horton, D.L., Abdo, Z., 2015. Field suppression of the peachtree borer, *Synanthedon exitiosa*, using *Steinernema carpocapsae*: effects of irrigation, a sprayable gel and application method. Biol. Control 82, 7–12.

Shapiro-Ilan, D.I., Cottrell, T.E., Mizell III, R.F., Horton, D.L., 2016. Efficacy of *Steinernema carpocapsae* plus fire gel applied as a single spray for control of the lesser peachtree borer, *Synanthedon pictipes*. Biol. Control 94, 33–36.

Siegel, J., Lacey, L.A., Fritts Jr., R., Higbee, B.S., Noble, P., 2004. Use of steinernematid nematodes for post-harvest control of navel orangeworm (Lepidoptera: Pyralidae, *Amyelois transitella*) in fallen pistachios. Biol. Control 30, 410–417.

Siegel, J., Lacey, L.A., Higbee, B.S., Noble, P., Fritts Jr., R., 2006. Effect of application rates and abiotic factors on *Steinernema carpocapsae* for control of overwintering navel orangeworm (Lepidoptera: Pyralidae, *Amyelois transitella*) in fallen pistachios. Biol. Control 36, 324–330.

Soliman, N.A., Ibrahim, A.A., El-Deen, M.S., Ramadan, N.F., Farag, S.R., 2014. Entomopathogenic nematodes and fungi as biocontrol agents for the peach fruit fly, *Bactrocera zonata* (Saunders) and the Mediterranean fruit fly, *Ceratitis capitata* (Wiedemann) soil borne-stages. Egypt. J. Biol. Pest Control 24, 497–502.

Sookar, P., Alleck, M., Ahseek, N., Bhagwant, S., 2014. Sterile male peach fruit flies, *Bactrocera zonata* (Saunders) (Diptera: Tephritidae), as a potential vector of the entomopathogen *Beauveria bassiana* (Balsamo) Vuillemin in a SIT programme. Afr. Entomol. 22, 488–498.

Sorensen, A.A., Falcon, L.A., 1980. Comparison of microdroplet and high volume application of *Bacillus thuringiensis* on pear: suppression of fruit tree leafroller (*Archips argyrospilus*) and coverage on foliage and fruit. Environ. Entomol. 9, 350–358.

Steinkraus, D.C., 2006. Factors affecting transmission of fungal pathogens of aphids. J. Invertebr. Pathol. 92, 125–131.

Suckling, D.M., Brockerhoff, E.G., 2010. Invasion biology, ecology, and management of the light brown apple moth (Tortricidae). Annu. Rev. Entomol. 55, 285–306.

Tedders, W.L., Weaver, D.J., Wehunt, E.J., Gentry, C.R., 1982. Bioassay of *Metarhizium anisopliae*, *Beauveria bassiana*, and *Neoaplectana carpocapsae* against larvae of the plum curculio, *Conotrachelus nenuphar* (Herbst) (Coleoptera: Curculionidae). Environ. Entomol. 11, 901–904.

Todorova, S.I., Coderre, D., Vincent, C., Côté, J.C., 2002. Screening of *Beauveria bassiana* (Hyphomycetes) isolates against *Choristoneura rosaceana* (Lepidoptera: Tortricidae). Can. Entomol. 134, 77–84.

Unruh, T.R., Lacey, L.A., 2001. Control of codling moth, *Cydia pomonella* (Lepidoptera: Tortricidae) with *Steinernema carpocapsae*: effects of supplemental wetting and pupation site on infection rate. Biol. Control 20, 48–56.

Unruh, T.R., Lacey, L.A., Headrick, H.L., Pfannenstiel, R.S., 2012. The effect of the granulovirus (PapyGV) on larval mortality and feeding behaviour of the Pandemis leafroller, *Pandemis pyrusana* (Lepidoptera: Tortricidae). Biocontrol Sci. Technol. 22, 981–990.

Vail, P.V., Barnett, W., Cowan, D.C., Sibbett, S., Beede, R., Tebbets, J.S., 1991. Codling moth (Lepidoptera: Tortricidae) control on commercial walnuts with a granulosis virus. J. Econ. Entomol. 84, 1448–1453.

van der Geest, L.P.S., Evenhuis, H.H. (Eds.), 1991. Tortricid Pests, Their Biology, Natural Enemies and Control. Elsevier Science Publishers, Amsterdam, The Netherlands.

van Emden, H.F., Harrington, R., 2007. Aphids as Crop Pests. CABI, Wallingford, U.K., 717 pp.

Vänninen, I., Hokkanen, H., 1997. Efficacy of entomopathogenic fungi and nematodes against *Argyresthia conjugella* (Lep.: Yponomeutidae). Entomophaga 42, 377–384.

Vickers, R.A., Rothschild, G.H.L., 1991. Use of sex pheromones for control of codling moth. In: van der Geest, L.P.S., Evenhuis, H.H. (Eds.), Tortricid Pests, Their Biology, Natural Enemies and Control. Elsevier Science Publishers, Amsterdam, The Netherlands, pp. 339–354.

Vincent, C., Belair, G., 1992. Biocontrol of the apple sawfly, *Hoplocampa testudinea*, with entomogenous nematodes. Entomophaga 37, 575–582.

Winstanley, D., Jarrete, P.J., Morgan, A.W., 1998. The use of transgenic biological control agents to improve their performance in the management of pests. In: Kerry, B.R. (Ed.), Biotechnology in Crop Protection: Facts and Fallacies, Proceedings of the British Crop Protection Council Symposium No. 71, pp. 37–44.

Wu, Z.W., Fan, J.B., Yu, H., Wang, D., Zhang, Y.L., 2015. Ultraviolet protection of the *Cydia pomonella* granulovirus using zinc oxide and titanium dioxide. Biocontrol Sci. Technol. 25, 97–107.

Yee, W.L., Lacey, L.A., 2003. Stage-specific mortality of *Rhagoletis indifferens* (Diptera: Tephritidae) exposed to three species of *Steinernema* nematodes. Biol. Control 27, 349–356.

Yee, W.L., Lacey, L.A., 2005. Mortality of different lifestages of *Rhagoletis indifferens* (Diptera: Tephritidae) exposed to the entomopathogenic fungus *Metarhizium anisopliae*. J. Entomol. Sci. 40, 167–177.

Zingg, D., Zuger, M., Bollhalder, F., Andermatt, M., 2011. Use of resistance overcoming CpGV isolates and CpGV resistance situation of the codling moth in Europe seven years after the first discovery of resistance to CpGV-M. IOBC/WPRS Bull. 66, 401–404.

Chapter 18

Entomopathogens Routinely Used in Pest Control Strategies: Orchards in Tropical Climate

N.K. Maniania[1], S. Ekesi[1], C. Dolinski[2]

[1]International Centre of Insect Physiology and Ecology (icipe), Nairobi, Kenya; [2]Universidade Estadual do Norte Fluminense/CCTA/LEF, Rio de Janeiro, Brazil

18.1 INTRODUCTION

Fruit and nut trees are important components of tropical ecosystems. They are nutritious as sources of vitamins and important minerals and carbohydrates. Tropical fruits have also great economic importance as sources of both household incomes and national revenues. They provide important adaptive values and tend to be more resilient to climate change due to their perennial nature and have the property of sequestering carbon (Pan et al., 1998). According to FAO (2003), world production and trade of fresh tropical fruit are expected to expand over the next decades, with developing countries accounting for about 98% of total production, while developed countries account for 80% of world import trade. The major tropical fruits (mango, pineapples, papaya, and avocado) account for approximately 75% of global fresh tropical fruit production. Insect pests and mites are often the most important constraint to the production of tropical fruit. They indirectly reduce yields by debilitating the plant and directly reduce the yield or quality of fruit before and after they are harvested. They range from esthetic problems that lower the marketability of the harvested product to lethal problems that devastate local or regional production. Some of them are quarantine pests and restrict export markets. The control of these insect pests and mites is through the application of broad-spectrum agents, which has generated a panoply of problems including safety risks, outbreaks of secondary pests, environmental contamination, decrease in biodiversity, and insecticide resistance (Lacey and Shapiro-Ilan, 2008). Microbial control agents (MCAs), including *Bacillus thuringiensis*, entomopathogenic fungi (EPF), viruses, and entomopathogenic nematodes (EPN), can be used as alternatives to synthetic chemical insecticides for the control of several orchard insect pests and mites as they are selective and safe. In this chapter, we review insect and mite species for which the use of entomopathogens has been attempted.

18.2 NUT CROPS

A number of nut crops are of importance in tropical climate orchard systems, with coconut and cashew being the most important.

18.2.1 Coconut and Palm

The coconut palm, *Cocos nucifera*, is an important tree in most tropical islands and along the coastal regions of tropical Africa. It originated from Melanesia, but Southeast Asia remains an important cultivation region today. A diversity of insects and mites attack different parts of the plant.

18.2.1.1 Rhinoceros Beetle

Oryctes rhinoceros (Coleoptera: Scarabaeidae) is endemic to the coconut-growing regions of South and Southeast Asia and has now spread to other regions of the world (EPPO, 2014). Although coconut and oil palm are the primary hosts, it has also been recorded on banana, sugarcane, papaya, sisal, and pineapple and raphia palm. Adult beetles bore through petiole bases into the central unopened leaves, causing significant feeding damage. Among the entomopathogens that are associated with *O. rhinoceros*, *Nudivirus* is the most studied. It attacks larval and adult stages. Originally discovered in 1963 in Malaysia (Hüger, 1966), it has been since then introduced in many countries where the virus has established and spread as epizootics through the beetle populations, resulting in spectacular declines in the treated populations (Jackson, 2009). A long-term reduction in damage by the beetle to coconut palms at the virus release sites in Fiji was reported 35 years after its introduction (Bedford, 2013). The most economical and effective method of application was found to be the release of infected adults (Bedford, 1986). The virus was also introduced into

Microbial Control of Insect and Mite Pests. http://dx.doi.org/10.1016/B978-0-12-803527-6.00018-4

related beetle species, *Oryctes monoceros*, population in Tanzania by releasing approximately 2000 infected adult beetles in four time intervals from November 1983 to June 1987. The infection rates in the release sites 1.5 years after the last release ranged between 40% and 60%, indicating that *Oryctes rhinoceros nudivirus* was already established in the wild population of the beetle (Purrini, 1989). Natural infection of *Metarhizium anisopliae* on *O. rhinoceros* was reported in 1912 at Western Samoa, and since then, attempts have been made to use them as biological control agents. Field applications using both aqueous formulation and dry spores onto breeding sites were reported to significantly reduce the beetle population, especially the larvae (Ramle et al., 1999). Application of *M. anisopliae* was also reported to reduce field population by 72–80% (Gopal et al., 2006; Ramle et al., 2006). Varma (2013) observed 100% mycosis of *O. rhinoceros* grubs after 3 weeks following the application of *M. anisopliae* in manure pits coconut basins and other breeding sites and a reduction of pest attack from 85% to 10% after 6 months. A trap for the autodissemination of *M. anisopliae* was recently developed and tested in the laboratory and field (Ramle et al., 2011). In laboratory, 67% of the trapped adult beetles escaping the trap were confirmed dead due to fungal infection and infected adults were able to disseminate conidia to the breeding site, causing mortality of 92% of larvae. In the field test, the percentage of trapped adult beetles leaving the trap was between 85% and 95%. The density of viable spores from the soil in the trapping region showed an increase of inoculum, suggesting establishment of the fungus in the breeding sites.

18.2.1.2 Red Palm Weevil

Rhynchophorus ferrugineus (Coleoptera: Curculionidae) is widely distributed in Southern Asia and Melanesia. It feeds on a broad range of palms including coconut, sago, date, and oil palms (Rajamanickam et al., 1995). Since the mid 1980s, red palm weevil (RPW) has caused serious damage to date palms in the Gulf region (Abraham et al., 1998). Eggs of the RPW are laid in the trunk of the palms, and the larval stages feed on the soft plant tissue within the trunk of palms, which leads to the formation of tunnels inside the palm. Adults are also vector of the nematode *Bursaphelenchus cocophilus*, which is responsible for red ring disease (Griffith, 1987). Several entomopathogens have been isolated and tested against RPW. Three spore-forming bacilli, *Bacillus sphaericus*, *B. megaterium*, and *B. laterosporus*, isolated from natural habitats associated with RPW in Egypt caused mortality of 40% and 60% of second-instar larvae (Salama et al., 2004). A cytoplasmic polyhedrosis virus was reported to infect all stages of RPW in India, and infected late-larval stages resulted in malformed adults and drastic suppression of the host population (Gopinadhan et al., 1990). EPNs *Steinernema riobrave*, *Steinernema*

carpocapsae, and *Heterorhabditis* sp. were found to be pathogenic to both larval and adult stages of RPW in the laboratory (Abbas and Hononik, 1999). Application of genetically transformed strains of *Steinernema* and *Heterorhabditis* resulted in 95–100% mortality of larvae of RPW in the laboratory and 50% in the field (Hanounik, 1998). Gindin et al. (2006) found that strains of *Metarhizium anisopliae* were more virulent than those of *Beauveria bassiana* against RPW, causing 100% of larval mortality within 6–7 days. Moreover, incubation of eggs in a substrate treated with *M. anisopliae* spores increased egg mortality and reduced their hatchability. When RPW adults were challenged with dry powder and aqueous suspension of the pathogen, mortality of 100% was achieved in 2–3 weeks for the dry rice–based formulation and in 4–5 weeks for aqueous suspension, resulting in decreased longevity. Treated females had a shorter oviposition period and 3 times lower fertility than the controls (Gindin et al., 2006). Three methods of application (injection of *B. bassiana* in naturally infested palm trees, periodical dusting application of fungal spores on palm trees, and release of contaminated males of RPW with fungal spores) of an isolate of *B. bassiana* were tested by Sewify et al. (2009), and all resulted in considerable reduction of RPW populations. Recently, Dembilio et al. (2010) evaluated the potential of a strain of *B. bassiana* obtained from a naturally infected RPW pupa against this weevil in both laboratory and semi-field assays. The strain was able to infect eggs, larvae, and adults in the laboratory. In addition to mortality, adults of either sex inoculated with the fungus efficiently transmitted the disease to untreated adults of the opposite sex, resulting also in fecundity reduction. Efficacy of up to 85.7% was observed on 5-year-old potted palms in semi-field preventive assays. Three applications, at 3-month intervals, of solid formulation of *B. bassiana* in two sites in Spain in 2009 resulted in 70–85% RPW mortality, suggesting that *B. bassiana* could be applied as a preventive as well as curative treatment for RPW control (Güerri-Agulló et al., 2011). The use of traps for spreading conidia of *B. bassiana* by adult RPW was tested with success in three date palm plantations in the Northern Region of the United Arab Emirates. Mortality of adults caused by the fungus ranged from 41.2% to 51.3% compared to 4.8–4.9% in the control (El-Sufty et al., 2010).

18.2.1.3 Coconut Leaf Beetle

Brontispa longissima (Coleoptera: Chrysomelidae) is a native of Indonesia and Papua New Guinea and is currently distributed in Australia, many Pacific Islands, and Southern Asia. The beetle attacks more than 20 palm species, with coconut being the most favored host. Natural infection by bacterial disease, *Metarhizium* spp. and *B. bassiana*, has been reported on coconut leaf beetle (CLB) (Froggatt and O'Connor, 1941; Waterhouse and Norris, 1987; Vögele and

Zeddies, 1990; Hosang et al., 2004). Larva and adult CLB have been reported to be susceptible to EPF in the laboratory and field conditions. Liu et al. (1989) observed that an indigenous strain of *Metarhizium anisopliae* var. *anisopliae* isolated from infected CLB was most virulent against CLB than another isolate of *M. anisopliae* and *B. bassiana*, causing mortalities of 100% of larvae, pupae, and adults. In field trials conducted in the Pingtung area, Taiwan, in 1986 and 1987 using different formulations of the same isolate, CLB could not be detected after three applications of the pathogen. Hosang et al. (2004) reported that larval stage of CLB was more susceptible to *M. anisopliae* var. *anisopliae* and *B. bassiana*, and spray application of both fungal isolates twice-yearly at 2-weeks interval could efficiently control the pest in the field, reducing palm damage up to 90.3–95.1% at 7 months after application. Loc et al. (2004) reported that *M. anisopliae* isolate from naturally infected CLB was most virulent to both stages of the CLB, but larvae were more susceptible than adults. In the field experiments, three isolates of *M. anisopliae* and one of *B. bassiana* effectively controlled CLB, and the efficacy could be observed from 7 days after treatment and persisted up to 21 days after treatment.

18.2.1.4 Coconut Bug

Pseudotheraptus wayi (Hemiptera: Coreidae) is the most important pest of coconuts in East Africa (Brown, 1955) and West Africa (Julia and Mariau, 1978), while *P. devastans* has been reported from West Africa (Douaho, 1984). *P. wayi* has also been reported on other crops such as cashew and avocado. Infestation by these insect pests can result in more than 75% shoot damage, and at early flowering stage they can cause up to 98% flower drop and 80% loss in yield. No attempts have been made so far to use entomopathogens against this pest in the field. However, N. K. Maniania (unpublished data) screened eight isolates each of *B. bassiana* and *Metarhizium anisopliae* against third-instar nymphs of *P. wayi* in the laboratory. At standard concentration of 1×10^7 conidia/mL, all the fungal isolates were pathogenic to the host. *B. bassiana* isolates caused mortality of between 28.0% and 74.0% with lethal time 50% mortality (LT_{50}) values ranging from 5.7 to 8.1 days. On the other hand, isolates of *M. anisopliae* were most virulent, causing mortality of between 98.0% and 100% with LT_{50} values of between 3.5 and 7.2 days. Three commercial isolates of *M. anisopliae* (ICIPE 62, ICIPE 69, and ICIPE 78) were subjected to dose–response 50% mortality (LD_{50}) bioassays. The LD_{50} values varied between 8.2 and 13.2×10^5 conidia/mL.

18.2.2 Cashew

The cashew tree *Anacardium occidentale* is originally native to northeastern Brazil and is now widely grown in tropical regions, with Vietnam, Nigeria, India, and Cote d'Ivoire being major producers (FAO, 2011). Cashew is susceptible to over 60 different insect pest species throughout its growth stage, and damage intensity varies with location, varieties, and management practices.

18.2.2.1 Mirid Bugs

The tea mosquito bugs (TMB) *Helopeltis anacardii* and *H. schoutedeni* (Hemiptera: Miridae) are considered the most important species in East and West Africa, while *H. antonii*, *H. bakeryi* and *H. theivora* are the most economic important species in Oriental and Australasian regions (Stonedahl, 1991). Both nymph and adult TMB suck sap from the tender flushes, young shoots, inflorescence, panicles, growing young nuts, and apples. This injects toxic saliva into the plant parts, which causes death of vascular tissues resulting in dieback symptom (Devasahayam and Nair, 1986). *Aspergillus* sp., *A. flavus*, and *Verticillium lecanii* were isolated from *Helopeltis* spp. infesting cocoa crop and were found to be pathogenic, causing mortalities of between 77% and 90% in the laboratory (Pasaru et al., 2014). Four isolates of *Metarhizium anisopliae* (ICIPE 18, ICIPE 30, ICIPE 62, and ICIPE 69) were assayed against third-instar nymphs of *H. Schoutedeni* and were found virulent, causing 100% mortality within 4.3–4.7 days with lethal dose mortality response (LD_{50}) values varying between 6.7 and 8.6×10^5 conidia/mL (N. K. Maniania, unpublished data).

18.2.2.2 Cashew Stem and Root Borer

Plocaederus ferrugineus (Coleoptera: Ceramycidae) is a major pest of cashew in all cashew-growing tracts of India (Pillai et al., 1976) and was recently reported in Nigeria (Asogwa et al., 2009). The grubs form irregular tunnels in the cashew bark of the stem and roots, thereby damaging vascular tissues and resulting in gradual yellowing of the foliage trees. The virus *Rhabdionvirus oryctes* was tested against cashew stem and root borer (CSRB) in Sri Lanka (Rajapakse and Jeevaratnam, 1982). Treatments consisted of pouring virus suspension into the galleries, releasing virus-infected adults, and placing virus-mixed sawdust at the base of the infested stem. The infection of the virus was able to establish in the beetle population, resulting in reduction in the beetle damage to the tree in certain pockets of the cashew plantation, with sawdust mixture giving the best results. The susceptibility of CSRB to EPF has also been reported under field conditions. Ambethgar (2010) conducted field trials with *B. bassiana* and *Metarhizium anisopliae* by treating cashew trees with early phase of borer infestation using different methods of application (swabbing conidial-mud slurry over tree trunk, pouring aqueous conidial suspension through entry holes and soil incorporation of fungal spawn). Although all the treatments reduced the borer infestation, pouring of conidial suspension was superior. Application of *M. anisopliae* and *B. bassiana* at different time intervals and in combinations with neem cake was found to be effective in reducing CSRB infestation, up to 7.4–11.1% (Sahu and Sharma, 2008).

18.3 TREE FRUITS

18.3.1 Avocado

The avocado *Persea americana* is a tree native to Mexico and Central America (Chen et al., 2008). It is commercially valuable and cultivated in tropical and Mediterranean climates throughout the world. In South Africa, for instance, 18 potential insect pests have been reported on avocados; however, the most important are the fruit fly, the false codling moth (FCM), and the coconut bug (De Villiers and Van den Berg, 1987).

18.3.1.1 False Codling Moth

Thaumatotibia leucotreta (Lepidoptera: Tortricidae) is native to Ethiopia and sub-Saharan Africa. It is not known to be established in North America (Touhey, 2010). FCM is a pest of economic importance to many crops throughout sub-Saharan Africa, South Africa, and the islands of the Atlantic and Indian Oceans and is classified as quarantine pest by many export markets. Important host crops include citrus, avocado, macadamia, cotton, peach, and plum (Newton, 1998). Microbial control is currently being developed and used against FCM targets in citrus (Moore and Hattingh, 2012) and will be covered in Chapter 19. The strategies developed for FCM in citrus could therefore be applied to avocado. The potential use of EPNs for control of FCM was reported by Malan et al. (2011). The soilborne stages of FCM (larvae, pupae, and emerging moths) were found to be susceptible to EPN infection by the same authors. Recently, Manrakhana et al. (2014) studied the impact of naturally occurring EPNs on FCM in a citrus orchard in South Africa. Fruit infestation by FCM was 58.6% lower in an orchard block where EPNs were conserved compared to a block where EPNs were suppressed by nematicide. The virulence of *Metarhizium anisopliae* and *B. Beauveria* on larvae of FCM has also been reported (Begemann, 2008; Coombes et al., 2015).

18.3.1.2 Fruit Flies (Diptera: Tephritidae)

Avocados, especially thin-skinned varieties, are susceptible to attack by various species of fruit flies. Some fruit flies lay eggs under the skin of the fruit that is just beginning to ripen, but others attack young and old fruit. The strategies developed for fruit flies in mango and guava could also be applied to avocado.

18.3.1.3 Thrips (Thysanoptera: Thripidae)

The greenhouse thrips, *Heliothrips haemorrhoidalis*, although limited to tropical regions, is found throughout the world in greenhouses where it causes damage to a wide range of ornamental and vegetable plants. On the other hand, avocado thrips *Scirtothrips perseae* is a relatively new pest of avocados in California. Adult thrips can feed on 11 plant species, but larvae have only been found on avocados in the field in both California and Mexico, suggesting that *S. perseae* has a highly restricted host range (Hoddle et al., 2002). The redbanded thrips *Selenothrips rubrocinctus* is a pest of many plants, but the favorite tropical fruit hosts in Florida are mango and avocado. Entomopathogens, especially EPF, have an important role to play as microbial insecticides for thrips control because they are associated with thrips in nature (Butt and Brownbridge, 1997) and their successful use has been reported in many cropping systems (Ekesi and Maniania, 2002; Ansari et al., 2007). Recently, Zahn and Morse (2013) evaluated *Bacillus thuringiensis* subsp. *israelensis* proteins (Cyt 1A and Cry 11A) and multiple strains of *B. bassiana*, including the commercial strain of GHA against avocado and citrus thrips. Both thrips species were not susceptible to Bt protein tested while they were susceptible to all strains of *B. bassiana*, with GHA being the most effective.

18.3.2 Mango (*Mangifera* spp.)

Mango is one of the most important fruit crops grown in the tropical and subtropical parts of the world. It is a major source of income generation and household food and nutritional security. However, its production is affected by a variety of factors including insect pests. A plethora of arthropod pests attack mango, but the most important are undoubtedly fruit flies and mango seed weevil. They cause direct damage to fruits leading to 40–80% losses depending on locality, variety, and season (Lux et al., 2003; Ekesi et al., 2016).

18.3.2.1 Fruit Flies (Diptera: Tephritidae)

Bactrocera dorsalis, *Ceratitis cosyra*, *C. rosa*, *C. fasciventris*, *C. quinaria*, and *C. capitata* are known to attack mango in many African countries, while *Anastrepha ludens* is in Mexico, most of Central America, and the southern United States. On the other hand, the Caribbean fruit fly *A. suspensa* causes extensive damage to mango in Greater Antilles, Bahamas, and Florida. In Australia and many Pacific countries, mangoes are attacked by varieties of fruit flies within the *Bactrocera* genus including *B. dorsalis* and *B. tryoni* with high invasion potentials. Bacterial pathogens such as Bt, *Serratia marcercens*, and *B. pumilus* have been found naturally associated with dead larvae and pupae of fruit flies including *B. dorsalis* and *C. capitata* (Gingrich and El-Abbassi, 1988; Grimont and

Grimont, 1978). EPNs have been evaluated against some important fruit fly pests of mango. In the laboratory, *A. ludens* was reported to be highly susceptible to a variety of EPNs (Lezama-Gutierrez et al., 1996; Toledo et al., 2005a), and in field trials, effective control was achieved at the rate of 2.5×10^2 IJs of *Heterorhabditis bacteriophora/* cm^2. Similar results were achieved with *C. capitata* at the rate of $5–50 \times 10^2$ IJs of *Steinernema carpocapsae/* cm^2 (Lindegren et al., 1990; Gazit et al., 2000; Laborda et al., 2003). In South Africa, pupating larvae and adult *C. capitata* and *C. rosa* were found to be susceptible to *H. bacteriophora*, *H. zealandica*, and *S. khoisanae* infection in laboratory; however, no infection was recorded for the pupae (Malan and Manrakhan, 2009). Langford et al. (2014) tested the capacity of three EPN species with different foraging strategies (*S. feltiae*, *S. carpocapsae*, and *H. bacteriophora*) to cause larval and pupal mortality in *B. tryoni* across a range of concentrations (50, 100, 200, 500, and 1000 IJs/cm^2), substrate moisture (10%, 15%, 20%, and 25% w/v), and temperatures (15, 20, 25, and 30 °C). They found that all the EPN species tested caused environment- and density-dependent mortality in the third larval instar while pupae were not affected. Ekesi et al. (2007a) gave a comprehensive review of the role of EPF in the management of fruit flies. Six isolates of *B. bassiana* and *Metarhizium anisopliae* were reported to cause mortalities of 87–100% in *C. capitata* and *C. fasciventris* and 72–79% in *C. cosyra* after 4 days (Dimbi et al., 2003a, 2009). Several authors have also reported 46–100% mortality caused by the same species in European populations of *C. capitata* after 5–14 days (Castillo et al., 2000; Quesada-Moraga et al., 2006; Gindin et al., 1994; Krasnoff and Gupta, 1994). Infection by EPF was also reported to reduce fecundity and fertility (>70%) in native and exotic species in Kenya (Ouna, 2012; Dimbi et al., 2013). To minimize cost, labor, and materials required to achieve adequate spray coverage in tree crops, autodissemination (Fig. 18.1) is recommended for fruit fly suppression (Ekesi et al., 2007a,b). This has been demonstrated in *C. capitata*, *C. fasciventris*, *C. cosyra*, and *B. dorsalis* with mortality of up to 100% after 4 days (Dimbi et al., 2003b; Ouna, 2012). Because fruit flies spend part of their life-cycle in soil as larvae and puparia (White and Elson-Harris, 1992), *M. anisopliae* and *B. bassiana* were tested in suppressing late third-instar larvae of *C. capitata*, *C. fasciventris*, and *B. dorsalis* with greater than 80% mortality on the target species (Ekesi et al., 2002, 2003; Ouna, 2012). When conidia of these fungi were formulated as oil:water (50:50) and as granules, they persisted in the soil for 667 days. A single application of granular formulation of *M. anisopliae*-based biopesticide at 100 kg/ha reduced fruit infestation by 38–45% compared to 52–60% in the control (Ekesi et al., 2011).

FIGURE 18.1 Autodissemination device on mango canopy for fruit fly suppression. The inner inside of the chamber is lined with a velvet material on which fungal conidia are applied. The food lure/bait is placed in a small container at the bottom.

18.3.2.2 Mango Seed Weevil, *Sternochetus mangifera*

Adult mango seedweevil (MSW) have been reported to be attacked by *B. bassiana* (Joubert and Labuschagne, 1995; CABI/EPPO, 1997) and Baculovirus (Shukla et al., 1984); but very limited studies have been conducted to exploit them for management of the pest. Several isolates of *B. bassiana* and *Metarhizium anisopliae* from the ICIPE's Arthropod Germplasm caused mortality of 32–96% to adult MSW (Ekesi et al., 2016). Weekly application of an oil formulation of Met 69 (an *M. anisopliae*–based biopesticide) applied as soil drench to the base of the canopy and trunk for a total spray of 10 sprays when combined with orchard sanitation resulted in 0.02% fruit infestation compared with 75.8% fruit infestation in the control plots (Ekesi et al., 2016).

18.4 BANANA

Several varieties of banana and plantain (*Musa* spp.) are grown throughout the tropics and into the subtropics. Although a wide variety of insects attack banana, only banana weevil borer (BWB), the West Indian sugarcane borer, and the Tamarind owlet will be covered in this chapter.

18.4.1 Banana Weevil Borer

Banana Weevil Borer (BWB) *Cosmopolites sordidus* (Coleoptera: Curculionidae) is reported as the most important insect pest of banana and plantain worldwide. Heavy infestations can result in crop failure in newly established stands and reduced yield and shortened life span of plants in established stands (Gold et al., 2001, 2002). Figueroa (1990) evaluated three EPNs against BWB in greenhouse tests in Puerto Rico.

The nematodes significantly reduced the number of tunnels made by larvae in plantain corms at 4×10^2, 4×10^3, and 4×10^4 IJs/4-month-old plant. Treverrow et al. (1991) reported 43–68% mortality of BWB larvae in banana rhizomes after applying *Steinernema carpocapsae* with a water thickener into cuts or holes made in residual rhizomes. Schmitt et al. (1992) in Brazil evaluated an imported formulation of *S. carpocapsae* by spraying onto split pseudostems and pseudostem stumps as a baiting technique and observed 70% mortality of BWB adults. Treverrow and Bedding (1993) assayed 32 EPN strains/species against larvae and adults of BWB and observed that BWB strain of *S. carpocapsae* had the greatest activity against adults. They also described a method for introducing *S. carpocapsae* IJs into banana corms that involves removing cones from residual corms with a desuckering gouge and adding 2.5×10^5 IJs/cavity. The cone is then reinserted to produce a protected cavity that is attractive to adult weevils. In contrast, bimonthly treatments with *H. zealandica* and *S. carpocapsae* applied in a thickened aqueous solution into 200-mm-deep incisions in the residual rhizomes failed to produce adequate control (Smith, 1995). Strains of *B. bassiana* have shown good potential to control adult BWB (reviewed by Gold et al., 2002, 2003). Godonou et al. (2000) evaluated two formulations of *B. bassiana* applied to the planting holes and suckers of banana. Both formulations resulted in 75% mortality in artificially released weevils. Under natural infestation conditions, the oil palm kernel cake performed better causing 42% mortality than the conidial powder (6%). In Brazil, Batista Filho et al. (1995) reported up to 61% reduction of BWB adults using baits with *B. bassiana* (4.5×10^8 conidia/mL) applied as a rice paste with mineral oil. In Uganda, maize, soil-based, and oil formulations of an indigenous isolate of *B. bassiana* were evaluated. Weekly trapping of weevils over a 9-month monitoring period showed significant reduction in unmarked and marked weevil populations in *B. bassiana*–treated plots. Application of maize formulation at 2×10^{15} conidia/ha proved most effective, reducing the weevil populations by 63–72% within 8 weeks after a single application (Nankinga and Moore, 2000). BWB aggregation pheromone has been used to enhance the dissemination of EPF. Tinzaara et al. (2007) reported high mortalities of weevils from plots where *B. bassiana* was applied in combination with the pheromone than from plots without pheromone. Infected weevils could transmit conidia to healthy individuals, thereby increasing the dissemination of the fungus within weevil populations (Schoeman and Schoeman, 1999; Tinzaara et al., 2007). Gold et al. (2003) reviewed the potential of endophyte fungi for BWB suppression. Akello et al. (2007) reported that *B. bassiana* could endophytically colonize banana tissue culture. In screenhouse experiments, no effect of 8-week-old *B. bassiana*-inoculated tissue-cultured banana plants infested with female BWB could be observed on oviposition and egg hatchability 5 days after plant infestation. However, after 15 weeks, between 53.4% and 57.7%

of the BWB adults died due to *B. bassiana* infection, resulting in a reduction of plant damage by 29.1–62.7% depending on plant part (Akello et al., 2008a). In another study, tissue-cultured banana plants were inoculated by dipping roots in a *B. bassiana* suspension of 1.5×10^7 conidia/mL for 2 h, after which BWB larvae were introduced 2 months later. Two weeks after larval infestation, endophytic *B. bassiana* significantly reduced larval survivorship (23.5–88.9%), resulting in 42.0–86.7% reduction of plant damage (Akello et al., 2008b).

18.4.2 West Indian Sugarcane Borer

Metamasius hemipterus sericeus (Coleoptera: Curculionidae), the West Indian sugarcane borer, is also known as the silky cane weevil, and rotten stalk borer of sugar cane. It can be an important pest of banana in certain areas of the Americas (Gold et al., 2002). Giblin-Davis et al. (1996) evaluated *Steinernema carpocapsae* and insecticides in a field test with weevil-infested in Canary Island date palms (*Phoenix canariensis*). Lindane (1.5 g a.i./palm) and imidacloprid (1.2 g a.i./palm) had the greatest effect on the percentage mortality of total weevils present per palm (>60%), followed by *S. carpocapsae* (8×10^6 IJs/palm) (51%) and acephate (2.9 g a.i./palm) (39%), which were statistically equal to the controls (14%).

18.4.3 Tamarind Owlet

Opsiphanes tamarindi (Lepidoptera: Brassolidae), the tamarind owlet, is a major defoliator of plantains during the dry season in the region south of Lake Maracaibo, Venezuela. Briceno (1997) described an IPM system that combined cultural practices, application of Bt against early larval stages, and relying on natural enemies (parasitoids and predators) to control late larval and pupal stages. The seasonal application of Bt helped to eliminate first instars without affecting natural enemies.

18.5 PAPAYA

Papaya, *Carica papaya*, is a major tropical fruit from the Central America origin. It is currently grown in diverse geographical regions of the world mainly for its edible fruit in addition to its culinary medical and industrial uses (Pantoja et al., 2002). Over 130 species of arthropod pests have been reported to affect papaya worldwide. However, fruit flies, mites, and mealybugs are considered to be the key pests (Pantoja et al., 2002).

18.5.1 Papaya Fruit Flies *Bactrocera papayae* and *Toxotrypana curvicauda* (Diptera: Tephritidae)

These are the most serious fruit fly species of papaya in the Asian region and the Americas, respectively (White and Elson-Harris, 1992). In East Africa, *B. invadens* and *C. rosa* have been recorded from papaya (S. Ekesi, personal communication). Entomopathogens such as viruses, fungi, and bacteria have been reported on Tephritidae in nature (Bashirudin et al., 1988; Hedstrom, 1994; Strongman et al., 1997). The susceptibility of various species of fruit flies to EPNs (Lindegren et al., 1990), Bt (Robacker et al., 1996), and EPF (see review by Ekesi et al., 2007a) has been reported. The potential of EPF for fruit fly suppression in the field was reported by Ekesi et al. (2007a, 2011, 2016) and in a recent review Vargas et al. (2015) suggest their integration with other biological approaches. (Refer to Section 18.3.2 for more details.)

18.5.2 Papaya Mealybug *Paracoccus marginatus* (Hemiptera: Pseudococcidae)

Papaya mealybug is a polyphagous pest of many tropical crops. A native of Central America, it spread to the Caribbean region and South America in the 1990s; since then it has been accidentally introduced to some islands in the Pacific region. *P. marginatus* has been reported spreading in Asia and West Africa (CABI/EPPO, 2012). Banu et al. (2010) reported that nymphs and adults of *P. marginatus* were susceptible to *Metarhizium anisopliae*, *Lecanicillium lecanii*, and *B. bassiana*, with nymphs incurring 100% mortality 9 days postinfection. *L. lecanii* recorded the highest mortality of 80% in adults at 7 days after treatment. The efficacy of Campaign (*M. anisopliae* isolate ICIPE 69–based biopesticide commercialized by the Real IPM Kenya) was demonstrated against *P. marginatus* in laboratory and field trials in Ghana (J. Ofosu-Anim and V. Y. Eziah, unpublished data). The application of Campaign at the rate of 3.0 mL/L reduced mealybug population by 82% after 1 month and was as effective as the synthetic insecticide, Consider 200 SL, at the concentration of 2.0 mL/L.

18.5.3 Mites

Several species of mites have been reported as main pests of papaya. They include the broad mite *Polyphagotarsonemus latus* (Acari: Tarsonemidae), the two-spotted spider mite *Tetranychus urticae* (Acari: Tetranychidae), and the false spider mite *Brevipalpus phoenicis* (Acari: Tenuipalpidae) (Haramoto, 1969). In addition to damage, the broad mites vector plant diseases such as *Citrus leprosis virus* (CiLV) (Rodrigues et al., 1997). Many reports have been published on natural incidence of EPF on mites including *T. urticae* (Van der Geest, 1985; Chandler et al., 2000; Van der Geest et al., 2000; Jeyarani et al., 2011). *Beauveria*, *Metarhizium*, and *Isaria* spp. have been reported to be pathogenic to mites in the laboratory and glasshouse and have been developed as biocontrol agents against a range of mite species (Peña

et al., 1996; Chandler et al., 2000, 2005). The role of EPF in the control of *T. urticae* as pests of horticultural crops was reviewed by Maniania et al. (2008).

18.6 OTHER FRUITS

18.6.1 Guava

Psidium guajava is native to the American tropics but is currently grown in more than 50 subtropical and tropical countries. Brazil is the principal red guava producer, followed by Mexico, whereas India and Pakistan are major producers of white guava (Gould and Raga, 2002; Pomar Brasil, 2015). Different insect pests attack guava. In the Americas, the main pests are the guava weevil and the fruit flies (*Anastrepha* spp. and *Ceratitis* spp.). Occasionally, the leaf-footed bug can also cause damage (Souza et al., 2003). In Asia, Africa, and western Pacific countries, the main pests are the fruit flies (*Bactrocera* spp.) (Sarwar, 2006).

18.6.1.1 The Guava Weevil Conotrachelus psidii (Coleoptera: Curculionidae)

Guava weevil directly affects fruit quality, causing great damage. Females lay eggs in immature fruit, and larvae progress through four instars as the fruit develops. Infestation leads to acceleration in fruit maturation and fruit drop when ripe. Subsequently, larvae crawl into the soil where they develop into prepupae. Application of *H. baujardi* LPP7 at the doses of 0.17, 0.35, or 0.7 IJ/cm² in 20-L potted guava trees resulted in 30% and 58% mortality at the two highest doses (Dolinski et al., 2006). Del Valle et al. (2008) assessed the susceptibility of the guava weevil to *H. baujardi* IJs in the greenhouse and under field conditions by applying *Galleria mellonella* larvae–infected nematode cadavers. Significant differences were observed in the field when six cadavers/0.25 m² were applied. IJs from the cadavers persisted 6 weeks after application in the field, but decreased greatly thereafter. The nematode *Heterorhabditis bacteriophora* LPP30 has been assessed against *C. psidii* on conversion to organic from conventional guava cultivation systems as another potential agent. In guava crops, the population fluctuation of the nematode and the occurrence of the pest were evaluated over one year. Different application methods were tested, with aqueous suspension being the best, causing mortality from 79% to 85%. A reduction of the weevil population was observed, especially in the conversion to organic area (70–90%). Also, better cycling and persistence of IJs in the area under conversion to organic were observed (Minas, 2012). Silva et al. (2010) demonstrated that *H. indica* IBCB n5 applied at the dosages of 1 and 10 IJ/cm² could control the weevil, causing mortality of between 33% and 50%. The combination of isolates of *B. bassiana* and *Metarhizium anisopliae* with Tween 80, sunflower oil, and imidacloprid (IMI) was reported to increase their efficacy against adult *C. psidii*, but the efficiency was superior when applied together with IMI, resulting in LT_{50} values of 5.3–10.3 days compared to LT_{50} values of 9.5–17 days in Tween alone. However, conidial production on the cadavers of adult *C. psidii* was reduced when fungal isolates were applied together with IMI (Brito et al., 2008).

18.6.1.2 Fruit Flies (Diptera: Tephritidae)

Fruit flies are very important pests in guava because the adults lay eggs in the fruit, and the resulting damage by larvae lowers their quality. Several species of *Anastrepha* spp., *Ceratitis* spp., and *Bactrocera* spp. have been reported infesting guava trees (Weems, 2001; Zucchi, 2008; CABI, 2015). Control of some of these species by entomopathogens has been attempted.

18.6.1.2.1 The South American Fruit Fly *Anastrepha fraterculus*

This species is considered the key pest in many fruit crops. Experiments carried out in greenhouse in a peach orchard using *Heterorhabditis bacteriophora* RS88 and *Steinernema riobrave* RS59 showed no differences in pupal mortality by either nematode at 250 and 500 IJs/cm² (Barbosa-Negrisoli et al., 2009). In the field, both strains sprayed on natural and artificially infested fruit resulted in *A. fraterculus* larval mortality of 51.3%, 28.1% and 20%, 24.3%, respectively. Moreover, *H. bacteriophora* RS88 showed better efficacy independently of the application method (aqueous suspension or infected cadavers). The virulence of isolates of *M. anisopliae* and *B. bassiana* against *Anastrepha* spp. has been reported under laboratory and semi-field conditions (Lezama-Gutiérrez et al., 2000; De la Rosa et al., 2002; Toledo et al., 2006). Lezama-Gutiérrez et al. (2000) reported a reduction of adult emergence of 22% and 43% in sandy loam soil and loam soil, respectively, compared with the controls following application of 200 mL of one isolate of *M. anisopliae* at the concentration of 2.5×10^6 CFU/mL in a field-cage experiment.

18.6.1.2.2 The Mediterranean Fruit Fly, or Medfly, *Ceratitis capitata*

Medfly has been reported to afflict guava and many other fruit worldwide. Prepupae of *C. capitata* incurred significant mortality in a papaya field when exposed to different concentrations of *S. feltiae* Mexican, and eradication was achieved with 500 IJ/cm² applied in soils (Lindegren et al., 1990). In a guava orchard, Silva et al. (2010) confirmed that the Medfly's most susceptible stages to EPNs were prepupae and 1-day-old pupae. *H. indica* IBCBn5 was considered the most virulent strain applied at the dosages of one and 10 IJs/cm², causing mortality ranging from 66% to 93%. The efficacy of *H. baujardi* LPP7 was evaluated

in cages constructed over well-grown guava trees for the control of *C. capitata* prepupae on two separate dates. In each cage, 100 prepupae and a suspension of 10^5 IJs/500 mL of water were distributed evenly on the soil. The average larval mortality in treated trees was significantly different over the control (7.7–30.4% and 58.6–87.4%, respectively) (Minas, 2008). *Ceratitis* spp. have also been reported to be susceptible to EPF in the laboratory and field conditions (Castillo et al., 2000; Lezama-Gutiérrez et al., 2000; Konstantopoulou and Mazomenos, 2005; Mochi et al., 2006; Toledo et al., 2006; Almeida et al., 2007; Evangelos et al., 2013; Bissolli et al., 2014). In the greenhouse trial, application of *B. bassiana* IBCB 66 and *M. anisopliae* IBCB 425 at the concentration of 5×10^8 conidia/mL in soil pots containing citrus seedlings caused prepupal mortality of 66.6% (Almeida et al., 2007). Ali et al. (2009) tested *B. bassiana* 412 against adult Medfly in semi-field conditions using 4×10^8 spores/mL and obtained mortality of 46% compared with 16% in the control.

18.6.1.2.3 The Oriental Fruit Fly *Bactrocera dorsalis*

The Oriental fruit fly, originally described from Taiwan, is one of the most destructive fruit fly pests of East Asia and the Pacific region. Recent outbreaks have occurred in southern California and Florida. Lin-Jin et al. (2005) tested *S. carpocapsae* All, *S. carpocapsae* A24, *S. feltiae* SN, and *H. bacteriophora* H06. Field results showed that when the All strain (300 IJ/cm²) was used to treat the soil, larval and pupal mortality after 9 days reached 86.3% and the index of the Oriental fruit fly population was 15.5, significantly lower than 105.9 found in the orchard without nematode treatment. The virulence of isolates of EPF against *Bactrocera* spp. is well documented (Soliman et al., 2014; Gul et al., 2015). Gul et al. (2015) tested the pathogenecity of isolates of *B. bassiana*, *I. fumosorosea*, and *M. anisopliae* against *B. zonata* and observed mortality of between 20% and 40% following oral application of the inoculum. No significant effect on the pupal mortality of the fruit fly by the fungal isolates following soil application was observed. However, *I. fumosorosea* significantly affected the adult emergence.

18.6.1.3 Leaf-Footed Bug *Leptoglossus zonatus* (Hemiptera: Coreidae)

This species is usually a secondary pest on fruits and flowers. Grim and Guharay (1998) tested three *B. bassiana* isolates and one isolate of *Metarhizium anisopliae* against adults and observed that *M. anisopliae* NB was the most efficient. In a field trial, mineral oil–based ultralow-volume controlled droplet, applications of *M. anisopliae* NB at 10^{10} conidia/tree caused 94% adult mortality. Application of *B. bassiana* resulted in 28% increase in fruit yield.

18.6.2 Pineapple

Ananas comosus is native to South America and is currently cultivated in most tropical countries. Some of insect pests of pineapple worldwide are the pink pineapple mealybug, the gray pineapple mealybug, and the fruit borer caterpillar (Petty et al., 2002).

18.6.2.1 Fruit Borer Caterpillar

Thecla basalides (Lepidoptera: Lycaenidae) is an important pest of pineapple in Brazil, mainly in the cultivar "Pérola." In northern Brazil, the dose of 600 g/ha of *B. thuringiensis* is recommended for control (Sanches, 2005). In southern Brazil, Lorenzato et al. (1997) suggested that Bt, azinphos ethyl, and carbaryl were efficient in controlling *T. basalides* borer caterpillar with 100% control.

18.6.3 Vines

18.6.3.1 Passion Fruit

Passion fruit, *Passiflora edulis*, is native to Brazil, Paraguay, and northern Argentina. It is cultivated commercially in tropical and subtropical areas and is now widely grown in several countries of South America, Central America, the Caribbean, Africa, Southern Asia, Israel, Australia, and the United States. Passion fruit is attacked by several insect pests. However, fruit flies and mealybugs are the most important.

18.6.3.1.1 Fruit Flies (Diptera: Tephritidae)

The guava fruit flies *Anastrepha striata* and *Bactrocera correcta* are considered major pests of passion fruit. *A. striata* occurs in the American tropics and subtropics while *B. correcta* in India, Pakistan, Nepal, Sri Lanka, Thailand, and China (CABI/EPPO, 2003). Although there are no specific reports on microbial control of *A. striata* and *B. correcta* on passion fruit, laboratory and field experiments have demonstrated the efficacy of EPNs *Heterorhabditis* against larvae of *Anastrepha* spp. (Toledo et al., 2005b). Similarly, *B. bassiana*, *M. anisopliae*, and *I. fumosorosea* have been reported to be pathogenic to *Anastrepha* spp. (De La Rosa et al., 2002; Destéfano et al., 2005; Toledo et al., 2006; Osorio-Fajardo et al., 2011; Gandarilla-Pacheco et al., 2012).

18.6.3.1.2 Mealybugs

The cotton mealybug *Phenacoccus solenopsis* (Hemiptera: Pseudococcidae) has become an important plant pest worldwide since its first description in 1897 (Williams and Granara de Willink, 1992). It has been reported from over 200 hosts. On the other hand, *P. kenyae* occurs only in the Afrotropical Region and has limited distribution. It has been reported to be an important insect pest on passion fruit in Kenya (Odienki, 1975). There are no available reports on the use of entomopathogens against *P. solenopsis* and *P. kenyae* in passion fruit cropping system. However, few attempts have been made to use EPF against *P. solenopsis* on cotton. For instance, isolates of *Metarhizium* spp. were found to be pathogenic to *P. solenopsis* in the laboratory (Ujjan et al., 2015). At the concentration of 6.3×10^{12} conidia/acre, selected isolate caused 50% adult mortality after 13.8 and 19.6 days under screen house and field conditions, respectively. Combination of the *Metarhizium* isolate with insecticide IMI resulted in short lethal time of 6.6 and 8.4 days under screen house and field conditions, respectively (Ujjan et al., 2015).

18.7 CHALLENGES AND FUTURE PROSPECTS

MCAs have been developed and applied using the insecticide paradigm, such as in an inundative approach (Lacey and Goettel, 1995). However, this approach requires a high amount of inoculum with short survival in the environment due to solar radiation (Jaronski, 2010), thereby requiring frequent applications, which in turn increase their costs. A new strategy is currently being considered for the delivery of MCAs in crop protection. It consists of disseminating the pathogen among target pest populations by using devices that attract insects to baited stations, where they are contaminated with the pathogen and then return to the environment where they can transmit the disease to healthy individuals (Vega et al., 2007). This strategy has been evaluated with success in the field against the Asian citrus psyllid *Diaphorina citri* (Patt et al., 2015), *C. capitata* (Navarro-Llopis et al., 2015), and tsetse fly (Maniania et al., 2006). Combination of an entomopathogen agent with a food bait and/or semiochemicals has also been suggested (Vargas et al., 2005). However, the success of this strategy depends on the use of powerful semiochemicals (kairomones/pheromones) and their compatibility with the pathogen (Mfuti et al., 2016).

Concern about insecticide resistance and human and environmental safety creates opportunities for the development and use of biocontrol agents as alternatives to synthetic chemical insecticides (Lacey et al., 2015). Although MCAs are by no means able to solve all major problems of insect pests in agriculture, they have a prominent place in insect control. The problems are to define precisely the ecosystem in which they can effectively play a positive role and to determine the necessary strategies for the expression of their full potentialities (Ferron, 1985). Identifying markets where expectations are in line with the actual performance of the insect pathogen is also an important trend. Moreover, MCAs should be viewed as components of integrated pest management programs rather than as the sole resource available (Gelernter and Lomer, 2000; Lacey et al., 2015).

REFERENCES

Abbas, M.S.T., Hanonik, S.P., 1999. Pathogenicity of entomopathogenic nematodes to red palm weevil, *Rhynchophorus ferrugineus*. Intern. J. Nematol. 9, 84–86.

Abraham, V.A., Al-Shuaibi, M.A., Faleiro, J.R., Abuzuhairah, R.A., Vidyasagar, P.S.P.V., 1998. An integrated management approach for red palm weevil, *Rhynchophorus ferrugineus* Oliv., a key pest of date palm in the Middle East. Sultan Qabus Univ. J. Sci. Res. Agric. Sci. 3, 77–84.

Akello, J., Dubois, T., Coyne, D., Gold, C.S., Kyamanywa, S., 2007. Colonization and persistence of entomopathogenic fungus, *Beauveria bassiana*, in tissue culture of banana. Afr. Crop Sci. Conf. Proc. 8, 857–861.

Akello, J., Dubois, T., Coyne, D., Kyamanywa, S., 2008a. Effect of endophytic *Beauveria bassiana* on populations of the banana weevil, *Cosmopolites sordidus*, and their damage in tissue-cultured banana plants. Entomol. Exp. Appl. 129, 157–165.

Akello, J., Dubois, T., Coyne, D., Kyamanywa, S., 2008b. Endophytic *Beauveria bassiana* in banana (*Musa* spp.) reduces banana weevil (*Cosmopolites sordidus*) fitness and damage. Crop Prot. 27, 1437–1441.

Ali, A., Sermann, H., Lerche, S., Büttner, C., 2009. Soil application of *Beauveria bassiana* to control *Ceratitis capitata* in semi field conditions. Commun. Agric. Appl. Biol. Sci. 74, 357–361.

Almeida, J.E.M., Batista Filho, A., Oliveira, F.C., Raga, A., 2007. Pathogenicity of the entomopathogenic fungi and nematode on medfly *Ceratitis capitata* (Wied.) (Diptera: Tephritidae). BioAssay 2, 1–7.

Ambethgar, V., 2010. Field assessment of delivery methods for fungal pathogens and insecticides against cashew stem and root borer, *Plocaederus ferrugineus* L. (Cerambycidae: Coleoptera). J. Biopest. 3, 121–125.

Ansari, M.A., Shah, F.A., Whittaker, M., Prasad, M., Butt, T.M., 2007. Control of western flower thrips (*Frankliniella occidentalis*) pupae with *Metarhizium anisopliae* in peat and peat alternative growing media. Biol. Control 40, 293–297.

Asogwa, E.U., Anikwe, J.C., Ndubuaku, T.C.N., Okelana, F.A., 2009. Distribution and damage characteristics of an emerging insect pest of cashew, *Plocaederus ferrugineus* L. (Coleoptera: Cerambycidae) in Nigeria: a preliminary report. Afr. J. Biotechnol. 8, 53–58.

Banu, J.G., Suruliveru, T., Amutha, M., Gapalakrishnan, N., 2010. Susceptibility of cotton mealy bug, *Paracoccus marginatus* to entomopathogenic fungi. Ann. Plant Prot. Sci. 18, 247–248.

Barbosa-Negrisoli, C.R., Garcia, M.S., Dolinski, C., Negrisoli Jr., A.S., Bernardi, D., Nava, D.E., 2009. Efficacy of indigenous entomopathogenic nematodes (Rhabditida: Heterorhabditidae, Steinernematidae), from Rio Grande do Sul Brazil, against *Anastrepha fraterculus* (Wied.) (Diptera: Tephritidae) in peach orchards. J. Invertebr. Pathol. 102, 6–13.

Bashirudin, J., Martin, J., Reinganum, C., 1988. Queensland fruit fly virus, a probable member of the Picornaviridae. Arch. Virol. 61–74.

Batista Filho, A., Leite, L.G., Raga, A., Sato, M.E., Oliviera, J.A., 1995. Utilização de *Beauveria bassiana* (Balls.) Vuill. no manejo de *Cosmopolites sordidus* Germar, 1824, em Miracatu, SP. Biológico 57, 17–19.

Bedford, G.O., 1986. Biological control of the rhiniceros beetle (*Oryctes rhinoceros*) in the South Pacific by baculovirus. Agric. Ecosyst. Environ. 15, 141–147.

Bedford, G.O., 2013. Long-term reduction in damage by rhinoceros beetle *Oryctes rhinoceros* (L.) (Coleoptera: Scarabaeidae: Dynastinae) to coconut palms at *Oryctes* Nudivirus release sites on Viti Levu, Fiji. Afr. J. Agric. Res. 8, 6422–6425.

Begemann, G.J., 2008. The mortality of *Thaumatotibia leucotreta* (Meyrick) final instar larvae exposed to the entomopathogenic fungus *Beauveria bassiana* (Balsamo) Vuillemin. Afr. Entomol. 16, 306–308.

Bissolli, G., Correia, A.C.B., Barbosa, J.C., 2014. Selection of pathogenic fungi for the control of *Ceratitis capitata* (fruit fly) larvae and pupae. Científica 42, 338–345.

Briceno, A.J., 1997. Perspectivas de un manejo integrado del gusano verde del platano, *Opsiphanes tamarindi* Felder (Lepidoptera: Brassolidae). Rev. Fac. Agron. Univ. del Zulia 14, 487–495.

Brito, E.S., Paula, A.R., Vieira, L.P., Dolinski, C., Samuels, R.I., 2008. Combining vegetable oil and sub-lethal concentrations of Imidacloprid with *Beauveria bassiana* and *Metarhizium anisopliae* against adult guava weevil *Conotrachelus psidii* (Coleoptera: Curculionidae). Biocontrol Sci. Technol. 18, 665–673.

Brown, E.S., 1955. *Pseudotheraptus wayi*, a new genus and species of coreid (Hemiptera) injurious to coconuts in East Africa. Bull. Entomol. Res. 46, 221–240.

Butt, T.M., Brownbridge, M., 1997. Fungal pathogens of thrips. In: Lewis, T. (Ed.), Thrips as Crop Pests. CAB International, Wallingford, pp. 339–433.

CABI, 2015. Invasive Species Compendium. http://www.cabi.org/isc/datasheet/5654.

CABI/EPPO, 1997. Quarantine Pests for Europe, second ed. CAB International, Wallingford (GB).

CABI/EPPO, 2003. Bactrocera correcta. Distribution Maps of Plant Pests, No. 640. CAB International, Wallingford, UK.

CABI/EPPO, June 2012. Paracoccus marginatus. Distribution Maps of Plant Pests. CAB International, Map 614, Wallingford, UK.

Castillo, M.A., Moya, P., Primo-Yúfera, E., 2000. Susceptibility of *Ceratitis capitata* Wiedemann (Diptera: Tephritidae) to entomopathogenic fungi and their extracts. Biol. Control 19, 274–282.

Chandler, D., Davidson, G., Pell, J.G., Ball, B.V., Shaw, K., Sunderland, K.D., 2000. Fungal biocontrol of Acari. Biocontrol Sci. Technol. 10, 357–384.

Chandler, D., Davidson, G., Jacobson, R.J., 2005. Laboratory and glasshouse evaluation of entomopathogenic fungi against the two-spotted spider mite, *Tetranychus urticae* (Acari: Tetranychidae), on tomato, *Lycopersicon esculentum*. Biocontrol Sci. Technol. 15, 37–54.

Chen, H., Morrell, P.L., Ashworth, V.E.T.M., De La Cruz, M., Clegg, M.T., 2008. Tracing the geographic origins of major avocado cultivars. J. Hered. 100, 56–65.

Coombes, C.A., Hill, M.P., Moore, S.D., Dames, J.F., Fullard, T., 2015. *Beauveria* and *Metarhizium* against false codling moth (Lepidoptera: Tortricidae): a step towards selecting isolates for potential development of a mycoinsecticide. Afr. Entomol. 23, 239–242.

De La Rosa, W., Lopez, F.L., Liedo, P., 2002. *Beauveria bassiana* as a pathogen of the Mexican fruit fly (Diptera: Tephritidae) under laboratory. J. Econ. Entomol. 95, 36–43.

De Villiers, E.A., Van den Berg, M.A., 1987. Avocado insects of South Africa. South Afr. Avocado Growers' Assoc. Yearb. 10, 75–79.

Del Valle, E.E., Dolinski, C., Barreto, E.L.S., Souza, R.M., Samuels, R.I., 2008. Efficacy of *Heterorhabditis baujardi* LPP7 (Nematoda: Rhabditida) applied in *Galleria mellonella* (Lepidoptera: Pyralidae) insect cadavers to *Conotrachelus psidii*, (Coleoptera: Curculionidae) larvae. Biocontrol Sci. Technol. 18, 33–41.

Dembilio, O., Quesada-Moraga, E., Santiago-Álvarez, C., Jacas, J.A., 2010. Potential of an indigenous strain of the entomopathogenic fungus *Beauveria bassiana* as a biological control agent against the red palm weevil, *Rhynchophorus ferrugineus*. J. Invertebr. Pathol. 104, 214–221.

Destéfano, R.H.R., Bechara, I.J., Messias, C.L., Piedrabuena, A.E., 2005. Effectiveness of *Metarhizium anisopliae* against immature stages of *Anastrepha fraterculus* fruit fly (Diptera: Tephritidae). Braz. J. Microbiol. 36, 94–99.

Devasahayam, S., Nair, C.P.R., 1986. The mosquito bug, *Helopeltis antonii* Sign. on cashew in India. J. Plant. Crops 14, 1–10.

Dimbi, S., Maniania, N.K., Lux, S.A., Ekesi, S., Mueke, J.K., 2003a. Pathogenicity of *Metarhizium anisopliae* (Metsch.) Sorokin and *Beauveria bassiana* (Balsamo) Vuillemin, to three adult fruit flies *Ceratitis capitata* (Weidemann), *C. Fasciventris* (Karsch) and *C. Cosyra* (Walker) (Diptera: Tephritidae). Mycopathologia 156, 375–382.

Dimbi, S., Maniania, N.K., Lux, S.A., Ekesi, S., Mueke, J.K., 2003b. Host species, age and sex as factors affecting the susceptibility of the African tephritid fruit fly species, *Ceratitis capitata*, *C. cosyra* and *C. fasciventris* to infection by *Metarhizium anisopliae*. J. Pest Sci. 76, 113–117.

Dimbi, S., Maniania, N.K., Ekesi, S., 2009. Effect of *Metarhizium anisopliae* inoculation on the mating behavior of three 3 species of African tephritid fruit flies, *Ceratitis capitata*, *Ceratitis cosyra* and *Ceratitis fasciventris*. Biol. Control 50, 111–116.

Dimbi, S., Maniania, N.K., Ekesi, S., 2013. Horizontal transmission of *Metarhizium anisopliae* in fruit flies and effect of fungal infection on egg laying and fertility. Insects 4, 206–216.

Dolinski, C., Del Valle, E.E., Stuart, R.J., 2006. Virulence of entomopathogenic nematodes to larvae of guava weevil, *Conotrachelus psidii* (Coleoptera: Curculionidae), in laboratory and greenhouse experiments. Biol. Control 38, 422–427.

Douaho, A., 1984. Pests and diseases of oil palm and coconut. Biological control of *Pseudotheraptus* and related species. Oléagineux 39, 257–262.

Ekesi, S., Maniania, N.K., 2002. *Metarhizium anisopliae:* An effective biological control agent for the management of thrips in horti- and floriculture in Africa. In: Upadhyay, R. (Ed.), Advances in Microbial Control, Kluwer Academic/Plenum Publishers, pp. 164–180.

Ekesi, S., Maniania, N.K., Lux, S., 2002. Mortality in three African tephritid fruit fly puparia and adults caused by the entomopathogenic fungi, *Metarhizium anisopliae* and *Beauveria bassiana*. Biocontrol Sci. Technol. 12, 7–17.

Ekesi, S., Maniania, N.K., Lux, S.A., 2003. Effect of soil temperature and moisture on survival and infectivity of *Metarhizium anisopliae* to four tephritid fruit fly puparia. J. Invertebr. Pathol. 83, 157–167.

Ekesi, S., Dimbi, S., Maniania, N.K., 2007a. The role of entomopathogenic fungi in the integrated management of fruit flies (Diptera: Tephritidae) with emphasis on species occurring in Africa. In: Ekesi, S., Maniania, N.K. (Eds.), Use of Entomopathogenic Fungi in Biological Pest Management. Research Signpost, Kerala, India, pp. 239–274.

Ekesi, S., Lux, S., Billah, M.K., 2007b. Field Comparison of Food-Based Synthetic Attractants and Traps for African Tephritid Fruit Flies. pp. 205–222 IAEA Technical Document 1574.

Ekesi, S., Maniania, N.K., Mohamed, S.A., 2011. Efficacy of soil application of *Metarhizium anisopliae* and the use of GF-120 spinosad bait spray for suppression of *Bactrocera invadens* (Diptera: Tephritidae) in mango orchards. Biocontrol Sci. Technol. 21, 299–316.

Ekesi, S., De Meyer, M., Mohamed, S.A., Virgilio, M., Borgemeister, C., 2016. Taxonomy, bioecology and management of native and exotic fruit fly species in Africa. Annu. Rev. Entomol. 61, 219–238.

El-Sufty, R., Al Bgham, S., Al-Awash, S., Shahdad, A., Al Bathra, A., 2010. A study on a trap for autodissemination of the entomopathogenic fungus *Beauveria bassiana* by red palm weevil adults in date palm plantations. J. Basic Appl. Mycol. 1, 61–65.

EPPO, 2014. PQR Database. European and Mediterranean Plant Protection Organization, Paris, France. http://www.eppo.int/DATABASES/pqr/pqr.htm.

Evangelos, I.B., Dimitrios, P.P., Anastasia, F., Spyridon, A.A., Dimitrios, C.K., 2013. Pathogenicity of three entomopathogenic fungi on pupae and adults of the Mediterranean fruit fly, *Ceratitis capitata* (Diptera: Tephritidae). J. Pest Sci. 86, 275–284.

FAO, 2003. Medium-Term Prospects for Agricultural Commodities. Projections to the Year 2010. 99 pp.

FAO, 2011. Major Food and Agricultural Commodities and Producers – Countries by Commodity. (Retrieved 02/02/2015).

Ferron, P., 1985. Fungal control. In: Kerkut, G.A., Gilbert, L.I. (Eds.), Comprehensive Insect Physiology, Biochemistry, and Pharmacology, 12. Pergamon Press, pp. 313–346.

Figueroa, W., 1990. Biocontrol of the banana root borer weevil, *Cosmopolites sordidus* (Germar), with steinernematid nematodes. J. Agric. Univ. P.R. 74, 15–19.

Froggatt, J.L., O'Connor, B.A., 1941. Insects associated with the coconut palm. Pt. II. New Guinea Agric. Gaz. 7, 125–133.

Gandarilla-Pacheco, F.L., Nava-González, H.D., Arévalo-Niño, K., Galán-Wong, L.J., Elías-Santos, M., Quintero-Zapata, I., 2012. Evaluation of native strains of *Isaria fumosorosea* (Wize) against *Anastrepha ludens* (Loew) (Diptera: Tephritidae). J. Life Sci. 6, 957–960.

Gazit, Y., Rossler, Y., Glazer, I., 2000. Evaluation of entomopathogenic nematodes for the control of Mediterranean fruit fly (Diptera: Tephritidae). Biocontrol Sci. Technol. 10, 157–164.

Gelernter, W.D., Lomer, C.J., 2000. Success in biological control of above-ground insects by pathogens. In: Gurr, G., Wratte, S. (Eds.), Biological Control: Measures of Success. Kluwer Academic Publishers, Dordrecht, pp. 297–322.

Giblin-Davis, R.M., Peña, J.E., Duncan, R.E., 1996. Evaluation of an entomopathogenic nematode and chemical insecticides for control of *Metamasius hemipterus sericeus* (Coleoptera: Curculionidae). J. Entomol. Sci. 31, 240–251.

Gindin, G., Barash, I., Harari, N., Raccah, B., 1994. Effect of endotoxic compounds isolated from Verticillium lecanii on the sweet potato whitefly, Bemisia tabaci. Phytoparasitica 22, 189–196.

Gindin, G., Levski, S., Glazer, I., Soroker, V., 2006. Evaluation of the entomopathogenic fungi *Metarhizium anisopliae* and *Beauveria bassiana* against the red palm weevil *Rhynchophorus ferrugineus*. Phytoparasitica 34, 370–379.

Gingrich, R.E., El-Abbassi, T.S., 1988. Diversity among *Bacillus thuringiensis* active against Mediterranean fruit fly *Ceratitis capitata*. In: Proc. International Symposium on Modern Insect Control: Nuclear Techniques and Biotechnology, Series. IAEA; International Atomic Energy Agency (IAEA), Vienna (Austria), pp. 77–84. November 16–20, 1987.

Godonou, I., Green, K.R., Oduro, K.A., Lomer, C.J., Afreh-Nuamah, K., 2000. Field evaluation of selected formulations of *Beauveria bassiana* for the management of the banana weevil (*Cosmopolites sordidus*) on plantain (*Musa* spp., AAB Group). Biocontrol Sci. Technol. 10, 779–788.

Gold, C.S., Peña, J.E., Karamura, E.B., 2001. Biology and integrated pest management for the banana weevil *Cosmopolites sordidus* (Germar) (Coleoptera: Curculionidae). Int. Pest Manage. Rev. 6, 79–155.

Gold, C.S., Pinese, B., Peña, J.E., 2002. Pests of banana. In: Peña, J.E., Sharp, J.L., Wysoki, M. (Eds.), Tropical Fruit Pests and Pollinators: Biology, Economic Importance, Natural Enemies, and Control. CABI Publishing, Wallingford, UK, pp. 13–56.

Gold, C.S., Nankinga, C., Niere, B., Godonou, I., 2003. IPM of banana weevil in Africa with emphasis on microbial control. In: Neuenschwander, P., Borgemeister, C., Langewald, J. (Eds.), Biological Control in IPM Systems in Africa. CABI Publishing, Wallingford, UK, pp. 243–257.

Gopal, M., Gupta, A., Thomas, G.V., 2006. Prospects of using *Metarhizium anisopliae* to check the breeding of insect pest, *Oryctes rhinoceros* L. in coconut leaf vermicomposting sites. Bioresour. Technol. 97, 1801–1806.

Gopinadhan, P.B., Mohandas, N., Nair, K.P.V., 1990. Cytoplasmic polyhedrosis virus infecting redpalm weevil of coconut. Curr. Sci. 59, 577–580.

Gould, W.P., Raga, A., 2002. Pests of guava. In: Peña, J.E., Sharp, J.L., Wysoki, M. (Eds.), Tropical Fruit Pests and Pollinators. CABI Publishing, Wallingford, UK, pp. 295–313.

Griffith, R., 1987. Red ring disease of coconut palm. Plant Dis. 71, 193–196.

Grim, C., Guharay, F., 1998. Control of leaf-footed bug *Leptoglossus zonatus* and shield-backed bug *Pachycoris klugii* with entomopathogenic fungi. Biocontrol Sci. Technol. 8, 356–376.

Grimont, P., Grimont, F., 1978. The genus *Serratia*. Ann. Rev. Microb. 32, 221–248.

Güerri-Agulló, B., López-Follana, R., Asensio, L., Barranco, P., Lopez-Llorca, L.V., 2011. Use of a solid formulation of *Beauveria bassiana* for biocontrol of the red palm weevil (*Rhynchophorus ferrugineus*) (Coleoptera: Dryophthoridae) under field conditions in SE Spain. Fla. Entomol. 94, 737–747.

Gul, H.T., Freed, S., Akmal, M., Malik, M.N., 2015. Vulnerability of different life stages of *Bactrocera zonata* (Tephritidae: Diptera) against entomogenous fungi. Pak. J. Zool. 47, 307–317.

Hanounik, S.B., 1998. Steinernematids and heterorhabditids as biological control agents for the red palm weevil (*Rhynchophorus ferrugineus* Oliv.). Sultan Qabus Univ. J. Sci. Res. Agric. S.C. 3, 95–102.

Haramoto, F.H., 1969. Biology and control of *Brevipalpus phoenicis* (Acari: Tenuipalpidae). Hawaii Agric. Exp. Stn. Tech. Bull. 68.

Hedstrom, I., 1994. *Stigmatomyces* species on the guava fruit flies in seasonal and non-seasonal neotropical forest environments. Mycol. Res. 98, 403–407.

Hoddle, M.S., Nakahara, S., Phillips, P.A., 2002. Foreign exploration for *Scirtothrips perseae* Nakahara (Thysanoptera: Thripidae) and associated natural enemies on avocado (*Persea americana* Miller). Biol. Control 24, 251–265.

Hosang, M.L.A., Alouw, J.C., Novarianto, H., 2004. Biological Control of *Brontispa longissima* (Gestro) in Indonesia. FAO, Report of the expert consultation on coconut beetle outbreak in APPPC member.

Hüger, A.M., 1966. A virus disease of the Indian rhinoceros beetle, *Oryctes rhinoceros* (Linnaeus), caused by a new type of insect virus, *Rhaddionvirus oryctes* gen.n. sp.n. J. Invertebr. Pathol. 8, 38–41.

Jackson, T.A., 2009. The use of *Oryctes* virus for control of rhinoceros beetle in the Pacific Islands. Prog. Biol. Control 6, 133–140.

Jaronski, S.T., 2010. Ecological factors in the inundative use of fungal entomopathogens. BioControl 55, 159–185.

Jeyarani, S., Banu, J.G., Ramaraju, K., 2011. First record of natural occurrence of *Cladosporium cladosporioides* (Fresenius) de Vries and *Beauveria bassiana* on two spotted spider mite, *Tetranychus urticae* Koch. J. Entomol. 8, 274–279.

Joubert, P.H., Labuschagne, T.I., 1995. Alternative measures for controlling mango seed weevil, *Sternochetus mangiferae* (Fab.) (Coleoptera: Curculionidae). S. Afr. Mango Growers' Assoc. Yearb. 15, 94–96.

Julia, J.F., Mariau, D., 1978. The coconut bug: *Pseudotheraptus* sp. in the Ivory Coast. I. – studies preliminary to the devising of a method of integrated control. Oleagineux 33, 65–67.

Konstantopoulou, M.A., Mazomenos, B.E., 2005. Evaluation of *Beauveria bassiana* and *B. brongniartii* strains and four wild-type fungal species against adults of *Bactrocera oleae* and *Ceratitis capitata*. BioControl 50, 293–305.

Krasnoff, S.B., Gupta, S., 1994. Identification of the antibiotic phomalactone from the entomopathogenic fungus, *Hirsutella thompsonii* var. *synnematosa*. J. Chem. Ecol. 20, 293–302.

Laborda, R., Bargues, L., Navarro, C., Barajas, O., Arroyo, M., Garcia, E.M., Montoro, E., Llopis, E., Martinez, A., Sayagues, J.M., 2003. Susceptibility of the Mediterranean fruit fly (*Ceratitis capitata*) to entomopathogenic nematode *Steinernema* spp. ("Biorend C"). Bull. OILB/SROP 26, 95–97.

Lacey, L.A., Goettel, M.S., 1995. Current developments in microbial control of insect pests and prospects for the early 21st century. Entomophaga 40, 3–27.

Lacey, L.A., Shapiro-Ilan, D.I., 2008. Microbial control of insect pests in temperate orchard systems: potential for incorporation into IPM. Annu. Rev. Entomol. 53, 121–144.

Lacey, L.A., Grzywacz, D., Shapiro-Ilan, D.I., Frutos, R., Brownbridge, M., Goettel, M.S., 2015. Insect pathogens as biological control agents: back to the future. J. Invertebr. Pathol. 132, 1–41.

Langford, E.A., Nielsen, U.N., Johnson, S.N., Riegler, M., 2014. Susceptibility of Queensland fruit fly, *Bactrocera tryoni* (Froggatt) (Diptera: Tephritidae), to entomopathogenic nematodes. Biol. Control 69, 34–39.

Lezama-Gutiérrez, R., Molina, O.L., Contreras-Ochoa, M., Gonzáles-Ramírez, A., Trujillo-De La Luz, A., Rebolledo-Domínguez, O., 1996. Susceptibilidad de larvas de *Anastrepha ludens* (Diptera: Tephritidae) a diversos nemátodos entomopatógenos (Steinernematidae y Heterorhabditidae). Vedalia 3, 31–33.

Lezama-Gutierrez, R., Trujillo-De La Luz, A., Molina-Ochoa, J., Rebolledo-Dominguez, O., Pescador, A.R., Lopez-Edwards, M., Aluja, M., 2000. Virulence of *Metarhizium anisopliae* (Deuteromycotina: Hyphomycetes) on *Anastrepha ludens* (Diptera: Tephritidae): laboratory and field trials. J. Econ. Entomol. 93, 1080–1084.

Lindegren, I.E., Wong, T.T., McInnis, D.O., 1990. Responses of Mediterranean fruit fly (Diptera: Tephritidae) to the entomogenous nematode *Steinernema feltiae* in field tests in Hawaii. Environ. Entomol. 19, 383–386.

Lin-Jin, T., Zeng, L., Liang, G., Lu-Yong, Y., Bin-Shu, Y., 2005. Effects of entomopathogenic nematodes on the oriental fruit fly, *Bactrocera* (*Bactrocera*) *dorsalis* (Hendel). Acta Entomol. Sin. 48, 736–741.

Liu, S.D., Lin, S.C., Shiau, J.F., 1989. Microbial control of coconut leaf beetle (*Brontispa longissima*) with green muscardine fungus, *Metarhizium anisopliae* var. *anisopliae*. J. Invertebr. Pathol. 53, 307–314.

Loc, N.T., Chi, V.T.B., Nhan, N.T.N., Ngu Thanh, N.D., Hong, T.T.B., Hung, P.Q., 2004. Biocontrol potential of *Metarhizium anisopliae* against coconut beetle, *Brontispa longissima*. Omonrice 12, 85–91.

Lorenzato, D., Chouene, E.C., Medeiros, J., Rodrigues, A.E.C., Pederzolli, R.C.D., 1997. Ocorrência e controle da broca-do-fruto-do-abacaxi *Thecla basalides* (Geyer, 1837). Pesq. Agropec. Gaúcha 3, 15–19.

Lux, S.A., Ekesi, S., Dimbi, S., Mohamed, S., Billah, M., 2003. Mango infesting fruit flies in Africa-perspectives and limitations of biological approaches to their management. In: Neuenschwander, P., Borgemeister, C., Langewald, J. (Eds.), Biological Control in Integrated Pest Management Systems in Africa. CABI, Wallingford, pp. 277–293.

Malan, A.P., Manrakhan, A., 2009. Susceptibility of the Mediterranean fruit fly (*Ceratitis capitata*) and the Natal fruit fly (*Ceratitis rosa*) to entomopathogenic nematodes. J. Invertebr. Pathol. 100, 47–49.

Malan, A.P., Knoetze, R., Moore, S.D., 2011. Isolation and identification of entomopathogenic nematodes from citrus orchards in South Africa and their biocontrol potential against false codling moth. J. Invertebr. Pathol. 108, 115–125.

Maniania, N.K., Ekesi, S., Odulaja, A., Okech, M.A., Nadel, D.J., 2006. Prospects of a fungus-contamination device for the control of tsetse fly *Glossina fuscipes fuscipes*. Biocontrol Sci. Technol. 16, 129–139.

Maniania, N.K., Bugeme, D.M., Wekesa, V.W., Delalibera Jr., I., Knapp, M., 2008. Role of entomopathogenic fungi in the control of *Tetranychus evansi* and *Tetranychus urticae* (Acari: Tetranychidae), pests of horticultural crops. Exp. Appl. Acarol. 46, 259–274.

Manrakhana, A., John-Henry Daneela, J.-H., Moore, S.D., 2014. The impact of naturally occurring entomopathogenic nematodes on false codling moth, *Thaumatotibia leucotreta* (Lepidoptera: Tortricidae), in citrus orchards. Biocontrol Sci. Technol. 24, 241–245.

Mfuti, D.K., Subramanian, S., van Tol, R.W.H.M., Wiegers, G.L., De Kogel, W.J., Niassy, S., Du Plessis, H., Ekesi, S., Maniania, N.K., 2016. Spatial separation of semiochemical Lurem-TR and entomopathogenic fungi to enhance their compatibility and infectivity in an autoinoculation system for thrips management. Pest Manag. Sci. 72, 131–139.

Minas, R.S., 2008. M.Sc. Dissertation presented at Universidade Estadual do Norte Fluminense Darcy Ribeiro, Campos dos Goytacazes, RJ, Brazil.

Minas, R.S., 2012. Caracterização biológica de uma linhagem de nematoide entomopatogênico visando o controle do gorgulho da goiaba (*Conotrachelus psidii*) em dois sistemas de cultivo. Doctor Thesis presented at Universidade Estadual do Norte Fluminense Darcy Ribeiro. Campos dos Goytacazes, RJ, Brazil.

Mochi, D.A., Monteiro, A.C., Bortoli, S.A., Dória, H.O.S., Barbosa, J.C., 2006. Pathogenicity of *Metarhizium anisopliae* for *Ceratitis capitata* (Wied.) (Diptera: Tephritidae) in soil with different pesticides. Neotrop. Entomol. 35, 382–389.

Moore, S., Hattingh, V., 2012. A review of current pre-harvest control options for false codling moth in citrus in southern Africa. S.A. Fruit. J. 11, 82–85.

Nankinga, C.M., Moore, D., 2000. Reduction of banana weevil populations using different formulations of the entomopathogenic fungus *Beauveria bassiana*. Biocontrol Sci. Technol. 10, 645–657.

Navarro-Llopis, V., Ayala, I., Sanchis, J., Primo, J., Moya, P., 2015. Field Efficacy of a *Metarhizium anisopliae*-based attractant–contaminant device to control *Ceratitis capitata* (Diptera: Tephritidae). J. Econ. Entomol. 108, 1570–1578.

Newton, P.J., 1998. False codling moth *Cryptophebia leucotreta* (Meyrick). In: Bedford, E.C.G., Van den Berg, M.A., de Villiers, E.A. (Eds.), Citrus Pests in the Republic of South Africa. Agricultural Research Council, Republic of South Africa, Nelspruit, pp. 192–200.

Odienki, J.J., 1975. Diseases and pests of passion fruit in Kenya. Acta Hortic. 49, 291–293.

Osorio-Fajardo, A., Canal, N.A., 2011. Selection of strains of entomopathogenic fungi for management of *Anastrepha obliqua* (Macquart, 1835) (Diptera: Tephritidae) in Colombia. Rev. Fac. Nal. Medellin 64, 6129–6139.

Ouna, E.A., 2012. Entomopathogenicity of Hyphomycetes Fungi to the Invasive Fruit Flies *Bactrocera invadens* (MSc dissertation). Kenyatta University, Kenya.

Pan, Q., Wang, Z., Quebedeaux, B., 1998. Responses of the apple plant to CO_2 enrichment: changes in photosynthesis, sorbitol, other soluble sugars, and starch. Aust. J. Plant Physiol. 25, 293–297.

Pantoja, A., Follett, P.A., Villanueva-Jiménez, J.A., 2002. Pest of papaya. In: Peña, E., Sharp, J.L., Wysoki, M. (Eds.), Tropical Fruit Pests and Pollinators: Biology, Economic Importance, Natural Enemies and Control. CAB International, pp. 131–156.

Pasaru, F., Anshary, A., Kuswinanti, T., Mahfudz, Shahabuddin, 2014. Prospective of entomopathogenic fungi associated with *Helopeltis* spp. (Hemiptera: Miridae) on cacao plantation. Int. J. Curr. Res. Acad. Rev. 2, 227–234.

Patt, J.M., Chow, A., Meikle, W.G., Gracia, C., Jackson, M.A., Flores, D., Sétamou, M., Dunlap, C.A., Avery, P.B., Hunter, W.B., Adamczyk, J.J., 2015. Efficacy of an autodisseminator of an entomopathogenic fungus, *Isaria fumosorosea*, to suppress Asian citrus psyllid, *Diaphorina citri*, under greenhouse conditions. Biol. Control 88, 37–45.

Peña, J.E., Osborne, L.S., Duncan, R.E., 1996. Potential of fungi as biocontrol agents of *Polyphagotarsonemus latus* (Acari: Tarsonemidae). Entomophaga 41, 27–36.

Petty, G.J., Stirling, G.R., Bartholomew, D.P., 2002. Pests of pineapple. In: Peña, J., Sharp, J., Wysoki, M. (Eds.), Tropical Fruit Pests and Pollinators: Biology, Economic Importance, Natural Enemies and Control. CABI Publishing, Wallingford, UK, pp. 157–195.

Pillai, G.B., Dubey, O.P., Singh, V., 1976. Pests of cashew and their control in India: a review of current status. J. Plant. Crops 4, 37–50.

Pomar Brasil., 2015. Available via: http://www.pomarbrasil.com/success/index.html. Cited May 2015.

Purrini, K., 1989. *Baculovirus oryctes* release into *Oryctes monoceros* population in Tanzania, with special reference to the interaction of virus isolates used in our laboratory infection experiments. J. Invertebr. Pathol. 53, 285–300.

Quesada-Moraga, E., Ruiz-García, A., Santiago-Álvarez, C., 2006. Laboratory evaluation of the entomopathogenic fungi *Beauveria bassiana* and *Metarhizium anisopliae* against puparia and adults of *Ceratitis capitata* (Diptera: Tephritidae). J. Econ. Entomol. 99, 1955–1966.

Rajamanickam, K., Kennedy, J.S., Christopher, A., 1995. Certain components of integrated management of red palm weevil, *Rhynchophorus ferrugineus* F. (Curculionidae: Coleoptera) on coconut. Meded. Fac. Landbouwk. Toegepaste Biol. Wet. 60, 803–805.

Rajapakse, R.H.S., Jeevaratnam, K., 1982. Use of a virus against the root and stem borer *Plocaederus ferrugineus* L. (Coleoptera:Cerambydiae) of the cashew. Int. J. Trop. Insect Sci. 3, 49–51.

Ramle, M., Mohd Basri, W., Norman, K., Sharma, M., Siti Ramlah, A.A., 1999. Impact of *Metarhizium anisopliae* (Deuteromycotina: Hyphomycetes) applied by wet and dry inoculums on oil palm rhinoceros beetles, *Oryctes rhinoceros* (Coleoptera: Scarabaeidae). J. Oil Palm. Res. 11, 25–40.

Ramle, M., Mohd Basri, W., Norman, K., Siti Ramlah, A.A., Hisham, N.H., 2006. Research into the commercialization of *Metarhizium anisopliae* (Hyphomycetes) for biocontrol of oil palm rhinoceros beetle, *Oryctes rhinoceros* (Scarabaeidae) in oil palm. J. Oil Palm. Res. (Special issue), 37–49.

Ramle, M., Norman, K., Mohd Basri, W., 2011. Trap for the autodissemination of *Metarhizium anisopliae* in the management of rhinoceros beetle, *Oryctes rhinoceros*. J. Oil Palm Res. 1011–1017.

Robacker, D.C., Martinez, A.J., Garcia, J.A., Diaz, M., Romero, C., 1996. Toxicity of *Bacillus thuringiensis* to Mexican fruit fly (Diptera: Tephritidae). J. Econ. Entomol. 89, 104–110.

Rodrigues, J.C., Nogueira, N.L., Freitas, D.S., Prates, H.S., 1997. Virus-like particles associated with *Brevipalpus phoenicis* Geijskes (Acari: Tenuipalpidae), vector of *Citrus leprosis virus*. Ann. Soc. Entomol. Bras. 26, 391–395.

Sahu, K.R., Sharma, D., 2008. Management of cashew stem and root borer, *Plocaederus ferrugineus* L. by microbial and plant products. J. Biopest. 1, 121–123.

Salama, H.S., Foda, M.S., El-Bendary, M.A., Abdel-Razek, A., 2004. Infection of red palm weevil, *Rhynchophorus ferrugineus* by spore-forming bacilli indigenous to its natural habitat in Egypt. J. Pest Sci. 77, 27–31.

Sanches, N.F., 2005. Manejo integrado da broca-do-fruto do abacaxi. Embrapa Mandioca e Fruticultura Tropical Bol. Tec. no 36, 37 pp.

Sarwar, M., 2006. Occurrence of insect pests on guava (*Psidium guajava*) tree. Pak. J. Zool. 38, 197–200.

Schmitt, A.T., Gowen, S.R., Hague, N.G.M., 1992. Baiting techniques for the control of *Cosmopolites sordidus* Germar (Coleoptera: Curculionidae) by *Steinernema carpocapsae* (Nematoda: Steinernematidae). Nematropica 22, 159–163.

Schoeman, P.S., Schoeman, M.H., 1999. Transmission of *Beuaveria bassiana* from infected adults of the banana weevil *Cosmopolites sordidus* (Coleoptera: Curculionidae). Afr. Plant Prot. 5, 53–54.

Sewify, G.H., Belal, M.H., Al-Awash, S.A., 2009. Use of the entomopathogenic fungus, *Beauveria bassiana* for the biological control of the red palm weevil, *Rhynchophorus ferrugineus* Olivier. Egypt J. Biol. Pest Control 19, 157–163.

Shukla, R.P., Tandon, P.L., Singh, S.J., 1984. Baculovirus – a new pathogen of mango nut weevil, *Sternochetus mangiferae* (Fabricius) (Coleoptera : Curculionidae). Curr. Sci. 53, 593–594.

Silva, A.C., Batista Filho, A., Leite, L.G., Tavares, F.M., Raga, A., Schmidt, F.S., 2010. Effect of entomopathogenic nematodes on the mortality of the fruit fly *Ceratitis capitata* and of the guava weevil *Conotrachelus psidii*. Nematol. Braz. 34, 31–40.

Smith, D., 1995. Banana weevil borer control in south-eastern Queensland. Aust. J. Exp. Agric. 35, 1165–1172.

Soliman, N.A., Ibrahim, A.A., El-Deen, M.M.-S., Ramadan, N.F., Farag, S.R., 2014. Entomopathogenic nematodes and fungi as biocontrol agents for the peach fruit fly, *Bactrocera zonata* (Saunders) and the Mediterranean fruit fly, *Ceratitis capitata* (Wiedemann) soil borne-stages. Egypt J. Biol. Pest Control 24, 497–502.

Souza, J.C., Haga, A., Souza, M.A., 2003. Pragas da goiabeira. Tech. Bull. no. 71. 60 pp.

Stonedahl, G.M., 1991. The oriental species of *Helopeltis* (Heteroptera: Miridae): a review of economic literature and guide to identification. Bull. Entomol. Res. 81, 465–490.

Strongman, D., MacKenzie, K., Dixon, P., 1997. Entomopathogenic fungi in the lowbush blueberry fields. Acta Hortic. 465–476.

Tinzaara, W., Gold, C.S., Dicke, M., Van Huis, A., Nankinga, C.M., Godfrey H. Kagezi, G.H., Ragama, P.E., 2007. The use of aggregation pheromone to enhance dissemination of *Beauveria bassiana* for the control of the banana weevil in Uganda. Biocontrol Sci. Technol. 17, 111–124.

Toledo, J., Pérez, C., Liedo, P., Ibarra, J., 2005a. Susceptibilidad de larvas de *Anastrepha obliqua* Macquart (Diptera: Tephritidae) a *Heterorhabditis bacteriophora* (Poinar) (Rhabditida: Heterorhabtididae) en condiciones de laboratorio. Vedalia 12, 11–21.

Toledo, J., Ibarra, J.E., Liedo, P., Gomez, A., Rasgado, M.A., Williams, T., 2005b. Infection of *Anastrepha ludens* (Diptera: Tephritidae) larvae by *Heterorhabditis bacteriophora* conditions. Biocontrol Sci. Technol. 15, 627–634.

Toledo, J., Liedo, P., Flores, S., Campos, S.E., Villasenor, A., Montoya, P., 2006. Use of *Beauveria bassiana* and *Metarhizium anisopliae* for fruit fly control: a novel approach. In: Proceedings of the 7th International Symposium on Fruit Flies of Economic Importance, Salvador, Brazil, pp. 127–132 Available at: http://www.moscamed.org.br/pdf/Cap_13.pdf.

Touhey, P., 2010. False Codling Moth. Personal Report to P. Michalak.

Treverrow, N., Bedding, R.A., 1993. Development of a system for the control of the banana weevil borer, *Cosmopolites sordidus*, with entomopathogenic nematodes. In: Bedding, R.A., Akhurst, R., Kaya, H.K. (Eds.), Nematodes and the Biological Control of Insect Pests. CSIRO Publishing, Melbourne, pp. 41–47.

Treverrow, N., Bedding, R.A., Dettmann, E.B., Maddox, C., 1991. Evaluation of entomopathogenic nematodes for control of *Cosmopolites sordidus* Germar (Coleoptera: Curculionidae), a pest of bananas in Australia. Ann. Appl. Biol. 119, 139–145.

Ujjan, A.A., Khanzada, M.A., Shahzad, S., 2015. Efficiency of *Metarhizium* spp. (Sorokin) strains and insecticides against cotton mealybug *Phenacoccus solenopsis* (Tinsley). Pak. J. Zool. 47, 351–360.

Van der Geest, L.P.S., 1985. Pathogens of spider mites. In: Helle, W., Sabelis, M.W. (Eds.), Spider Mites. Their Biology, Natural Enemies and Control. World Crop Pests, vol. 1B. Elsevier, Amsterdam, pp. 247–258.

Van der Geest, L.P.S., Elliot, S.L., Breeuwer, J.A.J., Beerling, E.A.M., 2000. Diseases of mites. Exp. Appl. Acarol. 24, 497–560.

Vargas, R.I., Stark, J.D., Mackey, B., Bull, R., 2005. Weathering trials of amulet cue-lure and amulet methyl eugenol "attract-and-kill" stations with male melon flies and oriental fruit flies (Diptera: Tephritidae) in Hawaii. J. Econ. Entomol. 98, 1551–1559.

Vargas, R.I., Piñero, J.C., Leblanc, L., 2015. An overview of pest species of *Bactrocera* fruit flies (Diptera: Tephritidae) and the integration of biopesticides with other biological approaches for their management with a focus on the Pacific region. Insects 6, 297–318.

Varma, C.K., 2013. Efficacy of ecofriendly management against rhinoceros beetle grubs in coconut. J. Biopest. 6, 101–103.

Vega, F.E., Dowd, P.F., Lacey, L.A., Pell, J.K., Jackson, D.M., Klein, M.G., 2007. Dissemination of beneficial microbial agents by insects. In: Lacey, L.A., Kaya, H.K. (Eds.), Field Manual of Techniques in Invertebrate Pathology, second ed. Springer, Dordrecht, The Netherlands, pp. 127–146.

Vögele, J.M., Zeddies, J., 1990. Economic analysis of classical biological pest control: a case study from Western Samoa. Dtsch. Landwirtschafts-Gesellschaft 1, 45–51.

Waterhouse, D.F., Norris, K.R., 1987. Biological Control: Pacific Prospects. VIII, 454 pp.

Weems Jr., H.V., 2001. In: Feature Creatures – *Anastrepha serpentina* Available via: http://entnemdept.ufl.edu/creatures/fruit/tropical/sapote_fruit_fly.htm updated on January 2015. Cited May 2015.

White, I.M., Elson-Harris, M.M. (Eds.), 1992. Fruit Flies of Economic Significance: Their Identification and Bionomics. CAB International, Wallingford, UK. 601 pp.

Williams, D.J., Granara de Willink, M.C. (Eds.), 1992. Mealybugs of Central and South America. CAB International, Wallingford, UK. 635 pp.

Zahn, D.K., Morse, J.G., 2013. Investigating alternatives to traditional insecticides: effectiveness of entomopathogenic fungi and *Bacillus thuringiensis* against citrus thrips and avocado thrips (Thysanoptera: Thripidae). J. Econ. Entomol. 106, 64–72.

Zucchi, R.A., 2008. Fruit Flies in Brazil – *Anastrepha* Species Their Host Plants and Parasitoids. Available via: www.lea.esalq.usp.br/anastrepha/ updated on March 11, 2015. Cited May 2015.

Chapter 19

Microbial Control of Insect and Mite Pests of Citrus

S.D. Moore[1,2], L.W. Duncan[3]

[1]*Citrus Research International, Port Elizabeth, South Africa;* [2]*Rhodes University, Grahamstown, South Africa;* [3]*University of Florida, Lake Alfred, FL, United States*

19.1 INTRODUCTION

19.1.1 Commercial Citrus Production

Most citrus is grown in broad belts of the subtropics between 15 and 35 degrees in both the northern and southern latitudes, where minimal temperatures remain above lethal limits—generally from −4 to 7°C. Trees are long lived (≈100 years), but commercial plantings continually replace trees that decline in productivity. Trees are grown on rootstocks adapted to specific local conditions. The principal species include *Citrus sinensis* (sweet orange), *C. paradisi* (grapefruit), *C. reticulata* (mandarin), *C. limon* (lemon) and *C. aurantiifolia* (lime), *Poncirus trifoliata* (trifoliate orange), and *Fortunella* spp. (kumquat). These species hybridize easily and are generally graft compatible (Jackson, 1999). Among the 10 major citrus-growing countries, China produced nearly one-third of the global citrus in 2013–2014, followed by Brazil (20%), the European Union (EU) (12%), the United States (USA) (9%), Mexico (8%), Turkey (4%), Egypt and South Africa (3% each), and Argentina and Morocco (2% each) (USDA-FAS, 2015).

19.1.2 Pests of Citrus

Citrus pest management in the past was heavily dependent on whether fruit were destined for fresh or processed consumption (Browning, 1999). Citrus that was processed for juice (primarily Brazil and Florida) required fewer pest management inputs than did that sold on the fresh market. Industries that rely heavily on exported fruit, such as South Africa and the Mediterranean countries, are especially vulnerable to market issues involving quarantine pests. However, the recent introduction of huanglongbing (HLB, or Asian citrus greening disease) into the new world caused a complete paradigm shift in the integrated pest management (IPM) programs of the world's largest producers. Orchards in Brazil and Florida that relied heavily on natural control of arthropod pests now apply chemical pesticides repeatedly throughout the year to reduce the population density of the HLB disease vector, the Asian citrus psyllid (*Diaphorina citri*), and will probably continue to do so until a sustainable means of disease management is discovered (Rogers et al., 2015). Although there is currently sufficient diversity in the mode of action among the chemical insecticides that are available and effective for control of the Asian citrus psyllid in order to delay evolution of insecticide resistance by the pest, this cannot be viewed as a long-term sustainable approach (Qureshi et al., 2014). Not only does it become cost prohibitive but also some degree of resistance to key insecticides has already been documented (Tiwari et al., 2011). Additionally, despite the proven high level of efficacy of many of the available insecticides (Qureshi et al., 2014), there has been a dramatic decline in citrus production as a result of HLB, particularly in Florida (Hodges and Spreen, 2012).

Although the exact species present in the pest complex and their relative importance differ between countries and citrus production regions, there are certain key pests that are common among most citrus-producing regions in the world. For example, armored scales, mealybugs, thrips, mites, and fruit flies are common to most citrus-producing countries, even though the exact species may differ. On citrus in California, around 53 different species of insect and mite pests are listed (Dreistadt, 2012). The most important of these include scale insects, citrus thrips, and certain mites. Futch (2011) lists 39 different insect and mite pests of citrus in Florida. However, this is not a comprehensive list. The University of Florida Citrus Extension Website (UF/IFAS Citrus Extension) probably provides the most comprehensive list of citrus pests in Florida, totaling 47. However, even here there is acknowledgment that the list is not exhaustive but focuses on the most economically important pests. Browning et al. (1995) claimed that only a few arthropods found on Florida citrus are economic pests that damage relatively small areas, and thus, they rarely justify pest control. However, this assertion was made before the arrival of the Asian citrus psyllid and the devastation caused by HLB (Halbert, 2005). On citrus in South Africa, Bedford et al. (1998) listed more than 100 insect pests. When mites were included, this total increased to more than 110 species. However, Grout and Moore (2015) only listed 63 insect pests of citrus, excluding a number of previously listed species that were rarely, if ever, recorded on citrus in recent years. Due to the emphasis on exports, the most serious pests in southern Africa are the phytosanitary pests, false codling moth (*Thaumatotibia leucotreta*) and various fruit flies. In China, citrus orchard insect pests include more than 74 species among 36 families in nine orders (Niu et al., 2014). However, only a few are widely distributed and occur sufficiently consistently to have a significant economic importance. Niu et al. (2014) list those arthropods considered to be major pests in the past 40 years. Smith et al. (1997) listed 84 insect and mite pests of citrus in Australia and a further 20 insect species that were of such minor importance they did not warrant any detailed mention. In the citrus-producing countries of the Mediterranean basin, 108 insect pests and 10 mite pests are reported (Franco et al., 2006). In Brazil, which is by far the largest producer of citrus in South and Central America, there are numerous pests that are considered as either major or occasional (Campanhola et al., 1995; Smith and Peña, 2002; Dolinski and Lacey, 2007). Around 10 years ago, the Asian citrus psyllid was added to this list (Teixeira et al., 2005).

As it is not the purpose of this review to comprehensively list all important citrus pests occurring in all notable citrus production regions, there are a number of citrus-producing countries that have not been covered here, such as Turkey, Morocco, Egypt, and Mexico. However, there is a high degree of overlap of citrus pests in proximal regions throughout the world. Additionally, there are a number of other publications that do cover this topic (eg, Morse et al., 1996).

It must be noted that not all pest species that occur on citrus are equally suitable targets for microbial control agents. This may be a result of their morphology, life cycle, or habitat. However, it is extremely rare to find an insect or mite pest that is not at least susceptible to attack by an entomopathogen, and therefore all citrus pests should be seen as potential targets and investigated as such.

19.2 MICROBIAL CONTROL OF CITRUS PESTS WITH ENTOMOPATHOGENIC FUNGI

19.2.1 Armored Scales

Large, stationary populations of scale insects would seem to be ideal targets for fungal attack (Evans and Prior, 1990). However, this may not be the case as although a number of species of entomopathogenic fungi (EPF) have been recorded naturally infecting various species of armored scales on citrus (Table 19.1), it is questionable how much population suppression actually occurs (McCoy et al., 2009).

The first recorded efforts at biological control of scale insects on citrus using EPF were in Florida with *Cosmospora flammea* against San Jose scale (*Quadraspidiotus perniciosus*) and with *Podonectria coccicola* and *Aschersonia aleyrodis* against other scale insects (Rolfs and Fawcett, 1908). This led to commercialization of the fungi by some private companies and, in the opinion of some, saved the Florida citrus industry in the 1930s (Berger, 1942). However, not everyone agreed, and subsequently, the initiative died (Fawcett, 1944) and has not been revived. Interestingly, Bedford (1954) reported a similar initiative with *C. flammea* against California red scale in South Africa in the early 1900s but concluded that field applications failed to produce epizootics. In Georgia, *Hypocrella* spp. were apparently used for biological control of Japanese citrus scale (*Lopholeucaspis japonica*) (Tabatadze and Yasnosh, 1999). Despite these efforts, there appear to be no recent or current cases of the applied use of fungi for control of armored scale in citrus (Dolinski and Lacey, 2007).

19.2.2 Soft Scales, Wax Scales, and Mealybugs

Fungi-attacking soft scales (Table 19.1) differ from those reported to attack armored scales, with the exception of a *Myiophagus* sp. attacking Caribbean black scale (*Saisettia neglecta*) (McCoy et al., 2009). In South Africa, Samways and Grech (1986) applied *Cladosporium oxysporum* at 1.5×10^8 spores/mL to the green soft scale (*Pulvinaria aethiopica*), among other hemipteran pests, in laboratory bioassays. They claimed that death and hyphal growth resulted but did not quantify these, as these trials were designed to simply test host range, as a precursor to field trials.

There appears to be a dearth of information on EPF attacking mealybugs on citrus (Table 19.1), possibly indicative of their lack of potential against pseudococcids in this environment. In laboratory trials, *C. oxysporum*, at 1.5×10^8 spores/mL, killed citrus mealybug and longtailed mealybug (*Pseudococcus longispinus*) (Samways and Grech, 1986). Chartier-Fitzgerald et al. (2016) had similar success with a *Metarhizium anisopliae* isolate and a *Beauveria bassiana* isolate, with estimated LC_{50} values against the crawlers of 5.29×10^5 and 4.25×10^6 conidia/mL, respectively.

As was the case with armored scales, no record could be found of the applied use or commercialization of fungi for control of soft scales, wax scales, or mealybugs on citrus. However, several years ago, it was reported that *Colletorichum gloeosporioides* was under development as a

TABLE 19.1 Entomopathogenic Fungi Recorded Attacking Citrus Pests

Pest Species	Fungal Species	Region	References
California red scale (*Aonidiella aurantii*)	*Myiophagus ucrainicus*	Florida, USA	McCoy et al. (2009)
		Australia	Dao et al. (2015)
	Microcera coccophila	Australia	Dao et al. (2015)
	Hirsutella besseyi		
	Cosmospora aurantiicola	South Africa	Annecke (1963)
	Fusarium coccinellum		
	Fusarium lateritium	South Africa	Catling (1971)
	Podonectria sp.		
Various (unspecified) armored scales	*Myiophagus ucrainicus*	Florida, USA	McCoy et al. (2009)
	M. coccophila		
	Hirsutella besseyi		
	Cosmospora flammea		
Louse scale (*Unaspis citri*) and Glover's scale (*Lepidosaphes gloverii*)	*M. coccophila*	Australia	Dao et al. (2015)
	Microcera larvarum		
Citrus mussel scale (*Lepidosaphes beckii*)	*C. aurantiicola*	South Africa	Annecke (1963)
	Podonectria sp.	India	Rao and Sohi (1980)
Circular purple scale (*Chrysomphalus aonidum*)	*Myiophagus* sp.	South Africa	Annecke (1963)
	Podonectria sp.	India	Rao and Sohi (1980)
	Podonectria coccicola	Taiwan	Yen and Tsai (1969)
	Pseudomicrocera henningsii		
	C. aurantiicola		
Unspecified citrus scales	*P. coccicola*	India	Marcelino (2007)
Carribean black scale (*Saisettia neglecta*)	*Myiophagus* sp.	Florida, USA	McCoy et al. (2009)
	Aschersonia cubensis	Costa Rica	Liu et al. (2006)
Citrus orthezia (*Orthezia praelonga*)	*Colletotrichum* spp.	Brazil	Robbs et al. (1991)
	Colletorichum gloeosporioides	Brazil	Cesnik and Oliveira (1993)
Hemispherical scale (*Saissetia coffeae*)	*Podonectria* sp.	India	Rao and Sohi (1980)
Black scale (*Parlatoria ziziphi*)	*Aschersonia aleyrodis*	Cuba	El-Choubassi et al. (2001)
	Aschersonia goldiana		
Rufous scale (*Selenaspidus articulates*)	*A. aleyrodis*	Brazil	Gravena et al. (1988)
Florida wax scale (*Ceroplastes floridensis*), green scale (*Coccus viridis*)	*A. cubensis*	Florida, USA	Liu et al. (2006)
Green soft scale (*Pulvinaria aethiopica*)	*Lecanicillium lecanii*	South Africa	Moore (2002)
Coffee green scale (*Coccus viridis*)	*Podonectria* sp.	India	Rao and Sohi (1980)
Various soft scales	*Aschersonia turbinata*	Florida, USA	Watson and Berger (1937)
Citrus mealybug (*Planococcus citri*)	*Cladosporium oxysporum*	South Africa	Samways and Grech (1986)
	Entomophthora fumosa	Florida, USA	Speare (1922)

Continued

TABLE 19.1 Entomopathogenic Fungi Recorded Attacking Citrus Pests—cont'd

Pest Species	Fungal Species	Region	References
White wax scale (*Ceroplastes destructor*) and Chinese wax scale (*C. sinensis*)	*L. lecanii, Fusarium* spp.	New Zealand	Lo and Chapman (1998)
Citrus whitefly (*Dialeurodes citri*)	*A. aleyrodis*	Florida, USA	Fawcett (1944)
		Texas, USA	Meyerdirk et al. (1980)
Cloudywinged whitefly (*Singhiella citrifolii*)	*A. aleyrodis*	Florida, USA	Fawcett (1944)
Citrus blackfly (*Aleurocanthus woglumi*)	*A. aleyrodis*	Cost Rica	Elizondo and Quesada (1990)
Woolly whitefly (*Aleurothrixus flocossus*)	*A. aleyrodis*	Nigeria	Umeh and Adeyemi (2011)
Various whitefly species	*A. aleyrodis*	Brazil	Alves (1998)
Giffard's whitefly (*Bemisia giffardi*)	*A. aleyrodis*	Taiwan	Yen and Tsai (1969)
Asian citrus psyllid (*Diaphorina citri*)	*Hirsutella citriformis*	Cuba, Indonesia, and Florida, USA	Rivero Aragón and Grillo Ravelo (2000), Subandiyah et al. (2000), and Hall et al. (2012)
	Isaria fumosorosea	Indonesia	
	L. lecanii	Cuba	
	Beauveria bassiana	Cuba	
Citrus measuring worm (*Ascotis selenaria reciprocaria*)	*B. bassiana*	South Africa	Schoeman (1960)
Cotton bollworm (*Helicoverpa armigera*)	*Entomophthora* spp.	Australia	Smith et al. (1997)
	Nomuraea rileyi		
False codling moth (*Thaumatotibia leucotreta*)	*B. bassiana*	South Africa	Begemann (2008)
Fuller's rose beetle (*Pantomorus cervinus*)	*B. bassiana*	Florida, USA	Gyeltshen and Hodges (2012)
	Beauveria sp. and *Metarhizium* sp.	Australia	Smith et al. (1997)
Various thrips species	Various species	Australia	Smith et al. (1997)
Various minor thrips species	Various species	Various	Butt and Brownbridge (1997)
Citrus rust mite (*Phyllocoptruta oleivora*)	*Hirsutella thompsonii*	Florida, USA	Fisher (1950) and McCoy et al. (2009)

mycoinsecticide for commercialization in Brazil (Marcelino, 2007), but no record of subsequent progress with this could be found.

19.2.3 Whiteflies and Blackflies

Around 30 species of whiteflies and blackflies have been reported as pests of citrus worldwide (Smith and Peña, 2002). However, most of them are minor pests. *Aschersonia aleyrodis* is the primary entomopathogen detected on whiteflies (Table 19.1). In addition to natural occurrence, *Aschersonia* spp. have been successfully applied against whiteflies in citrus orchards in Georgia, China, and Japan, achieving up to 90% control (Gao et al., 1985).

19.2.4 Liviids and Triozids (Psyllids)

Recently, the Asian citrus psyllid, *Diaphorina citri*, was moved from the family Psyllidae to Liviidae (Burckhardt and Ouvrard, 2012). Despite this, the common name of Asian citrus psyllid has remained in use. It is arguably the most devastating pest currently known in any citrus industry around the world due to its vectoring of HLB (Grafton-Cardwell et al., 2013). A wide range of

EPF have been reported to infect the Asian citrus psyllid (Table 19.1). Lezama-Gutiérrez et al. (2012) compared the efficacy of various strains of *M. anisopliae*, *B. bassiana*, and *Isaria fumosorosea* against the pest under field conditions, achieving between 22% and 60% control of adults and nymphs following four successive sprays at 2×10^{13} conidia/ha. Although a *B. bassiana* isolate was the most effective and *I. fumosorosea* was the least effective, most recent studies seem to have focused on the development of *I. fumosorosea*, with examples that follow, indicating that this species holds the most promise for biological control of psyllids.

In Mexico, both *B. bassiana* and *I. fumosorosea* caused 61% mortality of adult psyllids, when a 10^8 conidia/mL concentration was sprayed directly onto them (Gandarilla-Pacheco et al., 2013). However, when sprayed onto seedlings, *I. fumosorosea* was more efficacious than *B. bassiana*. Delalibera et al. (2014) reported up to 96% mortality of the Asian citrus psyllid in the field, with a maximum concentration of 5×10^6 conidia/mL of an *I. fumosorosea* isolate. A mycopesticide based on this isolate is currently in preparation for commercial registration (Delalibera et al., 2014). A novel autodissemination technique for *I. fumosorosea* is also under development (Avery et al., 2009; Patt et al., 2015), using a bright yellow color as the attractant for the device and demonstrated 55% mycosis of adults in a greenhouse trial and significant horizontal transmission.

In South Africa, *Cladosporium oxysporum* was applied against the African citrus triozid (*Trioza erytreae*) (previously the African citrus psyllid) in the field at 4×10^8 conidia/mL (Samways and Grech, 1986). Within 21 days after application, 100% mortality was recorded. However, very little hyphal growth was observed, leading to speculation that an unidentified toxin may have been more instrumental in causing death.

19.2.5 Aphids

Aphids are generally a minor pest on citrus, except on young nonbearing trees and where the aphid may vector a disease, such as the brown citrus aphid's (*Toxoptera citricida*) transmission of citrus tristeza virus. In South Africa, the same *C. oxysporum* isolate referred to for green soft scale, mealybugs. and the citrus triozid was applied against the brown citrus aphid (known in South Africa as the black citrus aphid) in the field (Samways and Grech, 1986). Results were not promising; however, Poprawski et al. (1999) had more success, recording overt mycosis to range from 23% to 78% with isolates of *I. fumosorosea*, *M. anisopliae*, and *B. bassiana* in laboratory assays. Field trials with the *B. bassiana*-based mycoinsecticde, Mycotrol ES (Mycotech, USA), achieved more than 94% control of the brown citrus

aphid. In bioassays with *Lecanicillium lecanii* against the brown citrus aphid on caged citrus seedlings in Trinidad, Balfour and Khan (2012) recorded almost 79% mortality after 12 days with a concentration of 1.49×10^9 spores/mL.

19.2.6 Fruit Flies

There are a number of studies that have investigated the efficacy of EPF (*M. anisopliae*, *B. bassiana*, and *I. fumosorosea*) in the laboratory, particularly against Mediterranean fruit fly (medfly; *Ceratitis capitata*) (Fig. 19.1). Some of these studies were targeted against soil-dwelling life stages (eg, Ekesi et al., 2002) and some were targeted against adults (eg, Dimbi et al., 2003), whereas others investigated and compared the susceptibility of pupal and adult flies (eg, Goble et al., 2011; Beris et al., 2013). All studies demonstrated potential for control of fruit flies, with better results against adults than pupae.

Despite promising laboratory results, McCoy et al. (2009) were of the opinion that it is highly unlikely that entomopathogens have potential as biological control agents of fruit flies, since regulatory programs required virtually 100% control of these pests, which fungi will not give. However, the single report of a field study that could be found appeared to contradict this expectation, producing very positive results (Navarro-Llopis et al., 2015). The efficacy of an *M. anisopliae*–based attractant-contaminant device to control *C. capitata* was evaluated in six replicates in a 40-ha orchard over a 2-year period, using a density of 24 devices per ha. Two fungus treatments per year reduced the fruit fly population by 71% and 37% relative to a reference field (several aerial and ground-based bait applications with malathion and protein hydrolysate) and a single fungus application field, respectively.

FIGURE 19.1 Mediterranean fruit fly (*Ceratitis capitata*), infected with *Metarhizium anisopliae*. *Photo courtesy of Tarryn Goble.*

19.2.7 Moths

Very few cases of fungal diseases of citrus Lepidoptera have been reported (*Lepidoptera*Table 19.1). *Beauveria bassiana* isolated from false codling moth pupae was able to induce a 97% mortality of larvae in laboratory assays (Begemann, 2008). However, field trials with this isolate were not successful, probably due to low relative humidity (Begemann, 2008). As well as investigating the effect of EPF against fruit flies, Goble et al. (2011) also tested the susceptibility of false codling moth prepupae and pupae to a range of fungal isolates from citrus orchards (Goble et al., 2010), with far more promising results than those achieved against fruit flies. Depending on fungal isolate, up to 95% mortality and 93% mycosis of pupae were recorded (Goble et al., 2011). Consequently, Coombes et al. (2016) conducted large-scale field trials in which they applied between 10^{12} and 10^{14} spores/ha, achieving more than 80% reduction in fruit infestation by false codling moth over a 32-week period after just a single application to the soil in spring. However, use of the higher rate may not be economically feasible.

19.2.8 Beetle

Probably more work has been conducted with EPF against the Diaprepes root weevil (*Diaprepes abbreviatus*) (Fig. 19.2) in Florida than with any other beetle pests of citrus. As with most citrus beetles, it is the soil-borne larval life stage that is targeted with fungi. As fungi such as *B. bassiana* and *M. anisopliae* generally offer inadequate control of the pest when applied at cost-effective rates, Quintela and McCoy (1997) investigated the use of imidacloprid (a neonicotinoid insecticide) as a synergist. At sublethal rates, the addition of imidacloprid increased larval mortality and mycosis (Quintela and McCoy, 1997). The activity of imidacloprid, in reducing larval movement and influencing

FIGURE 19.2 Diaprepes root weevil (*Diaprepes abbreviatus*), a target for control with entomopathogenic fungi and entomopathogenic nematodes. *Photo by Robin Stuart.*

the infection process, was so effective that mortality (and mycosis) was dramatically improved. When used in combination, mortality was almost 100% in bioassays, whereas on their own, mortality was always below 5%, 15%, and 35% for *B. bassiana*, *M. anisopliae*, and imidacloprid. No published reports of the combined use of imidacloprid and EPFs in the field could be found. However, McCoy et al. (2007) stated that the efficacy of Mycotrol, used as a stand-alone weevil control agent, was hindered by its poor persistence in the soil.

19.2.9 Thrips

Very little information could be found on EPF attacking thrips (Table 19.1). Zahn and Morse (2013) concluded that the commercially available GHA *B. bassiana* strain was effective against adult female citrus thrips (*Scirtothrips citri*) in laboratory bioassays (LC_{50} and LC_{95} of 8.6×10^4 and 4.8×10^6 conidia/mL, respectively) and that field trials were thus justified. Coombes et al. (2016) conducted some very preliminary bioassays against adult citrus thrips (*S. aurantii*) from South Africa and recorded 60% and 70% mortality with 10^7 conidia/mL of an *M. anisopliae* and a *B. bassiana* strain, respectively.

19.2.10 Mites

McCoy et al. (1971) applied the mycelia of *Hirsutella thompsonii* against a rust mite infestation on citrus trees, observing sporulation after about 48 h, followed by a decline in mite infestation, which only began to recover 10–14 weeks later. Consequently, this fungus became the first microbial product to be registered for use in the U.S. citrus industry and apparently in any citrus industry anywhere in the world (McCoy and Couch, 1982). This was, in 1976 under the tradename Mycar (Abbott Laboratories), formulated as a wettable powder. The product also proved effective in controlling rust mite, when applied with either oil or a wetter (McCoy and Couch, 1982). *Beauveria bassiana* was also shown in laboratory bioassays to be effective against the citrus rust mite (Alves et al., 2005). However, McCoy et al. (2009) were of the opinion that the use of fungal pathogens to control spider mites was not feasible as promising results in laboratory bioassays too often failed to translate into acceptable efficacy in the field. This may be why Mycar's survival in the market was so brief (McCoy, 1996).

Hirsutella thompsonii was also shown to be effective against the two-spotted spider mite (*Tetranychus urticae*) in laboratory bioassays, causing 96.5% mortality with unformulated conidia and up to 99% mortality with the formulated Mycar (Gardner et al., 1982). Shi et al. (2008) also demonstrated that *B. bassiana* had a significant ovicidal effect on the two-spotted spider mite, with up to 87.5%

egg mortality in laboratory bioassays. Additionally, Shi and Feng (2009) showed that in laboratory bioassays *B. bassiana*, *I. fumosorosea*, and *M. anisopliae* had a significant effect on reproductive potential of mites, reducing egg laying by 3- to 4-fold.

In China, Shi and Feng (2006) established an exciting synergy between *B. bassiana* and low rates of pyridaben for control of citrus red mite (*Panonychus citri*) in orange orchards. Two applications of around 1.5×10^{13} conidia/ha plus a low rate of pyridaben, with a 15-day spray interval, resulted in up to 92% control of citrus red mite for 35 days, which was significantly more effective and protracted than with the chemical treatment alone.

Studies with other mite species reported promising results for *B. bassiana* against broad (or silver) mite (*Polyphagotarsonemus latus*) (Peña et al., 1996) in laboratory bioassays and greenhouse trials and with various fungal isolates against the reddish-black flat mite (or false spider mite, *Brevipalpus phoenicis*) in laboratory bioassays (Rossi-Zalaf and Alves, 2006).

Although the commercial use of mycoinsecticides against mite pests on crops other than citrus is well documented (eg, Alves et al., 2002; Chandler et al., 2005; Maniania et al., 2008), other than the early use of Mycar, no further records of commercial use of fungi against mite pests on citrus could be found.

19.3 MICROBIAL CONTROL OF CITRUS PESTS USING ENTOMOPATHOGENIC NEMATODES

Entomopathogenic nematodes (EPNs) have been used in citrus IPM programs to varying degrees for nearly a quarter century (Dolinski et al., 2012; Campos-Herrera et al., 2015). The primary targets for EPN-based tactics have been false codling moth and the citrus mealybug, both in South Africa, and the Diaprepes root weevil in Florida, the Caribbean, and California. As in all cropping systems to date, nearly all of the applied research projects involving EPNs in citrus have focused on augmentation with commercially formulated EPNs. However, there is evidence that conservation biological control may be a feasible and profitable approach to pest management with EPNs in some citrus orchards (Duncan et al., 2013; Campos-Herrera et al., 2015).

19.3.1 Root Weevils

Managing root weevils in Florida (and elsewhere) has acquired new urgency because it was recently shown that HLB destroys nearly half of the citrus root system before aboveground symptoms become evident (Graham et al., 2013a). As a consequence, root herbivory by weevils is even more damaging to tree health than prior to the arrival of the disease.

The Diaprepes root weevil life cycle is asynchronous and occupies both the soil and tree canopy (Woodruff, 1985). A broad range of larval instars, pupae, and teneral adults occur in the soil simultaneously, and the continuous, season-long movement of weevil stages between the soil and canopy provides no time or location where the entire population is vulnerable to management. In the absence of soil pesticides, the only effective control tactic for subterranean larvae is application of EPNs. These nematodes function as nonpersistent biopesticides, and treatment efficacy is typically a matter of 1–2 weeks (Duncan et al., 2003, 2007). Augmentation with *Steinernema riobrave* twice annually in orchards on coarse sandy soil was shown to reduce emerging adult weevil populations by approximately half (Duncan et al., 2007, 2010). Increasing the EPNs in soil by augmentation or by cultural practices also reduced the population density of the plant pathogen *Phytophthora nicotianae*, which invades weevil-damaged roots (Duncan et al., 2010; Campos-Herrera et al., 2014).

Steinernema carpocapsae was the first nematode to be developed commercially for root weevil control (Schroeder, 1987). *Heterorhabidis bacteriophora* and *H. indica* have also been marketed for use in citrus. Shortly after its discovery, *S. riobrave* was shown to have greater efficacy to Diaprepes root weevil than the other species (Schroeder, 1994), with some field studies in Florida and California reporting from 60% to greater than 90% suppression of larvae following treatment (Duncan et al., 1996, 2007; Bender et al., 2014). Application rates in most studies were in the range of 20–100 infective juveniles (IJs)/cm^2 soil surface, applied with watering cans or pressurized spray equipment. Commercial applications were generally by injection into microsprinkler irrigation systems, twice annually at rates between 20 and 40 IJs/cm^2. Various commercial formulations of *S. riobrave* were used in citrus for more than 15 years. However, in 2011 production of formulated *S. riobrave* was discontinued due to insufficient sales, caused in part by new costs associated with HLB management (Dolinski et al., 2012). As the interaction between weevil herbivory and HLB root loss has become evident, there is an urgent need to reintroduce *S. riobrave* formulations (Graham et al., 2013b).

19.3.2 False Codling Moth

As with root weevils in Florida, EPNs are currently the only tactic commercially available in South Africa to manage false codling moth larvae in soil. Malan et al. (2011) surveyed citrus orchards throughout South Africa and isolated six EPN species, all of which were virulent to false codling moth. *Steinernema yirgalemense* and *H. zealandica* (Fig. 19.3) were the most effective, and a single soil application of *Heterorhabidis bacteriophora* (10–20 IJs/cm^2) in an orchard trial (Fig. 19.4) was recently shown to reduce false codling

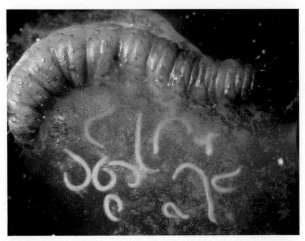

FIGURE 19.3 False codling moth (*Thaumatotibia leucotreta*) larva infected with *Heterorhabditis bacteriophora*. *Photomicrograph courtesy of Antoinette Malan.*

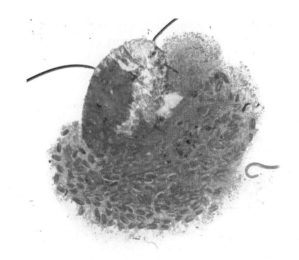

FIGURE 19.5 Citrus mealybug (*Planococcus citri*) infected with *Heterorhabditis bacteriophora*. *Photomicrograph courtesy of Sonnica van Niekerk.*

FIGURE 19.4 Application of entomopathogenic nematodes to citrus orchard soil. Entomopathogenic fungi are applied in the same manner. *Photo by Sean Moore.*

moth fruit infestation by as much as 81% (Moore et al., 2013). However, monthly application of these nematodes at just 5–10 IJs/cm^2 gave more consistent efficacy in these trials (Moore et al., 2014). The relatively shallow depth at which false codling moth inhabits the soil (Love, 2015) may facilitate EPN efficacy at such low application rates. Ongoing research is aimed at producing affordable commercial products via liquid fermentation with native isolates of *H. zealandica* and *H. bacteriophora* for use in IPM programs to manage false codling moth (Ferreira et al., 2014).

19.3.3 Citrus Mealybug

Mealybugs infest all parts of citrus trees except the root system, which would ordinarily make them unsuitable targets for biological control with EPNs. Although there are no reports yet of full-scale field trials to determine efficacy of EPNs against citrus mealybug, research indicates they hold promise (Fig. 19.5). South African isolates of

H. zealandica and *S. yirgalemense* were both highly virulent to adult mealybugs, particularly when applied with polymer-surfactant adjuvants. Use of adjuvants retarded desiccation for several hours following application of EPN suspensions to detached, infested leaves suspended in tree canopies and thereby improved the efficacy of *S. yirgalemense* against citrus mealybug by 53% (van Niekerk and Malan, 2014).

19.3.4 Conservation of Entomopathogenic Nematodes

Conservation biological control to exploit the services of subterranean natural enemies including EPNs remains primarily a matter of speculation (Barbercheck and Hoy, 2005; Stuart et al., 2008). However, there is evidence from both South Africa and Florida that cultural practices can be modified profitably to favor EPNs. Manrakhan et al. (2014) reported that the nematicide cadusafos (FMC Corporation, Philadelphia, PA, USA) suppressed the infection rate of sentinel false codling moth larvae by naturally occurring *H. zealandica* in a Navel orange orchard in South Africa. One month following treatment, the number of prematurely dropped fruit infested by false codling moth increased significantly in the plots receiving the nematicide, relative to untreated plots. Although previous surveys revealed most South African citrus orchards to be relatively depauperate of EPNs (Malan et al., 2011), the infection rate of false codling moth larvae buried in situ in this orchard was about 50%. Therefore, while management of false codling moth by EPNs requires augmentation in most situations (Moore et al., 2013), some orchards support robust EPN communities that can be profitably conserved and exploited for pest management by avoiding nematicide use during the time when fruit are susceptible to infestation.

In Florida, the natural rate of weevil larval infection by native EPNs was reported to be inversely related to the abundance of weevils in different regions (Duncan et al., 2003). The possibility that naturally occurring EPNs regulate Diaprepes root weevil spatial patterns suggests that soil properties can be modified in ways that conserve or enhance biological control of the pest. Surveys of orchards and natural areas in Florida's citrus growing regions revealed unusually rich and abundant communities of EPNs that vary regionally in community structure (*Diaprepes abbreviatus*-Campos-Herrera et al., 2013). All citrus orchards surveyed and 60% of natural areas had detectable EPN communities composed variously of nine EPN species. Soil properties that best explained the variation in EPN spatial patterns (depth to groundwater, organic matter and clay content, water-holding capacity) also affect soil water potential. Of the four major EPN species in the region, *S. diaprepesi*, *H. indica*, and *H. zealandica* dominated communities in areas with drier soils and fewer weevils, whereas *H. indica* and occasionally an undescribed *Steinernema* sp. dominated regions with poorly drained soils and abundant weevils. A 4-year field trial in a poorly drained orchard heavily infested with Diaprepes root weevil and augmented with four EPN species showed that when trees were planted in holes filled with coarse sand or in holes with the native loamy soil, the better drained, sandy sites supported greater EPN efficacy against buried sentinel weevil larvae and less emergence of adult weevils than did the loamy sites (Duncan et al., 2013). The two soil habitats created in this experiment supported EPN communities with virtually the same structure and biocontrol efficacy as those characterized in the different ecoregions surveyed by Campos-Herrera et al. (2013, 2015).

19.4 BACULOVIRUSES FOR MICROBIAL CONTROL OF LEPIDOPTERA IN CITRUS

There are very few reports of the use of viruses in citrus pests, and most reports on the use of baculoviruses are from South Africa. This is probably reflective of the relative lack of importance of lepidopteran pests in the different citrus industries around the world, as baculoviruses are predominantly isolated from and effective against Lepidoptera.

19.4.1 *Helicoverpa armigera* Nucleopolyhedrovirus

The cotton bollworm, *Helicoverpa armigera*, has historically only been considered as a pest in the southern African and Australian citrus industries. Recently, it was introduced into Brazil and was reported on citrus there (Paiva and Yamamoto, 2014). It is only a pest during spring, when blossoms and fruitlets are the primary feeding targets

FIGURE 19.6 Cotton bollworm (*Helicoverpa armigera*) feeding on a nucleopolyhedrovirus-infected *H. armigera* larval cadaver on a citrus tree. *Photo by Sean Moore.*

(Grout and Moore, 2015). In Australia, it is considered as a minor pest (Smith et al., 1997). However, in South Africa it has been known to damage over 80% of the setting fruit and reduce the total crop by over 80% (Moore et al., 2004).

Natural attack of bollworm larvae on citrus by a nucleopolyhedrovirus (NPV) and a granulovirus (GV) was reported in Australia but without any details (Smith et al., 1997). The first report of the use of the *H. armigera* NPV (HearNPV) on citrus was by Kunjeku et al. (1998), referring to an unpublished study conducted by Jones in Thailand. Very good control was demonstrated on tangerines, with some trials showing twice the yield in virus-treated plots. However, the details of this work were not disseminated. Subsequently, Moore et al. (2004) and Moore and Kirkman (2010) conducted a comprehensive range of field trials against bollworm on Navel oranges in South Africa. Moore et al. (2004) demonstrated that concentrations as low as 7.3×10^5 OBs/mL resulted in 100% reduction in bollworm infestation within 14 days (Fig. 19.6). Bollworm damage to fruit was reduced by up to 84%, yield was increased by up to 99% and rejection for export was reduced by up to 96%. Subsequently, a HearNPV product, Helicovir (River Bioscience, South Africa), was registered for use against bollworm on citrus (among other crops) in South Africa (Moore and Kirkman, 2010).

19.4.2 *Cryptophlebia leucotreta* Granulovirus

Fritsch (1988) was the first to report a field trial with the *Cryptophlebia leucotreta* granulovirus (CrleGV) against false codling moth. This was a small-scale trial in citrus on the Cape Verde Islands. Concentrations of 10^8 and 10^9 OBs/mL were formulated with skimmed milk powder and a wetting agent and applied at 4 L per tree (per ha rate not reported), resulting in reduction of false codling moth damage by 77%. In the last 15 years, extensive work on CrleGV has been conducted in South Africa, initially in the laboratory but subsequently in the field too. Dose-response and time-response bioassays conducted on artificial diet and on detached fruit demonstrated the virus to be highly virulent against false codling moth neonate larvae (Moore et al., 2011). Consequently, extensive field trials were conducted, using between 9×10^{12} and 1.2×10^{15} OBs/ha (Moore et al., 2015), leading to registration of the biopesticide Cryptogran recommended at a rate of 5×10^{13} OBs/ha (River Bioscience) (Moore et al., 2011). Two CrleGV products, Cryptex and Gratham (both Andermatt Biocontrol, Switzerland), have also been registered for use against false codling moth in South Africa.

Moore et al. (2015) reported that in excess of 50 field trials had been conducted with CrleGV against false codling moth on citrus in South Africa over a 13-year period (with 10 years of commercial use). In a representative sample of 13 field trials, using the application rates mentioned, efficacy as measured by reduction in larval infestation of fruit ranged between 30% and 92% and was shown to persist at a level of 70% for up to 17 weeks after application of CrleGV (Moore et al., 2015).

19.4.3 Viruses of Other Pests

There are several other minor reports of baculovirus infections recorded against citrus pests. For example, an unidentified GV, reported to be killing citrus leafroller (*Christoneura occidentalis*) larvae on citrus estates in Swaziland, was isolated and morphologically described (Smith et al., 1990). Swart et al. (1975) reported, without quoting any evidence, that bacterial or viral diseases often kill larvae of the fruit-piercing moth, *Serrodes partita*. Smith et al. (1997) reported that on citrus in Australia, the light brown apple moth (*Epiphyas postvittana*) can be affected by an NPV but provided no details. However, this was probably the *Epiphyas postvittana* NPV (EppoNPV), which was previously reported in Australia by MacCollum and Reed (1971).

Only one case of a nonoccluded virus infecting a citrus pest could be found in the literature, namely the citrus red mite. This was first discovered affecting populations of red mite on citrus in Florida (Muma, 1955) but surprisingly has not been found there subsequently (McCoy et al., 2009). However, the virus was observed in California in 1958 and until today is regarded as the cause of epizootics of red mite in citrus in California and Arizona (McCoy et al., 2009). Shaw et al. (1968) cultured the virus in vivo in a laboratory culture of red mite. They applied the virus in the field, as a macerated suspension of infected mites, and reported satisfactory results but achieved better results by harvesting and reapplying field-collected infected mites.

19.5 BACTERIA FOR MICROBIAL CONTROL OF CITRUS PESTS

One of the most commercially successful entomopathogens, *Bacillus thuringiensis*, remains the only bacterial species widely tested or applied for use against citrus pests, especially lepidopterans (Lacey and Shapiro-Ilan, 2008). However, *Serratia entomophila* was isolated from the citrus leafminer, *Phyllocnistis citrella,* in Peru (Meca et al., 2009) and *Bacillus subtilis* and *B. mycoides* were isolated from citrus leaf miner in Algeria (Saiah et al., 2011). A few others have been studied only in the context of their natural history in association with citrus pests (El-Borai et al., 2005; Ruiu et al., 2013).

19.5.1 *Bacillus thuringiensis*

Bacillus thuringiensis is remarkably ubiquitous and variable in most habitats. In a survey of Spanish citrus soils, Vidal-Quist et al. (2009) recovered 376 isolates from 150 environmental samples. Although *B. thuringiensis* is not widely used in citrus pest management, it has potential against a wide range of insect taxa, particularly within the Lepidoptera. This is especially true when protection of natural enemies and avoidance of pesticide resistance are concerns.

Most lepidopteran pests of citrus have been evaluated for susceptibility to *B. thuringiensis*. Although *B. thuringiensis* controlled cotton bollworm, HearNPV provided superior efficacy (Moore et al., 2004). When the pest status of the citrus flower moth (or lemon borer moth), *Prays citri*, suddenly escalated in South Africa, it was shown to be amenable for control by *B. thuringiensis* (Moore and Kirkman, 2014). Commercially formulated *B. thuringiensis* similarly reduced citrus flower moth by 90% during 2 years in field trials in Egypt and was recommended as a suitable replacement for synthetic insecticides (Shetata and Nasr, 1998). Li and Bouwer (2012) recently screened a variety of *B. thuringiensis* Cry proteins against false codling moth larvae and reported the potential utility of several for future use. Lime swallowtail larvae (*Papilio demoleus*) on citrus in India were reduced by more than 90%, comparable to chemical insecticide standards, when treated with *B. thuringiensis* (Dipel) at the rate of 1 kg/ha (Gopalakrishnan and Gangavisalakshy, 2005). "Orange dog" larvae of the giant swallowtail butterfly (*Papilio cresphontes*) are occasional problems on young trees in Florida and are readily controlled by formulated *B. thuringiensis*. Similarly, *B. thurigiensis* is registered in South Africa for use not only against swallowtail butterflies (*P. demodocus* and *P. nireus lyaeus*) on citrus but also against the citrus leafroller and the apple leafroller (*Lozotaenia capensana*).

Although citrus leaf miners are afforded some protection from nonsystemic insecticidal sprays by the leaf cuticle, Dias et al. (2005) achieved 27–57% mortality of citrus leaf miner with *B. thuringiensis* in field trials in the Azores, applied at between 2.5×10^7 and 2.5×10^9 spores/mL applied once at 1 L per tree. Certain Jordanian strains of the bacterium and reference strains caused additive effects against citrus leafminer when combined with spray oil in laboratory trials, and the lethal concentration of a pyrethroid insecticide was reduced by half in combination with *B. thuringiensis kurstaki* (Khyami-Horani and Ateyyat, 2002). No field trials were reported.

Many dipteran species are also susceptible to *B. thuringiensis*, although control tactics remain undeveloped. Preliminary screening of *B. thuringiensis* activity against medfly has provided a foundation to develop formulations to relieve the selection pressure against currently used pesticides (Vidal-Quist et al., 2009). Isolated toxins, from what appeared to be *B. thuringiensis* subsp. *kurstaki* and *B. thuringiensis* subsp. *israelensis*, based on sodium dodecyl sulfate–polyacrylamide gel electrophoresis protein profiles, were generally tested on adult flies, since larvae are inaccessible when in the fruit. Unfortunately, efficacy reported to date against tephritids tends to be low compared to other dipterans such as mosquitoes (Vidal-Quist et al., 2009).

Diaprepes root weevil larvae were fed spores and *Diaprepes abbreviatus* δ-endotoxins isolated from 19 isolates of coleopteran-active *B. thuringiensis* (Weathersbee et al., 2006). Several isolates expressing CryET33 and Cry ET34 were identified as especially lethal to the larvae. Despite the proposal by Weathersbee et al. (2006) that transformation of citrus rootstocks to express δ-endotoxins is the most promising approach to exploit *B. thuringiensis* and development of citrus germplasm shown to express a chimeric δ-endotoxin with coleopteran activity (Rhim et al., 2004), no transformed plants have yet been tested for tolerance to the pest.

19.6 PERSPECTIVES ON THE FUTURE

Reviews of the prospects for the future of microbial control are not always optimistic (Lord, 2005); however, the global citrus industry has a very rich and long legacy of biological control and IPM (McCoy et al., 2009; Niu et al., 2014; Grout and Moore, 2015). Moreover, regulatory and market pressure on industries that supply discerning western markets are increasing (Grout and Moore, 2015). This is having a strong influence in compelling, rather than incentivizing, the citrus industry and its suppliers of insecticides toward investigating nonchemical pest control options, including microbial control. Evidence of this is the purchasing of biocontrol companies by many of the multinational agricultural chemical companies in the past few years (Moore, 2015).

Nevertheless, as discerning and demanding as sophisticated markets have become for reduced chemical pesticide use (Glare et al., 2012), so too, have they become more demanding of cosmetically superior fruit and for adherence to strict regulations aimed to prevent importation of phytosanitary organisms, leading to a risk-aversion approach by farmers (Chandler et al., 2011). These factors mean more intensive pest management and often increased pesticide application. Similarly, a lack of sustainable tactics to manage HLB currently forces producers in Brazil and Florida to rely heavily on insecticides to manage the disease vector (Qureshi et al., 2014). Microbial products will therefore face the challenge of having to prove similar efficacy to currently available chemical insecticides. Innovative research will need to be intensified to achieve this through integration with other modes of control (Lacey et al., 2015). This could be through simple integration with other technologies such as semiochemicals, soft chemical pesticides, other natural enemies, and resistant plants (Lacey et al., 2001) or through synergism, either with chemicals (eg, Quintela and McCoy, 1997; Shi and Feng, 2006) or with other microbials (eg, Thurston et al., 1993; Koppenhöfer et al., 1999). Glare et al. (2012) stated that new technological advances will make the most difference in improvement of activity spectra, delivery options, persistence of pesticidal effect, and implementation strategies to ensure improved uptake. Use of pathogen genes, especially toxins, will also be very important (Lord, 2005). However, the problem of HLB and its vector will need to be solved; otherwise, the debate of microbial control and IPM as a whole will become superfluous as the viability of the citrus industry in a number of major production regions in the world will become severely compromised.

Conservation of aboveground natural enemies has been a mainstay of citrus IPM since vedalia beetles were released in California orchards in the late 19th century (Browning, 1999; McCoy et al., 2009). The associations between pesticide use and numerous secondary pests in the citrus canopy are clearly evident (Browning, 1999). By contrast, nontarget effects of cultural and pest management tactics on the pests and diseases of roots are not directly observable and remain unknown for most parts. Given the important information generated by a few studies using molecular probes to study belowground community dynamics (Campos-Herrera et al., 2013, 2015; Pathak et al., 2012), there is rich potential for progress in subterranean conservation tactics, as these research methods are more widely used. There was little awareness of the importance of native EPFs and EPNs in regulating pests until recently (Meyling and Eilenberg, 2007; Pell et al., 2010; Manrakhan et al., 2014; Campos-Herrera et al., 2015). Moreover, as optimal habitat conditions for native natural enemies are identified and exploited, biocontrol companies and their customers can also profit by improving the soil conditions prior to augmentation or by selecting species best adapted to local conditions (Meyling and Eilenberg, 2007; Stuart et al., 2008). This will benefit control of not only root pests but also arboreal pests that pupate in the soil.

Many of the needs experienced by the citrus industries worldwide for improved efficacy and accessibility of microbial control products are shared by other agricultural sectors and, hence, the chapters in this book dedicated to the specific groups of microbes (particularly fungi, nematodes, viruses, and bacteria) can be consulted.

REFERENCES

Alves, S.B., 1998. Fungos entomopatogênicos. In: Alves, S.B. (Ed.), Controle Microbiano de Insetos, Fundação de Estudos Agrários Luiz de Queiroz, pp. 289–381 Piracicaba, Brasil.

Alves, S.B., Pereira, R.M., Lopes, R.B., Tamai, M.A., 2002. Use of entomopathogenic fungi in Latin America. In: Upadhyay, R.K. (Ed.), Advances in Microbial Control of Insect Pests. Kluwer Academic/Plenum Publishers, New York, USA, pp. 193–211.

Alves, S.B., Tamai, M.A., Rossi, L.S., Castiglioni, E., 2005. *Beauveria bassiana* pathogenicity to the citrus rust mite *Phyllocoptruta oleivora*. Exp. Appl. Acarol. 37, 117–122.

Annecke, D.P., 1963. Observations on some citrus pests in Moçambique and Southern Rhodesia. J. Entomol. Soc. S. Afr. 26, 194–225.

Avery, P.B., Hunter, W.B., Hall, D.G., Jackson, M.A., Powell, C.A., Rogers, M.E., 2009. *Diaphorina citri* (Hemiptera: Psyllidae) infection and dissemination of the entomopathogenic fungus *Isaria fumosorosea* (Hypocreales: Cordycipitaceae) under laboratory conditions. Fla. Entomol. 92, 608–618.

Balfour, A., Khan, A., 2012. Effects of *Verticillium lecanii* (Zimm.) Viegas on *Toxoptera citricida* Kirkaldy (Homoptera: Aphididae) and its parasitoid *Lysiphlebus testaceipes* Cresson (Hymenoptera: Braconidae). Plant Prot. Sci. 48, 123–130.

Barbercheck, M.E., Hoy, C.W., 2005. A systems approach to conservation of nematodes. In: Grewal, P.S., Ehlers, R.-U., Shapiro-Ilan, D.I. (Eds.), Nematodes as Biocontrol Agents. CABI Publishing, Wallingford, Oxon, UK, pp. 331–347.

Bedford, E.C.G., 1954. Summary of the Biological Control of Insect Pests in South Africa 1895–1953. Department of Agriculture Technical Services, Pretoria. Special Report No. 2/53–54, 32 pp.

Bedford, E.C.G., Van den Berg, M.A., De Villiers, E.A., 1998. Citrus Pests in the Republic of South Africa, second ed. Dynamic Ad, South Africa (revised) 288 pp.

Begemann, G.J., 2008. The mortality of *Thaumatotibia leucotreta* (Meyrick) final instar larvae exposed to the entomopathogenic fungus *Beauveria bassiana* (Balsamo) Vuillemin. Afr. Entomol. 16, 306–308.

Bender, G.S., Bates, L.M., Bethke, J.A., Lewis, E., Tanizaki, G., Morse, J.G., Godfrey, K.E., 2014. Evaluation of insecticides, entomopathogenic nematodes, and physical soil barriers for control of *Diaprepes abbreviatus* (Coleoptera: Curculionidae) in citrus. J. Econ. Entomol. 107, 2137–2146.

Berger, E.W., 1942. Status of the friendly fungus parasites of armored scale-insects. Fla. Entomol. 26–29.

Beris, E.I., Papachristos, D.P., Fytrou, A., Antonatos, S.A., Kontodimas, D.C., 2013. Pathogenicity of three entomopathogenic fungi on pupae and adults of the Mediterranean fruit fly, *Ceratitis capitata* (Diptera: Tephritidae). J. Pest Sci. 86, 275–284.

Browning, H.W., 1999. Arthropod pests of fruit and foliage. In: Timmer, L.W., Duncan, L.W. (Eds.), Citrus Health Management. APS Press, St. Paul, Minnesota, USA, pp. 116–123.

Browning, H.W., Calvert, D.V., McGovern, R.J., Jackson, L.K., Wardowski, W.F., 1995. Florida Citrus Diagnostic Guide. Florida Science Source, Inc., Lake Alfred, Florida, USA.

Burckhardt, D., Ouvrard, D., 2012. A revised classification of the jumping plant-lice (Hemiptera: Psylloidea). Zootaxa 3509, 1–34.

Butt, T.M., Brownbridge, M., 1997. Fungal pathogens of thrips. In: Lewis, T. (Ed.), Thrips as Crop Pests. CABI Publishing, Wallingford, UK, pp. 399–433.

Campanhola, C., José de Moraes, G., De Sá, L.A.N., 1995. Review of IPM in South America. In: Mengech, A.N., Saxena, K.N., Gopalan, H.N.B. (Eds.), Integrated Pest Management in the Tropics: Current Status and Future Prospects. John Wiley & Sons Ltd., Chichester, England, pp. 121–151.

Campos-Herrera, R., El-Borai, F.E., Ebert, T.E., Schumann, A., Duncan, L.W., 2014. Management to control citrus greening alters the soil food web and severity of a pest-disease complex. Biol. Control 76, 41–51.

Campos-Herrera, R., El-Borai, F.E., Duncan, L.W., 2015. It takes a village: entomopathogenic nematode community structure and conservation biological control in Florida (US) orchards. In: Campos Herrera, R. (Ed.), Nematode Pathogenesis of Insects and Other Pests. Springer, New York, USA, pp. 329–351.

Campos-Herrera, R., Pathak, E., El-Borai, F.E., Stuart, R.J., Gutiérrez, C., Rodríguez-Martín, J.A., Graham, J.H., Duncan, L.W., 2013. Geospatial patterns of soil properties and the biological control potential of entomopathogenic nematodes in Florida citrus groves. Soil Biol. Biochem 66, 163–174.

Catling, H.D., 1971. Studies on the citrus red scale, *Aonidiella aurantii* (Mask.), and its biological control in Swaziland. J. Entomol. Soc. S. Afr. 34, 393–411.

Cesnik, R., Oliveira, G., 1993. *Colletotrichum gloeosporioides*, isolado de *Orthezia praelonga* causando patogenicidade em *Coccoloba* sp. In: Congresso Paulista de Fitopatologia, Campinas. Resumos, vol. 19. Summa Phytopathologica, Jaguariúna, p. 37.

Chandler, D., Davidson, G., Jacobson, R.J., 2005. Laboratory and glasshouse evaluation of entomopathogenic fungi against the two-spotted spider mite, *Tetranychus urticae* (Acari: Tetranychidae), on tomato, *Lycopersicon esculentum*. Biocontrol Sci. Technol. 15, 37–54.

Chandler, D., Bailey, A.S., Tatchell, G.M., Davidson, G., Greaves, J., Grant, W.P., 2011. The development, regulation and use of biopesticides for integrated pest management. Philos. Trans. R. Soc. Lond. B Biol. Sci 366, 1987–1998.

Chartier-Fitzgerald, V., Dames, J., Moore, S.D., Hill, M.P., 2016. Screening of entomopathogenic fungi against citrus mealybug *Plannococcus citri* (Hemiptera: Pseudococcidae). Afr. Entomol. (in press).

Coombes, C.A., Chartier-Fizgerald, V., Wiblin, D., Dames, J., Hill, M.P., Moore, S.D., 2016. The role of entomopathogenic fungi in the control of citrus pests in South Africa: cause for optimism. IOBC-WPRS Bull 113, 25–29.

Dao, H.T., Beattie, G.A.C., Rossman, A.Y., Burgess, L.W., Holford, P., 2015. Systematics and biology of two species of *Microcera* associated with armoured scales on citrus in Australia. Mycol. Prog. 14, 1–14.

Delalibera Jr., I., D'Alessandro, C.P., Conceschi, M.R., Ausique, J.J.S., 2014. A new mycopesticide developed especially for the control of the citrus greening vector *Diaphorina citri* (Hemiptera: Liviidae). In: Proceedings of the 47th Annual Meeting of the Society for Invertebrate Pathology and International Congress on Invertebrate Pathology and Microbial Control, 7–14 August 2014, Mainz, Germany, p. 44.

Dias, D., Garcia, P., Simões, N., Oliveira, L., 2005. Efficacy of *Bacillus thuringiensis* against *Phyllocnistis citrella* (Lepidoptera: Phyllocnistidae). J. Econ. Entomol. 98, 1880–1883.

Dimbi, S., Maniania, N.K., Lux, S.A., Ekesi, S., Mueke, J.K., 2003. Pathogenicity of *Metarhizium anisopliae* (Metsch.) Sorokin and *Beauveria bassiana* (Balsamo) Vuillemin, to three adult fruit fly species: *Ceratitis capitata* (Weidemann), *C. rosa* var. *fasciventris* Karsch and *C. cosyra* (Walker) (Diptera: Tephritidae). Mycopathologia 156, 375–382.

Dolinski, C., Lacey, L.A., 2007. Microbial control of arthropod pests of tropical tree fruits. Neotrop. Entomol. 36, 161–179.

Dolinski, C., Choo, H.Y., Duncan, L.W., 2012. Grower acceptance of entomopathogenic nematodes: case studies on three continents. J. Nematol. 44, 226–235.

Dreistadt, S.H. (University of California Integrated Pest Management Program), 2012. Integrated Pest Management for Citrus, vol. 3303. UCANR Publications, Richmond, USA. 270 pp.

Duncan, L.W., McCoy, C.W., Terranova, A.C., 1996. Estimating sample size and persistence of entomogenous nematodes in sandy soils and their efficacy against the larvae of *Diaprepes abbreviatus* in Florida. J. Nematol. 28, 56–67.

Duncan, L.W., Graham, J.H., Dunn, D.C., Zellers, J., McCoy, C.W., Nguyen, K., 2003. Incidence of endemic entomopathogenic nematodes following application of *Steinernema riobrave* for control of *Diaprepes abbreviatus*. J. Nematol. 35, 178–186.

Duncan, L.W., Graham, J.H., Zellers, J., Bright, D., Dunn, D.C., El-Borai, F.E., 2007. Food web responses to augmentation biological control using entomopathogenic nematodes in bare and composted-manure amended soil. J. Nematol. 39, 176–189.

Duncan, L.W., Dewdney, M., Graham, J.H., 2010. Remember *Diaprepes*?—It's still a problem. Citrus Ind. 91, 10–14.

Duncan, L.W., Stuart, R.J., El-Borai, F.E., Campos-Herrera, R., Pathak, E., Graham, J.H., 2013. Modifying orchard planting sites conserves entomopathogenic nematodes, reduces weevil herbivory and increases citrus tree growth, survival and fruit yield. Biol. Control 64, 26–36.

Ekesi, S., Maniania, N.K., Lux, S.A., 2002. Mortality in three African tephritid fruit fly puparia and adults caused by the entomopathogenic fungi, *Metarhizium anisopliae* and *Beauveria bassiana*. Biocontrol Sci. Technol. 12, 7–17.

El-Borai, F.E., Duncan, L.W., Preston, J.F., 2005. Bionomics of a phoretic association between *Paenibacillus* sp. and the entomopathogenic nematode *Steinernema diaprepesi*. J. Nematol. 37, 18–25.

El-Choubassi, W., Iparraguirre-Cruz, M.A., Sisne-Luis, M.L., Grillo-Ravelo, H., 2001. Incidencia del genero *Aschersonia* sobre la población *Parlatoria ziziphi* (Lucas) (Homoptera: Diaspididae) en naranjo Valencia (*Citrus sinensis*) de la provincia de Ciego de Avila. Cent. Agric. 28, 42–45.

Elizondo, J.M., Quesada, J.R., 1990. Identificacincia (*Citrus sinensis*) de la provincia es de la mosca prieta de los citricos *Aleurocanthus woglumi* Ashby (Hompotera: Aleyrodidae) en cuatro zonas citricolas de Costa Rica. Turrialba 40, 190–197.

Evans, H.C., Prior, C., 1990. Entomopathogenic fungi. In: Rosen, D. (Ed.), Armoured Scale Insects: Their Biology, Natural Enemies and Control, vol. B. Elsevier Press, Amsterdam, pp. 3–17.

Fawcett, H.S., 1944. Fungus and bacterial diseases of insects as factors in biological control. Bot. Rev. 10, 327–348.

Ferreira, T., Addison, M.F., Malan, A.P., 2014. In vitro liquid culture of a South African isolate of *Heterorhabditis zealandica* for the control of insect pests. Afr. Entomol. 22, 80–92.

Fisher, F.E., 1950. Two new species of *Hirsutella* Patouillard. Mycologia 42, 290–297.

Franco, J.C., García-Marí, F., Ramos, A.P., Besri, M., 2006. Survey on the situation of citrus pest management in Mediterranean countries. IOBC-WPRS Bull. 29, 335–346.

Fritsch, E., 1988. Biologische bekämpfung des falschen apfelwicklers, *Cryptophlebia leucotreta* (Meyrick) (Lep., Tortricidae), mit granuloseviren. Mitt. Dtsch. Ges. Allg. Angew. Ent. 6, 280–283.

Futch, H.S., 2011. Identification of Mites, Insects, Diseases, Nutritional Symptoms and Disorders on Citrus. University of Florida, Gainesville, USA. 152 pp.

Gandarilla-Pacheco, F.L., Galán-Wong, L.J., López-Arroyo, J.I., Rodríguez-Guerra, R., Quintero-Zapata, I., 2013. Optimization of pathogenicity tests for selection of native isolates of entomopathogenic fungi isolated from citrus growing areas of México on adults of *Diaphorina citri* Kuwayama (Hemiptera: Liviidae). Fla. Entomol. 96, 187–195.

Gao, R.X., Ouyang, Z., Gao, Z.X., Zheng, J.X., 1985. A preliminary report on the application of *Aschersonia aleyrodis* for the control of citrus whitefly. Chin. J. Biol. Control 1, 45–46.

Gardner, W.A., Oetting, R.D., Storey, G.K., 1982. Susceptibility of the two-spotted spider mite, *Tetranychus urticae* Koch, to the fungal pathogen *Hirsutella thompsonii* Fisher. Fla. Entomol. 65, 458–465.

Glare, T.R., Caradus, J., Gelernter, W., Jackson, T., Keyhani, N., Kohl, J., Marrone, P., Morin, L., Stewart, A., 2012. Have biopesticides come of age? Trends Biotechnol. 30, 250–258.

Goble, T.A., Dames, J.F., Hill, M.P., Moore, S.D., 2010. The effects of farming system, habitat type and bait type on the isolation of entomopathogenic fungi from citrus soils in the Eastern Cape Province, South Africa. Biocontrol 55, 399–412.

Goble, T.A., Dames, J.F., Hill, M.P., Moore, S.D., 2011. Investigation of native isolates of entomopathogenic fungi for the biological control of three citrus pests. Biocontrol Sci. Technol. 21, 1193–1211.

Gopalakrishnan, C., Gangavisalakshy, P.N., 2005. Field efficacy of commercial formulations of *Bacillus thuringiensis* var. kurstaki against *Papilio demoleus* L. on citrus. Entomon 30, 93–95.

Grafton-Cardwell, E.E., Stelinski, L.L., Stansly, P.A., 2013. Biology and management of Asian citrus psyllid, vector of the huanglongbing pathogens. Annu. Rev. Entomol. 58, 413–432.

Graham, J.H., Johnson, E.G., Gottwald, T., Irey, M., 2013a. Pre-symptomatic fibrous root decline in citrus trees caused by huanglongbing and potential interaction with *Phytophthora* spp. Plant Dis. 97, 1195–1199.

Graham, J.H., Johnson, E.G., Gottwald, T., Irey, M., 2013b. Integrated management of root health. Citrus Ind. 94 (2), 12–14.

Gravena, S., Leão-Neto, R.R., Moretti, F.C., Tozatti, G., 1988. Eficiência de inseticidas sobre *Selenaspidus articulates* (Morgan) (Homoptera, Diaspididae) e efeito sobre inimigos naturaes em pomar citrico. Científica 16, 209–217.

Grout, T.G., Moore, S.D., 2015. Citrus. In: Prinsloo, G.L., Uys, G.M. (Eds.), Insects of Cultivated Plants and Natural Pastures in Southern Africa. Entomological Society of Southern Africa, Pretoria, South Africa, pp. 447–501.

Gyeltshen, J., Hodges, A., 2012. Fuller Rose Beetle, *Naupactus godmanni* (Crotch) (Insecta: Coleoptera: Curculionidae). http://entomology.ifas.ufl.edu/creatures.

Halbert, S.E., 2005. Pest Alert: Citrus Greening/Huanglongbing. Florida Department of Agriculture and Consumer Services. Division of Plant Industry. Available: http://www.freshfromflorida.com/Divisions-Offices/Plant-Industry/Plant-Industry-Publications/Pest-Alerts/Pest-Alerts-Citrus-Greening.

Hall, D.G., Hentz, M.G., Meyer, J.M., Kriss, A.B., Gottwald, T.R., Boucias, D.G., 2012. Observations on the entomopathogenic fungus *Hirsutella citriformis* attacking adult *Diaphorina citri* (Hemiptera: Psyllidae) in a managed citrus grove. Biocontrol 57, 663–675.

Hodges, A.W., Spreen, T.H., 2012. Economic Impacts of Citrus Greening (HLB) in Florida. 2006/07–2010/11. FE 903 http://edis.ifas.ufl.edu/pdffiles/FE/FE90300.pdf.

Jackson, L.K., 1999. Citrus cultivation. In: Timmer, L.W., Duncan, L.W. (Eds.), Citrus Health Management. APS Press, St. Paul, Minnesota, USA, pp. 17–20.

Khyami-Horani, H., Ateyyat, M., 2002. Efficacy of Jordanian isolates of *Bacillus thuringiensis* against the citrus leafminer, *Phyllocnistis citrella* Stainton (Lepidoptera: Gracillariidae). Int. J. Pest Manag. 48, 297–300.

Koppenhöfer, A.M., Choo, H.Y., Kaya, H.K., Lee, D.W., Gelernter, W.D., 1999. Increased field and greenhouse efficacy against scarab grubs with a combination of an entomopathogenic nematode and *Bacillus thuringiensis*. Biol. Control 14, 37–44.

Kunjeku, E., Jones, K.A., Moawad, G.M., 1998. Africa, the near and Middle East. In: Hunter-Fujita, F.R., Entwistle, P.F., Evans, H.F., Crook, N.E. (Eds.), Insect Viruses and Pest Management. Wiley, Chichester, England, pp. 280–302.

Lacey, L.A., Shapiro-Ilan, D.I., 2008. Microbial control of insect pests in temperate orchard systems: potential for incorporation into IPM. Annu. Rev. Entomol 53, 121–144.

Lacey, L.A., Frutos, R., Kaya, H.K., Vail, P., 2001. Insect pathogens as biological control agents: do they have a future? Biol. Control 21, 230–248.

Lacey, L.A., Grzywacz, D., Shapiro-Ilan, D.I., Frutos, R., Brownbridge, M., Goettel, M.S., 2015. Insect pathogens as biological control agents: back to the future. J. Invertebr. Pathol. 132, 1–41.

Lezama-Gutiérrez, R., Molina-Ochoa, J., Chávez-Flores, O., Ángel-Sahagún, C.A., Skoda, S.R., Reyes-Martínez, G., Foster, J.E., 2012. Use of the entomopathogenic fungi *Metarhizium anisopliae*, *Cordyceps bassiana* and *Isaria fumosorosea* to control *Diaphorina citri* (Hemiptera: Psyllidae) in Persian lime under field conditions. Int. J. Trop. Insect Sci. 32 (01), 39–44.

Li, H., Bouwer, G., 2012. The larvicidal activity of *Bacillus thuringiensis* Cry proteins against *Thaumatotibia leucotreta* (Lepidoptera: Tortricidae). Crop Prot. 32, 47–53.

Liu, M., Chaverri, P., Hodge, K.T., 2006. A taxonomic revision of the insect biocontrol fungus *Aschersonia aleyrodis*, its allies with white stromata and their *Hypocrella* sexual states. Mycol. Res. 110, 537–554.

Lo, P.L., Chapman, R.B., 1998. The role of parasitoids and entomopathogenic fungi in mortality of third-instar and adult *Ceroplastes destructor* and *C. sinensis* (Hemiptera: Coccidae: Ceroplastinae) on citrus in New Zealand. Biocontrol Sci. Technol. 8, 573–582.

Lord, J.C., 2005. From Metchnikoff to Monsanto and beyond: the path of microbial control. J. Invertebr. Pathol. 89, 19–29.

Love, C., 2015. The Biology, Behaviour and Survival of Pupating False Codling Moth, *Thaumatotibia leucotreta* (Meyrick) (Lepidoptera: Tortricidae), a Citrus Pest in South Africa (MSc thesis). Rhodes University, Grahamstown, South Africa. 214 pp.

MacCollum, G.B., Reed, E.M., 1971. A nuclear polyhedrosis virus of the light brown apple moth, *Epiphyas postvittana*. J. Invertebr. Pathol. 18, 337–343.

Malan, A.P., Knoetze, R., Moore, S.D., 2011. Isolation and identification of entomopathogenic nematodes from citrus orchards in South Africa and their biocontrol potential against false codling moth. J. Invertebr. Pathol. 108, 115–125.

Maniania, N.K., Bugeme, D.M., Wekesa, V.W., Delalibera Jr., I., Knapp, M., 2008. Role of entomopathogenic fungi in the control of *Tetranychus evansi* and *Tetranychus urticae* (Acari: Tetranychidae), pests of horticultural crops. Exp. Appl. Acarol. 46, 259–274.

Manrakhan, A., Daneel, J.-H., Moore, S.D., 2014. The impact of naturally occurring entomopathogenic nematodes on false codling moth, *Thaumatotibia leucotreta* (Lepidoptera: Tortricidae), in citrus orchards. Biocontrol Sci. Tech. 24, 241–245.

Marcelino, J., 2007. Epizootiology and Phylogenetics of Entomopathogenic Fungi Associated with *Fiorinia externa* Ferris (Hemiptera: Diaspidiae) in the Northeastern USA (Ph.D. thesis). University of Vermont, Vermont, USA. 197 pp.

McCoy, C.W., 1996. Pathogens of eriophyid mites. In: Lindquist, E.E., Sabelis, M.M., Bruim, J. (Eds.), Eriophyid Mites: Their Biology, Natural Enemies and Control. Elsevier, Dordrecht, pp. 481–490.

McCoy, C.W., Couch, T.L., 1982. Microbial control of the citrus rust mite with the mycoacaricide, Mycar. Fla. Entomol. 116–126.

McCoy, C.W., Selhime, A.G., Kanavel, R.F., Hill, A.J., 1971. Suppression of citrus rust mite populations with application of fragmented mycelia of *Hirsutella thompsonii*. J. Invertebr. Pathol. 17, 270–276.

McCoy, C.W., Samson, R.A., Boucias, D.G., Osborne, L.S., Pena, J.E., Buss, L.J., 2009. Pathogens Infecting Insects and Mites of Citrus. LLC Friends of Microbes. Winter Park, Florida, USA, p. 193.

McCoy, C.W., Stuart, R.J., Shapiro-Ilan, D.I., Duncan, L.W., 2007. Application and evaluation of entomopathogens for citrus pest control. In: Lacey, L.A., Kaya, H.K. (Eds.), Field Manual of Techniques in Invertebrate Pathology: Application and Evaluation of Pathogens for Control of Insects and Other Invertebrate Pests, second ed. Springer Scientific Publishers, Dordrecht, pp. 567–581.

Meca, A., Sepúlveda, B., Ogoña, J.C., Grados, N., Moret, A., Morgan, M., Tume, P., 2009. In vitro pathogenicity of northern Peru native bacteria on *Phyllocnistis citrella* Stainton (Gracillariidae: Phyllocnistinae), on predator insects (*Hippodamia convergens* and *Chrisoperna externa*), on *Citrus aurantifolia* Swingle and white rats. Span. J. Agric. Res. 7, 137–145.

Meyerdirk, D.E., Kreasky, J.B., Hart, W.G., 1980. Whiteflies (Aleyrodidae) attacking citrus in southern Texas with notes on natural enemies. Can. Entomol. 112, 1253–1258.

Meyling, N.V., Eilenberg, J., 2007. Ecology of the entomopathogenic fungi *Beauveria bassiana* and *Metarhizium anisopliae* in temperate agroecosystems: potential for conservation biological control. Biol. Control 43, 145–155.

Moore, S.D., 2002. Entomopathogens and microbial control of citrus pestsin South Africa: a review. S. Afr. Fruit. J 1 (3), 30–32.

Moore, S.D., 2015. The future of microbial pesticides: fantasy or reality? In: Proceedings of the Joint 19th ESSA and 37th ZSSA Congress, p. 72 Grahamstown South Africa, 12–17 July 2015.

Moore, S.D., Kirkman, W., 2010. Helicovir: a virus for the biological control of bollworm. S. Afr. Fruit. J. 9 (4), 63–67.

Moore, S.D., Kirkman, W., 2014. The lemon borer moth = the citrus flower moth, Prays citri: its biology and control on citrus. S. Afr. Fruit. J 13, 86–91.

Moore, S.D., Pittaway, T., Bouwer, G., Fourie, J.G., 2004. Evaluation of *Helicoverpa armigera* nucleopolyhedrovirus, HearNPV, for control of *Helicoverpa armigera* Hübner (Lepidoptera: Noctuidae) on citrus in South Africa. Biocontrol Sci. Technol. 14, 239–250.

Moore, S.D., Hendry, D.A., Richards, G.I., 2011. Virulence of a South African isolate of the *Cryptophlebia leucotreta* granulovirus (CrleGV-SA) to *Thaumatotibia leucotreta* neonate larvae. Biocontrol 56, 341–352.

Moore, S., Coombes, C., Manrakhan, A., Kirkman, W., Hill, M.P., Ehlers, R.-U., Daneel, J.-H., De Waal, J., Dames, J., Malan, A.P., 2013. Subterranean control of an arboreal pest: EPNs and EPFs for FCM. IOBC-WPRS Bull. 90, 247–250.

Moore, S.D., Manrakhan, A., Gilbert, M., Kirkman, W., Daneel, J.-H., deWaal, J., Ehlers, R.-U., 2014. Conservation and augmentation of entomopathogenicnematodes on citrus for control of false codling moth. In: Proceedings of the 6th International Congress of Nematology, p. 14 Cape Town, South Africa, 4–9 May 2014.

Moore, S.D., Kirkman, W., Richards, G.I., Stephen, P., 2015. The *Cryptophlebia leucotreta* granulovirus – 10 years of commercial field use. Viruses 7, 1284–1312. http://dx.doi.org/10.3390/v7031284.

Morse, J.G., Luck, R.F., Gumpf, D.J., 1996. Citrus Pest Problems and Their Control in the Near East. FAO Plant Production and Protection Paper 135. Food and Agriculture Organization of the United Nations, Rome, Italy. 403 pp.

Muma, M.H., 1955. Factors contributing to the natural control of citrus insects and mites in Florida. J. Econ. Entomol. 48, 432–438.

Navarro-Llopis, V., Ayala, I., Sanchis, J., Primo, J., Moya, P., 2015. Field efficacy of a *Metarhizium anisopliae*-based attractant–contaminant device to control *Ceratitis capitata* (Diptera: Tephritidae). J. Econ. Entomol. 108, 1570–1578.

Niu, J.Z., Hull-Sanders, H., Zhang, Y.X., Lin, J.Z., Dou, W., Wang, J.J., 2014. Biological control of arthropod pests in citrus orchards in China. Biol. Control 68, 15–22.

Paiva, P.E.B., Yamamoto, P.T., 2014. Citrus caterpillars, with an emphasis on *Helicoverpa armigera*: a brief review. Citrus Res. Technol. 35, 11–17.

Pathak, E., El-Borai, F.E., Campos-Herrera, R., Johnson, E.G., Stuart, R.J., Graham, J.H., Duncan, L.W., 2012. Use of real-time PCR to discriminate parasitic and saprophagous behaviour by nematophagous fungi. Fungal Biol 116, 563–573.

Patt, J.M., Chow, A., Meikle, W.G., Gracia, C., Jackson, M.A., Flores, D., Sétamou, M., Dunlap, C.A., Avery, P.B., Hunter, W.B., Adamczyk, J.J., 2015. Efficacy of an autodisseminator of an entomopathogenic fungus, *Isaria fumosorosea*, to suppress Asian citrus psyllid, *Diaphorina citri*, under greenhouse conditions. Biol. Control 88, 37–45.

Pell, J.K., Hannam, J.J., Steinkraus, D.C., 2010. Conservation biological control using fungal entomopathogens. In: Roy, H.E., Vega, F.E., Goettel, M.S., Chandler, D., Pell, J.K., Wajnberg, E. (Eds.), The Ecology of Fungal Entomopathogens. Springer, Dordrecht, Netherlands, pp. 187–198.

Peña, J.E., Osborne, L.S., Duncan, R.E., 1996. Potential of fungi as biocontrol agents of *Polyphagotarsonemus latus* (Acari: Tarsonemidae). Entomophaga 41, 27–36.

Poprawski, T.J., Parker, P.E., Tsai, J.H., 1999. Laboratory and field evaluation of hyphomycete insect pathogenic fungi for control of brown citrus aphid. Environ. Entomol. 28, 315–321.

Quintela, E.D., McCoy, C.W., 1997. Pathogenicity enhancement of *Metarhizium anisopliae* and *Beauveria bassiana* to first instars of *Diaprepes abbreviatus* (Coleoptera: Curculionidae) with sublethal doses of imidacloprid. Environ. Entomol. 26, 1173–1182.

Qureshi, J.A., Kostyk, B.C., Stansly, P.A., 2014. Insecticidal suppression of Asian citrus psyllid *Diaphorina citri* (Hemiptera: Liviidae) vector of huanglongbing pathogens. PLoS One 9, e112331.

Rao, N.N.R., Sohi, H.S., 1980. Occurrence of *Podonectria coccicola* (Ellis and Everh.) Petch on citrus scale insects. Curr. Sci. 49, 359–360.

Rhim, S.L., Kim II, G., Jin, T.E., Lee, J.H., Kuo, C., Sun, S.C., Huang, L.C., 2004. Transformation of citrus with Coleopteran specific delta-endotoxin gene from *Bacillus thuringiensis* ssp. *tenebrionis*. J. Plant Biotechnol. 6, 21–24.

Rivero Aragón, A., Grillo Ravelo, H., 2000. Natural enemies of *Diaphorina citri* Kuwayama (Homoptera: Psyllidae) in the central region of Cuba. Cent. Agric. 27 (3), 87–88.

Robbs, C.F., Sá, L.A.N., Lucchini, F., Cesnik, R., Sadi, C.V.S., 1991. Utilização de estirpes do fungo *Colletotrichum gloeosporioides* no controle da *Orthezia praelonga*. In: 13 Congresso brasileiro de entomologia. Recife, Brazil, p. 245.

Rolfs, P.H., Fawcett, H.S., 1908. Fungus diseases of scale insects and white fly. Fla. Agric. Exp. Sta. Bull 94, 1–17.

Rogers, M.E., Stansly, P.A., Stelinski, L.L., 2015. Asian citrus psyllid and citrus leafminer. In: Rogers, M.E., Dewdney, M.M. (Eds.), Florida Citrus Pest Management Guide. Univ. of Fla., pp. 33–38. Coop. Ext. Serv., IFAS. SP-43.

Rossi-Zalaf, L.S., Alves, S.B., 2006. Susceptibility of *Brevipalpus phoenicis* to entomopathogenic fungi. Exp. Appl. Acarol. 40, 37–47.

Ruiu, L., Satta, A., Floris, I., 2013. Emerging entomopathogenic bacteria for insect pest management. Bull. Insectology 66, 181–186.

Saiah, F., Bendahmane, B.S., Benkada, M.Y., 2011. Isolement et identification de bactéries entomopathogènes à partir de *Phyllocnistis citrella* Stainton 1856 dans l'Ouest algérien. Entomol. Faun. 63, 121–123.

Samways, M.J., Grech, N.M., 1986. Assessment of the fungus *Cladosporium oxysporum* (Berk. and Curt.) as a potential biocontrol agent against certain Homoptera. Agric. Ecosyst. Environ. 15, 231–239.

Schoeman, O.P., 1960. The Biology, Ecology and Control of the Citrus Measuring Worm, *Ascotis Selenaria* Reciprocaria. Proefskrif, Universiteit van Pretoria.

Schroeder, W.J., 1987. Laboratory bioassays and field trials of entomogenous nematodes for control of *Diaprepes abbreviatus*. Environ. Entomol. 16, 987–989.

Schroeder, W.J., 1994. Comparison of two steinernematid species for control of the root weevil *Diaprepes abbreviatus*. J. Nematol. 26, 360–362.

Shaw, J.G., Chambers, D.L., Tashiro, H., 1968. Introducing and establishing the non inclusion virus of the citrus red mite in citrus groves. J. Econ. Entomol. 61, 1352–1355.

Shetata, W.A., Nasr, F.N., 1998. Laboratory evaluation and field application of bacterial and fungal insecticides on the citrus flower moth, *Prays citri* Miller (Lep., Hyponomeutidae) in lime orchards in Egypt. Anz. für Schädlingskd. Pflanzenschutz Umweltschutz 71, 57–60.

Shi, W.B., Feng, M.G., 2006. Field efficacy of application of *Beauveria bassiana* formulation and low rate pyridaben for sustainable control of citrus red mite *Panonychus citri* (Acari: Tetranychidae) in orchards. Biol. Control 39, 210–217.

Shi, W.B., Feng, M.G., 2009. Effect of fungal infection on reproductive potential and survival time of *Tetranychus urticae* (Acari: Tetranychidae). Exp. Appl. Acarol. 48, 229–237.

Shi, W.B., Zhang, L.L., Feng, M.G., 2008. Field trials of four formulations of *Beauveria bassiana* and *Metarhizium anisoplae* for control of cotton spider mites (Acari: Tetranychidae) in the Tarim Basin of China. Biol. Control 45, 48–55.

Smith, D., Peña, J.E., 2002. Tropical citrus pests. In: Peña, J.E., Sharp, J.L., Wysoki, M. (Eds.), Tropical Fruit Pests and Pollinators: Biology, Economic Importance, Natural Enemies and Control. CABI Publishing, Wallingford, UK, pp. 57–101.

Smith, D., Beattie, G.A.C., Broadley, R., 1997. Citrus Pests and Their Natural Enemies: IPM in Australia, p. 272 Queensland, Department of Primary Industries, Brisbane, Australia. Information Series Q, 197030.

Smith, D.H., Da Graça, J.V., Whitlock, V.H., 1990. Granulosis virus from *Cacoecia occidentalis*: isolation and morphological description of a granulosis virus of the citrus leafroller, *Cacoecia occidentalis*. J. Invertebr. Pathol. 55, 319–324.

Speare, A.T., 1922. Natural control of the citrus mealybug in Florida. US Dept. Agric. Bull. 1117, 1–18.

Stuart, R.J., El-Borai, F.E., Duncan, L.W., 2008. From augmentation to conservation of entomopathogenic nematodes: trophic cascades, habitat manipulation and enhanced biological control of *Diaprepes abbreviatus* root weevils in Florida citrus groves. J. Nematol. 40, 73–84.

Subandiyah, S., Nikoh, N., Sato, H., Wagiman, F., Tsuyumu, S., Fukatsu, T., 2000. Isolation and characterization of two entomopathogenic fungi attacking *Oiaphorina citri* (Homoptera, Psylloidea) in Indonesia. Mycoscience 41, 509–513.

Swart, P.L., De Kok, A.E., Rust, D.J., 1975. Fruit-piercing moths. Deciduous Fruit. Grow. 25, 97–102.

Tabatadze, E.S., Yasnosh, V.A., 1999. Population dynamics and biocontrol of the Japanese scale. Entomol. Bari 33, 429–434.

Teixeira, D.C., Ayers, J., Danet, L., Jagoueix-Eveillard, S., Saillard, C., Bové, J.M., 2005. First report of a huanglongbing-like disease of citrus in São Paulo State, Brazil and association of a new *Liberibacter* species, 'Candidatus Liberibacter americanus', with the disease. Plant Dis. 89, 107.

Thurston, G.S., Kaya, H.K., Burlando, T.M., Harrison, R.E., 1993. Milky disease bacterium as a stressor to increase susceptibility of scarabaeid larvae to an entomopathogenic nematode. J. Invertebr. Pathol. 61, 167–172.

Tiwari, S., Mann, R.S., Rogers, M.E., Stelinski, L.L., 2011. Insecticide resistance in field populations of Asian citrus psyllid in Florida. Pest Manag. Sci. 67, 1258–1268.

UF/IFAS Citrus Extension. Citrus Extension Entomology and Nematology. http://www.crec.ifas.ufl.edu/extension/entomology/index.shtml#.

Umeh, V.C., Adeyemi, A., 2011. Population dynamics of the woolly whitefly *Aleurothrixus floccosus* (Maskell) on sweet orange varieties in Nigeria and association of *A. floccosus* with the enotmopathogenic fungi *Aschersonia* spp. Fruit 66, 385–392.

USDA-FAS, July 2015. Citrus: World Markets and Trade. United States Department of Agriculture Foreign Agricultural Service. http://apps.fas.usda.gov/psdonline/circulars/citrus.pdf.

van Niekerk, S., Malan, A.P., 2014. Evaluating the efficacy of a polymer-surfactant formulation to improve control of the citrus mealybug, Planococcus citri (Risso) (Hemiptera: Pseudococcidae), using entomopathogenic nematodes under simulated natural conditions. Afr. Plant Prot 17, 1–8.

Vidal-Quist, J.C., Castañera, P., González-Cabrera1, J., 2009. Diversity of *Bacillus thuringiensis* strains isolated from citrus orchards in Spain and evaluation of their insecticidal activity against *Ceratitis capitata*. J. Microbiol. Biotechnol. 19, 749–759.

Watson, J.R., Berger, E.W., 1937. Citrus insects and their control. Fla. Coop. Ext. Serv. Bull. 88, 134.

Weathersbee III, A.A., Lapointe, S.L., Shatters Jr., R.G., 2006. Activity of *Bacillus thuringiensis* isolates against *Diaprepes abbreviatus* (Coleoptera: Curculionidae). Fla. Entomol. 89, 441–448.

Woodruff, R.E., 1985. Citrus weevils in Florida and the West Indies: preliminary report on systematics, biology, and distribution (Coleoptera: Curculionidae). Fla. Entomol. 68, 370–379.

Yen, D.F., Tsai, Y.T., 1969. Entomogenous fungi of citrus Homoptera in Taiwan. Plant Prot. Bull. 11, 1–10.

Zahn, D.K., Morse, J.G., 2013. Investigating alternatives to traditional insecticides: effectiveness of entomopathogenic fungi and *Bacillus thuringiensis* against citrus thrips and avocado thrips (Thysanoptera: Thripidae). J. Econ. Entomol. 106, 64–72.

Chapter 20

Microbial Control of Sugarcane Insect Pests

T. Goble[1], J. Almeida[2], D. Conlong[3,4]

[1]Cornell University, Ithaca, NY, United States; [2]Instituto Biológico, Laboratório de Controle Biológico, Campinas, Brazil; [3]South African Sugarcane Research Institute, Mount Edgecombe, South Africa; [4]University of KwaZulu-Natal, Pietermaritzburg, South Africa

20.1 INTRODUCTION

Sugarcane is cultivated on 20 million hectares in over 110 countries, mainly in emerging and developing countries, and accounts for 80% of the world sugar production while sugarbeet accounts for the remaining 20% (FAO, 2015). Brazil is the world's largest sugar producer with 739,267,000 metric tons (TMT) in 2013 and is the second largest sugar-based ethanol producer after the United States (FAO, 2015). Other major sugar producers in 2013 included India (341,200 TMT), China (125,536 TMT), Thailand (100,096 TMT), and Pakistan (63,750 TMT) (FAO, 2015). Sugar consumption is unlikely to decrease, and as ethanol and biomass production continue to increase, additional agricultural land will be converted to sugarcane production (Goebel and Sallam, 2011). This scenario has consequences in that there may be increased pest populations if they are not adequately controlled and increased biosecurity risks from incursion of invasive insect pests if proper phytosanitary measures are not in place (Goebel and Sallam, 2011). Further, the scenario suggests that indigenous insects may become pests, increasing the need for knowledge-based integrated pest management (IPM) tactics for pest regulation (Conlong and Rutherford, 2009). Many countries practice IPM using resistant sugarcane varieties, biocontrol/natural enemies (either as a combination of parasitoids, predators, and entomopathogens or each group separately), cultural control, habitat management, and mechanical control. IPM has emerged out of necessity because of the cryptic nature of lepidopteran stalk borers and scarabaeid larvae (white grubs), which make insecticide applications challenging. Insect pests represent major production constraints, and each geographic region has a distinctive fauna, mostly indigenous insects on continents, and exotic insects on islands, which feed on sugarcane as an adopted host, mainly due to its extensive cultivation (Pemberton and Williams, 1969).

Sugarcane arthropod pests can generally, for convenience be divided into three ecological categories: stalk borers; sap suckers and leaf feeders; and soil pests (Carnegie and Conlong, 1994). We review only the major pests of sugarcane, not insects that have the potential to become major pests, nor are minor or occasional pests reviewed; therefore, mites (Acari) will not be covered.

20.2 LEPIDOPTERAN STALK BORING PESTS

Stalk boring lepidopteran pests represent the most important and damaging group of insects in all sugarcane-growing countries, except for Australia and Fiji, where they are largely absent (Goebel and Sallam, 2011). Many stalk borer larvae attack the shoots, stalks, and roots of sugarcane, reducing plant biomass and sugar content, while maintaining their populations in wild grasses or sedges (Conlong and Way, 2015). Losses can be substantial: 51% yield reduction is reported for top shoot borer, *Scirpophaga excerptalis* (Crambidae), which is a major pest in India (Sallam and Allsopp, 2005), and *Eldana saccharina* (Pyralidae) causes losses of ZAR744 million (US$54 million) for South African sugarcane growers mainly because of the damage they cause as well as the requirement of early harvesting before full crop maturation (Rutherford, 2015). For this reason as well as the biosecurity risks they impose, control of stalk borers has been well documented. Sallam (2006) compiled a list of 276 records of parasitoids, predators, and pathogens of 18 key lepidopteran stalk borers in Asia and the Indian Ocean islands, highlighting an enormous amount of research on the use of natural enemies as control agents, especially parasitoids. Here, we report on research and development of entomopathogenic bacteria, viruses, fungi (EPF), and nematodes (EPN) to control important lepidopteran stalk borers of sugarcane worldwide.

Microbial Control of Insect and Mite Pests. http://dx.doi.org/10.1016/B978-0-12-803527-6.00020-2

20.2.1 *Bacillus thuringiensis* for Control of Stalk Boring Lepidoptera

Until the early 1990s, only three commercial *Bacillus thuringiensis* subspecies *kurstaki* (Btk) products—Spore-ine (first commercially available Bt product used in France in 1938), Dipel (Abbott, USA), and Thuricide (originally Thermo Trilogy, now Certis, USA)—were available to developing countries like Brazil and India, mainly from the United States and Europe (Sanahuja et al., 2011). Subsequently, more Btk products became available: Delfin (Certis, USA), Halt (Biostadt Limited, India), and Biobit (Valent, USA), which were field-tested or used augmentatively against various stalk borers in Brazil, China, and India (Table 20.1) and in some cases gave better or comparable results to chemical insecticides (Easwaramoorthy, 2009). In India, applications of 0.4% Thuricide at weekly intervals from July–October reduced *Acigona steniellus* (Crambidae) infestations and was more effective than the organophosphate Malathion and the organochloride Endrin (Easwaramoorthy, 2009). Other augmentative uses of Btk, as a wettable powder (WP, 16,000 international units/mg), diluted 1000 times, at the start of *Chilo infuscatellus* (Crambidae) borer hatching in South China are reported (Huang, 2002 cited in Zeng, 2004). Bt-based products used in combination with other substances such as molasses and corn bran reduced *Diatraea saccharalis* (Crambidae) infestations in Brazilian sugarcane by more than 50% (Gravena et al., 1980). There are no reports, however, of the natural infection or isolation of Bt strains from *Diatraea* spp. from the Americas, but various Bt strains and pure proteins have been evaluated against *D. saccharalis* in the laboratory (Hernández-Velázquez et al., 2012). There are reports indicating the ineffectiveness of Bt products against noctuid stalk borers in the field. For example, in Ramu, Papua New Guinea, Delfin was field-trialed against *Sesamia grisescens* (Noctuidae) with variable and low (<20%) field mortality, despite producing >80% mortality in laboratory bioassays (Kuniata, 2000).

The encountered environmental instability of formulations of Bt crystals and spores (West, 1984) has led to the development of cloning of insecticidal crystal protein genes and their expression in plant-associated bacteria or transgenic plants as alternative control strategies. A number of transgenic sugarcane lines have been developed with Bt genes expressing the *Cry* protein, proteinase inhibitor, or lectin resistance to lepidopteran stalk borers, hemipteran insects, or scarabaeid white grubs (Srikanth et al., 2011). Laboratory and field trials in Brazil (Braga et al., 2001) and China (LiXing et al., 2006) confirmed borer-resistant transgenic lines against the economically important species *D. saccharalis* and *Chilo sacchariphagus sacchariphagus* (Crambidae), respectively. The latter species resulted in 100% mortality within one week exposure to sugarcane

lines expressing high levels of synthetic *cry1Ac* gene proteins. Sugarcane giant lepidopteran borer *Telchin licus* (Castniidae) is a major native pest in Brazil, and transgenic sugarcane lines against this pest are showing promise (Alves et al., 2003). Further, approval for field trials using transgenic sugarcane was granted to Australia in 2009, followed by India for limited field trials (Srikanth et al., 2011). In South Africa, new control strategies for *E. saccharina* control have been developed using recombinants of the sugarcane endophyte *Gluconacetobacter diazotrophicus*, which expresses truncated Bt *cry1Ac* genes (Rapulana and Bouwer, 2013). Further, the Omegon-Km vector has been used to introduce *cry1Ac* from native Bt strains showing activity against *E. saccharina* into the chromosome of *Pseudomonas fluorescens*, which is capable of colonizing sugarcane. Glasshouse trials indicated that sugarcane treated with *P. fluorescens* carrying the Bt gene were more resistant to *E. saccharina* damage than untreated sugarcane (Herrera et al., 1997).

20.2.2 Baculovirus for Control of Stalk Borers

An endemic granulovirus (GV) obtained from *D. saccharalis* larvae (DisaGV) attacking sugarcane in the United States was subsequently introduced into Brazil (Alves, 1986), where it was applied as an experimental product, with two applications of DisaGV at 10^8 occlusion bodies (OB)/mL, 50 days after sugarcane planting (Alves, 1986). However, the program did not achieve the expected success because only 27% control of *D. saccharalis* larvae was achieved in the field (Alves, 1986). Given the large area under sugarcane cultivation at the time and the need for large spray volumes of the virus, for which production costs were too high, the program was deemed unviable (Alves, 1986). In India, the natural occurrence of two GVs that infect larvae of *C. infuscatellus* (ChinGV) (1.4–30% natural incidence in larvae) and *Chilo sacchariphagus indicus* (ChsaGV) (31.5% natural incidence in eight districts) (Crambidae) were widely distributed in Pondicherry and other agro-climatic locations in Tamil Nadu (Easwaramoorthy and Jayaraj, 1987). ChinGV has been used successfully to control *C. infuscatellus* in sugarcane in Tamil Nadu (Table 20.1); current augmentation is crude, and consists of foliar applications of ChinGV (10^7 OB/mL) in 500 L/ha water at 35 and 50 days with the sticker, Teepol 610S (Sigma Aldrich, USA) (Hunsigi, 2001). During a field trial in Tamil Nadu, two spray applications of ChinGV (10^9 OB/mL) at 30 and 45 days after planting gave equal control to Sevidol (carbaryl + lindane) 4:4G (Union Carbide Corporation, owned by Dow Chemical Co., USA) applied 30 days after planting (Patil et al., 1996). Commercial virus production in India is low but is being undertaken on a small scale by various IPM centers and state agricultural departments (Gupta and Dikshit, 2010). There are various virus isolations reported

TABLE 20.1 Examples of Some Augmented and Experimental Field Programs Using Microbial Control of Insect Pests in Sugarcane

Country	Microbial Species	Target Pest	Application	References
India	Granulovirus (ChinGV)	*Chilo infuscatellus* (Lepidoptera: Crambidae)	1 foliar spray application. of GV (10^7 OB/mL) in 500 L water, 30 days after planting (in combination with *Trichogramma chilonis* @ 50,000 ha^{-1}, applied at 45 days) showed 8.4% pest reduction by 60 days resulting in a higher number of millable stalks. Augmentation: GV (10^7 OB/mL) in 500 L water at 35 and 50 days with the sticker, Teepol (0.05%)	Rachappa et al. (2000), Hunsigi (2001), Tiwari and Tanwar (2001)
	Bacillus thuringiensis subsp. *kurstaki* (Btk)		Bt products used in two field trials in Tamil Nadu, (in order of efficacy); Delfin (Btk)> Biobit (Btk) > Halt (Btk) >Spicturin (Bt subsp. *galleriae*) increased sugar yields and commercial cane sugar (CCS) significantly compared to controls	Kesavan et al. (2003)
	Beauveria bassiana	*Chilo auricilius* (Lepidoptera: Crambidae)	Foliar spray of 2.3×10^7 conidia/mL in 750–1000 mL/ha, controls carry-over populations	Naidu (2009), Tiwari and Tanwar (2001)
		Cavelerius sweeti (Hemiptera: Lygaeidae)	Inoculative releases of fungal-infected adults @ 5000 adults/ha, to control carry-over populations from November to December	Naidu (2009), Tiwari and Tanwar (2001)
	Beauveria brongniartii	*Holotrichia serrata* (Coleoptera: Scarabaeidae)	Soil appl. of fungus-sorghum grains in furrow at 10^{14} and 10^{15} conidia/ha gave 46.2–68.6% infection respectively, in 2 years of appl.	Srikanth et al. (2010)
	Metarhizium anisopliae		Two field appl. of *M. anisopliae* in three formulations: talc, lignite, and liquid, made at 15-d intervals, appl. to root zone 5–10 cm, at 10^{12} conidia/ha. All treatments were effective (liquid best) with 76.87–81% mortality after 15 d	Thamarai-Chelvi et al. (2011)
		Pyrilla perpusilla (Homoptera: Lophopidae)	Field evaluation using a foliar spray of 1.5×10^6 conidia/mL, resulted in 42.8% adult and 60% nymph mortality. Field releases of *Metarhizium*-infected adults at 100 adults/ha.	Varma et al. (1992), Tiwari and Tanwar (2001)
		Ceratovacuna lanigera (Hemiptera: Aphididae)	Trials with oil-in-water formulations of *M. anisopliae* (1.65×10^8 conidia/mL); 1 g of conidia dust mixed in 5 mL sunflower oil + 95 mL water applied to five leaves, caused 42.3% mycosis of in situ adults	Nirmala et al. (2007)
Pakistan		*Odontotermes obesus and Microtermes obesi* (Isoptera: Termitidae)	Setts dipped in suspensions of 10^8 conidia/mL + diesel oil significantly reduced termite damage in the field (higher sett germination >55% and lower bud damage <5.50%) at two sites in two seasons in Punjab	Hussain et al. (2011)
Australia		*Dermolepida albohirtum* (Coleoptera: Scarabaeidae)	BioCane granules (2×10^9 conidia/g) applied at 33 kg/ha provided 50–60% control, with persistence effective for 2 years in ratoon crops	Samson et al. (1999)
Thailand		*Dorysthenes buqueti* (Coleoptera: Cerambycidae)	Soil treatment of 10^7–10^8 conidia/mL solution in greenhouse resulted in 80–100% mortality	Suasa-ard et al. (2008)

Continued

TABLE 20.1 Examples of Some Augmented and Experimental Field Programs Using Microbial Control of Insect Pests in Sugarcane—cont'd

Country	Microbial Species	Target Pest	Application	References
Indonesia	*Metarhizium flavoviride*	*Dorysthenes hydropicus*	Field appl. of formulated fungus (9.4 × 10^6 conidia/mL) at a rate of 74 kg/ha, resulted in 34% mortality and 15–17% more infected larvae at three sites	Pramono et al. (2001)
China	*S. glaseri*	*Alissonotum impressicolle* (Coleoptera: Scarabaeidae)	Appl. of suspension @ 1.5 × 10^6 IJs/ha, along plants bases, in early spring, resulted in 71.2–80% mortality of third instar grubs, 15 d after appl.	Wang and Li (1987)
Brazil, Cuba	*B. bassiana; M. anisopliae*	*Diatraea saccharalis* (Lepidoptera: Crambidae)	Aerial appl. of *B. bassiana* 6 × 10^12 conidia/ha to 15,000 ha sugarcane, gave 60% control in Brazil. In Cuba, *B. bassiana* applied at 1–3 × 10^9 conidia/ha in 10–40 L/ha of liquid formulation, gave 65% pest control. Appl. of *M. anisopliae* at 10^13 conidia/ha applied to 780 ha, gave 58% control in Brazil	Nicholls et al. (2002), Zappelini (2009), Zappelini et al. (2010), Estrada et al. (1997)
Brazil	*Steinernema braziliense*	*Sphenophorus levis* (Coleoptera: Curculionidae)	500 g of conidia/ha are impregnated in cane stem baits (0.4 g/ stem); applied at a rate of 200 stems/ha, yields 75% mortality after 30 d	Badilla and Alves (1991)
			Isolate IBCBn 06, applied at 1 × 10^8 IJs/ha or 1 kg/ha in 400 L of water (ground appl.) used with a sublethal conc. of 500 g of thiamethoxam 250 WG/ha, resulted in increased yields by 28 ton/ha with 70% control	Leite et al. (2012)
	Steinernema spp. and *Heterorhabditis* spp.	*Mahanarva fimbriolata* (Hemiptera: Cercopidae)	Greenhouse trials on nymphs, using 4 EPN species in sprays @ 200 IJs/nymph in 2 mL, caused 48–72% mortality. Field trials using *Heterorhabditis* spp. at various conc. 6.6 × 10^7–3.3 × 10^9 IJs/ ha applied at 2000 L/ha to mulch resulted in 43–74% control by 6 d	De Paula Batista et al. (2011)
	M. anisopliae		Ground appl. of isolate IBCB 425 @ 5 × 10^12 conidia/ha, sprayed with water 300 L/ha, close to the roots, or using air at flow rates of 30 L/ha, gave 70% control. Aerial appl. made using 7–10 kg/ha (1.8 × 10^11 conidia/ha) to 5000 ha, yields ≈80% control	Alves et al. (2003), Almeida and Batista Filho (2006)
		Mahanarva posticata	Appl. of the isolate PL43 @ 3.5 × 10^12 (200–500 g) conidia/ha in 50–200 L water/ha using ground equipment and 20–30 L/ha by aircraft (2–5 m above the sugarcane), yields ≈80% control after 15 d	Alves et al. (2003), Alves and Lopes (2008)
Venezuela		*Aeneolamia varia* (Hemiptera: Cercopidae)	In Venezuela, a formulated, wettable powder (10^10 conidia/g) product, Cobican, is applied to >50,000 ha of sugarcane with 72% control.	Mendonça (2005)
French Reunion Island	*Beauveria hoplocheli*	*Hoplochelus marginalis* (Coleoptera: Scarabaeidae)	The product, Betel, a clay-granule formulation, has been systematically applied since 1996, in-furrow at 50 kg/ha at plant. Density dependent range of 33–67% mycosis now occurs in different areas	Jeuffrault et al. (2004), Robène-Soustrade et al. (2015)

from *Sesamia calamistis* (Noctuidae) and *Sesamia grisescens,* which cause 40% larval mortality in sugarcane fields in Reunion Island and Papua New Guinea, respectively, but no augmented use has been reported yet (Jacquemard et al., 1985; Kuniata and Sweet, 1994).

20.2.3 Entomopathogenic Fungi for Control of Stalk Borers

In Latin America, the use of EPF to control stalk borers, mainly *Diatraea* spp., was first studied 30 years ago. However, the research was discontinued due to the success of the wasp larval endoparasitoid *Cotesia flavipes* (Braconidae) (Zappelini et al., 2010). New studies on the augmented application of EPF have now become necessary mainly due to the expansion of sugarcane areas, the growing demand for some parasitoid cultures, and the lack of establishment of some natural enemies, particularly in the warmer areas of Brazil (Zappelini et al., 2010). Small-scale field use is reported against *D. saccharalis* and *Chilo auricilius* (Crambidae) although limited to the ascomycota, *Beauveria bassiana s.l.* and *Metarhizium anisopliae s.l.* (Clavicipitaceae) in Brazil, Cuba, and India (Table 20.1). In Brazil, isolates of *B. bassiana* and *M. anisopliae,* produced by solid-state fermentation on rice, were experimentally applied at rates of 6×10^{12} conidia/ha (60% insect control) and 1×10^{13} conidia/ha (58% control) against *D. saccharalis,* respectively (Zappelini, 2009; Zappelini et al., 2010) (Table 20.1). According to Estrada et al. (1997), the application of 1×10^{12} conidia/ha of *B. bassiana* in Cuba gave an increased yield of 10 tons/ha. However, in Brazil there is no augmented use of these EPF due to higher production and application costs compared to registered insecticides despite similar efficacies (60% insect control) (Zappelini et al., 2010). Laboratory studies and natural infections of *D. saccharalis* by the fungi *Hirsutella nodulosa, Isaria tenuipes,* and *I. farinosa* (=*Paecilomyces*) are reported (Hernández-Velázquez et al., 2012), which have not led to augmentation. Likewise, *I. farinosa* is an important fungal pathogen of *S. inferens* in China (Li, 1992) and the fungus is reported to infect *C. s. sacchariphagus* in Mauritius (Ganeshan, 2000). In Taiwan, strains of *I. tenuipes* and *I. farinosa* were pathogenic to *S. inferens* larvae and a strain of *B. amorpha* killed larvae of *Tetramoera schistaceana* (Tortricidae) in the laboratory (Wang, 1996; ZuNan, 1996). In India, *H. nodulosa* showed 11.4% natural infection of *C. s. indicus* in sugarcane in Tamil Nadu. Natural field activity of this fungus was recorded for up to one year and could be grown on artificial media but later failed to infect host larvae in the laboratory (Easwaramoorthy et al., 1998). *B. bassiana* and *Cordyceps* spp. have been isolated from *E. saccharina* attacking indigenous host plants in Ethiopia

(Assefa et al., 2010) (Fig. 20.1C) and *B. bassiana* is a major mortality factor in *Cyperus papyrus* L. in South Africa (Conlong and Way, 2015). Research undertaken to establish *B. bassiana* as an endophyte within sugarcane in South Africa has been undertaken with variable results (Memela, 2014). In Mauritius, locally isolated strains of *M. anisopliae* have been field-tested against *C. s. sacchariphagus* but have not been commercialized (Behary Paray et al., 2012). *M. anisopliae* was recorded as a mortality factor on *C. s. indicus* larvae in sugarcane fields in India; when brought back to the laboratory, the fungus was able to cause 90% mortality of fourth instar larvae when they were treated with a 10^9 conidia/mL suspension (Easwaramoorthy et al., 2001). In China, 16 *M. anisopliae* isolates, of which eight displayed high virulence, were recovered from *Chilo venosatus* and tested (10^7 conidia/mL) with larval mortality ranging from 60% to 100% under laboratory conditions (Liu et al., 2012). *B. bassiana* at a concentration of 10^7 conidia/mL, caused 69–76% mortality in *C. infuscatellus* larvae in India (Sivasankaran et al., 1990), and *B. bassiana, Fusarium oxysporum,* and *M. anisopliae* were pathogenic to larvae of *S. inferens* in the laboratory (Varma and Tandan, 1996). Both *Beauvaria* spp. and *M. anisopliae* have been recorded as pathogens of *Sesamia grisescens* in Papua New Guinea, but no augmentative use is reported (Kuniata and Sweet, 1994; Young and Kuniata, 1995).

20.2.4 Entomopathogenic Nematodes for Control of Stalk Borers

In South Africa, locally isolated nematodes *Steinernema* and *Heterorhabditis* spp. were applied to *E. saccharina*–infested sugarcane with encouraging results (Spaull, 1991). During these field trials, an initial concentration of 130 billion infective juveniles per hectare (IJs/ha) of *Heterorhabditis* spp. in 26,000 L/ha was sprayed onto sugarcane stalks, which achieved 56% control of *E. saccharina* larvae. Thereafter, lower volume suspensions (87,000 IJs in 57 mL) achieved 45% control (Spaull, 1991). Nevertheless, the volume of water required to apply EPNs rendered this control option logistically untenable. However, given the advent of additional adjuvants and irrigation systems in some sugarcane areas in South Africa, this option should be reconsidered. More recently, a previously undescribed species, *Steinernema sacchari,* has been identified from *E. saccharina* larvae exposed in soil; however, the efficacy of this species has not been tested in the field (Nthenga et al., 2014). Hernández-Velázquez et al. (2012) believe that EPNs have good potential for biological control of *Diatraea* spp. because of their natural presence in sugarcane soils; moreover, *Steinernema feltiae, S. glaseri, S. rarum, Heterorhabditis heliothis,* and *H. bacteriophora* demonstrated high infectivity (90–100% mortality) within 5 days during

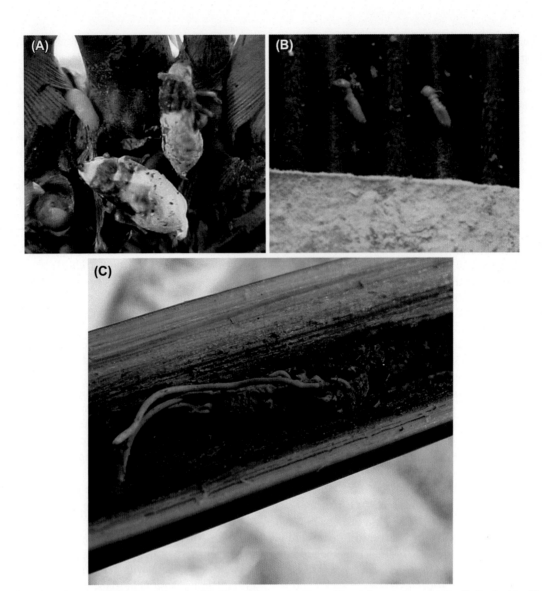

FIGURE 20.1 (A) *Mahanarva fimbriolata* infected with *Metarhizium anisopliae* collected in sugarcane in Brazil (J. de Almeida, IBLCB); (B) *Heterotermes tenuis* infected with *Beauevria bassiana* from a Termitrap device in Brazil (J. deAlmeida, IBLCB); (C) *Eldana saccharina* fifth instar larvae infected with unknown *Cordyceps* spp. from a *Cyperus papyrus* umbel collected from Lake Awasa in Ethiopia (D. Conlong, SASRI). **ARC-SGI**-Agricultural Research Council-Small Grains Institute, Bethlehem, South Africa; **IBLCB**-Instituto Biológico, Laboratório de Controle Biológico, Campinas, Brazil; **SASRI**-South African Sugarcane Research Institute, KwaZulu-Natal, South Africa; **CIRAD**-Centre de coopération Internationale en Recherche Agronomique pour le Développement, Saint-Pierre, La Réunion.

laboratory bioassays. Since 1987, *H. bacteriophora* has been mass produced in Cuba to control soil pests, such as *D. saccharalis*, however no successful field trials are reported (Hernández-Velázquez et al., 2012). In India, early research on EPNs that commenced in 1966 focused on imported nematode strains (i.e., *Steinernema carpocapsae* [strain DD 136]), but results were inconsistent (Divya and Sankar, 2009). Following exploratory diversity and field isolation studies, *H. bacteriophora*, *H. indica*, *S. siamkayai,* and *S. glaseri* were recovered from different geographical areas in Tamil Nadu, India, that are pathogenic to *C. s. indicus*; *C. infuscatellus,* and *S. excerptalis*, athough there is no augmentation reported (Sankaranarayanan and Easwaramoorthy, 2003; Umamaheswari et al., 2004).

20.3 HEMIPTERAN SAP-SUCKING PESTS AND THEIR MICROBIAL CONTROL WITH ENTOMOPATHOGENIC FUNGI

Sucking pests, mainly hemipterans, cause serious damage to sugarcane either by direct feeding or by transmitting plant viral diseases (Rabindra and Ramanujam, 2007). Because most sucking pests feed on plants via a stylet, augmented microbial control has focused on pathogens that do not need to be ingested to be pathogenic such as EPF and EPNs (Alves et al., 2003). The augmented use of *M. anisopliae s.l.* for control of the hemipteran families Cercopidae, Lophopidae, and Aphididae is well represented in Latin American countries and India (Table 20.1). Impressively,

FIGURE 20.2 (A) Unknown scarab infected with *Beauveria* spp. found on *Acacia mearnsii* trunk, this tree species is often found cultivated adjacent to sugarcane in KwaZulu-Natal, South Africa (KZN SA) (T. Goble, SASRI); (B) *Pegylis sommeri* grubs infected with *Beauveria* spp. found during an epizootic in KZN SA (J. Hatting, ARC-SGI); (C) *Hoplochelus marginalis larvae infected with Beauveria hoplocheli* showing long hyphasmata (B. Vercambre; CIRAD) (D) *Beauveria hoplocheli* (B. Vercambre; CIRAD); (E) *Beauveria brongniartii* growing out of *P. sommeri* obtained in KZN SA (M. Way SASRI); (F) *Orphiocordyceps* growing out of the head of an unknown *Anomala* spp. found in KZN SA; the scale is in mm (M. Way, SASRI).

there are 58 listed commercial products based on *M. aniso-pliae* of which 37 (63.8%) are reported to be active against cercopids, and most of these products are available for use in South and Central America (Faria and Wraight, 2007).

Currently, the largest single microbial control program uses *M. anisopliae*, grown by solid fermentation on rice, to control a complex of spittlebugs, including *Mahanarva postica* and *Mahanarva fimbriolata* (Cercopidae) in Brazilian sugarcane (Table 20.1). *M. postica* occurs in 800,000 ha of sugarcane in northeastern Brazil, causing 11% loss in crop weight and 15% loss in industrial yields (sugar weight)

(Alves et al., 2003). Between 1970 and 1998, the Sugar and Alcohol Institute (IAA/Planalsucar) and State of Pernanbuco Agricultural Research Enterprise (IPA) produced 40,000 kg of *M. anisopliae* conidia, which was applied to 500,000 ha of sugarcane, resulting in a 72% reduction in infestation rates by spittlebugs (Alves et al., 2003). The area under augmentation is now less than 250,000 ha in these regions because *M. anisopliae* is already present in sugarcane fields and the cost is approximately US$22 for an application of 3.5×10^{12} conidia/ha (Alves et al., 2003) (Table 20.1).

The root spittlebug *M. fimbriolata* became an important sugarcane pest in 1997, mainly in São Paulo State and thereafter in all sugarcane fields in Brazil (Almeida, 2014). *M. anisopliae* (IBCB 425) is mas -produced by 37 companies in this country to control this pest. During 2013/2014, the fungus was applied to ≈350,000 ha of sugarcane in São Paulo State, with 60% average control efficacy (Almeida, 2014) (Fig. 20.1A). The current augmentation strategies include ground applications of *M. anisopliae* at 5×10^{12} conidia/ha, either by spraying in water at 300 L/ha to sugarcane roots or ground applications using 5×10^{12} conidia/ha with air, at flow rates of 30 L/ha. Aerial applications using 7–10 kg/ha (1.8×10^{11} conidia/ha) to 5000 ha of sugarcane yields 80% insect control from November to February (Alves et al., 2003; Almeida and Batista Filho, 2006) (Table 20.1). Greenhouse and field trials using EPNs, mainly *Steinernema* and *Heterorhabditis* spp., have shown promising results for control of *M. fimbriolata* in Brazil (De Paula Batista et al., 2011) (Table 20.1).

In Columbia, augmented use of *M. anisopliae* (CCMa01) applied at 5×10^{13} conidia/ha controls *Aeneolamia varia* (Cercopidae) (Alves et al., 2003)—similarly in Venezuela (Table 20.1) and Peru (Alves and Lopes, 2008; Cuarán et al., 2012; Solarte et al., 2012). Since the 1980s, various Central American countries have initiated microbial control programs utilizing specifically selected *M. anisopliae* isolates applied aerially (5×10^{13} conidia/ha) to control the major cercopids (spittlebugs) in these regions. Mexico applies the product Meta-Sin (Agrobiologicos del Noroeste S.A. de C.V. [Agrobionsa], Mexico) to >50,000 ha of sugarcane to control *Aeneolamia postica* and against other cercopids (Torres et al., 2013). In Guatemala, Salivase (Produtos Ecológicos, Guatemala) is applied to control *A. postica*, although the efficiency is not as high when compared to Mexico, thus requiring development of more-virulent isolates (Melgar, 2014). In Nicaragua, a *M. anisopliae*–based product, Dextruxin 50 WP (Laverlam S.A., Colombia) imported from Columbia is applied to control cercopids (Alves et al., 2003; Faria and Wraight, 2007). In Trinidad and Tobago, *M. anisopliae* is applied to control *Aeneolamia* spp. and is applied to >5000 ha of sugarcane in an oil formulation (Alves et al., 2003). The Brazilian product, Biotech (Biotech, Brazil), is imported into Panama where it is applied (5×10^{12} conidia/ha) to sugarcane to control *Aeneolamia lepidior* (Cercopidae) (Alves et al., 2003). The *M. anisopliae* product, Metadieca (Liga Agricola Industrial de LaCanã de Azucar—LAICA), produced by the Extension Service in Costa Rica is applied to control the sugarcane pests: *A. postica*, *Zulia vilios* and *Prosapia simulans* (Cercopidae); applications of *M. anisopliae* increased from 350 ha of sugarcane treated, using 115 kg of product, to >6000 ha treated with 1650 kg of product in 1999 (Alves et al., 2003).

Other important hemipteran pests include the sugarcane plant hopper *Perkinsiella saccharicida* (Delphacidae) that vectors viral Fiji disease (Rheoviridae), which belongs to a species complex of Australian origin and occurs mostly in tropical Asia, parts of Africa, and the Middle East (Hughes and Robinson, 1961). In 1900, this pest was accidentally introduced into Hawaii and subsequently spread to Central and South America. In the United States, sugarcane damaged by *P. saccharicida* resulted in the reduction of sugar production by approximately 17% resulting in US$5 million/annum in revenue loss (Clausen, 1978). Classical biological control comprising parasitoids and predators is the main tactic in the management of *P. saccaricida* in the United States (including Hawaii), the Pacific Region, and China (Swezey, 1936; Huffaker and Caltagirone, 1986). Both *M. anisopliae s.l.* and *Beauveria* spp. have been recovered from *P. saccharicida* in Ecuador (Badilla et al., 2004), and greenhouse trials using a strain of *M. anisopliae* in a water-based suspension of 8.3×10^9 conidia/sugarcane plant caused 73.8% and 52.5% adult and nymph mortality, respectively. *B. bassiana* at the same concentration caused 62.8% and 63.6% mortality, respectively (Badilla et al., 2004). These findings resulted in the commercial production of the *M. anisopliae* strain on rice, which is applied at 10^{13} conidia/ha or 10 kg/ha by aircraft. Natural infections of *Hirsutella* spp. are also known to occur on *P. saccharicida* in Hawaii (Speare, 1920) and Ecuador (Badilla et al., 2004) during the rainy seasons, but there are no reports of augmented applications in sugarcane.

In India, small-scale field use of *M. anisopliae* and inoculative releases of *B. bassiana* are used to control plant hoppers *Pyrilla perpusilla* (Lophopidae) and black bugs, *Cavelerius sweeti* (Lygaeidae), respectively (Table 20.1). *Pyrilla perpusilla* causes 28% yield loss, poor growth of seed sets, and difficulty milling cane. Periodic epizootics of *M. anisopliae s.l.* occur under ideal conditions and Kulashreshtha and Gursahani (1961) recorded a 60–75% level of natural mortality of nymphs and adults in India from 1957 to 1959, when rainfall was high. Field evaluations of *M. anisopliae* against this pest, in Uttar Pradesh in India, are also reported (Varma et al., 1992; Tiwari and Tanwar, 2001) (Table 20.1). Waghulde et al. (1991) reported that *P. perpusilla* nymph and adult suppression by *M. anisopliae* was 61% and 55%, respectively, at 15 days postapplication and increased to 77% and 89% at 30 days without adverse effects on the moth ectoparasitoid *Epiricania melanoleuca* (Epipyropidae). Also in India, inoculative releases of adult *C. sweeti* infected with *B. bassiana* are made to sugarcane fields at the rate of 5000 adults/ha to control carry-over populations from November to December (Naidu, 2009) (Table 20.1).

Sugarcane wooly aphid *Ceratovacuna lanigera* (Hemiptera: Aphididae) lives in large colonies that are reported in outbreak proportions (25% infestation of total sugarcane grown) in India; it is also a serious pest of sugarcane in several parts of the Oriental region (Joshi and

Viraktamath, 2004). During field trials in India against *C. lanigera*, oil-in-water formulations (1.65×10^8 conidia/mL) of *B. bassiana* and *M. anisopliae* (100 mL/5 leaves of sugarcane) caused 19.8% and 42.3% mycosis, respectively (Nirmala et al., 2007) (Table 20.1). Further, conservation biological control—the modification of the environment or existing practices to protect and enhance natural enemies (Eilenberg et al., 2001)—is practiced in India to create ideal conditions for the growth of other fungi, *Cladosporium oxysporum*, *M. anisopliae*, *Lecanicillium longisporum* (=*lecanii*), and *B. bassiana,* that suppress *C. lanigera* (Satyagopal et al., 2014). *Acremonium zeylandicum* caused 90% mortality during natural epizootics in *C. lanigera* and under ideal conditions the fungus can spread at a rate of 100 acres (0.4 km^2) each month (Tippannavar et al., 2006). Further, *Fusarium moniliforme* and *Penicillium oxalicum* are reported natural pathogens of *C. lanigera* in China (Fang and Tan, 1986). Conservation biological control is also practiced in the northern regions of Argentina where research centers use irrigation as an environmental manipulation strategy in sugarcane to promote infection of cicadas *Proarna bergi* (Cicadidae) by *Cordyceps sobolifera* (=*Ophiocordyceps*) (Alves et al., 2003). In Venezuela, the *M. anisopliae* product, Cobican (Probioagro S.A., Venezuela), a wettable powder (10^{10} conidia/g), is applied to sugarcane to control yellow sugarcane aphid *Sipha flava* (Aphididae) (Alves et al., 2003). CREE (Centros de Reprodução de Entomófagos e Entomopatógenos) laboratories in Cuba produce Vertisav 57 (*Lecanicillium* spp.), applied at a rate of 1 kg/ha to control whiteflies, aphids, and mites (Nicholls et al., 2002; Alves et al., 2003). Interestingly, *Aschersonia placenta* occurs naturally with 3.4–87.3% infection level on white fly, *Aleurolobus barodensis* (Alyrodidae) in India, but there are no reports of augmentation (Rajak and Varma, 2001).

20.4 ROOT-FEEDING PESTS AND THEIR MICROBIAL CONTROL

Of the soil pests associated with sugarcane worldwide, endemic scarabaeid larvae (white grubs) are the dominant group in causing damage and in their distribution (Wilson, 1969). Third instar white grubs cause the most damage by feeding on the roots and underground stalks of sugarcane plants, reducing plant vigor, crop yield, and sugar content (Wilson, 1969).

20.4.1 Control of Scarabaeid Root Pests with Entomopathogenic Fungi

Worldwide, *Metarhizium* spp. have been used extensively (India, Australia, South America, Thailand, and Indonesia) to control a broad range of scarabs (dynastids, rutelines, and melolonthids), and other coleopteran pests

(Faria and Wraight, 2007), whereas *Beauveria brongniartii* and *Beauveria hoplocheli* (the latter previously thought to be *B. brongniartii* but recently recognized as a novel species that is closely related to *Beauveria malawiensis* [Robène-Soustrade et al., 2015]) (Fig. 20.2C, D) have been used specifically against melolonthids in India, Reunion Island, and South Africa (nothing is yet registered in SA). In Australia, BioCane (Becker Underwood Inc., USA-Australian division) granules (2×10^9 conidia/g), applied at 5 g/m per sugarcane row (33 kg/ha) resulted in 70% mortality of *Antitrogus consanguineus* (Melolonthinae) after 8 weeks in ratoon sugarcane with 2-year persistence (Samson et al., 1999). At the same rate, 50–60% control of *Dermolepida albohirtum* (Melolonthinae), a pest that causes economic losses of AUD$40 million in damage and management costs, can be achieved (Table 20.1). Despite this, the production of BioCane has been suspended due to low market demand (Sugar Research Australia, 2013). In India, *B. brongniartii* (Srikanth et al., 2010) and *M. anisopliae* (Thamarai-Chelvi et al., 2011) isolates were effective against *Holotrichia serrata* (Melolonthinae) under experimental field conditions (Table 20.1). On Reunion Island, the clay-based product Betel (Betel Reunion S.A., Reunion Island is a subsidiary of Natural Plant Protection, France strain: *B. hoplocheli*) is available commercially to control *Hoplochelus marginalis* (Melolonthinae) in sugarcane (Jeuffrault et al., 2004) (Table 20.1). *B. brongniartii* isolates have recently been recovered from the melolonthid *Pegylis sommeri* (=*Hypopholis* see: Harrison, 2014) in sugarcane fields in South Africa (Goble et al., 2012) (Fig. 20.2B, C). These isolates are also pathogenic towards adults and larvae of another sugarcane melolonthid pest, *Schizonycha affinis* (Goble et al., 2015). Field trials will commence soon in South African sugarcane to test formulations of two virulent isolates for registration purposes. In 2010, a multiparty consortium, funded under the EU-Africa, Caribbean and Pacific Economic Partnership Agreement (EU-ACP), was established for research and development of bioinsecticides against white grubs affecting sugarcane (Ngubane et al., 2012). To date, of the 1549 insect cadavers shipped to South Africa from Swaziland, Tanzania, Malawi, Zimbabwe and Mauritius, four species of fungi were identified: *M. anisopliae*, *B. bassiana*, *B. brongniartii,* and *Lecanicillium* spp. Notably the South African material yielded more *Beauveria* spp. (Fig. 20.2A, B), whereas cadavers from other countries were infected primarily with *Metarhizium* spp. (Ngubane et al., 2012). In Mauritius, strains of locally isolated *M. anisopliae* are being researched to control *Heteronychus licas* adults and larvae, and the trials are in the laboratory testing phase (Behary Paray et al., 2012). Development of mass production methods for strains of *M. anisopliae* have been completed for research and trial purposes against *H. licas* in Zimbabwe and Tanzania (Ngubane et al., 2012), but it is unknown whether commercialization will be realized (Mazodze and Zvoutete, 1999). An interesting occurrence are fungal pathogens of the *Ophiocordyceps*, which have been recovered from scarabs associated with sugarcane in South Africa (Goble, 2012) (Fig. 20.2F), Tanzania (Ngubane et al., 2012), and Somalia (Evans et al., 1999).

20.4.2 Control of Scarabaeid Root Pests with EPNs

There are no reports of extensive augmented use of EPNs to control white grubs in sugarcane; research has been limited to laboratory and small-scale field trials (Table 20.1). In India, EPN species *S. glaseri*, *S. riobrave*, *H. bacteriophora,* and *H. indica* were tested in laboratory bioassays against both larvae and pupae of *H. serrata* and *Holotrichia consanguinea* (Sankaranarayanan et al., 2008). All EPNs caused mortality in pupae; the lowest LD_{50} (dose required to kill 50% of individuals in a sample) were for *S. glaseri* (113 IJs/pupa) and *H. indica* (127 IJs/pupa) with LT_{50} (time taken for 50% of individuals in a sample to die) of 24.9h and 27.3h, respectively (Sankaranarayanan et al., 2008). In field trials in India, *S. glaseri* (10^5 IJs/m²) caused in 40% and 100% infection of *H. consanguinea* grubs 10 and 20days posttreatment (Vyas and Yadav, 1993). *Alissonotum impressicolle* is a major pest in Australia and China. Field applications of *S. glaseri* (1.5×10^9 IJs/ha) sprayed on the ground along the base of sugarcane plants to control third instar *A. impressicolle* in early spring resulted in 71–80% mortality 15days after application in China. Applications to sandy soil resulted in higher mortality than in clay, which was attributed to increased nematode migration (Wang and Li, 1987) (Table 20.1).

20.4.3 Control of Scarabaeids with Entomopathogenic Bacteria

The effectiveness of milky disease caused by the soil bacterium *Paenibacillus popilliae* against *H. consanguinea* was evaluated in the laboratory and in field applications using a bacterial spore powder in Gujarat, India, which resulted in 35–61% larval infection in 1976 and between 20% and 75% in 1977 (Vora and Ramakrishnan, 1978). However, the product was expensive to import and was not considered for large-scale use in sugarcane. Many strains of *B. thuringiensis* have had little or no effect against scarabs with the exception of the Bt subsp. *japonensis* Buibui strain, which carries the *Cry8Ca1* delta-endotoxin and is effective against Rutelinae larvae but less effective against Dynastinae and Melolonthinae (Bixby et al., 2007). During an extensive

survey for entomopathogens in South African sugarcane, the bacteria *Bacillus laevolacticus, Serratia marcescens, S. liquefaciens,* and *S. grimesii* were isolated from white grubs and tested in laboratory bioassays with variable results (Hatting, 2008); no augmentation is reported.

20.5 MICROBIAL CONTROL OF OTHER SOIL PESTS

Metarhizium flavoviride that was mass produced in larvae in the laboratory and *M. anisopliae* cultured on white rice are effective against the economically important species of sugarcane longhorn stem borers *Dorysthenes hydropicus* from Indonesia and *Dorysthenes buqueti* (Cerambycidae) from Thailand, respectively (Table 20.1). Further, *M. anisopliae* (ESALQ 259) and *B. bassiana* (ESALQ 447) are used to control the sugarcane root borer *Sphenophorus levis* (Curculionidae) in Brazil (Badilla and Alves, 1991) (Table 20.1). The field application of *H. indica* and *Steinernema* spp. applied alone, or in combination with *M. anisopliae* and *B. bassiana*, showed promise to control *S. levis* in Brazil (Leite et al., 2012). Simi (2014) showed that *M. anisopliae* IBCB 383 was antagonistic when applied with *Steinernema braziliense* IBCBn 06 for the control of *S. levis* in sugarcane, but when applied with *B. bassiana* IBCB 170 the combination was synergistic. The Brazilian company, Bio Controle—Métodos de Controle de Pragas Ltda, in partnership with the Brazilian Biological Institute, produced *S. braziliense* IBCBn 06 in vitro using the sponge method (Ehlers and Shapiro-Ilan, 2005), which is now registered in Brazil (Brasil, 2015). The beetle *Migdolus fryanus* (Vesperidae) causes serious damage to sugarcane roots and rhizomes in Sao Paulo State, Brazil. Machado et al. (2005) found that *S. glaseri* and *H. indica* applied at 600 IJs/larva, respectively, caused 76% mortality of *M. fryanus* larvae in the laboratory. However, Arrigoni et al. (1986) found that *S. glaseri* and *S. carpocapsea* applied at a rate of between 50,000 and 150,000 IJs/m² infected *M. fryanus* larvae in the field but <10% mortality was observed, which was not satisfactory. Control of subterranean termites *Heterotermes tenuis* (Isoptera: Rhinotermitidae) and *Procornitermes* spp. (Isoptera: Termitidae), which feed on sugarcane roots in Brazil, is achieved using a commercially developed attractive bait trap, called Termitrap loaded with the conidia of *B. bassiana*, or *M. anisopliae*. The current biological control strategy employs Termitraps® loaded with *B. bassiana* and 0.01% imidacloprid, which are disseminated at a rate of 50 bags per hectare, buried near the soil surface adjacent to sugarcane fields; at this rate, ≈70% of the nests in the area can be controlled after application (Almeida and Alves, 1999) (Fig. 20.1B).

20.6 CONCLUSIONS

Current trends in microbial control of sugarcane pests indicate that commercialization of EPF will continue to grow in developing countries because (1) production of these microbes is simple and culture media are inexpensive compared to those for viruses and nematodes; (2) fungi do not need to be ingested but infect directly through the cuticle, thus phloem-feeding insect pests can be controlled; (3) fungi can infect all insect life stages; (4) a single fungal product may control several sugarcane pests; and (5) tropical climates in which sugarcane is produced facilitates fungal infection and epizootics. Augmented use and commercialization of viruses against lepidopteran stalk borers have been limited because viruses are very specific and are not always found in all species, as is the case with *E. saccharina,* and where they are found and developed, multinational companies, who have the production capabilities, may not be interested in the low market potential. Further, sometimes smaller, local producers have had issues with production scale-up, quality control, formulation, and field efficacy (Ramanujam et al., 2014). Use of Bt products in developing countries is still relatively limited probably due to the cost of imported products from Europe and the United States although Cuba and Mexico have begun cheaper production using inexpensive media such as whey to culture Bt which reduces costs. Other limitations to Bt use include competition with chemicals and reduced action spectrum (i.e., *Cry* toxins are more active against young larval stages) (Ramanujam et al., 2014). There is a need to develop virulent indigenous Bt strains against sugarcane pests and affordable formulations for large-scale production in developing countries (Ramanujam et al., 2014). Some EPNs, such as *H. indica, S. braziliense, S. carpocapsae, S. glaseri,* and *S. thermophilum,* are now commercialized and mass produced in India (Divya and Sankar, 2009) and Brazil (Brasil, 2015). The reasons for low adoption of EPN technology include that they are expensive to rear via in vitro media because there are separate media and fermenter requirements for both nematode growth and their associated bacteria (Divya and Sankar, 2009); application of EPNs above ground can be challenging unless formulations are stable to reduce desiccation of nematodes; and timing of applications to target stalk boring neonates before they bore into sugarcane stalks is challenging.

There is considerable scope for the development and commercialization of bioinsecticides in developing countries because subtropical and tropical climates enhance infections in insects; there is usually a rich biodiversity of virulent species and isolates to serve as active ingredients for products; traditional farming knowledge provide valuable skills for developing novel effective products; and regulatory systems in developing countries are often more relaxed, with cheaper registrations of products, compared with First World countries. Major reasons for the lack of adoption by end-users in developing countries are unequivocal evidence that microbial insecticides have efficacy to control crop damage and increase yields, availability of high-quality products that are at affordable, lack of education and literacy, and reliable supply chains.

REFERENCES

Almeida, J.E.M., 2014. Cigarrinha-da-raiz da cana: Como controlar? Rev. Cultiv. Grand. Cult. 177, 30–31.

Almeida, J.E.M., Alves, S.B., 1999. Controle de *Heterotermes tenuis* (Hagen, 1858) (Isoptera: Rhinotermitidae) em cana-de-açúcar com iscas Termitrap® associadas ao fungo entomopatogênico *Beauveria bassiana* (Bals.) Vuill. e/ou a inseticidas em época seca. Arq. Inst. Biol. 66, 85–90.

Almeida, J.E.M., Batista Filho, A., 2006. Controle biológico da cigarrinha-da-raiz da cana-de-açúcar com o fungo *Metarhizium anisopliae.* Bol. Téc. Inst. Biol. 16, 19.

Alves, S.B., 1986. In: Controle Microbiano de Insetos. Manole, São Paulo, Brazil, p. 407.

Alves, S.B., Lopes, R.B., 2008. In: Controle Microbiano de Pragas na América Latina. FEALQ, Piracicaba, Brazil, p. 414.

Alves, S.B., Pereira, R.M., Lopes, R.B., Tamai, M.A., 2003. Use of entomopathogenic fungi in Latin America. In: Upadhyay, R.K. (Ed.), Advances in Microbial Control of Insect Pests. Kluwer Academic Plenum Publishers, NY, pp. 193–211.

Arrigoni, E.B., Dinardo, L.L., Conde, A.J., Terán, F.O., 1986. Aplicação de *Neoplectana carpocapsea* (Weiser, 1955) em condições de campo para o controle de *Migdolus* spp. (Coleóptera: Cerambycidae). Nemat. Bras. 10, 81–189.

Assefa, Y., Conlong, D.E., van den Berg, J., Mitchell, A., 2010. Distribution of sugarcane stem borers and their natural enemies in small-scale farmers' fields, adjacent margins and wetlands of Ethiopia. Int. J. Pest Manage. 56, 233–241.

Badilla, F.F., Alves, S.B., 1991. Controle do gorgulho da cana-de-açúcar *Sphenophorus levis* Vaurie, 1978 (Coleoptera: Curculionidae) com *Beauveria* spp. em condições de laboratório e campo. An. Soc. Entomol. Brasil 20, 251–253.

Badilla, F.F., Jara, W., Gordillo, W., 2004. Control of the leaf-hopper *Perkinsiella saccharicida* with the fungi *Beauveria bassiana* and *Metarhizium anisopliae.* M. Integ. Plagas. Agroecosyst. 73, 29–34.

Behary Paray, N., Mmanga, S., Hatting, J.L., Conlong, D.E., Ganeshan, S., 2012. Detection, isolation and characterization of white grub (Coleoptera: Scarabaeidae) pathogens in Mauritius and Tanzania. Proc. South Afr. Sugar Technol. Ass. 85, 123–128.

Bixby, A., Alm, S.R., Power, K., Grewal, P., Swier, S.R., 2007. Susceptibility of four species of turfgrass infesting scarabs (Coleoptera: Scarabaeidae) to *Bacillus thuringiensis* serovar *japonensis* strain Buibui. J. Econ. Entomol. 100, 1604–1610.

Braga, D.P.V., Arrigoni, E.D.B., Burnquist, W.L., Silva Filho, M.C., Ulian, E.C., 2001. A new approach for control of *Diatraea saccharalis* (Lepidoptera: Crambidae) through the expression of an insecticidal CryIa(b) protein in transgenic sugarcane. Proc. Int. Soc. Sugar Technol. 24, 331–336.

Brasil, 2015. Ministério da Agricultura, Pecuária e Abastecimento. Agrofit: Sistema de agrotóxicos fitossanitários. Available at: http://extranet.agricultura.gov.br/agrofit_cons/principal_agrofit_cons (accessed 07.05.15.).

Carnegie, A.J.M., Conlong, D.E., 1994. In: Biology, Pest Status and Control Measure Relationships of Sugarcane Insect Pests. Proc. Second Sugarcane Entomology Workshop. International Soc. Sugar. Technol. SASA Experiment Station. ISBN: 1-874903-10-7.

Clausen, C.P., 1978. Delphacidae. In: Clausen, C.P. (Ed.), Introduced Parasites and Predators of Arthropod Pests and Weeds. US Department of Agriculture Handbook 480, p. 545.

Conlong, D.E., Rutherford, R.S., 2009. Biological and habitat interventions for integrated pest management systems. Proc. South Afr. Sug. Technol. Ass. 82, 486–494.

Conlong, D.E., Way, M.J., 2015. Sugarcane. In: Prinsloo, G.L., Uys, V.M. (Eds.), Insects of Cultivated Plants and Natural Pastures in Southern Africa. Entomol. Soc. South Afr, Hatfield, South Africa. ISBN: 978-0-620-60841-1, pp. 156–176.

Cuarán, V.L., Valderrama, U.C., Pardey, A.E.B., Cobo, N.C.N., Sanchez, G.D.R., Gil, C.A.M., Laverde, L.A.G., 2012. Metodo para evaluar el daño de los salivazos (Hemiptera: Cercopidae) sobre la caña de azúcar. Rev. Col. Entomol. 38, 171–176.

De Paula Batista, E.S., Auad, A.M., De Resende, T.T., De Oliveira Monteiro, C.M., 2011. Screening of entomopathogenic nematodes to control Mahanarva fimbriolata (Hemiptera: Cercopidae). Rev. Colom. Entomol. 37, 198–202.

Divya, K., Sankar, M., 2009. Entomopathogenic nematodes in pest management. Indian J. Sci. Technol. 2, 53–60.

Easwaramoorthy, S., 2009. Biological control of sugarcane pests: present status and future thrusts. In: Upadhyay, R.K., Mukerji, K.G., Chamola, B.P., Dubey, O.P. (Eds.), Integrated Pest and Disease Management. APH Publishing Co, New Delhi, India, pp. 361–381.

Easwaramoorthy, S., Jayaraj, S., 1987. Survey of granulosis virus infection in sugarcane borers, Chilo infuscatellus Snellen and C. sacchariphagus indicus (Kapur) in India. Trop. Pest. Manage. 33, 200–201.

Easwaramoorthy, S., Strongman, D.B., Santhalakshmi, G., 1998. Record of Hirsutella nodulosa Petch from Chilo sacchariphagus indicus (Kapur), sugarcane internode borer in India. J. Biol. Control 11, 79–80.

Easwaramoorthy, S., Nirmala, R., Santhalakshmi, G., 2001. Occurrence of Metarhizium anisopliae var. anisopliae on sugarcane internode borer, Chilo sacchariphagus indicus (Kapur). J. Biol. Control 15, 81–84.

Ehlers, R.U., Shapiro-Ilan, D.I., 2005. Mass production. In: Grewal, P.S., Ehlers, R.U., Shaprio-Ilan, D.I. (Eds.), Nematodes as Biocontrol Agents. Cabi Publishing, Cambridge, MA, USA, pp. 65–78.

Eilenberg, J., Hajek, A., Lomer, C., 2001. Suggestions for unifying the terminology in biological control. BioControl 46, 387–400.

Estrada, M.E., Romero, M., Snowball, M., 1997. Aplicación de Beauveria bassiana en la lucha biológica contra Diatraea saccharalis. Caña Azúcar 15, 39–43. Available at: http://www.ceniap.gov.ve/bdigital/cana/canal1501/texto/aplicacion.htm (accessed 20.05.15.).

Evans, H.C., Smith, S.S., Katundu, J.M., Kapama, J.T., 1999. A Cordyceps pathogen of sugar-cane white grubs in Tanzania. Mycologist 13, 11–14.

Fang, H.M., Tan, S.M., 1986. Pathogenic fungi of several insect pests on sugarcane. Microbiol. China 13, 97–100.

Faria, M.R., Wraight, S.P., 2007. Mycoinsecticides and mycoacaricides: a comprehensive list with worldwide coverage and international classification of formulation types. Biol. Control 43, 237–256.

Food and Agriculture Organization of the United Nations (FAO), 2015. Crop Production. Available at: http://www.fao.org/cfs/cfs-home/list-events/en/ (accessed 27.01.15.).

Ganeshan, S., 2000. Biological control of sugarcane pests in Mauritius: current status and future prospects. In: Proc. IV Sugarcane Entomol. ISSCT, Thailand, pp. 3–9.

Goble, T.A., 2012. Towards the Development of a Mycoinsecticide to Control White Grubs (Coleoptera: Scarabaeidae) in South African Sugarcane (Ph.D. thesis of Rhodes University Grahamstown, South Africa). p. 267.

Goble, T.A., Costet, L., Robene, I., Nibouche, S., Rutherford, R.S., Conlong, D.E., Hill, M.P., 2012. Beauveria brongniartii on white grubs attacking sugarcane in South Africa. J. Invertebr. Pathol. 111, 225–236.

Goble, T.A., Conlong, D.E., Hill, M.P., 2015. Virulence of Beauveria brongniartii and B. bassiana against Schizonycha affinis white grubs and adults (Coleoptera: Scarabaeidae). J. Appl. Entomol. 139, 134–145.

Goebel, F.R., Sallam, N., 2011. New pest threats for sugarcane in the new bioeconomy and how to manage them. Curr. Opin. Environ. Sustainability 3, 81–89.

Gravena, S., Sanguino, J.R., Bara, J.R., 1980. Controle biológico da broca da cana Diatraea saccharalis (Fabricius, 1794) por predadores de ovos e Bacillus thuringiensis Berliner. An. Soc. Entomol. Brasil 9, 87–95.

Gupta, S., Dikshit, A.K., 2010. Biopesticides: an ecofriendly approach for pest control. J. Biopestic. 3, 186–188.

Harrison, J., Du, G., 2014. Review of the South African species of Pegylis Erichson, 1847 (Coleoptera: Scarabaeidae: Melolonthinae) commonly known as large wattle chafers. Afr. Entomol. 22, 685–713.

Hatting, J.L., 2008. Final Project Report on Microbial Control of White Grubs and Sugarcane Borer (Collaborative project between SASRI and ARC-SGI). South African Sugar. Res. Institute Internal report. , pp. 1–51.

Hernández-Velázquez, V.M., Lina-García, L.P., Obregon-Barboza, V., Trejo-Loyo, A.G., Peña-Chora, G., 2012. Pathogens associated with sugarcane borers, Diatraea spp. (Lepidoptera: Crambidae): a review. Int. J. Zool. 2012, 1–20.

Herrera, G., Snyman, S.J., Thomson, J.A., Mihm, J.A., 1997. Construction of a bioinsecticidal strain of Pseudomonas fluorescens active against sugarcane borer. In: Insect Resistant Maize: Recent Advances and Utilisation. Proceedings of an International Symposium Held at the International Maize and Wheat Improvement Center (CIMMYT), 27 November-3 December 1994, Mexico City, Mexico, pp. 159–162.

Huffaker, C.B., Caltagirone, L.E., 1986. The impact of biological control on the development of the Pacific. Agr. Ecosyst. Environ. 15, 95–107.

Hughes, C.G., Robinson, P.E., 1961. Fiji disease. In: Martin, J.P., Abbott, E.V., Hughes, C.G. (Eds.), Sugarcane Diseases of the World. Elsevier, Amsterdam, Netherlands, pp. 389–405.

Hunsigi, G., 2001. In: Sugarcane in Agriculture and Industry. Prism Books Pvt Ltd, Bangalore, India. ISBN: 81-7286-149-4, p. 269.

Hussain, A., Ahmed, S., Shahid, M., 2011. Laboratory and field evaluation of Metarhizium anisopliae var. anisopliae for controlling subterranean termites. Neotrop. Entomol. 40, 244–250.

Jacquemard, P., Croizier, G., Amargier, A., Veyrunes, J.C., Croizier, L., Bordat, D., Vercambre, B., 1985. Presence of three viruses in Sesamia calamistis Hampson (Lepidoptera Agrotidae) on the island of Reunion. Agron. Trop. 40, 66–71.

Jeuffrault, E., Rolet, A., Reynaud, B., Manikom, R., Georger, S., Taye, T., Chiroleu, F., Fouillaud, M., Vercambre, B., 2004. Vingt ans de lutte contre le ver blanc de la canne à sucre à la Réunion: Un succès, mais il reste des questionnements scientifiques pour confirmer la durabilité de la lutte biologique. Phytoma 573, 16–19.

Joshi, S., Viraktamath, C.A., 2004. The sugarcane woolly aphid, Ceratovacuna lanigera Zehntner (Hemiptera: Aphididae): its biology, pest status and control. Curr. Sci. India 87, 307–316.

Kesavan, R., Easwaramoorthy, S., Santhalakshmi, G., 2003. Evaluation of different formulations of Bacillus thuringiensis against sugarcane early shoot borer Chilo infuscatellus Snellen. Sugar. Tech. 5, 51–55.

Kulashreshtha, R.C., Gursahani, K.A., 1961. An observation on an ento-mogenous fungus, *Metarhizium anisopliae* (Metsch.) Sor. Indian J. Sugar Res. Dev. Work 5, 163–164.

Kuniata, L.S., 2000. Integrated management of sugarcane stemborers in Papua New Guinea. In: Allsopp, P.G., Suasa-ard, W. (Eds.), Sugarcane Pest Management Strategies in the New Millennium. Int. Soc. Sugar. Technol, Brisbane, Australia, pp. 37–50.

Kuniata, L.S., Sweet, C.P.M., 1994. Management of *Sesamia grisescens* Walker (Lepidoptera: Noctuidae), a sugar-cane borer in Papua New Guinea. Crop Prot. 13, 488–493.

Leite, L.G., Tavares, F.M., Batista Filho, A., 2012. Eficiência de nematoides entomopatogênicos e inseticidas químicos contra *Sphenophorus levis* e *Leucothyreus* sp. em cana-de-açúcar. Pesq. Agrop. Trop. 42, 40–48.

Li, H.K., 1992. A preliminary study on *Isaria farinosa*, an important pathogen of pyralid larvae. Chin. J. Biol. Control 8, 93.

LiXing, W., Haihua, D., JinLing, X., Qi, L., LianHui, W., Zide, J., HaiBao, Z., Qiwei, L., LianHui, Z., 2006. Regeneration of sugarcane elite breeding lines and engineering of stem borer resistance. Pest Manage. Sci. 62, 178–187.

Liu, L., Zhan, R., Yang, L., Liang, C., Zeng, D., Huang, J., 2012. Isolation and identification of *Metarhizium anisopliae* from *Chilo venosatus* (Lepidoptera: Pyralidae) cadaver. Afr. J. Biotechnol. 11, 7609–7617.

Machado, L.A., Habib, M., Leite, L.G., Calegari, L.C., Goulart, R.M., Tavares, F.M., 2005. Patogenicidade de nematoides entomopatogêni-cos a ovos e larvas de *Migdolus fryanus* (Westwood, 1863) (Coleop-tera; Vesperidae). Arq. Inst. Biol. 72, 221–226.

Mazodze, R., Zvoutete, P., 1999. Efficacy of *Metarhizium anisopliae* against *Heteronychus licas* (Scarabaedae: Dynastinae) in sugarcane in Zimbabwe. Crop Prot. 18, 571–575.

Melgar, M., 2014. Desarrollo tecnológico de la agroindústria azucarera y perspectivas. In: Melgar, M., Meneses, A., Orozco, H., Pérez, O., Espi-nosa, R. (Eds.), El cultivo de la caña de azucar em Guatemala. Artemis Editer, Guatemala. Available at: http://pt.slideshare.net/mmelgar0506/libro-el-cultivo-de-la-caa-de-azcar-16-febdoc (accessed 20.05.15.).

Memela, N.S., 2014. Isolation of *Beauveria Bassiana* Strains From Kwa-Zulu-natal as Biocontrol Agents against the African Sugarcane Stem Borer *Eldana Saccharina* Walker (Lepidoptera: Pyralidae) MSc the-sis. University of KwaZulu-Natal, South Africa, p. 170.

Mendonça, A.F., 2005. Cigarrinhas da cana-de-açúcar. Insetos & Cia, Maceió, Brazil, p. 317.

Naidu, P., 2009. IPM in Sugarcane. Module B.XII. Available at: http://tsri.com.vn/uploads/files/ipm_in_sugarcane.pdf (accessed 30.05.15.).

Ngubane, N.P., Hatting, J.L., Truter, M., 2012. Entomopathogens associ-ated with African and Mauritian Scarabaeidae affecting sugarcane. Proc. South Afr. Sug. Technol. Ass. 85, 114–117.

Nicholls, C.I., Pérez, N., Vasquez, L., Altieri, M.A., 2002. The develop-ment and status of biologically based integrated pest management in Cuba. Int. Pest Manage. Rev. 7, 1–16.

Nirmala, R., Harlapur, S.I., Ramanujam, B., Rabindra, R.J., Rao, N.S., 2007. Effect of entomofungal pathogens on sugarcane woolly aphid, (*Ceratova-cuna lanigera* Zehntner) and its predators. J. Biol. Control 21, 179–182.

Nthenga, I., Knoetze, R., Berry, S., Tiedt, L.R., Malan, A.P., 2014. *Steiner-nema sacchari* n. sp. (Rhabditida: Steinernematidae), a new entomo-pathogenic nematode from South Africa. Nematology 16, 475–494.

Patil, S.B., Jayarao, K., Khot, R.S., 1996. Bioefficacy of granulosis virus in the management of early shoot borer *Chilo infuscatellus* Snell, of sugarcane. Bharatiya Sugar 22, 7–10.

Pemberton, C.E., Williams, J.R., 1969. Distribution, origins and spread of sugarcane insect pests. In: Williams, J.R., Metcalfe, R.W.,

Mungomery, R.W., Mathes, R. (Eds.), Pests of Sugar Cane. Elsevier, London, UK, pp. 1–9.

Pramono, D., Thoharisman, A., Putranto, D.P., Juliadi, D., Achadian, E.M., 2001. Effectiveness of an indigenous entomopathogenic fungus as a biocontrol agent of *Dorysthenes* sp. (Coleoptera: Cerambycidae). In: Proc. XXIV Cong. Int. Soc. Sugar. Technol. Brisbane, Australia, 17–21 September 2001, pp. 401–403.

Rabindra, R.J., Ramanujam, B., 2007. Microbial control of sucking pests using entomopathogenic fungi. J. Biol. Control 21, 21–28.

Rachappa, V., Krishna Naik, L., Goud, J.V., 2000. Efficacy of different IPM modules against early shoot borer, *Chilo infuscatellus* (Snellen). Kamataka J. Agric. Sci. 13, 878–881.

Rajak, D.C., Varma, A., 2001. Natural enemy complex of insect-pest of sugarcane in the southern zone Andhra Pradesh. Indian J. Sugarcane. Tech. 16, 114–116.

Ramanujam, B., Rangeshwaran, R., Sivakmar, G., Mohan, M., Yandigeri, M.S., 2014. Management of insect pests by microorganisms. Proc. Indian Natl. Sci. Acad. 80, 455–471.

Rapulana, T., Bouwer, G., 2013. Toxicity to *Eldana saccharina* of a recombinant *Gluconacetobacter diazotrophicus* carrying a trun-cated *Bacillus thuringiensis* cry1Ac gene. Afr. J. Microbiol. Res. 7, 1207–1214.

Robène-Soustrade, I., Jouen, E., Pastou, D., Payet-Hoarau, M., Goble, T., Linderme, D., Lefeuvre, P., Calmès, C., Reynaud, B., Nibouche, S., Costet, L., 2015. Description and phylogenetic placement of *Beauve-ria hoplocheli* sp. nov. used in the biological control of the sugarcane white grub, *Hoplochelus marginalis*, in Reunion Island. Mycologia 107, 1221–1232, http://dx.doi.org/10.3852/14-344.

Rutherford, R.S., 2015. IPM for Eldana Control. An Integrated Pest Man-agement (IPM) Approach for the Control of the Stalk Borer *Eldana saccharina* Walker (Lepidoptera: Pyralidae). SA Sugarcane Research Institute, Mount Edgecombe, South Africa. ISBN: 1-874903-41-7, p. 80.

Sallam, N.M., 2006. A review of sugarcane stem borers and their natural enemies in Asia and Indian Ocean Islands: an Australian perspective. Ann. Soc. Entomol. Fr. 42, 263–283.

Sallam, M.S., Allsopp, P.G., 2005. Our home is girt by sea – but how well are we prepared in Australia for exotic cane borers. In: Proc. Australian Soc. Sugar Technol. Bundaberg, Australia, 3–6 May, pp. 358–366.

Samson, P.R., Milner, R.J., Bullard, G.K., 1999. Development of *Metarhi-zium*-based biopesticides for sugarcane pest management-current progress and future prospects. Proc. South Afr. Sugar Technol. Ass. 21, 156–163.

Sanahuja, G., Banakar, R., Twyman, R.M., Capell, T., Christou, P., 2011. *Bacillus thuringiensis*: a century of research, development and com-mercial applications. Plant Biotechnol. J. 9, 283–300.

Sankaranarayanan, C., Easwaramoorthy, S., 2003. Bio-efficacy of ento-mopathogenic nematodes isolates against sugarcane internode borer *Chilo sacchariphagus* (Kapur) at two constant temperatures. In: Ignacimuthu, S., Jayaraj, S. (Eds.), Biological Control of Insect Pests. Phoenix Publishing House, New Delhi, India, pp. 73–78.

Sankaranarayanan, C., Somasekhar, N., Singaravelu, B., Shanmugasun-daram, M., 2008. Pathogenicity of entomopathogenic nematodes to sugarcane internode borer, *Chilo sacchariphagus indicus* Kapur (Lep-idoptera: Crambidae). J. Biol. Control 22, 167–171.

Satyagopal, K., Sushil, S.N., Jeyakumar, P., Shankar, G., Sharma, O.P., Boina, D.R., Sain, S.K., Reddy, M.N., Rao, N.S., Sunanda, B.S., Asre, R., Kapoor, K.S., Arya, S., Kumar, S., Patni, C.S., Chattopadhyay, C.,

Badgujar, M.P., Choudhary, A.K., Varshney, S.K., Tippannavar, P.S., Basavraj, M.K., Thakare, A.Y., Halepyati, A.S., Patil, M.B., Sreenivas, A.G., 2014. AESA Based IPM Package for Sugarcane. Available at: http://farmer.gov.in/imagedefault/ipm/Sugarcane.pdf (accessed 01.06.15.).

Simi, L.D., 2014. Controle de *Sphenophorus levis* e *Conotrachelus humeropictus* pelo uso combinado de nematoides e fungos entomopatogênicos (Ph.D. thesis). UNESP, Botucatu. 107 pp.

Sivasankaran, P., Easwaramoorthy, S., David, H., 1990. Pathogenicity and host range of *Beauveria* nr. *bassiana*, a fungal pathogen of *Chilo infuscatellus* Snellen. J. Biol. Control 4, 48–51.

Solarte, J.A.M., Bustillo, A.E.P., Castro, U.V., Mesa, N.C.C., Moreno, C.A.G., 2012. Eficacia de *Metarhizium anisopliae* para controlar *Aeneolamia varia* (Hemiptera: Cercopiae), en caña de azúcar. Rev. Colomb. Entomol. 38, 177–181.

Spaull, V.W., 1991. *Heterorhabditis* and *Steinernema* species (Nematoda: Rhabditida) for the control of a sugarcane stalk borer in South Africa. Phytophylactica 23, 213–215.

Speare, A.T., 1920. On certain entomogenous fungi. Mycologia 12, 62–76.

Srikanth, J., Easwaramoorthy, S., Santhalakshmi, G., 2010. Field efficacy and persistence of *Beauveria brongniartii* (Sacc.) Petch applied against *Holotrichia serrata* F. (Coleoptera: Scarabaeidae) infesting sugarcane in southern India. Sugarcane Intern 28, 151–156.

Srikanth, J., Subramonian, N., Premachandran, M.N., 2011. Advances in transgenic research for insect resistance in sugarcane. Trop. Plant Biol. 4, 52–61.

Suasa-ard, W., Sommartya, P., Buchatian, P., Puntongcum, A., Chiangsin, R., 2008. Effect of *Metarhizium anisopliae* on infection of sugarcane stems borer, *Dorysthenes buqueti* Guerin (Coleoptera: Cerambycidae) in laboratory. In: Proc. 46th Kasetsart Uni. Ann. Conference, Kasetsart, Thailand, 29 January 2008, pp. 155–160.

Sugar Research Australia, 2013. Greyback Canegrub (*Dermolepida Albohirtum*). Information sheet ISI 3039. Available at: http://www.sugar-research.com.au/icms_docs/164162_IS13039_ Greyback_canegrub.pdf (accessed 14.07.15.).

Swezey, O.H., 1936. Biological Control of the Sugar Cane Leafhopper in Hawaii, 40. Plrs. Rec, Hawaii, pp. 57–101.

Thamarai-Chelvi, C., Thilagaraj, W.R., Nalini, R., 2011. Field efficacy of formulations of microbial insecticide *Metarhizium anisopliae* (Clavicipitaceae) for the control of sugarcane white grub *Holotrichia serrata* F (Coleoptera: Scarabaeidae). J. Biopest. 4, 186–189.

Tippannavar, P.S., Mallapur, C.P., Kulkarni, S., Patil, S.B., Yalmali, 2006. Record of a new entomopathogenic fungus on sugarcane woolly aphid. Curr. Sci. India 91, 56.

Tiwari, N.K., Tanwar, R.K., 2001. Biocontrol agents of sugarcane pests: their bioecology, mass production and field application. In: Upadhyay, R.K., Mukerji, K.G., Chamola, B.P. (Eds.), Biocontrol Potential and its Exploitation in Sustainable Agriculture, vol. 2. Kluwer Academic. Plenum Publishers, NY, pp. 189–213.

Torres, M.C., Cortez, H.M.M., Ortiz, C.F.G., Capello, S.G., Cruz, A.P., 2013. Caracterización de aislamientos nativos de *Metarhizium anisopliae* y su patogenicidad hacia *Aeneolamia postica*, en Tabasco, México. Rev. Colomb. Entomol. 39, 40–46.

Umamaheswari, R., Sivakumar, M., Subramanian, S., 2004. Host range of native entomopathogenic nematodes from Tamil Nadu. Insect. Environ. 10, 151–152.

Varma, A., Tandan, B.K., 1996. Pathogenicity of three entomogenous fungi against insect pests of sugarcane. J. Biol. Control 10, 87–91.

Varma, A., Tandan, B.K., Singh, K., 1992. Field evaluation of the entomogenous fungus, *Metarhizium anisopliae* (Deuteromycotina: Hyphomycetes) of *Pyrilla perpusilla* (Homoptera: Lophopidae). 42. Indian Sug, pp. 463–470.

Vora, V.J., Ramakrishanan, N., 1978. Studies on the milky disease of white grub, *Holotrichia consanguinea* Blanchard (Coleoptera: Scarabaeidae). J. Entomol. Res. 2, 136–141.

Vyas, R.V., Yadav, D.N., 1993. *Steinernema glaseri* (Steiner) Travassor, a biological control agents of root grub *Holotrichia consanguinea* blan. Pak. J. Nematol. 11, 41–44.

Waghulde, J.K., Patil, A.S., Hapase, D.G., 1991. Suppression of *Pyrilla perpusilla* Wlk. in sugarcane with *Metarhizium anisopliae*. Bhartiya Sugar 53–58.

Wang, J.X., Li, L.Y., 1987. Entomogenous nematode research in China. Rev. Nematol. 10, 483–489.

Wang, Z.N., 1996. Three Entomogenous Fungi From Taiwan and Their Pathogenicity to Some Insects, 154. Sugar Res. Institute, Taiwan, pp. 15–30.

West, A.W., 1984. Fate of the insecticidal, proteinaceous parasporal crystal of *Bacillus thuringiensis* in soil. Soil Biol. Biochem. 16, 357–360.

Wilson, G., 1969. White grubs as pests of sugarcane. In: Williams, J.R., Metcalfe, J.R., Mungomery, R.W., Mathes, R. (Eds.), Pests of Sugar Cane. Elsevier, London, UK, pp. 237–258.

Young, G.R., Kuniata, L.S., 1995. The population dynamics of the borer, *Sesamia grisescens* Walker (Lepidiota: Noctuidae), on sugarcane in the Ramu valley of Papua New Guinea. P. N. G. J. Agric. For. Fish. 38, 94–101.

Zappelini, L.O., 2009. Seleção de isolados dos fungos entomopatogênicos *Beauveria bassiana* e *Metarhizium anisopliae* visando ao controle da broca da cana-de-açúcar, *Diatraea saccharalis* (Lepidoptera: Crambidae) MSc thesis. Instituto Biológico, São Paulo, p. 50.

Zappelini, L.O., Almeida, J.E.M., Batista Filho, A., Giometti, F.H.C., 2010. Seleção de isolados do fungo entomopatogênico *Metarhizium anisopliae* (Metsch.) Sorok. visando ao controle da broca da cana-de-açúcar *Diatraea saccharalis* (Fabr. 1794). Arq. Inst. Biológico 77, 75–82.

Zeng, T., 2004. Control of insect pests in Sugarcane: IPM approaches in China. Sugar Tech. 6, 273–279.

ZuNan, W., 1996. Three Entomogenous Fungi From Taiwan and Their Pathogenicity to Some Insects. Report of the Taiwan Sugar Research Institute 154, pp. 15–30.

Chapter 21

Use of Entomopathogens Against Forest Pests

A.E. Hajek[1], K. van Frankenhuyzen[2]

[1]Cornell University, Ithaca, NY, United States; [2]Natural Resources Canada, Sault Ste. Marie, ON, Canada

21.1 INTRODUCTION

The topic of this chapter is the use of entomopathogens and nematodes to control native or exotic invasive insect pests on or in trees in natural forests, plantations, woodlots, and urban environments. Some great successes in the use of microbes for the control of arthropod pests have been achieved in these different types of forests, where the application of chemical insecticides is not always practical or acceptable and where tolerance to pest infestations is often higher than in agricultural crops. The microbial control of forest insects embraces a variety of strategies that differ in both key management objectives and pathogen deployment. The first strategy is classical biological control, an approach that has been used extensively to obtain long-term if not permanent control of established invasive arthropod pests through the introduction of parasitoids, predators, and herbivorous arthropods. Entomopathogens have also been used in classical biological control programs as reviewed by Hajek et al. (2007, 2009) and Hajek and Delalibera (2010). Cases of successful classical biological control of forest insects through the release of inoculative levels of nonindigenous pathogens with the goal of permanent establishment include the introduction of *Entomophaga maimaiga* (Zygomycota: Entomophthorales) for the control of gypsy moth (*Lymantria dispar*; Lepidoptera: Erebidae) in the United States and Europe, nucleopolyhedroviruses (NPVs) for the control of European spruce sawfly (*Gilpinia hercyniae*) and European pine sawfly (*Neodiprion sertifer*; Hymenoptera: Diprionidae) in North America, and introduction of *Deladenus* (=*Beddingia*) *siricidicola* (Nematoda: Neotylenchidae) for the control of the pine-killing wood wasp *Sirex noctilio* (Hymenoptera: Siricidae) in Australia, South America, and South Africa.

A second approach is the augmentation of specialist pathogens to increase their impact on the pest's generational survival (inoculative augmentation). Specialist pathogens are applied to initiate or accelerate epizootics that typically terminate local or regional outbreaks. Examples include the use of NPVs for the suppression of gypsy moth in North America and Europe and for the suppression of Douglas-fir tussock moth (*Orgyia pseudotsugata*; Lepidoptera: Erebidae), balsam fir sawfly (*Neodiprion abietis*; Hymenoptera: Diprionidae), and redheaded pine sawfly (*Neodiprion lecontei*; Hymenoptera: Diprionidae) in North America and the release of generalist hypocrealean fungi (Ascomycota: Hypocreales) to suppress pine caterpillars (*Dendrolimus* spp.; Lepidoptera: Lasiocampidae) in China and wood-boring beetles (*Anoplophora glabripennis* and *Monochamus alternatus*; Coleoptera: Cerambycidae and *Agrilus planipennis*; Coleoptera: Buprestidae) in Asia and North America.

A third approach is inundative augmentation, when entomopathogens are used more as a biological insecticide than a biological control agent. The best known example is the use of *Bacillus thuringiensis* var. *kurstaki* (Btk) for the broad-scale suppression of outbreak populations of defoliating forest insects. Because Btk is a generalist pathogen that does not play a role in driving forest insect population cycles, suppression is temporary, and repeated application during the same outbreak event is often required. Cases of the inundative release of Btk discussed in this chapter include its use for the suppression of spruce budworm (*Choristoneura fumiferana*; Lepidoptera: Tortricidae), other coniferous defoliators, and gypsy moth in North America and pine processionary caterpillar (*Thaumetopoea pityocampa*; Lepidoptera: Notodontidae) and other defoliators in Europe, Asia, and Africa. Btk has also been used successfully to eradicate invasive forest pests before they become established. Prominent cases include the eradication of new introductions of Asian and European gypsy moths in North America and tussock moths in New Zealand.

21.2 CONTROL OF FOREST INSECTS THROUGH CLASSICAL BIOLOGICAL CONTROL

Classical biological control involves the release of natural enemies from the pest's origin into the area where the pest

Microbial Control of Insect and Mite Pests. http://dx.doi.org/10.1016/B978-0-12-803527-6.00021-4

has been introduced to achieve permanent control. It has predominantly been used for the control of invasive invertebrate pests (through the release of parasitoids or predators) and invasive weeds (through the release of herbivorous invertebrates) (Cock et al., 2010). Entomopathogens or parasitic nematodes have been used in 131 cases in addition to eight instances of accidental introductions that resulted in establishment (Hajek et al., 2007). We will describe major introductions of pathogens against forest insect pests, some of which provide dramatic examples of exotic entomopathogens providing control. Surprisingly, some of the most successful examples are for pathogens that may have been introduced accidentally.

21.2.1 Biological Control of Gypsy Moth With Exotic Pathogens

The gypsy moth, *L. dispar*, is a polyphagous hardwood defoliator that is native to Eurasia. It was accidentally introduced from France to the Boston area in North America in 1868 and has been spreading in western and southern directions since then. In North America outbreaks are more severe than in the gypsy moth's native range (Alalouni et al., 2013) and earlier in outbreaks exhibit complex cyclical behavior with outbreak episodes occurring at roughly 10-year intervals in more mesic areas and at roughly 5-year intervals in more susceptible forest types (Johnson et al., 2006). Models suggest that both predators and pathogens influence the occurrence of cyclical outbreaks (Bjørnstad et al., 2010), although this analysis is based on the presence of only the gypsy moth virus and not the more recently introduced fungal pathogen, *E. maimaiga*, as discussed in the following.

21.2.1.1 Lymantria dispar Multiple Nucleopolyhedrovirus

The *Lymantria dispar multiple nucleopolyhedrovirus* (LdMNPV) is highly host specific and occurs throughout the gypsy moth's native range. In Europe it is the dominant pathogen in the regulation of outbreak populations (Novotny, 1989). A classical biological control program against the invasive gypsy moth through introductions of parasitoids began in North America in 1905 (Hoy, 1976). By 1907, a "wilt" disease was reported, caused by LdMNPV, which was thought to have arrived with introduced parasitoids (Hajek et al., 2005) and which naturally spread throughout established gypsy moth populations. Although infection generally does not prevent defoliation in lower density populations, LdMNPV often caused rapid crashes of outbreak populations. However, after the establishment of *E. maimaiga* (see below), both *E. maimaiga* and LdMNPV caused infections during epizootics, with *E. maimaiga* often being more abundant (Liebhold et al., 2013; Hajek et al.,

2015). Methods for the mass production of this virus were developed, and it is now being used successfully in inoculative augmentation approaches for gypsy moth control in both the United States and Canada (see Section 21.3.1).

21.2.1.2 Entomophaga maimaiga in North America

Around the time that the classical biological control with parasitoids against gypsy moth had started, scientists learned of a fungal pathogen in the Entomophthorales (*E. maimaiga*) that killed gypsy moth larvae in Japan. Two larval cadavers containing spores were brought from Japan to the Boston area. Studies with the fungus were conducted in lab and field in 1910–11, but the fungus did not establish. It was never found infecting gypsy moth larvae in pathogen surveys made between 1912 and 1989 as the gypsy moth spread in North America. The fungus was again released in 1985 and 1986 in New York State and Virginia (Hajek et al., 1995), but once again it was thought not to have established.

In the very wet spring of 1989, epizootics caused by *E. maimaiga* occurred in seven northeastern states (Hajek, 1999). The fungus subsequently spread throughout the contiguous gypsy moth distribution over the next 3 years. Molecular studies suggest that the fungus did not originate from the 1910–11 introductions (Nielsen et al., 2005). A model based on weather and larval and fungal development suggests that *E. maimaiga* had been introduced into Connecticut sometime after 1971 (Weseloh, 1998), which implies that the currently established isolate resulted from an accidental introduction. Host specificity studies showed that while *E. maimaiga* could infect numerous species under optimal conditions in the laboratory, in the field very few lepidopteran larvae in the foliage or leaf litter aside from *L. dispar* became infected during epizootics (Hajek, 2007).

As gypsy moth continues to spread in North America, *E. maimaiga* has moved into low-density populations near the edge of the spread (Hajek and Tobin, 2011). The prevalence of infection is associated with environmental moisture (Reilly et al., 2014). *E. maimaiga* at times responds density independently to host populations (eg, Liebhold et al., 2013) and at other times is density dependent (Hajek et al., 2015). In at least some of the more northern areas where gypsy moth occurs, outbreaks have not occurred since 1992, while *E. maimaiga* infections occur every year (AEH, unpublished data). *E. maimaiga* appears to have replaced LdMNPV as the dominant pathogen causing epizootics ending outbreaks and, while cyclical outbreaks still occur in the mid-Atlantic area, damage caused by outbreaks and lengths of outbreaks are thought to have decreased since the fungus became established.

21.2.1.3 Entomophaga maimaiga in Eurasia

L. dispar populations periodically increase to defoliating outbreak densities across Europe and Asia with a periodicity and intensity that shift with latitude and longitude (Alalouni et al., 2013). Outbreaks are more intensive and cause more damage in the Mediterranean and Balkan regions. *E. maimaiga* did not occur in Europe, and so it was sent from the United States to Bulgaria for release in 1996, 1999, and 2000 (Hajek et al., 2005). The 1999 releases resulted in establishment, and cadavers containing resting spores were collected during epizootics and released at additional sites in Bulgaria. In 2005 the fungus was found in Georgia, but the origin of that isolate, which slightly differs at the molecular level from Bulgarian samples, is not known. Beginning in 2011, *E. maimaiga* was reported from additional sites in Eastern Europe and Turkey (Table 21.1; Zúbrik et al., 2016). To what extent *E. maimaiga* spread on its own (most probably due to the airborne dispersal of conidia) or was moved via soil (ie, on soles of shoes or by other means) is not known. When gypsy moth populations increase again it is expected that *E. maimaiga* will have spread further. It is too early to understand how *E. maimaiga* will influence the outbreak dynamics of gypsy moth in Europe.

21.2.1.4 Microsporidia

Extensive studies conducted in Europe identified five species of microsporidia infecting gypsy moth in this area of endemism, and host specificity of these microsporidia was evaluated (eg, Solter and Hajek, 2009). Between 1986 and 2010, four species were released within 5 years in three US states (Hajek et al., 2005; Solter, personal communication). While evaluations of the early introductions in Maryland have not been conducted, the later introductions (in Michigan and Illinois) were considered unsuccessful due to environmental conditions or because recipient populations were killed by an epizootic of *E. maimaiga*.

TABLE 21.1 Years When *Entomophaga maimaiga* was First Reported as it Spread in Western Asia and Eastern Europe After Introduction to Bulgaria (Zúbrik et al., 2016)

Year	Countries
1999[a]	Bulgaria
2005	Georgia
2011	Turkey, Serbia
2012	Greece, Macedonia
2013	Slovakia, Hungary, Croatia, Bosnia and Hercegovina

[a]*Purposefully successfully introduced in Bulgaria in 1999 while all other records were not purposeful introductions.*

21.2.2 Biological Control of Sawflies With Exotic Nucleopolyhedroviruses

21.2.2.1 European Spruce Sawfly Nucleopolyhedrovirus

The European spruce sawfly, *G. hercyniae*, became a serious defoliator of spruce in eastern Canada and nearby US states in the 1930s (Cunningham, 1998). Introductions of parasitoids from Europe and Japan started in 1933. In 1936, an NPV (GiheNPV) was reported in New Brunswick and adjacent US states (Hajek et al., 2005). It is hypothesized that the virus was introduced along with some of the parasitoids. A natural dispersal of the virus was facilitated by intentional releases in Quebec and Ontario. Collapse of the spruce sawfly outbreak between 1938 and 1942 was attributed largely to the virus (Balch and Bird, 1944; Bird and Elgee, 1957). Today, *G. hercyniae* is no longer a problem, and the combined action of virus and parasitoids continues to hold spruce sawfly populations at endemic levels (Neilson and Morris, 1964). This is considered a stellar example of successful long-term control by an exotic pathogen acting in concert with two of the parasitoid species introduced during the classical biological control program.

21.2.2.2 European Pine Sawfly Nucleopolyhedrovirus

The European pine sawfly, *N. sertifer*, is a serious pest of pine in continental Europe and the British Isles (Huber, 1998). *N. sertifer* was first reported in New Jersey in 1925 but was not recognized as exotic until 1937, by which time it had spread west and north, with the first reports from Canada in 1939 (McGugan and Coppel, 1962). In North America, it became a problem in Christmas tree plantations and ornamental pines (Cunningham, 1998). Successful control of the pest with its NPV (NeseNPV) in Europe provided an impetus to investigate the use of NeseNPV for control in North America. *N. sertifer* larvae feed gregariously, and the virus, which replicates in the midgut, is released via defecation when infected gut cells are sloughed off (called "infectious diarrhea"). So, once the virus is introduced to a colony, transmission within the colony as well as to colonies lower down on trees can be very rapid (Cunningham and Entwistle, 1981). The virus was first introduced from Sweden to southern Ontario in 1950 and to many sites in Canada and the United States during the two decades that followed (Hajek et al., 2005). Few results of releases were recorded, but one example from 1951 reported that an introduction of the virus controlled an infestation over 40 ha within 3 years.

The USDA Forest Service conducted host specificity testing, and in 1983 Neochek-S was registered by the Environmental Protection Agency (EPA) as a biopesticide for use in the United States. A registration petition was also submitted in Canada for a product named Sertifervirus. The virus was used to treat small areas in the United States and Canada between 1975 and 1993 (Cunningham, 1998). Those products are no longer available, because the European pine sawfly did not continue to be a serious pest in North America, and registrations were not renewed due to declining demand. NeseNPV products were also developed in Finland, the Czech Republic, and the United Kingdom and were used operationally in Scandinavia (Huber, 1998), but registrations of those products have also expired.

21.2.3 Biological Control of Sirex Woodwasp With Exotic Nematodes

Sirex noctilio is native to Eurasia and North Africa where it attacks pines. This woodwasp is obligately associated with the basidiomycete white rot fungus *Amylostereum areolatum*, which females transport in mycangia and deposit in trees when they oviposit. Pines are killed by the joint activity of the fungus and the phytotoxic venom noctilisin, also deposited by ovipositing females. Although *S. noctilio* is not considered a pest where it is native, in areas of introduction in the Southern Hemisphere it is very aggressive and has killed large plantations of introduced pines. *S. noctilio* was first found in the Southern Hemisphere in New Zealand around 1900. This species subsequently spread to Australia (first found in Tasmania in 1952), South America (Uruguay in 1980), South Africa (1994), North America (New York State in 2004) (Hajek and Morris, 2014), and most recently northeastern China, where it is not considered native (J. Xi, personal communication). In many of these introduced locations, *S. noctilio* populations are still spreading. In the Southern Hemisphere, pines are not native, and communities associated with pines largely consist of introduced species. In contrast, in the Northern Hemisphere, pines are native, hosting communities that include native *Sirex* species and their associates.

During classical biological control foreign exploration in the 1960s and 1970s (Hajek and Morris, 2014), several hundred strains of seven species of nematodes in the genus *Deladenus*, parasitizing several species of siricids, were also collected (Bedding and Akhurst, 1974). The potential for control by nematodes is due to sterilizing adult females, because all eggs in a parasitized *S. noctilio* female can be dead and full of nematodes. All of the siricid-associated species of *Deladenus* have an unusual life cycle, with morphologically different forms either feeding on *Amylostereum* or parasitizing siricids. This dimorphic life cycle has been advantageous, because the nematodes could easily be cultured in the mycophagous form, injected into appropriate trees, and would increase within the tree until encountering *S. noctilio* larvae, when they change form and parasitize. Natural transmission is possible, because when high densities of *S. noctilio* occur, both parasitized and healthy eggs are laid in the same tree by different mothers, and nematodes disperse within the tree to locate new hosts. After an evaluation of many strains, the nematode chosen for use was a strain of *D.* (=*Beddingia*) *siricidicola* from Sopron, Hungary. This strain yielded high levels of *S. noctilio* parasitism, sterilized adult females, and fed on *A. areolatum* (the symbiotic fungus carried by *S. noctilio*, instead of other *Amylostereum* species). It did not parasitize the hymenopteran parasitoids, and it minimally decreased the size of parasitized *S. noctilio* (so parasitism would not negatively impact nematode dispersal by *S. noctilio*).

The Sopron strain of *S. noctilio* was found to be very effective for the control of *S. noctilio* in Australia, with up to 100% parasitism (Bedding and Akhurst, 1974), and by 1984 the nematode was considered superior to any parasitoids (Neumann and Morey, 1984), so subsequent emphasis for the biological control of *S. noctilio* has been on the nematodes. Methods for mass producing the nematodes and releasing them in pine plantations infested by *S. noctilio* were developed. The use of this nematode for the control of *S. noctilio* in Australia has been considered a stellar example of classical biological control, but, with time and continued spread of *S. noctilio*, the picture has become more complicated. In 1987–90, an outbreak of *S. noctilio* affected a large area of *Pinus radiata* plantations in southeastern Australia and did not seem to be controlled by the Sopron strain that was already present. With extended nematode culture in the mycophagous form, the nematode had lost the ability to become parasitic. This discovery resulted in field reisolation of a strain that made parasitics (the Kamona strain, isolated from an original testing site in Tasmania) and developing methods for in vivo cryopreservation so that continuing production of virulent *D. siricidicola* was possible.

An Australian company (Ecogrow) now mass produces and sells *D. siricidicola*, which is used for classical biological control but is also used for inoculative augmentation when a fast response to curtail a new outbreak is needed. Australian pine plantations are surveyed by air and, if *S. noctilio*-struck trees are present, *S. noctilio* are reared from trees to evaluate existing nematode parasitism. Based on results as well as the density of struck trees, a new inoculative release of *D. siricidicola* may be made to bolster the existing nematode population (Carnegie and Bashford, 2012).

After *D. siricidicola* was developed for release in Australia, it was subsequently released in Brazil, Argentina, Chile, and South Africa (Slippers et al., 2012). However, efficacy has often been variable. Problems encountered include the following: (1) in Australia, trap trees used for release become infested with a competing bark beetle and blue stain fungus, so that numbers and sizes of *S. noctilio* emerging from inoculated trap trees to spread the nematodes are reduced (Yousuf et al., 2014); (2) in Brazil, the attenuated Sopron strain had been initially released, although the Kamona strain was subsequently used, and parasitism levels of >70% have subsequently been documented (Iede et al., 2012); (3) in South African areas where rain falls during the summer and not the winter, low parasitism has principally been attributed to low wood moisture (Hurley et al., 2012); (4) the potential for variable success of the genetically homogeneous Kamona strain (Mlonyeni et al., 2011) against different genotypes of *S. noctilio* introduced in different areas of the world (Boissin et al., 2012; Bedding, 1972); and (5) the presence of "nonsterilizing" strains of *D. siricidicola* in the field in New Zealand, Argentina, and North America (Zondag, 1975; Kroll et al., 2013; Klasmer and Botto, 2012) with the potential for competition and/or hybridization with Kamona (Akhurst, 1975; Williams, personal communication), resulting in unknown impacts on sterilization levels. Researchers in the United States are presently evaluating the potential impact of *D. siricidicola* Kamona on the *S. noctilio* genotypes introduced to North America as well as the potential for nontarget impacts on native siricid communities. Results will be used to inform decisions about whether and/or how *D. siricidicola* might be used for control in North America.

21.3 SUPPRESSION OF FOREST INSECTS THROUGH INOCULATIVE AUGMENTATION

21.3.1 Use of Nucleopolyhedrovirus for Control of Gypsy Moth

The *L. dispar* NPV has been present in North America since at least the start of the 20th century (see Section 21.2.1). Although suggested as a microbial control agent in the early 1900s, efforts to develop the virus for gypsy moth control did not begin until the 1960s with two decades of field testing in Europe, the Soviet Union, and the United States (Lewis, 1981). In the United States, ground spray trials in the early 1960s were followed by aerial spray trials in the 1970s and culminated in the USDA Forest Service obtaining registration of a product called Gypchek in 1978. Since its registration, Gypchek has been the subject of intense research to maximize its production and field efficacy. Various application rates and volumes were tested using various types of aircraft and nozzles (Reardon and Podgwaite, 1994; Reardon et al., 1996). Efficacy improved with the introduction of a new tank mix in 1987. Aqueous flowable carriers were evaluated in the early 1990s and permitted the application of recommended dose rates in 9.7 L per ha as compared to 19.4 L per ha for the previous tank mix. The Canadian Forest Service obtained acceptable infection and mortality at even lower application volumes (2.5–5.0 L per ha) and received registration for Disparvirus in 1997 (Cunningham, 1998). However, high costs of in vivo production and market limitations resulting from high target specificity

combined with the cyclical nature of gypsy moth outbreaks and multiyear treatment efficacy made LdMNPV unattractive for private investment, and repeated efforts to establish commercial production failed (van Frankenhuyzen et al., 2015). The sole source of product was a 6000-ha equivalent that was produced annually by the USDA Forest Service and USDA APHIS, a production that has now been contracted out to Sylvar Technologies Inc (Fredericton, New Brunswick, Canada). The acquisition of Sylvar by an international biological control company, Swiss-based Andermatt Biocontrol AG, is hoped to secure the future availability of Gypchek as well as other viral biopesticides mentioned in this chapter. Gypchek has been used for the suppression of gypsy moth populations on a cumulative total of ~32,000 ha since 1988, mostly in environmentally sensitive areas, in addition to its use in eradication programs and the Slow-the-Spread management program (see Section 21.5.1) (USDA Forest Service Gypsy Moth Digest; http://www.na.fs.fed/fhp/gm). The current recommendation for population suppression is two applications 3 days apart soon after egg hatch is complete, using 5×10^{11} occlusion bodies in 4.7–9.7 L per ha (van Frankenhuyzen et al., 2007). In Canada, the virus has not been used for operational suppression.

The same virus that is present in Gypchek and Disparvirus was also produced under the name Biolavirus-LD in the Czech Republic. In the former Soviet Union, the development of a different NPV of gypsy moth with a broader host range was pioneered in the early 1960s. It was registered under the product name Virin-ENSh, and, in 1998, it was reported as being the most important viral product in terms of production volume and area of application in the former Soviet Union (Lipa, 1998). These products are no longer available as their registrations have expired. The commercialization of gypsy moth NPV was also attempted in Germany in response to large-scale outbreaks (Huber, 1998).

21.3.2 Use of Nucleopolyhedrovirus for Control of Sawflies

21.3.2.1 Balsam Fir Sawfly

The balsam fir sawfly (*N. abietis*) is an eruptive defoliator of firs and spruces in North America. Periodic outbreaks have been reported from Central to Atlantic Canada, usually lasting 3–4 years (Cunningham, 1984). When an outbreak in precommercially thinned stands in western Newfoundland in the early 1990s did not collapse as anticipated, research was undertaken to explore options for biological control (Moreau and Lucarotti, 2007). An NPV, NeabNPV, was isolated from a local population and mass produced in the field. Evaluation between 2001 and 2005 on ~22,000 ha demonstrated that aerial applications of NeabNPV to increasing or peaking populations at rates as low as 1×10^9

polyhedral occlusion bodies per ha can initiate or hasten the decline of an outbreak (Moreau et al., 2005). NeabNPV was granted conditional registration in 2006 and full registration in 2009 for the control of balsam fir sawfly in Canada under the trade name Abietiv. A licensing agreement with Sylvar Technologies Inc led to the first operational use of a commercially produced baculovirus product in Canada: Abietiv was used for the control of balsam fir sawfly in Newfoundland on 15,000 ha annually from 2006 to 2009 and in New Brunswick on 10,000 ha in 2011 (National Forestry Database, http://nfdp.ccfm.org).

21.3.2.2 Redheaded Pine Sawfly

The redheaded pine sawfly (*N. lecontei*) is a serious defoliator of red pine (*Pinus resinosa*) plantations in eastern Canada. Heavy defoliation reduces height growth and causes branch mortality and growth deformity. In 1950, an NPV (NeleNPV) was found in Ontario. Field trials demonstrated that the virus can be introduced into outbreak populations to initiate epizootics that carry over into subsequent generations (Bird, 1971). An intensive research effort was initiated in Canada to develop NeleNPV as a biological control agent, which resulted in its registration as a pest control product under the trade name Lecontvirus in 1983. The virus was not available commercially but was produced by the Canadian Forestry Service through harvesting virus-killed larvae from treated infestations. The recommended dosage of 5×10^9 occlusion bodies per ha was obtainable from about 50 field-collected larvae. Lecontvirus was the only entomopathogenic virus that was ever used routinely in Canadian forests, albeit on a limited scale: between 1983 and 1994 it was used on a total of ~6000 ha of plantations, mostly by private landowners (Cunningham, 1998). Surveys of treated plantations showed that one application can suppress populations for several years. Field production was discontinued in the mid-1990s due to a change in mandate of the Canadian Forest Service. About 10 years later Lecontvirus was included in the Abietiv licensing agreement with Sylvar Technologies and is now awaiting renewed production and commercialization.

21.3.3 Use of Nucleopolyhedrovirus for Control of Douglas-Fir Tussock Moth

The Douglas-fir tussock moth (*O. pseudotsugata*) is an important defoliator in the interior dry-belt forests of southern British Columbia and the western United States. The primary host tree is Douglas fir, *Pseudotsuga menziesii*. Outbreaks occur periodically, at intervals of about 8–14 years, with noticeable defoliation usually lasting from 2–5 years. Douglas-fir tussock moth infestations in the western United States were targeted by extensive chemical spray programs, primarily with DDT in the 1940s to 1960s,

and various organophosphates after that. Outbreaks are generally terminated by epizootics of naturally occurring NPVs but not before serious host tree damage has already occurred (Otvos et al., 1995). Field trials during the 1970s resulted in the registration of OpMNPV (the multicapsid strain) by the USDA Forest Service under the trade name TM-BioControl in 1976. The same virus received registration in Canada as Virtuss in 1983.

Field testing in the early 1980s demonstrated that the virus can be introduced into Douglas-fir tussock moth populations during the early outbreak phase, before naturally occurring virus increases its prevalence, and application causes collapse of the outbreaks before tree mortality occurs (Otvos et al., 1995; Cunningham, 1998). OpMNPV was used for operational suppression of Douglas-fir tussock moth outbreaks in British Columbia during the early 1990s (1850 ha in 1991–93) and on a larger scale in 2009–10 (6350 ha; National Forestry Database, http://nfdp.ccfm.org). The treatment of 15,840 ha in Oregon in 2000 represented the first large-scale operational use of the virus in the United States. It was subsequently used in Washington (6400 ha in 2001 and 4800 ha in 2010) and New Mexico (450 ha in 2007; USDA Forest Service, Region 6 reports, http://www.fs.usda.gov/detail/r6/forest-grasslandhealth/insects-diseases). The material used in these programs came from a stockpile that was produced by the USDA Forest Service between 1985 and 1995. Production of the virus has now been licensed to Sylvar Technologies for use in Canada and the United States.

21.3.4 Use of *Beauveria bassiana* Against Pine Caterpillars in China

Caterpillars in the genus *Dendrolimus* are common forest defoliators from northeastern to southern China. Masson's pine caterpillar, *D. punctatus*, the most common species in the south, is of particular concern as it increases to outbreak densities that can lead to tree mortality every 3–5 years. *D. punctatus* larvae are susceptible to *Beauveria bassiana*, and this fungus has been used extensively for their control. Beginning in the 1970s, *B. bassiana* was mass produced by many small plants throughout China for application against *D. punctatus* (Li, 2007). The fungus is locally produced cheaply on bran or peat, and by 1980 there were 60 factories in Hunan Province alone. Since then, the production methodology has improved but is still not centralized in large industries. Although not technically sophisticated, the use of *B. bassiana* against *D. punctatus* has been and continues to be a very successful biological control program in pine forests in China and has been called the "largest use of a biocontrol agent"; *B. bassiana* has been used on up to a million ha of pine forest (Lord, 2005).

The application of *B. bassiana* conidia as sprays or mists was tested, but the optimal application is as a dust.

Nonformulated, dry conidia are preferred for application because of their smaller volume and light weight, which make this type of application material more convenient for transport. Conidia applied as dry dusts also are carried by the wind over longer distances and up into tree canopies. Results from the field vary from poor levels of infection to over 80% mortality of *D. punctatus*. Epizootiological studies conducted have shown that at least 32 native *B. bassiana* genotypes reside in Chinese pine forests (Li, 2007). Studies suggest that the *B. bassiana* that is applied persists in forests but is aided in control by the native strains.

Inundative treatment of forests with ground sprayers was very difficult, because forests needing treatment are often rugged, thorny, and difficult to access. A variety of methods were tested for inoculative application, including the release of field-infected insects and the use of explosives to spread conidia to the tops of trees and over larger areas (Li, 2007). Although both inundative and inoculative applications reduced populations dramatically, only frequent inoculative releases resulted in long-term suppression, making it the recommended method for *D. punctatus* suppression (Li et al., 2010).

21.3.5 Use of Fungi Against Wood-Boring Beetle

Hypocrealean fungi have been investigated for the inoculative control of invasive wood-boring beetles in Japan, China, and the United States during the last several decades. In all cases, control is primarily based on infecting emerging adults, which inoculate themselves during dispersal, although in a few instances some earlier stages might also become infected.

21.3.5.1 Monochamus alternatus

In Japan and China, the pinewood nematode, *Bursaphelenchus xylophilus*, has caused extensive mortality of pine trees. This nematode is vectored by the cerambycid *M. alternatus*, and the vector is the primary target for control programs. When *M. alternatus* ecloses, pinewood nematodes in the surrounding wood move into the tracheae of the beetles. After *M. alternatus* adults emerge from dead trees, they fly to the canopy of a healthy tree for maturation feeding, and it is during this activity that the nematodes move into the new tree. Subsequent nematode feeding in the tree disrupts the water transport of susceptible pines, leading to the death of the pine. To control these beetles, bioassays demonstrated that *B. bassiana* was virulent against adults. Cultures of *B. bassiana* were grown within long strips of a loosely woven material so that the fungus grew within the material, and the surface of the material became covered with conidia. A commercialized product for the control of cerambycids in Japanese orchards was based on these "fungal bands"

(Higuchi et al., 1997). For the control of *M. alternatus* in Japan, the conidiated strips of material were placed on top of stacks of wood inhabited by immature stages of *M. alternatus*. Then the stacks of wood were covered with tarps so that when adults emerged, they self-inoculated when walking across the bands in order to disperse (Shimazu, 2009). Although 10–15 days were required before emerging adults died from fungal infection, infected adults were immobile and seldom fed, so transmission of the nematodes to new trees was considered minimal. In China the strategy for the use of fungal bands was somewhat different as fungal bands made with *B. bassiana* were wrapped around infested pine trees, and *M. alternatus* adults were attracted to these trees using baited traps (Li et al., 2010). The adult beetles became infected when contacting the fungal bands, and this resulted in epicenters of disease and suppression of the beetles over large areas. In 2006, the control program, covering 6000 ha of pine plantations, resulted in population decline, and the treated area expanded in 2007.

21.3.5.2 Asian Longhorned Beetle

In the United States yet another approach to the use of fungal bands was tested but this time against an invasive polyphagous cerambycid with the common name Asian longhorned beetle (*A. glabripennis*). This aggressive pest has been introduced to five states in the United States, one province in Canada, and numerous European countries, and at all locations it is under eradication based on the extensive potential damage possible to hardwood forests if it becomes established. As is typical of cerambycids, the detection of *A. glabripennis* is difficult. Fungal isolates were screened, and the F52 strain of *Metarhizium brunneum*, which is already registered with the US EPA, was shown to be virulent. Research continued with this strain since much of the testing necessary for registration would not be necessary (Hajek and Bauer, 2009). F52 was grown in fungal bands and placed around trunks where scaffold branches left the trunk, because *A. glabripennis* commonly walk on tree trunks and branches in the shaded canopy of trees. Field trials conducted in Anhui Province, China, demonstrated adult mortality and decreased oviposition in treated areas. Fungal bands in Queens, New York, retained high densities of infective conidia for >3 months. Although fungal bands appeared promising as a preventive approach around areas where infested trees had been removed, application against *A. glabripennis* would require climbing individual trees, and for this reason the US federal authority responsible for eradication was not interested in using fungal bands. In recent years, research turned to growing *M. brunneum* F52 as longer-lived microsclerotia and then formulating them with materials to both retain moisture as well as stick them to the tree trunk. In this case, the formulations can be sprayed onto tree trunks from the ground, which makes application much easier.

21.3.5.3 Emerald Ash Borer

The emerald ash borer, *A. planipennis*, is native to China but has been introduced to North America and Russia. It attacks and kills ash trees, *Fraxinus* spp., and is a huge threat to both natural and urban forests. It was first found in Michigan (United States) and Ontario (Canada) in 2002 and around Moscow, Russia, in 2005 and has been spreading from these sites. The US EPA-registered *B. bassiana* strain GHA was found to be virulent to *A. planipennis*, and different methods for application were tested. Ground-based foliar and trunk applications reduced numbers of *A. planipennis* feeding in and emerging from infested ash trees and reduced new infestations of healthy ash trees (Hajek and Bauer, 2009). An oil-based emulsifiable concentrate of GHA sprayed onto trees provided infection with subsequent mortality for at least 14 days and could be used as a preemergent spray (Castrillo et al., 2010a,b). However, the longevity of conidia on trees was not long compared to the period that *A. planipennis* adults are active, and the broad spectrum activity of *B. bassiana* was considered a concern in light of the potential for nontarget impacts, so a more directed approach using fungal cultures followed.

Native isolates of *B. bassiana* virulent to *A. planipennis* were identified and tested using autodissemination traps, which trapped live beetles. Beetles contacted conidiated *B. bassiana* cultures within traps, and after becoming inoculated, beetles could disperse with the goal of further transmission. *B. bassiana* within autocontamination chambers could retain pathogenicity to adults over 43 days of outdoor exposure; adult beetles exposed to a culture from a trap that had been in the field for 29 days became abundantly contaminated with conidia in the laboratory (Lyons et al., 2012). However, results in terms of increasing the prevalence of fungal infections in emerald ash borer outbreaks were disappointing, and work on autodissemination traps has been discontinued.

21.4 SUPPRESSION OF DEFOLIATORS THROUGH INUNDATIVE RELEASE OF PATHOGENS

21.4.1 Role of Forestry in the Commercialization of *Bacillus thuringiensis*

Btk is the main pathogen that is being used in inundative release approaches for the suppression of forest insect outbreak populations. It is currently the most successful microbial pest control product used around the world for the protection of not only forests but also crops (Entwistle et al., 1993). That success is rooted in its development as a commercial product for the suppression of defoliating forest insects in North America during the second half of the previous century. Cyclical outbreaks of various defoliators provided an arena that was attractive for commercial development, because it offered a market that was easy to access (single user) and economically attractive (large-scale, high-product volume). Another factor was that a considerable degree of uncertainty in efficacy could be tolerated, because in forests, only a proportion of the current year's foliage needs to be saved to prevent tree mortality and to minimize impacts on tree growth and yield. The forestry experience demonstrated that Btk could be developed as a commercially viable alternative to synthetic insecticides, at a time when there was widespread uncertainty regarding how competitive Btk could be on the market, versus other control options. Success in the North American forestry market provided the experience and confidence that were critical for the subsequent development in other pest control markets around the world. How Btk became the mainstay in forest protection programs was reviewed in detail by others (Reardon et al., 1994; van Frankenhuyzen, 1995; van Frankenhuyzen et al., 2015), but a brief summary here will set the stage for sections that follow.

The development of Btk for forest insect control occurred along two parallel and interconnected tracks. Main testing arenas were provided by massive outbreaks of spruce budworm in Canada and of gypsy moth in the United States. Outbreaks of both species became targets for large-scale aerial spray programs in the late 1940s throughout northeastern North America to protect threatened timber resources, using DDT for the first decade and mostly organophosphates after that. As evidence for negative environmental impacts of those operations accumulated, research was initiated to develop environmentally safe alternatives. The bacterium Bt quickly became a key focus for that research as it had been known as an entomopathogen since the start of the century and had already been tested for insect control during the late 1920s and early 1930s. Initial field tests against spruce budworm and gypsy moth in the early 1960s yielded poor results due to poor formulations and the use of ineffective strains. Test results improved during the 1970s through the use of products based on the superior HD-1 *kurstaki* isolate with standardized potencies. Research on spray atomization, optimum droplet sizes, and spray deposition in softwood and hardwood canopies guided the development of better formulations and more sophisticated application technologies. For ease in quantification and making comparisons, application rates are expressed in billions of international units (BIU), which is a measure of the activity or potency of the formulation rather than an expression of mass. The advent of high-potency formulations that could be applied undiluted in ultra-low spray volumes made the most significant contribution to reducing constraints of high costs and unreliable efficacy. Operational use in both markets in the early 1980s proved critical by providing the practical experience that was needed to further improve operational field efficacies, while competitive bidding by suppliers forced reductions in product cost. By the mid-1980s, Btk was generally accepted as a viable alternative to synthetic insecticides for the control of spruce budworm and gypsy moth. Between 1983 and 1990, the use of Btk products skyrocketed (Table 21.2), eventually replacing all or much of the synthetic insecticides in forest protection programs across North America for the suppression of budworms, gypsy moth, and other defoliators. This was followed by similar trends in Europe and other parts of the world. During the 35 years since its first operational use against spruce budworm in 1980, Btk has been used to suppress outbreak populations of native forest defoliators and eradicate invasive forest pests on a cumulative total of 11.3 million ha in Canada, 6.2 million ha in the United States, and at least 1.8 million ha in Europe.

21.4.2 Microbial Control of Forest Defoliators in Europe

Improvements in formulation and application of Btk resulting from more than a decade of operational experience in North America, together with the introduction of differential Global Positioning Systems and other electronic guidance systems to aid operational navigation of spray aircrafts (van Frankenhuyzen et al., 2015), set the stage for a sharp increase in the use of Btk in Europe's forestry market during the 1990s. Btk use in Europe had grown steadily between the early 1970s and mid-1980s for the control of various defoliators (van Frankenhuyzen, 2000). For example, the use of Btk-based products in Bulgaria rose from 2% of the total area treated in 1975 to 53% in 1984 for the treatment of up to 40,000 ha annually; treatment typically involved the application of diluted products in high volumes. In the early 1990s, high-potency products and undiluted applications in ultra-low volumes were introduced by the Btk manufacturers and quickly became standard practice. High-potency products at 20–50 BIU per ha are now routinely used for control of the pine processionary caterpillar, *Thaumetopoea pityocampa,* in the Mediterranean region. In Poland, high-potency products were applied at 50 BIU in 4.0 L per ha on ~148,000 ha in 1994 to control an outbreak of nun moth, *Lymantria monacha.* The ultra-low-volume application technology was introduced in Russia by Abbott Laboratories in 1996 in a cooperative program for control of the Siberian moth, *Dendrolimus superans sibericus,* on ~120,000 ha using 38 BIU in 3 L per ha. Btk was used for the first time in Sweden in 1997 on ~4000 ha against pine looper, *Bupalus piniara,* and again in 1998 on 1000 ha against nun moth. The total use in Europe for the control of defoliators is estimated to have involved the treatment of ~1.8 million ha of deciduous and coniferous forests between 1990 and

1998 alone (van Frankenhuyzen, 2000), but quantitative data for use after 1998 are not readily available.

21.4.3 Suppression of Gypsy Moth in North America

The gypsy moth has defoliated more than 37 million ha of hardwood forests in North America since 1924 (USDA Forest Service Gypsy Moth Digest; http://www.na.fs.fed/fhp/gm). The main impacts are on oak (*Quercus* spp.), the preferred host. Timber losses estimated in 1981 had a value of US\$ 72 million in the state of Pennsylvania alone (Montgomery and Wallner, 1988). Because of such losses, gypsy moth populations have been subjected to massive aerial applications of insecticides since the early 1940s in attempts to suppress or eradicate local outbreaks. Between 1949 and 2014, ~6.6 million ha of gypsy moth infested-forest lands have been treated with synthetic insecticides, initially DDT (~3.7 million ha between 1949 and 1960), followed by carbaryl and trichlorfon (~0.85 million ha between 1967 and 1987) and most recently the more selective diflubenzuron (~2.0 million ha between 1982 and 2014; Liebhold and McManus, 1999; USDA Forest Service Gypsy Moth Digest; http://www.na.fs.fed/fhp/gm). Microbial control products have featured prominently in gypsy moth suppression programs since the early 1980s, in particular Btk.

In the United States, gypsy moth infestations provided the main arena for the development of Btk as a viable forest insect control product, parallel to and interconnected with its development for the control of spruce budworm in Canada, as described earlier. The use of ultra-low-volume applications of undiluted high-potency Btk-based products took a decade longer to gain general acceptance in gypsy moth control programs. Currently, doses of 60–95 BIU per ha are applied undiluted in volumes of 3.0–4.7 L, usually in a single application. In Canada, the recommended prescription is two applications of 30–50 BIU in 2.4–3.9 L per ha. Spray applications are typically conducted when oak leaves are 40–50% expanded or when the majority of larvae are second instars (van Frankenhuyzen et al., 2007). Such applications provide routinely acceptable foliage protection but highly variable reductions in egg mass densities (Smitley and Davis, 1993). In Canada, Btk is the only product that has been used for gypsy moth suppression since the mid-1980s, with use on a total of ~0.6 million ha (Table 21.2). In the United States, the market is split between Btk and the insect-growth regulator diflubenzuron (Dimilin). Since 1980, Btk has been applied to a cumulative total of ~3.2 million ha for the suppression of gypsy moth outbreaks on federal and state-owned land, in addition to extensive use in specific eradication programs and the Slow-the-Spread management program (see Section 21.5.1; USDA Forest Service Gypsy Moth Digest; http://www.na.fs.fed/fhp/gm).

TABLE 21.2 Ha[a] (in Thousands) Treated With Commercial *Bacillus thuringiensis* var. *kurstaki* Products for the Suppression of Major Forest Defoliators[b] in North America Between 1980 and 2014

Year	ESBW Canada	ESBW United States	WSBW Canada	WSBW United States	JPBW	EHL	GM Canada	GM United States[c]	Total
1980	61.0	56.6	0.0	0.0	0.0	0.0	0.0	6.8	124.4
1981	53.5	50.6	0.0	0.0	0.0	0.0	0.0	8.9	113.0
1982	46.1	35.8	0.0	0.0	0.0	0.0	0.0	26.9	108.8
1983	59.5	47.7	0.0	0.0	0.0	0.0	0.0	190.3	297.5
1984	360.6	88.4	0.0	0.0	0.0	0.0	0.0	87.3	536.3
1985	675.7	133.1	0.0	16.5	220.0	2.4	0.17	108.1	1,155.8
1986	356.8	0.0	0.0	0.0	482.0	5.4	216.27	87.7	1,148.2
1987	404.8	0.0	0.0	64.3	105.4	5.1	80.50	127.5	787.6
1988	434.5	0.0	1.8	251.3	0.0	25.6	27.57	110.4	851.3
1989	432.7	0.0	0.5	4.9	14.3	9.1	25.90	165.3	652.8
1990	1,061.9	0.0	0.0	29.0	0.0	44.9	67.91	340.6	1,544.4
1991	526.7	0.0	3.0	46.1	0.0	16.9	73.15	295.3	961.2
1992	261.1	0.0	35.6	74.5	0.0	0.9	80.00	263.8	716.0
1993	195.2	0.0	34.2	25.9	0.1	45.2	2.62	148.9	453.4
1994	47.9	0.0	21.0	0.0	21.5	17.7	2.08	138.3	260.2
1995	204.4	0.0	0.0	11.3	51.0	89.6	1.06	108.8	479.3
1996	213.5	0.0	0.0	8.1	25.6	145.5	0.36	80.6	481.3
1997	112.8	0.0	16.1	0.0	0.0	14.4	0.00	18.7	169.6
1998	201.5	0.0	21.2	6.5	0.0	7.2	0.00	36.7	286.9
1999	283.0	0.0	21.7	0.0	0.0	16.3	32.42	61.3	417.7
2000	102.3	0.0	8.4	0.0	0.0	107.2	0.00	91.1	342.7
2001	103.6	0.0	26.8	0.0	0.0	68.2	0.92	109.6	334.1
2002	160.3	0.0	30.9	0.0	0.0	141.2	0.00	59.9	403.9
2003	79.1	0.0	22.1	0.0	0.0	42.0	0.00	27.2	198.5
2004	0.00	0.0	25.5	0.0	0.0	0.0	0.60	29.4	108.0
2005	60.4	0.0	30.4	0.0	0.0	0.0	0.00	2.9	137.2
2006	52.6	0.0	43.9	0.0	104.2	0.0	0.00	58.0	297.1
2007	0.2	0.0	57.3	0.0	173.0	0.0	0.63	64.8	319.0
2008	21.2	0.0	61.9	0.0	0.0	0.0	0.00	180.2	281.6
2009	63.4	0.0	72.9	0.0	80.8	29.7	2.10	116.6	380.1
2010	105.4	0.0	47.9	0.0	0.0	31.6	0.00	2.3	210.8
2011	161.7	0.0	49.6	0.0	0.0	4.5	0.00	1.0	234.1
2012	178.0	0.0	117.2	0.0	0.0	0.0	0.69	1.1	309.5
2013	241.1	0.0	78.0	0.0	0.0	18.6	0.00	21.7	370.5

TABLE 21.2 Ha[a] (in Thousands) Treated With Commercial *Bacillus thuringiensis* var. *kurstaki* Products for the Suppression of Major Forest Defoliators[b] in North America Between 1980 and 2014—cont'd

Year	ESBW Canada	ESBW United States	WSBW Canada	WSBW United States	JPBW	EHL	GM Canada	GM United States[c]	Total
2014	289.7	0.0	56.7	0.0	0.0	0.0	0.00	3.0	357.8
Total	**7,621.5**	**412.1**	**885.2**	**538.5**	**1,278.0**	**889.5**	**614.9**	**3,181.2**	**15,830.5**

[a]*Number of ha sprayed × number of applications, except GM United States data [see [c]].*
[b]*Main target insects: EHL, eastern hemlock looper; ESBW, eastern spruce budworm; GM, gypsy moth; JPBW, jack pine budworm; WSBW, western spruce budworm.*
[c]*Surface area treated regardless of number of applications, not including eradication programs (1980–2014: 1.66 million ha, Asian and European gypsy moth) or Slow-the-Spread program (1993–2014: 0.42 million ha).*
WSBW United States: Sheehan, K.A., 1996. Effects of Insecticide Treatments on Subsequent Defoliation by Western Spruce Budworm in Oregon and Washington: 1982–1992. General Technical Report PNW-GTR-367. Pacific Northwest Research Station, USDA Forest Service. Portland, OR; GM United States: Gypsy Moth Digest, USDA Forest Service, Northeastern Area, http://na.fs.fed.us/fhp/gm; Other species: Canadian National Forestry Database (http://nfdp.ccfm.org) supplemented by Annual Pest Forum reports, available at http://cfs.nrcan.gc.ca/publications, Canadian Forestry Service.

21.4.4 Suppression of Coniferous Defoliators in North America

21.4.4.1 Spruce Budworm

The eastern spruce budworm, *Choristoneura fumiferana*, a native defoliator of spruce (*Picea* spp.) and fir (*Abies* spp.) in boreal and mixed coniferous forests across North America, is considered the most destructive pest of these forests and remains the primary target for control with Btk. Outbreaks are cyclical, and every 30–40 years populations increase to outbreak levels, where they remain for 10 years or longer (Royama, 1984). The last outbreak began around 1970 and persisted for more than 20 years. At its peak, defoliation affected 58 million ha and caused an estimated loss of 44 million m³ per year of timber volume (MacLean et al., 2002). Severe defoliation sustained for 5–7 years causes extensive mortality of fir and spruce hosts. To keep forests alive and maintain associated socioeconomic benefits, most jurisdictions in eastern Canada and northeastern United States conducted extensive and often controversial aerial spray programs.

Population cycles of the spruce budworm are mostly driven by a complex array of natural enemies (Royama, 1984; Eveleigh et al., 2007), which interact with defoliation-induced and density-dependent effects of host tree conditions to influence larval survival (Régnière and Nealis, 2008). Diseases play a limited role in outbreak dynamics: pathogens causing lethal infections typically occur at low frequencies (Lucarotti et al., 2004) or in isolated epizootics (Vandenberg and Soper, 1978) and are believed to have generally low or, at best, localized population impacts (Neilson, 1963). This probably explains the limited success of the many attempts over the years to use spruce budworm pathogens for population management. Early efforts to initiate epizootics by entomophthoralean fungi through the inoculative release of infected larvae or the inundative release of mass-produced hyphal bodies failed to increase field prevalence (Lim and Perry, 1983; Soper, 1985), while aerial spray trials with spruce budworm viruses (ie, entomopoxvirus, granulovirus, and nucleopolyhedrovirus) on a cumulative total of thousands of ha did not yield satisfactory results (Cunningham and Kaupp, 1995). Only Btk showed promise and eventually became the mainstay in spruce budworm suppression programs.

As a biological insecticide, Btk is applied the same way and using the same application technologies as synthetic insecticides. Because Btk is a pathogen that does not naturally cause epizootics in insect populations, application has to be repeated each pest generation to sustain protection. For protection against spruce budworm and coniferous defoliators in general, the most commonly used products contain 12.7–16.9 BIU per liter and are usually applied undiluted at 30 BIU in 1.6–2.4 L per ha, using rotary atomizers to produce a finely dispersed spray cloud with a median droplet diameter of 50–80 μm (Bauce et al., 2004; van Frankenhuyzen et al., 2007). To maximize foliage protection, Btk is applied after buds have flushed and shoots are starting to elongate. Between 1980 and 2014, Btk was applied to a cumulative total of 8.0 million ha of spruce budworm-infested forests, mostly in Canada (Table 21.2). Its use has declined since the early 1990s and shifted toward western provinces as the spruce budworm outbreak in eastern Canada collapsed. The steep increase in the areas sprayed after 2007 reflects the recurrence of outbreak spruce budworm populations in Québec, which is believed to be the onset of a new outbreak that may once again sweep across eastern North America.

21.4.4.2 Jack Pine Budworm

Another species in the genus *Choristoneura* that exhibits destructive cyclical outbreaks in central and eastern

North America is the jack pine budworm, *Choristoneura pinus*. Outbreaks occur throughout the range of its primary host (jack pine; *Pinus banksiana*) at intervals of 10–15 years and can cause severe defoliation over millions of ha (McCullough, 2000; van Frankenhuyzen et al., 2011). Driven primarily by a reciprocal interaction between severe defoliation and low pollen cone production coupled with increased parasitism of late-instar larvae, outbreaks are short-lived and typically persist for only 2–4 years (Nealis et al., 2003). However, significant losses in merchantable volume of jack pine can result from a single outbreak episode, because severe defoliation during one season is sufficient to kill trees and reduce growth (Gross, 1992). Extensive foliage protection programs with Btk, using the same application prescriptions as for spruce budworm, have been successfully conducted in Ontario during three outbreak episodes since the mid-1980s on a cumulative total of ~1.2 million ha (Table 21.2).

21.4.4.3 Western Spruce Budworm

Western spruce budworm, *Choristoneura freemani* (formerly *Choristoneura occidentalis*), occurs west of the Continental Divide where it is most commonly associated with Douglas fir (*P. menziesii*). Outbreaks are intermediate to those of the previous two species, causing severe defoliation on hundreds of thousands of ha with a variable periodicity of 20–35 years and an average duration of 10–15 years (Axelson et al., 2015). The three species share similar mechanisms for population regulation, which involve lagged density-dependent effects of defoliation-induced changes in host condition that interact with mortality caused by natural enemies. Differences between *C. freemani*, *C. pinus*, and *C. fumiferana* in periodicity and spatial extent of outbreaks result from differences in intensity of feedbacks between larval survival and the host tree (ie, defoliation, tree mortality, or flower production, depending on the budworm species; Nealis and Régnière, 2009). The western spruce budworm differs from its eastern counterparts in that pathogens appear to play a more prominent role in outbreaks: an NPV was widespread and caused up to 20% mortality at high larval densities (Nealis et al., 2015). Experimental application of various baculoviruses in the early 1980s reduced populations by up to 50% in the year of treatment and caused measurable increases in larval mortality for two years after but failed to exert population control through epizootics (Shepherd et al., 1995), and none of these baculoviruses were pursued for commercial development.

Western spruce budworm has been a target for large-scale spray programs with synthetic insecticides since the late 1940s in the western United States, with a primary objective to reduce populations so that resurgence during an outbreak episode will not occur (Mason and Paul, 1996). In the mid-1980s, Btk started to replace those products in the suppression of outbreak populations (van Frankenhuyzen, 2000) and has since been used on a cumulative total of ~0.54 million ha (Table 21.2). In British Columbia, no spray interventions were made until the mid-1990s, when Btk had been validated for foliage protection on overstory and understory Douglas fir (Shepherd et al., 1995). Since then, Btk has been used in annual foliage protection programs on ~0.88 million ha cumulatively (Table 21.2), using similar application prescriptions as for eastern spruce budworm.

21.4.4.4 Hemlock Looper

The eastern hemlock looper, *Lambdina fiscellaria fiscellaria* (Lepidoptera: Geometridae), is a native species in North America, distributed from northeastern to northcentral regions. It feeds on a variety of trees, but outbreaks occur predominantly in stands of balsam fir (*Abies balsamea*). Periodic outbreaks occur throughout its distribution and vary in size over time and space from a few thousand to a few hundred thousand ha. Larger scale outbreaks cycle at 10–15 year intervals and last from 3–6 years. Since larvae feed on both new and old foliage, severe defoliation can kill trees within 1 or 2 years. The largest recorded outbreak (1966–71) killed 12 million m³ of merchantable wood in Newfoundland (van Frankenhuyzen et al., 2002). Foliage protection programs with Btk have been undertaken since the early-1980s on a cumulative total of ~0.89 million ha (Table 21.2), mostly in Newfoundland and Québec. The same products and application prescriptions are used as for spruce budworms.

21.5 ERADICATION OF INVASIVE FOREST INSECTS WITH ENTOMOPATHOGENS

The following successful programs have all targeted invasive forest defoliators, and Btk has been the product of choice for these eradication programs based on its specificity and high safety to humans.

21.5.1 Eradication of Gypsy Moths in North America

21.5.1.1 Asian Gypsy Moth

Lymantria dispar asiatica is a subspecies that is known as the Asian gypsy moth. It is distributed mostly east of the Ural Mountains through China to Korea. Asian gypsy moth has been flagged worldwide as a species that requires eradication as soon as it has been detected in locations outside its natural distribution. It is a species of main concern because it has a broader host range than its European counterpart, and its females are capable of sustained flight, which would likely facilitate their spread (females of European gypsy moth are incapable of sustained flight). Early detection of new incursions is accomplished through extensive pheromone trapping programs throughout North America, and many eradication programs have been undertaken in response to positive trap catches of either Asian gypsy moths or Asian-European gypsy moth hybrids. Because Asian gypsy moths typically arrive on the continent via grain or cargo ships, eradication programs are most often conducted in urban areas near ports of entry.

Eradication programs have varied from small-scale, ground-based operations to aerial treatment of a few hundred to tens of thousands of ha with two or three applications of Btk at 60–95 BIU per ha. Notable examples of large-scale eradication programs include aerial treatment of ~20,000 ha in Vancouver (British Columbia) and ~46,600 ha in Tacoma (Washington) in 1992 and treatment of ~55,800 ha in North Carolina in 1994. These programs have been discussed in detail by Hajek and Tobin (2009). Numerous smaller scale programs have been conducted as well: for example, in the state of Washington, ~1300 ha have been treated from air or ground between 1996 and 2000 (http://agr.wa.gov/PlantsInsects/InsectPests/GypsyMoth/ControlEfforts/docs/GMTreatmentsSince1979.pdf).

21.5.1.2 European Gypsy Moth

Because of negative ecological and socioeconomic impacts associated with the spread of European gypsy moth in eastern North America during the past century, much effort is being spent to prevent or slow establishment in new areas. Gypsy moth currently occupies about one-quarter of susceptible habitat that is available in North America and continues to spread west and south. New populations that arise in the transition zone between infested and noninfested areas grow and coalesce with the population front, which greatly enhances the rate of gypsy moth spread. Under the Slow-the-Spread Program (Tobin and Blackburn, 2007), these newly established populations are delimited and aggressively targeted for eradication to limit their influence on the rate of spread. The program involves all levels of government and many universities and is currently conducted on a belt of ~20 million ha that stretches from Wisconsin to North Carolina. Depending on density, size, and location, leading-edge populations are treated with Btk, Dimilin, or Gypchek, sometimes followed by the use of pheromones for mating disruption. Btk is the product of choice and has been used on ~0.42 million ha since the pilot stage of the program in 1993, as compared to ~21,000 ha that were treated with Dimilin (USDA Forest Service Gypsy Moth Digest; http://www.na.fs.fed/fhp/gm). Btk doses generally range from 60 to 95 BIU per ha using either one or two applications during the period of early instar activity. A lesser used tactic is the application of Gypchek at 5×10^{12} occlusion bodies in 9.4 L per ha, which has been used between 1993 and 2014 on ~67,000 ha of mostly environmentally sensitive lands.

Gypsy moths are also detected on a regular basis across North America far from established infestations as a result of human-mediated dispersal, and eradication programs are conducted on a regular basis across the continent. Between 1980 and 2014, Btk was used on ~1.66 million ha to eradicate isolated infestations of both gypsy moths across the United States (USDA Forest Service Gypsy Moth Digest; http://www.na.fs.fed/fhp/gm). Of particular concern are frequent introductions and potential establishment along the Pacific Coast. Large-scale eradication programs were conducted in Oregon in 1985–86, involving treatment of ~168,000 ha of urban area with multiple applications of Btk. In Canada, small-scale eradication programs with Btk are frequently conducted in British Columbia. The largest program took place in the greater Victoria region on Vancouver Island in 1999 when ~13,400 ha were treated with multiple applications of Btk (de Amorim et al., 2001).

21.5.2 Eradication of Tussock Moths in New Zealand

As an island nation with an agricultural-based economy, New Zealand places a high priority on preventing invasion by exotic pest species. The discovery of established populations of two tussock moth species in 1996 and 1999 precipitated eradication programs that are well known for their thoroughness and success and which were described in detail by Glare (2009). Because both species are polyphagous and produce multiple generations a year, risks of colonization and subsequent economic losses and environmental damages were considered high. The white spotted tussock moth (*Orgyia thyellina*, Erebidae), a native of eastern Asia where it is occasionally a pest, was found in Auckland in 1996. Surveys established that the infestation was established over about 7 km². An eradication program was conducted in late 1996 and early 1997 on ~4000 ha of suburbs, with an estimated 86,000 people living within the treatment zone. An undiluted, high-potency Btk product was applied at 65 BIU in 5 L per ha in nine applications at weekly intervals, which was a high treatment intensity compared to the eradication of gypsy moths in North America. Intensive pheromone trapping failed to detect any males over ensuing years, and by 1998 the white spotted tussock moth was declared eradicated. One year later, another tussock moth species was found in Auckland, triggering serious concerns. The painted apple moth (*Teia anartoides*, Erebidae) is a minor pest of urban gardens in southeastern Australia but was considered a high risk due to its wide host range and spread potential. Limited ground spraying with synthetic insecticides was complemented with targeted aerial application of Btk on 600 ha commencing in 2002, which led to a spray program over a period of 9 months that eventually covered ~12,000 ha treated with up to nine applications. In 2003, the final stages of spraying were complemented with the release of sterile males to disrupt the reproduction cycle of any survivors. Eradication was declared a success a few years later. The success of the tussock moth eradications was attributed to the rapid response when established populations were still localized and the pests' restricted ability to disperse due to the fact that females are flightless. A third eradication program was launched after the capture of a single Asian gypsy moth in Hamilton in 2003. Because no ground populations were detected, a target area of ~1250 ha was defined based on moth flight behavior. The area was treated with eight weekly aerial applications of Btk at 65–90 BIU (5–7 L per ha), and no further trap catches were made after that. The New Zealand experience thus confirmed the experience in North America that the eradication of invasive lepidopteran pests with Btk is possible and feasible.

21.6 THE FUTURE OF MICROBIAL CONTROL IN FORESTRY

The emerging promise of insect pathogens for the ecologically acceptable control of insect pests around the middle of the previous century led to active research programs in insect pathology in many countries (Steinhaus, 1975). This effort has resulted in a pathogen-based control of a broad range of insect pests in agriculture, forestry, and human and animal health. Pathogen-based control for forest pests currently includes commercial products for inundative and inoculative release, based on Bt (Charles et al., 2000), baculoviruses (Hunter-Fujita et al., 1998), hypocrealean fungi (de Faria and Wraight, 2007), and a nematode (Slippers et al., 2012). Much of the effort was driven by the desire to develop products that could replace broad spectrum synthetic insecticides. Investigations therefore focused predominantly on pathogens that fit the "chemical control paradigm," that is, products that are suitable for mass production, long-term storage, formulation into sprayable products, and commercialization. Several factors are now necessitating a shift in research away from developing such products and toward more biologically based control approaches. Increasing costs of spray product development and registration, waning public support for broad-scale applications of any pesticide, be it synthetic or microbial, and the desire to adopt more proactive suppression approaches earlier in outbreak cycles rather than continuing current reactive "fire-fighting" approaches in response to full-blown outbreaks all point to the need for developing more biologically based control approaches. These new directions will require a shift in research investment that is driven more by public good than strictly commercial considerations.

Current successes clearly demonstrate that insect pathogens can be successfully used in classical biological control and other ecologically based control approaches such as inoculative augmentation. Promising target pests are established invasive exotic pests and native forest pests with outbreak cycles that are often governed by host–pathogen interactions. Current successes pertain mostly to defoliating forest insects. Future efforts should build on our knowledge of the importance of specific system parameters to building successful programs. For example, target pest biology and habitat can strongly influence pathogen success, eg, baculoviruses have been especially successful against sawfly species for which transmission is enhanced due to gregarious behavior, and application technologies must be optimized to reach intended targets with correct concentrations of viable inocula. A key focus for future research in insect pathology is the control of cryptic pests that spend much of their life within trees. The control of cryptic species will require more creative approaches than simply applying a pathogen. Invasive forest pests will continue to be a key focus for scientists working in microbial control as the inadvertent introduction of invasive arthropods that pose a threat to our forests will continue with the relentless advance in the globalization of trade.

While there are new types of pests for which control methods need to be developed, research methods are changing and improving to allow scientists to gain new types of data. Until recently, the search for suitable pathogens has often been constrained by the limitations of microscopy-based methods. The advent of affordable genome sequencing and other molecular tools for pathogen identification and characterization is now offering new and powerful ways to reexamine the natural diversity of entomopathogens in forest insect populations in order to identify new biological control candidates. In addition, changes in research focus and methodology are allowing greater abilities to understand interactions among entomopathogens, host insects, and their host trees. Scientists now have greater abilities to investigate when methods for pathogen production, formulation, and transmission in the field are not optimal, with a goal toward improving overall efficacy.

The examples provided in this chapter outline specifically how achievements in the use of entomopathogens for pest control are contributing to the sustainable protection of forests around the world. Future challenges lie in store, but forest ecosystems, usually biodiverse areas that have relatively high economic or aesthetic injury levels (ie, some level of damage from pests can be tolerated), are excellent systems for effective control using entomopathogens.

ACKNOWLEDGMENTS

We thank Drs. Mitsuaki Shimazu, Zengzhi Li, and Johannes Jehle for information about biopesticide products and Dr. Tonya Bittner for help with the manuscript.

REFERENCES

Akhurst, R.J., 1975. Cross-breeding to facilitate the identification of *Deladenus* spp., nematode parasites of woodwasps. Nematologica 21, 267–272.

Alalouni, U., Schädler, M., Brandl, R., 2013. Natural enemies and environmental factors affecting the population dynamics of the gypsy moth. J. Appl. Entomol. 137, 721–738.

Axelson, J.N., Smith, D.J., Daniels, L.D., Alfaro, R.I., 2015. Multicentury reconstruction of western spruce budworm outbreaks in central British Columbia, Canada. For. Ecol. Manag. 335, 235–248.

Balch, R.E., Bird, F.T., 1944. A disease of the European spruce sawfly, *Gilpinia hercyniae* (Htg.), and its place in natural control. Sci. Agric. 25, 65–80.

Bauce, E., Carisey, N., Dupont, A., van Frankenhuyzen, K., 2004. *Bacillus thuringiensis* subsp. *kurstaki* (Btk) aerial spray prescriptions for balsam fir stand protection against spruce budworm (Lepidoptera: Tortricidae). J. Econ. Entomol. 97, 1624–1634.

Bedding, R.A., 1972. Biology of *Deladenus siricidicola* (Neotylenchidae) an entomophagous-mycetophagous nematode parasitic in siricid woodwasps. Nematologica 18, 482–493.

Bedding, R.A., Akhurst, R.J., 1974. Use of the nematode *Deladenus siricidicola* in the biological control of *Sirex noctilio* in Australia. J. Aust. Entomol. Soc. 13, 129–135.

Bird, F.T., 1971. *Neodiprion lecontei* (Fitch), red-headed pine sawfly (Hymenoptera: Diprionidae). In: Biological Control Programmes against Insects and Weeds in Canada, 1959–1968. Commonwealth Agricultural Bureaux, Farnham Royal, Bucks, England, pp. 148–150.

Bird, F.T., Elgee, D.E., 1957. A virus disease and introduced parasites as factors controlling the European spruce sawfly, *Diprion hercyniae* (Htg.), in central New Brunswick. Can. Entomol. 89, 371–378.

Bjørnstad, O.N., Robine, C., Liebhold, A.M., 2010. Geographic variation in North American gypsy moth cycles: subharmonics, generalist predators, and spatial coupling. Ecology 91, 106–118.

Boissin, E., Hurley, B., Wingfield, M.J., Vasaitis, R., Stenlid, J., Davis, C., de Groot, P., Ahumada, R., Carnegie, A., Goldarazena, A., Klasmer, P., Wermelinger, B., Slippers, B., 2012. Retracing the routes of introduction of invasive species: the case of the *Sirex noctilio* woodwasp. Mol. Ecol. 21, 5728–5744.

Carnegie, A.J., Bashford, R., 2012. Sirex woodwasp in Australia: current management strategies, research and emerging issues. In: Slippers, B., de Groot, P., Wingfield, M.J. (Eds.), The Sirex Woodwasp and its Fungal Symbiont: Research and Management of a Worldwide Invasive Pest. Springer, New York, pp. 175–201.

Castrillo, L.A., Griggs, M.H., Liu, H., Bauer, L.S., Vandenberg, J.D., 2010. Assessing deposition and persistence of *Beauveria bassiana* GHA (Ascomycota: Hypocreales) applied for control of the emerald ash borer, *Agrilus planipennis* (Coleoptera: Buprestidae), in a commercial tree nursery. Biol. Control 54, 61–67.

Castrillo, L.A., Bauer, L.S., Liu, H., Griggs, M.H., Vandenberg, J.D., 2010. Characterization of *Beauveria bassiana* (Ascomycota: Hypocreales) isolates associated with *Agrilus planipennis* (Coleoptera: Buprestidae) populations in Michigan. Biol. Control 54, 135–140.

Charles, J.-F., Delécluse, A., Nielsen-LeRoux, C., 2000. Entomopathogenic Bacteria: From Laboratory to Field Application. Kluwer Academic Publishers, Dordrecht, The Netherlands.

Cock, M.J.W., van Lenteren, J.D., Brodeur, J., Barratt, B.I.P., Bigler, F., Bolckmans, K., Cônsoli, F.L., Haas, F., Mason, P.G., Parra, J.R.P., 2010. Do new Access and Benefit Sharing procedures under the Convention on Biological Diversity threaten the future of biological control? BioControl 55, 199–218.

Cunningham, J.C., 1984. *Neodiprion abietis* (Harris), balsam fir sawfly (Hymenoptera: Diprionidae). In: Kelleher, J.S., Hulme, M.A. (Eds.), Biological Control Programs against Insects and Weeds in Canada 1969–1980. Commonwealth Agricultural Bureaux, Slough, UK, pp. 321–322.

Cunningham, J.C., 1998. North America. In: Hunter-Fujita, F.R., Entwistle, P.F., Evans, H.F., Crook, N.E. (Eds.), Insect Viruses and Pest Management. Wiley, Chichester, UK, pp. 313–331.

Cunningham, J.C., Entwistle, P.F., 1981. Control of sawflies by baculovirus. In: Burges, H.D. (Ed.), Microbial Control of Pests and Plant Diseases 1970–1980. Academic Press, New York, pp. 379–407.

Cunningham, J.C., Kaupp, W.J., 1995. Insect viruses. In: Armstrong, J.A., Ives, W.G.H. (Eds.), Forest Insect Pests in Canada. Natural Resources Canada, Science and Sustainable Development Directorate, Ottawa, Canada, pp. 327–340.

de Amorim, G.V., Whittome, B., Shore, B., Levin, D.B., 2001. Identification of *Bacillus thuringiensis* subspecies *kurstaki* strain HD1-like bacteria from environmental and human samples after aerial spraying of Victoria, British Columbia, Canada with Foray 48B. Appl. Environ. Microbiol. 67, 1035–1043.

de Faria, M.R., Wraight, S.P., 2007. Mycoinsecticides and mycoacaracides: a comprehensive list with worldwide coverage and international classification of formulation types. Biol. Control 43, 237–256.

Entwistle, P.F., Cory, J.S., Bailey, M.J., Higgs, S., 1993. *Bacillus thuringiensis*, an Environmental Biopesticide: Theory and Practice. John Wiley & Sons, Chicester, UK.

Eveleigh, E.S., McCann, K.S., McCarthy, P.C., Pollock, S.J., Lucarotti, C.J., Morin, B., McDougal, G.A., Strongman, D.B., Huber, J.T., Umbanhowar, J., Faria, L.D.B., 2007. Fluctuations in density of an outbreak species drive diversity cascades in food webs. Proc. Natl. Acad. Sci. U.S.A. 104, 16976–16981.

Glare, T.R., 2009. Use of pathogens for eradication of exotic lepidopteran pests in New Zealand. In: Hajek, A.E., Glare, T.R., O'Callaghan, M. (Eds.), Use of Microbes for Control and Eradication of Invasive Arthropods. Springer, Dordrecht, The Netherlands, pp. 49–70.

Gross, H., 1992. Impact analysis for a jack pine budworm infestation in Ontario. Can. J. For. Res. 22, 193–196.

Hajek, A.E., 1999. Pathology and epizootiology of the lepidoptera-specific mycopathogen *Entomophaga maimaiga*. Microbiol. Mol. Biol. Rev. 63, 814–835.

Hajek, A.E., 2007. Introduction of a fungus into North America for control of gypsy moth. In: Vincent, C., Goettel, M., Lazarovits, G. (Eds.), Biological Control: International Case Studies. CABI Publ, UK, pp. 53–62.

Hajek, A.E., Bauer, L.S., 2009. Use of entomopathogens against invasive wood boring beetles in North America. In: Hajek, A.E., Glare, T.R., O'Callaghan, M. (Eds.), Use of Microbes for Control and Eradication of Invasive Arthropods. Springer, Dordrecht, The Netherlands, pp. 159–179.

Hajek, A.E., Delalibera Jr., I., 2010. Fungal pathogens as classical biological control agents against arthropods. BioControl 55, 147–158 Also published in Roy, H.E., Vega, F.E., Chandler, D., Goettel, M.S., Pell, J.K., Wajnberg, E. (Eds.), The Ecology of Fungal Entomopathogens. Springer, pp. 147–158.

Hajek, A.E., Morris, E.E., 2014. Biological control of *Sirex noctilio*. In: Van Driesche, R.G., Reardon, R. (Eds.), The Use of Classical Biological Control to Preserve Forests in North America. FHTET-2013-02. USDA Forest Service, Forest Health Technology Enterprise Team, Morgantown, West Virginia, pp. 331–346.

Hajek, A.E., Tobin, P.C., 2009. North American eradications of Asian and European gypsy moth. In: Hajek, A.E., Glare, T.R., O'Callaghan, M. (Eds.), Use of Microbes for Control and Eradication of Invasive Arthropods. Springer, Dordrecht, The Netherlands, pp. 50–71.

Hajek, A.E., Tobin, P.C., 2011. Introduced pathogens follow the invasion front of a spreading alien host. J. Anim. Ecol. 80, 1217–1226.

Hajek, A.E., Humber, R.A., Elkinton, J.S., 1995. The mysterious origin of *Entomophaga maimaiga* in North America. Am. Entomol. 41, 31–42.

Hajek, A.E., McManus, M.L., Delalibera Jr., I., 2005. Catalogue of Introductions of Pathogens and Nematodes for Classical Biological Control of Insects and Mites. FHTET-2005-05, USDA Forest Service, Forest Health Technology Enterprise Team, Morgantown, West Virginia.

Hajek, A.E., McManus, M.L., Delalibera Jr., I., 2007. A review of introductions of pathogens and nematodes for classical biological control of insects and mites. Biol. Control 41, 1–13.

Hajek, A.E., Glare, T.R., O'Callaghan, M., 2009. Use of Microbes for Control and Eradication of Invasive Arthropods. Springer, Dordrecht, The Netherlands.

Hajek, A.E., Tobin, P.C., Haynes, K.J., 2015. Replacement of a dominant viral pathogen by a fungal pathogen does not alter the synchronous collapse of a forest insect outbreak. Oecologia 177, 785–797.

Higuchi, T., Saika, T., Senda, S., Mizobata, T., Kawata, Y., Nagai, J., 1997. Development of biorational pest control formulation against longicorn beetles using a fungus, *Beauveria brongniartii* (Sacc.) Petch. J. Ferment. Bioeng. 84, 236–243.

Hoy, M.A., 1976. Establishment of gypsy moth parasitoids in North America: an evaluation of possible reasons for establishment or non-establishment. In: Anderson, J.F., Kaya, H.K. (Eds.), Perspectives in Forest Entomology. Academic Press, New York, pp. 215–232.

Huber, J., 1998. Western Europe. In: Hunter-Fujita, F.R., Entwistle, P.F., Evans, H.F., Crook, N.E. (Eds.), Insect Viruses and Pest Management. Wiley, Chichester, UK, pp. 201–215.

Hunter-Fujita, F.R., Entwistle, P.E., Evans, H.F., Crook, N.E. (Eds.), 1998. Insect Viruses and Pest Management. J. Wiley & Sons, Chichester, UK.

Hurley, B.P., Hatting, H.J., Wingfield, M.J., Klepzig, K.D., Slippers, B., 2012. The influence of *Amylostereum areolatum* diversity and competitive interactions on the fitness of the *Sirex* parasitic nematode *Deladenus siricidicola*. Biol. Control 61, 207–214.

Iede, E.T., Penteado, S.R.C., Filho, W.R., 2012. The woodwasp *Sirex noctilio* in Brazil: monitoring and control. In: Slippers, B., de Groot, P., Wingfield, M.J. (Eds.), The Sirex Woodwasp and its Fungal Symbiont: Research and Management of a Worldwide Invasive Pest. Springer, Dordrecht, Netherlands, pp. 217–228.

Johnson, D.M., Liebhold, A., Bjørnstad, O.N., 2006. Geographical variation in the periodicity of gypsy moth outbreaks. Ecography 29, 367–374.

Klasmer, P., Botto, E., 2012. The ecology and biological control of the woodwasp *Sirex noctilio* in Patagonia, Argentina. In: Slippers, B., de Groot, P., Wingfield, M.J. (Eds.), The Sirex Woodwasp and its Fungal Symbiont: Research and Management of a Worldwide Invasive Pest. Springer, Dordrecht, The Netherlands, pp. 202–215.

Kroll, S.A., Hajek, A.E., Morris, E.E., Long, S.J., 2013. Parasitism of *Sirex noctilio* by non-sterilizing *Deladenus siricidicola* in northeastern North America. Biol. Control 67, 203–211.

Lewis, F.B., 1981. Control of gypsy moth by a baculovirus. In: Burges, H.D. (Ed.), Microbial Control of Pests and Plant Diseases 1970–1980. Academic Press, New York, pp. 363–377.

Li, Z., 2007. *Beauveria bassiana* for pine caterpillar management in the People's Republic of China. In: Vincent, C., Goettel, M.S., Lazarovits, G. (Eds.), Biological Control: A Global Perspective. CABI Publ, Wallingford, UK, pp. 300–310.

Li, Z., Alves, S.B., Roberts, D.W., Fan, M., Delalibera Jr., I., Tang, J., Lopes, R.B., Faria, M., Rangel, D.E.N., 2010. Biological control of insects in Brazil and China: history, current programs and reasons for their successes using entomopathogenic fungi. BioControl Sci. Technol. 20, 117–136.

Liebhold, A., McManus, M., 1999. The evolving use of insecticides in gypsy moth management. J. For. 97, 20–23.

Liebhold, A.M., Plymale, R.C., Elkinton, J.S., Hajek, A.E., 2013. Emergent fungal entomopathogen does not alter density dependence in a viral competitor. Ecology 94, 1217–1222.

Lim, K.P., Perry, D.F., 1983. Field Test of the Entomopathogenic Fungi *Zoophthora radicans* and *Entomophthora egressa* against the Spruce Budworm in Western Newfoundland, 1982. Newfoundland Forestry Centre Trial 3281 (1982–1983). Canadian Forest Service, St John's, Newfoundland.

Lipa, J.J., 1998. Eastern Europe and the former Soviet Union. In: Hunter-Fujita, F.R., Entwistle, P.F., Evans, H.F., Crook, N.E. (Eds.), Insect Viruses and Pest Management. Wiley, Chichester, UK, pp. 216–231.

Lord, J.C., 2005. From Metchnikoff to Monsanto and beyond: the path of microbial control. J. Invertebr. Pathol. 89, 19–29.

Lucarotti, C.J., Eveleigh, E.S., Royama, T., Morin, B., McCarthy, P., Ebling, P.M., Kaupp, W.J., Guertin, C., Arella, M., 2004. Prevalence of baculoviruses in spruce budworm (Lepidoptera: Tortricidae) populations in New Brunswick. Can. Entomol. 136, 255–264.

Lyons, D.B., Lavalle, R., Kyei–poku, G., van Frankenhuyzen, K., Johny, S., Guertin, C., Francese, J.A., Jones, J.C., Blais, M., 2012. Towards the development of an autocontamination trap system to manage populations of emerald ash borer (Coleoptera: Buprestidae) with the native entomopathogenic fungus, *Beauveria bassiana*. J. Econ. Entomol. 105, 1929–1939.

MacLean, D.A., Beaton, K.P., Porter, K.B., MacKinnon, W.E., Budd, M., 2002. Potential wood supply losses to spruce budworm in New Brunswick estimated using the spruce budworm decision support system. For. Chron. 78, 739–750.

Mason, R.R., Paul, H.G., 1996. Case History of Population Change in a *Bacillus thuringiensis* – Treated vs. Untreated Population of the Western Spruce Budworm. PNW-RN-521. USDA Forest Service, Pacific Northwest Research Station, Portland, OR.

McCullough, D.G., 2000. A review of factors affecting the population dynamics of jack pine budworm (*Choristoneura pinus pinus* Freeman). Popul. Ecol. 42, 243–256.

McGugan, B.M., Coppel, H.C., 1962. Part II. Biological control of forest insects, 1910–1958. In: A Review of the Biological Control Attempts against Insects and Weeds in Canada. Commonwealth Institute of Biological Control, Trinidad, Technical Communication 2, CAB, Farnham Royal, England, pp. 35–216.

Mlonyeni, X.O., Wingfield, B.D., Wingfield, M.J., Ahumada, R., Klasmer, P., Leal, I., deGroot, P., Slippers, B., 2011. Extreme homozygosity in Southern Hemisphere populations of *Deladenus siricidicola*, a biological control agent of *Sirex noctilio*. Biol. Control 59, 348–353.

Montgomery, M.E., Wallner, W.E., 1988. The gypsy moth: a westward migrant. In: Berryman, A.A. (Ed.), Dynamics of Forest Insect Populations: Patterns, Causes, Implications. Plenum, New York, pp. 353–375.

Moreau, G., Lucarotti, C.J., 2007. A brief review of the past use of baculoviruses for the management of eruptive forest defoliators and recent developments on a sawfly virus in Canada. For. Chron. 83, 105–112.

Moreau, G., Lucarotti, C.J., Kettela, E.G., Thurston, G.S., Holmes, S., Weaver, C., Levin, D.B., Morin, B., 2005. Aerial application of nucleopolyhedrovirus induces decline in increasing and peaking populations of *Neodiprion abieitis*. Biol. Control 30, 65–73.

Nealis, V.G., Régnière, J., 2009. Risk of dispersal in western spruce budworm. Agric. For. Entomol. 11, 213–223.

Nealis, V.G., Magnussen, S., Hopkin, A.A., 2003. A lagged, density-dependent relationship between jack pine budworm *Choristoneura pinus pinus* and its host tree, *Pinus banksiana*. Ecol. Entomol. 28, 183–192.

Nealis, V.G., Turnquist, R., Morin, B., Graham, R.I., Lucarotti, C.J., 2015. Baculoviruses in populations of the western spruce budworm. J. Invertebr. Pathol. 127, 76–80.

Neilson, M.M., 1963. Disease and the spruce budworm. In: Morris, R.F. (Ed.), The Dynamics of Epidemic Spruce Budworm Populations, vol. 31. Entomological Society of Canada Memoirs, pp. 288–310.

Neilson, M.M., Morris, R.F., 1964. The regulation of European spruce sawfly numbers in the Maritime Provinces of Canada from 1937 to 1963. Can. Entomol. 96, 773–784.

Nielsen, C., Milgroom, M.G., Hajek, A.E., 2005. Genetic diversity in the gypsy moth fungal pathogen *Entomophaga maimaiga* from founder populations in North America and source populations in Asia. Mycol. Res. 109, 941–950.

Neumann, F.G., Morey, J.L., 1984. Influence of natural enemies on the *Sirex* wood wasp in herbicide-treated trap trees of radiata pine in north-eastern Victoria. Austral. For. 47, 218–224.

Novotny, J., 1989. Natural disease of gypsy moth in various gradation phases. In: Wallner, W.E., McManus, K.A. (Eds.), Proceedings. Lymantriidae: A Comparison of Features of New and Old World Tussock Moths. USDA Forest Service, Gen. Tech. Rpt. NE-123, pp. 101–111.

Otvos, I.S., Cunningham, J.C., Shepherd, R.F., 1995. Douglas-fir tussock moth, *Orgyia pseudotsugata*. In: Armstrong, J.A., Ives, W.G.H. (Eds.), Forest Insect Pests in Canada. Natural Resources Canada, Science and Sustainable Development Directorate, Ottawa, Canada, pp. 127–132.

Reardon, R., Podgwaite, J., 1994. Summary of efficacy evaluations using aerially applied Gypchek against gypsy moth in the USA. J. Environ. Sci. Health B 29, 739–756.

Reardon, R., Dubois, N., McLane, W., 1994. *Bacillus thuringiensis* for Managing Gypsy Moth: A Review. FHM-NC-01-94, USDA Forest Service, National Center of Forest Health Management, Hamden, CT.

Reardon, R., Podgwaite, J., Zerillo, R., 1996. Gypchek – The Gypsy Moth Nucleopolyhedrosis Virus Product. FHTET-96–16, Forest Health Technology Enterprise Team, USDA Forest Service, Morgantown, WV.

Régnière, J., Nealis, V.G., 2008. The fine-scale population dynamics of spruce budworm: survival of early instars related to forest condition. Ecol. Entomol. 33, 362–373.

Reilly, J.R., Hajek, A.E., Liebhold, A.M., Plymale, R.S., 2014. The impact of *Entomophaga maimaiga* on outbreak gypsy moth population: the role of weather. Environ. Entomol. 43, 632–641.

Royama, T., 1984. Population dynamics of the spruce budworm *Choristoneura fumiferana*. Ecol. Monogr. 54, 429–462.

Sheehan, K.A., 1996. Effects of Insecticide Treatments on Subsequent Defoliation by Western Spruce Budworm in Oregon and Washington: 1982–1992. General Technical Report PNW-GTR-367, Pacific Northwest Research Station, USDA Forest Service, Portland, OR.

Shepherd, R.F., Cuningham, J.C., Otvos, I.S., 1995. Western spruce budworm, *Choristoneura occidentalis*. In: Armstrong, J.A., Ives, W.G.H. (Eds.), Forest Insect Pests in Canada. Natural Resources Canada, Science and Sustainable Development Directorate, Ottawa, Canada, pp. 119–121.

Shimazu, M., 2009. Use of microbes for control of *Monochamus alternatus*, vector of the invasive pinewood nematode. In: Hajek, A.E., Glare, T.R., O'Callaghan, M. (Eds.), Use of Microbes for Control and Eradication of Invasive Arthropods. Springer, Dordrecht, The Netherlands, pp. 141–157.

Slippers, B., de Groot, P., Wingfield, M.J. (Eds.), 2012. The Sirex Woodwasp and its Fungal Symbiont. Springer, Dordrecht, The Netherlands.

Smitley, D.R., Davis, T.W., 1993. Aerial application of *Bacillus thuringiensis* for suppression of gypsy moth in *Populus-Quercus* forests. J. Econ. Entomol. 86, 1178–1184.

Solter, L.F., Hajek, A.E., 2009. Control of gypsy moth, *Lymantria dispar*, in North America since 1878. In: Hajek, A.E., Glare, T.R., O'Callaghan, M. (Eds.), Use of Microbes for Control and Eradication of Invasive Arthropods. Springer, Dordrecht, The Netherlands, pp. 181–212.

Soper, R.S., 1985. *Erynia radicans* as a mycoinsecticide for spruce budworm control. In: Grimble, D.G., Lewis, F.B. (Eds.), Proceedings Symposium on Microbial Control of Spruce Budworms and Gypsy Moths. USDA Forest Service, Northeastern Forest Experiment Station GTR-NE-100, Broomall, PA, pp. 69–76.

Steinhaus, E.A., 1975. Disease in a Minor Chord. Ohio State University Press, Columbus, Ohio.

Tobin, P.C., Blackburn, L.M., 2007. Slow the Spread: A National Program to Manage the Gypsy Moth. Gen. Tech. Rep. NRS-6. USDA Forest Service, Northern Research Station, Newtown Square, PA.

van Frankenhuyzen, K., 1995. Development and current status of *Bacillus thuringiensis* for control of defoliating forest insects. In: Armstrong, J.A., Ives, W.G.H. (Eds.), Forest Insect Pests in Canada. Natural Resources Canada, Science and Sustainable Development Directorate, Ottawa, Canada, pp. 315–325.

van Frankenhuyzen, K., 2000. Application of *Bacillus thuringiensis* in forestry. In: Charles, J.F., Delécluse, A., Nielsen-LeRoux, C. (Eds.), Entomopathogenic Bacteria: From Laboratory to Field Application. Kluwer Academic Publishers, Dordrecht, The Netherlands, pp. 371–382.

van Frankenhuyzen, K., West, R.J., Kenis, M., 2002. *Lambdina fiscellaria fiscellaria* (Guenée), hemlock looper (Lepidoptera: Geometridae). In: Mason, P.G., Huber, J.T. (Eds.), Biological Control Programmes in Canada, 1981–2000. CABI Publ, Wallingford, UK, pp. 141–144.

van Frankenhuyzen, K., Reardon, R.C., Dubois, N.R., 2007. Forest defoliators. In: Lacey, L.L., Kaya, H.K. (Eds.), Field Manual of Techniques in Invertebrate Pathology. Springer, Dordrecht, The Netherlands, pp. 481–504.

van Frankenhuyzen, K., Ryall, K., Liu, Y., Meating, J., Bolan, P., Scarr, T., 2011. Prevalence of *Nosema* sp. (Microsporidia: Nosematidae) during an outbreak of the jack pine budworm in Ontario. J. Invertebr. Pathol. 108, 201–208.

van Frankenhuyzen, K., Lucarotti, C.J., Lavallée, R., 2015. Canadian contributions to forest insect pathology and the use of pathogens in forest pest management. Can. Entomol. http://dx.doi.org/10.4039/tce.2015.20.

Vandenberg, J.D., Soper, R.S., 1978. Prevalence of Entomophthorales mycoses in populations of spruce budworm, *Choristoneura fumiferana*. Environ. Entomol. 7, 847–853.

Weseloh, R.M., 1998. Possibility for recent origin of the gypsy moth (Lepidoptera: Lymantriidae) fungal pathogen *Entomophaga maimaiga* (Zygomycetes: Entomophthorales) in North America. Environ. Entomol. 27, 171–177.

Yousuf, F., Gurr, G.M., Carnegie, A.J., Bedding, R.A., Bashford, R., Gitau, C.W., Nicol, H.I., 2014. The bark beetle, *Ips grandicollis*, disrupts biological control of the woodwasp, *Sirex noctilio*, via fungal symbiont interactions. FEMS Microbiol. Ecol. 88, 38–47.

Zondag, R., 1975. A non-sterilising strain of *Deladenus siricidicola*. In: Forestry Research Institute Report 1974. Rotorua, New Zealand, pp. 51–52.

Zúbrik, M., Hajek, A.E., Pilarska, D., Špilda, I., Georgiev, G., Hrašovec, B., Csóka, G., Goertz, D., Hoch, G., Barta, M., Saniga, M., Kunca, A., Nikolov, C., Vakula, J., Galko, J., Hirka, A., 2016. Potential of *Entomophaga maimaiga* in regulation of gypsy moth *Lymantria dispar* (L.) (Lepidoptera: Erebidae) in Europe: a review. J. Appl. Entomol. (in press).

Chapter 22

Microbial Control of Insect Pests of Turfgrass

A.M. Koppenhöfer, S. Wu

Rutgers University, New Brunswick, NJ, United States

22.1 THE TURFGRASS SYSTEM

Worldwide about 50 grass species are amenable to use in turfgrass systems due to their abilities to form a mat of intertwined plants with an extensive root mass, regenerate from the crown after defoliation, and form a high shoot density under the continuous mowing regimes characteristic for turfgrass systems (Christians, 1998). Turfgrasses can provide a hard-wearing permanent or semipermanent ground cover that can be used for various recreational spaces in urban and suburban environments, including lawns, parks, golf courses, and athletic fields. Other areas in which turfgrasses are grown include cemeteries, roadsides, and sod farms. Turfgrasses also control soil erosion, capture and clean runoff water from urban areas, provide soil improvement and restoration, moderate temperature, reduce glare and noise, reduce pests, pollen, and human disease exposure, create good wildlife habitats, and improve the physical and mental health of urban populations (Beard and Green, 1994).

The turfgrass market in the United States is by far the largest in the world. Turfgrass areas cover about 20 million ha, and the size of the turfgrass industry is estimated at $40 billion per year (Breuninger et al., 2012). Home lawns represent the largest part of the total turf area (66%). Golf courses only represent about 2% of the total area covered by turfgrasses in the United States, but, due to their much higher maintenance and use intensity, contribute about 20% to the total economic impact. In most other countries, turf hectarages primarily consist of golf courses followed by parks and athletic fields, with residential turf being insignificant. Of the ~30,000 golf courses worldwide, the highest numbers are found in the United States (~15,000), the United Kingdom (~2630), Japan (~2300), Canada (~1725), and Australia (1560; Breuninger et al., 2012).

Permanent turf provides habitat for many invertebrate species, most of which feed on vegetation and detritus without causing obvious damage or loss of productivity. However, amenity turf is under constant critical scrutiny from the public, and playing and safety standards on athletic fields and golf courses are very high. Consequently, damage thresholds are generally low, and therefore, a large number of insect species are regarded as pests.

22.2 IMPORTANT TURFGRASS INSECT PESTS

The most important groups of turfgrass insect pests are white grubs, weevils, and caterpillars, because they can do extensive damage and are found worldwide. Other groups such as crane flies, March flies, mole crickets, and mites can cause significant damage, but the damage tends to be more limited in distribution (Table 22.1).

22.2.1 Coleoptera and Diptera

The larvae of weevils (Coleoptera: Curculionidae) [billbugs (*Sphenophorus* spp.), annual bluegrass weevil (*Listronotus maculicollis*)] and flies (Diptera) [crane flies (*Tipula* spp., Tipulidae), frit fly (*Oscinellafrit*, Chloropidae), March flies (*Bibio* spp., *Dilophus* spp., Bibionidae)] feed on roots and/or bore into stems and/or crowns, killing tillers or entire plants (Potter, 1998; Vittum et al., 1999). White grubs, the larvae of scarabaeid beetles (Coleoptera: Scarabaeidae), are among the most damaging turfgrass and pasture pests in different parts of the world with numerous indigenous and exotic species causing severe problems. The grubs feed on the grass roots near the soil surface, which at high larval densities and under warm and dry conditions can lead to wilting of plants, gradual thinning of the turf, and death of large turf areas. In addition, vertebrate predators (eg, skunks, raccoons, wild pigs, crows, etc.) can tear up the turf to feed on the grubs even at larval densities that by themselves would not cause damage (Potter, 1998; Vittum et al., 1999).

Microbial Control of Insect and Mite Pests. http://dx.doi.org/10.1016/B978-0-12-803527-6.00022-6

TABLE 22.1 Major Turf Pests, Part of the Plant They Attack, Geographical Problem Areas, and Potential Microbial Controls

Plant Part Attacked	Pests	Pest Life Stage	Location	Potential Microbial Controls
Root: chewing	White grubs	Larva	Worldwide	Bacteria, fungi, nematodes
	Mole crickets	Nymph, adult	s.e. United States, s. Europe	Fungi, nematodes
Stem/crown: chewing	Annual bluegrass weevils	Larva	n.e. United States, s.e. Canada	Fungi, nematodes
	Billbugs	Larva	United States, Japan, New Zealand, Australia	Fungi, nematodes
	Crane flies	Larva	United States, Canada, Europe	Bacteria, nematodes
Leaf/stem: chewing	Armyworms	Larva	Worldwide	Bacteria, nematodes
	Cutworms	Larva	Worldwide	Nematodes
	Sod webworms	Larva	United States	Bacteria, nematodes
Leaf/stem: sucking	Chinch bugs	Nymph, adult	United States, s.e. Canada, Japan	Fungi
	Greenbug aphids	Nymph, adult	United States	Fungi
	Mealybugs	Nymph, adult	s. United States, New Zealand	–
	Mites	Nymph, adult	United States	–
	Scales	Nymph, adult	s. United States, Japan	–
	Spittlebugs	Nymph, adult	e. United States, Brazil	Fungi

e., eastern; n.e., northeastern; s., southern; s.e., southeastern.
After Klein, M.G., Grewal, P.S., Jackson, T.A., Koppenhöfer, A.M., 2007. Lawn, turf and grassland pests. In: Lacey, L.A., Kaya, H.K. (Eds.), Field Manual of Techniques for the Application and Evaluation of Entomopathogens, second ed. Springer, Dordrecht, The Netherlands. pp. 655–675.

22.2.2 Lepidoptera

Most caterpillar pests (Lepidoptera: Noctuidae and Pyralidae) initially skeletonize foliage, but as they get larger they create burrows in the soil from which they emerge at night to feed on the grass shoots (webworms, hepialids, cutworms), or they live on the surface feeding on grass foliage and stems (armyworms). The black cutworm, *Agrotis ipsilon*, is a perennial problem on golf course greens and tees throughout the world (Potter, 1998; Vittum et al., 1999). The larvae dig burrows in the thatch or soil and emerge at night to eat the grass blades and stems, which creates dead patches, sunken areas, or pockmarks, disrupting the uniformity and smoothness of the putting surface.

22.2.3 Orthoptera

Nymphs and adults of mole crickets (Orthoptera: Gryllotalpidae) cause damage by feeding on grass roots and shoots and through their extensive tunneling activity. *Scapteriscus* spp. mole crickets, accidentally introduced into the United States from South America, have become major turfgrass insect pests throughout the coastal plain region of the southeastern United States (Potter, 1998; Vittum et al., 1999).

22.2.4 Hemiptera and Acarina

Nymphs and adults of several species each of chinch bugs (*Blissus* spp.), aphids, scales, mealybugs, spittlebugs (all Hemiptera), and mites (Acarina) can cause damage by piercing grass stems or foliage with their syringe-like mouthparts and sucking out plant sap (Potter, 1998; Vittum et al., 1999). The damage potential is greater for species that also inject salivary secretions that are toxic to the grass during feeding. Most of these pests live primarily on stems and foliage, but some species, like chinch bugs, live mainly in the more protected thatch zone.

22.3 MICROBIAL CONTROL AGENTS OF TURFGRASS INSECT PESTS

22.3.1 Baculoviruses

To date no virus-based insecticides are labeled for any turfgrass usage. However, baculoviruses have been isolated

TABLE 22.2 Some Microbial Products for Turfgrass Pest Management

Pathogenic Microbes	Target Insect	Product Name
Bacteria		
Bacillus thuringiensis (Bt)		
Bt subsp. *aizawai*	Sod webworms, armyworms	XenTari
Bt subsp. *galleriae*	White grubs	grubGONE!
Bt subsp. *kurstaki*	Sod webworms, armyworms	CoStar, Deliver, Dipel, Javelin, Crymax
Paenibacillus popilliae	*Popillia japonica*	Milky Spore
Serratia entomophila	*Costelytra zealandica*	Bioshield
Fungi		
Beauveria bassiana	Chinch bugs	BotaniGard, Naturalis L
Beauveria brongniartii	*Melolontha melolontha,* *Hoplochelus marginalis*	BeauPro, Engerlingspilz, Betel, Melocont
Metarhizium anisopliae	*Adoryphorus couloni* *Dermolepida albohirtum*	Biogreen Bio-Cane
Metarhizium brunneum	White grubs, weevils	Met52, GranMet
Nematodes		
Heterorhabditis bacteriophora	Billbugs, white grubs	Heteromask, Nemasys G, Nema-green, Terranem
Steinernema carpocapsae	Annual bluegrass weevils, armyworms, billbugs, cutworms, fleas, European crane flies, mole crickets, sod webworms	Bionema C, Biosafe, Carponem, Capsanem, Millenium, Nemastar, Ecomask, NEMAgräs
Steinernema glaseri	White grubs	Biotopia

After Klein, M.G., Grewal, P.S., Jackson, T.A., Koppenhöfer, A.M., 2007. Lawn, turf and grassland pests. In: Lacey, L.A., Kaya, H.K. (Eds.), Field Manual of Techniques for the Application and Evaluation of Entomopathogens, second ed. Springer, Dordrecht, The Netherlands. pp. 655–675.

from several insect species that can be major turfgrass pests. On golf courses in Kentucky, United States, a baculovirus of *A. ipsilon*, the AgipMNPV, was found, causing epizootics of this pest (Prater et al., 2006). Sprayed suspensions of AgipMNPV (5×10^8–6×10^9 occlusion bodies/m^2) provided 75–93% control of third to fourth instar *A. ipsilon* in field trials and also persisted on mowed and irrigated bentgrass field plots for at least 4 weeks. AgipMNPV thus holds promise as a preventive bioinsecticide with some long-term effects (Prater et al., 2006; Bixby-Brosi and Potter, 2010a), especially if it can suppress cutworms in the surrounds of greens and tees, thereby reducing reinfestation in these areas.

The presence of the *Neotyphodium lolii* fungal endophyte in perennial ryegrass (Bixby-Brosi and Potter, 2010b) and the treatment of creeping bentgrass plots with the chitinsynthesis-inhibiting fungicide polyoxin-d (Bixby-Brosi and Potter, 2011) had a slightly negative effect on AgipMNPV activity against *A. ipsilon* larvae, likely by reducing consumption of the virus-treated grass. The addition of the optical brightener M2R, meant to synergize the virus and protect it from UV degradation, had no effect under field conditions (Bixby-Brosi and Potter, 2010a).

22.3.2 Fungi

The Hypocreales fungi, *Beauveria* spp. and *Metarhizium* spp., naturally infect white grubs, mole crickets, chinch bugs, and other insects in turfgrass. *Beauveria bassiana* has been associated with the large-scale natural mortality of chinch bugs, especially under hot, humid conditions (Vittum et al., 1999). In the United States, products containing *B. bassiana* (Table 22.2) are or have been labeled for use against white grubs, mole crickets, sod webworm, and chinch bugs.

Against *L. maculicollis* adults, which are active on the soil surface and in the turf canopy, *B. bassiana* (strain GHA; two applications of 0.14×10^{15} conidia/ha) provided 0–42% suppression (Koppenhöfer, unpublished data). Against mature *L. maculicollis* larvae, which feed on or near the soil surface, *B. bassiana* (strain GHA; two applications of 0.14×10^{15} conidia/ha) had no effect (Koppenhöfer, unpublished data), and *Metarhizium brunneum* (strain F52; 2.5–11×10^{15} conidia/ha) gave 31–46% suppression (Ramoutar et al., 2010). Because of the destructive nature of *L. maculicollis* larval feeding, these levels of suppression are not acceptable. The use of fungi for golf course pests may also be limited by the often heavy use of fungicides on the higher profile areas.

European crane fly, *Tipula paludosa*, larval populations were reduced by 63–83% after an application of *B. bassiana* (strain GHA; 0.5×10^{15} conidia/ha) in the fall targeting first and second instars but only by 29% after application in the spring targeting fourth instars (Peck et al., 2008). In greenhouse pots with grass, *Metarhizium robertsii* (strain V1005) applied at 0.2, 2, and 20×10^{15} conidia/ha gave 50%, 90%, and 100% control, respectively, of third instar *T. paludosa* at 4 weeks after application (Ansari and Butt, 2011).

The use of mycoinsecticides against soil insects in turfgrass is also limited by the difficulty of delivering the fungal conidia through the dense turf canopy and thatch layer into the upper soil layers, resulting in often low and highly variable pest suppression. For example, commercial (based on conidia) and experimental (based on microsclerotia) formulations of *M. brunneum* (strain F52) provided 14–69% and 39–69% suppression when applied against first to second instars and second to third instars of the Japanese beetle, *Popillia japonica*, respectively (Behle et al., 2015). Neither the emulsifiable nor wettable powder formulation of *B. bassiana* (strain GHA) provided adequate control of southern masked chafer, *Cyclocephala lurida*, larvae in the field (Wu, 2013) or even in greenhouse pots with grass (Wu et al., 2014). Subsurface applications may improve efficacy and reduce variability but require highly specialized equipment, like modified seed drills, which are usually not available for turfgrass applications. Also, using a slit seeder for the application of *Metarhizium anisopliae sensu lato* against *Phyllopertha horticola* larvae in golf course roughs resulted in only around 15% lower larval populations after 8 weeks, whether granular formulations were directly placed into the slits or sprayable formulations were washed into the slits with water. However, the turf in this study had a thick thatch layer, and no comparison was made to applications without mechanical disruption of the soil surface (Strasser et al., 2005). Whether an application of entomopathogenic fungi (EPF) in conjunction with soil aeration or dethatching could improve efficacy needs to be tested. In addition, mole crickets, and possibly white grubs, can actively avoid soil contaminated with EPF (Villani et al., 1994, 2002; Thompson and Brandenburg, 2005).

In grass pastures where subsurface applications are more feasible, particularly at establishment, EPFs have shown good suppression of some pests, including *Beauveria brongniartii* for the control of *Melolontha melolontha*

larvae in pastures in Europe (Enkerli et al., 2004) and *M. anisopliae sensu lato* against *Adoryphorus couloni* grubs in Australian pastures (Rath et al., 1995). *M. anisopliae sensu lato* is commercially available for the control of froghoppers and spittlebugs on pasture turf in Brazil.

22.3.3 Bacteria

22.3.3.1 Bacillus thuringiensis

Products containing *Bacillus thuringiensis* (Bt) subsp. *kurstaki* or Bt subsp. *aizawai* or their insecticidal crystal proteins (ICPs) are registered for the control of turf pests such as armyworms and sod webworms (Table 22.2). However, they are not very effective against this group of pests in turf because of the rapid photo degradation of the toxins, the slow activity of the ICP, and the lack of contact activity. The weak performance of these products against late-instar larvae makes successful use of these products dependent on the detection of early instars of the pests. If egg hatch occurs over an extended period, multiple treatments of Bt may be needed. Some turf Lepidoptera, such as black cutworm larvae, are also generally not very susceptible to the commercial Bt strains.

The Buibui strain of Bt subsp. *japonensis* (Btj) has shown good activity against the larvae of various scarab species, including the cupreous chafer (*Anomala cuprea*; Suzuki et al., 1994), *P. japonica*, oriental beetle (*Anomala orientalis*), and the green June beetle (*Cotinis nitida*; Alm et al., 1997; Bixby et al., 2007). To be effective, Btj has to be applied against early instars in a preventive mode. Its commercialization in the United States in the 1990s was hindered by formulation difficulties and competition from highly effective and reduced-risk synthetic insecticides that are also applied preventively. The commercial development of this strain had recently been revived in the United States but was halted once again due to the relatively high rates required and the low efficacy against some important white grub species. Thus a rate of 100–120 g δ-endotoxin/ha targeting first and second instars provided around 80% control of *A. orientalis*, but to achieve this level of control reliably for *P. japonica*, around 600 g toxin/ha were necessary. European chafer (*Rhizotrogus majalis*) and especially Asiatic garden beetle (*Maladera castanea*) larvae could not be controlled with Btj (Bixby et al., 2007). Whether the finding that other soil bacteria can improve the virulence (Mashtoly et al., 2010) and persistence of Btj needs to be further examined.

The SDS-502 strain of Bt subsp. *galleriae* has also shown high toxicity to various scarab larvae (Asano et al., 2003; Yamaguchi et al., 2008) and *P. japonica* adults (Yamaguchi et al., 2008) but may face the same issues from competition and differences in susceptibility among different scarab species (Koppenhöfer, unpublished data). Nonetheless, it is now available in the United States for the control of white grub larvae and adults (Table 22.2).

Bt subsp. *israelensis* applied against the first and second instars of the crane fly, *T. paludosa*, has provided 74% and 83% control, respectively, with 13 kg/ha of 5700 International Toxic Units (ITU) and 20 kg/ha of 3000 ITUs (Oestergaard et al., 2006) and 25–58% control at 280 g toxin/ha (Peck et al., 2008); to date there has been no commercial product for crane fly control.

The use of Bt products in turfgrass may increase in the future if biotechnology can provide more versatile and persistent products. For example, the creation of recombinant ICPs could enhance the toxicity or the target spectrum of Bt toxins as has already been shown with hybrid ICPs against the otherwise Bt-resistant black cutworm (de Maagd et al., 2003).

22.3.3.2 Paenibacillus Species

Paenibacillus popilliae and *Paenibacillus lentimorbus* are obligate pathogens that cause "milky disease" in scarab larvae in the subfamilies Melolonthinae, Rutelinae, Aphodinae, and Dynastinae (Klein, 1992; Garczynski and Siegel, 2007; Jurat-Fuentes and Jackson, 2012). Strains of these bacteria are highly specific, showing little to no cross-infectivity to species other than the one they were isolated from. Only the strain infecting *P. japonica* has been commercialized (Table 22.2) and has been used for more than 60 years in the United States. Due to the absence of an effective mass production system, the product has to be applied in an inoculative approach, relying on recycling in infected larvae to spread the disease. Spores can persist in the soil for several years; however, limited availability combined with inconsistent performance and establishment has curbed the use of *P. popilliae* (Klein, 1992; Redmond and Potter, 1995). Strains that infect grubs of species other than *P. japonica* have not been commercially developed. Until effective in vitro production or more virulent strains can be developed, the use of *P. popilliae* will remain very limited.

22.3.3.3 Serratia Species

Amber disease is caused by certain strains of *Serratia* spp. and may be a significant natural mortality factor in white grub populations in turfgrass soils. In New Zealand, *Serratia entomophila* is mass produced in vitro and is used commercially against *Costelytra zealandica* in pastures (Jackson et al., 1992). A more recent improved formulation allows storage at room temperature for several months and is applied with sowing machines (Jackson, 2007). Similar bacteria have been isolated from grubs of *P. japonica* and *Cyclocephala* spp. in the United States, but no commercial development has taken place in the United States.

22.3.4 Nematodes

Nematodes with the potential for the biological control of insects are found in 23 nematode families. Because of problems with culture and/or limited virulence of other nematodes, presently only entomopathogenic nematodes (EPNs) in the families Heterorhabditidae and Steinernematidae (Order: Rhabditida) are used as microbial control agents (MCAs). Several EPN-based products are labeled for the control of turfgrass pests (Table 22.2). In most situations and against most turfgrass pests, EPNs are recommended to be applied at a rate of $2.5–5.0 \times 10^9$ infective juvenile nematodes (IJs)/ha (Grewal et al., 2005; Georgis et al., 2006).

22.3.4.1 Mole Crickets

In broadcast applications at 2.5×10^9 IJs/ha, *Steinernema carpocapsae* provided 58% control of *Scapteriscus* spp. mole crickets on average whereas *Steinernema scapterisci* and *Steinernema riobrave* averaged 75% control (Georgis and Poinar, 1994). However, only the mole cricket-specific *S. scapterisci* can effectively reproduce in *Scapteriscus* spp. and provide long-term suppression of mole cricket populations. It is more effective against the southern mole cricket (*Scapteriscus borellii*) than against the tawny mole cricket (*Scapteriscus vicinus*) and is least effective against the short-winged mole cricket (*Scapteriscus abbreviatus*). It is most effective against adult mole crickets and is ineffective against small nymphs (Parkman and Frank, 1992).

The introduction of *S. scapterisci* (Rhabditida: Steinernematidae) into the United States for mole cricket control is the first successful use of an EPN in classical biological control (Parkman and Smart, 1996; Frank and Walker, 2006). This species was isolated from mole crickets in Uruguay and Argentina and successfully established after the inundative application of *S. scapterisci*-infected cricket cadavers and by using electronic mating callers to attract mole crickets to the site of application (Parkman et al., 1993).

S. scapterisci is an ideal control agent for pastures and turfgrass areas that can tolerate some mole cricket damage (Frank and Walker, 2006). In more damage-sensitive areas, *S. scapterisci* use has been limited due to the competition from insecticides and by the need for the application of nematicides for the control of plant-parasitic nematodes. In pastures, the potentially biggest market for *S. scapterisci*, it has been applied using slit injectors in strips covering 12.5% of the area from which it then spreads over a period of several years (Frank and Walker, 2006). This approach reduced the cost to well below that of chemical insecticides that provide only short-term suppression.

In areas of Florida where both *S. scapterisci* and a parasitoid wasp, *Larra bicolor* (Hymenoptera: Crabronidae), also introduced from South America, have become established, mole cricket populations are significantly suppressed and cause much fewer problems (Frank and Walker,

2006). Because of the nematode's slow spread from inoculation sites, the widespread use of a commercial product or a large-scale inoculation program would be necessary to accelerate its spread. However, even though the *S. scapterisci*-based product Nematac-S appeared to become a success story, especially for uses in low-maintenance turf and in pastures, it was discontinued in 2012 due to insufficient demand.

22.3.4.2 White Grubs

As soil-dwelling insects, white grubs share their natural soil and rhizosphere habitats with EPNs. At least five EPN species, *Steinernema arenarium*, *Steinernema glaseri*, *Steinernema kushidai*, *Steinernema scarabaei*, and *Heterorhabditis megidis*, were originally collected and described from naturally infected white grubs, and many more species have been documented to use white grubs as natural hosts (Peters, 1996). However, as a result of their coevolution with soil pathogens, white grubs have developed defense mechanisms, including infrequent carbon dioxide output, sieveplates over their spiracles, frequent defecation, defensive and evasive behaviors, a dense peritrophic membrane, and a strong immune response that make them relatively resistant to infection by EPNs (Grewal et al., 2005 and references therein).

Early field trials with white grubs in the United States found that *Heterorhabditis* spp. and *S. glaseri* were generally more effective than *Steinernema feltiae* and *S. carpocapsae* (Klein, 1993). An analysis of 82 field trials against *P. japonica* indicated that *Heterorhabditis bacteriophora* (at 2.5×10^9 IJs/ha) applied under the right conditions was as effective as the standard chemical insecticides used at the time (organophosphates and carbamates), whereas *S. carpocapsae* was ill-adapted for white grub control (Georgis and Gaugler, 1991). Studies have shown that white grub species differ in their susceptibility to EPNs and that the relative virulence of different EPN species also varies among white grub species (Grewal et al., 2002; Koppenhöfer et al., 2004, 2006).

Grewal et al. (2005) provide an extensive summary of field studies on EPN efficacy against white grubs using as a qualifier for "good" control at least 70% control (at 2.5×10^9 IJs/ha). Good control of *P. japonica* has been achieved with *S. scarabaei* AMK001 (100%; Koppenhöfer and Fuzy, 2003a), *H. bacteriophora* GPS11 (34–97%; Grewal et al., 2004), *H. bacteriophora* TF (65–92%; Koppenhöfer and Fuzy, 2003a; Koppenhöfer et al., 2000a,b), and *H. zealandica* X1 (73–98%; Grewal et al., 2004). *S. scarabaei* is the only nematode species that has provided high field control of *A. orientalis* (87–100%; Koppenhöfer and Fuzy, 2003a,b, 2009), *Maladera castanea* (71–86%; Koppenhöfer and Fuzy, 2003b), and *Rhizotrogus majalis* (89%; Cappaert and Koppenhöfer, 2003). Good control of the northern

masked chafer, *Cyclocephala borealis*, has been observed with *H. zealandica* X1 (72–96%), *S. scarabaei* (84%), and *H. bacteriophora* GPS11 (47–83%; Grewal et al., 2004; Koppenhöfer and Fuzy, 2003a).

Larval stages of white grubs may also differ in their susceptibility, but the effect varies with white grub and EPN species. Against *P. japonica*, efficacy is higher against second instars than third instars for *H. bacteriophora* (Grewal et al., 2004; Koppenhöfer and Fuzy, 2004; Power et al., 2009), but there is no significant effect for *S. scarabaei* (Koppenhöfer and Fuzy, 2004) or *H. zealandica* (Grewal et al., 2004). For *A. orientalis*, susceptibility decreases from second to third instars for *H. bacteriophora* (Koppenhöfer and Fuzy, 2004, 2008) and *Heterorhabditis* sp., *S. carpocapsae*, *S. glaseri*, and *Steinernema longicaudum* (Lee et al., 2002) and from young third instars to older third instars for *H. bacteriophora* (Koppenhöfer and Fuzy, 2004). However, there is no difference in susceptibility between second and third instars for *S. scarabaei* or *S. glaseri* (Koppenhöfer and Fuzy, 2004). *M.castanea* susceptibility to *Steinernema scarabaei* increases from second to third instar (Koppenhöfer and Fuzy, 2004).

As soil temperature drops below 20°C, EPNs, especially *H. bacteriophora*, become increasingly ineffective for white grub control (Georgis and Gaugler, 1991). IJ downward movement is restricted by thatch, an accumulation of organic matter between the soil and turfgrass foliage, and thatch thickness is negatively related to EPN efficacy (Georgis and Gaugler, 1991). Irrigation volume and frequency and soil moisture are positively related to efficacy (Georgis and Gaugler, 1991; Grewal et al., 2004) with a minimum of 7.4 mm of postapplication irrigation required for the establishment of the IJs in turfgrass (Shetlar et al., 1988). Georgis and Gaugler (1991) concluded that *H. bacteriophora* was more effective against *P. japonica* in fine-textured soils, probably because finer soils retain moisture better and restrict IJ movement to the upper soil layers where most of the white grubs can be found.

The timing of EPN applications should consider not only the presence of the most susceptible white grub stages but also other environmental conditions, particularly soil temperatures. For example, in the northeastern United States, *H. bacteriophora* applications against *P. japonica* or *A. orientalis* will tend to be more effective if applied between mid-August and early September due to the presence of the more susceptible younger larvae and a longer period after application before soil temperatures become too cool for good IJ activity. The efficacy of application before mid-August could be limited by hot and dry conditions, especially where irrigation is limited and due to limited recycling of the EPNs in the smallest larval stages (Power et al., 2009).

EPN persistence beyond a season following application against third instar white grubs has been reported for *H. bacteriophora* and *S. scarabaei*, suggesting the potential impact of EPNs on multiple generations of white grubs (eg, Klein and Georgis, 1992; Koppenhöfer and Fuzy, 2009). Due to the excellent adaptation of *S. scarabaei* to white grubs as hosts (high virulence and recycling in hosts) and the outstanding persistence of individual IJs, this species provided high control of *A. orientalis* within 1 month of application at low application rates (77–100% control at 0.25–2.5 × 10^9 IJs/ha). It also provided additional control of overwintered larvae in the following spring (86–100% control at 0.1–2.5 × 10^9 IJs/ha) and persisted and suppressed *A. orientalis* for up to 4 years after application (Koppenhöfer and Fuzy, 2009). Thus this species could be an excellent candidate for the long-term suppression of white grubs with periodic reapplication at low rates.

22.3.4.3 Weevils

No detailed studies on billbug–EPN interaction have been published. In field tests targeting the larvae in the soil, control of the bluegrass billbug, *Sphenophorus parvulus*, by *S. carpocapsae* (average 78%) and *H. bacteriophora* (average 74%; 2.5 × 10^9 IJs/ha) was similar to that by standard insecticides at the time (organophosphates, carbamate; Georgis and Poinar, 1994). But the use of EPN products against billbugs is limited in the United States due to the availability of several newer insecticides that are easier to use, more effective, and cheaper. On golf courses in Japan, *S. carpocapsae* (2.5 × 10^9 IJs/ha) was very effective for the control of the hunting billbug, *S. venatus vestitus* (average 84%; Smith, 1994; Kinoshita and Yamanaka, 1998), in part due to favorable environmental conditions (temperature and rainfall) and the adoption of EPN friendly application protocols. However, *S. carpocapsae* sales for billbug control in Japan significantly declined after the registration of imidacloprid for turfgrass uses.

As for *L. maculicollis*, McGraw and Koppenhöfer (2009) observed that indigenous populations of *S. carpocapsae* and *H. bacteriophora* were common on insecticide-free fairways but failed to generate a functional response capable of increasing mortality with increasing weevil larval densities, suggesting their unreliability in reducing densities in a conservation biological-control approach. Despite the promising control of fourth and fifth instar larvae by *S. carpocapsae*, *S. feltiae*, and *H. bacteriophora* under laboratory conditions (McGraw and Koppenhöfer, 2008), numerous field trials with multiple species, rates, and combinations showed variable performance (0–94% control at 2.5 × 10^9 IJs/ha) and limited persistence (McGraw et al., 2010). However, ongoing research indicates that split applications of *S. carpocapsae* (about 1 week apart) improve control and that the simultaneous application of *S. carpocapsae* or *H. bacteriophora* with imidacloprid (applied for white grub control) has an additive effect on larval mortality (Wu, unpublished data). Even though the destructive

nature of *L. maculicollis* necessitates high levels of control, the dearth of effective insecticides for insecticide-resistant populations (Ramoutar et al., 2009) may increase opportunities for EPN use.

22.3.4.4 Caterpillars

In field efficacy testing of EPN products in the 1980s and early 1990s, *S. carpocapsae* was highly effective in controlling *A. ipsilon* larvae (average 95%), but *H. bacteriophora* did not provide satisfactory control (average 62%; Georgis and Poinar, 1994). In detailed laboratory studies that included seven EPN species (*H. bacteriophora*, *Heterorhabditis megidis*, *H. indica*, *S. carpocapsae*, *S. riobrave*, *S. feltiae*, and *S. kraussei*), Ebssa and Koppenhöfer (2012) found that *S. carpocapsae* tended to cause the highest larval mortality in pots with grass followed by *H. bacteriophora*, *H. megidis*, and *S. riobrave*, and that fourth and/or fifth instar larvae were the most susceptible stage to most EPN species. Pupae were the least susceptible.

In field studies using commercial products against fourth instar larvae at 2.5 × 10^9 IJs/ha (Ebssa and Koppenhöfer, 2011), *S. carpocapsae* was the best performing species due to a combination of high (average 83%) and most consistent (70–90%) control rates (at 7 days after treatment) and fastest kill (average 68% at 4 days after treatment); *S. feltiae* (average 68%) and *H. bacteriophora* (average 62%) were less consistent, whereas *S. riobrave* was ineffective (average 32%). In additional field studies (Koppenhöfer, unpublished data), (1) combinations of two EPN species at half rate each did not provide significantly better control than either of the two species alone at full rate; (2) syringing (ie, twice daily small amounts of irrigation) provided some limited improvement of *S. carpocapsae* efficacy under warm, sunny conditions; and (3) a split application (two applications at half rate 3 days apart) improved *S. carpocapsae* efficacy by around 20%. Despite this high efficacy, EPNs are not widely used for *A. ipsilon* control, because damage thresholds on golf course tees and especially greens are so low that golf course superintendents prefer to use chemical insecticides that provide even better and more consistent control at a lower cost than *S. carpocapsae*.

Studies with other turfgrass caterpillars are more limited. In field studies with fifth instars of the armyworm *Mythimna* (*Pseudaletia*) *separata*, Korean strains of *S. carpocapsae* (GSN1), *S. longicaudum* (Nosan), *Heterorhabditis* sp. (HG), and *Steinernema monticolum* (Jiri) provided 65%, 60%, 50%, and 45% control, respectively, at a rate of 1 × 10^9 IJs/ha (Jung et al., 2013). Against sixth instar larvae of the common armyworm, *Mythimna* (*Pseudaletia*) *unipuncta*, local Azorean EPN strains provided only limited control (*H. bacteriophora* 43%; *S. carpocapsae* 32%) in a field study despite using a high application rate (10 × 10^9 IJs/ha); however, efficacy might have been higher if younger stage larvae would have been used and if the larvae had been exposed for more than 3 days in the field (Rosa and Simões, 2004). Against fifth and sixth instars of the armyworm *Spodoptera cilium* on sod arenas in the laboratory, Turkish isolates of *S. carpocapsae* and *H. bacteriophora* provided 77% and 29% control at 2.5×10^9 IJs/ha (Gulcu et al., 2014). In greenhouse pots with grass infested with larvae of the tropical sod webworm, *Herpetogramma phaeopteralis*, a commercial *S. carpocapsae* product applied at 2.5×10^9 IJs/ha provided 80%, 86%, and 90% control of first and second, third and fourth, and fifth and sixth instars, respectively (Tofangsazi et al., 2014).

22.3.4.5 Crane Flies

Larvae of *T. paludosa* are susceptible to heterorhabditid nematodes and particularly to *S. feltiae* (Ehlers and Gerwien, 1993; Simard et al., 2006). Larval susceptibility, at least to *S. feltiae*, decreases with larval development (Peters and Ehlers, 1994). In field trials using a rate of 5×10^9 IJs/ha, *S. carpocapsae* (All strain) provided 75–82% control when applied against first and second instars but no control when applied against second to third instars (Oestergaard et al., 2006). The lack of control in the later application was likely due to a combination of decreasing susceptibility of older larvae and decreasing soil temperatures; *S. feltiae* (strain OBSIII) was ineffective (0–16%) irrespective of application timing.

22.4 CONCLUSIONS AND RECOMMENDATIONS

The use of MCAs in turfgrass is very limited. A large proportion of them are used by homeowners who value reduced exposure to synthetic insecticides more than they value product efficacy, ease of use, and price. While concerns about health risks and environmental hazards of pesticides have led to pesticide legislation in the United States and Canada that have resulted in the loss of many insecticides, especially organophosphates, for turfgrass uses (Bélair et al., 2010), they have been replaced by new active ingredients from several new insecticide classes that are considered low-risk insecticides (eg, neonicotinoids, oxadiazines, anthranilic diamides). Nonetheless, in Canada, the use of pesticides for cosmetic purposes is banned in two provinces (Québec and Ontario), and more than 152 municipalities have adopted by-laws restricting or banning the use of landscape pesticides (Bélair et al., 2010). Canadian golf courses may still use pesticides but have to comply with new, stricter regulations.

In the United States the regulatory process for pesticides is less cumbersome and faster than in Canada and the European Union, and the turf market is much larger and legislatively more uniform. As a result, more insecticides, particularly the newer types, are available (Bélair et al., 2010), and synthetic insecticides remain the mainstay for insect control in turfgrass. Nonetheless, public concerns about insecticide use persist. To take advantage of the growing interest in alternatives to synthetic insecticides and organic land care, MCAs will have to be improved with respect to efficacy (more virulent strains from field populations or through biotechnology, improved application technologies), costs (better production technologies), and especially ease of use (formulations with extended shelf life and tolerance to temperature extremes). Most such advances would probably be only gradual, but they may suffice to significantly increase the use of microbials, at least where pesticide regulations, local ordinances, and public opinion already impinge on the use of synthetic insecticides.

The use of MCAs could also increase in many situations by more systematically exploiting their ability to recycle in hosts and provide long-term pest suppression. Most of the research on MCAs in turfgrass has concentrated on inundative application with the emphasis on quick and effective pest control. However, many of the agents developed for ease of production and commercialization may have lost virulence and fitness during laboratory culture (Wang and Grewal, 2002) and with that would be less likely to persist in the environment. In addition, species that have a potential for long-term suppression generally appear to be somewhat host-specific, but if their hosts are key pests, they can still be successful [eg, *B. brongniartii* against *Melolontha* (Enkerli et al., 2004), *S. entomophila* against *C. zealandica* (Jackson et al., 1992), or *S. scapterisci* against mole crickets (Parkman and Smart, 1996)]. Two other organisms that may lend themselves to long-term pest suppression are the *Agip*M-NPV for *A. ipsilon* suppression in surrounds of golf course greens (Prater et al., 2006) and *S. scarabaei* for white grub suppression (Koppenhöfer and Fuzy, 2009). However, since these specific pathogens depend on the presence of hosts for recycling and may become patchy in distribution over time, their use may only be feasible in areas with some tolerance for occasional pest damage.

Ultimately, significant increases in MCA use, whether as biopesticides or inoculative agents, will likely happen only through education and legislation. Major changes in insecticide use patterns will have to be encouraged through legislative incentives, regulations, and restrictions (Bélair et al., 2010).

REFERENCES

Alm, S.R., Villani, M.G., Yeh, T., Shutter, R., 1997. *Bacillus thuringiensis japonensis* strain Buibui for control of Japanese and oriental beetle larvae (Coleoptera: Scarabaeidae). Appl. Entomol. Zool. 32, 477–484.

Ansari, M.A., Butt, T.M., 2011. Evaluation of entomopathogenic fungi and a nematode against the soil-dwelling stages of the crane fly *Tipula paludosa*. Pest Manage. Sci. 68, 1337–1344.

Asano, S., Yamashita, C., Iizuka, T., Takeuchi, K., Yamanaka, S., Cerf, D., Yamamoto, T., 2003. A strain of *Bacillus thuringiensis* subsp. *galleriae* containing a novel *cry8* gene highly toxic to *Anomala cuprea* (Coleoptera: Scarabaeidae). Biol. Control 28, 191–196.

Beard, J.B., Green, R.L., 1994. The role of turfgrass in environmental protection and their benefits to humans. J. Environ. Qual. 23, 452–460.

Behle, R.W., Richmond, D.S., Jackson, M.A., Dunlap, C.A., 2015. Evaluation of *Metarhizium brunneum* F52 (Hypocreales: Clavicipitaceae) for control of Japanese beetle larvae in turfgrass. J. Econ. Entomol. 108, 1587–1595.

Bélair, G., Koppenhöfer, A.M., Dionne, J., Simard, L., 2010. Current and potential use of pathogens in the management of turfgrass insects as affected by new pesticide regulations in North America. Int. J. Pest Manage. 56, 51–60.

Bixby, A., Alm, S.R., Power, K., Grewal, P.S., Swier, S., 2007. Susceptibility of four species of turfgrass-infesting scarabs (Coleoptera: Scarabaeidae) to *Bacillus thuringiensis* serovar *japonensis* strain Buibui. J. Econ. Entomol. 100, 1604–1610.

Bixby-Brosi, A.J., Potter, D.A., 2010a. Evaluating a naturally occurring baculovirus for extended biological control of the black cutworm (Lepidoptera: Noctuidae) in golf course habitats. J. Econ. Entomol. 103, 1555–1563.

Bixby-Brosi, A.J., Potter, D.A., 2010b. Influence of endophyte (*Neotyphodium lolii*) infection of perennial ryegrass on susceptibility of the black cutworm (Lepidoptera: Noctuidae) to a baculovirus. Biol. Control 54, 141–146.

Bixby-Brosi, A.J., Potter, D.A., 2011. Can a chitin-synthesis-inhibiting turfgrass fungicide enhance black cutworm susceptibility to a baculovirus? Pest Manage. Sci. 68, 324–329.

Breuninger, J.M., Welterlen, M.S., Augustin, B.J., Cline, V., Morris, K., 2012. The turfgrass industry. In: Stier, J.C., Horgan, B.P., Bonos, S.A. (Eds.), Turfgrass: Biology, Use, and Management. Aronomy Monograph, vol. 56. Am. Soc. Agron, Crop Sci. Soc. Am., Soil Sci. Soc. Am, Madison, WI., USA, pp. 933–1006.

Cappaert, D.C., Koppenhöfer, A.M., 2003. *Steinernema scarabaei*, an entomopathogenic nematode for control of the European chafer. Biol. Control 28, 379–386.

Christians, N.E., 1998. Fundamentals of Turfgrass Management. Ann Arbor Press, Chelsea, MI.

de Maagd, R.A., Weemen-Henriks, M., Molthoff, J.W., Naimov, S., 2003. Activity of wild type and hybrid Bacillus thuringiensis δ-endotoxins against Agrotis ipsilon. Arch. Microbiol. 179, 363–367.

Ebssa, L., Koppenhöfer, A.M., 2011. Efficacy and persistence of entomopathogenic nematodes for black cutworm control in turfgrass. Biocontrol Sci. Technol. 21, 779–796.

Ebssa, L., Koppenhöfer, A.M., 2012. Entomopathogenic nematodes for black cutworm management: effect of instar, nematode species, and nematode production method. Pest Manage. Sci. 68, 947–957.

Ehlers, R.-U., Gerwien, A., 1993. Selection of entomopathogenic nematodes (Steinernematidae and Heterorhabditidae, Nematoda) for the biological control of cranefly larvae *Tipula paludosa* (Tipulidae, Diptera). Z. Pflanzenk. Pflanz. 100, 343–353.

Enkerli, J., Widmer, F., Keller, S., 2004. Long-term field persistence of *Beauveria brongniartii* strains applied as biocontrol agents against European cockchafer larvae in Switzerland. Biol. Control 29, 115–123.

Frank, J.H., Walker, T.J., 2006. Permanent control of pest mole crickets (Orthoptera: Gryllotalpidae) in Florida. Am. Entomol. 52, 139–144.

Garczynski, S.F., Siegel, J.P., 2007. Bacteria. In: Lacey, L.A., Kaya, H.K. (Eds.), Field Manual of Techniques in Invertebrate Pathology. Kluwer, Dordrecht, The Netherlands, pp. 175–198.

Georgis, R., Gaugler, R., 1991. Predictability in biological control using entomopathogenic nematodes. J. Econ. Entomol. 84, 713–720.

Georgis, R., Poinar Jr., G.O., 1994. Nematodes as bioinsecticides in turf and ornamentals. In: Leslie, A. (Ed.), Integrated Pest Management for Turf and Ornamentals. CRC Press, Boca Raton, FL, pp. 477–489.

Georgis, R., Koppenhöfer, A.M., Lacey, L.A., Bélair, G., Duncan, L.W., Grewal, P.S., Samish, M., Tan, L., Torr, P., van Tol, R.W.H.M., 2006. Successes and failures in the use of parasitic nematodes for pest control. Biol. Control 37, 103–123.

Grewal, P.S., Grewal, S.K., Malik, V.S., Klein, M.G., 2002. Differences in the susceptibility of introduced and native white grub species to entomopathogenic nematodes from various geographic localities. Biol. Control 24, 230–237.

Grewal, P.S., Power, K.T., Grewal, S.K., Suggars, A., Haupricht, S., 2004. Enhanced consistency in biological control of white grubs (Coleoptera: Scarabaeidae) with new strains of entomopathogenic nematodes. Biol. Control 30, 73–82.

Grewal, P.S., Koppenhöfer, A.M., Choo, H.Y., 2005. Lawn, turfgrass, and pasture applications. In: Grewal, P.S., Ehlers, R.-U., Shapiro-Ilan, D.I. (Eds.), Nematodes as Biocontrol Agents. CABI Publishing, Wallingford, UK, pp. 115–146.

Gulcu, B., Ulug, D., Hazir, C., Karagoz, M., Hazir, S., 2014. Biological control potential of native entomopathogenic nematodes (Steinernematidae and Heterorhabditidae) against *Spodoptera cilium* (Lepidoptera: Noctuidae) in turfgrass. Biocontrol Sci. Technol. 24, 965–970.

Jackson, T.A., 2007. A novel bacterium for control of grass grub. In: Vincent, C., Goettel, M.S., Lazarovits, G. (Eds.), Biological Control: A Global Perspective. CABI Publishing, Wallingford, UK, pp. 160–168.

Jackson, T.A., Pearson, J.F., O'Callaghan, M.O., Mahanty, H.K., Willocks, M.J., 1992. Pathogen to product-development of *Serratia entomophila* (Enterobacteriaceae) as a commercial biological control agent for the New Zealand grass grub (*Costelytra zealandica*). In: Jackson, T.A., Glare, T.R. (Eds.), Use of Pathogens in Scarab Pest Management. Intercept Ltd., Andover, UK, pp. 191–198.

Jung, Y.H., Kim, J.J., You, E.J., Lee, C.M., Choo, H.Y., Lee, D.W., 2013. Evaluation of entomopathogenic nematodes against armyworm, *Pseudaletia separata* on tall fescue, *Festuca arundinacea*. Weed Turf. Sci. 2, 312–317.

Jurat-Fuentes, J.L., Jackson, T.A., 2012. Bacterial entomopathogens. In: Vega, F.E., Kaya, H.K. (Eds.), Insect Pathology, second ed. Elsevier, Academic Press, San Diego, CA, pp. 265–349.

Kinoshita, M., Yamanaka, S., 1998. Development and prevalence of entomopathogenic nematodes in Japan. Japn. J. Nematol. 28, 42–45.

Klein, M.G., 1992. Use of *Bacillus popilliae* in Japanese beetle control. In: Jackson, T.A., Glare, T.R. (Eds.), Use of Pathogens in Scarab Pest Management. Intercept Ltd., Andover, UK, pp. 179–189.

Klein, M.G., 1993. Biological control of scarabs with entomopathogenic nematodes. In: Bedding, R., Akhurst, R., Kaya, H. (Eds.), Nematodes and the Biological Control of Insect Pests. CSIRO Press, East Melbourne, Australia, pp. 49–58.

Klein, M.G., Georgis, R., 1992. Persistence of control of Japanese beetle (Coleoptera: Scarabaeidae) larvae with steinernematid and heterorhabditid nematodes. J. Econ. Entomol. 85, 727–730.

Klein, M.G., Grewal, P.S., Jackson, T.A., Koppenhöfer, A.M., 2007. Lawn, turf and grassland pests. In: Lacey, L.A., Kaya, H.K. (Eds.), Field Manual of Techniques for the Application and Evaluation of Entomopathogens, second ed. Springer, Dordrecht, The Netherlands, pp. 655–675.

Koppenhöfer, A.M., Fuzy, E.M., 2003a. *Steinernema scarabaei* for the control of white grubs. Biol. Control 28, 47–59.

Koppenhöfer, A.M., Fuzy, E.M., 2003b. Biological and chemical control of the Asiatic garden beetle, *Maladera castanea* (Coleoptera: Scarabaeidae). J. Econ. Entomol. 96, 1076–1082.

Koppenhöfer, A.M., Fuzy, E.M., 2004. Effect of white grub developmental stage on susceptibility to entomopathogenic nematodes. J. Econ. Entomol. 97, 1842–1849.

Koppenhöfer, A.M., Fuzy, E.M., 2008. Early timing and new combinations to increase the efficacy of neonicotinoid-entomopathogenic nematodes (Rhabditida: Heterorhabditidae) combinations against white grubs (Coleoptera: Scarabaeidae). Pest Manage. Sci. 64, 725–735.

Koppenhöfer, A.M., Fuzy, E.M., 2009. Long-term effects and persistence of *Steinernema scarabaei* applied for suppression of *Anomala orientalis* (Coleoptera: Scarabaeidae). Biol. Control 48, 63–72.

Koppenhöfer, A.M., Brown, I.M., Gaugler, R., Grewal, P.S., Kaya, H.K., Klein, M.G., 2000a. Synergism of entomopathogenic nematodes and imidacloprid against white grubs: greenhouse and field evaluation. Biol. Control 19, 245–251.

Koppenhöfer, A.M., Wilson, M.G., Brown, I., Kaya, H.K., Gaugler, R., 2000b. Biological control agents for white grubs (Coleoptera: Scarabaeidae) in anticipation of the establishment of the Japanese beetle in California. J. Econ. Entomol. 93, 71–80.

Koppenhöfer, A.M., Fuzy, E.M., Crocker, R., Gelernter, W., Polavarapu, S., 2004. Pathogenicity of *Steinernema scarabaei*, *Heterorhabditis bacteriophora* and *S. glaseri* to twelve white grub species. Biocontrol Sci. Technol. 14, 87–92.

Koppenhöfer, A.M., Grewal, P.S., Fuzy, E.M., 2006. Virulence of the entomopathogenic nematodes *Heterorhabditis bacteriophora*, *H. zealandica*, and *Steinernema scarabaei* against five white grub species (Coleoptera: Scarabaeidae) of economic importance in turfgrass in North America. Biol. Control 38, 397–404.

Lee, D.W., Choo, H.Y., Kaya, H.K., Lee, S.M., Smitley, D.R., Shin, H.K., Park, C.G., 2002. Laboratory and field evaluation of Korean entomopathogenic nematode isolates against the oriental beetle *Exomala orientalis* (Coleoptera: Scarabaeidae). J. Econ. Entomol. 95, 918–926.

Mashtoly, T.A., Abolmaaty, A., Thompson, N., El-Zemaity, M.E., Hussien, M.I., Alm, S.R., 2010. Enhanced toxicity of *Bacillus thuringiensis japonensis* strain Buibui toxin to oriental beetle and northern masked chafer (Coleoptera: Scarabaeidae) larvae with *Bacillus* sp. NFD2. J. Econ. Entomol. 103, 1547–1554.

McGraw, B.A., Koppenhöfer, A.M., 2008. Evaluation of two endemic and five commercial entomopathogenic nematode species (Rhabditida: Heterorhabditidae and Steinernematidae) against annual bluegrass weevil (Coleoptera: Curculionidae) larvae and adults. Biol. Control 46, 467–475.

McGraw, B.A., Koppenhöfer, A.M., 2009. Population dynamics and interactions between endemic entomopathogenic nematodes and annual bluegrass weevil populations in golf course turfgrass. Appl. Soil Ecol. 41, 77–89.

McGraw, B.A., Vittum, P.J., Cowles, R.S., Koppenhöfer, A.M., 2010. Field evaluation of entomopathogenic nematodes for the biological control of the annual bluegrass weevil, *Listronotus maculicollis* (Coleoptera: Curculionidae) in golf course turfgrass. Biocontrol Sci. Technol. 20, 149–163.

Oestergaard, J., Belau, C., Strauch, O., Ester, A., van Rozen, K., Ehlers, R.-U., 2006. Biological control of *Tipula paludosa* (Diptera: Nematocera) using entomopathogenic nematodes (*Steinernema* spp.) and *Bacillus thuringiensis* supsp. *israelensis*. Biol. Control 39, 525–531.

Parkman, J.P., Frank, J.H., 1992. Infection of sound–trapped mole crickets, *Scapteriscus* spp., by *Steinernema scapterisci*. Fla. Entomol. 75, 163–165.

Parkman, J.P., Smart Jr., G.C., 1996. Entomopathogenic nematodes, a case study: introduction of *Steinernema scapterisci* in Florida. Biocontrol Sci. Technol. 6, 413–419.

Parkman, J.P., Hudson, W.G., Frank, J.H., Nguyen, K.B., Smart Jr., G.C., 1993. Establishment and persistence of *Steinernema scapterisci* (Rhabditida: Steinernematidae) in field populations of *Scapteriscus* spp. mole crickets (Orthoptera: Gryllotalpidae). J. Entomol. Sci. 28, 182–190.

Peck, D.C., Olmstead, D., Morales, A., 2008. Application timing and efficacy of alternatives for the insecticidal control of *Tipula paludosa* Meigen (Diptera: Tipulidae), a new invasive pest of turf in the northeastern United States. Pest Manage. Sci. 64, 989–1000.

Peters, A., 1996. The natural host range of *Steinernema* and *Heterorhabditis* spp. and their impact on insect populations. Biocontrol Sci. Technol. 6, 389–402.

Peters, A., Ehlers, R.-U., 1994. Susceptibility of leatherjackets (*Tipula paludosa* and *T. oleracea*; Tipulidae: Nematocera) to the entomopathogenic nematode *Steinernema feltiae*. J. Invertebr. Pathol. 63, 163–171.

Potter, D.A., 1998. Destructive Turfgrass Insects: Biology, Diagnosis, and Control. Ann Arbor Press, Chelsea, MI. 344 pp.

Power, K.T., An, R., Grewal, P.S., 2009. Effectiveness of *Heterorhabditis bacteriophora* strain GPS11 applications targeted against different instars of the Japanese beetle *Popillia japonica*. Biol. Control 48, 232–236.

Prater, C.A., Redmond, C.T., Barney, W., Bonning, B.C., Potter, D.A., 2006. Microbial control of black cutworm (Lepidoptera: Noctuidae) in turfgrass by using *Agrotis ipsilon* multiple nucleopolyhedrovirus. J. Econ. Entomol. 99, 1129–1137.

Ramoutar, D., Alm, S.R., Cowles, R.S., 2009. Pyrethroid resistance in populations of *Listronotus maculicollis* (Coleoptera: Curculionidae) from southern New England golf courses. J. Econ. Entomol. 102, 388–392.

Ramoutar, D., Legrand, A.I., Alm, S.R., 2010. Field performance of *Metarhizium anisopliae* against *Popillia japonica* (Coleoptera: Scarabaeidae) and *Listronotus maculicollis* (Coleoptera: Curculionidae) larvae in turfgrass. J. Entomol. Sci. 45, 1–7.

Rath, A.C., Worledge, D., Koen, T.B., Rowe, B.A., 1995. Long-term field efficacy of the entomogenous fungus *Metarhizium anisopliae* against the subterranean scarab, *Adoryphorus couloni*. Biocontrol Sci. Technol. 5, 439–451.

Redmond, C.T., Potter, D.A., 1995. Lack of efficacy of in vivo- and putatively in vitro-produced *Bacillus popilliae* against field populations of Japanese beetle (Coleoptera: Scarabaeidae) grubs in Kentucky. J. Econ. Entomol. 88, 846–854.

Rosa, J.S., Simões, N., 2004. Evaluation of twenty-eight strains of *Heterorhabditis bacteriophora* isolated in Azores for biological control of the armyworm, *Pseudaletia unipunctata* (Lepidoptera: Noctuidae). Biol. Control 29, 409–417.

Shetlar, D.J., Suleman, P.E., Georgis, R., 1988. Irrigation and use of entomogenous nematodes, *Neoaplectana* spp. and *Heterorhabditis heliothidis* (Rhabditida: Steinernematidae and Heterorhabditidae), for control of Japanese beetle (Coleoptera: Scarabaeidae) grubs in turfgrass. J. Econ. Entomol. 81, 1318–1322.

Simard, L., Bélair, G., Gosselin, M.E., Dionne, J., 2006. Virulence of entomopathogenic nematodes (Rhabditida: Steinernematidae, Heterorhabditidae) against *Tipula paludosa* (Diptera: Tipulidae), a turfgrass pest on golf courses. Biocontrol Sci. Technol. 16, 789–801.

Smith, K.A., 1994. Control of weevils with entomopathogenic nematodes. In: Smith, K.A., Hatsukade, M. (Eds.), Control of Insect Pests with Entomopathogenic Nematodes. Food and Fertilizer Technology Center, Republic of China in Taiwan, pp. 1–13.

Strasser, H., Zelger, R., Pernfuss, B., Längle, T., Seger, C., 2005. EPPO based efficacy study to control *Phyllopertha horticola* in golf courses. IOBC WPRS Bull. 28, 189–192.

Suzuki, N., Hori, H., Tachibana, M., Asano, S., 1994. *Bacillus thuringiensis* strain Buibui for control of cupreous chafer, *Anomala cuprea* (Coleoptera: Scarabaeidae), in turfgrass and sweet potato. Biol. Control 4, 361–365.

Thompson, S.R., Brandenburg, R.L., 2005. Tunneling responses of mole crickets (Orthoptera: Gryllotalpidae) to the entomopathogenic fungus, *Beauveria bassiana*. Environ. Entomol. 34, 140–147.

Tofangsazi, N., Cherry, R.H., Arthurs, S.P., 2014. Efficacy of commercial formulations of entomopathogenic nematodes against tropical sod webworm, *Herpetogramma phaeopteralis* (Lepidoptera: Crambidae). J. Appl. Entomol. 138, 656–661.

Villani, M.G., Krueger, S.R., Schroeder, P.C., Consolie, F., Consolie, N.H., Preston-Wilsey, L.M., Roberts, D.W., 1994. Soil application effects of *Metarhizium anisopliae* on Japanese beetle (Coleoptera: Scarabaeidae) behavior and survival in turfgrass microcosms. Environ. Entomol. 23, 502–513.

Villani, M.G., Allee, L.L., Preston-Wilsey, L., Consolie, N., Xia, Y., Brandenburg, R.L., 2002. Use of radiography and tunnel castings for observing mole cricket (Orthoptera: Gryllotalpidae) behavior in soil. Am. Entomol. 48, 42–50.

Vittum, P.J., Villani, M.G., Tashiro, H., 1999. Turfgrass Insects of the United States and Canada. Cornell University Press, Ithaca, NY. 422 pp.

Wang, X., Grewal, P.S., 2002. Rapid deterioration of environmental tolerance and reproductive potential of an entomopathogenic nematode during laboratory maintenance. Biol. Control 23, 71–78.

Wu, S., 2013. Efficacy of Entomopathogenic Nematodes and Entomopathogenic Fungi against Masked Chafer with Grubs, Cyclocephala Spp. (Coleoptera: Scarabaeidae) (Ph.D. dissertation). Virginia Polytechnic Institute and State University, Blacksburg, VA.

Wu, S., Youngman, R.R., Kok, L.T., Laub, C.A., Pfeiffer, D.G., 2014. Interaction between entomopathogenic nematodes and entomopathogenic fungi applied to third instar southern masked chafer white grubs, *Cyclocephala lurida* (Coleoptera: Scarabaeidae), under laboratory and greenhouse conditions. Biol. Control 76, 65–73.

Yamaguchi, T., Sahara, K., Bando, H., Asano, S., 2008. Discovery of a novel *Bacillus thuringiensis* Cry8D protein and the unique toxicity of the Cry8D-class proteins against scarab beetles. J. Invertebr. Pathol. 99, 257–262.

Chapter 23

LUBILOSA: The Development of an Acridid-Specific Mycoinsecticide

R. Bateman[1], N. Jenkins[2], C. Kooyman[3], D. Moore[4], C. Prior[5]

[1]Imperial College, Ascot, United Kingdom; [2]Pennsylvania State University, University Park, PA, United States; [3]Éléphant Vert, Nanyuki, Kenya;
[4]Commonwealth Agricultural Bureaux International Europe – UK, Surrey, United Kingdom; [5]Mount Pleasant, Bampton, Devon, United Kingdom

23.1 HISTORY OF THE LUBILOSA PROGRAM

The LUBILOSA (Lutte Biologique contre les Locustes et Sauteriaux) research program ran throughout the late 1980s and early 1990s and successfully developed a formulation of the entomopathogenic fungus *Metarhizium acridum* suitable for the control of acridids and potentially some other insect pests under hot and arid conditions. Although the program drew on several previous scientific studies, two were central. Prior et al. (1988) had shown that at high humidity, spores of *Beauveria bassiana* were about 30 times more infective to a weevil pest of cocoa when formulated in oil as opposed to water, or, to put a different emphasis on the results, only 1/30 of the inoculum was needed to achieve mortality if the spores were formulated in oil, which is a very large difference. Marcandier and Khachatourians (1987) had shown in an independent study, not yet conducted at the time of Prior et al.'s investigations, that *B. bassiana* spores in water applied topically, as if by spraying, infected a grasshopper pest independent of the ambient humidity. This was surprising, because up to that time it had generally been thought that high humidity was required for *B. bassiana* infection.

Prior had been working in Papua New Guinea in the late 1970s/early 1980s on fungal diseases of cocoa and by way of diversion investigated fungi that infect pest insects; these fungi are common in the humid tropics. He carried out some preliminary experiments on the infection of *Pantorhytes plutus*, a major weevil pest of cocoa, with *B. bassiana*. At this point the resident spray application technologist, Pete Jollands, suggested using ultra-low volume (ULV) sprayers for application. These need pesticides formulated in diluents of low volatility, such as oils, and he and Prior tested coconut oil formulations of *B. bassiana* instead of aqueous ones. The work showed that the oil formulations were much more infective than the water ones (Prior et al., 1988).

How could these studies be applied to locust control? In the 1980s, locusts were controlled by the application of ULV formulations of chemical insecticide. This is because conventional high-volume sprays, which are low concentrations of pesticide in large volumes of water, are applied by ground-based equipment, but such formulations weigh too much to be carried in spray planes. Aerial application is the only realistic way of delivering pesticides to the large areas that need spraying, so to support this technology it is necessary to provide very high concentrations of pesticide in low volumes of diluent. Application rates are in the range 0.25–2.0 L/ha. Diluents are usually liquids, but powders can also be used as dust diluents.

Because ULV pesticides are concentrated, the lethal dose per insect that is required is very small. To achieve this with minimal waste, ULV formulations are broken up into very small droplets. The formulations must therefore be in oils of low volatility rather than water, because at such a very small droplet size, the water diluent would evaporate immediately, and the residual drop would be so small that it would drift away under Brownian motion and never hit the target. Prior knew not only that fungal pathogens of insects could be formulated in oils but also that such formulations were much more virulent than water ones. This insight led to the suggestion that an environmentally benign fungal pathogen might be formulated to replace chemical pesticides for locust control.

These ideas developed around 1988 at the end of a period of extensive locust control. A number of international aid donors had just funded a massive chemical spray operation costing millions of dollars and using huge volumes of fenitrothion. Dieldrin, an insecticide that could be laid down in narrow, persistent bands and remain active for months, killing the hopper bands as they crossed them, had been the preferred control agent. It was effective but caused concerns about damage to people and the environment, especially because the dieldrin bands were often laid down along roads where people and domestic animals were likely to become contaminated. Dieldrin was eventually banned for being too persistent and environmentally damaging, so in the 1980s, when the next plague erupted, the donors were forced to fund the use of another insecticide, fenitrothion, which is not as persistent.

This meant that instead of laying down low dosages of pesticide in narrow bands, the control teams were forced to blanket spray large areas; ironically, the ban on the unacceptably persistent dieldrin led to the massively greater use of an alternative chemical, quite possibly with even greater environmental damage. The donors received much criticism from the environmental lobby and were anxious to find an alternative. They were therefore interested in a suggestion to make a ULV formulation of an entopathogenic fungus that could be put through the existing equipment, was host specific, would kill locusts, and was environmentally benign (Prior and Greathead, 1989).

Prior became interested in locust control while working on an unrelated project on cocoa pests in Uganda. Hearing radio reports of a very large locust swarm in Algeria, he held discussions on his return with David Greathead, the director of CIBC [Commonwealth Institute of Biological Control, now absorbed into Commonwealth Agricultural Bureaux International (CABI) Bioscience], who had formerly done fieldwork on locusts. He wrote a consultation document outlining the ideas and circulated an executive summary to the donor community. The donor response was very positive, and the generous level of financial support they provided allowed the assembly of a well-resourced team. However, the donors were taking a gamble. It was known that *B. bassiana* could be formulated in vegetable oil, and under high humidity, in plastic boxes in an air-conditioned laboratory, these oil formulations would kill insects at much lower doses than water suspensions. This was very different from the very hot and arid conditions in which locusts were sprayed. And there was not a proven isolate; there actually was a *Metarhizium* sp. from a locust in the CABI collection, but its virulence was unknown. There were a lot of potential problems.

However, with encouragement from the locust control community and generous financial support from Canadian, American, British, Dutch, and later Swiss aid agencies, the LUBILOSA team was able to begin a program to investigate formulations of entopathogenic fungi for locust control. The program had four phases of approximately 3 years each. In the first phase, the project explored for pathogenic fungi, screened them for virulence under laboratory conditions, and developed a formulation suitable for field use. Having achieved this, the program moved into a second phase to field test formulations on locusts in cages at Niamey, Niger, and against grasshoppers at sites in Benin, followed by a third phase of more extensive field testing and finally a fourth phase where activities were transferred to national programs, with LUBILOSA providing advice and support.

The laboratory research and field trials will be described in detail. Various aspects of the LUBILOSA program, including isolate collection and screening, formulation, and field testing have been described elsewhere (Lomer and Prior, 1992; Lomer et al., 1997, 1999, 2001; Van der Valk, 2007).

23.2 DEVELOPMENTAL RESEARCH

Because so few potentially suitable isolates were available, LUBILOSA adopted two approaches to seeking more. The exploratory entomologist, Paresh Shah, equipped a lorry as a mobile laboratory and shipped it to Pakistan, then Oman, to form a base for field exploration. In West Africa, Christiaan Kooyman set up a regional network among the agricultural extension staff of several countries offering a financial reward for suitable infected insects. It became very clear that although the field exploration yielded some interesting material, the network was the more productive approach, simply because when searching for rare items in the field, many pairs of eyes are better than one. The West Africa search yielded many isolates for the United Kingdom-based screening program.

Isolates were screened for virulence against Desert Locusts bred in cages at the CIBC center, maintained so that they could provide a continuous supply of adult insects. A standard dose was topically applied beneath the pronotum, and the insects were kept in plastic boxes at 30°C and a 12-h day of artificial light (Bateman et al., 1996). It was quickly established that the most virulent isolates were of a *Metarhizium* sp. and derived from acridids. The fungus was initially thought to be a morphologically distinct strain of *Metarhizium flavoviride* but is now recognized as a distinct species, *M. acridum* (Bischoff et al., 2009; Driver et al., 2000). Isolates of *Metarhizium* spp. from nonacridids, or other insect pathogenic fungi, were of lesser or no virulence. One isolate, IMI 330,189, was selected for further development, including submission for mandatory and very expensive mammalian safety testing, which yielded no cause for concern. Having selected an isolate for field testing, the next stage before large-scale field tests such as described by Kooyman et al. (1997) was a field bioassay where insects were sprayed in field "arenas" (Bateman et al., 1991). This allowed testing under real field conditions at very little expense.

Initially, the fungus was stored on slopes of potato carrot agar; single spore isolates were deposited in the culture collection at the International Mycological Institute at Kew. Various media were tested for conidial production, and Molisch's medium (Speare, 1920) was chosen for routine use. Conidia were harvested in cottonseed oil for use in bioassays. When larger quantities of conidia were required the fungus was grown in medical flats, or, in West Africa, old whisky bottles. An unexpected problem was that extraction from agar culture in oil also removed considerable quantities of water, which reduced the viability of the suspension if it was not used immediately. The addition of silica gel to the formulation solved this problem (Moore et al., 1995, 1996).

To scale up production still further, a two-stage production system was developed using a combination of techniques developed in Brazil and China, whereby fungal inoculum grown in liquid culture on waste brewery yeast and table sugar was used to inoculate a solid substrate. Many substrates were tested, but autoclaved white rice proved to be the most reliable. Cereal grains (rice, barley, or sorghum) seem to be universally favored for the purpose. Because the program was focused on African countries, these substrates were accepted to be appropriate for the region. Further details of the production system are given in Jenkins and Goettel (1997), Jenkins et al. (1998), and Cherry et al. (1999).

During the development of the liquid medium, the standard isolate of *M. acridum* was observed to produce conidia in submerged culture (Jenkins and Prior, 1993). Unfortunately, the conidia were hydrophilic and could not be suspended in the oils used for ULV spraying. In any case, a comparison of aerial and submerged conidia proved that the former had superior field efficacy (Jenkins and Thomas, 1996), though submerged conidia may have a value as yet unrealized.

Attempts were made to develop a production system in which suspended cellulose fiber cloths were soaked in the liquid inoculum in a sterile environment (Jenkins and Lomer, 1994). However, yields of conidia were lower than from rice, and the system was very prone to contamination.

The ULV formulations required a very high standard of product quality, having uniform particle size, high purity, and strictly controlled moisture content. This led to the development of the "Mycoharvester" and considerable work on shelf life and moisture content (Hong et al., 1997, 2000). No information was available on quality control, and the project developed a protocol for "Green Muscle," as the product was called, that ensured a satisfactory product (Jenkins and Grzywacz, 2000, 2003).

After selecting an appropriate system for producing conidia and enforcing the necessary quality controls, the next stage was to develop formulations and test them against adult locusts. A spray rig was built to which an adult locust could be attached with a rubber band and passed at a controlled speed along a fixed track; this allowed captive locusts to be passed through a cloud of spray droplets: tracer dye, visible under UV light, revealed the distribution of spray on the insect. Prior's original study on *Pantorhytes* had used coconut oil, but LUBILOSA tested a variety of mineral oil mixtures of appropriate viscosity, including cottonseed oil (Fig. 23.1). Formulation in oil had a striking effect on the dose–mortality relationship (Bateman et al., 1993). Other aspects of formulation that were studied included drying conidia to improve storage stability,

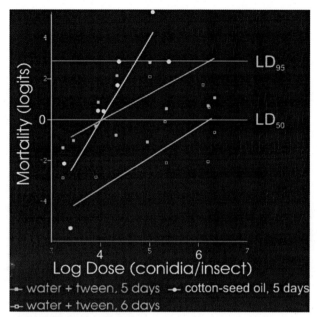

FIGURE 23.1 Dose–mortality relationships for water and oil formulations of *Metarhizium acridum* conidia.

packaging, effects of temperature on spore longevity, and UV protectant chemicals; see Moore et al. (1995, 1996), Hong et al. (1997, 2000), Griffiths and Bateman (1997, 1997), and Moore et al. (1993) for a detailed discussion. Once dried to a low moisture content, the conidia proved surprisingly long-lived. Subsequent work using a modified seed storage protocol indicates that shelf lives of more than 3 years are achievable at 5–8°C. Chemicals added to protect against UV light were relatively ineffective and not cost-effective.

A "spin-off" technology was the development of a device employing cyclone technology to collect separated conidia in a pure form to aid both drying and formulation stability. The initial work was done by Sylvia Mermelstein (Mermelstein, 1995) and was driven by the need for a strict particle size specification to maximize the duration of suspension and prevent blockages in ULV sprayers fitted with very narrow restrictors. Although there were socioeconomic arguments for developing a simple sieving technique appropriate for use in low-technology systems, such techniques could not deliver the specification required: the sieving allowed contamination by residual substrate, which impaired storage and clogged spray nozzles. The cyclone device was later further modified and named the "Mycoharvester" (Bateman, 2003). It is now used by others working on fungal pathogens, and at the time of writing (2015) more than 130 units have been sold worldwide.

Because the environment in which the fungus was to be used is hot and arid, with high levels of UV light, attention was given to the effect of these environmental factors on conidial survival and infection. Temperature tolerance

FIGURE 23.2 Aerial spraying operations during field tests in West Africa. *OF*, oil-miscible flowable concentrate.

increased dramatically at low moisture contents, whereas fresh conidia died within minutes at 50°C, those dried to 5% moisture content could survive for several days; the effects of UV were also reduced at low moisture contents. For long-term storage, it was found that with appropriate packaging, dried conidia could be kept under refrigeration for several years, though care was needed not to rehydrate too rapidly or imbibition damage occurred (Moore et al., 1997). Further studies in collaboration with Reading University, United Kingdom, resulted in an equation that could predict conidial longevity in storage and the effects of drying regimes and aspects of mass production (Hong et al., 2001).

Following the end of the LUBILOSA project, further developments have occurred. Direct exposure to UV remains a problem, and the addition of protectant chemicals has not proved helpful. There are signs that genetic modification techniques may be of value here, for example, by transferring dihydroxynapthalene genes into the fungus, as has been done for *Metarhizium anisopliae* (Tseng et al., 2011). Imbibition damage can be largely avoided by using water warmed to 33°C, and slow rehydration prevented damage, as did formulating in various oils (Faria et al., 2009; Xavier-Santos et al., 2011).

Modified atmospheric packaging (MAP) has resulted in remarkable survival at very high temperatures: *B. bassiana* conidia showed >80% survival after 2 months at 50°C (Faria et al., 2012). Isolate selection was a very important aspect of these studies, and the LUBILOSA isolate IMI 330,189, used in "Green Muscle," appears to be very robust, such that MAP may not even be necessary for routine use. With hindsight, it is very important that at the outset of such a project, a very precise isolate specification is available. "Green Muscle" became increasingly sophisticated as field operations were scaled up. For aerial spraying (Fig. 23.2), a proprietary oil-miscible flowable concentrate (OF) formulation was developed, which included surfactants and structuring agents to maintain the conidia in suspension. This enabled rapid tank mixing under field conditions.

23.2.1 Descriptions of Field Trials

M. acridum has been field tested in numerous trials throughout Africa, Australia, North America, and elsewhere. Target species have been, among others, the Desert Locust *Schistocerca gregaria*, the Brown Locust *Locustana pardalina*, the Red Locust *Nomadacris septemfasciata*, the Sahelian Tree Locust *Anacridium melanorhodon*, the Australian Plague Locust *Chortoicetes terminifera*, the Migratory Locust *Locusta migratoria*, the Senegalese Grasshopper *Oedaleus senegalensis*, the Rice Grasshopper *Hieroglyphus daganensis*, and the Variegated Grasshopper *Zonocerus variegatus*. Five selected trials are described in the following sections.

23.2.1.1 Trial Against Desert Locust in Mauritania

In November, 2006, a trial was conducted with the OF formulation produced in South Africa. The targets were hopper bands of the Desert Locust in an area of low dunes near Benishab, 220 km northeast of Nouakchott, Mauritania. The formulation was diluted 1:9 in diesel and sprayed at a Volume Application Rate (VAR) of 1 L/ha by means of a motorized knapsack sprayer with a Micron AU8000 attached. Thus the nominal dose of the active ingredient was 50 g/ha. At the moment of application, the dominant hopper stage was third instar. A total of 10 hopper bands were treated, while nine others at a distance of several kilometers away were left untreated.

Eight days after the treatments, no intact hopper bands remained in the treated area, and only a few scattered hoppers could still be found. In contrast, only one of the untreated bands had been eliminated by brown-necked ravens (*Corvus ruficollis*), while the other bands were still intact or had merged. As can be seen from Fig. 23.3, some bands lasted only 4 days after application. That was even before clear signs of infection were visible among the hoppers. All these bands had been relatively small, and since predation pressure was high, it was assumed that they were

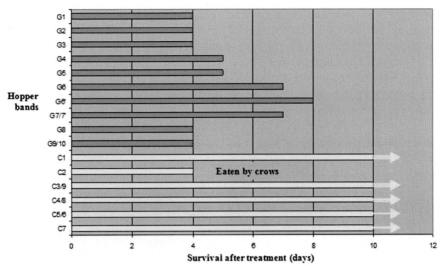

FIGURE 23.3 Survival of hopper bands after treatment. The *green bars* represent treated bands and the *yellow bars* untreated bands. An *arrow* indicates that the band was intact on the last day of observation. The double numbers (eg, G9/G10) indicate that the bands had joined.

eliminated by predators. It was not clear to what extent the treatments had contributed to this. The fact that six small untreated bands were also eliminated by predators before and after the start of the trial shows that predation alone can be sufficient to prevent hopper bands from developing until adulthood.

Close observations on the larger bands revealed that predators quickly discovered that hoppers had become sick. From about 4 days after treatment, predation pressure on the treated bands became higher than on the untreated bands. The impression was that most hoppers disappeared during the night. Several nocturnal visits to the site revealed a number of mammalian predators, especially several rodent species, but wild cats as well, all of which are known to eat locusts. During the day, the most important predators were birds and lizards. Observations on sampled hoppers in cages confirmed that the hoppers from treated bands had been infected.

23.2.1.2 Trial Against Red Locust in Tanzania

An operational trial on red locusts (*N. septemfasciata*) was conducted with the South African OF formulation in Tanzania in 2003. Swarms of adult locusts were targeted in the Katavi National Park in the southwest of the country. Two dose rates were tested, so the formulation was diluted 1:9 for a dose rate of 50 g/ha and 1:19 for a rate of 25 g/ha using a VAR of 1 L/ha in both cases. The locusts were concentrated in a number of swarms of varying sizes that did not move much. It was therefore possible to delineate blocks to include these swarms. A block of 1400 ha was sprayed with Green Muscle at 50 g/ha, two blocks of 800 and 400 ha,

respectively, were sprayed at 25 g/ha, one block of 600 ha was sprayed with Fenitrothion 96% ULV, and one block of 400 ha was kept untreated.

Samples of locusts were taken from all blocks and kept in large outdoor cages 1 m^3 made of a wooden frame covered with mosquito wire mesh. Some cages were kept in the shade, but most were exposed to the sun. To prevent the temperature in the cages from rising too much, they were covered with grass thatch in such a way that locusts could choose to bask in the sun.

Mortality in the field was almost impossible to check. Only eight intact cadavers were found, and these proved all to be infected on incubation. Apart from these, only bits and pieces of dead locusts were found. A number of predators were suspected of catching sick locusts, including storks, rollers, crows, falcons, servals, civet cats, and mongooses. More than half of 40 cadavers placed in the field had disappeared after 24 h, and all had gone in 3 days, which shows that scavengers, especially ants and beetles, were active in removing cadavers.

There were also other signs that the treatments had been effective. From about 10 days after application, many locusts were seen basking on the ground in groups. Soon afterward, the locusts became increasingly reluctant to fly to the extent that it became relatively easy to catch them by hand. Population reduction became evident after about 3 weeks. Especially during the first 2 weeks of the trial, movement of locusts within the blocks was such that simple transect counts were not possible. The decline in populations could therefore not be graphically represented. It is estimated that the populations in the 25 g/ha blocks had declined by more than 70% after 4 weeks. Decline in the

50 g/ha block was more difficult to estimate because of the large number of swarmlets, but more than 50% reduction was definitely achieved.

Mortality in the cages could of course be quantified easily and is shown in Fig. 23.4. One has to be a bit cautious in interpreting the high infection rates in the samples of treated locusts, because during the collection of samples, insect nets will be contaminated with spores from the vegetation and will transfer some to uninfected locusts. It is, however, reasonable to assume that more than half of the locusts had been infected at the time of sampling (1 day after application). In 2009, 10,000 ha were sprayed in a successful control campaign against *N. septemfasciata*. In Tanzania, further use of LUBILOSA technology has been very sporadic.

23.2.1.3 Trial Against Senegalese Grasshopper in Senegal

In 2008, the Senegalese government decided to treat about 7000 ha of savannah with scattered crop fields that had been infested with high densities of grasshoppers (up to 120 nymphs/m^2). This time, the OF formulation produced in Senegal itself was used. The applications were carried out with a spray aircraft fitted with four Micronair AU5000 spray heads. The intended dose rates were 25 and 50 g/ha, but one area was sprayed at 100 g/ha by mistake.

This trial turned out to be one of the few during which large numbers of dying grasshoppers were observed suspended from the vegetation (Fig. 23.5). Many cadavers could therefore be collected and incubated. Because of the logistical challenges related to the large plot sizes, the

first population estimates after the treatments could only be made on day 14. At that moment, there were no significant differences in population reduction between the dose rates. The results conclusively show that the lowest dose rate of 25 g/ha is as effective as the higher ones. This makes the use of Green Muscle also financially attractive. At 2009 prices, the cost of product per hectare at 25 g/ha would be €6.25 (CFA4100), which compares favorably with the cost of recommended chemical products.

FIGURE 23.5 Dying Senegalese Grasshopper.

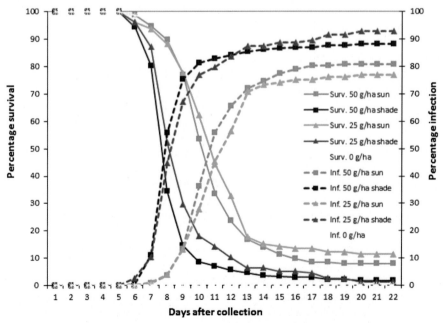

FIGURE 23.4 Mortality in cages following field application.

23.2.1.4 Trials Against Variegated Grasshopper in Benin and Ghana

Isolate I91-609, which is particularly effective against pyrgomorphid grasshoppers, has been tested several times against *Z. variegatus* L., most often in Benin and Ghana. The active ingredient for these trials was produced in the experimental production unit of the IITA in Cotonou. Some trials compared different dose rates either by varying the concentration in the spray tank or by varying the distance between spray passes (Douro-Kpindo et al., 1995).

During a trial carried out in 1996 near Azovè, 140 km northwest of Cotonou, Benin, the fungus was applied at four dose rates each replicated 3 times: 10^{11}, 4.65×10^{11}, 2.15×10^{12}, and 10^{13} conidia/ha (equivalent to 2, 9.3, 43, and 200 g/ha) at a VAR of 2 L/ha. Population densities were estimated every 3 days, and they declined at approximately the same rate for all doses. Fig. 23.6 shows that even the populations in the control plots declined and that those in some treated plots started declining before the treatments. This phenomenon was caused by the movement of hoppers out

of the plots, which, at 1 ha each, were rather small. Unfortunately, this made it impossible to evaluate the differences between the various dose rates. However, it is clear that at the end of the trial, all dose rates had achieved roughly the same population decline.

Differences between the dose rates did become apparent in the grasshopper samples taken from the field and incubated in cages in the laboratory and the field. Average survival times dropped at increasing dose rates, though not all differences were significant (Table 23.1). It can be concluded from this trial that very low dose rates of I91-609 are sufficient to reduce populations of *Z. variegatus* within a reasonable time period. At 2009 prices, a dose rate of 10 g/ha would cost only €2.50 (CFA 1650) of product per hectare treated. Another trial, conducted in Ghana, has shown that a further cost reduction can be achieved by doubling the distance between spray passes, eg, from 5 to 10 m, when using a handheld ULV sprayer. This cuts in half the time needed to spray a certain area or doubles the area that can be treated within a certain amount of time.

In all field trials using "Green Muscle" there was skepticism, principally among experienced field control workers, about efficacy, due to the slow speed of kill compared to chemical insecticides. However, studies showed that infection had important prelethal effects. After 24 h at the highest doses used, infected *S. gregaria* consumed significantly less food than uninfected controls, and this effect strengthened as the insects became more ill (Moore et al., 1992; Seyoum et al., 1994). This was also observed in studies on *Z. variegatus* (Thomas et al., 1997). Similar effects have also been observed on mosquitoes infected by entomopathogenic fungi (Blanford et al., 2012).

23.3 EFFECTS OF HUMIDITY AND FORMULATION ON EFFICACY

When locusts are inoculated and kept under humid or dry conditions, the Median Lethal Time (MLT) decreases as the dose increases (Fig. 23.7). MLT is lower under humid conditions and lower for oil formulations than aqueous ones. Further information on the effect of oil formulation on dose response is given in Fig. 23.1, which shows how oil alters

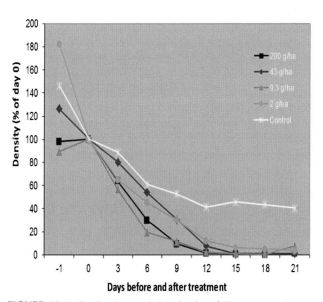

FIGURE 23.6 Decline in population density of *Zonocerus variegatus* following field application.

TABLE 23.1 Average Survival Times (Days) of *Zonocerus variegatus* Treated With *Metarhizium acridum* and Incubated in the Laboratory and in the Field as a Function of the Dosage Rate

	Dosage Rate (g/ha)				
	0	**2**	**9.3**	**43**	**200**
Laboratory	18.33 ± 0.88 a (A)	12.33 ± 0.88 b (A)	9.33 ± 0.33 c (A)	8.00 ± 0.56 cd (A)	6.33 ± 0.33 d (A)
Field	20.33 ± 1.67 a (A)	15.33 ± 2.33 b (A)	12.33 ± 0.67 bc (B)	10.67 ± 1.45 bc (A)	9.67 ± 1.20 c (A)

The means followed by the same small letters within the lines and by the same capital letters within the columns are not significantly different.

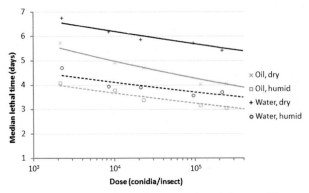

FIGURE 23.7 Effect of formulation and inoculation conditions on Median Lethal Time.

the slope of the dose–response relationship, bringing it closer to that for chemical insecticides; see also Bateman et al. (1993).

23.4 REASONS FOR SUCCESS

First, this program was based on an original idea, derived as described in Section 23.1. This allowed the scientific team a clear lead in the subject, devoid of the stresses of competition. Second, LUBILOSA assembled a team of scientists with a diverse range of skills: not only insect pathology but also entomology, spray application and formulation technology, and field experimentation, backed up by a project manager who combined scientific knowledge with financial and managerial insights. This was a team that could not only develop and test an original idea but also take it and field test it under operational conditions, thus turning a laboratory idea into a practical reality. The team was based initially at a unique site: Silwood Park, an outstation of London University's Imperial College, housed many academics who provided valuable consultations. It was also the home of the International Pesticide Application and Research Center, the base of the team's application specialist, Roy Bateman.

The third reason for success was undoubtedly the generous financial support provided by the donor agencies. This was far greater than the funding provided for previous CIBC projects and enabled the LUBILOSA team to make rapid progress. There was some skepticism expressed at the start of the program that the 3-year time frame for the first phase, laboratory screening, was too short, but with adequate funding to support the work, this did not prove to be the case, and by the second phase the project was ready to move into field testing, confident from preliminary tests in the first phase that these tests were likely to be successful.

Another factor in the project's favor was that as well as no significant problems with mammalian safety, the chosen fungus had very little environmental impact (Peveling and Demba, 2003), confirming Zimmermann's contention

(Zimmermann, 2007) that *Metarhizium* spp. have good safety characteristics, making them very suitable for development as alternatives to chemical insecticides.

23.5 BARRIERS TO OVERCOME

At the time of writing (2015), there has been locust control activity in the Central Region of Africa and Madagascar but very little use of Green Muscle. Reasons for this may include difficulties with registration and limited marketing of the product. While commercial aspects of product marketing are outside the authors' competence for detailed comment, it seems that there is room for improvement.

Other barriers to overcome are the formidable bureaucratic hurdles that must be surmounted to register "Green Muscle" for use in the United States and also many African countries. Work is in progress to achieve this, but there are major difficulties in Africa, because the holder of the registration is no longer in business and cannot issue the required authorization to allow other organizations to take over. In the United States, a major difficulty is likely to be approval for field use of a nonnative organism.

Two other points must be made. First, many people in positions of influence and with long experience of locust control had little or no knowledge of insect pathology, and to such people the ideas behind LUBILOSA may have seemed to be simply too unusual to offer any chance of success, despite abundant evidence from field tests to the contrary. Second, it seems likely that the commercial world of well-established chemical insecticides produced by large industrial companies was not entirely comfortable with a radical new alternative product that offered a challenge in an area where one was not expected.

23.6 SPIN-OFF/COMMERCIALIZATION

Richard Milner, insect pathologist with CSIRO in Canberra, Australia, collaborated with Prior to test an Australian isolate of *M. acridum* against the Australian Plague Locust, *C. terminifera* (Milner, 2002). This was successfully field tested in trials in collaboration with the Australian Plague Locust Commission (Hunter et al., 2001) and led to the registration of a product, "Green Guard," which is used for control operations on a limited scale in organic agriculture and in environmentally sensitive areas. Interest has also been shown in China (Bao-Ping et al., 2000).

The use of mycoinsecticides has often been frustrated by the fact that while very effective in the laboratory, results in the field are variable. A major task for LUBILOSA was to understand more clearly the field environment and its effects on control and the role of the insect itself in resisting infection. Environmental temperature and thermal biology profoundly affect insect–pathogen interactions (Thomas et al., 1996; Inglis et al., 1997; Blanford and Thomas, 1999). The ability of locusts to thermoregulate, especially raising their body temperature when infected by a pathogen to obtain a behavioral fever and hence to slow the growth of the pathogen (Blanford et al., 1998; Thomas and Blanford, 2003), means that time to kill can vary greatly. Thermoregulation depends on many factors that influence environmental temperature, such as season, vegetation, and cloud cover. Aspects of the pathogen's biology also affect the ability of the insect to resist infection; some isolates grow within a narrower temperature range than others, and some are much more effective than others at resuming growth after exposure to very high or low temperatures.

LUBILOSA investigated these environment–temperature interactions by the use of models (Thomas et al., 1995, 1999), which were vital to link laboratory and field studies and assist in the interpretation of field results. The models clarified infection mechanisms postapplication (ie, following inundation with a pathogen) and later helped elucidate the further benefit of secondary cycling (Thomas et al., 1995); infected insects die and sporulate, infecting other insects, thus maintaining and even increasing the fungal inoculum in the environment after the initial application. This contrasts with control using chemical insecticides and helps explain why a mycoinsecticide can be far more effective (Langewald et al., 1999), giving a long-term management using conventional application methods.

Because LUBILOSA was funded by aid agencies with a strong focus on developing countries, the project concentrated on using technologies that were appropriate but could still produce a high-quality, reliable product. An analysis by economist Dan Swanson compared the LUBILOSA production system developed in West Africa with the high-technology system used by Mycotech (now Lavalam) in Butte, Montana, to produce *B. bassiana* (Swanson, 1997). When outputs to treat up to 20,000 ha per annum were required, the low-tech system was more cost-effective; when more than 80,000 ha per annum were to be treated, the reverse was the case. This posed a dilemma, because locust control operations require very large areas to be treated, and it appeared that the low-tech approach was inappropriate. However, discussions with national program representatives involved in locust control showed that they believed production in Africa using high-tech methods was desirable. For this reason, the way forward had to involve commercial entities. This was one of the first examples of product commercialization; as an example of a Development Assistance Project, see Dent and Lomer (2001) for further discussion. After of a series of commercial takeovers, the license for production of "Green Muscle" currently (2015) resides with BASF. Commercial production by a plant in Senegal resulted in considerable sales of "Green Muscle," but this plant has now closed; some commercial production also occurred in Madagascar (Luke, personal communication).

Despite the scientific success of LUBILOSA, only limited commercialization of Green Muscle has occurred to date. Reasons for this are discussed by Lisansky and Sy (2002). In a constructively critical review, these authors note that the project was strong on scientific expertise but weak in knowledge of commerce. In particular, the project scientists and management concentrated on the scientific aspects, assuming that if success was achieved and was what the sponsors wanted, the technology would automatically be adopted. This understandable but mistaken view fails to take into account the fact that if a new technology is to be adopted, it will displace existing ones, and this may not be met with enthusiasm by the commercial interests that provide the current technology. Lisansky and Sy (2002) comment that especially in the final phase, LUBILOSA would have benefited from advice and guidance from those more familiar with the commercial world. This is an important point to be noted by those embarking on future projects of this nature.

ACKNOWLEDGMENTS

The authors gratefully acknowledge the help, advice, and encouragement given by friends and colleagues in the field of insect pathology and elsewhere; these are too numerous to name individually. We also acknowledge the generous funding provided by the international aid agencies of Canada, the Netherlands, Switzerland, the United Kingdom, and the United States. We dedicate this chapter to the memory of three key program scientists who are sadly no longer with us: David Greathead, the director of the International Institute of Biological Control during the early part of the project; Chris Lomer, entomologist and program leader during the later stages; and Bill Steele, agronomist and program manager from 1990 to 1998.

REFERENCES

Bao-Ping, L., Bateman, R., Guo-You, L., Ling, M., Yun, Z., Ainiwar, 2000. Field trial on the control of grasshoppers in mountain grassland by oil formulation of *Metarhizium flavoviride*. Chin. J. Biol. Control 16, 145–147.

Bateman, R.P., 1997. Methods of application of microbial pesticide formulations for the control of locusts and grasshoppers. In: Goettel, M.S., Johnson, D.L. (Eds.), Microbial Control of Grasshoppers and Locusts. Mem. Entomol. Soc. Can. 171, 67–79.

Bateman, R.P., 2003. The 'MycoHarvester': Cleaning Up Locust Control. International Pest Control. See also http://www.mycoharvester.info.

Bateman, R.P., Godonou, I., Kpindu, D., Lomer, C.J., Paraiso, A., 1991. Development of a novel "field bioassay" technique for assessing mycopesticide ULV formulations. In: Lomer, C.J., Prior, C. (Eds.), Biological Control of Locusts and Grasshoppers. CABI International and International Institute of Tropical Agriculture, pp. 255–262.

Bateman, R.P., Carey, M., Moore, D., Prior, C., 1993. The enhanced infectivity of *Metarhizium flavoviride* in oil formulations to desert locusts at low humidities. Ann. Appl. Biol. 122, 145–152.

Bateman, R.P., Carey, M., Batt, D., Prior, C., Abraham, Y., Moore, D., Jenkins, N.E., Fenlon, J., 1996. Screening for virulent isolates of entomopathogenic fungi against the desert locust, *Schistocera gregaria* (Forskal). Biocontrol Sci. Technol. 6, 549–560.

Bischoff, J.F., Rehner, S.A., Humber, R.A., 2009. A multilocus phylogeny of the *Metarhizium anisopliae* lineage. Mycologia 101, 512–530.

Blanford, S., Thomas, M.B., 1999. Host thermal biology: the key to understanding host-pathogen interactions and microbial pest control. Agric. For. Biol. 1, 195–202.

Blanford, S., Thomas, M.B., Langewald, J., 1998. Behavioural fever in a population of the Senegalese grasshopper *Oedaleus senegalensis*, and its implications for biological control using pathogens. Ecol. Entomol. 23, 9–14.

Blanford, S., Jenkins, N.E., Read, A.F., Thomas, M.B., 2012. Evaluating the lethal and pre-lethal effects of a range of fungi against adult *Anopheles stephens*i mosquitoes. Malar. J. 11, 365. http://dx.doi.org/10.1186/1475-2875-11-365 (10 pages).

Cherry, A., Jenkins, N.E., Heviefo, G., Bateman, R.P., Lomer, C.J., 1999. A West African pilot scale production plant for aerial conidia of *Metarhizium* spp. for use as a mycoinsecticide against locusts and grasshoppers. Biocontrol Sci. Technol. 9, 35–51.

Dent, D.R., Lomer, C., 2001. The Convention on Biological Diversity and Product Commercialisation and Development Assistance Projects: A Case Study of LUBILOSA Biopesticides. Series 3. CABI Publishing.

Douro-Kpindo, O.K., Godonou, I., Houssou, A., Lomer, C.J., Shah, P., 1995. Control of *Zonocerus variegatus* with a ULV formulation of *Metarhizium flavoviride* conidia. Biocontrol Sci. Technol. 5, 131–139.

Driver, F., Milner, R.J., Trueman, J.W.H., 2000. A taxonomic revision of *Metarhizium* based on a phylogenetic analysis of rDNA sequence data. Mycol. Res. 104, 134–150.

Faria, M., Hajek, A.E., Wraight, S.P., 2009. Imbibitional damage in conida of the entomopathogenic fungi *Beauveria bassiana*, *Metarhizium acridum* and *Metarhizium anisopliae*. Biol. Control 51, 346–354.

Faria, M., Hajek, A.E., Wraight, S.P., 2012. Application of modified atmosphere packaging (gas flushing and active packaging) for extending the shelf life of *Beauveria bassiana* conidia. Biol. Control 61, 78–88.

Griffiths, J., Bateman, R.P., 1997. Evaluation of the Francome Mark 2 Exhaust Nozzle Sprayer to apply oil-based formulations of *Metarhizium flavoviride* for locust control. Pest. Sci. 51, 176–184.

Hong, T.D., Ellis, R.H., Moore, D., 1997. Development of a model to predict the effect of temperature and moisture on fungal spore longevity. Ann. Bot. 79, 121–128.

Hong, T.D., Jenkins, N.E., Ellis, R.H., 2000. The effects of duration of development and drying regime on the longevity of conidia of *Metarhizium flavoviride*. Mycol. Res. 106, 662–665.

Hong, T.D., Gunn, J., Ellis, R.H., Jenkins, N.E., Moore, D., 2001. The effects of storage environments on the longevity of conidia of *Beauveria bassiana*. Mycol. Res. 105, 597–602.

Hunter, D., Milner, R., Spurgin, P., 2001. Aerial treatment of the Australian plague locust *Chortoicetes terminifera* (Orthoptera: Acrididae) with *Metarhizium anisopliae* (Deuteromycotina: Hyphomycetes). Bull. Entomol. Res. 91, 93–100.

Inglis, G.D., Johnson, D.L., Goettel, M.S., 1997. Use of pathogen combinations to overcome constraints of temperature on entomopathogenic Hyphomycetes against grasshoppers. Biol. Control 8, 143–152.

Jenkins, N.E., Goettel, M.S., 1997. Mass production of microbial control agents for the control of locusts and grasshoppers. In: Goettel, M.S., Johnson, D.L. (Eds.), Microbial Control of Grasshoppers and Locusts. Mem. Entomol. Soc. Can. 71, 37–48.

Jenkins, N.E., Grzywacz, D., 2000. Quality control of fungal and viral biocontrol agents – assurance of product performance. Biocontrol Sci. Technol. 10, 753–777.

Jenkins, N.E., Grzywacz, D., 2003. Towards the understanding of quality control of fungal and viral biocontrol agents. In: van Lenteren, J.C. (Ed.), Quality Control and Production of Biological Control Agents: Theory and Testing Procedures. CAB International, Wallingford, Oxon, pp. 247–263.

Jenkins, N.E., Lomer, C.J., 1994. Development of a new procedure for the mass production of *Metarhizium flavoviride*. 10BC/WPRS Bull. 17, 181–184.

Jenkins, N.E., Prior, C., 1993. Growth and formation of true conidia by *Metarhizium flavoviride* in a simple liquid medium. Mycol. Res. 97, 1489–1494.

Jenkins, N.E., Thomas, M.B., 1996. Effect of formulation and application method on the efficacy or aerial and submerged conidia of *Metarhizium flavoviride* for locust and grasshopper control. Pest. Sci. 46, 299–306.

Jenkins, N.E., Heviefo, G., Langewald, J., Cherry, A.J., Lomer, C.J., 1998. Development of mass production technology for aerial conidia of mitosporic fungi for use as mycopesticides. Biocontrol News Inf. 19, 21N–31N.

Kooyman, C., Bateman, R.P., Langewald, J., Lomer, C.J., Oambama, Z., Thomas, M.B., 1997. Operational-scale application of entomopathogenic fungi for the control of Sahelian grasshoppers. Proc. R. Soc. Lond. B 264, 541–546.

Langewald, J., Ouambama, Z., Mamadou, A., Peveling, R., Stolz, I., Bateman, R., Attignon, S., Blanford, S., Arthur, S., Lomer, C., 1999. Comparison of an organophosphate insecticide with a mycoinsecticide for the control of *Oedaleus senegalensis* (Orthoptera: Acrididae) and other Sahelian grasshoppers at an operational scale. Biocontrol Sci. Technol. 9, 199–214.

Lisansky, S., Sy, A.-A., 2002. Review of LUBILOSA. Use of *Metarhizium* in Africa. CPL Business Consultants, The Manor House, Howbery Park, Wallingford, Oxfordshire OX10 8BA, UK. 27 pp. Available by negotiation from CPL Business Consultants.

Lomer, C.J., Prior, C. (Eds.), 1992. Biological control of locusts and grasshoppers. In: Proceedings of a Workshop Held at the International Institute of Tropical Agriculture, Cotonou, Republic of Benin, April 29–May 1, 1991. CAB International in Association with the International Institute Tropical Agriculture. xii + 394 pp.

Lomer, C.J., Thomas, M.B., Douro-Kpindu, O.-K., Gbongboui, C., Godonou, I., Langewald, J., Shah, P., 1997. Control of grasshoppers, particularly *Hieroglyphus daganensis*, in northern Benin using *Metarhizium flavoviride*. In: Goettel, M.S., Johnson, D.L. (Eds.), Microbial Control of Grasshoppers and Locusts. Mem. Entomol. Soc. Can. 171, 301–311.

Lomer, C.J., Bateman, R.P., Dent, D., De Groote, H., Douro-Kpindou, O.-K., Kooyman, C., Langewald, J., Peveling, R., Thomas, M., 1999. Development of strategies for the incorporation of biological insecticides into the integrated management of locusts and grasshoppers. Agric. For. Entomol. 1, 71–78.

Lomer, C.J., Bateman, R.P., Johnson, D.L., Langewald, J., Thomas, M., 2001. Biological control of locusts and grasshoppers. Annu. Rev. Entomol. 46, 667–702.

Marcandier, S., Khachatourians, G.G., 1987. Susceptibility of the migratory grasshopper *Melanopus sanguipes* (Fab.) (Orthoptera: Acrididae) to *Beauveria bassiana* (Bals.) Vuillemin (Hyphomycete): influence of relative humidity. Can. Entomol. 119, 901–907.

Mermelstein, S.P., 1995. The Design, Manufacture and Evaluation of a Fungal Spore Collector (MSc thesis). Department of Architecture and Design, London South Bank University. 194 pp. plus annexe.

Milner, R., 2002. See: http://pubs.rsc.org/en/Content/ArticlePDF/2002PO/B200948J?page=Search.

Moore, D., Reed, M., Le Patourel, G., Abraham, Y.J., Prior, C., 1992. Reduction of feeding by the desert locust *Schistocerca* gregaria after infection with *Metarhizium flavoviride*. J. Invertebr. Pathol. 60, 304–307.

Moore, D., Bridge, P.D., Higgins, P.M., Bateman, R.P., Prior, C., 1993. Ultra-violet radiation damage to *Metarhizium flavoviride* conidia and the protection given by vegetable and mineral oils and chemical sunscreens. Ann. Appl. Biol. 122, 605–616.

Moore, D., Bateman, R.P., Carey, M., Prior, C., 1995. Long term storage of *Metarhizium flavoviride* conidia in oil formulations for the control of locusts and grasshoppers. Biocontrol Sci. Technol. 5, 193–199.

Moore, D., Douro-Kpindou, O.K., Jenkins, N.E., Lomer, C.J., 1996. Effects of moisture content and temperature on storage of *Metarhizium flavoviride* conidia. Biocontrol Sci. Technol. 6, 51–61.

Moore, D., Langewald, J., Obognon, F., 1997. Effects of rehydration on the conidial viability of *Metarhizium flavoviride* mycopesticide formulations. Biocontrol Sci.Technol. 7, 87–94.

Peveling, R., Demba, S.A., 2003. Toxicity and pathogenicity of *Metarhizium anisopliae* var. *acridum* (Deuteromycotina, Hyphomycetes) and fipronil to the fringe-toed lizard *Acanthodactylus dumerili* (Squamata: Lacertidae). Environ. Toxicol. Chem. 22, 1437–1447.

Prior, C., Greathead, D.J., 1989. Biological control of locusts: the potential for the exploitation of pathogens. FAO Plant Prot. Bull. 37, 37–48.

Prior, C., Jollands, P., le Patourel, G., 1988. Infectivity of oil and water formulations of *Beauveria bassiana* (Deuteromycotina: Hyphomycetes) to the cocoa weevil pest *Pantorhytes plutus* (Coleoptera: Curculionidae). J. Invertebr. Pathol. 52, 66–72.

Seyoum, E., Moore, D., Charnley, A.K., 1994. Reduction in flight activity and food consumption by the desert locust *Schistocerca gregaria* Forskal (Orthoptera: Cyrtacanthacrinae) after infection with *Metarhizium flavoviride*. J. Appl. Entomol. 118, 310–315.

Speare, A.T., 1920. Further studies of *Sorosporella uvella*, a fungus parasite of noctuid larvae. J. Agic. Res. 18, 399–447.

Swanson, D., 1997. Economic feasibility of two technologies for production of a mycopesticide in Madagascar. In: Goettel, M.S., Johnson, D.L. (Eds.), Microbial Control of Grasshoppers and Locusts. Mem. Entomol. Soc. Can. 171, 101–113.

Thomas, M.B., Blanford, S., 2003. Thermal biology in insect-parasite interactions. Trend. Ecol. Evol. 18, 344–350.

Thomas, M.B., Wood, S.N., Lomer, C.J., 1995. Biological control of locusts and grasshoppers using a fungal pathogen: the importance of secondary cycling. Proc. R. Soc. Lond. B 259, 265–270.

Thomas, M.B., Gbongboui, C., Lomer, C.J., 1996. Between-season survival of the grasshopper pathogen *Metarhizium flavoviride* in the Sahel. Biocontrol Sci. Technol. 6, 569–574.

Thomas, M.B., Blanford, S., Lomer, C.J., 1997. Reduction of feeding by the variegated grasshopper *Zonocerus variegatus* following infection by the fungal pathogen *Metarhizium flavoviride*. Biocontrol Sci. Technol. 7, 327–334.

Thomas, M.B., Wood, S.N., Solorzano, V., 1999. Application of insect-pathogen models to biological control. In: Hawkins, B.A., Cornell, H.V. (Eds.), Theoretical Approaches to Biological Control. University Press, Cambridge, pp. 368–384.

Tseng, M.N., Chung, P.C., Tzean, S.S., 2011. Enhancing the stress tolerance and virulence of an entomopathogen by metabolic engineering of dihydroxynapthalene melaninbiosynthesis genes. Appl. Environ. Microbiol. 77, 4508–4519.

Van der Valk, H., 2007. Review of the Efficacy of *Metarhizium anisopliae* var. *acridum* against the Desert Locust. Desert Locust Technical Series. FAO Plant Prod. Prot. Div. No. AGP/DL/TS/34/77 pp.

Xavier-Santos, S., Lopes, R.B., Faria, M., 2011. Emulsifiable oils protect *Metarhizium robertsii* and *Metarhizium pingshaense* conidia from imbibitional damage. Biol. Control 59, 261–267.

Zimmermann, G., 2007. Review on the safety of the entomopathogenic fungus *Metarhizium anisopliae*. Biocontrol Sci. Technol. 17, 879–920.

Chapter 24

Microbial Control of Nursery Ornamental and Landscape Plant Pests

S.P. Arthurs[1], D.J. Bruck[2]

[1]University of Florida, Apopka, FL, United States; [2]DuPont Pioneer, Johnston, IA, United States

24.1 INTRODUCTION

Ornamental trees, shrubs, annuals, grasses, and other plants are cultivated and planted for aesthetic and utility purposes in urban landscapes, including yards, parks, gardens, sports fields, cemeteries, medians, and roadsides. The landscape "green" industry is among the largest of all agricultural enterprises in the United States and consists of many components, including plant production in nurseries as well as landscape design, establishment, and maintenance. Total gross sales of nursery crops exceeded $15 billion and comprised over 1.6 billion acres for operations with $100,000 or more in sales among 17 surveyed states (USDA-NASS, 2009; Fig. 24.1).

These ornamental plants are infested by many different pests, including aphids, scales, thrips, mealybugs, weevils, and many others. Over 2500 species of insects and mites are destructive to nursery crops in the United States (Johnson and Lyon, 1991; Dreistadt, 2004; Alford, 2012). Many of these pests and some additional ones infest ornamental plants grown in landscape settings in different geographic regions (Raupp et al., 2010). In many states, this situation is exacerbated by the influx of invasive species. Although economic data for individual introductions is scarce, the combined effects of invasive insects on ornamental plants run into billions of dollars annually (Pimentel et al., 2005).

Pest management on ornamental plants in many regions is largely accomplished by the application of insecticides, often on a calendar basis (Grewal, 2012; LeBude et al., 2012). Ornamental plants now represent a significant and growing market for the agrochemical industry in the United States. In 2007, the home and garden market sector of the pesticide industry comprised 38% of user expenditure for insecticides/miticides (EPA, 2011). While chemical insecticides provide useful tools for managing pests on ornamental plants, overreliance on this approach has caused several problems, including insecticide resistance, a resurgence of pest populations due to the elimination of beneficial species, and removal of pesticides from the marketplace

(Gao et al., 2012; Raupp et al., 2012; Szczepaniec et al., 2013). Neonicotinoids, the most common insecticides used on ornamental plants, have been cited as a problem for beekeepers (Cresswell, 2011) and may contaminate urban water bodies (Van Dijk et al., 2013).

Microbial-based pesticides containing entomopathogens and their derivatives offer an alternative to chemical insecticides with reduced environmental impact. However, nursery and landscape plants represent a complex challenge for pest control due to the large number of plant species, insect pests, growing conditions, potting media, and production techniques. In addition, feeding injury on many ornamental plants directly affects plant marketability or aesthetic value, and the threshold for damage is often relatively low. In this chapter, we review recent progress on the use of entomopathogens to manage the major pests of ornamental plants of nurseries and landscapes. We consider major pest guilds separately, based on their feeding damage, habitat, and associated control strategies (Table 24.1). In general, methods for evaluating insect pathogens or microbial insecticides will reflect the pest habitat (eg, roots, leaves, stems) and crop situation, ie, nursery or landscape (Fig. 24.2). Related pests of ornamental plants in greenhouses as well as lawn and turf are covered elsewhere in this book (Chapters 16 and 22).

24.2 PHYTOPHAGOUS MITES

Several families of mites (Acari) are significant pests of ornamental plants; the major ones are spider (Tetranychidae), tarsonemid (Tarsonemidae), and eriophyid mites (Eriophyidae). Adult mites generally have eight legs and piercing/sucking mouthparts used to remove fluids from plant cells. Spider mites spin webs and are persistent invaders in greenhouses and nurseries and rapidly reach high populations under hot and dry conditions (Gerson and Weintraub, 2012). Eriophyid mites are tiny (with only four legs in all stages) and cause plant distortion, including galls, and also transmit several important diseases of

FIGURE 24.1 Nursery production of ornamental trees in the Pacific Northwest showing the scale and arrangement of planting blocks. *Photographs by Denny Bruck.*

ornamental plants. Many different miticides are registered and used for ornamental plants (Stamps and Osborne, 2009). Several arthropod biological control agents (usually predatory mites) are also commercially produced and used to control phytophageous mites on nursery-grown ornamental plants (McMurtry et al., 2013). Thus the compatibility of these predators with insecticides and miticides is an important consideration for integrated mite management.

24.2.1 Pathogens of Mites

All major pathogen groups (bacteria, rickettsiae, fungi, protozoans, nematodes, and viruses) have been associated with mite disease (Poinar and Poinar, 1998). However, to

date few pathogens have been developed as commercial miticides due to issues in mass production or poor performance under field conditions.

24.2.2 Fungi as Mycoacaricides

Mites and ticks are susceptible to pathogenic fungi, and there are opportunities to develop these pathogens as mycoacaricides. There are also opportunities to apply fungal pathogens as classical biological control agents against mites, as reviewed by Wekesa et al. (2015). Chandler et al. (2000) collated records of 58 species of fungi infecting at least 73 species of mites in nature or via experiments. Two mite-specific fungi, *Hirsutella*

TABLE 24.1 Pest Guilds of Ornamental Nursery and Landscape Plants Targeted With Different Entomopathogens

Guild	Bacteria	Nematodes	Fungi	Selected References
Phytophagous mites Tetranychidae, Eriophyidae, Tarsonemidae			*Beauveria bassiana, Metarhizium brunneum, Hirsutella thompsonii*[b], *Lecanicillium* spp.[a]	Chandler et al. (2005), Seyed-Talebi et al. (2014), and Wekesa et al. (2015)
Wood borers Peachtree borer, spruce bark beetle, ambrosia beetle, emerald ash borer, Asian longhorned beetle		*Steinernema* spp., *Heterorhabditis* spp.	*B. bassiana, Beauveria brongniartii, Isaria fumosorosea, M. brunneum*	Castrillo et al. (2010), Shapiro-Ilan et al. (2010a), Lyons et al. (2012), Tanabe et al. (2012), Castrillo et al. (2013), Kepenekci and Atay (2014), Carrillo et al. (2015), and Goble et al. (2015)
Root feeders Black vine weevil, Diaprepes root weevil		*Steinernema* spp., *Heterorhabditis* spp.	*B. bassiana, M. brunneum*	Kakouli-Duarte et al. (1997), Shapiro et al. (2000), Bruck et al. (2005), Bruck (2007), Stuart et al. (2008), Ansari et al. (2008b), and Ansari and Butt (2011)
Defoliators Cottonwood leaf beetle, elm leaf beetle, tent caterpillar, fall webworm, bagworm, spruce budworm, gypsy moth	*Bacillus thuringiensis* subspp. *aizawai, kurstaki, tenebrionis*	*Steinernema carpocapsae, Heterorhabditis megidis*		Pinkham et al. (1984), Gill and Raupp (1994), Thurston (2001), Arthurs et al. (2004), Coyle et al. (2005), Bauce et al. (2006), Summerville and Crist (2008), and Blackburn et al. (2011)
Sap feeders Aphid, whiteflies (*Bemisia tabaci, Trialeurodes vaporariorum*), thrips (western flower, chilli)		*Steinernema feltiae*	*B. bassiana, M. brunneum, I. fumosorosea, Lecanicillium muscarium*[a], *Lecanicillium longisporum*[b]	Arthurs and Heinz (2006), Olson and Oetting (1999a, 1999b), Maniania et al. (2002), Ansari et al. (2008a), Lacey et al. (2008), Shan and Feng (2010), Arthurs et al. (2013), and Mascarin et al. (2014, 2015)

[a]*Available in Europe.*
[b]*No longer commercially available.*

thompsonii and *Neozygites floridana*, are obligate pathogens of pestiferous eriophyid and tetranychid mites, respectively. *H. thompsonii* was developed as the product "Mycar" for the control of eriophyoid mites on citrus but was withdrawn from sale in the 1980s. Members of the Hypocreales (*Beauveria, Metarhizium, Isaria,* and *Lecanicillium*) appear to hold the most promise for future development as mycoacaricides.

24.2.3 Fungi for Control of Spider Mites

Several Hypocreales fungi have been successfully tested to control spider mites. *Lecanicillium muscarium* is registered in Europe as "Mycotal" for the control of whitefly larvae, thrips, and spider mites. Bugeme et al. (2009) assessed 23 isolates of *Metarhizium anisopliae* and three isolates of *Beauveria bassiana* against the two-spotted spider

mite, *Tetranychus urticae*, and reported that several were pathogenic at constant temperatures between 20 and 35°C. Chandler et al. (2005) demonstrated that a commercial formulation of *B. bassiana* (Naturalis-L) controlled *T. urticae* on greenhouse tomato plants. The authors reported that foliar applications of *B. bassiana* at 10^5 conidia/mL (equivalent to 0.5 g/L water) with and without the predatory mite *Phytoseiulus persimilis* reduced the numbers of *T. urticae* adults, nymphs, and eggs by up to 98%. Seyed-Talebi et al. (2014) studied the compatibility of *B. bassiana* with the acaricide spirodiclofen. Spider mites exposed to both spirodiclofen and the fungus showed significantly increased mortality, suggesting a synergistic effect.

Mycoacaricides are likely to be the most successful in greenhouses where environmental conditions can be manipulated to enhance infectivity and when efforts are taken to ensure compatibility between any supplemental

FIGURE 24.2 Experimental application of microbial insecticides on ornamental plants in a landscape (upper plate) and nursery situation. *Photographs by Steve Arthurs.*

pesticides and biological control agents. Additional studies are needed to evaluate commercial mycoacaricides in integrated pest management programs for spider mites infesting ornamental plants. Few studies have reported good activity of fungi against immature mites; the short developmental period and rapid molting of mites may be a hindrance to obtaining good control. Investigations to enhance mycopathogens with selective growth inhibitors might thus be warranted.

24.3 WOOD BORERS

The larvae of several insects develop internally in the trunks, stems, and major limbs of trees and woody shrubs. Wood-boring pests of nursery and landscape ornamental plants include several moths (Lepidoptera; Solomon, 1977) and beetles (Coleoptera; Brockerhoff

et al., 2006; Haack, 2006). Wood-boring Lepidoptera include clearwings (Sesiidae), pyralids (Pyralidae), and carpenter moths (Cossidae), while wood-boring beetles include flathead borers (Buprestidae), roundhead borers (Cerambycidae), and bark and ambrosia beetles (Scolytidae). Young trees may be killed by girdling injury, while older infested trees may become vulnerable to secondary pests and diseases. Healthy plants tend not to be attacked by most wood-boring insects, and many of these pests are secondary problems associated with stressed, damaged, or dying plants and trees (Drees et al., 1994). In general, the control of pests of severely damaged plants is impractical for various reasons, including poor uptake of systemic insecticides, and sanitation or removal is the preferred option. However, there are cases where insect pathogens have been used successfully to control important wood-boring pests.

24.3.1 Entomopathogenic Fungi for Control of Wood-Boring Insects

Residual bark sprays of insecticides, such as pyrethroids, are used to target the adult stages of several wood-boring insects in nurseries before they penetrate the tree and lay eggs. The use of entomopathogenic fungi has been investigated as an environmentally benign replacement of such chemicals. Carrillo et al. (2015) investigated the use of three commercial strains of entomopathogenic fungi (*Isaria fumosorosea* 3581 and PFR97) and *B. bassiana* (GHA) against the redbay ambrosia beetle, *Xyleborus glabratus*. Approximately 70% of beetles dipped or exposed to avocado bolts treated with $\approx 2 \times 10^6$ spores/mL became infected and died inside the galleries before egg laying. In another laboratory study, the female granulate ambrosia beetle, *Xylosandrus crassiusculus*, exposed to beech stems treated with *B. bassiana* GHA and *Metarhizium brunneum* F52 applied at 600 conidia/mm^2 showed up to 95% mortality 5 days after inoculation, with fewer galleries, offspring, and symbiont growth compared with controls (Castrillo et al., 2013).

There are practical challenges in targeting adult wood-boring insects with entomopathogenic fungi in the field over large areas, including high application costs and limited environmental persistence. Castrillo et al. (2010) observed a significant decline in the recovery of colony-forming units of *B. bassiana* GHA applied to target the adult emerald ash borer *Agrilus planipennis* within 1 week of application. In this study, the decline was more pronounced on leaves than on bark. However, sufficient inoculum persisted on bark and leaves for 7–14 days to make trunk and foliar sprays a possible means to target adults during emergence, preoviposition feeding, or oviposition. Lyons et al. (2012) describe an attempt to contaminate adult *A. planipennis* populations with *B. bassiana* using an autocontamination trapping system. Insects entering a multifunnel trap baited with (Z)-3-hexenol acquired fungal conidia from cultures growing on pouches in the chambers. Since only 6.5% of beetles became contaminated, better methods to contaminate and distribute the fungus are probably needed.

Several methods for improving the environmental persistence of fungal applications to trees have been proposed. The Asian longhorned beetle, *Anoplophora glabripennis*, is an invasive pest of urban street trees. In the biological control of *A. glabripennis*, the use of *M. brunneum* strain F52 and *Beauveria brongniartii* formulated on agar or nonwoven fiber fungal bands (which extended the period of sporulation) has been tested successfully in the laboratory (Ugine et al., 2013) and field (Hajek et al., 2006). A sprayable "hydromulch" formulation of fungus (which removes the need to apply bands) has also been developed. This hydromulch consists of hydrated wheat straw with natural glue and contains hyphal aggregates (microsclerotia)

of *M. brunneum* F52, which sporulate over time as water becomes available (Goble et al., 2015).

24.3.2 Entomopathogenic Nematodes for Control of Wood-Boring Insects

Entomopathogenic nematodes (EPNs) have been used to control the larval stages of several wood-boring Lepidoptera that attack ornamental shade and fruit trees. Clearwing moths, including the lesser peachtree borer, *Synanthedon pictipes*, and banded ash borer *Podosesia aureocincta* have been targeted with *Steinernema carpocapsae* and *Steinernema feltiae* (Smith-Fiola et al., 1996; Shapiro-Ilan et al., 2010a). Additional studies suggest that several carpenter moths that infest hardwood shade trees are similarly susceptible to EPNs (Forschler and Nordin, 1988; Huaiwen et al., 1993; Ashtari et al., 2011). In general, the successful control of clearwing and carpenter moth larvae requires that high volume applications deliver the infective juvenile (IJ) stage close to the feeding galleries, since there is limited time for host location before the IJs desiccate. Although more labor intensive, the direct injection of EPN suspensions into larval feeding galleries with a syringe or fine spray stream often results in a higher mortality rate of larvae. Application during overcast conditions and additional light irrigation (wetting) before and after EPNs are applied may also be used to enhance IJ survival and infectivity in aboveground locations (Arthurs et al., 2004). Shapiro-Ilan et al. (2010a) investigated the combined use of *S. carpocapsae* with a sprayable fire-retardant gel formulation (Barricade). The gel enhanced postapplication IJ survival in exposed locations and significantly improved the subsequent control of the lesser peachtree borer when compared with other treatments. Bark beetles may be similarly targeted with EPN (Kepenekci and Atay, 2014) although there are few good field studies on this topic. Several parasitic and phoretic nematodes are naturally associated with scolytid beetles (Kanzaki and Kosaka, 2009) although these have not been developed as inundative biological control agents. Few studies have reported the use of steinernematids against ambrosia beetles although there are prospects to investigate this further. One study reported that *S. carpocapsae* was attracted to the ambrosia beetle, *Platypus quercivorus*, the vector of Japanese oak wilt, in the laboratory (Tanabe et al., 2012).

24.4 ROOT FEEDERS

Several root-feeding weevils are serious pests in nurseries and landscape plantings in North America. The black vine weevil, *Otiorhynchus sulcatus*, which is a parthenogenetic flightless female, infests nearly 150 plant species (Moorhouse et al., 1992). This insect lays 500–800 eggs between May and September, and emerging larvae move through the soil in search of roots. Large larvae consume feeder roots and bark on larger roots in the spring, causing reduced plant vigor and death (Moorhouse et al., 1992). EPNs and entomopathogenic fungi have been widely and successfully studied for the control of *O. sulcatus* in nursery stock, especially as some of the more persistent and toxic insecticides have been phased out. Although more commonly associated with citrus, the Diaprepes root weevil, *Diaprepes abbreviatus*, is an important pest of nursery ornamentals that has been targeted by EPNs (Stuart et al., 2008). This insect is invasive in North America where it not only infests 100,000 acres of citrus in Florida alone but it also attacks woody ornamental plants (Mannion et al., 2003; Diaz et al., 2006). The adult stages feed and lay egg clusters on leaves, while the emerging larvae drop to the ground, enter the soil, and feed on roots.

24.4.1 Entomopathogenic Fungi Against Black Vine Weevil

M. brunneum (F52) is registered by the US Environmental Protection Agency for the control of black vine weevil. In general the fungus may be applied as a curative drench or a dried formulation incorporated prophylactically into the potting media at the time of transplanting. Bruck (2007) tested fungal applications to control black vine weevil in container-grown plant material maintained outdoors in the Pacific Northwest. Results showed that drench applications to soilless potting media were nearly 100% effective 28 days postapplication in the spring but were less effective in the fall, possibly due to cooler temperatures (10–12°C) over the course of the experiment, since fungal growth significantly declined below 20°C. Another study showed that the same fungus incorporated into the media persisted for at least 133 days (Bruck, 2006). A greenhouse study (Ansari et al., 2008b) evaluated the combined and supplemental use of EPNs (*Heterorhabditis bacteriophora*) and the fungus *M. brunneum*. Results suggested that the third instar black vine weevil targeted by both agents suffered >97% mortality even at reduced rates (ie, 10^8 conidia/L and 1 to 2 IJs/cm^2), suggesting that the EPN and fungus might work together synergistically.

24.4.2 Entomopathogenic Nematodes for Control of Black Vine Weevil

In general, good success (ie, >90% control) has been achieved from applications of *S. carpocapsae*, *S. feltiae*, *Heterorhabditis marelatus* (OH10 strain), *Heterorhabditis megidis*, *H. bacteriophora*, and *Heterorhabditis downesi* against black vine weevil larvae and pupae on container-grown nursery plants and landscape ornamentals (reviewed by Bruck et al., 2007). There are reports that heterorhabditids provide better control of black vine weevil when

compared with steinernematids, likely due to their different host penetration and foraging strategies (Stimmann et al., 1985; Shanks and Agudelo-Silva, 1990). The use of EPN on nursery plants with very low damage thresholds, such as cyclamen (one larva/pot) and rhododendron (three larvae/plant), may require more frequent applications at higher rates, or the EPN may need to be integrated with additional chemical treatments (Kakouli-Duarte et al., 1997). Another study also confirmed that the type and composition of potting media (eg, peat, bark, coir, and peat blended with compost green waste) significantly affected the efficacy and dispersal of IJs applied for black vine weevil control (Ansari and Butt, 2011). For example, *Heterorhabditis* spp. caused 100% larval mortality in all media whereas *Steinernema* spp. only caused 100% larval mortality in the peat blended with 20% compost green waste.

24.4.3 Entomopathogenic Nematodes for Control of Diaprepes Root Weevil

Comparatively, little work has evaluated EPNs for the control of *Diaprepes abbreviates* in ornamental plants where *D. abbreviates* is a pest (Shapiro-Ilan et al., 2010b). However, several lessons may be learned from research evaluating EPNs for the control of *D. abbreviates* in citrus (see Chapter 19).

24.5 DEFOLIATORS

Caterpillars and leaf beetles are the most common arthropod pests that consume leaves and other aboveground plant structures, including flowers, fruits, needles, and buds, in nurseries and landscapes (Alford, 2012). These insects often feed gregariously and can consume vast quantities, defoliating entire trees in outbreak years (Wallner, 1987). The bacterium *Bacillus thuringiensis* (Bt) and its insecticidal toxins are the most widely used entomopathogens for control of this feeding guild (Sanchis, 2010). Strains of Bt are readily mass produced and are commercially available for the grower, landscaper, and homeowner.

24.5.1 Baculoviruses for Control of Defoliators

Several baculoviruses (nucleopolyhedroviruses and granuloviruses) are used against several defoliating Lepidoptera in crops and forestry (Chapters 3 and 21). A major advantage of baculoviruses over Bt products is their relative specificity; thus their use is favored in environmentally sensitive areas to protect nontarget Lepidoptera (Peacock et al., 1998). However, due to their much higher costs of production, baculoviruses are used far less frequently than Bt formulations.

24.5.2 *Bacillus thuringiensis* for Control of Defoliators

Bt subsp. *kurstaki* (Btk), and to a lesser extent Bt subsp. *aizawai* (Bta), are used for the control of Lepidoptera infesting forest and urban shade trees, including tent caterpillars *Malacosoma* spp., fall webworms, *Hyphantria cunea*, bagworms, *Thyridopteryx ephemeraeformis*, spruce budworms, *Choristoneura occidentalis*, and gypsy moths *Lymantria dispar* (Pinkham et al., 1984; Gill and Raupp, 1994; Coyle et al., 2005; Bauce et al., 2006; Summerville and Crist, 2008; Blackburn et al., 2011). Bt subsp. *tenebrionis* (Btt) has also been used against several leaf beetles that infest ornamental and shade trees (Cranshaw et al., 1989; Wells et al., 1994; Coyle et al., 2000; Beveridge and Elek, 2001; Tenczar and Krischik, 2006). In cases where products are successfully applied, mortality rates of 90% or more with a significant reduction of defoliation are typically observed. The commercial availability of Btt products tends to be lower compared with Btk and Bta products.

Bt products are available in various formulations, including suspension concentrates, wettable powders, and water-dispersible granules. Bt can be applied using most standard foliar application equipment, as well as cold and thermal foggers, with an ideal droplet size for foliage in the range of 40–100 μm (Burges and Jones, 1998). Since the crystal toxins need to be ingested, a good spray coverage at the target site is critical, and wetting or spreading agents are often used. Bt residues tend to break down under sunlight and rainfall (Gindin et al., 2007) and hence should be timed to coincide with the pest already at the target and actively feeding. In general, Bt products are more effective against small larval stages, with larger instars having progressively higher LC_{50} values, a trend observed in both Lepidoptera (Shanmugam et al., 2006) and Coleoptera (James et al., 1999). Unfortunately, arborists and nurserymen often overlook small, early instars, and pesticide applications are made to late instars (Gill and Raupp, 1994). Variation in susceptibility to Bt products has also been noted among different field populations (Moar et al., 2008).

24.5.3 Entomopathogenic Nematodes for Control of Defoliators

EPNs, especially *S. carpocapsae*, have been tested against defoliating caterpillars and leaf beetles attacking ornamental plants in the nursery and landscape. Under certain conditions, EPNs may provide significant control of foliar pests. Gill and Raupp (1994) reported that foliar applications of *S. carpocapsae* at 200 IJ/cm² of plant foliage plus folicote (spreader/sticker) and 0.25% v/v horticultural oil provided acceptable levels of control (>90% reduction) of mid-instar bagworms when applications were made during cloudy and rainy conditions. In this case, the favorable environmental

conditions and protection of the IJs inside the larval bags may have enhanced activity. In general, however, applications of EPNs typically do not result in high levels of control of foliar pests due to issues with rapid desiccation of IJs on exposed leaf surfaces (Arthurs et al., 2004). Targeting a soil dwelling stage of leaf-feeding pests may thus be more productive. For example, Tomalak (2004) reported that applications of *H. megidis* conducted under the canopy of urban trees at 10⁵ IJs/m² were highly effective against two common urban tree leaf beetles, *Altica quercetorum* and *Agelastica alni*, pupating in the soil. The formulation of EPNs may help overcome abiotic constraints when targeting aboveground pests. For example, Kaya et al. (1981) demonstrated that elm leaf beetle, *Xanthogaleruca luteola*, life stages were highly susceptible to *S. carpocapsae* in the laboratory but ineffective when applied as foliar sprays. Later, Thurston (2001) reported that *S. carpocapsae* added to tree bands containing cellulose mulch at a rate of 2 × 10⁵ IJs in 200 mL water/liter infected >90% of *X. luteola* larvae migrating down the tree trunk to locate pupation sites. The cellulose wood fiber mulch (hydromulch) placed in the tree band provided a suitable habitat for nematode survival for at least 1 month.

24.6 SAP FEEDERS

Hemipteran insects that suck or remove plant sap, including aphids, thrips, lacebugs, scales, and whiteflies, constitute the largest pest-feeding guild affecting ornamental nursery and landscape plants (Johnson and Lyon, 1991; Dreistadt, 2004; Alford, 2012). Many of these pests are highly invasive, polyphageous, reproductively prolific, or have some combination of these factors. Damage symptoms vary according to the group but often include tissue distortion at growing terminals, silvering, bronzing, or chlorosis due to the removal of chlorophyll. Later symptoms include wilting or tip dieback of small branches, sometimes followed by crown dieback due to the removal of plant nutrients. Some members of this pest-feeding guild also vector viruses of ornamental plants and are thus of particular concern as pests in nursery and greenhouse crops (Whitfield et al., 2015). Due to their internal plant-feeding method that limits per os entry of candidate pathogens, entomopathogenic fungi, which infect through the host cuticle, are the most promising microbial control agents for hemipteran pests. Below we briefly review studies to evaluate microbial control agents for major sap-feeding pests of nursery and landscape plants.

24.6.1 Entomopathogenic Fungi for Control of Whiteflies

Whiteflies (Hemiptera: Aleyrodidae), such as the silverleaf whitefly, *Bemisia tabaci*, cause major damage to agricultural crops worldwide through phloem feeding and the injection of toxins and indirectly through the transmission of geminiviruses and the excretion of honeydew (De Barro et al., 2011). *B. bassiana* and *I. fumosorosea* (Ascomycota: Cordycipitaceae) are promising insect pathogenic fungi for use against whiteflies (Lacey et al., 2008; Zimmermann, 2008). Commercial products containing *B. bassiana* strain GHA (eg, Botanigard and Naturalist-O) applied at $\approx 10^7$ spores/mL have been tested with mixed success against *B. tabaci* infesting ornamental plants (Olson and Oetting, 1999a; Liu and Stansly, 2009). Commercial blastospore formulations of *I. fumosorosea* (PFR-97 WDG Preferal WDG, and NoFly WP containing $\approx 10^9$ CFU/g) are registered for use against whiteflies, psyllids, and other soft-bodied pests in North America and Europe. Blastospores can be produced inexpensively and germinate more quickly than conidia and thus can be quite effective but have a more limited shelf life compared with conidial formulations (Mascarin et al., 2015). *L. muscarium* isolate (marketed as "Mycotal" by Koppert in the Netherlands) has also been registered in the European Union for the control of whitefly nymphs and thrips, while *Lecanicillium longisporum* was marketed as "Vertalec" for the management of aphids.

Because whiteflies live on the underside of leaves, getting good coverage is important for achieving high infection rates. Air-assisted and electrostatic sprayers may enable deposition in these places (Saito, 2005). Several authors have quantified the compatibility of *I. fumosorosea* with chemical insecticides (Zou et al., 2014) and adjuvants (Mascarin et al., 2014) that might be concurrently used to enhance the control of these pests, with mostly additive and synergistic effects reported. Products based on entomopathogenic fungi are thus useful components in organic production, insecticide-resistant management strategies, or to target insecticide-resistant whitefly populations. However, the limited residual activity and sensitivity of fungal spores to sunlight and rainfall may limit their application in some situations, especially outdoors.

24.6.2 Entomopathogenic Fungi for Control of Thrips

Several thrips species, such as the western flower thrips, *Frankliniella occidentalis* (Thysanoptera: Thripidae), are global pests of ornamental plants that are difficult to control with chemical insecticides (Gao et al., 2012). Several studies have evaluated commercially produced and experimental entomopathogenic fungi for use against *F. occidentalis* and other species. Studies have either evaluated foliar sprays targeting feeding stages or applied/incorporated conidia into potting media to target prepupal and pupal stages in the soil. Maniania et al. (2002) reported that an African strain of *M. brunneum* (ICIPE 69) applied at 10^{13} conidia/ha achieved good control of adult thrips infesting

chrysanthemum although the level of control of larval populations was much lower than for adults. Combined application of *M. anisopliae* and methomyl (Lannate) provided improved control of both the larval and adult stages (Maniania et al., 2002). Arthurs et al. (2013) evaluated the pathogenicity of commercial products against chilli thrips, *Scirtothrips dorsalis*, and determined LC_{50} values in the range of 10^4–10^5 spores/mL for adults but higher values (10^6–10^8 spores/mL) against second instars, ostensibly due to the molting process, allowing immature stages to shed conidia before they are able to penetrate the cuticle. In nursery tests on roses during hot sunny conditions, four applications of the mycoinsecticides reduced *S. dorsalis* populations by 44–81% over 10 weeks, compared with >90% for spinosad. These studies show that mycoinsecticides can be useful in management strategies for low to moderate populations of *S. dorsalis* but were not as effective as the current standards (Arthurs et al., 2013). The application of entomopathogenic fungi though soil drenches or premixing with potting compost can be an effective strategy against *F. occidentalis* infesting potted ornamental plants. The species/strain of fungus may influence the level of control achieved. Ansari et al. (2008a) reported that *M. brunneum* strains V275 and ERL700 were the most effective, causing 85–96% mortality of *F. occidentalis* larvae and pupae 11 days after inoculation in the soil, compared with 51–84% mortality in four other *M. brunneum* strains, 54–84% mortality from two *B. bassiana* strains, 63–75% mortality from two *I. fumosorosea* strains, and 15–54% for the insecticide fipronil.

24.6.3 Entomopathogenic Fungi for Control of Aphids

Aphids (Hemiptera: Aphididae) are major pests of ornamental plants that rapidly reach pest levels. Natural epizootics caused by several obligate entomophthoralean fungi (Entomophthoromycota) are responsible for the decimation of high-density aphid populations worldwide (Latgé and Papierok, 1988). Entomophthoralean fungi tend not to be amenable for mass production, however, and there are comparatively few successful reports of aphid control with commercial or experimental entomopathogenic Hypocreales fungi, other than *Lecanicillium* spp. In general, the rapid molting of aphids limits successful control for many fungi. Some studies report that *B. bassiana* alone or with sublethal rates of insecticides might be an effective strategy for some aphids that infest ornamental plants under some situations, such as in glasshouses (Olson and Oetting, 1999b; Ye et al., 2005). The potential for some *Metarhizium* strains to infect aphids at higher temperatures also exists (Shan and Feng, 2010). However, more research is needed to develop and optimize the potential of entomopathogenic fungi for the successful biological control of aphids on nursery and landscape plants.

24.6.4 Entomopathogenic Nematodes for Control of Thrips

Several EPN species have been tested against thrips (primarily western flower thrips) that infest ornamental plants. In general, applications are targeted against the prepupal and pupal stages residing in the substrate (Ebssa et al., 2004; Pozzebon et al., 2014) although high-volume foliar applications have also been tested (Arthurs and Heinz, 2006). A commercial formulation of *S. feltiae* (Nemasys) is currently marketed in Europe and Canada as a high-volume application for thrips control (Brownbridge et al., 2014). The latest research suggests that individual nematode treatments have a measurable (although variable) suppressive effect on thrips populations. A combined nematode/fungus treatment (ie, Nemasys + Met52) may also provide superior control compared with either agent alone.

24.6.5 Entomopathogenic Nematodes for Control of Lacebugs

There are few reports of EPNs as biocontrol agents of Hemiptera. Shapiro-Ilan and Mizell (2012) evaluated five EPN species against sycamore lacebug *Corythucha ciliata* in the laboratory. *H. indica* (HOM1) exhibited the highest virulence, killing ≈70% of adult *C. ciliata* on sycamore leaf discs at a relatively low rate (25 IJs/cm^2) and also produced the most IJ stages (≈300 per insect) compared with other tested species. The possibility of successfully targeting *C. ciliata* overwintering in cryptic locations under loose bark or cracks and crevices remains to be proven.

24.7 CONCLUSIONS

There are significant opportunities for controlling important pests of ornamental plants using entomopathogens, but much remains to be done before they are widely adopted for practical use in nursery and landscape systems. Most studies with entomopathogens of ornamental plant pests focused on greenhouse-produced crops (Chapter 16), with relatively few studies conducted outdoors, with the exception of lawn and turf systems (Chapter 22). Given the high value of the nursery and landscape industry and concerns about chemical inputs in environmentally sensitive areas and near housing, the lack of research on microbial pest control in urban landscapes is a major oversight and is in need of targeted research. Commercial products supported by simple robust recommendations are needed for the successful adoption of microbial insecticides used on nursery and ornamental plants. Practical and economical application techniques have also not yet been developed for many pest systems. However, we note that advances in "smart" application technologies may significantly reduce the quantities of pesticides applied to nursery stock

(Zhu et al., 2012), thus reducing the potential costs of microbial applications. Microbial control products are also more sensitive to environmental conditions and may not be effective under all conditions. Finally, continual improvements in the commercial production, distribution, and storage of viable products are needed.

REFERENCES

Alford, D.V., 2012. Pests of Ornamental Trees, Shrubs and Flowers: A Colour Handbook. CRC Press. 480 pp.

Ansari, M.A., Butt, T.M., 2011. Effect of potting media on the efficacy and dispersal of entomopathogenic nematodes for the control of black vine weevil, *Otiorhynchus sulcatus* (Coleoptera: Curculionidae). Biol. Control 58, 310–318.

Ansari, M.A., Brownbridge, M., Shah, F.A., Butt, T.M., 2008a. Efficacy of entomopathogenic fungi against soil-dwelling life stages of western flower thrips, *Frankliniella occidentalis*, in plant-growing media. Entomol. Exp. Appl. 127, 80–87.

Ansari, M.A., Shah, F.A., Butt, T.M., 2008b. Combined use of entomopathogenic nematodes and *Metarhizium anisopliae* as a new approach for black vine weevil, *Otiorhynchus sulcatus*, control. Entomol. Exp. Appl. 129, 340–347.

Arthurs, S., Heinz, K.M., 2006. Evaluation of the nematodes *Steinernema feltiae* and *Thripinema nicklewoodi* as biological control agents of western flower thrips *Frankliniella occidentalis* infesting chrysanthemum. Biocontrol Sci. Technol. 16, 141–155.

Arthurs, S., Heinz, K.M., Prasifka, J.R., 2004. An analysis of using entomopathogenic nematodes against above-ground pests. Bull. Entomol. Res. 94, 297–306.

Arthurs, S.P., Aristizábal, L.F., Avery, P.B., 2013. Evaluation of entomopathogenic fungi against chilli thrips, *Scirtothrips dorsalis*. J. Insect Sci. 13, 31. Available online: http://jinsectscience.oxfordjournals.org/content/13/1/31.

Ashtari, M., Karimi, J., Rezapanah, M.R., Hassani-Kakhki, M., 2011. Biocontrol of leopard moth, *Zeuzera pyrina* L. (Lep.: Cossidae) using entomopathogenic nematodes in Iran. IOBC/WPRS Bull. 66, 333–335.

Bauce, E., Kumbasli, M., Van Frankenhuyzen, K., Carisey, N., 2006. Interactions among white spruce tannins, *Bacillus thuringiensis* subsp. *kurstaki*, and spruce budworm (Lepidoptera: Tortricidae), on larval survival, growth, and development. J. Econ. Entomol. 99, 2038–2047.

Beveridge, N., Elek, J.A., 2001. Insect and host-tree species influence the effectiveness of a *Bacillus thuringiensis* ssp. *tenebrionis*-based insecticide for controlling chrysomelid leaf beetles. Aust. J. Entomol. 40, 386–390.

Blackburn, L.M., Leonard, D.S., Tobin, P.C., 2011. The use of *Bacillus thuringiensis kurstaki* for managing gypsy moth populations under the Slow the Spread Program, 1996–2010, relative to the distributional range of threatened and endangered species. U.S. For. Serv. Res. Pap. NRS 18, 1–20.

Brockerhoff, E.G., Jones, D.C., Kimberley, M.O., Suckling, D.M., Donaldson, T., 2006. Nationwide survey for invasive wood-boring and bark beetles (Coleoptera) using traps baited with pheromones and kairomones. For. Ecol. Manage. 228, 234–240.

Brownbridge, M., Saito, T., Cote, P., 2014. Considerations and combinations to improve control of pupating western flower thrips in chrysanthemums. IOBC/WPRS Bull. 102, 29–35.

Bruck, D.J., 2006. Effect of potting media components on the infectivity of *Metarhizium anisopliae* against the black vine weevil (Coleoptera: Curculionidae). J. Environ. Hortic. 24, 91.

Bruck, D.J., 2007. Efficacy of *Metarhizium anisopliae* as a curative application for black vine weevil (*Otiorhynchus sulcatus*) infesting container-grown nursery crops. J. Environ. Hortic. 25, 150–156.

Bruck, D.J., Shapiro-Ilan, D.I., Lewis, E.E., 2005. Evaluation of application technologies of entomopathogenic nematodes for control of the black vine weevil. J. Econ. Entomol. 98, 1884–1889.

Bruck, D.J., Berry, R.H., DeAngelis, J.D., 2007. Insect and mite control on nursery and landscape plants with entomopathogens. In: Lacey, L.A., Kaya, H.K. (Eds.), Field Manual of Techniques in Invertebrate Pathology, second ed. Springer, Dordrecht, The Netherlands, pp. 609–626.

Bugeme, D.M., Knapp, M., Boga, H.I., Wanjoya, A.K., Maniania, N.K., 2009. Influence of temperature on virulence of fungal isolates of *Metarhizium anisopliae* and *Beauveria bassiana* to the two-spotted spider mite *Tetranychus urticae*. Mycopathologia 167, 221–227.

Burges, H.D., Jones, K.A., 1998. Formulation of bacteria, viruses and protozoa to control insects. In: Burges, H.D. (Ed.), Formulation of Microbial Biopesticides. Kluwer Academic Publishers, Dordrecht, The Netherlands, pp. 33–127.

Carrillo, D., Dunlap, C.A., Avery, P.B., Navarrete, J., Duncan, R.E., Jackson, M.A., Behle, R.W., Cave, R.D., Crane, J., Rooney, A.P., Peña, J.E., 2015. Entomopathogenic fungi as biological control agents for the vector of the laurel wilt disease, the redbay ambrosia beetle, *Xyleborus glabratus* (Coleoptera: Curculionidae). Biol. Control 81, 44–50.

Castrillo, L.A., Griggs, M.H., Liu, H., Bauer, L.S., Vandenberg, J.D., 2010. Assessing deposition and persistence of *Beauveria bassiana* GHA (Ascomycota: Hypocreales) applied for control of the emerald ash borer, *Agrilus planipennis* (Coleoptera: Buprestidae), in a commercial tree nursery. Biol. Control 54, 61–67.

Castrillo, L.A., Griggs, M.H., Vandenberg, J.D., 2013. Granulate ambrosia beetle, *Xylosandrus crassiusculus* (Coleoptera: Curculionidae), survival and brood production following exposure to entomopathogenic and mycoparasitic fungi. Biol. Control 67, 220–226.

Chandler, D., Davidson, G., Pell, J.K., Ball, B.V., Shaw, K., Sunderland, K.D., 2000. Fungal biocontrol of Acari. Biocontrol Sci. Technol. 10, 357–384.

Chandler, D., Davidson, G., Jacobson, R.J., 2005. Laboratory and glasshouse evaluation of entomopathogenic fungi against the two-spotted spider mite, *Tetranychus urticae* (Acari: Tetranychidae), on tomato, *Lycopersicon esculentum*. Biocontrol Sci. Technol. 15, 37–54.

Coyle, D.R., McMillin, J.D., Krause, S.C., Hart, E.R., 2000. Laboratory and field evaluations of two *Bacillus thuringiensis* formulations, Novodor and Raven, for control of cottonwood leaf beetle (Coleoptera: Chrysomelidae). J. Econ. Entomol. 93, 713–720.

Coyle, D.R., Nebeker, T.E., Hart, E.R., Mattson, W.J., 2005. Biology and management of insect pests in North American intensively managed hardwood forest systems. Annu. Rev. Entomol. 50, 1–29.

Cranshaw, W.S., Day, S.J., Gritzmacher, T.J., Zimmermann, R.J., 1989. Field and laboratory evaluations of *Bacillus thuringiensis* strains for control of elm leaf beetle. J. Arboric. 15, 31–43.

Cresswell, J.E., 2011. A meta-analysis of experiments testing the effects of a neonicotinoid insecticide (imidacloprid) on honey bees. Ecotoxicol 20, 149–157.

De Barro, P.J., Liu, S.S., Boykin, L.M., Dinsdale, A.B., 2011. *Bemisia tabaci*: a statement of species status. Annu. Rev. Entomol. 56, 1–19.

Diaz, A.P., Mannion, C., Schaffer, B., 2006. Effect of root feeding by *Diaprepes abbreviatus* (Coleoptera: Curculionidae) larvae on leaf gas exchange and growth of three ornamental tree species. J. Econ. Entomol. 99, 811–821.

Drees, B.M., Jackman, J.A., Merchant, M.E., 1994. Wood-Boring Insects of Trees and Shrubs. Texas Agricultural Extension Service. Publication B-5086.

Dreistadt, S.H., 2004. Pests of Landscape Trees and Shrubs: An Integrated Pest Management Guide, vol. 3359. UCANR Publications. 501 pp.

Ebssa, L., Borgemeister, C., Poehling, H.M., 2004. Effectiveness of different species/strains of entomopathogenic nematodes for control of western flower thrips (*Frankliniella occidentalis*) at various concentrations, host densities, and temperatures. Biol. Control 29, 145–154.

EPA (Environmental Protection Agency), 2011. Pesticides Industry Sales and Usage, 2006 and 2007: Market Estimates. Office of Pesticide Programs, U.S. Environmental Protection Agency, Washington, DC 20460. 33 pp.

Forschler, B.T., Nordin, G.L., 1988. Suppression of carpenterworm, *Prionoxystus robiniae* (Lepidoptera: Cossidae), with the entomophagous nematodes, *Steinernema feltiae* and *S. bibionis*. J. Kans. Entomol. Soc. 61, 396–400.

Gao, Y., Lei, Z., Reitz, S.R., 2012. Western flower thrips resistance to insecticides: detection, mechanisms and management strategies. Pest Manage. Sci. 68, 1111–1121.

Gerson, U., Weintraub, P.G., 2012. Mites (Acari) as a factor in greenhouse management. Annu. Rev. Entomol. 57, 229–247.

Gill, S.A., Raupp, M.J., 1994. Using entomopathogenic nematodes and conventional and biorational pesticides for controlling bagworm. J. Arboric. 20, 318–322.

Gindin, G., Navon, A., Saphir, N., Protasov, A., Mendel, Z., 2007. Environmental persistence of *Bacillus thuringiensis* products tested under natural conditions against *Thaumetopoea wilkinsoni*. Phytoparasitica 35, 255–263.

Goble, T.A., Hajek, A.E., Jackson, M.A., Gardescu, S., 2015. Microsclerotia of *Metarhizium brunneum* F52 applied in hydromulch for control of Asian longhorned beetles (Coleoptera: Cerambycidae). J. Econ. Entomol. 108, 433–443.

Grewal, P.S., 2012. From integrated pest management to ecosystem management: the case of urban lawn. In: Abrol, D.P., Shankar, U. (Eds.), Integrated Pest Management: Principles and Practice. CABI Bioscience, Wallingford, UK, pp. 450–488.

Haack, R.A., 2006. Exotic bark- and wood-boring Coleoptera in the United States: recent establishments and interceptions. Can. J. For. Res. 36, 269–288.

Hajek, A.E., Huang, B., Dubois, T., Smith, M.T., Li, Z., 2006. Field studies of control of *Anoplophora glabripennis* (Coleoptera: Cerambycidae) using fiber bands containing the entomopathogenic fungi *Metarhizium anisopliae* and *Beauveria brongniartii*. Biocontrol Sci. Technol. 16, 329–343.

Huaiwen, Y., Gangying, Z., Shanago, Z., Heng, J., 1993. Biological control of tree borers (Lepidoptera: Cossidae) in China with the nematode *Steinernema carpocapsae*. In: Bedding, R.A., Akhurst, R.J., Kaya, H.K. (Eds.), Nematodes and the Biological Control of Insect Pests. CSIRO Publishing, Melbourne, Australia, pp. 33–40.

James, R.R., Croft, B.A., Strauss, S.H., 1999. Susceptibility of the cottonwood leaf beetle (Coleoptera: Chrysomelidae) to different strains and transgenic toxins of *Bacillus thuringiensis*. Environ. Entomol. 28, 108–115.

Johnson, W.T., Lyon, H.H., 1991. Insects That Feed on Trees and Shrubs: An Illustrated Practical Guide, second ed. Cornell University Press, Ithaca. 560 pp.

Kakouli-Duarte, T., Labuschagne, T.L., Hague, N.G.M., 1997. Biological control of black vine weevil, *Otiorhynchus sulcatus* (Coleoptera: Curculionidae) with entomopathogenic nematodes (Nematoda: Rhabditida). Ann. Appl. Biol. 131, 11–27.

Kanzaki, N., Kosaka, H., 2009. Relationship between the nematodes and bark and the ambrosia beetles. J. Jpn. For. Soc. 91, 446–460.

Kaya, H.K., Hara, A.H., Reardon, R.C., 1981. Laboratory and field evaluation of *Neoaplectana carpocapsae* (Rhabditida: Steinernematidae) against the elm leaf beetle (Coleoptera: Chrysomelidae) and the western spruce budworm (Lepidoptera: Tortricidae). Can. Entomol. 113, 787–793.

Kepenekci, I., Atay, T., 2014. Evaluation of aqueous suspension and entomopathogenic nematodes infected cadaver applications against the great spruce bark beetle *Dendroctonus micans* (Kugelann), (Coleoptera: Scolytidae). Egypt. J. Biol. Pest Control 24, 335.

Lacey, L.A., Wraight, S.P., Kirk, A.A., 2008. Entomopathogenic fungi for control of *Bemisia tabaci* biotype B: foreign exploration, research and implementation. In: Gould, J., Hoelmer, K., Goolsby, J. (Eds.), Classical Biological Control of *Bemisia tabaci* in the United States – a Review of Interagency Research and Implementation. Progress in Biological Control, vol. 4. Springer, Dordrecht, The Netherlands, pp. 33–69.

Latgé, J.P., Papierok, B., 1988. Aphid pathogens. In: Minks, A.K., Harrewijn, P. (Eds.), Aphids, Their Biology, Natural Enemies and Control, vol. B. Elsevier, Amsterdam, pp. 323–433.

LeBude, A.V., White, S.A., Fulcher, A.F., Frank, S., Klingeman III, W.E., Chong, J.H., Chappell, M.R., Windham, A., Braman, K., Hale, F., Dunwell, W., Williams-Woodward, J., Ivors, K., Adkins, C., Neal, J., 2012. Assessing the integrated pest management practices of southeastern US ornamental nursery operations. Pest Manage. Sci. 68, 1278–1288.

Liu, T.X., Stansly, P.A., 2009. Effects of relative humidity on efficacy of BotaniGuard™ (*Beauveria bassiana*) on nymphs of sweetpotato whitefly, *Bemisia tabaci* (Hemiptera: Aleyrodidae), on hibiscus in greenhouses. Southwest Entomol. 34, 189–191.

Lyons, D.B., Lavallée, R., Kyei-Poku, G., van Frankenhuyzen, K., Johny, S., Guertin, C., Francese, J.A., Jones, G.C., Blais, M., 2012. Towards the development of an autocontamination trap system to manage populations of emerald ash borer (Coleoptera: Buprestidae) with the native entomopathogenic fungus, *Beauveria bassiana*. J. Econ. Entomol. 105, 1929–1939.

Maniania, N.K., Ekesi, S., Lohr, B., Mwangi, F., 2002. Prospect for biological control of the western flower thrips, *Frankliniella occidentalis*, with the entomopathogenic fungus, *Metarhizium anisopliae*, on chrysanthemum. Mycopathologica 155, 229–235.

Mannion, C., Hunsberger, A., Peña, J.E., Osborne, L., 2003. Oviposition and larval survival of *Diaprepes abbreviatus* (Coleoptera: Curculionidae) on select host plants. Fla. Entomol. 86, 165–173.

Mascarin, G.M., Kobori, N.N., Quintela, E.D., Arthurs, S.P., Delalibera Jr., Í.D., 2014. Toxicity of non-ionic surfactants and interactions with fungal entomopathogens toward *Bemisia tabaci* biotype B. BioControl 59, 111–123.

Mascarin, G.M., Jackson, M.A., Kobori, N.N., Behle, R.W., Delalibera Jr., Í.D., 2015. Liquid culture fermentation for rapid production of desiccation tolerant blastospores of *Beauveria bassiana* and *Isaria fumosorosea* strains. J. Invertebr. Pathol. 127, 11–20.

McMurtry, J.A., De Moraes, G.J., Sourassou, N.F., 2013. Revision of the lifestyles of phytoseiid mites (Acari: Phytoseiidae) and implications for biological control strategies. Syst. Appl. Acarol. 18, 297–320.

Moar, W., Roush, R., Shelton, A., Ferré, J., MacIntosh, S., Leonard, B.R., Abel, C., 2008. Field-evolved resistance to Bt toxins. Nat. Biotechnol. 26, 1072–1074.

Moorhouse, E.R., Charnley, A.K., Gillespie, A.T., 1992. A review of the biology and control of the vine weevil, *Otiorhynchus sulcatus* (Coleoptera: Curculionidae). Ann. Appl. Biol. 121, 431–454.

Olson, D.L., Oetting, R.D., 1999a. The efficacy of mycoinsecticides of *Beauveria bassiana* against silverleaf whitefly (Homoptera: Aleyrodidae) on poinsettia. J. Agric. Urban Entomol. 16, 179–185.

Olson, D.L., Oetting, R.D., 1999b. Compatibility of insect growth regulators and *Beauveria bassiana* (Balsamo) Vuillemin in controlling green peach aphid (Homoptera: Aphididae) on greenhouse chrysanthemums. J. Entomol. Sci. 34, 286–294.

Peacock, J.W., Schweitzer, D.F., Carter, J.L., Dubois, N.R., 1998. Laboratory assessment of the effects of *Bacillus thuringiensis* on native Lepidoptera. Environ. Entomol. 27, 450–457.

Pimentel, D., Zuniga, R., Morrison, D., 2005. Update on the environmental and economic costs associated with alien-invasive species in the United States. Ecol. Econ. 52, 273–288.

Pinkham, J.D., Frye, R.D., Carlson, R.B., 1984. Toxicities of *Bacillus thuringiensis* isolates against the forest tent caterpillar (Lepidoptera: Lasiocampidae). J. Kans. Entomol. Soc. 672–674.

Poinar Jr., G., Poinar, R., 1998. Parasites and pathogens of mites. Annu. Rev. Entomol. 43, 449–469.

Pozzebon, A., Boaria, A., Duso, C., 2014. Single and combined releases of biological control agents against canopy- and soil-dwelling stages of *Frankliniella occidentalis* in cyclamen. BioControl 60, 341–350.

Raupp, M.J., Shrewsbury, P.M., Herms, D.A., 2010. Ecology of herbivorous arthropods in urban landscapes. Annu. Rev. Entomol. 55, 19–38.

Raupp, M.J., Shrewsbury, P.M., Herms, D.A., 2012. Disasters by design: outbreaks along urban gradients. In: Barbosa, P., Letournea, D.K., Agrawal, A. (Eds.), Insect Outbreaks Revisited. Wiley-Blackwell, Chichester, West Sussex, UK, pp. 311–340.

Saito, T., 2005. Preliminary experiments to control the silverleaf whitefly with electrostatic spraying of a mycoinsecticide. Appl. Entomol. Zoo. 40, 289–292.

Sanchis, V., 2010. From microbial sprays to insect-resistant transgenic plants: history of the biospesticide *Bacillus thuringiensis*. A review. Agron. Sustainable Dev. 31, 217–231.

Seyed-Talebi, F.S., Kheradmand, K., Talaei-Hassanloui, R., Talebi-Jahromi, K., 2014. Synergistic effect of *Beauveria bassiana* and spirodiclofen on the two-spotted spider mite (*Tetranychus urticae*). Phytoparasitica 42, 405–412.

Shan, L.T., Feng, M.G., 2010. Evaluation of the biocontrol potential of various *Metarhizium* isolates against green peach aphid *Myzus persicae* (Homoptera: Aphididae). Pest Manag. Sci. 66, 669–675.

Shanks Jr., C.H., Agudelo-Silva, F., 1990. Field pathogenicity and persistence of heterorhabditid and steinernematid nematodes (Nematoda) infecting black vine weevil larvae (Coleoptera: Curculionidae in cranberry bogs. J. Econ. Entomol. 83, 107–110.

Shanmugam, P.S., Ballagurunathan, R., Sathiah, N., 2006. Susceptibility of *Helicoverpa armigera* Hubner instars to *Bacillus thuringiensis* insecticidal crystal proteins. Pestic. Res. J. 18, 186–189.

Shapiro, D.I., McCoy, C.W., Fares, A., Obreza, T., Dou, H., 2000. Effects of soil type on virulence and persistence of entomopathogenic nematodes in relation to control of *Diaprepes abbreviatus* (Coleoptera: Curculionidae). Environ. Entomol. 29, 1083–1087.

Shapiro-Ilan, D.I., Cottrell, T.E., Mizell, R.F., Horton, D.L., Behle, R.W., Dunlap, C.A., 2010a. Efficacy of *Steinernema carpocapsae* for control of the lesser peachtree borer, *Synanthedon pictipes*: improved above ground suppression with a novel gel application. Biol. Control 54, 23–28.

Shapiro-Ilan, D.I., Morales-Ramos, J.A., Rojas, M.G., Tedders, W.L., 2010b. Effects of a novel entomopathogenic nematode-infected host formulation on cadaver integrity, nematode yield, and suppression of *Diaprepes abbreviatus* and *Aethina tumida*. J. Invertebr. Pathol. 103, 103–108.

Shapiro-Ilan, D.I., Mizell III, R.F., 2012. Laboratory virulence of entomopathogenic nematodes to two ornamental plant pests, *Corythucha ciliata* (Hemiptera: Tingidae) and *Stethobaris nemesis* (Coleoptera: Curculionidae). Fla. Entomol. 95, 922–927.

Solomon, J.D., 1977. Frass characteristics for identifying insect borers (Lepidoptera: Cossidae and Sesiidae; Coleoptera: Cerambycidae) in living hardwoods. Can. Entomol. 109, 295–303.

Smith-Fiola, D.C., Gil, S.A., Way, R.G., 1996. Evaluation of entomopathogenic nematodes as biological control against the banded ash clearwing borer. J. Environ. Hortic. 14, 67–71.

Stamps, R.H., Osborne, L.S., 2009. Selected Miticides for Use on Ornamental Plants. University of Florida. IFAS Extension, EDIS document #ENH1118 https://edis.ifas.ufl.edu/ep383.

Stuart, R.J., El-Bopai, F.E., Duncan, L.W., 2008. From augmentation to conservation of entomopathogenic nematodes: trophic cascades, habitat manipulation and enhanced biological control of *Diaprepes abbreviatus* root weevils in Florida citrus groves. J. Nematol. 40, 73–84.

Stimmann, M.W., Kaya, H.K., Burlando, T.M., Studdert, J.P., 1985. Black vine weevil management in nursery plants. Calif. Agric. 39, 25–26.

Summerville, K.S., Crist, T.O., 2008. Structure and conservation of lepidopteran communities in managed forests of northeastern North America: a review. Can. Entomol. 140, 475–494.

Szczepaniec, A., Raupp, M.J., Parker, R.D., Kerns, D., Eubanks, M.D., 2013. Neonicotinoid insecticides alter induced defenses and increase susceptibility to spider mites in distantly related crop plants. PLoS One 8 (5), e62620.

Tanabe, H., Sanada, Y., Takeuchi, Y., Futai, K., 2012. Host-finding behavior and insecticidal activity of steinernematid entomopathogenic nematodes against the ambrosia beetle, *Platypus quercivorus* (Coleoptera : Platypodidae). Nematol. Res. 42, 15–25.

Tenczar, E.G., Krischik, V.A., 2006. Management of cottonwood leaf beetle (Coleoptera: Chrysomelidae) with a novel transplant soak and biorational insecticides to conserve coccinellid beetles. J. Econ. Entomol. 99, 102–108.

Thurston, G.S., 2001. *Xanthogaleruca luteola* (Muller), elm leaf beetle (Coleoptera: Chrysomelidae). In: Mason, P.G., Huber, J.T. (Eds.), Biological Control Programmes in Canada, 1981–2000. CABI Publishing, Wallingford, UK, pp. 272–275.

Tomalak, M., 2004. Infectivity of entomopathogenic nematodes to soil-dwelling developmental stages of the tree leaf beetles *Altica quercetorum* and *Agelastica alni*. Entomol. Exp. Appl. 110, 125–133.

Ugine, T.A., Jenkins, N.E., Gardescu, S., Hajek, A.E., 2013. Conidial acquisition and survivorship of adult Asian longhorned beetles exposed to flat versus shaggy agar fungal bands. J. Invertebr. Pathol. 113, 247–250.

USDA-NASS, February 2009. U.S. Department of Agriculture, National Agricultural Statistics Service, 2007 Census of Agriculture, vol. 1.

Van Dijk, T.C., Van Staalduinen, M.A., Van der Sluijs, J.P., 2013. Macroinvertebrate decline in surface water polluted with imidacloprid. PLoS One. 8 (5), e62374.

Wallner, W.E., 1987. Factors affecting insect population dynamics: differences between outbreak and non-outbreak species. Annu. Rev. Entomol. 32, 317–340.

Wekesa, V.W., Hountondji, F.C., Dara, S.K., 2015. Mite pathogens and their use in biological control. In: Carrillo, D., Moraes, G.J., Peña, J.E. (Eds.), Prospects for Biological Control of Plant Feeding Mites and Other Harmful Organisms. Springer International Publishing, Dordrecht, pp. 309–328.

Wells, A.J., Kwong, R.M., Field, R., 1994. Elm leaf beetle control using the biological insecticide, Novodor® (*Bacillus thuringiensis* subsp. *tenebrionis*). Plant Prot. Q. 9, 52–55.

Whitfield, A.E., Falk, B.W., Rotenberg, D., 2015. Insect vector-mediated transmission of plant viruses. Virology 479, 278–289.

Ye, S.D., Dun, Y.H., Feng, M.G., 2005. Time and concentration dependent interactions of *Beauveria bassiana* with sublethal rates of imidacloprid against the aphid pests *Macrosiphoniella sanborni* and *Myzus persicae*. Ann. Appl. Biol. 146, 459–468.

Zhu, H., Ozkan, E., Derksen, R.D., Reding, M.E., Ranger, C.M., Canas, L., Krause, C.R., Locke, J.C., Ernst, S.C., Zondag, R.H., Fulcher, A., Rosetta, R., Jeon, H., Chen, Y., Gu, J., Liu, H., Shen, Y., 2012. Development of smart spray systems to enhance delivery of pesticides in field nursery production. Phytopathology 102, 144–145.

Zimmermann, G., 2008. The entomopathogenic fungi *Isaria farinosa* (formerly *Paecilomyces farinosus*) and the *Isaria fumosorosea* species complex (formerly *Paecilomyces fumosoroseus*): biology, ecology and use in biological control. Biocontrol Sci. Technol. 18, 865–901.

Zou, C., Li, L., Dong, T., Zhang, B., Hu, Q., 2014. Joint action of the entomopathogenic fungus *Isaria fumosorosea* and four chemical insecticides against the whitefly *Bemisia tabaci*. Biocontrol Sci. Technol. 24, 315–324.

Chapter 25

Microbial Control of Black Flies (Diptera: Simuliidae) With *Bacillus thuringiensis* subsp. *israelensis*

E.W. Gray[1], R. Fusco[2]

[1]*University of Georgia, Athens, GA, United States;* [2]*Valent BioSciences Corporation (Retired), Mifflintown, PA, United States*

25.1 INTRODUCTION

Larval black flies (Diptera: Simuliidae) develop in a broad spectrum of flowing waters ranging from the smallest creeks to the largest rivers. The adult fly emerges into the terrestrial environment and can be a vector of the causal agents of human and animal diseases or cause a significant biting and swarming nuisance to humans and domestic animals across vast areas. As a result of their wide range of pestiferous activity and the diversity of their larval habitats, black flies have long been a target for suppression programs. The daytime biting activity of the adults and the lack of effectiveness and efficiency of adulticides require most black fly suppression work to focus on the larval stage.

Early black fly suppression programs were conducted using conventional broad-spectrum insecticides (Brown, 1962; Jamnback, 1973). Eventually, the negative environmental impacts and the development of resistance in pest populations were recognized. These factors resulted in a tremendous research effort through the mid-20th century that peaked after the biological control agent *Bacillus thuringiensis* subsp. *israelensis* (Bti) was discovered. This discovery was followed by the rapid commercial development of highly effective formulations that provided a solution to the environmental, resistance, and efficacy issues related to suppression programs. Many of the scientists who were involved in the early black fly and Bti research were victims of their own success. After a safe and effective larvicide was developed, many researchers were forced to focus on other areas of research as funding for black fly work became difficult to secure. We were fortunate to remain active in black fly research and control for the past 30 years. This chapter summarizes some of the key developmental efforts relating to the effectiveness of today's larvicide formulations and describes the factors that we believe are most relevant to the microbial control of black flies with *B. thuringiensis* subsp. *israelensis* in the 21st century.

25.2 PEST STATUS

Black flies are widely recognized as significant pests of humans and animals (Adler et al., 2004). In large parts of West Africa, and to a lesser degree in Central and South America, black flies are vectors of the filarial nematode *Onchocerca volvulus*, the causative agent of human onchocerciasis or river blindness. Black flies are also pests of livestock, causing significant losses to producers in Canada (Fredeen, 1985) and South Africa (Palmer, 1998). Wildlife also suffer from black flies, with the common loon (*Gavia immer*), the critically endangered whooping crane (*Grus americanna*), and the snowy owl (*Bubo scandiacus*), all having their nesting behavior disrupted by black flies (McIntyre and Barr, 1977; Urbanek et al., 2010; Solheim et al., 2013). Black flies also are vectors of the hemosporidean genus *Leucocytozoon*, which has caused significant losses among ducks, geese, chickens, and turkeys (Horosko and Noblet, 1986). In addition to these severe biting and vector examples, the nuisance caused by black flies swarming about the head and shoulders has caused economic losses for outdoor recreation facilities (Gray et al., 1996) and tourism (Sariozkan et al., 2014).

25.3 BLACK FLY BIOLOGY

25.3.1 Adult Black Flies

Adults emerge from the larval habitat and subsequently mate, blood feed (females only), and oviposit. Female simuliids are, with some exceptions, obligate blood feeders on warm-blooded vertebrates. Males do not blood feed. For detailed presentation of the biology of the adult flies, see Adler et al. (2004) and Adler and McCreadie (2009).

25.3.2 Immature Black Flies

Only the larval stage is targeted for suppression using microbial control agents. All stages of immature black flies (egg, larvae, and pupae) are found in lotic habitats. Eggs are deposited on the water's surface or on rocks of the larval habitat, trailing vegetation or other objects just at or slightly below the surface of the water. Upon hatching and drifting, the larvae attach to a variety of substrates (rocks, sticks, leaf packs, trailing vegetation, and debris) using a silk attachment pad and their posterior hooklets. They orient downstream with the current, twist 90–180 degrees, and filter particles from the current with their labral fans (Fig. 25.1).

Larval black flies are commonly described as indiscriminant filter feeders or "passive" suspension feeders requiring currents in their larval habitats to move suspended particles to their labral fans (Wotton, 1994). Larvae ingest particles 0.091 to 350 μm in size, with the majority of the particles being less than 30 μm (Ross and Craig, 1980). Typical particles ingested include bacteria, diatoms, leaf fragments and similar detritus, pollen, fecal pellets, protozoa, and minute arthropods (Ross and Craig, 1980; Merritt et al., 1996; Adler et al., 2004; Wotton, 2009). In addition to sieving and direct-impact filtration that occurs on the labral fan, larvae can also remove colloidal and dissolved materials from the water column (Wotton, 2009). The turbulence of water passing through and around the labral fans appears to cause the flocculation of colloid and dissolved materials that are then captured by the fans. The effective capture of small particles is relevant to Bti-based larvicides because the particles in today's most effective formulations are in the 2- to 9-μm range (Valent BioSciences, 2003).

Pupation takes place in a silken cocoon on the same substrates used by larvae. On completing development, adult flies emerge from the pupa and fly a short distance to a temporary resting site before mating and host seeking.

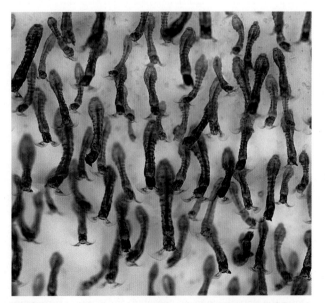

FIGURE 25.1 *Simulium vittatum* larvae from the University of Georgia colony.

25.4 DISCOVERY OF *BACILLUS THURINGIENSIS* SUBSP. *ISRAELENSIS* AND ITS DEVELOPMENT AS A MICROBIAL CONTROL AGENT OF BLACK FLIES

To find an effective alternative to the conventional pesticides used in the Onchocerciasis Control Programme (OCP), an extensive screening effort was undertaken. Lacey and Mulla (1977) evaluated 13 strains of *Bacillus thuringiensis* against larval black flies, and while some demonstrated efficacy, the results did not compare favorably with the organophosphate-based products of the time. In 1976, a *B. thuringiensis* isolate was discovered in Israel by Goldberg and Margalit (1977), which demonstrated a high level of larvicidal activity to several mosquito species. This new strain was described as *B. thuringiensis* subsp. *israelensis* (Bti) serotype H-14 (de Barjac, 1978). Margalit and Dean (1985) provided a short review on the discovery and initial development of Bti. Undeen and Nagel (1978) conducted the first laboratory bioassays on the susceptibility of simuliid larvae to Bti. Laboratory-reared larvae of *Simulium verecundum* and field-collected larvae of *Cnephia ornithophilia*, *Stegopterna mutata*, *Simulium vittatum*, and *Prosimulium mixtum* demonstrated exceptional susceptibility to varying concentrations (3×10^4 to 1×10^5 cells/mL) of a bacterial suspension. Subsequently, Undeen and Colbo (1980) reported on the efficacy of Bti against larvae of *S. verecundum* complex, *P. mixtum*, *C. ornithophilia*, *S. vittatum*, *S. tuberosum* complex, and *S. mutata* in small Newfoundland streams. One-minute dosages of 10^5 viable cells/mL caused up to 100% mortality.

25.4.1 Formulation Development

The discovery of this agent initiated a new era in black fly research and control. During this time, researchers developed and modified a variety of assay techniques to evaluate the efficacy of the various formulations and products that had become available (Wilton and Travis, 1965; Lacey and Mulla, 1977; Gaugler et al., 1980; Lacey et al., 1982; Molloy et al., 1984; Barton et al., 1991; Riley and Fusco, 1990; O'Callaghan et al., 2012). Techniques included bubblers, flushing-funnels, gyratory shakers, magnetic stir bars, mini-gutters, in-stream evaluations, and field collected samples. All of these techniques are based on creating moving water that black fly larvae require to enable feeding or are some version of an actual in-stream evaluation.

25.4.1.1 Criteria for Acceptable Bti Formulations

Some of the challenges encountered in developing a cost-effective and efficacious Bti formulation included formulating a product with the optimum physical properties that did not change from batch to batch, that was stable under tropical conditions, and that could be applied easily through all application systems. Thus, a list of requisites was given to industry to follow in the development process. These criteria for formulation development for effective black fly control were reviewed by multiple researchers (Kurtak et al., 1987; Devisetty et al., 1998; Molloy, 1990; Guillet et al., 1990).

25.4.1.2 The Influence of Formulation on Downstream Carry

In lotic environments, the efficacy of Bti formulations is usually expressed as the distance downstream at which the treatment produces high black fly mortality (85–99%, depending on the program). Carry is directly proportional to the discharge of the river and has a direct relationship to the concentration and physical properties of the formulation. The carry requirements established for an acceptable formulation in the Onchocerciasis Control Programme (OCP) were greater than 10 km under high-discharge river conditions and 2 km for low-discharge conditions (Guillet et al., 1990).

25.4.1.3 Dispersion

The formulation should, when applied diluted or undiluted, by hand or through spray equipment, have rapid and even dispersion throughout the water column to ensure effective carry down river and uniform contact with all larval substrates.

25.4.1.4 Aqueous Bti Formulation Development

The starting point in Bti formulation development was primarily wettable powders and low-potency aqueous suspensions. The powdered formulations were generally high potency with a larger particle size. These were more efficacious in laboratory tests; however, field trials showed them to have little carry, the greatest limiting factor to downstream efficacy, and thus not suitable in the lotic environment. Aqueous formulations, when applied at the proper concentration, were effective for a greater distance downstream. Therefore, industry concentrated on developing these aqueous formulations to meet the criteria in the WHO/OCP (Lacey et al., 1982; Lacey and Undeen, 1984; Molloy et al., 1984; Guillet et al., 1990).

The two leading companies that partnered with WHO in the development of an improved Bti formulation for the OCP were Abbott Laboratories (VectoBac AS) and Sandoz, Inc. (Teknar). Both companies were leaders in microbial fermentation and the production of biopharmaceuticals. At the start of collaboration in 1981, they had registered low-potency (600 ITU/mg) formulations that had to be diluted with a minimum 20% water for application, and they generally had poor biological stability in tropical climates. To be effective, these products had to be applied at a rate of 1.2 L per m³/s of river discharge, which was about 8 times higher than the rate for temephos. This was economically unacceptable in the large aerial application program being operated by the OCP. In 1985, both companies registered aqueous formulations with potencies of 1200 ITU/mg, such as Teknar HPD, VectoBac 12AS (Cibulsky and Fusco, 1987). These formulations could be applied undiluted and were biologically stable for at least 1 year in tropical climates. Most importantly, they had the physical properties necessary to produce excellent black fly control for long distances downstream from the application sites.

The results of this work led to the development of today's high-potency, aqueous formulations of Bti-based larvicides. These formulations have been refined over the course of 35 years of research and development, field evaluation, and operational use. However, one factor remains the same. To be effective, the insecticidal proteins of the Bti-based larvicides must be ingested during larval feeding. Once ingested, a cascade of events occurs to convert the proteinaceous protoxins to toxic units. The act of ingestion is critical to the effectiveness of a Bti-based larvicide.

25.4.2 Methods of Application

Application of Bti-based larvicides is usually made by helicopters or fixed-wing aircraft in large rivers or by hand from boats or bridges or via "walking delivery" in smaller rivers or streams. In large rivers, the Bti is applied undiluted, and in smaller rivers, either undiluted or, preferably diluted with water, using a pressurized sprayer or poured from gardening watering cans, graduated cylinders, jugs, or carboys.

25.4.3 Treatment Sites and Location

Due to the extensive flight capabilities of many black fly species (Fredeen, 1969; Moore and Noblet, 1974), suppression programs are usually conducted over broad areas. When applications are exclusively ground or boat based, access to the various sites on the waterway becomes increasingly important. Aerial applications usually eliminate this consideration. No matter the application technique, the ideal application site would be located upstream of a riffle area that would aid in mixing the Bti-based larvicide into the water column. Applications from bridges also are effective in enhancing mixing. The location of individual treatment sites and the distance between them can vary, depending on many factors. The characteristics of the individual waterway (stream bed composition/topography), the flow rate, and points of access are critical in determining the application site and the concentration of the formulation applied to a waterway and the downstream carry. In addition, the suspended solids of a river, determined by river-side assays (Palmer, 1997), can play a critical role in application-site selection and larvicide concentration.

The carry of each larvicide application is important. Carry is described as the distance downstream that significant mortality is observed after a larvicide application (Undeen et al., 1984). Carry is primarily affected by the characteristics of the waterway. Small streams and creeks produce the least carry due to the relative amount of surface area that the flow passes over and the typical slow stretches that are present. Areas with many rocky ledges often have rapids, interspersed with large pools. In these circumstances, individual rapids might require treatment. The areas of slow flow between the rapids allow the insecticidal proteins to settle from the water column and not be available to downstream larvae. In larval habitats where the flows are more uniform, carry can be expected for greater distances (5–12 km). Dense beds of aquatic vegetation in any type of waterway should be noted. These areas reduce the flow, thereby increasing the probability that the insecticidal proteins will settle from the water column and provide additional surface area for the *Bti*-based larvicide to adhere. All of these factors emphasize the need for actual in-stream surveys to assess waterway characteristics and not just flow calculations, as proposed by Colbo and O'Brien (1984). In addition, although mortality and carry increase with concentration, the increase is not proportional (Lacey and Undeen, 1984). As a result, the economical balance between increased concentration and increased application sites will always be a consideration.

25.4.4 Flow Calculations

Once the larval habitats have been identified, the flow of the waterways to be treated must be determined. In the United States, many larger waterways have United States Geological Survey (USGS) flow-gauging stations. Depending on where the gauge station is located in relation to the targeted larval habitat, calculations can be conducted under base flow conditions to estimate the percent change in flow based on river mile. Otherwise, the waterway's flow should be measured using standard hydrological techniques (Amrine, 1983; Skovmand et al., 2007). Flow rates produced by water control structures can often be obtained from the operating agency.

25.4.5 Application Rate

After the waterway's flow is determined, the amount of larvicide required to achieve a specified concentration is calculated based on the label specifications. Typically, operational concentrations range from 10 to 25 ppm for a 1-min application. Higher rates are used when the suspended

organic solids of river water would have a direct effect on black fly feeding behavior and susceptibility. For example, rate calculations in the Susquehanna River in Pennsylvania, where applications are made perpendicular to the river banks, are of 1-min duration (Pennsylvania Department Environmental Protection, 2015). In the Orange River, South Africa, good results are achieved when applications are applied across and up the river in a serpentine pattern for durations of less than 30 s (Palmer, 1997).

Longer application times have been described in the past but operationally are seldom used today. Lacey and Undeen (1984) proposed that shorter application time produces a higher initial concentration that has a tighter inoculum front and carries farther downstream. They also mention the reduced cost of labor associated with the shorter application times; this factor should not be discounted. Suppression programs require multiple applications conducted as succinctly as possible when conditions are conducive for effective and efficient larvicide applications.

25.4.6 Timing of Application

Timing of Bti applications is critical for integrated pest management (IPM)-oriented black fly management programs. For best results, it is paramount to initiate Bti treatments when the black fly populations are in the larval stage and before larvae enter the final instar. Larvae in the final instar can be recognized by the large, dark spots (gill histoblasts) on the sides of the thorax in which the individual filaments can clearly be observed. Larval instar is important, as just before pupation the larva stops feeding and the pupa is not affected by Bti-based larvicides. Consequently, Bti applications must be conducted before pupation takes place, or adult emergence will occur following the larvicide application. However, to minimize the number of larvicide applications over a season, it is also important to target later instar larvae. This practice allows the longest possible interval between larvicide applications and is crucial economically and ecologically. The final decision to initiate a larvicide application is often based on the "best professional judgment" of the applicator after careful consideration of larval surveillance, flow conditions, water temperatures, and weather forecasts.

25.5 PRESCRIPTION SITE APPLICATIONS OF BTI-BASED LARVICIDES

The cost of conducting black fly control programs has increased over the years due to increased costs for the Bti-based products and their application, especially aerially. Increases in worker wages, price of raw materials, shipping, and energy costs have all increased the final costs. Consequently, operators have begun to optimize methods to reduce costs but still maintain programs at levels necessary for control and without sacrificing the integrity of the program. Prescription site applications use all available prespray data at individual sites or segments of rivers to prescribe the rate and method of treatment for those areas.

The South African program on the Orange River demonstrated that, depending on the river discharges, larvicide volumes could be reduced by 20–55% using optimized application techniques (Rivers-Moore et al., 2008). To reduce Bti volumes and cost, the Orange River program managers avoided making applications in low-flow conditions and used test applications at different concentrations the day before operational applications. Applying Bti only in high-flow conditions produced longer carry of the material and fewer treatment sites. Using test sites before large-scale operations allowed the managers to assess the effect of seston and other factors on black fly larval mortality and make decisions on the rate to use for treatments.

In the Pennsylvania program, similar optimization practices are being implemented. Regional managers are beginning to recognize the value of the prescription-site application approach that allows lower rates than the standard 12 ppm applied over 1 min in many of the 45 targeted rivers and streams. Significant rate reductions have been implemented on some of the more pristine rivers and streams, and effective larval mortality has still been observed. However, true prescription-site applications, which address seston levels and other factors affecting larval susceptibility, require additional operational effort and have been difficult to fully implement. Initiating these changes will continue to be a challenge.

25.5.1 Biological and Environmental Factors

Black fly suppression programs today target a multitude of species under a wide range of environmental conditions. Research conducted in the laboratory with early Bti-based larvicides demonstrated varying susceptibilities among species (Molloy et al., 1981). Today's operational applications have proven effective across a broad spectrum of targeted species; consequently, varying species susceptibility does not appear to be a significant operational consideration.

Water temperature is another aspect of the aquatic environment that has been evaluated in relation to Bti-based larvicides and their effectiveness. The general conclusion is that there is a positive correlation between water temperature and Bti effectiveness against larval black flies (Lacey et al., 1978; Lacoursiere and Charpentier, 1988; de Barjac and Southerland, 1990; Atwood et al., 1992). However, the effect of water temperature is a complicated combination of behavioral and physiological parameters affecting feeding rates and enzymatic activity (Lacoursiere and Charpentier, 1988). Operationally, we have achieved excellent mortality at water temperatures ranging from 1 to 2°C (Gray et al., 2012), and black fly suppression is routinely conducted in cold water environments. Operators should initiate applications using

maximum label rates under cold water conditions and then reduce the rates as mortality evaluations dictate.

Once the larvicide has been applied to a waterway, larval feeding and the ingestion of the insecticidal proteins become critical. Under ideal conditions, mortality occurs rapidly (≤30 min, personal observation) when larvae are exposed to a high concentration of a Bti-based larvicide. This scenario is often demonstrated by larvae immediately downstream of a larvicide application. In these cases, when larvae are feeding actively, they rapidly ingest enough insecticidal proteins to induce irreversible damage to the larval gut epithelium. In contrast, larvae farther downstream are exposed to lower concentrations of the insecticidal proteins and require significantly longer for mortality to occur.

25.5.2 Water Quality and Feeding Inhibition

A factor that continues to vex black fly control is water quality and turbidity and its effect on larval feeding. Turbidity has been a concern for black fly suppression specialists, and algal blooms in particular have been recognized as problematic (Palmer, 1998; Stephens et al., 2004). These observations raise questions of whether normal feeding activity was disturbed, or if the presence of additional particles merely reduced the probability that the insecticidal proteins would be ingested. Gaugler and Molloy (1980) first addressed this topic, describing feeding inhibition as "any atypical feeding behavior resulting in reduced ingestion." The typical feeding posture for filter feeding is the full extension of both labral fans (Fig. 25.1). Normal feeding activity involves the rapid abduction of the labral fans, typically alternating one fan after the other, and then immediately fully extending the fan (Merritt et al., 1996). Labral fans retracted for extended periods or only partially extended would result in feeding inhibition and in reduced larval susceptibility to Bti-based larvicides.

Kurtak (1978) determined that particles larger in diameter than the space between the labral fan rays (~35 μm) are captured more efficiently than smaller particles. Gaugler and Molloy (1980) demonstrated that a variety of materials would cause feeding inhibition and that any differences were probably due to the various particle sizes of the materials and the feeding efficiency of the larvae in capturing that size. The percentage of larvae exhibiting feeding inhibition increased with increasing particle concentration and, as a result, feeding inhibition was attributed to rapid gut filling. Normal feeding activity could be restored when the particulate matter was flushed from the bioassay system and removed from the larval medium.

25.5.3 Effects of Seston

Research related to the Pennsylvania program has provided the impetus for some of the most recent studies related to water quality and the effect of seston on larval susceptibility. Initial studies focused on the algal populations during certain times of the suppression season and their effects on larval susceptibility (Stephens et al., 2004). This work demonstrated the varying effects of different species of algae on larval susceptibility to the Bti-based larvicide.

Laboratory studies identified the green alga *Scenedesmus quadricauda* as particularly problematic to larval susceptibility to Bti at levels as low as 16,000 cells/mL. This alga commonly occurs in four-cell colonies that approach 100 μm in diameter, a size that Kurtak (1978) thought to be ideal for larval capture. The other three algal types used in these studies had much smaller cells (3–12 μm). Consequently, these cells would not have been captured as efficiently as the larger algal colonies and had no effect on larval susceptibility at ≥250,000 cells/mL. This work would appear to support Gaugler's and Molloy's idea that feeding inhibition was related to rapid gut filling and that continual ingestion is required to maintain the full status of the digestive tract with the smaller cells. This could also explain why small clay particles have little effect on larval susceptibility even when turbidity is high (Iburg et al., 2011; Gray et al., 2012).

Overmeyer et al. (2006) demonstrated that the reduction in larval susceptibility caused by the green alga *Scenedesmus quadricauda* could be overcome by increased concentration of the insecticidal proteins or by allowing the larvae longer to feed in the medium containing the algal cells and proteins. This work could suggest that the larvae are more efficient at capturing the larger particles and consequently do not have to feed as long as when in a medium containing smaller particles. Palmer (1998) and Palmer et al. (1996) recommend increasing the larvicide concentration 25–150% during algal blooms to overcome the reduced larval susceptibility associated with these conditions.

Iburg et al. (2011) provided additional insight into the effect of seston on larval susceptibility. River water from an area where less than optimal larval mortality was observed was analyzed with multiple techniques. Filtration experiments demonstrated that the filtrate and the dissolved cations did not alter larval susceptibility to the Bti-based larvicide. This finding supported Gaugler and Molloy (1980) who determined that nonparticulate matter did not discernibly affect the larval feeding behavior. However, these experiments did clearly demonstrate that the seston in the river water negatively affected larval susceptibility.

Scanning electron microscopy identified significant populations of radial centric and pennate diatoms in the river water during the times of reduced larval susceptibility. Two species of diatoms were cultured in the laboratory, *Cyclotella meneghiniana* and *Nivicula pelliculosa*, both of which are common in the microflora of the Susquehanna River. Further experiments demonstrated that live diatoms and diatom frustules, 5–15 μm in diameter, significantly reduced larval susceptibility at low levels when little

turbidity was observed in the larval media. Consequently, feeding inhibition caused by small diatom cells and frustules is not likely related to a "rapid gut filling" event, as described by Gaugler and Molloy (1980). This type of feeding inhibition might be related to the irritant effect of the silicon dioxide particles on the feeding apparatus and digestive system.

Iburg et al. (2015) built on these findings while investigating the length of time required for larval mortality after exposure to Bti insecticidal proteins. This work demonstrated that full mortality occurred within 4 h in *Simulium vittatum* larvae after exposure to an operational concentration. Lesser concentrations achieved similar mortality but took nearly twice as long. Inclusion of 50 ppm of kaolinite clay produced a turbidity of 41 NTUs and significantly delayed mortality at 1 and 2 h postexposure. However, by 3 h post-exposure, mortality was not different than in medium with no particulates. These results again indicate that in the presence of competing particles, fewer proteins come in contact with binding sites and mortality is delayed. These observations support the work of Lacoursiere and Charpentier (1988) who stated that "the efficacy of an insecticide requiring ingestion is not only dependent on the probability of encounter between the toxic particles and the insect, but is also dependent on the feeding behavior of the larva, as dictated by prevailing environmental conditions".

25.6 BLACK FLY SUPPRESSION PROGRAMS—OVERVIEW

Black fly control programs are initiated in response to a problem associated with the adult stage. Like most types of pest management conducted today, black fly suppression specialists try to use an IPM approach. This effort requires adult and larval surveillance to help determine when and where larvicide applications should be conducted. Once the adult pest species is identified, larval surveillance is targeted to find the waterways closest to the area where the flies are causing the problem. This effort is important from an operational, economic, and regulatory aspect. The flight range of the pest species, the objective of the suppression effort and the resources provided by the entity initiating the work will determine what larval habitats will be targeted. The planning and surveillance aspect of the work requires the best available maps of the area. USGS maps (1:24,000) are a good example, but any map or image of the area will be an asset with modern technology providing many new and useful tools.

While black fly suppression programs are initiated to target a pest population, Malmqvist et al. (2004) reminds us that black flies serve important roles in both the aquatic and terrestrial environments. Using an IPM approach to suppress the pest population will eliminate unnecessary larvicide applications and target only larval habitats producing the pest species. The costs associated with black fly suppression also serve to target larvicide applications where the greatest efficiencies are achieved related to the goals of the operators.

25.6.1 Onchocerciasis Control Programme

Bti-based larvicides are used worldwide in various black fly management programs against many species. The OCP was the first program that used large volumes of Bti-based larvicides in one of the world's largest human-health vector management programs. The OCP involved the highest levels of international collaboration of multilateral agencies, private industry, governments, and nongovernment organizations to break disease transmission in 11 disease-endemic countries (Benin, Burkina Faso, Côte d'Ivoire, Ghana, Guinea Bissau, Guinea, Mali, Niger, Senegal, Sierra Leone, and Togo). Between 1974 and 2002, the program covered more than 1.2 million km^2 and protected 40 million people from the scourge of insect-borne disease, prevented blindness in 600 thousand people, and ensured that 18 million children were born free from the threat of river blindness. Prior to the program, along some river breeding sites, up to 50% of adults were blind and the disease burden caused people to abandon land in fertile river valleys for fear of contracting onchocerciasis, which led to even higher levels of poverty and famine. After the program, according to WHO (2015a,b), approximately 25 million hectares of the abandoned land were reclaimed for settlement and agricultural production, which was capable of feeding 17 million people annually.

25.6.2 Pennsylvania Black Fly Control Program

The Pennsylvania program is the longest continuous program using *Bti* in a lotic environment. The Pennsylvania Department of Environmental Protection (PA-DEP) treats 45 rivers and streams in three major drainage basins (Susquehanna, Delaware, and Allegheny Rivers) totaling c. 3000 km (Arbegast, 1994; PA DEP, 2015). The program targets four black fly species in the *S. jenningsi* species group, which are a nuisance to humans. The Bti-based larvicide VectoBac 12AS is applied every 10–14 days with helicopters (Fig. 25.2) and some ground applications using motorized backpack sprayers. This product has been proven to be highly efficacious in Pennsylvania with minimal effects on nontarget organisms (Brancato, 1996; Jackson et al., 2002).

25.6.3 Other Notable Programs

Another notable program is conducted in South Africa on the Orange River where *Simulium chutteri* is considered the primary pest species; 10^4–3×10^4 L of VectoBac 12AS is used

FIGURE 25.2 Aerial application over the Susquehanna River, Pennsylvania, USA.

annually for its control. Reviews of the Orange River Black Fly Control Program have been written by Palmer et al. (1996), Palmer (1997, 1998), and Myburg and Nevill (2003).

In the United States, suppression programs are conducted against a variety of species. Two of the larger programs include work conducted on the Colorado River against *S. tribulatum* (formerly "cytoform IIIL-1" of *Simulium vittatum*). This program is unique with the pest causing nuisance problems in both Arizona and Nevada requiring a multistate collaboration. The state of West Virginia targets species from the *S. jenningsi* group in the Greenbrier, Lower Bluestone and New Rivers. Other programs include the Metropolitan Mosquito Control District in Minnesota which targets *S. luggeri*, *S. meridionale*, and *S. johannseni* in the Mississippi, Minnesota, and Rum Rivers. Smaller programs are conducted in Idaho, South Carolina, New York, Nevada, and occasionally Wisconsin and North Carolina. A summary of all North American programs through 2004 is included in Adler et al. (2004).

25.7 EFFECTS OF BTI ON NONTARGET AQUATIC INVERTEBRATES

The worldwide use of Bti for mosquito control in public health programs has increased yearly since its commercial availability from various companies, beginning in the early 1980s. However, widespread acceptance of Bti for black fly control has been slower because of possible effects on community food-web structure and fisheries in flowing water. Since many consumers in this environment are polyphagous and exhibit considerable overlap in their diets, the conventional fear has been potential trophic cascades (Lacey and Mulla, 1990). Many studies have demonstrated Bti's lack of direct, short-term toxic effect on nontarget aquatic invertebrates and fish. The majority of the literature reports little, if any, direct mortality or density changes for these nontarget organisms (Back et al., 1985; Lacey and Mulla, 1990; Dejoux and Elouard, 1990; Molloy, 1992; Boisvert and Boisvert, 2000; Jackson et al., 2002; Lacey and Merritt, 2003).

Fewer studies exist on the indirect, long-term effects due to the complexity of trophic relationships in river and stream ecosystems and the in-stream nature of the studies. The questions most asked have focused on the ecological implications of continuous applications of Bti in systems where simuliids are abundant and support higher guilds, both aquatic and terrestrial. The major concerns have been the predators, competitors, and scavengers: more than 20 aquatic families including the EPT groups (Ephemeroptera, Plecoptera, Trichoptera).

25.7.1 Predators

The predators are likely the first trophic level affected by continuous Bti use, particularly specialist predators in a simuliid-poor environment. Generalist predators were least affected. Wipfli and Merritt (1994) showed changes in feeding habits of some species of stoneflies (Plecoptera: Perlidae), following a Bti application; however, some stoneflies switch to alternate prey after reduction of larval black flies. In some species, indirect effects occurred through loss of black fly biomass, but the effect was short. Also, in this study, the predatory odonates fed more on mayflies than on larval black flies and did not seem to be affected by the Bti treatments.

25.7.2 Scavengers

In a study of the feeding habits of nontarget aquatic insects in response to live and Bti-killed Simuliidae, Merritt et al. (1991) found decreased feeding on dead black fly larvae by *Acroneuria lycorias* (Plecoptera:Perlidae) and *Nigronia serricornis* (Megaloptera: Corydalidae). In the same study, the detritivore *Prostoia completa* (Plecoptera: Nemouridae) fed on dead black fly larvae after the treatment, illustrating that *Bti* treatments might be beneficial to some species.

In the Pennsylvania program, which was initiated in 1983, numerous studies have investigated effects on fish and other predators and scavengers. A 5-year study by the Academy of Natural Sciences of Philadelphia and Stroud

Water Research Center found no significant impacts to fish and aquatic macroinvertebrates (Jackson et al., 2002). There were some initial concerns about small darters (*Etheostoma* and *Etheostorna* spp.), which are abundant in many Pennsylvania streams (Brancato, 1996). A comprehensive study conducted by the Pennsylvania Department of Environmental Protection, however, showed that the concerns were not warranted and that darter growth, condition, and abundance were not affected by repeated Bti applications. Gut contents from a large sample of darters showed that larval black flies represented <1% of their diet; chironomids were their main food source. Empirical observations by the PA-DEP over a 29-year period has shown no evidence of significant impacts to aquatic macroinvertebrates, especially the EPT group, which makes up a large portion of the aquatic insect fauna of the Bti-treated streams and rivers in Pennsylvania.

25.7.3 Competitors

Some black fly competitors have been shown to benefit from Bti treatments. Reduction in competition by black fly larvae following applications of Bti might provide, to some of the predators, food, and vacated niches that were occupied by black fly larvae (Hemphill and Cooper, 1983). In the WHO/OCP Program, although it is difficult to make definitive conclusions due to the alternation of insecticides in the wet and dry seasons, Dejoux and Elouard (1990) reported changes in the macroinvertebrate assemblages in some African rivers but found no evidence of deleterious effects in lotic environments after weekly dry-season applications of Bti. After 2 months, population reductions occurred in Tipulidae, Baetidae, and Hydropsychidae, but densities increased for Hydroptilidae, Ceratopogonidae, Chironomidae, and a filter-feeding Trichoptera.

Other studies have shown an inverse correlation between simuliid and hydropsychid caddisfly densities suggesting competition (Chutter, 1968; Hemphill and Cooper, 1983; McAuliffe, 1984). Lower benthic densities after Bti applications have been reported, particularly of some chironomid species, but no researchers report permanent changes (Back et al., 1985; Boisvert and Boisvert, 2000; Simmons, 1991; Molloy, 1992; Palmer and Palmer, 1995). Many of these studies were not repeated at different times or at multiple sites, suggesting that observed differences might reflect natural spatial and temporal variability.

The application of Bti-based larvicides targeting mosquito populations has also been investigated extensively. In some of the most recent work, Lagadic et al. (2014) determined that Bti-based larvicide applications conducted over 7 years had no influence on the taxonomic structure and abundance of nontarget aquatic invertebrate communities. They also concluded that the number of invertebrates that could be used as food resources for birds was maintained in the environment. This work directly addresses earlier papers by

Poulin et al. (2010) and Poulin (2012) that raised questions about the long-term impacts of Bti applications. These studies reiterate the importance of conducting larvicide applications in an IPM fashion to prevent unnecessary larvicide applications.

25.8 CONCLUSIONS AND FUTURE DIRECTIONS

Larvicides based on *Bacillus thuringiensis* subsp. *israelensis* are the instruments of choice for the suppression of nuisance and vector black fly populations. Today's formulations are highly effective in a wide range of habitats and have been repeatedly shown to be safe to nontarget organisms. Multiple studies have demonstrated that long-term black fly suppression programs do not cause significant deleterious effects on higher trophic levels or ecosystem structure. The success of Bti-based larvicides has led to a significant decline in research into other materials that could be used against black fly populations. There are no other active ingredients registered for larval black fly applications in the United States today. Recent research has demonstrated the importance of multiple factors related to larval feeding inhibition and the resulting reduction in larval susceptibility. Continued study of these factors will help guide operational personnel as to when to apply Bti-based larvicides for maximum effectiveness and efficiency.

REFERENCES

Adler, P.H., McCreadie, J.W., 2009. Black flies (Simuliidae). In: Mullen, G., Durden, L. (Eds.), Medical and Veterinary Entomology, second ed. Academic Press, San Diego, pp. 183–200.

Adler, P.H., Currie, D.C., Wood, D.M., 2004. The Black Flies (Simuliidae) of North America. Cornell University Press, Ithaca, NY, USA. 941 pp.

Amrine, J.W., 1983. Measuring stream discharge and calculating treatment of rates of *Bacillus thuringiensis* (H-14) for black fly control. Mosq. News 43, 17–21.

Arbegast, D.H., 1994. Black fly suppression, the Pennsylvania experience. Proc. Annu. Meet. N.J. Mosq. Control Assoc. 81, 107–113.

Atwood, D.W., Robinson, J.V., Meisch, M.V., Olson, J.K., Johnson, D.R., 1992. Efficacy of *Bacillus thuringi*ensis var. *israelensis* against larvae of the southern buffalo gnat, *Cnephia pecuarum* (Diptera: Simuliidae) and the influence of water temperature. J. Am. Mosq. Control Assoc. 8, 126–130.

Back, C., Boisvert, J., Lacoursière, J.O., Charpentier, G., 1985. High-dosage treatment of a Quebec stream with *Bti*: efficacy against black fly larvae and impact on non-target insects. Can. Entomol. 117, 1523–1534.

de Barjac, H., 1978. Une nouvelle variété de *Bacillus thuringiensis* très toxique pour les moustiques: *B. thuringiensis* var. *israelensis* sérotype H14. C.R. Acad. Sci. Paris t. 286, sér. D 797–800.

de Barjac, H., Sutherland, D., 1990. Bacterial Control of Mosquitoes and Black Flies. Rutgers University Press, New Brunswick, NJ, USA. 349 pp.

Barton, W., Noblet, R., Kurtak, D.C., 1991. A simple technique for determining relative toxicities of *Bacillus thuringniensis* var. *israelensis* formulations against larval black flies (Diptera: Simuliidae). J. Am. Mosq. Control Assoc. 7, 313–315.

Boisvert, M., Boisvert, J., 2000. Effects of *Bti* on target and nontarget organisms: a review of laboratory and field experiments. Biocontrol Sci. Technol. 10, 517–561.

Brancato, J.C., 1996. Environmental impact of Pennsylvania black fly control programs on fish & invertebrates. Proc. Annu. Meet. N.J. Mosq. Control Assoc. 83, 64–71.

Brown, A.W.A., 1962. A survey of *Simulium* control in Africa. Bull. World Health Organ. 27, 511–527.

Chutter, F.M., 1968. On the ecology of fauna on stones in the current in a South African river supporting a very large *Simulium* (Diptera) population. J. Appl. Ecol. 5, 531–561.

Cibulsky, R.J., Fusco, R.A., 1987. Recent experiences with Vectobac for black fly control: an industrial perspective on future developments. In: Kim, K.C., Merritt, R.W. (Eds.), Black Flies, Ecology, Population Management and Annotated World List. The Pennsylvania State University. University Park and London, pp. 419–424.

Colbo, M.H., O'Brien, H., 1984. A pilot black fly (Diptera: Simuliidae) control program using *Bacillus thuringiensis* var. *israelensis* in Newfoundland. Can. Entomol. 116, 1085–1096.

Dejoux, C., Elouard, J.M., 1990. Potential impact of microbial insecticides on the freshwater environment, with special reference to the WHO/UNDP/World Bank, Onchocerciasis Control Programme. In: Laird, M., Lacey, L.A., Davidson, E.W. (Eds.), Safety of Microbial Insecticides. CRC Press, Boca Raton, FL, pp. 65–83.

Devisetty, B.N., Wang, Y., Sudershan, P., Kirkpatrick, B.L., Cibulsky, R.J., Birkhold, D., 1998. Formulation and delivery systems for enhanced and extended activity of biopesticides. ASTM Spec. Tech. Publ. 1347, 242–272.

Fredeen, F.J.H., 1969. Outbreaks of the black fly *Sumulium arcticum* Malloch in Alberta. Quaest. Entomol. 5, 341–372.

Fredeen, F.J.H., 1985. Some economic effects of outbreaks of black flies (*Simulium luggeri* Nicholson and Mickel) in Saskatchewan. Quaest. Entomol. 21, 175–208.

Gaugler, R., Molloy, D., 1980. Feeding inhibition in black fly larvae (Diptera: Simuliidae) and its effect on the pathogenicity of *Bacillus thuringiensis* var. *israelensis*. Environ. Entomol. 9, 704–708.

Gaugler, R., Molloy, D., Haskins, T., Rider, G., 1980. A bioassay system for the evaluation of black fly (Diptera: Simuliidae) control agents under simulated stream conditions. Can. Entomol. 112, 1271–1276.

Goldberg, L.J., Margalit, J.A., 1977. Bacterial spore demonstrating rapid larvicidal activity against *Anopheles sergentii, Uranotaenia unguiculata, Culex univitattus, Aedes aegypti* and *Culex pipiens*. Mosq. News 37, 355–358.

Gray, E.W., Adler, P.H., Noblet, R., 1996. Economic impact of black flies (Diptera: Simuliidae) in South Carolina and development of a localized suppression program. J. Am. Mosq. Control Assoc. 12, 676–678.

Gray, E.W., Wyatt, R.D., Adler, P.H., Smink, J., Cox, J.E., Noblet, R., 2012. The lack of effect of low temperature and high turbidity on operational *Bacillus thuringiensis* subsp. *israelensis* activity against larval black flies (Diptera: Simuliidae). J. Am. Mosq. Control Assoc. 28, 134–136.

Guillet, P., Kurtak, D.C., Philippon, B., Meyer, R., 1990. Use of *Bacillus thuringiensis israelensis* for onchocerciasis control in West Africa. In: de Barjac, H., Sutherland, D.J. (Eds.), Bacterial Control of Mosquitoes and Black Flies. Rutgers University Press, New Brunswick, NJ, USA, pp. 187–201.

Hemphill, N., Cooper, S.D., 1983. The effect of physical disturbance on the relative abundances of two filter-feeding insects in a small stream. Oecologia 58, 378–382.

Horosko, S., Noblet, R., 1986. Black fly control and suppression of Leucocytozoonosis in turkeys. J. Agric. Entomol. 3, 10–24.

Iburg, J.P., Gray, E.W., Wyatt, R.D., Cox, J.E., Fusco, R.A., Noblet, R., 2011. The effect of seston on mortality of *Simulium vittatum* (Diptera: Simuliidae) from insecticidal proteins produced by *Bacillus thuringiensis* subsp. *israelensis*. Environ. Entomol. 40, 1417–1426.

Iburg, J.P., Gray, E.W., Noblet, R., 2015. Mortality patterns of *Simulium vittatum* larvae (Diptera: Simuliidae) following exposure to insecticidal proteins produced by *Bacillus thuringiensis* var. *israelensis*. J. Am. Mosq. Control Assoc. 31, 44–51.

Jackson, J.K., Horwitz, R.J., Sweeney, B.W., 2002. Effects of *Bti* on black flies and nontarget macroinvertebrates and fish in a large river. Trans. Am. Fish. Soc. 131, 910–930.

Jamnback, H., 1973. Recent developments in control of blackflies. Annu. Rev. Entomol. 8, 281–304.

Kurtak, D.C., 1978. Efficiency of filter feeding of black fly larvae (Diptera: Simuliidae). Can. J. Zool. 56, 1608–1623.

Kurtak, D., Jamnback, H., Meyer, R., Ocran, M., Renaud, P., 1987. Evaluation of larvicides for control of *Simulium damnosum s.l.*, (Diptera: Simuliidae) in West Africa. J. Am. Mosq. Control Assoc. 3, 201–210.

Lacey, L.A., Merritt, R.W., 2003. The safety of bacterial microbial agents used for black fly and mosquito control in aquatic environments. In: Hokkanen, H.M.T., Hajek, A.E. (Eds.), Environmental Impacts of Microbial Insecticides: Need and Methods for Risk Assessment. Kluwer Academic Publishers, Dordrecht, The Netherlands, pp. 151–168.

Lacey, L.A., Mulla, M.S., 1977. Evaluation of *Bacillus thuringiensis* as a biocide of blackfly larvae (Diptera: Simuliidae). J. Invertebr. Pathol. 30, 46–49.

Lacey, L.A., Mulla, M.S., 1990. Safety of *Bacillus thuringiensis* ssp. and *Bacillus sphaericus* to non-target organisms in the aquatic environment. In: Laird, M., Lacey, L.A., Davidson, W.W. (Eds.), Safety of Microbial Insecticides. CRC Press, Boca Raton, FL, pp. 169–188.

Lacey, L.A., Undeen, A.H., 1984. Effect of formulation, concentration, and application time on the efficacy of *Bacillus thuringiensis* (H-14) against black fly (Diptera: Simuliidae) larvae under natural conditions. J. Econ. Entomol. 77, 412–418.

Lacey, L.A., Mulla, M.S., Dulmage, H.T., 1978. Some factors affecting the pathogenicity of *Bacillus thuringiensis* Berliner against black flies. Environ. Entomol. 7, 583–588.

Lacey, L.A., Escaffre, H., Philippon, B., Sékétéli, A., Guillet, P., 1982. Large river treatment with *Bacillus thuringiensis* (H-14) for the control of *Simulium damnosum s.l.* in the Onchocerciasis Control Programme. Tropenmed. Parasitol. 33, 97–101.

Lacoursiere, J.O., Charpentier, G., 1988. Laboratory study of the influence of water temperature and pH on *Bacillus thuringiensis* var. *israelensis* efficacy against black fly larvae (Diptera: Simuliidae). J. Am. Mosq. Control Assoc. 4, 64–72.

Lagadic, L., Roucaute, M., Caquet, T., 2014. *Bti* sprays do not adversely affect non-target aquatic invertebrates in French Atlantic coastal wetlands. J. Appl. Ecol. 51, 102–113.

Malmqvist, B., Adler, P., Kuusela, K., Merritt, R.W., Wotton, R.S., 2004. Black flies in the boreal biome, key organisms in both terrestrial and aquatic environments: a review. Ecoscience 11, 187–200.

Margalit, J., Dean, D., 1985. The story of *Bacillus thuringiensis* var. *israelensis* (*B.t.i.*). J. Am. Mosq. Control Assoc. 1, 1–7.

McAuliffe, J.R., 1984. Competition for space, disturbance, and the structure of a benthic stream community. Ecology 65, 894–908.

McIntyre, J.W., Barr, J.F., 1997. Common loon (*Gavia immer*). In: Poole, A., Gill, F. (Eds.), Birds of North America – Online. http://dx.doi.org/10.2173/bna.313.

Merritt, R.W., Wipfli, M.W., Wotton, R.S., 1991. Changes in feeding habits of nontarget aquatic insects in response to live and *Bti*-killed black fly larvae (Diptera: Simuliidae). Can. Entomol. 123, 179–185.

Merritt, R.W., Craig, D.A., Wotton, R.S., Walker, E.D., 1996. Feeding behavior of aquatic insects: case studies on black fly and mosquito larvae. Invertebr. Biol. 115, 206–217.

Molloy, D.P., 1990. Progress in the biological control of black flies with *Bacillus thuringiensis israeliensis* with emphasis on temperate climates. In: de Barjac, H., Southerland, D.J. (Eds.), Bacterial Control of Mosquitoes and Black Flies. Rutgers University Press, New Brunswick, NJ, pp. 161–189.

Molloy, D.P., 1992. Impact of the black fly control agent *Bti* on chironomids & other non-target insects: results of ten field trials. J. Am. Mosq. Control Assoc. 8, 24–31.

Molloy, D.P., Gaugler, R., Jamnback, H., 1981. Factors influencing efficacy of *Bacillus thuringiensis* var. *israelensis* as a biological control agent of black fly larvae. J. Econ. Entomol. 74, 61–64.

Molloy, D.P., Wraight, S.P., Kaplan, B., Gerardi, J., Peterson, P., 1984. Laboratory evaluation of commercial formulations of *Bacillus thuringiensis israeliensis* against mosquito and black fly larvae. J. Agric. Entomol. 1, 161–168.

Moore, H.S., Noblet, R., 1974. Flight range of *Simulium slossonae*, the primary vector of *Leucocytozoon smithi* of turkeys in South Carolina. Environ. Entomol. 3, 365–369.

Myburgh, E., Neville, E.M., 2003. Review of black fly (Diptera: Simuliidae) control in South Africa. Onderstepoort J. Vet. Res. 70, 307–317.

O'Callaghan, M., Glare, T.R., Lacey, L.A., 2012. Bioassay of bacterial entomophatogens against insect larvae. In: Lacey, L.A. (Ed.), Manual of Techniques in Invertebrate Pathology, second ed. , pp. 101–126San Diego.

Overmyer, J.P., Stephens, M.S., Gray, E.W., Noblet, R., 2006. Mitigating the effects of the green alga *Scenedesmus quadricauda* on the efficacy of *Bacillus thuringiensis* var. *israelensis* against larval black flies. J. Am. Mosq. Control Assoc. 22, 135–139.

Palmer, R.W., 1997. Principles of Integrated Control of Black Flies (Diptera; Simuliidae) in South Africa. WRC Report No. 650/1/97. 307 pp.

Palmer, R.W., 1998. An overview of black fly (Diptera: Simuliidae) control in the Orange River, South Africa. Isr. J. Entomol. 32, 99–110.

Palmer, R.W., Palmer, A.R., 1995. Impacts of repeated applications of *Bacillus thuringiensis* var. *israelensis* de Barjac and temephos used in black fly (Diptera: Simuliidae) control on macroinvertebrates in the middle Orange River, South Africa. South. Afr. J. Aquat. Sci 21, 35–55.

Palmer, R.W., Edwardes, M., Nevill, E.M., 1996. Control of pest black flies (Diptera: Simuliidae) along the Orange River, South Africa: 1990–1995. Onderstepoort J. Vet. Res. 63, 289–304.

Pennsylvania Department Environmental Protection. , 2015. http://www.depweb.state.pa.us/portal/server.pt/community/black_fly/13774.

Poulin, B., 2012. Indirect effects of bioinsecticides on the nontarget fauna: the Camargue experiment calls for future research. Acta Oecol. 44, 28–32.

Poulin, B., Lefebvre, G., Paz, L., 2010. Red flag for green spray: adverse trophic effects of Bti on breeding birds. J. Appl. Ecol. 47, 884–889.

Riley, C.M., Fusco, R., 1990. Field efficacy of Vectobac®-12AS and Vectobac®-24AS against black fly larvae in New Brunswick streams (Diptera: Simuliidae). J. Am. Mosq. Control Assoc. 6, 43–46.

Rivers-Moore, N.A., Bangay, S., Palmer, R.W., 2008. Optimisation of *Bacillus thuringiensis* var. *israelensis* (Vectobac®) applications for black fly control programme on the Orange River, South Africa. Water SA 34, 193–198.

Ross, D.H., Craig, D.A., 1980. Mechanisms of fine particle capture by larval black flies. Can. J. Zool. 58, 1186–1192.

Sariozkan, S., Inci, A., Yildirim, A., Duzlu, O., Gray, E.W., Adler, P.H., 2014. Economic losses during an outbreak of *Simulium* (*Wilhelmia*) species (Diptera: Simuliidae) in the Cappadocia region of Turkey. Turk. Soc. Parasitol. 38, 116–119.

Simmons, K.R., 1991. Effects of Multiple Applications of the Bacterial Insecticide *Bti* on Black Fly & Non-target Invertebrates in the Mississippi River: Results of 1989 Studies (Metropolitan MCD. Techn. Rept. St. Paul, MN).

Skovmand, O., Kerwin, J., Lacey, L.A., 2007. Microbial Control of Mosquitoes and Black Flies. Field Manual of Techniques in Invertebrate Pathology. Springer, Dordrecht, pp. 735–750.

Solheim, R., Jacobsen, K.O., Oien, I.J., Aarvak, T., Polojarvi, P., 2013. Snowy owl nest failures caused by black fly attacks on incubating females. Ornis Nor. 36, 1–5.

Stephens, M.S., Overmyer, J.P., Gray, E.W., Noblet, R., 2004. Effect of algae on the efficacy of *Bacillus thuringiensis* var. *israelensis* against larval black flies. J. Am. Mosq. Control Assoc. 20, 171–175.

Undeen, A.H., Colbo, M.H., 1980. The efficacy of *Bacillus thuringiensis* var. *israelensis* against blackfly larvae (Diptera: Simuliidae) in their natural habitat. Mosq. News 40, 181–184.

Undeen, A.H., Nagel, W.L., 1978. The effect of *Bacillus thuringiensis* ONR-60A strain (Goldberg) on *Simulium* larvae in the laboratory. Mosq. News 38, 524–527.

Undeen, A.H., Lacey, L.A., Avery, S.W., 1984. A system for recommending dosage of *Bacillus thuringiensis* (H-14) for control of simuliid larvae in small streams based upon stream width. Mosq. News 44, 553–559.

Urbanek, R.P., Zimorski, S.E., Fasoli, A.M., Szyszkoski, E.K., 2010. Nest desertion in a reintroduced population of migratory whooping cranes. Proc. North Am. Crane Workshop 11, 133–141.

Valent BioSciences Corporation, 2003. Technical Use Bulletin for Vectobac® 12AS Mosquito and Black Fly Larvicide (Libertyville, Ill).

Wilton, D.P., Travis, B.V., 1965. An improved method for simulated stream tests of black fly larvicides. Mosq. News 25, 118–123.

Wipfli, M.S., Merritt, R.W., 1994. Disturbance to a stream food web by a bacterial larvicide to black flies: feeding responses of predatory macroinvertebrates. Freshwater Biol. 32, 91–103.

World Health Organization, 2015a. Onchocerciasis Control Programme, Prevention of Blindness and Visual Impairment. http://who.int/blindness/partnerships/onchocerciasis_OCP/en/.

World Health Organization, 2015b. Onchocerciasis. Fact Sheet 374. http://who.int/blindness/partnerships/onchocerciasis_OCP/en/.

Wotton, R.S., 1994. Methods for capturing particles in benthic animals. In: Wotton, R.S. (Ed.), The Biology of Particles in Aquatic Systems. Lewis Publ., Boca Raton, pp. 183–204.

Wotton, R.S., 2009. Feeding in black fly larvae (Diptera: Simuliidae) – the capture of colloids. Acta Zool. Lith. 19, 64–67.

Chapter 26

Mosquito Control With Entomopathogenic Bacteria in Europe

N. Becker[1,2], P. Lüthy[3]

[1]German Mosquito Control Association (KABS), Speyer, Germany; [2]University of Heidelberg, Heidelberg, Germany; [3]Swiss Federal Institute of Technology, ETH Zürich, Zürich, Switzerland

26.1 INTRODUCTION

Mosquitoes are the cause of major nuisances in temperate climates, and several species are potential vectors, predominantly of arboviruses. In Europe, the two mass breeding species causing disturbances and a reduction of the quality of life are the floodwater mosquitoes *Aedes vexans* and *Ochlerotatus sticticus*. Their breeding sites are temporarily flooded zones along rivers and shallow borders of lakes. Frequently, the biting rates are so high that staying outdoors becomes virtually impossible, especially during the evening hours. Outbreaks may last for 4–6 weeks, the life span of floodwater mosquitoes. The massive disturbance by mosquitoes during the summer season can have negative economic impacts. Restaurants, beer gardens, hotels, and camping sites remain empty, and tourists stay away or cut their holidays short. It has even happened that residents have sold their houses because of mosquitoes. It is estimated that the economic loss due to plagues of mosquitoes in Europe is in the range of 100 million Euros. In Germany, the yearly loss in revenue is estimated at 20 million Euros. In the Upper Rhine Valley alone more than 12 million Euros are lost per year (Hirsch and Becker, 2009). In the plain of Magadino, an important tourist area in the Canton Ticino (Switzerland), the losses amounted to 2 to 5 million Euros during the mosquito season (Lüthy, personal communication). With the onset of mosquito control in 1988, no more losses occurred. As a reaction to the regional mosquito problems, about 20 abatement districts have been founded in Europe during the past six decades, dealing with more than 2 million ha of breeding sites.

The halophilic or halotolerant species, *Ochlerotatus caspius* and *Ochlerotatus detritus*, breed particularly in shallow lagoons found along the coasts of Southern Europe, Asia Minor, and North Africa. *Oc. caspius* also occurs in masses in freshly flooded rice fields in Mediterranean areas where they can also cause serious nuisance problems.

Along lakes with littoral vegetation, eg, by *Phragmites australis* or *Typha* spp., *Coquillettidia richiardii* finds ideal conditions for mass development. In swampy woodlands

and tundra areas we find the snow-melt mosquitoes, eg, *Ochlerotatus communis*, *Ochlerotatus rusticus*, *Ochlerotatus punctor*, and *Ochlerotatus hexodontus*, which may also cause nuisance problems and limit outdoor activities during certain periods. The rock-pool mosquito *Ochlerotatus mariae* are found along parts of the Mediterranean rocky coasts, where mass occurrences can become a great nuisance. Species of the *Culex pipiens* complex, known as the "house mosquito" because of its presence close to human settlements, can likewise cause problems in temperate zones.

With the invasion of exotic species, the mosquito problem for Europe has taken on a new dimension. Mosquitoes are passively shipped by long-distance transport within hours or days from their places of origin to Europe. Increasingly favorable climatic conditions allow their ultimate establishment. Among these mosquitoes, there are vectors of human infectious diseases, representing a severe threat to public health.

In Europe, established invasive species that are distributed by international trade are *Aedes albopictus*, *Aedes aegypti*, *Ochlerotatus j. japonicus*, *Ochlerotatus koreicus*, *Ochlerotatus atropalpus*, and *Ochlerotatus triseriatus*. Since its first introduction to Europe, the Asian Tiger mosquito, *Ae. albopictus*, has rapidly spread along the Mediterranean coast and even to central European countries like Germany (Adhami and Reiter, 1998; Becker et al., 2012a). It is now present in almost 20 European countries (Becker et al., 2012b). It is not only a major nuisance due to its aggressive and daytime biting activity but also *Ae. albopictus* is a vector of more than 20 arboviruses, especially dengue, chikungunya, and the Zika-Virus. Thus new diseases, and those thought to be eradicated, including malaria, have started to appear or reemerge in Europe. Autochthonous transmitted cases have been diagnosed in the Mediterranean area (Jetten and Takken, 1994; Santa-Olalla Peralta, 2010). Thus the general risk for the transmission of mosquito-borne diseases has increased again in Europe (ECDC, 2008). A new potential threat has arisen with the Zika-Virus, which

Microbial Control of Insect and Mite Pests. http://dx.doi.org/10.1016/B978-0-12-803527-6.00026-3

has spread in Brazil and to more than 20 countries in the Americas, the suspected cause of microcephaly.

The trigger of the autochthonous transmission of dengue, chikungunya, and Zika-Virus is infected persons entering disease-free regions. Their number is rapidly increasing. For example, *Ae. albopictus* was the vector of a local chikungunya epidemic, started by a visitor from India, affecting 200 persons in the area of Rimini (Italy) in 2007 (Angelini et al., 2007). The autochthonous transmission of dengue and chikungunya by *Ae. albopictus* has become a common incident in southern France during late summer/fall (Delisle et al., 2015). More cases where *Ae. albopictus* populations are high will emerge in the years to come. Not only does globalization favor the spread and establishment of exotic mosquitoes but also higher temperatures and extreme precipitation associated with climate change (Becker, 2008). The effect of global warming, which favors the completion of the sporogonic cycle of *Plasmodium* spp. in anophelines, and the hundreds of malaria cases, mostly caused by *P. falciparum* acquired in the tropics and imported to Europe, could enhance the risk of indigenous malaria transmission.

26.2 STRATEGIC CONSIDERATIONS OF MOSQUITO CONTROL PROGRAMS IN EUROPE

Most mosquito control programs in Europe aim to protect humans from mosquitoes in an effective and economically acceptable manner without impact on the environment, maintaining biodiversity and avoiding toxicological and ecotoxicological effects. Already by protecting and promoting antagonists of mosquitoes, the natural balance of ecosystems can be maintained. This can reinforce the sustainable control of mosquitoes by natural enemies. Furthermore, when selective control tools are applied and only the number of mosquitoes is reduced to an acceptable level, all remaining insects can serve as food sources for bats, birds, amphibians, and invertebrates (Becker et al., 2003).

These are the challenges of a modern control strategy, which has been established as a model in the Upper Rhine region.

It makes sense to take action against the developmental stages of mosquitoes, because these are concentrated in the breeding sites. They can be controlled by currently available biological means, especially microbial control agents (MCAs). In contrast, the adult mosquitoes, especially most of the floodwater mosquitoes, usually migrate into large areas at least 10 times more vast than the breeding sites. Therefore, to control the larval stages is ecologically well accepted and cost-effective.

Public opinion has become more and more critical toward the unbridled use of broad spectrum chemical pesticides due to their negative impacts. This has enhanced the search and development of selective, environmentally friendly measures, in particular, biological control agents such as pathogens and parasites of mosquitoes, which can be used in the context of integrated mosquito management. A great variety of organisms, including mermithids (eg, *Romanomermis culicivorax* and *Romanomermis iyengari*), fungi such as *Coelomomyces* spp., *Beauveria bassiana*, and *Metarhizium anisopliae*, and the Oomycete *Lagenidium giganteum* have been isolated and tested. However, the mass rearing and proliferation of these organisms in an economically feasible way and their sustainable establishment in the ecosystem have posed major problems (Becker et al., 2003).

The discovery of efficacious, easily and relatively cheaply produced bacteria [*Bacillus thuringiensis* subsp. *israelensis* (Bti) and *Lysinibacillus sphaericus*] within the past four decades has opened a new chapter in the biological control of mosquitoes and black flies (Singer, 1973; Weiser, 1984; Lacey, 2007).

26.3 CONTROL OF MOSQUITOES WITH BACTERIA

26.3.1 *Bacillus thuringiensis* Subsp. *israelensis*

The discovery of the gram-positive, endospore-forming soil bacterium, Bti, in the Negev desert of Israel in 1976 (Goldberg and Margalit, 1977; Barjac, 1978) has opened the door for MCAs as effective and environmentally safe tools for mosquito control. They are in their efficacy comparable to chemical larvicides. The outstanding feature of these MCAs is their specificity. The target insects are exclusively mosquito and black fly larvae and few other Nematocera. Thus the environmental impact is negligible.

During sporulation Bti produces a so-called parasporal body (Fig. 26.1) comprising four protein toxins, which passing through a cascade of activation steps become highly toxic to mosquito and black fly larvae. The four protein components have the following designations and molecular weights: Cry4A (125 kDa), Cry4B (135 kDa), Cry10A (58 kDa), and Cry11A (68 kDa; Delecluse et al., 1996). In

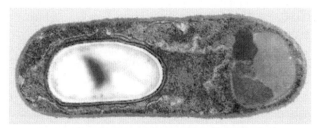

FIGURE 26.1 *Bacillus thuringiensis* subsp. *israelensis* with spore (left) and parasporal body, so-called protein crystal (right). *Micrograph courtesy of J.-F. Charles, Pasteur Institute, Paris.*

addition, Bti produces the so-called Cyt proteins, toxins of lesser specificity reacting with membrane lipids. Cyt toxins support the larvicidal activity of the Cry-proteins but differ in their mode of action (Höfte and Whiteley, 1989; Federici et al., 1990; Priest, 1992; see also Chapter 4). It is assumed that the four Cry toxins recognize different receptors on the intestinal epithelium, which has so far prevented the appearance of resistance (Wirth et al., 2004; Wirth, 2010). Details of the molecular mode of action of *Bacillus thuringiensis* are covered in Chapter 4. Nontarget organisms remain undamaged, because they are unable to activate the protoxin or the receptors are missing. This selective mode of action is the major reason for the high degree of biosafety of Bti.

26.3.2 *Lysinibacillus sphaericus*

This is basically a water-borne species with only a few strains exhibiting mosquitocidal properties. New strains of *Lysinibacillus sphaericus*, such as 2362 (serotype 5a5b), isolated from an adult black fly in Nigeria (Weiser, 1984), and 2297 (Serotype 25), isolated in Sri Lanka (Wickramasinghe and Mendis, 1980), are much more potent than the first isolates. *L. sphaericus* is particularly active against larvae of *Culex* species and *Anopheles* spp. (Ragoonanansingh et al., 1992; Fillinger et al., 2003; Lacey, 2007). New strains are still being isolated, which show at least similar toxicity as the already commercially used strain 2362 (Park et al., 2009). *L. sphaericus* has become increasingly important in mosquito control programs. The potential of *L. sphaericus* as an MCA lies in its narrow spectrum of activity and its ability to recycle for weeks. Thus long-term control can be achieved in many cases (Lacey, 1990; Ludwig et al., 1994; Silva-Filha et al., 2001). The time span between retreatments can be extended to over 1 month, providing more cost-efficient control of *Culex* species, particularly of *Cx. pipiens* as a vector of West Nile virus and *Culex quinquefasciatus*, which is the most important vector of lymphatic filariasis. *Cx. quinquefasciatus* prefers highly polluted waterbodies found in urban areas and associated with animal husbandry (eg, dairy drains and the like) in which to breed.

L. sphaericus can easily be identified by its round spore located terminally within a swollen sporangium. Its larvicidal efficacy is based on a binary (Bin) toxin (BinA, molecular weight 42-kDa, and BinB, molecular weight 51 kDa) The toxins are bound within the spore coat (exosporium). The formation of parasporal bodies adjacent to the spore is very likely due to an overproduction of the toxins. Both the binding domains of BinA and BinB are required for a high level of larvicidal activity (Baumann et al., 1991; Priest, 1992). The mode of action and receptor-binding of the Bin toxins are similar to that in Bti (Davidson and Youston, 1990; Charles et al., 2000). Exotoxins with larvicidal activity have also been identified (Mtx1: 100 kDa; Mtx2: 31 kDa; Mtx3: 36 kDa).

L. sphaericus kills only mosquito larvae. At higher concentrations, larvae of certain Psychodidae can also be affected (Becker et al., 2010). Certain mosquito species, such as *Cx. quinquefasciatus* and *Anopheles gambiae*, are highly susceptible to *L. sphaericus* toxins whereas larvae of *Ae. aegypti* and several other *Aedes* species are more than 100 times less susceptible. Black fly larvae, other insects, mammals, and other nontarget organisms are not susceptible to *L. sphaericus*. A particularly attractive feature of *L. sphaericus* is its potential to persist and recycle under certain field conditions. Appropriate formulations have shown a significant residual activity against larvae of *Culex p. pipiens* and *Cx. quinquefasciatus* in highly polluted breeding habitats (Davidson and Yousten, 1990; Lacey, 1990).

26.3.3 Factors Affecting Activity and Persistence of *Bacillus thuringiensis* Subsp. *israelensis* and *Lysinibacillus sphaericus*

In addition to the different susceptibility of various mosquito species to the bacterial toxins of Bti and *L. sphaericus*, a variety of factors can influence their efficacy (Becker et al., 1993; Ludwig et al., 1994; Puchi, 2005). The efficacy depends on the developmental stage of the target organisms, their feeding behavior, organic content of the water, the filtration effect of target larvae as well as that of other nontarget organisms, photosensitivity and other abiotic factors such as water temperature and depth, the sedimentation rate, and the shelf life of Bti and *L. sphaericus* formulations (Mulla et al., 1990; Becker et al., 1992). The long-term effect is also strongly influenced by the recycling capacity of the agent (Aly, 1985; Becker et al., 1995).

26.3.3.1 Species and Instar Sensitivity

Larvae become less sensitive to bacterial toxins in the course of their development (Becker et al., 1992). For instance, second instars *Ae. vexans* are about 11 times more sensitive than fourth instar larvae at a water temperature of 25°C (second instars: LC$_{90}$ 0.014±0.007 mg/L; fourth instars: LC$_{90}$ 0.149±0.004 mg/L). The differences in sensitivity are less at temperatures between 8 and 15°C. Nonetheless, the second instars of *Ae. vexans* are more than twice as sensitive as fourth instars at a water temperature of 15°C (second instars: LC$_{90}$ 0.062±0.025 mg/L; fourth instars: LC$_{90}$ 0.145±0.004 mg/L). In field experiments, only half of the concentration is required to kill third instar larvae compared to the fourth instar larvae. If fourth instar larvae are dominant, then the dosage must be doubled. It is therefore recommended to start control measures while the larvae are at an early developmental stage.

Large differences in toxin sensitivity exist between mosquito species due to differences in their feeding habits and their ability to activate the protoxin and binding to midgut cell receptors. For example, larvae of *Cx. pipiens* were found to be 2–4 times less susceptible to Bti than Aedini larvae of the same instar. By contrast, larvae of *Cx. pipiens* are much more sensitive to the toxins of *L. sphaericus* than *Aedes/Ochlerotatus*.

26.3.3.2 Temperature

The feeding rates of mosquito larvae depend on the water temperature. For instance, the feeding rate of *Ae. vexans* is reduced as the temperature decreases, and therefore the intake of bacterial toxins is less. In bioassays, second instar larvae of *Ae. vexans* are more than 10 times less sensitive at 5°C than at 25°C (5°C: LC_{90} 0.145±0.065 mg/L; 25°C: LC_{90} 0.0142±0.007 mg/L; Becker et al., 1992). The same effect was found when fourth instar larvae were tested. Applications of bacterial formulations should be conducted at a temperature above 8°C when Aedini species have to be controlled.

26.3.3.3 Size of the Water Body

Because bacterial toxins spread throughout the entire body of water, deep water requires higher rates than shallow water. Since larvae of many mosquito species feed near or at the surface (eg, *Anopheles* spp.), the effectiveness depends on the concentration and durability of the toxins in the upper layers of the water body. Floating formulations (eg, ice granules, Becker, 2003) will prolong the availability of toxin to surface-feeding larvae.

26.3.3.4 Larval Density

Bioassays with *Ae. vexans* have shown that higher densities of larvae require increased amounts of Bti (Becker et al., 1992). At 10 fourth instar larvae/150 mL water, the LC_{50} value was 0.0162±0.004 mg/L; at 75 larvae, the LC_{50} was ~7 times higher (LC_{50} 0.1107±0.02 mg/L). The presence of other filter-feeding organisms such as Cladocera causes similar effects. In tests with Bti, the LC_{50} and LC_{90} values were 5 and 6 times higher at a density of 90 *Daphnia* sp./150 mL than were those without *Daphnia* sp.

26.3.3.5 State of Nutrition

The state of nutrition and the amount of available food compete with Bti uptake. In laboratory studies 2 or 3 times more Bti was required for an equal level of mortality in the presence of added food (or polluted water) compared to clean water (Mulla et al., 1990).

26.3.3.6 Sunlight

Although radiation, such as that from a Cobalt[60] source, is well suited for the sterilization of Bti products without a significant reduction of toxicity (Krieg, 1986; Becker, 2002), strong sunshine appears to reduce larvicidal effects. For example, *L. sphaericus* preparations were active more than 3 times longer in shaded water than exposed to full sunlight (Sinegre et al., 1994). The LC_{90} values obtained with Bti [6000 International Toxic Units (ITUs)/mg] in bioassays with third instar *Cx. p. pipiens* in sunny sites (6000–12,000 lux for 7 h) and in shaded sites (<150 lux) at the same time and under identical conditions ($t = 25 \pm 1$°C) were very different (Becker et al., 1992). While the LC_{90} value was 0.054±0.008 ppm under shaded conditions, it was c. 4 times higher at sunny sites (LC_{90} 0.235±0.036 ppm).

26.3.4 Recycling Processes

A particularly attractive feature of *L. sphaericus* is its high efficacy against *Culex* species and its potential to persist and recycle under certain field conditions (Des Rochers and Garcia, 1984). In laboratory tests, it was shown that the presence of mosquito larval cadavers in the water contributes to the maintenance of toxic levels. Larval cadavers seem to contain all the nutrients necessary, both for multiplication of the bacteria and for toxin synthesis associated with the sporulation process. Aly (1985) was also able to demonstrate experimentally the germination of Bti in the gut of Aedini larvae. When compared with other environmental conditions, it seems that larval cadavers are of crucial importance for recycling processes (Becker et al., 1995).

It is important to understand the impact of all these factors for routine treatments, particularly because it allows the correct calculation of the optimal dosage, the selection of the right formulation under various environmental situations, and the optimal timing for application against different mosquito species (Becker and Rettich, 1994).

26.3.5 Environmental Safety

The exceptional environmental safety of the two bacterial control agents has been confirmed in many laboratory and field tests as well as in the course of worldwide application by thousands of tons of formulated Bti and *L. sphaericus* products over more than three decades without any harm to the environment. The US Environmental Protection Agency approved the use of Bti as early as 1981. In safety tests on representative aquatic organisms it was shown that in addition to plants and Mammalia none of the taxa tested such as Cnidaria, Turbellaria, Rotatoria, Mollusca, Annelida, Acari, Crustacea, Ephemeroptera, Odonata, Heteroptera, Coleoptera, Trichoptera, Pisces, and Amphibia were affected when exposed to water containing concentrations far above the recommended dosages of the bacterial preparations (Becker and Margalit, 1993; Becker et al., 2003). The World Health Organization (WHO) states that Bti can also be applied to reservoirs of drinking water for the control of container-breeding mosquitoes (WHO, 1999).

In Germany, Switzerland, and Italy as well as in the tropical countries of Thailand, Philippines, and Indonesia, Bti tablets are in use against larvae of *Cx. pipiens*, *Ae. aegypti*, and *Ae. albopictus* (Becker et al., 2003; Mahilum et al., 2005). The Vectobac DT/Culinex tablets used in drinking water are sterilized by gamma radiation and do not contain living Bti cells (Becker, 2002).

Within the order Diptera, the toxicity of Bti is restricted to mosquitoes and a few other nematocerous families (Colbo and Undeen, 1980; Miura et al., 1980; Ali, 1981; Margalit and Dean, 1985; Lacey and Merritt, 2003). In addition to mosquito and black fly larvae, only species in the families Psychodidae, Chironomidae, Dixidae, Sciaridae, and Tipulidae are sensitive to Bti, however, generally far less so than mosquitoes or black flies. In contrast to Bti, the range of the *L. sphaericus* insecticidal activity is even more restricted. Black fly larvae as well as other insects (except Psychodidae), mammals, and other nontarget organisms are not susceptible to *L. sphaericus*. Another important aspect is the widespread natural occurrence of both bacilli, *Bacillus thuringiensis* in soil and the phylloplane, and *L. sphaericus* in soil and water. They are natural components of the microecosystem. Thus residues after application are integrated into the natural cycle (Becker et al., 2010).

26.3.6 Commercial Development and Implementation

Bti and *L. sphaericus* were rapidly developed by industry with close support by universities and national and international organizations such as WHO. The WHO Pesticide Evaluation Scheme (WHOPES), which investigates and coordinates pesticides for public health, has approved MCAs as safe and efficient pesticides to support the protection of public health (WHO/CDS/WHOPES/2004.8).

Following production in bioreactors, Bti and *L. sphaericus* can be formulated into stable products with shelf lives comparable to chemical insecticides. The high degree of biosafety make Bti and *L. sphaericus* products suitable for integrated mosquito management programs, including community participation. Thanks to the different Cry proteins of Bti, recognizing separate receptors, supported by Cyt components, the risk for the development of resistance is minimal. The resistant pressure by *L. sphaericus* products is relatively high because of a single toxin unit (Nielsen-LeRoux et al., 1995; Charles and Nielsen-LeRoux, 2000; Wirth, 2010). Beside numerous reports of resistance in *Cx. quinquefasciatus* to *L. sphaericus* in the tropics (see Chapter 28), it has also been reported in Europe along the Mediterranean coast of France.

26.3.7 Formulations

Basic requirements for the successful use of bacterial control agents are effective formulations suited to the biology and habitats of the target organisms. Formulating pesticides is considered an art. Formulated products are available as water-dispersible granules (eg, Vectobac WDG and Vectolex WDG), wettable powders, fluid concentrates (eg, Vectobac 12AS and Aquabac), corncob granules (Vectobac G and Vectolex G), pellets, tablets (eg, Vectobac DT/Culinex tablets), briquettes, and ice granules (Table 26.1).

TABLE 26.1 *Bacillus thuringiensis* Subsp. *israelensis* and *Lysinibacillus sphaericus* Products Registered in Europe

Brand	Formulation	Potency (ITU/mg)	Active Substance
VectoBac 12AS	Aqueous suspension concentrate	1200	*Bacillus thuringiensis* subsp. *israelensis* strain AM65-52
VectoBac G	Granule	200	*Bacillus thuringiensis* subsp. *israelensis* strain AM65-52
VectoBac WG	Water-dispersible granule	3000	*Bacillus thuringiensis* subsp. *israelensis* strain AM65-52
VectoBac DT	Tablet	2200	*Bacillus thuringiensis* subsp. *israelensis* strain AM65-52
VectoMax FG	Granule	50 *Lysinibacillus sphaericus*	*Lysinibacillus sphaericus* strain ABTS 1743 *Bacillus thuringiensis* subsp. *israelensis* strain AM65-52
Teknar SC	Suspension concentrate	1200	*Bacillus thuringiensis* subsp. *israelensis* strain SA3A
Culinex Tab Plus	Tablet	1000	*Bacillus thuringiensis* subsp. *israelensis* strain AM65-52

ITU, International Toxic Units.

Products containing a combination of Bti and *L. sphaericus* such as VectoMax exhibit two specific advantages: fast killing is caused by Bti, and the presence of *L. sphaericus* provides persistence. This is especially important for the control of mosquitoes, which show a continuous reproduction over the whole season, as in the case of container breeders. VectoMax is also suitable for the control of mixed populations of culicine and/or *Anopheles*, allowing the application of a single product. In addition, the likelihood of resistance against *L. sphaericus* can be significantly reduced or even avoided (Zahiri et al., 2002).

Genetic engineering can be promising by improving mosquito larvicides based on both bacteria Bti and *L. sphaericus*, reducing, for example, the risk of resistance to *L. sphaericus* toxins. The best of these recombinants contain all major Bti endotoxins, specifically, Cry4A, Cry4B, Cry11A, and Cyt1A, plus the Bin endotoxin of *L. sphaericus*. The presence of Cyt1A in these recombinants, which synergizes Cry toxicity and delays resistance to these proteins and *L. sphaericus* Bin, should enable long-term use with little if any risk of resistance (Federici et al., 2007).

The activity of Bti and *L. sphaericus* products is expressed in ITUs per mg based on standard powders, originally produced by the Pasteur Institute in Paris (de Barjac, 1983; Dulmage et al., 1990; Skovmand and Becker, 2000) The procedure for bioassays is described in Chapter 8. Per hectare of floodwater about 3×10^9 ITUs are required. Each formulation differs in ITUs. Following the selection of a suitable formulation adapted to the application equipment and habitats to be treated, the quantity required per ha can be reliably determined. As an example, a few hundred grams or even less of powder, 1.5 to 2 L of liquid concentrate, or a few kg of granules/ha are usually enough to kill all mosquito larvae. In some situations, a long-term effect is achieved if larger amounts are used (Becker and Margalit, 1993; Becker and Rettich, 1994; Russell et al., 2003; Rydzanicz et al., 2009). With the production of tablets, water-soluble pouches, or briquette formulations, progress has been made toward achieving long-term effects (Kahindi et al., 2008; Su, 2008). In Germany, tablet formulations, based on Bti or *L. sphaericus*, have to be sterilized by γ-radiation to prevent the presence of living bacterial spores in drinking water. The tablets can be safely used for the control of container-breeding mosquitoes such as *Cx. p. pipiens* or *Ae. aegypti/Ae. albopictus* or even against *An. gambiae* (Becker et al., 1991; Kroeger et al., 1995; Mahilum et al., 2005).

In addition to commercially available granules, based on ground corncobs, sand granules can also serve as carriers for wettable powder formulations: 50 kg fire-dried quartz sand (grain size 1–2 mm) with 0.8–1.4 L of vegetable oil as a binding material and 1.8 kg of Bti powder with 5000 ITUs/mg is homogenized in a cement mixer. This mixture is sufficient to treat 2–3 ha of mosquito breeding sites. Recently, a very cost-effective formulation based on ice pellets was developed (Becker, 2003). Ice granules can be easily produced by transforming water suspensions containing the bacterial toxins into small ice cubes or pearls (3–5 mm) and kept in cold storage until used. The advantages of ice granules are as follows:

1. The toxins are incorporated into the ice pellet. A loss of active material by friction during application is avoided.
2. As the specific weight of ice is less than that of water, the ice pellets float in the upper water layer, releasing the toxins into the feeding zone of mosquito larvae as they melt.
3. The ice pellets fall through dense vegetation and do not stick to leaves even when it is raining. They show little drift by wind.
4. Increased swath is achieved because the friction is reduced due to the physical properties of ice.
5. The production is cost-effective.
6. The "carrier" is water.

The amount of active material/ha can thus be significantly reduced when compared with granules based on sand.

When appropriately stored, most preparations based on bacterial toxins can be kept for a long period without losing activity. Experience has shown that powder or corncob formulations lose little of their activity even after many years in storage. On the other hand, the activity of fluid concentrates may be more labile. Such formulations should therefore be retested in bioassays according to WHO guidelines (Skovmand and Becker, 2000) when they have been stored for more than a year.

26.4 SUITABILITY OF MICROBIAL CONTROL AGENT-INTEGRATED MOSQUITO CONTROL PROGRAMS WITH COMMUNITY PARTICIPATION

Bacterial control agents are particularly well suited for use in integrated programs because their toxic effect is selective and does not affect predatory organisms. Predators can therefore be included as additional elements in an integrated control strategy. The beneficial effect of predators can continue after the bacterial control agents have been applied. This indirectly produces a sustained suppressive effect (Mulla, 1990; Becker, 1992).

The WHO "Primary Health Care Concept" increasingly seeks to involve local residents in the search for solutions to healthcare problems. Bacterial control agents have a considerable safety advantage over synthetic insecticides, because neither the operator nor the residents within the treated perimeters are exposed to potentially dangerous chemicals. For this reason, such preparations are particularly well suited for use by volunteers (Becker, 1992).

Applications of Bti or *L. sphaericus* do not harm beneficial organisms such as honeybees, silkworms, or aquatic animals such as fish, shrimp, or oysters. These formulations can therefore be used in ecologically sensitive areas. Because they are biodegradable, no toxic residues remain. Their environmental safety permits bacterial control agents to be accepted both by registration authorities and by the public.

26.5 OPERATIONAL USE OF MICROBIAL CONTROL AGENTS

Many control programs aim at integrated biological control (IBC), allowing the protection of humans against mosquitoes and the safeguarding of biodiversity. When the ecosystem is compared with a web and each group of organisms represents one mesh, the strategy is to effectively reduce only the single mesh, representing the excessive quantity of target mosquitoes without affecting other meshes within the food web.

This goal can only be achieved in an optimum way if MCAs are used as a component within the overall IBC strategy. The conservation and enhancement of predators are important elements of IBC. Therefore microbial agents and biological methods go hand in hand with environmental management, eg, improving ditch systems for regulating water levels and creating permanent habitats for aquatic predators such as fish. IBC programs have been successfully implemented in Switzerland, Sweden, France, and particularly in Germany.

For the successful implementation and use of MCAs, the following prerequisites are necessary: entomological investigations, precise mapping and logging data of all major breeding sites, assessment of effective dosages in bioassays and conducting field tests, adaptations of application techniques compatible to the requirements in the field, design of the control strategy as well as training of the field staff, and last but not least dealing with governmental regulations.

Mosquitoes breed in a great variety of habitats. Almost everywhere, where stagnant water is present, the probability of finding mosquito larvae is high. Where the control of larvae is necessary, formulations and dosages have to be compatible with the ecology of breeding sites and factors affecting the efficacy of the MCAs.

Aerial and ground applications are the two modes of intervention. Fixed-wing aircraft and/or helicopters equipped with conventional buckets, booms, and nozzles or rotary mist atomizers are commonly used to treat densely vegetated and/or large surface areas. For ground treatments, knapsack sprayers or motorized backpack blowers are the conventional equipment used with liquid or granular formulations.

Various commercial products of Bti and *L. sphaericus* are used in Europe and worldwide (Russell et al., 2003; Boisvert, 2005; Kahindi et al., 2008) (Table 26.1).

26.5.1 Mosquito Control in Germany

The control of mosquitoes in Germany has a long history. In the 1920s and 1930s, breeding sites were treated with petroleum oils (Becker and Ludwig, 1983). During the 1950s and 1960s, adulticides were favored. In the early 1970s, the mosquito populations were extremely high because of frequent fluctuations of the water level of the Rhine. The attack by mosquitoes was so severe that residents in villages adjacent to temporarily flooded areas could not spend any length of time outside their houses. As a reaction to this unbearable situation, 44 towns and communities on both sides of the Rhine merged in a united mosquito control program, the Kommunale Aktionsgemeinschaft zur Bekämpfung der Stechmückenplage e.V. (KABS), which was founded in 1976. At present, 100 cities and municipalities along a 310-km stretch of the Upper Rhine River, with a total population of 2.7 million people, have joined forces to control the mosquitoes, mainly *Ae. vexans* and *Oc. sticticus*, over a breeding area of some 600 km² of the Rhine's flood plain. The budget of the program is approximately 3.4 million Euros a year, which results in overall costs per person per year of approximately 1.3 Euros.

The overall concept follows the abovementioned IBC (Becker, 1997). This includes microbial and biological methods integrated within environmental management (eg, improving the ditch system for regulation of the water level and providing permanent habitats for aquatic predators such as fish). The mosquito control program of KABS is one of the largest in Europe and worldwide. A wealth of experience and scientific knowledge has been acquired over the past four decades.

Before the implementation and use of MCAs, the design of the control strategy was based on entomological studies, mapping, numbering, and characterization of breeding sites by employing geographic information system (GIS), assessment of the optimum effective dosage of the product in use, and applying for governmental approval. For four decades, Bti and *L. sphaericus* have been successfully used in Germany as biological control agents against floodwater mosquitoes (eg, *Ae. vexans*) and *Culex* mosquitoes (eg, *Cx. p. pipiens* biotype *molestus*). Over 1500 km² of breeding areas have been treated with Bti, resulting in a reduction of the mosquito population each year by more than 90%.

The flood plains of the Rhine are usually inundated 2 to 4 times each summer. Due to climate change and heavy precipitation events, the number of peaks increased significantly. The extent of the flooding depends on the snowmelt in the Alps and on the intensity of rainfall. These parameters are constantly monitored. During flooding, *Ae. vexans* and other floodwater mosquito larvae hatch within minutes or hours at temperatures exceeding 8°C. Before control measures are initiated, the larval density and the larval stages are checked by means of 10 samples at representative breeding sites as the most important for the decision-making process for interventions; 1 day postapplication, spot samples are taken at reference breeding sites to check the efficacy of the treatment.

According to the extent of the flooding, 10–20% of the potential breeding areas of 600 km² have to be treated regularly by the 400 collaborators of the KABS. For treating first and second larval instars, 250 g of powder formulation (Vectobac WG, 3000 ITU/mg) or 1 L of liquid concentrate (Vectobac 12AS, 1200 ITU/mg) are suspended in 9–10 L of filtered pond water for each ha treated and applied by knapsack sprayers (Fig. 26.2). For deeper breeding sites or when late instar larvae are present, the dosage is doubled. During severe flooding, usually one-third of the area has to be treated with Bti granules, which are dispensed with the aid of a helicopter (dosages: 10–20 kg/ha). From 1981 to 2015, 140 tons of Bti powder and fluid concentrate and more than 3500 tons of Bti ice and sand granules mixed with quartz sand, vegetable oil, and Bti powder were used to treat over 3500 km² (350,000 ha) of breeding area. Since 1997, the more cost-effective Bti ice granules have been applied by helicopters to thousands of ha of floodwater mosquito breeding sites (Fig. 26.3).

Intervention against urban mosquito species is mainly carried out by residents. They are assisted by KABS providing advice and information on the biology of *Cx. p. pipiens* biotype *molestus* and the appropriate control measures. Culinex tablets have been particularly successful. They kill *Culex* larvae in water containers over a period of several weeks. In drainage systems and large cesspools with eutrophic water bodies, *L. sphaericus* as a liquid or powder formulation is applied against *Culex* larvae. Each year about 1 million Culinex-Bti tablets are applied against *Cx. p. pipiens*, especially in rainwater containers.

Monitoring of the program: Some 8% of the KABS budget is invested in monitoring mosquito populations, mosquito resistance, and investigating possible environmental impacts. All the studies carried out to date have shown that the introduction of Bti and *L. sphaericus* has reduced the number of nuisance mosquitoes to a tolerable level. The diversity of the ecosystem as a whole has not been affected.

Monitoring mosquito abundance: To monitor the abundance of adult mosquitoes, a sufficient number of traps are placed at comparable sites throughout the entire inundation area. Monitoring is done twice a month from April to September. From sunset to dawn the mosquitoes are caught by means of carbon dioxide light traps. Catches in nontreated areas serve as reference to determine the percentage reduction in the treated zones. It has been shown that since the

FIGURE 26.2 Ground application of microbial control agents by means of knapsack sprayers.

FIGURE 26.3 Aerial application of Bti ice granules.

widespread application of Bti in 1981, mass occurrences of mosquitoes were successfully suppressed. Naturally, the sustained control measures are greatly appreciated by the local people.

Monitoring the environmental impact: It has been essential to document the environmental impact by Bti and *L. sphaericus* applications in order to provide a scientific basis to rebut arguments commonly cited against mosquito control by opponents. Ahead of large-scale applications of MCAs, key members of various aquatic species groups

(*Cnidaria* to *Amphibia*) were screened in the laboratory and in small-scale field trials for their susceptibility to MCAs. These studies showed that in addition to mosquitoes (Culicidae) and black flies (Simuliidae), only a few species of midges (Chironomidae) were affected by Bti. For the most part, these midges were much less susceptible to Bti than the target organisms. *L. sphaericus* is toxic to an even narrower range of insects: only *Culex* species are highly susceptible, whereas *Aedes/Ochlerotatus* species respond much less. Black fly larvae as well as other insects (exception: Psychodidae) and nontarget organisms are not susceptible.

The development of insects in treated and untreated water is regularly monitored using emergence traps. The occurrence and abundance of insects in treated areas is assessed by light trap catches. All investigations have shown that while the numbers of *Aedes/Ochlerotatus* mosquitoes are drastically reduced, all other aquatic insects continue to develop normally as winged adults, providing a food resource for birds, amphibians, and bats.

Monitoring resistance: Mosquito populations are checked at regular intervals for the development of resistance (Becker and Ludwig, 1993). No resistance has been detected after 40 years of treatment with Bti. Bti and *L. sphaericus* products are alternated in control programs against *Culex* spp. in order to prevent resistance to *L. sphaericus*.

26.5.2 Mosquito Control in Switzerland

Switzerland has several wetland conservation areas at the foothills on both sides of the Alps. They are located at the end of valleys where the rivers have formed plains or deltas before flowing into lakes. These wetland areas are flooded periodically during the spring and summer months by snowmelt and by increasingly heavy precipitations, caused by climate change, along the rim of both sides of the Alps.

Although relatively small in size, these wetland areas exhibit a unique fauna and flora due to special climatic conditions. Many wetlands are under federal protection, especially the natural reserve Bolle di Magadino in the south of Switzerland, which is located along one of the main routes for migratory birds. Another area of interest exhibiting an extraordinary dynamic is situated at the upper end of the Lac de la Gruyère in the western part of Switzerland.

The frequently flooded wetland zones have become major breeding places for floodwater mosquitoes such as *Ae. vexans* and *Oc. sticticus*. The adult mosquitoes with a flight range of 10 km or more have created a serious nuisance problem for the nearby residential areas and for tourism, which is a vital economic source of income for Switzerland. In 1988 the first long-term mosquito control project in the plain of Magadino was initiated (Lüthy, 2001). Before, masses of mosquitoes in the search for blood meals severely affected the quality of life and likewise the economy. Since the mosquito control program using products exclusively based on Bti has been so successful, a second project followed in 1995 at the Lac de la Gruyère. It is likely that other areas such as the natural reserve "Les Grangettes" at the upper end of Lake Geneva may eventually follow.

Over the past decade, control practices have been refined and optimized. Although we would like to promote the integrated approach, the use of Bti has remained so far the main viable option. It is the safest mosquito control agent that is commercially produced. Other practices of mosquito control were either not efficient enough or would have interfered with the ecosystem (WHO, 1999).

The aerial application of Bti by helicopter has proved to be the only efficient method to reach the larval breeding sites. In the Swiss program, granular formulations of Bti are ideal for use in the flooded zones, which are predominantly covered by dense vegetation. Granular formulations (eg, Vectobac G) are applied at a dosage of about 15 kg/ha to reach a mortality rate approaching 100%. The granular formulation has an excellent shelf life over a period of at least 2–3 years. With enough stock on hand, we can always react very fast to control mass breeding of floodwater mosquitoes.

Mapping of the breeding sites and timing of applications are the most important elements for successful larviciding. Timing includes the climatic factors, the forecasted movements of the water levels in the wetland zones, and the standby of manpower and equipment.

The benefit of mosquito control is financially measurable. Most importantly, it improves the quality of life of the residents living within the range of mosquitoes. In a cost–benefit analysis, it was demonstrated that the cost–benefit factor is at least 1 to 5.

Wetlands deserve special protection. However, these areas need mutual acceptance and support by the local population as well as by the authorities. Emissions in both directions, into and from wetlands, should be treated on a status of equal rights. Complaints by residents about mosquito disturbance have to be taken seriously, and the responsible authorities should agree to control the breeding sites inside the wetlands. Residents should be exposed to no more than six bites by floodwater mosquitoes per evening. We cannot demand daily protection by mosquito repellents, which is especially undesirable in the case of children. If governmental approval is granted, we will be able to protect residents living adjacent to wetlands using Bti. This will ease local social conflicts. Furthermore, protected wetlands are accepted by residents and even become an attraction for tourists.

26.5.3 Mosquito Control in Sweden

The first Swedish abatement district (Biologisk Myggkontroll-Nedre Dalälven) was created in September 2000 as a response to many years of complaints about mosquito nuisance by the local population of the lower portion of the Dalälven river (Nedre Dalälven) in Central Sweden. A first estimate of the size of the potential mosquito breeding areas was 10 km². It was not possible, within a reasonable time frame, to use traditional ecological mapping for such a large and often inaccessible area. Thus from the very beginning, high-resolution orthophoto maps in combination with direct differential global positioning system were used to create precise maps of breeding site locations and sizes.

Approximately 10 km² of the Dalälven flood plain between the city of Avesta and the Sea of Bothnia consist of temporary wetland. These areas are usually flooded once or twice during spring and summer. The extent of flooding depends on the snowmelt in the Scandinavia Mountain Range and on rainfall in the very large catchment area for the River Dalälven. It is constantly necessary to monitor the water level in the river and in the flood plain from the beginning of snowmelt.

Some 24 different species of mosquitoes have been recorded in the Nedre Dalälven area, and the majority of species attack people. Determining the seasonal and spatial distribution of mosquito populations is an integral component of providing solutions to the periodic molestation by mosquitoes in the Nedre Dalälven area. Distribution patterns and seasonal dynamics of adult mosquitoes in this area have shown that the predominant species is *Oc. sticticus* (80%) followed by *Aedes rossicus* (8%) and *Aedes cinereus* (7%). These three species constitute the key nuisance species during spring and summer floods. In spring, *Oc. communis*, *Oc. punctor*, and *Ochlerotatus intrudens* can also be found in high numbers in the wetlands. *Oc. sticticus* is causing most of the problems due to its mass occurrence and dispersal ability. It feeds viciously on humans and restricts outdoor activities. Up to 60,000 mosquitoes per trap per night have been recorded, showing that the mosquito populations can be very large. The large flight range of *Oc. sticticus* increases the affected area to about 100 km² or about 10 times the size of the wetland breeding areas. The seasonal dynamics of mosquitoes in the area are mainly influenced by the flooding regime of the Dalälven River, which is controlled by a large number of power stations.

The control of *Ochlerotatus* and *Aedes* mosquitoes is based on the use of Bti products. As a prerequisite the following activities were conducted in 2000 and 2001: (1) mapping of 1000 ha of potential breeding areas for floodwater mosquitoes, (2) demand of governmental approval for the aerial application of Bti by helicopter, (3) calibration of the helicopter spraying equipment, (4) biweekly adult mosquito monitoring covering the whole Nedre Dalälven area, (5) training of the local fire brigade to perform field work with larval sampling, (6) creation of an often updated homepage to disseminate accurate information to the public (www.sandviken.se/mygg), and (7) performing 100% successful treatment against *Oc. sticticus* larvae in 70 ha of flooded wetlands.

All major nonprofit organizations concerned about eventual negative impacts on nontarget organisms have been invited to discuss the development of the mosquito control operations in order to minimize ecological side effects. Several research projects have been set up to analyze the patterns of mosquito interactions with the fauna in the wetland areas as well as long-term studies of any side effects of Bti on nontarget organisms.

26.5.4 The Use of Microbial Control Agents in Other European Countries

26.5.4.1 Croatia

Ae. vexans, mainly associated with *Oc. sticticus*, *Ae. cinereus*, and *Anopheles messeae*, is the major nuisance species in the inundation areas of the rivers Sava, Drava, and Danube, whereas along the Mediterranean coast *Oc. caspius* is abundant. Within human settlements *Cx. pipiens* is the main nuisance species. Mosquito control operations are usually organized and financed at the local level. Aerial spraying against adult mosquitoes is restricted by the Croatian government. For ground application in addition to Malathion, Baygon, Icon, and AquaReslin products based on Bti are increasingly being used.

26.5.4.2 Czech and Slovak Republics

Ochlerotatus cantans causes discomfort, and during the summer floodwater mosquitoes, *Ae. vexans*, *Oc. sticticus*, and *Ae. rossicus* are a severe nuisance to the inhabitants in years with regular flooding of the river basins of the Morava and Danube. Although adulticiding with pyrethroids (permethrin) is mainly practiced in some years, additional larvicides such as Bti granules are applied.

26.5.4.3 France

Mosquito control has a long history in France due to public health (eg, West-Nile Fever) and nuisance problems. Five independent organizations are responsible for the control operations, which are financed by the local (departments and communities) and regional administrations. Entente Interdepartementale pour la Demoustication (EID) Méditerranée, based in Montpellier, covers the French part of the Mediterranean basin (control area: 350,000 ha); EID Atlantique, based in St-Crépin, is responsible for the Atlantic coast (100,000 ha). In both areas, the major nuisance species along the coasts is *Oc. caspius*. EID Ain Isère-Rhône-Savoie, based in Chindrieux, covers the Rhone-Alps region (250,000 ha). The abatement district in the Alsace protects the people on the French side of the Upper Rhine Valley dealing with the floodwater mosquitoes *Ae. vexans* and *Oc. sticticus* as well as snowmelt mosquitoes such as *Oc. cantans*. Larval control is achieved by the management of wetlands encompassing the whole fluvial hydrosystem and by the use of larvicides directed against larval instars in specific habitats. From the mid-1980s, chemical insecticides were partly replaced with biological larvicides. By 1997, some of the organizations, such as EID Rhone-Alpes and in the Alsace region, based their larval control exclusively on biological larvicides, mainly Bti. Usually about 10,000 ha of breeding sites are treated with Bti products.

26.5.4.4 Greece

The largest projects in Greece are concerned with mosquito control in rice fields in the areas of Thessaloniki and the prefectures of Serres, Chalkidiki (14 municipalities), Larissa (2 municipalities), Kavala (3 municipalities), Pieria (11 municipalities), Imathia (3 municipalities), Pella (7 municipalities), and Lamia. Altogether more than 100,000 ha of breeding sites are treated with Diflubenzuron [Insect growth regulator (IGR), DU-DIM 10WP], and in ecologically sensitive areas products based on Bti (Vectobac and Teknar) are mainly used against *Oc. caspius*, *Anopheles sacharovi*, and *Cx. p. pipiens*.

26.5.4.5 Hungary

Major mosquito problems exist in two regions in Hungary: the regularly flooded areas along the Danube, the Tisza river, and their tributaries, and the shores and marsh area around Lake Balaton and Lake Velence. In addition to *Aedes/Ochlerotatus* species, which are prolific in all flood plains, *Cq. richiardii* is abundant at Lake Balaton in the densely reeded marsh areas. Current mosquito control in Hungary is carried out by several private abatement or pest control organizations, which are performing their activities in collaboration with urban pest control authorities. The main products used are organophoshates or pyrethroids and to a lesser extent some IGRs and products based on Bti.

26.5.4.6 Italy

Several programs have been implemented in Italy, mainly based on regional and local public financial support. One of the first programs was established in 1987 in the Bologna province. In the region of Comacchio, a tourist resort in the Po delta and a natural protected area, the mosquito control program was started in 1991. The strategy is based on a combination of both larvicides and adulticides. An important nuisance is caused by *Oc. caspius*, *Cx. p. pipiens* biotype *molestus*, *Oc. detritus*, and *Culex modestus*.

From 1996, following the approval of Regional Law n.75/1995, several programs were started in the Piedmont, which rapidly became the most organized region in Italy. From the beginning, major attention was devoted to larval control amended by occasional adult control in defined areas. As a result of specific research in surveys and monitoring, the control program has adopted larviciding as the only means of intervention. In the campaigns against the two major target species, *Oc. caspius* and *Cx. pipiens*, 95% of the total larvicide products used are based on Bti, while the remaining 5% consists of IGRs, used only in catch basins. More programs are expected in the near future, especially in central and southern Italy to protect tourism and in northern Italy to cope with *Ae. albopictus*, which is rapidly becoming the main noxious species.

26.5.4.7 Poland

Modern mosquito control operations started in Poland with the catastrophic flood of the Odra River in 1997 followed by huge mosquito problems, which increased the public interest in mosquito control. As a result, large-scale control activities have been reinitiated in several regions such as in the towns of Wroclaw, Szczecin, Gorzow Wielkopolski, and the region of Krynica Morska. Although the control of adult mosquitoes is emphasized with pyrethroids, more and more larval control is carried out with Bti formulation (Rydzanicz et al., 2000).

26.5.4.8 Spain

Organized mosquito control started in Spain in the early 1900s following the discovery of mosquitoes as vectors of malaria. Nevertheless, interest in mosquito studies ceased as soon as the disease was eradicated in this country in 1963 (Pletsch, 1965). Spain's economic development resulted in an improved standard of living. It was increasingly recognized by the public and by administrators that mosquito nuisance should be considered as a factor interfering with the development of the country. This applied especially to towns and cities close to mosquito breeding places and in areas where tourism was the most important source of income. The first mosquito control service (MCS) was created in 1982, in Roses Bay and Ter River, one of the tourist areas in Catalonia exposed to 34,000 ha of breeding sites. In the following years, three more services appeared under the direction of public administrations. In 1983, the MCS of the Baix Llobregat (Catalonia) was created, being responsible for 15 municipalities stretching south of Barcelona, including 25,000 ha of breeding sites. The MCS of Huelva in Andalusia was formed in 1985, covering 11 municipalities with more than 130,000 ha, and the MCS of the Ebro Delta (Catalonia) was formed in 1991, covering seven municipalities with 32,000 ha of breeding sites. *Oc. caspius*, *Oc. detritus*, and *Ae. vexans* are the most important species in natural breeding places while *Anopheles atroparvus* is the major species in rice fields. Methods used by the Spanish MCS are based on an integrated control strategy. Some of the control problems are similar across all the MCS, but there are also important local differences due to topography, the presence of different breeding habitats, and of course, different species. Natural parks are exclusively treated with Bti while the most important *Cx. pipiens* breeding places, septic tanks spread over residential areas, are treated with *L. sphaericus* or Pyriproxyfen.

26.5.4.9 Serbia

In Serbia, especially in the province of Vojvodina, a continuous mosquito surveillance and partially control programs have been conducted for the past 40 years. The flood plains of the rivers Danube, Sava, and Tisza provide ideal breeding conditions for floodwater mosquitoes such as *Ae. vexans*, *Oc. sticticus*, *Ae. cinereus*, and *Ae. rossicus* (Petrić et al., 1988). Oxbow swamps and deep depressions in meanders of the rivers are breeding sites of *An. messeae*, *Anopheles maculipennis* s.s., *Cx. modestus*, *Cx. pipiens* s.l., and *Cq. richiardii*. The stagnant and slowly flowing water on the alkaline and saline soils with a relatively high concentration of dissolved salts offers good conditions for *An. atroparvus*, *Oc. caspius*, and *Oc. dorsalis*. These species can also be found in the marches along the Adriatic coast. In the Danube flood plains, the most abundant species are *Ae. vexans* and *Oc. sticticus*, which appear in very large numbers 2 to 5 times per year, depending on the discharge pattern of the Danube.

Mosquito control in Vojvodina has been implemented since 1976 under the umbrella of the Province government with various degrees of treatment magnitudes and different achievements regarding effectiveness, environmental impact, and people protection (Zgomba and Petrić, 2008). The Faculty of Agriculture (University of Novi Sad) was the initiator and organizer of the control program until 1985, when insecticide purchase and pest control organizations were transferred to the local mosquito abatement district. The Faculty of Agriculture remained involved in education, monitoring, and the evaluation of insecticide properties, dosages, and the effectiveness of control measures.

The main goal is to provide the prerequisites for a biological mosquito control program based on the use of Bti in various formulations. Biological larval control is carried out in 19 out of 40 municipalities in Vojvodina. Six municipalities assert that they practice through other measures like terrain modification and public involvement/participation through distributed tablets with Bti or *L. sphaericus*. Other seasonal outbreaks of floodwater mosquitoes are suppressed by combined methods (Bti for larviciding and ultra-low volume application of pyrethroids for adulticiding) using air or ground equipment. GIS software serves as a database for the implementation and documentation of control activities.

REFERENCES

Adhami, J., Reiter, P., 1998. Introduction and establishment of *Aedes* (*Stegomyia*) *albopictus* Skuse (Dipetera: Culicidae) in Albania. J. Am. Mosq. Control Assoc. 14, 340–343.

Ali, A., 1981. *Bacillus thuringiensis* serovar *israelensis* (ABG-6108) against chironomids and some non-target aquatic invertebrates. J. Invertebr. Pathol. 38, 264–272.

Aly, C., 1985. Germination of *Bacillus thuringiensis* var. *israelensis* spores in the gut of *Aedes* larvae (Diptera: Culicidae). J. Invertebr. Pathol. 45, 1–8.

Angelini, R., Finarelli, A.C., Angelini, P., Po, C., Petropulacos, K., Silvi, G., Macini, P., Fortuna, C., Venturi, G., Magurano, F., Fiorentini, C., Marchi, A., Benedetti, E., Bucci, P., Boros, S., Romi, R., Majori, G., Ciufolini, M.G., Nicoletti, L., Rezza, G., Cassone, A., November 22, 2007. Chikungunya in north-eastern Italy: a summing up of the outbreak. Euro Surveill. 12 (11). euro-surveillance.org.

Barjac, H.D., 1978. A new variety of *Bacillus thuringiensis* very toxic for mosquitoes: *B. thuringiensis* var. *israelensis* serotype 14. C. R. Acad. Sci. Hebd. Seances. Acad. Sci. D. 286, 797–800.

Baumann, P.M., Clark, A., Baumann, L., Broadwell, A.H., 1991. *Bacillus sphaericus* as a mosquito pathogen: properties of the organism and its toxins. Microbiol. Rev. 55, 425–436.

Becker, N., 1992. Community participation in the operational use of microbial control agents in mosquito control programs. Bull. Soc. Vector Ecol. 17, 114–118.

Becker, N., 1997. Microbial control of mosquitoes: management of the Upper Rhine mosquito population as a model programme. Parasitol. Today 13, 485–487.

Becker, N., 2002. Sterilization of *Bacillus thuringiensis israelensis* products by gamma radiation. J. Am. Mosq. Control Assoc. 18, 57–62.

Becker, N., 2003. Ice granules containing endotoxins of microbial control agents for the control of mosquito larvae-a new application technique. J. Am. Mosq. Control Assoc. 19, 63–66.

Becker, N., 2008. Influence of climate change on mosquito development and mosquito-borne diseases in Europe. Parasitol Res. 103 (Suppl. 1), 19–28.

Becker, N., Ludwig, H.W., 1983. Mosquito control in West Germany. Bull. Soc. Vector Ecol. 8, 85–93.

Becker, N., Margalit, J., 1993. Use of *Bacillus thuringiensis israelensis* against mosquitoes and black flies. In: Entwistle, P.F., Cory, J.S., Baily, M.J., Higgs, S. (Eds.), *Bacillus thuringiensis* an Environmental Biopesticide: Theory and Practice, pp. 147–170.

Becker, N., Rettich, F., 1994. Protocol for the introduction of new *Bacillus thuringiensis israelensis* products into the routine mosquito control program in Germany. J. Am. Mosq. Control Assoc. 10, 527–533.

Becker, N., Djakaria, S., Kaiser, A., Zulhasril, O., Ludwig, H.W., 1991. Efficacy of a new tablet formulation of an asporogenous strain of *Bacillus thuringiensis israelensis* against larvae of *Aedes aegypti*. Bull. Soc. Vector Ecol. 16, 176–182.

Becker, N., Zgomba, M., Ludwig, M., Petric, D., Rettich, F., 1992. Factors influencing the activity of *Bacillus thuringiensis* var. *israelensis* treatments. J. Am. Mosq. Control Assoc. 8, 285–289.

Becker, N., Ludwig, M., Beck, M., Zgomba, M., 1993. The impact of environmental factors on the efficacy of *Bacillus sphaericus* against *Culex pipiens*. Bull. Soc. Vector Ecol. 18, 61–66.

Becker, N., Zgomba, M., Petric, D., Beck, M., Ludwig, M., 1995. Role of larval cadavers in recycling processes of *Bacillus sphaericus*. J. Am. Mosq. Control Assoc. 11, 329–334.

Becker, N., Petrić, D., Zgomba, M., Boase, C., Dahl, C., Lane, J., Kaiser, A., 2003. Mosquitoes and Their Control. Kluwer Academic/Plenum Publishers, New York. 498 pp.

Becker, N., Petric, D., Zgomba, M., Boase, C., Madon, M., Dahl, C., Kaiser, A., 2010. Mosquitoes and Their Control. Springer, Dordrecht, York (sec. edition), 577 pp.

Becker, N., Pluskota, B., Kaiser, A., Schaffner, F., 2012a. Exotic mosquitoes conquer the world. In: Mehlhorn, H. (Ed.), Arthropods as Vectors of Emerging Diseases. Springer, Dordrecht, pp. 31–60.

Becker, N., Geier, M., Balczun, C., Bradersen, U., Huber, K., Kiel, E.A., Lühken, R., Orendt, C., Plenge-Bönig, A., Rose, A., Schaub, G., Tannich, E., 2012b. Repeated introduction of *Aedes albopictus* into Germany, July to October 2012. Parasitol. Res. 112, 1787–1790.

Boisvert, M., 2005. Utilization of *Bacillus thuringiensis* var. *israelensis* (*Bti*)-based formulations for the biological control of mosquitoes in Canada. In: 6th Pacific Rim Conference on the Biotechnology of *Bacillus thuringiensis* and its Environmental Impact, Victoria BC, pp. 87–93.

Charles, J.F., Nielsen-LeRoux, C., 2000. Mosquitocidal bacterial toxins: diversity, mode of action and resistance phenomena. Mem. Inst. Oswaldo Cruz 95, 201–206.

Colbo, A.H., Undeen, A.H., 1980. Effect of *Bacillus thuringiensis* var. *israelensis* on non-target insects in stream trials for control of Simuliidae. Mosq. News 40, 368–371.

Davidson, E.W., Yousten, A.A., 1990. The mosquito larval toxin of *Bacillus sphaericus*. In: de Barjac, H., Sutherland, D. (Eds.), Bacterial Control of Mosquitoes and Black Flies: Biochemistry, Genetics and Applications of *Bacillus thuringiensis israelensis* and *Bacillus sphaericus*. Rutgers Univ. Press, New Brunswick, N J, pp. 237–255.

de Barjac, H., 1983. Bioassay Procedure for Samples of *Bacillus thuringiensis israelensis* Using IPS-82 Standard. WHO Report TDR/VED/SWG (5) (81.3), Geneva.

Delecluse, A., Barloy, F., Rosso, M.L., 1996. Les bacteries pathogenes des larves de dipteres: structure et spécificité des toxines. Ann. Inst. Pasteur/Actual. 7 (4), 217–231.

Delisle, E., Rousseau, C., Broche, B., Leparc-Goffart, I., L'Ambert, G., Cochet, A., Prat, C., Foulongne, V., Ferré, J.B., Catelinois, O., Flusin, O., Tchernonog, E., Moussion, I.E., Wiegandt, A., Septfons, A., Mendy, A., Moyano, M.B., Laporte, L., Maurel, J., Jourdain, F., Reynes, J., Paty, M.C., Golliot, F., 2015. Chikungunya outbreak in Montpellier, France, September to October 2014. Euro Surveill. 20 (17).

Des Rochers, B., Garcia, R., 1984. Evidence for persistence and recycling of *Bacillus sphaericus*. Mosq. News 44, 160–165.

Dulmage, H.T., Correa, J.A., Gallegos-morales, G., 1990. Potential for improved formulations of *Bacillus thuringiensis israelensis* through standardization and fermentation development. In: De Barjac, H., Sutherland, D. (Eds.), Bacterial Control of Mosquitoes and Black Flies: Biochemistry, Genetics and Applications of *Bacillus thuringiensis israelensis* and *Bacillus sphaericus*. Rutgers Univ Press, New Brunswick, NJ, pp. 16–44.

ECDC, 2008. Europe Faces Heightened Risk of Vector-Borne Disease Outbreaks Such as Chikungunya Fever. ECDC, Stockholm.

Federici, B., Lüthy, P., Ibarra, J.E., 1990. Parasporal body of *Bacillus thuringiensis israelensis*: structure, protein composition, and toxicity. In: Bacterial Control of Mosquitoes and Black Flies: Biochemistry, Genetics and Applications of *Bacillus thuringiensis israelensis* and *Bacillus sphaericus*. Rutgers Univ. Press, New Brunswick, NJ, pp. 45–65.

Federici, B.A., Park, H.W., Bideshi, D.K., Wirth, M.C., Johnson, J.J., Sakano, Y., Tang, M., 2007. Developing recombinant bacteria for control of mosquito larvae. J. Am. Mosq. Control Assoc. 23, 164–175.

Fillinger, U., Knols, B.G.J., Becker, N., 2003. Efficacy and efficiency of new *Bacillus thuringiensis* var. *israelensis* and *Bacillus sphaericus* formulations against afrotropical anophelines in western Kenya. Trop. Med. Int. Health 8 (1), 37–47.

Goldberg, L.H., Margalit, J., 1977. A bacterial spore demonstrating rapid larvicidal activity against *Anopheles sergenti*, *Uranotaenia unguiculata*, *Culex univittatus*, *Aedes aegypti* and *Culex pipiens*. Mosq. News 37, 355–358.

Hirsch, H., Becker, N., 2009. Cost-benefit analysis of mosquito control operations based on microbial control agents in the Upper Rhine Valley (Germany). Eur. Mosq. Bull. 27, 47–55.

Höfte, H., Whiteley, H.R., 1989. Insecticidal crystal proteins of *Bacillus thuringiensis*. Microbiol. Rev. 53, 242–255.

Jetten, T.H., Takken, W., 1994. Anophelism Without Malaria in Europe. A Review of the Ecology and Distribution of the Genus *Anopheles* in Europe. Wagening. Agric. Univ. 69 pp.

Kahindi, S.C., Midega, J.T., Mwangangi, J.M., Kibe, L.W., Nzovu, J., Luethy, P., Githure, J., Mbogo, C., 2008. Efficacy of Vectobac DT and Culinex combi against mosquito larvae in unused swimming pools in Malindi, Kenya. J. Am. Mosq. Control Assoc. 24, 538–542.

Krieg, A., 1986. *Bacillus thuringiensis*, ein mikrobielles Insektizid. Grundlagen und Anwendung. Parey Verlag. 191 pp.

Kroeger, A., Dehlinger, U., Burkhardt, G., Anaya, H., Becker, N., 1995. Community based dengue control in Columbia: people's knowledge and practice and the potential contribution of the biological larvicide *B. thuringiensis israelensis* (*Bacillus thuringiensis israelensis*). Trop. Med. Parasitol. 46, 241–246.

Lacey, L.A., 1990. Persistence and formulation of *Bacillus sphaericus*. In: Barjac de, H., Sutherland, D. (Eds.), Bacterial Control of Mosquitoes and Black Flies: Biochemistry, Genetics and Applications of *Bacillus thuringiensis israelensis* and *Bacillus sphaericus*. Rutgers Univ. Press, New Brunswick, NJ, pp. 284–294.

Lacey, L.A., 2007. *Bacillus thuringiensis* serovariety *israelensis* and *Bacillus sphaericus* for mosquito control. Bull. Am. Mosq. Control Assoc. 7, 133–163.

Lacey, L.A., Merritt, R.W., 2003. The safety of bacterial microbial agents used for black fly and mosquito control in aquatic environments. In: Hokkanen, H.M.T., Hajek, A.E. (Eds.), Environmental Impacts of Microbial Insecticides: Need and Methods for Risk Assessment. Kluwer Academic Publishers, Dordrecht, pp. 151–168.

Ludwig, M., Beck, M., Zgomba, M., Becker, N., 1994. The impact of water quality on the persistence of *Bacillus sphaericus*. Bull. Soc. Vector Ecol. 19, 43–48.

Lüthy, P., 2001. La lotta biologica contro le zanzare alle Bolle di Magadino. In: Patocchi, N. (Ed.), Contributo alla conoscenza delle Bolle di Magadino, pp. 139–145.

Mahilum, M.M., Ludwig, M., Madon, M.B., Becker, N., 2005. Evaluation of the present dengue situation and control strategies against *Aedes aegypti* in Cebu City, Philippines. J. Vector Ecol. 30, 277–283.

Margalit, J., Dean, D., 1985. The story of *Bacillus thuringiensis israelensis* (B.t.i.). J. Am. Mosq. Control Assoc. 1, 1–7.

Miura, T., Takahashi, R.M., Mulligan, F.S., 1980. Effects of the bacterial mosquito larvicide, *Bacillus thuringiensis* serotype H-14 on selected aquatic organisms. Mosq. News 40, 619–622.

Mulla, M.S., 1990. Activity, field efficacy, and the use of *Bacillus thuringiensis israelensis* against mosquitoes. In: de Barjac, H., Sutherland, D.J. (Eds.), Bacterial Control of Mosquitoes and Black Flies: Biochemistry, Genetics and Applications of *Bacillus thuringiensis israelensis* and *Bacillus sphaericus*. Rutgers Univ. Press, New Brunswick, NJ, pp. 134–160.

Mulla, M.S., Darwazeh, H.A., Zgomba, M., 1990. Effect of some environmental factors on the efficacy of *Bacillus sphaericus* 2362 and *Bacillus thuringiensis* (H-14) against mosquitoes. Bull. Soc. Vector Ecol. 15, 166–175.

Nielsen-LeRoux, C., Charles, J.F., Thiery, I., Georghiou, G.P., 1995. Resistance in a laboratory population of *Culex quinquefasciatus* (Diptera: Culicidae) to *Bacillus sphaericus* binary toxin is due to a change in the receptor on midgut brush-border membranes. Eur. J. Biochem. 228, 206–210.

Park, H.W., Sabrina, R., Hayes, S.R., Stout, G.M., Day-Hall, G., Latham, M.D., John, P., Hunter, J.P., 2009. Identification of two mosquitocidal *Bacillus cereus* strains showing different host ranges. J. Invertebr. Pathol. 100, 54–56.

Petrić, D., Zgomba, M.I., Srdić, Ž., 1988. Mosquito species (Dip.: Culicidae) abundance in the Danube inundation area of Vojvodina (YU). In: Proc. XVIII International Congress of Entomology, pp. 286–287 (Vancouver, Canada).

Pletsch, D., 1965. Informe sobre una misión efectuada en España en Septiembre-Noviembre de 1963 destinada a la certificación de la erradicación del paludismo. Rev. Sanid. Hig. Publica. 7,8,9, 309–335.

Priest, F.G., 1992. Biological control of mosquitoes and other biting flies by *Bacillus sphaericus* and *Bacillus thuringiensis*. J. Appl. Bacteriol. 72, 357–369.

Puchi, N.D., 2005. Factors affecting the efficiency and persistance of *Bacillus thuringiensis* var. *israelensis* on *Anopheles aquasalis* Curry (Diptera: Culicidae), a malaria vector in Venezuela. Entomotropica 20, 213–233.

Ragoonanansingh, R.N., Njunwa, K.J., Curtis, C.F., Becker, N., 1992. A field study of *Bacillus sphaericus* for the control of culicine and anopheline mosquito larvae in Tanzania. Bull. Soc. Vector Ecol. 17 (1), 45–50.

Russell, T.L., Brown, M.D., Purdie, D.M., Ryan, P.A., Brian, H., Kay, B.H., 2003. Efficacy of VectoBac (*Bacillus thuringiensis* variety *israelensis*) formulations for mosquito control in Australia. J. Econ. Entomol. 96, 1786–1791.

Rydzanicz, K., Lonc, E., 2000. Proby integrowanej kontroli komarow na terenie miasta i okolic Wroclawia. Biul. PSPDDD 1, 8–10.

Rydzanicz, K., Lonc, E., Kiewra, D., Dechant, P., Krause, S., Becker, N., 2009. Evaluation of two application techniques of three microbial larvicide formulations against *Culex p. pipiens* in irrigation fields in Wroclaw. Pol. J. Am. Mosq. Control Assoc. 25, 140–148.

Santa-Olalla Peralta, P., Vazquez-Torres, M.C., Latorre-Fandos, E., Mairal-Claver, P., Cortina-Solano, P., Puy-Azon, A., Adiego Sancho, B., Leitmeyer, K., Lucientes-Curdi, J., Sierra-Moros, M.J., October 14, 2010. First autochthonous malaria case due to *Plasmodium vivax* since eradication, Spain. Euro. Surveill. 15 (41), 19684.

Silva-Filha, M.H., Regis, L., Oliveira, C.M.F., Furtado, A.F., 2001. Impact of a 26-month *Bacillus sphaericus* trial on the preimaginal density of *Culex quinquefasciatus* in an urban area of Recife. Braz. J. Am. Mosq. Control Assoc. 17, 45–50.

Sinegre, G., Babino, M., Quermel, J.M., Gavon, B., 1994. First field occurrence of *Culex pipiens* resistance to *Bacillus sphaericus* in southern France. In: Abstr VIIIth Europ Meet Soci Vector Ecol, Barcelona, p. 17.

Singer, S., 1973. Insecticidal activity of recent bacterial isolates and their toxins against mosquito larvae. Nature 244, 110–111.

Skovmand, O., Becker, N., 2000. Bioassays of *Bacillus thuringiensis* subsp. *israelensis*. In: Navon, A., Ascher, K.R.S. (Eds.), Bioassays of Entomopathogenic Microbes and Nematodes. CABI Publishing, Oxon, UK, pp. 41–47.

Su, T.S., 2008. Evaluation of water-soluble pouches of *Bacillus sphaericus* applied as prehatch treatment against *Culex* mosquitoes in simulated catch basins. J. Am. Mosq. Control Assoc. 24, 54–60.

Weiser, J., 1984. A mosquito-virulent *Bacillus sphaericus* in adult *Simulium damnosum* from Northern Nigeria. Zbl. Mikrobiol. 139, 57–60.

WHO, 1999. *Bacillus thuringiensis*, Environmental Health Criteria. 217 pp.

WHO, 2004. Review of VectoBac WG, PermaNet and Gokilaht 5EC. WHO/CDS/WHOPES/2004.8, Geneva.

Wickramasinghe, B., Mendis, C.L., 1980. *Bacillus sphaericus* spores from Sri Lanka demonstrating rapid larvicidal activity on *Culex quinquefasciatus*. Mosq. News 40, 387–389.

Wirth, M.C., 2010. Mosquito resistance to bacterial larvicidal toxins. Open J. Toxinol. 3, 126–140.

Wirth, M.C., Jiannino, J.A., Federici, B.A., Walton, W.E., 2004. Synergy between toxins of *Bacillus thuringiensis* subsp. *israelensis* and *Bacillus sphaericus*. J. Med. Entomol. 41, 935–941.

Zahiri, N.S., Su, T., Mulla, M.S., 2002. Strategies for the management of resistance in mosquitoes to the microbial control agent *Bacillus sphaericus*. J. Med. Entomol. 39, 513–520.

Zgomba, M., Petrić, D., 2008. Risk assessment and management of mosquito-born diseases in the European region. In: Robinson, W.H., Bajomi, D. (Eds.), Proceedings of the 6th International Conference on Urban Pest. pp. 29–39, ISBN: 978-963-06-5326-8.

Chapter 27

Microbial Control of Pest and Vector Mosquitoes in North America North of Mexico

T. Su

West Valley Mosquito and Vector Control District, Ontario, CA, United States

27.1 INTRODUCTION

The need for mosquito control by biorational approaches, particularly a reliance on bacterial larvicides based on *Bacillus thuringiensis* (Bt) subsp. *israelensis* (Bti) and *Lysinibacillus sphaericus*, in North America has intensified for the following reasons. Besides being nuisances, some of the over 350 mosquito species in North America pose a significant threat of transmitting disease-causing agents, both indigenous and exotic, to humans and animals. These include Saint Louis encephalitis, Western equine encephalomyelitis, Eastern equine encephalitis, West Nile fever and encephalitis, La Crosse encephalitis, Dengue, Chikungunya fever, and most recently the Zika virus infection and others. Among these, the incidence of certain mosquito-borne arboviruses has been on the rise. Increasing human and animal population movements and global freight exchange facilitate the spread of mosquitoes and mosquito-borne pathogens. Furthermore, fast demographic growth, economic development, and subsequent environmental impact promote the need for sustainable mosquito control. Finally, strict governmental regulations and high environmental sensitivities demand ecologically sound mosquito control based on biorational approaches, especially the use of bacterial larvicides for their high efficacy, target specificity, and environmental and nontarget safety profile. The solids and liquids from the fermentations of Bti and *L. sphaericus* can be formulated alone or in combination with each other to optimize efficacy against mosquitoes and mitigate insecticide resistance development.

27.2 DEVELOPMENTAL RESEARCH FOR USE OF *BACILLUS THURINGIENSIS* SUBSP. *ISRAELENSIS* FOR CONTROL OF MOSQUITO LARVAE

27.2.1 Formulations, Applications, and Efficacy

Many unique properties make Bti an excellent bacterial larvicide against mosquitoes, black flies, and other nematoceran species. Since its discovery, formulations of this bacterium have increasingly become the predominant nonchemical means to combat mosquito larvae by local governmental vector control agencies, private pest control operations, and, to a lesser extent, even some home owners. The efficacy of Bti formulations has been demonstrated in a wide variety of habitats against a multitude of mosquito species, either being used alone or as part of an integrated mosquito management with other biorational approaches (Lacey and Lacey, 1990; Lacey, 2007). The most comprehensive research and development on the fermentation, formulation, and application of Bti has occurred in North America since 1980. Leading industries have successfully developed a wide variety of formulations, ranging from onsite formulation technical powders, fast-release aqueous suspension concentrates, controlled release granules, and water soluble pouches (WSPs) to long-lasting dunks and briquets. These formulations are customized for different habitats, application methods, and desired residual efficacy. Some of the formulations are Organic Materials Review Institute (OMRI) listed for application on organic farms.

Microbial Control of Insect and Mite Pests. http://dx.doi.org/10.1016/B978-0-12-803527-6.00027-5

Products based on Bti are suitable for virtually any habitat associated with natural, agricultural, and urban environments, although efficacy can be reduced in the habitats with high organic matter and very high larval densities, regardless of formulations (Mulla, 1990). Applications by hand-tossing WSPs and briquets, ground or aerially applying aqueous suspensions and water dispersible granules (WDGs), or broadcasting dry granules are widely used to combat different species of mosquitoes. Along with the advancement of formulation technology, diverse formulations for different purposes and by various application methods are readily available.

Laboratory and field evaluation on Bti products started in the early 1980s, and early studies were reviewed by Mulla (1990). The results of field trials since 1980 are summarized in Table 27.1. Early attempts were made to evaluate the field efficacy of Bti (Mulligan et al., 1980), where the residual action of Bti technical material was short-lived in the catch basins. However, the treatment of pasture plots resulted in the effective control of *Aedes* spp. and *Culex tarsalis* at 1 kg/ha. An aerial application on a duck club pond at the same rate produced almost complete control of *Cx. tarsalis*. Purcell (1981) conducted laboratory and field studies on the activity and efficacy of Bti wettable powder against *Aedes taeniorhynchus* and realized that salt marsh mud reduced Bti activity, and high-level larval mortality was achieved at 4.5 International Toxic Unit (ITU)/mL. Moderate to high larval mortality was noted after treatment with Bti wettable powder against *Culex peus* and *Culex pipiens* in log ponds (Eldridge and Callicrate, 1982). Mulla et al. (1982) evaluated four experimental formulations of Bti in the laboratory and field against larvae of the floodwater mosquitoes *Aedes nigromaculis* and *Psorophora columbiae* with ground and aerial application methods. Both species were similar in susceptibility, and a high rate was needed for the same level of control when applied by aerial application rather than by ground application. The field efficacy of wettable powder of Bti was extended by formulating it into floating pellets and briquets against *Culex quinquefasciatus* by Lacey et al. (1984). One notable example is the use of the Beecomist ultra-low volume generator for application of Bti liquid formulations for control of *Anopheles quadrimaculatus* larvae in rice fields in Louisiana. This method provided effective control at very low application rates (Table 27.1; Sandoski et al., 1985). The small droplets (<80 μm) produced with the device penetrated dense foliage and remained in the feeding zone of *An. quadrimaculatus* larvae longer than larger droplets or granules that sank after application. Mulla et al. (1985) conducted field trials on granular formulations against *Ps. columbiae* breeding on irrigated pastures; excellent control of larvae was obtained at various rates, depending on the potency and type of the formulations. Variable field efficacy among granules, flowable concentrates, and briquets of Bti was obtained against *Aedes vexans* and *Culex* spp. larvae

by Berry et al. (1987). Among the wide variety of Bti formulations, WDGs possess some unique desired properties of high potency, low minimum effective doses, and ease of use as dry, conventional, or low volume spray. Excellent control was obtained when applied at a fairly low rate against *Culex* spp. in microcosms (Su and Mulla, 1999a). During the screening and evaluation of new formulations, Su and Mulla (1999b) noted that when achieving high levels of mosquito control, productivity of some algal species was noticeably suppressed and that certain water quality parameters were enhanced by the suppression of primary productivity by Bti and *L. sphaericus*. Products such as VectoBac CG and VectoBac 12AS based on Bti performed fairly well under highly challenging conditions associated with underground storm drain systems in urban areas, which are considered significant breeding sources of *Culex* spp. (Su et al., 2003). Longevity was extended when the products were applied at 3 times the maximum label rate (Anderson et al., 2011). In contrast, Mulla et al. (1993) found that excessive dosages at 5 and 10 times the minimum effective rate (1.0 and 2.0 kg/ha) of fast-release formulations such as VectoBac 12AS did not significantly extend the larvicidal longevity against *Culex* spp. in outdoor microcosms. Products of Bti were also highly effective against *Culiseta* spp. When Bti was applied at 15 ppm to shaded tires for the control of *Culiseta incidens*, great initial and residual efficacy was observed (Kramer, 1990). Lawler et al. (1999) evaluated the efficacy of Bti granules (VectoBac G) against larval *Ae. taeniorhynchus*. Good overall efficacy was indicated, but occasional failures occurred.

27.2.2 Factors Influencing Larvicidal Activity and Efficacy

There are a number of biological, technological, and environmental factors that impact the larvicidal activity and efficacy of Bti against mosquitoes. For example, different strains of Bti such as AM65-52, BMP144, and SA3A are expected to perform differently in toxin production during culturing. Fermentation technology, including the media used and the culture conditions, results in significant differences in toxin production. Formulation technology is another key factor for preserving the active ingredients for the longest shelf life possible and delivering the active ingredients to the feeding zone at a preferred rate. Studies indicated that formulation technology, rather than the particle behavior, is the determining factor for active ingredient distribution in aquatic environments (Clark et al., 2007). Often, the type of formulations has a major impact on efficacy. Great efforts have been made to deliver the active ingredients to the feeding zone of various genera of mosquitoes, as Bti particles naturally only exist at the surface layer of water for a short period of time and then reach equilibrium throughout the entire water body after being released from the formulations (Mulla, 1990; Nguyen

TABLE 27.1 Examples of *Bacillus thuringiensis* Subsp. *israelensis* Used for Control of Mosquito Larvae in North America

Targeted Mosquito Species	Habitats	Formulations	Application Rate and Efficacy	References
Aedes spp., *Culex tarsalis*	Pastures, duck club ponds, catch basins	Technical materials	Effective control of *Aedes* spp. and *Cx. tarsalis* at 1 kg/ha in pasture, 99% reduction of *Cx. tarsalis* at the same rate in duck club ponds, short-lived residual action in catch basins	Mulligan et al. (1980)
Aedes taeniorhynchus	Salt marshes	WP[a](3000 ITU/mg)	99% larval mortality at 4.5 ITU/mL[b]	Purcell (1981)
Culex peus, Culex pipiens	Log ponds	WP (2000 ITU/mg)	73–99% larval mortality for 1 week at 0.4–1.63 kg/ha	Eldridge and Callicrate (1982)
Psorophora columbiae, Aedes nigromaculis	Irrigated fields	WP and water dispersible concentrates (400–6000 ITU/mg)	83–100% larval reduction at 0.44–1.12 kg/ha, aerial application 2–3-fold less effective	Mulla et al. (1982)
Culex quinquefasciatus	Organically enriched containers (8–15 L)	WP, floating pellets (0.17 g WP/pellet)	Complete control by WP at 1 mg WP/L, limited control for 6 weeks by 1–2 pellets/container	Lacey et al. (1984)
Anopheles quadrimaculatus	Rice fields	FC	71–98% control for <7 d at 0.07–0.54 L/ha	Sandoski et al. (1985)
Ps. Columbiae	Flooded pastures	Corncob granules (100–300 ITU/mg), aqueous suspension (600 ITU/mg)	64–100% reduction by corncob granules at 2.8–11.2 kg/ha and 87–98% by suspension at 0.28–1.12 kg/ha at 24 h	Mulla et al. (1985)
Aedes trivittatus, Cx. spp.	Artificial pools, ditches, culverts, large lagoons	Granules, FC, and briquet	90–100% control by granules and flowable concentrate but 12–76% by briquet, no residual efficacy noted	Berry et al. (1987)
Cx. quinquefasciatus, Culex stigmatosoma, Culex tarsalis	Organically enriched microcosms	VectoBac 12AS (1200 ITU/mg)	>65% for up to 28 d at 1.0–2.0 kg/ha	Mulla et al. (1993)
Ae. taeniorhynchus	Mangrove swamps	VectoBac G (200 ITU/mg)	90% control for 14 d at 5.61 kg/ha	Lawler et al. (1999)
Cx. quinquefasciatus, Cx. stigmatosoma, Cx. tarsalis	Organically enriched microcosms	VectoBac WDG (3000 ITU/mg)	85% control for 14 d at 0.30–1.23 kg/ha	Nguyen et al. (1999a)
Cx. quinquefasciatus, Cx. stigmatosoma, Cx. tarsalis	Organically enriched microcosms	VectoBac G (200 ITU/mg)	73–93% reduction for 7 d at 5.92–11.84 kg/ha	Nguyen et al. (1999b)
Cx. quinquefasciatus, Cx. stigmatosoma, Cx. tarsalis	Organically enriched microcosms	VectoBac WDG (4000 ITU/mg)	80% control for 7–12 d at 1.2–3.0 kg/ha	Su and Mulla (1999a)
Cx. quinquefasciatus, Cx. stigmatosoma, Cx. tarsalis	Organically enriched microcosms	VectoBac G (200 ITU/mg) and VectoBac WDG (4000 ITU/mg)	85% control for 21 d by VectoBac G at 48.1 kg/ha, 65% control for 17 d by VectoBac WDG at 0.6 kg/ha	Su and Mulla (1999b)
Culiseta incidens	Used tires	VectoBac AS (600 ITU/mg)	>90% efficacy for 2 weeks and >50% for 4 weeks at 15 ppm	Kramer (1990)
Ochlerotatus squamiger	Tide pools	Sand granules made with VectoBac TP	98.4–99.4% larval mortality at 24–72 h at 22.39 kg/ha	Christiansen et al. (2004)
Cx. pipiens, Culex restuans, and Aedes japonicus	Underground storm drains	VectoBac CG, VectoBac 12AS	Almost 100% control for 1 week by VectoBac CG at 1.8 g or VectoBac 12AS at 0.193 mL per catch basin, efficacy extended to 2–4 weeks at 3 times doses	Anderson et al. (2011)

[a]AS, aqueous suspension; FC, flowable concentrate; TP, technical powder; WDG, water dispersible granules; WP, wettable powder.
[b]ITU/mg, International Toxic Unit/mg.

et al., 1999a,b). Innovative formulation methods have been developed to overcome the short persistence of Bti products. Briquets may result in more prolonged control than liquid formulations, as they have greater persistence through the slow-release process. Special challenges exist when formulating the solids from fermentation, which consist of culture media, spores, crystals, and other components, as compared with formulating pure compounds.

Biological factors associated with target mosquitoes such as species, feeding behavior, instars, and density also significantly influence the larvicidal activity and efficacy of the Bti products. For instance, *Culex* spp. forage through the water column even though they spend a great amount of time just below the water surface. *Aedes* spp. are considered substrate/bottom feeders, while *Anopheles* spp. are primarily surface feeders. Additionally, the filtration rate varies among mosquito genera and species. The mosquitoes with different feeding patterns respond to Bti treatments differently. Younger larvae with more active feeding activity tend to ingest more active ingredients within a given time under the same conditions as compared with older larvae; hence a higher larval mortality can be expected. In all species studied, increasing age of the larvae resulted in reduced susceptibility in mosquitoes (Mulla et al., 1982). The age-dependent differences to Bti from second to fourth instars can vary up to 10-fold (Mulla, 1990). As all larvae that dwell in the same aquatic habitat share the available toxins, larval densities impact activity and efficacy. For example, it is estimated that 1.5–2-fold more inoculum would be needed against 50–100 larvae/dip than 5–20 larvae/dip to yield the same mortality (Mulla, 1990).

Water quality and water temperature in mosquito habitats are among the most important environmental factors that govern the outcome of Bti treatment. Among the water quality aspects, organic matters not only bind the Bti particles and make them unavailable to mosquito larvae but they also provide mosquito larvae with food sources in competition with Bti toxins (Mulla, 1990). However, Bti still provides operationally acceptable levels of control against *Cx. peus* and *Cx. pipiens* in log ponds (Eldridge and Callicrate, 1982) or urban storm drains (Anderson et al., 2011) with poor water quality. Water temperature is an important environmental factor that not only impacts the release process of the active ingredients from the formulations but also controls the feeding rate and therefore the ingestion of the Bti particles. The data from the feeding and susceptibility experiments suggest that decreased efficacy of Bti at low temperatures may occur because the rate of larval feeding decreases (Walker, 1995). The similar phenomenon was observed by Christiansen et al. (2004). Laboratory bioassays were conducted to determine the efficacy of VectoBac TP at different temperatures against the salt marsh mosquito *Ochlerotatus squamiger*. Bioassays on late third and early fourth stage larvae, 72 h posttreatment and at 14°C, produced an LD_{90} of 0.223 ppm, whereas more

than double this dose was required to produce a similar mortality at 6°C. It seems that Bti can tolerate certain pH ranges. No effect of pH in the range of 6.3–8.6 (27°C) on the efficacy of VectoBac AS in the bioassay against third instar larvae of *Cx. quinquefasciatus* was observed (Floore et al., 1987). Other factors such as solar irradiation also adversely impact the larvicidal efficacy of Bti (Kramer, 1990).

27.2.3 Potential for Resistance and Its Management

Generally, the risk of resistance development to wild type Bti is very low (Su, 2016). Reports on laboratory selection and field occurrence of resistance are rare. Multiple attempts to select resistance in laboratory colonies of *Cx. quinquefasciatus* or *Aedes aegypti* for various generations only resulted in low level and unstable resistance (Vasquez-Gomez, 1983; Goldman et al., 1986). Susceptibility baseline to Bti varied within 3–4-fold among 31 populations of *Cx. pipiens* in California (Wirth et al., 2001). One report from New York showed a low-level resistance in a wild population of *Cx. pipiens* (Paul et al., 2005); however, more follow-ups are necessary to verify the one time sampling and bioassay.

Four synergistic endotoxins, including Cry4A, Cry4B, Cry11A, and Cyt1A, are produced during the sporulation of Bti. Exposures to individual toxins are conducive to resistance development, where *Cx. quinquefasciatus* developed 91–900-fold resistance to individual toxins in the absence of Cyt1A toxin (Georghiou and Wirth, 1997). Cross-resistance occurs among the Cry toxins of Bti (Wirth and Georghiou, 1997). Cyt1A itself from Bti does not possess significant larvicidal activity, but it plays a critical role in overcoming, preventing, and delaying resistance development to Cry toxins in Bti (Wirth et al., 1997; 2005b; Pérez et al., 2007). It was discovered that the mosquitocidal toxin (Mtx) from some *L. sphaericus* strains not only enhances the larvicidal activity of Bti Cry toxins but also mitigates resistance development to Cry toxins (Wirth et al., 2007, 2013). There is a lack of cross-resistance to wild types of other Bt subspecies, such as Bt subsp. *jegathesan*, Bt subsp. *kyushuensis*, and Bt subsp. *fukuokuensis* in *Cx. quinquefasciatus*, which are resistant to individual toxins from Bti (Wirth, 2010). From the perspective of resistance prevention in target species based on the findings on the importance of Cyt1A toxin, it is considered inadvisable to express only Cry toxins in transgenic microbial organisms or algae for mosquito control.

27.2.4 Safety to Nontarget Organisms

Overall, the findings indicate that Bti is specific to mosquitoes, black flies, some chironomid midge larvae, and some other nematoceran insects. Profound safety profiles have been well established for other invertebrates, fish, mammals, and the environment, as comprehensively reviewed by Lacey and Mulla (1990), Siegel and Shadduck (1990), Lacey and Siegel (2000), Lacey and Merritt (2003), and Siegel (2012). The principal reasons for the benign environmental and nontarget profile of Bti are attributable to the binding site specificity and secondarily its short persistence and lack of recycling as well as the requirement of a high gut pH for protoxins to be activated. Short persistence is confirmed by a quick resurgence of young larvae after initial treatment (Mulla et al., 1985; Mulla, 1990). Detailed laboratory evaluation (Wipfli and Merritt, 1994) and field studies in Florida (Purcell, 1981), New York (Molloy, 1992) and southcentral Minnesota (Hershey et al., 1995) all indicated the lack of negative effects of larvicidal treatment on the richness and biomass of nontarget invertebrate groups. However, a detrimental effect of Bti in streams and rivers can occur for specialist predatory species through the loss of prey. Generalist predators, on the other hand, may be least affected if alternative prey are abundant (Lacey and Merritt, 2003).

Generally, the susceptibility of chironomid midge larvae to Bti is much lower than that of mosquitoes. Early field studies demonstrated a greater safety margin to this group of benthic dwellers (Miura et al., 1980). However, chironomid midge larvae can be one group of the aquatic organisms that are adversely impacted by Bti if high dosages are applied, which sometimes can be within the dose range for mosquito control. Species in the subfamily Chironominae are apparently the most susceptible to direct toxicity, relating perhaps to their feeding mode, particularly ones with filter feeding mechanisms (Pont et al., 1999). As in mosquito larvae, younger midge larvae are more susceptible than older ones. Other factors such as water temperature, the presence of algae and vegetation, organic pollution, and larval densities also affect some of the potential negative impacts to chironomid larvae (Charbonneau et al., 1994). Short-term studies on nontarget effects may not be adequate to detect potential long-term impacts from pesticide use (Hershey et al., 1995). A long-term study by Hershey et al. (1998) did indicate adverse impacts on chironomid midge populations in wetlands as a result of disruption of the invertebrate food web. On the other hand, an elevated concentration of Bti can be used as a tool to control midges in places with high midge populations (Ali, 1980; Rodcharoen et al., 1991) without an adverse impact on other nontargets (Ali, 1980). Lawler et al. (1999) evaluated the efficacy and nontarget effects of Bti to control larval *Ae. taeniorhynchus* in mangrove swamps on Sanibel Island, FL. No detectable mortality of amphipods or flying insects was attributed to the application of Bti (VectoBac G at 5.6 kg/ha). Lawler et al. (2000) tested the nontarget effects of a combined formulation (Duplex) of Bti and methoprene against immature mosquitoes *Ae. dorsalis* and water boatmen (*Trichocorixa reticulata*). There were no detectable effects of either pesticide on the survival or maturation of the nontarget species. Because of the excellent nontarget safety profile,

Bti products are compatible with other control agents such as freshwater crustacean tadpole shrimp (TPS) *Triops newberryi* used as a mosquito predator (Su and Mulla, 2005). The exposure of TPS to VectoBac TP (7000 ITU/mg) at 3 ppm on a weekly basis in the laboratory or 11.2–22.4 kg/ha of VectoBac G (200 ITU/mg) in the field did not have significant adverse effects on growth, longevity, and fecundity, which indicates the compatibility of TPS with Bti and that they can be used jointly in practical mosquito control programs.

27.3 DEVELOPMENTAL RESEARCH FOR USE OF *LYSINIBACILLUS SPHAERICUS* FOR CONTROL OF MOSQUITO LARVAE

27.3.1 Formulations, Application, and Efficacy

Various granular formulations or WSPs and their efficacy are listed in Table 27.2. Formulations that contain *L. sphaericus* alone are less popular because of concerns of resistance development, even though high efficacy has been well documented against *Culex* species breeding in habitats with high organic content, such as dairy wastewater lagoons and septic ditches. *Lysinibacillus sphaericus* has been shown to persist longer than Bti in polluted habitats and, under certain circumstances, can recycle in larval cadavers (Hertlein et al., 1979; Davidson et al., 1984; Lacey, 1990, 2007; Correa and Yousten, 1995). Several formulations are OMRI listed for application on organic farms.

The earliest field evaluations on *L. sphaericus* preparations were reported in the late 1970s (Ramoska et al., 1978; Mulligan et al., 1978), followed by more studies on various technical materials and later commercial formulations (Table 27.2). A comprehensive review on early field evaluations of *L. sphaericus* preparations was made by Yap (1990). Some highlights of early studies were mentioned in this chapter to reflect the continuous history. Mulligan et al. (1980) evaluated strain 1593-4 in urban catch basins and realized that the active ingredient settled to the bottom of the habitats where prolonged residual activity occurred. The next generation of larvae, however, was not affected; apparently, they did not feed from the bottom. Lacey et al. (1984) compared the efficacy of *L. sphaericus* powder, floating pellets, and briquets in buckets containing water and organic enrichment against *Cx. quinquefasciatus*. Operationally significant control was achieved under field conditions. Mulla et al. (1984a) evaluated unformulated materials such as primary powder and cream of strain 2362 (IF-117) against *Culex* spp. in organically enriched earthen ponds. Both preparations yielded a high initial control and a long residual efficacy at fairly low rates in the ponds without or with low levels of organic enrichment. Spores of both preparations settled quickly and accumulated in the mud out of the reach of feeding mosquito larvae. At the same time,

TABLE 27.2 Examples of *Lysinibacillus sphaericus* Used for Control of Mosquito Larvae in North America

Targeted Mosquito Species	Habitats	Formulations	Application Rate and Efficacy	References
Culex nigripalpus	Roadside ditches	SSII-1, 1404-9, 1593-4 culture	89% control at 48 h at $3.1-19 \times 10^{-4}$ cell/mL	Ramoska et al. (1978)
Culex tarsalis, Culex quinquefasciatus	Ponds, pastures, dairy drains	1593-4 lyophilized powder and wettable powder	90–100% control by LP at 0.67 kg/ha and 40–60% by WP at 0.84–2.24 kg/ha at 48 h	Mulligan et al. (1978)
Culex spp.	Catch basins	PP strain 1593	95–100% for 1–4 weeks at 500 ppm	Mulligan et al. (1980)
Cx. tarsalis	Small field plots	PP strains 1593 and 2362	83–100% for 2 d and 94–100% for 4 d at 0.122 and 0.244 kg/ha	Davidson et al. (1984)
Cx. quinquefasciatus	Organically enriched 15-L containers	Acetone precipitated spore powder, floating pellets (0.21 mg powder/pellet)	Complete control by powder at 1 mg/L, good to excellent control for 8 weeks by 2 pellets/container	Lacey et al. (1984)
Culex spp.	Organically enriched earthen ponds	PP and cream of strain 2362 (IF-117)	90% initial control and lasts about 3 weeks at no and low pollution at 0.22 kg/ha	Mulla et al. (1984a)
Culex peus, Cx. tarsalis,	Field mesocosms	PP of strains 2362 (IF-97) and 1593 (IF-94)	82% reduction in *Cx. peus* and 96% reduction in *Cx. tarsalis* for up to 14 d at 0.12 and 0.22 kg/ha	Mulla et al. (1984b)
Aedes melanimon, Ps. columbiae	Irrigated pastures	PP of strains 2362 and 1593	66–99% reduction at the rates of 0.06–0.56 kg/ha against *Ps. columbiae*, 4–94% at 0.11–0.56 kg/ha against *Ae. melanimon* at 48 h	Mulla et al. (1985)
Anopheles quadrimaculatus	Rice fields	FC	71–82% reduction of *An. quadrimaculatus* at 48 h at 0.58–1.17 L/ha, 50–98% reduction for *Ps. columbiae* at 0.29–0.58 L/ha	Lacey et al. (1986)
Psorophora columbiae				
Culex pipiens, Aedes trivittatus	Natural sites and artificial pools	Granular formulations	100% reduction of *Cx. pipiens*, 84–98% reduction of a mixed population of *Ae. trivittatus* and *Cx. pipiens* for 96 h at 2.78–8.42 kg/ha	Berry et al. (1987)
Ps. columbiae and Aedes nigromaculis	Irrigated fields	PP, liquid formulation (BSP-2), corncob granules of strain 2362	83–100% control of both species at 0.28–0.56 kg/ha by PP at 48 h; 40–100% control against the same species at 0.56–2.23 kg/ha by BSP-2; 91% and 98% reduction of *Ae. nigromaculis* by high spore count granules (1.5×10^9 spores/g) at 2.8 and 5.6 kg/ha, complete control by lower spore count granules (7.6×10^8 spores/g) at 11.2 kg/ha but poor results at 5.6 kg/ha	Mulla et al. (1988a)
Culex spp.	Dairy wastewater lagoons	PP, FC (BSP-2), and granules	Short-term control of *Culex* spp. by PP at 0.26 and 0.56 kg/ha, control and persistence increased at 1.12–4.48 kg/ha, almost 100% control by 1.12 and 2.24 kg/ha for 4 weeks and 99% control at 4.48 kg/ha for 7 weeks or longer; complete initial and persistent control by BSP-2 at 2.24–5.6 kg/ha for 14–21 d; >80% control by granules at 2.8–22.4 kg/ha for 14–21 d	Mulla et al. (1988b)

Culex spp., Anopheles spp.	Rice fields, tires, woodland pools	FC, granules, pellets, and briquets	93–100% reduction of Culex spp. by FC (12.8% PP) at 0.25 kg/ha;	Lacey et al. (1988)
			97% and 99% reduction of Cx. quinquefasciatus by granules (5% PP) at 10 kg/ha in polluted tanks and at 2.5 kg/ha to sod-lined potholes; 68% and 92–100% reduction of An. and Ps. columbiae at 5 kg/ha in rice fields	
			Elimination of Culex restuans by sustained-release pellets (30% PP) in small woodland pools at 4 pellets/pool for >8 d; 88–95% reduction of Cx. quinquefasciatus by sustained-release briquets (5% PP) at 1/2 briquet/1.8 m^2 sod-lined potholes for up to 2 weeks in open sunlight	
Culex stigmatosoma	Dairy wastewater lagoons	BSP-2 and ABG-6184	About 90% reduction for up to 4 weeks by BSP-2 at 4.48 kg/ha and ABG-6184 at 2.24 kg/ha	Matanmi et al. (1990)
Culiseta incidens	Waste tires	Liquid formulation of strain 2362 (300 ITU/mg)	20–100% reduction for 10 weeks at 3.75–15 ppm	Kramer (1990)
Culex spp., Ps. columbiae	Sod-lined pools	Sustained release pellets	80–100% control for 84 d against Culex spp., elimination of breeding when applied as prehatch treatment against Ps. columbiae	Lord (1991)
Cx. quinquefasciatus, Cx. stigmatosoma, Cx. tarsalis	Organically enriched microcosms	VectoLex WDG, VectoLex CG	Approximately 60–80% reduction for 28 d at 0.05–0.10 kg/ha for VectoLex CG and 1.6–3.0 kg/ha for VectoLex CG	Nguyen et al. (1999a)
Cx. quinquefasciatus, Cx. stigmatosoma, Cx. tarsalis	Organically enriched microcosms	VectoLex WDG (350 ITU/mg)	90% or greater control for 25 d at 1.2–3.0 kg/ha	Su and Mulla (1999a)
Cx. quinquefasciatus, Cx. stigmatosoma, Cx. tarsalis	Organically enriched microcosms	VectoLex WDG (350 ITU/mg)	85–100% control for 25 d at 3.1 kg/ha	Su and Mulla (1999b)
Cx. quinquefasciatus, Cx. stigmatosoma	Simulated catch basins	VectoLex WSP (50 ITU/mg)	90% or greater for 203 d at 1 WSP/catch basin	Su (2008)
Cx. pipiens, Culex restuans, Aedes japonicus	Catch basins	VectoLex CG	Almost 100% control for 1 week at 1.8 g/catch basin, 3 times dose for 5 weeks	Anderson et al. (2011)

FC, flowable concentrate; PP, primary powder; WDG, water dispersible granule; WSP, water soluble pouch.

Mulla et al. (1984b) tested several strains of *L. sphaericus* in the laboratory against second and fourth instar larvae of five mosquito species. All strains exhibited high levels of activity against *Culex* larvae. Strains 2362 (IF-97), 2013-4, and 2013-6 were as effective as the standard 1593-4 (RB-80), causing 90% mortality in fourth instar larvae of *Cx. quinquefasciatus* at 0.04–0.05 ppm. Strain 1593 (IF-94) was somewhat less active, with an LC$_{90}$ of 0.11 ppm. All the strains tested displayed lower activity against larvae of *Anopheles*, while *Ae. aegypti* was the least susceptible species, with an LC$_{90}$ of more than 40 ppm. Second instar larvae were in general more susceptible than fourth instar larvae, and the maximum mortality at a given concentration was obtained with an exposure period of 48 h; there was little or no mortality after exposure for 24 h. Under field conditions in California, strains 2362 (IF-97) and 1593 (IF-94) yielded excellent initial and residual control of *Cx. tarsalis* and *Cx. peus* at fairly low rates. In another study by Mulla et al. (1985), *L. sphaericus* were evaluated in the laboratory and field against *Psorophora columbiae* and *Aedes melanimon*. *Ps. columbiae* was slightly more susceptible than *Cx. quinquefasciatus* to active strains. The LC$_{90}$ for active strains ranged from 0.013 to 0.069 ppm. *Ae. melanimon* was slightly less susceptible than *Ps. columbiae*. In field trials, aqueous suspensions of primary powder of strains 2362 and 1593 yielded high control at low application rates (Table 27.2).

The Beecomist ultra-low volume generator was used to apply *L. sphaericus* liquid formulations for the control of *An. quadrimaculatus* larvae in rice fields in Louisiana. This method provides effective control at very low application rates (Lacey et al., 1986; Table 27.2). The small droplets demonstrated high penetration through dense vegetation and delivered the active ingredients to the feeding zone of *An. quadrimaculatus*. A high efficacy of *L. sphaericus* against *Cx. pipiens* and *Aedes trivittatus* was achieved by a granular formulation applied at a range of doses in Iowa (Berry et al., 1987).

Control efficacy evaluation against field populations of floodwater mosquitoes was continued and published in 1988 on four new formulations of strain 2362, which yielded excellent control of *Ps. columbiae* and *Ae. nigromaculis* in irrigated fields in California, where control levels depended on species, formulations, and potency (Mulla et al., 1988a). Mulla et al. (1988b) evaluated the initial efficacy and longevity of strain 2362 formulations for the control of *Culex* spp. larvae under more challenging conditions in dairy wastewater lagoons. Test materials included two primary powder preparations, a flowable concentrate (BSP-2), and granular formulations that were applied at various doses. Some formulations were more active than others, yielding great initial and persistent control with one treatment. Field studies on a wide variety of formulations, including granules, flowable concentrate, briquettes, and pellets conducted by Lacey et al. (1988)

in Florida, Louisiana, and Tennessee indicated good efficacy against *Culex* spp. and other species (Table 27.2). Evaluation on dairy wastewater lagoons was continued by Matanmi et al. (1990) against *Culex stigmatosoma*. Two commercial formulations, BSP-2 and ABG-6184, provided high levels of control for an extended period of time at suggested rates. Results indicated that the extended control obtained was due primarily to the ingestion of spores from bottom water and mud by larvae (Table 27.2).

In controlling *Cs. incidens* breeding in waste tires, products applied at 3.75–15 ppm yielded great initial and residual efficacy under sunlit or shaded conditions (Kramer, 1990). Lord (1991) reported that the residual activity of *L. sphaericus* against *Culex* spp. larvae was extended by formulating pellets with partially hydrogenated vegetable oil, talc, and a starch-based superabsorbent polymer. When the pellets were applied as a prehatch treatment 5 days prior to flooding, the *Ps. columbiae* that hatched at the later flooding were eliminated. After the pools were dried and reflooded, *L. sphaericus* were present at 611 CFU/mL in the surface water (Table 27.2).

Upon new developments of highly potent formulations, Su and Mulla (1999a) evaluated WDG formulations in outdoor microcosms against natural populations of *Culex* mosquitoes. The minimum effective dosage of the WDGs with 350–630 ITU/mg for initial and extended efficacy was quite low. Su (2008) conducted an outdoor test to evaluate the initial efficacy and longevity of VectoLex WSP applied as a prehatch treatment against *Culex* species in simulated catch basins. VectoLex WSP applied at one pouch (10 g) per basin (single treatment) yielded significant immediate and long-term control of late instars. Spore counts in water and sludge samples verified the presence of *L. sphaericus* spores on day 196 posttreatment after 28 weekly flushes. The results strongly indicate that the WSP formulation of *L. sphaericus* could be one of the best candidates for controlling the larvae of *Culex* mosquitoes developing in catch basins. Studies on efficacy to control *Cx. pipiens*, *Culex restuans*, and *Aedes japonicus* in urban catch basins in the northeastern United States were conducted by Anderson et al. (2011). VectoLex CG applied at a label rate significantly reduced the numbers of larvae for 1 week. When applied at 3 times the maximum label rate, VectoLex CG significantly reduced the numbers of larvae for 5 weeks (Table 27.2).

Lacey et al. (1987) studied the delayed effects of *L. sphaericus* on *Cx. quinquefasciatus* after a sublethal exposure. Treated individuals delayed their pupation, and adults showed lower survival rates and lower stored nutrients. Research by Mulla et al. (1991) showed similar results of delayed mortality in larvae, pupae, and adults. Mortality increased in succeeding cohorts and developmental stages, resulting from the reduced surviving larvae. Only 10% and 25% overall emergence of viable adults occurred in the sublethal treatments of *L. sphaericus* preparations. A wide range of external morphogenetic aberrations in dead larvae, pupae, and adults was noted.

27.3.2 Factors Influencing Larvicidal Activity and Efficacy

Strains that are isolated from various invertebrate hosts with various geographical origins may carry different biological and genetic attributes, which may lead to different toxin production during fermentation, thus affecting the larvicidal activity and efficacy. Mulla et al. (1984b) investigated the laboratory activity of various strains against different mosquito species. Strain 2362 (IF-97), 2013-4, and 2013-6 were as effective as strain 1593-4 (RB-80), while strain 1593 (IF-94) was somewhat less active than the other three strains. The fermentation conditions, especially medium, and processes in different laboratories or industrial facilities may result in variable biological potency and quality (Yousten et al., 1984). The long-term stability of a given strain under fermentation and subsequent storage needs to be monitored and maintained. As in Bti with the same consideration of solids harvested from fermentation, formulation technology is always a challenge to maintain the viability of the highly uneven active ingredients such as spores, crystal inclusions, and others in a given formulation and to deliver them to the feeding zone of the target species within a proper time frame (Mulligan et al., 1980; Nguyen et al., 1999a). Studies by Clark et al. (2006) indicated that the declines in particle availability in the field may not be due to the particle itself but rather to the formulations.

In contrast to that of Bti, the target species spectrum of *L. sphaericus* is narrower with high activity against all *Culex* spp., *Ps. columbiae*, and most *Anopheles* species but low to no larvicidal activity against some *Aedes* species (Mian and Mulla, 1983; Davidson et al., 1984; Mulla et al., 1984b, 1985; Ali and Nayar, 1986). Species-dependent susceptibility is related to the nature of the target sites of the epithelial midgut cells but not to uptake or processing of the toxins (Baumann et al., 1991). Cyt1A produced by Bti can promote the insertion of the Bin B from *L. sphaericus* into midgut epithelium cells, hence significantly increasing their toxicity against *Aedes* spp. (Wirth et al., 2000a).

Larval feeding behavior is a biological factor that significantly impacts the larvicidal activity and efficacy of *L. sphaericus* against *Cx. quinquefasciatus*, *Anopheles albimanus*, and *Ae. aegypti* (Ramoska and Hopkins, 1981). Additionally, the feeding activity in young larvae is more active than in old larvae, and therefore younger larvae exhibit higher mortality than older larvae in *Ae. aegypti* and *Cx. quinquefasciatus* (Mian and Mulla, 1983; Mulla et al., 1984b). Starved larvae showed higher mortality as compared with unstarved larvae under the same conditions for the same species (Mian and Mulla, 1983). In an unsynchronized larval population, some late instars may survive

treatment of *L. sphaericus* products. This phenomenon can be more profound than that in Bti, as the latter acts faster than *L. sphaericus* against *Cx.* spp. (Su and Mulla, 1999a, 1999b). Larval density can play some role in larvicidal efficacy, as in *Cx. tarsalis* (Mulla et al., 1984a), especially when released active ingredients in the habitat are at the threshold concentration.

By nature, *L. sphaericus* can tolerate higher organic matter in aquatic environments than Bti, even though better activity and efficacy were observed in clear water than in polluted water, as seen in *Aedes* and *Culex* spp. (Mulligan et al., 1980; Mian and Mulla, 1983; Mulla et al., 1984b). The presence of certain organic matter may be conducive to the recycling of spores if adequate numbers of intact larval cadavers of *Cs. incidens* exist (Kramer, 1990). At the same time, mosquito larvae, as opportunistic feeders in aquatic environments, may ingest fewer *L. sphaericus* spores and inclusions in the presence of overwhelming organic matter as food for *Cx. quinquefasciatus* and *An. albimanus* (Ramoska and Pacey, 1979). Water temperature, as one of the most important abiotic factors associated with mosquito habitats, impacts the efficacy against *Ae. melanimon* and *Ps. columbiae* (Mulla et al., 1985) by governing the feeding efficiency of target larvae as well as affecting the release of the active ingredient from the formulations through physical and chemical processes. High pH in water resulted in a loss of larvicidal activity of strain 1593-4 against *Cx. quinquefasciatus* (Mulligan et al., 1980). Finally, high solar irradiation reduced larvicidal efficacy of *L. sphaericus* products against *Cx. quinquefasciatus* and *Cs. incidens* (Mulligan et al., 1980; Kramer, 1990).

27.3.3 Potential for Resistance and Its Management

Resistance to *L. sphaericus* in the *Cx. pipiens* complex has been reported in different countries since 1994 (Wirth, 2010; Su, 2016, Chapter 28). Laboratory populations in North America developed resistance to strain 2362 at different levels in response to selection for numerous generations (Rodcharoen and Mulla, 1994; Wirth et al., 2000b; Zahiri et al., 2002; Zahiri and Mulla, 2003). The resistance development depends on the genetic background, size of the population used, selection pressure, and so on. Resistance level is also dependent on the susceptibility of the reference population against which the resistance ratio is calculated. Resistance to *L. sphaericus* in wild populations of the *Cx. pipiens* complex was first reported in 1994 in southern France (Sinègre et al., 1994). Thereafter, various levels of resistance to different strains in response to field applications have been reported in a number of countries (Wirth, 2010) but not in North America. The monitoring of susceptibility to *L. sphaericus* in California indicated some temporal and spatial variability (Wirth et al., 2001). The resistance levels vary greatly, depending on prior exposure to naturally existing strains, population genetic background, gene exchange with untreated populations, and product application strategies. Furthermore, once mosquitoes develop resistance to a given strain, they are also often resistant to other strains (Rodcharoen and Mulla, 1996) because of the similarity of the Bin toxins in most strains. This cross-resistance among strains, however, is moderate between the strains that also produce Mtx toxins, and there is a lack of cross-resistance between Bin toxins and Mtx toxins, which indicates that Mtx could be a potential tool to manage resistance to Bin toxins. Fortunately, mosquitoes that have developed resistance to various strains of *L. sphaericus* remain susceptible to Bti (Rodcharoen and Mulla, 1996; Wirth et al., 2000b; Zahiri et al., 2002; Zahiri and Mulla, 2003). It is mostly believed that a recessive genetic mechanism is involved in the resistance to *L. sphaericus*, and a lack of binding of the Bin toxins to the brush border membrane is the main reason associated with resistance (Wirth, 2010). Behavioral modifications such as reduced ingestion on toxins (Rodcharoen and Mulla, 1995) and other unknown mechanisms are also involved.

Bti can be used as a powerful tool to mitigate resistance to *L. sphaericus* because of the synergism between Bti and *L. sphaericus* toxins (Wirth et al., 2004, 2010). After resistance to *L. sphaericus* Bin toxins has occurred, Bti alone, a mixture, or a rotation of Bti and *L. sphaericus* can restore the susceptibility to *L. sphaericus* in *Cx. quinquefasciatus*, even though this process can take many generations (Zahiri et al., 2002). Resistance to *L. sphaericus* in susceptible *Cx. quinquefasciatus* can be delayed or prevented by the mixture of Bti and *L. sphaericus* (Zahiri and Mulla, 2003) because of the synergistic action among four toxins, particularly the presence of Cyt1A (Wirth et al., 2005a, 2010). Furthermore, Cyt1A from Bti restores the toxicity of *L. sphaericus* against resistant *Cx. quinquefasciatus* (Wirth et al., 2000c). The rotation of Bti and *L. sphaericus* seemed to accelerate the development of resistance to *L. sphaericus* in susceptible *Cx. quinquefasciatus* (Zahiri and Mulla, 2003), which is not recommended for delaying or preventing resistance evolution to *L. sphaericus* in mosquitoes. The recombinant that produces toxins from both Bti and *L. sphaericus*, even at greater amounts than the wild type of bacteria (Federici, 2003; Federici et al., 2007, 2010), provides another path for the mitigation of resistance and the enhancement of larvicidal activity and efficacy.

27.3.4 Safety for Nontarget Organisms

As a bacterial larvicide with a high target specificity, its impact on nontarget organisms and the environment is negligible. The available data indicate a high degree of specificity of *L. sphaericus* binary toxins for mosquitoes, with no demonstrated toxicity to nontargets such as chironomid larvae and others at mosquito control application rates (Mulla et al., 1984b; Ali and Nayar, 1986; Lacey and Mulla, 1990). Studies by Yousten et al. (1991) indicated that the spores of strain 2362 were rapidly eliminated after ingestion by midge larvae. However, spores remained up to 21 days in oysters and up to 49 days in snails, but there was no indication of actively growing *L. sphaericus* in the animals. Passage through the oyster gut detoxified the mosquitocidal toxins. A 3-year study (2000–02) by Merritt et al. (2005) in southeastern Wisconsin indicated that when VectoLex was applied for six treatments at the labeled dose, no detrimental effects to nontarget organisms could be attributed to this microbial insecticide. After a comprehensive review on published papers, Lacey and Mulla (1990) as well as Lacey and Merritt (2003) concluded that *L. sphaericus* appears to pose little direct or indirect threat to nontarget benthic invertebrate species or fish. Su and Mulla (2005) indicated the high compatibility of *L. sphaericus* with a predatory biological control agent, tadpole shrimp, *T. newberryi*. As in Bti, the exposure of TPS to 8.5 ppm of *L. sphaericus* (VectoLex TP, 2000 ITU/mg) on a weekly basis did not have significant adverse effects on growth, longevity, and fecundity. This also held true when this organism was treated by VectoLex CG (50 ITU/mg) at 11.2–22.4 kg/ha in the field. To predict potential detrimental effects over the long term, a thorough understanding of the basic ecological interactions among *L. sphaericus*, target, and nontarget organisms is warranted. Long-term studies of the effects of *L. sphaericus* use on food resource loss for predators and the effects on fisheries and other wildlife are needed to understand their impact on overall aquatic ecosystems.

27.4 COMBINATION OF *BACILLUS THURINGIENSIS* SUBSP. *ISRAELENSIS* AND *LYSINIBACILLUS SPHAERICUS*

Bti and *L. sphaericus* possess many opposite biological traits such as speed of action, species spectrum, tolerance to organic matter in water, and risk of resistance development. Therefore the combination of both species has the advantages of quick action, broad species spectrum, and low risk of resistance development from Bti and long residual efficacy and tolerance to high organic pollution from *L. sphaericus*. Numerous granular, WSP, and briquet formulations of the combination of Bti and *L. sphaericus* have been developed and commercialized. Some products are formulated by combining technical powders of Bti and *L. sphaericus*, while others are produced by BioFuse technology, which ensures the simultaneous ingestion of active ingredients from both bacteria by mosquito larvae.

Quantitative data on field evaluations of formulations that contain both Bti and *B. sphaericus* are summarized in Table 27.3. Su (2008) conducted the first outdoor evaluation

TABLE 27.3 Examples of Combination of *Bacillus thuringiensis* Subsp. *israelensis* and *Lysinibacillus sphaericus* Used for Control of Mosquito Larvae in North America

Targeted Mosquito Species	Habitats	Formulations	Application Rate and Efficacy	References
Culex quinquefasciatus, Culex stigmatosoma	Simulated catch basins	WSP[a]	90% or greater for 203 d	Su (2008)
Culex pipiens, Culex restuans, Aedes japonicus	Catch basins	VectoMax WSP	70–100% control at 1 WSP/catch basin for 3–4 weeks	Anderson et al. (2011)
Culex spp.	Catch basins	FourStar 180-day briquet	1 briquet/catch basin, 8 consecutive weeks	Harbison et al. (2014)
Cx. pipiens	Septic tanks	VectoMax WSP	1–2 pouches (10–20 g) per septic tank, >96% control of larvae for 24 d	Cetin et al. (2015)

[a]WSP, *water dispersible pouch.*

on a WSP formulation (VBC-60035, later VectoMax WSP) based on the combination of Bti and *L. sphaericus* using simulated catch basins. This formulation contains a combination of Bti strain AM65-52 and *L. sphaericus* strain ABTS 1743. The formulation was applied as a prehatch treatment at one WSP per basin against *Culex* species. Immediate and long-term control of late instars of *Culex* mosquitoes was achieved. Anderson et al. (2011) evaluated the efficacy of the same formulation in urban storm drain systems in northeastern United States, and excellent efficacy was obtained against *Culex* spp. and *Ae. japonicus*. Cetin et al. (2015) evaluated the residual effectiveness of VectoMax WSP when applied to septic tanks against third and fourth stage larvae of *Cx. pipiens*. This formulation was evaluated at operational application rates of one pouch (10 g) and two pouches (20 g) per septic tank. Both application rates resulted in >96% control of larvae for 24 days (Table 27.3). Harbison et al. (2014) evaluated a new slow-release formulation of Bti and *L. sphaericus*, FourStar 180-day briquette, in storm water catch basins in the northern suburbs of Chicago. Over a 17-week period, FourStar reduced immature numbers in treated catch basins for 8 consecutive weeks. These results suggest that if effectively timed, a single application of the formulation tested may last an entire season (Table 27.3). The recombinant of Bti and *L. sphaericus*, which produces toxins from both, provides a new industrial development opportunity in the near future for highly potent bacterial larvicides (Federici, 2003; Federici et al., 2007, 2010; Park et al., 2003, 2005; Park and Federici, 2009). Bacterial larvicides based on the combination of both bacteria have the advantages of prolonged activity and efficacy enhancement as well as prevention of resistance development and restoration of susceptibility after resistance occurrence to *L. sphaericus* (Zahiri et al., 2002, 2004; Zahiri and Mulla, 2003).

27.5 INTEGRATED MOSQUITO CONTROL

The role of biological control of mosquitoes in integrated vector management was reviewed by Lacey and Orr (1994). Biological control has focused on the use of predators (fish and predatory invertebrates), parasites (nematodes), and pathogens (fungus, Bti, *L. sphaericus*, and others). The successful implementation of these organisms will largely depend on an in-depth understanding of the ecology of both the targeted species and the biological control agents to be used. Bti or *L. sphaericus* alone or in combination have been demonstrated as highly effective, target specific, nontarget safe, and environmentally sound tools to combat wide varieties of nuisance and vector mosquito species. The active ingredients from these two bacteria, particularly ones from Bti, have also been recognized as potential tools in integrated mosquito management by enhancing efficacy or mitigating resistance development.

As early as the late 1980s, field trials attempted to use a combination of Bti and methoprene to control *Ps. columbiae* and *An. quadrimaculatus* breeding in rice fields (Bassi et al., 1989). Lawler et al. (2000) evaluated the efficacy of the same combination against caged larval *Ae. vexans* in salt marsh ponds. The compatibility of these two pesticides and high efficacy were observed. *Culex quinquefasciatus* that are highly resistant to spinosad in the laboratory remain very susceptible to Bti and a combination of Bti and *L. sphaericus* (Su and Cheng, 2012, 2014a,b) and permethrin (Su et al., unpublished data). Bti restored the susceptibility to spinosad after a high-level resistance developed in response to long-term laboratory selection (Su et al., unpublished data). Susceptibility to methoprene in field populations of *Ae. nigromaculis* in central California was significantly restored by Bti applications after a high-level resistance developed in response to many years of Altosid applications

(Cornel et al., 2002). The earlier mentioned scenarios indicate that Bti could be a great tool to mitigate resistance in other biorational mosquito larvicides. Recently, a commercial formulation (eg, VectoPrime) that contains methoprene and Bti has become available for mosquito control operations. Additional biological and environmental methods of integrated vector control are presented in Lacey and Orr (1994) and in Chapter 28.

27.6 FUTURE PERSPECTIVES

Considering emerging and reemerging mosquito-borne diseases and numerous nuisance mosquito species in North America, the demand on control agents with high efficacy, target specificity, nontarget, and environmental safety remains. Among the biorational larvicides against mosquitoes ranging from microbial ingredients, botanical components, and hormone mimics, Bti and *L. sphaericus* are among the very few with great promise. In view of the narrow target spectrum and the potential for resistance development in response to repeated applications, *L. sphaericus* has limited prospects for control of some *Aedes* and *Anopheles* spp. However, with some resistance management, *L. sphaericus* can still be used against *Culex* mosquitoes. On the other hand, Bti formulations, which have a broader spectrum of activity against all mosquito species, have not showed a significant development of resistance in mosquitoes (Wirth, 2010). Formulations of these bacteria have become the predominant nonchemical means employed for the control of mosquito larvae in the United States and other countries. Due to their efficacy and relative specificity, both Bti and *L. sphaericus* can be ideal control agents in integrated programs, especially where other biological control agents, environmental management, personal protection, and the judicious use of insecticides are combined (Lacey, 2007). The increased use of Bti during recent years by mosquito abatement districts is due to strict regulations and the pressure from the public to decrease the use of adulticides.

Both Bti and *L. sphaericus* will continue to play an irreplaceable role in the foreseeable future in combating mosquitoes and mosquito-borne diseases. In order to keep the sustainability of these bacterial larvicides, the following aspects warrant attention from academic research, industrial development, and field application. It is believed that numerous new strains of bacteria remain to be isolated and identified, which might possess more desired activity and efficacy. Advanced genetic and molecular technology will enable the modification and customization of microbial organisms isolated from nature for toxin production under fermentation. Advancements in understanding the biology and ecology of microbial organisms intended for larvicidal development will facilitate successful fermentation and toxin production. Improved formulation technology will enhance the persistence and delivery of the active ingredients for the effective control of mosquito populations. It is equally important that precise application techniques and equipment are developed and applied for a customized delivery of the products used.

REFERENCES

Ali, A., 1980. *Bacillus thuringiensis* serovar. *israelensis* (ABG-6108) against chironomids and some non-target aquatic invertebrates. J. Invertebr. Pathol. 38, 264–272.

Ali, A., Nayar, J.K., 1986. Efficacy of *Bacillus sphaericus* Neide against larval mosquitoes (Diptera: Culicidae) and midges (Diptera: Chironomidae) in the laboratory. Fla. Entomol. 69, 685–690.

Anderson, J.F., Ferrandino, F.J., Dingman, D.W., Main, A.J., Andreadis, T.G., Becnel, J.J., 2011. Control of mosquitoes in catch basins in Connecticut with *Bacillus thuringiensis israelensis*, *Bacillus sphaericus*, and spinosad. J. Am. Mosq. Control Assoc. 27, 45–55.

Bassi, D.G., Weathersbee, A.A., Meisch, M.V., Inman, A., 1989. Efficacy of Duplex and Vectobac against *Psorophora columbiae* and *Anopheles quadrimaculatus* larvae in Arkansas ricefields. J. Am. Mosq. Control Assoc. 5, 264–266. http://www.ncbi.nlm.nih.gov/pubmed/2568399?dopt=Abstract - comments.

Baumann, P., Clark, M.A., Baumann, L., Broadwell, A.H., 1991. *Bacillus sphaericus* as a mosquito pathogen: properties of the organism and its toxins. Microbiol. Rev. 55, 425–436.

Berry, W.J., Novak, M.G., Khounlo, S., Rowley, W.A., Melchior, G.L., 1987. Efficacy of *Bacillus sphaericus* and *Bacillus thuringiensis* var. *israelensis* for control of *Culex pipiens* and floodwater *Aedes* larvae in Iowa. J. Am. Mosq. Control Assoc. 3, 579–582.

Cetin, H., Oz, E., Yanikoglu, A., Cilek, J.E., 2015. Operational evaluation of Vectomax® WSP (*Bacillus thuringiensis* subsp. *israelensis* + *Bacillus sphaericus*) against larval *Culex pipiens* in septic tanks. J. Am. Mosq. Control Assoc. 31, 193–195.

Charbonneau, C.S., Drobney, R.D., Rabeni, C.F., 1994. Effects of *Bacillus thuringiensis* var. *israelensis* on non-target benthic organisms in a lentic habitat and factors affecting the efficacy of the larvicide. Environ. Toxicol. Chem. 13, 267–279.

Christiansen, J.A., McAbee, R.D., Stanich, M.A., DeChant, P., Boronda, D., Cornel, J., 2004. Influence of temperature and concentration of VectoBac on control of the salt-marsh mosquito, *Ochlerotatus squamiger*, in Monterey County, California. J. Am. Mosq. Control Assoc. 20, 165–170.

Clark, J.D., Devisetty, B.N., Krause, S.C., Novak, R.J., Warrior, P., 2006. Particle distributional behavior of a spray-dried technical concentrate and a water-dispersible granule formulation of *Bacillus sphaericus* in an aqueous column. J. Am. Mosq. Control Assoc. 22, 718–724.

Clark, J.D., Devisetty, B.N., Krause, S.C., Novak, R.J., Warrior, P., 2007. A novel method for evaluating the particle distributional behavior of a spray-dried technical concentrate and a water-dispersible granule formulation of *Bacillus thuringiensis* subsp. *israelensis* in an aqueous column. J. Am. Mosq. Control Assoc. 23, 60–65.

Cornel, A.J., Stanich, M.A., McAbee, R.D., Mulligan III, F.S., 2002. High level methoprene resistance in the mosquito *Ochlerotatus nigromaculis* (Ludlow) in central California. Pest Manag. Sci. 58, 791–798.

Correa, M., Yousten, A.A., 1995. *Bacillus sphaericus* spore germination and recycling in mosquito larval cadavers. J. Invertebr. Pathol. 66, 76–81.

Davidson, E.W., Urbina, M., Payne, J., Mulla, M.S., Darwazeh, H., Dulmage, H.T., Correa, J.A., 1984. Fate of *Bacillus sphaericus* 1593 and 2362 spores used as larvicides in the aquatic environment. Appl. Environ. Microbiol. 47, 125–129.

Eldridge, B.F., Callicrate, J., 1982. Efficacy of *Bacillus thuringiensis* var. *israelensis* de Barjac for mosquito control in a western Oregon log pond. Mosq. News 42, 102–105.

Federici, B.A., 2003. Recombinant bacteria for mosquito control. J. Exp. Biol. 206, 3877–3885.

Federici, B.A., Park, H.W., Bideshi, D.K., Wirth, M.C., Johnson, J.J., Sakano, Y., Tang, M., 2007. Developing recombinant bacteria for control of mosquito larvae. J. Am. Mosq. Control Assoc. 23 (Suppl. 2), 164–175.

Federici, B.A., Park, H.W., Bideshi, D.K., 2010. Overview of the basic biology of *Bacillus thuringiensis* with emphasis on genetic engineering of bacterial larvicides for mosquito control. Open Toxinol. J. 3, 154–171.

Floore, T.G., Rathbun Jr., C.B., Boike Jr., A.H., Masters, H.M., Puckett, J.M., 1987. The effect of water with three different pH values on the efficacy of VectoBac AS. J. Fla. Anti-Mosq. Assoc. 58, 15–16.

Georghiou, G.P., Wirth, M.C., 1997. Influence of exposure to single versus multiple toxins of *Bacillus thuringiensis* var. *israelensis* on the development of resistance in the mosquito *Culex quinquefasciatus* (Diptera: Culicidae). Appl. Environ. Microbiol. 63, 1095–1101.

Goldman, I., Arnold, J., Carlton, B.C., 1986. Selection for resistance to *Bacillus thuringiensis* subspecies *israelensis* in field and laboratory populations of the mosquito *Aedes aegypti*. J. Invertebr. Pathol. 47, 317–324.

Harbison, J.E., Henry, M., Xamplas, C., Berry, R., Bhattacharya, D., Dugas, L.R., 2014. A comparison of FourStar™ briquets and Natular™ XRT tablets in a north shore suburb of Chicago, IL. J. Am. Mosq. Control Assoc. 30, 68–70.

Hershey, A.E., Shannon, L., Axler, R., Ernst, C., Mickelson, P., 1995. Effects of methoprene and Bti (*Bacillus thuringiensis* var. *israelensis*) on non-target insects. Hydrobiologia 308, 219–227.

Hershey, A.E., Lima, A.R., Niemi, G.J., Regal, R.R., 1998. Effects of *Bacillus thuringiensis israelensis* (Bti) and methoprene on non-target macroinvertebrates in Minnesota wetlands. Ecol. Appl. 8, 41–60.

Hertlein, B.C., Levy, R., Miller Jr., T.W., 1979. Recycling potential and selective retrieval of *Bacillus sphaericus* from soil in a mosquito habitat. J. Invertebr. Pathol. 33, 217–221.

Kramer, V.L., 1990. Efficacy and persistence of *Bacillus sphaericus*, *Bacillus thuringiensis* var. *israelensis*, and methoprene against *Culiseta incidens* (Diptera: Culicidae) in tires. J. Econ. Entomol. 83, 1280–1285.

Lacey, L.A., 1990. Persistence and formulation of *Bacillus sphaericus*. In: de Barjac, H., Sutherland, D.J. (Eds.), Bacterial Control of Mosquitoes and Black Flies. Rutgers University Press, New Brunswick, NJ, pp. 284–294.

Lacey, L.A., 2007. *Bacillus thuringiensis* serovariety *israelensis* and *Bacillus sphaericus* for mosquito control. J. Am. Mosq. Control Assoc. 23 (Suppl. 2), 133–163.

Lacey, L.A., Lacey, C.M., 1990. The medical importance of riceland mosquitoes and their control using alternatives to chemical insecticides. J. Am. Mosq. Control Assoc. 6 (Suppl. 2), 1–93.

Lacey, L.A., Merritt, R.W., 2003. The safety of bacterial microbial agents used for black fly and mosquito control in aquatic environments. In: Hokkanen, H.M.T., Hajek, A.E. (Eds.), Environmental Impacts of Microbial Insecticides: Need and Methods for Risk Assessment. Kluwer Academic Publishers, Dordrecht, The Netherlands, pp. 151–168.

Lacey, L.A., Mulla, M.S., 1990. Safety of *Bacillus thuringiensis* (H-14) and *Bacillus sphaericus* to non-target organisms in the aquatic environment. In: Laird, M., Lacey, L.A., Davidson, E.W. (Eds.), Safety of Microbial Insecticides. CRC Press, Boca Raton, FL, pp. 169–188.

Lacey, L.A., Orr, B.K., 1994. The role of biological control of mosquitoes in integrated vector control. Am. J. Trop. Med. Hyg. 50 (Suppl), 97–115.

Lacey, L.A., Siegel, J.P., 2000. Safety and ecotoxicology of entomopathogenic bacteria. In: Charles, J.F., Delecluse, A., Nielsen-LeRoux, C. (Eds.), Entomopathogenic Bacteria: From Laboratory to Field Application. Kluwer Academic Publishers, Dordrecht, The Netherlands, pp. 253–273.

Lacey, L.A., Urbina, M.J., Heitzman, C., 1984. Sustained release formulations of *Bacillus sphaericus* and *Bacillus thuringiensis* (H-14) for the control of container breeding *Culex quinquefasciatus*. Mosq. News 44, 26–32.

Lacey, L.A., Heitzman, C.M., Meisch, M.V., Billodeaux, J., 1986. Beecomist® application of *Bacillus sphaericus* for control of riceland mosquitoes. J. Am. Control Assoc. 2, 548–551.

Lacey, L.A., Day, J., Heitzman, C.M., 1987. Long-term effects of *Bacillus sphaericus* on *Culex quinquefasciatus*. J. Invertebr. Pathol. 49, 116–123.

Lacey, L.A., Ross, D.H., Lacey, C.M., Inman, A., Dulmage, H.T., 1988. Experimental formulations of *Bacillus sphaericus* for the control of anopheles and culicine larvae. J. Ind. Microbiol. 2, 39–47.

Lawler, S.P., Jensen, T., Dritz, D.A., Wichterman, G., 1999. Field efficacy and nontarget effects of the mosquito larvicides temephos, methoprene, and *Bacillus thuringiensis* var. *israelensis* in Florida mangrove swamps. J. Am. Mosq. Control Assoc. 15, 446–452.

Lawler, S.P., Dritz, D.A., Jensen, T., 2000. Effects of sustained-release methoprene and a combined formulation of liquid methoprene and *Bacillus thuringiensis israelensis* on insects in salt marshes. Arch. Environ. Contam. Toxicol. 39, 177–182.

Lord, J.C., 1991. Sustained release pellets for control of *Culex* larvae with *Bacillus sphaericus*. J. Am. Mosq. Control Assoc. 7, 560–564.

Matanmi, B.A., Federici, B.A., Mulla, M.S., 1990. Fate and persistence of *Bacillus sphaericus* used as a mosquito larvicide in dairy wastewater lagoons. J. Am. Mosq. Control Assoc. 6, 384–389.

Merritt, R.W., Lessard, J.L., Wessell, K.J., Hernandez, O., Berg, M.B., Wallace, J.R., Novak, J.A., Ryan, J., Merritt, B.W., 2005. Lack of effects of *Bacillus sphaericus* (Vectolex) on non-target organisms in a mosquito-control program in southeastern Wisconsin: a 3-year study. J. Am. Mosq. Control Assoc. 21, 201–212.

Mian, L., Mulla, M.S., 1983. Factors influencing activity of the microbial agent *Bacillus sphaericus* against mosquito larvae. Bull. Soc. Vector Ecol. 8, 128–134.

Miura, T., Takahashi, R.M., Mulligan III, F.S., 1980. Effects of the bacterial mosquito larvicide, *Bacillus thuringiensis* serotype H-14 on selected aquatic organisms. Mosq. News 40, 619–622.

Molloy, D.P., 1992. Impact of the black fly (Diptera: Similiidae) control agent *Bacillus thuringiensis* var. *israelensis* on chironomids (Diptera: Chironomidae) and other non-target insects: results of ten field trials. J. Am. Mosq. Control Assoc. 8, 24–31.

Mulla, M.S., 1990. Activity, field efficacy, and use of *Bacillus thuringiensis israelensis* against mosquitoes. In: de Barjac, H., Sutherland, D.J. (Eds.), Bacterial Control of Mosquitoes and Black Flies. Rutgers University Press, New Brunswick, NJ, pp. 134–160.

Mulla, M.S., Federici, B.A., Darwazeh, H.A., Ede, L., 1982. Field evaluation of the microbial insecticide *Bacillus thuringiensis* serotype H-14 against floodwater mosquitoes. Appl. Environ. Microbiol. 43, 1288–1293.

Mulla, M.S., Darwazeh, H.A., Davidson, E.W., Dulmage, H.T., 1984a. Efficacy and persistence of the microbial agent *Bacillus sphaericus* against mosquito larvae in organically enriched habitats. Mosq. News 44, 166–173.

Mulla, M.S., Darwazeh, H.A., Davidson, E.W., Dulmage, H.T., Singer, S., 1984b. Larvicidal activity and field efficacy of *Bacillus sphaericus* strains against mosquito larvae and their safety to non-target organisms. Mosq. News 44, 336–342.

Mulla, M.S., Darwazeh, H.A., Ede, L., Kennedy, B., Dulmage, H.T., 1985. Efficacy and field evaluation of *Bacillus thuringiensis* (H-14) and *B. sphaericus* against floodwater mosquitoes in California. J. Am. Mosq. Control Assoc. 1, 310–315.

Mulla, M.S., Darwazeh, H.A., Tietze, N.S., 1988a. Efficacy of *Bacillus sphaericus* 2362 formulations against floodwater mosquitoes. J. Am. Mosq. Control Assoc. 4, 172–174.

Mulla, M.S., Axelrod, H., Darwazeh, H.A., Matanmi, B.A., 1988b. Efficacy and longevity of *Bacillus sphaericus* 2362 formulations for control of mosquito larvae in dairy wastewater lagoons. J. Am. Mosq. Control Assoc. 4, 448–452.

Mulla, M.S., Singh, N., Darwazeh, H.A., 1991. Delayed mortality and morphogenetic anomalies induced in *Culex quinquefasciatus* by the microbial control agent *Bacillus sphaericus*. J. Am. Mosq. Control Assoc. 7, 412–419.

Mulla, M.S., Chaney, J.D., Rodcharoen, J., 1993. Elevated dosages of *Bacillus thuringiensis* var. *israelensis* fail to extend control of *Culex* larvae. Bull. Soc. Vector Ecol. 18, 125–132.

Mulligan III, F.S., Schaefer, C.M., Miura, T., 1978. Laboratory and field evaluation of *Bacillus sphaericus* as a mosquito control agent. J. Econ. Entomol. 71, 774–777.

Mulligan III, F.S., Schaefer, C.M., Wilder, W.H., 1980. Efficacy and persistence of *Bacillus sphaericus* and *B. thuringiensis* H-14 against mosquitoes under laboratory and field conditions. J. Econ. Entomol. 73, 684–688.

Nguyen, T.T.H., Su, T., Mulla, M.S., 1999a. Mosquito control and bacterial flora in water enriched with organic matter and treated with *Bacillus thuringiensis* subsp. *israelensis* and *Bacillus sphaericus* formulations. J. Vector Ecol. 24, 138–153.

Nguyen, T.T.H., Su, T., Mulla, M.S., 1999b. Bacterial and mosquito abundance in microcosms enriched and unenriched with organic matter and treated with *Bacillus thuringiensis* subsp. *israelensis* formulation. J. Vector Ecol. 24, 191–201.

Park, H.W., Federici, B.A., 2009. Genetic engineering of bacteria to improve efficacy using the insecticidal proteins of *Bacillus* species. In: Stock, S.P., Glaser, I., Boemare, N., Vandenberg, J. (Eds.), Molecular Approaches and Techniques for the Study of Insect Pathogens. CABI Publishing, Rothamsted, England, pp. 275–305.

Park, H.W., Bideshi, D., Federici, B.A., 2003. Recombinant strain of *Bacillus thuringiensis* producing Cyt1A, Cry11B, and the *Bacillus sphaericus* binary toxin. Appl. Environ. Microbiol. 69, 1331–1334.

Park, H.W., Bideshi, D.K., Wirth, M.C., Johnson, J.J., Walton, W.E., Federici, B.A., 2005. Recombinant larvicidal bacteria with markedly improved efficacy against *Culex* vectors of West Nile virus. Am. J. Trop. Med. Hyg. 72, 732–738.

Paul, A., Harrington, L.C., Zhang, L., Scott, J.G., 2005. Insecticide resistance in *Culex pipiens* in New York. J. Am. Mosq. Control Assoc. 21, 305–309.

Pérez, C., Munoz-Garcia, C., Portugal, L.C., Sánchez, J., Gill, S.S., Soberón, M., Bravo, A., 2007. *Bacillus thuringiensis* subsp. *israelensis* Cyt1Aa enhances activity of Cry11Aa toxin by facilitating the formation of a pre-pore oligomeric structure. Cell. Microbiol. 9, 2931–2937.

Pont, D., Franquet, E., Tourenq, J.N., 1999. Impact of different *Bacillus thuringiensis* variety *israelensis* treatments on a chironomid (Diptera: Chironomidae) community in a temporary marsh. J. Econ. Entomol. 92, 266–272.

Purcell, B.H., 1981. Effects of *Bacillus thuring*iensis var. *israelensis* on *Aedes taeniorhynchus* and some non-target organisms in the salt marsh. Mosq. News 41, 476–484.

Ramoska, W.A., Hopkins, T.L., 1981. Effects of mosquito feeding behavior on *Bacillus sphaericus* efficacy. J. Invertebr. Pathol. 37, 269–272.

Ramoska, W.A., Pacey, C., 1979. Food availability and period of exposure as factors of *Bacillus sphaericus* efficacy on mosquito larvae. J. Econ. Entomol. 72, 523–525.

Ramoska, W.A., Burgess, J., Singer, S., 1978. Field application of a bacterial insecticide. Mosq. News 38, 57–60.

Rodcharoen, J., Mulla, M.S., 1994. Resistance development in *Culex quinquefasciatus* (Diptera: Culicidae) to *Bacillus sphaericus*. J. Econ. Entomol. 87, 1133–1140.

Rodcharoen, J., Mulla, M.S., 1995. Comparative ingestion rates of *Culex quinquefasciatus* (Diptera: Culicidae) susceptible and resistant to *Bacillus sphaericus*. J. Invertebr. Pathol. 66, 242–248.

Rodcharoen, J., Mulla, M.S., 1996. Cross-resistance to *Bacillus sphaericus* strains in *Culex quinquefasciatus*. J. Am. Mosq. Control Assoc. 12, 247–250.

Rodcharoen, J., Mulla, M.S., Chaney, J.D., 1991. Microbial larvicides for the control of nuisance aquatic midges (Diptera: Chironomidae) inhabiting mesocosms and man-made lakes in California. J. Am. Mosq. Control Assoc. 7, 56–62.

Sandoski, C.A., Yates, M.M., Olson, J.K., Meisch, M.V., 1985. Evaluation of Beecomist-applied *Bacillus thuringiensis* (H-14) against *Anopheles quadrimaculatus* larvae in rice fields. J. Am. Mosq. Control Assoc. 2, 316–319.

Siegel, J.P., 2012. Testing the pathogenicity and infectivity of entomopathogens to mammals. In: Lacey, L.A. (Ed.), Manual of Techniques in Invertebrate Pathology, second ed. Academic Press/Elsevier, San Diego, CA, pp. 441–450.

Siegel, J.P., Shadduck, J.A., 1990. Mammalian safety of *Bacillus thuringiensis israelensis*. In: de Barjac, H., Sutherland, D.J. (Eds.), Bacterial Control of Mosquitoes and Black Flies. Rutgers University Press, New Brunswick, NJ, pp. 202–217.

Sinègre, G., Babinot, M., Quermel, J.M., Gavon, B., 1994. First field occurrence of *Culex pipiens* resistance to *Bacillus Sphaericus* in southern France. In: Proc 8th European Meet Soc Vector Ecol, September 5–8, 1994; Barcelona, Spain, p. 17.

Su, T., 2008. Evaluation of water soluble pouches of *Bacillus sphaericus* applied as pre-hatch treatment against *Culex* mosquitoes in simulated catch basins. J. Am. Mosq. Control Assoc. 24, 54–60.

Su, T., 2016. Resistance and its management to microbial and insect growth regulator larvicides in mosquitoes. In: Trdan, S. (Ed.), Insecticides Resistance. InTech, Europe, Rijeka, Croatia, pp. 135–154.

Su, T., Cheng, M.L., 2012. Resistance development in *Culex quinquefasciatus* to spinosad: a preliminary report. J. Am. Mosq. Control Assoc. 28, 263–267.

Su, T., Cheng, M.L., 2014a. Laboratory selection of resistance to spinosad in *Culex quinquefasciatus* (Diptera: Culicidae). J. Med. Entomol. 50, 421–427.

Su, T., Cheng, M.L., 2014b. Cross resistances in spinosad – resistant *Culex quinquefasciatus* (Diptera: Culicidae). J. Med. Entomol. 50, 428–435.

Su, T., Mulla, M.S., 1999a. Field evaluation of new water-dispersible granular formulations of *Bacillus thuringiensis* ssp. *israelensis* and *Bacillus sphaericus* against *Culex* mosquitoes in microcosms. J. Am. Mosq. Control Assoc. 15, 356–365.

Su, T., Mulla, M.S., 1999b. Microbial agents *Bacillus thuringiensis* ssp. *israelensis* and *Bacillus sphaericus* suppress eutrophication, enhance water quality, and control mosquitoes in microcosms. J. Environ. Entomol. 28, 761–767.

Su, T., Mulla, M.S., 2005. Toxicity and effects of microbial mosquito larvicides and larvicidal oil on the development and fecundity of the tadpole shrimp *Triops newberryi* (Packard) (Notostraca: Triopsidae). J. Vector Ecol. 30, 107–114.

Su, T., Webb, J.P., Meyer, R.P., Mulla, M.S., 2003. Spatial and temporal distribution of mosquitoes in underground storm drain systems in Orange County, California. J. Vector Ecol. 28, 79–89.

Vasquez-Gomez, M., 1983. Investigations of the Possibility of Resistance to *Bacillus thguringiensis* Ser. H-14 in *Culex quinquefasciatus* through Accelerated Selection Pressure in the Laboratory. University of California 1983, Riverside.

Walker, E.D., 1995. Effect of low temperature on feeding rate of *Aedes stimulans* larvae and efficacy of *Bacillus thuringiensis* var. *israelensis* (H-14). J. Am. Mosq. Control Assoc. 11, 107–110.

Wipfli, M.S., Merritt, R.W., 1994. Effects of *Bacillus thuringiensis* var. *israelensis* on nontarget benthic insects through direct and indirect exposure. J. North Am. Benthol. Soc. 13, 190–205.

Wirth, M.C., 2010. Mosquito resistance to bacterial larvicidal toxins. Open Toxinol. J. 3, 126–140. http://dx.doi.org/10.2174/187541470 1003010126.

Wirth, M.C., Georghiou, G.P., 1997. Cross-resistance among CryIV toxins of *Bacillus thuringiensis* subsp. *israelensis* in *Culex quinquefasciatus* (Diptera: Culicidae). J. Econ. Entomol. 90, 1471–1477.

Wirth, M.C., Georghiou, G.P., Federici, B.A., 1997. CytA enables CryIV endotoxins of *Bacillus thuringiensis* to overcome high levels of CryIV resistance in the mosquito, *Culex quinquefasciatus*. Proc. Natl. Acad. Sci. USA 94, 10536–10540.

Wirth, M.C., Federici, B.A., Walton, W.E., 2000a. Cyt1A from *Bacillus thuringiensis* synergizes activity of *Bacillus sphaericus* against *Aedes aegypti* (Diptera: Culicidae). Appl. Environ. Microbiol. 66, 1093–1097.

Wirth, M.C., Georghiou, G.P., Malik, J.I., Hussain, G., 2000b. Laboratory selection for resistance to *Bacillus sphaericus* in *Culex quinquefasciatus* (Diptera: Culicidae) from California, USA. J. Med. Entomol. 37, 534–540.

Wirth, M.C., Walton, W.E., Federici, B.A., 2000c. Cyt1A from *Bacillus thuringiensis* restores toxicity of *Bacillus sphaericus* against resistant *Culex quinquefasciatus* (Diptera: Culicidae). J. Med. Entomol. 37, 401–407.

Wirth, M.C., Ferrari, J.A., Georghiou, G.P., 2001. Baseline susceptibility to bacterial insecticides in populations of *Culex pipiens* complex (Diptera: Culicidae) from California and from the Mediterranean Island of Cyprus. J. Med. Entomol. 94, 920–928.

Wirth, M.C., Jiannino, J.J., Federici, B.A., Walton, W., 2004. Synergy between toxins of *Bacillus thuringiensis* subsp. *israelensis* and *Bacillus sphaericus*. J. Med. Entomol. 41, 935–941.

Wirth, M.C., Jiannino, J.A., Federici, B.A., Walton, W.E., 2005a. Evolution of resistance toward *Bacillus sphaericus* or a mixture of *B. sphaericus*+Cyt1A from *Bacillus thuringiensis*, in the mosquito, *Culex quinquefasciatus* (Diptera: Culicidae). J. Invertebr. Pathol. 88, 154–162.

Wirth, M.C., Park, H.W., Walton, W.E., Federici, B.A., 2005b. Cyt1A of *Bacillus thuringiensis* delays resistance to Cry11A in the mosquito *Culex quinquefasciatus*. Appl. Environ. Microbiol. 71, 185–189.

Wirth, M.C., Yang, Y., Walton, W.E., Federici, B.A., Berry, C., 2007. Mtx toxins synergize *Bacillus sphaericus* and Cry11Aa against susceptible and insecticide-resistant *Culex quinquefasciatus* larvae. Appl. Environ. Microbiol. 73, 6066–6071.

Wirth, M.C., Walton, W.E., Federici, B.A., 2010. Evolution of resistance to the *Bacillus sphaericus* Bin toxin is phenotypically masked by combination with the mosquitocidal proteins of *Bacillus thuringiensis* subspecies *israelensis*. Environ. Microbiol. 12, 1154–1160.

Wirth, M.C., Berry, C., Walton, W.E., Federici, B.A., 2013. Mtx toxins from *Lysinibacillus sphaericus* enhance mosquitocidal cry-toxin activity and suppress cry-resistance in *Culex quinquefasciatus*. J. Insect Pathol. 115, 62–67.

Yap, H.H., 1990. Field trials of *Bacillus sphaericus* for mosquito control. In: de Barjac, H., Sutherland, D.J. (Eds.), Bacterial Control of Mosquitoes and Black Flies. Rutgers University Press, New Brunswick, NJ, pp. 307–320.

Yousten, A.A., Benfield, E.F., Campbell, R.P., Foss, S.S., Genthner, F.J., 1991. Fate of *Bacillus sphaericus* 2362 spores following ingestion by non-target invertebrates. J. Invertebr. Pathol. 58, 427–435.

Yousten, A.A., Madhekar, N., Wallis, D.A., 1984. Fermentation conditions affecting growth, sporulation, and mosquito larval toxin formation by *Bacillus sphaericus*. Dev. Ind. Microbiol. 25, 757–762.

Zahiri, N.S., Mulla, M.S., 2003. Susceptibility profile of *Culex quinquefasciatus* (Diptera: Culicidae) to *Bacillus sphaericus* on selection with rotation and mixture of *B. sphaericus* and *B. thuringiensis israelensis*. J. Med. Entomol. 40, 672–677.

Zahiri, N.S., Su, T., Mulla, M.S., 2002. Strategies for the management of resistance in mosquito to the microbial control agent *Bacillus sphaericus*. J. Med. Entomol. 39, 513–520.

Zahiri, N.S., Federici, B.A., Mulla, M.S., 2004. Laboratory and simulated field evaluation of a new recombinant of *Bacillus thuringiensis* ssp. *israelensis* and *Bacillus sphaericus* against *Culex* mosquito larvae (Diptera: Culicidae). J. Med. Entomol. 41, 423–429.

Chapter 28

Microbial Control of Medically Important Mosquitoes in Tropical Climates

L.A. Lacey

IP Consulting International, Yakima, WA, United States

28.1 INTRODUCTION

Vector-borne diseases account for more than 17% of all infectious human diseases, causing more than one million deaths annually (WHO, 2014, 2016b). Hundreds of species of insects, mites, and ticks are involved in disease transmission, but mosquitoes are by far the most medically important (Foster and Walker, 2009). Malaria and dengue alone pose a risk for half the world's population in over 100 countries in the tropics and sub-tropics (Stanaway et al., 2016; WHO, 2016a, 2016b).

Malaria, which is caused by protozoan parasites in the genus *Plasmodium* and transmitted by *Anopheles* mosquitoes (Warrell and Gilles, 2002; Hay et al., 2004), is responsible for more than 600,000 deaths every year globally, most of them children under 5 years of age (Hay et al., 2004; WHO, 2015c, 2016a). Ninety percent of all malaria deaths occur in sub-Saharan Africa. Increasing worldwide travel and trade, unplanned urbanization, and environmental changes due to shifting rainfall patterns or agricultural practices are having a significant impact on mosquito-borne disease transmission (WHO, 2015c).

Arthropod-borne viruses (arboviruses) such as dengue, chikungunya, Rift Valley fever, and West Nile and Zika viruses are transmitted by various mosquito species and are emerging in countries or regions where they were previously unknown (WHO, 2014, 2016b). The global incidence of dengue has grown dramatically in recent decades (WHO, 2012; Bhatt et al., 2013). The dengue virus is transmitted through bites from *Aedes* mosquitoes with an estimated 390 million infections per annum and approximately 96 million clinical manifestations (dengue fever, dengue hemorrhagic fever; WHO, 2012, 2015a; Bhatt et al., 2013).

Mosquitoes are also involved in the transmission of parasitic nematodes causing lymphatic filariasis in 58 tropical or subtropical countries. It is estimated that over 120 million people are currently infected, with about 40 million disfigured and incapacitated by the disease often referred to as elephantiasis (WHO, 2015b). Most of these mosquito-borne

diseases are preventable through informed protective measures, including the control of the mosquito vectors. The prevention and control of some of these diseases, for example dengue, solely depend on effective vector control measures.

The traditional means of mosquito vector control has been with broad spectrum residual chemical insecticides targeting adult mosquitoes. The drawbacks of this strategy have led to the development of the concept of integrated vector management (IVM), which aims to maximize alternative means of control along with the judicious use of insecticides. Among these alternatives is the biological control of mosquito larvae (Lacey and Orr, 1994; Cameron and Lorenz, 2013).

The use of microbial control agents (MCAs) is increasingly becoming a predominant form of biological control of insects, including mosquitoes in Europe (Chapter 26), North America (Chapter 27), and to a more limited extent in the tropics. A variety of MCAs that have been used or are being actively studied for mosquito control include the water mold *Lagenidium giganteum* (Oomycetes; Kerwin, 2007); nematodes in the family Mermithidae (Petersen, 1985; Platzer, 2007); viruses (Becnel and White, 2007); the fungi, *Culicinomyces clavisporus*, *Tolypocladium cylindrosporum*, *Coelomomyces* spp. (Federici, 1981; Roberts and Panter, 1985; Lacey and Undeen, 1986; Scholte et al., 2004a), *Metarhizium anisopliae*, and *Beauveria bassiana* (Scholte et al., 2004a, 2005; Blanford et al., 2005; Howard et al., 2010); and Microsporidia (Becnel et al., 2005; Becnel and Andreadis, 2014). However, the entomopathogenic bacteria, *Bacillus thuringiensis* subspecies *israelensis* (Bti) and *Lysinibacillus* (*Bacillus*) *sphaericus*, are the predominant MCAs used in IVM (Lacey, 2007; Silva-Filha et al., 2014). In this chapter, nematodes, *Romanomermis* spp. (Mermithidae), the entomopathogenic fungi *M. anisopliae* and *B. bassiana*, and the bacteria, Bti and *L. sphaericus*, will be emphasized for their use in tropical and subtropical habitats for the control of mosquitoes.

28.2 NEMATODA: MERMITHIDAE

Nematode species in the family Mermithidae are parasitic in several insect species but are best known from mosquitoes. Mermithid species in seven genera (*Culicidermis, Empidomermis, Hydromermis, Octomyomermis, Perutilimermis, Romanomermis,* and *Strelkovimermis*) are known to parasitize mosquito larvae. The most well known and best studied are *Romanomermis* species (Petersen, 1985; Platzer, 2007; Platzer et al., 2005). Fourteen of the 15 species in the genus attack mosquitoes (Platzer, 2007). The mermithid life cycle consists of free-living aquatic and obligatory parasitic stages. After penetration of early instar mosquito larvae by the preparasitic infective stage (L1), the nematodes molt and feed within the host larva (Fig. 28.1). When the fully developed stage of the nematode (L3) leaves the host insect by breaking through the integument or gut wall, the mosquito larva invariably dies (Fig. 28.2). This stage settles into the substratum, molts to the adult stage (L4), mates,

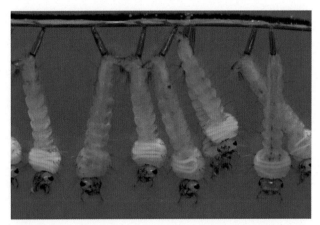

FIGURE 28.1 *Romanomermis culicivorax* L3 larvae coiled within *Culex quinquefasciatus* larvae. *Photograph courtesy of Edward Platzer.*

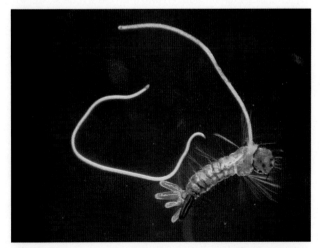

FIGURE 28.2 *Romanomermis iyengari* L3 larvae emerged and emerging from *Aedes aegypti* larva. *Photograph by R. Pérez-Pacheco courtesy of Edward Platzer.*

and lays eggs. The duration of development from egg to adult depends on the species of mermithid, temperature, moisture, and other factors. This information is presented in Sections 28.2.1 and 28.2.2 for *Romanomermis culicivorax* and *Romanomermis iyengari*. Although there are fewer publications on *Strelkovimermis spiculatus* compared to *Romanomermis* spp., good efficacy, tolerance for salinity and organic pollution, and ability to cause epizootics and become established in application sites have been demonstrated (Camino and Garcia, 1991; Campos et al., 1993; Micieli and García, 1999; Campos and Sy, 2003; Achinelly and Micieli, 2013).

28.2.1 *Romanomermis culicivorax*

Since its efficacy was initially demonstrated (Petersen, 1985), *R. culicivorax* has been extensively studied in the subtropical southern United States and subsequently in other locations in the United States, Cuba, Mexico, El Salvador, Colombia, Azerbaijan, and Iran and shown to be an effective control agent with the capability of becoming established in various mosquito habitats (Petersen and Willis, 1974, 1975; Brown-Westerdahl et al., 1982; Kerwin and Washino, 1985; Petersen, 1985; Rojas et al., 1987; Walker et al., 1988; Zaim et al., 1988; Alirzaev et al., 1990; Santamarina et al., 1992; Santamarina and Pérez-Pacheco, 1997; Platzer et al., 2005; Pérez-Pacheco et al., 2009).

28.2.1.1 Factors Affecting Efficacy of Romanomermis culicivorax

The species and age of mosquito larvae significantly affect infection. *Anopheles* species are more susceptible than Culicine larvae due to the common feeding sites of larvae and the aggregation of infective preparasites at the surface of the water (Petersen, 1985). First instars are the most susceptible to infection by preparasites with susceptibility declining with age. The life cycle from egg hatch to ovipositing adults depends on the temperature and habitat. The parasitic phase of *R. culicivorax* requires 7–8 days for development (Petersen, 1985). All of the acquisition of nutrient and development to L3 occurs in the larva during this phase. Platzer et al. (2005) observed that the period of time from emergence from the host until egg hatch was 21 days at 27°C.

Despite demonstrated efficacy, there are caveats to the microbial control success of *R. culicivorax*: it is ineffective in polluted (Petersen, 1985) and saline habitats (Brown and Platzer, 1977), pH outside of the range of 3.6–8.6 (Brown and Platzer, 1978; Petersen, 1985), and limited by low and high temperatures (Hughes and Platzer, 1977; Levi and Miller, 1977; Brown and Platzer, 1978). The optimal range of temperature for the development of *R. culicivorax* is 21–33°C (Platzer et al., 2005; Platzer, 2007; Petersen, 1985).

A minimum of 15°C is required for infectivity (Brown and Platzer, 1977), and temperatures above 37–44°C can be detrimental to free-living stages of the nematode (Levy and Miller, 1977). Petersen et al. (1978) observed drastically different results in the control of *Anopheles albimanus* when *R. culicivorax* preparasites were applied to lake margins in El Salvador when there was wind-caused wave action compared to when there were no waves (Petersen et al., 1978). Other limiting factors can be the presence of predators and pathogens of nematodes, use of particular pesticides (especially nematicides), and drying of temporary habitats before the nematode's life cycle can be completed.

28.2.1.2 Field Testing of Romanomermis culicivorax

Several field trials were conducted in Louisiana, United States, that demonstrated the efficacy of *R. culicivorax* for the control of *Anopheles* spp. (Petersen and Willis, 1974). Treatment rates of 1200 and 2400 preparasites/m² produced 76% and up to 90% infection rates, respectively. The first large-scale field test of *R. culicivorax* was conducted in El Salvador for the control of *An. albimanus* and *Anopheles punctipennis*. At treatment rates of 2400–4800 preparasites/ha, the nematode produced up to 94% larval reduction (Petersen et al., 1978). Santamarina and Pérez-Pacheco (1997) reported on the reduction of *An. albimanus*, *Culex atratus*, *Culex quinquefasciatus*, and *Psorophora confinnis* larvae for 9 weeks following the treatment of natural breeding sites in Cuba with 1000 preparasites/m². Mean levels of parasitism ranged from 82% to 83% for *Cx. quinquefasciatus*, 89–95% for *An. albimanus*, and 78–80% for *Cx. atratus* and *P. confinnis*. Pérez-Pacheco et al. (2009) reported good control of *An. albimanus* and *Culex nigripalpus* in several natural habitats in Oaxaca, Mexico. Inoculation rates of 500 and 1000 preparasites/m² produced 74–88% infection and 77–97% infection, respectively. They demonstrated significantly greater infection rates in *An. albimanus* (88–97%) than *Cx. nigripalpus* (78–82%). After the introduction of *R. culicivorax* eggs and adults into all breeding sites of *An. albimanus* surrounding two towns along the Pacific north coast of Colombia, Rojas et al. (1987) reported a progressive reduction in the prevalence of malaria in school children over a 2-year period. Additional information on *R. culicivorax* for mosquito control and the biotic and abiotic factors that affect efficacy in temperate climates is presented by Petersen (1985), Lacey and Lacey (1990), and Platzer et al. (2005).

28.2.2 Romanomermis iyengari

R. iyengari is indigenous to India and has received more international attention than *R. culicivorax*, especially in the tropics. The first discovery, taxonomic history, and earliest experiments were summarized by Platzer (2007). The strain of *R. iyengari* used for all published research since 1978 was isolated in Pondicherry, India, by Gajanana et al. (1978). It has been shown to be effective in controlling mosquitoes breeding in natural habitats, rice fields, flooded grasslands, tree holes, and other habitats (Table 28.1). The documented natural and experimental host range in India includes the following: in natural and agricultural habitats: *Anopheles subpictus*, *Anopheles vagus*, *Anopheles barbirostris*, *Anopheles hyrcanus*, *Culex tritaeniorhyncus*, *Culex bitaeniorhyncus*, *Culex vishnui*, *Cx. quinquefasciatus*, and *Aedes albopictus* (Gajanana et al., 1978; Chandrahas and Rajagopalan, 1979; Paily et al., 1994, 2013; Chandhiron and Paily, 2015); and in laboratory bioassays only: *Culex sitiens*, *Aedes aegypti*, *Armigeres subalbatus*, *Mansonia annulifera*, *Anopheles stephensi*, and *Anopheles culicifacies* (Paily and Balaraman, 2000).

28.2.2.1 Factors Affecting Efficacy of Romanomermis iyengari

The effect of mosquito species, age, and environmental limitations of *R. iyengari* are nearly the same as those demonstrated for *R. culicivorax*. Although similar, the effects of temperature on developmental time and limitations are marginally different. The development of *R. iyengari* is significantly lengthened at temperatures below the optimal range of 25–30°C (Paily and Balaraman, 1990, 1994). Other than the slight effect of temperature on development, the life cycle of *R. iyengari* is identical to that of *R. culicivorax*. From egg to adult requires 9 days at 35°C to 14 days at 20°C. However, the higher temperature significantly shortens the survival of preparasites; LT_{50} at 35°C was 10.6h versus 105h at 20°C (Paily and Balaraman, 1994). In tree holes, Paily et al. (2013) observed that a high parasite to host ratio is detrimental to parasite recycling due to premature mortality of the host. They determined that a 3:1 concentration (parasite to host) could be more amenable to recycling and therefore persistence. Other effects of temperature and parasite to host ratio include the effect of parasite load on sex and the survival of emerging preadults (Paily and Balaraman, 1990).

Gajanana et al. (1978) and Paily and Jayachandran (1987) demonstrated the negative effect of highly polluted habitats on the ability of *R. iyengari* to infect *Cx. quinquefasciatus* and *Culex tritaeniorhynchus* larvae, respectively. However, Paily et al. (1994) demonstrated moderate levels of control for *An. subpictus* and *Cx. tritaeniorhynchus* in grasslands irrigated with sewage contaminated water and treated with high concentrations of *R. iyengari*. Similarly, Santamarina et al. (1993) reported moderate control of *Cx. quinquefasciatus* in spillways of oxidation ponds (Table 28.1). The negative effect of salinity on *R. iyengari* infectivity was observed by Pridantseva et al. (1990). However, Santamarina et al. (1996) demonstrated control of *Aedes taeniorhynchus* in saline

TABLE 28.1 Selected References on Field Research of *Romanomermis iyengari* for Control of Mosquitoes

Targeted Species	Habitat	Application Rate[a] and % Mortality/Infection	Country	References
Culex (fatigans) quinquefasciatus	Cesspools and soak pits	1.2×10^3–4.6×10^4 (cesspool) and 1.2×10^4–3.1×10^5 (soak pit), 1.2%	India	Gajanana et al. (1978)
Anopheles subpictus	Grasslands irrigated with sewage-contaminated water	5×10^3 pp/m², 37–54%		Paily et al. (1994)
Culex tritaeniorhyncus		5×10^3 pp/m², 70–72%, infected larvae found for 9 + months		
Aedes albopictus	Tree holes	3–4 pp/larva, 86–97%		Paily et al. (2013)
Aedes taeniorhynchus	Carbonate rock depressions	1000 pp/m², 80–100%	Cuba	Santamarina et al. (1996)
Anopheles albimanus, Culex nigripalpus	Small temporary natural sites	1000 pp/m², 80–100%, recycling noted		Santamarina et al. (1992, 1993) and Santamarina (1994)
Cx. quinquefasciatus	Spillways of oxidation ponds	1000 pp/m², 75–80%		Santamarina et al. (1992, 1993)
Anopheles pseudopunctipennis	Various natural breeding sites	2–3×10^3 pp/m², 85–100%	Mexico	Santamarina et al. (1999)
	Natural breeding sites in Oaxaca	3×10^3 pp/m², 39–100%		Pérez-Pacheco et al. (2005)
An. pseudopunctipennis	Reservoirs	10^3 pp/m², 85–95%		Pérez-Pacheco et al. (2009)
Anopheles albitarsis, Anopheles rondoni	Various natural breeding sites	2000 pp/m², 85–97%	Brazil	Santamarina and Bellini (2000)
Anopheles gambiae	Natural habitats	3.5–5.0×10^3 pp/m², 89–94%	Benin	Abagli et al. (2012) and Alavo et al. (2015)
Anopheles martinius, Anopheles hyrcanus, Culex modestus	Various natural habitats	0.6–6.5×10^3 pp/m², 67% infection	Uzbekistan	Pridantseva et al. (1990)

[a]*Preparasitic (pp) stage.*

water in carbonate rock depressions. The effects of conductivity and pH on larvicidal activity are reported by Bheema Rao et al. (1979) and Paily et al. (1991). Drying of mosquito breeding sites and the nematode's continued persistence and subsequent infection of mosquito larvae in reflooded breeding sites attests to the desiccation tolerance of *R. iyengari* (Chandhiron and Paily, 2015). Paily and Balaraman (1993) demonstrated the effect of moisture on oviposition. As habitats become dry, adult females will oviposit lower in the substratum following a moisture gradient.

28.2.2.2 Field Testing of Romanomermis iyengari

The results of efficacy testing in a variety of tropical habitats in the Western and Eastern Hemispheres are presented in Table 28.1. Of particular interest for using *R. iyengari* as

a biological control agent is its ability to recycle and persist in a variety of natural and agricultural larval habitats in India, the putative center of origin of the species. For example, Paily et al. (1994) observed persistence of the nematode in irrigated grasslands over 3.5 months after treatment. Chandhiron and Paily (2015) reported on the natural appearance of *R. iyengari* in a periodically dried and cultivated rice field habitat after over 35 years since earlier observations of natural occurrence of the nematode. The level of parasitization was higher than those originally reported. Samples of *R. iyengari* were imported into Cuba, and initial field studies revealed its potential as a biological control agent for the management of a number of species, including vectors of the causal agent of malaria (Santamarina et al., 1993, 1996). A wide range of mosquito species, including, but not limited to, *An. albimanus*, *Cx. quinquefasciatus*, *Cx. nigripalpus*, and *Ae. taeniorhynchus*, have been successfully controlled in Cuba

(Table 28.1). Santamarina (1994) also demonstrated recycling of the nematode for several months following its introduction into natural breeding sites of *An. albimanus*. In this regard, *R. iyengari* can be considered as a successfully introduced classical biological control agent. The nematode was subsequently introduced into Mexico (Santamarina et al., 1999), Brazil (Santamarina and Bellini, 2000), and Benin (Abagli et al., 2012; Alavo et al., 2015). Evidence of recycling and persistence was reported in Mexico by Santamarina et al. (1999) and Pérez-Pacheco et al. (2009).

28.2.3 Positive and Negative Factors Affecting Widespread Use and Commercialization of *Romanomermis* Species

They are relatively easy to produce locally at low cost, can become established, and control larvae in more permanent habitats, resulting in extended control and providing an insecticide-free intervention that could be added to other tools for IVM and insecticide resistance management. Methods for the mass production of *Romanomermis* spp. (Petersen and Willis, 1972; Gajanana et al., 1978; Paily, 1990; Santamarina, 1996; Santamarina and Bellini, 2000; Perez-Pacheco and Flores, 2005; Alavo et al., 2015) have enabled larger scale testing and research in Cuba, Mexico, El Salvador, Brazil, India, Benin, and the US subtropics (Petersen et al., 1978; Petersen, 1985; Paily et al., 1994; Santamarina et al., 1996, 1996; Pérez-Pacheco et al., 2005; Santamarina and Bellini, 2000; Abagli et al., 2012). The principal obstacles for the widespread use and commercialization of mermithid parasites of mosquitoes are the requirement for in vivo production, short shelf life, thermal limitations, detrimental effects of saline, and limited use in many organically enriched habitats (cesspools, septic ditches, dairy drains, etc.).

28.3 *METARHIZIUM ANISOPLIAE S.L.* AND *BEAUVERIA BASSIANA* (FUNGI: ASCOMYCOTA: HYPOCREALES)

Both of these entomopathogenic fungi have proved to be efficacious for the control of several insect and mite pests (Chapters 5, 12–24, and 29). In addition to efficacy, the attractiveness of *M. anisopliae* and *B. bassiana* is increased by other desirable traits: inexpensive methods for mass production (Chapter 9), good shelf life under appropriate conditions of humidity and temperature (Chapter 9), and safety for vertebrates and most invertebrate nontarget organisms (Zimmermann, 2007a,b). Additionally, their ubiquitous distribution in nature would not encounter the obstacles of importation of exotic species if isolated within the country or region where they will be used.

28.3.1 Evaluation of *Metarhizium anisopliae* as a Mosquito Larvicide

This fungus was proposed as a mosquito larvicide in the early 1980s and was investigated for efficacy by Daoust and Roberts (1982, 1983) and others since then (Roberts and Panter, 1985; Lacey et al., 1988a; Alves et al., 2002; Bukhari et al., 2011). Although *M. anisopliae* was shown to have larvicidal activity, the successes and ease of application of Bti and *L. sphaericus* for larval control have eclipsed the practicality of further development for larval control.

28.3.2 Development of *Metarhizium anisopliae* and *Beauveria bassiana* for Control of Adult *Anopheles* spp.

Renewed interest in *M. anisopliae* as an MCA for mosquito control began with the demonstration of adulticidal activity for malaria vectors and the potential for a reduction in malaria transmission (Scholte et al., 2003, 2005, 2008; Blanford et al., 2005; Thomas and Reed, 2007). The time from exposure of female mosquitoes to *M. anisopliae* until death is dependent on isolate and dosage but not duration of exposure (Scholte et al., 2003). The LT_{50} for *Anopheles gambiae s.s.* exposed to 1.6×10^7 or 1.6×10^{10} conidia/m^2 is 9.7 and 5.9 days, respectively (Scholte et al., 2003). Mnyone et al. (2009) reported similar findings for *M. anisopliae* and *B. bassiana* against *An. gambiae*. Conidia of both species remained infective for up to 28 days postapplication.

In addition to mortality, temporarily surviving female mosquitoes infected with *M. anisopliae* reduce blood and nectar feeding and have lower fecundity (Scholte et al., 2006; Blanford et al., 2012; Ondiaka et al., 2015) and a reduced ability to transmit the sporozoites responsible for causing malaria (Blanford et al., 2005). Fang et al. (2011) improved the ability of the fungus to block transmission by genetically modifying its genome to express scorpine (an antimicrobial toxin), resulting in an up to 90% reduction of sporozoites.

Farenhorst et al. (2009, 2010) demonstrated the beneficial effect that *M. anisopliae* has on the potential to counter the development of insecticide resistance to permethrin. Similarly, Howard et al. (2010) showed that resistance to pyrethroids in *An. gambiae* led to increased susceptibility to *M. anisopliae* and *B. bassiana*.

28.3.3 Development of Delivery Systems to Improve the Mosquitocidal Efficacy of Entomopathogenic Fungi

Bed nets have been used for hundreds of years to protect humans from mosquito bites and mosquito-borne disease. They were further improved by treating them with pyrethroid insecticides (eg, deltamethrin or permethrin), resulting in the death of adult mosquitoes, reducing their population densities, and lowering human morbidity and mortality due to malaria (Bermejo and Veeken, 1992; Greenwood, 1995; Nevill et al., 1996; Rowland et al., 1997; Hawley et al., 2003; Russell et al., 2010). Similarly, Forenhorst et al. (2011) demonstrated good mosquitocidal activity for *B. bassiana* against *An. gambiae*. After exposing female mosquitoes for 30 min to polyester or cotton netting (400 cm^2) that had been sprayed with 5 mL of a Shellsol solvent-formulated suspension of 5×10^{11} conidia per mL, >90% of the mosquitoes died within 10 days.

The treatment of other surfaces used as entry points to domiciles (eaves) and indoor resting sites (walls, curtains) for endophilic mosquitoes has been shown to be an effective method for transferring conidia to mosquitoes. However, the type of fungus-treated surfaces upon which mosquitoes alight result in significantly different infection rates. Mnyone et al. (2010) evaluated *M. anisopliae* and *B. bassiana* applied 2×10^{10} conidia/m^2 on mud panels, black cotton cloth, and polyester netting against adult *An. gambiae s.s.* The mosquitoes were exposed to the treated surfaces 2, 14, and 28 days after conidia were applied. Both fungal treatments caused significantly higher mortality in the mosquitoes exposed to *M. anisopliae* conidia on mud panels and had a higher mortality rate than those on netting or cotton cloth. Mosquitoes exposed to *B. bassiana* conidia on mud panels or cotton cloth responded in a similar manner. Although adulticidal activity of both fungi declined over time, enough conidia remained infective 28 days postapplication to kill 73–82% of exposed mosquitoes, respectively, within 14 days of exposure. Lwetoijera et al. (2010) constructed stations baited with carboxylic acid, ammonia, and carbon dioxide as a method of delivering *M. anisopliae* conidia to *Anopheles arabiensis* females. By spraying eave baffles and panels with 3×10^{10} conidia/m^2, 95% of mosquitoes that entered and exited the station died within 14 days and subsequently produced conidia on the surfaces of 86% of the cadavers. Extending the adulticidal activity beyond the initially treated mosquitoes, Scholte et al. (2004b) demonstrated that *M. anisopliae*-contaminated *An. gambiae* females could disseminate fungal inoculum between mosquitoes during mating activity under laboratory conditions.

Forenhurst et al. (2008) evaluated *M. anisopliae* in clay water pots used as resting sites by *Anopheles* spp. adults. An application of 10^{10} conidia/m^2 infected an average of 92% of *An. gambiae* and *Anopheles funestus*, demonstrating the potential utility of this exposure tactic. Brooke et al. (2012) also proposed clay pots to deliver *M. anisopliae* and *B. bassiana* to *An. arabiensis* and other *Anopheles* spp. in Northern Kwazulu-Natal, South Africa.

28.3.4 Control of Adults of Other Mosquito Species With Entomopathogenic Fungi

Culex spp. and *Aedes* spp. are often found in peridomestic habitats (water tanks, small containers, etc.) and are important vectors of the causal agents of lymphatic filariasis and dengue, respectively. Scholte et al. (2003, 2007) demonstrated infection of two principal vectors of dengue virus, *Ae. aegypti* and *Ae. albopictus*, and *Cx. quinquefasciatus*, an important vector of *Wuchereria bancrofti*, the causal agent of Bancroftian filariasis, with *M. anisopliae*. Subsequently, Paula et al. (2011) showed that a combination of a 10^9 conidia/mL suspension of *M. anisopliae* and sublethal amounts of imidacloprid synergized the adulticidal activity of the fungus against *Ae. aegypti*. Snetselaar et al. (2014) developed a device that can lethally contaminate adult *Ae. aegypti* with *B. bassiana* at breeding sites.

28.4 BACTERIA: BACILLALES

Hundreds of bacterial species and varieties have been isolated from mosquitoes, closely related Nematocera, and mosquito habitats, but only two species, Bti and *L. sphaericus*, have been commercialized and developed for large-scale operational programs for the control of mosquitoes in Europe and North America. In many of these programs Bti and *L. sphaericus* are the main, if not only, larvicides that are employed. However, with some exceptions, their use as the predominant means of mosquito control in large operational control programs in the tropics is not common. The biotic and abiotic factors (ie, mosquito species and feeding habits, temperature, water depth and quality, vegetative cover, and formulation and application methodology) affecting the efficacy of Bti and *L. sphaericus* for the control of tropical mosquitoes are basically the same as those reported for mosquito larvae from temperate regions. For greater detail, see Lacey and Undeen (1986), Lacey (2007), and Chapters 26 and 27.

28.4.1 *Bacillus thuringiensis* subspecies *israelensis*

The efficacy of Bti as a larvicide of mosquitoes (Culicidae) and black flies (Simuliidae) and its safety in sensitive aquatic environments, both in terms of nontarget invertebrates and vertebrates, have been the major factors responsible for its rapid registration and commercialization (Goldberg and Margalit, 1977; Lacey and Undeen, 1986; Siegel and Shadduck, 1990; Boisvert and Boisvert, 2000; Lacey and Siegel, 2000; Lacey and Merritt, 2003). Soon after its discovery, developmental research on Bti led to the accelerated adoption of the bacterium as a principal means of control in many western countries (Chapters 26 and 27).

Good efficacy of experimental and commercial preparations has been demonstrated against dozens of tropical and subtropical mosquito species exposed to low concentrations of the bacterium in a myriad of habitats, ranging from small containers, used tires, ditches, and temporary puddles to rice fields and pond and lake margins (Tables 28.2 and 28.3). Some of the more difficult species to control are those with desiccation-resistant eggs that hatch soon after they are inundated with water (used tires, ditches), species in cryptic or unseen habitats (catch basins, tree holes), sites with dense vegetation (rice fields), and in highly polluted habitats (dairy drains, septic ditches, etc.). Formulation and application methods tailored to particular habitats and species of mosquitoes have helped overcome problems associated with reaching targeted species. For example, cryptic breeding sites and tires that are periodically inundated by rain can be treated with granular or tablet formulations of Bti that provide larvicidal toxins to larvae already present in such habitats as well as larvae that hatch soon after eggs are immersed in water (Siegel and Novak, 1999; Batra et al., 2000; Harwood et al., 2015). Granular formulations that penetrate foliage and float at the surface of the water or release active ingredients at the surface enable toxins to stay in the feeding zone of mosquito larvae longer. This is especially important for anopheline species. Another notable example is the use of the Beecomist UVL (ultralow volume) generator for aerial applications of Bti and *L. sphaericus* liquid formulations for the control of *Anopheles quadrimaculatus* larvae in subtropical rice fields in the United States. This method provides effective control at very low application rates (0.29–0.58 L/ha; Sandoski et al., 1985; Lacey et al., 1986). The small droplets (<80 μm) produced with the device penetrate dense foliage and remain in the feeding zone of *An. quadrimaculatus* larvae longer than larger droplets or granules that sink. The efficacy of Bti formulations against several tropical mosquitoes in a variety of habitats is presented in Table 28.2 for anopheline species and Table 28.3 for other species of medical importance.

Several large field trials in sub-Saharan Africa, Central and South America, and Western and Southeast Asia have demonstrated the efficacy of Bti as a larvicide for anopheline and culicine mosquitoes and a reduction in biting rates and in some cases a decrease in disease transmission (Perich et al., 1990; Karch et al., 1991; Xu et al., 1992; Lee et al., 1994; Kroeger et al., 1995b; Kumar et al., 1998; Fillinger et al., 2003, 2008, 2009; Fillinger and Lindsay, 2006; Geissbühler et al., 2009; Mwangangi et al., 2010; Maheu-Giroux and Castro, 2013; Nartey et al., 2013). In Tanzania, the microbial larvicides Bti and *L. sphaericus* have been used operationally since 2006 in the Urban Malaria Control Program in Dar es Salaam (Fillinger et al., 2008; Geissbühler et al., 2009). Bti can also be a substantial complementary component of IVM programs along with other biological control agents (naturally occurring and introduced), larval breeding

site reduction and the use of insecticide-treated bed nets (ITNs; see Section 28.5).

A disadvantage of Bti is the necessity for frequent reapplication due to denaturing of the proteinaceous toxins in organically polluted habitats and/or settling of the toxins from the feeding zone of some species (eg, *Anopheles* spp.; Hougard et al., 1983; Silapanuntakul et al., 1983; Mulla et al., 1993). However, in some habitats, such as small containers with relatively clear shallow water, Bti may persist for a considerable amount of time (see Section 28.5.2).

28.4.2 *Lysinibacillus sphaericus*

The second most used MCA for the control of mosquito larvae is *L. sphaericus* (serotype 5a5b: isolate 2362 used for most commercial products). It is not active against other families of Nematocera and has a narrower range of susceptible species within the Culicidae (ie, *Ae. aegypti* and other *Stegomya* species are not affected). Nevertheless, it is and has been successfully used for the control of *Culex* spp., *Anopheles* spp., *Psorophora* spp., and other mosquito species (Table 28.4 and Chapters 26 and 27). A binary endotoxin (Bin A Bin B), produced at the time of sporulation, is responsible for the larvicidal activity, and the end result is similar to that observed with Bti (ie, rupture of epithelial cells in the midgut). The molecular biology and mode of action of the binary toxin were reviewed by Charles et al. (1996) and Silva-Filha et al. (2014). The safety of *L. sphaericus* to nontarget organisms, including mosquito predators, is well documented (Lacey and Mulla, 1990; Blanco Castro et al., 2002; Lacey and Merritt, 2003).

The efficacy of *L. sphaericus* for the control of anopheline vectors of the causal agents of malaria (*Plasmodium* spp.) has been reported from several studies, ranging from southern Mexico (Arredondo-Jimenez et al., 1990), Honduras (Blanco Castro et al., 2002), Guatemala (Blanco Castro et al., 2000), Nicaragua (Rivera et al., 1997), Brazil (Galardo et al., 2013), Venezuela (Moreno et al., 2014), across sub-Saharan Africa (Karch et al., 1991, 1992; Ragoonanansingh et al., 1992; Skovmand and Baudin, 1997; Barbazan et al., 1998; Skovmand and Sanogo, 1999; Fillinger et al., 2003; Kandyata et al., 2012) and India (Sundararaj and Raghunatha-Rao, 1993; Kumar et al., 1994; Das and Amalraj, 1997) to Malaysia (Yap et al., 1991). In some of the larger, longer term studies, a decrease of malaria incidence has been reported. An even greater number of studies and programs are reported for the control of the ubiquitous *Cx. quinquefasciatus* with *L. sphaericus* (Table 28.4, Chapters 26 and 27).

A strong advantage of *L. sphaericus* over Bti is the ability to work well in polluted water (Fig. 28.3) (Silapanuntakul et al., 1983; Mulla et al., 1997, 2001) and to persist in a variety of habitats (Nicolas et al., 1987b; Paily et al., 1987; Karch et al., 1988; Pantuwatana et al., 1989; Hougard,

TABLE 28.2 *Bacillus thuringiensis* subsp. *israelensis* Used for Control of *Anopheles* Species in Tropical Countries

Targeted Species	Habitat	Application Rate and % Larval Reduction	Country	References
Anopheles gambiae	Various natural and man-made urban breeding sites	0.5 mL/m², 100%, initial mortality, reappearance of first instars at 28 h	Burkina Faso	Hougard et al. (1983)
	Rainwater pools	0.25–0.5 kg/ha 82–95%		Majori et al. (1987)
	Plastic tubs	0.2–1.0 kg/ha, 100% initial mortality, treated weekly, short residual		Dambach et al. (2014a)
	Rice fields	1 kg/ha, 92%	Liberia	Bolay and Trpis (1989)
	Irrigation ponds	2–8 L/ha, 90–100%,	Zaire	Karch et al. (1991)
	Artificial ponds	0.2 kg/ha, 95–100% mortality	Ghana	Nartey et al. (2013)
	Simulated pools	0.2–5 kg/ha, 100% initial reduction, 7-day intervals, no residual activity	Kenya	Fillinger et al. (2003)
An. gambiae, *Anopheles arabiensis*, *Anopheles funestus*	Various permanent and semipermanent manmade and natural breeding sites	Bti: 0.2 kg water dispersible granules (WDG) 5.0 corncob granules (CG) kg/ha, or *Lysinibacillus sphaericus* 1.0 kg WDG or 15 kg CG/ha weekly for 28 months, 95% larval and 92% adult reductions		Fillinger and Lindsay (2006)
	Various permanent and semipermanent manmade and natural breeding sites	0.4 kg/ha WDG or 10 kg/ha applied weekly, 96.5% larval reduction	Tanzania	Fillinger et al. (2008)
An. gambiae, *An. arabiensis*, *Anopheles melas*	Natural habitats in floodplains of the River Gambia	4 kg/ha CG or 0.2 kg/ha WDG, weekly application for 9 weeks, 100% control in first 24–48 h, no pupae produced during trial	The Gambia	Majambere et al. (2007)
	Rice fields and floodwater habitats covered with grass and sedge	0.2 kg/ha or 5 kg/ha, weekly during dry season for 2 years, 91–98% of sites free of larvae		Majambere et al. (2010)
An. Arabiensis	Pools, rice fields, ditches	0.6–1.0 L/ha, 90–100%, <5 days	Madagascar	Romi et al. (1993)
Anopheles stephensi	Tanks, construction sites	10 kg/ha, 90–100%, 3–7 days, 2–3 weeks	India	Kumar et al. (1995)
	Barrow pits, drains	0.5 g/m², 90–100%, 1–3 weeks		Mittal (2003)
Anopheles subpictus, *Anopheles nigerrimus*	Rice fields	0.5–1.0 kg/ha, 93–96%		Kramer (1984)
Anopheles culicifacies	Pools	0.5 g/m², 90–100%, 3–7 days		Mittal (2003)
Anopheles sinensis	Ponds, ditches	41–80 kg/ha, 89–95%	China	Xu et al. (1992)
Anopheles darling	Not specified	1.0 L/ha, 100%	Brazil	Vilarinhos et al. (1998)
Anopheles albimanus Anopheles spp.	Ditches, ponds, rice fields	1.17 L/ha, 90–100%, 10 days	Honduras	Perich et al. (1990)
	Rice fields, ponds	1.0–2.0 kg/ha, weekly for 7–10 weeks, good control, but rapid rebound, 50–70% reduction in adults and biting	Peru and Ecuador	Kroeger et al. (1995b)
An. albimanus, *Anopheles vestitipennis*	Various permanent habitats	10 mL "active ingredient" m⁻², 95–100%	Cuba	Pineda et al. (2005)
Anopheles quadrimaculatus	Rice fields	0.07–0.5 L/ha, 71–98%	Subtropical United States	Sandoski et al. (1985)

TABLE 28.3 *Bacillus thuringiensis* subsp. *israelensis* for Control of Culicine Mosquito Larvae

Targeted Species	Habitat	Application Rate and % Larval Reduction	Country	References
Culex quinquefasciatus	Roadside ditches	0.65 kg/ha 91–97%	Burkina Faso	Majori et al. (1987)
	Concrete channels	1.5 kg/ha 90–100%		
	Pools, drains	0.5 g/m², 90–100% 1–2 weeks	India	Mittal (2003)
Culex spp.	Dambos	1–1.5 briquets m⁻², reduced survival rate to 64% and 25%	Kenya	Logan and Linthicum (1992)
Culex melanoconion, *Culex coronator*	Artificial ponds	2 kg/ha, 98% reduction initially and significant reduction relative to controls until week 8 postapplication	Mexico	Marina et al. (2014)
Culex saltanensis	Polluted habitats	2 L/ha, 100% initial control, subsequent applications every 15 days	Brazil	Zequi and Lopes (2007)
Culex spp., *Psorophora* spp., *Mansonia titillans*	Various permanent habitats	10 mL a.i./m², 95–100%	Cuba	Pineda et al. (2005)
Aedes aegypti[a]	Various natural habitats	1.0–2.0 kg/ha, 100%, 1–2 days after treatment, weekly application	Colombia	Kroeger et al. (1995a)
	Small containers	1–2 g 50 L⁻¹, 96–100%, 12 weeks	Brazil	Vilarinhos and Monserrat (2004)
	Peridomestic containers	1 briquet/container, 100% (except at one site)	Colombia	Ocampo et al. (2009)
	Tires	0.75 and 0.375 g VectoBac tablets provided 100% control of late instars for 3 weeks	India	Batra et al. (2000)
	Industrial scrap	0.5 g/m², 90–100%, 5 weeks		Mittal (2003)
	200-L earthen jars, filled with 100 L of water	0.53 mL liquid concentrate (LC) 100 L⁻¹; 100% control for 4 weeks; 87–95% for 3 more weeks	Thailand	Chansang et al. (2004)
	Small containers	8 mg/L, 90–100%, 4 weeks	Australia	Ritchie et al. (2010)
Aedes spp.	Dambos	One briquette per 1.5–9.0/m², 13 days postflooding one briquette added per 4.6–9 m², reduced survival rate to 25% and 68%	Kenya	Logan and Linthicum (1992)

[a]*Numerous other field trials are reported on the use and evaluation of Bti for the control of container-breeding peridomestic Aedes aegypti by Boyce, R., Lenhart, A., Kroeger, A., Velayudhan, R., Roberts, B., Horstick, O., 2013. Bacillus thuringiensis israelensis (Bti) for the control of dengue vectors: systematic literature review. Trop. Med. Int. Health. 18, 564–577.*

1990). Persistence is likely influenced by the protection of toxin inclusions within the exosporium (Fig. 28.4). Recycling of *L. sphaericus* in larval cadavers may also influence longer term persistence (Des Rochers and Garcia, 1984; Karch and Coz, 1986; Nicolas et al., 1987b; Karch et al., 1990; Becker et al., 1995; Correa and Yousten, 1995; Yuan et al., 1999).

Due to the efficacy of *L. sphaericus* and its recycling and persistence in larval habitats, it has been successfully used in research and control efforts in both old and new world tropics

(Table 28.4). For example, Regis et al. (1995, 2000) reported good control of *Cx. quinquefasciatus* in an area where Bancroftian filariasis is endemic. Subsequently, Silva-Filha et al. (2001) reported good suppression of *Cx. quinquefasciatus* larval density throughout and following a 26-month *L. sphaericus* trial in urban habitats in Recife, Brazil, when applications of the bacterium were made on a monthly basis. Hougard et al. (1993) documented a significant decline of *Cx. quinquefasciatus* adult populations in Yaoundé, South Cameroon, after treatment of virtually every breeding site in

TABLE 28.4 Selected Examples of the Use of *Lysinibacillus sphaericus* for Control of Mosquito Larvae in the Tropics and Subtropics

Targeted Species	Habitat	Application Rate and % Larval Reduction	Country	References
Culex quinquefasciatus[a]	Cesspits	30 kg sustained release granules ha^{-1}, 99% reduction for 28 days	Burkina Faso	Skovmand and Sanogo (1999)
	Urban wastewater	10 g/m^2, every 3 months over 1 year, 41–55% decline in ♀♀ capture rate	Cameroon	Hougard et al. (1993) and Barbazan et al. (1997)
	Treated sewage water in containers, cesspools, ponds	30 L flowable concentrate (FC) ha^{-1}; or 30 kg sustained release granules ha^{-1} 95% reduction 4 days; >7 weeks in closed containers		Skovmand and Baudin (1997)
	Sewage ponds	10–30 kg/ha 5–7 days	Zaire	Karch et al. (1991)
	Pools, blocked streams, "more or less" polluted	Two or more weeks complete control with 2.5 ppm of liquid concentrate (LC)	Tanzania	Ragoonanansingh et al. (1992)
	Polluted drains, cesspits, septic tanks, etc.	1 g/m^2 weekly for one year, initially 68% positive, posttreatment mean of 5.9% positive	India	Kumar et al. (1996)
	Polluted drains, cesspits	1.0 g/m^2, 90–100%, 1–3 weeks	India	Mittal (2003)
	Organically polluted canals with standing water and other manmade water accumulations	50–200 mg of water dispersible granules (WDG) m^{-2} for 5 months resulted in 18–95% control with variable persistence depending on habitat; concomitantly biting rate was lowered	Thailand	Mulla et al. (2001)
	Manmade urban habitats	Treated monthly yielded "good control" over 26 months	Brazil	Silva-Filha et al. (2001)
	Polluted tanks, potholes	granules at 10 kg/ha in polluted tanks; 2.5 kg/ha to sod-lined potholes, 97–98% control, respectively	Subtropical United States	Lacey et al. (1988b)
Culex tritaeniorhynchus	Rice fields	87–100% reduction at the 4.3 kg microgel ha^{-1} for at least 5 weeks	India	Sundararaj and Reuben (1991)
		0.5–1.0 kg/ha, 62–69%		Kramer (1984)
Mansonia spp.	Swampy ditches, impounded paddy ditches	1.45 kg technical powder (TP) ha^{-1} 90% no residual; 1.0 kg TP ha^{-1}, initial 93% 2 week residual 80%	Malaysia	Yap et al. (1991)
Anopheles gambiae	Rice fields and swamps	10 kg/ha every 15 days, 98%	Zaire	Karch et al. (1991)
		10 kg/ha, 15-day interval, 98% initial control		Karch et al. (1992)
	Artificial pools	0.2 kg/ha, 50–100%	Tanzania	Ragoonanansingh et al. (1992)
	Ponds	30 kg granules ha^{-1}, 90–100%, 15 days 30 L FC ha^{-1}, 90–100%, 5 days	Senegal	Skovmand and Baudin (1997)
	Artificial ponds	0.24 kg/ha 2362 primary powder, 95–100% control	Burkina Faso	Majori et al. (1987)
	Small plots	1 mg/m^2, 100% initial control, 3–10day residual		Nicolas et al. (1987a)
	Puddles	50 L LC, 60–97% control for 10 days		Skovmand and Sanogo (1999)
	Artificial pools	5 kg corncob granules (CG) ha^{-1}, 100% initial reduction, 11-day residual activity	Kenya	Fillinger et al. (2003)

Species	Habitat	Dosage and results	Location	Reference
An. gambiae, Anopheles funestas, Anopheles arabiensis	Mostly human made	1 kg WDG ha^{-1}–15 kg CG ha^{-1} 95–97% initial reduction		Fillinger and Lindsay (2006)
An. arabiensis	Pools, rice fields, ditches	2.5–18 kg/ha, 90–100%, <5 days persistence	Madagascar	Romi et al. (1993)
An. gambiae, An. arabiensis, Anopheles melas	Rice fields and floodwater habitats covered with grass and sedge	0.2 kg/ha, 100% initial reductions, no residual activity in the wet season; up to 10 days control in the wet season	The Gambia	Majambere et al. (2007)
Anopheles fluviatilis/Anopheles culicifacies	Ponds	1.0 g/m^2, 90–100% for 3 weeks	India	Shukla et al. (1997)
Anopheles subpictus	Riverbed pools	1.0 g/m^2, 90–100% for 2 weeks		Mittal (2003)
	Rice fields	2.2–4.3 kg/ha, reduction of pupae 83–100% for 5 weeks		Sundararaj and Ruben (1991)
Anopheles stephensi	Masonry and sump tanks	1 g/m^2, weekly application, significant decrease in malaria incidence		Kumar et al. (1994)
Anopheles darlingi	Excavation pools	20 kg/ha; 78–93% control	Brazil, Amazon	Galardo et al. (2013)
An. darlingi, Anopheles marajoara, Anopheles nuneztovari	Various	20 kg/ha, for 48 months, 80% reduction in malaria	Venezuela	Moreno et al. (2014)
Anopheles albimanus, Culex coronator, Cx. Quinquefasciatus	Experimental plots in pastures	0.125 mL LC m^{-2} for Culex spp. and 2 g granules m^{-2} for An. albimanus for 70% mean reduction	Mexico	Arredondo-Jimenez et al. (1990)
An. Albimanus	Various	10 mL/m^2, for 4 months in 46 locations, 95% control of larvae, 50% reduction in malaria	Guatemala	Blanco Castro et al. (2000)
	Various	10 mL/m^2, 100% control up to 4 months	Honduras	Blanco Castro et al. (2002)
An. albimanus, Cx. quinquefasciatus, Aedes taeniorhynchus	Numerous natural and manmade sites	10 mL/m^2, 100%, up to 5 months	Cuba	Montero Lago et al. (1991)
Anopheles quadrimaculatus	Rice fields	0.58–1.17 LC ha^{-1}, 71–82% reduction	Subtropical United States	Lacey et al. (1986)
Psorophora spp.		0.3–0.6 L/ha 50–98%		Lacey and Inman (1985)
Anopheles spp, Psorophora columbiae		5 kg granules ha^{-1} larvae reduced by 68 and 92–100%, respectively		Lacey et al. (1988b)

a Results of numerous additional field trials of Lysinibacillus sphaericus for the control of Culex quinquefasciatus in organically polluted habitats are reported by Lacey, L. A. 2007. Bacillus thuringiensis serovariety israelensis and Bacillus sphaericus for mosquito control. In: Floore, T. G. (Ed.), Biorational Control of Mosquitoes. Am. Mosq. Control Assoc. Bull. 7, 133–163. and Silva-Filha, M. H. N. L, Berry, B., Regis, L., 2014. Lysinibacillus sphaericus: toxins and mode of action, applications for mosquito control and resistance management. In: Dhadialla, T.S., Gill, S.S., (Eds.), Advances in Insect Physiology, vol. 47, Academic Press, Oxford, pp. 89–176.

FIGURE 28.3 Application of *Lysinibacillus sphaericus* for the control of *Culex quinquefasciatus* in a polluted ditch. *Photograph from World Health Organization.*

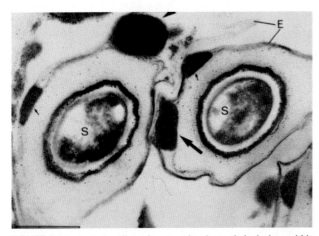

FIGURE 28.4 *Lysinibacillus sphaericus* showing toxin inclusions within the exosporium. *Micrograph from Karch, S., Charles, J.-F., 1987. Toxicity, viability and ultrastructure of* Bacillus sphaericus *2362 spore/crystal complex used in the field. Ann. Inst. Pasteur Microbiol. 138, 485–492.*

a 200-ha zone with *L. sphaericus* liquid concentrate applied at $10\,g/m^2$ every 3 months for 1 year. Similarly, Mulla et al. (2001) documented a reduction of adult *Cx. quinquefasciatus* after the treatment of larval breeding sites in low-income communities in Nonthaburi Province, Thailand.

An additional benefit of *L. sphaericus* on *Cx. quinquefasciatus* is the sublethal effects on vector potential (Lacey et al., 1987). Larvae that had survived an LC_{60} of the bacterium developed more slowly than untreated larvae, and a significant portion died as pupae. Emerging adults had lower stores of glucose, glycogen, and lipids than controls. These deficiencies would reduce vagility and the energy required to find nectar, mates, and blood sources.

A major drawback of *L. sphaericus* is the rapid development of high levels of resistance in some *Cx. quinquefasciatus* populations (Wirth et al., 2000; Nielsen-Laroux et al., 2002; Wirth, 2010; Silva-Filha et al., 2014) in India (Rao et al., 1995), Thailand (Mulla et al., 2003; Su and Mulla, 2004), China (Yuan et al., 2000; Olivera et al., 2004), Tunisia

(Nielsen-LeRoux, 2002), and Southern France (Chevillon et al., 2001). Mulla et al. (2003) reported one of the highest rates of resistance in *Cx. quinquefasciatus* in the Nonthanburi Province, Thailand (>125,000-fold). On an encouraging note, despite prolonged usage (26 months) of *L. sphaericus* for *Cx. quinquefasciatus* control in Recife, Brazil, only low levels of resistance were observed (Silva-Filha et al., 2008).

The mechanisms responsible for the development of resistance were reviewed by Charles et al. (1996), Romão et al. (2006), Wirth (2010), and Silva-Filha et al. (2014). Strategies for managing resistance have been proposed by Regis and Nielsen-Leroux (2000), Wirth et al. (2000), Mulla et al. (2003), and Wirth (2010). These include alternation with soft pesticides (eg, methoprene, spinosad, etc.) or Bti; combining *L. sphaericus* and Bti in a single formulation; leaving untreated refugia in which to maintain susceptible genes; and using environment management methods. Charles et al. (1996) suggested incorporating genes encoding for larvicidal toxins from other species and varieties of bacteria into the *L. sphaericus* genome (eg, *Clostridium bifermentans*, *L. sphaericus* serotype 25 [var. 2297], Bti). Wirth et al. (2000, 2004) observed synergistic activity between the toxins of *L. sphaericus* and Bti when bioassayed against resistant and nonresistant *Cx. quinquefasciatus*. The combination of toxins increased both endotoxin complexity and synergistic interactions and enhanced larvicidal activity.

Park et al. (2003) produced a recombinant Bti expressing both Bti endotoxins (Cyt 1A, Cry 11B) and *L. sphaericus* bin toxins that was as or more effective than the combined formulation (Bti/*L. sphaericus*) with the potential to persist longer in organically enriched habitats, broaden the range of susceptible mosquito species, and provide a means of avoiding resistance (Zahiri et al., 2004; Park et al., 2005; Federici et al., 2007) (Fig. 28.5).

28.5 ROLE OF MICROBIAL CONTROL AGENTS IN INTEGRATED VECTOR MANAGEMENT STRATEGIES FOR THE SUPPRESSION OF MOSQUITO-BORNE DISEASE

28.5.1 Microbial Control Agents for Malaria Control

According to Enayati and Hemingway (2010), the cessation of malaria transmission in Africa will depend on the development of a reliable vaccine and better drugs and insecticides. Traditionally, the control of vectors of the causal agents of malaria in Africa (*Plasmodium falciparum*) has depended on the use of insecticides (eg, spraying of walls with residual insecticides and the treatment of bed nets with pyrethroids). Killeen et al. (2002) contend that larval control should be prioritized for further development, evaluation, and implementation as an integral part of reducing malaria. Walker and Lynch (2007) suggest that targeting

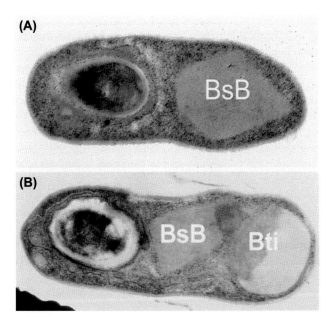

FIGURE 28.5 Binary toxin (Bin) of *Lysinibacillus sphaericus* (Ls) strain 2362 engineered for high levels of synthesis in *Bacillus thuringiensis* subspecies *israelensis* (Bti). (A) Transmission electron micrograph of a sporulated cell of an acrystalliferous strain of Bti showing a cross-section through the cell. Note the large size of the Ls crystal (BsB, the Bs standing for the previous name of Ly, ie, *Bacillus sphaericus*). (B) Electron micrograph of a recombinant strain of Bti that produces the normal parasporal body containing its endotoxins, the major ones being Cry4Aa, Cry4Ab, Cry11Aa, and Cyt1Aa, and additionally the Ls Bin endotoxin engineered for high levels of synthesis in Bti. The latter recombinant Bti/Ls Bin strain produces approximately 10-fold more international toxic units per mL of fermentation medium than the wild type strains of either Bti or Ls (see Park et al., 2005 for details). *Micrographs from Park, H.W., Bideshi, D.K., Wirth, M.C., Johnson, J.J., Walton, W.E., Federici, B.A., 2005. Recombinant larvicidal bacteria with markedly improved efficacy against* Culex *vectors of West Nile virus. Am. J. Trop. Med. Hyg. 72, 732–738 courtesy of Brian Federici.*

larvae, especially in human-made habitats, can significantly reduce malaria. They concurred with Killeen et al. (2002) that larval control is especially suitable for urban areas, where larval habitats are limited. As alternatives or supplements to chemical pesticides, MCAs are increasingly being developed and used as larvicidal components in IVM. Bti and *L. sphaericus* are the most used MCAs that can reduce larval mosquito populations (Lacey, 2007).

The use of MCAs for reducing mosquito-borne disease usually does not provide a standalone means of control due to the inability to find and adequately treat all mosquito breeding and resting sites. A notable exception was reported by Fillinger et al. (2008) in Dar es Salam, Tanzania. The reliance on ITNs for preventing indoor (endophilic) blood feeding by *An. gambiae s.l.* was failing due to a change in the host-seeking behavior of female mosquitoes from endophilic to exophilic feeding. Through a community-based program, breeding sites were mapped in detail and treated with Bti on a weekly basis over a 2-year period. Numbers of *An. gambiae* were reduced by 96%, and the incidence of malaria declined by 31%.

The combined use of MCAs and other elements of IVM, such as the removal or modification of larval breeding sites; protection of humans from being bitten by mosquitoes; and public education regarding malaria prevention (eg, use of antimalarial prophylaxes, early diagnosis and treatment, and community acceptance of mosquito control methods), can result in a significant reduction in the incidence of malaria (Fillinger et al., 2008, 2009; Geissbühler et al., 2009). Specific IVM strategies will depend on a variety of biological (mosquito species, environmental and human factors) and abiotic factors. Manpower, including community participation, is often inadequate, and the cost of MCAs can also be a constraint, especially when international financial support is unavailable.

28.5.1.1 Combined Use of Microbial Control Agents and Other Interventions and Their Effect on Malaria Transmission

28.5.1.1.1 Microbial Control Agents and Other Natural Enemies

Bti and *L. sphaericus* are compatible with both vertebrate and invertebrate natural enemies of *Anopheles* and other mosquito species (Lacey and Mulla, 1990; Siegel and Shadduck, 1990; Walton and Mulla, 1991; Lacey and Siegel, 2000; Lacey and Merritt, 2003). A broad spectrum of generalist invertebrate predators are unaffected when occupying habitats in which *L. sphaericus* or Bti have been applied (Lacey and Mulla, 1990; Lacey and Orr, 1994). When mosquito prey become scarce, other prey species can provide sustenance for generalist predators. However, specialized predators that feed only on targeted mosquitoes could be adversely affected when control is prolonged through residual larvicidal activity or when breeding sites are treated at regular intervals.

Evaluations of larvivorous fish (eg, *Gambusia affinis, Aplocheilus blocki, Poecilia reticulata, Oreochromis* [*Tilapia*] *nilotica*) combined with larvicidal bacteria have demonstrated good efficacy that can be greater than that of either natural enemy when used alone (Kramer et al., 1989; Bolay and Trpis, 1989; Lacey and Lacey, 1990; Walton and Mulla, 1991; Kumar et al., 1998). A caveat to the use of generalist larvivorous fish, such as *G. affinis*, in natural habitats is the negative environmental impact of the displacement of native species of fish and eutrophication due to the fish's preferential feeding on zooplankton. On the other hand, larvivorous edible fish, like *O. nilotica*, provide additional incentive for community participation (Howard et al., 2007).

28.5.1.1.2 Larval Control Using Habitat Management and Microbial Control Agents

Larval control strategies against malaria vectors could be highly effective in reducing the population density of emerging adults and complementary to adult control interventions, such as ITNs and insecticide application, including the application of *M. anisopliae* conidia, to resting sites of adults. Walker and Lynch (2007) reviewed the role of controlling larvae of *Anopheles* spp. in malaria suppression in tropical Africa through the removal or draining of larval habitats and larviciding, including the use of MCAs. Targeting larvae, particularly in human-made habitats, can significantly reduce malaria transmission and is especially suitable for urban areas, where larval habitats are limited (Utziner et al., 2001; Kileen et al., 2002; Castro et al., 2004, 2010; Fillinger et al., 2004; Majambere et al., 2010; Imbahale et al., 2012; Dambach et al., 2014b).

28.5.1.1.3 Microbial Control Agents and Insecticide-Treated Bed Nets

Since the implementation of ITNs in Africa, there has been a marked reduction in child mortality and morbidity due to malaria (Bermejo and Veeken, 1992; Hawley et al., 2003). Fillinger et al. (2009) demonstrated the combined effect of using ITNs and implementing a large-scale treatment of breeding sites with weekly applications of water-dispersible and granule formulations of Bti over a 1.5-year period in the western Kenyan highlands. As indices they used the incidence of new *Plasmodium* infections in children as the primary outcome and anopheline larval and adult abundance as secondary outcomes. In association with Bti treatments they reported a 91% reduction in the mean number of late instar anopheline larvae and a 65% reduction in adult mosquitoes resting indoors. Similar reductions were seen with the average biting rate. Vector control with microbial larvicides and ITNs combined resulted in a two-fold reduction in new malaria infections compared with ITNs alone.

28.5.2 Dengue Control

Dengue is the most rapidly spreading mosquito-borne viral disease, with a 30-fold increase in global incidence over the past 50 years (WHO, 2012). Each year an estimated 50–100 million dengue infections occur throughout the tropics and subtropics (WHO, 2012; Bhatt et al., 2013; Stanaway et al., 2016). Effective vector control measures are critical to achieving and sustaining a reduction of morbidity attributable to dengue (WHO, 2012). The principal vector of the dengue viruses, *Ae. aegypti*, breeds in virtually any collection of water in and around housing. Emptying, covering, or removing peridomestic breeding sites can have a significant effect on lowering biting rates and the prevalence of dengue (Rosenbaum et al., 1995; Wang et al., 2000; Winch et al., 2002). Eggs of *Ae. aegypti* and *Ae. albopictus*, another potential vector of the dengue virus, can remain dry and dormant for months and hatch when inundated with water. In some cases, the treatment of breeding sites with larvicides may be required. For example, sites that are temporarily dry, cryptic (unseen), or are almost impossible to empty (eg, used tires) can be treated preemptively with granular formulations of

Bti to target neonate *Aedes* spp. larvae when the breeding site is inundated with water. Chansang et al. (2004), Vilarinhos and Monnerat (2004), and Ritchie et al. (2010) demonstrated that Bti can provide residual control of *Ae. aegypti* in small to large containers for several weeks. Armengol et al. (2006) demonstrated prolonged control of *Ae. aegypti* in containers in full sunlight using a tablet formulation of Bti. Jacups et al. (2013) used a backpack mist blower to treat cryptic breeding sites in typical Far North Queensland, Australia, bushland with Bti (water-dispersible granules) at 400–800 g/ha. They recorded effective control (>80%) of *Ae. aegypti* for 11 weeks up to 16 months from the application point. Chansang et al. (2004) reported on the efficacy of using the predator copepod, *Mesocyclops thermocyclopoides*, and/or Bti in drinking water containers for the control of *Ae. aegypti*. The combination of the two resulted in better sustained control than either of the agents used individually. Zero larvae were detected during the first 8 weeks of the trial, and their numbers were kept at a very low level for the following 4 weeks.

28.6 CONCLUSIONS

The most successful IVM strategies target all stages of the mosquito life cycle. The MCA component of IVM has been shown to provide significant reductions of larvae and biting rates of *Culex* spp., *Anopheles* spp., *Aedes* spp., and other vector mosquitoes in the large-scale control trials and operational programs reported in this chapter. Evidence of the reduction of biting rates of *An. gambiae* and other *Anopheles* spp. and a lower incidence of malaria has been reported by several researchers across sub-Saharan Africa, tropical Asia, and the neotropics. An essential component for reducing the incidence of malaria and dengue fever is community participation with regard to knowledge and acceptance of larviciding, breeding site reduction, and early diagnosis and treatment of disease (Swaddiwudhipong et al., 1992; Kroeger et al., 1995a; Rosenbaum et al., 1995; Geissbühler et al., 2009; Ocampo et al., 2009; Gubler, 2011; Bhatt et al., 2013; Mboera et al., 2014; Stanaway et al., 2016). Sustainable control of malaria and other vector-borne diseases using MCAs within an IVM strategy in the poorest countries will require continued financial support from the more affluent countries and international organizations.

Bti and *L. sphaericus* will continue to be the most reliable MCAs for vector and disease control for the foreseeable future. The development of improved formulations and innovative application methods can increase their efficacy and spray coverage of a variety of habitats, including cryptic habitats. Classical biological control using mermithid nematodes and some of the other MCAs with the capability of becoming established in mosquito habitats will provide some control but should be integrated with bacterial MCAs or soft chemical larvicides (methoprene, spinosad, etc.) and the use of ITNs. In the future, with further development, *M. anisopliae* and *B. bassiana* could contribute to the significant control of adult mosquitoes.

ACKNOWLEDGMENTS

I am grateful to the several colleagues who furnished literature, micrographs, photographs, and constructive comments. Special thanks to Cynthia Lacey for reviewing the manuscript and providing grammatical advice. I appreciate Ulrike Fillinger's review of my first draft and for providing comments and suggested literature on the microbial control of vector mosquitoes in eastern Africa. Thanks also to Brian Federici for microgrpahs, literature, and for review of sections of the manuscript.

REFERENCES

Abagli, A.Z., Alavo, T.B.C., Platzer, E.G., 2012. Efficacy of the insect parasitic nematode, *Romanomermis iyengari*, for malaria vector control in Benin West Africa. Malar. World J. 11 (Suppl. 1), P5.

Achinelly, M.F., Micieli, M.V., 2013. Host range of the parasite *Strelkovimermis spiculatus* (Nematoda: Mermithidae) in Argentina mosquitoes. J. Vector Ecol. 38, 69–73.

Alavo, T.B.C., Abagli, A.Z., Pérez-Pacheco, R., Platzer, E.G., 2015. Large-scale production of the malaria vector biological control agent *Romanomermis iyengari* (Nematoda: Mermithidae) in Benin, West Africa. Malar. World J. 6, 1–5.

Alirzaev, G.U., Pridantseva, E., Vladimirova, V.V., Alekseev, A.U., 1990. The prospects for using *Romanomermis culicivorax* and *R. iyengari* (Nematoda: Mermithidae) for mosquito control in Azerbaijan. Med. Parazitol. Parazit. Bol. (1) (In Russian).

Alves, S.B., Alves, L.F.A., Lopes, R.B., Pereira, R.M., Vieira, S.A., 2002. Potential of some *Metarhizium anisopliae* isolates for control of *Culex quinquefasciatus* (Dipt., Culicidae). J. Appl. Entomol. 126, 504–509.

Armengol, G., Hernandez, J., Velez, J.G., Orduz, S., 2006. Long-lasting effects of a *Bacillus thuringiensis* serovar *israelensis* experimental tablet formulation for *Aedes aegypti* (Diptera: Culicidae) control. J. Econ. Entomol. 99, 1590–1595.

Arredondo-Jimenez, J.I., Lopez, T., Rodriguez, M.H., Bown, D.N., 1990. Small scale field trials of *Bacillus sphaericus* (strain 2362) against anopheline and culicine mosquito larvae in southern Mexico. J. Am. Mosq. Control Assoc. 6, 300–305.

Barbazan, P., Baldet, T., Darriet, F., Escaffre, H., Haman Djoda, D., Hougard, J.-M., 1997. Control of *Culex quinquefasciatus* (Diptera: Culicidae) with *Bacillus sphaericus* in Maroua, Cameroon. J. Am. Mosq. Control Assoc. 13, 263–269.

Barbazan, P., Baldet, T., Darriet, F., Escaffre, H., Haman Djoda, D., Hougard, J.-M., 1998. Impact of treatments with *Bacillus sphaericus* on *Anopheles* populations and the transmission of malaria in Maroua, a large city in a savannah region of Cameroon. J. Am. Mosq. Control Assoc. 14, 33–39.

Batra, C.P., Mittal, P.K., Adak, T., 2000. Control of *Aedes aegypti* breeding in desert coolers and tires by use of *Bacillus thuringiensis* var. *israelensis* formulation. J. Am. Mosq. Control Assoc. 16, 321–323.

Becker, N., Zgomba, M., Petric, D., Beck, M., Ludwig, M., 1995. Role of larval cadavers in recycling processes of *Bacillus sphaericus*. J. Am. Mosq. Control Assoc. 11, 329–334.

Becnel, J.J., Andreadis, T.G., 2014. Microsporida in insects. In: Weiss, L.M., Becnel, J.J. (Eds.), Microsporidia: Pathogens of Opportunity. John Wiley and Sons, West Sussex, Oxford, UK, pp. 521–570.

Becnel, J.J., White, S.E., 2007. Mosquito pathogenic viruses – the last 20 years. In: Floore, T.G. (Ed.), Biorational Control of Mosquitoes. Am. Mosq. Control Assoc. Bull. 7, 36–49.

Becnel, J.J., White, S.E., Shapiro, A.M., 2005. Review of microsporidia-mosquito relationships: from the simple to the complex. Folia Parasitol. 52, 41–50.

Bermejo, A., Veeken, H., 1992. Insecticide-impregnated bed nets for malaria control: a review of the field trials. Bull. World Health Org. 70, 293–296.

Bhatt, S., Gething, P.W., Brady, O.J., Messina1, J.P., Farlow, A.W., Moyes, C.L., Drake, J.M., Brownstein, J.S., Hoen, A.G., Sankoh, O., Myers, M.F., George, D.B., Jaenisch, T., Wint, G.R.W., Simmons, C.P., Scott, T.W., Farrar, J.J., Hay, S.I., 2013. The global distribution and burden of dengue. Nature 496, 504–507.

Bheema Rao, U.S., Gajanana, A., Rajagopalan, P.K., 1979. A note on the tolerance of the mermithid nematode *Romanomermis* sp. to different pH and salinity. Indian J. Med. Res. 69, 423–427.

Blanco Castro, S.D., Martínez Arias, A., Cano Velásquez, O.R., Tello Granados, R., Mendoza, I., 2000. Introducción del *Bacillus sphaericus* cepa-2362 (GRISELESF) para el control biológico de vectores maláricos en Guatemala. Rev. Cuba. Med. Trop. 52, 37–43.

Blanco Castro, S.D., Colombi, E., Flores, L.N., Canales, D., 2002. Aplicación del biolarvicida *Bacillus sphaericus*-2362 (GRISELESF) para el control de la malaria en un área de salud de la República de Honduras. Rev. Cuba. Med. Trop. 54, 134–141.

Blanford, S., Chan, B.H.K., Jenkins, N., Sim, D., Turner, R.J., Read, A.F., Thomas, M.B., 2005. Fungal pathogen reduces potential for malaria transmission. Science 308, 1638–1641.

Blanford, S., Jenkins, N.E., Read, A.F., Thomas, M.B., 2012. Evaluating the lethal and pre-lethal effects of a range of fungi against adult *Anopheles stephensi*. Malar. J. 11, 365.

Boisvert, M., Boisvert, J., 2000. Effects of *Bacillus thuringiensis* var. *israelensis* on target and nontarget organisms: a review of laboratory and field experiments. Biocontrol Sci. Technol. 10, 517–561.

Bolay, F.K., Trpis, M., 1989. Control of mosquitoes with *Bacillus thuringiensis* var. israelensis and *larvivorous* fish, *Tilapia nilotica*, in rice fields in Liberia, West Africa. Isr. J. Entomol. 23, 77–82.

Boyce, R., Lenhart, A., Kroeger, A., Velayudhan, R., Roberts, B., Horstick, O., 2013. *Bacillus thuringiensis israelensis* (Bti) for the control of dengue vectors: systematic literature review. Trop. Med. Int. Health 18, 564–577.

Brooke, B., Hargreaves, K., Spillings, B., Raswiswi, E., Koekemoer, L., Coetzee, M., 2012. Novel malaria control: can traditional clay pots be used to deliver entomopathogenic fungi to malaria vectors in Northern Kwazulu-Natal? Commun. Dis. Surv. Bull. 10, 13–17.

Brown, B.J., Platzer, E.G., 1977. The effects of temperature on the infectivity of *Romanomermis culicivorax*. J. Nematol. 9, 166–172.

Brown, B.J., Platzer, E.G., 1978. Salts and the infectivity of *Romanomermis culicivorax*. J. Nematol. 10, 53–64.

Brown-Westerdahl, B., Washino, R.K., Platzer, E.G., 1982. Successful establishment and subsequent recycling of *Romanomermis culicivorax* (Mermithidae: Nematoda) in a California rice field following postparasite application. J. Med. Entomol. 19, 34–41.

Bukhari, T., Takken, W., Koenraadt, C.J.M., 2011. Development of *Metarhizium anisopliae* and *Beauveria bassiana* formulations for control of malaria mosquito larvae. Parasites Vectors 4, 23.

Cameron, M.M., Lorenz, L.M. (Eds.), 2013. Biological and Environmental Control of Disease Vectors. CABI, Wallingford, Oxford.

Camino, N.B., Garcia, J., 1991. Influencia de salinidad y el pH en el parasitismo de *Strelkovimermis spiculatus* Poinar and Camino, 1986 (Nematoda: Mermithidae) en larvas de *Culex pipiens* Weid (Diptera: Culicidae). Neotropica 37, 107–112.

Campos, R.E., Sy, V.E., 2003. Mortality in immatures of the floodwater mosquito *Ochlerotatus albifasciatus* (Diptera: Culicidae) and effects of parasitism by *Strelkovimermis spiculatus* (Nematoda: Mermithidae) in Buenos Aires Province, Argentina. Mem. Inst. Oswaldo Cruz 98, 199–208.

Campos, R.E., Maciá, A., García, J.J., 1993. Fluctuaciones estacionales de culícidos (Diptera) y sus enemigos naturales en zonas urbanas de los alrededores de La Plata, Provincia de Buenos Aires. Geotrópica 39, 55–66.

Castro, M.C., Yamagata, Y., Mtasiwa, D., Tanner, M., Utzinger, J., Keiser, J., Singer, B.H., 2004. Integrated urban malaria control: a case study in Dar es Salaam, Tanzania. Am. J. Trop. Med. Hyg. 71 (Suppl. 2), 103–117.

Castro, M.C., Kanamori, S., Kannady, K., Mkude, S., Killeen, G.F., Fillinger, U., 2010. The importance of drains for the larval development of lymphatic filariasis and malaria vectors in Dar es Salaam, United Republic of Tanzania. PLoS Negl. Trop. Dis. 4, e693.

Chandhiron, K., Paily, K.P., 2015. Natural parasitism of *Romanomermis iyengari* (Welch) (Nematoda: Mermithidae) on various species of mosquitoes breeding in rice fields in Pondicherry, India. Biol. Control 83, 1–6.

Chandrahas, R.K., Rajagopalan, P.K., 1979. Mosquito breeding and the natural parasitism of larvae by a fungus *Coelomomyces* and mermithid nematode *Romanomermis* in paddy fields in Pondicherry. Indian J. Med. Res. 69, 63–70.

Chansang, U., Bhumiratana, A., Kittayapong, P., 2004. Combination of *Mesocyclops thermocyclopoides* and *Bacillus thuringiensis* var. *israelensis*: a better approach for the control of *Aedes aegypti* larvae in water containers. J. Vector Ecol. 29, 218–226.

Charles, J.F., Nielsen-LeRoux, C., Delécluse, A., 1996. *Bacillus sphaericus* toxins: molecular biology and mode of action. Annu. Rev. Entomol. 41, 451–472.

Chevillon, C., Bernard, C., Marquine, M., Pasteur, N., 2001. Resistance to *Bacillus sphaericus* in *Culex pipiens* (Diptera: Culicidae): interaction between recessive mutants and evolution in southern France. J. Med. Entomol. 38, 657–664.

Correa, M., Yousten, A.A., 1995. *Bacillus sphaericus* spore germination and recycling in mosquito larval cadavers. J. Invertebr. Pathol. 66, 76–81.

Dambach, P., Louis, V.R., Kaiser, A., Ouedraogo, S., Sié, A., Sauerborn, R., Becker, N., 2014a. Efficacy of *Bacillus thuringiensis* var. *israelensis* against malaria mosquitoes in northwestern Burkina Faso. Parasites Vectors 7, 371–378.

Dambach, P., Traoré, I., Becker, N., Kaiser, A., Sié, A., Sauerborn, R., 2014b. EMIRA: ecologic malaria reduction for Africa – innovative tools for integrated malaria control. Glob. Health Action. 7, 25908. http://dx.doi.org/10.3402/gha.v7.25908.

Daoust, R.A., Roberts, D.W., 1982. Virulence of natural and insect passaged strains of *Metarhizium anisopliae* to mosquito larvae. J. Invertebr. Pathol. 40, 107–117.

Daoust, R.A., Roberts, D.W., 1983. Studies on the prolonged storage of *Metarhizium anisopliae* conidia: effect of temperature and relative humidity on conidial viability and virulence against mosquitoes. J. Invertebr. Pathol. 41, 143–150.

Das, P.K., Amalraj, D.D., 1997. Biological control of malaria vectors. Indian J. Med. Res. 106, 174–197.

Des Rochers, B., Garcia, R., 1984. Evidence for persistence and recycling of *Bacillus sphaericus*. Mosq. News 44, 160–165.

Enayati, A., Hemingway, J., 2010. Malaria management: past, present, and future. Annu. Rev. Entomol. 55, 569–591.

Fang, W., Vega-Rodriguez, J., Ghosh, A.K., Jacobs-Lorena, M., Khang, A., St Leger, R.J., 2011. Development of transgenic fungi that kill human malaria parasites in mosquitoes. Science 331, 1074–1077.

Farenhorst, M., Farina, D., Scholte, E.-J., Takken, W., Hunt, R.H., Coetzee, M., Knols, B.G.J., 2008. African water storage pots for the delivery of the entomopathogenic fungus *Metarhizium anisopliae* to the malaria vectors *Anopheles gambiae s.s.* and *Anopheles funestas*. Am. J. Trop. Med. Hyg. 78, 910–916.

Farenhorst, M., Mouatcho, J.C., Kikankie, C.K., Brooke, B.D., Hunt, R.H., Thomas, M.B., Koekemoer, L.L., Knols, B.G.J., Coetzee, M., 2009. Fungal infection counters insecticide resistance in African malaria mosquitoes. Proc. Natl. Acad. Sci. U.S.A. 106, 17443–17447.

Farenhorst, M., Knols, B.G.J., Thomas, M.B., Howard, A.F.V., Takken, W., Rowland, M., N'Guessan, R., 2010. Synergy in efficacy of fungal entomopathogens and permethrin against West African insecticide-resistant *Anopheles gambiae* mosquitoes. PLoS One 5, e12081. http://dx.doi.org/10.1371/journal.pone.0012081.

Farenhorst, M., Hilhorst, A., Thomas, M.B., Knols, B.G.J., 2011. Development of fungal applications on netting substrates for malaria vector control. J. Med. Entomol. 48, 305–313.

Federici, B.A., 1981. Mosquito control by the fungi *Culicinomyces*, *Lagenidium*, and *Coelomomyces*. In: Burges, H.D. (Ed.), Microbial Control of Pests and Plant Diseases: 1970–1980. Academic Press, London, UK, pp. 555–572.

Federici, B.A., Park, H.-W., Bideshi, D.K., Wirth, M.C., Johnson, J.J., Sakano, Y., Tang, M., 2007. Developing recombinant bacteria for control of mosquito larvae. Bull. Am. Mosq. Control Assoc. 7, 164–175.

Fillinger, U., Lindsay, S.W., 2006. Suppression of exposure to malaria vectors by an order of magnitude using microbial larvicides in rural Kenya. Trop. Med. Int. Health 1, 1–14.

Fillinger, U., Knols, B.G., Becker, N., 2003. Efficacy and efficiency of new *Bacillus thuringiensis* var *israelensis* and *Bacillus sphaericus* formulations against Afrotropical anophelines in western Kenya. Trop. Med. Int. Health 8, 37–47.

Fillinger, U., Sonye, G., Killeen, G.F., Knols, B.G., Becker, N., 2004. The practical importance of permanent and semipermanent habitats for controlling aquatic stages of *Anopheles gambiae sensu lato* mosquitoes: operational observations from a rural town in western Kenya. Trop. Med. Int. Health 9, 1274–1289.

Fillinger, U., Kannady, K., William, G., Vanek, M.J., Dongus, S., Nyika, D., Geissbühler, Y., Chaki, P.P., Govella, N.J., Mathenge, E.M., Singer, B.H., Mshinda, H., Lindsay, S.W., Tanner, M., Mtasiwa, D., de Castro, M.C., Killeen, G.F., 2008. A tool box for operational mosquito larval control: preliminary results and early lessons from the Urban Malaria Control Programme in Dar es Salaam, Tanzania. Malar. J. 7, 20. http://dx.doi.org/10.1186/1475-2875-7-20.

Fillinger, U., Ndenga, B., Githeko, A., Lindsay, S.W., 2009. Integrated malaria vector control with microbial larvicides and insecticide treated nets in the western Kenyan highlands: a controlled study. Bull. WHO 87, 655–665.

Foster, W.A., Walker, E.D., 2009. Mosquitoes. In: Mullen, G., Durden, L. (Eds.), Medical and Veterinary Entomology, second ed. Academic Press, San Diego, pp. 201–253.

Gajanana, A., Kazmin, S.J., Bheema Rao, U.S., Suguna, S.G., Chandrahas, R.K., 1978. Studies on a nematode parasite (*Romanomermis* sp.: Mermithidae) of mosquito larvae isolated in Pondicherry. Indian J. Med. Res. 68, 242–247 Need rates of application.

Galardo, A.K.R., Zimerman, R., Galardo, C.D., 2013. Larval control of *Anopheles* (*Nyssorhinchus*) *darlingi* using granular formulation of *Bacillus sphaericus* in abandoned gold-miners excavation pools in the Brazilian Amazon rainforest. Rev. Soc. Bras. Med. Trop. 46, 172–177.

Geissbühler, Y., Kannady, K., Chaki, P., Emidi, B., Govella, N.J., Mayagaya, V., Mtasiwa, D., Mshinda, H., Lindsay, S.W., Fillinger, U., Tanner, M., Castro, M.C., Killeen, G.F., 2009. Microbial larvicide application by a large scale, community-based program reduces malaria infection prevalence in Dar es Salaam, United Republic of Tanzania. PLoS One 4, 5107.

Goldberg, L.H., Margalit, J., 1977. A bacterial spore demonstrating rapid larvicidal activity against *Anopheles sergenti*, *Uranotaenia unguiculata*, *Culex univittatus*, *Aedes aegypti* and *Culex pipiens*. Mosq. News 37, 355–358.

Greenwood, B., 1995. Mortality and morbidity from malaria in Gambian children after introduction of an impregnated bednet programme. Lancet 345, 479–483.

Gubler, D.J., 2011. Prevention and control of *Aedes aegypti*-borne disease: lesson learned from past successes and failures. Asia Pac. J. Mol. Biol. Biotechnol. 19, 111–114.

Harwood, J.F., Farooq, M., Turnwall, B.T., Richardson, A.G., 2015. Evaluating liquid and granular *Bacillus thuringiensis* var. *israelensis* broadcast applications for controlling vectors of dengue and chikungunya viruses in artificial containers and tree holes. J. Med. Entomol. 52, 663–671.

Hawley, W.A., Phillips-Howard, P.A., Kuile, F.O., Terlouw, D.J., Vulule, J.M., Ombok, M., Nahlen, B.L., Gimnig, J.E., Kariuki, S.K., Kolczak, M.S., Hightower, A.W., 2003. Community-wide effects of permethrin-treated bed nets on child mortality and malaria morbidity in western Kenya. Am. J. Trop. Med. Hyg. 68, 121–127.

Hay, S.I., Guerra, C.A., Tatem, A.J., Noor, A.M., Snow, R.W., 2004. The global distribution and population at risk of malaria: past, present, and future. Lancet Infect. Dis. 4, 327–336.

Hougard, J.M., 1990. Formulations and persistence of *Bacillus sphaericus* in *Culex quinquefasciatus* larval sites in tropical Africa. In: de Barjac, H., Sutherland, D. (Eds.), Bacterial Control of Mosquitoes and Black Flies: Biochemistry, Genetics, and Applications of *Bacillus thuringiensis israelensis* and *Bacillus sphaericus*. Rutgers University Press, New Brunswick, pp. 295–306.

Hougard, J.M., Darriet, F., Bakayoko, S., 1983. Evaluation en milieu naturel de l'activité larvicide de *Bacillus thuringiensis* serotype H-14 sur *Culex quinquefasciatus* Say, 1823 et *Anopheles gambiae* Giles, 1902 s. l. (Diptera: Culicidae) en Afrique de L'Ouest. Cah. ORSTOM Entomol. Méd Parasitol. 21, 111–117.

Hougard, J.M., Mbentengam, R., Lochouarn, L., Escaffre, H., Darriet, F., Barbazan, P., Quillévéré, D., 1993. Lutte contre *Culex quinquefasciatus* par *Bacillus sphaericus*: résultats d'une campagne pilote dans une grande agglomération urbaine d'Afrique equatorialé. Bull. World Health Org. 71, 367–375.

Howard, A.F., Zhou, G., Omlin, F.X., 2007. Malaria mosquito control using edible fish in western Kenya: preliminary findings of a controlled study. BMC Pub. Health 7, 199.

Howard, A.F.V., Koenraadt, C.J.M., Farenhorst, M., Knols, B.G.J., Takken, W., 2010. Pyrethroid resistance in *Anopheles gambiae* leads to increased susceptibility to entomopathogenic fungi *Metarhizium anisopliae* and *Beauveria bassiana*. Malar. J. 9, 168.

Hughes, D.S., Platzer, E.G., 1977. Temperature effects on the parasitic phase of *Romanomermis culicivorax* on *Culex pipiens*. J. Nematol. 9, 173–175.

Imbahale, S.S., Githeko, A., Mukabana, W.R., Takken, W., 2012. Integrated mosquito larval source management reduces larval numbers in two highland villages in western Kenya. BMC Pub. Health 12, 362.

Jacups, S.P., Rapley, L.P., Johnson, P.H., Benjamin, S., Ritchie, S.A., 2013. *Bacillus thuringiensis* var. *israelensis* misting for control of *Aedes* in cryptic ground containers in north Queensland, Australia. Am. J. Trop. Med. Hyg. 88, 490–496.

Kandyata, A., Mbata, K.J., Shinondo, C.J., Katongo, C., Kamuliwo, R.M., Nyirenda, F., Chanda, J., Chanda, E., 2012. Impacts of *Bacillus thuringiensis* var. *israelensis* and *Bacillus sphaericus* insect larvicides on mosquito larval densities in Lusaka, Zambia. Med. J. Zamb. 39, 12.

Karch, S., Charles, J.-F., 1987. Toxicity, viability and ultrastructure of *Bacillus sphaericus* 2362 spore/crystal complex used in the field. Ann. Inst. Pasteur Microbiol. 138, 485–492.

Karch, S., Coz, J., 1986. Recycling of *Bacillus sphaericus* in dead larvae of *Culex pipiens* (Diptera, Culicidae). Cah. ORSTOM Entomol. Méd Parasitol. 24, 41–43.

Karch, S., Monteny, N., Coz, J., 1988. Persistance de *Bacillus sphaericus* dans une gîte à moustiques 4 ans après son introduction en vue de lutte biologique. C. R. Acad. Sci. III Life Sci. 307, 289–292.

Karch, S., Monteny, N., Jullien, J.L., Sinégre, G., Coz, J., 1990. Control of *Culex pipiens* by *Bacillus sphaericus* and role of nontarget arthropods in its recycling. J. Am. Mosq. Control Assoc. 6, 47–54.

Karch, S., Manzambi, Z.A., Salaun, J.J., 1991. Field trials with Vectolex (*Bacillus sphaericus*) and Vectobac (*Bacillus thuringiensis* (H-14) against *Anopheles gambiae* and *Culex quinquefasciatus* breeding in Zaire. J. Am. Mosq. Control Assoc. 7, 176–179.

Karch, S., Asidi, N., Manzambi, Z.M., Salaun, J.J., 1992. Efficacy of *Bacillus sphaericus* against the malaria vector *Anopheles gambiae* and other mosquitoes in swamps and rice fields in Zaire. J. Am. Mosq. Control Assoc. 8, 376–380.

Kerwin, J.L., 2007. *Oomycetes: Lagenidium giganteum*. In: Floore, T.G. (Ed.), Biorational Control of Mosquitoes. Am. Mosq. Control Assoc. Bull. 7, 50–57.

Kerwin, J.L., Washino, R.K., 1985. Recycling of *Romanomermis culicivorax* (Mermithidae: Nematoda) in rice fields in California, USA. J. Med. Entomol. 22, 637–643.

Killeen, G.F., Fillinger, U., Bart, G.J., 2002. Advantages of larval control for African malaria vectors: low mobility and behavioural responsiveness of immature mosquito stages allow high effective coverage. Malar. J. 1, 8.

Kramer, V.L., 1984. Evaluation of *Bacillus sphaericus* and *Bacillus thuringiensis* (H-14) for mosquito control in rice fields. Indian J. Med. Res. 80, 642–648.

Kramer, V.L., Garcia, R., Colwell, A.E., 1989. An evaluation of *Gambusia affinis* and *Bacillus thuringiensis* var. *israelensis* as mosquito control agents in California wild rice fields. J. Am. Mosq. Control Assoc. 4, 87–92.

Kroeger, A., Dehlinger, U., Burkhardt, G., Atehortua, W., Anaya, H., Becker, N., 1995a. Community based dengue control in Colombia: people's knowledge and practice and the potential contribution of the biological larvicide Bti (*Bacillus thuringiensis israelensis*). Trop. Med. Parasitol. 46, 241–246.

Kroeger, A., Horstick, O., Riedl, C., Kaiser, A., Becker, N., 1995b. The potential for malaria control with the biological larvicide *Bacillus thuringiensis israelensis* (Bti) in Peru and Ecuador. Acta Trop. 60, 47–57.

Kumar, A., Sharma, V.P., Sumodan, P.K., Thavaseluam, D., Kamat, R.H., 1994. Malaria control utilising *Bacillus sphaericus* against *Anopheles stephensi* in Pinaji, Goa. J. Am. Mosq. Control Assoc. 10, 534–539.

Kumar, A., Sharma, V.P., Thavaselvam, D., Sumodan, P.K., 1995. Control of *Anopheles stephensi* breeding in construction sites and abandoned overhead tanks with *Bacillus thuringiensis* var. *israelensis*. J. Am. Mosq. Control Assoc. 11, 86–89.

Kumar, A., Sharma, V.P., Thavaselvam, D., Sumodan, P.K., Kamat, R.H., Audi, S.S., Surve, B.N., 1996. Control of *Culex quinquefasciatus* with *Bacillus sphaericus* in Vasco city, Goa. J. Am. Mosq. Control Assoc. 12, 409–413.

Kumar, A., Sharma, V.P., Sumodan, P.K., Thavaselvam, D., 1998. Field trials of biolarvicide *Bacillus thuringiensis* var. israelensis strain 164 and the larvivorous fish *Aplocheilus blocki* against *Anopheles stephensi* for malaria control in Goa, India. J. Am. Mosq. Control Assoc. 14, 457–462.

Lacey, C.M., Lacey, L.A., Roberts, D.W., 1988a. Route of invasion and histopathology of *Metarhizium anisopliae* in *Culex quinquefasciatus*. J. Invertebr. Pathol. 52, 108–118.

Lacey, L.A., 2007. *Bacillus thuringiensis* serovariety *israelensis* and *Bacillus sphaericus* for mosquito control. In: Floore, T.G. (Ed.), Biorational Control of Mosquitoes. Am. Mosq. Control Assoc. Bull. 7, 133–163.

Lacey, L.A., Inman, A., 1985. Efficacy of granular formulations of *Bacillus thuringiensis* (H-14) for the control of *Anopheles* larvae in rice fields. J. Am. Mosq. Control Assoc. 1, 38–42.

Lacey, L.A., Lacey, C.M., 1990. The medical importance of riceland mosquitoes and their control using alternatives to chemical insecticides. J. Am. Mosq. Control Assoc. 2 (Suppl. 6), 1–93.

Lacey, L.A., Merritt, R.W., 2003. The safety of bacterial microbial agents used for black fly and mosquito control in aquatic environments. In: Hokkanen, H.M.T., Hajek, A.E. (Eds.), Environmental Impacts of Microbial Insecticides: Need and Methods for Risk Assessment. Kluwer Academic Publishers, Dordrecht, pp. 151–168.

Lacey, L.A., Mulla, M.S., 1990. Safety of *Bacillus thuringiensis* (H-14) and *Bacillus sphaericus* to non-target organisms in the aquatic environment. In: Laird, M., Lacey, L.A., Davidson, E.W. (Eds.), Safety of Microbial Insecticides. CRC Press, Boca Raton, pp. 169–188.

Lacey, L.A., Orr, B.K., 1994. The role of biological control of mosquitoes in integrated vector control. Am. J. Trop. Med. Hyg. 50 (Suppl.), 97–115.

Lacey, L.A., Siegel, J.P., 2000. Safety and ecotoxicology of entomopathogenic bacteria. In: Charles, J.-F., Delécluse, A., Nielsen-LeRoux, C. (Eds.), Entomopathogenic Bacteria: From Laboratory to Field Application. Kluwer Academic Publishers, Dordrecht, pp. 253–273.

Lacey, L.A., Undeen, A.H., 1986. Microbial control of black flies and mosquitoes. Annu. Rev. Entomol. 31, 265–296.

Lacey, L.A., Heitzman, C.M., Meisch, M.V., Billodeaux, J., 1986. Beecomist® application of *Bacillus sphaericus* for control of riceland mosquitoes. J. Am. Control Assoc. 2, 548–551.

Lacey, L.A., Day, J., Heitzman, C.M., 1987. Long-term effects of *Bacillus sphaericus* on *Culex quinquefasciatus*. J. Invertebr. Pathol. 49, 116–123.

Lacey, L.A., Ross, D.H., Lacey, C.M., Inman, A., Dulmage, H.T., 1988b. Experimental formulations of *Bacillus sphaericus* for the control of anopheline and culicine larvae. J. Indus. Microbiol. 3, 39–47.

Lee, H.L., Seleena, P., Singh, A., Matusop, A., Bakri, M., 1994. Evaluation of the impact of *Bacillus thuringiensis* serotype H-14, a mosquito microbial control agent on malaria vector and incidence in Peninsular Malaysia. Trop. Biomed. 11, 31–38.

Levy, R., Miller Jr., T.W., 1977. Thermal tolerance of *Romanomermis culicivorax* a nematode parasite of mosquitoes. J. Nematol. 9, 259–260.

Logan, T.M., Linthicum, K.J., 1992. Evaluation of a briquette formulation of *Bacillus thuringiensis* var. *israelensis* (H-14) against *Aedes* spp. and *Culex* spp. larvae in dambos in Kenya. Biocontrol Sci. Technol. 2, 257–260.

Lwetoijera, D.W., Sumaye, R.D., Madumla, E.P., Kavishe, D.R., Mnyone, L.L., Russell, T.L., Okumu, F.O., 2010. An extra-domiciliary method of delivering entomopathogenic fungi, *Metarhizium anisopliae* IP 46 against malaria vectors, *Anopheles arabiensis*. Parasites Vectors 3, 18.

Maheu-Giroux, M., Castro, M.C., 2013. Impact of community-based larviciding on the prevalence of malaria infection in Dar es Salaam, Tanzania. PLoS One 8, e71638.

Majambere, S., Lindsay, S.W., Green, C., Kandeh, B., Fillinger, U., 2007. Microbial larvicides for malaria control in the Gambia. Malar. J. 6, 76. http://dx.doi.org/10.1186/1475-2875-6-76.

Majambere, S., Pinder, M., Fillinger, U., Ameh, D., Conway, D.J., Green, C., Jeffries, D., Jawara, M., Milligan, P.J., Hutchinson, R., Lindsay, S.W., 2010. Is mosquito larval source management appropriate for reducing malaria in areas of extensive flooding in the Gambia? A cross-over intervention trial. Am. J. Trop. Med. Hyg. 82, 176–184.

Majori, G., Ali, A., Sabatinelli, G., 1987. Laboratory and field efficacy of *Bacillus thuringiensis* var. *israelensis* and *B. sphaericus* against *Anopheles gambiae* s.l. and *Culex quinquefasciatus* in Ouagadougou, Burkina Faso. J. Am. Mosq. Control Assoc. 3, 20–25.

Marina, C.F., Bond, J.G., Muñoz, J., Valle, J., Novelo-Gutiérrez, R., Williams, T., 2014. Efficacy and non-target impact of spinosad, Bti and temephos larvicides for control of *Anopheles* spp. in an endemic malaria region of southern Mexico. Parasites Vectors 7, 55.

Mboera, L.E.G., Kramer, R.A., Miranda, M.L., Kilima, S.P., Shayo, E.H., Lesser, A., 2014. Community knowledge and acceptance of larviciding for malaria control in a rural district of East-Central Tanzania. Int. J. Environ. Res. Public Health 11, 5137–5154.

Micieli, M.V., García, J.J., 1999. Estudios epizootiológicos de *Strelkovimermis spiculatus* Poinar y Camino, 1986 (Nematoda, Mermithidae) en una población natural de *Aedes albifasciatus* Macquart (Diptera, Culicidae) en la Argentina. Misc. Zool. 22, 31–37.

Mittal, P.K., 2003. Biolarvicides in vector control: challenges and prospects. J. Vector Borne Dis. 40, 20–32.

Mnyone, L.L., Matthew, K.J., Kirby, M.J., Lwetoijera, D.W., Mpingwa, M.W., Knols, B.G.J., Takken, W., Russell, T.L., 2009. Infection of the malaria mosquito, *Anopheles gambiae*, with two species of entomopathogenic fungi: effects of concentration, co-formulation, exposure time and persistence. Malar. J. 8, 309. http://dx.doi.org/10.1186/1475-2875-8-309.

Mnyone, L.L., Koenraadt, C.J.M., Lyimo, I.N., Mpingwa, M.W., Takken, M.W., Russell1, T.L., 2010. Anopheline and culicine mosquitoes are not repelled by surfaces treated with the entomopathogenic fungi *Metarhizium anisopliae* and *Beauveria bassiana*. Parasites Vectors 3, 80. http://dx.doi.org/10.1186/1756-3305-3-80.

Montero Lago, G., Diaz Perez, M., Marrero Figueroa, A., Castillo Gonzalez, F.A., 1991. Results of the pilot applications of the biolarvicide *Bacillus sphaericus* 2362 in mosquito breeding sites of the Santa Cruz del Norte municipality Havana Province. Rev. Cuba. Med. Trop. 43, 39–44.

Moreno, J.E., Martínez, A., Acevedo, P., Sánchez, V., Amaya, W., Petterson, L., Guevara, J., Ascanio, Y., 2014. Evaluación preliminar de la eficiencia de *Bacillus sphaericus* en un área endémica a malaria del estado Bolívar, Venezuela. Bol. Malariol. Salud Amb. 54, 47–57.

Mulla, M.S., Chaney, J.D., Rodcharoen, J., 1993. Elevated dosages of *Bacillus thuringiensis* var. *israelensis* fail to extend control of *Culex* larvae. Bull. Soc. Vector Ecol. 18, 125–132.

Mulla, M.S., Rodcharoen, J., Ngamsuk, W., Tawatsin, A., Pan-Urai, P., Thavara, U., 1997. Field trials with *Bacillus sphaericus* formulations against polluted water mosquitoes in a surburban area of Bangkok, Thailand. J. Am. Mosq. Control Assoc. 13, 297–304.

Mulla, M.S., Thavara, U., Tawatsin, A., Kong-Ngamsuk, W., Chompoosri, J., Su, T., 2001. Mosquito larval control with *Bacillus sphaericus*: reduction in adult populations in low-income communities in Nonthaburi Province, Thailand. J. Vector Ecol. 26, 221–231.

Mulla, M.S., Thavara, U., Tawatsin, A., Chomposri, J., Su, T.Y., 2003. Emergence of resistance and resistance management in field populations of tropical *Culex quinquefasciatus* to the microbial control agent *Bacillus sphaericus*. J. Am. Mosq. Control Assoc. 19, 39–46.

Mwangangi, J.M., Kahindi, S.C., Kibe, L.W., Nzovu, J.G., Luethy, P., Githure, J.I., Mbogo, C.M., 2010. Wide-scale application of *Bti/Bs* biolarvicide in different aquatic habitat types in urban and peri-urban Malindi, Kenya. Parasitol. Res. 108, 1355–1363.

Nartey, R., Owusu-Dabo1, E., Kruppa, T., Baffour-Awuah1, S., Annan, A., Oppong, S., Becker, N., Obiri-Danso, K., 2013. Use of *Bacillus thuringiensis* var *israelensis* as a viable option in an integrated malaria vector control programme in the Kumasi Metropolis, Ghana. Parasites Vectors 6, 116–125.

Nevill, C., Some, E., Mungála, V., Mutemi, W., New, L., Marsh, K., Lengeler, C., Snow, R., 1996. Insecticide-treated bednets reduce mortality and severe morbidity from malaria among children on the Kenyan coast. Trop. Med. Int. Health 1, 139–146.

Nicolas, L., Darriet, F., Hougard, J.M., 1987a. Efficacy of *Bacillus sphaericus* 2362 against larvae of *Anopheles gambiae* under laboratory and field conditions in West Africa. Med. Vet. Entomol. 1, 157–162.

Nicolas, L., Dossou-Yovo, J., Hougard, J.M., 1987b. Persistence and recycling of *Bacillus sphaericus* 2362 spores in *Culex quinquefasciatus* breeding sites in West Africa. Appl. Microbiol. Biotechnol. 25, 341–345.

Nielsen-LeRoux, C., Pasteur, N., Pretre, J., Charles, J.-F., Ben Sheikh, H., Chevillon, C., 2002. High resistance to *Bacillus sphaericus* binary toxin in *Culex pipiens* (Diptera: Culicidae): the complex situation of West Mediterranean countries. J. Med. Entomol. 39, 729–735.

Ocampo, C.B., González, C., Morales, C.A., Pérez, M., Wesson, D., Apperson, C.S., 2009. Evaluation of community-based strategies for *Aedes aegypti* control inside houses. Biomédica 29, 282–297.

Oliveira, C.M., Silva-Filha, M.H., Nielsen-LeRoux, C., Pei, G., Yuan, Z., Regis, L., 2004. Inheritance and mechanism of resistance to *Bacillus sphaericus* in *Culex quinquefasciatus* (Diptera: Culicidae) from China and Brazil. J. Med. Entomol. 41, 58–64.

Ondiaka, S.N., Masinde, E.W., Koenraadt, C.J.M., Takken, M., Mukabana, W.R., 2015. Effects of fungal infection on feeding and survival of *Anopheles gambiae* (Diptera: Culicidae) on plant sugars. Parasites Vectors 8, 35. http://dx.doi.org/10.1186/s13071-015-0654-3.

Paily, K.P., 1990. An improved method of mass culturing of *Romanomermis iyengari*, a mermithid nematode parasite of mosquito larvae. Indian J. Med. Res. 91, 298–302.

Paily, K.P., Balaraman, K., 1990. Effect of temperature and host parasite ratio on sex differentiation of *Romanomermis iyengari* (Welch) a mermithid parasite of mosquitoes. Indian J. Exp. Biol. 128, 470–474.

Paily, K.P., Balaraman, K., 1993. Influence of soil moisture on survival and oviposition of *Romanomermis iyengari* a mermithid nematode parasite of mosquitoes. Indian J. Malariol. 30, 221–225.

Paily, K.P., Balaraman, K., 1994. Effect of temperature on different stages of *Romanomermis iyengari*, a mermithid nematode parasite of mosquitoes. Mem. Inst. Oswaldo Cruz 89, 635–642.

Paily, K.P., Balaraman, K., 2000. Susceptibility of ten species of mosquito larvae to the parasitic nematode *Romanomermis iyengari* and its development. Med. Vet. Entomol. 14, 426–429.

Paily, K.P., Jayachandran, S., 1987. Factors inhibiting parasitism of mosquito larvae by the mermithid nematode (*Romanomermis iyengari*) in a polluted habitat. Indian J. Med. Res. 86, 469–474.

Paily, K.P., Virmani, A.K., Jayachandra, S., 1987. Utility of *Bacillus sphaericus* in controlling *Culex tritaeniorhynchus* breeding. FEMS Microbiol. Ecol. 45, 313.

Paily, K.P., Arunachalam, N., Somachary, N., Balaraman, K., 1991. Infectivity of a mermithid nematode *Romanomermis iyengari* (Welch) in different conductivity levels under laboratory and field conditions. Indian J. Exp. Biol. 29, 579–581.

Paily, K.P., Arunachalam, M., Reddy, C.M.R., Balaraman, K., 1994. Effect of field application of *Romanomermis iyengari* (Nematoda: Mermithidae) on the larvae of *Culex tritaeniorhynchus* and *Anopheles subpictus* breeding in grassland. Trop. Biomed. 11, 23–29.

Paily, K.P., Chandhiran, K., Vanamail, P., Pradeep Kumar, N., Jambulingam, P., 2013. Efficacy of a mermithid nematode *Romanomermis iyengari* (Welch) (Nematoda: Mermithidae) in controlling tree hole breeding mosquito *Aedes albopictus* (Skuse) (Diptera: Culicidae) in a rubber plantation area of Kerala, India. Parasitol. Res. 112, 1299–1304.

Pantuwatana, S., Maneeroj, R., Upatham, E.S., 1989. Long residual activity of *Bacillus sphaericus* 1593 against *Culex quinquefasciatus* larvae in artificial pools. Southeast Asian J. Trop. Med. Pub. Health 20, 421–427.

Park, H.-W., Bideshi, D.K., Federici, B.A., 2003. Recombinant strain of *Bacillus thuringiensis* producing Cyt1A, Cry11B, and the *Bacillus sphaericus* binary toxin. Appl. Environ. Microbiol. 69, 1331–1334.

Park, H.W., Bideshi, D.K., Wirth, M.C., Johnson, J.J., Walton, W.E., Federici, B.A., 2005. Recombinant larvicidal bacteria with markedly improved efficacy against *Culex* vectors of West Nile virus. Am. J. Trop. Med. Hyg. 72, 732–738.

Paula, A.R., Carolino, A.T., Paula, C.O., Samuels, R.I., 2011. The combination of the entomopathogenic fungus *Metarhizium anisopliae* with the insecticide Imidacloprid increases virulence against the dengue vector *Aedes aegypti* (Diptera: Culicidae). Parasites Vectors 4, 8.

Perez-Pacheco, R., Flores, G., 2005. Mass production of mermithid nematode parasites of mosquito larvae in Mexico. J. Nematol. 37, 388.

Pérez-Pacheco, R., Rodríguez-Hernández, C., Lara-Reyna, J., Montez-Belmont, R., Ruiz-Vega, J., 2005. Control of the mosquito *Anopheles pseudopunctipennis* (Diptera: Culicidae) with *Romanomermis iyengari* (Nematoda: Mermithidae) in Oaxaca, Mexico. Biol. Control 32, 137–142.

Pérez–Pacheco, R., Santamarina–Mijares, A., Vásquez–López, A., Martínez–Tomás, S.H., Suárez–Espinosa, J., 2009. Efectividad y supervivencia de *Romanomermis culicivorax* en criaderos naturales de larvas de mosquitos. Agrociencia 43, 861–868.

Perich, M.J., Boobar, L.R., Stivers, J.C., Rivera, L.A., 1990. Field evaluation of four biorational larvicide formulations against *Anopheles albimanus* in Honduras. Med. Vet. Entomol. 4, 393–396.

Petersen, J.J., 1985. Nematode parasites. In: Chapman, H.C. (Ed.), Biological Control of Mosquitoes. Am. Mosq. Control Assoc. Bull. 6, 110–122.

Petersen, J.J., Willis, O.R., 1972. Procedures for the mass rearing of a mermithid parasite of mosquitoes. Mosq. News 32, 226–230.

Petersen, J.J., Willis, O.R., 1974. Experimental release of a mermithid nematode to control *Anopheles* mosquitoes in Louisiana. Mosq. News 34, 316–319.

Petersen, J.J., Willis, O.R., 1975. Establishment and recycling of a mermithid nematode for the control of larval mosquitoes. Mosq. News 35, 526–532.

Petersen, J.J., Chapman, H.C., Willis, O.R., Fukuda, T., 1978. Release of *Romanomermis culicivorax* for control of *Anopheles albimanus* in El Salvador. II. Application of nematode. Am. J. Trop. Med. Hyg. 27, 1268–1273.

Pineda, C.A.C., Lago, G.M., Ortega, A.N., Martín, P.L.M., 2005. Control de culícidos con el empleo de *Bacillus thuringiensis* SH-14 var. *israelensis* en criaderos permanentes de la localidad de Fomento, Provincia Sancti Spíritus, Cuba. Rev. Cuba. Med. Trop. 57.

Platzer, E.G., 2007. Mermithid nematodes. In: Floore, T.G. (Ed.), Biorational Control of Mosquitoes. Am. Mosq. Control Assoc. Bull. 7, 58–64.

Platzer, E.G., Mullens, B.A., Shamseldean, M.M., 2005. Mermithid nematodes. In: Grewal, P.S., Ehlers, R.-U., Shapiro-Ilan, D.I. (Eds.), Nematodes as Biocontrol Agents. CABI Publishing, Oxfordshire, UK, pp. 411–418.

Pridantseva, E.A., Lebeneva, N.I., Shcherban, Z.P., Kadyrovam, M.K., 1990. Assessment of the possibility of using the mermithid *Romanomermis iyengari* for the control of mosquitoes in Uzbekistan. Med. Parazitol. Parazit. Bol. 1, 15–17.

Ragoonanansingh, R.N., Njunwa, K.J., Curtis, C.F., Becker, N., 1992. A field study of *Bacillus sphaericus* for the control of culicine and anopheline mosquito larvae in Tanzania. Bull. Soc. Vector Ecol. 17, 45–50.

Rao, D.R., Mani, T.R., Rajendran, R., Josesph, A.S., Gajanana, A., Reuben, R., 1995. Development of a high level of resistance to *Bacillus sphaericus* in a field population of *Culex quinquefasciatus* from Kochi, India. J. Am. Mosq. Control Assoc. 11, 1–5.

Regis, L., Nielsen-LeRoux, C., 2000. Management of resistance to bacterial vector control. In: Charles, J.-F., Délécluse, A., Nielsen-LeRoux, C. (Eds.), Entomopathogenic Bacteria: From Laboratory to Field Application. Kluwer Academic Publishers, Dordrecht, The Netherlands, pp. 419–438.

Regis, L., Silva-Filha, M.H.N.L., Oliveira, C.M.F., Rios, E.M., Silva, S.B., Furtado, A.F., 1995. Integrated control measures against *Culex quinquefasciatus*, the vector of filariasis in Recife. Mem. Inst. Oswaldo Cruz 90, 115–119.

Regis, L., Oliveira, C.M.F., Silva-Filha, M.H., Silva, M.B., Maciel, A., Furtado, A.F., 2000. Efficacy of *Bacillus sphaericus* in control of the filariasis vector *Culex quinquefasciatus* in an urban area of Olinda, Brazil. Trans. R. Soc. Trop. Med. Hyg. 94, 488–492.

Ritchie, S.A., Rapley, L.P., Benjamin, S., 2010. *Bacillus thuringiensis* var. *israelensis* (Bti) provides residual control of *Aedes aegypti* in small containers. Am. J. Trop. Med. Hyg. 82, 1053–1059.

Rivera, P., Lugo, E., Lopez, M., Valle, S., Delgado, M., Lopez, D., Larios, F., 1997. Evaluación de la efectividad biolarvicida y residualidad de *Bacillus sphaericus* (Cepa 2362) para el control de *Anopheles albimanus* en la costa del Lago Xolotlan, Managua, Nicaragua, 1995. Rev. Nicarag. Entomol. 42, 7–14.

Roberts, D.W., Panter, C., 1985. Fungi other than *Coelomomyces* and *Lagenidium*. In: Chapman, H.C. (Ed.), Biological Control of Mosquitoes. Am. Mosq. Control Assoc. Bull. 6, 99–109.

Rojas, W., Northup, J., Gallo, O., Montoya, E., Restrepo, M., Nimnich, G., Arango, M., Echavarria, M., 1987. Reduction of malaria prevalence after introduction of *Romanomermis culicivorax* (Mermithidae: Nematoda) in larval *Anopheles* habitats in Colombia. Bull. World Health Org. 63, 331–337.

Rowland, M., Hewitt, S., Durrani, N., Saleh, P., Bouma, M., Sondorp, E., 1997. Sustainability of pyrethroid-impregnated bednets for malaria control in Afghan communities. Bull. World Health Org. 75, 23–29.

Romão, T.P., de Melo Chalegre, K.D., Key, S., Ayres, C.F.J., Fontes de Oliveira, C.M., de-Melo-Neto, O.P., Silva-Filha, M.H.N.L., 2006. A second independent resistance mechanism to *Bacillus sphaericus* binary toxin targets its a-glucosidase receptor in *Culex quinquefasciatus*. FEBS J. 273, 1556–1568.

Romi, R., Ravoniharimelina, B., Ramiakajato, M., Majori, G., 1993. Field trails of *Bacillus thuringiensis* H-14 and *Bacillus sphaericus* (Strain 2362) formulations against *Anopheles arabiensis* in the central highlands of Madagascar. J. Am. Mosq. Control Assoc. 9, 325–329.

Rosenbaum, J., Nathan, M.B., Ragoonanansingh, R., Rawlins, S., Gayle, C., Chadee, D.D., Lloyd, L.S., 1995. Community participation in dengue prevention and control: a survey of knowledge, attitudes, and practice in Trinidad and Tobago. Am. J. Trop. Med. Hyg. 53, 111–117.

Russell, T.L., Lwetoijera1, D.W., Maliti, D., Chipwaza, B., Kihonda1, J., Charlwood, J.D., Smith, T.A., Lengeler, C., Mwanyangala, M.A., Nathan, R., Knols, B.G.J., Takken, W., Killeen, G.F., 2010. Impact of promoting longer-lasting insecticide treatment of bed nets upon malaria transmission in a rural Tanzanian setting with pre-existing high coverage of untreated nets. Malar. J. 9, 187.

Sandoski, C.A., Yates, M.M., Olson, J.K., Meisch, M.V., 1985. Evaluation of Beecomist-applied *Bacillus thuringiensis* (H-14) against *Anopheles quadrimaculatus* larvae in rice fields. J. Am. Mosq. Control Assoc. 2, 316–319.

Santamarina, A.M., 1994. Actividad parasitaria de *Romanomermis iyengari* (Nematoda, Mermithidae) en criaderos naturales de larvas de mosquito. Misc. Zool. 17, 59–65.

Santamarina, A.M., 1996. Mass breeding of *Romanomermis culicivorax* (Nematoda: Mermithidae) in the tropical conditions of Cuba (in Spanish). Rev. Cuba. Med. Trop. 48, 26–33.

Santamarina, A.M., Bellini, A.C., 2000. Producción masiva de *Romanomermis iyengari* (Nematoda: Mermithidae) y su aplicación en criaderos de anofelinos en Boa Vista (Roraima), Brasil. Rev. Panam. Salud Publica 7, 155–161.

Santamarina, A.M., Pérez-Pacheco, R.P., 1997. Reduction of mosquito larval densities in natural sites after introduction of *Romanomermis culicivorax* (Nematoda: Mermithidae) in Cuba. J. Med. Entomol. 34, 1–4.

Santamarina, A.M., García, I.A., González, R.B., 1992. Capacidad infestiva del nematodo parásito *Romanomermis iyengari* (Welch, 1964) (Nematoda: Mermithidae) en larvas de mosquitos en condiciones naturales. Rev. Cuba. Med. Trop. 44, 92–97.

Santamarina, A.M., García, I.A., González, R.B., 1993. Valoración de la capacidad infectiva del nemátodo parásito *Romanomermis iyengari* (Nematoda: Mermithidae) en criaderos naturales de larvas de mosquitos. Rev. Cuba. Med. Trop. 45, 128–131.

Santamarina, A.M., García, I.A., Rosabal, J.R., Solís, A.M., 1996. Release of *Romanomermis iyengari* (Nematoda: Mermithidae) to control *Aedes taeniorhynchus* (Diptera: Culicidae) in Punta del Este, Isla de la Juventud. J. Med. Entomol. 33, 680–682.

Santamarina, A.M., Pérez-Pacheco, R., Martinez, S.H., Cantón, L.E., Flores-Ambrosio, G.F., 1999. The *Romanomermis iyengari* parasite for *Anopheles pseudopunctipennis* suppression in natural habitats in Oaxaca State, Mexico. Rev. Panam. Salud Publica 5, 23–28.

Scholte, E.J., Njiru, B.N., Smallegange, R.C., Takken, W., Knols, B.G.J., 2003. Infection of malaria (*Anopheles gambiae s.s.*) and filariasis (*Culex quinquefasciatus*) vectors with the entomopathogenic fungus *Metarhizium anisopliae*. Malar. J. 2, 29.

Scholte, E.J., Knols, B.G.J., Samson, R., Takken, W., 2004a. Entomopathogenic fungi for mosquito control: a review. J. Insect Sci. 4, 19.

Scholte, E.L., Knols, B.G.J., Takken, W., 2004b. Autodissemination of the entomopathogenic fungus *Metarhizium anisopliae* amongst adults of the malaria vector *Anopheles gambiae s.s.* Malar. J. 3, 45.

Scholte, E.J., Ng'habi, K., Kihonda, J., Takken, W., Paaijmans, K., Abdulla, S., Killeen, G.F., Knols, B.G.J., 2005. An entomopathogenic fungus for control of adult African malaria mosquitoes. Science 308, 1641–1642.

Scholte, E.J., Knols, B.G.J., Takken, W., 2006. Infection of the malaria mosquito *Anopheles gambiae* with the entomopathogenic fungus *Metarhizium anisopliae* reduces blood feeding and fecundity. J. Invertebr. Pathol. 91, 43–49.

Scholte, E.J., Takken, W., Knols, B.G.J., 2007. Infection of adult *Aedes aegypti* and *Ae. albopictus* mosquitoes with the entomopathogenic fungus *Metarhizium anisopliae*. Acta Trop. 102, 151–158.

Scholte, E.J., Knols, B.G.J., Takken, W., 2008. An entomopathogenic fungus (*Metarhizium anisopliae*) for control of the adult African malaria vector *Anopheles gambiae*. Entomol. Berich. 68, 21–26.

Shukla, R.P., Kohli, V.K., Ojha, V.P., 1997. Larvicidal efficacy of *Bacillus sphaericus* H5a, 5b and *B. thuringiensis* var *israelensis* H-14 against malaria vectors in Bhabar area, District Nainital, U.P. Indian J. Malariol. 34, 208–212.

Siegel, J.P., Novak, R.J., 1999. Duration of activity of the microbial larvicide Vectolex CG® (*Bacillus sphaericus*) in Illinois catch basins and waste tires. J. Am. Mosq. Control Assoc. 15, 366–370.

Siegel, J.P., Shadduck, J.A., 1990. Safety of microbial insecticides to vertebrate-humans. In: Laird, M., Lacey, L.A., Davidson, E.W. (Eds.), Safety of Microbial Insecticides. CRC Press, Boca Raton, pp. 101–113.

Silapanuntakul, A., Pantuwatana, S., Bhumiuratana, A., Charoensiri, K., 1983. The comparative persistence of toxicity of *Bacillus sphaericus* Strain 1593 and *Bacillus* thuringiensis serotype H-14 against mosquito larvae in different kinds of environments. J. Invertebr. Pathol. 42, 387–392.

Silva-Filha, M.H., Regis, L., Oliveira, C.M.F., Furtado, A.F., 2001. Impact of a 26-month *Bacillus sphaericus* trial on the preimaginal density of *Culex quinquefasciatus* in an urban area of Recife, Brazil. J. Am. Mosq. Control Assoc. 17, 45–50.

Silva-Filha, M.H.N.L., Chalegre, K.D.D.M., Anastacio, D.B., Fontes de Oliveira, C.M., Batista da Silva, S., Acioli, R.V., Hibi, S., Cardoso de Oliveira, D., Parodi, E.S.M., Marques Filho, C.A.M., Furtado, A.F., Regis, L., 2008. *Culex quinquefasciatus* field populations subjected to treatment with *Bacillus sphaericus* did not display high resistance levels. Biol. Control 44, 227–234.

Silva-Filha, M.H.N.L., Berry, B., Regis, L., 2014. *Lysinibacillus sphaericus*: toxins and mode of action, applications for mosquito control and resistance management. In: Dhadialla, T.S., Gill, S.S. (Eds.), Advances in Insect Physiology, vol. 47. Academic Press, Oxford, pp. 89–176.

Skovmand, O., Baudin, S., 1997. Efficacy of a granular formulation of *Bacillus sphaericus* against *Culex quinquefasciatus* and *Anopheles gambiae* in West African countries. J. Vector Ecol. 22, 43–51.

Skovmand, O., Sanogo, E., 1999. Experimental formulation of *Bacillus sphaericus* and *B. thuringiensis israelensis* against *Culex quinquefasciatus* and *Anopheles gambiae* (Diptera: Culicidae) in Burkina Faso. J. Med. Entomol. 36, 62–67.

Snetselaar, J., Andriessen, R., Suer, R.A., Osinga, A.J., Knols, G.J., Farenhorst, M., 2014. Development and evaluation of a novel contamination device that targets multiple life-stages of *Aedes aegypti*. Parasites Vectors 7, 200.

Stanaway, J.D., Shepard, D.S., Undurraga, E.A., Halasa, Y.A., Coffeng, L.E., Brady, O.J., Hay, S.I., Bedi, N., Bensenor, I.M., Castañeda-Orjuela, C.A., Chuang, T.-W., Gibney, K.B., Memish, Z.A., Rafay, A., Kingsley, N., Ukwaja, K.N., Yonemoto, N., Murray, C.J.L., 2016. The global burden of dengue: an analysis from the Global Burden of Disease Study 2013. Lancet Infect. Dis. http://dx.doi.org/10.1016/S1473-3099(16)00026-8.

Su, T., Mulla, M.S., 2004. Documentation of high-level *Bacillus sphaericus* 2362 resistance in field populations of *Culex quinquefasciatus* breeding in polluted water in Thailand. J. Am. Mosq. Control Assoc. 20, 405–411.

Sundararaj, R., Raghunatha-Rao, D., 1993. Field evaluation of a microgel droplet formulation of *Bacillus sphaericus* 1593M (Biocide-S) against *Anopheles culicifacies* and *Anopheles subpictus* in South India. Southeast Asian J. Trop. Med. Pub. Health 24, 363–368.

Sundararaj, R., Reuben, R., 1991. Evaluation of a microgel droplet formulation of *Bacillus sphaericus* 1593 M (Biocide-S) for control of mosquito larvae in rice fields in southern India. J. Am. Mosq. Control Assoc. 7, 556–559.

Swaddiwudhipong, W., Chaovakiratipong, C., Nguntra, P., Koonchote, S., Khumklam, P., Lerdlukanavonge, P., 1992. Effect of health education on community participation in control of dengue hemorrhagic fever in an urban area of Thailand. Southeast Asian J. Trop. Med. Pub. Health 23, 200–206.

Thomas, M.B., Read, A.F., 2007. Can fungal biopesticides control malaria? Nat. Rev. Microbiol. 5, 377–383.

Utzinger, J., Tozan, Y., Singer, B.H., 2001. Efficacy and cost-effectiveness of environmental management for malaria control. Trop. Med. Int. Health 6, 677–687.

Vilarinhos, P.T., Monnerat, R., 2004. Larvicidal persistence of formulations of *Bacillus thuringiensis* var. *israelensis* to control larval *Aedes aegypti*. J. Am. Mosq. Assoc. 20, 311–314.

Vilarinhos, P.T., Dias, J.C.M.S., Andrade, C.F.S., Araújo-Coutinho, C.J.P.C., 1998. Uso de bactérias para o controle de culicídeos e simulídeos. In: Alves, S.B. (Ed.), Controle Microbiano de Insetos, second ed. Fundação de Estudos Agrários Luiz de Queiros – FEALQ, Piracicaba, Brazil, pp. 447–480.

Walker, K., Lynch, M., 2007. Contributions of *Anopheles* larval control to malaria suppression in tropical Africa: review of achievements and potential. Med. Vet. Entomol. 21, 2–21.

Walker, T.W., Meek, C.L., Wright, V.L., 1988. Establishment and recycling of *Romanomermis culicivorax* in Louisiana ricelands. J. Am. Mosq. Control Assoc. 1, 468–473.

Walton, W.E., Mulla, M.S., 1991. Integrated control of *Culex tarsalis* larvae using *Bacillus sphaericus* and *Gambusia affinis*: effects on mosquitoes and nontarget organisms in field mesocosms. Bull. Soc. Vector Ecol. 16, 203–221.

Wang, C.H., Chang, N.T., Wu, H.H., Ho, C.M., 2000. Integrated control of the dengue vector *Aedes aegypti* in Liu-Chiu Village, Ping-Tung county, Taiwan. J. Am. Mosq. Control Assoc. 16, 93–99.

Warrell, D.A., Gilles, H.M., 2002. Essential Malariology, fourth ed. CRC Press, Boca Raton, FL. 352 pp.

WHO, 2012. Global Strategy for Dengue Prevention and Control 2012–2020. 35 pp.

WHO, 2014. Vector-Borne Diseases Fact Sheet 387.

WHO, 2015a. Dengue and Severe Dengue Fact Sheet 117.

WHO, 2015b. Lymphatic Filariasis Fact Sheet 102.

WHO, 2015c. World Malaria Report 2015.

WHO, 2016a. Malaria.

WHO, 2016b. Vector-Borne Disease.

Winch, P.J., Leontsini, E., Rigau-Pérez, J.G., Mervin Ruiz-Pérez, M., Clark, G.G., Gubler, D.J., 2002. Community-based dengue prevention programs in Puerto Rico: impact on knowledge, behavior, and residential mosquito infestation. Am. J. Trop. Med. Hyg. 67, 363–370.

Wirth, M.C., 2010. Mosquito resistance to bacterial larvicidal toxins. Open Toxinol. J. 3, 126–140.

Wirth, M.C., Walton, W.E., Federici, B.A., 2000. Cyt1A from *Bacillus thuringiensis* restores toxicity of *Bacillus sphaericus* against *Culex quinquefasciatus* (Diptera; Culicidae). J. Med. Entomol. 37, 401–407.

Wirth, M.C., Jiannino, J.A., Federici, B.A., Walton, W.E., 2004. Synergy between toxins of *Bacillus thuringiensis* subsp. *israelensis* and *Bacillus sphaericus*. J. Med. Entomol. 41, 935–941.

Xu, B.Z., Becker, N., Xianqi, X., Ludwig, H.W., 1992. Microbial control of malaria vectors in Hubei Province, People's Republic of China. Bull. Soc. Vector Ecol. 17, 140–149.

Yap, H.H., Tan, H.T., Yahaya, A.M., Baba, R., Chong, N.L., 1991. Small-scale field trials of *Bacillus sphaericus* (strain 2362) formulations against *Mansonia* mosquitoes in Malaysia. J. Am. Mosq. Control Assoc. 7, 24–29.

Yuan, Z.M., Zhang, Y.M., Chen, Z.S., Cai, Q.X., Lieu, E.Y., 1999. Recycling of *Bacillus sphaericus* in mosquito larval cadavers and its effect on persistence. Chin. J. Biol. Control 15, 23–26.

Yuan, Z.M., Zhang, Y.M., Cai, Q.X., Liu, E.Y., 2000. High level field resistance to *Bacillus sphaericus* C3-41 in *Culex quinquefasciatus* from southern China. Biocontrol Sci. Technol. 10, 41–49.

Zahiri, N.S., Federici, B.A., Mulla, M.S., 2004. Laboratory and field evaluation of a new recombinant of *Bacillus thuringiensis* ssp. *israelensis* and *Bacillus sphaericus* against *Culex* mosquito larvae (Diptera: Culicidae). J. Med. Entomol. 41, 423–429.

Zaim, M., Ladonni, H., Ershadi, M.R.Y., Manouchehri, A.V., Sahabi, Z., Nazari, M., Shahmohammadf, H., 1988. Field application of *Romanomermis culicivorax* (Mermithidae: Nematoda) to control anopheline larvae in southern Iran. J. Am. Mosq. Control Assoc. 4, 351–355.

Zequi, J.A.C., Lopes, J., 2007. Biological control of *Culex* (*Culex*) *saltanensis* Dyar (Diptera, Culicidae) through *Bacillus thuringiensis israelensis* in laboratory and field conditions. Rev. Bras. Zool. 24, 164–168.

Zimmermann, G., 2007a. Review on safety of the entomopathogenic fungi *Beauveria bassiana* and *Beauveria brongniartii*. Biocontrol Sci. Technol. 17, 553–596.

Zimmermann, G., 2007b. Review on safety of the entomopathogenic fungus *Metarhizium anisopliae*. Biocontrol Sci. Technol. 17, 879–920.

Chapter 29

Microbial Control of Structural Insect Pests

R.M. Pereira[1], D.H. Oi[2], M.V. Baggio[3], P.G. Koehler[1]

[1]University of Florida, Gainesville, FL, United States; [2]USDA-ARS, Gainesville, FL, United States; [3]UNESP, Jaboticabal, São Paulo, Brazil

29.1 INTRODUCTION TO THE ECONOMIC PROBLEM CAUSED BY STRUCTURAL INSECT PESTS

As the world population concentrates in large and small cities, the importance of structural and domestic insect pests continues to grow. The concomitant growth in commerce and travel serves as a contributing factor in the spread of pests. The recent resurgence of bed bugs as major urban pests in different areas is an example of the rapid spread and internationalization of urban pests. As different cultures adopt similar housing structures, habits, and construction techniques, including the use of temperature control and similar living arrangements, the most common structural domestic pests become prevalent worldwide. Prime examples are German cockroaches, house flies, termites, such as Formosan and Asian subterranean termites, several ant species, such as the red imported fire ant, and bed bugs.

In contrast to agricultural pest control, which depends more directly on local climate and agricultural practices, structural pests survive and cause problems that are very similar in different regions. Therefore the control of these pests, either through the use of conventional chemical insecticides or through the use of microbial biopesticides, can be achieved with minimal modification to products or application protocols across global regions.

29.1.1 Physical Damage, Annoyance, and Disease Vectoring Caused by Structural Insect Pests

29.1.1.1 Cockroaches

There are more than 3500 cockroach species described worldwide, but only a few are considered pests [eg, *Periplaneta americana* (American cockroach) and *Blattella germanica* (German cockroach); Koehler et al., 2008; Rafael et al., 2008]. Many of the most significant pest species are especially abundant in food preparation areas, sewage systems, and storage areas; accordingly, they have great potential as disease carriers (Bell et al., 2007). Although cockroaches are well-known carriers of several human pathogens (Mariconi, 1999; Miranda and Silva, 2008; Rust, 2008; Kassiri and Kazemi, 2012), they are feared more for their annoying presence. Additionally, cockroach allergens are the primary cause of asthma cases among low-income populations in large cities (Arruda et al., 2001).

29.1.1.2 Ants

Ants can be significant pests of structures when they invade buildings and contaminate food and supplies, potentially transmit disease, damage electrical equipment, or are a maddening nuisance simply by their presence (Beatson, 1972a, 1973; Lard et al., 1999; Groden et al., 2005; Wetterer and Keularts, 2008). Stinging ants (eg, *Solenopsis invicta*) are a health and legal concern because of the potential for anaphylaxis that can lead to death (Lockey, 1974) and the associated litigation (deShazo et al., 1999, 2004; Goddard et al., 2002). In medical facilities and other situations, ants are vectors of diseases (Moreira et al., 2005), spreading pathogenic bacteria and other microorganisms (Beatson, 1972b; Rodovalho et al., 2007; Oi, 2008), causing disease transmission and spoilage of food (Zarzuela et al., 2005, 2007). As structural pests, carpenter ants cause considerable damage to wood structures (Klotz et al., 1995; Hansen and Klotz, 2005).

29.1.1.3 Termites

Termites damaging structures are generally divided into soil termites, which require contact with soil where they have access to moisture required for their survival, and drywood termites, which survive within wood without the need to access moisture beyond what is provided by the wood or wood digestion (Potter, 2011). These insects can weaken structures to the point of collapse if control measures are not taken (Grace and Yates, 1999). Beyond the structural damage, termites also cause loss in the insulating capacity of buildings (Koehler et al., 2008)

Microbial Control of Insect and Mite Pests. http://dx.doi.org/10.1016/B978-0-12-803527-6.00029-9

due to the elimination of insulating material and the addition of soil and moisture into the structure (Tucker et al., 2008). Termites can also be of great importance in museums and other historical preservation sites as they consume wood and paper objects of historical value (Ferrari and Marini, 1999).

29.1.1.4 Bed Bugs and Kissing Bugs

Although kissing bugs (*Rodnius* spp. and *Triatoma* spp.) have been transmitting serious diseases for many years in American countries (Schofield, 1994), their significance as structural pests has been limited to more rural areas. Although bed bugs (*Cimex lectularius* and *Cimex hemipterus*) have not been linked to the large-scale transmission of disease-causing pathogens (Harlan et al., 2008), studies have demonstrated the ability of bed bugs to transmit the parasite that causes Chagas disease (Salazar et al., 2014) and the bacteria that causes trench fever (Leulmi et al., 2015). These insects have reached considerable notoriety in the 21st century due to outbreaks of this pest in many large urban areas in western countries after almost a complete absence as widespread urban pests since the middle of the 20th century (Davies et al., 2012). Large bed bug infestations are quite unsanitary for humans (Whitney, 2012) as the bloody bed bug feces are considered to be a health hazard due to the possibility of contamination with human pathogens and other microorganisms (Delauney et al., 2011). Also, the large consumption of blood by increasing the bed bug infestations (Pereira et al., 2013) can cause anemia in humans, which can lead to more severe heart problems (Paulke-Korinek et al., 2012).

29.1.1.5 Flies

Several nonbiting and biting flies are important pests in medical facilities, food-handling establishments, and human habitations (Pereira et al., 2014) due to their ability to transmit diseases. Structural pest flies can be divided into large flies (house, flesh, blow, bottle, and soldier flies), which normally breed on decaying organic matter or animal feces outside of structures, entering them for shelter and food, and small flies (filter, moth, or drain flies, fruit flies, eye gnats, phorid flies), which usually breed on bacterial and fungal growth on decaying organic matter inside structures (Pereira et al., 2014). Flies develop or feed in unsanitary locations and move pathogens to clean areas (Förster et al., 2007; Butler et al., 2010), degrading human food and spreading pathogens. Flies are involved in numerous human deaths, especially among children in poverty-stricken areas who acquire dysentery (Levine and Levine, 1991), and

the spread of antibiotic-resistant microorganisms within medical facilities (Rady et al., 1992).

29.1.1.6 Other Structural and Domestic Pests

Fleas, mites, wood destroying and stored food beetles, crickets, and many other insects and arthropods such as spiders and ticks can also be important domestic pests, causing material damage and the deterioration of food items and other goods and being involved in human disease transmission (Koehler et al., 2013).

29.2 MICROBIAL CONTROL OF STRUCTURAL INSECT PESTS

Like many other areas, the microbial control of domestic and structural pests has not reached a level of commercial development and practical use seen with some agricultural pest. Structural pests are unique in their close contact with humans, food, and their homes, which forces the control of these pests to be different from that of field crops and other outdoor settings where human contact is less. Much of the management of structural pests has to be done directly in or around human-occupied structures, so the chance for human contact with entomopathogens is high. Despite the proven safety of most entomopathogens to humans and their pets, the application of large quantities of microbial control agents (MCAs) to human-occupied structures carries a certain stigma due to fears associated with microorganisms and disease transmission. However, because many of the structural pests come from outside, the application of control measures can in some instances occur outdoors, where health and other concerns are minimized. Also, in both industrial and residential areas, many cockroaches, scorpions, and other arthropods inhabit sewage systems where the use of microbial control may be more acceptable.

29.2.1 Cockroaches

Although microbial products are not widely used for the control of cockroaches, studies show the pathogenic action and the potential of MCAs such as bacteria, entomopathogenic nematodes (EPNs), and fungi. A commercial product formulated as a gel containing a densonucleosis virus has been available in China for the control of cockroaches (Bergoin and Tijssen, 2010). The pathogenicity of *Bacillus thuringiensis* subsp. *israelensis* (Singh and Gill, 1985) and *B. thuringiensis* subsp. *kurstaki* (Lonc et al., 1997) to *P. americana* and *B. germanica* was demonstrated. However, despite the encouraging results, there are few reports of the use of these microbial agents in field experiments.

A reduction of up to 67% of *B. germanica* in apartments was observed after the use of stations containing EPNs of the genus *Steinernema* (Appel et al., 1993). The use of food baits with *Steinernema carpocapsae* for indoor application also caused a relatively high mortality of American and German cockroaches, especially among nymphs (Maketon et al., 2010).

Entomopathogenic fungi, such as *Metarhizium anisopliae* and *Beauveria bassiana*, have produced high cockroach mortalities in laboratory experiments. These entomopathogenic fungi can infect cockroach oothecae (Mohan et al., 1999; Lopes and Alves, 2011; Hubner-Campos et al., 2013) and reduce nymphal hatching (Quesada-Moraga et al., 2004), but effects can vary with relative humidity and fungal isolates (Hernanez-Ramirez et al., 2008). Oil suspensions of *B. bassiana* and *M. anisopliae* conidia can be used against oothecae, which are not susceptible to chemical insecticides applied topically, and cause a high mortality of *P. americana* adults (Hubner-Campos et al., 2013; Fig. 29.1A and B). Another fungus genus, *Aspergillus*, has not been seriously considered as an MCA against cockroaches due to its production of toxins; however, natural infection of *Aspergillus flavus* on *B. germanica* has been reported (Kulshrestha and Pathak, 1997).

The main difficulties in the use of MCAs for cockroach control concerns safety during domestic application due to the risk of human allergies. Also, the time required for pest elimination, the formulation choice, and the application time to ensure the survival of the MCA in the environment represent further problems for their use against cockroaches. An attempt to insert these organisms in household cleaning procedures was the commercial formulation of *M. anisopliae*, which the company Ecoscience Corporation (United States) patented in the 1990s, but it had a short life in the market. The product was placed in stations distributed at strategic points inside homes and business buildings (Andis, 1994).

The strategy for the use of entomopathogens against cockroaches depends on the target species. Food bait stations with EPNs, *B. thuringiensis*, *M. anisopliae*, or other MCAs can be used for managing *B. germanica* (Appel et al., 1993; Lonc et al., 1997; Lopes and Alves, 2011),

because these pests are mainly inside of homes, in protected environments where UV radiation and large temperature fluctuations, which normally cause problems for the use of entomopathogens on crops, are not important considerations. Bait formulations have less environmental impact and are safer to use, because they are implemented with exact placement and amount and are not likely to have use restrictions in homes, food shops, and other establishments (Anaclerio and Molinari, 2012).

Feeding baits containing *S. carpocapsae* and *B. thuringiensis* have been used with low efficacy for *P. americana* control (Maketon et al., 2010). Although *B. thuringiensis* works by ingestion, *S. carpocapsae* penetrates by spiracles and intersegmental membranes, and the use of baits may not provide the best conditions for infective juveniles' penetration into the host insect. Considering the habitat of *P. americana*, which may spend a great part of their lives outside structures, it is possible to eliminate American cockroach infestations outside structures. Leaf litter and other areas where this cockroach lives outdoors provide favorable environmental conditions for the viability and growth of fungi (Vieira et al., 2007) and other pathogens. These favorable conditions can extend the effect of the biological products, and the outdoor use reduces risks of allergies and undesirable contamination of domestic areas. Other formulations that could be used are liquid applications of fungal conidia and adjuvants such as oil and Tween (Hernanez-Ramirez et al., 2008; Hubner-Campos et al., 2013) or dusting with *M. anisopliae* and *B. bassiana* conidia, which has provided efficient control of grain beetles (Shafighi et al., 2014). Another possibility is the combination of chemical insecticides with pathogenic microbial agents in food baits, liquid, or dust applications.

Although entomopathogenic fungi and EPNs have been emphasized in studies of cockroach microbial control, more studies are needed on the susceptibility of different insect stages to entomopathogens and the feasibility of different formulations and application methods. The chosen

(A) **(B)**

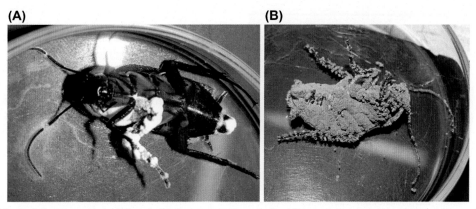

FIGURE 29.1 *Periplaneta americana* infected with the entomopathogenic fungi *Beauveria bassiana* (A) and *Metarhizium anisopliae* (B).

formulations should enhance survival of the microbial agents, with a minimum of 94% viability (Lopes et al., 2013), and be appropriate for the type of environment in which each cockroach species occurs. Given the environments where MCAs should be used to control structural cockroaches, it may be ideal to use baits indoors and reserve spraying or dusting applications for outdoors and in sewage systems.

29.2.2 Ants

Ant management with entomopathogens has mainly focused on their utilization as classical biological control agents to suppress invasive ant populations in landscapes surrounding buildings or in other outdoor habitats, which serve as reservoirs for invading ants (Oi and Valles, 2009). While many pathogens and nematodes infecting ants have been reported, only a few have been the subject of extensive efforts to utilize them as MCAs (Evans, 1974; Oi and Pereira, 1993; Milner and Pereira, 2007; Evans et al., 2010, 2011). Nematodes and entomopathogens isolated from the red imported fire ant (*S. invicta*) have been the most extensively evaluated for their microbial control potential. Several microorganisms have been discovered and evaluated as microbial and biological control agents (Allen and Buren, 1974; Allen and Silveira-Guido, 1974; Broome, 1974; Jouvenaz et al., 1977, 1981; Williams et al., 2003). These include the microsporidia *Kneallhazia* (formerly *Thelohania*) *solenopsae* and *Vairimorpha invictae*. Other findings and evaluations include the fungus *B. bassiana*, the neogregarine protozoan *Mattesia* sp. (Pereira et al., 2002) (Fig. 29.2), and the nematodes *Tetradonema solenopsis* and *Allomermis solenopsi* (Nickle and Jouvenaz, 1987; Poinar et al., 2007). Molecular screenings have facilitated the discovery of several viruses of *S. invicta* (Valles, 2012; Valles et al., 2013b).

FIGURE 29.2 *Solenopsis invicta* cadaver pile with yellow-head disease-infected cadaver at the center. Ant cadaver piles can be a source of pathogens for future research and a source of infection for ants in the field.

The inability to transmit the nematode *A. solenopsi* (Poinar et al., 2007), the protozoan *Mattesia* sp. (Pereira et al., 2002), and other pathogens and nematodes precluded further evaluation of some potential ant biocontrol agents. In contrast, other pathogens have been studied in detail. An isolate (447) of *B. bassiana* was obtained from and shown to be virulent to red imported fire ants (Alves et al., 1988; Stimac et al., 1989; Stimac and Alves, 1994) before being directly applied to individual fire ant nests using different techniques and rates, but colony mortality was less than 50% (Oi et al., 1994). Bait formulations containing *B. bassiana* 447 or other isolates (Collins et al., 1999; Stimac and Pereira, 2001; Bextine and Thorvilson, 2002; Thorvilson et al., 2002; Milner and Pereira, 2007) have produced inconsistent results attributed to the limited intracolony transmission of fungus, microbial antagonism, and fungistatic action of fire ant venom (Storey et al., 1991; Pereira and Stimac, 1992; Oi and Pereira, 1993). Currently, *B. bassiana* is no longer being considered as an MCA for fire ants.

Major efforts have been made to utilize *K. solenopsae* for the classical biological control of red imported fire ants in the United States. First reported from *S. invicta* in Brazil in 1974 and described in 1977 as *T. solenopsae*, but later placed in a new genus, *Kneallhazia* (Allen and Buren, 1974; Knell et al., 1977; Sokolova and Fuxa, 2008), this pathogen was extensively studied on the black imported fire ant, *Solenopsis richteri* (Briano et al., 1995, 1996, 2012). Natural infections of *K. solenopsae* in fire ant populations in the United States (Williams et al., 1998, 2003) allowed studies on host specificity, transmission, impact, and releases of this pathogen in the United States (Oi and Valles, 2009).

K. solenopsae causes a chronic, debilitative disease in fire ants with significant declines in brood production and queen longevity (Williams et al., 1999; Oi and Williams, 2002). While infected colonies are smaller, they often are not eliminated (Cook, 2002; Oi and Williams, 2002; Fuxa et al., 2005), thus the control for a stinging ant is insufficient. However, when *K. solenopsae* and a fire ant decapitating parasitic fly, *Pseudacteon tricuspis*, were established around plots cleared of fire ants by insecticide, reinfestation was delayed, and the need for retreatment was delayed over 1 year where the biological control agents were established (Oi et al., 2008). Possibly contributing to the slower reinfestation is the impairment of colony founding by *K. solenopsae*-infected queens, which could be attributable to reduced lipid reserves (Cook et al., 2003; Oi and Williams, 2003; Overton et al., 2006; Preston et al., 2007).

Transmission of *K. solenopsae* can be initiated by the introduction of live infected brood into uninfected fire ant colonies (Fig. 29.3; Williams et al., 1999; Oi et al., 2001). *K. solenopsae* has a host range of six species of *Solenopsis*, in the *Solenopsis saevissima* species group (Oi and Valles, 2009, 2012), and *Solenopsis geminata*

FIGURE 29.3 *Knellhazia solenopsae*-infected *Solenopsis invicta* brood ready to be used as an inoculum for application into uninfected fire ant colonies. The ants in the inoculated colony may attack and kill any adult ants added to the colony but will adopt the brood, introducing the microsporidian pathogen into the ant colony.

and the *S. geminata* × *Solenopsis xyloni* hybrid, in the *S. geminata* group (Ascunce et al., 2010). The establishment and spread of *K. solenopsae* was documented in six states in the United States (Williams et al., 1999; Fuxa et al., 2005; Oi, unpublished data). *K. solenopsae* is naturally widespread in populations of polygyne fire ants (Streett et al., 2004; Mitchell et al., 2006; Milks et al., 2008; Allen et al., 2010; Oi, unpublished data).

Another microsporidium under consideration for the biological control of fire ants is *V. invictae*, which was described in 1986 (Jouvenaz et al., 1980; Jouvenaz and Ellis, 1986). Natural infections of *V. invictae* in Argentina were associated with fluctuating declines in field populations of red imported fire ants, which included 100% reductions for over 31 months (Briano, 2005). Transmission has been documented with the introduction of live, infected brood or dead, *V. invictae*-infected adults into small fire ant colonies; infected colonies had reductions in growth of greater than 80% (Oi et al., 2005), dead workers with higher infection prevalence than live workers, and faster mortality of starved infected workers (Briano and Williams, 2002).

Using the metatranscriptomics approach (Valles et al., 2008), four RNA viruses, *Solenopsis invicta* virus 1 (SINV-1; Valles et al., 2004); *Solenopsis invicta* virus 2 (SINV-2; Valles et al., 2007); *Solenopsis invicta* virus 3 (SINV-3; Valles and Hashimoto, 2009); and *Solenopsis invicta* densovirus (SiDNV; Valles et al., 2013b) have been discovered and characterized from fire ants. SINV-1, -2, and -3 have been detected in both the United States and Argentina, while SiDNV has been found in Argentina. SINV-3 is most consistently associated with the demise of fire ant colonies (Valles and Hashimoto, 2009), and it

is being evaluated as a classical biological control agent and a biopesticide, although production is still an unresolved obstacle. SINV-3 was found to infect all stages of fire ants, including the eggs (Valles and Hashimoto, 2009; Valles et al., 2014). Host range studies of SINV-3 revealed that infections occurred only in red imported fire ants and the *S. invicta* × *S. richteri* hybrid (Porter et al., 2013, 2015), and symptoms include the cessation of feeding and the retrieval of solid food, brood and worker mortality, decreased egg production, ovary degeneration, and, finally, colony collapse. This disease progression stems from altered worker behavior affecting the acquisition and/or distribution of food (Valles et al., 2014).

Baits consisting of homogenized live and frozen infected fire ants mixed with a food lure, such as a 5–10% sucrose solution, have effectively transmitted SINV-3 (Valles et al., 2013a; Valles and Oi, 2014). Inoculative releases of the sucrose solution formulation resulted in infections in the urban, desert habitat of the Coachella Valley in California (Valles and Oi, unpublished data). The impact of SINV-3 on field populations of fire ants is yet to be determined. Unfortunately, a major limitation of using RNA viruses, like SINV-3, is the lack of large-scale virus production capability.

The cooccurrence of combinations of SINV-1, -2, -3, *K. solenopsae*, and even species of fire ant decapitating flies (parasitoids) has been documented from individual fire ant colonies in the field (Valles et al., 2010). The effects of the interactions among the pathogens and the species of decapitating flies vary, with some flies acquiring *K. solenopsae* but not the SINVs (Valles and Porter, 2007; Oi et al., 2009; Mészáros et al., 2014). Thus far there is no evidence of the flies vectoring *K. solenopsae* (Oi, unpublished data). A realistic goal is that the establishment of several fire ant pathogens and parasites may weaken, instead of eliminate, the majority of fire ant colonies (Fuxa et al., 2005).

Research programs targeting the microbial control of other invasive ant species that infest structures and urban landscapes have also considered pathogens. Metagenomic surveys for entomopathogens of the tawny crazy ant, *Nylanderia fulva*, currently plaguing the southern United States, have identified a virus that is being characterized (Valles et al., 2012). In addition, a new microsporidium was also described from this invasive ant (Plowes et al., 2015). Surveys conducted in Europe and North America for natural enemies of the red ant, *Myrmica rubra*, an ant that has become an invasive pest in the northeast United States, have yielded four new fungal pathogens (Groden et al., 2005). Biological control with pathogens is considered to be one of the few long-term strategies for suppressing invasive ants that can dominate large areas and inhabit diverse habitats. The appeal of biological control with entomopathogens is the potential for host specificity, self-sustaining suppression, and natural spread into untreated or inaccessible populations.

29.2.3 Termites

Because termites are serious pests in some agricultural systems, a great deal of attention has been dedicated to the control of these pests in rural settings, but relatively little attention has been dedicated to the control of termites in urban areas. One of the main reasons for this disparity is probably the relative cost/benefit of termite control in rural and urban settings, with differences in construction techniques and materials used also playing important roles in the lower enthusiasm for the microbial control of termites in urban settings. While wood and wood products in urban construction continue to dominate in certain areas of the world, in other areas, the use of mineral products has eliminated termites as a major threat to structures. However, certain termite species continue to be and in some cases are increasing their importance as a threat for urban trees and landscape plants. The spread of termite species to new areas (eg, *Coptotermes gestroi* and *Coptotermes formosanus* spread in the Americas) has raised the interest in new control techniques, including microbial control.

Several authors have raised doubts as to whether the microbial control of termites in structures can be successful (Rath, 2000; Verma et al., 2009; Rust, 2014). The possibility of the microbial control of termites has been severely dismissed by Chouvenc et al. (2011) after these authors reinterpreted data on studies published over half a century and concluded that few field study results should be considered positive. Reevaluation of several study results by these authors indicated that no virus or bacterium studies conducted in the field produced clearly positive results. Among the field tests of nematodes used in the control of termites, only two cases of nematodes being injected within bush stems (Dantharanarayana and Vitharana, 1987) or tree cavities (Lenz and Runko, 1992) were considered positive. Chouvenc et al. (2011) evaluated fungal pathogens and determined nine cases of positive field trials using entomopathogenic fungi: (1) three with injection of *M. anisopliae* into trees against drywood termites in the family Kalotermitidae (Vitharana, 1988; Lenz and Runko, 1992), (2) two with the same fungus applied directly into mounds of the Termitidae *Conitermes cumulans* (Fernandes and Alves, 1992; Neves and Alves, 1999), and (3) four with *B. bassiana* also applied to mounds of the same termite species (Fernandes and Alves, 1991, 1992; Alves et al., 1995; Neves and Alves, 1999).

In terms of urban pests, these evaluations seem to indicate that EPNs and fungi could have limited potential as control agents for termites found in easily located, contained nests, in landscape trees or mounds. Most likely, these conditions favor greater contact between the pathogens and the hosts and avoid the usual pitfalls that affect field applications of entomopathogens (eg, UV radiation and extreme temperatures). Past attempts to use *M. anisopliae* for the treatment of structural infestations of soil termites have not succeeded in establishing this or even other entomopathogens as control agents for urban termite pests (Rath, 2000).

More recent works have raised new possibilities in terms of the microbial control of structural termites by combining entomopathogenic fungi and chemical termiticides (Wright and Lax, 2013) or testing an attractive cellulose bait formulation containing *M. anisopliae* (Wang and Powell, 2004). However, these new approaches have not produced consistent microbial termiticide use in infested structures. For the near future, no commercially viable microbial products can be used for the control of structural termite infestations.

29.2.4 Bed Bugs

The recent resurgence of bed bugs as urban pests in the most developed countries around the world has sparked an interest in the microbial control of these insects. Little research has been conducted on the potential role entomopathogens may have in controlling bed bug populations (Strand, 1977). As expected, first attempts at the microbial control of these insects were with the entomopathogenic fungi *B. bassiana* (Barbarin et al., 2012), which was determined to cause rapid mortality of bed bugs, and *M. anisopliae* (Ulrich et al., 2014), which was judged as "a poor pathogen for control of bed bugs." Whether the two diverse conclusions represent a true nature of the relationship between these fungi and the insect host or if it is just a reflection of different approaches to the possibility of the microbial control of bed bugs remains to be seen.

Regardless of laboratory assays with these and other pathogens, the microbial control of bed bugs is bound to face similar obstacles as the microbial control of other urban pests. An important characteristic of bed bug infestations is the close proximity of the insect harborage to the human hosts on which the insects feed. This proximity between target bed bug harborages and humans is cause for concern, especially when fungal conidia, which may cause allergic reactions, are used in the dissemination of the insect pathogens. Beyond this problem, human structures that are air-conditioned usually have humidity levels well below optimal for entomopathogenic fungi.

An interesting opportunity for the introduction of MCAs in bed bug populations is represented by the sexual behavior of these insects. Because bed bug mating occurs through the process of traumatic insemination, with the male penetrating and causing a wound on the body of the female (Pfiester et al., 2009), there is potential for the introduction of pathogens into the female body (Reinhardt et al., 2005). The development of pathogens that could be used in a way to exploit this vulnerability in the bed bug life cycle represents an opportunity to be investigated.

29.2.5 Flies

The efficacy of MCAs for the control of flies in rural and outdoor environments has been demonstrated in several habitats

(Geden, 2012). However, several factors probably prevent these controls from being applied in urban areas, including: (1) flies are produced in one area but end up as pests in a different location after dispersal, (2) controlling adult flies without controlling larval development at the breeding sites provides only temporary effects, and (3) entomopathogens are normally too slow in producing adult fly mortality.

29.2.6 Fleas

Although several products are available for the control of flea larva with EPNs (Grewal and Georgis, 1999; Georgis et al., 2006), these applications are only practical when the flea larvae are developing on soil, lawns, and other areas outdoors, where the EPNs can be applied in water suspensions. For fleas that develop entirely indoors, the use of nematodes is impractical.

29.2.7 Other Structural Pests

Unless there is significant progress in the microbial control of the major pests mentioned previously, it is very unlikely that entomopathogens will become commercially available for the control of other structural pests. Label restrictions, the cost of development, and complications associated with marketing MCAs for structural pests restrict the development of any commercial product to the major structural pests. Use for biopesticides against other minor pests is a consequence of the potential success that a control product may have against at least one of the major pests. Without clear commercial microbial product winners for cockroaches, ants, or termites, the use of pathogens against minor structural pests cannot be seriously considered.

29.3 CONCLUSIONS AND RECOMMENDATIONS FOR FUTURE DEVELOPMENT OF MICROBIAL CONTROL AGENTS OF STRUCTURAL INSECT PESTS

The three major pest groups affecting urban structures (ants, termites, and peridomestic cockroaches) are potentially the most attractive for the development of microbial controls, not only because of their economic importance, but mainly because of aspects of their biology and ecology that make them more susceptible and amenable to control by entomopathogens. These three pest groups usually spend a considerable part of their life cycle outside human structures, which offers better opportunities for microbial control.

Moisture levels outdoors can be more adequate to provide optimal environmental conditions for the development and survival of MCAs. Ants, termites, and peridomestic cockroaches spend considerable time in the soil, leaf litter, mulch, or similar habitat, either in their nest or foraging, where moisture levels

are high and temperatures are moderate and usually within the optimal range for entomopathogenic activity.

The application of microbes outdoors, even those with a proven record of safety in relation to humans, does not have the same perceived risk to human health as indoors applications. Efficacy and regulatory testing as well as the implementation of the microbial control of structural pests can be much less costly and require less stringent requirements if restricted to applications outside homes, offices, and other human-occupied structures.

Normally the control of ants, termites, and peridomestic cockroaches already occurs outdoors as a preventive measure to avoid their entry into homes and structures. In the case of termites, preventive control is the norm to avoid the high costs associated with repairs. The control of peridomestic cockroaches and most structure-invading ants can follow the same model.

Both ants and termites are social insects that live in colonies. Cockroaches and bed bugs do not present true social behavior, but large numbers of individuals are found within harborage sites. These behaviors and ecological characteristics can facilitate the spread of pathogens (Table 29.1). Social behaviors have been described as having both negative and positive effects on disease occurrence (Oi

TABLE 29.1 Behaviors That May Increase (+) or Decrease (−) Infection Rates, Dissemination, and Dispersal of Entomopathogens Used for Control of Social and Subsocial Urban Pests

Behavior	Infection Process	Intracolony Dissemination	Out of Nest Dispersal
Grooming	−	+	−/+
Use of secretions	−	−	−/+
Food sharing	+	+	ne
Avoidance of pathogens	−	−	−/+
Nest hygiene/ necrophoresis	ne	−/+	−/+
Summit disease syndrome	ne	−	+
Altered activity time or place	ne	−	−/+
Brood care	−/+	+	ne
Brood sharing	+	+	+
Colony movement	−	−	+

ne, no effect.
Modified from Oi, D.H., Pereira, R.M., 1993. Ant behavior and microbial pathogens (Hymenoptera: Formicidae). Fla. Entomol. 76, 63–73, with permission from authors.

and Pereira, 1993). Close proximity, food, and secretion exchange are some of the behaviors that may facilitate the exchange and transmission of disease organisms within social insect nests and harborages of subsocial insects such as cockroaches. The colony and harborage areas can be targeted for the application of MCAs, either in direct application or in the form of baits that are carried back to the nest and shared among the target pest population, either directly or by trophallaxis, coprophagy, or other behaviors. These areas offer extra protection from variations in environmental factors such as temperature, humidity, and exposure to ultraviolet radiation.

Ants, termites, and cockroaches can be targeted with bait applications. The use of bait formulations for the distribution of entomopathogens is still not widely used but does show some promise as a way to reach the large proportion of the social insect population that never leaves the nest. The use of attractive powder bait formulations containing *B. bassiana* spores cause a fungal infection on fire ants living in almond orchards in California and decrease damage to nuts (Pereira, unpublished results). Food attraction has also been used to draw termites to areas treated with entomopathogenic fungi (Wang and Powell, 2004).

The three pest groups highlighted here, as well as other structural pests, are a great concern to an increasing population, which worries about the application of toxic chemicals around their living spaces. Natural products have an increasing following among urban and suburban populations around the world. The cost of using alternatives to conventional insecticides becomes less important when consumers consider the elimination of pests in structures they occupy. With increased urbanization, greater concerns regarding diseases and damage caused by structural pests, continuous spread of pests at an international level, and a certain fear of insects among the human population, structural pest management with MCAs is poised for strong growth in the future relative to other areas of pest management. Despite the difficulties in fitting microbial products into structural pest management, this area may yet see increased interest. However, this can only occur if monetary and time investments are made into the search and implementation of microbial control solutions for the structural pest market, which can lead to the selection of efficient entomopathogens and practical application methods.

REFERENCES

Allen, G.E., Buren, W.F., 1974. Microsporidan and fungal diseases of *Solenopsis invicta* Buren in Brazil. J. N.Y. Entomol. Soc. 82, 125–130.

Allen, G.E., Silveira-Guido, A., 1974. Occurrence of microsporidia in *Solenopsis richteri* and *Solenopsis* sp. in Uruguay and Argentina. Fla. Entomol. 57, 327–329.

Allen, H.R., Valles, S.M., Miller, D.M., 2010. Characterization of *Solenopsis invicta* (Hymenoptera: Formicidae) populations in Virginia: social form genotyping and pathogen/parasitoid detection. Fla. Entomol. 93, 80–88.

Alves, S.B., Stimac, J.L., Camargo, M.T.V., 1988. Suscetibilidade de *Solenopsis invicta* Buren e *Solenopsis saevissima* FR. Smith. a isolados de *Beauveria bassiana* (Bals.) Vuill. An. Soc. Entomol. Brasil 17, 379–387.

Alves, S.B., Almeida, J.E.M., Moino, A., Stimac, J.L., Pereira, R.M., 1995. Uso de *Metarhizium anisopliae* e *Beauveria bassiana* no controle de *Cornitermes cumulanus* (Kollar, 1832) em pastagens. Ecossistema 20, 50–57.

Anaclerio, M., Molinari, F., 2012. Intra and inter-specific attraction of cockroach faecal extracts: studies for improving bait activity. Bull. Insectol. 65, 113–118.

Andis, M., 1994. The BioPath cockroach control chamber uses nature to control nature's pests. Pest Control 62, 44–48.

Appel, A.G., Benson, E.P., Ellenberger, J.M., Manweileri, S.A., 1993. Laboratory and field evaluations of an entomogenous nematode (Nematoda: Steinernematidae) for German cockroach (Dictyoptera: Blattellidae) control. J. Econ. Entomol. 86, 777–784.

Arruda, L.K., Vailes, L.D., Ferriani, V.P.L., Santos, A.B.R., Pomés, A., Chapman, M.D., 2001. Cockroach allergens and asthma. J. Allergy Clin. Immunol. 107, 419–428.

Ascunce, M.S., Valles, S.M., Oi, D.H., Shoemaker, D., Plowes, R., Gilbert, L., Lebrun, E.G., Sánchez-Arroyo, H., Sanchez-Peña, S., 2010. Molecular diversity of the microsporidium *Kneallhazia solenopsae* reveals an expanded host range among fire ants in North America. J. Invertebr. Pathol. 105, 279–288.

Barbarin, A.M., Jenkins, N.E., Rajotte, E.G., Thomas, M.B., 2012. A preliminary evaluation of the potential of *Beauveria bassiana* for bed bug control. J. Invertebr. Pathol. 111, 82–85.

Beatson, S.H., 1972a. Pharoah's ants as pathogen vectors in hospitals. Lancet (London) 299, 425–427.

Beatson, S.H., 1972b. Pharoah's ants as pathogen vectors in hospitals. Lancet 1, 425–427.

Beatson, S.H., 1973. Pharoah's ants enter giving sets. Lancet 301, 606.

Bell, W.J., Louis, M.R., Nalepa, C.A., 2007. Cockroaches: Ecology, Behavior, and Natural History. Johns Hopkins University Press, Baltimore, MD.

Bergoin, M., Tijssen, P., 2010. Densoviruses: a highly diverse group of arthropod parvoviruses. In: Ascari, S., Johnson, K. (Eds.), Insect Virology. Caister Academic Press, Norfolk, UK, pp. 59–82.

Bextine, B.R., Thorvilson, H.G., 2002. Field applications of bait-formulated *Beauveria bassiana* alginate pellets for biological control of the red imported fire ant (Hymenoptera: Formicidae). Environ. Entomol. 31, 746–752.

Briano, J.A., 2005. Long-term studies of the red imported fire ant, *Solenopsis invicta*, infected with the microsporidia *Vairimorpha invictae* and *Thelohania solenopsae* in Argentina. Environ. Entomol. 34, 124–132.

Briano, J.A., Williams, D.F., 2002. Natural occurrence and laboratory studies of the fire ant pathogen *Vairimorpha invictae* (Microsporida: Burenellidae) in Argentina. Environ. Entomol. 31, 887–894.

Briano, J., Patterson, R., Cordo, H., 1995. Long term studies of the black imported fire ant (Hymenoptera: Formicidae) infected with a microsporidium. Environ. Entomol. 24, 1328–1332.

Briano, J.A., Patterson, R.S., Becnel, J.J., Cordo, H.A., 1996. The black imported fire ant, *Solenopsis richteri*, infected with *Thelohania solenopsae*: intracolonial prevalence of infection and evidence for transovarial transmission. J. Invertebr. Pathol. 67, 178–179.

Briano, J.A., Calcaterra, L., Varone, L., 2012. Fire ants (*Solenopsis* spp.) and their natural enemies in southern South America. Psyche:198084 19 p.

Broome, J.R., 1974. Microbial Control of the Imported Fire Ant, *Solenopsis richteri* Forel (Hymenoptera: Formicidae) (Ph.D. dissertation). Mississippi State University. vii, 66 pp.

Butler, J.F., Garcia-Maruniak, A., Meek, F., Maruniak, J.E., 2010. Wild Florida house flies (*Musca domestica*) as carriers of pathogenic bacteria. Fla. Entomol. 93, 218–223.

Chouvenc, T., Su, N.-Y., Grace, J.K., 2011. Fifty years of attempted biological control of termites – analysis of a failure. Biol. Control 59, 69–82.

Collins, H., Callcott, A.M., Mcanally, L., Ladner, A., Lockley, T., Wade, S., 1999. Field and Laboratory Efficacy of *Beauveria bassiana* Alone or in Combination with Imidacloprid Against RIFA. 1999 Accomplishment Report. Gulfport Plant Protection Station, CPHST, PPQ, USDA, Gulfport, Mississippi, pp. 96–99.

Cook, T.J., 2002. Studies of naturally occurring *Thelohania solenopsae* (Microsporida: Thelohaniidae) infection in red imported fire ants, *Solenopsis invicta* (Hymenoptera: Formicidae). Environ. Entomol. 31, 1091–1096.

Cook, T.J., Lowery, M.B., Frey, T.N., Rowe, K.E., Lynch, L.R., 2003. Effect of *Thelohania solenopsae* (Microsporida: Thelohaniidae) on weight and reproductive status of polygynous red imported fire ant, *Solenopsis invicta* (Hymenoptera: Formicidae), alates. J. Invertebr. Pathol. 82, 201–203.

Danthanarayana, W., Vitharana, S.I., 1987. Control of the live-wood tea termite *Glyptotermes dilatatus* using *Heterorhabditis* sp. (Nemat.). Agric. Ecosyst. Environ. 19, 333–342.

Davies, T.G.E., Field, L.M., Williamson, M.S., 2012. The re-emergence of the bed bug as a nuisance pest: implications of resistance to the pyrethroid insecticides. Med. Vet. Entomol. 26, 241–254.

Delaunay, P., Blanc, V., Del Giudice, P., Levy-Bencheton, A., Chosidow, O., Marty, P., Brouqui, P., 2011. Bedbugs and infectious diseases. Clin. Infect. Dis. 52, 200–210.

deShazo, R.D., Williams, D.F., Moak, E.S., 1999. Fire ant attacks on residents in health care facilities: a report of two cases. Ann. Intern. Med. 131, 424–429.

deShazo, R.D., Kemp, S.F., deShazo, M.D., Goddard, J., 2004. Fire ant attacks on patients in nursing homes: an increasing problem. Am. J. Med. 116, 843–884.

Evans, H.C., 1974. Natural control of arthropods, with special reference to ants (Formicidae), by fungi in the tropical high forest of Ghana. J. App. Ecol. 11, 37–49.

Evans, H.C., Groden, E., Bischoff, J.F., 2010. New fungal pathogens of the red ant, *Myrmica rubra*, from the UK and implications for ant invasions in the USA. Fungal Biol. 114, 451–466.

Evans, H.C., Elliot, S.L., Hughes, D.P., 2011. Hidden diversity behind the zombie-ant fungus *Ophiocordyceps unilateralis*: four new species described from carpenter ants in Minas Gerais, Brazil. PLoS One 6, e17024.

Fernandes, P.M., Alves, S.B., 1991. Controle de *Cornitermes cumulans* (Kollar, 1832) (Isoptera: Termitidae) com *Beauveria bassiana* (Bals.) Vuill. e *Metarhizium anisopliae* (Metsch.) Sorokin em condições de campo. An. Soc. Entomol. Brasil 20, 46–49.

Fernandes, P.M., Alves, S.B., 1992. Seleção de isolados de *Beauveria bassiana* (Bals.) Vuill. e *Metarhizium anisopliae* (Metsch.) Sorok. para controle de *Cornitermes cumulans* (Kollar, 1832) (Isoptera: Termitidae). An. Soc. Entomol. Brasil 21, 319–328.

Ferrari, R., Marini, M., 1999. Hidden diversity behind the zombie-ant fungus *Ophiocordyceps unilateralis*: four new species described from carpenter ants in Minas Gerais. In: Robinson, W.H., Rettich, F., Rambo, G.W. (Eds.), Proceed. 3rd Int. Conf. on Urban Pests. Czech University of Agriculture, Prague, pp. 357–365.

Förster, M., Klimpel, S., Mehlhorn, H., Sievert, K., Messler, S., Pfeffer, K., 2007. Pilot study on synanthropic flies (e.g. *Musca, Sarcophaga, Calliphora, Fannia, Lucilia, Stomoxys*) as vectors of pathogenic microorganisms. Parasitol. Res. 101, 243–246.

Fuxa, J.R., Sokolova, Y.Y., Milks, M.L., Richter, A.R., Williams, D.F., Oi, D.H., 2005. Prevalence, spread, and effects of the microsporidium *Thelohania solenopsae* released into populations with different social forms of the red imported fire ant (Hymenoptera: Formicidae). Environ. Entomol. 34, 1139–1149.

Geden, C.J., 2012. Status of biopesticides for control of house flies. J. Biopest. 5 (Suppl.), 1–11.

Georgis, R., Koppenhöfer, A.M., Lacey, L.A., Bélair, G., Duncan, L.W., Grewal, P.S., Samish, M., Tan, L., Torr, P., van Tol, R.W.H.M., 2006. Successes and failures in the use of parasitic nematodes for pest control. Biol. Control 38, 103–123.

Goddard, J., Jarratt, J., deShazo, R.D., 2002. Recommendations for prevention and management of fire ant infestation of health care facilities. South. Med. J. 95, 627–633.

Grace, J.K., Yates III, J.R., 1999. Termite resistant construction and building materials. In: Robinson, W.H., Rettich, F., Rambo, G.W. (Eds.), Proc. 3rd Int. Conf. Urban Pests. Czech University of Agriculture, Prague, pp. 399–406.

Grewal, P.S., Georgis, R., 1999. Entomopathogenic nematodes. In: Hall, F.R., Menn, J.J. (Eds.), Methods in Biotechnology. Biopesticides: Use and Delivery, vol. 5. Humana Press, Totowa, NJ, pp. 271–299.

Groden, E., Drummond, F.A., Garnas, J., Francoeur, A., 2005. Distribution of an invasive ant, *Myrmica rubra* (Hymenoptera: Formicidae), in Maine. J. Econ. Entomol. 98, 1774–1784.

Hansen, L.D., Klotz, J.H., 2005. Carpenter Ants of the United States and Canada. Cornell Univ. Press, Ithaca, NY.

Harlan, H.J., Faulde, M.K., Baumann, G.J., 2008. Bedbugs. In: Bonnefoy, X., Kampen, H., Sweeney, K. (Eds.), Public Health Significance of Urban Pests. World Health Organization, Copenhagen, Denmark, pp. 131–153.

Hernanez-Ramirez, G., Sanchez-Arroyo, H., Alatorre-Rosas, R., 2008. Pathogenicity of *Metarhizium anisopliae* and *Beauveria bassiana* to the American cockroach (Dictyoptera: Blattidae). In: Proceedings of the 6th International Conference on Urban Pests. OOK Press Kft, Hungary, pp. 143–144.

Hubner-Campos, R.F., Leles, R.N., Rodrigues, J., Luz, C., 2013. Efficacy of entomopathogenic hypocrealean fungi against *Periplaneta americana*. Parasitol. Int. 62, 517–521.

Jouvenaz, D.P., Ellis, E.A., 1986. *Vairimorpha invictae* N. sp. (Microspora: Microsporida), a parasite of the red imported fire ant, *Solenopsis invicta* Buren (Hymenoptera: Formicidae). J. Protozool. 33, 457–461.

Jouvenaz, D.P., Allen, G.E., Banks, W.A., Wojcik, D.P., 1977. A survey for pathogens of fire ants, *Solenopsis* spp., in the Southeastern United States. Fla. Entomol. 60, 275–279.

Jouvenaz, D.P., Banks, W.A., Atwood, J.D., 1980. Incidence of pathogens in fire ants, *Solenopsis* spp., in Brazil. Fla. Entomol. 63, 345–346.

Jouvenaz, D.P., Lofgren, C.S., Banks, W.A., 1981. Biological control of imported fire ants: a review of current knowledge. Bull. Entomol. Soc. Am. 27, 203–208.

Kassiri, H., Kazemi, S., 2012. Cockroaches (*Periplaneta americana* (L.), Dictyoptera; Blattidae) as carriers of bacterial pathogens, Khorramshahr county, Iran. Jundishapur J. Microbiol. 5, 320–322.

Klotz, J.H., Mangold, J.R., Vail, K.M., Davis Jr., L.R., Patterson, R.S., 1995. A survey of the urban pest ants (Hymenoptera: Formicidae) of peninsular Florida. Fla. Entomol. 78, 109–118.

Knell, J.D., Allen, G.E., Hazard, E.I., 1977. Light and electron microscope study of *Thelohania solenopsae* N. sp. (Microsporida: Protozoa) in the red imported fire ant, *Solenopsis invicta*. J. Invertebr. Pathol. 29, 192–200.

Koehler, P.G., Kern Jr., W.H., Pereira, R.M., 2008. Pests on and near food. In: Koehler, P.G., Kern Jr., W.H., Pereira, R.M. (Eds.), General Household Pest Control: Applicator Training Manual, second ed. UF/IFAS Florida Cooperative Extension Service, Gainesville, pp. 68–81.

Koehler, P.G., Bayer, B.E., Branscome, D., 2013. Cockroaches and their management. In: Koehler, P.G., Buss, E.A., Kern Jr., W.H., Pereira, R.M. (Eds.), Pests in and Around the Florida Home, third ed. UF/IFAS Florida Cooperative Extension Service, Gainesville, p. 119.

Kulshrestha, V., Pathak, S.C., 1997. Aspergillosis in German cockroach *Blattella germanica* (L.) (Blattoidea: Blatellidae). Mycopathologia 139, 75–78.

Lard, C.F., Hall, C.R., Salin, V., Vinson, S.B., Cleere, K.H., Purswell, S., 1999. The Economic Impact of the Red Imported Fire Ant on the Homescape, Landscape, and the Urbanscape of Selected Metroplexes of Texas: A Part of the Texas Fire Ant Initiative 1997–1999. Fire Ant Economic Research Report # 99-08, 63 p.. Texas A&M Univ., Department of Agricultural Economics.

Lenz, M., Runko, S., 1992. Use of Microorganisms to Control Colonies of the Coconut Termite *Neotermes rainbowi* (Hill) on Vaitupu. Commonwealth Scientific and Industrial Research Organization, Division of Entomology, Tuvalu, p. 47. Termite group report No. 92/16.

Leulmi, H., Bitam, I., Berenger, J.M., Lepidi, H., Rolain, J.M., Almeras, L., Raoult, D., Parola, P., 2015. Competence of *Cimex lectularius* bed bugs for the transmission of *Bartonella quintana*, the agent of trench fever. PLoS Negl. Trop. Dis. 9 (5), e0003789. http://dx.doi.org/10.1371/journal.pntd.0003789.

Levine, O.S., Levine, M.M., 1991. Houseflies (*Musca domestica*) as mechanical vectors of shigellosis. Rev. Infect. Dis. 13, 688–696.

Lockey, R.F., 1974. Systemic reactions to stinging ants. J. Allergy Clin. Immunol. 54, 132–146.

Lonc, E., Lecadet, M.-M., Lachowicz, T.M., Panek, E., 1997. Description of *Bacillus thuringiensis wratislaviensis* (H-41), a new serotype originating from Wroclaw (Poland), and other *Bt* soil isolates form the same area. Lett. Appl. Microbiol. 24, 467–473.

Lopes, R.B., Alves, S.B., 2011. Differential susceptibility of adults and nymphs of *Blattella germanica* (L.) (Blattodea: Blattellidae) to infection by *Metarhizium anisopliae* and assessment of delivery strategies. Neotrop. Entomol. 40, 368–374.

Lopes, R.B., Martins, I., Souza, D.A., Faria, M., 2013. Influence of some parameters on the germination assessment of mycopesticides. J. Invertebr. Pathol. 112, 236–242.

Maketon, M., Hominchan, A., Hotaka, D., 2010. Control of American cockroach (*Periplaneta americana*) and German cockroach (*Blattella germanica*) by entomopathogenic nematodes. Rev. Colomb. Entomol. 36, 249–253.

Mariconi, F.A.M., 1999. As baratas. In: Mariconi, F.A.M., Fontes, L.R., Araújo, R.L., Zamith, A.P.L., Neto, C.C., Bueno, O.C., Campos-Farinha, A.E.C., Matthiesen, F.A., Taddei, V.A., Oliveira Filho, A.M.O., Ferreira, W.L.B. (Eds.), Insetos e outros invasores de residências. FEALQ, Piracicaba, pp. 13–33.

Mészáros, A., Oi, D.H., Valles, S.M., Beuzelin, J.M., Reay-Jones, F.P.F., Johnson, S.J., 2014. Distribution of *Pseudacteon* spp. (Diptera: Phoridae), biological control agents of *Solenopsis* spp. (Hymenoptera: Formicidae), in Louisiana and associated prevalence of *Kneallhazia solenopsae* (Microsporidia: Thelohaniidae). Biol. Control 77, 93–100.

Milks, M.L., Fuxa, J.R., Richter, A.R., 2008. Prevalence and impact of the microsporidium *Thelohania solenopsae* (Microsporidia) on wild populations of red imported fire ants, *Solenopsis invicta*, in Louisiana. J. Invertebr. Pathol. 97, 91–102.

Milner, R., Pereira, R.M., 2007. Microbial control of urban pests – cockroaches, ants and termites. In: Lacey, L.A., Kaya, K.H. (Eds.), Field Manual of Microbial Control of Insects, second ed. Springer, Dordrecht, The Netherlands, pp. 695–711.

Miranda, R.A., Silva, J.P., 2008. Enterobactérias isoladas de *Periplaneta americana* capturadas em um ambiente hospitalar. Ciência Praxis 1, 21–24.

Mitchell, F.L., Snowden, K., Fuxa, J.R., Vinson, S.B., 2006. Distribution of *Thelohania solenopsae* (Microsporida: Thelohaniidae) infecting red imported fire ant (Hymenoptera: Formicidae) in Texas. Southwest. Entomol. 31, 297–306.

Mohan, C.H.M., Lakshmi, A., Devi, U., 1999. Laboratory evaluation of the pathogenicity of three isolates of the entomopathogenic fungus *Beauveria bassiana* (Bals.) Vuillemin on the American cockroach (*Periplaneta americana*). Biocontrol Sci. Technol. 9, 29–33.

Moreira, D.D.O., Morais, V., Vieira da Motta, O., Campos-Farinha, A.E.C., Tonhasca Jr., A., 2005. Ants as carrier of antibiotic-resistant bacteria in hospitals. Neotrop. Entomol. 34, 999–1006.

Neves, P.O.J., Alves, S.B., 1999. Controle associado de *Cornitermes cumulans* (Kollar 1832) (Isoptera: Termitidae) com *Metarhizium anisopliae*, *Beauveria bassiana* e imidacloprid. Sci. Agric. 56, 313–319.

Nickle, W.R., Jouvenaz, D.P., 1987. *Tetradonema solenopsis* N. sp. (Nematoda: Tetradonematidae) parasitic on the red imported fire ant *Solenopsis invicta* Buren from Brazil. J. Nematol. 19, 311–313.

Oi, D.H., 2008. Pharaoh ants and fire ants. In: Bonnefoy, X., Kampen, H., Sweeney, K. (Eds.), Public Health Significance of Urban Pests. World Health Organization, Copenhagen, Denmark, pp. 175–208.

Oi, D.H., Pereira, R.M., 1993. Ant behavior and microbial pathogens (Hymenoptera: Formicidae). Fla. Entomol. 76, 63–73.

Oi, D.H., Valles, S.M., 2009. Fire ant control with entomopathogens in the USA. In: Hajek, A.E., Glare, T.R., O'Callaghan, M. (Eds.), Use of Microbes for Control and Eradication of Invasive Arthropods. Springer Science+Business Media B.V., pp. 237–257.

Oi, D.H., Valles, S.M., 2012. Host specificity testing of the *Solenopsis* fire ant (Hymenoptera: Formicidae) pathogen, *Kneallhazia* (=*Thelohania*) *solenopsae* (Microsporidia: Thelohaniidae), in Florida. Fla. Entomol. 95, 509–512.

Oi, D.H., Williams, D.F., 2002. Impact of *Thelohania solenopsae* (Microsporidia: Thelohaniidae) on polygyne colonies of red imported fire ants (Hymenoptera: Formicidae). J. Econ. Entomol. 95, 558–562.

Oi, D.H., Williams, D.F., 2003. *Thelohania solenopsae* (Microsporidia: Thelohaniidae) infection in reproductives of red imported fire ants (Hymenoptera: Formicidae) and its implication for intercolony transmission. Environ. Entomol. 32, 1171–1176.

Oi, D.H., Pereira, R.M., Stimac, J.L., Wood, L.A., 1994. Field applications of *Beauveria bassiana* for control of the red imported fire ant (Hymenoptera: Formicidae). J. Econ. Entomol. 87, 623–630.

Oi, D.H., Becnel, J.J., Williams, D.F., 2001. Evidence of intracolony transmission of *Thelohania solenopsae* (Microsporidia: Thelohaniidae) in red imported fire ants (Hymenoptera: Formicidae) and the first report of spores from pupae. J. Invertebr. Pathol. 78, 128–134.

Oi, D.H., Briano, J.A., Valles, S.M., Williams, D.F., 2005. Transmission of *Vairimorpha invictae* (Microsporidia: Burenellidae) infections between red imported fire ant (Hymenoptera: Formicidae) colonies. J. Invertebr. Pathol. 88, 108–115.

Oi, D.H., Williams, D.F., Pereira, R.M., Horton, P.M., Davis, T.S., Hyder, A.H., Bolton, H.T., Zeichner, B.C., Porter, S.D., Hoch, A.L., Boswell, M.L., Williams, G., 2008. Combining biological and chemical controls for the management of red imported fire ants (Hymenoptera: Formicidae). Am. Entomol. 54, 46–55.

Oi, D.H., Porter, S.D., Valles, S.M., Briano, J.A., Calcaterra, L.A., 2009. *Pseudacteon* decapitating flies (Diptera: Phoridae): are they potential vectors of the fire ant pathogens *Kneallhazia* (=*Thelohania*) *solenopsae* (Microsporidia: Thelohaniidae) and *Vairimorpha invictae* (Microsporidia: Burenellidae)? Biol. Control 48, 310–315.

Overton, K., Rao, A., Vinson, S.B., Gold, R.E., 2006. Mating flight initiation and nutritional status (protein and lipid) of *Solenopsis invicta* (Hymenoptera: Formicidae) alates infected with *Thelohania solenopsae* (Microsporidia: Thelohaniidae). Ann. Entomol. Soc. Am. 99, 524–529.

Paulke-Korinek, M., Széll, M., Laferl, H., Auer, H., Wenisch, C., 2012. Bed bugs can cause severe anaemia in adults. Parasitol. Res. 110, 2577–2579.

Pereira, R.M., Stimac, J.L., 1992. Transmission of *Beauveria bassiana* within nests of *Solenopsis invicta* (Hymenoptera: Formicidae) in the laboratory. Environ. Entomol. 21, 1427–1432.

Pereira, R.M., Williams, D.F., Becnel, J.J., Oi, D.H., 2002. Yellow-head disease caused by a newly discovered *Mattesia* sp. in populations of the red imported fire ant, *Solenopsis invicta*. J. Invertebr. Pathol. 81, 45–48.

Pereira, R.M., Taylor, A.S., Lehnert, M.P., Koehler, P.G., 2013. Potential population growth and harmful effects on humans from bed bug populations exposed to different feeding regimes. Med. Vet. Entomol. 27, 147–155.

Pereira, R., Cooksey, J., Baldwin, R., Koehler, P., 2014. Filth fly management in urban environments. In: Dhang, P. (Ed.), Urban Insect Pests: Sustainable Management Strategies. AB International, Reading, UK, pp. 43–64.

Pfiester, M., Pereira, R.M., Koehler, P.G., 2009. Effect of population structure and size on aggregation behavior of the bed bug *Cimex lectularius* L. J. Med. Entomol. 46, 1015–1020.

Plowes, R.M., Becnel, J.J., Lebrun, E.G., Oi, D.H., Valles, S.M., Jones, N.T., Gilbert, L.E., 2015. *Myrmecomorba nylanderiae* gen. et sp. nov., a microsporidian parasite of the tawny crazy ant *Nylanderia fulva*. J. Invertebr. Pathol. 129, 45–56.

Poinar Jr., G.O., Porter, S.D., Tang, S., Hyman, B.C., 2007. *Allomermis solenopsi* N. sp. (Nematoda: Mermithidae) parasitising the fire ant *Solenopsis invicta* Buren (Hymenoptera: Formicidae) in Argentina. Syst. Parasitol. 68, 115–128.

Porter, S.D., Valles, S.M., Oi, D.H., 2013. Host specificity and colony impacts of the fire ant pathogen, *Solenopsis invicta* virus 3. J. Invertebr. Pathol. 114, 1–6.

Porter, S.D., Valles, S.M., Wild, A.L., Dieckmann, R., Plowes, N.J.R., 2015. *Solenopsis invicta* virus 3: further host-specificity tests with native *Solenopsis* ants (Hymenoptera: Formicidae). Fla. Entomol. 98, 122–125.

Potter, M.F., 2011. Termites. In: Mallis, A. (Ed.), Handbook of Pest Control: The Behavior, Life History, and Control of Household Pests, tenth ed. Mallis Handbook Company, Richfield, OH, pp. 293–441.

Preston, C.A., Fritz, G.N., Vander Meer, R.K., 2007. Prevalence of *Thelohania solenopsae* infected *Solenopsis invicta* newly mated queens within areas of differing social form distributions. J. Invertebr. Pathol. 94, 119–124.

Quesada-Moraga, E., Santos-Quiros's, R., Valverde-García, P., Santiago-Álvarez, C., 2004. Virulence, horizontal transmission, and sublethal reproductive effects of *Metarhizium anisopliae* (Anamorphic fungi) on the German cockroach (Blattodea: Blattellidae). J. Invertebr. Pathol. 87, 51–58.

Rady, M.H., Abdel-Raouf, N., Labib, I., Merdan, A.I., 1992. Bacterial contamination of the housefly *Musca domestica*, collected from 4 hospitals at Cairo. J. Egypt. Soc. Parasitol. 22, 279–288.

Rafael, J.A., Silva, N.M., Dias, R.M.N.S., 2008. Baratas (Insecta, Blattaria) sinantrópicas na cidade de Manaus, Amazonas, Brasil. Acta Amaz. 38, 173–178.

Rath, A.C., 2000. The use of entomopathogenic fungi for control of termites. Biocontrol Sci. Technol. 10, 563–581.

Reinhardt, K., Naylor, R.A., Siva-Jothy, M.T., 2005. Potential sexual transmission of environmental microbes in a traumatically inseminating insect. Ecol. Entomol. 30, 607–611.

Rodovalho, C.M., Santos, A.L., Marcolino, M.T., Bonetti, A.M., Brandeburgo, M.A.M., 2007. Urban ants and transportation of nosocomial bacteria. Neotrop. Entomol. 36, 454–458.

Rust, M.K., 2008. Cockroaches. In: Bonnefoy, X., Kampen, H., Sweeney, K. (Eds.), Public Health Significance of Urban Pests. World Health Organization, Copenhagen, Denmark, pp. 53–84.

Rust, M.K., 2014. Management strategies for subterranean termites. In: Dhang, P. (Ed.), Urban Insect Pests: Sustainable Management Strategies. AB International, Reading, UK, pp. 114–129.

Salazar, R., Castillo-Neyra, R., Tustin, A.W., Borrini-Mayorí, K., Náquira, C., Levy, M.Z., 2014. Bed bugs (*Cimex lectularius*) as vectors of *Trypanosoma cruzi*. Am. J. Trop. Med. Hyg. 92, 331–335.

Schofield, D.C., 1994. Triatominae: Biology & Control. Eurocommunica Publ, West Sussex, UK.

Shafighi, Y., Ziaee, M., Ghosta, Y., 2014. Diatomaceous earth used against insect pests, applied alone or in combination with *Metarhizium anisopliae* and *Beauveria bassiana*. J. Plant Prot. Res. 54, 62–66.

Singh, G.J.P., Gill, S.S., 1985. Myotoxic and neurotoxic activity of *Bacillus thuringiensis* var. *israelensis* crystal toxin. Pestic. Biochem. Physiol. 24, 406–414.

Sokolova, Y.Y., Fuxa, J.R., 2008. Biology and life-cycle of the microsporidium *Kneallhazia solenopsae* Knell Allan Hazard 1977 gen. n., comb. n., from the fire ant *Solenopsis invicta*. Parasitology 135, 903–929.

Stimac, J.L., Alves, S.B., 1994. Ecology and biological control of fire ants. In: Rosen, D., Bennett, F.D., Capinera, J.L. (Eds.), Pest Management in the Subtropics: Biological Control – A Florida Perspective. Intercept Ltd., Andover, UK, pp. 353–380.

Stimac, J.L., Pereira, R.M., 2001. Methods and Formulations for Control of Pests. U.S. Patent No. 6,254,864.

Stimac, J.L., Alves, S.B., Camargo, M.T.V., 1989. Controle de *Solenopsis* spp. (Hymenoptera: Formicidae) com *Beauveria bassiana* (Bals.) Vuill. em condições de laboratório e campo. An. Soc. Entomol. Brasil 181, 95–103.

Storey, G.K., Vander Meer, R.K., Boucias, D.G., McCoy, C.W., 1991. Effect of fire ant (*Solenopsis invicta*) venom alkaloids on the in vitro germination and development of selected entomogenous fungi. J. Invertebr. Pathol. 58, 88–95.

Strand, M.A., 1977. Pathogens of cimicidae (bedbugs). Bull. World Health Org. 55 (Suppl. 1), 313–315.

Streett, D.A., Freeland Jr., T.B., Pranschke, A.M., 2004. Distribution of *Thelohania solenopsae* in red imported fire ant populations in Mississippi. In: Pollet, D., Johnson, S., Beckley, P., Clayton, S. (Eds.), Proceed. Ann. Red Imported Fire Ant Conf., March 21–23, 2004, Baton Rouge, LA, pp. 150–153.

Thorvilson, H., Wheeler, D., Bextine, B., San Francisco, M., 2002. Development of *Beauveria bassiana* formulations and genetically marked strains as a potential biopesticide for imported fire ant control. Southwest. Entomol. 25 (Suppl.), 19–29.

Tucker, C.L., Koehler, P.G., Pereira, R.M., 2008. Development of a method to evaluate the effects of eastern subterranean termite damage to the thermal properties of building construction materials (Isoptera: Rhinotermitidae). Sociobiology 51, 589–600.

Ulrich, K.R., Feldlaufer, M.F., Kramer, M., St Leger, R.J., 2014. Exposure of bed bugs to *Metarhizium anisopliae* at different humidities. J. Econ. Entomol. 107, 2190–2195.

Valles, S.M., 2012. Positive-strand RNA viruses infecting the red imported fire ant, *Solenopsis invicta*. Psyche:821591, 14 p.

Valles, S.M., Hashimoto, Y., 2009. Isolation and characterization of *Solenopsis invicta* virus 3, a new positive-strand RNA virus infecting the red imported fire ant, *Solenopsis invicta*. Virology 388, 354–361.

Valles, S.M., Oi, D.H., 2014. Successful transmission of *Solenopsis invicta* virus 3 to field colonies of *Solenopsis invicta* (Hymenoptera: Formicidae). Fla. Entomol. 97, 1244–1246.

Valles, S.M., Porter, S.D., 2007. *Pseudacteon* decapitating flies: potential vectors of a fire ant virus? Fla. Entomol. 90, 282–283.

Valles, S.M., Strong, C.A., Dang, P.M., Hunter, W.B., Pereira, R.M., Oi, D.H., Shapiro, A.M., Williams, D.F., 2004. A picorna-like virus from the red imported fire ant, *Solenopsis invicta*: initial discovery, genome sequence, and characterization. Virology 328, 151–157.

Valles, S.M., Strong, C.A., Hashimoto, Y., 2007. A new positive-strand RNA virus with unique genome characteristics from the red imported fire ant, *Solenopsis invicta*. Virology 365, 457–463.

Valles, S.M., Strong, C.A., Hunter, W.B., Dang, P.M., Pereira, R.M., Oi, D.H., Williams, D.F., 2008. Expressed sequence tags from the red imported fire ant, *Solenopsis invicta*: annotation and utilization for discovery of viruses. J. Invertebr. Pathol. 99, 74–81.

Valles, S.M., Oi, D.H., Porter, S.D., 2010. Seasonal variation and the co-occurrence of four pathogens and a group of parasites among monogyne and polygyne fire ant colonies. Biol. Control 54, 342–348.

Valles, S.M., Oi, D.H., Yu, F., Tan, X.X., Buss, E.A., 2012. Metatranscriptomics and pyrosequencing facilitate discovery of potential viral natural enemies of the invasive Caribbean crazy ant, *Nylanderia pubens*. PLoS One 7 (2), e31828 9 p.

Valles, S.M., Porter, S.D., Choi, M.Y., Oi, D.H., 2013a. Successful transmission of *Solenopsis invicta* virus 3 to *Solenopsis invicta* fire ant colonies in oil, sugar, and cricket bait formulations. J. Invertebr. Pathol. 113, 198–204.

Valles, S.M., Shoemaker, D., Wurm, Y., Strong, C.A., Varone, L., Becnel, J.J., Shirk, P.D., 2013b. Discovery and molecular characterization of an ambisense densovirus from South American populations of *Solenopsis invicta*. Biol. Control 67, 431–439.

Valles, S.M., Porter, S.D., Firth, A.E., 2014. *Solenopsis invicta* virus 3: pathogenesis and stage specificity in red imported fire ants. Virology 460, 66–71.

Verma, M., Sharma, S., Prasad, R., 2009. Biological alternatives for termite control: a review. Int. Biodeter. Biodegr. 63, 959–972.

Vieira, P.D.S., Silva, W.M.T., Paiva, L.M., Lima, E.A.L.-A., Cavalcanti, V.L.B., 2007. Estudo da caracterização morfológica, esporulação e germinação dos conídios de *Metarhizium anisopliae* var. *acridum* em diferentes temperaturas. Biológico 69, 17–21.

Vitharana, S.I., 1988. Feasability of biological control of the low-contry live wood termite (*Glyptotermes dilatatus*) of tea (*Camellia sinensis*) in Sri Lanka. In: Proc. Regional Tea Conf. Colombo, pp. 99–112.

Wang, C.L., Powell, J.E., 2004. Cellulose bait improves the effectiveness of *Metarhizium anisopliae* as a microbial control of termites (Isoptera: Rhinotermitidae). Biol. Control 30, 523–529.

Wetterer, J.K., Keularts, J.L.W., 2008. Population explosion of the hairy crazy ant, *Paratrechina pubens* (Hymenoptera: Formicidae), on St. Croix, US Virgin Islands. Fla. Entomol. 91, 423–427.

Whitney, D.W., 2012. Application of OSHA's Blood-Borne Pathogen Standard to Bed Bugs. Tox. Law Rep. 27 TXLR 1186.

Williams, D.F., Knue, G.J., Becnel, J.J., 1998. Discovery of *Thelohania solenopsae* from the red imported fire ant, *Solenopsis invicta*, in the United States. J. Invertebr. Pathol. 71, 175–176.

Williams, D.F., Oi, D.H., Knue, G.J., 1999. Infection of red imported fire ant (Hymenoptera: Formicidae) colonies with the entomopathogen *Thelohania solenopsae* (Microsporidia: Thelohaniidae). J. Econ. Entomol. 92, 830–836.

Williams, D.F., Oi, D.H., Porter, S.D., Pereira, R.M., Briano, J.A., 2003. Biological control of imported fire ants (Hymenoptera: Formicidae). Amer. Entomol. 49, 150–163.

Wright, M.S., Lax, A.R., 2013. Combined effect of microbial and chemical control agents on subterranean termites. J. Microbiol. 51, 578–583.

Zarzuela, M.F.M., Campos-farinha, A.E.C., Peçanha, M.P., 2005. Evaluation of urban ants (Hymenoptera: Formicidae) as carriers of pathogens in residential and industrial environments: I. Bacteria. Sociobiology 45, 9–14.

Zarzuela, M.F.M., Campos-farinha, A.E.C., Russomanno, O.M.R., Kruppa, P.C., Gonçalez, E., 2007. Evaluation of urban ants (Hymenoptera: Formicidae) as vectors of muicroorganisms in residential and industrial environments: II. Fungi. Sociobiology 50, 653–658.

Part V

Registration of Microbial Control Products: A Comparative Overview

Chapter 30

Registration of Microbial Control Agents (A Comparative Overview)

R. Gwynn

Biorationale Limited, Duns, United Kingdom

30.1 INTRODUCTION

Registration is the process whereby the responsible national government or regional authority approves the sale and use of a pesticide following the evaluation of comprehensive scientific data demonstrating that the product is effective for the intended purposes and does not pose an unacceptable risk to human or animal health or the environment (FAO, 2002). Depending on the country or region, registration may be straightforward and take a few months or complex and take many years. When microorganisms are used for plant (crop) protection, they need to be registered: while microorganism-based plant protection products (PPPs) can represent technologies with low risk profiles generally, they are usually still required to be registered if they are intended to be used on crops. This is to protect users, consumers, and the environment.

Crop protection is still based predominantly on the use of conventional (synthesized) chemical pesticides. However, there continue to be concerns over the impact of some of these products on human health and the environment. In response, many of the more toxic active ingredients have been withdrawn from use, and there has been the development of alternative crop protection approaches such as Integrated Pest Management (IPM), including the use of microorganism-based PPPs. In general, these types of products are viewed as a desirable technology which, when compared to many conventional chemical pesticides, have a lower toxicity and hence offer greater food safety and environmental protection. This more favorable risk profile has contributed to creating market opportunities for microorganism PPPs and has therefore increased demand for their registration.

Long and complex registration processes are often a barrier to the availability of products for farmers, and it is easy to be frustrated with the process. But it is important to consider that the risks associated with the technology may not always be fully realized.

Most PPP registration systems are mature systems that have been developed over many decades for evaluating conventional chemical pesticides, whereas microorganisms are usually viable and can be whole cells with or without the fermentation materials and with or without the secondary compounds produced by the microorganisms. These differences in types of active substance mean microorganism-based PPPs are not usually a good fit into the existing registration data requirements, and often regulators have no, or minimal, expertise in them. This does not mean that data requirements or registration for microorganisms are less stringent than for conventional chemical pesticides but rather that the data requirements for microorganisms should be appropriate to the technology and consider both the microorganism itself and any relevant secondary compounds present. For example, information about chemical structure is not relevant whereas correct taxonomy is critical to microorganism registration. However, it is worth noting that *Bacillus thuringiensis*-based products are still the most sold microorganism-based biopesticides, and they have been mostly registered following the relevant conventional chemical pesticide data requirements. However, these were sometimes adapted on a case-by-case basis.

That biopesticides have characteristics requiring particular consideration for registration is shared by some countries including the United States, European Union (EU), and some Organization for Economic and Cooperative Development (OECD) member countries through support for the work of its Biopesticide Steering Group. Further, specific biopesticide registration guidelines have been developed by certain regions and countries (Brazil, China, Ghana, Kenya, and Southeast Asia).

An in-depth review of regulatory procedures in different jurisdictions worldwide is provided by Kabaluk et al. (2010). This chapter describes approaches to microorganism registration in principle and explores different approaches taken by some countries or regions.

Microbial Control of Insect and Mite Pests. http://dx.doi.org/10.1016/B978-0-12-803527-6.00030-5

30.2 REGISTRATION PRINCIPLES: DATA REQUIREMENTS AND PROCESSES

For registration purposes, "microorganism" usually means microbial pest control agents that are microorganisms (eg, protozoan, fungus, bacterium, virus, or other microscopic self-replicating biotic entity; FAO, 2005) and any associated metabolites, to which the effects of pest control are attributed (OECD, 2006). A microorganism active substance may contain viable and nonviable microorganisms. It can contain relevant secondary compounds/metabolites/toxins produced during cell proliferation (growth), material from the growth medium, and sometimes microbial contaminants, provided none of these components have been intentionally altered, including no genetic modification. Most countries do not consider genetically manipulated organisms as biopesticides.

It is important to note that generally, microorganism registration is *strain* specific, and therefore the isolate or strain code is an integral part of the active substance definition, eg, *Metarhizium brunneum* strain 52. There are some exceptions, such as for baculoviruses, which, due to the specificity of their relationship with their insect host, are regulated according to their host specificity, eg, *Cydia pomonella* granulovirus.

Registration is a two-stage process: stage one is to address issues related to the *active substance* and stage two for the *product(s)*. These stages may happen together or separately, and if separately, there may be several years between the two stages (eg, in the United States they are concurrent, but in the EU it can take 1–3 years for the active substances and 1–2 years for the products). To register an active substance and product, the applicant provides "information" to the regulator following a specified set of questions or data requirements. This information is compiled into one or more "dossiers." All the data requirements have to be answered, even when the applicant may consider them nonapplicable or irrelevant. If they are nonapplicable or irrelevant then the rationale for this must be explained and justified based on good reasoned arguments supported by good quality information (eg, scientific papers, review texts).

30.2.1 Data Requirements for Registration

The data requirements can be addressed by submitting new or unpublished data/studies, published studies from scientific journals or other sources in the public domain, and/or providing a reasoned case (also called scientific justification or a waiver) for why the data requirement does not need to be addressed by a new study. Where information is provided from a study, the study is usually conducted following Good Laboratory Practice (GLP) or Good Experimental Practice Standard as appropriate. Information for a reasoned case should be collated from good quality information such as (but not only) papers published in good quality peer-reviewed scientific journals or textbooks or reviews by experts. Information from studies generated by applicants is usually data protected

for a period (eg, 10 years in the EU) whereas information from public domain sources or reasoned cases is not.

While there may be some exceptions in countries where data requirements for microorganisms have been simplified, the applicant must address the following data requirements:

1. Active substances: identity and purity; physical and chemical and/or biological properties; further information (use, production, etc.); analytical methods; human health; residues; fate and behavior in the environment; effects on nontarget organisms.
2. Products: identity of the active substance and composition of the formulation; physical and chemical and/or biological properties; application, labeling, and packaging; analytical methods; efficacy data; toxicology and exposure; residues; fate and behavior in the environment; effects on nontarget organisms.

Studies and reasoned cases are prepared, and the information is compiled into a "dossier" plus the copies of the study reports and other relevant papers. The applicant then submits one or more copies of this entire dossier to his or her chosen country's regulatory agency or agencies. Some countries have different parts of the dossier evaluated by experts located in different organizations. The commonly followed data requirements and dossier format for microorganisms that many countries use is the OECD (2003). Following submission, the first step is for the regulatory agency to check that the dossier is complete; that is, it is confirmed that all the data requirements have been addressed. At this stage the information in the dossier is not evaluated. Once a dossier is confirmed as "complete" the regulatory agency makes a detailed evaluation of the information provided by the applicant. Depending on the regulatory agency, there may be ongoing dialogue with the applicant to clarify issues. In some case this does not happen, but best practice is considered to be when there is a good two-way dialogue. There may also be pauses in the evaluation process, so-called "stop-the-clock," if insufficient information has been provided by the applicant, giving the applicant time to provide new information or consult an expert. Once the evaluation is completed, the regulatory agency summarizes its findings and conclusions in a report. This registration report usually makes a recommendation for approval or nonapproval and justifies this decision. In some countries a vote is then taken to approve or reject the application at the government level.

In registration, especially for microorganisms, there are certain aspects of regulatory practice that facilitate the registration process. These include the following:

For the applicant:

- good knowledge of registration procedures in the country of application,
- knowledge of microbiology,
- familiarity with relevant guidance documents,
- availability for presubmission meetings and development of clear regulatory strategy,

- sufficient expertise in his or her regulatory area to be able to prepare a good quality dossier and to have good quality scientific discussions with the regulators on the technology to ensure that the data requirements are sufficiently and appropriately addressed,
- sufficient funds to conduct the relevant studies and the ability to collate reasoned cases and prepare the dossier, and
- timeliness in preparing the dossier and responding to regulatory questions.

For the regulatory agency:

- knowledge of microbiology,
- dedicated biopesticide or microorganism evaluators,
- familiarity with all relevant guidance documents,
- availability for presubmission meetings (preferably free),
- availability for clarification of minor points with the applicant prior to dossier submission,
- openness to discuss nontypical features during dossier development,
- all specialists have sufficient expertise in their regulatory area to have good quality scientific discussions on the technology to ensure that the data requirements are sufficiently and appropriately addressed,
- facilitating approach,
- 1-day completeness check of the completed dossier,
- timeliness in evaluating the dossier and responding to regulatory questions, and
- proportional registration fees.

This section has outlined the principles of registration; however, approaches, processes, and data requirements differ by country, as discussed in the following sections.

30.3 GLOBAL REGISTRATION AND HARMONIZATION ACTIVITIES FOR MICROORGANISMS

30.3.1 Food and Agriculture Organization Guidelines for Registration of Pesticides, Including Microorganisms

The Food and Agriculture Organization (FAO) has prepared the "Guidelines on data requirements for the registration of pesticides" (FAO, 2013). These guidelines generally focus on the scientific data and other information that may be required to determine what products can be permitted for use and for what purposes. The data and other information provided can be used to register all types of pesticides, including public health pesticides.

In 1988, the FAO published guidelines for the registration of biological pest control agents (FAO, 1988), which sit alongside the guidance for pesticides, following the same principles and processes. This document provides guidance on the adaptation of data requirements that may be needed for

microorganism-based PPPs. Currently, this guideline is being updated to provide advice on the principles and processes for the registration of microorganism-based biopesticides, accounting for their differences from conventional chemical pesticides. This guideline should be available by late 2016. Harmonization of data requirements and of procedures for registration is recognized by the FAO as an important step to facilitate the availability of microorganisms.

The guidance provided considers that microorganisms are distinguished from conventional chemical pesticides by a combination of their active substance material and/ or nature. This specialized guidance is being developed because, in many countries, microorganisms are evaluated and registered following the same system as for conventional chemical pesticides; this approach can pose an unnecessarily high regulatory burden by having to satisfy inappropriate requirements. The guidance document aims to provide a framework for individual countries to integrate relevant procedures into their existing system for PPPs. The implementation of the guidelines is carried out at the national level of the country choosing to adopt them.

30.3.2 Organization for Economic Cooperation and Development Guidelines for Registration of Microorganisms

Like the FAO, the OECD aims to harmonize data requirements for pesticide registration and has a project specifically for biological pesticides: the BioPesticide Steering Group (BPSG), initiated in 1999. The OECD BPSG considers microorganisms to include, but not necessarily limited to, bacteria, algae, protozoa viruses, and fungi. This program aims to help governments work together to assess biological pesticide risks to humans and the environment. The assessment of risks provides the basis for governments' decisions on whether to register new biological pesticides and whether to renew the registration of old ones. OECD BPSG considers that by working together, governments can evaluate a biological pesticide's risks more quickly and thoroughly, thereby hastening the approval of safer pesticides and disqualifying others.

The OECD BPSG has been responsible for developing some important and useful guidance documents specifically for microorganism based-PPPs. These documents are used globally to support the development and assessment of microorganism dossiers. OECD BPSG holds regular seminars and workshops with regulators, invited experts, and industry to consider biopesticide-specific aspects of registration and to develop guidance documents. Information on their work can be accessed online (http://www.oecd.org/chemicalsafety/pesticides-biocides/biologicalpesticideregistration.htm#OECD_Work). Their most important and relevant guidance documents include those for the submission and evaluation of biological pesticide test data: OECD Guidance for Industry Data Submissions for Microbial Pest Control Products and their Microbial Pest Control Agents (Dossier Guidance for Microbials), Series on Pesticides No. 23 and OECD Guidance for Country Data Review Reports on Microbial Pest Control Agents (Monograph Guidance for Microbials), Series on Pesticides No. 22. These two documents provide the framework followed by many countries for microorganism dossiers and assessments.

30.3.3 European and Mediterranean Plant Protection Organization Guidance Documents

It is also useful to note that for efficacy testing of microorganisms, the European and Mediterranean Plant Protection Organization (EPPO) develops standards or guidance documents on how to design, run, assess, and analyze efficacy trials (http://www.eppo.int/). In particular, they have developed a standard for microorganisms: PP1/276(1) Principles of efficacy evaluation for microbial PPPs. This document outlines some principles of good practice for conducting efficacy assessment with microorganisms. So far there are no other EPPO guidance documents that are specific to microorganism-based PPPs.

30.4 REGISTRATION SYSTEMS WITHIN INDIVIDUAL COUNTRIES OR REGIONS

30.4.1 European Union Registration System for Microorganisms

The EU regulatory system is probably one of the most complex and takes the longest. In the EU, all biological pesticides, including microorganisms, are regulated under the same Regulation as the conventional chemical pesticides. However, there are adapted data requirements for microorganisms. As for many countries, microorganisms are regulated at strain level. In the EU, there is no distinction between indigenous and nonindigenous strains; however, the geographical location and ecosystem of the originally isolated microorganism is considered in the dossier. This may influence, for example, the environmental risk assessment, depending on the species and source. It should also be noted that the EU does not accept dossiers developed outside of the EU, although certain data and study results might be used.

The current system (Regulation EC/1107/2009) is one in which active substances are evaluated separately to the products: the active substance dossiers are submitted to an EU representative Member State (country) who conducts an assessment on behalf of the other 27 Member States. This assessment is presented as a draft Registration Report and is then peer reviewed in a process coordinated by the European Food Standards Agency (EFSA). Once the peer review process is completed, the EFSA present their findings and recommendations in a report. This report is then discussed by the European Commission, and a decision is made for approval or nonapproval.

At the same time, or more usually after active substance approval, the product dossier is submitted for approval by zone ("zonal approach"). There are three zones (north, center, and south), which are mainly political but are also related to geography. For products to be used in greenhouses or on seeds, one Member State can make an assessment for all the other EU Member States, across all zones. For products to be used in the field, a Member State in either/or each of three zones conducts an assessment for the designated countries within their zone. Following successful completion of this process the product is authorized for use in each designated country. At the beginning of the process the applicant will have decided in which countries product authorizations are wanted; this decision is usually based on the market opportunity. This newer approach is an improvement on Council Directive 91/414/EEC, which was described in detail by Gwynn and Dale (2010). Under this previous system, the timeline was open and often caused evaluation to take more than 10 years. Under the new Regulation EC/1107/2009 implemented in June 2011, the timelines for each step are prescribed, and the whole process should be completed within 4–5 years plus some extra months if it has been necessary to "stop-the-clock" for an expert meeting.

In Regulation EC/1107/2009 there is a provision for "Low Risk" active substances. If an active substance can be categorized as Low Risk, the process for the product authorization is shorter: the product evaluation should be completed within 120 days (instead of about 18 months), data protection is for 13 years (instead of the usual 10 years), and the active substance is allowed to stay on the approved list, without renewal, for 15 years. The criteria for microorganism active substances to be classified as Low Risk are currently being finalized, but it is expected that many, if not most, microorganisms will be classified as Low Risk.

Provisions for microorganisms were made for the first time in Regulation EC/1107/2009. The data requirements for microorganisms (active substances) were established for the first time in Commission Regulation (EU) No. 283/2013 and for products in Commission Regulation (EU) No. 284/2013. These regulations broadly follow the OECD format (OECD Series on Pesticides No. 23). In addition, the EU has provided details of the testing methods that could be used for microorganisms. These changes represent an important step in that for the first time in the EU, there is a transparent and agreed approach that applicants and evaluators can follow for the registration of microorganism active substances and products.

In addition, Regulation EC/1107/2009 prescribes the development of guidance documents for microorganisms. The EU has so far developed various guidance documents, which can be found on the EU Commissions website (http://ec.europa.eu/food/plant/pesticides/eu-pesticides-database/public/?event=homepage&language=EN). The EU is also an active member of the OECD BPSG, organizing, hosting, and participating in seminars and workshops. Within the EU there are now certain Member States' regulatory agencies who have developed particular expertise in microorganism registration, notably The Netherlands, France, Sweden, Germany, and the United Kingdom. All these activities contribute greatly to increasing facilitation for the registration of microorganism based-PPPs.

30.4.2 Ghana Registration System for Microorganisms

In 1994, the government of Ghana set a clear goal for economic development based on the sustainable use of the country's natural resources. To enable this they have strengthened institutional capacity, including the development of the "Guide to registration of biological control agents" (EPA, Ghana, 2011). This guideline recognizes that microorganisms are different from conventional chemical pesticides and need a different approach for their registration. The approach is to simplify the data requirements and for the evaluation to be undertaken by specialists who understand the technology, particularly in the following areas: human toxicology/ecotoxicology, labeling and advertising, and bioefficacy (www.epa.gov.gh).

The guideline indicates that all data requirements must be addressed, even if only a reasoned case for nonapplicability. Data supplied as dossiers must be generated using Collaborative International Pesticides Analytical Council (CIPAC), EU, or OECD methods, as appropriate. The active substance and product information is evaluated in one dossier following the mandatory, but limited, data requirements, set as follows:

Background information on applicant and product: identification, designation, composition, origin, use, registrations.

Active ingredient information: designation of the active ingredient, physical and chemical properties, purity, toxicology, residues on the plant, ecotoxicology, behavior in the environment, other relevant studies.

Formulation information: physical and chemical properties, toxicology, emergency measures for accident or fire, labeling, packaging, trials (site of trial, object, layout, treatments, observations and results, assessments of trials).

The provisions for biopesticides, including microorganisms, allow for a provisional clearance, allowing products in the market for up to 1 year during which the national regulatory authority (Environmental Protection Agency [EPA]) evaluates the relevant dossiers submitted. Full approvals need to be renewed every 3 years.

30.4.3 Kenya Registration System for Microorganisms

In 2002, Kenya developed a biopesticide-specific regulatory pathway, comprehensively described in Gwynn and Maniania (2010). Under these regulations, microorganisms include naturally occurring bacteria, protozoa, fungi, viruses, and Rickettsia. In Kenya, microorganisms are regulated at strain level, and there is a distinction between indigenous and nonindigenous microorganisms. Nonindigenous organisms must be approved by the Kenya Standing Technical Committee on Imports and Exports on live organisms prior to initiating any in-country work with them.

When applying for microorganism registration, it is considered good practice and encouraged to have a preregistration consultation with the relevant experts with Kenya's regulatory organization, the Pest Control Products Board (PCPB). The active substance and the product are evaluated concurrently, and the time from preconsultation with PCPB to registration varies with the complexity of the product, ranging from 2–4 years.

In addition to providing the dossier, the applicant must also submit samples of the technical grade of the active ingredient, the laboratory standard of the active ingredient, the pest control product, and potentially other forms/formulations as appropriate. In other countries, such as the EU, this is in theory required, but in practice it is never requested, although evidence that such samples are available should be provided.

PCPB will consider allowing data in the dossier that has been generated outside of Kenya, provided it was carried out to the required standards such as GLP. However, it is necessary to carry out at least two seasons of efficacy trials of the product in Kenya.

There is evidence that these specific biopesticide, including microorganisms, provisions are effective, as Kenya has the highest number of biopesticides registered and available out of all the African countries.

30.4.4 New Zealand Registration System for Microorganisms

New Zealand has developed a new guidance for "Microbial Agricultural Chemicals" to provide specific provisions for microorganism-based biopesticides when they are used in agriculture and veterinary medicine. This document considers microorganism PPPs to be microorganisms that are living organisms such as bacteria, protozoa, Rickettsia, fungi, and viruses or the genetically modified or naturally occurring mutants of any of these microbes. This inclusion of genetically modified microorganisms goes further than most countries. It should be noted that, in advance of making an application to register a product, it is compulsory that nonindigenous microorganisms have a biosecurity permit from the import authority, plus each import of the product must comply with the relevant Import Health Standard that has been decided for the product.

In New Zealand, the active substance and product are assessed together in a combined application. The data requirements are similar to those outlined in Section 30.2, but are, importantly, more simplified, pragmatic, and congruent with the risks that most microorganism-based products represent.

30.4.5 Southeast Asia Registration System for Microorganisms

Pesticides are applied widely in Southeast Asia, not only to ensure food security for a growing population but also to protect people from vector-borne diseases. There is good awareness of the pesticide risk and safety. Since 1982, the FAO has played an important role in assisting and supporting countries in Southeast Asia to have legislation in place to regulate the use of these pesticides. As part of this effort, guidelines on biopesticide registration harmonization were developed and adopted in November, 2011 (FAO, 2012). The main objective of the FAO was to assist countries in Southeast Asia in harmonizing their respective country's biopesticide regulatory process with the FAO Code of Conduct on Pesticides. The guidance details the data requirements for microorganism-based biopesticides and importantly indicates what data is not required. It also provides guidance on regulatory study methods that are appropriate. Implementation of this guidance is the responsibility of individual countries.

30.4.6 United States Registration System for Microorganisms

In 1994, the EPA established the Biopesticides and Pollution Prevention Division (BPPD) in the Office of Pesticide Programs to facilitate the registration of biopesticides. The BPPD aims were to promote the use of safer pesticides, including biopesticides, as components of IPM programs. For the EPA, microbial pesticides are considered to consist of a microorganism as the active ingredient that produces a pesticidal effect; they are eukaryotic microorganisms, including, but not limited to, protozoa, algae, and fungi, and prokaryotic microorganisms, including, but not limited to, bacteria or autonomous replicating microscopic elements, including, but not limited to, viruses.

The EPA developed specific provisions, as biopesticides are considered to be less toxic than conventional pesticides, to generally affect only the target pest and closely related organisms, and are often effective in very small quantities and often decompose quickly, lowering exposure and avoiding pollution problems. Further, when used as a component of IPM

programs, biopesticides are considered to greatly reduce the use of conventional pesticides, while crop yields remain high.

To facilitate the availability of biopesticides, the EPA generally requires less data for registration compared to conventional pesticides. For microorganisms, the EPA indicates that specific protocols should be used in studies such as those contained in Series 885: Microbial Pesticide Test Guidelines, which is a comprehensive set of guidelines specifically adapted for microorganisms. While biopesticides generally require less data, the EPA considers the standards for assessments to be equally rigorous as those for conventional chemical pesticides. Proportional data requirements together with skilled evaluators mean that new biopesticides are often registered in less than a year, compared with an average of more than 3 years for conventional pesticides.

The EPA requires that efficacy data is developed by the applicants, but they do not usually require it to be presented or evaluated. This reduced efficacy requirement reduces the costs and time for applicants prior to registration. However, in practice, postregistration, the applicant usually needs to have a number of good quality efficacy trials to demonstrate to farmers that the products are effective PPPs.

The success of a biopesticide-specific registration program is demonstrated by the United States having the highest number of active substances and products available worldwide. As of September, 2015, there are 436 registered biopesticide active ingredients and 1401 active biopesticide product registrations.

REFERENCES

EPA, Ghana, 2011. Guideline to Registration of Biological Control Agents. Environmental Protection Agency, Ministries Area, Accra, Ghana.

FAO, 1988. Guidelines for the Registration of Biological Control Agents. FAO.

FAO, 2002. International Code of Conduct on the Distribution and Use of Pesticides. FAO.

FAO, 2005. International Standards for Phytosanitary Measures. Publication No. 3, 2005.

FAO, 2012. Guidance for Harmonizing Pesticide Regulatory Management in Southeast Asia. FAO.

FAO, January 2013. International FAO/WHO Code of Conduct on Pesticide Management, Guidelines on Data Requirements for the Registration of Pesticides. FAO.

Gwynn, R.L., Dale, J., 2010. European Union with special reference to the United Kingdom. In: Kabaluk, J.T., Svircev, A.M., Goettel, M.S., Woo, S.G. (Eds.), The Use and Regulation of Microbial Pesticides in Representative Jurisdictions Worldwide. IOBC Global, pp. 24–34.

Gwynn, R.L., Maniania, N., 2010. Africa with special reference to Kenya. In: Kabaluk, J.T., Svircev, A.M., Goettel, M.S., Woo, S.G. (Eds.), The Use and Regulation of Microbial Pesticides in Representative Jurisdictions Worldwide. IOBC Global, pp. 94–99.

Kabaluk, J.T., Svircev, A.M., Goettel, M.S., Woo, S.G. (Eds.), 2010. The Use and Regulation of Microbial Pesticides in Representative Jurisdictions Worldwide. IOBC Global, p. 99.

OECD, 2003. OECD Series on Pesticides Number 18: Guidance for Registration Requirements for Microbial Pesticides.

OECD, 2006. OECD Guidance for Industry Data Submissions for Microbial Pest Control Products and Their Microbial Pest Control Agents.

Index

'*Note*: Page numbers followed by "f" indicate figures and "t" indicate tables.'

Printed in the United States
By Bookmasters